MICROFUNGI ON LAND PLANTS

MICROFUNGI ON LAND PLANTS

An Identification Handbook

MARTIN B. ELLIS BSc., PhD. (Lond.)

and

J. PAMELA ELLIS BSc., Dip. syst. Mycol.

CROOM HELM
London & Sydney

© 1985 Martin B. Ellis & J. Pamela Ellis
Croom Helm Ltd, Provident House, Burrell Row,
Beckenham, Kent BR3 1AT
Croom Helm Australia Pty Ltd, First Floor, 139 King Street,
Sydney, NSW 2001, Australia

British Library Cataloguing in Publication Data

Ellis, Martin B.
  Microfungi on land plants.
  1. Fungi, Phytopathogenic
  I. Title II. Ellis, J. Pamela
  589.2'045249   SB733

ISBN 0-7099-0950-0

Typeset by Columns of Reading
Printed and bound in Great Britain by
Biddles Ltd, Guildford and King's Lynn

CONTENTS

Introduction	vii
Plurivorous Wood and Bark Fungi	1
Plurivorous Leaf-litter Fungi	68
Fungi Specific to Trees, Shrubs and Woody Climbers	75
Plurivorous Fungi on Herbaceous Plants	275
Fungi Specific to Herbaceous Plants other than Grasses, Rushes, Sedges, Bur-reeds and Reedmaces	300
Plurivorous Fungi on Grasses	451
Fungi Specific to Grasses	474
Fungi on Rushes, Sedges, Bur-reeds and Reedmaces	521
Fungi on Ferns, Horsetails and Clubmosses	562
Fungi Parasitic on Rusts and Powdery Mildews	571
Some Recommended Books	573
Glossary	574
Plates	580
Fungus Index	787
Host Index	813

INTRODUCTION

Microfungi grow in most habitats: in the soil, in fresh and salt water, on dung, on insects and other animals, on food and on textiles. By far the greatest number are found on living and dead plants, and it is these with which we are concerned here.

At the present time there is no general up-to-date book available to which a naturalist can turn when he or she needs to name microscopic fungi, and the task of doing so, especially for the amateur, is indeed a daunting one. We try in this book to make identification simpler, and hope by so doing to encourage more people to look at and learn about microscopic fungi. Little is known about the distribution of even the very common species, and valuable contributions to our knowledge in this field can be made.

Some fungi are plurivorous but more are either host limited or show marked preference for particular plants in either a living or dead state. Rusts, smuts, downy and powdery mildews and many other microfungi grow on living plants but others are found only on moribund tissues, the age and condition of which can be important. In this book, species that show preference for particular substrata are described on the host plant, and illustrations of many of them are placed together for comparison. Where a very large number of different fungi occur, as for example on an oak tree, they are divided into those growing on leaves, fruits, and wood and bark, and separated further taxonomically when this is thought to be helpful. By checking a fungus first of all against descriptions of species known to occur on the host plant, much time and effort can be saved.

Plurivorous species are grouped under four headings: wood and bark, leaf litter, herbaceous plants other than grasses, and grasses. Keys are provided where many genera or species are involved.

Host plants are arranged alphabetically under genera and are divided into five main sections: (1) trees, shrubs and woody climbers, (2) herbaceous plants other than grasses, sedges and similar plants, (3) grasses, (4) rushes, sedges, bur-reeds and reedmaces, (5) ferns, horsetails and clubmosses. At the end are described fungi found growing on rusts and powdery mildews.

Descriptions of fungi vary in length, and some are very brief. Technical terms have been avoided where possible and those used are explained in the glossary. Measurements, except where stated otherwise, are always in thousandths of a millimetre (microns) and much space has been saved by not constantly repeating the μm sign. A short list of books which we recommend and to which reference can be made for fuller descriptions is given after the Glossary.

Although we have collected many of the specimens needed in the preparation of this book, others have been loaned to us by the Commonwealth Mycological Institute and the Royal Botanic Gardens, Kew, and we are very grateful to the directors and staff of these establishments. We have been helped enormously by our friends Malcolm and Marjorie Clark and Douglas Graddon, who have not only sent us named specimens of discomycetes about which we knew very little, but have helped us in many other ways. We have learnt much also from Dr E.A. Ellis, the brother of one of us, whose unrivalled field experience and numerous collections, especially of marsh and fen fungi, have been always at our disposal. We owe a great deal to our mentors of the past, in particular Dr B.F. Barnes, Professor C.G.C. Chesters and Mr E.W. Mason, who stimulated and helped us in our early days. Reprints and books have been given or lent to us by our generous friends and former colleagues including Drs G.C. Ainsworth, Colin Booth, R.W.G. Dennis, David L. Hawksworth, Douglas M. Henderson, Lennart and Kerstin Holm, Paul M. Kirk, and David W. Minter, Professor J.A. Nannfeldt, and Drs E. Punithalingam, A. Sivanesan, Brian M. Spooner, Brian C. Sutton, and Grace M. Waterhouse, and Professor John Webster.

PLURIVOROUS WOOD AND BARK FUNGI

DISCOMYCETES

KEY TO GENERA

    Apothecia superficial or erumpent becoming superficial 1
    Apothecia immersed 29
1. Apothecia translucent, waxy, paraphyses usually abruptly swollen to a knob at the apex .... *Orbilia*
    Apothecia and paraphyses not so 2
2. Apothecia scarlet or orange-red 3
    Apothecia purple or purplish, gelatinous 4
    Apothecia some other colour 5
3. Apothecia up to 4cm diam., smooth .... *Sarcoscypha*
    Apothecia up to 1cm, fringed with dark hairs .... *Scutellinia*
4. Ascospores septate .... *Ascocoryne*
    Ascospores non-septate, with striate walls .... *Ascotremella*
    Ascospores non-septate, with 2 guttules .... *Neobulgaria*
5. Apothecia stalked, with heads becoming spherical and powdery .... *Roesleria*
    Apothecia stalked with discs becoming convex 6
    Apothecia stalked or sessile, discs plane or concave 7
6. Discs bright orange, stalks blackish .... *Vibrissea*
    Discs whitish or dull coloured .... *Cudoniella*
7. Paraphyses slender with swollen, fusiform, septate apices .... *Diplocarpa*
    Paraphyses not so 8
8. Apothecia distinctly hairy 9
    Apothecia smooth or slightly downy 13
9. Apothecia on a cobwebby subiculum .... *Arachnopeziza*
    No cobwebby subiculum 10
10. Hairs brown, setose, often forked at base .... *Trichophaeopsis*
    Hairs highly refractive, 'glassy' .... *Unguicularia*
    Hairs encrusted with orange granules .... *Perrotia*
    Hairs tapered, often to a fine point, smooth 11
    Hairs mostly cylindrical or clavate and granular 12
11. Apothecia white or pale .... *Hyaloscypha*
    Apothecia brown .... *Dematioscypha*
12. Paraphyses as a rule lanceolate .... *Dasyscyphus*

	Paraphyses filiform .... *Cistella*
13.	Ascospores coloured when mature 14
	Ascospores hyaline (sometimes pale brown in *Phaeohelotium*) 18
14.	Ascospores purple or purplish with striate walls .... *Ascobolus*
	Ascospores brown or olivaceous brown 15
15.	Ascospores 1-septate 16
	Ascospores without septa 17
16.	Ascospores dark brown to black, asci bitunicate .... *Poetschia*
	Apothecia black, asci unitunicate .... *Rhizodiscina*
17.	Ascospores kidney-shaped, apothecia blackish .... *Bulgaria*
	Ascospores dumb-bell-shaped, discs olive .... *Catinella*
	Ascospores elliptical, discs dark olive .... *Velutarina*
18.	Asci operculate 19
	Asci not operculate 20
19.	Apothecia cup-shaped or flattened, ascus walls staining deep blue with iodine .... *Peziza*
	Apothecia pulvinate, ascus walls turning only very pale blue with iodine .... *Pachyella*
20.	Apothecia funnel-shaped, often one-sided .... *Chlorencoelia*
	Not so 21
21.	Apothecia green or dark green .... *Claussenomyces*
	Apothecia white pruinose, mostly on Diatrypaceae .... *Polydesmia*
	Apothecia pale brown, gelatinous .... *Ascocoryne solitaria*
	Apothecia black or becoming black 22
	Apothecia some other colour, not gelatinous 24
22.	Apothecia having anchoring hyphae encrusted with red granules .... *Patinellaria*
	Not so 23
23.	Asci unitunicate .... *Durella*
	Asci bitunicate .... *Patellaria*
24.	Excipulum of apothecia composed of brown, rounded, polygonal or clavate cells 25
	Excipulum not so 27
25.	Apothecia on a dark hyphal subiculum .... *Tapesia*
	Apothecia with slight or no subiculum 26
26.	Ascospores broadly ellipsoidal, 8–9 wide .... *Graddonia*
	Ascospores not more than 4 wide .... *Mollisia* and *Niptera*
27.	Apothecia clustered on a stroma .... *Pezicula*
	Apothecia not clustered on a stroma 28
28.	Excipulum tough, composed of interwoven rather thick-walled hyphae; often bright yellow .... *Bisporella*
	Excipulum composed of large thin-walled cells, paraphyses often branched .... *Phaeohelotium*
	Excipulum usually downy with protuberant cells, apothecia often with ring of brown cells at the base .... *Calycellina*

	Excipulum easy to cut, composed of thin-walled hyphae. Apothecia often stalked .... *Hymenoscyphus* and *Pezizella*
29.	Ascospores brown, muriform .... *Melittosporium*
	Ascospores hyaline, not muriform 30
30.	Ascospores allantoid, 22–26 × 6–7 .... *Propolomyces*
	Ascospores very long and narrow .... *Stictis*
	Ascospores shorter and broader 31
31.	Apothecia white or creamy .... *Cryptodiscus*
	Apothecia dark green to black with pale brown rims .... *Melittosporiella*

Arachnopeziza aurata Fuckel (Fig. 1)
Colonies effuse with yellowish, cobwebby subiculum. Apothecia sessile, hairy, gold to orange, 0.2–1mm diam. Ascospores hyaline, 6- or 7-septate, 50–75 × 1.5–3. On rotten wood and sometimes inside bark of *Acer, Betula, Quercus*, etc., Jan.–July.

Arachnopeziza candido-fulva (Schw.)Korf
Subiculum whitish, cobwebby. Apothecia 0.5–3mm diam., yellow to orange, hairy, hairs sometimes sticking together to form teeth. Ascospores hyaline, 1-septate, subfusiform, more tapered at base, 11–17 × 2.5–3.5. On rotten wood, uncommon.

Ascobolus lignatilis Alb. & Schw.
Apothecia short-stalked, 5–12mm diam., yellow to yellowish green, with black protuberant ascus tips. Ascospores ellipsoidal, purple or purplish brown, striate with some striae anastomosing, 16–20 × 9–12. On rotting branches, etc., Jan.–May.

Ascocoryne cylichnium, described on *Fagus*, is found sometimes on wood of other deciduous trees, e.g. *Acer* and *Salix*.

Ascocoryne sarcoides, described on *Fagus*, its commonest host, is, however, also found frequently on dead branches of other trees including *Betula, Carpinus, Pinus, Quercus* and *Ulmus*, often in its *Coryne* conidial state. The large reddish-purple gelatinous clusters can be recognised easily in the field.

Ascocoryne solitaria (Rehm)Dennis
Apothecia turbinate, pale brown and gelatinous when fresh, drying darker. Ascospores hyaline, 0- to 3-septate, 9–15 × 3.5–4. On decorticated dead wood, Nov., rare.

Ascotremella faginea (Peck)Seaver (Fig. 2)
Apothecia gelatinous, up to 2cm diam., short-stalked, reddish yellow to purplish brown or reddish purple, convoluted, *Tremella*-like. Ascospores hyaline with striate walls, 6–10 × 4. On fallen trunks and branches of *Acer, Fagus, Sambucus*, etc., July–Oct.

Bisporella, apothecia with tough walls composed of interwoven, rather thick-walled hyphae.

KEY

    Apothecia whitish or cream with ochraceous discs and ascospores
        without septa .... *subpallida*
    Apothecia yellow at least when dry, some ascospores 1-septate 1
1.   Apothecia whitish or pale yellow when fresh, found usually with a black
       hyphomycete, *Bispora antennata* .... *pallescens*
    Apothecia bright yellow when fresh, not found with *Bispora antennata* .. 2
2.   Ascospores 3–4 wide, asci up to 140 long .... *citrina*
    Ascospores 2 wide, asci up to 70 long .... *sulfurina*

Bisporella citrina (Batsch)Korf & Carpenter (Fig. 3)
Apothecia erumpent, up to 3mm diam., yellow with discs drying orange. Ascospores hyaline, some 1-septate, 9–14 × 3–4. On dead branches and decorticated wood of *Alnus, Betula, Corylus, Fagus, Fraxinus, Quercus*, etc., Sept.–Dec., common.

Bisporella pallescens (Pers.)Carpenter & Korf
Apothecia stalked or subsessile, up to 1.5mm diam., whitish or pale yellow with discs drying yellow. Ascospores hyaline, rather pointed at each end, 1-septate, 9–10 × 2.5–3. On cut surfaces of stumps, usually with a black hyphomycete, *Bispora antennata*, Dec.–Feb.

Bisporella subpallida (Rehm)Dennis (Fig. 4)
Apothecia up to 1.5mm diam., whitish or cream with ochraceous discs. Ascospores hyaline, without septa, 6–9 × 2.5–3. Usually on decorticated wood, Nov.–Jan.

Bisporella sulfurina (Quélet)Carpenter (Fig. 5)
Apothecia 0.5–1mm diam., sulphur-yellow. Ascospores hyaline, mostly 1-septate, 9–10 × 2. Common on fallen twigs and branches of *Acer, Cornus, Crataegus, Fagus, Fraxinus, Quercus, Tilia, Ulex* and *Ulmus*, often overgrowing stromatic pyrenomycetes such as *Diatrype stigma* and *Diatrypella favacea*; Sept.–Feb.

Bulgaria inquinans, described on *Quercus*. The large blackish-brown gelatinous apothecia are found occasionally on fallen trunks and branches of other trees, e.g. *Carpinus, Castanea, Fagus* and *Ulmus*.

Calycellina ochracea (Grelet & Croz.)Dennis (Fig. 6)
Apothecia up to 0.5mm diam., rather pale yellow, downy. Ascospores hyaline, becoming 3-septate, 16–20 × 3. On bark of small, rotten branches in very damp places, Jan.–Aug.

Catinella olivacea (Batsch)Boud. (Fig. 7)
Apothecia up to 1cm diam., brown with dark olive discs rimmed with yellow ochre. Ascospores dumb-bell-shaped, olivaceous brown when mature, 8–13 × 3–5. On damp logs, May–Nov.

Chlorencoelia versiformis (Pers.)Dixon (Fig. 8)
Apothecia 1–3mm diam., funnel-shaped, often one-sided, with short stalks, olivaceous, darkening with age. Ascospores slightly curved, hyaline, occasionally 1-septate, 9–15 × 3. On rotting wood of conifers and some deciduous trees, e.g. *Fraxinus, Quercus, Salix*, Aug.–Nov.

Cistella, apothecia sessile or subsessile, white to pale ochraceous, covered with short cylindrical to clavate hyaline hairs often granulate towards their tips. Paraphyses filiform, about the same length as the asci. Ascopores hyaline, without septa.

KEY

Ascospores not more than 6 long .... *geelmuydenii*
Ascospores 7–11 long .... *perparvula*
Ascospores 12–13 long .... *dentata*

Cistella dentata (Pers.)Quélet
Apothecia 1–2mm diam., pale ochraceous. Ascospores 12–13 × 3–4. Hairs 25–35 × 4, granulate.

Cistella geelmuydenii Nannf.
Apothecia 0.1–0.3mm diam., white or greyish white. Ascospores 6 × 2.5. Hairs about 30 × 4, granulate.

Cistella perparvula (P. Karsten)Nannf.
Apothecia 0.2–0.4mm diam., whitish. Ascospores 7–11 × 2–2.5. Hairs about 20 × 3–4, smooth.

Claussenomyces atrovirens (Pers.)Korf & Abawi (Fig. 9)
Apothecia 0.5–1.5mm diam., dark green drying almost black, gelatinous. Ascospores hyaline, at first 6- to 10-septate, 16–29 × 3.5–4, but these soon form innumerable tiny rod-shaped ascoconidia which come to fill the asci. On damp rotting wood of *Acer, Fagus, Sorbus, Ulex*, etc., mostly Apr.–June.

Claussenomyces prasinulus (P. Karsen)Korf & Abawi (Fig. 10)
Apothecia 0.2–0.4mm diam., green, gelatinous. Ascospores hyaline, 3-septate, 10–13 × 3. Apothecia accompanied by green synnemata of the *Dendrostilbella* state, which produces minute rod-shaped conidia. On damp rotting wood of *Betula, Fagus, Sambucus*, etc., June–Oct.

Cryptodiscus pallidus (Pers.)Corda (Fig. 11)
Apothecia with rich cream discs deeply sunken in wood. Ascospores hyaline, 3-septate, 11–17 × 3–6. On thin decorticated twigs of deciduous trees, uncommon.

Cryptodiscus rhopaloides Sacc. (Fig. 12)
Apothecia 0.3–0.5mm diam., deeply sunken in embedded white stromata which split into teeth to expose creamy discs. Ascospores clavate, hyaline, 6-

to 8-septate, 30–40 × 4–5. On dead branches of *Clematis*, *Pinus*, *Rubus*, *Salix*, etc., June–Nov.

Cudoniella acicularis, described on *Quercus*, has been found on stumps of other trees.

Cudoniella clavus and its var. *grandis*, described on *Alnus*, are found occasionally on other woody debris in very wet places, usually where there is shallow running water.

Cudoniella tenuispora (Cooke & Massee)Dennis
Similar to *C. clavus* var. *grandis* but more tawny, white and downy underneath, and with slightly smaller spores, which are often 1-septate. On dead twigs, chips and other woody debris in damp places, May–June.

Dasyscyphus, apothecia covered with hairs which are mostly granulate; paraphyses usually lanceolate.

KEY

	Apothecia fasciculate, paraphyses septate .... *fascicularis*
	Apothecia not fasciculate, paraphyses not septate 1
1.	Apothecia white when fresh with cream or yellowish discs 2
	Apothecia yellowish, buff, grey or brown 4
2.	Paraphyses lanceolate, 5 wide .... *virgineus*
	Paraphyses more slender 3
3.	Ascospores 4–6 × 1, paraphyses with yellow sap .... *papyraceus*
	Ascospores 6–8 × 1.5–2, paraphyses hyaline .... *brevipilus*
4.	Apothecia less than 1mm diam. 5
	Apothecia 2mm diam. or more 6
5.	Ascospores 1-septate, 11–18 × 2–3.5 .... *corticalis*
	Ascospores without septa, 6–9 × 2–3 .... *pulveraceus*
6.	Paraphyses slender, ascospores 4–6 × 2–2.5 .... *cerinus*
	Paraphyses lanceolate, 5 wide, ascospores 7–11 × 1.5–2.5 .... *pygmaeus*

Dasyscyphus brevipilus, described on *Fagus*, is not uncommon also on rotten branches of other deciduous trees, including *Acer*, *Betula*, *Crataegus* and *Prunus*.

Dasyscyphus cerinus (Pers.)Fuckel (Fig. 13)
Apothecia up to 2mm diam., short-stalked, brown, covered with yellow or golden-brown hairs. Ascospores hyaline, 4–6 × 2–2.5. On mainly decorticated wood of deciduous trees and shrubs including *Alnus*, *Betula*, *Crataegus*, *Fagus*, *Fraxinus*, *Malus*, *Quercus* and *Salix*, Mar.–Sept.

Dasyscyphus corticalis (Pers.)Massee (Fig. 14)
Apothecia up to 0.7mm diam., sometimes short-stalked, buff or flesh-coloured, hairy. Hairs long, yellowish-brown below, covered with minute spicules. Ascospores hyaline, 1-septate, 11–18 × 2–3.5. Paraphyses slender.

On dead twigs and branches of *Betula, Fraxinus, Hedera, Lonicera, Quercus, Viburnum*, etc., Mar.–Dec.

Dasyscyphus fascicularis (Velen.)Le Gal (Fig. 15)
Apothecia fasciculate, 1.5–2mm diam., white, drying russet. Hairs short. Ascospores hyaline, 5–6 × 1–1.5. Paraphyses with brownish contents, septate. On dead branches of deciduous trees and shrubs, including *Crataegus*, Apr.–Sept., uncommon.

Dasyscyphus papyraceus (P. Karsten)Sacc. (Fig. 16)
Apothecia 0.5–1mm diam., white, hairy including stalks, with yellowish discs. Ascospores hyaline, 4–6 × 1. Paraphyses slender with yellowish contents. On decorticated branches of *Betula, Larix, Pinus* and *Salix*, Oct.–Nov.

Dasyscyphus pulveraceus (Alb. & Schw.)Höhnel (Fig. 17)
Apothecia up to 0.5mm diam., brown but covered with pale grey incrustation, discs pale yellow with fringe of white hairs. Marginal hairs closely septate. Ascospores hyaline, 6–9 × 2–3. Paraphyses slender. On dead branches of *Betula, Fagus, Fraxinus, Hedera, Ilex, Sorbus* and *Ulex*, June–Sept.

Dasyscyphus pygmaeus, described on *Ulex* but found also on twigs and branches of *Acer, Betula, Fraxinus, Quercus, Salix*, etc., lying in grass or partly buried in soil.

Dasyscyphus virgineus (Batsch)Gray (Fig.18)
Apothecia up to 1mm diam., white, very hairy including stalks, discs creamy. Hairs cylindrical, up to 120 × 4–5, some swollen a little at the tip. Ascospores 6–10 × 1.5–2.5. Paraphyses lanceolate, 5 wide. Very common on dead twigs and debris of *Alnus, Betula, Fagus, Quercus, Rubus, Salix*, etc., Feb.–Oct.

Dematioscypha dematiicola, described on *Quercus*, is common also, especially in its *Haplographium* state, on dead branches of many other trees and shrubs, including *Acer, Alnus, Betula, Castanea, Corylus, Fagus, Fraxinus, Ilex, Prunus, Rhododendron, Salix, Sorbus* and *Ulex*.

Diplocarpa bloxamii (Berk. ex Phill.)Seaver (Fig. 19)
Apothecia stalked, 2–3mm diam., dark brown with olivaceous discs. Ascospores hyaline, 6–8 × 2.5–3. Paraphyses slender with swollen, fusiform, septate apices. On underside of rotting logs of *Quercus*, etc., May–Sept., uncommon.

Durella atrocyanea (Fr.)Höhnel (Fig. 20)
Apothecia superficial or partly immersed, 0.5–1mm diam., black, leathery or wrinkled when dry. Ascospores hyaline, mostly 3-septate, 18–21 × 4–5. Mainly on decorticated parts of branches of *Cytisus, Fraxinus, Quercus, Salix* and *Ulex*, Dec.–Feb.

Durella connivens, described on *Fagus*, and *D. commutata* and *D. macrospora*, both on *Quercus*, are found occasionally on wood of other trees and shrubs.

Graddonia coracina (Bres.)Dennis (Fig. 21)
Apothecia 1–2.5mm diam., greyish brown with buff discs when fresh, turning reddish brown or black on drying. Ascospores 0- or 1-septate, hyaline, mostly 18–20 × 8–9. On fallen twigs and rotten wood in damp places, Feb.–May.

Hyaloscypha hyalina, described on *Quercus*, has been found also on dead branches of *Aesculus*, *Betula*, *Crataegus* and *Populus*.

Hyaloscypha leuconica, described on *Larix*, has been collected also on dead wood of *Fagus*, *Pinus*, *Rhododendron*, *Salix*, *Sorbus* and *Ulex*.

Hymenoscyphus, with apothecia usually stalked and smooth; flesh easy to cut, made up of thin-walled hyphae.

KEY

	Ascospores 8–11 long	1
	Ascospores over 15 long	2
1.	Apothecia white, drying reddish brown	*imberbis*
	Apothecia yellow or yellowish brown, with almost black stalks	*vernus*
2.	Apothecia reddish brown	*subferrugineus*
	Apothecia white, yellow or ochraceous	3
3.	Ascospores 15–22 × 3–4.5	*calyculus*
	Ascospores 16–20 × 5–6	*vitigenus*
	Ascospores mostly 20–25 × 5–6	*laetus*

Hymenoscyphus calyculus (Sow.)Phill. (Fig. 22)
Apothecia about 2mm diam., stalked, yellow, drying somewhat ochraceous, downy especially near base of stalk. Ascospores hyaline, 15–22 × 3–4.5. On fallen dead twigs and branches of *Acer*, *Alnus*, *Betula*, *Castanea*, *Corylus*, *Fraxinus*, *Rubus*, etc., June–Dec.

Hymenoscyphus imberbis (Bull.)Dennis (Fig. 23)
Apothecia 1–2mm diam., stalked, white, drying reddish brown, with short, slightly curved hairs especially on the stalks. Ascospores hyaline, 8–11 × 3–4. On fallen dead twigs and branches of *Alnus*, *Corylus*, *Crataegus*, *Salix*, *Sambucus*, etc., in wet places, Aug.–Dec.

Hymenoscyphus laetus (Boud.)Dennis (Fig. 24)
Apothecia about 2mm diam., short-stalked, yellow, drying yellowish brown. Ascospores hyaline, occasionally 1-septate, 18–25 × 5–6. On fallen dead branches of *Alnus*, *Betula*, *Fraxinus*, *Quercus*, etc. in wet places, Sept.–Mar.

Hymenoscyphus subferrugineus, described on *Salix*, has been found also on dead wood and bark of other deciduous trees.

Hymenoscyphus vernus (Boud.)Dennis (Fig. 25)
Apothecia up to 7mm diam., yellow to yellowish brown with stalks very dark brown to black especially near the base. Ascospores hyaline, 8–11 × 3–3.5.

On fallen dead twigs and branches of *Alnus, Quercus, Populus, Salix*, etc., especially in wet places, Feb.–Sept.

Hymenoscyphus vitigenus (de Not.)Dennis
Apothecia up to 1mm diam., stalked, white becoming ochraceous. Ascospores hyaline, biguttulate, 16–20 × 5–6. On rotten wood.

Melittosporiella pulchella Höhnel (Fig. 26)
Apothecia 1–2mm diam., dark green to black with pale brown rims, deeply immersed in bleached parts of old decorticated wood, opening by teeth. Ascospores hyaline, 3-septate, 15–18 × 4.5–5.

Melittosporium propolidoides (Rehm)Rehm (Fig. 27)
Apothecia up to 1mm diam., almost black, immersed in decorticated wood, seen through oval slits. Ascospores 1 to each ascus, muriform, becoming brown, 25–35 × 8–10. Mostly on conifers, uncommon.

Mollisia, with apothecia saucer-shaped or plane, sessile, soft-fleshed with outer cells brown and mostly rounded, polygonal or clavate. See also *Niptera ramincola*.

KEY

	Ascospores often septate	1
	Ascospores not septate	2
1.	Ascospores 20–30 × 2–2.5	*ramealis*
	Ascospores up to 21 × 3.5	*ventosa*
2.	Ascospores 5–6 × 2.5–3.5	*aquosa*
	Ascospores 4–7 × 1	*caespiticea*
	Ascospores larger	3
3.	Ascospores mostly straight	4
	Ascospores often curved	5
4.	Ascospores mostly narrowly clavate with a guttule at each end, 7–10 × 2–2.5	*cinerea*
	Ascospores not narrowly clavate, 8–11 × 2.5	*cinerella*
5.	Apothecia erumpent in clusters, drying yellow	*discolor* var. *longispora*
	Apothecia not in clusters drying yellow	6
6.	Apothecia drying like three-cornered hats, ascospores 8–10 × 2	*ligni*
	Apothecia not so, ascospores 8–14 × 2–2.5	*melaleuca*

Mollisia aquosa (Berk. & Br.)Phill. (Fig. 28)
Apothecia up to 0.7mm diam., olivaceous brown with pale-rimmed watery grey discs. Ascospores hyaline, 5–6 × 2.5–3.5. On rotten wood including that of *Salix*, accompanied by *Lasiosphaeria hirsuta* and hyphomycetes such as *Sporoschisma mirabile* and *Bactrodesmium abruptum*, favouring stumps and the underside of rotting logs, Aug.–Jan.

Mollisia caespiticea (P. Karsten)P. Karsten (Fig. 29)
Apothecia up to 1.5mm diam., dark grey to greyish brown, in dense clusters erumpent from bark. Ascospores hyaline, 4–7 × 1. On dead branches of *Betula, Crataegus, Quercus, Ribes, Salix* and *Sambucus*, Sept.–Apr.

Mollisia cinerea (Batsch)P. Karsten (Fig. 30)
Apothecia 1–3mm diam., often with wavy edges, greyish brown below, with anchoring hyphae, discs watery grey or pale yellowish grey. Ascospores mostly narrowly clavate, hyaline, with a guttule at each end, 7–10 × 2–2.5. Very common throughout the year on dead wood and branches of *Acer, Betula, Castanea, Crataegus, Fagus, Fraxinus, Pinus, Quercus, Rhododendron, Rubus, Salix, Ulex*, etc.

Mollisia cinerella Sacc. (Fig. 31)
Apothecia up to 1mm diam., blackish brown with greyish olive to dark, rather bluish discs, marginal cells of excipulum much elongated. Ascospores hyaline, 8–11 × 2.5. On decorticated dead wood of *Acer, Betula, Castanea, Fagus, Ilex*, etc., Jan.–Aug.

Mollisia discolor var. *longispora*, described on *Quercus*, is seen occasionally on branches of other trees, e.g. *Corylus, Fagus, Salix* and *Tilia*. Ascospores 9–16 × 2–2.5.

Mollisia ligni (Desm.)P. Karsten (Fig. 32)
Apothecia about 1mm diam., mid- to dark brown, downy, with dark greyish discs, drying like little three-cornered hats. Protruding excipular cells dark brown, distinctly clavate; ascospores somewhat curved, mostly 8–10 × 2. Very common all the year round on dead branches of *Betula, Carpinus, Castanea, Crataegus, Fagus, Fraxinus, Quercus, Salix, Tilia, Ulex, Ulmus*, etc.

Mollisia melaleuca (Fr.)Sacc. (Fig. 33)
Apothecia up to 2mm diam., blackish brown with whitish discs. Ascospores slightly curved, hyaline, with guttules at each end, 8–14 × 2–2.5. On rotten wood and dead branches of *Betula, Cytisus, Fraxinus, Rubus, Ulmus*, etc., Oct.–May.

Mollisia ramealis, described on *Betula*, has been found also occasionally on *Alnus* and *Castanea*.

Mollisia ventosa P. Karsten (Fig. 34)
Apothecia 1–2.5mm diam., greyish brown, discs yellow or yellowish grey with pale margins. Ascospores hyaline, often 1-septate, occasionally 2- or 3-septate, 12–21 × 2–3.5. On rotten wood of *Alnus, Salix, Ulex*, etc., Mar.–May.

Neobulgaria lilacina (Wulfen)Dennis (Fig. 35)
Apothecia 2–3mm diam., purplish grey, gelatinous. Ascospores hyaline, biguttulate, 6–8 × 3–3.5. On decorticated wood, especially cut ends of stumps, and logs of *Fagus, Populus* and *Salix* in wet places, Oct.–May.

Niptera ramincola Rehm (Fig. 36)
Apothecia up to 1mm diam., disc becoming ochraceous. Ascospores 1-septate, hyaline, 9–15 × 2–3. On dead wood of *Fagus, Lonicera*, etc.

Orbilia, with translucent, waxy-looking apothecia. Most have slender paraphyses swollen to form a knob at the apex.

KEY

	Apothecia distinctly cyathiform	*cyathea*
	Apothecia not cyathiform	1
1.	Ascospores 11–16 × 1–1.5	*vinosa*
	Ascospores up to 15 × 3	*occulta*
	Ascospores not more than 11 long	2
2.	Ascospores spherical, 2–3 diam.	*euonymi*
	Ascospores not more than 6 long	3
	Ascospores more than 6 long	6
3.	Ascospores curved, allantoid 3–4 × 1–1.5	4
	Ascospores straight	5
4.	Apothecia golden yellow	*xanthostigma*
	Apothecia white or very pale cream	*leucostigma*
5.	Ascospores 3–5 × 1–1.5, apothecia orange or peachy orange	*alnea*
	Ascospores 4–6 × 0.5–1, apothecia yellow with white anchoring hyphae	*auricolor*
	Ascospores 4–5 × 2.5, apothecia red	*coccinella*
6.	Ascospores curved, 10–11 × 0.5	*curvatispora*
	Ascospores 6–9 × 0.5, apothecia pink to coppery	*sarraziniana*
	Ascospores 7–11 × 1–1.5, apothecia tangerine, drying paler	*luteorubella*

Orbilia alnea Velen. (Fig. 37)
Apothecia 0.2–0.5mm diam. Ascospores with polar guttules. On dead branches of *Alnus, Betula, Fagus, Salix, Tilia*, etc., all the year round.

Orbilia auricolor (Bloxam ex Berk.)Sacc. (Fig. 38)
Apothecia up to 1.5mm diam. Anchoring hyphae distinctive. On fallen dead branches of *Acer, Corylus, Cupressus, Fraxinus*, etc., all the year round.

Orbilia coccinella (Sommerf.)Fr.
Apothecia up to 0.25mm diam. Ascospores biguttulate. On rotten wood of *Betula, Populus, Salix*, etc., Sept.–Oct.

Orbilia curvatispora Boud. (Fig. 39)
Apothecia up to 1mm diam., translucent, white becoming somewhat yellowish. On rotten wood and fallen branches of *Acer, Fagus, Hedera, Salix, Ulmus*, etc., all the year round.

Orbilia cyathea Velen. (Fig. 40)
Apothecia permanently cyathiform, the deep cups 1–2mm diam., tapered to narrow stalks, translucent fawn to wine-coloured. Ascospores 4–8 × 1. On dead wood of *Acer, Crataegus, Salix,* etc., Aug.–Oct.

Orbilia euonymi Velen.
Apothecia 0.3–0.4mm diam., pale yellowish or apricot-coloured. Small spherical spores distinctive. On dead wood.

Orbilia leucostigma, described on *Pinus,* is found occasionally on wood of other trees.

Orbilia luteorubella (Nyl.)P. Karsten (Fig. 41)
Apothecia up to 1.5mm diam., tangerine when fresh, drying paler and somewhat yellowish. On rotten wood of *Fraxinus, Populus, Salix, Ulex,* etc., Apr.–Sept.

Orbilia occulta (Rehm)Sacc.
Apothecia 0.3–0.7mm diam., flesh-coloured to blood red. Ascospores oval, drawn out into long threads, up to 15 × 3. On dead wood including that of *Salix.*

Orbilia sarraziniana Boud. (Fig. 42)
Apothecia up to 1mm diam. On bark and wet wood of *Acer, Crataegus, Fraxinus, Salix,* etc., May–Nov.

Orbilia vinosa (Alb. & Schw.)P. Karsten
Apothecia 1–2mm diam., flesh-coloured. Ascospores often slightly curved. On wood of *Castanea, Quercus, Ulmus,* etc.

Orbilia xanthostigma (Fr.)Fr. (Fig. 43)
Apothecia up to 1mm diam. Probably the commonest species of *Orbilia,* found throughout the year on decorticated wood and on bark of *Alnus, Betula, Fagus, Pinus, Quercus, Salix, Ulex, Ulmus,* etc.

Pachyella babingtonii (Berk.)Boud. (Fig. 44)
Apothecia 5–9mm diam., raw sienna to reddish brown, gelatinous. Ascospores hyaline, 19–21 × 11–12. Paraphyses with brown contents. On rotten logs, etc., e.g. those of *Salix* lying in water, May–Nov.

Patellaria atrata Fr. (Fig. 45)
Apothecia superficial, 0.5–1.5mm diam., almost black. Asci bitunicate. Ascospores hyaline, 7- to 10-septate, 35–46 × 8–9. On decorticated wood of *Acer, Quercus, Tilia,* etc., Mar.–May.

Patinellaria sanguinea (Pers.)P. Karsten (Fig. 46)
Apothecia up to 0.2mm diam., reddish at first, soon becoming black with anchoring hyphae encrusted with red granules which join up with similar hyphae forming a subiculum. Paraphyses with swollen terminal brown cells agglutinated to form an epithecium. Ascospores hyaline, 7–10 × 3. On

decorticated wood of *Corylus, Quercus, Populus, Rosa, Salix*, etc.

Perrotia flammea (Alb. & Schw.)Boud. (Fig. 47)
Apothecia 1–2mm diam., reddish orange, covered with hairs up to 200 × 5 encrusted with orange granules. Ascospores hyaline, often curved, some septate, 10–14 × 2.5–3. On dead twigs of *Corylus, Fagus, Ligustrum, Salix*, etc., Aug.–Sept.

Pezicula cinnamomea, described on *Quercus*, is found also on *Acer, Aesculus, Fagus, Fraxinus* and *Ulmus*.

Peziza, operculate asci have walls which turn blue in Melzer's fluid.

KEY

	Ascospores verruculose with an appendage at each end .... *apiculata*
	Ascospores smooth without appendages 1
1.	Flesh of apothecia stratified 2
	Flesh not stratified 3
2.	Paraphyses with rows of inflated cells .... *varia*
	Paraphyses not so .... *repanda*
3.	Flesh thick, watery, containing very large cells linked together by short hyphae .... *ampliata*
	Flesh thin, without these very large cells .... *micropus*

Peziza ampliata Pers.
Apothecia 1–3cm diam., pale brown or yellowish brown, margin often dentate. Flesh containing very large round cells linked together by short hyphae. Ascospores hyaline, smooth, 18–20 × 10–11, without guttules. On very rotten wood, Apr.–May.

Peziza apiculata, described on *Fagus*, has been found also on wet rotten wood of other trees.

Peziza micropus Pers. (Fig. 48)
Apothecia up to 4cm diam., often short-stalked, fawn or ochraceous brown, with dentate margins. Ascospores hyaline, smooth, 15–20 × 9–11. On rotten wood, often emerging through cracks in bark of *Alnus, Fagus, Salix, Ulmus*, etc., Sept.–Nov.

Peziza repanda Pers. (Fig. 49)
Apothecia 5–12cm diam., hazel or chestnut brown, with crenate margins; flesh stratified. Ascospores hyaline, smooth, 15–16 × 9–10. On rotten wood, the base of fallen trees or logs of *Salix, Ulmus*, etc., Oct.–May.

Peziza varia (Hedw.)Fr. (Fig. 50)
Apothecia 1–5cm diam., often short-stalked, greyish brown; flesh stratified. Ascospores hyaline, smooth, 14–16 × 9–11. Paraphyses with rows of inflated cells. On rotten wood and sawdust, June–Oct. Often found on half-buried

14 Plurivorous Wood and Bark Fungi

wood around stumps of *Fagus*, etc.

Pezizella, apothecia pale, often downy on the outside; flesh as in *Hymenoscyphus*.

KEY

	Ascospores 1-septate .... *parilis*
	Ascospores without septa 1
1.	Apothecia clustered, ascospores 1–2 wide .... *vulgaris*
	Apothecia solitary, ascospores 4 wide .... *leucostigma*

Pezizella leucostigma (Fuckel)Sacc.
Apothecia 1mm diam., sessile, white, diaphanous, disc drying yellowish. Ascospores cylindric oblong, slightly curved, hyaline, biguttulate, 10–12 × 4. Recorded on branches of *Prunus spinosa*, described originally on *Betula* and *Fagus*.

Pezizella parilis (P. Karsten)Dennis (Fig. 51)
Apothecia up to 2mm diam., stalked, creamy white, drying honey-coloured. Ascospores hyaline, 1-septate, 10–14 × 2. On dead twigs of *Acer*, *Alnus*, *Betula*, *Quercus*, *Salix*, etc., May–Jan., but mostly Sept.–Nov.

Pezizella vulgaris (Fr.)Sacc (Fig. 52)
Apothecia erumpent, clustered, 1mm diam., short-stalked, white to cream or flushed pink. Ascospores hyaline, slightly curved, 6–10 × 1–2. On dead twigs of *Corylus*, *Rosa*, *Rubus*, *Salix*, etc., July–Dec.

Phaeohelotium, apothecia turbinate or sessile and cup-shaped, cream, yellow or pinkish when fresh with excipular cells usually large and thin-walled, and paraphyses in a number of species branched.

KEY

	Ascospores not more than 11 long 1
	Ascospores often more than 11 long 4
1.	Ascospores 1-septate .... *italicum*
	Ascospores without septa 2
2.	Ascospores straight .... *flexuosum*
	Ascospores slightly curved 3
3.	Overgrowing stromatic pyrenomycetes .... *extumescens*
	On dead wood only .... *trabinellum*
4.	Apothecia pale lilac or purplish pink 5
	Apothecia yellow 6
5.	Ascospores 9–13 × 2.5–4 .... *subcarneum*
	Ascospores 14–20 × 5–8 .... *lilacinum*
6.	Ascospores becoming brown, 1-septate 12–18 × 4–5 .... *monticola*
	Ascospores becoming 1-septate, hyaline 11–14 × 3–4 .... *nobilis*

Ascospores with groups of guttules at each end, 12–20 × 3–4.5 ....
.... *umbilicatum*

Phaeohelotium extumescens (P. Karsten)Dennis (Fig. 53)
Apothecia 0.2–0.5mm diam., cream, drying yellow ochre. Ascospores 8–11 × 2.5–3. Overgrowing *Diatrype stigma*, but also on surrounding wood of *Sambucus* and on *Eutypa* on *Fagus*, Oct.–Mar.

Phaeohelotium flexuosum (Crossl.)Dennis (Fig. 54)
Apothecia up to 0.7mm diam., yellow ochre drying reddish brown. Ascospores 8–11 × 3. On rotten wood and bark of *Betula*, *Salix*, also on cupules of *Castanea*, Oct.–July.

Phaeohelotium italicum (Sacc.)Dennis (Fig. 55)
Apothecia 0.2–0.5mm diam., pale yellowish. Ascospores 8–10 × 2–3, 1-septate. On dead wood of *Carpinus*, *Fagus*, *Fraxinus*, *Quercus*, etc., Aug.–Nov.

Phaeohelotium lilacinum (Bres.)Dennis (Fig. 56)
Apothecia up to 4mm diam., pale lilac or greyish. On rotten wood including that of *Alnus*, Oct.–Nov.

Phaeohelotium monticola (Berk.)Dennis (Fig. 57)
Apothecia 1–2mm diam. Brown 1-septate spores minutely verruculose. On wood, Sept.–Oct.

Phaeohelotium nobilis (Velen.)Dennis
Apothecia 1–2mm diam. On rotten wood of *Quercus* and on woody debris, Aug.–Oct.

Phaeohelotium subcarneum (Schum.)Dennis
Apothecia 0.3–0.7mm diam., pale purplish pink. Paraphyses branched with swollen apices. On damp rotten wood of *Carpinus*, *Betula*, etc., Mar.–Apr.

Phaeohelotium trabinellum (P. Karsten)Dennis (Fig. 58)
Apothecia 0.3–0.9mm diam., honey-coloured or pale burnt sienna, paraphyses swollen towards the apex. Ascospores 9–11 × 3–4. On rotten wet wood of *Alnus*, *Rosa*, etc., and on *Fagus* cupules, May–Oct.

Phaeohelotium umbilicatum (Le Gal)Dennis
Apothecia over 1mm diam. On rotten wood, described originally on *Quercus*.

Poetschia cratincola (Rehm)Hafellner (Fig. 59)
Apothecia superficial, up to 0.3mm diam., dark brown to black. Ascospores 1-septate, 30–40 × 10–14, dark brown when mature. On decorticated wood, described originally on *Salix*.

Polydesmia pruinosa (Berk. & Br.)Boud. (Fig. 60)
Apothecia sessile, white, pruinose, growing mostly on stromatic pyrenomycetes such as *Diatrype* and *Diatrypella*. Ascospores hyaline, kidney-shaped, mostly 15–20 × 4–6. Found all the year round but especially in the winter,

when it is very common.

Propolomyces versicolor (Fr.)Dennis (Fig. 61)
Apothecia immersed, up to 5mm diam., with white pruinose discs exposed by peeling-back or shedding of part of periderm of the host. Ascospores hyaline, curved, mostly 22–26 × 6–7. On dead branches of *Acer, Betula, Corylus, Fagus, Fraxinus, Quercus, Rosa, Salix, Sambucus, Sorbus, Tilia,* etc.; found throughout the year but in best condition Feb.–Mar.

Rhizodiscina lignyota (Fr.)Hafellner (Fig. 62)
Apothecia superficial, sessile, up to 1mm diam., black or almost so. Ascospores brown, 1-septate, 9–13 × 4–5. On decorticated wood of *Fagus, Fraxinus, Quercus,* etc., Oct.–July.

Roesleria pallida (Pers.)Sacc. (Fig. 63)
Apothecia stalked, up to 1mm diam., at first cup-shaped, soon disintegrating to form spherical pale-grey powdery spore masses; stalks synnema-like, pale yellow, the hyphae splaying out in heads and some of them protruding. Ascus walls diffluent, spores soon free, hyaline or pale olivaceous, 5–6 × 4–5. On bark including that of dead roots of *Acer, Aesculus, Castanea, Rosa,* etc., Oct.–Mar.

Sarcoscypha coccinea (Scop.)Lambotte (Fig. 64)
Apothecia sometimes stalked, up to 4cm diam., or occasionally even larger, stalk and outer surface white, downy, inside of cup usually deep scarlet, rarely white. Ascospores hyaline, 28–40 × 11–15. On mossy twigs especially of *Corylus* on the ground in damp places; also on fallen branches of *Alnus* and *Salix* in places subject to flooding, Jan.–Mar.

Scutellinia scutellata, described on *Salix,* occurs occasionally on wet wood of other trees, e.g. *Fagus, Fraxinus* and *Quercus.*

Stictis friabilis (Phill. & Plowr.)Sacc. & Traverso (Fig. 65)
Apothecia up to 0.5mm diam., erumpent becoming superficial, discs sunken, reddish, margins broad creamy, pruinose. Ascospores multiseptate, hyaline, 55–70 × 2.5–3.5. On dead wood and bark including that of *Castanea* and *Rubus idaeus.*

Stictis radiata (L.)Pers.
Apothecia up to 0.5mm diam., with yellow discs deeply sunken and surrounded by broad white margins split into lobes. Ascospores hyaline, multiseptate, about 160 × 2–3. Upper part of hymenium stains deep blue with iodine. On dead twigs and branches of *Corylus, Hedera, Quercus, Sambucus, Ulmus,* etc., all the year round.

Tapesia fusca, described on *Alnus,* has been found also on fallen dead branches of *Betula, Corylus, Crataegus, Quercus, Rhododendron, Salix* and *Ulex.*

Tapesia lividofusca (Fr.)Rehm
Rather similar to *T. fusca* with whitish or pale yellowish or grey apothecia seated on an extensive brown subiculum, but the ascospores measure 10–12 × 4. On dead branches of *Betula*, *Carpinus*, etc., Jan.–Aug.

Trichophaeopsis bicuspis (Boud.)Korf & Erb (Fig. 66)
Apothecia 2–4mm diam., white with brown rims, with the outside covered with dark-brown septate setae up to 500 × 15, many of which have a downwardly pointed prong or branch arising near the base. Asci operculate. Ascospores hyaline, 15–20 × 10–11. On rotting sticks on damp soil. We have found this in great abundance in one area on rotting twigs and leaves of *Populus nigra* × Serotina, June–Oct.

Unguicularia scrupulosa (P. Karsten)Höhnel (Fig. 67)
Apothecia 0.1–0.2mm diam., urceolate, brown, covered with highly refractive 'glassy' hairs 15–40 × 4–6. Ascospores hyaline, 5–6 × 1–2. On dead twigs and branches of *Alnus*, *Fagus*, *Populus*, *Rubus idaeus*, *Salix*, etc., Oct.–June.

Velutarina rufo-olivacea (Alb. & Schw.)Korf (Fig. 68)
Apothecia erumpent, sessile, 2–3mm diam., buff often powdered reddish brown, discs dark olivaceous. Swollen cells in the excipulum characteristic. Ascospores becoming pale brown, 10–15 × 7–8. On dead twigs and branches of *Acer*, *Fagus*, *Fraxinus*, *Rosa*, *Rubus*, *Ulex*, *Ulmus*, etc., all the year round, in very good condition usually in Aug.

Vibrissea truncorum (Alb. & Schw.)Fr. (Fig. 69)
Apothecia with orange convex discs up to 5mm diam. and blackish stalks up to 1cm high, partly covered with short black hairs. Ascospores hyaline, multiseptate, thread-like, about 200 × 1.5. On dead twigs and branches half submerged in running streams, May.

OTHER ASCOMYCETES

KEY TO GENERA

	Ascomata pale or brightly coloured 1
	Ascomata dark, mostly brown, blackish brown or black 7
	Ascomata (perithecia) with walls purple by transmitted light, ascospores hyaline, 3-septate .... *Gibberella*
1.	Perithecia not embedded in stromata 2
	Perithecia embedded in stromata, ascospores each dividing to form 2 part-spores 5
2.	Perithecia setose, ascospores with more than 1 septum .... *Trichonectria*
	Perithecia without setae 3
3.	Ascospores with more than 1 septum .... *Calonectria*
	Ascospores with 1 septum 4

4. Ascospores each dividing to form 2 part-spores, perithecia yellow-ochre, seated on a felted subiculum .... *Protocrea*
    Ascospores not dividing to form 2 part-spores, perithecia often red or reddish, solitary or clustered on stromata .... *Nectria*
5. Stromata erect, club-shaped, yellow .... *Podostroma*
    Stromata pulvinate or flattened 6
6. Part-spores hyaline .... *Hypocrea*
    Part-spores green .... *Creopus* and *Chromocrea*
7. Ascomata elongated with slit-like openings (hysterothecia) 8
    Ascomata with laterally compressed necks 11
    Ascomata not so 12
8. Ascospores with 1 transverse septum 9
    Ascospores with several transverse septa .... *Hysterium*
    Ascospores with transverse and longitudinal septa 10
9. Ascospores hyaline, no subiculum .... *Glonium*
    Ascospores becoming brown, hysterothecia nestling among brown hyphae .... *Byssolophis*
    Ascospores with large brown cell and small hyaline one .... *Farlowiella*
10. Ascospores pale yellow .... *Gloniopsis*
    Ascospores brown .... *Hysterographium*
11. Ascospores with transverse septa only .... *Lophiostoma*
    Ascospores with transverse and longitudinal septa .... *Platystomum*
12. Ascomata with walls which split up into radiating polygonal plates 13
    Ascomata not so 14
13. Ascospores kidney-shaped .... *Fragosphaeria*
    Ascospores elliptical .... *Cephalotheca*
14. Ascospores with 1 or more cells darker than the others 15
    Ascospores colourless or evenly coloured 19
15. Mature ascospores with a short, fat, brown cell and a long, often curved, colourless cell 16
    Mature ascospores with 2 large brown cells and a small colourless basal cell .... *Apiorhynchostoma*
    Ascospores with median cells darker than end cells 17
16. Perithecia superficial, clustered, without setae .... *Bombardia*
    Perithecia mostly immersed, solitary, setose around ostiole ... *Cercophora*
17. Ascomata (perithecia) superficial with black velvety subiculum .... *Chaetosphaerella*
    Ascomata mostly or partly immersed 18
18. Ascomata (perithecia) long-necked with stroma .... *Melogramma*
    Ascomata (pseudothecia) short-necked without stroma, thick-walled .... *Trematosphaeria*
19. Ascospores colourless or straw-coloured in mass 20
    Ascospores individually coloured, usually brown 38
20. Ascomata (perithecia) small, black, superficial, nestling in a forest of brown conidiophores .... *Chaetosphaeria* (see also *Trichosphaeria*)

	Ascomata not so 21
21.	Ascomata (perithecia) superficial, black, collapsing from above when old in a cupulate manner, often seated on a dark subiculum 22
	Ascomata not so 23
22.	Perithecia surrounded by black setae .... *Acanthonitschkea*
	Perithecia not surrounded by setae .... *Nitschkia* (see also *Melanopsamma*)
23.	Ascospores 2-septate .... *Melomastia*
	Ascospores without septa 24
	Ascospores 1-septate 33
	Ascospores with 3 or more septa 35
24.	Ascospores straight 25
	Ascospores curved, often allantoid 27
25.	Ascospores large (over 25 long) .... *Botryosphaeria*
	Ascospores medium sized (7–13 long) 26
	Ascospores very small (2.5–3 long) .... *Linostomella*
26.	Perithecia superficial, setose .... *Trichosphaeria*
	Perithecia immersed, smooth, with long necks .... *Ceratostomella*
27.	Perithecia superficial, smooth or setose, ascospores mostly more than 20 long .... *Lasiosphaeria*
	Perithecia immersed, ascospores less than 20 long 28
28.	Asci polysporous 29
	Asci with not more than 8 spores 30
29.	Perithecia solitary or in clusters beneath bark, without ostioles, no stroma .... *Coronophora*
	Perithecia embedded in pulvinate erumpent stromata, necks protuberant .... *Diatrypella*
30.	Necks of perithecia not or scarcely protruding, stroma effuse or discoid .... *Diatrype*
	Necks of perithecia markedly protuberant interspersed with setae or brown synnemata .... *Peroneutypa*
	Necks protuberant but without setae or synnemata 31
31.	Perithecia often arranged in a ring with convergent necks emerging through a stromatic disc .... *Valsa*
	Perithecia more irregularly arranged, ostioles often sulcate 32
32.	Stroma effuse .... *Eutypa*
	Stroma pulvinate, erumpent .... *Eutypella*
33.	Perithecia immersed in groups, often delimited by black stromatic lines, necks protuberant .... *Diaporthe*
	Perithecia superficial 34
34.	Perithecia warted, looking like little black mulberries, ascospores 30 or more long, curved .... *Bertia*
	Perithecia flask-shaped, smooth, spores falcate .... *Ceratocystiopsis*
	Perithecia in dense swarms, tending to collapse in the centre, spores constricted at septum .... *Melanopsamma*
35.	Perithecia immersed, with long necks 36

	Perithecia superficial 37
36.	Ascospores more than 40 long .... *Ceratosphaeria*
	Ascospores less than 15 long .... *Ceratostomella*
37.	Perithecia greenish yellow to almost black with yellow hairs ...... *Tubeufia*
	Perithecia warted like black mulberries .... *Bertia*
	Perithecia black, smooth .... *Zignoella*
38.	Ascospores without septa, often with germ slits, sometimes appendaged 39
	Ascospores septate 42
39.	Perithecia not embedded in a stroma 40
	Perithecia embedded in a stroma 41
40.	Perithecia setose .... *Coniochaeta*
	Perithecia not setose .... *Rosellinia*
41.	Stromata pulvinate, very large, in section distinctly zonate .... *Daldinia*
	Stromata erect, club-shaped or branched .... *Xylaria*
	Stromata erumpent, host tissues often stained yellow .... *Camarops*
	Stromata superficial, pulvinate or effuse .... *Hypoxylon* and *Ustulina*
42.	Ascomata (pseudothecia) overlaid by and nestling among brown hyphae .... *Herpotrichia*
	Ascomata not so 43
43.	Ascospores 1-septate 44
	Ascospores with 3 or more transverse septa 48
	Ascospores with transverse and longitudinal septa 53
44.	Perithecia in groups immersed in a stroma .... *Valsaria*
	No stroma 45
45.	Ascomata (perithecia) setose .... *Neopeckia*
	Ascomata not setose 46
46.	Ascomata (pseudothecia) superficial or only in part immersed .... *Microthelia*
	Ascomata immersed 47
47.	Ascomata (perithecia) mostly solitary .... *Amphisphaeria*
	Ascomata (pseudothecia) in groups seated on an immersed black subiculum .... *Otthia*
48.	Ascospores each dividing into two 1-septate part-spores inside the ascus .... *Ohleria*
	Ascospores not dividing 49
49.	Perithecia solitary or in pairs beneath a clypeus .... *Clypeosphaeria*
	No clypeus 50
50.	Ascospores curved .... *Lasiosphaeria*
	Ascospores not curved 51
51.	Pseudothecia interspersed with masses of orange–red granules and immersed in a stroma .... *Thyridaria*
	Pseudothecia not so 52
52.	Pseudothecia setose .... *Herpotrichiella*
	Pseudothecia not setose .... *Melanomma*

53. Pseudothecia setose, not immersed in a stroma .... *Dictyotrichiella*
    Stroma present 54
54. Stroma superficial, small, pseudothecia bearing short setae ..... *Berlesiella*
    Stroma erumpent, large, pseudothecia not setose .... *Fenestella*

Acanthonitschkea tristis, described on *Acer*, is found occasionally also on fallen dead branches of *Fagus*, *Prunus* and *Ulmus*.

Amphisphaeria millepunctata (Fuckel)Petrak (Fig. 70)
Perithecia immersed, 0.3–0.4mm diam., with short just-protruding necks appearing as black dots on the surface. Ascospores brown, 1-septate, 12–17 × 5–7. On dead twigs of *Acer*, *Carpinus*, *Corylus*, *Quercus*, *Salix*, *Sorbus*, etc.

Apiorhynchostoma curreyi (Rabenh.)E. Müller (Fig. 71)
Perithecia partly immersed, up to 0.4mm diam. Ascospores 2-septate, small basal cell hyaline, other cells brown, 22–28 × 6–8. On decorticated conifer wood. Figured from material on a fence post.

Berlesiella nigerrima (Bloxam ex Currey)Sacc. (Fig. 72)
Pseudothecia crowded 0.05–0.1mm diam., black, with very short projecting dark-brown spines, partly immersed in dark-grey pulvinate stromata about 1mm across. Ascospores pale olivaceous brown, 15–21 × 5–6, with 4 to 6 (usually 5) transverse septa and 1 longitudinal septum. Mostly on old stromata of *Diatrype stigma*, *Eutypa acharii*, and *Hypoxylon multiforme* on wood of *Acer*, *Corylus*, *Fagus*, *Fraxinus*, *Hedera*, *Ilex*, *Prunus* and *Salix*, Sept.–Apr.

Bertia moriformis (Tode)de Not. (Fig. 73)
Perithecia superficial, up to 1.5mm high, black, thick-walled, without ostioles, warted on the outside and resembling little mulberries. Ascospores hyaline, 1-septate, 30–50 × 4.5–5.5. On dead wood of *Acer*, *Alnus*, *Fagus*, *Fraxinus*, *Picea*, *Sambucus*, *Tilia*, *Ulmus*, etc., all the year round. It seems to favour wood lying on mossy banks.

Bertia moriformis var. *multiseptata* A. Sivan. (Fig. 74)
In this variety the ascospores tend to be more fusiform, develop 3 to 8 septa and often form little phialides. On dead wood, May.

Bombardia bombarda (Batsch)Schröter (Fig. 75)
Perithecia clustered, superficial, black, shining, up to 1.5mm high. Ascospores at first hyaline, cylindrical, 30–40 × 3–4, with an appendage at each end, then swelling at one end, a septum cutting off a dark-brown cell 12–15 × 8–10. On dead wood, especially stumps, of *Fagus*, *Fraxinus*, etc., Sept.–Nov.

Botryosphaeria obtusa (Schw.)Shoemaker (Fig. 76)
Pseudothecial stromata immersed, up to 3mm diam., dark-brown to black, thick-walled. Ascospores hyaline, 26–33 × 7–12. Pycnidia thick-walled, up to 0.5mm diam.; conidia brown, with thin verruculose walls, 20–26 × 9–12. On

branches of *Acer, Alnus, Betula, Crataegus, Malus, Rhamnus, Salix, Ulmus* and *Viburnum*, causing cankers and dieback.

Byssolophis sphaerioides (P. Karsten)E. Müller (Fig. 77)
Pseudothecia up to 1.5 × 0.5mm, black, seated on a thin brown hyphal subiculum. Ascospores 1-septate, 16–20 × 5–6, hyaline in asci but finally mid–pale brown. On wood and bark, uncommon, several records on *Rubus*, seen also on dead stems of *Erica*.

Calonectria pseudopeziza (Desm.)Sacc. (Fig. 78)
Colonies sometimes covering large areas. Perithecia superficial, smooth, 0.3–0.35mm diam., pale tangerine with creamy spore tendrils when fresh and moist. Ascospores hyaline, 2- to 8-septate, 28–50 × 4–6. On dead branches of *Berberis, Cytisus, Fagus, Ilex, Lupinus arboreus, Populus, Sambucus, Ulex, Ulmus*, etc., Sept.–Apr.

Camarops lutea, described on *Buxus*, has been found also on dead branches and stumps of *Acer, Alnus, Corylus, Fagus, Fraxinus, Quercus* and *Sambucus*. Surrounding wood is often stained bright yellow.

Cephalotheca sulfurea Fuckel (Fig. 79)
Cleistothecia 0.25–0.5mm diam., black, with walls made up of polygonal radiating plates covered when young with sulphur-coloured hyphae. Ascospores pale brown, 4–6 × 3–4. On rotting wood including that of *Quercus*.

Ceratocystiopsis falcata (Wright & Cain)Upadhyay
Perithecia superficial, dark brown or black, up to 0.08mm diam., with short conical necks. Asci evanescent, spores falcate, faintly 1-septate, 22–28 × 1–1.5. Accompanied by a *Chalara* state. On cut stumps of *Betula, Fagus, Quercus*, etc.

Ceratosphaeria lampadophora (Berk. & Br.)Niessl (Fig. 80)
Perithecia deeply immersed, up to 0.8mm diam., black, necks variable: some long and curved, others shorter. Ascospores hyaline, multiseptate, 40–60 × 4–4.5. In very rotten wood of *Fagus, Fraxinus, Salix*, etc.

Ceratostomella ampullasca (Cooke)Sacc. (Fig. 81)
Perithecia immersed except for long necks, black, up to 0.5mm diam. Ascospores hyaline with 2 large guttules, 9–13 × 3.5–4.5. On rotten wood of *Acer, Betula, Fagus, Fraxinus, Ilex, Quercus, Ulmus*, etc., usually in good condition Apr.–May.

Ceratostomella cirrhosa (Pers.)Sacc. (Fig. 82)
Perithecia immersed except for long necks, black, up to 0.4mm diam. Ascospores hyaline, 9–11 × 4–5, often becoming rather faintly 1- to 3-septate. On rotten wood of *Fagus, Fraxinus, Populus, Prunus spinosa, Ulmus*, etc.

Cercophora caudata (Currey)Lundq. (Fig. 83)
Perithecia 0.5–0.7mm diam., brown, thinly hairy, immersed, with dark protuberant necks which bear short thick-walled setae or processes. Ascospores at first cylindrical, flexuous, hyaline, 40–60 × 5–6 with a small appendage at each end, later swelling to 9–11 at the upper end, which becomes brown and cut off by a septum. Additional septa occasionally develop in very old spores. On rotten wood of *Fagus, Quercus, Sorbus*, etc., Aug.–Mar.

Chaetosphaerella fusca (Fuckel)E. Müller & Booth (Fig. 84)
Perithecia up to 0.4 × 0.3mm, black. Ascospores 3-septate, 20–26 × 7–8.5, median cells golden brown, end cells hyaline. Perithecia formed on brown to black hyphal mats, which also bear branched conidiophores of the *Oedemium* state. Conidia dumb-bell-shaped, 12–20 × 9–14, 1-septate, brown in the middle, paler at each end. Associated with diatrypaceous fungi on fallen branches of *Acer, Fagus, Ilex, Salix* and *Tilia*, Sept.–Feb., uncommon.

Chaetosphaerella phaeostroma (Durieu & Mont.)E. Müller & Booth (Fig. 85)
Perithecia up to 0.35 × 0.5mm, dark grey to black, seated on black velvety mycelial mats, which also bear dark-brown to black setae and branched nodose conidiophores. Ascospores curved, 3-septate, 32–35 × 6–9, with brown median cells and hyaline end cells. Conidia similarly coloured but straight 20–35 × 10–15. Commonly growing over or associated with other fungi such as *Diatrype stigma* and *Eutypa flavovirens* on fallen branches of *Acer, Alnus, Corylus, Fagus, Fraxinus, Hedera, Ilex, Populus, Prunus spinosa, Quercus, Ribes, Salix, Sambucus, Tilia* and *Ulmus*, Sept.–Apr.

Chaetosphaeria, mostly superficial, small black perithecia accompanied by conidial states.

KEY

	Ascospores becoming more than 1-septate	1
	Ascospores not more than 1-septate	2
1.	Ascospores 1- to 5-septate up to 32 long, with *Catenularia* conidial state	*cupulifera*
	Ascospores 1- to 3-septate up to 22 long, with *Menispora* conidial state	*pulviscula*
2.	Ascospores more than 12 long	3
	Ascospores less than 12 long	4
3.	Perithecia always seated among long dark setae; *Codinaea* conidial state	*callimorpha*
	Perithecia without surrounding setae; *Catenularia* conidial state	*innumera*
4.	Ascospores not dividing into 2 part-spores; *Catenularia* conidial state	*myriocarpa*
	Ascospores each dividing into 2 part-spores	5

5. Part-spores unequal in size, *Gonytrichum* conidial state .... *inaequalis*
    Part-spores equal; *Chloridium* conidial state 6
6. Ascospores 5–7 × 1.5–2 .... *preussii*
    Ascospores 7–10 × 2–3 .... *vermicularioides*

Chaetosphaeria callimorpha, described on *Rubus*, occurs occasionally also on fallen dead branches of *Castanea, Corylus, Quercus*, etc.

Chaetosphaeria cupulifera (Berk. & Br.)Sacc. (Fig. 86)
Perithecia about 0.2mm diam., black, verrucose. Ascospores often slightly curved, hyaline, 1- to 5-septate, 17–32 × 4–5. *Catenularia* conidial state has brown conidiophores up to 300 × 6–8 with open cups at apex 10–11 wide, and wedge-shaped brown conidia 9–14 × 9–10 (5–7 at base). On dead fallen branches of *Fagus, Fraxinus, Hedera, Ilex, Quercus, Taxus* and *Ulmus*, Dec.–June.

Chaetosphaeria inaequalis (Grove ex Berl. & Vogl.)W. Gams & Hol.-Jech. (Fig. 87)
Perithecia about 0.2mm diam., black. Ascospores hyaline, at first 1-septate 6–9 × 2–3 but soon splitting into 2 part-spores, the upper shorter than the lower. *Gonytrichum* state often present, colonies grey when sporing freely, later dark blackish brown. Conidiophores up to 500 × 2–3, dark brown with a very complex branching system, the branches often anastomosing and difficult to tease apart. Conidia hyaline 2–3 × 1–1.5. Very common on dead branches lying on the ground, especially where covered by leaf mould; found on *Acer, Betula, Buxus, Corylus, Crataegus, Fagus, Fraxinus, Hedera, Larix, Quercus, Rhamnus, Rhododendron, Ribes, Rubus, Sambucus, Ulex* and *Ulmus*, all the year round.

Chaetosphaeria innumera Tul. & C. Tul. (Fig. 88)
Perithecia 0.15–0.2mm diam., smooth. Ascospores hyaline, 1-septate, 13–18(20) × 3–4. Associated *Catenularia* state has brown conidiophores up to 150 × 3–6 which often proliferate percurrently, and hyaline conidia 3–5 × 2–3. On fallen dead branches of *Acer, Alnus, Betula, Castanea, Fagus, Fraxinus, Juglans, Quercus, Sambucus, Sorbus* and *Ulmus*, Sept.–May.

Chaetosphaeria myriocarpa (Fr.)Booth (Fig. 89)
Perithecia 0.1–0.15mm diam. Ascospores hyaline 0- or 1-septate, 5–7 × 2–3. Associated *Catenularia* state has brown conidiophores up to 80 × 3–4 with hyaline conidia 2–2.5 × 1.5–2. Very common all the year round on fallen dead branches of *Acer, Alnus, Betula, Carpinus, Castanea, Corylus, Crataegus, Cytisus, Fagus, Fraxinus, Ilex, Prunus spinosa, Quercus, Rubus, Tilia, Ulex* and *Ulmus*.

Chaetosphaeria preussii W. Gams & Hol.-Jech. (Fig. 90)
Perithecia 0.15–0.2mm diam. Ascospores hyaline, 1-septate, 5–7 × 1.5–2 readily breaking into almost equal part-spores. Associated *Chloridium* state has brown conidiophores up to 80 × 3–5 with frequent percurrent proliferation,

conidia 2.5–3.5 × 1.5–2, in slimy heads. On fallen dead branches of *Acer, Corylus, Cytisus, Quercus, Rubus* and *Taxus*, Nov.–May.

Chaetosphaeria pulviscula (Currey)Booth (Fig. 91)
Perithecia erumpent, 0.15–0.25mm diam. Ascospores hyaline, 1- to 3-septate, 18–22 × 3.5–4. Associated *Menispora* state has brown setae up to 500 × 3–5 and branched conidiophores up to 350 × 3–4. Conidia slightly curved, hyaline, 16–22 × 3–4. On dead, often rotten wood of *Acer, Alnus, Betula, Carpinus, Castanea, Corylus, Fagus, Fraxinus, Hedera, Quercus, Salix, Sambucus, Ulex* and *Ulmus*, Oct.–Apr.

Chaetosphaeria vermicularioides (Sacc. & Roum.)W. Gams & Hol.-Jech. (Fig. 92)
Perithecia 0.1–0.3mm diam. Ascospores hyaline, smooth to minutely verruculose, 1-septate, 7–10 × 2–3, breaking readily into equal-sized part-spores. *Chloridium* conidial state has brown conidiophores 30–60 × 2–4 and conidia 2–3 × 2–2.5, in greenish masses, often in columns. On rotten wood of *Alnus, Fraxinus* and *Quercus*.

Chromocrea aureoviridis (Plowr. & Cooke)Petch (Fig. 93)
Stromata about 2.5mm diam., yellow, at first pulvinate then pezizoid. Part-spores dark green in mass, verruculose, globose or subglobose 4–4.5 × 4. *Trichoderma* state has smooth, yellowish-green conidia 3–4.5 × 2–3. On dead wood of *Acer, Corylus, Fagus* and *Sambucus*, Oct.–Mar.

Clypeosphaeria mamillana (Fr.)Lambotte, which closely resembles *C. notarisii* described on *Rubus*, is found occasionally on dead twigs of *Acer, Cornus* and *Quercus*.

Coniochaeta, perithecia black, shortly setose, superficial or partly immersed, ascospores brown with germ slits.

KEY

Ascospores 6–10 × 4–6 .... *velutina*
Ascospores 9–12 × 7–8 .... *pulveracea*
Ascospores 12–18 × 8–12 .... *ligniaria*

Coniochaeta ligniaria, described on *Ulex*, has been found also on conifer wood and on dead branches of *Fagus, Quercus* and *Ulmus*.

Coniochaeta pulveracea (Ehrenb. ex Pers.)Munk (Fig. 94)
Perithecia partly immersed or superficial, 0.2–0.4mm diam., black, with setae 5–11 × 3–4. Ascospores brown, mostly 9–12 × 7–8. On bark and decorticated wood of *Fraxinus, Quercus*, etc., Jan.–Mar.

Coniochaeta velutina (Fuckel)Cooke
Perithecia superficial or partly immersed, 0.15–0.2mm diam., black, with setae

20–30 × 3–4. Ascospores brown, 6–10 × 4–6. On fallen dead branches of *Acer, Aesculus, Corylus, Fagus, Quercus, Ulex*, etc.

Coronophora angustata Fuckel (Fig. 95)
Perithecia 0.5–1mm diam., black, often in small groups, without any stroma, immersed in bark. Asci polysporous; spores hyaline curved, 4–6 × 1. On twigs and branches of *Alnus, Betula, Corylus, Fagus, Fraxinus, Tilia*, etc.

Coronophora gregaria, described on *Sorbus*, occurs also on branches of other deciduous trees; it has perithecia up to 2mm diam. and ascospores 7–9 × 1–2.

Creopus gelatinosus (Tode)Link (Fig. 96)
Stromata pulvinate, 1–3mm diam., pale yellow, translucent, with dark contents of perithecia clearly visible through walls. Part-spores dark green verruculose, unequal, upper one globose, 4 diam., lower one 5–6 × 3–4. On dead branches of *Acer, Betula, Corylus, Fagus, Fraxinus, Salix*, etc., June–Dec.

Daldinia concentrica, described on *Fraxinus*, occurs also occasionally on *Acer, Alnus, Betula, Fagus, Salix*, etc.

Diaporthe, black perithecia immersed, usually in groups, with necks emerging through the bark and surrounded by stromatic tissue which appears in sections through the wood in the form of thin black lines. Ascospores hyaline, 1-septate with 4 guttules, 10–15 × 2.5–4.

KEY

    Ascospores mostly constricted at septum .... *eres*
    Ascospores not markedly constricted at septum 1
1.   Causing dark-brown 'leopard's' spots, each surrounded by a
    black, slightly raised line .... *pardalota*
    Not so, perithecial necks long and sinuous .... *rudis*

Diaporthe eres (Fig. 97), described on *Ulmus* its type host, has been recorded on many other trees and shrubs including *Abies, Acer, Aesculus, Cedrus, Corylus, Cupressus, Cytisus, Euonymus, Fraxinus, Hedera, Ilex, Ligustrum, Lonicera, Malus, Prunus, Pyrus, Rhododendron, Rosa, Rubus, Salix, Sorbus, Symphoricarpus, Tilia* and *Ulex*.

Diaporthe pardalota, described on *Lonicera*, is common also on *Cornus, Ilex* and *Rubus*.

Diaporthe rudis, described on *Laburnum*, has been found also on *Crataegus, Fagus, Fraxinus, Lupinus, Rubus* and *Ulex*.

Diatrype disciformis, described on *Fagus*, has been recorded occasionally also on *Acer, Betula* and *Carpinus*.

Diatrype stigma (Hoffm.)Fr. (Fig. 98)
Stromata often forming widely effused purplish-brown to black crusts over the surface of decorticated wood, or erumpent through periderm causing it to roll back in a characteristic manner. Perithecia immersed, about 0.3mm diam., with short necks only very slightly protruding above the stroma. Ascospores allantoid, mostly 6–8 × 1.5, pale straw-coloured in mass. Very common all the year round on dead branches of *Acer, Aesculus, Betula, Carpinus, Castanea, Crataegus, Corylus, Fagus, Fraxinus, Ilex, Malus, Populus, Prunus, Quercus, Salix, Sorbus, Syringa, Tilia, Ulex, Ulmus,* etc.

Diatrypella favacea, described on *Betula*, is found almost equally commonly on attached and fallen branches of *Fagus* and less frequently on *Alnus, Corylus, Crataegus* and *Quercus*.

Dictyotrichiella pulcherrima Munk (Fig. 99)
Pseudothecia erumpent, black, 0.15mm diam., covered with dark brown septate setae 20–40 × 3–4. Ascospores pale greyish brown, with up to 5 transverse septa and 1 longitudinal septum, 10–13 × 3–4.5. On rotten wood.

Eutypa, widely effused blackish stromata, somewhat protuberant necks and small curved ascospores coloured in mass.

KEY

    Cut stromata greenish yellow .... *flavovirens*
    Cut stromata not greenish yellow 1
1.   Ascospores 5–7 × 1 .... *acharii*
    Ascospores 8–12 × 2 .... *lata*

Eutypa acharii, described on *Acer*, has been found also on dead branches of *Betula, Castanea* and *Salix*.

Eutypa flavovirens (Fr.)Tul. & C. Tul. (Fig. 100)
Stromata pulvinate but rather thin, often elongated, dull black outside but greenish yellow inside (seen when sliced through with a knife). This is a good diagnostic character in the field. Perithecia immersed, 0.4–0.5mm diam., necks just protruding. Ascospores curved, hyaline, golden brown in mass, 7–10 × 2. On dead wood of *Acer, Betula, Corylus, Crataegus, Fagus, Fraxinus, Prunus spinosa, Quercus, Tilia* and *Ulmus* throughout the year, very common.

Eutypa lata (Pers.)Tul. & C. Tul. (Fig. 101)
Stromata effuse, dark grey to black. Perithecia immersed, up to 0.5mm diam., necks protruding, black, shining. Ascospores curved, hyaline, golden brown in mass, 8–12 × 2. On dead twigs and branches of *Crataegus, Fraxinus, Hedera, Ulmus,* etc.

Eutypella acericola, described on *Acer*, is found occasionally on dead branches of other trees, e.g. *Fraxinus* and *Sambucus*.

Farlowiella carmichaeliana (Berk.)Sacc. (Fig. 102)
Hysterothecia up to 3 × 0.25mm, dark grey to black, smooth. Ascospores 19–26 × 9–12, 1-septate, upper cell large, very dark brown, lower cell small, hyaline. *Acrogenospora* state with dark-brown conidiophores up to 400 × 5–9 (9–12 at base), conidia dark brown, 20–40 × 15–25. On dead wood and bark of *Alnus, Betula, Corylus, Cytisus, Fagus, Fraxinus, Larix, Pinus, Prunus, Quercus, Sorbus, Taxus,* Feb.–Apr.

Fenestella fenestrata (Berk. & Br.)Schröter (Fig. 103)
Pseudothecia embedded in large, pulvinate, erumpent blackish-brown stromata. Ascospores brown, muriform, with a protruding hyaline cell at each end, mostly 40–60 × 17–22. On dead twigs and branches of *Alnus, Corylus, Quercus,* etc.

Fenestella vestita, described on *Acer,* has been found also on *Corylus, Fagus, Tilia* and *Ulmus.* Ascospores 18–30 × 9–13.

Fragosphaeria reniformis (Sacc. & Therry)Malloch & Cain (Fig. 104)
Cleistothecia black, 0.25–0.35mm diam., with walls made up of radiating polygonal plates, covered when young with blackish-brown hairs. Ascospores kidney-shaped, olivaceous brown, 4–5 × 3–3.5. In cavities in rotten wood of *Fagus, Quercus, Populus,* etc.

Gibberella pulicaris, described on *Lupinus,* occurs also on dead branches of *Clematis, Cytisus, Fraxinus, Humulus, Juniperus, Populus, Sambucus, Ulmus,* etc.

Gloniopsis praelonga, described on *Rubus,* is found commonly throughout the year also on dead twigs and branches of many trees and shrubs including *Acer, Betula, Corylus, Fagus, Fraxinus, Hedera, Ilex, Ligustrum, Lonicera, Malus, Populus, Prunus, Quercus, Rhamnus, Rhododendron, Rosa, Salix, Sambucus, Ulex* and *Ulmus.*

Glonium lineare (Fr.)de. Not. (Fig. 105)
Hysterothecia black, very narrow, up to 2mm long. Ascospores hyaline, 1-septate, 14–20 × 5–7. On decorticated wood, uncommon. We have seen especially good material on twigs of *Ilex.*

Herpotrichia herpotrichoides, described on *Rubus,* has been found on dead twigs of *Acer.*

Herpotrichia macrotricha, more commonly found on dead herbaceous stems, has been recorded on fallen twigs and branches of *Acer, Fagus, Rosa, Rubus* and *Salix.*

Herpotrichiella moravica Petrak (Fig. 106)
Pseudothecia superficial, 0.1–1.2mm diam., black, covered with dark brown pointed hairs 20–30 × 2–3. Ascospores olivaceous, 3-septate, 10–14 × 3–4. We have found this only on wet wood of *Salix* in March, but it has been recorded on other trees, e.g. *Fagus.*

Herpotrichiella pilosella (P. Karsten)Munk ex Barr (Fig. 107)
Pseudothecia superficial or immersed, 0.1–0.2mm diam., blackish brown to black, with short setae forming a compact ring around the ostiole and a few others scattered over the surface. Ascospores olivaceous, 3-septate, 14–17 × 4.5–5. On dead twigs and on rotten wood including that of *Betula*.

Hypocrea, stromata superficial, often pulvinate or discoid, usually brightly coloured, with embedded perithecia; ascospores each dividing to form two part-spores, which are hyaline.

KEY

	Stromata rusty brown with broad white margins .... *rufa*
	Stromata gamboge becoming bright reddish orange .... *splendens*
	Stromata dark olive to black .... *schweinitzii*
	Stromata at first white or pale yellow, becoming clay-coloured or brownish 1
1.	Stromata pulvinate, part-spores unequal, upper 3–3.5 diam., lower 4–4.5 × 3 .... *argillacea*
	Stromata effuse, part spores unequal, upper 4–5 diam., lower 4.5–7 × 3–4 .... *citrina*

Hypocrea argillacea Phill. & Plowr. (Fig. 108)
Stromata pulvinate, 1–4mm diam., at first lemon yellow, becoming clay-coloured and collapsing when old. Part-spores verruculose, unequal. On dead branches and rotten wood of *Betula*, *Castanea*, etc., Oct.–Mar.

Hypocrea citrina (Pers.)Fr. (Fig. 109)
Stromata widely effused, whitish to pale lemon or creamy yellow, becoming brownish when old but remaining pale at byssoid margin, dotted with darker brown ostioles. Perithecia 0.2mm diam. Spores hyaline, very minutely echinulate (for measurements see key). On rotting wood especially of conifers and on debris, Aug.–Sept.

Hypocrea rufa (Pers.)Fr. (Fig. 110)
Stromata discoid or pulvinate, 2–10mm diam., rusty brown often with a broad white margin of mycelium. Part-spores hyaline, minutely verruculose, 3.5–4.5 diam. or 5–6 × 3.5–4. Accompanied or preceded by its *Trichoderma* state with much-branched hyaline conidiophores, the branching almost at right angles, and verruculose, globose or subglobose, blue–green conidia 3.5–4.5 diam. Common on rotten wood of *Acer*, *Betula*, *Corylus*, *Fagus*, *Quercus*, *Pinus*, *Populus*, etc., July–Nov.

Hypocrea schweinitzii (Fr.)Sacc.
Stromata undulating, up to 1cm diam., very dark olive to black. Part-spores equal, globose, almost smooth, 3.5 diam., with large guttules. On dead wood including very rotten trunks of *Quercus*, Sept.–Nov.

30 Plurivorous Wood and Bark Fungi

Hypocrea splendens Phill. & Plowr. (Fig. 111)
Stromata pulvinate to hemispherical, 4–6mm diam., gamboge, becoming bright reddish orange. Part spores 4–5 diam. or 4.5–6 × 3.5–4, minutely verruculose. On dead branches of *Laurus* and *Rubus*, Sept.–Oct.

Hypoxylon: a few species are strictly host limited, six show a distinct preference for one particular host but are found from time to time on dead branches of other trees and these are *H. confluens* (*Quercus*), *H. fuscum* and *H. howeianum* (*Corylus*), *H. multiforme* (*Betula*), *H. rubiginosum* (*Fraxinus*) and *H. rutilum* (*Fagus*); they are often accompanied or preceded by *Nodulisporium* conidial states. The only species without any special host preference is *H. serpens*.

KEY

	Ascospores not more than 10 × 4.5 1
	Ascospores often more than 10 × 4.5 2
1.	Ostioles umbilicate .... *howeianum*
	Ostioles papillate .... *rutilum*
2.	Ascospores 8–12 wide .... *confluens*
	Ascospores not more than 8 wide 3
3.	Ascospores not more than 5 wide .... *multiforme*
	Ascospores often more than 5 wide 4
4.	Ostioles papillate .... *serpens*
	Ostioles umbilicate 5
5.	Stromata widely effused, thin, reddish purple .... *rubiginosum*
	Stromata mostly pulvinate or hemispherical .... *fuscum*

Hypoxylon confluens, described on *Quercus*, found also on *Betula*, *Castanea*, *Fagus*, *Fraxinus* and *Ulmus*.

Hypoxylon fuscum, described on *Corylus*, found also on *Alnus*, *Betula*, *Fagus*, *Prunus padus*.

Hypoxylon howeianum, described on *Corylus*, found also on *Acer*, *Castanea*, *Crataegus*, *Fraxinus*, *Prunus spinosa* and *Quercus*.

Hypoxylon multiforme, described on *Betula*, found also on *Acer*, *Alnus*, *Corylus*, *Crataegus*, *Fagus*, *Prunus*, *Pyrus*, *Salix*, *Sambucus* and *Sorbus*.

Hypoxylon rubiginosum, described on *Fraxinus*, found also on *Acer*, *Fagus*, *Malus*, *Salix* and *Ulmus*.

Hypoxylon rutilum, described on *Fagus*, found also on *Betula*.

Hypoxylon serpens (Pers.)Fr. (Fig. 112)
Stromata superficial or erumpent, irregular in size and shape, often elongated, when mature, dark purplish brown to black; perithecia prominent with distinctly papillate ostioles. Ascospores dark brown, 10–17 × 5–8. *Geniculosporium* state forms effuse greyish colonies with pale brown conidiophores up

to 300 × 2–3 and conidia 2.5–4.5 × 2–3. On dead branches of *Acer, Betula, Castanea, Corylus, Crataegus, Fagus, Fraxinus, Hedera, Picea, Populus, Prunus laurocerasus, Quercus, Salix, Sorbus, Ulex* and *Ulmus*.

Hysterium: four species occur on wood and bark but only one, *H. angustatum*, is at all common. All have longitudinally furrowed hysterothecia.

KEY

	Ascospores 4- to 9-septate .... *insidens*
	Ascospores 3-septate 1
1.	End cells of ascospores much paler than median cells .... *pulicare*
	Cells uniformly golden brown 2
2.	Ascospores 12–18 × 5–6, on conifers .... *acuminatum*
	Ascospores 17–24 × 6–7, on hardwoods .... *angustatum*

Hysterium acuminatum Fr.
An uncommon species on bark of *Pinus* and *Larix*.

Hysterium angustatum Alb. & Schw. (Fig. 113)
Hysterothecia about 1mm long. Common, especially on dead branches of *Acer, Betula, Cornus, Crataegus, Fagus, Fraxinus, Populus, Prunus spinosa, Quercus, Rhododendron, Rubus, Ulex, Ulmus* and *Viburnum* which have lost their bark, Mar.–May.

Hysterium insidens Schw. (Fig. 114)
We have not found hysterothecia of this species but the pulvinate black sporodochia of its *Coniosporium* state with long chains of brown, verrucose conidia with transverse and longitudinal septa mostly 20–30 × 7–11 are found occasionally on fence posts and other decorticated wood.

Hysterium pulicare Pers (Fig. 115)
Ascospores with median cells golden brown and end cells subhyaline, 20–30 × 7–9. Usually on bark of *Betula, Hedera, Fraxinus* and *Quercus*, Mar.–May.

Hysterographium mori (Schw.)Rehm (Fig. 116)
Hysterothecia 0.5–2mm long, dark grey. Ascospores pale to mid golden brown, with transverse and longitudinal septa, mostly 15–25 × 7–10. On old fence posts and dead branches of *Lonicera, Populus* and *Salix*, uncommon.

Lasiosphaeria, perithecia superficial, ascospores more or less cylindrical, often bent or flexuous.

KEY

	Perithecia appearing smooth 1
	Perithecia obviously setose or hairy 2
1.	Perithecia black, shining .... *spermoides*
	Perithecia white with small black ostioles .... *ovina*

2. Ascospores over 45 long, becoming septate 3
   Ascospores not more than 40 long, without septa 4
3. Ascospores 3–5 wide, with basal appendage when young .... *caudata*
   Ascospores 6–7 wide, without appendages .... *hirsuta*
4. Setae flexuous, thin-walled, clumped together .... *phyllophila*
   Setae straight, pointed, very thick-walled .... *canescens*

Lasiosphaeria canescens (Pers.)P. Karsten (Fig. 117)
Perithecia brown, about 0.5mm diam. with very thick-walled, straight, pointed setae. Ascospores hyaline or slightly yellowish, 30–40 × 4–5. On rotten wood of *Acer, Alnus, Fagus, Fraxinus, Quercus, Salix, Sorbus, Ulmus* and *Viburnum*, Nov.–June.

Lasiosphaeria caudata (Fuckel)Sacc. (Fig. 118)
Perithecia 0.3–0.5mm diam., brown, covered with dark brown, thick-walled setae. Ascospores hyaline with a small appendage at the base when young, becoming pale brown and 3-septate, 45–50 × 3–5. On dead branches of *Acer, Alnus, Crataegus, Larix, Quercus, Rubus* and *Salix*, July–Aug.

Lasiosphaeria hirsuta (Fr.)Ces. & de Not. (Fig. 119)
Perithecia 0.4–0.5mm diam., brown, covered with dark brown setae. Ascospores becoming pale brown and 7-septate when mature, 55–75 × 6–7, sometimes germinating to form small phialides. Very common on old wood of *Fagus* and *Fraxinus*, frequent also on *Acer, Alnus, Crataegus, Hedera, Quercus, Salix, Sorbus* and *Ulmus*, Sept.–Apr.

Lasiosphaeria ovina (Pers.)Ces. & de Not. (Fig. 120)
Perithecia white with dot-like dark ostioles, appearing smooth but in fact coated with a short dense white felt. Ascospores hyaline, 40–50 × 4. Very common especially on wood which has been decayed by *Armillaria mellea*, the perithecia frequently being found near its rhizomorphs. On *Acer, Betula, Cornus, Fagus, Fraxinus, Ilex, Quercus, Rubus, Salix, Sambucus, Tilia*, etc., Sept.–Apr.

Lasiosphaeria phyllophila Mouton (Fig. 121)
Perithecia dark brown to black, about 0.5mm diam., smooth around the ostiole, long hairs clumped and radiating. Ascospores hyaline, 24–30 × 3.5–4.5. On fallen dead twigs and debris of *Acer, Populus* and *Salix*, Nov.–Apr.

Lasiosphaeria spermoides (Hoffm.)Ces. & de Not. (Fig. 122)
Perithecia up to 0.6mm diam., black, shining, usually closely packed together in large clusters. Ascospores hyaline, 17–25 × 3–4. On rotting wood of *Acer, Alnus, Betula, Crataegus, Fagus, Fraxinus, Prunus, Quercus, Salix, Sorbus* and *Ulmus*, Nov.–Apr., very often associated with *Armillaria mellea*, common.

Linostomella sphaerosperma (Fuckel)Petrak (Fig. 123)
Perithecia about 0.25mm diam., with slender necks protruding from

superficial black stromata. Ascospores hyaline, 2.5–3 × 2–2.5. On decorticated wood, Apr.–May, uncommon.

Lophiostoma nucula (Fr.)Ces. & de Not. (Fig. 124)
Pseudothecia 0.3–0.4mm diam., immersed except for thinly compressed necks. Ascospores hyaline, 1- to 3-septate, 17–26 × 4–6. On wood of *Acer*, *Populus*, *Quercus*, *Salix* and *Ulmus*, Apr.–July.

Melanomma fuscidulum (Sacc.)Sacc. (Fig. 125)
Pseudothecia almost entirely immersed or erumpent, blackish-brown to black, 0.4–0.6mm diam. Ascospores 3-septate, pale to mid golden brown, 12–16 × 3–4.5. *Aposphaeria* state has hyaline conidia 3–4.5 × 1.5. Common on fallen dead branches of *Acer*, *Cytisus*, *Fagus*, *Fraxinus*, *Hedera*, *Lonicera*, *Quercus*, *Ulex* and *Ulmus*, Mar.–May.

Melanomma pulvis-pyrius (Pers.)Fuckel (Fig. 126)
Pseudothecia superficial, 0.4–0.5mm diam., black, often in very large clusters, looking like grains of gunpowder. Ascospores rather pale brown, 3-septate, 14–18 × 4–6. *Aposphaeria* state has hyaline conidia 4–5 × 1–2. Very common, especially on dry hard wood of decorticated branches of *Fagus*, *Fraxinus* and *Ulmus*, but found also on many other trees and shrubs including *Acer*, *Alnus*, *Betula*, *Carpinus*, *Clematis*, *Corylus*, *Crataegus*, *Cytisus*, *Ilex*, *Lonicera*, *Populus*, *Prunus*, *Quercus*, *Rhamnus*, *Rhododendron*, *Ribes*, *Rosa*, *Rubus*, *Salix*, *Sorbus* and *Ulex*, Sept.–May.

Melanopsamma pomiformis (Pers.)Sacc. (Fig. 127)
Perithecia superficial, thickly clustered, black, 0.25–0.35mm diam., tending to collapse in the centre when old. Ascospores hyaline, 1-septate, 12–18 × 5–8. *Stachybotrys* state has hyaline conidiophores up to 250 long, and greenish- or greyish-brown smooth conidia 6–11 × 4.5–7. On wood of *Aesculus*, *Alnus*, *Carpinus*, *Fagus*, *Fraxinus*, *Juglans*, *Malus*, *Populus* and *Ulmus*, Nov.–Apr.

Melogramma campylosporum, described on *Carpinus*, has been recorded also on *Corylus* and *Fagus*.

Melomastia mastoidea (Fr.)Schröter (Fig. 128)
Perithecia immersed or erumpent, 0.5–0.8mm diam., black. Ascospores hyaline, 2-septate, 14–18 × 4–6. On dead twigs and branches of *Cornus*, *Fraxinus*, *Hedera*, *Lonicera*, *Populus*, *Sambucus*, *Symphoricarpus* and *Viburnum*.

Microthelia incrustans (Ell. & Ev.)Corlett & Hughes (Fig. 129)
Pseudothecia mostly superficial, conical, black, 0.5–0.8mm diam. Ascospores brown, 1-septate, very variable in size, mostly 25–35 × 8–11 but occasionally up to 45 × 15. *Dendryphiopsis* state has branched, blackish-brown conidiophores up to 500 × 8–11 and 2- to 5-septate, smoky or olivaceous brown conidia 40–80 × 12–25. On decorticated branches of *Alnus*, *Buxus*, *Carpinus*, *Fagus*, *Populus*, *Quercus*, *Tilia*, etc., Dec.–Apr.

Nectria species are very common. They are brightly coloured, mostly red or yellow, at least when young, and have 1-septate ascospores. They grow mainly on deciduous trees. For species which occur exclusively on conifer wood, see *N. fuckeliana*, described on *Picea* (found also on *Abies* and *Pseudotsuga*) and *N. pinea*, described on *Pinus* (found also on *Abies*, *Cupressus*, *Larix*, *Picea* and *Pseudotsuga*).

KEY

    Ascospores not more than 11 long 1
    Ascospores more than 11 long 3
1. Perithecia not showing pinched collapse on drying, ascospores remaining smooth .... *purtonii*
    Perithecia showing pinched collapse on drying, ascospores becoming verruculose 2
2. Always growing on other pyrenomycetes, pinched collapse involving apical disc .... *episphaeria*
    Growing directly on wood and bark, pinched collapse not involving apical disc .... *viridescens*
3. Ascospores, longitudinally striate walls soon apparent, perithecia showing cupulate collapse on drying .... *peziza*
    Ascospore walls not longitudinally striate, perithecia seldom showing cupulate collapse 4
4. Growing on *Diatrypella*, ascospores smooth, 10–15 × 4.5–6 .... *magnusiana*
    Growing on *Melanconis* and *Hapalocystis*, ascospores becoming verrucose and thick-walled, 15–19 × 8–10 .... *wegeliniana*
    Growing mainly on wood and bark 5
5. Perithecia red, encrusted. Accompanied by pink cushions of *Tubercularia* state .... *cinnabarina*
    Perithecia creamy yellow to orange, encrusted. Accompanied by black spore masses of *Myrothecium* state .... *ralfsii*
    No *Tubercularia* or *Myrothecium* states 6
6. Perithecia staying yellow, ascospores 10–14 × 3.5–5 .... *pallidula*
    Perithecia not staying yellow, ascospores longer 7
7. Perithecia warted except for dark disc, ascospores 16–24 × 7–8.5 .... *veuillotiana*
    Perithecia not warted, ascospores mostly smaller 8
8. Perithecia 0.4–0.7mm diam., orange to reddish brown with black papillate ostioles .... *mammoidea*
    Perithecia not more than 0.35mm diam., bright red, at least when young 9
9. No stroma. Perithecia ampulliform, red with very short dark necks. Associated with yellow or yellowish green synnemata .... *flavo-viridis*

Stroma often present. Perithecia mostly somewhat oval. No synnemata 10
10. Ostiolar papilla pointed, asci with chitinoid apical ring .... *coccinea*
    Ostiolar papilla blunt, asci without chitinoid apical ring .... *galligena*

Nectria cinnabarina, described on *Acer*, is easily the commonest of all species and even has a common name, 'coral spot'. It has been collected on dead branches of many trees and shrubs including *Aesculus, Alnus, Betula, Chamaecyparis, Corylus, Cytisus, Fagus, Fraxinus, Ilex, Malus, Populus, Prunus, Rhododendron, Ribes, Salix, Sambucus, Sorbus, Tilia* and *Ulmus*.

Nectria coccinea, described on *Fagus*, occurs also on trunks and dead branches of many other trees and shrubs, including *Acer, Aesculus, Alnus, Betula, Carpinus, Corylus, Cytisus, Fraxinus, Hedera, Ilex, Morus, Myrica, Picea, Populus, Quercus, Rubus, Salix, Sambucus, Sorbus, Taxus, Tilia* and *Ulmus*.

Nectria episphaeria (Tode)Fr. (Fig. 130)
Perithecia 0.1–0.2mm diam., red, translucent, when dry showing pinched collapse which involves the apical disc. Ascospores mostly 7–10 × 3–4, becoming pale brown and verruculose. Growing on other pyrenomycetes, especially on stromata of *Diatrype stigma*. Recorded also on *Anthostoma, Diatrypella, Eutypa, Hypoxylon, Melanconis, Melogramma* and *Quaternaria* on *Acer, Betula, Corylus, Crataegus, Cytisus, Fagus, Fraxinus, Hedera, Prunus, Quercus, Ribes, Salix* and *Ulmus*, Mar.–May.

Nectria flavo-viridis (Fuckel)Wollenw. (Fig. 131)
Perithecia 0.25–0.35mm diam., red with short, very dark necks and some lateral collapse on drying. Ascospores 10–16 × 5–7, becoming pale brown and sometimes minutely verruculose. Associated conidial state has yellow or yellowish green synnemata up to 600 high with spherical pale heads bearing 1 to 5 septate, hyaline *Fusarium*-type conidia. On dead branches of *Acer, Fraxinus, Larix, Pinus, Quercus, Salix*, sometimes overgrowing effete stromatic pyrenomycetes, Feb.–May.

Nectria galligena, described on *Malus*, is found almost as frequently on *Fraxinus* and occasionally on *Betula, Fagus, Salix* and *Sorbus*.

Nectria magnusiana Rehm ex Sacc. (Fig. 132)
Perithecia 0.25–0.35mm diam., yellowish red with flat discs, collapsing when old. Ascospores smooth, becoming pale brown, 10–15 × 4.5–6. *Dendrodochium* state usually forms first with hyaline allantoid conidia 4.5–6 × 1.5–2. On *Diatrypella favacea* on *Betula* and *D. quercina* on *Quercus*, Aug.–Apr.

Nectria mammoidea Phill. & Plowr. (Fig. 133)
Perithecia 0.4–0.7mm diam., orange to reddish brown with almost black papillate ostioles; often with brown sterile hyphae between them. Ascospores sometimes pale brown, 16–19 × 6–8. On fallen dead branches of *Acer, Carpinus, Cornus, Corylus, Cytisus, Fagus, Fraxinus, Populus, Quercus, Rosa,*

Sorbus, Ulex and *Ulmus*, Sept.–May.

Nectria pallidula Cooke (Fig. 134)
Perithecia 0.2–0.3mm diam., yellow with short brownish papillate ostioles; seated on yellowish stromata. Ascospores becoming pale brown and slightly verruculose when mature, 10–14 × 3.5–5. On dead twigs and branches of *Acer, Aesculus, Fagus, Quercus, Robinia* and *Taxus*, Aug.–Jan., not common.

Nectria peziza (Tode)Fr. (Fig. 135)
Perithecia 0.25–0.35mm diam., yellow becoming brownish orange, somewhat crystalline and looking like little tangerine lozenges when fresh, collapsing from above to form cups. Ascospores hyaline to pale straw-coloured, with longitudinally striate walls, 12–16 × 5–7. On dead, often rotten branches and stumps of *Acer, Aesculus, Betula, Fagus, Malus, Salix, Sambucus* and *Ulmus*, Aug.–Dec.

Nectria purtonii (Grev.)Berk. (Fig. 136)
Perithecia 0.15–0.25mm diam., yellow to red, short-necked with flat or concave discs. Ascospores 8–11 × 3–4, smooth, becoming pale brown. Mostly on effete, often valsoid pyrenomycetes on dead branches of *Alnus, Betula, Fagus, Pinus, Prunus* and *Ulmus*, Mar.–July.

Nectria ralfsii Berk. & Br. (Fig. 137)
Perithecia 0.3–0.4mm diam., creamy yellow, often orange in the middle, covered with granules, collapsing from above to form cups, seated on stromata and accompanied by black masses of conidia. Ascospores hyaline, smooth, 19–23 × 6–7. *Myrothecium* state has limoniform, grey or greenish-grey to black conidia, 12–17 × 7–9. On dead cut or fallen branches of *Acer, Aesculus, Taxus, Ulex* and *Ulmus*, Sept.–Jan., most collections from Cornwall.

Nectria veuillotiana Roum. & Sacc. (Fig. 138)
Perithecia 0.3–0.5mm diam., at first yellow, then red becoming brownish, warted except for smooth dark ostiolar discs. Ascospores hyaline, 16–24 × 7–8.5. *Cylindrocarpon* state has mostly 3-septate hyaline conidia 45–55 × 5–7. On dead branches of *Cytisus, Fagus, Populus, Salix*, Oct.–Mar.

Nectria viridescens Booth (Fig. 139)
Perithecia 0.15–0.2mm diam., yellow when quite young, soon becoming red; pinched collapse on drying not involving apical disc. Ascospores hyaline becoming pale brown and verruculose when old, 7.5–10 × 4–5. Unlike the rather similar *N. episphaeria*, this grows directly on wood and bark, not on other pyrenomycetes. It has been recorded on *Abies, Acer, Betula, Cytisus, Fagus, Fraxinus, Ilex, Picea, Pinus, Salix* and *Sorbus*, Sept.–May.

Nectria wegeliniana (Rehm)Höhnel (Fig. 140)
Perithecia 0.2–0.3mm diam., red, darkening with age. Ascospores at first hyaline but becoming reddish brown, verrucose and thick-walled, 15–19 × 8–10. We have collected this only on old stromata of *Melanconis modonia* on

Castanea in Feb., but it was found originally on *Hapalocystis bicaudata* on *Ulmus*.

Neopeckia fulcita (Bucknall)Sacc. (Fig. 141)
Perithecia becoming superficial, rough-walled, 0.4mm diam., blackish brown, setose. Ascospores 1-septate, brown, 12–15 × 5–6. On dead branches of *Betula, Fagus, Fraxinus, Prunus* and *Quercus*.

Nitschkia: perithecia mostly superficial, black, sometimes warted, tending to collapse in a cupulate manner when old; often seated on a dark subiculum. *N. confertula* is described on *Fraxinus* on which it is found with *Hypoxylon rubiginosum*.

KEY

	Perithecia bearing short spines .... *brevispina*
	Perithecia without spines 1
1.	Ascospores mostly straight, fusiform 2
	Ascospores mostly slightly curved, not fusiform 3
2.	Ascospores 6–9 × 1.5–2.5 .... *grevillei*
	Ascospores 12–18 × 4–6 .... *collapsa*
3.	Ascospores subreniform, 3.5–5 wide, coloured .... *confertula*
	Ascospores not subreniform, 2–3 wide, hyaline 4
4.	Perithecia 0.2–0.3 mm diam., on *Nectria cinnabarina* .... *parasitans*
	Perithecia 0.3–0.45mm diam., not on *Nectria* .... *cupularis*

Nitschkia brevispina (Munk)Nannf. (Fig. 142)
Perithecia 0.3–0.4mm diam., bearing blackish-brown spines 8–20 × 5–8. Ascospores hyaline, curved, 9–15 × 3–4. On *Acer, Fagus* and *Ligustrum*, associated with other pyrenomycetes, Feb.–Apr.

Nitschkia collapsa (Romell)Chenant. (Fig. 143)
Perithecia 0.5–0.75mm diam. Ascospores often 1-septate, becoming pale greyish brown, 12–18 × 4–6. On *Acer, Corylus, Crataegus, Fraxinus* and *Prunus spinosa*, usually with *Diatrype stigma*, Aug.–Mar.

Nitschkia cupularis (Pers.)P. Karsten (Fig. 144)
Perithecia 0.3–0.45mm diam. Ascospores hyaline with 4 large guttules, 12–19 × 2–2.5. On dead branches of *Acer, Aesculus, Fagus, Fraxinus, Prunus spinosa, Tilia, Ulmus*, often conspicuous on wood where loose bark has come away, Oct.–Apr.

Nitschkia grevillei (Rehm)Nannf. (Fig. 145)
Perithecia 0.3–0.5mm diam. Ascospores hyaline with 2 to 4 guttules, occasionally 1-septate, 6–9 × 1.5–2.5. On dead branches of *Acer, Aesculus, Carpinus, Fagus, Ilex, Populus*, and *Ulmus*, sometimes on other fungi such as *Peroneutypa heteracantha*, Oct.–Mar.

Nitschkia parasitans (Schw.)Nannf. (Fig. 146)
Perithecia 0.2–0.3mm, growing on stromata of *Nectria cinnabarina*. Ascospores hyaline, 9–16 × 2–3. On dead branches of *Acer, Aesculus, Fagus, Malus, Tilia* and *Ulmus*, Sept.–May.

Ohleria rugulosa Fuckel (Fig. 147)
Perithecia gregarious, partly immersed, 0.5–0.8mm diam., globose with small papillate ostioles, black, rough-walled; with thin effuse stroma. Ascospores brown, 14–18 × 4.5–5, at first 3-septate but soon breaking at the median septum to form two 1-septate part-spores. On decorticated dead wood of *Alnus, Carpinus, Tilia* and *Ulmus*.

Otthia spiraeae (Fuckel)Fuckel (Fig. 148)
Pseudothecia immersed, 0.3–0.5mm diam., dark brown to black, in small groups seated on a blackish-brown hyphal subiculum, with short necks showing through cracks in bark. Ascospores dark brown, 1-septate, 23–28 × 10–14. *Diplodia* state has brown 1-septate conidia 25–30 × 10–14. On dead twigs and branches of *Acer, Clematis, Crataegus, Fraxinus, Prunus spinosa, Salix, Ulmus*, etc., Nov.–Apr.

Peroneutypa heteracantha (Sacc.)Berl. (Fig. 149)
Black stromata formed under the bark, each containing 2 to 8 perithecia with long slender necks which protrude through cracks and are accompanied by spine-like groups of sterile brown hyphae or by synnemata of the *Harpographium* state. Ascospores allantoid, hyaline, mostly 4–6 × 1. *Harpographium* conidia hyaline, falcate, 11–15 × 1–2. On dead branches of *Acer, Alnus, Crataegus, Cytisus, Euonymus, Fagus, Fraxinus, Lupinus, Rubus, Sambucus, Tilia, Ulmus*, etc., Jan.–Apr.

Platystomum compressum (Pers.)Trevisan (Fig. 150)
Pseudothecia 0.5–0.8mm diam., black, usually immersed with just the large, laterally compressed necks protruding. Ascospores rather pale brown, with 3 to 9 transverse septa and 1 longitudinal septum, 20–35 × 7–10. On decorticated wood of *Hedera, Fraxinus, Populus, Rubus* and *Salix*, Oct.–Feb.

Podostroma alutaceum (Pers.)Atk. (Fig. 151)
Stromata clavate, erect, 2–4cm high, pale yellow; perithecia immersed in upper part with ostioles showing on the surface as dots. Ascospores hyaline, verruculose, each dividing in the ascus to form 2 part-spores 4 diam. or 4–5 × 3. On decaying stumps of half-buried rotten wood of conifers, Sept.–Nov.

Protocrea delicatula (Tul.) Petch, which grows on coniferous wood, has slightly smaller perithecia than *P. farinosa*, and ascospores 6–8 × 2.5–3.

Protocrea farinosa (Berk. & Br.)Petch (Fig. 152)
Perithecia 0.2–0.25mm diam., hyaline becoming yellow–ochre, seated on a cottony, closely felted, creamy hyphal subiculum. Ascospores hyaline, minutely verruculose, 8–10 × 3–4, dividing at septum into 2 part-spores. On dead

wood of deciduous trees, e.g. *Fraxinus*, and on decaying polypores, May–Sept.

Rosellinia, perithecia rather large, dark-brown to black with papillate ostioles and sometimes a subiculum. Ascospores dark-brown with germ slits and in some species with hyaline appendages.

KEY

Ascospores without appendages .... *mammiformis*
Appendages short .... *aquila*
Appendages long and slender .... *thelena*

Rosellinia aquila (Fr.)de Not. (Fig. 153)
Perithecia 1–1.5mm diam., dark brown to black with papillate ostioles, gregarious, seated on a dense brown hyphal subiculum. Ascospores bean-shaped, dark brown with a small hyaline appendage at each end, 20–28 × 7–9. Often accompanied or preceded by its pale brown to grey *Geniculosporium* state with much-branched conidiophores and hyaline conidia 7–15 × 3–4.5. On dead branches, especially common near the bottom of piles of brushwood. Found on *Acer, Cornus, Corylus, Crataegus, Fagus, Fraxinus, Lupinus, Rhododendron, Rubus, Salix, Sambucus, Tilia* and *Ulmus*, Feb.–May.

Rosellinia mammiformis, described on *Hedera*, has been found occasionally also on dead branches of *Acer, Fraxinus* and *Ilex*.

Rosellinia thelena (Fr.)Rabenh. (Fig. 154)
Perithecia 1–1.5mm diam., black with short, acutely pointed necks, seated on a brown subiculum. Ascospores dark brown, 20–26 × 7.5–8.5, with slender hyaline appendages 5–9 long at each end. The *Geniculosporium* state has conidia 5–8 × 2.5–3. On fallen dead branches and debris of *Fagus, Larix, Pinus, Quercus, Ribes*, etc., Sept.–May.

Thyridaria rubronotata, described on *Acer*, has been collected also on *Fagus, Juglans* and *Ulmus*.

Trematosphaeria, pseudothecia partly or wholly immersed, thick-walled, ascospores septate, brown with end cells usually paler than median ones.

KEY

Ascospores more than 70 long, 5- to 7-septate .... *callicarpa*
Ascospores shorter, 3-septate 1
1. Ascospores less than 30 long .... *pertusa*
Ascospores more than 30 long .... *anglica*

Trematosphaeria anglica (Sacc.)Sacc. (Fig. 155)
Pseudothecia immersed, 0.35–0.5mm diam., black. Ascospores 3-septate, reddish brown with end cells paler than median ones, 32–38 × 8–9. On rotten wood, uncommon.

Trematosphaeria callicarpa (Sacc.)Sacc.
Pseudothecia partly immersed in wood, somewhat conical, up to 0.7mm diam. Ascospores broadly fusiform, constricted in the middle, becoming 5- to 7-septate, 75–90 × 15–25. On dead wood and bark of *Populus*, etc.

Trematosphaeria pertusa Fuckel (Fig. 156)
Pseudothecia partly immersed in wood, 0.35–0.45mm diam., black. Ascospores 3-septate, golden brown with end cells slightly paler than median ones, 22–29 × 7–8. On decorticated branches of *Acer, Fagus, Quercus*, etc., Mar.–May.

Trichonectria hirta (Bloxam ex Currey)Petch (Fig. 157)
Perithecia superficial, up to 0.25mm diam., salmon pink, setose; setae thick-walled, white, up to 160 long. Ascospores hyaline, 8- to 15-septate, 50–90 × 5–6. On logs and decorticated branches, sometimes overgrowing lichens, Dec.–Jan.

Trichosphaeria notabilis Mouton (Fig. 158)
Perithecia superficial, 0.2–0.4mm diam., black, growing among conidiophores of *Brachysporium*. Ascospores hyaline, guttulate, 15–18 × 6–7. On rotten wood of *Acer, Castanea, Quercus*, etc., with *Brachysporium britannicum* and *B. nigrum*, Apr.–Aug.

Trichosphaeria pilosa (Pers.)Fuckel (Fig. 159)
Perithecia superficial, gregarious, 0.2–0.25mm diam., black, setose. Setae brown, up to 50 × 3–4.5. Ascospores hyaline, 7–9 × 3–4. On dead branches of *Fagus, Fraxinus*, etc.

Tubeufia cerea (Berk. & Curtis)Booth (Fig. 160)
Perithecia in groups, superficial, greenish yellow to almost black, warted, hairy; hairs yellow, sometimes branched, up to 100 × 3–6. Ascospores hyaline, 7- to 11-septate, 30–50 × 3–5. Preceded and accompanied by bright lemon or greenish-yellow colonies of its *Helicosporium* state with dark brown conidiophores up to 350 × 4–5 and conidia helically coiled 2 or 3 times in one plane, 10–20 diam., with filaments about 1 thick. On dead wood and bark of *Acer, Betula, Corylus, Fagus, Fraxinus, Quercus*, etc., also on beech cupules, birch seeds and inside acorn cups, often with or growing over *Diatrype stigma, Lasiosphaeria hirsuta* and *Polydesmia pruinosa*, Apr.–Oct.

Ustulina deusta, described on *Fagus*, is found occasionally on stumps and dead roots of *Aesculus, Betula, Taxus* and *Ulmus*.

Valsa, perithecia in clusters or circles embedded in bark, with necks converging, their tips protruding through small stromatic discs. Ascospores curved, allantoid, colourless. Preceded and often accompanied by *Cytospora* conidial states.

KEY

   Ascospores not more than 10 × 2 .... *ceratophora*
   Ascospores up to 15 × 3 or more 1
1. *Cytospora* state with olive or dark-green spore tendrils .... *pruinosa*
   Cytospora state with white or yellow spore tendrils 2
2. Conidia of *Cytospora* state 5–7 × 1 .... *ambiens*
   Conidia of *Cytospora* state 4–5 × 1 .... *pustulata*

Valsa ambiens (Pers.)Fr. (Fig. 161)
Perithecia immersed, 0.4–0.5mm diam., black, clustered or arranged in circles beneath bark with necks protruding slightly through greyish discs. Ascospores hyaline, cylindrical, curved, mostly 15–18 × 3–4. Conidiomata of *Cytospora* state multilocular with chambers arranged in a circle or confluent forming a lobed cavity, with a single ostiole in the centre of a prominent greyish disc. Conidia allantoid, 5–7 × 1, exuding in white or yellowish tendrils. Very common and often extending uniformly along the whole length of twigs and branches of *Acer, Betula, Corylus, Crataegus, Fagus, Fraxinus, Malus, Populus, Prunus, Quercus, Rosa, Rubus, Sorbus, Ulmus*, etc., Oct.–May.

Valsa ceratophora Tul. & C. Tul.
Perithecia 0.2–0.4mm diam., black, in groups of about 10 below bark, with long necks protruding through raised, greyish-brown discs. Ascospores slightly curved, hyaline, 7–10 × 1.5–2. *Cytospora* state has allantoid conidia 4–5 × 1. On branches of *Acer, Betula, Castanea, Corylus, Fagus, Fraxinus, Ilex, Malus, Quercus, Rosa, Rubus, Salix, Sorbus, Ulex*, etc., Nov.–Mar.

Valsa pruinosa (Fr.)Défago
Perithecia immersed, 3 to 8 per small stroma, 0.35–0.6mm diam., necks 0.25–0.35mm long scarcely protruding. Asci 8-spored, 45–70 × 7–11. Ascospores allantoid, hyaline, 9–18 × 1.5–4. *Cytospora* state with necks erumpent on a black disc, conidia 3–9 × 0.8–2.5, exuding in olivaceous or dark green tendrils. On dead branches of *Fraxinus, Ligustrum, Prunus, Quercus, Salix* and *Ulmus*.

Valsa pustulata Auersw. ex Nitschke
Perithecia immersed, usually in groups of 3 to 8, black, 0.7–0.9mm diam., with rather short necks protruding through raised greyish-brown discs. Ascospores hyaline, slightly curved, 12–16 × 3. *Cytospora* state with allantoid conidia 4–5 × 1. On branches of *Crataegus, Fagus, Prunus, Rosa*, Nov.–Apr.

Valsaria foedans (P. Karsten)Sacc. (Fig. 162)
Perithecia 0.4–0.6mm diam., in groups of 20 to 30 immersed in stromata, with long necks only just visible on dark surface of discs. Ascospores olivaceous brown, 1-septate, 12–15 × 4. On dead branches of *Alnus, Betula, Castanea*, etc., Apr.–May.

Valsaria insitiva (Tode)Ces. & de Not. (Fig. 163)

Perithecia 0.2–0.3mm diam., mostly in groups of 7 to 10 immersed in stromata with tips of long necks emerging through brown discs. Ascospores dark brown, 1-septate, 16–20 × 8–10. On dead branches of *Fagus, Prunus, Quercus* and other deciduous trees, Oct.–Nov.

Xylaria hypoxylon (L.)Grev. (Fig. 164)
Stag's-horn or candle-snuff fungus. Stromata up to 7cm high, black and hairy near the base, divided above into a number of flat branches which are white and covered with a powdery layer of hyaline conidia 10–11 × 2.5–3 early in the year. Later protuberant perithecia, ripe by Mar.–Apr., develop above the hairy base. Ascospores bean-shaped, brown, 12–14 × 5–6. Very common on dead branches partly buried in the ground and on old stumps. Found on *Betula, Corylus, Fraxinus, Hedera, Malus, Salix* and *Ulmus*.

Xylaria polymorpha (Pers.)Grev. (Fig. 165)
Dead-man's fingers. Stromata club-shaped, short-stalked, up to 8 × 2cm, dark chocolate brown to black. Ascospores brown, 20–31 × 5–8. Usually in small clusters at about soil level on dead stumps or branches of *Acer, Prunus, Quercus*, etc., Sept.–Nov.

Zignoella ovoidea (Fr.)Sacc. (Fig. 166)
Perithecia superficial, black, about 0.3mm diam., with papillate ostioles. Ascospores hyaline, 3-septate, 24–28 × 4–5. On decorticated wood including that of *Fraxinus*.

HYPHOMYCETES

KEY TO GENERA

	Conidia unbranched	1
	Conidia branched	80
1.	Conidia colourless or brightly coloured in mass	2
	Conidia dark or dull coloured	33
2.	Conidia without septa	3
	Conidia 1-septate	24
	Conidia with more than 1 transverse septum	26
	Conidia helicoid, spirally coiled in one plane or in several planes	31
3.	Conidia allantoid	*Dendrodochium*
	Conidia not regularly allantoid	4
4.	Conidia in slimy heads capping green synnemata	*Dendrostilbella*
	Conidia borne on purple gelatinous conidiomata	*Coryne*
	Conidia in dry heads capping rust synnemata	*Pachnocybe*
	Colonies pink	5
	Colonies yellow or orange	6
	Colonies white or some other colour	8
5.	Sporodochial or effuse, conidiophores penicillately branched	*Gliocladium*

	Sporodochial, conidiophores not penicillately branched ...... *Tubercularia*
6.	Conidia in slimy heads capping erect conidiophores .... *Gliocladium*
	Not so 7
7.	Conidia in chains, mycelium without clamps .... *Alysidium*
	Conidia not in chains, mycelium with clamps .... *Sporotrichum*
8.	Sporodochia with branched twisted brown setae .... *Sarcopodium*
	Not so 9
9.	Conidiophores hyaline or subhyaline 10
	Conidiophores brown or black 15
10.	Conidiophores dichotomously branched, conidia in chains, with disjunctors .... *Amblyosporium*
	Conidiophores verticillately branched 11
	Conidiophores irregularly branched, like mycelium, conidia on flat-topped pegs 14
11.	Conidia on little pegs or denticles at ends of branches 12
	Conidia emerging in slime from phialides 13
12.	Conidia spherical, ends of branches resembling cocks' combs .... *Costantinella*
	Conidia 4–6 × 1.5–2 .... *Calcarisporium*
13.	Conidia in columns .... *Clonostachys*
	Conidia often in slipped chains .... *Mariannaea*
	Conidia in spherical masses, colourless or green, often rough-walled .... *Trichoderma*
14.	Conidia solitary .... *Acladium*
	Conidia in chains .... *Alysidium*
15.	Conidiophores synnematous 16
	Conidiophores mononematous 18
16.	Conidia falcate .... *Harpographium*
	Conidia not falcate 17
17.	Conidia aggregated in slimy heads .... *Graphium*
	Conidia dry, on little pegs .... *Phaeoisaria*
18.	Conidiophores unbranched 19
	Conidiophores branched 20
19.	Conidia in chains, wedge-shaped, formed inside little cups at apex of conidiophore .... *Catenularia*
	Conidia fusiform, issuing from little cups along the sides as well as at the apex of the conidiophore .... *Codinaea*
	Conidia ellipsoidal, formed in slimy masses or long columns at apex of conidiophore .... *Chloridium*
20.	Branching very complex, branches anastomosing and difficult to tease apart .... *Gonytrichum*
	Branching near apex of conidiophore only 21
	Branching loose and irregular 22
21.	Branching penicillate .... *Haplographium*
	Branches and phialides in verticils .... *Phaeostalagmus*

44 Plurivorous Wood and Bark Fungi

22. Conidia developing in slime at tips of short lateral phialides .... *Menispora*
    Conidia formed on very short pegs which fracture leaving a little frill at the base of each conidium 23
23. Branches geniculate along much of their length .... *Geniculosporium*
    Branches tending to be nodose or swollen at their ends .... *Nodulisporium*
24. Conidiophores brown with lateral branches and setiform apices .... *Chaetopsis*
    Conidiophores hyaline 25
25. Colonies soon turning pink, conidia held together in clusters in slime .... *Trichothecium*
    Colonies remaining white, conidia dry on flat-topped pegs .... *Diplorhinotrichum*
26. Conidiophores hyaline 27
    Conidiophores brown 28
27. Conidia cylindrical without foot-cells .... *Cylindrocarpon*
    Conidia tapered towards each end, mostly curved, often with foot-cells .. .. *Fusarium*
28. Conidiophores loosely branched with short lateral phialides ... *Menispora*
    Conidiophores unbranched, no lateral phialides 29
29. Conidia leaving scars at and below apex of conidiophore when they secede .... *Pseudospiropes*
    No scars on conidiophores 30
30. Conidia in chains .... *Pleurotheciopsis*
    Conidia in groups in slime .... *Cacumisporium*
31. Conidia ellipsoidal, filaments coiled in more than one plane ..... *Helicoon*
    Conidia with filaments coiled in one plane 32
32. Colonies lemon yellow .... *Helicosporium*
    Colonies effuse, white becoming pink .... *Helicomyces*
    Colonies discrete sporodochia, watery white, gelatinous .... *Hobsonia*
33. Conidia without septa 34
    Conidia 1-septate 48
    Conidia with several septa or pseudosepta 55
    Conidia with longitudinal and transverse septa 74
    Conidia helicoid 78
34. Conidiophores colourless 35
    Conidiophores brown or black 36
35. Colonies effuse, conidia in black slimy heads .... *Stachybotrys*
    Colonies discrete, sporodochial, conidia in black slimy masses .... *Myrothecium*
36. Conidiophores synnematous, black .... *Graphium*
    Conidiophores mononematous, brown 37
37. Conidiophores not branched 38
    Conidiophores branched 43

38.	Conidia in chains	39
	Conidia not in chains	41
39.	Conidia at and below the apex of the conidiophore leaving scars when they secede	*Cladosporium*
	Conidia at apex only	40
40.	Conidia wedge-shaped, formed inside little cups	*Catenularia*
	Conidia not wedge-shaped, no cups	*Xylohypha*
41.	Conidia at apex only	*Acrogenospora*
	Conidia at and below apex	42
42.	Conidia formed through small pores in the conidiophore wall	*Spadicoides*
	Conidia leaving small scars	*Veronaea*
	No pores or obvious scars	*Virgariella*
43.	Conidia in chains	44
	Conidia not in chains	46
44.	Conidiophores dichotomously branched	*Aegerita*
	Conidiophores tree-like with branches almost at right-angles	*Oidiodendron*
	Branching irregular	45
45.	Conidia all spherical	*Periconia*
	Conidia limoniform, oval and spherical	*Alysidium*
46.	Conidia kidney-shaped	*Virgaria*
	Conidia not kidney-shaped	47
47.	Branches geniculate along much of their length	*Geniculosporium*
	Branches nodose or swollen at ends	*Nodulisporium*
48.	Conidiophores like mycelium, conidia solitary at ends of short branches	*Trichocladium*
	Conidiophores quite different from mycelium	49
49.	Conidiophores mostly branched	50
	Conidiophores not branched	51
50.	Conidia in long chains terminal and lateral, developing through pores	*Diplococcium*
	Conidia solitary or in very short chains on terminal and intercalary swellings	*Oedemium*
51.	Conidia in long terminal chains	52
	Conidia not in chains	53
52.	Conidiophores short, conidia with very dark broad band at septum	*Bispora*
	Conidiophores longer, no dark bands	*Heteroconium*
53.	Conidia terminal	*Endophragmia* and *Endophragmiella*
	Conidia terminal and lateral	54
54.	Conidiophores not nodose, conidia formed through tiny pores in the wall	*Spadicoides*
	Conidiophores often nodose, conidia on short thin denticles	*Cordana*

55.	Conidiophores synnematous .... *Arthrobotryum*	
	Conidiophores mononematous 56	
56.	Conidiophores like mycelium, conidia solitary at ends of short branches .... *Trichocladium*	
	Conidiophores quite different from mycelium 57	
57.	Conidiophores branched 58	
	Conidiophores mostly unbranched or sparingly branched near base .. 59	
58.	Short lateral branches near apex with conidia at their tips .... *Dendryphiopsis*	
	Branching loose, often dichotomous, conidia on terminal and intercalary swellings .... *Oedemium*	
59.	Conidia in long chains 60	
	Conidia not in chains (occasionally short chains in *Corynespora*) 63	
60.	Conidia formed inside large brown phialides .... *Sporoschisma*	
	Conidia formed at ends of conidiophores 61	
61.	Conidia and conidiophores forming sporodochia .... *Septotrullula*	
	Colonies effuse 62	
62.	Conidiophores short .... *Taeniolella*	
	Conidiophores longer .... *Heteroconium*	
63.	Conidia terminal, conidiophores often proliferating straight on through a previous apex (percurrently) 64	
	Conidia beneath as well as at the apex, conidiophores proliferating sympodially or not at all 68	
64.	Conidia developing through a large pore .... *Corynespora*	
	Conidia developing as blown-out ends 65	
65.	Conidia hanging together in slimy clumps .... *Cacumisporium*	
	Conidia dry, separate 66	
66.	Conidia long .... *Sporidesmium*	
	Conidia short 67	
67.	Conidiophores very slender, splayed out, conidia large forming small sporodochia .... *Bactrodesmium*	
	Conidiophores broader sometimes proliferating through cups .... *Endophragmia* and *Endophragmiella*	
68.	Conidia developing through small pores in walls of conidiophores which do not proliferate 69	
	Conidia not developing through pores, conidiophores proliferating sympodially 70	
69.	Conidia long, mostly obclavate .... *Helminthosporium*	
	Conidia short, not obclavate .... *Spadicoides*	
70.	Conidia hanging together in slimy clumps .... *Pleurothecium*	
	Conidia dry, separate 71	
71.	Conidia leaving well-defined scars on the conidiophore when they secede 72	
	Conidia formed at ends of long slender pedicels 73	
72.	Conidia 2-septate, limoniform, with very large dark central cell and	

	small pale end cells .... *Brachydesmiella*
	Conidia not so .... *Pseudospiropes*
73.	Conidiophores short, conidia long, cylindrical, multiseptate .... *Camposporium*
	Conidiophores long, conidia short, 2 to 4 septa .... *Brachysporium*
74.	Conidia in chains, forming very dark pulvinate sporodochia 75
	Conidia not in chains, colonies effuse 76
75.	Chains branched .... *Trimmatostroma*
	Chains not branched .... *Coniosporium*
76.	Conidiophores stout, 9–12 thick at base .... *Dactylosporium*
	Conidiophores slender, mostly 2–3 thick 77
77.	Conidia bearing short conical horns .... *Oncopodiella*
	Conidia without conical horns .... *Monodictys*
78.	Conidia ellipsoidal with filaments coiled in more than one plane .... *Helicoon*
	Conidia coiled in one plane 79
79.	Conidia borne on short conidiophores .... *Cirrenalia*
	Conidia borne on pegs along the sides of long brown conidiophores .... *Helicoma*
80.	Conidia often in chains .... *Taeniolina*
	Conidia not in chains 81
81.	Conidiophores erect, long, stout, dark brown, conidia terminal with radiating or splayed out arms .... *Actinocladium* and *Triposporium*
	Conidiophores short or slender, sometimes lateral pegs on mycelium. Arms of conidia tending to be parallel and sometimes close together like the fingers of a hand 82
82.	Branches of conidia growing downwards from a cap-like plate .... *Cryptocoryneum*
	Branches growing upwards 83
83.	Conidia on pulvinate sporodochia .... *Digitodesmium*
	Colonies effuse 84
84.	Branches up to 230 × 22, widely spaced .... *Ceratosporium*
	Branches up to 45 × 7, touching each other ....*Dictyosporium*

Acladium state of *Botryobasidium conspersum* J. Eriksson (Fig. 167)
Colonies effuse, cottony and pale at first, later velvety and fulvous or snuff-coloured. Conidiophores up to 350 × 6–9, subhyaline with numerous cylindrical denticles. Conidia subhyaline or straw-coloured, 15–20 × 9–14. Very common on dead wood and bark of *Acer*, *Alnus*, *Betula*, *Fagus*, *Fraxinus*, *Larix*, *Picea*, *Pinus*, *Prunus*, *Quercus*, *Rhododendron*, *Salix*, *Ulex*, etc.

Acrogenospora sphaerocephala (Berk. & Br.)M.B. Ellis (Fig. 168)
Colonies effuse, dark blackish brown to black, hairy. Conidiophores dark brown, up to 380 long, 9–11 thick at base, 5–8 at apex. Conidia spherical or subspherical, mid to dark brown by transmitted light, black and shining by reflected light, 15–33 × 14–33 (28 × 27); truncate base 5–7 wide. On rotten

wood of *Acer, Alnus, Betula, Cornus, Prunus spinosa, Quercus, Sambacus* and *Taxus*. The *Acrogenospora* state of *Farlowiella carmichaeliana* (v.s.) is similar but with broadly ellipsoidal conidia averaging 28 × 20.

Actinocladium rhodosporum Ehrenb. (Fig. 169)
Colonies effuse, hairy, dark brown to black. Conidiophores brown, up to 130 long. Conidia brown, branched, with septate branches up to 140 long, 7–10 thick at base. On rotten wood of *Betula, Buxus, Carpinus, Corylus, Fraxinus, Picea, Quercus, Sorbus, Ulex* and *Ulmus*.

Aegerita viridis, described on *Alnus*, has been found also on *Betula* and *Fagus*.

Alysidium resinae (Fr.)M.B. Ellis (Fig. 170)
Colonies effuse, black. Hyphae and conidiophores blackish brown, 5–11 thick. Conidia in simple and branched chains, spherical and 7–12 diam. or limoniform to ellipsoidal and 11–20 × 7–14, blackish brown. On rotten wood of *Acer, Betula, Fagus, Quercus, Pinus*, etc.

Alysidium state of *Botryobasidium aureum* Parm. (Fig. 171)
Colonies effuse, thin, cottony, white to yellowish or almost orange when sporulating freely. Hyphae hyaline, 4–9 thick. Conidia in branched chains, 18–28 × 10–12. On rotten wood of *Betula, Corylus, Fagus, Fraxinus, Ilex, Pinus, Quercus, Sorbus*, etc.

Alysidium state of *Botryobasidium candicans* J. Eriksson (Fig. 172)
Colonies effuse, cottony, greyish white. Hyphae 4–10 thick. Conidia in short chains, mostly 15–17 × 8–9. On rotten wood of *Betula, Fagus, Quercus, Rhododendron*, etc., all the year round.

Amblyosporium botrytis Fresen. (Fig. 173)
Colonies effuse, yellowish white to buff. Conidiophores up to 5mm long, 12–20 thick, repeatedly branched at apex, hyaline or pale straw-coloured. Conidia in chains, separated by disjunctors, commonly barrel-shaped, verruculose to echinulate, mostly 15–25 × 6–8. On very rotten wood.

Arthrobotryum stilboideum, described on *Quercus*, occurs also on *Ulmus*.

Bactrodesmium, sporodochia scattered, punctiform, brown to black, usually about 0.1–0.3mm diam., occasionally up to 0.5mm. Large, dark, shining conidia formed at the ends of very narrow conidiophores can be seen quite distinctly under a binocular dissecting microscope.

KEY

Conidia with thick black bands at one or more of the septa 1
Conidia without black bands at septa 2
1. Conidia 12–17 wide, with the penultimate cell always much longer than the others .... *abruptum*
Conidia 15–25 wide, with the penultimate cell slightly longer than the

          terminal cell .... *obovatum*
2.   Conidia more than 20 wide .... *atrum*
     Conidia less than 20 wide 3
3.   Conidia pale to mid brown, usually 4-septate .... *spilomeum*
     Conidia very pale brown, 5- to 6-septate .... *pallidum*

Bactrodesmium abruptum (Berk. & Br.)Mason & Hughes (Fig. 174)
Conidia clavate, mid to dark reddish brown, paler towards the basal cell, 3- to 7-septate, 32–70 × 12–17. On dead wood and bark; most collections on *Acer, Fraxinus* and *Quercus*, Sept.–Apr.

Bactrodesmium atrum M.B. Ellis (Fig. 175)
Conidia obovoid, when mature almost black except near base, 3- to 5-septate, 43–72 × 22–38. Inside bark of *Betula* and on rotten wood of *Fagus*, Mar.–June.

Bactrodesmium obovatum (Oudem.)M.B. Ellis (Fig. 176)
Conidia clavate, upper cells mid to dark brown, lower cells subhyaline or pale brown, 4- or 5-septate, 25–58 × 15–25. The commonest species on wood and bark of *Alnus, Betula, Fagus, Fraxinus, Populus* and *Quercus*, all the year round.

Bactrodesmium pallidum M.B. Ellis (Fig. 177)
Conidia more or less ellipsoid, pale brown, 5- or 6-septate, 35–55 × 9–12. On rotten wood and bark of *Fagus, Fraxinus* and *Quercus*, Oct.–Apr.

Bactrodesmium spilomeum (Berk. & Br.)Mason & Hughes (Fig. 178)
Conidia ellipsoid to clavate, mostly 4-septate, pale to mid brown, 24–43 × 8–12. on *Betula, Fagus, Fraxinus, Ulmus*, etc., Apr.–June.

Bispora antennata (Pers.)Mason (Fig. 179)
Colonies effuse, black. Conidiophores inconspicuous, pale brown, 5–30 × 2–5. Conidia in long chains, doliiform, 1-septate, mid to dark brown with almost black band at septum, 13–20 × 7–8. Common on the cut ends of stumps and on chips of *Carpinus, Corylus, Fagus, Quercus, Ulmus*, etc.

Bispora betulina (Corda)Hughes (Fig. 180)
Colonies punctiform or effuse, blackish brown. Conidia 1-septate or occasionally 2-septate, cylindrical, mid to dark brown with almost black bands at septa, 9–14 × 4–5. On dead wood of *Betula, Castanea, Fagus, Ilex, Populus, Quercus*, etc.

Brachydesmiella biseptata, described on *Fraxinus*, has been found also on *Fagus*.

Brachysporium, colonies thin, effuse, brown or dark brown, hairy, on rotten wood and bark. Conidia attached to the tops of their conidiophores by long, sometimes twisted, narrow, cylindrical pedicels or separating cells.

KEY

	Conidia 2-septate	1
	Conidia 3-septate	2
	Conidia 4-septate	*masonii*
1.	Conidia pale brown	*obovatum*
	Conidia dark brown	*britannicum*
2.	Conidia ellipsoidal or oval	*nigrum*
	Conidia obovoid or clavate	3
3.	Conidial measurements averaging 29 × 11.7	*bloxami*
	Conidial measurements averaging 23.5 × 13.3	*dingleyae*

Brachysporium bloxami (Cooke)Sacc. (Fig. 181)
Conidiophores up to 380 long. Conidia 22–37 × 9–14, basal cell small and very pale, other cells much larger and mid pale to rather dark brown. Common on *Acer, Alnus, Betula, Castanea, Fagus, Fraxinus, Pinus, Prunus* and *Quercus*.

Brachysporium britannicum Hughes (Fig. 182)
Conidiophores up to 500 long. Conidia 18–24 × 11–14, basal cell small, subhyaline, other cells dark or very dark brown. On *Betula, Buxus, Castanea, Fagus, Fraxinus, Quercus* and *Sambucus*.

Brachysporium dingleyae Hughes, described on *Rhododendron*, has been recorded also on *Fraxinus* and *Quercus*.

Brachysporium masonii Hughes (Fig. 183)
Conidiophores up to 400 long. Conidia nearly all 4-septate, 26–34 × 10–14, cell at each end hyaline or very pale, median cells brown. On *Castanea, Fagus, Prunus, Quercus*.

Brachysporium nigrum (Link)Hughes (Fig. 184)
Conidiophores up to 300 long. Conidia 17–24 × 8–11, cell at each end hyaline or very pale, median cells brown. Very common on *Betula, Castanea, Fagus, Fraxinus, Quercus, Ulmus*, etc.

Brachysporium obovatum (Berk.)Sacc. (Fig. 185)
Conidiophores up to 380 long. Conidia 19–28 × 11–14, basal cell very small, subhyaline, other cells much larger and rather pale brown. Common on *Acer, Betula, Fagus, Fraxinus, Quercus, Populus, Prunus* and *Sambucus*.

Cacumisporium capitulatum (Corda)Hughes (Fig. 186)
Colonies effuse, hairy, brown or greyish brown. Conidiophores up to 300 long, dark brown, paler towards the apex where there are up to 13 growth rings or annellations. Conidia aggregated in slimy masses, slightly curved, at first colourless with large guttules, later 3-septate, brown with end cells paler than median ones, 15–22 × 4.5–5. On dead wood of *Acer, Betula, Fagus, Pinus, Populus* and *Quercus*.

Calcarisporium arbuscula Preuss (Fig. 187)
Colonies white, growing over the surface of other fungi such as *Dasyscyphus virgineus* and *Rosellinia aquila* on rotten wood. Conidiophores hyaline, up to 150 × 2–5, verticillately branched. Conidia borne on short pegs at the ends of branches, where there is usually a slight swelling, hyaline, 4–6 × 1.5–2.

Camposporium cambrense, described on *Fagus* cupules, has been found also on rotten wood of *Betula, Ilex, Lonicera, Quercus* and *Ulmus*.

Camposporium pellucidum, described on *Fagus* cupules, occurs quite frequently also on dead wood and bark of *Acer, Aesculus, Quercus, Rosa, Rubus, Ulex*, etc.

Catenularia, v.s. under *Chaetosphaeria cupulifera* and *C. myriocarpa*.

Ceratosporium fuscescens, described on *Ilex*.

Chaetopsis grisea (Ehrenb.)Sacc. (Fig. 188)
Colonies effuse, at first grey, later blackish brown, hairy. Conidiophores brown, up to 1mm long, 5–7 thick, upper part sterile, setiform, with branches near middle terminating in polyphialides. Conidia aggregated in slimy bundles, hyaline, mostly 1-septate, 8–12 × 2. On fallen dead branches of *Fagus, Fraxinus, Populus* and *Ulmus*.

Chloridium, v.s. under *Chaetosphaeria preussii* and *C. vermicularioides*.

Cirrenalia lignicola P.M. Kirk (Fig. 189)
Colonies inconspicuous, dark brown, punctiform. Conidia helicoid, olivaceous brown, 15–20 diam. On rotten wood and bark of *Fagus* and *Quercus*.

Cladosporium britannicum M.B. Ellis (Fig. 190)
Colonies effuse, dark brown or blackish brown, hairy. Conidiophores dark brown, up to 350 × 3–8. Conidia in simple or branched chains, without septa, pale to mid brown, 10–20 × 5–6. On dead wood of *Acer, Fagus, Quercus* and *Ulmus*.

Clonostachys compactiuscula, described on *Fagus*, has been found also on dead twigs of *Quercus* and *Salix*.

Codinaea, v.s. under *Chaetosphaeria callimorpha*.

Coniosporium, v.s. under *Hysterium insidens*.

Cordana pauciseptata Preuss (Fig. 191)
Colonies effuse, dark brown to black, thinly hairy. Conidiophores brown, up to 170 long. Conidia mid to dark brown, 1-septate, often with a thick very dark band at the septum, 8–12 × 5–7. On dead wood of *Betula, Fagus* and *Quercus*.

Coryne, v.s. under *Ascocoryne*.

Corynespora biseptata, described on *Fagus*, has been found also on rotten wood of *Acer*.

52   Plurivorous Wood and Bark Fungi

Corynespora smithii, described on *Ilex*, is not uncommon also on *Carpinus*, *Fagus* and *Hedera*.

Costantinella, characterised by the recurved tips of the spore-bearing branches which are covered with short denticles and look like little cocks' combs. Two species occur on fallen twigs and woody debris; both have hyaline spherical conidia 3.5–4.5 diam.

KEY

Conidiophores smooth, hyaline, without a setiform apex .... *micheneri*
Conidiophores pale yellowish brown, verrucose with a setiform apex .... *terrestris*

Costantinella micheneri (Berk. & Curt.)Hughes (Fig. 192)
Colonies white, loose, cottony. Conidiophores up to 250 long, 8–10 thick at base. Recorded on *Crataegus*, *Fagus*, *Fraxinus* and *Pinus*.

Costantinella terrestris (Link)Hughes (Fig. 193)
Colonies becoming fawn or greyish brown. Conidiophores up to 1mm long, 9–18 thick at base. Found on *Acer*, *Fraxinus*, *Larix*, *Pinus* and *Quercus*.

Cryptocoryneum, sporodochia small, pulvinate, flat-topped, dark blackish brown to black. Conidia cheiroid, each made up of a small number of dark cap-cells from which septate, subhyaline or pale brown arms grow downwards towards the substratum.

KEY

Conidial arms with 3 to 9 septa .... *rilstonii*
Conidial arms with up to 17 septa .... *condensatum*

Cryptocoryneum condensatum (Wallr.)Mason & Hughes (Fig. 194)
Conidia 40–85 × 20–30, arms 3–5 thick. Very common on fallen dead wood and bark of *Acer*, *Betula*, *Cornus*, *Corylus*, *Fagus*, *Fraxinus*, *Hedera*, *Ilex*, *Populus*, *Prunus*, *Quercus*, *Salix*, *Sorbus*, *Taxus* and *Ulmus*.

Cryptocoryneum rilstonii M.B. Ellis (Fig. 195)
Conidia 20–45 × 12–30, black cap cells firmly united and arms 4–6 thick. On dead branches of *Fraxinus* and *Rubus*, not common.

Cylindrocarpon, v.s. under *Nectria veuillotiana*.

Dactylosporium macropus (Corda)Harz (Fig. 196)
Colonies effuse, black, hairy. Conidiophores dark brown, closely septate near base, paler at apex, up to 1mm long, 9–12 thick below, tapering to 4–7. Conidia subhyaline to brown, smooth, 18–36 × 7–13, with longitudinal and transverse septa. On dead branches of *Acer*, *Fagus*, *Ilex*, *Quercus*, etc.

Dendrodochium, v.s. under *Nectria magnusiana*.

Dendrostilbella, v.s. under *Claussenomyces prasinulus.*

Dendryphiopsis, v.s. under *Microthelia incrustans.*

Dictyosporium toruloides (Corda)Guéguen (Fig. 197)
Colonies effuse, black, granular. Conidia cheiroid, 38–45 × 25–34, flattened in one plane, rather like hands with fingers held close together. Cells mostly olive to brown but the terminal one in each row is often hyaline. On rotten wood of *Acer, Alnus, Betula, Carpinus, Castanea, Corylus, Fagus, Fraxinus, Juglans, Quercus, Rubus, Sambucus* and *Sorbus*, all the year round.

Digitodesmium elegans P.M. Kirk (Fig. 198)
Sporodochia punctiform, pulvinate, pale to mid brown. Conidiophores short, 3–4 wide. Conidia digitate, pale brown, 45–60 long, basal cells 3–5 wide, 2 to 6 arms, each 5–6 wide, 5- to 13-septate, some with gelatinous caps. On rotten wood of *Fagus, Quercus*, etc., Sept.–May.

Diplococcium spicatum Grove (Fig. 199)
Colonies effuse, dark brown, cottony. Conidiophores up to 900 long, 2.5–4 thick, brown, branched, with minute pores in walls. Conidia in chains, very pale brown, 1-septate, 6–9 × 3–4. On dead, often rotting, wood of *Betula, Fagus, Pinus, Prunus, Quercus* and *Sorbus*.

Diplorhinotrichum candidulum Höhnel (Fig. 200)
Colonies effuse, thin white. Conidiophores hyaline, up to 50 × 4–5, the upper half bearing numerous denticles. Conidia 1-septate, 16–22 × 3–4. On rotten wood of *Acer, Fraxinus* and *Sorbus*, Jan.–Apr.

Endophragmia and *Endophragmiella*: species belonging to these genera have brown, septate, often ellipsoidal or pyriform conidia and frequently a little cup near the apex of the conidiophore through which proliferation has taken place. Species are placed in one genus or the other according to whether the conidium breaks away from the conidiophore through or below a septum. If fragmentation takes place below the septum, the conidium has a little frill at its base. This character is not always clear when the frill is very short or there is a lot of pigment present. For this reason the two genera are treated together in one key.

KEY

	Conidia cylindrical or oblong rounded at apex	1
	Conidia obclavate or ovoid	2
	Conidia mostly broadly ellipsoidal	3
	Conidia pyriform, obovoid, turbinate or clavate	5
1.	Conidia 6- to 8-septate	*alternata*
	Conidia mostly 2-septate	*fallacia*
	Conidia mostly 1-septate	*pallescens*
2.	Conidia 1-septate, 14–16 long	*ovoidea*

	Conidia 1- to 3-septate, 14–42 long	*corticola*
3.	Conidia 2- to 3-septate, no black bands at septa	*subolivacea*
	Conidia with 4 or 5 septa and broad black bands at some of them	4
4.	Conidia 4-septate; spherical secondary conidia formed	*prolifera*
	Conidia 5-septate; no secondary conidia	*elliptica*
5.	Conidia all or mostly 1-septate	6
	Conidia all or mostly 2-septate	8
	Conidia all or mostly 3-septate	9
	Conidia all or mostly 4-septate, very pale	*hyalosperma*
6.	Conidia becoming verruculose	*verruculosa*
	Conidia remaining smooth	7
7.	Conidiophores branched	*cambrensis*
	Conidiophores not branched	*uniseptata*
8.	Conidia 8–12 wide	*biseptata*
	Conidia 12–20 wide	*nannfeldtii*
9.	Conidia 10–17 × 9–11.5	*glanduliformis*
	Conidia 22–30 × 11–14	*boothii*

Endophragmia alternata, described on leaf litter, has been found also on dead twigs and branches of *Cytisus*, *Fagus*, *Pinus*, *Rubus* and *Salix*.

Endophragmia biseptata M.B. Ellis (Fig. 201)
Colonies effuse, dark blackish brown. Conidiophores brown, up to 180 × 3–7. Conidia 15–24 × 8–12, mostly 2-septate, uppermost cell dark brown, middle cell brown, basal cell subhyaline. On dead wood of *Betula*, *Cytisus*, *Fagus*, *Ilex*, *Sorbus* and *Ulex*, Sept.–May.

Endophragmia boothii, described on *Rubus*, occurs also on dead wood of *Acer*, *Fagus* and *Prunus spinosa*.

Endophragmia elliptica, a very common species described on herbaceous stems, has been found occasionally also on dead twigs and branches of *Aesculus*, *Fagus*, *Populus*, *Rubus* and *Salix*.

Endophragmia glanduliformis, described on *Betula*, has been found also on *Fagus*.

Endophragmia hyalosperma, described on herbaceous stems, is common also on dead branches of *Cytisus*, *Fagus*, *Picea*, *Quercus*, *Rubus* and *Sambucus*.

Endophragmia nannfeldtii M.B. Ellis (Fig. 202)
Colonies effuse, black. Conidiophores mid to dark brown, up to 550 × 7–10. Conidia 18–33 × 12–20, upper cells mid to dark brown, basal cell hyaline. On fallen twigs and branches of *Castanea*, *Pinus* and *Quercus*.

Endophragmia prolifera, described on *Filipendula*, has been found also on rotting branches of *Populus*.

Endophragmia uniseptata M.B. Ellis (Fig. 203)
Colonies effuse, hairy, black. Conidiophores mid to dark brown, up to 220 × 4–7. Conidia 13–27 × 9–13, upper cell mid to dark brown, lower cell paler. On fallen dead branches of *Castanea*, *Fagus*, *Quercus*, *Ulex* and *Ulmus*, Sept.–Apr.

Endophragmia verruculosa, described on *Fagus*, has been found also on rotten wood which could not be identified.

Endophragmiella cambrensis M.B. Ellis (Fig. 204)
Colonies shortly hairy, greyish brown. Branched conidiophores pale brown, up to 100 × 3–4. Conidia 13–18 × 8–10, dark brown. On unidentified rotten wood.

Endophragmiella corticola P.M. Kirk (Fig. 205)
Colonies effuse, hairy, dark brown. Conidiophores pale to mid brown, 50–60 × 2.5–4. Conidia 14–42 × 5–6.5, 1- to 4-septate, mid to dark brown, paler towards the apex. On rotten wood of conifers and on *Quercus*, Sept.–Apr.

Endophragmiella fallacia P.M. Kirk (Fig. 206)
Colonies effuse, hairy, blackish brown to black. Conidiophores mid to dark brown, up to 300 × 4–6. Conidia with protuberant base, brown, 16–30 × 9–13. On rotten wood of *Betula*, *Corylus* and *Quercus*.

Endophragmiella ovoidea P.M. Kirk (Fig. 207)
Colonies effuse, hairy, blackish brown to black. Conidiophores brown, up to 200 × 3–4. Conidia 14–16 × 5–6.5, apical cell pale, basal cell mid to dark brown. On dead wood of *Alnus*, *Castanea*, *Fagus* and *Quercus*, Sept.–Apr.

Endophragmiella pallescens Sutton (Fig. 208)
Colonies reddish brown, hairy. Conidiophores up to 200 × 3–4, pale brown. Conidia 15–24 × 7–8, rather pale brown. On rotten wood of *Populus* and *Quercus*.

Endophragmiella subolivacea (Ell. & Ev.)Hughes (Fig. 209)
Colonies hairy, pale brown. Conidiophores up to 120 × 4–6. Conidia 27–37 × 15–17, pale brown, sometimes minutely verruculose. On rotten wood.

Fusarium, v.s. under *Nectria flavo-viridis*.

Geniculosporium, v.s. under *Hypoxylon serpens* and *Rosellinia aquila*.

Gliocladium luteolum Höhnel (Fig. 210)
Colonies thinly hairy, yellowish. Conidiophores up to 850 × 10, pale straw-coloured, penicillately branched at apex. Conidia agglutinated to form yellow slimy heads, individually appearing smooth, 5–7 × 2–3. On rotten wood.

Gliocladium roseum Bainier (Fig. 211)
Colonies variable in size, sometimes effuse, at first creamy, soon becoming salmon pink. Conidiophores hyaline, penicillately branched, 3–5 thick.

Conidia hyaline, pink in mass, mostly 5–7 × 2–3.5, often flattened on one side or slightly curved, held together in slimy masses. Commonly found on dead branches of *Rubus* and stacked logs of *Acer* but recorded also on *Fagus, Picea, Pinus, Robinia* and *Rosa*.

Gonytrichum, v.s. under *Chaetosphaeria inaequalis.*

Graphium calicioides (Fr.)Cooke & Massee (Fig. 212)
Colonies effuse with erect black synnemata up to 5mm high and 40–100 thick clearly visible under a low-power dissecting microscope. Conidia aggregated in olivaceous brown to black slimy heads, individually hyaline or olivaceous, 1.5–3 × 1–2. On rotten wood of *Acer, Alnus, Betula, Cytisus, Fagus, Fraxinus, Juglans, Prunus, Quercus, Sorbus* and *Ulmus*, common, Sept.–May.

Haplographium, v.s. under *Dematioscypha dematiicola.*

Harpographium, v.s. under *Peroneutypa heteracantha.*

Helicoma state of *Thaxteriella pezicula* (Berk. & Curtis)Petrak (Fig. 213)
Colonies effuse, hairy or velvety, buff, olivaceous brown or brown. Conidiophores simple or occasionally branched, brown or olivaceous brown, up to 180 × 5–8, apex pale and sometimes setiform. Conidia on pegs, helicoid, $1\frac{1}{2}$–$1\frac{3}{4}$ times coiled, 17–21 diam., becoming pale olivaceous brown, filaments 5- to 12-septate, 3.5–6 thick. Not uncommon on wet decaying wood and bark.

Helicomyces roseus Link (Fig. 214)
Colonies superficial, effuse, pale pink. Conidia helically coiled, filaments multiseptate, hyaline, 160–180 × 6, borne on short conidiophores. On rotten wood.

Helicoon ellipticum (Peck)Morgan (Fig. 215)
Colonies olivaceous brown, cottony or velvety. Conidia on pegs attached to hyphae, coiled in more than one plane to form ellipsoidal, pale straw-coloured spore-bodies 32–40 × 22–30 with 7 to 9 coils; filaments multiseptate, 3–5 thick. On rotten, mainly coniferous, wood.

Helicosporium, v.s. under *Tubeufia cerea.*

Helminthosporium velutinum Link (Fig. 216)
Colonies effuse, black, hairy. Conidiophores dark brown, up to 950 long, 8.5–12 thick near apex. Conidia developing through small pores at apex and in whorls beneath septa, obclavate, rather pale golden brown, 6- to 16-pseudoseptate, 40–118 × 11–20. Very common on fallen dead twigs and branches of *Acer, Alnus, Betula, Cornus, Corylus, Crataegus, Cytisus, Euonymus, Fagus, Fraxinus, Hedera, Ilex, Ligustrum, Myrica, Prunus, Quercus, Rhododendron, Ribes, Salix, Sambucus, Tilia, Ulex* and *Ulmus*, all the year round. Occasionally parasitised by *Letendraea helminthicola* (Berk. & Br.)Weese (Fig. 217) with minute yellow–ochre or flesh-coloured perithecia and pale brown, 1-septate

ascospores 12–18 × 5–6.

Heteroconium chaetospira (Grove)M.B. Ellis (Fig. 218)
Colonies effuse, scarcely visible to the naked eye. Conidiophores pale brown, up to 50 × 3–4. Conidia in long, simple, often spirally twisted chains, very pale brown, mostly 1- to 3-septate, thin-walled, 20–35 × 3–4. On rotten wood, usually overgrowing other fungi such as *Chaetosphaeria cupulifera* and *Sporoschisma mirabile*.

Heteroconium tetracoilum (Corda)M.B. Ellis (Fig. 219)
Colonies effuse, olivaceous, velvety, often forming glistening tufts around the ostioles of diatrypaceous fungi. Conidiophores pale olivaceous brown, up to 65 × 2.5–5. Conidia in long, unbranched chains, pale olive, thick-walled, 15–65 × 4–7. Usually found growing on or close to fructifications of Diatrypaceae on logs and fallen branches, especially where these have been kept damp by a covering of dead leaves. Most commonly on *Diatrype stigma* but recorded also on *Anthostoma*, *Eutypa flavovirens* and *Peroneutypa heteracantha* on *Acer*, *Corylus*, *Fagus*, *Fraxinus*, *Hedera*, *Ilex*, *Populus*, *Prunus* and *Quercus*.

Hobsonia mirabilis (Peck)Linder (Fig. 220)
Sporodochia watery white, gelatinous, 2mm diam. Conidia helicoid, multiseptate, hyaline, 25–40 diam.; filaments 9–12 thick, guttulate. On rotten wood, uncommon.

Mariannaea elegans (Corda)Samson (Fig. 221)
Colonies effuse, whitish to pale golden brown. Conidiophores hyaline, 200–750 × 10–14, verticillately branched at the apex. Conidia formed in slipped chains or slimy heads, hyaline, fusiform or angular, 4–6 × 2–3.5. On wood and bark of *Acer*, *Betula*, *Corylus*, *Fagus*, *Fraxinus*, *Larix*, *Quercus*, *Pinus* and *Rhododendron*.

Menispora: four species found on wood and bark. One is the conidial state of *Chaetosphaeria pulviscula* (*v.s.*). All form colonies which are effuse, hairy or velvety, at first pale greyish brown, later olivaceous and sometimes blackish brown. Conidia hyaline, slightly curved, aggregated in slimy masses at tips of lateral phialides.

KEY

	Conidia 3-septate with a setula at each end 1
	Conidia without septa 2
1.	Phialides recurved at tip .... *glauca*
	Phialides not recurved at tip .... *tortuosa*
2.	Conidia with a setula at each end .... *ciliata*
	Conidia without setulae .... *Chaetosphaeria pulviscula*

Menispora ciliata Corda (Fig. 222)
Conidiophores occasionally branched, flexuous, frequently anastomosing, up

to 900 × 3–5, upper part sterile, setose. Conidia 12–21 (mostly 14–18) × 3–3.5; setulae up to 12 long. Very common on dead wood and bark of *Fagus, Fraxinus, Hedera, Ilex, Prunus, Quercus, Rosa, Rubus, Sambucus* and *Ulex*, all the year round.

Menispora glauca Pers. (Fig. 223)
Conidiophores unbranched or with a few long, upwardly directed branches, up to 800 × 3–5, upper part sterile, setose, twisted or loosely coiled. Conidia mostly 18–23 × 4–5; setulae up to 14 long. On wood and bark of *Acer, Betula, Fagus, Populus*, etc., fairly common, most collections made in winter.

Menispora tortuosa Corda, found on *Betula, Fagus* and *Fraxinus*, is similar to *M. glauca* but has phialides which are clustered and not bent over at the tip.

Monodictys, conidia brown or black, muriform, formed at the ends of very short conidiophores; colonies effuse, rather powdery or granular.

KEY

```
   Conidia verrucose ............................................................... 1
   Conidia smooth .................................................................. 2
1. Conidia deeply lobed ................................................. antiqua
   Conidia not deeply lobed ........................................ castaneae
2. Conidia mostly pyriform or subspherical ................... putredinis
   Conidia mostly cylindrical or oblong ........................... lepraria
```

Monodictys antiqua (Corda)Hughes (Fig. 224)
Colonies black. Conidia often in erect groups, dark brown, up to 75 × 25. On dead wood.

Monodictys castaneae (Wallr.)Hughes (Fig. 225)
Colonies lavender to dark grey or black. Conidia mid to dark reddish brown, 14–40 × 10–25. On rotten wood of *Quercus, Sambucus*, etc.

Monodictys lepraria (Berk.)M.B. Ellis (Fig. 226)
Colonies dark blackish brown to black. Conidia mid to dark brown or olivaceous brown, sometimes shading from dark to pale, up to 100 × 50. On dead branches of *Quercus, Pinus, Tilia*, etc., often associated with or growing over lichens.

Monodictys putredinis (Wallr.)Hughes (Fig. 227)
Colonies blackish brown to black. Conidia dark reddish brown to almost black, 20–30 × 15–25. On rotten wood including that of *Prunus spinosa*.

Myrothecium, v.s. under *Nectria ralfsii*.

Nodulisporium, v.s. under *Hypoxylon*.

Oedemium, v.s. under *Chaetosphaerella*.

Oidiodendron, conidiophores look like tiny trees with their branches often almost at right-angles to the main axis and to each other, fragmenting to form chains of conidia.

KEY

    Surrounding wood often stained red .... *rhodogenum*
    Surrounding wood not stained red 1
1.   Conidia oblong or ellipsoidal .... *griseum*
    Conidia spherical or subspherical .... *tenuissimum*

Oidiodendron griseum Robak (Fig. 228)
Colonies grey or olivaceous brown. Conidiophores olivaceous to blackish brown, smooth, up to 100 × 1.5–2. Conidia mostly smooth, rarely minutely verruculose, 2–3.5 × 1.5–2. On rotten wood of *Betula, Fagus, Pinus,* etc., Sept.–Jan.

Oidiodendron rhodogenum Robak (Fig. 229)
Colonies grey, greenish grey or brown. Conidiophores up to 150 × 2–3, brown, frequently verrucose especially towards the apex. Conidia minutely verruculose, 2.5–3.5 × 1.5–2. On rotten decorticated wood of *Pinus, Quercus,* etc.

Oidiodendron tenuissimum (Peck)Hughes (Fig. 230)
Colonies pale grey to blackish brown. Conidiophores brown or olivaceous brown, smooth or verrucose, up to 300 × 1.5–2.5. Conidia 2–4 × 2–3, with dark and distinctly verruculose walls, linked in chains by narrow connectives and resembling beads on a string. Common on wood and bark of *Betula, Carpinus, Castanea* and *Fagus,* all the year round.

Oncopodiella trigonella, described on *Hedera,* has been found also on rotten wood of *Pinus* and *Ulmus.*

Pachnocybe albida Berk. (Fig. 231)
This is not congeneric with the lectotype species of *Pachnocybe, P. ferruginea,* and needs taxonomic revision. Colonies effuse. Synnemata erect, white or cream, up to 400 × 40–80, composed of hyphae 2–3 thick. Conidia hyaline, smooth, 15–25 × 11–15. On rotting logs of *Buxus, Fraxinus* and *Quercus.*

Pachnocybe ferruginea (Sow.)Berk. (Fig. 232)
Colonies effuse, rust-coloured, the scattered, rather dark reddish-brown synnemata up to 1mm high and 10–20 thick, each capped by a pale, splayed-out head. Conidia 4–7 × 2–3.5. Mostly on sawn timber.

Periconia cambrensis Mason & M.B. Ellis (Fig. 233)
Conidiophores widely scattered and not forming well defined colonies, pale brown, 150–300 × 3–5. Conidia in short chains, spherical, brown, smooth, 5–8 diam. On dead wood and bark of *Betula, Fagus, Ilex* and *Quercus,* Aug.–Apr.

Phaeoisaria species form effuse black hairy colonies made up of erect synnemata the upper parts of which are covered with a white or pale grey powdery mass of conidia.

KEY

Conidia subspherical 1–2 diam. .... *clavulata*
Conidia fusiform or narrowly ellipsoid 4–10 × 1.5–2.5 .... *clematidis*

Phaeoisaria clavulata (Grove)Mason & Hughes (Fig. 234)
Synnemata brown, up to 450 × 10–20. On rotten wood of *Fraxinus, Laurus, Pinus, Prunus* and *Sambucus*.

Phaeoisaria clematidis (Fuckel)Hughes (Fig. 235)
Synnemata brown, up to 1.5mm high, 20–80 thick at base. Common on fallen dead branches of *Acer, Betula, Clematis, Rubus, Sambucus, Ulmus*, etc., Nov.–Apr.

Phaeostalagmus species have erect, brown, tree-like conidiophores which bear near their apices either whorls of short branches bearing phialides or whorls of phialides. Conidia hyaline.

KEY

Conidia 5–8 × 2 .... *peregrinus*
Conidia not more than 4 long 1
1. Phialides only or very few branches, conidia 2–4 long .... *tenuissimus*
   Branches and phialides, conidia less than 2 long .... *cyclosporus*

Phaeostalagmus cyclosporus (Grove)W. Gams (Fig. 236)
Colonies pale grey and frosted when fresh, becoming dark blackish brown, hairy. Conidiophores up to 500 long, dark brown and 3–7 thick near base, tapering to 2–3 and paler at the apex, with usually 4 to 10 whorls of branches. Very common at all times of the year on fallen dead branches of *Alnus, Betula, Castanea, Corylus, Fagus, Fraxinus, Ilex, Populus, Prunus, Quercus, Rhododendron, Rosa, Rubus, Salix, Sambucus* and *Ulex*.

Phaeostalagmus peregrinus, described on pine cones, has been found inside bark of *Eucalyptus* and on *Castanea*.

Phaeostalagmus tenuissimus (Corda)W. Gams (Fig. 237)
Colonies grey to dark brown, hairy. Conidiophores usually about 200 long, 4–8 thick and dark brown at base, tapering and becoming paler towards the apex. Phialides formed directly on stipe in 1 to 5 whorls. On fallen branches and litter of *Alnus, Castanea, Fagus, Ilex, Picea, Pinus, Quercus, Rhododendron, Rubus* and *Ulmus*, common.

Pleurotheciopsis bramleyi, described on *Alnus*, has been found also on fallen branches of *Betula, Ilex, Quercus* and *Salix*.

Pleurothecium recurvatum (Morgan)Höhnel (Fig. 238)
Colonies effuse, hairy, brown or greyish brown. Conidiophores brown, up to 260 × 4–7. Conidia formed on flat-topped pegs, slightly curved, 3-septate, at first hyaline becoming brown, with median cells darker than end ones, 16–27 × 5–7, aggregated in slimy heads. On dead branches of *Acer*, *Betula*, *Fagus*, *Quercus*, *Tilia* and *Ulmus*.

Pseudospiropes species form effuse, often velvety or hairy colonies and have unbranched conidiophores with a number of thickened, often dark conidial scars near their apices.

KEY

    Conidia pseudoseptate, fusiform or navicular 1
    Conidia septate, usually other shapes 2
1. Conidia 12–18 wide, 5–7 at scar .... *nodosus*
    Conidia 9–13 wide, 2–3 at scar .... *simplex*
2. Conidia obclavate .... *obclavatus*
    Conidia not obclavate 3
3. Conidia 12–18 long, mostly ellipsoidal .... *hughesii*
    Conidia often more than 20 long, not ellipsoidal 4
4. Conidia mostly somewhat clavate, 3.5–5.5 wide .... *subuliferus*
    Conidia never clavate, 5–8 wide .... *rousselianus*

Pseudospiropes hughesii, described on *Fagus*, has been found also on *Fraxinus* and *Sambucus*.

Pseudospiropes nodosus (Wallr.)M.B. Ellis (Fig. 239)
Colonies black. Conidiophores 100–350 × 8–10, dark blackish brown. Conidia pale to dark golden brown, with 6 to 10 pseudosepta, 32–50 × 12–18, 5–7 wide at scar. Common on dead wood and bark of *Acer*, *Betula*, *Corylus*, *Fagus*, *Fraxinus*, *Hedera*, *Populus*, *Rubus* and *Salix*.

Pseudospiropes obclavatus M.B. Ellis (Fig. 240)
Colonies olivaceous brown to dark blackish brown. Conidiophores olivaceous brown or dark brown, up to 60 × 3.5–5. Conidia 3- to 10-septate, pale to mid olivaceous brown, smooth, rugulose or verruculose, 16–38 × 3.5–4.5. On fallen branches of *Betula*, *Castanea*, *Corylus*, *Fagus*, *Hedera*, *Pinus*, *Quercus* and *Sambucus*.

Pseudospiropes rousselianus, described on herbaceous stems, is found occasionally on wood and bark of deciduous trees including *Fagus*; conidia 3- to 7-septate, 20–37 × 5–8.

Pseudospiropes simplex (Kunze)M.B. Ellis (Fig. 241)
Colonies dark olivaceous brown or blackish brown. Conidiophores dark brown, up to 400 × 4.5–6. Conidia pale to mid golden brown, with 6 to 11 pseudosepta, 26–44 × 9–13, 2–3 wide at scar. Very common on dead

branches of *Acer, Betula, Corylus, Crataegus, Fagus, Fraxinus, Hedera, Ilex, Malus, Prunus, Quercus, Ruscus, Salix, Ulex* and *Ulmus*.

Pseudospiropes subuliferus (Corda)M.B. Ellis (Fig. 242)
Colonies dark blackish brown to black. Conidiophores dark brown, subulate, closely septate, up to 500 long, 7–10 thick just above base. Conidia hyaline or subhyaline, 1- to 6- (mostly 3-)septate, 12–29 × 3.5–5.5. On dead wood and bark of *Acer, Carpinus, Crataegus, Fagus, Fraxinus, Salix* and *Sorbus*.

Sarcopodium tortuosum (Wallr.)Hughes (Fig. 243)
Sporodochia yellowish or reddish brown, hairy, often with a whitish margin. Setae branched, twisted, pale to mid golden brown, smooth or indistinctly verruculose, up to 400 × 2–3. Conidia aggregated in slimy orange masses when fresh, hyaline, 3–8 × 1–1.5. On dead wood and bark of *Fagus, Populus, Quercus* and *Rhamnus*.

Septotrullula bacilligera, described on *Betula*, has been found occasionally also on *Fagus, Quercus* and *Salix*.

Spadicoides species form effuse, blackish-brown to black, hairy or velvety colonies and have dark brown or blackish-brown conidiophores with minute pores in their walls through which short, brown, 0- to 3-septate conidia develop.

KEY

Conidia without septa ....*atra*
Conidia 1-septate .... *bina*
Conidia 3-septate .... *grovei*

Spadicoides atra, described on *Larix*, has been found on dead branches of other trees, mainly conifers.

Spadicoides bina (Corda)Hughes (Fig. 244)
Conidiophores 20–550 × 2.5–4.5. Conidia mid pale to dark reddish brown, mostly 1-septate, with a wide, almost black band at the septum, 7–12 × 3–5. On dead wood and bark of *Betula, Fagus, Quercus, Sambucus* and *Sorbus*, all the year round.

Spadicoides grovei M.B. Ellis (Fig. 245)
Conidiophores 70–300 × 4–6. Conidia mid to dark brown with almost black bands at septa, 16–26 × 8–13. On dead wood of *Fagus* and on rotten unidentified wood, May–Sept.

Sporidesmium species form effuse, hairy or velvety usually brown or black colonies and bear septate, brown conidia singly at the apex of erect, brown conidiophores.

KEY

	Conidia cylindrical, pseudoseptate .... *folliculatum*
	Conidia conical to obpyriform, septate 1
	Conidia fusiform or obclavate, septate or pseudoseptate 3
1.	Conidia 5- to 8-septate, 14–25 wide .... *altum*
	Conidia 2- to 4-septate, 7–12 wide 2
2.	Conidia conico-truncate and protruding at base .... *cookei*
	Conidia rounded at base .... *aturbinatum*
3.	Conidia narrowly obclavate, 5–8 wide .... *leptosporum*
	Conidia more than 8 wide 4
4.	Conidia pseudoseptate 5
	Conidia septate 6
5.	Conidia not more than 70 long .... *coronatum*
	Conidia 145–300 long .... *vagum*
6.	Conidia 30–40 long, fusiform .... *socium*
	Conidia over 50 long, obclavate to fusiform 7
7.	Conidia 50–130 long, obclavate .... *pedunculatum*
	Conidia up to 270 long, fusiform to obclavate .... *hormiscioides*
	Conidia up to 600 long, obclavate, rostrate .... *anglicum*

Sporidesmium altum, described on *Sambucus*, has been found also on *Acer*, *Buxus*, *Clematis*, *Hedera*, *Prunus* and *Ulmus*.

Sporidesmium anglicum (Grove)M.B. Ellis (Fig. 246)
Conidiophores 20–100 × 4–8. Conidia mid to dark reddish brown, 90–600 × 12–16, with 24–142 very dark septa. On dead wood, rare.

Sporidesmium aturbinatum, described on *Sambucus*, has been found also on *Clematis vitalba*.

Sporidesmium cookei, described on *Sambucus*, has been found occasionally also on *Clematis*, *Hedera* and *Ulex*.

Sporidesmium coronatum Fuckel (Fig. 247)
Colonies dark olivaceous to chocolate brown. Conidiophores brown, 10–70 × 4–7. Conidia pale brown, with 8 to 14 pseudosepta, 35–70 × 9–12. On dead wood and bark of *Acer*, *Cytisus*, *Fagus* and *Sambacus*; sometimes growing on conidiophores of *Helminthosporium velutinum*.

Sporidesmium folliculatum (Corda)Mason & Hughes (Fig. 248)
Colonies black. Conidiophores dark reddish brown, 40–98 × 5–6.5. Conidia pale brown when young, rather dark reddish brown when mature, with 6 to 12 pseudosepta, 38–81 × 8–11. On rotten stumps and fallen branches of *Fagus*, *Fraxinus*, *Hedera*, *Quercus*, *Salix*, *Sorbus* and *Ulmus*, Mar.–Nov.

Sporidesmium hormiscioides, described on *Fagus*, is found also occasionally on *Quercus*.

Sporidesmium leptosporum, described on *Sambucus*, is found quite commonly also on dead branches of many other trees and shrubs, including *Acer*, *Cupressus*, *Fagus*, *Fraxinus*, *Quercus*, *Rhododendron* and *Ulmus*.

Sporidesmium pedunculatum (Peck)M.B. Ellis (Fig. 249)
Colonies grey to black. Conidiophores dark reddish brown to black, 25–130 × 5–8. Conidia rather dark brown, 9- to 26-septate, 50–130 × 11–13. On rotting wood of *Populus*, *Taxus* and an unidentified conifer.

Sporidesmium socium, described on *Hedera*, has been found also on *Tilia*.

Sporidesmium vagum C.G. & T.F.L. Nees (Fig. 250)
Colonies black. Conidiophores 17–31 × 3.5–6. Conidia pale to mid brown, with 19 to 38 pseudosepta, 145–300 × 11–14. On unidentified rotten wood.

Sporoschisma species form effuse, black colonies and have capitate setae scattered or in groups mixed with large, dark brown conidiophores. Conidia cylindrical with truncate ends, brown, septate, formed inside conidiophores.

KEY

Conidia dark brown, smooth .... *mirabile*
Conidia pale brown, verruculose .... *juvenile*

Sporoschisma juvenile Boud. (Fig. 251)
Conidiophores up to 240 long. Conidia 20–42 × 7–11, 3-septate when mature. On wood and bark of *Alnus*, *Fagus*, *Fraxinus*, *Hedera*, *Quercus*, *Pinus*, *Ribes*, *Sorbus* and *Ulmus*, common.

Sporoschisma mirabile Berk. & Br. (Fig. 252)
Conidiophores in groups of up to 20, up to 320 long. Conidia mostly 3-septate, 23–45 × 10–15. On rotten wood and bark of *Alnus*, *Betula*, *Fagus*, *Fraxinus*, *Quercus*, *Salix*, *Sorbus*, *Ulmus*, etc., common.

Sporotrichum state of *Ceriporiopsis aneirina* (Sommerf.)Dom. (Fig. 253)
Colonies effuse, white when young becoming golden yellow to bright orange when sporulating freely. Mycelium yellow, with clamp connections. Conidia yellow, orange in mass, 8–12 × 6–8. On very rotten wood of *Populus*, *Quercus*, etc.

Stachybotrys, v.s. under *Melanopsamma pomiformis*.

Taeniolella breviuscula (Berk. & Curtis)Hughes (Fig. 254)
Colonies effuse, black, thin. Conidiophores dark blackish brown, 6–10 thick. Conidia mostly 2- or 3-septate, rarely with up to 5 transverse septa and often with 1 or 2 longitudinal septa, smooth, dark blackish brown, 17–45 × 10–13. On wood and bark.

Taeniolella rudis (Sacc.)Hughes (Fig. 255)
Colonies effuse, black, velvety. Conidia in long, unbranched chains, mid to

Plurivorous Wood and Bark Fungi

dark brown, mostly 40–60 × 10–13, with 5–8 thick black septa. On wood lying on the ground.

Taeniolella stilbospora, described on *Salix*, has been found occasionally on *Alnus* and *Corylus*.

Taeniolina scripta (P. Karsten)P.M. Kirk (Fig. 256)
Colonies pulvinate and discrete or effuse, dark blackish brown to black. Conidia branched; branches 2- to 17-septate, 12–70 × 5–7, mid to dark brown or reddish brown. On dead wood and bark of *Betula*, *Corylus*, *Fagus*, *Pinus*, *Quercus* and *Sorbus*.

Trichocladium species form effuse, cottony, grey to black colonies, and their dark brown, septate conidia are formed at the ends of short branches.

KEY

Conidia verrucose, mostly 1-septate .... *asperum*
Conidia smooth, mostly with 3 or more septa .... *opacum*

Trichocladium asperum Harz (Fig. 257)
Conidia 15–30 × 10–15. A plurivorous species occasionally found on rotten wood.

Trichocladium opacum (Corda)Hughes (Fig. 258)
Conidia 1- to 5-septate but mostly with 4 septa, 25–40 × 11–17. Recorded on wood and bark of *Castanea*, *Fagus*, *Fraxinus* and *Quercus* and on petioles of *Acer*.

Trichoderma species form either effuse or compact and pulvinate colonies, many of them green when sporulating freely, and have much branched conidiophores, the branches often almost at right-angles to the main axis and to each other.

KEY

    Conidiophores often terminating in flexuous sterile setae 1
    Conidiophores not terminating in sterile setae 2
1. Conidia hyaline, 2.5–4 × 2 .... *polysporum*
    Conidia green, 4–6 × 2–3 .... *hamatum*
2. Conidia smooth 3.5–4.5 × 2–3 .... *Chromocrea aureoviridis (v.s.)*
    Conidia verruculose, 3.5–4.5 diam. .... *Hypocrea rufa (v.s.)*

Trichoderma hamatum (Bon.)Bainier (Fig. 259)
Colonies at first white, later greyish green, tufted. On rotten wood, occasional.

Trichoderma polysporum (Link)Rifai (Fig. 260)
Colonies remaining pure white, as scattered, raised velvety patches. Common on wood of *Acer*, *Fagus*, *Fraxinus* and *Pinus*.

Trichothecium roseum Link (Fig. 261)
Colonies effuse, at first white but soon turning rosy pink. Conidiophores up to 150 × 3–4, hyaline, often slightly swollen at their tips. Conidia hyaline, pink in mass, 1-septate, thick-walled, each with a flattened protuberance at the base, 13–27 × 7–11, often clustered and held together in slime. Common on felled trunks and fallen branches of *Acer, Corylus, Fagus, Prunus, Quercus* and *Ulmus*, Oct.–Mar.

Trimmatostroma betulinum, described on *Betula*, is common also on attached and fallen branches of *Corylus, Ilex, Salix* and *Sorbus*.

Triposporium elegans, described on *Rubus*, fairly common also on dead twigs and branches of *Alnus, Corylus, Fagus, Fraxinus, Ilex, Prunus, Quercus, Rhododendron* and *Ulex*.

Tubercularia, v.s. under *Nectria cinnabarina*.

Veronaea parvispora M.B. Ellis (Fig. 262)
Colonies effuse, dark blackish brown to black, hairy. Conidiophores brown, up to 250 × 1.5–2, with numerous minute scars near apex. Conidia pale brown, smooth, 2–3 × 1.5–2. On dead wood of *Quercus, Salix*, etc., and overgrowing *Chaetosphaeria callimorpha*.

Virgaria nigra (Link)Nees (Fig. 263)
Colonies effuse, dark olivaceous brown or almost black, often thick and felted. Conidiophores branched, pale to mid brown, 60–250 × 2–3. Conidia reniform, pale to mid brown, smooth, 4–6 × 2.5–4, borne on cylindrical denticles. On wood and bark of *Acer, Betula, Corylus, Fagus, Fraxinus, Juglans, Quercus, Rubus* and *Sorbus*.

Virgariella species form effuse, velvety or hairy, blackish-brown to black colonies, and have simple, erect, brown conidiophores without apparent scars or denticles, on which are borne dark brown non-septate conidia.

KEY

Conidia ellipsoidal to spherical, 11–15 × 10–13 .... *atra*
Conidia ovoid, 8–9.5 × 5–6 .... *ovoidea*

Virgariella atra Hughes (Fig. 264)
Conidiophores up to 200 × 4–7. On rotten wood of *Fagus, Fraxinus* and *Quercus*.

Virgariella ovoidea Kirk (Fig. 265)
Conidiophores 25–60 × 2.5–5. On rotten wood including that of *Quercus*.

Xylohypha species form effuse or pulvinate and scattered, powdery, dark colonies and long acropetal chains of brown, non-septate conidia formed at the ends of short brown conidiophores.

KEY

Colonies reddish brown, conidia 5–7 × 3.5–5 .... *ferruginosa*
Colonies brown to black, conidia 7–15 × 4–6 .... *nigrescens*

Xylohypha ferruginosa (Corda)Hughes (Fig. 266)
Conidiophores very short. Conidia reddish brown. On wood of *Fagus*, *Quercus* and *Salix*.

Xylohypha nigrescens (Pers.)Mason (Fig. 267)
Conidiophores 15–35 × 2.5–4.5. Conidia pale to mid brown. Conidial chains break up very readily and hands become covered with brown powder when it is being collected. Common on wood of *Acer*, *Cornus*, *Corylus*, *Fagus*, *Fraxinus*, *Hedera*, *Juglans*, *Ligustrum*, *Populus*, *Salix*, *Sambucus*, *Tilia* and *Viburnum*.

COELOMYCETES

Aposphaeria, v.s. under *Melanomma fuscidulum* and *M. pulvis-pyrius*.

Cytospora, v.s. under *Valsa*.

Dichomera saubinetii, described on *Acer*, occurs also on twigs of *Corylus* and *Rhamnus*.

Diplodia, v.s. under *Otthia spiraeae*.

Microsphaeropsis olivacea (Bon.)Höhnel (Fig. 268)
Pycnidia at first immersed becoming erumpent, olivaceous brown, thin-walled, 0.2–0.3mm diam., darker around the papillate ostiole. Conidia oval or oblong–ellipsoid, pale olivaceous brown, 4–8 × 2.5–5. On twigs and branches of *Cytisus*, *Hedera*, *Laurus*, *Lycium* and *Sambucus*.

Truncatella angustata (Pers.)Hughes (Fig. 269)
Acervuli subepidermal, up to 0.35–0.4mm diam., black. Conidiophores hyaline, up to 25 × 2. Conidia 16–20 × 7–8, 3-septate, median cells brown, cell at each end hyaline, apical appendage often branched, 12–28 long. On wood including rootstocks of *Malus*, *Prunus* and *Ribes*.

BASIDIOMYCETES

Cyphellopsis anomala (Pers.)Donk (Fig. 270)
This little basidiomycete is included because the clustered snuff-coloured, white-edged fructifications are cup-shaped and sometimes mistaken in the field for discomycetes. Hairs long, pale yellow, verruculose, swollen at their hyaline tips. Basidia 18–30 × 5–6. Spores hyaline, mostly 8–10 × 4. On logs, branches and twigs of *Alnus*, *Betula*, *Fagus*, *Fraxinus* and *Quercus*, Aug.–Feb.

PLURIVOROUS LEAF-LITTER FUNGI

DISCOMYCETES

Betulina fuscostipitata, described on *Betula*, has been found also on dead *Rubus* leaves and *Castanea* cupules.

Botryotinia fuckeliana (de Bary)Whetzel (Fig. 271)
Apothecia 0.5–3mm diam., pale brown, downy, with finely hairy stalks blackish brown at base, arising from black sclerotia. Ascospores hyaline, 8–10 × 4–5. Preceded or accompanied by a *Botrytis* state (*B. cinerea* type). Inside *Castanea* cupules and on dead leaves and other debris of *Quercus*, *Rubus*, *Rosa* and *Salix*.

Calycellina populina, described on *Populus*, occurs commonly also on dead leaves of *Betula*, *Quercus* and *Rubus*.

Coccomyces coronatus, described on *Quercus*, is found also on *Betula*, *Fagus* and *Sorbus* leaves.

Crocicreas subhyalinum, described on *Acer*, has been collected also on dead leaves and cupules of *Aesculus* and on leaves of *Alnus* and *Castanea*.

Dasyscyphus ciliaris, with ascospores 15–23 × 2–2.5, described on *Quercus*, is found occasionally also on *Castanea* and *Fagus*.

Dasyscyphus coruscatus, with ascospores 4–6 × 1.5–2 and thin-walled hairs capped by crystals, described on *Quercus* leaves, has been collected also on *Castanea* and *Rubus*.

Dasyscyphus deflexus, with ascospores 4–5 × 1.5–2 and distinctly uncinate, granulate hairs; described on *Ulmus* leaves, has been seen also on *Acer* and *Salix*.

Dasyscyphus virgineus S.F.Gray (Fig. 18)
Apothecia up to 1mm diam., long-stalked, all white or with creamy discs, very hairy both on the outside of the cup and on the stalk. Hairs hyaline, up to 100 × 4–5, some slightly swollen at the tip, variable, minutely granulate. Ascospores hyaline, 6–10 × 1.5–2.5. Lanceolate paraphyses up to 5 wide. The commonest of all litter species on dead leaves, stems, twigs, cupules, cones and bracken fronds, Mar.–Aug.

Hyalopeziza ciliata, described on *Acer*, is not uncommon also on dead leaves of *Carpinus*, *Fagus* and *Quercus*, Oct.–Nov.

Hyaloscypha lachnobrachya, described on decaying leaves of *Acer*, where dense swarms occur from Sept. to Nov., is found quite frequently also on *Aesculus, Alnus, Fagus, Populus, Quercus* and *Ulmus*.

Hyaloscypha lachnobrachya var. *araneocincta*, described on *Betula*, occurs also on dead leaves of *Quercus*.

Hymenoscyphus caudatus (P. Karsten)Dennis (Fig. 272)
Apothecia stalked, white or creamy, about 1mm diam. Ascospores hyaline, more pointed at base than apex, 17–22 × 4–5. A very common and truly plurivorous species found on decaying leaves of deciduous trees, mainly on veins and petioles, June–Dec. We have found it on *Acer, Alnus, Betula, Castanea, Corylus, Fagus, Populus, Quercus, Tilia* and *Salix*. There are six other species of *Hymenoscyphus* found in litter which show preference for but are not strictly confined to particular host plants: *H. albopunctus* (described on *Alnus*), *H. epiphyllus* (*Quercus*), *H. fructigenus* (*Corylus*), *H. immutabilis* and *H. phyllogenus* (*Populus*), and *H. phyllophilus* (*Fagus*).

KEY

	Apothecia bright yellow when fresh	*epiphyllus*
	Apothecia white or pale cream when fresh	1
1.	Ascospores not more than 15 long	2
	Ascospores more than 15 long	4
2.	Apothecia 1–1.5mm diam.	*immutabilis*
	Apothecia not more than 0.5mm diam.	3
3.	Ascospores 1-septate	*phyllophilus*
	Ascospores without septa	*phyllogenus*
4.	Apothecia on nuts and cupules, up to 3mm diam.	*fructigenus*
	Apothecia on rotting leaves	5
5.	Ascospores 4–5 wide	*caudatus*
	Ascospores 2.5–3 wide	*albopunctus*

Additional hosts include:
for *H. albopunctus*: *Fagus, Rosa, Rubus, Quercus, Salix*.
for *H. epiphyllus*: *Betula, Carpinus, Castanea, Rosa, Rubus*.
for *H. fructigenus*: *Carpinus, Fagus, Quercus*.
for *H. immutabilis*: *Betula, Quercus, Ulmus*.
for *H. phyllogenus*: *Acer, Castanea, Quercus*.
for *H. phyllophilus*: *Quercus*.

Mollisina acerina, described on *Acer*, has been recorded also on *Populus* and *Rubus*.

Mollisina rubi, a minute species with short, thread-like, often branched, processes from marginal cells, described on *Rubus* leaves, occurs also on *Acer, Quercus, Salix* and *Ulmus*.

Phialea sp.1 Dennis (Fig. 273)
Apothecia about 0.3mm diam., like little eggcups, greyish white or creamy, dark brown at the base. Ascospores hyaline, 16–18 × 2. On dead leaves of *Betula, Castanea, Fagus, Quercus* and *Salix*, Sept.–Nov.

Pilatia foliorum Velen.
Apothecia scattered, 0.5–0.7mm diam., deeply cup-shaped, stalked, white, with hyaline, thick-walled, acute marginal setae 120–200 × 3. Asci 25–30 × 5–6. Ascospores 3–5 long. Described originally on *Prunus avium*, has been recorded in Britain on *Betula* and *Salix*.

Pyrenopeziza petiolaris, described on *Acer*, is found also on petioles of *Aesculus* and *Populus*.

Trichodiscus virescentulus (Mouton)Dennis (Fig. 274)
Apothecia sessile, up to 0.2mm diam., pale greenish yellow, hairy. Hairs about 80 × 4, encrusted with greenish matter. Ascospores hyaline, 8–13 × 3, occasionally becoming 1-septate. On decaying leaves of *Acer, Carpinus*, etc.

Uncinia foliicola, described on *Alnus*, has been found also on dead leaves of *Betula, Rubus* and *Salix*.

OTHER ASCOMYCETES

Apiognomonia errabunda, described on *Fagus*, occurs also quite commonly on dead leaves of other deciduous trees, such as *Betula, Carpinus* and *Quercus*.

Arachnocrea stipata (Lib.)Moravec (Fig. 275)
Perithecia golden brown, 1–2mm diam., embedded in a pale cobwebby subiculum. Ascospores hyaline, acutely pointed at each end, 9–11 × 2.5–3, at first 1-septate later dividing to form 16 part-spores. On leaf litter and debris from deciduous trees, e.g. *Fagus* and *Quercus*, Oct.–Jan.

Aulographum hederae, described on *Hedera*, is quite common also on dead leaves of *Ilex* and *Rhododendron*.

Discosphaerina fagi, described on *Fagus*, occurs also on rotting leaves of *Aesculus, Castanea, Quercus* and *Ulmus*.

Gnomonia setacea, described on *Betula*, is found quite commonly also on *Acer, Alnus, Crataegus, Populus* and *Quercus*.

Microthyrium ciliatum var. *hederae*, described on *Hedera*, has been found also on dead leaves of *Buxus, Quercus, Rubus* and *Salix*. The thyriothecia have no dark collar around the ostiole.

Microthyrium microscopicum, described on *Quercus*, but found also on dead leaves of *Castanea, Fagus, Hedera, Ilex* and debris of *Ulex*, forms thyriothecia each with a dark collar around the ostiole.

Mycosphaerella punctiformis, described on *Quercus,* is common also on overwintered fallen leaves of *Acer, Alnus, Castanea, Corylus, Fagus, Salix, Tilia* and *Ulmus.*

Sphaerognomonia carpinea, described on *Carpinus,* occurs also on overwintered dead leaves of *Acer, Alnus, Betula, Castanea, Corylus, Fraxinus* and *Quercus.*

Trichothyrina nigroannulata, more commonly found on grasses, rushes and sedges, has been collected also on dead leaves of *Acer, Betula, Populus, Quercus* and *Salix.*

Xylaria filiformis (Fr.)Fr.
Stromata black, thread-like, up to 6cm long, with scattered protuberant perithecia. Ascospores blackish brown, 11–14 × 5–6. On rotting leaves of deciduous trees, e.g. *Acer, Betula, Castanea, Fagus* and *Salix,* not common.

HYPHOMYCETES

Anavirga laxa, described on *Castanea* cupules, occurs also on rotting leaves of *Betula, Fraxinus* and *Quercus.*

Anungitea fragilis Sutton (Fig. 276)
Colonies effuse or discrete, pale greyish brown, hairy. Conidiophores up to 80 × 2.5–4, pale to mid brown, sometimes extended to form a setiform portion up to 400 long. Conidia in short, simple, fragile chains at tips of cicatrised denticles, cylindrical, 1-septate, pale olivaceous brown, with flat thickened scars at one or both ends, 10–14 × 1.5–2. On dead leaves and other litter of *Fagus, Hedera* and *Ilex,* May–Sept.

Belemnospora epiphylla P.M.Kirk (Fig. 277)
Colonies effuse, pale to mid brown, minutely hairy. Conidiophores up to 35 × 2–3, brown. Conidia bullet-shaped, 1-septate, very pale brown, 9–16 × 2–2.5. On dead leaves of *Eucalyptus, Rhododendron,* etc.

Botrytis, v.s. under *Botryotinia fuckeliana.*

Camposporium cambrense and *C. pellucidum,* both described on *Fagus* cupules, have long, cylindrical, brown, multiseptate conidia. They are very common also on litter of *Acer, Aesculus, Betula, Ilex, Quercus, Rosa, Rubus, Ulmus,* etc.

Candelabrum spinulosum, described on *Castanea* cupules, has hyaline, complex, lobed conidia. It is common also on cupules of *Fagus,* and on dead leaves of *Aesculus, Fagus* and *Quercus.*

Chalara aurea (Corda)Hughes (Fig. 278)
Colonies orange or golden yellow. Conidiophores brown, up to 90 long, venter 6–7 wide, neck 3 wide. Conidia in long chains, 0- or 1-septate, hyaline, 12–16 × 2. On dead leaves of *Acer, Alnus, Fraxinus* and *Salix,* and on old cupules of *Castanea.*

Three other species of *Chalara* are common on leaf litter: *C.affinis* and *C. cylindrosperma*, described on *Fagus* cupules, and *C.cylindrica* on *Picea* leaves.

KEY

    Colonies creamy .... *affinis*
    Colonies orange or golden yellow .... *aurea*
    Colonies dark brown 1
1.   Conidia often 1-septate, 3–6 × 1 .... *cylindrica*
    Conidia without septa, 4–11 × 1–1.7 .... *cylindrosperma*

C. affinis has been found on *Aesculus*, *Alnus*, *Fagus*, *Pinus* and *Quercus*; *C. cylindrica* on *Castanea*, *Picea* and *Taxus*; *C.* cylindrosperma on *Aesculus*, *Fagus*, *Picea* and *Pinus*.

Chromelosporium carneum (Pers.)Hennebert (Fig. 279)
Colonies of plumose or isarioid, at first white later flesh-coloured, groups of synnemata; branching dichotomous, branches 7–14 thick. Conidia spherical, hyaline, verruculose, 5–8 diam. On rotting leaves especially of *Fagus* and *Quercus*.

Chromelosporium ochraceum Corda (Fig. 280)
This species occurs much more commonly on dead herbaceous plants but is seen occasionally in leaf litter. The colonies become ochraceous, conidiophores are dichotomously branched but not synnematous, and the spherical, verruculose conidia, 4–5.5 diam., are clustered along the sides of the terminal branches.

Clathrosphaerina state of *Hyaloscypha zalewskii* described on *Castanea* cupules is found also occasionally on rotting leaves of *Fagus*, *Quercus* and *Ulmus*. The hyaline, clathrate conidia are 12–24 diam.

Codinaea britannica M.B.Ellis (Fig. 281)
Colonies effuse, black, hairy. Setae blackish brown, up to 400 long, not fertile at tip. Conidiophores pale brown, up to 120 × 2–5. Conidia hyaline, slightly curved, 15–20 × 2–3, with a setula 5–9 long at each end. On *Acer* petioles and cupules of *Aesculus*, *Castanea* and *Fagus*, common.

Codinaea fertilis, described on *Fagus* cupules has been found also on *Aesculus*. Setae often fertile at tip.

Codinaea simplex Hughes & Kendrick (Fig. 282)
Colonies effuse, olivaceous or grey. No setae. Conidiophores brown, up to 100 × 3.5–4. Conidia hyaline, 14–20 × 2–3 with a setula 6–8 long at each end. On fallen dead leaves of *Corylus*, *Laurus* and *Quercus ilex*, Apr.–June.

Cylindrium elongatum, described on *Quercus*, is found also on cupules and leaves of *Fagus* and occasionally other deciduous trees. Colonies white, effuse.

Endophragmia alternata Tubaki & Saitô (Fig. 283)
Colonies pulvinate or effuse, dark olive to black. Conidiophores brown, up to 150 × 4–7. Conidia cylindrical, pale brown, 6- to 8-septate, 30–50 × 5–8. Common on *Alnus* catkins and on dead leaves of *Betula, Cytisus, Fagus, Hedera, Ilex, Quercus* and *Rubus*.

Fusidium aeruginosum, described on *Quercus*, is common also on rather newly dead fallen leaves of *Betula* and *Castanea*. It forms velvety greenish-yellow colonies.

Fusidium griseum Link (Fig. 284)
Colonies effuse, pale grey, velvety. Conidiophores forming a superficial palisade, hyaline, mostly 10–20 × 2–3. Conidia hyaline, mostly 15–20 × 2, in long, sometimes branched chains. On dead leaves of *Betula, Castanea* and *Quercus*, Nov.–Jan., common.

Haplariopsis fagicola, described on *Fagus* cupules, has been found occasionally on cupules and dead leaves of *Aesculus* and *Quercus*. Conidia 1-septate, 14–20 × 3.5–5.

Henicospora minor, described on *Rubus*, has been found also on dead leaves of *Ilex, Quercus* and *Rhododendron*. Conidia pale olivaceous brown with 5 pseudosepta, 16–26 × 4–5.

Polyscytalum fagicola, described on *Fagus*, occurs also on *Quercus*.

Polyscytalum fecundissimum Riess (Fig. 285)
Colonies effuse, white or greenish. Conidiophores pale olivaceous brown, up to 75 × 2–5, often swollen at the base to 7. Conidia in branched chains, hyaline or pale brown, 0- or 1-septate, 13–18 × 2. On rotting leaves of *Betula, Fagus, Laurus, Quercus* and *Salix* and on old fruits of *Castanea*.

Tricladium castaneicola, described on *Castanea* cupules, has been found also on *Quercus* cupules and other litter.

COELOMYCETES

Coleophoma cylindrospora, described on *Ilex*, has been found also on dead leaves of *Hedera, Laurus, Juniperus, Mahonia, Prunus* and *Quercus ilex*. Conidia hyaline, 16–30 × 2.5–3.

Coleophoma empetri, described on *Rhododendron*, occurs also on dead leaves of *Betula, Eucalyptus, Fraxinus, Lonicera* and *Prunus laurocerasus*. Conidia 13–18 × 2–3.

Coniella castaneicola, described on *Castanea*, is recorded also on dead leaves of *Betula, Quercus* and *Tilia*. Conidia pale brown, 15–30 × 2.5–3.5.

Discosia artocreas, described on *Betula*, occurs also on dead leaves of *Carpinus*,

Castanea, Fagus, Quercus and *Tilia*. Conidia 3-septate, 15–20 × 2–3.5 with a setula near each end.

Pilidium acerinum Kunze (Fig. 286)
Conidiomata 0.2–0.3mm diam., dark brown. Conidia falcate, hyaline, 14–16 × 2. On fallen dead leaves of *Abies, Acer, Betula, Carpinus, Castanea, Fagus* and *Quercus*, Sept.–May.

FUNGI SPECIFIC TO TREES, SHRUBS AND WOODY CLIMBERS

ABIES: including Giant Fir, Noble Fir and Silver Fir

ON LEAVES

UREDINALES

Four rusts are recorded in Great Britain as having pycnia and aecia on *Abies*.

KEY

	Aecia reddish yellow, causing witches' brooms .... *Melampsorella caryophyllacearum*
	Aecia white or whitish, not causing brooms 1
1.	Aeciospores 15–21 × 10–14 .... *Pucciniastrum epilobii*
	Aeciospores larger 2
2.	Aecia in 2 irregular rows on yellowed areas, spores minutely verruculose .... *Milesina kriegeriana*
	Aecia in 2 regular rows not confined to yellowed areas, spores with warts smaller on one side .... *Milesina blechni*

Melampsorella caryophyllacearum Schröter, 0,1
On *A.alba*, *A. cephalonica*, *A. lowiana*, *A. nordmanniana* and *A. pinsapo*. Mycelium perennial causing witches' brooms. Infected shoots pale yellow and very conspicuous in summer; they are thicker and somewhat shorter than other shoots, grow vertically upwards and bear short thick leaves arranged spirally. Aecia hypophyllous, 0.5–1mm long, reddish yellow, in 2 rows, 1 on each side of the midrib; spores verruculose, 16–30 × 15–20 with orange–yellow contents, June–Aug. 11, 111 on various species of *Cerastium* and *Stellaria*.

Milesina blechni Syd., 0,1
On *A. alba* and *A. cephalonica*. Aecia hypophyllous, white, cylindrical, 0.3–0.4mm diam., arranged in 2 rows, 1 on each side of the midrib; spores white, verruculose with warts smaller on one side, 28–35 × 21–27. Rare. 11, 111 on *Blechnum spicant*.

Milesina kriegeriana (Magn.)Magn.,0,1
On *A. alba*, *A. cephalonica*, *A. grandis* and *A. nordmanniana*. Aecia hypophyllous, white, cylindrical, 0.3–0.8mm diam., up to 1.3mm high, in 2 irregular rows on yellowed parts of leaves; spores white, minutely verruculose,

24–48 × 20–35, June–Sept. 11, 111 on species of *Dryopteris* including *D. carthusiana*, *D. dilatata* and *D. filix-mas*.

Pucciniastrum epilobii Otth, 0,1
On *A. grandis*. Aecia hypophyllous, white or whitish, cylindrical, 0.25mm diam., 1mm high, in 2 rows, spores hyaline, minutely verruculose with a smooth area on one side, 15–21 × 10–14, June–July. 11, 111 on species of *Epilobium* including *E. angustifolium*, *E. montanum* and *E. palustre*.

OTHER FUNGI

Botryosphaeria abietina (Prill. & Delacr.)Maubl.
Pseudothecia immersed, 0.25–0.35mm diam., dark brown to black, thick-walled. Ascospores hyaline, 21–30 × 8–11.

Delphinella abietis (Rostr.)E. Müller (Fig. 287)
Pseudothecia at first covered by epidermis, then erumpent, 0.15–0.2mm diam., black, thick-walled. Asci 16-spored; spores hyaline, mostly 1-septate, occasionally 3-septate, 12–20 × 4–6.5. Parasitic on living leaves, mostly on the upper surface, Mar.–May.

Phacidium abietinum Schm.
Apothecia subepidermal, round, about 1mm diam. with greyish disc exposed by splitting of the covering epidermis into 3 or 4 teeth which fold back. Ascospores hyaline, biguttulate, 9–11 × 3.5–4.5. On decaying leaves, May–Aug.

Thysanophora penicillioides (Roum.)Kendrick (Fig. 288)
Colonies effuse, olivaceous, dark blackish brown or black. Conidiophores dark brown below, pale at apex, branches 11–22 × 3–10, pale olivaceous brown. Conidia in dry basipetal chains, colourless or pale brown, smooth or minutely verruculose, 2–5 × 1.5–3. Common on rotting fallen leaves of *A. grandis*. Found occasionally also on *Larix*, *Picea* and *Pseudotsuga*.

ON WOOD AND BARK

Antennatula pinophylla (Nees)Strauss (Fig. 289)
Colonies effuse, often very large, up to 10cm long, dark blackish brown to black, spongy. Hyphae anastomosing, brown, 9–20 thick. Conidia pale to mid brown, 6 to 11 septate, 50–130 × 10–20. Sooty mould, sometimes associated with *Tripospermum* and other fungi on twigs.

Camarops tubulina (Alb. & Schw.)Shear
Stromata oblong or oval, superficial or partly immersed, reddish brown to black. Perithecia large, crowded, becoming angular through pressure, with long necks. Asci long-stalked. Ascospores 6–7 × 3–4, blackish brown. On rotten wood of *A. alba*, May–Nov.

Cucurbidothis pithyophila (Schm. & Kunze)Petrak (Fig. 290)
Pseudothecia blackish brown to black, up to 0.45mm diam., clustered on an erumpent, effuse, encrusting stroma which is black on the surface, white inside. Ascospores muriform, golden brown, smooth, 18–25 × 6–8. On twigs and branches of *A. alba*, *A. nordmanniana* and *A. procera*, uncommon. It has been reported as being responsible for disease in nurseries but in general it does not seem to cause much injury and is chiefly found on poorly developed trees.

Herpotrichia parasitica (Hartig)Rostrup (Fig. 291)
Pseudothecia seated on an extensive dark brown to black subiculum, setose, 0.1–0.2mm diam. Setae dark brown, 50–60 × 5–6. Ascospores when mature brownish, 3-septate, 15–20 × 4–5. On twigs of *A. alba*, causing dieback, June.

Sarea resinae (Fr.)Kuntze (Fig. 292)
Apothecia up to 1.5mm diam., orange. Asci multispored; spores smooth, hyaline, yellowish in mass, 2–3 diam. *Pycnidiella* conidial state with yellow, thick-walled conidiomata up to 0.45mm diam. and spherical conidia 2 diam. On resinous exudate from stumps of *A. alba* and other conifers throughout the year.

Valsa kunzei, described on *Larix*, has been collected occasionally also on *Abies*.

ACER: Including Field Maple, Norway Maple and Sycamore
ON LEAVES

DISCOMYCETES

KEY

	Apothecia with minute ciliate processes arising from marginal cells .... *Mollisina acerina*
	Apothecia hairy 1
	Apothecia without hairs or processes 5
1.	Apothecia brown with white fringe .... *Dasyscyphus acerinus*
	Apothecia white or yellowish 2
2.	Hairs with crystals 3
	Hairs without crystals 4
3.	Ascospores 4–5 × 1–1.5 .... *Dasyscyphus rhytismatis*
	Ascospores 5–8 × 2.5–3 .... *Dasyscyphus radotinensis*
4.	Hairs up to 200 long, ascospores 4–5 × 2 .... *Hyalopeziza ciliata*
	Hairs up to 40 long, ascospores 12–14 × 2–2.5 .... *Hyaloscypha lachnobrachya*
5.	Apothecia stalked 6
	Apothecia not stalked 8

6. Apothecia greenish yellow .... *Rutstroemia luteovirescens*
    Apothecia white or cream 7
7. Excipulum tough, ascospores 14–16 long .... *Crocicreas subhyalinum*
    Excipulum not tough, ascospores 17–22 long ..... *Hymenoscyphus caudatus*
8. Apothecia erumpent from large black tarspot stromata ....
    .... *Rhytisma acerinum*
    Apothecia erumpent from petioles, hysterioid .... *Pyrenopeziza petiolaris*

Crocicreas subhyalinum (Rehm)S.Carp. (Fig. 293)
Apothecia up to 0.6mm diam., ivory white, drying yellowish, short-stalked. Excipulum tough and elastic, made up of elongated cells with highly refractive walls. Ascospores hyaline, mostly 14–16 × 3–4. Common, especially on petioles of dead fallen leaves of *A. pseudoplatanus*, Oct.–Nov.

Dasyscyphus acerinus (Cooke & Ellis)Cash (Fig. 294)
Apothecia 0.2–0.4mm diam., golden brown, white fringed. Hairs up to 40 × 4, those at the margin hyaline, those clothing the sides pale to mid brown. Ascospores hyaline, 5–7 × 1.5, clavate. Few paraphyses. On dead leaves of *A. pseudoplatanus*, Oct., uncommon.

Dasyscyphus radotinensis (Velen.)Dennis (Fig. 295)
Apothecia 0.2–0.4mm diam., white, short-stalked. Hairs hyaline, smooth, 60–80 × 4, with large scattered crystals and sometimes also bundles of smaller crystals. Ascospores hyaline, 5–8 × 2.5–3. On dead leaves of *A. pseudoplatanus*, Oct.–Nov.

Dasyscyphus rhytismatis (Phill.)Sacc. (Fig. 296)
Apothecia white, 0.2mm diam. Hairs about 40 × 3–4, hyaline, each capped by a mass of small crystals. Ascospores hyaline, 4–5 × 1–1.5. Paraphyses 50 × 3–4. On the underside of fallen dead leaves of *A. pseudoplatanus*, Oct.–Apr., mostly associated with *Rhytisma acerinum*.

Hyalopeziza ciliata Fuckel (Fig. 297)
Apothecia 0.2–0.3mm diam., white drying pale yellow, with spaced-out, stiff, tapered, hyaline hairs up to 200 × 6–7. Ascospores hyaline, mostly 4–5 × 2. Typically on old fallen leaves of *A. campestre* and *A. pseudoplatanus*, Oct.–Jan.

Hyaloscypha lachnobrachya (Desm.)Nannf. (Fig. 298)
Apothecia up to 0.3mm diam., white or yellowish with paler rim, sessile on a narrow brown base. Hairs hyaline, up to 40 long, basal swelling 4 wide. Asci usually 4-spored but occasionally 8-spored; spores hyaline, mostly 12–14 × 2–2.5. Often in dense swarms on decaying leaves of *A. campestre* and *A. pseudoplatanus*, Sept.–Nov.

Hymenoscyphus caudatus, described on leaf litter, is common on decaying leaves of *Acer*, especially on petioles and midribs.

Mollisina acerina (Mouton)Höhnel (Fig. 299)
Apothecia 0.2–0.3mm diam., obconical with flat base, white drying pale greyish brown, with a white margin where the small cells bear hyaline ciliate processes, 5–10 × 0.5–1 (best seen in water mounts). Ascospores 5–8 × 2.5–4. On decaying leaves of *A. pseudoplatanus*, Nov.–Feb.

Pyrenopeziza petiolaris (Alb. & Schw.)Nannf. (Fig. 300)
Apothecia erumpent, hysterioid, usually 0.4–0.8mm long, although occasionally up to 2mm dark grey or yellowish grey when fully exposed, often with a pale rim, excipulum brown. Ascospores hyaline, mostly 6–9 × 2, rarely up to 12 × 3. Common on petioles of fallen leaves of *A. pseudoplatanus*, May–Oct.

Rhytisma acerinum (Pers.)Fr. (Fig. 301)
This 'tarspot' fungus produces black, at first shiny, later wrinkled, stromata up to 2cm diam. on living leaves from July onwards. In Aug. and Sept. these contain conidiomata of the *Melasmia* state with hyaline conidia 8–10 × 1. The leaves fall to the ground and it is not until Mar. or Apr. of the following year and then only after rain that the black crusts split open to expose grey or cream apothecial discs. Ascospores hyaline, 60–80 × 1.5–2. Very common on *A. pseudoplatanus* and seen occasionally on *A. campestre*; found everywhere except in towns where there is high sulphur dioxide atmospheric pollution.

Rutstroemia luteovirescens (Rob. ex Desm.)White (Fig. 302)
Apothecia bright greenish yellow when fresh, up to 5mm diam., with yellow stalks arising from blackened patches on petioles of old fallen leaves of *A. pseudoplatanus*, Oct.–Nov., common. A good place to look for them is in woodland rides where dead leaves are covered by long, damp grass. Ascospores hyaline, mostly 12–14 × 5–6.

OTHER ASCOMYCETES

Gnomonia cerastris (Riess)Ces. & de Not. (Fig. 303)
Perithecia immersed, black, 0.3–0.4mm diam., with protuberant necks usually about 0.35mm but sometimes up to 1mm long. Ascospores 1-septate, hyaline, 14–18 × 2–3.5. On dead leaves and petioles of *A. pseudoplatanus*, Feb.–Apr.

Gnomonia setacea, with small appendaged ascospores, described on *Betula*, is sometimes found also on *Acer* leaves.

Guignardia acerifera (Cooke)Lindau
Pseudothecia hypophyllous, immersed, minute, black. Asci broadly cylindrical, spores navicular, hyaline with 2 guttules, about 20 × 6–7. On rotting leaves of *A. campestre*.

Plagiostoma inclinata (Desm.)Barr (Fig. 304)
Perithecia immersed, black, 0.3–0.4mm diam., with protuberant necks about 1mm long. Ascospores 1-septate, hyaline, 16–20 × 2–2.5, with a filiform

appendage at each end. Common on petioles of rotting leaves of *A. pseudoplatanus*, Feb.–July.

Uncinula bicornis (Wallr.)Lév. (Fig. 305)
White powdery mildew mostly on lower leaf surface. Conidia of *Oidium* state hyaline, 25–35 × 15–20. Cleistothecia superficial, dark brown, about 0.2mm diam., appendages hyaline, at first straight and simple, then forked at tip, finally twice forked with the ultimate branches curled round. Asci usually 6 to 8 with 8 spores per ascus; spores hyaline, 17–27 × 7–14. On living leaves of *A. campestre* and *A. pseudoplatanus*, fairly common; asci ripe Oct.

HYPHOMYCETES

Aureobasidium pullulans (de Bary)Arnaud (Fig. 306)
Colonies very extensive, effuse, black. Mycelial cells dark brown, sometimes rounding off and separating. Conidia aggregated in slimy masses, hyaline, mostly 4–6 × 2–3. Together with *Cladosporium* and sometimes other fungi this forms a complex often called '*Fumago vagans*', which grows in honeydew secreted by aphids on the upper surface of living leaves of *A. pseudoplatanus* and other trees.

Cristulariella depraedens (Cooke)Höhnel (Fig. 307)
Conidiophores hyaline, about 200 × 10–15 with shining white spherical heads 100 diam. Budded-off conidia spherical, hyaline, 7–10 diam. Causes round or irregular pale buff spots 2–12mm diam. with narrow brown borders on green leaves of *A. pseudoplatanus* and *A. platanoides*, Aug.–Sept. and sometimes is responsible for premature defoliation.

Trisulcosporium acerinum Hudson & Sutton (Fig. 308)
Mycelium hyaline at first becoming brown. Conidia hyaline, branched, main axis 50–60 × 2–3.5, second and third branches up to 60 long. On rotten fallen leaves of *A. pseudoplatanus*.

COELOMYCETES

Chaetomella acutiseta Sutton & Sarbhoy
Pycnidia superficial, dark brown to black, setose, thick-walled, up to 0.8 × 0.5mm. Setae thick-walled, acutely pointed, rather pale brown, 4- to 6-septate, up to 750 × 12–15. Conidia hyaline, guttulate, 7.5–10 × 1.5–2. On dead leaves of *A. pseudoplatanus*.

Phloeospora aceris (Lib.)Sacc. (Fig. 309)
Acervuli amphigenous, brown or reddish brown, about 0.2mm diam. Conidia hyaline, guttulate, becoming 3-septate and constricted at the septa, 20–50 ×

2–5. On living and fading leaves of *A. campestre* and *A. pseudoplatanus*, especially on seedlings, sometimes causing small brown spots, from July on.

ON WOOD AND BARK

DISCOMYCETES

Allophylaria crystallifera Graddon (Fig. 310)
Apothecia greyish white turning brown when old, up to 0.5mm diam. Ascospores 0- or 1-septate, hyaline, 12–16 × 3–3.5. Accompanied by a small *Chalara*. On decorticated wood of *A. campestre*, Feb.

Hymenoscyphus subpallescens Dennis (Fig. 311)
Apothecia about 0.5mm diam., yellowish white to pale ochraceous, shortly stipitate. Ascospores fusiform, hyaline, 12–15 × 4–4.5. On bark associated with *Cryptodiaporthe lebiseyi*, Nov.

Incrupila viridipilosa Graddon (Fig. 312)
Apothecia up to 0.3mm diam., bright green when fresh, becoming dull ochraceous. Hairs about 35 × 3, heavily encrusted with greenish crystals. Ascospores hyaline, 3–4 × 1.5–2. On wood, Jan.–July.

Pezicula acericola (Peck)Sacc. (Fig. 313)
Apothecia in clusters erumpent through cracks in bark, often short-stalked, yellowish becoming cinnamon, pruinose. Ascospores hyaline, some eventually becoming multiseptate, 23–35 × 8–10. On branches of *A. campestre*, Oct.–Mar.

Pezicula carnea (Cooke & Ellis)Rehm
Similar to the previous species but apothecia sessile and less yellow. Ascospores 18–26 × 6–9. On bark of *A. pseudoplatanus*.

OTHER ASCOMYCETES

KEY

```
    Ascocarps setose or with dark setae arising from the subiculum ............ 1
    Ascocarps and subiculum without setae ................................. 3
1.  Ascospores hyaline, without septa ................. Acanthonitschkea tristis
    Ascospores 3-septate, median cells brown ............................. 2
2.  Ascospores 12–17 long ..................... Chaetosphaerella fusispora
    Ascospores 20–26 long ......................... Chaetosphaerella fusca
    Ascospores 32–35 long ................... Chaetosphaerella phaeostroma
3.  Ascocarps immersed in large, erect, club-shaped stromata ............ 4
    Ascocarps immersed in round, reddish-brown, superficial stromata ........
           ................................................ Hypoxylon howeianum
    Ascocarps not so ...................................................... 5
```

4. Brown ascospores 12–15 long .... *Xylaria longipes*
   Brown ascospores 20–31 long .... *Xylaria polymorpha*
5. Ascocarps superficial, red .... *Nectria cinnabarina*
   Ascocarps immersed or erumpent, dark 6
6. Ascospores without septa, allantoid 7
   Ascospores septate 8
7. Ascospores 5–7 long .... *Eutypa acharii*
   Ascospores 10–11 long .... *Eutypella acericola*
8. Ascospores 1-septate, hyaline 9
   Ascospores 3-septate, hyaline .... *Calospora platanoides*
   Ascospores 3-septate, brown 10
   Ascospores muriform 11
9. Ascospores 9–11 × 2–2.5 .... *Cryptodiaporthe lebiseyi*
   Ascospores 15–20 × 2.5–3, no black line in wood ....
   .... *Cryptodiaporthe hystrix*
   Ascospores 12–16 × 2–4, black line .... *Diaporthe pustulata*
   Ascospores 14–19 × 4–6.5, black line .... *Diaporthe varians*
10. Ascospores 14–20 × 5–6.5 .... *Thyridaria rubronotata*
    Ascospores 40–50 × 12–18 .... *Splanchnonema pupula*
    Ascospores 80–95 × 18–21 .... *Massaria inquinans*
11. Ascospores hyaline, 18–20 × 8–9 .... *Rhamphoria bevanii*
    Ascospores golden brown, 18–30 × 8–13 .... *Fenestella vestita*

Acanthonitschkea tristis (Pers.)Nannf. (Fig. 314)
Perithecia black, 0.3–0.35mm diam., seated on a black subiculum from which arise dark blackish-brown to black, thick-walled setae up to 200 × 8, grouped mainly around the perithecia. Ascospores hyaline, 6–9 × 1.5–2. On wood of *A. pseudoplatanus*, Mar.–May.

Calospora platanoides (Pers.)Niessl ex Sacc. (Fig. 315)
Perithecia 0.5mm diam., in groups of up to 20 with black necks protruding through a grey disc. Ascospores hyaline, mostly 3-septate and 20–29 × 7–8, occasionally with a small appendage at each end. Very common on twigs and small branches of *A. pseudoplatanus*, Jan.–Mar.

Chaetosphaerella fusispora Sivan. (Fig. 316)
Perithecia black, setose, about 0.3mm diam., seated on a thin subiculum. Setae blackish brown, up to 220 × 6. Ascospores 3-septate, 12–17 × 4–6, with 2 median brown cells, end cells hyaline. On dead branches of *A. pseudoplatanus*, Sept.–July. Two plurivorous species of *Chaetosphaerella* occur on *Acer*, *C. fusca* is seldom found but *C. phaeostroma* is extremely common.

Cryptodiaporthe hystrix (Tode)Petrak (Fig. 317)
Perithecia about 0.5mm diam. in loose clusters, immersed, with black, sinuous, flattened necks protruding through a pale disc. Ascospores hyaline, 1-septate, 15–20 × 2.5–3, sometimes with a small appendage at each end.

Diplodina conidial state has erumpent conidiomata which split to expose a pale orange spore mass. Conidia 1-septate, hyaline, 12–17 × 3–3.5. On branches of *A. pseudoplatanus*; conidia formed Mar.–Apr., perithecia May–July.

Cryptodiaporthe lebiseyi (Desm.)Wehmeyer (Fig. 318)
Perithecia about 0.3mm diam., solitary or in loose clusters with small, slender protuberant necks. Ascospores hyaline, 1-septate, 9–11 × 2–2.5. *Phomopsis* conidial state has biguttulate hyaline conidia 8–10 × 3. On branches of *A. campestre* and *A. pseudoplatanus*.

Diaporthe pustulata (Desm.)Sacc.
Perithecia up to 0.6mm diam., in clusters with stout necks protruding through a black disc; black line in wood. Ascospores hyaline, slightly constricted at the septum, 12–16 × 2–4. *Phomopsis* state with A-conidia guttulate, 7–12 × 2.5–3.5 and B-conidia 15–20 × 1–1.5. On branches of *A. pseudoplatanus*, Sept.–Oct.

Diaporthe varians (Currey)Sacc. (Fig. 319)
Perithecia up to 0.65mm diam. Ascospores hyaline, 1-septate, 14–19 × 4–6.5. On branches of *A. campestre*.

Eutypa acharii Tul & C.Tul. (Fig. 320)
Grey stroma widespread on surface of wood with perithecial necks just protruding. Perithecia about 0.5mm diam. Ascospores dark straw colour in mass, 5–7 × 1. On branches of *A. pseudoplatanus*, Mar.–Apr., common.

Eutypella acericola (de Not.)Berl. (Fig. 321)
Stromata dark brown, cushion-shaped or somewhat irregular; protruding perithecial necks with sulcate ostioles. Ascospores straw-coloured or pale brown in mass, 10–11 × 1.5–2. On branches of *A. campestre* and *A. pseudoplatanus*, Feb.–Mar.

Fenestella vestita (Fr.)Sacc. (Fig. 322)
Pseudothecia up to 0.8mm diam., immersed, clustered in groups of up to 15 with black necks protruding through a 2–3mm disc. Ascospores muriform, golden brown, 18–30 × 9–13. Fairly common on small dead branches of *A. pseudoplatanus*, Feb.–Mar.

Hypoxylon howeianum, described on *Corylus*, is found occasionally on *A. campestre* and *A. pseudoplatanus*.

Massaria inquinans (Tode)de Not. (Fig. 323)
Pseudothecia immersed, black, about 1mm diam., usually found singly or occasionally in small groups, raising the bark slightly but inconspicuous and with the short necks only just showing. Ascospores 3-septate, dark brown, mostly 80–90 × 18–21. On dead branches of *A. campestre* and *A. pseudoplatanus*, Mar.–May.

Nectria cinnabarina (Tode)Fr. (Fig. 324)
The 'coral spot' fungus. Perithecia in groups of up to 15 clustered on a stroma, about 0.5mm diam., tending to collapse in the middle, red with encrusted walls. Ascospores hyaline, 1-septate, 12–20 × 4.5–6. Preceded or accompanied by pink cushions up to 3mm diam. of the *Tubercularia* state which has hyaline conidia mostly 5–7 × 2. A plurivorous species especially common on cut and fallen branches of *A. pseudoplatanus*, Sept.–Mar.

Rhamphoria bevanii Sivan. (Fig. 325)
Perithecia up to 0.3mm diam., partly immersed, black. Ascospores muriform, hyaline or subhyaline, 18–20 × 8–9. On dead branches of *A. pseudoplatanus*, July.

Splanchnonema pupula (Fr.)O. Kuntze (Fig. 326)
Pseudothecia up to 0.8mm diam. Ascospores 3-septate, 40–50 × 12–18, brown with thick gelatinous sheaths. On dead branches of *A. pseudoplatanus*, Jan.–Mar.

Thyridaria rubronotata (Berk. & Br.)Sacc. (Fig. 327)
Pseudothecia about 0.3mm diam. in groups immersed in pulvinate stromata erumpent through cracks in bark. Characterised by orange–red granular material which fills spaces between the pseudothecia. Ascospores brown, 3-septate, 14–20 × 5–6.5. Accompanying *Cyclothyrium* state has pale olivaceous brown conidia 4–6 × 2–3. A plurivorous species in our experience most commonly found on *A. pseudoplatanus*, Dec.–Mar.

Xylaria longipes Nitschke (Fig. 328)
Perithecia immersed in the swollen upper part of large, dark brown, long-stalked, club-shaped stromata (up to 10cm × 4–7mm). Ascospores brown, with germ slits, 12–15 × 5–6. On rotten fallen branches of *A. pseudoplatanus*, Sept.–Dec., common. The plurivorous *X. polymorpha*, with short-stalked clubs and much larger ascospores, is also not uncommon on sycamore.

HYPHOMYCETES

Cryptostroma corticale Gregory & Waller (Fig. 329)
The cause of 'sooty bark' disease of sycamore. The outer layers of bark of standing trees peel off, exposing enormous dark blackish-brown masses of conidia which often extend over a very large area. Conidia brown, 4–6 × 3.5–4.

Tubercularia, *v.s.* under *Nectria cinnabarina*.

COELOMYCETES

Camarosporium ambiens (Cooke)Grove (Fig. 330)
Pycnidia dark brown, shortly papillate, raising and eventually rupturing the epidermis. Conidia brown, with transverse and often also longitudinal septa, 16–25 × 6–8. On fallen small dead branches of *A. campestre* and *A. pseudoplatanus*, Jan.–July.

Cyclothyrium, *v.s.* under *Thyridaria rubronotata*.

Dichomera saubinetii (Mont.)Cooke (Fig. 331)
Conidiomata immersed becoming erumpent, up to 2mm diam., dark brown, multilocular, thick-walled. Conidia brown, with transverse and longitudinal septa, 11–17 × 7–10. Causes diamond-shaped cankers on trunks of *A. pseudoplatanus*.

Diplodina, *v.s.* under *Cryptodiaporthe hystrix*.

Phomopsis, *v.s.* under *Cryptodiaporthe lebiseyi* and *Diaporthe pustulata*.

Phomopsis platanoidis (Cooke)Died.
Conidiomata unilocular, densely gregarious, erumpent and conical. Conidia fusiform, hyaline, biguttulate, 6–8 × 1.5–2.5. On twigs of *A. pseudoplatanus*.

Stegonosporium pyriforme (Hoffm.)Corda (Fig. 332)
Acervuli up to 1mm diam., with conidia oozing out and forming a compact black mass. Conidia greyish olive to brown, muriform, 35–42 × 17–20. Common on dead, still-attached branches of *A. pseudoplatanus*, Oct.–June.

AESCULUS: Horse Chestnut

Apart from plurivorous leaf-litter fungi recorded on fallen fruits and leaves, few species have been found on *A. hippocastanum*.

Cryptodiaporthe aesculi (Fuckel)Petrak (Fig. 333)
Perithecia 0.3–0.4mm diam., immersed usually in groups of 2 to 5 with black necks protruding through a small white stromatic disc. Ascospores hyaline, 1-septate, mostly 18–24 × 4.5–7, sometimes with a small appendage at each end. *Diplodina* state with hyaline, 1-septate fusiform conidia 17–20 × 3–3.5. On dead branches and twigs, Nov.–Mar.

Gnomonia setacea, described on *Betula*, is not uncommon on petioles of fallen dead leaves.

Guignardia aesculi (Peck)Stewart
Pycnidia epiphyllous, small, black, scattered or crowded. Some conidia about 9 × 3, others 10–16 × 6.5–10, hyaline. Causes leaf-blotch disease. Lesions irregular, beginning at the tip and along the sides, extending inwards, at first

water-soaked then red to brown with yellow margins. Pseudothecia form on overwintered fallen leaves in May. Ascospores hyaline with granular contents, 12–18 × 7–9.

Septoria hippocastani Berk. & Br.
Pycnidia few, epiphyllous, dark brown, about 0.2mm diam. Conidia curved, hyaline, 3-septate, 30–60 × 2.5–3.5. On living leaves causing round or somewhat angular spots, at first brown, later whitish with narrow, dark brown borders.

ALNUS: Alder

ON LEAVES

UREDINALES

Melampsoridium betulinum, described on *Betula*, has been recorded on *Alnus* in Scotland.

DISCOMYCETES AND TAPHRINALES

Hymenoscyphus albopunctus (Peck)O. Kunze (Fig. 334)
Apothecia up to 2mm diam., white drying somewhat ochraceous when old. Ascospores hyaline, mostly 15–25 × 2.5–3, often hooked at the tip. Common on midribs and petioles of rotten leaves, Oct.–Nov. *H. caudatus*, a very common plurivorous species described on leaf litter, has wider spores (4–5.5).

Pyrenopeziza foliicola (P.Karsten)Sacc. (Fig. 335)
Apothecia 0.15–0.35mm diam., mid pale brown, with white rims and greyish discs. Ascospores hyaline, 6–9 × 1.5–2. On rotting fallen leaves of previous year, Apr.–July.

Rutstroemia conformata (P.Karsten)Nannf. (Fig. 336)
Apothecia 1–3mm diam., pale brown with stalks usually about 5mm long but sometimes up to 1.5cm when deeply buried in litter. Ascospores hyaline, 10–12 × 4.5–5.5. Arising from small blackened areas on midribs of fallen rotting leaves, Apr.–June, very common.

Taphrina sadebackii Johansson
Causes bright yellow unthickened spots up to 1cm diam. on the lower surface of leaves.

Taphrina tosquinetii (Westend.)Magn. (Fig. 337)
Asci 18–40 × 7–13, spores 2.5–6 × 2.5–5, often budding in the ascus. Infected leaves up to twice as large as normal, thickened, rather brittle and incurved. Large blisters formed with a whitish bloom on both surfaces. Common, especially on sucker shoots, June–Sept.

Uncinia foliicola Graddon (Fig. 338)
Apothecia up to 0.2mm diam., white when fresh, pale yellow when dry, with ring of brown cells at base. Hairs hooked, up to 15 long. Ascospores hyaline, 9–12 × 2–2.5. On fallen dead leaves, June–Dec.

OTHER ASCOMYCETES

Gnomonia setacea, described on *Betula*, occurs also on *Alnus*.

Gnomoniella tubiformis (Tode)Sacc. (Fig. 339)
Perithecia 0.4–0.5mm diam., brown, immersed but raising the leaf surface, with protuberant sometimes curved necks. Ascospores hyaline, olivaceous in mass, 13–17 × 5–6. On fallen dead leaves, Mar.–Apr. Acervuli of the *Asteroma* state are found on dark chestnut-coloured spots on the upper surface of living leaves in autumn; conidia fusiform, hyaline, 11–14 × 3.

Microsphaera penicillata (Wallr.)Lév.
Powdery mildew. White or greyish colonies mostly on lower surface of leaves, forming few conidia. Cleistothecia about 0.1mm diam., dark brown; appendages hyaline, sometimes yellowish at base, dichotomously branched 3 to 6 times at tips with ultimate branchlets recurved. Asci 3 to 8, 2- to 8-spored; spores 18–24 × 10–12.

HYPHOMYCETES

Passalora bacilligera (Mont. & Fr.)Mont. & Fr. (Fig. 340)
Colonies hypophyllous, effuse, olivaceous, velvety. Conidiophores olivaceous brown, up to 180 × 3–6. Conidia pale olivaceous, 1-septate, 21–68 × 4.5–8.5. On pale green or yellowish-green angular areas 1–2mm wide, limited by small veins and without definite margins.

Sporidesmium wroblewskii (Bubák)M.B.Ellis (Fig. 341)
Colonies brown, mycelium superficial. Conidia golden brown, smooth or verruculose, 3- or 4-septate, 30–46 × 9–11. On living leaves and also on catkins.

ON CATKINS

Ciboria amentacea (Balb.)Fuckel (Fig. 342)
Apothecia 4–10mm diam., pale brown, often long-stalked. Ascospores hyaline, 7–11 × 4–5. Common on old fallen male catkins, Feb.–Apr.

Ciboria viridifusca (Fuckel)Höhnel (Fig. 343)
Apothecia up to 4mm diam., disc yellowish olive, receptacle and stalk dark olivaceous brown. Ascospores hyaline, 6–9 × 2.5–3.5. Common on fallen female catkins, Oct.–Dec.

Mollisia amenticola (Sacc.)Rehm (Fig. 344)
Apothecia brown, disc pale, 0.2–0.4mm diam. Ascospores hyaline, 4–7 × 1.5–2. On old fallen female catkins, Aug.–Feb., common.

Pezizella alniella (Nyl.)Dennis (Fig. 345)
Apothecia up to 0.5mm diam., downy, cream drying yellow. Ascospores hyaline, mostly 9–10 × 2.5–3. The *Acleistia* state has spherical hyaline conidia 2–2.5 diam. On old female catkins, Oct.–Apr., common and sometimes in great abundance.

Taphrina amentorum (Sadeb.)Rostr.
Asci 30–50 × 10–20; spores 3–5 × 4–5. Infects scales of female catkins causing them to protrude as tongue-shaped outgrowths 1–2cm long, at first pale cream then bright red. Uncommon.

ON WOOD AND BARK

DISCOMYCETES

KEY

	Apothecia stalked 1
	Apothecia sessile or almost so 2
1.	Ascospores 1-septate, 10–14 × 2 .... *Pezizella parilis*
	Ascospores 0-septate, 10–16 × 3–5 .... *Cudoniella clavus*
2.	Apothecia on brown subiculum .... *Tapesia fusca*
	No extensive brown subiculum 3
3.	Asci becoming filled with hundreds of tiny secondary spores .... *Tympanis alnea*
	No secondary spores 4
4.	Ascospores with appendage at each end .... *Thecotheus rivicola*
	Ascospores not appendaged 5
5.	Ascospores verruculose, 18–25 × 11–12 .... *Miladina lecithina*
	Ascospores quite smooth 6
6.	Apothecia cream or pale yellow .... *Pezizella albohyalina*
	Apothecia dark 7
7.	Ascospores more than 20 long .... *Mollisia ramealis*
	Ascospores 13–17 long .... *Cenangium graddonii*
	Ascospores less than 13 long 8
8.	Ascospores distinctly curved .... *Encoelia furfuracea*
	Ascospores straight or slightly curved .... *Pyrenopeziza benesuada*

Cenangium graddonii Dennis (Fig. 346)
Apothecia superficial, 1–2mm diam., sessile, urn-shaped with incurved margin, disc ochraceous, receptacle black. Ascospores straight or slightly curved, hyaline, biguttulate becoming eventually 1- to 3-septate, 13–17 × 4–5. Paraphyses filiform, swollen at apex to 5. On decorticated dry branches, May.

Cudoniella clavus (Alb. & Schw.)Dennis (Fig. 347)
Apothecia stalked, usually less than 1cm diam., disc flat becoming convex, off-white or somewhat flesh-coloured. Ascospores hyaline, 10–16 × 3–5. The var. *grandis* (Boud.)Dennis (Fig. 347A) has short-stalked apothecia mostly 2cm diam. On fallen dead twigs lying in boggy ground and often immersed in water, May–July. It seems to prefer sloping ground where there is water movement, and may colonise debris other than that of *Alnus* on occasion.

Encoelia furfuracea, described on *Corylus*, has been recorded occasionally on attached branches of *Alnus*, Dec.–May.

Miladina lecithina (Cooke)Svrček (Fig. 348)
Apothecia single or clustered, 1–2mm diam., yellow, surrounded by white hyphae. Ascospores minutely verruculose, hyaline, 18–25 × 11–12. Paraphyses swollen to 8 at apex, with orange pigment. On wet wood, May–Oct.

Mollisia ramealis, described on *Betula*, is found occasionally on *Alnus*.

Pezizella albohyalina (P.Karsten)Rehm (Fig. 349)
Apothecia 0.5–0.6mm diam., cream or pale yellow, downy. Ascospores hyaline, 9–12 × 2–3. On decorticated wood, Aug.–Oct.

Pezizella parilis, a plurivorous wood and bark species with 1-septate spores, is not uncommon on small branches of *Alnus*.

Pyrenopeziza benesuada (Tul.)Gremmen (Fig. 350)
Apothecia up to 1.5mm diam., erumpent often in clusters, greyish brown with off-white to cream discs when fresh, becoming darker. Ascospores hyaline, 9–12 × 2(–2.5). On fallen dead twigs and branches, Mar.–June.

Tapesia fusca (Pers.)Fuckel (Fig. 351)
Apothecia up to 2mm diam., with pale bluish grey or sometimes yellowish discs, seated on an extensive brown hyphal subiculum. Ascospores hyaline, 9–14 × 2–2.5. Common all the year round on fallen branches.

Thecotheus rivicola (Vaček)Kimbrough & Korf (Fig. 352)
Apothecia up to 5mm diam., grey, pulvinate. Asci up to 200 × 12 with walls blueing in iodine; spores hyaline, 17–18 × 7.5–8.5, with a hyaline appendage up to 3 long at each end, which stains deeply in lactic cotton blue. On water-soaked twigs and wood, July–Aug., rare.

Tympanis alnea (Pers.)Fr.
Apothecia almost black, erumpent in clusters through cracks in bark of dead branches. Asci soon filled with hundreds of hyaline, slightly curved, secondary spores 2–4 × 1–1.5, which have replaced the evanescent primary ascospores, Feb.–June.

90   Fungi Specific to Trees, Shrubs and Woody Climbers

OTHER ASCOMYCETES

KEY

	Asci 16-spored .... *Ditopella ditopa*
	Asci polysporous .... *Diatrypella favacea*
	Asci generally 8-spored 1
1.	Ascospores without septa 2
	Ascospores with septa 3
2.	Ascospores not more than 6 long, olivaceous .... *Camarops*
	Ascospores 50–60 long, hyaline .... *Ophiovalsa suffusa*
3.	Ascospores not appendaged .... *Phragmoporthe conformis*
	Ascospores appendaged 4
4.	Ascospores pale brown, 3-septate .... *Prosthecium auctum*
	Ascospores hyaline, 1-septate 5
5.	Ascospores 20–24 × 4–7 .... *Melanconis alni*
	Ascospores 35–45 × 8–11 .... *Melanconis thelebola*

Camarops microspora (P.Karsten)Shear (Fig. 353)
Perithecia flask-shaped, 0.5–0.8mm diam., immersed in a brittle black stroma up to 5mm long. Ascospores pale olivaceous, 4.5–5.5 × 2–2.5. On dead branches.

Camarops polysperma (Mont.)Miller (Fig. 354)
Perithecia long, cylindrical, closely packed together, below an irregular blackish-brown stromatic crust often several cm diam. Ascospores pale olivaceous, 5–6 × 2–3. On rotten logs, Aug.–Nov., uncommon.

Diatrypella favacea, described on *Betula*, is common also on dead branches of *Alnus*.

Ditopella ditopa (Fr.)Schröter (Fig. 355)
Perithecia 0.5–0.8mm diam., immersed, usually solitary, occasionally in pairs, forming pustules on the surface of the bark, with black ostioles just showing. Asci 16-spored; spores hyaline often appearing to have a thin septum in the middle when mature, 17–23 × 2.5–4. Very common on attached and fallen twigs, Jan.–Mar.

Melanconis alni Tul. & C.Tul. (Fig. 356)
Perithecia 0.6–0.7mm diam., immersed in groups with black ostioles protruding through a pale disc. Ascospores hyaline, appendaged, 1-septate, 20–24 × 4–7. Often accompanied by its *Melanconium* state with rather pale brown conidia 10–12 × 6–8. Common on fallen dead twigs and branches, Oct.–Mar.

Melanconis thelebola (Fr.)Sacc. (Fig. 357)
Perithecia up to 1mm diam. in groups of up to 10 with black necks protruding slightly through a brown disc. Ascospores 1-septate, hyaline, 35–45 × 8–11,

some with a long appendage at each end. This species has a *Hendersoniopsis* state with two kinds of conidia (a) pale brown, 2- to 7-septate, 30–50 × 10–12, and (b) hyaline, allantoid, 7–8 × 1.5–2. On dead twigs and branches, July–Dec.

Ophiovalsa suffusa (Fr.)Petrak (Fig. 358)
Perithecia black, 0.4–0.5mm diam., in groups of 6 to 10, forming pustules. Ascospores hyaline, 50–60 × 4–5. The *Disculina* state often found on the same dead twigs and branches, especially in Mar. and Apr., has hyaline curved conidia 25–45 × 4–4.5. Perithecia ripe Mar.–July.

Phragmoporthe conformis (Berk. & Br.)Petrak (Fig. 359)
Perithecia immersed, solitary, black, up to 1mm diam. Ascospores hyaline, with 2 to 4 (mostly 3) septa, 20–25 × 6–9. On dead twigs Jan.–Mar.

Prosthecium auctum (Berk. & Br.)Petrak (Fig. 360)
Perithecia immersed, black, 0.5–0.7mm diam., in groups of 3 to 7, with necks erumpent through cracks in bark. Ascospores with a small appendage at each end (best seen in water mounts), 30–40 × 11–15, remaining for a long time 1-septate and hyaline but eventually becoming pale brown and 3-septate. On dead branches.

HYPHOMYCETES

Aegerita viridis Bayliss Elliott (Fig. 361)
Colonies made up of a number of discrete green or grey granular masses up to 0.5mm across, with conidiophores which branch dichotomously many times, the branches terminating in chains of spherical or subspherical, smooth, olivaceous conidia about 4 diam. This species has now been renamed *Pseudaegerita viridis* (Bayliss Elliott)Abdullah & Webster.

Bactrodesmium longisporum M.B.Ellis (Fig. 362)
Sporodochia punctiform, black. Conidia pale to mid brown, multiseptate, 50–80 × 7–8. On dead wood, May, rare.

Cordana crassa Tóth (Fig. 363)
Colonies effuse, dark blackish brown, shortly hairy. Conidiophores up to 300 × 5–9. Conidia 18–27 × 11–15, with dark brown basal cell and paler apical one. On dead wood, Apr.–June, uncommon.

Pleurotheciopsis bramleyi Sutton (Fig. 364)
Colonies effuse, brown, hairy, inconspicuous. Conidiophores up to 300 × 5–7. Conidia catenate, 3-septate, hyaline, 15–25 × 4–6. Very common all the year round on fallen, wet dead twigs and branches.

Taeniolella alta (Ehrenb.)Hughes (Fig. 365)
Colonies pulvinate or effuse, dark blackish brown. Conidia 2- to 5- (mostly 3-)

septate, dark brown or olivaceous brown, 20–50 × 10–13. On dead branches.

COELOMYCETES

Cytospora occulta Sacc.
Conidiomata about 1mm diam., conical. Conidia emerging in golden tendrils, curved, 5–6 × 1.5. On dead branches, Mar.–Apr.

Disculina, v.s. under *Ophiovalsa suffusa*.

Hendersoniopsis, v.s. under *Melanconis thelebola*.

Melanconium, v.s. under *Melanconis alni*.

Phomopsis alnea (Sacc.)Höhnel
Conidiomata erumpent through a short slit, up to 0.4mm long. A-conidia fusiform, hyaline, biguttulate, 7–10 × 2–3, B-conidia curved, filiform, 20–25 long. On dead twigs, May–June.

Phragmotrichum rivoclarinum (Peyronel)Sutton & Pirozynski (Fig. 366)
Conidiomata up to 300 diam., dark brown. Conidia catenate, with 2 to 7 (mostly 3) septa, pale golden brown, 18–30 × 6–8. On twigs and sometimes also on catkins.

Prosthemium stellare Riess (Fig. 367)
Conidiomata mostly separate, round, dark brown. Conidia compound, brown, with septate radiating arms each 25–35 × 7–8. On dead twigs and branches.

ANDROMEDA: Marsh Andromeda

Placuntium andromedae (Pers.)Ehrenb.
We have not seen this conspicuous fungus, which apparently forms black stromata on leaves and stems of *A. polifolia*. Pycnidial cavities containing conidia 4–8 × 1 are found in June–July. Mature ascigerous stromata should be sought May–July.

ARCTOSTAPHYLOS: Bearberry

Coccomyces arctostaphyli (Rehm)B. Eriksson
Apothecia subepidermal, up to 0.5mm diam., black, shining, opening by a slit or by 4 or 5 teeth to expose the disc. Ascospores hyaline, 45–55 × 2–2.5, with gelatinous sheaths. On upper surface of dead leaves.

Propolis phacidioides (Fr.)Corda
Apothecia up to almost 1mm diam. Ascospores thread-like, twisted around

one another spirally in the ascus, often 1-septate, 70–85 × 2–2.5. Common on the underside of grey leaves of *A. uva-ursi* in the Scottish Highlands.

BERBERIS: Barberry

UREDINALES

Puccinia brachypodii Otth var. *arrhenatheri* (Kleb.)Cummins & Greene, 0,1
Few records.

Puccinia graminis Pers., 0,1
Aecia found on *B. vulgaris* usually belong to this species.

OTHER FUNGI

Cucurbitaria berberidis (Pers.)Gray (Fig. 368)
Pseudothecia clustered in cracks in bark, up to 0.7mm diam. Ascospores golden brown, muriform, 24–30 × 10–13. On dead twigs and branches of *B. vulgaris*; also on *Mahonia*, Nov.–Apr.

Diaporthe detrusa (Fr.)Fuckel (Fig. 369)
Perithecia up to 0.5mm diam. in groups of 8 to 16 with necks pushing up through the bark. Ascospores hyaline, 1-septate, 14–17 × 5–6. On dead twigs and branches of *B. vulgaris*; also on *Mahonia*, Apr.–May.

Microsphaera berberidis (DC.)Lév. (Fig. 370)
Powdery mildew. Few conidia formed. Cleistothecia on both leaf surfaces, up to 0.15mm diam.; asci 6 to 10, each with 3 to 6 spores. Ascospores about 24 × 15. On *B. thunbergii*, *B. vulgaris* and *Mahonia*.

BETULA: Downy Birch and Silver Birch

ON LEAVES

UREDINALES

Melampsoridium betulinum (Fr.)Kleb.,11,111
Very common, especially on living leaves of young trees where, in Aug.–Oct., the lower surface of nearly every leaf may be covered with powdery yellow uredinia, the upper surface showing corresponding yellow spots; spores 24–38 × 9–15, echinulate except near apex. Telia orange or brownish, hypophyllous, subepidermal, spores in palisade-like layer, 30–50 × 8–15, smooth.

DISCOMYCETES AND TAPHRINALES

Betulina fuscostipitata Graddon (Fig. 371)
Apothecia up to 0.2mm diam., hairy, with pale greyish white discs and dark brown stalks. Hairs 25–35 × 2–3, subhyaline. Ascospores hyaline, biguttulate, 4–7 × 1. On dead fallen leaves, July–Oct.

Calycellina leucella (P.Karsten)Dennis ex E.Müller (Fig. 372)
Apothecia 0.2–0.35mm diam., creamy white with a brown ring at the base. Ascospores hyaline, 12–17 × 2–2.5. On fallen dead leaves, Oct.–Dec.

Hyaloscypha lachnobrachya (Desm.)Nannf. var. *araneocincta* (Phill.)Dennis (Fig. 373)
Apothecia sessile, up to 0.4mm diam., white to pale yellow, with marginal hairs 50–110 long, 4–5 wide at base. Asci mostly 4-spored, spores 10–13 × 1.5–2. On fallen dead leaves, Sept.–Oct., common.

Leucoscypha erminea (Bomm. & Rouss.)Boud. (Fig. 374)
Apothecia sessile, up to 5mm diam. with hairs 200–900 × 12–18. Ascospores hyaline, minutely spinulose, 20–30 × 9–13. On dead leaves among wet *Sphagnum*. Mature in June.

Taphrina betulae (Fuckel)Johansson
Asci cylindrical, 18–45 × 8–18, spores 4–6 × 3.5–5. Causes yellow or reddish spots up to 1cm diam. on living leaves, June–Sept., uncommon.

Taphrina betulina Rostr.
Asci 25–70 × 10–25; spores 4–6.5 × 2.5–5. On small, swollen, pale leaves on short erect shoots formed in witches' brooms, June.

Trichodiscus diversipilus, described on *Quercus*, has been found also on dead leaves of *Betula*.

OTHER ASCOMYCETES

Atopospora betulina (Fr.)Petrak (Fig. 375)
Stromata 0.5mm diam., black, arranged in clusters, multilocular. Ascospores yellowish in mass, with single septum near upper end, 10–14 × 4–5. On dead leaves.

Gnomonia setacea (Pers.)Ces. & de Not. (Fig. 376)
Perithecia immersed or erumpent, 0.2–0.4mm diam., with protuberant necks up to 1mm long. Ascospores hyaline, 1-septate, mostly 11–16 × 1–2 with appendage up to 9 long at each end. On dead fallen leaves, Jan.–June, very common and often in great abundance.

Venturia ditricha (Fr.)P.Karsten (Fig. 377)
Pseudothecia immersed, up to 0.15mm diam., brown, darker around the

ostiole, setae up to 80 long. Ascospores olivaceous to brown, septate below the middle, 10–17 × 4–7. On fallen dead leaves, Jan.–June, very common. The *Fusicladium* state on living leaves in late summer has pale olivaceous, 1-septate conidia 15–23 × 5–8.

COELOMYCETES

Asteroma microspermum (Peck)Sutton
Acervuli subcuticular, sometimes confluent. Conidia hyaline, fusiform, straight or slightly curved, 6.5–9.5 × 2–2.5. Causes brown lesions on living leaves.

Discosia artocreas (Tode)Fr. (Fig. 378)
Conidiomata amphigenous, flat, black, 0.1–0.3mm diam. Conidia hyaline or very pale brown, usually with 3, occasionally 4, septa, mostly 15–20 × 2–3.5, setulae 7–10 long. Common both on newly fallen and older dead leaves, Oct.–Apr.

Fusicladium, v.s. under *Venturia ditricha*.

Pilidium acerinum, described on leaf litter, is quite common on dead leaves.

ON SEEDS

Ciboria betulae (Woronin)White (Fig. 379)
Apothecia mostly 2–3mm diam., with stalks up to 1cm long, pale brown. Ascospores hyaline, 10–14 × 3.5–5. On stromatised fallen seeds, Apr.–May. We have found hundreds of apothecia at a time on seeds lying in wet *Sphagnum*.

ON WOOD AND BARK

DISCOMYCETES

Bulgariella pulla (Fr.)P. Karsten (Fig. 380)
Apothecia 1–3mm diam., dark olivaceous to almost black, gelatinous. Ascospores dark brown, 10–15 × 7–8.5. On rotten decorticated wood, Sept.–Oct.

Clavidisculum microsporum Graddon (Fig. 381)
Apothecia gregarious, up to 0.5mm diam., shortly hairy; hairs hyaline, granulate, up to 30 long, swollen at tips to 6. Ascospores 3–4 × 1.5–2. On bark, Feb.

Dasyscyphus brevipilus, described on *Fagus*, is not uncommon on decorticated branches.

Dencoeliopsis johnstonii (Berk.)Korf
Apothecia cup-shaped, up to 2mm diam., on short fat stalks, dark brown, scurfy with yellowish discs when fresh. Ascospores hyaline, becoming 1-septate, 18–33 × 6–8. On dead branches, rare.

Godronia urceolus (Schm.)P.Karsten
Apothecia solitary, subglobose, ventricose, blackish brown. Ascospores hyaline, multiseptate, 50–75 × 1.5, parallel in asci. On small fallen branches, May.

Hymenoscyphus cortisedus (Karst.)Dennis
Apothecia almost sessile, pale yellow. Asci 65–90 × 6–8. Paraphyses slender with brownish oil drops. Ascospores hyaline, 6–9 × 3–4. On rotten logs, Oct.

Mollisia ramealis (P.Karsten)P.Karsten (Fig. 382)
Apothecia about 1mm diam., dark brown outside with pale rims and discs orange when fresh drying raw-sienna. Ascospores hyaline, occasionally 1-septate, mostly 20–25 × 2–2.5. On dead twigs and branches, May–Dec., fairly common.

Pezizella parilis, a plurivorous 'wood and bark' species not uncommon on fallen twigs.

OTHER ASCOMYCETES

KEY

```
     Asci polysporous ........................................................................ 1
     Asci with not more than 8 spores ................................................ 2
1.   Asci long-stalked ............................................. Diatrypella favacea
     Asci short-stalked ............................................... Valsella adhaerens
2.   Ascospores hyaline ..................................................................... 3
     Ascospores brown ..................................................................... 10
3.   Ascospores 0-septate, somewhat curved ................................... 4
     Ascospores 1-septate, straight .................................................... 7
4.   Ascospores 30–40 long ........................................ Ophiovalsa betulae
     Ascospores not more than 12 long ............................................. 5
5.   Ascospores 9–12 × 2–3 ............................................. Enchnoa lanata
     Ascospores not more than 1.5 wide ........................................... 6
6.   Perithecia in a ring with long necks pointing inwards ..............
                                                      .......... Calosphaeria wahlenbergii
     Perithecia embedded in effuse stroma with tooth-like protuberances ......
                                                      .................. Xenotypa aterrima
7.   Ascocarps superficial ..................................... Stomiopeltis betulae
     Ascocarps immersed ................................................................. 8
8.   Septum near base of spore ........................... Anisogramma virgultorum
```

Septum at or near centre of spore 9
9. Ascospores 18–27 × 8–11 .... *Sydowiella ambigua*
   Ascospores 17–20 × 5–6 .... *Melanconis stilbostoma*
10. Ascospores without septa or pseudosepta 11
    Ascospores with septa or pseudosepta 12
11. Ascospores 9–12 × 4–5 .... *Hypoxylon multiforme*
    Ascospores 12–15 × 5–7.5 .... *Hypoxylon fuscum*
    Ascospores 6–9 × 3–4 .... *Hypoxylon howeianum*
12. Ascospores 1-septate, cells unequal .... *Pteridiospora scoriadea*
    Ascospores with more than 1 septum 13
13. Ascospores 3-septate .... *Melanomma subdispersum*
    Ascospores with more than 3 septa or pseudosepta 14
    Ascospores with longitudinal septa .... *Pleomassaria siparia*
14. Ascospores encapsuled .... *Splanchnonema argus*
    Ascospores not encapsuled .... *Pseudovalsa lanciformis*

Anisogramma virgultorum (Fr.)Theiss. & H. Syd. (Fig. 383)
Perithecia 0.3–0.5mm diam., in large groups immersed in a blackish-brown stroma. Ascospores hyaline, 1-septate towards the base, 8–12 × 4–6. On dead thin twigs, Mar., rare.

Calosphaeria wahlenbergii (Desm.)Nitschke (Fig. 384)
Perithecia 0.3–0.4mm diam., black, clustered in rings beneath bark, with long necks pointing inwards. Ascospores allantoid, hyaline, 8–11 × 1–1.5. On dead branches, Nov.–Mar.

Diatrypella favacea (Fr.)de Not. (Fig. 385)
Perithecia clustered in large erumpent grey to blackish-brown stromata. Asci polysporous, spores allantoid, pale golden brown in mass, 5–6 × 1. On attached and newly fallen branches which may be completely colonised by it, Nov.–Mar., extremely common everywhere.

Enchnoa lanata (Fr.)Fr. ex Ces. & de Not. (Fig. 386)
Perithecia without necks, up to 1.5mm diam., black, in clusters nestling beneath the bark among a feltwork of dark brown hyphae. The bark surface is slightly raised and sometimes cracked. Ascospores hyaline, curved, 9–12 × 2–3. On dead branches Apr.–May.

Hypoxylon multiforme (Fr.)Fr. (Fig. 387)
Stromata pulvinate when erumpent from bark, more effuse on bare wood, reddish brown darkening with age to almost black; perithecia prominent with papillate necks. Ascospores dark brown, mostly 9–12 × 4–5. *Nodulisporium* state greenish yellow, conidia 5–8 × 1.5–2.5. On dead branches and logs, Oct.–Apr., very common. *H. fuscum* and *H. howeianum*, both described on *Corylus*, occur also on *Betula* but less frequently.

Melanconis stilbostoma (Fr.)Tul. & C.Tul. (Fig. 388)
Perithecia 0.4–0.5mm diam., immersed and arranged in a circle with tips of black necks protruding through a prominent pale grey disc. Ascospores hyaline, 1-septate, mostly 17–20 × 5–6. The closely associated conidial state forms prominent conical black pustules, each with a white eye. Conidia, which ooze out in a black mass, are individually smoky grey, 11–12 × 7–8. On attached and newly fallen branches especially those of overcrowded, self-sown saplings, Oct.–Feb., very common everywhere.

Melanomma subdispersum (P.Karsten)Berl. & Vogl. (Fig. 389)
Pseudothecia superficial, black, 0.2–0.4mm diam. Ascospores 3-septate, with brown median cells and paler end cells, mostly 20–25 × 6–9. On periderm of logs, Sept.–Mar. *Pseudospiropes* state, common all the year round, forms effuse black, hairy colonies with dark blackish-brown conidiophores up to 1mm long, 6–9 thick, and usually 3-septate brown conidia, 16–26 × 6–11.

Ophiovalsa betulae (Tul. & C.Tul.)Petrak (Fig. 390)
Perithecia about 0.4mm diam. in small groups, immersed with necks just protuberant. Ascospores hyaline, 30–40 × 3.5–4. Conidia falcate, hyaline, 30–40 × 3–4. On small, dead, often attached twigs, Feb.–Mar.

Pleomassaria siparia (Berk. & Br.)Sacc. (Fig. 391)
Pseudothecia 1–1.5mm diam., black, flattened, just visible through very small slits in bark. Ascospores golden brown, encapsuled, with 5 to 8 transverse septa and a few longitudinal ones, 50–75 × 15–22. On dead branches, Dec.–Apr.

Pseudovalsa lanciformis (Fr.)Ces. & de Not. (Fig. 392)
Perithecia 0.5mm diam., in groups of up to 10 immersed in dark brown to black stromata with narrow elongated discs appearing through transverse slits in bark. Ascospores olivaceous brown with hyaline ends, pseudoseptate, 37–47 × 14–18, usually not mature before March. *Coryneum* state with pale brown conidia, mostly 45–60 × 15–18 found earlier in the year. Common on attached twigs and small branches.

Pteridiospora scoriadea (Fr.)Dennis (Fig. 393)
Pseudothecia black, 1mm diam., erumpent through bark. Ascospores each composed of a long, dark brown cell and a short, pale brown one, 55–67 × 20–25. On twigs, Sept.–Nov., uncommon.

Splanchnonema argus (Berk. & Br.)O. Kuntze (Fig. 394)
Pseudothecia up to 1mm diam., in small groups with rudimentary stroma, lifting bark and just exposing short necks. Ascospores brown, pseudoseptate, with thick gelatinous coats, 55–65 × 14–17. On dead branches, uncommon.

Stomiopeltis betulae J.P.Ellis (Fig. 395)
Thyriothecia superficial, round, 0.18–0.25mm diam., dark brown to black; scutellum thick, cells irregularly lobed and meandering. Ascospores hyaline,

nearly all 1-septate, just occasionally with 2 or 3 septa, guttulate, 16–22 × 3.5–5. On bark of dead twigs and branches, common; in good condition Mar.–May.

Sydowiella ambigua (Mouton)Mont. (Fig. 396)
Perithecia about 1 × 0.6mm, black, crowded, in groups of 2 to 5 under loose periderm. Ascospores hyaline, 1-septate, 18–27 × 8–11. On dead twigs and branches, Dec.–July.

Valsella adhaerens Fuckel (Fig. 397)
Perithecia immersed, in groups of 3 to 5, with a small stroma and black necks just protruding through a very pale grey or white disc. Asci polysporous, spores hyaline, 5–6 × 1. On dead branches.

Xenotypa aterrima (Fr.)Petrak (Fig. 398)
Stroma dark blackish brown to black, effuse, extending over large areas, surface rough with black, tooth-like, protuberant, sterile, stromatic processes. Perithecia immersed, 0.3–0.5mm diam. with necks up to 0.4mm long. Ascospores allantoid, hyaline, mostly 8–10 × 1.5 but occasionally up to 15 long. On dead branches, uncommon.

HYPHOMYCETES

Bactrodesmium betulicola M.B.Ellis (Fig. 399)
Sporodochia black, shining, 0.1–0.3mm diam. Conidiophores up to 40 × 1–3. Conidia 4-septate, median cells mid to dark brown, cell at each end hyaline or pale brown, 26–40 × 9–15. On dead branches, May–Nov.

Corynespora cespitosa (Ellis & Barth.)M.B.Ellis (Fig. 400)
Colonies dark blackish brown, 0.5–1.5mm diam. Conidiophores arising from stroma, brown, 15–45 × 8–11. Conidia brown, with 3 to 9 pseudosepta, 55–85 × 18–29. On dead twigs, Apr., uncommon.

Endophragmia glanduliformis (Höhnel)M.B.Ellis (Fig. 401)
Colonies effuse, thin, brown inconspicuous. Conidiophores brown, 15–40 × 2–3. Conidia 1- to 3-septate, 10–17 × 9–11.5, upper cell dark brown to black, other cells paler or subhyaline. On bark of logs, Sept., uncommon.

Endophragmiella suttonii P.M.Kirk (Fig. 402)
Colonies effuse, blackish brown to black, hairy. Conidiophores pale to mid brown, 50–100 × 2–4. Conidia pyriform, mostly 3-septate, 16–18 × 10–11, apical cell mid to dark brown, other cells paler. On rotten bark, Apr.–May.

Monodictys paradoxa (Corda)Hughes (Fig. 403)
Colonies effuse with very small black shining clusters of conidia. Conidiophores short, hyaline, often inflated. Conidia muriform, dark olivaceous or blackish olive, smooth, often with a protuberant hyaline cell at the base, 20–43

× 17–30. On bark of felled trunks and thick branches, common all the year round.

Nodulisporium, v.s. under *Hypoxylon multiforme*.

Pseudospiropes, v.s. under *Melanomma subdispersum*.

Septonema secedens Corda (Fig. 404)
Colonies effuse, hairy or velvety, pale olivaceous to dark blackish brown. Conidiophores brown, up to 200 × 4–6. Conidia in long, branched chains, 3-septate, olivaceous or reddish brown, 17–23 × 5–7. Common on felled trunks and dead fallen branches.

Septotrullula bacilligera Höhnel (Fig. 405)
Sporodochia pulvinate, 1–2mm diam., olivaceous buff or brown. Conidiophores 10–30 × 2–3. Conidia catenate, pale olivaceous, smooth, 1- to 5- (mostly 3-) septate, 19–29 × 2–3.5. Conidial mass slimy, becoming encrusted, often cracks on drying. Quite common on bark of dead branches.

Trimmatostroma betulinum (Corda)Hughes (Fig. 406)
Sporodochia pulvinate, sometimes confluent, or effuse, dark blackish brown to black. Conidia in branched chains, massed over the surface of large stromata, very variable in shape and septation, brown, smooth or verruculose, 5–20 × 5–14. Extremely common on attached and fallen twigs and branches, Feb.–May.

COELOMYCETES

Coryneum, v.s. under *Pseudovalsa lanciformis*.

Libertella betulina Desm.
Acervuli subepidermal, conidiophores branched. Conidia issuing from the bark in long golden tendrils, curved, pointed at the ends, 13–16 × 1–1.5. Common on dead branches.

Myxocyclus polycistis (Berk. & Br.)Sacc. (Fig. 407)
Acervuli scattered, erumpent, up to 0.8mm diam., dark brown to black. Conidiophores very pale brown, verruculose, 100–200 × 4–6. Conidia muriform, smoky brown, verruculose, 45–65 × 20–25, with gelatinous sheaths. On dead twigs and branches, Mar.–May.

BUXUS: Box

UREDINALES

Puccinia buxi DC., 111
Telia amphigenous, dark rather purplish brown; spores brown, smooth, 60–90

× 20–35, with long pedicels. Sept.–Oct. and overwintering, quite common.

ASCOMYCETES

Camarops lutea (Alb. & Schw.)Nannf. (Fig. 408)
Stromata erumpent, round or elliptical, up to 2.5cm diam., yellowish, becoming blackened on the surface, with black perithecia immersed at different levels. Ascospores olivaceous, 6–7 × 3–4. On stumps and thick branches, often staining the surrounding wood bright yellow, Aug.–Apr.

Gibberella buxi (Fuckel)Winter (Fig. 409)
Perithecia superficial, in small clusters, 0.2mm diam., black, walls purplish by transmitted light. Ascospores 1- to 3-septate, mostly 15–18 × 5–7. On dead twigs, Oct.–Jan.

Hyponectria buxi (DC.)Sacc. (Fig. 410)
Perithecia immersed, crowded, 0.2–0.35mm diam., with yellowish walls, appearing on the lower leaf surface as slightly depressed, at first pinkish then olivaceous brown spots. Ascospores hyaline, mostly 13–16 × 4–5. On fading and dead leaves, May–Dec.

Microthyrium macrosporum (Sacc.)Höhnel (Fig. 411)
Thyriothecia amphigenous, brown, 0.2mm diam., with cells of scutellum in radiating rows, rectangular and only slightly longer than wide. Asci 4-spored, spores 13–16 × 3.5–4.5, hyaline, 1-septate below the middle, with tufts of delicate setulae near apex and septum. On bleached dead leaves, Dec.–Jan.

Mycosphaerella buxicola (DC.)Tomilin
Pseudothecia subepidermal, 0.1–0.15mm diam., golden brown, darker around the ostiole. Ascospores hyaline, 1-septate, 22–30 × 3–5, the upper cell slightly wider than the lower one. Causes milk-white spots with brown borders mostly around the edges of living leaves, May–June.

Nectria desmazieresii Becc. & de Not. (Fig. 412)
Perithecia superficial, less than 0.2mm diam., orange. Ascospores hyaline, 1-septate, 12–17 × 5–7. On twigs and branches suffering from dieback, usually clustered on leaf scars, Feb.–Apr.

Pseudonectria rousseliana (Mont.)Wollenw. (Fig. 413)
Perithecia superficial, straw-coloured to pale orange, 0.2–0.3mm diam., with hyaline setae 40–100 × 4–6. Ascospores hyaline, mostly 11–17 × 3–4.5. Associated *Volutella* state with small pink, setose sporodochia, pale pink setae up to 120 × 4 and conidia 10–11 × 3–4. On lower side of dead leaves, Nov.–Mar.

Rosellinia buxi Fabre (Fig. 414)
Perithecia superficial, about 0.5mm diam., almost black, nestling in an effuse,

blackish brown, bristly, byssoid subiculum. Ascospores dark brown, 20–35 × 6–8. *Geniculosporium* state with brown conidiophores and hyaline conidia 3.5–7 × 1.5–3.5. On wood at base of bushes.

HYPHOMYCETES

Geniculosporium, v.s. under *Rosellinia buxi*.

Sesquicillium buxi (Schm.)K.W. Gams (Fig. 415)
Colonies on dead leaves, effuse, pale. Conidiophores hyaline, smooth, verticillately branched, 2.5–3.5 broad. Conidia in slipped chains, hyaline, smooth, 6–8 × 2–2.5.

Sporidesmium adscendens Berk. (Fig. 416)
Colonies black, hairy, effuse. Conidiophores dark reddish brown, 20–45 × 7–10. Conidia 110–375 × 14–20, with 16–62 pseudosepta. On rotten wood, Jan., uncommon.

Volutella, v.s. under *Pseudonectria rousseliana*.

COELOMYCETES

Blennoria buxi Fr. (Fig. 417)
Conidiomata brown, about 1mm diam., pushing up and splitting epidermis. Conidia hyaline, 11–13 × 2–3. On both surfaces of fading and dead leaves and on small twigs, Dec.–June.

Dothiorella candollei (Berk. & Br.)Petrak (Fig. 418)
Conidiomata immersed then splitting epidermis, dark brown, 0.3–0.4mm diam. Conidia hyaline, cloudy or with 2 guttules, 27–40 × 8–11. On fading and dead leaves and on small twigs.

Phomopsis stictica (Berk. & Br.)Trav. (Fig. 419)
Conidiomata blackish brown, about 0.25mm diam. Conidia hyaline, biguttulate, 7–10 × 2.5–3.5. On dead leaves and twigs, Nov.–Apr.

Sarcophoma state of *Guignardia miribelii* van der Aa (Fig. 420)
Conidiomata immersed, dark brown, somewhat flattened, up to 0.3mm diam., without an ostiole. Conidia hyaline, 9.5–11 × 4.5–5.5. On fading and dead leaves.

CARPINUS: Hornbeam

ON LEAVES

Mamiania fimbriata (Pers.)Ces. & de Not. (Fig. 421)
Perithecia amphigenous, 0.3mm diam., embedded in black stromata, with long

curved protuberant necks. Ascospores hyaline, 9–11 × 4, with septum near base. Appearing in summer on living leaves, maturing the following spring on fallen dead leaves, uncommon.

Sphaerognomonia carpinea (Fr.)Potebnia ex Höhnel (Fig. 422)
Perithecia 0.1–0.25mm diam., immersed, with small clypeus. Ascospores hyaline, 10–17 × 3.5–7. On fallen overwintered leaves.

Taphrina carpini Rostrup
Asci 20–30 × 7–15; spores 3.5–5 × 3–4.5. Infected leaves pale, on long shoots, May–July. Causes stubby witches' brooms with thick short branchlets.

ON WOOD AND BARK

Anthostoma decipiens (DC.)Nitschke (Fig. 423)
Perithecia 0.3–0.4mm diam., in off-white, black-coated stromata with tips of necks just protruding. Ascospores pale brown, 6–8 × 2.5–3.5. On bark.

Ceratosporella stipitata (Goid.)Hughes (Fig. 424)
Colonies effuse, dark brown to black, velvety. Conidiophores dark brown, up to 110 × 6–10. Branched brown conidia with branches up to 130 × 6–9. On dead branches.

Diaporthe carpini (Fr.)Fuckel (Fig. 425)
Perithecia crowded in pale stromata; necks black, protuberant. Ascospores hyaline, 1-septate, 13–16 × 3.5–4. On dead branches, Feb.–Mar., common.

Discosporina deplanata (Speg. & Roum.)Höhnel (Fig. 426)
Acervuli up to 2mm diam., olivaceous with black borders, at first covered, then erumpent. Conidia hyaline, 10–12 × 2.5–3. On dead twigs and branches, May–Oct.

Encoelia glaberrima (Rehm)Kirschst. (Fig. 427)
Apothecia erumpent, solitary or clustered, about 1cm diam., reddish brown. Ascospores hyaline, 6–8 × 1–1.5. On dead branches, Sept.–Oct.

Melanconis chrysostroma (Fr.)Tul. (Fig. 428)
Perithecia immersed, with necks emerging together through a yellow disc. Ascospores hyaline, 1-septate, 19–25 × 5–7, often with a small appendage at each end. On dead branches, Dec.–July.

Melanconis spodiae Tul. & C.Tul. (Fig. 429)
Perithecia 0.4mm diam., grouped in rudimentary stromata with necks emerging through small grey discs. Ascospores brown, 1-septate, 16–21 × 7–9, some with a small appendage at each end. On dead branches, Mar.–May.

Melanconium stromaticum Corda (Fig. 430)
Acervuli up to 1.5mm diam., with black spore masses flowing out like lava

from a volcano. Conidia greyish brown, 12–17 × 7–9. On dead twigs and branches.

Melogramma campylosporum Fr.
Perithecia immersed in black erumpent stromata 1–3mm diam., necks protuberant. Ascospores fusoid, curved, about 40 × 5–6, 3-septate, guttulate, median cells brown, end cells subhyaline. On dead branches.

Pezicula carpinea (Pers.)Tul. ex Fuckel (Fig. 431)
Apothecia erumpent, clustered, pale buff, discs raw sienna. Ascospores hyaline, mostly 20–30 × 10–12. *Cryptosporiopsis* state with reddish brown conidiomata 1–3mm diam., conidia 16–21 × 8–9. On dead twigs and branches.

Rutstroemia bolaris (Batsch)Rehm
Apothecia stalked, yellowish, drying pale reddish brown, with almost orange discs drying darker, stalks somewhat hairy at base. Ascospores hyaline, becoming 1- to 4-septate, 15–20 × 6–8. On small fallen twigs, uncommon.

Stilbospora macrosperma Pers. (Fig. 432)
Acervuli dark brown, up to 1.5mm diam. Conidia dark olivaceous brown, 3-septate, 32–46 × 11–15. On small dead branches, Feb.–Apr.

CASTANEA: Sweet Chestnut

ON LEAVES

Coniella castaneicola (Ellis & Ev.)Sutton (Fig. 433)
Pycnidia immersed or partly immersed, spherical, pale brown, thin-walled. Conidia pale brown, 15–30 × 2.5–3.5.

Discohainesia oenotherae (Cooke & Ellis)Nannf. (Fig. 434)
Apothecia hypophyllous, pulvinate, up to 1mm diam. Disc cream or flushed pale pink with the amber receptacle showing as a narrow darker border. Ascospores hyaline, 8–12 × 2–3. Paraphyses slender, branched near apex. On dead fallen leaves, Sept.–Oct. Found also on herbaceous plants, usually in its *Hainesia* state with pale brown, cup-shaped acervuli and hyaline conidia 5–7.5 × 1.5–2.

Microthyrium ilicinum, described on *Quercus*, is not uncommon also on *Castanea*.

Pyrenopeziza nervicola (Desm.)Boud. (Fig. 435)
Apothecia erumpent, 0.2–0.45mm diam., dark brown, disc creamy when quite fresh. Ascospores hyaline, 6–10 × 1.5–2.5. On dead fallen leaves, May–July.

Septoria castaneicola Desm.
Pycnidia hypophyllous, partly immersed, blackish brown, less than 0.1mm

diam. Conidia curved, 3-septate, 30–40 × 3–4.5, forming white tendrils. On living and fading leaves causing numerous small tawny spots, Sept.–Oct.

Discomycetes described and found more commonly on leaves of *Quercus* which occur also on *Castanea* include: *Arachnopeziza eriobasis, Ciborinia hirtella, Coccomyces dentatus, Dasyscyphus coruscatus, Pezizella roburnea* and *Rutstroemia sydowiana*. Plurivorous leaf-litter fungi commonly found include *Fusidium griseum* and *Mycosphaerella punctiformis*.

ON FRUIT INCLUDING CUPULES AND CUPULE SPINES

DISCOMYCETES

Arachnoscypha aranea (de Not.)Boud. ex Dennis (Fig. 436)
Apothecia up to 0.3mm diam., white becoming creamy yellow, fringed with hyaline hairs up to 50 × 2; seated on a thin, white, cobwebby subiculum. Ascospores hyaline, 7–8 × 2–2.5. On inner surface of decaying cupules, Sept.–Nov.

Dasyscyphus castaneicola Graddon (Fig. 437)
Apothecia up to 0.35mm diam., white, stalked, hairy; hairs hyaline, granulate, about 100 × 3–3.5, with clusters of crystals. Ascospores hyaline, 8–10 × 2–2.5. On spines on old rotting fallen cupules, Oct.–Dec.

Gorgoniceps charnwoodensis Graddon (Fig. 438)
Apothecia up to 0.5mm diam., sessile, white drying ochraceous, with a basal ring of white radiating hyphae. Ascospores hyaline, 7-septate, 40–45 × 1.5. Paraphyses branched near base. On old fallen cupules.

Helotium humile Sacc. (Fig. 439)
Apothecia sessile, 0.25–1mm diam., creamy yellow drying brownish. Ascospores hyaline, 11–15 × 2–3. On cupules and cupule stalks, Nov.

Hyalopeziza spinicola Graddon (Fig. 440)
Apothecia up to 0.3mm diam., short-stalked, lower part dark olivaceous, upper part hyaline, covered with hyaline hairs of two kinds, many tapered ones up to 250 × 3 and a few cylindrical ones up to 50 × 3.5. Ascospores 5–6 × 2. On dead cupule spines, Nov.

Rutstroemia americana (Durand)White (Fig. 441)
Apothecia up to 5mm diam., dark golden brown, long-stalked. Ascospores hyaline, 8–11 × 3–4.5. On old fallen fruits and cupules, Sept.–Nov.

Rutstroemia echinophila (Bull.)Höhnel (Fig. 442)
Apothecia up to 11mm diam., disc dark reddish brown, receptacle and stalk paler, arising from blackened areas. Ascospores curved, hyaline, mostly 15–18 × 5–6, becoming 3-septate and budding off small spherical conidia. On old fallen cupules, Oct.–Nov.

HYPHOMYCETES

Acrospeira mirabilis Berk. & Br. (Fig. 443)
Colonies effuse, powdery, chocolate brown, filling the space between the fleshy cotyledons. Conidia coiled, with 2 pale cells and a darker brown, verrucose, globose, terminal cell 20–30 diam. In fruits, sometimes causing considerable losses.

Anavirga laxa Sutton (Fig. 444)
Colonies effuse, dark brown to black, cobwebby or hairy. Conidia dark brown, branched, mostly triradiate or V-shaped, branches 50–180 × 11–14. On decaying fruit, cupules and spines.

Candelabrum spinulosum van Beverwijk (Fig. 445)
Colonies effuse, hyaline. Conidia complex, lobed, 12–17 × 7–10 in side view, 12–17 diam. in top-surface view. Fairly common on decaying cupules.

Chalara species, although not showing marked preference for this substrate, are not uncommon. Species seen include *C. aurea*, described on leaf litter, and *C. cylindrica*, described on *Picea* leaves.

Clathrosphaerina state of *Hyaloscypha zalewskii* Descals & Webster (Fig. 446)
Colonies effuse, hyaline. Conidia clathrate, hyaline, 12–24 diam. On cupules and spines. We have not seen the *Hyaloscypha* state but it was found by Descals & Webster on cupule spines and they described the apothecia as white, drying cream, up to 0.3mm diam., hairy; hairs up to 40 × 5; ascospores elliptical sometimes curved, 6.5–9 × 1.5–3.

Phialocephala fumosa (Ellis & Ev.)Sutton (Fig. 447)
Colonies effuse, brown. Conidiophores 70–120 × 5–6, dark brown. Conidia hyaline, 3–4.5 × 1. Common on cupules and spines.

Phialocephala truncata Sutton (Fig. 448)
Similar in many ways to *P. fumosa* but with shorter conidiophores, conidia 5–6.5 × 1 and up to 30 phialides in the head. On cupule spines.

Pleurotheciopsis pusilla Sutton (Fig. 449)
Colonies effuse, hairy, pale brown. Conidiophores up to 170 × 4–6. Conidia hyaline, 1-septate, 6–12 × 1.5–2.5. On rotting cupules and spines.

Polyscytalum fecundissimum, described on leaf litter, is common on rotting cupules.

Pseudomicrodochium aciculare Sutton (Fig. 450)
Colonies white or salmon pink. Conidia produced from short lateral phialides on the hyphae in glistening slimy masses, hyaline, acicular, 1-septate, 22–32 × 1. Common on rotting cupules, July–Oct.

Pseudomicrodochium cylindricum Sutton (Fig. 451)
Colonies white, glistening. Similar to the previous, much commoner species

but with rather differently shaped phialides and cylindrical conidia 16–24 × 1.5–2. On rotting cupules.

Scolecobasidium echinophilum Sutton
Colonies effuse, pinkish white, powdery. Sterile hyphae 20–66 × 2–2.5 swollen at apex to 3–4.5. Conidiophores up to 45 × 3.5–4, 0- to 2-septate, tapered towards apex, sometimes branched at base; 3 to 12 denticles. Conidia dry, hyaline to very pale brown, cylindrical to clavate, verruculose, truncate at base with marginal frill, 14–30 × 3.5–4.5, with 0 to 4 (mostly 3) septa. On cupule spines.

Tricladium castaneicola Sutton (Fig. 452)
Colonies effuse, thin cobwebby, white. Conidiophores hyaline, 15–35 × 3–5. Conidia hyaline, multiseptate, branched, main axis 80–130 × 3–3.5, lateral branches up to 50 long. On rotting cupules. We have found this also on *Quercus* cupules and on a dead rachis of *Dryopteris dilatata*.

ON WOOD AND BARK

Cryptodiaporthe castanea (Tul.)Wehmeyer (Fig. 453)
Perithecia black, up to 0.8mm diam., clustered, with short necks just erumpent through pale brown disc. Ascospores hyaline, appendaged, 1-septate, 13–18 × 2–4. Causing angular cracks in bark of dead branches, Oct.–Nov. *Diplodina* state found on dead twigs and branches May–Oct. Conidia issuing in flesh-coloured tendrils, mostly 8–12 × 2–3, long remaining without a septum, which tends to develop in longer conidia just before germination.

Didymosphaeria superapplanata Sivan. (Fig. 454)
Pseudothecia solitary, erumpent, black, up to 0.4mm diam., thick-walled. Ascospores brown, 1-septate, 20–27 × 10–12. On dead branches.

Melanconis modonia Tul. & C.Tul. (Fig. 455)
Perithecia in groups deeply immersed in a stroma, with necks just erumpent through a brown disc. Ascospores hyaline becoming yellowish brown when mature, 1-septate or rarely 3-septate, 26–38 × 8–12. More commonly found in its *Coryneum* state with pulvinate dark brown acervuli up to 0.5mm diam. and rather pale brown conidia usually 50–70 × 14–20, which have 5 to 8 pseudosepta. On dead twigs and branches, Feb.–Mar.

Psilachnum auranticolor Graddon (Fig. 456)
Apothecia up to 0.6mm diam., pale golden with shining white hairs 60 × 3–5. Ascospores hyaline, 8–10 × 2–2.5. Paraphyses lanceolate, septate, 5 broad. On wood, Feb.

CHAMAECYPARIS: Lawson's Cypress

Morenoina chamaecyparidis J.P.Ellis (Fig. 457)
Thyriothecia brown, up to 0.45 × 0.1mm. Asci 14–20 × 8–9. Ascospores 1-septate, 8–10 × 3–4, hyaline to pale brown, minutely verruculose. On dead leaves, May.

Pestalotiopsis monochaetioides (Doyer)Steyaert (Fig. 458)
Acervuli punctiform, black. Conidia 26–33 × 8–10 with 3 median brown cells and a hyaline cell at each end, smooth, with 1 apical setula up to 60 long or occasionally with 2 or 3 apical setulae, basal appendage 2–6 long. On dead leaves.

Trichothyrina fimbriata, described on *Cupressus*, occurs on bleached or brown, usually fallen, leaves.

CLEMATIS: Traveller's Joy or Old Man's Beard

Broomella vitalbae (Berk. & Br.)Sacc. (Fig. 459)
Perithecia subepidermal, solitary or in short rows, 0.35–0.4mm diam., dark brown, rather inconspicuous. Ascospores 3-septate, 25–40 × 5–7, median cells pale brown, end cells hyaline, with setulae 7–12 long. Conidia rather similar to ascospores in shape, colour and septation, 35–45 × 5–7. On dead stems, Oct.–Jan.

Excipularia fusispora (Berk. & Br.)Sacc. (Fig. 460)
Sporodochia cupulate, black, up to 0.15mm diam. Setae rather pale brown, branched near base, 50–70 × 2–4, marginal ones sometimes with apical cell dark brown and thick-walled. Conidia mostly 8-septate with the cell at each end hyaline and all the median cells rather pale brown, mostly 40–45 × 4–5. Inside bark of *C. vitalba*, Jan.–Sept.

Leptosphaeria haematites (Rob. ex Desm.)Niessl
Pseudothecia immersed, 0.35mm diam., causing red staining of host tissues. Ascospores pale brown, 3-septate, with the second cell slightly swollen, 20–27 × 4–5. On dead stems.

Phomopsis demissa (Sacc.)Trav.
Conidiomata immersed in wood, under bark, 0.25mm diam., black. Conidia hyaline, biguttulate, 6–8 × 2.5. On dead branches, July.

Pleospora vitalbae (de Not.)Berl.
Perithecia immersed, often in rows, black, 0.3–0.4mm diam. Ascospores mostly with 5 transverse septa and a longitudinal septum in one or both median cells, golden brown, 19–23 × 6–7. *Hendersonia* state with 3-septate, pale brown conidia 12–15 × 3–5. On dead stems, Feb.–Mar.

Rebentischia unicaudata (Berk. & Br.)Sacc.
Pseudothecia minute, immersed. Ascospores clavate, 3-septate, pale brown, upper cell palest, 18–21 × 6, with a 1-septate setula about 6 long at the base. On dead stems.

Septoria clematidis Rob. & Desm.
Pycnidia amphigenous, immersed, very small, pale brown. Conidia mostly curved, sometimes thicker at one end, 4- to 6-septate, hyaline, 50–80 × 4. On living leaves causing round or irregular greyish spots with dark brown borders, June–Oct.

CORNUS: Dogwood

Cheirospora botryospora, described on *Hedera*, has been found also on dead twigs of *C. alba*.

Diaporthe pardalota, described on *Lonicera*, is common also on twigs of *C. sanguinea*.

Didymella corni (Sow.)Sacc. (Fig. 461)
Pseudothecia black, immersed, with ostioles just showing on the surface. Ascospores hyaline, 1-septate, 11–14 × 3–4. On dead branches of *C. sanguinea*.

Diplodia mamillana Fr. (Fig. 462)
Pycnidia immersed, black, 0.3mm diam., prominent. Conidia golden brown, 1-septate, 21–23 × 9–12. On dead twigs of *C. sanguinea*, June-Oct.

Erysiphe tortilis (Wallr.)Fr. (Fig. 463)
White powdery mildew common on leaves of *C. sanguinea*. Cleistothecia, formed Sept.–Nov., mostly contain 3 to 5 asci, each with 3 to 5 spores; appendages long, often all turned in one direction.

Griphosphaeria corticola, described on *Rubus*, is commonly found in its *Seimatosporium* state on dead attached twigs of *C. sanguinea*.

Leiosphaerella vexata (Sacc.)E.Müller (Fig. 464)
Perithecia immersed, 0.25–0.3mm diam. Ascospores hyaline to olivaceous yellow, 1-septate, often with many small guttules, 20–30 × 9–14. On dead twigs of *C. sanguinea*, Oct.–Nov.

Mollisia discolor (Mont.)Phill. (Fig. 465)
Apothecia up to 1.5mm diam., greyish brown with discs drying pale yellow, erumpent in clusters through bark, often lobed. Ascospores 7–10 × 2–3. On dead twigs of *C. sanguinea*, June–Oct.

Phomopsis corni (Fuckel)Trav.
Conidiomata up to 0.25mm diam., erumpent. A-conidia hyaline, often curved,

with 2 or 3 guttules, 7–10 × 2–3, B-conidia curved or hamate, 20–30 × 1. On dead attached twigs of *C. sanguinea*, Mar.–Nov.

Pseudomassaria corni (Sow.)v. Arx (Fig. 466)
Perithecia immersed, up to 0.25mm diam. Ascospores hyaline, 1-septate with septum near base, 18–24 × 6–8. On dead twigs of *C. alba*.

Septoria cornicola Desm. (Fig. 467)
Pycnidia less than 0.1mm diam. Conidia hyaline, 2- to 4-septate, 30–40 × 2–3. Causes greyish spots with dark purple borders up to 1cm diam. on living leaves of *C. sanguinea*, June–Oct.

CORYLUS: Hazel

ON LEAVES

Asteroma coryli (Fuckel)Sutton (Fig. 468)
Acervuli hypophyllous, subcuticular, black, about 0.1mm diam., often in large irregular groups. Conidia curved, hyaline, 12–19 × 1–1.5. On living and fading leaves causing large, rather indeterminate, brown spots, Sept.–Nov.

Gnomonia gnomon (Tode)Schröter (Fig. 469)
Perithecia up to 0.25mm diam., dark brown, with long necks. Ascospores hyaline, 1-septate, slightly curved, tapered at the ends, with small appendages, 16–22 × 1.5. On fallen dead leaves, Mar.–May.

Mamiania coryli (Batsch)Ces. & de Not. (Fig. 470)
Perithecia about 0.4mm diam., embedded singly or in pairs in black stromata which are often arranged in rings and are visible on both leaf surfaces, necks protuberant with frilly white collar at base. Ascospores hyaline, 8–9 × 3. Usually immature when found on living leaves, maturing on fallen dead ones.

Phyllactinia guttata (Wallr.)Lév. (Fig. 471)
Cleistothecia up to 0.25mm diam., containing up to 20 asci each with usually 2 spores, developing Oct.–Dec. on the lower surface of leaves, especially on new growth following cutting earlier in the year. *Ovulariopsis* state usually poorly developed, conidia solitary, clavate, 50–90 × 10–20.

Piggotia coryli (Desm.)Sutton (Fig. 472)
Acervuli subcuticular, 0.1–0.15mm diam., pale brown. Conidia very pale brown, 9–16 × 4–7. Causing irregular brown spots, July–Nov.

ON FRUIT

Hymenoscyphus fructigenus (Bull.)Gray (Fig. 473)
Nut cup. Apothecia up to 3mm diam., stalked, creamy. Ascospores hyaline, sometimes 1-septate, mostly 18–22 × 3–4. On fallen nuts, very common, Sept.–Nov.

ON WOOD AND BARK

DISCOMYCETES

Dasyscyphus calyculiformis (Schm.)Rehm (Fig. 474)
Apothecia 1.5–2mm diam., short-stalked, brown with yellowish discs. Hairs golden brown, up to 160 × 4, multiseptate, granulate, often tipped with small crystals. Ascospores hyaline, 8–12 × 2. Lanceolate paraphyses 4–5 wide. On dead twigs, uncommon.

Encoelia furfuracea (Roth)P. Karsten (Fig. 475)
Apothecia erumpent, up to 1.5cm diam., pale tan, leathery, furfuraceous, with reddish-brown smooth discs. Ascospores allantoid, hyaline, 7–10 × 2–2.5. Paraphyses clavate. Mainly on attached dead or dying branches, Jan.–May.

Encoelia glauca Dennis (Fig. 476)
Apothecia erumpent, up to 2mm diam., bluish green, pruinose with yellow discs. Ascospores allantoid, hyaline, 4–5 × 1. On branches. We have seen only one collection of this beautiful species from Mull, sent to us by Malcolm Clark.

Pezicula coryli (Tul.)Tul. & C.Tul. (Fig. 477)
Apothecia erumpent, creamy yellow, 0.5–0.75mm diam. Ascospores hyaline, 22–30 × 7–9, becoming 3-septate. Commonly found in its *Cryptosporiopsis* state. The pulvinate acervuli develop on living branches in May and exude slimy golden yellow or pale orange masses of conidia which are mostly 25–30 × 9–10.5.

Pezicula corylina Groves
Apothecia erumpent, 0.2–0.5mm diam., pale sulphur yellow. Asci 4- or 8-spored. Ascospores cylindric clavate, 0- or 1-septate, 15–27 × 6.5–10. Paraphyses swollen at tips to 3–5, simple or branched. The *Cryptosporiopsis* state forms somewhat conical olivaceous or black acervuli and has conidia 17–24 × 7.5–8.5.

Pezicula paradoxa Dennis
Described as having ascospores mostly 28–40 × 8–11. We have not seen this species.

OTHER ASCOMYCETES

Cryptodiaporthe pyrrhocystis (Berk. & Br.)Wehmeyer (Fig. 478)
Perithecia 0.3–0.4mm diam., immersed in groups, with necks just protruding through discs which rupture the periderm and are about the same colour or slightly paler. Ascospores hyaline, appendaged, 1-septate, 22–28 × 7–9. On small dead branches, May.

Diaporthe decedens (Fr.)Fuckel (Fig. 479)
Perithecia 0.5–0.7mm diam., immersed, with necks emerging separately or through small discs. Ascospores hyaline, 1-septate, mostly 17–22 × 4–5 when mature. On dead branches, Oct.–June.

Diatrypella favacea, described on *Betula*, is very common also on *Corylus*, and both attached and newly fallen branches may be completely colonised by it.

Hypoxylon fuscum (Pers.)Fr. (Fig. 480)
A very common early primary colonist on living branches where, on the bark, it forms hemispherical 2–4mm diam., purplish-brown or pale greyish-brown stromata in which the perithecia are embedded with their ostioles flush with the surface or slightly sunken, umbilicate. Ascospores dark brown, 12–15 × 5–7.5. On bare wood the stroma becomes widely effused and much thinner. *Nodulisporium* state has conidia 3–6.5 × 1–4. On living and dead branches, Feb.–May.

Hypoxylon howeianum Peck (Fig. 481)
Stromata small, about 2–3mm diam., round, reddish brown, becoming darker with age; perithecia raised above the surface with reddish granules surrounding them but ostioles umbilicate. Ascospores pale to mid brown, 6–9 × 3–4. *Geniculosporium* state with conidia 3.5–5 × 1–2.5. Very common on dead branches, Jan.–Apr.

Melanconis flavovirens (Otth)Wehmeyer (Fig. 482)
Perithecia black, up to 0.5mm diam., grouped in circles under yellowish discs, with slightly protuberant necks. Ascospores rather thick-walled, broadly fusiform, hyaline, 1-septate with a small pointed appendage at each end, 22–30 × 7–8. On dead branches, Dec.–Apr.

Ophiovalsa corylina (Tul. & C.Tul.)Petrak (Fig. 483)
Perithecia 0.4–0.5mm diam., immersed, with necks showing through cracks in bark. Ascospores slightly curved, hyaline or pale olivaceous, mostly 60–85 × 4–5. On dead branches, Dec.–Feb.

Sillia ferruginea (Pers.)P.Karsten (Fig. 484)
Stromata erumpent through transverse slits in bark, up to 4mm across, mid pale reddish brown, with numerous black necks of perithecia projecting above the surface. Ascospores hyaline, mostly 5- or 6-septate, 60–90 × 2.5–3. On the basal part of standing branches between 3 and 10cm above soil level; Oct.–Dec., locally common.

HYPHOMYCETES AND COELOMYCETES

Camarosporium propinquum (Sacc.)Sacc. (Fig. 485)
Pycnidia immersed, up to 0.4mm diam., rather thick-walled. Conidia dark reddish brown, muriform, 12–18 × 6–8.5. On branches, Oct.–May.

Chalara inflatipes (Preuss)Sacc. (Fig. 486)
Colonies effuse, brown, hairy or velvety. Conidiophores dark brown, up to 250 long, stalk 10–12, sac 12–19 and neck 8–10 wide. Conidia in long chains, hyaline, smooth, 3- to 7-septate, 25–48 × 5–7. On bark.

Cryptosporiopsis, v.s. under *Pezicula*.

Geniculosporium, v.s. under *Hypoxylon howeianum*.

Nodulisporium, v.s. under *Hypoxylon fuscum*.

Phomopsis revellens (Sacc.)Höhnel
Conidiomata erumpent. A-conidia hyaline with 2 or 3 guttules, 6–9 × 2–3, B-conidia curved, 20–28 × 1–1.5. On dead twigs, branches and occasionally nuts, Feb.–May.

CRATAEGUS: Hawthorn

ON LEAVES

UREDINALES

Gymnosporangium clavariiforme (Pers.)DC., 0,1 (Fig. 487)
Aecia found from July–Sept., mainly on veins on lower surface of leaves but also on stems and fruits causing orange or reddish-brown swellings up to 2cm long, yellowish with very deeply cut white margins made up of chains of peridial cells. Peridial cells curved, up to 120 × 20–30, with scattered small warts. Aeciospores up to 30 × 26, pale yellowish brown. Alternate host *Juniperus communis*.

Gymnosporangium confusum Plowr., 0,1
Aecia on thickened brownish spots with yellowish margins, mostly hypophyllous. The peridial cells differ from those of *G. clavariiforme* in having oblique parallel ridges or elongated warts on their walls and the aeciospores are only up to 26 × 22. Alternate host *Juniperus sabina*.

OTHER FUNGI

Entomosporium state of *Diplocarpon mespili* (Sorauer)Sutton (Fig. 488)
Acervuli epiphyllous, subcuticular, up to 0.2mm diam. Conidia complex, hyaline, appendaged, 15–26 × 6–9. Causing small reddish spots on living leaves, not common.

Lophodermium hysterioides (Pers.)Sacc. (Fig. 489)
Apothecia black, hysterioid, up to 1mm long. Ascospores hyaline, mostly 40–45 × 1. On bleached areas on brown fallen leaves, May.

Monilia state of *Monilinia johnsonii* (*v.i.* on fruit)
Colonies effuse, silvery grey or buff. Conidia hyaline, spherical, 9–16 diam. in chains with narrow connectives, looking like beads on a string. Causes large brown or black blotches on living leaves, May–June.

Mycosphaerella crataegi (Fuckel)Johanson ex Oud.
Pseudothecia hypophyllous, immersed, 0.15mm diam., black. Ascospores hyaline, fusiform, medianly 1-septate, 35–40 × 3. On dead leaves, Apr.

Podosphaera clandestina (Wallr.)Lév. (Fig. 490)
Powdery mildew, common especially on seedlings and soft shoots in hedges. Cleistocarps about 0.1mm diam., with mature asci formed Oct.–Nov. Appendages dichotomously branched at tips. Ascospores 15–25 × 10–12.

Taphrina crataegi Sadeb.
Causes reddish-yellow blisters up to 1cm diam. along margins which become inrolled.

Venturia crataegi Aderh. (Fig. 491)
Pseudothecia immersed, 0.1–0.15mm diam., with dark brown setae up to 60 long around the papillate ostioles. Ascospores pale olivaceous brown, slightly verruculose, 1-septate, 11–14 × 4–6. On dead fallen leaves, Apr.–June.

ON FRUIT

Monilinia johnsonii (Ellis & Ev.)Honey (Fig. 492)
Apothecia up to 9mm diam., long-stalked, reddish brown with paler discs. Ascospores hyaline, 10–15 × 5–6. On mummified fallen fruits, Mar.–Apr.

Xylaria oxyacanthae Tul. & C.Tul.
Stromata emerging from soil, simple, cylindrical then expanding, compressed and variously divided above, black and hairy at base. Upper part at first white or yellowish with conidia, later black with crowded, immersed perithecia. Ascospores dark brown, about 10 × 4. On buried fallen fruits.

ON WOOD AND BARK

Cytospora oxyacanthae Rabenh.
Conidiomata erumpent, discs at first whitish then dark grey, with numerous radiating and labyrinthiform locules seen when sliced across with a razor blade. Conidia curved, 6–7 × 1–2. On dead twigs, common, Feb.–Mar.

Diaporthe crataegi Nitschke ex Fuckel (Fig. 493)
Perithecia 0.4–0.5mm diam., in groups of 4 to 9, immersed in small erumpent stromata. Ascospores hyaline, 1-septate, 14–19 × 5–6. On dead twigs, Dec.–Mar.

Mycosphaerella slaptoniensis D.Hawksw. & Sivan.
Pseudothecia immersed, black, up to 50 diam. Asci 20–24 × 9–13.

Ascospores hyaline, 1-septate, 10–15 × 2–4. On dead twigs.

Patellariopsis atrovinosa (Bloxam)Dennis (Fig. 494)
Apothecia sessile, up to 1mm diam., purplish brown, pruinose, with black discs, solitary or in small groups. Ascospores hyaline, 3-septate, 20–30 × 3–4. On dead branches, Feb.–Mar.

Pezicula sepium (Desm.)Dennis (Fig. 495)
Apothecia 0.5mm diam., erumpent, solitary or several together, brownish orange, pruinose, with dark reddish-brown discs. Ascospores hyaline, 20–30 × 9–12, becoming 3-septate. On dead branches, Nov.–Mar.

Pleospora shepherdiae Peck (Fig. 496)
Pseudothecia solitary or in groups, immersed, up to 0.6mm diam., black. Ascospores uniseriate, pale brown, mostly with 3 transverse and 1 or 2 longitudinal septa, 15–22 × 6.5–9. *Coniothyrium* state has pycnidia up to 0.3mm diam. and straw-coloured conidia 4–6 × 2.5–3. On dead twigs, May.

CUPRESSUS: Monterey Cypress

Chloroscypha seaveri, described on *Thuja*, has been recorded on dead leaves of *Cupressus*.

Microthyrium pinophyllum, described on *Pinus*, occurs also on dead leaves of *Cupressus*.

Pestalotiopsis funerea (Desm.)Steyaert (Fig. 497)
Acervuli punctiform, black, with spore masses sometimes holding together in thick erect columns. Conidia 25–33 × 8–13, 4-septate, smooth, with mid to dark brown median cells and colourless end cells; apical setulae 3 to 6, often recurved, 10–27 long, basal setula 5–14 long. On living and dead leaves, Dec.–Apr., common.

Pestalotiopsis monochaetioides, described on *Chamaecyparis*, is found also on *Cupressus*.

Seiridium cardinale (Wagener)Sutton & Gibson (Fig. 498)
Acervuli black, numerous, found on cankers and wounds near the base of twigs and branches. Conidiophores up to 30 × 1.5–2, hyaline. Conidia smooth, 5-septate, 18–30 × 8–11, median cells mid to dark brown, end cells hyaline.

Stomiopeltis cupressicola J.P.Ellis (Fig. 499)
Mycelium superficial, abundant. Thyriothecia round, up to 0.22mm diam., ostiole small. Scutellum of brown, sinuous, irregularly lobed cells. Ascospores hyaline, 1-septate, with 4 small guttules, 8.5–9.5 × 1.5–3. On dead leaves, Feb.–May.

Trichothyrina fimbriata J.P.Ellis (Fig. 500)
Thyriothecia round, 60–80 diam., brown, dark around ostiole, with distinctly fimbriate margins of narrow cells up to 20 long. Ascospores smooth, hyaline or pale yellow with dark septum, 6.5–8 × 1.5–3. On bleached or brown, usually fallen, leaves, Feb.–Aug. With a hand lens the thyriothecia can be seen as tiny black shining spots at the tips of the leaves.

Trichothyrina pinophylla, described on *Pinus* but found also on *Cupressus*, has thyriothecia with entire margins.

CYDONIA: Quince

Gymnosporangium confusum, 0,1, described on *Crataegus*, has been recorded on *Cydonia*.

Entomosporium state of *Diplocarpon mespili*, described on *Crataegus*, is found occasionally on leaves of *Cydonia*.

Monilia state of *Monilinia linhartiana* (Prill. & Delacr.)Dennis
Conidia hyaline, in chains, separated by disjunctors, 10–20 × 9–14. Causes blackish-brown blotches on leaves.

CYTISUS: Broom

UREDINALES

Uromyces pisi-sativi, 11, 111, described on *Pisum*, has been recorded on *Cytisus*.

DISCOMYCETES

Crocicreas complicatum, described on *Ulex*, has been found several times on dead wood of *Cytisus*, Jan.–Mar.

Durella atrocyanea, described on 'wood and bark', has been seen in great abundance on *Cytisus*, Jan.–Feb.

Pezicula scoparia (Cooke)Dennis
Apothecia erumpent, mostly solitary, sessile, up to 0.8mm diam., buff. Ascospores fusiform, hyaline, 0- to 5-septate, 19–25 × 6–10. Paraphyses narrow, enlarged to 5 at tip. On dead branches, uncommon.

OTHER ASCOMYCETES

Cucurbitaria spartii (Nees ex Schm. & Kunze)Ces. & de Not. (Fig. 501)
Pseudothecia 0.3–0.6mm diam., black or blackish brown, erumpent in clusters seated on a scanty brown subiculum. Ascospores muriform, golden brown, 25–30 × 11–12. On dead twigs and branches, Jan.–Apr., common.

Diaporthe inaequalis, described on *Ulex*, with ascospores 12–18 × 7–9, occurs also on *Cytisus*.

Diaporthe sarothamni Auersw. ex Nitschke
Stroma effuse with black line deeply embedded in wood. Perithecia solitary or in groups of 2 or 3, 0.3–0.6mm diam., with short necks just protruding. Ascospores hyaline, constricted at septum, 14–17 × 3–4.5. On dead twigs and branches, Nov.–Apr.

Erysiphe trifolii, white powdery mildew described on *Trifolium* is seen occasionally on *Cytisus*.

Gibberella pulicaris, described on *Lupinus*, is very common also on dead branches of *Cytisus*.

Kalmusia sarothamni Feltg. (Fig. 502)
Perithecia embedded in blackened areas of decorticated wood. Asci long-stalked. Ascospores brown, 3-septate, 13–15 × 5–6 inside the asci but sometimes enlarging up to 22 × 8 when free. On dead branches.

Pleospora cytisi Fuckel (Fig. 503)
Pseudothecia scattered, small, black, immersed. Ascospores muriform, golden brown, 30–35 × 12–15. On dead twigs and branches, Jan.–Feb.

HYPHOMYCETES

None specific to this host but several interesting species including *Endophragmia alternata*, *Henicospora minor*, *Sporidesmium coronatum* and *Volutella ciliata* have been found.

COELOMYCETES

Camarosporium spartii Trail
Pycnidia erumpent, black, 0.2mm diam. Conidia oblong, rounded at ends, dark brown, smooth, with 3 transverse septa and some with a longitudinal septum, 13–15 × 6–7. On dead branches.

Selenophoma state of *Guignardia cytisi* (Fuckel)v. Arx & E. Müller (Fig. 504)
Pycnidia immersed, 0.1–0.2mm diam., dark blackish brown. Conidia falcate,

truncate at base, hyaline, smooth, 20–30 × 4–5.5. On dead twigs and leaves.

DAPHNE: Mezereon and Spurge Laurel

Dothidea puccinioides, described on *Ulex*, occurs also on dead branches of *D. laureola*.

Marssonina daphnes Magn.
Acervuli amphigenous, small, pale brown. Conidia ovoid to pyriform, slightly curved, 12–20 × 4–5, hyaline, smooth, with a septum towards the base. Causes irregular brownish spots on fading leaves of *D. mezereum*, Sept.–Oct.

DRYAS: Mountain Avens

Cainiella johansonii (Rehm)E.Müller (Fig. 505)
Perithecia immersed, 0.2mm diam., with thick brown necks protruding. Ascospores kidney-shaped, 1-septate, full of oil drops, smooth, hyaline, 30–40 × 18–27. On petioles, June–Aug.

Chaetapiospora islandica (Johanson)Petrak (Fig. 506)
Perithecia immersed, 0.15mm diam., brown, with short dark protuberant necks and blackish-brown setae up to 180 × 10 around the ostioles. Ascospores hyaline or pale yellow, smooth, with septum near base, 20–27 × 9–12. On leaves, June–Sept.

Grahamiella dryadis (Nannf. ex Holm)Spooner (Fig. 507)
Apothecia urn-shaped, up to 0.2mm diam., olivaceous brown, with marginal fringe of thick-walled hairs 30–40 × 4–6. Excipulum of thick-walled cells. Ascospores hyaline, 1- to 3-septate, 10–17 × 2. On hairs on lower surface of leaves, June–Aug.

Isothea rhytismoides (Babington ex Berk.)Fr. (Fig. 508)
Perithecia about 0.25mm diam., sunken in leaf palisade, each surrounded by a small pseudostroma and provided with a black shining clypeus on the upper surface. Ascospores hyaline, 10–15 × 5–7. Causes dark red to black areas up to 5mm diam., on living leaves.

Mycosphaerella octopetalae (Oudem.)Lind (Fig. 509)
Pseudothecia epiphyllous, immersed, black, up to 0.1mm diam. Ascospores hyaline or pale straw-coloured, 1-septate, 20–25 × 7–10. On dead leaves, June–July.

Sphaerotheca volkartii Blumer
Powdery mildew with few conidia but abundant hypophyllous brown cleistothecia less than 1mm diam., on scanty brown mycelium. Asci solitary, 8-spored; spores 18–20 × 16–18. June-Aug.

Stomiopeltis dryadis (Rehm)Holm (Fig. 510)
Thyriothecia epiphyllous, black, round, about 0.15mm diam., with upper cells of scutellum irregularly lobed and with a fringe of superficial brown hyphae 3–4 thick. Ascospores hyaline, 1-septate, 10–12 × 2.5–4. On living and dead leaves, May–June.

Wettsteinina dryadis (Rostrup)Petrak (Fig. 511)
Pseudothecia amphigenous, erumpent, pyriform, 0.15mm diam. Ascospores 30–40 × 12–16, with thick gelatinous sheaths, at first hyaline, smooth 1-septate, becoming 3-septate and sometimes pale brown and verruculose. On leaves and pedicels, July.

EMPETRUM: Crowberry

Arwidssonia empetri (Rehm)Eriksson (Fig. 512)
Apothecia dark brown, with tops splitting into 3 or 4 lobes to expose discs. Ascospores hyaline, 3-septate, 15–18 × 5–6. Paraphyses with pale brown, slightly swollen apices. On dead attached leaves.

Botryosphaeria hyperborea Barr
Similar to *Physalospora empetri* (v.i.) but asci without apical ring.

Chrysomyxa empetri Cummins, 11 (Fig. 513)
Uredinia epiphyllous, orange, resembling aecia; spores orange, closely verrucose to echinulate, 25–50 × 20–30, commonly about 30 diam. On leaves, July–Oct.

Phaeangellina empetri (Phill.)Dennis (Fig. 514)
Apothecia erumpent, up to 0.5mm diam., dark purplish brown with dark olivaceous discs. Ascospores at first hyaline, without septa, later dark brown and 1-septate, 19–24 × 10–12. Common on dead attached leaves, June–Sept.

Physalospora empetri Rostrup
Perithecia epiphyllous, scattered, about 0.2mm diam., black, immersed with short protuberant necks. Asci with refractive apical ring. Ascospores ellipsoid, hyaline, 20–25 × 6–9. On dead, still attached leaves.

EUCALYPTUS: Eucalyptus

Aulographina eucalypti (Cooke & Massee)v. Arx & E. Müller (Fig. 515)
Hysterothecia black, crowded on pale areas on dead leaves of *E. pauciflora* × *coccifera* lying on the ground, June–Sept. Mature ascospores hyaline, 1-septate, mostly 10–11 × 3–4.

Parapleurotheciopsis inaequiseptata (Matsushima)P.M.Kirk (Fig. 516)
Characterised by the single terminal ramoconidium with many denticles, and

conidia each with its one septum well above the centre. Conidia pale olivaceous, up to 22 × 5. On dead fallen leaves, Sept.

Polyscytalum hareae (Sutton)P.M.Kirk (Fig. 517)
Colonies amphigenous, pale brown. Conidiophores up to 35 × 3–4, pale brown with 1 to 3 denticles at apex. Conidia catenate, hyaline, smooth, mostly 3-septate, 15–24 × 1.5–2, sometimes separating readily, at other times remaining attached, the chains then resembling longer, multiseptate conidia. On fallen dead leaves, Aug.–Nov.

Polyscytalum truncatum Sutton & Hodges (Fig. 518)
Colonies effuse, inconspicuous, buff or pale brown. Conidiophores pale brown, up to 95 × 4–4.5, with up to 10 denticles at apex. Conidia catenate, hyaline, 1-septate, 10–18 × 2–2.5. On fallen dead leaves, Sept.–Nov.

Readeriella mirabilis H. & P. Sydow (Fig. 519)
Pycnidia epiphyllous, immersed, up to 0.15mm diam., dark brown, thick-walled. Conidia trigonous, pale brown, 6.5–9.5 diam.

Zoellneria eucalypti (Berk.)Dennis (Fig. 520)
Apothecia erumpent, 0.5–2mm diam., stalked, olivaceous or yellow becoming dark reddish brown, with yellow discs, setose; setae dark brown, up to 250 × 7–8. Ascospores hyaline, 12–17 × 3–5. On fallen, dead leaves and on small, dead, attached branches, Sept.

EUONYMUS: Spindle-tree and Japanese Spindle-tree

UREDINALES

Melampsora epitea Thüm. var. *epitea*, 0,1
Aecia mostly hypophyllous, bright orange, on yellowish spots. On leaves of *E. europaeus*, Apr.–June. Alternate hosts *Salix* spp.

OTHER FUNGI

Ceuthospora euonymi Grove (Fig. 521)
Conidiomata up to 0.5mm diam., black, shining, visible on both leaf surfaces, with 1 to 3 ostioles. Conidia hyaline, 13–17 × 2–2.5; a small gelatinous apical appendage can be seen on them in water mounts. On fallen leaves and branches of *E. japonicus*, Aug.–Mar.

Cytospora euonymi Cooke
Conidiomata multilocular, conical, erumpent, dark brown. Conidia allantoid, curved, hyaline, 4–7 × 1–1.5. On dead branches of *E. japonicus*, June–Sept.

Microsphaera euonymi (DC.)Sacc. (Fig. 522)
Few conidia. Colonies mostly hypophyllous. Cleistothecia chestnut brown, up to 0.15mm diam.; appendages hyaline. Asci 5 to 10 with 3 to 5 spores each. On leaves of *E. europaeus*.

Microsphaera euonymi-japonici Viennot-Bourgin
Colonies amphigenous, effuse or compact, white, powdery. Conidia abundant, solitary or in short chains, 21–36 × 7–13. Cleistothecia 0.1–0.5mm diam., with long, dichotomously branched appendages. Very common but only in *Oidium* conidial state on living leaves of *E. japonicus*.

Pestalotiopsis neglecta (Thümen)Steyaert (Fig. 523)
Acervuli epiphyllous, about 0.1mm diam. Conidia 22–28 × 6–8, 4-septate, end cells hyaline, intermediate cells pale brown, the upper two often slightly darker than the lower one. Apical setulae 2 to 3, 10–23 long, basal appendage 4–7 long. On leaves of *E. japonicus*, June.

Phomopsis ramealis (Desm.)Died.
Conidiomata about 0.3mm diam., crowded and often covering large areas. A-conidia fusiform, biguttulate, 7–10 × 2.5–3, B-conidia filiform, hooked, 25–30 × 1–1.5. On thin twigs of *E. europaeus* and on twigs and leaves of *E. japonicus*, May–Sept.

Septogloeum carthusianum (Sacc.)Sacc. (Fig. 524)
Acervuli pale brown, about 0.1mm diam. Conidia hyaline, somewhat curved, 0- to 3-septate, 20–36 × 8–12. Causing white spots about 3mm diam., on leaves of *E. europaeus*.

Septoria euonymi Rabenh.
Pycnidia very small, dark brown to black. Conidia slightly curved, hyaline, 0- to 5-septate, 20–40 × 1.5–3. Causes pale greyish-brown spots on leaves of *E. japonicus*, Apr.–May.

FAGUS: Beech

ON LEAVES

DISCOMYCETES

Coccomyces dentatus, described on *Quercus*, is found also on fallen leaves of *Fagus*.

Hymenoscyphus albopunctus, described on *Alnus*, occurs also on decaying leaves of *Fagus*.

Hymenoscyphus phyllophilus (Desm.)O.Kuntze (Fig. 525)
Apothecia up to 0.4mm diam., thick-stalked, white drying yellowish.

Ascospores fusiform, hyaline, 1-septate, 12–14 × 3–4. On veins of fallen dead leaves, Sept.–Oct.

Pezizella fagi (Jaap)Matheis (Fig. 526)
Apothecia white or creamy, 0.2–1.2mm diam., with stalks 1–9mm long. Ascospores hyaline, 6–10 × 2–3. Found only on old bud scales lying on damp ground, May.

Pyrenopeziza petiolaris, described on *Acer*, is found also on petioles of dead fallen leaves of *Fagus*.

Rutstroemia petiolorum (Rob. ex Desm.)White (Fig. 527)
Apothecia up to 4mm diam., hazel brown, with dark teeth around edge, usually long-stalked. Ascospores curved, hyaline, 12–16 × 3–5. On petioles and midribs of rotting fallen leaves, Oct.

Scutoscypha fagi Graddon (Fig. 528)
Apothecia up to 0.2mm diam., pale greyish brown, developing beneath and pushing up and to one side small, dark brown shields. Hairs hooked, up to 20 long. Ascospores hyaline, sometimes 1-septate, 14–17 × 2–3.5. On upper side of very thin dead leaves, especially near midribs, Oct.

OTHER ASCOMYCETES

Apiognomonia errabunda (Rob. ex Desm.)Höhnel (Fig. 529)
Hypophyllous, perithecia mostly immersed, black, 0.2–0.3mm diam., with short protuberant necks. Ascospores hyaline, with septum near base, 14–17 × 3–4. On fallen dead leaves, Mar.–May. The *Discula* state with honey-coloured acervuli about 0.15mm diam., and hyaline conidia 10–13 × 4–6 is found on fading attached leaves, Aug.–Sept.

Discosphaerina fagi (Hudson)Barr (Fig. 530)
Inconspicuous, pseudothecia about 0.1mm diam., immersed becoming slightly erumpent through small slits. Ascospores hyaline, 13–22 × 5–7.5. It has an *Aureobasidium* conidial state. On fallen dead leaves, Apr.–May.

Microthyrium fagi J.P.Ellis (Fig. 531)
Thyriothecia 0.1mm diam.; cells surrounding ostioles thickened and forming raised dark collars. Asci 20–26 long. Ascospores hyaline, 1-septate, 7–8.5 × 1.5–2, without cilia. On decaying leaves, Oct.–Nov.

Microthyrium inconspicuum J.P.Ellis (Fig. 532)
Thyriothecia 0.1–0.14mm diam., without dark collars around ostioles. Asci 26–40 long. Ascospores 1-septate, hyaline, 8.5–10 × 2–2.5, without cilia. On decaying leaves, Aug.–Apr.

Microthyrium microscopicum, described on *Quercus*, has been recorded also on rotting leaves of *Fagus*; its ascospores are up to 14 long and have 2 trailing apical cilia.

Mycosphaerella fagi (Auersw.)Lindau
Pseudothecia mostly hypophyllous, scattered or gregarious, up to 0.1mm diam., black, shining. Asci 30–45 × 4–5. Ascospores hyaline, constricted at septum, 7–8 × 2–2.5. On fading leaves.

Mycosphaerella punctiformis, described on *Quercus* but found also on *Fagus*, has ascospores 8–14 × 2–3.

Phyllactinia guttata, powdery mildew of *Corylus*, has been seen also on *Fagus*.

HYPHOMYCETES

Ampulliferina fagi M.B.Ellis (Fig. 533)
Colonies effuse, grey. Mycelium superficial, brown, hyphopodiate. Conidia in chains, mostly 3- or 4-septate, dark brown, 21–30 × 7–10. On fallen dead leaves.

Brachysporiella laxa (Hudson)M.B.Ellis (Fig. 534)
Colonies effuse dark blackish brown to black, hairy. Conidiophores brown, up to 250 × 6–9. Conidia solitary or catenate, brown, with dark bands at septa, end cells often paler, 16–30 × 8–12. On fallen dead leaves.

Lobatopedis foliicola, described on *Quercus*, has been found also on dead leaves of *Fagus*.

Polyscytalum fagicola P.M.Kirk (Fig. 535)
Colonies amphigenous, effuse or compact, white. Conidiophores pale to mid brown, up to 150 × 3–4.5. Conidia formed in short chains on denticles which arise from the surface of slightly swollen ends of short lateral branches, cylindrical, subhyaline, 5–11 × 1–1.5. On dead leaves, Apr.–Sept.

ON FRUIT, INCLUDING CUPULES

DISCOMYCETES

Dasyscyphus fuscescens (Pers.)Rehm var. *fagicola* (Phill.)Dennis (Fig. 536)
Apothecia up to 1mm diam., rather short-stalked, brown, hairy. Hairs brown with paler end cells, capped by groups of small crystals. Ascospores hyaline, 7–11 × 1.5. On fallen dead cupules, May–June.

Dasyscyphus virgineus, described on 'wood and bark'. This plurivorous, white, very hairy species is extremely common on fallen cupules in spring.

Helotium humile, described on *Castanea* cupules, has been found on several occasions in great abundance on fallen cupules of *Fagus*, Oct.

Hymenoscyphus fagineus (Pers.)Dennis (Fig. 537)
Apothecia sessile or almost sessile, about 1mm diam., white or cream when fresh, drying reddish brown. Ascospores hyaline, 9–15 × 4–5. On fallen cupules, Sept.–Nov.

Hymenoscyphus rokebyensis Svrček (Fig. 538)
Apothecia 1–2mm diam., stalked, creamy white. Ascospores hyaline, 10–15 × 3–4. On fallen cupules, Sept.–Oct.

OTHER ASCOMYCETES

Trichothyrina cupularum J.P.Ellis (Fig. 539)
Thyriothecia up to 0.14mm diam., chestnut brown; ostiolar collars with small obtuse or truncate projecting teeth, margin not fringed. Ascospores hyaline, 1-septate, 13–16 × 3–4, with 6 cilia attached to mid-point of upper cell. Inside and outside decaying cupules, Nov.

Xylaria carpophila (Pers.)Fr. (Fig. 540)
Stromata 2–6cm long, 1–2mm thick, hairy at base, white-tipped when forming conidia, otherwise all black; perithecia formed in upper part, protuberant. Ascospores brown, 10–12 × 4–5. Very common on fallen cupules; perithecia mature Sept.–Oct.

HYPHOMYCETES

KEY

	Conidia complex with claw-like processes .... *Arachnophora fagicola*
	Conidia simple 1
1.	Conidia without septa 2
	Conidia with septa 6
2.	Conidia curved, more or less fusiform 3
	Conidia straight, cylindrical 5
3.	Conidia with a setula at each end 4
	Conidia without setulae .... *Codinaea hughesii*
4.	Conidia 10–15 long .... *Codinaea fertilis*
	Conidia 15–20 long .... *Codinaea britannica*
5.	Conidiophores about 50 long .... *Chalara affinis*
	Conidiophores up to 180 long .... *Chalara cylindrosperma*
6.	Conidia 1-septate 7
	Conidia 2-septate 8
	Conidia with 4 or 5 septa 9
	Conidia with 9 or more septa 10
7.	Conidia 15–25 × 3 .... *Chalara spiralis*

Conidia 14–20 × 3.5–5 .... *Haplariopsis fagicola*
8. Conidia hyaline, 3–4 wide .... *Spondylocladiopsis cupulicola*
    Conidia brown, 8–12 wide .... *Bactrodesmiella masonii*
9. Conidia with black bands at septa .... *Endophragmia catenulata*
    Conidia without black bands .... *Endophragmiella fagicola*
10. Conidia cylindrical, 9 or 10 septa .... *Camposporium cambrense*
     Conidia often somewhat tapered, up to 16 septa ....
     .... *Camposporium pellucidum*

Arachnophora fagicola Hennebert (Fig. 541)
Colonies small, brown or blackish. Conidiophores up to 150 × 4–6, brown. Conidia complex, pale brown, 16–20 × 10–13. On fallen cupules, July–Sept.

Bactrodesmiella masonii (Hughes)M.B.Ellis (Fig. 542)
Sporodochia scattered, punctiform, dark brown, up to 0.2mm diam. Conidiophores pale brown, up to 40 long, 2–4 thick at base, 4–7 at apex. Conidia pale to mid brown, mostly 2-septate, 16–26 × 8–12. On fallen cupules, Oct.–Apr.

Camposporium cambrense Hughes (Fig. 543)
Colonies effuse, grey, brown or olivaceous brown, sometimes glistening. Conidiophores brown, usually short, but can be up to 80 × 6–7. Conidia cylindrical, pale brown, end cells often subhyaline and the apical one occasionally prolonged into a long filiform appendage, 60–105 × 8–10, mostly 9- or 10-septate; attached to conidiophore apex by narrow cylindrical denticles. On fallen dead cupules, all the year round.

Camposporium pellucidum (Grove)Hughes (Fig. 544)
This differs from *C. cambrense* in having generally larger conidia, up to 140 × 12, with up to 16 septa, often tapered gradually towards the apex.

Chalara affinis Sacc. & Berl. (Fig. 545)
Colonies effuse, creamy. Conidiophores pale brown, 30–80 but mostly about 50 long, venter 4–7 and neck 2–2.5 wide. Conidia catenate, hyaline, smooth, 6–18 × 1.5–2.5. On fallen cupules, Oct.–July.

Chalara cylindrosperma (Corda)Hughes (Fig. 546)
Colonies blackish brown, hairy. Conidiophores brown, up to 180 long, venter 6–7 and neck 2–3 wide. Conidia hyaline, 4–11 × 1–1.7. On fallen cupules and other debris, Oct.–Feb.

Chalara spiralis Nag Raj & Kendrick (Fig. 547)
Colonies effuse, velvety, pale yellow or creamy. Conidiophores twisted, pale brown, up to 130 long, venter 6–8 and neck 2.5–4.5 wide. Conidia hyaline, 1-septate, mostly 15–25 × 3. On fallen cupules, Nov., uncommon.

Codinaea britannica, described on leaf litter, is common on cupules.

Codinaea fertilis Hughes & Kendrick (Fig. 548)
Colonies effuse, greyish brown to blackish brown, hairy. Conidiophores pale brown, smooth, up to 100 × 3–5, with a number of rather thick-walled, funnel-shaped collarettes at and below the apex. Setae brown, up to 300 × 4–6, mostly fertile with several collarettes. Conidia hyaline, curved, 10–15 × 2–3, with a setula 5–10 long at each end. Common on fallen cupules all the year round.

Codinaea hughesii M.B.Ellis (Fig. 549)
Colonies effuse, black, hairy. Conidiophores pale brown, up to 90 × 4–6. Setae dark brown, up to 350 long. Conidia hyaline, without setulae, 20–26 × 2.5–4. On fallen cupules and on dead wood.

Endophragmia catenulata M.B.Ellis (Fig. 550)
Colonies dark brown to black, hairy. Conidiophores brown, up to 230 × 6–8. Conidia in chains, 2- to 5- (mostly 4-)septate, brown, with pale end cells and thick blackish bands at one or more of the septa, 20–45 × 14–19. On fallen cupules, Sept.–Oct.

Endophragmiella fagicola P.M.Kirk (Fig. 551)
Colonies effuse, brown to dark blackish brown, hairy. Conidiophores brown, up to 280 × 6–10. Conidia 4- or 5-septate, brown with pale end cells, 60–105 × 15–18. On fallen cupules, Aug.–Oct.

Haplariopsis fagicola Oud. (Fig. 552)
Colonies effuse, velvety, grey or olivaceous grey. Conidiophores branched, lower part olivaceous, upper part hyaline or subhyaline, up to 100 × 5–6. Conidia hyaline, smooth, 1-septate, 14–20 × 3.5–5. Mainly on fallen cupules, Oct.–Nov.

Spondylocladiopsis cupulicola M.B.Ellis (Fig. 553)
Colonies effuse, velvety, olivaceous grey. Conidiophores up to 300 × 6–8, reddish brown near base, other parts subhyaline. Conidia mostly 2-septate, hyaline, smooth, 18–23 × 3–4. On fallen cupules, Oct.–Nov.

ON WOOD AND BARK

DISCOMYCETES

KEY

	Apothecia purple or pinkish lilac 1
	Apothecia dark brown to black 2
	Apothecia pale but not lilac 4
1.	Ascospores 6–9 long .... *Neobulgaria pura*
	Ascospores 12–19 long, 1 to 3 septa .... *Ascocoryne sarcoides*
	Ascospores 18–30 long, more septa .... *Ascocoryne cylichnium*
2.	Ascospores brown, kidney-shaped .... *Bulgaria inquinans*

	Ascospores hyaline 3
3.	Apothecia hysterioid, ascospores without septa or appendages .... *Ascodichaena rugosa*
	Apothecia erumpent, ascospores septate .... *Durella connivens*
	Apothecia superficial, ascospores appendaged at each end .... *Peziza apiculata*
4.	Apothecia long-stalked 5
	Apothecia sessile or short-stalked 6
5.	Apothecia white or pale yellow .... *Dasyscyphus brevipilus*
	Apothecia golden brown .... *Hymenoscyphus serotinus*
6.	Ascospores 3 wide, up to 3 septa .... *Calycellina ochracea*
	Ascospores 4–5 wide, up to 7 septa .... *Strossmayeria basitricha*
	Ascospores 6 or more wide 7
7.	Apothecia pale buff, spores 10–12 wide .... *Pezicula carpinea*
	Apothecia cinnamon, spores 6–9 wide .... *Pezicula cinnamomea*

Ascocoryne cylichnium (Tul.)Korf (Fig. 554)
Apothecia up to 1.5cm diam., purple or reddish purple, gelatinous, solitary or in clusters. Ascospores hyaline, mostly 18–30 × 4–6, becoming multiseptate and budding off spherical conidia about 2 diam. On rotting logs and stumps, Oct.–Nov.

Ascocoryne sarcoides (Jacq.)Groves & Wilson (Fig. 555)
Apothecia up to 1cm diam., purple or reddish purple, gelatinous, often in large clusters. Ascospores hyaline, 12–19 × 4–5, becoming 1- to 3-septate. On wet logs and stumps, Oct.–Dec. Often accompanied by, or found at other times of the year in, its *Coryne* state, which is superficially similar but forms rod-like conidia 4 × 1.

Ascodichaena rugosa Butin (Fig. 556)
Apothecia in dense groups, erumpent, hysterioid, sometimes with forked or stellate openings, up to 0.45mm long, blackish brown. Ascospores hyaline, 18–24 × 13–16. Most frequently seen in its *Polymorphum* state in which the rugose black stromata are often large and conspicuous and the conidiomata are polymorphous; conidia hyaline, 15–30 × 10–16. On bark of living trees, especially the basal part of trunks, Apr.–July, very common.

Bulgaria inquinans, described on *Quercus*, occurs also on *Fagus*.

Calycellina ochracea, described on 'wood and bark', has been seen on very rotten wood of *Fagus*.

Dasyscyphus brevipilus Le Gal (Fig. 557)
Apothecia white, drying pale yellow, up to 1.5mm diam., often long-stalked. Hairs granulate, hyaline, not more than 50 × 3, swollen to 4–5 at tips. Ascospores hyaline, mostly 6–8 × 1.5–2. Paraphyses filiform or narrowly lanceolate. On rotten, usually decorticated wood, Jan.–Sept.

Durella connivens (Fr.)Rehm (Fig. 558)
Apothecia erumpent, up to 0.5mm diam., black, with grey disc when fresh. Ascospores hyaline, with 3 to 7 septa, mostly 25–40 × 4–5. On hard decorticated wood, Sept.–Mar.

Hymenoscyphus serotinus (Pers.)Phill.
Apothecia up to 5mm diam., long-stalked, golden brown, with yellow or orange discs. Ascospores hyaline, clavate, pointed at base, sometimes hooked at apex, 18–28 × 3–4. On fallen dead twigs, Sept.–Nov.

Neobulgaria pura (Fr.)Petrak (Fig. 559)
Apothecia rather pale pinkish or lilac, up to 2mm diam., clustered, gelatinous, becoming horny on drying. Ascospores hyaline, biguttulate, 6–9 × 3–4. On fallen trunks, June–Jan.

Pezicula carpinea, described on *Carpinus*, and *P. cinnamomea*, described on *Quercus*, both occur also on *Fagus*.

Peziza apiculata Cooke (Fig. 560)
Apothecia 1.5–2cm diam., very dark olivaceous brown. Ascospores hyaline, verruculose, with an appendage at each end which is occasionally forked, 15–25 × 9–10; mature spores may be golden brown and up to 27 long. On wet, partly buried wood, Nov.–May.

Strossmayeria basitricha, described on *Quercus*, has been found also on *Fagus*.

OTHER ASCOMYCETES

KEY

	Ascospores curved allantoid, often straw-coloured in mass	1
	Ascospores not curved allantoid	9
1.	Asci polysporous	2
	Asci 8-spored	3
2.	Asci long-stalked	*Diatrypella favacea*
	Asci short-stalked	*Valsella amphoria*
3.	Ascospores 2–3 long, walls of ascocarps breaking into polygonal plates	*Fragosphaeria purpurea*
	Ascospores 3.5–4.5 long, no plates	*Calosphaeria parasitica*
	Ascospores more than 5 long	4
4.	Ascospores 15–18 long	*Quaternaria quaternata*
	Ascospores not more than 12 long	5
5.	Cut surface of stroma greenish yellow	*Eutypa flavovirens*
	Cut surface of stroma not greenish yellow	6
6.	Ascospores 8–12 × 2	*Eutypa leioplaca*
	Ascospores not more than 8 long	7
7.	Stromata small, round, flat-topped, brown	*Diatrype disciformis*

Stromata widely effused, dark brown to black 8
8. Surface of stroma almost smooth .... *Diatrype stigma*
   Surface of stroma coarsely roughened by knobby protuberant necks ......
   .... *Eutypa spinosa*
9. Ascomata bright yellow or red 10
   Ascomata not yellow or red 11
10. Perithecia at first yellow .... *Nectria ditissima*
    Perithecia never yellow .... *Nectria coccinea*
11. Ascospores hyaline 12
    Ascospores mid to dark brown 15
12. Ascospores 1-septate .... *Cryptodiaporthe galericulata*
    Ascospores 3-septate .... *Massarina eburnea*
    Ascospores 15- to 25-septate .... *Tubeufia helicoma*
    Ascospores without septa 13
13. Ascospores 3–6 long, ascomata superficial, long-necked ....
    .... *Ceratocystis moniliformis*
    Ascospores more than 15 long 14
14. Ascospores 16–28 × 8–16 .... *Cryptosporella compta*
    Ascospores 25–35 × 14–17 .... *Botryosphaeria quercuum*
    Ascospores up to 40 × 20 .... *Botryosphaeria hoffmanni*
15. Ascospores 3-septate .... *Asteromassaria macrospora*
    Ascospores 7-septate .... *Melogramma spiniferum*
    Ascospores without septa 16
16. Perithecia in small, valsoid clusters in bark 17
    Perithecia in large, mostly superficial, brown stromata 18
17. Ascospores 10–12 × 5–6 .... *Lopadostoma turgidum*
    Ascospores 25–30 × 11–13 .... *Anthostoma amoenum*
18. Ascospores 25–35 × 7–10 .... *Ustulina deusta*
    Ascospores less than 20 long 19
19. Ostioles umbilicate, spores 11–15 × 5–8 .... *Hypoxylon fragiforme*
    Ostioles papillate 20
20. Spores 10–16 × 6–10 .... *Hypoxylon nummularium*
    Spores averaging 10.2 × 4.3 .... *Hypoxylon cohaerens*
    Spores averaging 8.5 × 3.8 .... *Hypoxylon rutilum*

Anthostoma amoenum Sacc. (Fig. 561)
Stromata erumpent, 2–3mm diam., with white or creamy discs. Ascospores brown, mostly 25–30 × 11–13. On dead branches, Mar.–Apr., uncommon.

Asteromassaria macrospora (Desm.)Höhnel (Fig. 562)
Pseudothecia black, 0.5–0.7mm diam., erumpent through bark in small groups. Ascospores at first hyaline, 1-septate, becoming brown and 3-septate, 43–56 × 15–20, with gelatinous sheaths. On fallen dead branches, Dec.–Feb.

Botryosphaeria hoffmanni Höhnel (Fig. 563)
Stromata seen through triangular fissures in bark, black, 1–1.5mm diam., each

containing 4 to 12 pseudothecia. Ascospores hyaline, 25–40 × 12–20. Associated pycnidial state has hyaline conidia 35–50 × 10–13. On fallen dead branches.

Botryosphaeria quercuum (Schw.)Sacc. (Fig. 564)
Stromata similar to those of *B. hoffmanni*. Ascospores mostly 25–35 × 14–17. On fallen dead branches.

Calosphaeria parasitica Fuckel
Perithecia 0.3–0.35mm diam., black, long-necked, associated with and often embedded in the stromata or perithecial cavities of *Quaternaria quaternata*. Asci long-stalked; spores allantoid, 3.5–4.5 × 0.5.

Ceratocystis moniliformis (Hedgc.)C.Moreau (Fig. 565)
Perithecia superficial, dark brown to black, up to 0.25mm diam.; walls ornamented with short brown conical spines, necks up to 1mm long with splayed-out hyaline hyphae at tips. Ascospores hyaline, 3–6 × 2–3. *Thielaviopsis* state with catenate brown arthroconidia 7–15 × 4–9 and phialides which produce hyaline cylindrical conidia 4–18 × 1.5–3. On rotten wood.

Cryptodiaporthe galericulata (Tul. & C.Tul.)Wehmeyer (Fig. 566)
Perithecia black, 0.4mm diam., immersed in groups of up to 6, with ostioles just showing through small creamy white discs. Ascospores hyaline, 1-septate, 19–29 × 5–8, sometimes with an appendage at each end. Conidiomata of *Diplodina* state pulvinate, somewhat conical, 1–2mm diam., multilocular, eventually splitting open. Conidia hyaline, 15–20 × 4–5. On dead twigs, Sept.–Mar.

Cryptosporella compta (Tul. & C.Tul.)Sacc.
Perithecia 6 to 9 in each stroma, with necks just showing through powdery white or grey dics. Ascospores ovoid or occasionally cylindrical, hyaline, 16–28 × 8–16. On dead branches, uncommon.

Diatrype disciformis (Hoffm.)Fr. (Fig. 567)
Stromata usually numerous, erumpent through stellate fissures, 1.5–3mm diam., round, flat-topped, brown, dotted with very slightly protuberant black perithecial necks. Perithecia up to 60 per stroma. Ascospores pale straw-coloured in mass, curved allantoid, 6–8 × 1–2. On bark of attached and fallen branches, Oct.–Apr., very common.

Diatrype stigma, a plurivorous 'wood and bark' species, is extremely common on both bark and decorticated wood.

Diatrypella favacea, described on *Betula*, sometimes completely colonises attached and freshly fallen branches.

Eutypa flavovirens, another plurivorous 'wood and bark' fungus commonly seen on *Fagus*.

Fungi Specific to Trees, Shrubs and Woody Climbers 131

Eutypa leioplaca (Fr.)Cooke (Fig. 568)
Stromata effuse, irregular, very dark grey, dotted with black protuberant perithecial necks. Ascospores straw-coloured in mass, curved, mostly 8–12 × 2. On fallen dead branches.

Eutypa spinosa (Pers.)Tul. & C.Tul (Fig. 569)
Stromata effuse, black, the surface coarsely roughened by numerous sulcate, knobby, protuberant necks. Ascospores straw-coloured in mass, curved, 6–8 × 2. On dead wood, Sept.–May.

Fragosphaeria purpurea Shear (Fig. 570)
Cleistothecia spherical, black, 0.3–0.5mm diam., the walls fracturing along well-defined sutures to form polygonal plates; invested with purplish hyphae when young. Ascus walls diffluent; spores curved, pale brown in mass, 2.5–3 × 1.5–2. Embedded in decaying wood.

Hypoxylon cohaerens (Pers.)Fr. (Fig. 571)
Stromata 2–4mm diam., 1–2mm thick, gregarious, dark chestnut to blackish brown, with 6–50 perithecia per stroma; ostioles papillate. Ascospores dark brown, 9–12 × 4–5. *Nodulisporium* state pale fawn or creamy; conidia hyaline 3–7 × 1–3.5. On dead branches, erumpent from bark and on decorticated wood, Oct.–May.

Hypoxylon fragiforme (Scop.)Kickx (Fig. 572)
Stromata 2–10mm diam., hemispherical, brick-red becoming black or blackish brown when old; perithecia prominent but ostioles umbilicate. Ascospores dark brown, 11–15 × 5–8. *Nodulisporium* state yellowish, conidia subhyaline, 3–6 × 1–3. On dead branches, Sept.–May.

Hypoxylon nummularium Bull. (Fig. 573)
Stromata erumpent through bark, often large, elliptical, flat and crust-like, black and shiny when mature; ostioles small, papillate. Ascospores dark brown or blackish brown, 10–16 × 6–10. *Nodulisporium* state off-white to pale brown, at first forming palisade beneath periderm which later peels back. Conidia 3.5–7 × 1.5–2.5. On dead branches, uncommon.

Hypoxylon rutilum Tul & C.Tul.
Stromata when erumpent from bark 2–3mm diam., 1–2mm thick, on bare wood thinner and more effuse, pale reddish brown at first, darkening and eventually almost black; ostioles papillate. Ectostroma blood red beneath surface. Ascospores 7–10 × 3.5–4.5. On dead branches, Nov.–Dec., occasional.

Lopadostoma turgidum(Pers.)Traverso (Fig. 574)
Stromata 1–2mm diam., immersed, rudimentary, raising the periderm in a series of hemispherical bumps each with a small black papilla in the middle where the necks of 3 to 6 perithecia converge. Ascospores very dark brown, mostly 10–12 × 5–6. Very common and often covering large areas of fallen

branches, which have a characteristic bright reddish-brown colour.

Massarina eburnea (Tul. & C.Tul.)Sacc. (Fig. 575)
Pseudothecia immersed, about 0.5mm diam., each with a small black clypeus raising the epidermis slightly but rather inconspicuous. Ascospores 3-septate, hyaline, with gelatinous sheaths, 30–35 × 8–10. On thin newly dead twigs, Apr.–May.

Melogramma spiniferum (Wallr.)de Not (Fig. 576)
Stromata 2–3mm diam., erumpent, often confluent, forming black crusts roughened by long protuberant necks of perithecia. Ascospores 7-septate, 50–75 × 7.5–8, pale to mid brown with cell at each end hyaline. At base of trunks and on exposed roots.

Nectria coccinea (Pers.)Fr. (Fig. 577)
Perithecia about 0.3mm diam., mostly oval with pointed ostiolar papillae, bright red when young, darkening with age, usually found on bark in groups of 5 to 30 seated on an erumpent stroma but occasionally scattered, with no stroma, on bare wood. Ascospores hyaline to very pale brown, 1-septate, smooth or minutely verruculose, 12–17 × 5–7. Very common on dead branches and on trunks of standing trees, Sept.–May, sometimes accompanied by its *Cylindrocarpon* state with 1- to 5-septate, curved, cylindrical, hyaline conidia 30–80 × 6–7.

Nectria ditissima Tul. & C.Tul.
Perithecia 0.2–0.3mm diam., yellow becoming dark red, darker around the ostioles, in groups on scanty erumpent stromata. Ascospores hyaline, smooth, 1-septate, 15–21 × 5–8. Associated with dieback or canker, Oct.–May, uncommon.

Quaternaria quaternata (Pers.)Schröter (Fig. 578)
Stromata about 0.5mm diam., immersed, raising the periderm in a series of hemispherical bumps, each containing 2 to 5 perithecia, with short necks converging on a small black disc. Ascospores curved, straw-coloured in mass, mostly 15–18 × 3. Very common on dark grey areas of freshly fallen branches and on trunks of trees blown down by gales, sometimes completely colonising them, Nov.–May.

Tubeufia helicoma (Phill. & Plowr.)Piroz. (Fig. 579)
Colonies effuse, felty, brown or olivaceous brown. Pseudothecia superficial, scattered among conidiophores, about 0.3mm diam., golden brown to black, sometimes coarsely tuberculate. Ascospores hyaline to pale brown, 15- to 25-septate, 55–95 × 6–8. *Helicosporium* state has dark brown, setiform conidiophores up to 300 × 6–8, and coiled, multiseptate, hyaline conidia mostly 50–60 diam., with filaments 6–8 thick. On dead branches, Dec.

Ustulina deusta (Hoffm.)Lind (Fig. 580)
Stromata up to 10cm long, flat and thin when growing on bare wood but

pulvinate and irregular on bark, pale greyish brown to almost black. Perithecia immersed in stromata, up to 1.5 × 1mm. Ascospores dark brown, mostly 25–35 × 7–10. Found usually at the base of old stumps.

Valsella amphoria (Nitschke)Sacc. (Fig. 581)
Stromata erumpent through bark, each with the black tips of necks of 3 to 6 perithecia visible on the surface of small, white discs. Asci polysporous; spores hyaline or pale straw-coloured in mass, 6–7 × 1–1.5. On dead branches, uncommon.

HYPHOMYCETES

KEY

	Conidia of two kinds, fat brown ones and cylindrical hyaline ones 1
	Conidia of one kind only 2
1.	Brown conidia subglobose .... *Chalara ovoidea*
	Brown conidia obovate or barrel-shaped .... *Thielaviopsis*
2.	Fructifications purple, gelatinous .... *Coryne*
	Fructifications not purple 3
3.	Conidia helicoid .... *Helicosporium*
	Conidia not helicoid 4
4.	Conidia with a setula at each end 5
	Conidia without setulae 6
5.	Conidia 3-septate .... *Menispora tortuosa*
	Conidia without septa .... *Menispora ciliata*
6.	Conidia pale yellow, 100–200 × 40–50 .... *Bactridium flavum*
	Conidia not yellow, narrower and mostly shorter 7
7.	Conidiophores hyaline 8
	Conidiophores brown 9
8.	Branching verticillate, conidia 0-septate .... *Clonostachys compactiuscula*
	Branching not verticillate, conidia septate .... *Cylindrocarpon*
9.	Brown branched setae present .... *Ceratocladium microspermum*
	No brown setae 10
10.	Conidia cylindrical, catenate .... *Septotrullula bacilligera*
	Conidia not cylindrical and not catenate 11
11.	Conidiophores branched, especially near apex .... *Nodulisporium*
	Conidiophores unbranched or sparingly so near base 12
12.	Conidia without septa, conidiophores with terminal and intercalary swellings .... *Gonatobotryum fuscum*
	Conidia septate, conidiophores different 13
13.	Conidia formed sympodially 14
	Conidia formed only at conidiophore apex 15
14.	Conidia 2-septate .... *Brachydesmiella biseptata*
	Conidia with 3 or more septa or pseudosepta .... *Pseudospiropes*

134 Fungi Specific to Trees, Shrubs and Woody Climbers

15. Conidia in long chains .... *Taeniolella faginea*
    Conidia solitary or in very short chains 16
16. Conidia developing through a pore at the apex of the conidiophore ....
    .... *Corynespora biseptata*
    Conidia developing as blown-out ends 17
17. Conidia mostly less than 30 long .... *Endophragmia*
    Conidia mostly more than 30 long .... *Sporidesmium*

Bactridium flavum, described on *Salix*, is not uncommon also on wet rotten wood of *Fagus*.

Brachydesmiella biseptata, described on *Fraxinus*, has been found also on *Fagus*.

Ceratocladium microspermum Corda (Fig. 582)
Colonies effuse, velvety, pale olivaceous to dark brown. Brown setae up to 300 × 4–5, branched at apex. Conidia hyaline, 4–7 × 0.5–1. On wood and inside bark, uncommon.

Chalara ovoidea Nag Raj & Kendrick (Fig. 583)
Colonies effuse, black. Conidiophores hyaline, variable in length, 4–6 thick. Conidia of 2 kinds: (1) subglobose or pyriform, brown, mostly 8–14 × 6–10 formed at ends of branches; (2) cylindrical, hyaline or pale brown, 5–20 × 2.5–4. On dead wood and bark.

Clonostachys compactiuscula (Sacc.)Hawksw. & W.Gams (Fig. 584)
Colonies effuse, powdery, white or tinged with pink. Conidiophores hyaline, 3–5 thick and up to 1mm long, verticillately branched towards the apex, lower part slightly verruculose. Conidia in long columns, hyaline, mostly 6–8 × 1.5–3. On dead twigs among leaf litter.

Coryne, v.s. under *Ascocoryne*.

Corynespora biseptata M.B.Ellis (Fig. 585)
Colonies effuse, dark blackish brown to black. Conidiophores brown, 80–150 × 5–7. Conidia mostly 2-septate, brown, 20–35 × 7–9. On rotten wood.

Cylindrocarpon, v.s. under *Nectria coccinea*.

Endophragmia glanduliformis, described on *Betula*, has been found also on bark of fallen branches of *Fagus*.

Endophragmia stemphylioides (Corda)M.B.Ellis (Fig. 586)
Colonies black, effuse. Conidiophores up to 110 × 5–7, brown. Conidia mostly 4-septate, brown or dark brown with hyaline or pale brown end cells and black bands at septa, 15–32 × 10–15. Pale brown spherical secondary conidia formed at tips of some primary conidia. Described originally on *Taxus* but in Britain found only on rotten wood of *Fagus*.

Endophragmia verruculosa M.B.Ellis (Fig. 587)
Colonies effuse, dark blackish brown to black. Conidiophores pale brown, up to 150 × 4–7. Conidia 1-septate, verruculose, 16–21 × 10–13, upper cell mid to dark brown, lower cell subhyaline or pale brown. On rotten wood, Apr.–June.

Gonatobotryum fuscum, described on *Quercus*, has been found also on logs of *Fagus*.

Helicosporium, v.s. under *Tubeufia helicoma*.

Menispora ciliata and *M. tortuosa*, plurivorous 'wood and bark' species, are found commonly on *Fagus*.

Nodulisporium, v.s. under *Hypoxylon*.

Pseudospiropes hughesii M.B.Ellis (Fig. 588)
Colonies effuse, dark blackish brown to black, velvety or hairy. Conidiophores dark brown, up to 170 × 4–6. Conidia mid to dark brown, smooth, mostly 3-septate, 12–18 × 4.5–6. On dead wood. Five plurivorous species of *Pseudospiropes*, *P. nodosus*, *P. obclavatus*, *P. rousselianus*, *P. simplex* and *P. subuliferus* occur also, and *P. simplex* is extremely common.

Septotrullula bacilligera, described on *Betula*, has been found also on *Fagus*.

Sporidesmium hormiscioides Corda (Fig. 589)
Colonies effuse, dark blackish brown to black. Conidiophores reddish brown, 10–25 × 5–11. Conidia 65–270 × 10–17, mid to dark reddish brown, with many black septa. On dead wood. Three plurivorous species of *Sporidesmium* occur also, *S. coronatum*, *S. folliculatum* and *S. leptosporum*.

Taeniolella faginea (Fuckel)Hughes (Fig. 590)
Colonies effuse, dark blackish brown to black. Conidiophores pale to mid brown, 2–5 thick. Conidia brown, verruculose, mostly 3- to 5-septate, 15–37 × 6–9. On bark.

Thielaviopsis, v.s. under *Ceratocystis moniliformis*.

COELOMYCETES

Asterosporium asterospermum (Pers.)Hughes (Fig. 591)
Acervuli 1–2mm diam., subepidermal, splitting irregularly to expose the dark spore mass. Conidia 4-armed, smoky brown, mostly 45–50 from tip to tip of the branches which are about 15 thick at the base. Most commonly on dead twigs and small branches, Oct.–Feb.

Camarosporium ambiens, described on *Acer*, occurs also on *Fagus*.

Diplodina, v.s. under *Cryptodiaporthe galericulata*.

Libertella faginea Desm. (Fig. 592)
Acervuli subepidermal. Conidia curved, hyaline, mostly 15–19 × 1–1.5, issuing in long, twisted, glistening honey-yellow tendrils or sometimes broad ribbon-like bands. Very common on trunks and branches, July–Oct. Similar but bright orange spore tendrils which contain allantoid conidia 4–6 × 1–1.5 are found also very commonly on *Fagus* bark; these belong to the *Naemospora* state of *Diatrype stigma*.

Neohendersonia kickxii (Westend.)Sutton & Pollack (Fig. 593)
Conidiomata immersed, 0.5–1mm diam., dark brown to black, with papillate ostioles and 1 or several locules. Conidia emerge in black tendrils; they are clavate, 25–35 × 14–17 when mature, with 2 or rarely 3 septa, smoky brown, the basal cell paler or subhyaline. On twigs and branches, Mar.–Apr.

Polymorphum, *v.s.* under *Ascodichaena*.

Scolicosporium macrosporium (Berk.)Sutton (Fig. 594)
Acervuli immersed becoming erumpent, dark brown to black, up to 0.25mm diam. Conidia fusiform, up to 190 × 12–15, with 7 to 12 septa, smoky brown except for the end cells which are subhyaline. On twigs and branches, Dec.–Mar.

BASIDIOMYCETES

Phleogena faginea (Fr.)Link (Fig. 595)
Fructifications with pale brown stalks and almost spherical, off-white to snuff-coloured heads 1–3mm diam., composed of radiating, branched and twisted hyphae which bear 3-septate basidia 25–35 × 3–5. Basidiospores pale brown, 8–10 × 7–9. Common on bark.

FRANGULA: Alder Buckthorn

UREDINALES

Puccinia coronata Corda, 0,1
The orange aecia of crown rust of grasses, found in May and June, often cause distortion, especially of the petioles.

OTHER FUNGI

Diaporthe syngenesia (Fr.)Nitschke ex Fuckel (Fig. 596)
Perithecia immersed, in clusters of 7 to 15, with necks erumpent through discs which are up to 1mm diam. Ascospores hyaline, 1-septate, 12–14 ×

2–3.5, occasionally with a short appendage at each end. On dead twigs, June–Feb.

Karstenula rhodostoma (Alb. & Schw.)Speg. (Fig. 597)
Pseudothecia 0.3–0.4mm diam., immersed and surrounded by dark brown hyphae, short necks just showing on surface of slightly raised bark. Ascosphores mostly 20–25 × 7–9, dark reddish brown with 3 transverse septa and sometimes a longitudinal septum. On dead twigs, Jan.–Apr.

Microsphaera divaricata (Wallr.)Lév.
Powdery mildew mainly on leaves of young shoots, amphigenous; conidia abundant. Cleistothecia with 4 to 10 appendages which are 3 or 4 times dichotomously branched with recurved tips. Asci 3 to 7, each with 3 to 6 spores.

Pezicula frangulae (Fr.)Fuckel (Fig. 598)
Apothecia solitary or in small groups, up to 0.7mm diam., at first somewhat reddish but drying almost black. Ascospores 18–22 × 7–9, at first hyaline without septa, eventually brownish and 3-septate. On dead branches, Sept.–Nov.

Valsa auerswaldii Nitschke
Perithecia in groups of 2 to 8 per stroma, with necks erumpent through small whitish discs. Ascospores hyaline, curved, mostly 12–14 × 2–3. On dead branches.

FRAXINUS: Ash

ON LEAVES

Ascochyta metulispora Berk. & Br.
Pycnidia epiphyllous, immersed, pale brown, 0.15mm diam. Conidia hyaline, becoming 1-septate, 8–11 × 2.5–3. Causes brown spots up to 1cm diam.

Crocicreas dolosellum (P.Karsten)S. Carp. (Fig. 599)
Apothecia stalked, about 1mm diam., white or creamy, with short teeth around edges of discs. Ascospores hyaline 12–14 × 2–2.5. Common on dead petioles, Oct.–Nov. *C. coronatum* with much larger teeth, described on 'herbaceous stems', is found also quite commonly on *Fraxinus* petioles.

Hymenoscyphus albidus (Rob. ex Desm.)Phill. (Fig. 600)
Apothecia up to 2mm diam., cream drying pale brown, arising from blackened areas on petioles, with stalks the bases of which have short black sheaths of host tissue. Ascospores hyaline, 14–18 × 3–5. Common on fallen leaves, July–Oct.

Venturia fraxini Aderh. (Fig. 601)
Pseudothecia immersed, up to 0.1mm diam., setose; setae brown, up to 70

long. Ascospores 1-septate, pale brown, 11–14 × 3–4.5. On fallen dead leaves. The *Spilocaea* state with olivaceous brown, 1-septate conidia, mostly 15–20 × 4–6, causes rather indefinite pale brown spots on living and withering leaves in late summer and autumn.

ON FRUIT

Phoma samararum Desm.
Pycnidia immersed, black, up to 0.15mm diam. Conidia hyaline, biguttulate, 6–7 × 2.5. Common on wings of fallen samaras, Jan.–Apr.

Phomopsis pterophila (Nits.)Died.
Conidiomata erumpent, black, up to 0.3mm diam., each on a whitish spot surrounded by a pale brown halo. A-conidia biguttulate, 7–8 × 2–3, B-conidia curved or hooked, 20–25 × 1. On thick part of samaras containing the seed, Oct.–May, common.

ON WOOD AND BARK

DISCOMYCETES

Dactylospora stygia (Berk. & Curtis)Hafellner (Fig. 602)
Apothecia superficial scattered, sessile, black, up to 0.6mm diam. Tips of paraphyses encrusted with yellowish brown pigment, adhering to form epithecia. Ascospores 1-septate, olivaceous brown, smooth but sometimes appearing striate, 13–18 × 3.5–5. On decorticated wood, Apr.–May.

Dermea tulasnei Groves
Apothecia erumpent, solitary or in small clusters, dark brown with black discs. Ascospores hyaline, 0- to 3-septate, 15–22 × 6–8. On dead branches, Aug.

Encoelia fascicularis, described on *Populus*, has been found also on branches of *Fraxinus*.

Laquearia sphaeralis (Fr.)Fr. (Fig. 603)
Apothecia up to 0.4mm diam., at first immersed, eventually breaking through the bark to expose a brown disc with narrow white pruinose margin. Ascospores hyaline, 6–9 × 1.5–2. On dead branches, uncommon.

Triblidium octosporum D.Hawksw. & Coppins (Fig. 604)
Apothecia 0.5–1mm diam., black, thick-rimmed, superficial but each one narrowed below to a short foot immersed in bark. Ascospores muriform, hyaline, 16–23 × 8–10. On bark, Apr.

OTHER ASCOMYCETES

KEY

	Ascomata red .... *Nectria galligena*
	Ascomata brown or black 1
1.	Asci polysporous 2
	Asci 8-spored 3
2.	Ascospores 7–9 × 1 .... *Cryptosphaerella annexa*
	Ascospores 9–11 × 2–3 .... *Cryptovalsa protracta*
3.	Stromata large, zoned in section .... *Daldinia concentrica*
	No zonation 4
4.	Ascospores curved, allantoid or reniform 5
	Ascospores not curved, not allantoid or reniform 8
5.	Ascospores without septa 6
	Ascospores septate 7
6.	Perithecia superficial, becoming cupulate .... *Nitschkia confertula*
	Perithecia immersed, scattered .... *Cryptosphaeria eunomia*
	Perithecia immersed, clustered .... *Valsa pruinosa*
7.	Ascospores hyaline .... *Zignoella rhytidodes*
	Ascospores golden brown .... *Cryptosphaerina fraxini*
8.	Ascospores hyaline, without septa .... *Botryosphaeria stevensii*
	Ascospores brown, without septa 9
	Ascospores muriform 10
9.	Perithecia immersed in host .... *Anthostoma melanotes*
	Perithecia in superficial stromata .... *Hypoxylon*
10.	Ascospores hyaline .... *Myriangium duriaei*
	Ascospores brown 11
11.	Ascospores 20–30 × 8–12 .... *Teichospora obducens*
	Ascospores 40–48 × 16–20 .... *Hysterographium fraxini*

Anthostoma melanotes, described on *Ulmus*, has been found also on *Fraxinus* branches.

Botryosphaeria stevensii Shoemaker
Stromata erumpent, black, up to 1cm diam., containing numerous pseudothecia. Ascospores hyaline, mostly 30–36 × 13–14 and without septa. Conidiomata of *Diplodia* state 0.2–0.5mm diam. Conidia long remaining hyaline with rather thick glassy walls, eventually some becoming brown and 1-septate, 25–30 × 10–14. On dead branches, Mar.–Oct.

Cryptosphaerella annexa, described on *Salix*, occurs also on *Fraxinus* branches.

Cryptosphaeria eunomia (Fr.)Fuckel (Fig. 605)
Perithecia 0.4–0.5mm diam., immersed, with a stroma seen in sections as a thick black line beneath them; necks short. The surface of the bark often

appears slightly pimply over very large areas, each pimple with a black dot in its centre. Ascospores curved allantoid, straw-coloured, 11–16 × 2–2.5. On dead attached and fallen twigs and branches, Feb.–May, very common.

Cryptosphaerina fraxini Lambotte & Fautrey ex Sacc. & Syd.
This species superficially resembles *Cryptosphaeria eunomia* but has 3- to 7-septate, pale golden brown, allantoid ascospores 10–30 × 3.5–6. It is much less common.

Cryptovalsa protracta (Pers.)Ces. & de Not. ex Fuckel (Fig. 606)
Stromata broadly effused causing extensive blackening of host tissues. Asci long-stalked, polysporous, spores straw-coloured in mass, curved, 9–11 × 2–3. On dead branches, July, uncommon.

Daldinia concentrica (Bolt.)Ces. & de Not. (Fig. 607)
Stromata called cramp balls or King Arthur's Cakes, superficial, hemispherical, dark brown to black, up to 4cm diam., showing characteristic pale and dark concentric zones when cut vertically through the middle; fertile perithecia just below surface. Ascospores black, 12–18 × 6–10. A primary colonist, commonly seen on quite thick attached as well as fallen branches.

Hypoxylon

KEY

	Ostioles papillate, ascospore walls striate .... *chestersii*
	Ostioles umbilicate, ascospore walls not striate 1
1.	Stroma effuse, ascospores 9–13 × 4–7 .... *rubiginosum*
	Stroma discrete, round or hemispherical 2
2.	Ascospores mostly 20–22 × 9–10 .... *fraxinophilum*
	Ascospores 6–9 × 3–4 .... *howeianum*

Hypoxylon chestersii Rogers & Whalley
Stromata effuse, 1–2mm thick, blackish brown to black, with papillate ostioles. Ascospores elliptical to navicular with one end flattened, brown, walls distinctly striate, mostly 14–17 × 6–7, rarely up to 23 × 8. On decorticated wood, Dec., uncommon.

Hypoxylon fraxinophilum Pouzar (Fig. 608)
Stromata hemispherical, up to 5mm diam., clay-coloured, with umbilicate ostioles. Ascospores blackish brown, mostly 20–22 × 9–10. On dead branches.

Hypoxylon howeianum, described on *Corylus*, has been found also on *Fraxinus*.

Hypoxylon rubiginosum (Pers.)Fr. (Fig. 609)
Stromata widely effused, usually forming a thin reddish purple or purplish brown crust, sometimes surrounded by a black line and, when young, by a

pale greenish yellow zone; ostioles umbilicate. Ascospores brown, 9–13 × 4–7. Very common especially on decorticated wood.

Hysterographium fraxini (Pers.)de Not. (Fig. 610)
Hysterothecia mostly superficial, black, up to 2 × 0.5mm. Ascospores muriform, golden brown, sometimes with hyaline sheaths, usually 40–48 × 16–20. On bark of twigs and branches. Uncommon now except in parts of Wales.

Myriangium duriaei Mont. & Berk. (Fig. 611)
Stromata pulvinate, black, up to 5mm diam. Ascospores muriform, hyaline, when mature mostly 25–35 × 10–15. On bark associated with scale insects, uncommon.

Nectria galligena, described on *Malus*, also forms cankers on *Fraxinus* branches.

Nitschkia confertula (Schw.)Nannf. (Fig. 612)
Perithecia 0.3–0.5mm diam., black, collapsing to become cupulate, in groups seated on a blackish-brown subiculum. Ascospores reniform, pale brown or greyish brown, biguttulate, 7–11 × 3.5–5. On very rotten wood, usually with effete *Hypoxylon rubiginosum*.

Teichospora obducens (Schum.)Fuckel (Fig. 613)
Pseudothecia 0.2–0.4mm diam., black, smooth or rough, thick-walled, superficial or with base immersed. Ascospores muriform, rather pale golden brown, mostly 20–30 × 8–12. On blackened areas of decorticated wood.

Valsa pruinosa, described on 'wood and bark', is quite commonly found on *Fraxinus*.

Zignoella rhytidodes (Berk. & Br.)Sacc.
Perithecia superficial, black, rugulose, with papillate ostioles, seated on a black subiculum. Ascospores curved, hyaline, 3- to 6-septate, constricted at septa, 25–30 × 4. On decorticated wood of dead branches.

HYPHOMYCETES

Brachydesmiella biseptata Arnaud ex Hughes (Fig. 614)
Colonies effuse, black, shining. Conidiophores very pale brown, up to 70 long. Conidia 42–48 × 19–22, 2-septate, central cell large and smooth, brown to almost black, end cells small, very pale and usually verrucose or echinulate. On dead wood, uncommon.

Pseudospiropes hughesii, described on *Fagus*, has been found also on *Fraxinus*.

Xylohypha nigrescens, described on 'wood and bark', is very common on dead wood of *Fraxinus*.

COELOMYCETES

Camarosporium orni Henn. (Fig. 615)
Pycnidia immersed, dark brown to black, thick-walled, 0.3–0.5mm diam. Conidia chestnut brown, with 3 transverse septa and usually 1 longitudinal septum, 12–18 × 6–8. On branches.

Diplodia, v.s. under *Botryosphaeria.*

Dothiorella fraxinea Sacc. & Roum.
Conidiomata clustered, erumpent, black, up to 0.5mm diam. Conidia oblong–ellipsoid to subclavate, often tapering at ends, hyaline, 8–10 × 2–2.5. On bark, Feb.–Mar.

Gloeosporidiella turgida (Berk. & Br.)Sutton (Fig. 616)
Acervuli discoid, dark brown to black, up to 0.5mm diam. Conidia falcate, hyaline, 20–30 × 4–5. On dead twigs.

Phomopsis controversa (Sacc.)Traverso
Conidiomata gregarious, often confluent, becoming erumpent, 0.2–0.3mm diam. Conidia fusiform, hyaline, biguttulate, 7–8 × 2–3. On dead twigs, Mar.–May.

Phomopsis scobina Höhnel
Conidiomata often clustered near nodes, black, up to 0.5mm diam. A-conidia fusiform, hyaline, 7–10 × 2–2.5, B-conidia curved or uncinate, 20–25 × 1. On dead twigs and petioles, Mar.–Sept., common.

GENISTA: Dyer's Greenweed and Petty Whin

Uromyces pisi-sativi, described on *Pisum*, has been recorded occasionally on *G. anglica* and *G. tinctoria*.

Microthyrium cytisi Fuckel var. *cytisi* (Fig. 617)
Mycelium superficial, abundant. Thyriothecia 0.15–0.2mm diam., with dark conical collar around ostiole. Ascospores fusiform, hyaline, 1-septate, 11–13 × 2–3. On dead stems of *G. tinctoria*, Aug.–Oct.

HEDERA: Ivy

ON LEAVES

DISCOMYCETES

Hypoderma hederae (Martius)de Not. (Fig. 618)
Apothecia black, up to 1mm long. Ascospores hyaline with many guttules,

20–23 × 3.5–4.5. Accompanied by round shining black conidiomata 0.1–0.4mm diam. of the *Leptothyrina* state with conidia 2–2.5 × 1. On bleached areas on dead leaves and petioles, Oct.–Apr.

Lophodermium hedericola Ahmad (Fig. 619)
Apothecia subcuticular becoming erumpent, up to 0.7mm long, black. Ascospores hyaline, 55–90 × 1.5–2. On bleached areas on dead leaves, Sept.–Apr.

Trochila craterium (DC.)Fr. (Fig. 620)
Apothecia hypophyllous, about 0.4mm diam., immersed, the dark umber or olivaceous discs exposed by a splitting of the thin epidermis into teeth. Ascospores hyaline or pale olive, 7–9 × 4–5. On fallen dead leaves, Dec.–Sept., fairly common.

OTHER ASCOMYCETES

Aulographum hederae Lib. (Fig. 621)
Hysterothecia up to 0.7 × 0.15mm, black, with cells of the scutellum irregularly arranged and each with a hypostroma consisting of an extensive plate of subcuticular hyphae. Ascospores hyaline, 1-septate, 10–14 × 2.5–3.5. On fallen dead leaves, Jan.–July.

Calonectria hederae Arnaud ex Booth & Murray (Fig. 622)
Perithecia about 0.3mm diam., orange to red, rough-walled, seated on a small stroma. Ascospores hyaline, 3-septate, mostly 50–65 × 6–8 but extruded ones up to 80 long. Conidia of the accompanying *Cylindrocladium* state usually 3-septate and 60–90 × 5–7. On brown patches on fallen leaves, Oct.–Nov.

Guignardia philoprina, with a *Phyllosticta* conidial state described on *Ilex*, occurs also on fallen leaves of *Hedera*.

Microthyrium ciliatum Gremmen & de Kam var. *hederae* J.P.Ellis (Fig. 623)
Superficial mycelium abundant. Thyriothecia about 0.1mm diam., brown with no darkening around ostiole. Ascospores hyaline, 1-septate, 8–10 × 3–3.5, with 2 tufts of 3 or 4 cilia attached to the longer upper cell. On dead and decaying leaves and on trailing stems, Sept.–May.

Mycosphaerella hedericola Lindau
Pseudothecia minute, black, mostly epiphyllous. Ascospores narrowly elliptical, hyaline, 1-septate, about 10 long. On round or irregular whitish spots with broad, dark brown margins on living leaves.

COELOMYCETES

Ceuthospora hederae Grove (Fig. 624)
Conidiomata about 1mm diam., black, discoid, leathery, visible on both leaf

surfaces. Conidia hyaline, 12–14 × 2–2.5. On dead leaves, Sept.–Oct.

Colletotrichum helicis (Desm.)Morgan-Jones
Conidia truncate at base, wider (6–7) than those of the much commoner *C. trichellum*.

Colletotrichum trichellum (Fr.)Duke (Fig. 625)
Acervuli setose, about 0.1mm diam. Setae brown, up to 150 × 3–7. Conidia falcate, hyaline, 15–25 × 4–6. On living and dead leaves causing pale brown, dark-bordered spots, Oct.–Feb., very common everywhere.

Cryptocline paradoxa (de Not.)von Arx (Fig. 626)
Acervuli amphigenous, in clusters, discoid, 0.15–0.2mm diam., amber or orange-brown, opening irregularly. Conidia hyaline or very pale brown, 7–9 × 4–5. On leaves, Apr.–Aug.

Leptothyrina, v.s. under *Hypoderma hederae*.

Phoma hedericola (Dur. & Mont.)Boerema
Pycnidia epiphyllous, up to 0.2mm diam., brown, darker around the ostiole. Conidia hyaline, 4–7 × 2–3. On living leaves, causing whitish or pale spots with wide brown borders. Common at all times of the year.

Phomopsis, v.i. under *Diaporthe pulla*.

Phyllosticta, v.s. under *Guignardia philoprina*.

Septoria hederae Desm.
Pycnidia epiphyllous, 0.1mm diam., black. Conidia hyaline, 30–40 × 1–2. On living leaves causing whitish, brown-bordered spots up to 1cm diam., Apr.–Nov., common.

ON WOOD AND BARK

PYRENOMYCETES

Diaporthe hederae Wehmeyer (Fig. 627)
Stromata widely effused, blackening decorticated wood. Perithecia 0.4–0.7mm diam., immersed, with protuberant black necks. Ascospores hyaline, constricted at septum, 10–12.5 × 3.5–5. On dead twigs, Sept.–July.

Diaporthe pulla Nitschke (Fig. 628)
Perithecia immersed, about 0.5mm diam., with long slender black necks protruding through bark and through blackened areas of decorticated wood. Ascospores hyaline, 1-septate, 9–13 × 2–2.5. *Phomopsis* conidiomata with hyaline, biguttulate conidia, 7–8 × 2–3, precede and sometimes accompany the perithecia and are found also on leaves; Dec.–Aug., common.

Nectria hederae Booth (Fig. 629)
Perithecia 0.2–0.3mm diam., red, with conidiophores on the surface which

produce long chains of hyaline fusiform conidia 6–9 × 2–3. Ascospores 1-septate, hyaline and smooth or pale brown and verruculose, 12–16 × 4.5–6. On cut ends of stems and on bark, Dec.–Jan., uncommon.

Nectria sinopica (Fr.)Fr. (Fig. 630)
Perithecia 0.25–0.3mm diam., pale red when young, darkening and showing pomiform collapse when old, often covered with yellow granules, seated on erumpent yellow to red stromata up to 0.7mm diam. Ascospores hyaline, 1-septate, smooth, 9–13 × 4–6. *Zythiostroma* state has red multilocular conidiomata 0.4–0.5mm diam., with hyaline conidia 2–3 × 1; these precede and quite often also accompany the perithecia. On dead wood and bark of recently cut stems, Aug.–Mar., common.

Rosellinia mammiformis (Pers.)Ces. & de Not. (Fig. 631)
Perithecia superficial, scattered, about 1mm diam., black, papillate, with no subiculum. Ascospores broadly fusiform or limoniform without appendages, dark reddish brown, 25–30 × 8–10. On dead fallen branches, May–Sept., common.

HYPHOMYCETES

Oncopodiella trigonella (Sacc.)Rifai (Fig. 632)
Colonies blackish brown but inconspicuous. Conidiophores up to 30 × 1–3, pale brown. Conidia muriform, 14–19 × 12–16, reddish brown, with 2 to 4 conical, hyaline projecting horns 2–3 long. On rotten wood.

Sporidesmium socium M.B.Ellis (Fig. 633)
Colonies effuse, grey or black. Conidiophores brown, up to 300 × 5–7. Conidia 6- to 8-septate, 30–40 × 9–13, rather dark brown except for the hyaline or pale terminal cells. Grows with and sometimes on *Helminthosporium velutinum* on dead wood.

COELOMYCETES

Cheirospora botryospora (Mont.)Berk. (Fig. 634)
Acervuli topped by dark olivaceous brown to black, pulvinate, tremelloid masses of conidia about 1mm diam. Conidiophores hyaline, up to 200 × 2–4. Conidia complex, brown, up to 35 across, each enclosed in a gelatinous sheath and composed of a number of spherical cells 5 in diameter. On dead twigs, Mar.–Sept.

Melanconium hederae Preuss
Acervuli up to 1mm diam., black. Conidia oval or obovoid, becoming brown or dark olivaceous, with 1 guttule, 6–8 × 3–5. On dead twigs all the year round.

Phomopsis, v.s. under *Diaporthe pulla.*

Zythiostroma, v.s. under *Nectria sinopica.*

HIPPOPHAË: Sea Buckthorn

Lepteutypa hippophaes (Sollm.)v. Arx (Fig. 635)
Pseudothecia immersed with epidermis raised to form small bumps. Ascospores 3-septate, olivaceous brown, 15–26 × 7–9. On dead branches, Aug.–Sept.

ILEX: Holly

ON LEAVES

DISCOMYCETES

Lophodermium neesii Duby
Apothecia hypophyllous, on pale spots, black, 1 × 0.5mm. Asci 120–130 × 8–9; spores parallel, hyaline, the length of the ascus and 1 broad. Paraphyses thread-like, slightly swollen at apex. On fallen dead leaves, June–Aug.

Microscypha enrhiza Graddon (Fig. 636)
Apothecia short-stalked, up to 0.5mm diam., pale yellow, hairy, solitary or in clusters, erumpent from hyaline pseudostromata. Hairs hyaline, up to 60 × 2. Ascospores 6–9 × 2. On fallen dead leaves, usually deeply buried in litter under hedges, Dec.–Jan.

Phacidiostroma multivalve (DC.)Höhnel (Fig. 637)
Apothecia 1–2mm diam., each seated in a round black stroma the thickness of the leaf, opening by teeth to expose pale greyish discs. Ascospores hyaline, 9–11 × 3–4. The *Ceuthospora* conidial state has black, shining, leathery, multilocular conidiomata and cylindrical hyaline conidia, mostly 12–15 × 2–3. The conidial state is very common on fallen dead leaves, Jan.–Apr.; the apothecia, which develop Apr.–May, are seldom encountered.

Trochila ilicina (Nees)Greenhalgh & Morgan-Jones (Fig. 638)
Apothecia up to 1mm diam., immersed, each opening by a little pale lid to expose an olive-green disc. Ascospores pale olivaceous, 10–12 × 4–5. Common on the upper surface of fallen dead leaves, usually in good condition Dec.–Feb., but found also at other times of the year.

OTHER ASCOMYCETES

Calonectria erubescens (Rob. ex Desm.)Sacc. (Fig. 639)
Perithecia superficial, up to 0.3mm diam., reddish orange, tending to collapse.

Ascospores hyaline, 3-septate, 17–27 × 3–4. On the underside of fallen dead leaves, Oct.–Apr.

Guignardia philoprina (Berk. & Curtis)van der Aa (Fig. 640)
Pseudothecia epiphyllous, immersed, 0.2–0.3mm diam., black, thick-walled. Ascospores hyaline, 15–22 × 5–8. On dead leaves, Dec. *Phyllosticta* state has unilocular pycnidia 0.1–0.25mm diam., sometimes in a stroma, often arranged in concentric rings on leaf spots or dead leaves. Conidia variable in shape, often subglobose or pyriform, hyaline, frequently with a large central guttule, mostly 10–15 × 6–10, with gelatinous envelopes; some have an apical appendage up to 30 long. There is also a *Dothiorella* state with dumb-bell-shaped conidia 6–15 × 1.5–3.

Microthyrium ciliatum Gremmen & de Kam (Fig. 641)
Thyriothecia amphigenous, dark brown to black, up to 0.18mm diam., with fringed margins. Cells of scutellum 3–5 × 4–6, not darker around ostiole. Ascospores hyaline, 1-septate, 12.5–14.5 × 2.5–3.5, the upper cell with 2 tufts of 3 to 8 cilia. On dead and skeletal leaves, Oct.–May, common.

Niesslia ilicifolia (Cooke)Winter
Perithecia superficial, mostly epiphyllous, about 0.1mm diam., dark brown to black, setose; setae rigid, brown, 50–60 × 5. Ascospores hyaline, 1-septate, 6–8 × 1.5–2. On dead leaves, Mar.–June.

HYPHOMYCETES

Chaetochalara bulbosa Sutton & Pirozynski (Fig. 642)
Colonies effuse, dark brown to black, hairy. Setae subulate, dark brown, 60–90 × 2.5–3.5, bulbous base 7–8 wide. Conidiophores brown, 20–30 long, basal part 7–8 wide, cylindrical neck 3 thick. Conidia in chains, hyaline, 6–12 × 1.5–2. On rotting fallen leaves, Jan.–Feb.

Codinaea setosa Hughes & Kendrick (Fig. 643)
Colonies effuse, brown to blackish brown, hairy. Setae dark brown, up to 300 × 3.5–4, often swollen to 5 at apex. Conidiophores up to 100 × 3–4. Conidia hyaline, mostly 1-septate, 20–30 × 2–3.5. On fallen dead leaves.

Oidiodendron flavum Szilvinyi (Fig. 644)
Colonies effuse, bright yellow. Conidiophores pale yellow, up to 100 × 1.5–2.5. Conidia yellowish, finely verruculose or echinulate, 3.5–6 × 2.5–3.5. On fallen rotting leaves in litter, Dec.–Jan.

COELOMYCETES

Ceuthospora, v.s. under *Phacidiostroma multivalve*.

Coleophoma cylindrospora (Desm.)Höhnel (Fig. 645)
Pycnidia immersed, dark brown, flattened, 0.1–0.5mm wide, unilocular, thick-walled. Conidia hyaline with several large guttules, 16–30 × 2.5–3. On dead leaves, Apr.–May.

Coniothyrium ilicis A.L.Smith & Ramsb.
Pycnidia epiphyllous, immersed, 0.15–0.2mm diam., blackish brown. Conidia pale brown, 3–5 × 2–3. On fading leaves, causing whitish spots, Mar.–July.

Diplodia ilicicola Desm. (Fig. 646)
Pycnidia immersed but raising the epidermis, black, shining. Conidia becoming brown and 1-septate, 22–25 × 9–10. On dead fallen leaves and also sometimes on twigs, Oct.–Apr.

Dothiorella, v.s. under *Guignardia philoprina*.

Phomopsis crustosa Trav. (Fig. 647)
Conidiomata immersed, black, up to 0.5mm diam., on dark brown spots each surrounded by a narrow black line. A-conidia hyaline, 7–9 × 2.5–3. B-conidia formed occasionally, about 20 × 1. On dead leaves and twigs, Dec.–Feb.

Phyllosticta, v.s. under *Guignardia philoprina*.

Pyrenochaeta ilicis M.Wilson (Fig. 648)
Pycnidia erumpent, up to 0.8mm diam., brown to black, setose. Setae around ostiole, tapered, septate, dark brown, up to 250 × 10. Conidia hyaline, 7–9 × 1–1.5. On dead leaves, Oct.–May, common.

ON WOOD AND BARK

PYRENOMYCETES

Nectria aquifolii (Fr.)Berk. (Fig. 649)
Perithecia in clusters on erumpent yellow or red stromata, globose, about 0.3mm diam., red or purplish, often covered with greenish-yellow granules, showing pomiform collapse on drying. Ascospores hyaline, 1-septate, 9–12 × 3–5, forming allantoid ascoconidia, 3–4.5 × 1, which eventually fill the asci. On dead attached and fallen branches, Jan.–Apr., fairly common but inconspicuous.

Nectria peristomialis (Berk. & Br.)Samuels
Perithecia up to 0.4mm high and 0.2mm wide, pale yellow with apical disc surrounded by white triangular teeth. Ascospores fusiform, hyaline, 3- to 5-septate, 20–27 × 4–5. On dead bark, rare.

Nectria punicea (Schm.)Rabenh. var. *ilicis* Booth (Fig. 650)
Perithecia crowded on erumpent stromata, red or dark reddish brown, 0.25mm diam. Ascospores 1-septate, mostly 15–20 × 7–10, becoming pale

brown and minutely verruculose when old. On trunks and larger branches of standing, often dying trees, Oct.–May.

Vialaea insculpta (Fr.)Sacc. (Fig. 651)
Perithecia immersed, up to 0.4mm diam., olive green when fresh, becoming black, with slightly protuberant necks, surrounded by pale haloes. Ascospores dumb-bell-shaped, hyaline, 1-septate, 70–80 long, 7–8 wide near ends, 1–2 wide in the middle. Common on long, drooping, attached twigs, Sept.–Apr.

HYPHOMYCETES

Ceratosporium fuscescens Schw. (Fig. 652)
Colonies effuse, dark brown to black. Conidia branched, dark brown, each composed of a small pyriform basal cell and 2 or 3 divergent, tapered, septate arms up to 230 long and 14–22 thick. On dead wood and bark, Feb.–June.

Corynespora smithii (Berk. & Br.)M.B.Ellis (Fig. 653)
Colonies dark brown or black, velvety or spongy, round or oval on bark, effuse on wood. Conidiophores brown, up to 500×6–12. Conidia solitary or in short chains formed through a broad pore at the apex of the conidiophore, subhyaline to golden brown, 70–400×12–19, with 7 to 45 pseudosepta. Sept.–June, common.

COELOMYCETES

Coniothyrium slaptoniense D.Hawksw. & Punithalingam
Pycnidia immersed, solitary, 0.3–0.5mm diam., dark brown to black, thick-walled. Conidia irregularly subglobose, pale brown or raw sienna, verruculose, 5–8×3–4.5. On dead twigs, Aug.

Cytospora aquifolii Fr.
Conidiomata conical, black, up to 0.5mm diam., with dark discs, plurilocular. Conidia hyaline, allantoid, 4–6×1. On dead stems, Mar.–Apr. The plurivorous *Cytospora* state of *Valsa ambiens* with conidia 5–7×1 occurs also.

JUGLANS: Walnut

Gnomonia leptostyla (Fr.)Ces. & de Not. (Fig. 654)
Perithecia immersed in fallen dead leaves, black, with slender necks protruding. Ascospores hyaline, 1-septate, 17–22×3–4.5. Mar.–Apr. Acervuli of the *Marssonina* state are found in Sept.–Oct. causing large greyish-brown spots on the lower surface of attached fading leaves; conidia curved, hyaline, 1-septate, 15–25×3–4.5.

Microstroma juglandis (Bereng.)Sacc.
Acervuli forming large white patches on lower surface of living leaves in July, pulvinate, erumpent. Conidiophores clavate, hyaline, 12–20 × 5–7. Conidia hyaline, 5.5–7.5 × 3.5–5.

JUNIPERUS: Juniper

UREDINALES

Four species of *Gymnosporangium* form 1-septate teliospores on *Juniperus*, 2 on *J. communis* and 2 on *J. sabinae*.

KEY

	On *J. communis* 1
	On *J. sabinae* 2
1.	Teliospores more than 50 long .... *clavariiforme*
	Teliospores less than 50 long .... *cornutum*
2.	Teliospores with upper cell broadly rounded .... *confusum*
	Teliospores with upper cell conical .... *sabinae*

Gymnosporangium clavariiforme (Pers.)DC., 111
Telia on branches, from perennial mycelium, small, brown and inconspicuous when dry but swollen, bright orange and very conspicuous when wet. Spores either brown and 50–60 × 15–20 or pale yellow, 100–120 × 10–12. Common, Apr.–May. Aecial hosts *Crataegus* spp.

Gymnosporangium confusum Plowr., 111
Telia on branches, dark brown. Some spores rounded at apex with thick dark brown walls and orange contents, 30–50 × 20–25, others paler, thin-walled, rather narrower and longer. Uncommon, Apr.–May. Aecial hosts *Crataegus*, *Cydonia*, *Mespilus* and *Pyrus*.

Gymnosporangium cornutum Kern, 111
Telia on branches and occasionally leaves, brown when dry, orange when wet. Spores dark cinnamon brown, 30–48 × 18–28. Apr.–June. Aecial host *Sorbus aucuparia*.

Gymnosporangium sabinae (Dicks.)Wint., 111
Telia on branches becoming yellowish brown, gelatinous. Spores pale brown, 40–48 × 20–30, the upper cell bluntly conical. Apr.–May, uncommon. Aecial host *Pyrus communis*.

DISCOMYCETES

Chloroscypha sabinae (Fuckel)Dennis (Fig. 655)
Apothecia epiphyllous, erumpent, solitary or several together, short-stalked, up to 0.5mm diam., dark olive green to black, gelatinous when moist, drying horny. Ascospores hyaline, 17–21 × 7–8. On living and dead leaves of *J. communis* and *J. sabinae*, Sept.–Mar.

Coccomyces juniperi P. Karsten (Fig. 656)
Apothecia erumpent, black, 1–3mm long, splitting open irregularly to expose the pale yellow discs. Ascospores 40–50 × 1–1.5. On dead twigs and branches of *J. communis*, June–Aug.

Didymascella tetraspora (Phill. & Keith)Maire (Fig. 657)
Apothecia epiphyllous, 1mm diam., immersed, opening by teeth to expose blackish-brown discs. Asci 4-spored; spores olivaceous brown with septum near lower end, 21–24 × 13–16. On living leaves of *J. communis*, uncommon.

Lophodermium juniperinum (Fr.)de Not. (Fig. 658)
Apothecia black, hysterioid, about 1mm long. Ascospores hyaline, mostly 60–90 × 2. Common on dead leaves of *J. communis* and *J. nana*, Aug.–Oct.

Pithya cupressina (Pers.)Fuckel (Fig. 659)
Apothecia short-stalked, up to 1mm diam., with orange discs, excipulum pale, base downy. Ascospores spherical or subspherical, hyaline, 9–10 diam. On dead leaves, uncommon, June–Aug.

Rutstroemia juniperi K.Holm & Holm (Fig. 660)
Apothecia up to 2mm diam., dark reddish brown. Ascospores hyaline, becoming 3-septate, 12–18 × 5–5.5. On recently dead leaves and small twigs of *J. communis*, mostly seated on old stromata of other fungi, May–Sept.

Velutarina juniperi (Dennis)K.Holm & Holm
This does not seem to be morphologically distinct from the plurivorous *V. rufo-olivacea* described on 'wood and bark'.

OTHER ASCOMYCETES

Actidium nitidum (Ellis)Zogg (Fig. 661)
Hysterothecia black, 0.2–0.5mm long, superficial. Ascospores 1-septate, pale olivaceous brown, 11–17 × 2–3. On dead leaves and bark, Oct.–Feb.

Herpotrichia juniperi (Duby)Petrak (Fig. 662)
Pseudothecia erumpent, 0.2–0.5mm diam., blackish brown, with a weft of dark brown hyphae. Ascospores at first hyaline and 1-septate, finally brown and 3-septate, 25–33 × 7–11. Parasitic on leaves and branches, the so-called black snow mould. Pseudothecia formed on leaves.

Kriegeriella minuta (Barr)v.Arx & E.Müller (Fig. 663)
Pseudothecia superficial, blackish brown to black, 0.1mm diam. Ascospores hyaline to brown, 3-septate, 20–25 × 8–9. On dead leaves of *J. communis*, in litter, Aug.–Mar.

Metacapnodium juniperi (Phill. & Plowr.)Speg. (Fig. 664)
Colonies spongy, pulvinate, sometimes coalescing and covering large areas, at first rusty brown, becoming almost black. Hyphae moniliform, 15–20 thick tapering to 5–6. Pseudothecia black, 0.2mm diam. Ascospores brown with 1 to 3 septa, 20–34 × 8–14. Conidia of *Capnophialophora* state hyaline, 1 diam. On twigs of *J. communis*, Scotland.

Mytilinidion acicola Winter (Fig. 665)
Hysterothecia scattered, superficial, up to 0.5mm long, black, each with its base expanded to form a black disc. Ascospores 3-septate, pale brown, 19–22 × 6–8.5. On twigs and dead attached leaves.

Mytilinidion decipiens (P.Karsten)Sacc.
Hysterothecia small, shell-shaped. Ascospores pale brown, 3-septate, 13–15 × 3–4. Mostly on dead bark.

Seynesiella juniperi (Desm.)Arnaud (Fig. 666)
Thyriothecia 0.25mm diam., black, conical, mycelium subcuticular. Ascospores 1-septate, hyaline to mid pale brown, 20–22 × 7–8. On living and dead leaves of *J. communis*.

Stomiopeltis juniperina (Grove)K.Holm & Holm (Fig. 667)
Thyriothecia superficial, 0.25mm diam., dark brown. Ascospores hyaline, 1-septate, 12–14 × 3–4. On old decaying leaves.

Sydowia polyspora, described on *Pinus*, occurs also on *Juniperus*. The *Sclerophoma* conidial state causes leaf blight and dieback.

HYPHOMYCETES

Stigmina glomerulosa (Sacc.)Hughes (Fig. 668)
Sporodochia punctiform, up to 0.25mm diam., dark olivaceous brown to black. Conidia brown, verruculose, almost always 7-septate, 27–50 × 6–7. On leaves of *J. communis*.

Troposporella monospora, described on *Pinus*, is found also on dead leaves of *J. communis*.

COELOMYCETES

Phomopsis juniperovora Hahn
Conidiomata dark brown, thick-walled, up to 0.4mm diam., erumpent. A-

conidia hyaline, biguttulate, fusiform, 8–10 × 2–3. B-conidia curved or hooked, 20–30 × 0.5–1. On leaves and twigs, causing blight and dieback.

Seimatosporium foliicola (Berk.)Shoemaker (Fig. 669)
Acervuli epiphyllous, 0.15mm diam., dark brown to black. Conidia 5-septate, 19–23 × 7.5–9, median cells pale brown, end cells hyaline, apical and basal appendages up to 9 long. On dead leaves of *J. communis*.

LABURNUM: Laburnum, Golden Rain

Uromyces pisi-sativi, 11,111, described on *Pisum*, has been recorded occasionally on leaves of *Laburnum*.

Ascochyta kabatiana Trotter
Pycnidia epiphyllous, honey-coloured, 0.1–0.15mm diam. Conidia hyaline, often bent, becoming 1-septate, 7–15 × 3–4. Causing ochraceous, sinuous spots up to 2cm wide with thin brown margins, visible on both surfaces of living leaves, Sept.

Cucurbitaria laburni (Pers.)de Not.
Pseudothecia 0.5–0.7mm diam., black, papillate, usually in large groups seated on a black hyphal subiculum. Ascospores golden brown, 25–35 × 9–15, muriform, with 5 to 7 transverse septa, constricted at the middle one, and 1 or 2 longitudinal septa. Preceded and sometimes accompanied by its *Camarosporium* state with groups of black pycnidia about 1mm diam., and muriform golden-brown conidia 15–25 × 8–9. On dead and dying branches.

Diaporthe rudis (Fr.)Nitschke (Fig. 670)
Perithecia immersed, with necks protruding through cracks in bark, surrounded by a black line. Ascospores 1-septate, hyaline, 14–16 × 3.5–4. The *Phomopsis* state has conidiomata about 0.5mm diam. and conidia 7–9 × 2–2.5. On small dead twigs.

Diplodia rudis Desm. & Kickx
Pycnidia black, up to 0.5mm diam., in groups on a blackish-brown subiculum. Conidia remaining subhyaline for a long time, eventually becoming brown and 1-septate, 26–34 × 10–11. On dead twigs.

Hymenoscyphus infarciens (Ces.)Dennis
Apothecia erumpent, short-stalked, up to 1mm diam., ochraceous, externally paler and downy. Ascospores broadly fusiform, hyaline, some with 4 guttules, 17–25 × 5–8. On decorticated parts of dead twigs.

LARIX: European Larch and Japanese Larch

UREDINALES

Four species of *Melampsora* and one of *Melampsoridium* form aecia on leaves of *Larix*; the spores in all of them are in the range 15–25 × 10–20. *Melampsora capraearum* Thüm. and *M. epitea* Thüm. have as their alternate hosts species of *Salix*; *M. larici-populina* Kleb. and *M. populnea* (Pers.)Karst. form their uredinia and telia on *Populus*; and *Melampsoridium betulinum* (Fr.)Kleb. forms them on *Betula*.

KEY

    Aecia with well-defined peridia, often formed in longitudinal rows on one or both sides of the midrib, spore walls verruculose but with a smooth spot on one side .... *Melampsoridium betulinum*
    Aecia caeomoid (without well-defined peridia), scattered or on spots, spore walls verruculose all over 1
1. Aecia on yellowish spots, spore walls not more than 1.5 thick 2
    Aecia scattered or in groups, spore walls often 2 or more thick 3
2. Aecia orange .... *Melampsora larici-populina*
    Aecia pale yellow .... *Melampsora populnea*
3. Aecia pale orange, spore walls 2 thick .... *Melampsora capraearum*
    Aecia bright orange, spore walls 1.5–5 thick .... *Melampsora epitea*

DISCOMYCETES

Arachnopeziza obtusipila Grelet (Fig. 671)
Apothecia 0.2–0.3mm diam., white, hairy, seated on a thin whitish subiculum. Hairs about 100 × 3–4. Ascospores hyaline, mostly 3-septate, 15–27 × 3–3.5. On dead cones and small twigs, Nov.–Mar.

Ciliolarina laricina, described on *Pinus*, is fairly common also on dead branches of *Larix*.

Hyaloscypha leuconica (Cooke)Nannf. (Fig. 672)
Apothecia up to 0.5mm diam., white drying yellowish, each with a fringe of tapered hairs up to 140 long, 5–7 thick at base. Ascospores 6–7 × 2–2.5. On dead cones and branches all the year round, common.

Hyaloscypha velenovskii Graddon (Fig. 673)
Apothecia about 0.5mm diam., pale peach yellow; hairs up to 50 × 4, with amber-coloured globules between them. Ascospores hyaline, 7–10 × 2–2.5. On wood, July–Nov.

Lachnellula occidentalis (Hahn & Ayers)Dharne (Fig. 674)
Apothecia erumpent, often clustered, up to 3mm diam., with bright orange discs surrounded by thick rims of white hairs. Ascospores hyaline, elliptic oblong, 15–20 × 5–6. On dead twigs and branches, found throughout the year but most commonly Oct.–Mar.

Lachnellula willkommii (Hartig)Dennis (Fig. 675)
Apothecia 1–6mm diam., similar to those of *L. occidentalis* but with discs not always such a bright orange, often becoming buff or straw-coloured. Ascospores tapered to a point at one end or both ends, 18–25 × 7–9. Parasitic on branches, causing canker and dieback.

Sarcotrochila alpina (Fuckel)Höhnel (Fig. 676)
Apothecia erumpent, pushing to one side little lids, up to 1mm diam., yellow or flesh-coloured. Ascospores hyaline, biguttulate, occasionally 1-septate, 10–12 × 3.5–4. On fallen dead leaves, Nov.–Mar.

Tympanis laricina (Fuckel)Sacc.
Apothecia erumpent, black, 0.5–1mm diam. Asci becoming filled with rod-like ascoconidia 2–2.5 × 1. On dry dead branches.

OTHER ASCOMYCETES

Leucostoma curreyi (Nitschke)Défago (Fig. 677)
Stromata erumpent through periderm stellately fissured, each containing 3 to 15 perithecia with short necks protruding through white to pale brownish discs. Ascospores hyaline, 12–16 × 2.5–3 in 8-spored asci, up to 20 × 3 in 4-spored asci. Often accompanied by *Cytospora* state with allantoid conidia, 3–5 × 1. On dead branches about 1cm thick, Jan.–Mar.

Mytilinidion gemmigenum Fuckel (Fig. 678)
Hysterothecia black, about 1mm long. Ascospores mostly 7-septate, 30–40 × 5–7, straw-coloured. On bark, uncommon.

Valsa kunzei (Fr.)Fr.
Stromata erumpent with dark greyish-brown discs, about 0.7mm diam., each containing up to 20 perithecia. Ascospores 5–7 × 1.5. On dead branches about 2cm thick, Dec.–Jan.

HYPHOMYCETES

Meria laricis Vuill. (Fig. 679)
Fructifications on lower surface of leaves, emerging through stomata. Conidia mostly 5–11 × 2–3. Causes discoloration and browning of living needles from the tips downwards, and eventually defoliation. First symptoms about the

beginning of May.

Spadicoides atra (Corda)Hughes (Fig. 680)
Colonies effuse, dark blackish brown to black, velvety. Conidiophores brown, 60–300 × 2.5–4. Conidia formed through many small pores in conidiophore walls, brown, 4–6.5 × 3–4. On dead wood.

Trimmatostroma scutellare (Berk. & Br.)M.B.Ellis (Fig. 681)
Sporodochia pulvinate, black, shining. Conidiophores brown, up to 30 × 1–4. Conidia in chains which fragment readily, brown, many-celled, lobed, 10–30 × 8–25. On dead branches.

Zalerion arboricola Buczaki (Fig. 682)
Conidiophores brown. Conidia when coiled 10–20 diam., filaments mid to dark brown, often verruculose, 4–7 thick. On cankers caused by *Lachnellula willkommii*.

COELOMYCETES

Cytospora abietis Sacc.
Conidiomata 0.5–0.8mm diam., multilocular, with grey or brownish discs. Conidia allantoid, 3–6 × 1, yellowish in mass as they are extruded. On dead twigs, Dec.–Apr.

LAURUS: Sweet Bay or Poets' Laurel

ASCOMYCETES

Diaporthe nobilis Sacc. & Speg.
Perithecia immersed, surrounded by a black line, necks just protuberant. Ascospores hyaline, slightly constricted at septum, 11–14 × 2–3.5. Conidiomata of *Phomopsis* state immersed, up to 0.3mm diam., raising the epidermis. A-conidia hyaline, fusiform, 8–10 × 2.5–3, B-conidia curved or hooked, 16–18 × 1. On dead twigs and fading leaves, Apr.–Sept.

Microthyrium lauri Dennis & Spooner (Fig. 683)
Thyriothecia up to 0.2mm diam., brown, with cells darker around ostiole. Ascospores hyaline, 1-septate, 9.5–12 × 3–4, the upper cell larger than the lower one, biguttulate and bearing medianly 2 tufts of cilia. On fallen dead leaves, June.

Nectriella consolationis (Sacc.)E.Müller (Fig. 684)
Perithecia erumpent, solitary, pale orange. Ascospores hyaline, 1-septate, mostly 20–30 × 7–8. On dead leaves, June.

Fungi Specific to Trees, Shrubs and Woody Climbers

Stegopeziza lauri (Caldesi)Höhnel (Fig. 685)
Apothecia 0.3–0.8mm diam., opening by little lids to expose cream or olivaceous discs. Ascospores 4.5–5.5 × 1–1.5. On the upper surface of dead leaves, May–June.

HYPHOMYCETES

Beltraniella pirozynskii P.M.Kirk (Fig. 686)
Colonies effuse, pale, hairy. Conidiophores setiform, up to 400 × 4–6, brown, arising from radially lobed basal cells and bearing conidiogenous cells in 1 to 4 verticils. Conidia pale olivaceous brown, 20–24 × 6–7. On dead leaves, June.

Circinotrichum britannicum P.M.Kirk (Fig. 687)
Colonies effuse, hairy, blackish brown. Setae dark brown, up to 150 × 3–5. Conidiophores hyaline or very pale brown, up to 20 × 3–4.5. Conidia hyaline, 12–16 × 1.5–2. On dead leaves, June.

Corynesporopsis uniseptata P.M.Kirk (Fig. 688)
Colonies effuse, blackish brown, hairy. Conidiophores up to 100 × 3–4. Conidia in short chains, 1-septate, brown, 12–16 × 5–7. On dead leaves, June.

Cylindrotrichum clavatum W.Gams (Fig. 689)
Colonies effuse, hairy, black. Conidiophores brown, up to 110 × 3.5–4.5. Conidia clumped, hyaline, 1-septate, 9.5–12 × 3.5–4.5. On dead leaves.

Dactylaria obtriangularia Matsushima (Fig. 690)
Colonies effuse, inconspicuous. Conidiophores up to 15 × 4.5, hyaline. Conidia hyaline, 1-septate, 20–35 × 2–3. On dead leaves, Sept.

Endophragmiella lauri P.M. & C.M.Kirk (Fig. 691)
Colonies effuse, pale brown, hairy. Two types of conidiophore and conidia: (1) Simple conidiophores up to 200 × 3–4.5, brown, with numerous percurrent proliferations and ellipsoid or obclavate, 2-septate, pale brown conidia 25–30 × 7–8. (2) Conidiophores shorter, branched at apex, conidia curved, 10–15 × 1. On dead leaves.

Isthmolongispora minima Matsushima (Fig. 692)
Colonies inconspicuous. Conidiophores hyaline, short, with denticles. Conidia hyaline, 14–24 × 2–3. On dead leaves.

Polyscytalum gracilisporum (Matsushima)B. Sutton & Hodges (Fig. 693)
Colonies effuse, hairy, pale buff. Conidiophores pale brown, simple or umbellately branched, 2.5–3.5 wide. Conidia catenate, 1-septate, hyaline or subhyaline, 15–20 × 2. On dead leaves.

158 Fungi Specific to Trees, Shrubs and Woody Climbers

Subulispora minima P.M.Kirk (Fig. 694)
Colonies effuse, inconspicuous. Conidiophores hyaline, 4–15 × 1.5–2. Conidia 1- to 3-septate, hyaline, 16–25 × 1.5–2. On dead leaves.

Wiesneriomyces javanicus Koorders (Fig. 695)
Sporodochia pulvinate, with golden-yellow slimy conidial masses encircled by dark brown setae which are up to 600 × 9–15. Conidia hyaline, 10–12 × 3–4.5, remaining attached to one another quite firmly by short isthmi in chains of up to 15. On fallen dead leaves.

Zygosporium echinosporum Bunting & Mason (Fig. 696)
Colonies effuse, greyish brown, thin. Conidiophores brown, mostly 50–150 × 2–3, with vesicles 7–10 thick. Conidia spherical, hyaline, verruculose, 6–9 diam. On dead leaves.

Zygosporium gibbum (Sacc., Rouss. & Bomm.)Hughes (Fig. 697)
Short-stalked vesicles 10–15 × 7–9. Conidia spherical, hyaline, 4.5–6 diam. On dead leaves.

COELOMYCETES

Camarosporium lauri (Sacc.)Grove (Fig. 698)
Pycnidia erumpent, dark olivaceous brown. Conidia muriform, smoky brown, 15–20 × 7–9. On dead twigs, May–June.

Cytospora lauri Grove
Conidiomata conico-truncate, 0.5–1mm diam., with white furfuraceous discs. Conidia allantoid, hyaline, 3–5 × 1. On dead twigs and leaves.

Diplodia laurina Sacc.
Pycnidia immersed, black, 0.2–0.3mm diam. Conidia 1-septate, dark brown, 18–25 × 7–10. On dead twigs, May–June.

Gloeosporidiella nobilis (Sacc.)Sutton
Acervuli erumpent, up to 0.3mm diam. Conidia falcate, hyaline, 17–20 × 4–4.5. On dead leaves.

Phomopsis, v.s. under *Diaporthe nobilis*.

LIGUSTRUM: Privet

Diplodia ligustri Westend.
Pycnidia immersed, 0.4–0.5mm diam., dark brown. Conidia remaining hyaline a long time, eventually becoming brown and 1-septate, 20–26 × 8–10. On thin dead twigs, Apr.–June.

Mycosphaerella ligustri (Rob. ex Desm.)Lindau
Pseudothecia mostly epiphyllous, numerous, about 0.1mm diam., black,

collapsing. Ascospores hyaline, 1-septate, mostly about 10 × 4. On leaves.

Phomopsis brachyceras Grove
Conidiomata immersed becoming erumpent, 0.3–0.4mm diam. A-conidia hyaline, biguttulate, often slightly curved, 7–9 × 1.5–2. B-conidia curved or flexuous, 20–24 × 1. On dead attached twigs, Apr.–July.

Thedgonia ligustri (Boerema)Sutton (Fig. 699)
Conidiophores and conidia hyaline; conidia cylindrical, truncate at ends, 1- or 2-septate, 25–65 × 4–5.5, in chains. Causes zonate brown spots with grey centres and purplish-brown raised borders on living leaves of *L. ovalifolium*.

Tympanis ligustri Tul. & C. Tul. (Fig. 700)
Apothecia 0.5–0.8mm diam., black, erumpent in small groups from cracks in bark. Asci soon filled with tiny secondary spores about 2.5–3 long. Paraphyses forked, swollen to 6 at their brownish tips, which are glued together. On dead branches, Jan.–May.

Valsa cypri (Tul.)Tul. & C.Tul.
Perithecia in groups of 3 to 8 nestling in the cortex, with short necks converging to a small disc. Ascospores allantoid, hyaline, 11–12 × 2. *Cytospora* state has conidia 5–6 × 1–1.5. On dead branches.

LONICERA: Honeysuckle

UREDINALES

Puccinia festucae Plowr., 0,1
Aecia hypophyllous on yellow or pale brown spots, June–Aug. Alternate hosts *Festuca* spp.

DISCOMYCETES

Dasyscyphus barbatus (Kunze)Massee
Apothecia brown, hairy, up to 2mm diam. Hairs up to 250 × 4–5, reddish brown, smooth, thick-walled except at the paler, sometimes slightly swollen, tip. Ascospores hyaline, some 1-septate, 9–11 × 2. Lanceolate paraphyses 3–4 thick. On dead stems, uncommon.

Unguiculella robergei (Desm.)Dennis (Fig. 701)
Apothecia up to 1mm diam., downy, with dark red discs. Hairs up to 55 long, 3–4 thick at base which is swollen, upper part 1 thick, without a lumen and curved or hooked. Ascospores hyaline, biguttulate, 5–8 × 2.5. On dead stems, uncommon.

OTHER ASCOMYCETES

Amphisphaerella xylostei (Pers.)Munk (Fig. 702)
Perithecia immersed, 0.5–0.75mm diam., each with a short neck protruding through a small brown clypeus. Ascospores dark brown with equatorial pores, mostly 15–20 × 10–14. On dead branches, Aug.–May.

Diaporthe pardalota (Mont.)Nitschke ex Fuckel (Fig. 703)
Perithecia immersed, 0.3–0.5mm diam., with short necks just erumpent through bark. Ascospores hyaline, 1-septate, 10–15 × 2.5–4. On dead attached twigs causing dark brown 'leopard's' spots each surrounded by a black, slightly raised line, Jan.–Aug., very common.

Glyphium elatum (Grev.)Zogg (Fig. 704)
Hysterothecia superficial, hatchet-shaped, 1–2mm high, up to 0.5mm wide, seated on a brown subiculum. Ascospores pale olive, multiseptate, 150–330 × 2–3. We have found this rare fungus only on old bark of *L. periclymenum* in the Channel Islands, Sept., but it has been recorded on other plants, mainly deciduous trees.

Lasiobotrys lonicerae (Fr.)Kunze (Fig. 705)
Brown pseudothecia formed around the edges of small black sclerotia which cover the surface of round, dark mycelial mats 3–4mm diam., and are initially anchored to them by groups of dark brown hyphae. At maturity the sclerotia are shot off. Ascospores hyaline, 1-septate, 8–10 × 4–5. On living leaves, Aug.–Oct.

Melomastia mastoidea, a plurivorous species described on 'wood and bark', is fairly common on *Lonicera*.

Microsphaera lonicerae (DC.)Winter
Powdery mildew. Colonies amphigenous, patchy, thin, with few conidia. Cleistothecia with appendages 4 times dichotomously branched, the tips often coralloid and sometimes recurved; 3 to 7 asci, each with 3 to 5 spores. On living and fading leaves.

Mycosphaerella clymenia (Sacc.)Johanson ex Oudem. (Fig. 706)
Pseudothecia amphigenous, immersed, 60–80 diam., black, surrounded by thick brown hyphae. Ascospores hyaline, 1-septate, 10–16 × 2.5–3. On pale brown spots 4–15mm diam., on living leaves, Aug.–Oct.

COELOMYCETES

Ascochyta vulgaris Kabat & Bubák
Pycnidia brown, 0.2mm diam., darker around ostiole. Conidia hyaline, becoming 1-septate, guttulate, 6–14 × 2.5–4.5. Causes greyish brown to white, purple-bordered spots, up to 1cm diam., on living leaves, common.

Diplodia lonicerae Fuckel
Pycnidia erumpent in elongated clusters, up to 1mm diam., black. Conidia dark brown, 1-septate, 20–30 × 10–12. On thin dead branches, Feb.–June.

Kabatia periclymeni (Desm.)Morelet (Fig. 707)
Conidiomata epiphyllous, flat, black, subcuticular, dehiscing by radial splitting from centre outwards. Conidia hyaline, curved, obliquely tapered at one end, rounded at the other, 20–28 × 7–9. Causes pale olivaceous, brown-bordered spots on living and fading leaves, July–Sept.

Phoma minutula Sacc.
Pycnidia black, about 0.1mm diam. Conidia hyaline, 4 × 1. On small dead branches, Feb.–July.

Phomopsis cryptica Höhnel
Conidiomata gregarious, subepidermal. A-conidia fusiform, hyaline, biguttulate, 7–9 × 2.5–3, B-conidia bent or hamate, 20–30 × 1. On living and dead branches, Jan.–Apr.

LUPINUS: Lupin and Tree Lupin

Diaporthe rudis, described on *Laburnum*, is common also on dead branches of *Lupinus arboreus*.

Erysiphe trifolii, powdery mildew described on *Trifolium*, is seen occasionally on leaves of *L. polyphyllus*.

Gibberella pulicaris (Fr.)Sacc. (Fig. 708)
Perithecia about 0.2mm diam., in clusters, dark chocolate brown, often appearing warted; when mounted in lactic acid a purple or reddish-purple pigment diffuses out from their walls. Ascospores hyaline, 3-septate, 20–30 × 6–10. Often accompanied by its *Fusarium* state which has curved, fusiform, 2- to 5-septate macroconidia 35–55 × 4–5.5, with foot-cells. Parasitic on *L. arboreus*, Jan.–Apr.

Pleiochaeta setosa (Kirchn.)Hughes (Fig. 709)
Colonies effuse, grey, olivaceous brown or black. Conidiophores up to 150 × 8–13, with broad, thin scars. Conidia mostly 5-septate, 60–90 × 14–22, with hyaline end cells and golden brown median cells; hyaline appendages up to 100 × 3–4. Parasitic on leaves and pods of *L. polyphylla*.

Rhabdospora lupini Buchw.
Pycnidia 0.1–0.15mm diam., brown. Conidia falcate, hyaline, 20–24 × 1.5–2. On dead stems, Oct.

LYCIUM: Duke of Argyll's Tea Plant

Microsphaera mougeotii Lév. (Fig. 710)
Powdery mildew. Colonies amphigenous, Cleistothecia 0.1–0.15mm diam., with numerous dichotomously branched appendages; each containing 12–20 asci, often with 2 spores each.

Phomopsis importata Died.
Conidiomata slightly erumpent, blackish. Conidia hyaline, biguttulate, 7–10 × 2.5–3. On dead branches of *L. barbarum* and *L. chinense*, Mar.–May.

MAHONIA: Oregon Grape

UREDINALES

The aecia of *Puccinia graminis* occur on leaves of *Mahonia* as well as on *Berberis* but the rust most commonly found is a *Cumminsiella*.

Cumminsiella mirabilissima (Peck)Nannf., 0,1,11,111 (Fig. 711)
The peridia of the aecia are yellow, not white as they are in *P. graminis*. Uredinia and telia formed on red or purplish spots. Teliospores 30–35 × 20–25, brown, thick-walled, with very long, hyaline pedicels.

OTHER FUNGI

Ceuthospora mahoniae Grove (Fig. 712)
Conidiomata multilocular, black, with a central pore. Conidia hyaline, 10–14 × 1.5–2. On dead leaves, Mar.–Apr.

Cucurbitaria berberidis and *Diaporthe detrusa*, which occur on dead branches, and the powdery mildew *Microsphaera berberidis* on leaves are described and figured on *Berberis*.

MALUS: Apple

DISCOMYCETES

Monilinia fructigena Honey ex Whetzel (Fig. 713)
Apothecia long-stalked, golden brown or greyish. Ascospores hyaline, 9–12 × 5–7. On overwintered mummified fruit. The *Monilia* conidial state is extremely common and widespread, especially on apples lying on the ground. It causes brown rot, and the pale fawn, cushion-like sporodochia, arranged in circles, are very conspicuous. Conidia hyaline, ellipsoidal, mostly 15–20 × 9–11.

Pezicula. Three species occur on apple, causing branch canker or fruit rot, but usually only their conidial states are found.

KEY

Ascospores becoming 3- to 6-septate .... *alba*
Ascospores 0- to 3-septate 1
1. Ascospores 13–22 × 4.5–8 .... *malicorticis*
Ascospores 17–30 × 7–9 .... *corticola*

Pezicula alba Guthrie
Apothecia sessile, grey, about 1mm diam. Ascospores hyaline, becoming 3- to 6-septate, 20–30 × 7–10. Conidiomata of *Phlyctema* state yellowish brown, pulvinate, up to 0.6mm diam., conidia hyaline, curved, 15–18 × 2.5–3, mixed with sterile hyphae. A cause of fruit rot.

Pezicula corticola Nannf. (Fig. 714)
Apothecia erumpent, clustered, yellow, up to 0.8mm diam. Ascospores hyaline, 17–30 × 7–9. Conidiomata of the *Cryptosporiopsis* state 1–2mm diam., white, pustular; conidia hyaline, 30–33 × 9.5–10.5. A cause of bark canker.

Pezicula malicorticis (H.S.Jackson)Nannf.
Apothecia erumpent, becoming dark brown. Ascospores hyaline, 0- to 3-septate, often flattened on one side, 13–22 × 4.5–8. Conidiomata of *Cryptosporiopsis* state pulvinate, yellowish, about 1mm diam., but sometimes confluent, macroconidia 11.5–16 × 3–4, microconidia 5–6 × 1–1.5. A cause of branch canker and fruit rot.

Potebniamyces pyri, described on *Pyrus*, also causes a bark canker of apple.

Rutstroemia rhenana (Kirscht.)Dennis (Fig. 715)
Apothecia erumpent, 2–5mm diam., short-stalked, dark reddish brown. Ascospores hyaline, slightly curved, 12–14 × 2.5–3.5. Paraphyses with brown contents in apical cells. On dead branches, Oct.

Tympanis conspersa (Fr.)Fr. (Fig. 716)
Apothecia up to 0.8mm diam., black, with inrolled margins, sometimes with mealy white granules between them, erumpent through the bark, often in large clusters. Ascospores hyaline, 4–8 × 3–6; these are soon replaced by rod-shaped secondary spores 2–4 × 1–1.5 which fill the asci. On dead twigs, Jan.–July.

OTHER ASCOMYCETES

Diaporthe perniciosa, described on *Prunus*, sometimes causes branch canker and dieback of apple.

Nectria galligena Bres. (Fig. 717)
Perithecia 0.25–0.35mm diam., bright red, dark around the ostiole, when old becoming darker and somewhat warted. Ascospores hyaline, 1-septate, smooth to minutely verruculose, mostly 15–20 × 6–8. Often preceded or accompanied by its *Cylindrocarpon* state with hyaline, slightly curved, 1- to 5-septate conidia, 20–65 × 4–7. On the surface or around the edge of brown fissured cankers on branches, all the year round.

Podosphaera leucotricha (Ellis & Ev.)Salmon (Fig. 718)
White powdery mildew. Conidia in chains, hyaline, 23–30 × 15–20. Cleistothecia with appendages only occasionally dichotomously branched at tips, each containing one 8-spored ascus. On living leaves and thin twigs.

Valsa ambiens, described on 'wood and bark', is very common on dead branches of *Malus*.

Venturia inaequalis (Cooke)Winter (Fig. 719)
Pseudothecia amphigenous, immersed, about 0.2mm diam., brown, with or without short blackish-brown setae around ostioles. Ascospores olivaceous brown, 1-septate, 12–16 × 5–7. On blackened fallen leaves, Feb.–Mar. The *Spilocaea* state, which forms olivaceous brown, velvety colonies on living leaves, shoots and fruit, has 1-septate conidia 16–24 × 7–10. This is the very common scab fungus.

HYPHOMYCETES

Cylindrocarpon, v.s. under *Nectria galligena*.

Monilia, v.s. under *Monilinia fructigena*.

Monodictys melanopa (Ach.)M.B.Ellis (Fig. 720)
Colonies effuse, black. Conidia muriform, lower half pale brown, upper half blackish brown, 27–45 × 18–27. On bark.

Penicillium expansum Link
Colonies bluish green, velvety or powdery, with a strong musty smell. Conidiophores up to 500 × 3–3.5, hyaline, penicillately branched at apex, the branches terminating in groups of 5 to 9 phialides. Conidia blue–green in mass, smooth, 4–5 × 2.5–3.5. Causes a soft brown rot of fruit, especially in storage, colonies appearing on watery, yellowish-brown areas which often originate at either the stem or the calyx end. Very common.

Spilocaea, v.s. under *Venturia inaequalis*.

COELOMYCETES

Cryptosporiopsis, v.s. under *Pezicula corticola* and *P. malicorticis*.

Phlyctema, *v.s.* under *Pezicula alba*.

MESPILUS: Medlar

Gymnosporangium confusum Plowr., 0,1
Aecia on thickened brown spots with yellowish margins on leaves, calyces and fruit. Alternate host *Juniperus sabina*.

Monilinia mespili Whetzel
Apothecia up to 9mm diam., brown, paler at margin, with stalks up to 5mm long. Ascospores hyaline, elliptical or broader at one end, 12–18 × 6–8. On mummified fruit, Apr. *Monilia* state causes leaf blotch in May; colonies white, powdery, sweet-smelling; conidia hyaline, 12–25 × 10–20, in long chains.

MORUS: Mulberry

Mycosphaerella mori (Fuckel)Wolf (Fig. 721)
Found mostly in its *Phloeospora* state with acervuli up to 0.2mm diam., and hyaline, 2- to 4-septate conidia, 30–55 × 4–5; parasitic on leaves of *M. alba* and *M. nigra*. Leaf spots 1–2mm diam., off-white or ochraceous with reddish-brown borders, Sept.–Oct.

MYRICA: Bog Myrtle or Sweet Gale

Ciboria acerina Whetzel & Buchwald (Fig. 722)
Apothecia long-stalked, 1–4mm diam., buff to brown, finely pubescent. Asci 4 spored. Ascospores hyaline, 10–15 × 4.5–6. On overwintered, stromatised male catkins lying on the ground, Apr.–May.

Cryptodiaporthe aubĕrtii (Westend.)Wehmeyer (Fig. 723)
Perithecia about 0.3mm diam., with short necks, immersed in bark in small clusters. Ascospores hyaline or pale olivaceous, 1-septate, 14–18 × 4–5, some with a minute appendage at each end. On dead branches.

Dasyscyphus sulphurellus (Peck)Sacc. (Fig. 724)
Apothecia mostly less than 0.5mm diam., very pale lemon yellow. Hairs topped by crystals. Ascospores hyaline, 7–10 × 1–1.5. On dead branches, Apr.–Aug.

Hyalotricha corticicola Dennis (Fig. 725)
Apothecia about 0.2mm diam., with olivaceous discs but appearing white to shining silvery grey owing to a coating of long, thick-walled hairs. Ascospores hyaline, 5–7 × 1. On dead twigs, May–Sept.

Pestalotiopsis oxyanthi (Thüm.)Steyaert
Acervuli punctiform, black. Conidia 24–29 × 6–8, 4-septate, end cells hyaline, other cells brown, the middle one darkest, with 2 to 4 apical setulae mostly 20–30 long and a basal setula 6–8 long. On dead branches.

Ramularia destructiva Plowr. & Phill.
Colonies pale flesh-coloured. Conidia ovoid, 14–20 × 8–10, in short chains. Causes reddish-brown spots, 1–3cm diam., on twigs and on the lower side of leaves.

MYRTUS: Myrtle

Pseudocercospora myrticola (Speg.)Deighton (Fig. 726)
Colonies amphigenous. Conidiophores mid pale brown or olivaceous brown, 20–50 × 3–5. Conidia pale straw-coloured, 3- to 6-septate, 20–70 × 2.5–4. Causes pale grey or brown spots, often with dark purple borders, on living leaves.

PICEA: Spruce

ON LEAVES AND BUDS

UREDINALES

Chrysomyxa abietis Unger, 111
Telia hypophyllous, orange or reddish brown, up to 1cm long; spores with smooth hyaline walls, 20–30 × 10–15, in short chains. On yellowish spots on leaves of *P. abies* and *P. sitchensis*, Mar.–May, uncommon.

Chrysomyxa rhododendri de Bary, 0,1
Aecia amphigenous, up to 3mm long, yellow with white, torn peridia, spores yellow, verruculose, 25–28 × 20–22. On transverse yellow bands on leaves on *P. abies* and *P. sitchensis*, July–Sept. Alternate hosts *Rhododendron* spp.

DISCOMYCETES

Dasyscyphus acuum, described on *Pinus*, is sometimes found on fallen dead leaves of *Picea*.

Heyderia abietis (Fr.)Link (Fig. 727)
Apothecia up to 3cm high, with swollen, reddish-brown heads and blackish stalks. Ascospores hyaline, somewhat curved, 12–15 × 1.5–2.5, multiguttulate. On fallen dead leaves, Sept.–Nov.

Lirula macrospora (Hartig)Darker
Apothecia hysterioid, black, shining. Ascospores filiform, 60 long, hyaline, with mucilaginous sheaths. On leaves of *P. abies*.

Lophodermium piceae (Fuckel)Höhnel (Fig. 728)
Apothecia up to 1.8 × 0.4mm, with black lips. Ascospores 60–95 × 1.5–2, each with a sheath up to 2 thick. On fallen dead leaves of *P. sitchensis*, May.

Pezizella subtilis (Fr.)Dennis (Fig. 729)
Apothecia stalked, up to 1mm diam., white, becoming creamy on drying. Ascospores hyaline, 5–8 × 1–1.5. On fallen dead leaves of *P. excelsa*, Oct.–Nov.

OTHER ASCOMYCETES

Gemmamyces piceae (Borthwick)Casagrande (Fig. 730)
Pseudothecia crowded, 0.3–0.6mm diam., dark brown to black. Ascospores muriform, mid pale golden brown, 40–50 × 13–19. Causes bud blight of *P. engelmannii*, *P. pungens* and var. *glauca*, June–Dec.; buds swollen and characteristically twisted.

Stomiopeltis pinastri and *Trichothyrina pinophylla*, both described on *Pinus*, are found also on dead leaves of *Picea*.

HYPHOMYCETES

Chalara cylindrica Karsten (Fig. 731)
Colonies dark brown, hairy. Conidiophore stalks up to 65 × 4, brown. Conidia hyaline, 0- or 1-septate, 3–6 × 1, catenate. On dead leaves and also on dry scales of *P. excelsa*.

Dactylaria lepida Minter (Fig. 732)
Conidiophores hyaline to pale brown, flexuous, with many denticles, 20–30 × 4–5. Conidia cylindrical, hyaline, 1-septate, very thick-walled at each truncate end, 25–35 × 2. On fallen dead leaves of *P. abies*.

Sesquicillium candelabrum, described on *Pinus*, and *Thysanophora penicillioides*, described on *Abies*, both occur also on dead leaves of *Picea*.

COELOMYCETES

Megaloseptoria mirabilis Naumov (Fig. 733)
Pycnidia clustered, superficial, dark brown to black, 0.4–0.6mm diam. Conidia septate, hyaline to very pale brown, mostly 150–200 × 5–7. On buds,

associated with *Gemmamyces*.

Rhizosphaera kalkoffii Bubák (Fig. 734)
Pycnidia minute, black, superficial, in rows above stomata. Conidia hyaline, 5–10 × 3–5. On dead leaves of *P. abies* and *P. pungens*.

ON CONES

Pucciniastrum areolatum (Fr.)Otth, 0,1
Aecia pale reddish brown, cupulate; spores verrucose, 20–30 × 16–22. On cone scales of *P. abies*, Aug.–Nov. Alternate host *Prunus padus*, few records.

Phialea strobilina (Fr.)Gillet
Apothecia up to 0.5mm diam., olivaceous brown with pale rims and black stalks. Ascospores hyaline, 8–10 × 1.5–2. On fallen cones of *P. excelsa*.

Phomopsis conorum, described on *Pinus* leaves, is found also on cones of *Picea*.

ON WOOD AND BARK

DISCOMYCETES

Agyrium rufum (Pers.)Fr. (Fig. 735)
Apothecia up to 1mm diam., pulvinate, reddish brown, gelatinous. Ascospores hyaline, 10–14 × 6–8. On decorticated wood, Apr.–Aug.

Claussenomyces olivaceus (Fuckel)Sherwood
Apothecia 0.5–1mm diam., dark olivaceous to black, on a dark brown hyphal subiculum. Tips of paraphyses surrounded by brown gelatinous substance forming epithecia. Ascospores hyaline, 0- to 7-septate, occasionally with a longitudinal septum, 11–28 × 3.5–7, forming spherical ascoconidia 2–2.5 diam. which soon replace them and fill the asci. On resinous exudates.

Colpoma crispum (Pers.)Sacc. (Fig. 736)
Apothecia 2–5 × 1–2mm, dark brown, becoming black, opening by thick white lips to expose greenish discs. Ascospores hyaline, 40–50 × 2–2.5. On bark of dead branches of *P. abies*, June.

Lachnellula resinaria (Cooke & Phill.)Rehm
Apothecia short-stalked, 0.5–1.5mm diam., with orange discs, covered by white, granulate hairs up to 90 × 2.5–3. Ascospores hyaline, 2–3 × 1–1.5. On resin exudates.

Pezicula livida, described on *Pinus*, is found occasionally also on dead twigs of *P. abies*.

Pseudophacidium piceae E.Müller (Fig. 737)
Apothecia erumpent, 1–1.5mm diam., black, opening irregularly to expose pale grey to brown discs. Ascospores hyaline, 11–16 × 5–9. On old but not

decayed logs of *P. sitchensis*, May.

OTHER ASCOMYCETES

Cucurbidothis pithyophila, described on *Abies*, has been recorded also on *Picea*.

Lophium mytilinum, described on *Pinus*, has been found also on old decorticated logs of *P. sitchensis*.

Nectria fuckeliana Booth (Fig. 738)
Perithecia 0.3–0.4mm diam., papillate, bright red becoming maroon, in large groups on erumpent stromata. Ascospores hyaline to pale brown, smooth to verruculose, 13–17 × 5–6. On dead branches of *P. excelsa* and *P. sitchensis*, Mar.–Aug.

Nectria pinea, described on *Pinus*, and *Valsa kunzei*, described on *Larix*, have both been found on *Picea*.

HYPHOMYCETES

Catenularia piceae M.B.Ellis (Fig. 739)
Colonies effuse, dark blackish brown to black, velvety. Conidiophores brown, up to 300 × 4–6. Conidia catenate, hyaline to pale brown, 8–11 × 3.5–5. On dead twigs, May.

Dictyopolyschema pirozynskii M.B.Ellis (Fig. 740)
Colonies effuse, dark blackish brown to black. Mycelium forming a brown superficial network. Conidia spherical, muriform, brown, 13–23 diam. On dead twigs, May.

Endophragmiella resinae P.M.Kirk (Fig. 741)
Colonies effuse, hairy, dark blackish brown to black. Conidiophores brown, up to 200 × 4.5. Conidia obovoid to pyriform, 1-septate, pale brown, basal cell very pale, 17–22 × 9–10.5. On old wounds on branches of *P. sitchensis*, Sept.

Sterigmatobotrys macrocarpa, described on *Taxus*, is found occasionally also on dead wood of *Picea*.

PINUS: Pines

ON LEAVES

UREDINALES

Coleosporium tussilaginis (Pers.)Berk.,0,1
Aecia amphigenous, up to 3mm long, yellow; spores verruculose, 20–40 ×

17–27. On living leaves of *P. sylvestris* and *P. nigra*, alternate hosts various Compositae, Campanulaceae and Scrophulariaceae.

DISCOMYCETES

KEY

	Apothecia clavate with long stalks ..... *Heyderia pusilla*
	Apothecia not club-shaped 1
1.	Apothecia hairy 2
	Apothecia not hairy 5
2.	Hairs dark brown, long, pointed ..... *Desmazierella acicola*
	Hairs pale brown, discs orange ..... *Melastiza asperula*
	Hairs yellow 3
	Hairs white or hyaline 4
3.	Hairs tipped with red exudate ..... *Dasyscyphus pulverulentus*
	Hairs tipped with yellow exudate ..... *Dasyscyphus mughonicola*
4.	Hairs 20–30 long, swollen at tip ..... *Dasyscyphus acuum*
	Hairs up to 50 long, uncinate ..... *Hyaloscypha laricionis*
	Hairs up to 150 long, thick-walled ..... *Urceolella trichodea*
5.	Apothecia opening by teeth ..... *Phacidium*
	Apothecia opening by slits 6
	No teeth or slits 8
6.	Discs exposed by folding back of flaps of epidermis ..... *Naemacyclus*
	Apothecia hysterioid 7
7.	Ascospores narrowly clavate ..... *Lophodermella*
	Ascospores not clavate, long ..... *Lophodermium*
	Ascospores not clavate, short ..... *Meloderma*
8.	Apothecia white or cream ..... *Pezizella subtilis*
	Apothecia orange or reddish ..... *Phialea fumosella*
	Apothecia golden brown or reddish brown 9
9.	Ascospores without septa, 12–15 long ..... *Cenangium acuum*
	Ascospores 3-septate, 22–25 long ..... *Pseudohelotium pineti*

Cenangium acuum Cooke & Peck (Fig. 742)
Apothecia erumpent, solitary, turbinate, golden brown, 0.5–3mm diam., leathery, margin inrolled when dry. Ascospores hyaline, 12–15 × 3.5–5. Paraphyses swollen to 3–5 at brownish apices. On rotting leaves.

Dasyscyphus acuum (Alb. & Schw.)Sacc. (Fig. 743)
Apothecia 0.2–0.3mm diam., white when fresh, becoming buff, hairs hyaline, 20–30 long, swollen to 4–5 at their tips. Ascospores 4–5 × 1–1.5. On dead leaves of *P. nigra* var. *maritima*, *P. radiata* and *P. sylvestris*, Sept.–Mar., common, especially on leaves still attached to cut-off branches.

Dasyscyphus mughonicola Svrček
Apothecia bright yellow. Hairs similar to those of *D. pulverulentus* but with yellow not red exudate at their tips. Acospores hyaline, 10–15 × 2–3.

Dasyscyphus pulverulentus (Lib.)Sacc. (Fig. 744)
Apothecia up to 1mm diam., yellow, hairs with yellow contents many of them tipped with red exudate, seen clearly under a lens or binocular dissecting microscope. Ascospores 4–5 × 1. On fallen dead leaves of *P. nigra* var. *maritima* and *P. sylvestris*, Mar.–Sept., uncommon.

Desmazierella acicola Lib. (Fig. 745)
Apothecia up to 5mm diam., brown and bristly with pale or almost white discs and dark brown pointed hairs up to 1mm long. Ascospores hyaline, mostly 17–22 × 9–11, some with thin hyaline sheaths or apiculi. On fallen blackened decaying leaves of *P. sylvestris*, Mar.–Oct., best in spring. Conidiophores of the *Verticicladium* state, looking like little brown telegraph poles under a lens or dissecting microscope, can be found throughout the year and are extremely common. Conidiophores 300–800 × 8–10, conidia hyaline 4–6 × 2.5–3.5.

Heyderia pusilla (Alb. & Schw.)Link (Fig. 746)
Apothecia mostly about 1cm high, with swollen, straw-coloured heads and reddish-brown stalks. Ascospores hyaline, 12–15 × 2–3. On fallen dead leaves of *P. sylvestris*, Sept.–Nov.

Hyaloscypha laricionis (Velen.)Nannf. (Fig. 747)
Apothecia white, glistening, 0.2–0.3mm diam., covered with uncinate, hyaline hairs, 1- or 2-septate, up to 50 long, 2–3 wide at base. Ascospores hyaline 4–6 × 1–1.5. Very common on fallen dead leaves of *P. nigra* var. *maritima* and *P. sylvestris*, Sept.–June.

Lophodermella conjuncta (Darker)Darker (Fig. 748)
Apothecia up to 3mm long, grey, opening by slits. Ascospores hyaline, 75–100 long, 3–3.5 thick at their tips, tapered below, with gelatinous sheaths. On discoloured areas of living leaves of *P. nigra* var. *maritima* and *P. sylvestris*.

Lophodermella sulcigena (Rostrup)Höhnel
Similar to *L. conjuncta* but with ascospores 30–35 × 4–5.

Lophodermium

KEY

    Transverse stromatic lines thin and black 1
    Transverse stromatic lines absent or broad and brown 2
1. Stromatic lines few, apothecial lips grey .... *pini-excelsae*
    Stromatic lines many, apothecial lips red or orange ...... *pinastri*
2. Centre quarter of apothecium black, remainder grey .... *conigenum*
    Apothecium all pale grey, all subepidermal .... *seditiosum*

Lophodermium conigenum, described below on cones but found sometimes on leaves.

Lophodermium pinastri (Schrad.)Chev. (Fig. 749)
Dry apothecia black over more than half the surface, the remainder grey surrounded by a black line. Stromatic lines across leaves numerous, thin, black. Apothecia about 1mm long, usually with red or orange lips, covered by host epidermis on either side but above it in the centre. Ascospores hyaline, 70–110 × 2, with sheaths. Very common on detached leaves in litter.

Lophodermium pini-excelsae Ahmad
Similar to *L. pinastri* but apothecia with grey lips and ascospores 50–70 long. On leaves of *P. excelsa* and less commonly on *P. sylvestris*.

Lophodermium seditiosum Minter, Staley & Millar (Fig. 750)
Apothecia totally subepidermal, when dry appearing pale grey, 1–1.5mm long, lips hyaline, green or blue. Ascospores mostly 90–120 × 2, with sheaths. On leaves attached to living or dead branches of *P. sylvestris* and *P. nigra* var. *maritima*.

Apothecia of *L. conigenum*, *L. pinastri* and *L. seditiosum* are commonly parasitised by a small white tremellaceous fungus *Pseudostypella translucens* (Gordon)Reid & Minter (Fig. 751) with hyaline basidiospores 7–9 × 3–4.

Melastiza asperula Spooner
Apothecia clustered, up to 1cm diam., pale brown with orange discs, hairy. Hairs pale brown, septate, some tapered, up to 270 × 16–18, others obtuse, 50–120 × 15–28. Ascospores hyaline, biguttulate, minutely verruculose, 17–21 × 8–10.5. On fallen dead leaves and other debris, Nov.

Meloderma desmazieresii (Duby)Darker (Fig. 752)
Apothecia subepidermal, up to 1mm long, black, opening by slits, each with a grey halo enclosed by a narrow black line. Ascospores hyaline, 22–32 × 4–5, with thick gelatinous sheaths. On leaves of *P. strobus* and *P. sylvestris*.

Naemacyclus minor Butin (Fig. 753)
Apothecial discs 0.5mm long, pale grey, exposed when flaps of epidermis at the sides fold back. Ascospores hyaline, 75–85 × 2.5–3, occasionally 1- or 2-septate. Paraphyses branched and swollen at tips. On fallen leaves of *P. radiata* and *P. sylvestris*, Nov.–May.

Pezizella subtilis, described on *Picea*, occurs also on *Pinus*.

Phacidium infestans P.Karsten
Apothecia immersed, about 1mm diam., disc grey, exposed when epidermis splits radially into teeth which fold back. Ascospores hyaline, 15–28 × 5–8. Parasitic on leaves of *P. sylvestris*, May–Aug.

Phacidium lacerum Fr. (Fig. 754)
Apothecial discs grey, 1mm diam., exposed when epidermis splits and teeth

fold back. Ascospores hyaline, 8–12 × 3–4. On decaying leaves of *P. sylvestris*, especially those on cut branches lying on the ground. It has a *Ceuthospora* conidial state.

Phialea fumosella (Cooke & Ellis)Sacc.
Apothecia up to 0.5mm diam., cup-shaped on a broad base, orange or reddish. Excipulum tough, elastic, made up of refractive, thick-walled cells. Ascospores in bundles in asci, hyaline, 1- to 3-septate, about 10 × 1.5–2. On fallen dead leaves, Sept.–Nov. *Pseudocenangium* state has cup-shaped, pale brown conidiomata about 0.3mm diam. and 1-septate hyaline conidia 12–18 × 2.

Pseudohelotium pineti (Batsch)Fuckel (Fig. 755)
Apothecia superficial, up to 1mm diam., reddish brown with pale discs. Ascospores hyaline, 3-septate, 22–25 × 1.5–2. On rotting leaves of *P. sylvestris*, especially those attached to cut-off branches, Sept.–Nov. The *Linodochium* state forms yellowish gelatinous cushions, 1–2mm diam., which flatten down and become brown and horny when dry. Conidia hyaline, 2-septate 30–50 × 1. It should be looked for in July–Aug. on whitened leaves.

Urceolella trichodea (Phill. & Plowr.)Dennis (Fig. 756)
Apothecia up to 0.6mm diam., golden brown with hyaline refractive thick-walled hairs about 150 × 4. Ascospores hyaline, 6–7 × 1.5. On black, decaying leaves, Apr.–June.

OTHER ASCOMYCETES

Anthostomella. Four species have been recorded by Dr S. Francis and others but we find only one of them to be common.

KEY

    Ascospores 1-celled 1
    Ascospores 2-celled 2
1.   Ascospores 7–10 wide .... *conorum*
    Ascospores 6–7 wide .... *pedemontana*
2.   Ascospores asymmetrical with oblique dwarf cells .... *sabiniana*
    Ascospores symmetrical .... *formosa*

Anthostomella conorum, described below on cones; occurs also on dead leaves but without a clypeus.

Anthostomella formosa Kirschst. (Fig. 757)
Perithecia immersed with small protuberant papilla surrounded usually by a black mass of ascospores; no clypeus. Ascospores symmetrical, 13–16 × 6–8 with large brown cell and small hyaline one. On fallen dead leaves of *P. nigra* var. *maritima* and *P. sylvestris*, Feb.–Sept., fairly common.

Anthostomella pedemontana Ferr. & Sacc.
Perithecia as in *A. formosa*. Ascospores brown, 1-celled, 13–16 × 6–7. On fallen leaves of *P. sylvestris*, uncommon.

Anthostomella sabiniana S.Francis
Perithecia with black clypeus. Ascospores asymmetrical, 14–15 × 6–7, brown, with obliquely placed hyaline dwarf cell. On dead leaves of *P. nigra* var. *maritima*.

Klasterskya acuum (Mouton)Petrak (Fig. 758)
Perithecia superficial, surrounded by a mat of greyish-brown hyphae, 0.1–0.15mm diam., dark brown, with necks up to 1mm long, fringed at their apices. Asci evanescent. Ascospores hyaline, 1-septate, 15–30 × 5.5–8. On rotting fallen leaves.

Kriegeriella mirabilis Höhnel (Fig. 759)
Pseudothecia superficial, black, cone-shaped or like halma pieces, 0.1–0.2mm high. Ascospores hyaline to pale brown, 4- to 6- (mostly 5-)septate, 30–37 × 8–10. On fallen dead leaves of *P. nigra* var. *maritima* and *P. sylvestris*, Nov.–Apr.

Melanospora chionea (Fr.)Corda (Fig. 760)
Perithecia superficial, clustered, 0.5mm diam., white cottony, with curved yellow necks. Ascospores dark brown, 11–15 × 9–12 × 5–6. On fallen dead leaves of *P. sylvestris*, Apr.–Oct.

Microthyrium pinophyllum (Höhnel)Petrak (Fig. 761)
Thyriothecia up to 0.15mm diam., not dark around ostiole, with marginal cells of scutellum elongated to form an irregular fringe. Ascospores hyaline, 1-septate, upper cell larger and bearing medianly 2 tufts of 3 cilia. On decaying needles of *P. nigra* var. *maritima*, *P. radiata* and *P. sylvestris*, Feb.–Apr., quite common.

Mytilinidion mytilinellum (Fr.)Zogg (Fig. 762)
Hysterothecia like minute mussel shells set on edge, black or blackish brown. Ascospores pale straw-coloured to brown, 1- to 3-septate, 17–25 × 2.5–3. On decaying leaves and cones.

Niesslia exilis (Fr.)Schröter (Fig. 763)
Perithecia superficial, 0.1–0.15mm diam., black, setose. Setae dark brown, 50–60 × 4. Ascospores hyaline, 1-septate, 10–13 × 2–2.5. On decaying leaves and occasionally on cones, Feb.–Apr.

Scirrhia pini Funk & Parker (Fig. 764)
Ascomata erumpent, up to 0.6mm diam., black, multilocular. Ascospores hyaline, 1-septate, 11–16 × 3–4. *Dothistroma* state with 1- to 5- (mostly 3-)septate, hyaline conidia 20–60 × 2–3.5. Parasitic on *P. nigra*, *P. radiata*, etc.; leaves reddish with necrotic bands.

Stomiopeltis pinastri (Fuckel)v.Arx (Fig. 765)
Thyriothecia up to 0.17mm diam., with irregular but not fimbriate margins. Cells of scutellum irregularly branched and lobed. Ascospores hyaline, 1-septate below middle, 6–8 × 1–1.5. Pycnidia of *Sirothyriella* state up to 0.4mm diam.; conidia hyaline, 4.5–7 × 1–1.5. On fallen rotting leaves of *P. nigra* var. *maritima*, *P. radiata* and *P. sylvestris*, Sept.–Mar.

Trichothyrina pinophylla (Höhnel)Petrak (Fig. 766)
Thyriothecia 80–90 diam., margin entire, ostiolar collar of 3 or 4 rings of thick-walled, dark brown cells. Ascospores slipper-shaped, hyaline, 1-septate, 8–11 × 1.5–3. On fallen dead leaves of *P. nigra* var. *maritima*, *P. radiata*, etc., Feb.–July.

HYPHOMYCETES

Belemnospora pinicola P.M.Kirk (Fig. 767)
Colonies effuse, shortly hairy, pale brown, inconspicuous. Conidiophores pale to mid brown, up to 12 × 1.5–2. Conidia bullet-shaped, very pale brown, 6.5–9 × 1.5–2. On dead leaves of *P. sylvestris*, May–July.

Bloxamia bohemica Minter & Hol.–Jech. (Fig. 768)
Sporodochia up to 2 × 1mm, amber becoming blackish brown on drying. Conidiophores forming a palisade. Conidia hyaline, 3–5.5 × 1. On rotting leaves of *P. sylvestris*, Oct.–Nov.

Chalara affinis and *C. cylindrosperma*, described on *Fagus* cupules, are quite common on rotting needles of *P. sylvestris*.

Chalara fusidioides (Corda)Rabenh. (Fig. 769)
Colonies effuse, pale yellowish. Conidiophores mostly without stalk cells, consisting of just a very pale brown phialide 12–25 long, 4–8 wide at base, with neck 1.5–3 thick. Conidia catenate, mostly 5–8 × 1.5–2.5. On dead leaves of *P. nigra* var. *maritima* and *P. sylvestris*.

Cladosporium staurophorum (Kendrick)M.B.Ellis (Fig. 770)
Colonies effuse, velvety, hazel or pale brown. Conidiophores 20–30 × 2–4. Ramoconidia 1-septate, up to 30 long. Conidia in long chains, pale brown, 5–18 × 1.5–3. Chlamydospores dark brown, mostly 5-celled, 12–17 × 10–16. On decaying fallen leaves of *P. sylvestris*.

Clonostachys compactiuscula, described on *Fagus* wood, has been found also on dead leaves of *P. sylvestris*.

Dactylaria lepida, described on *Picea* leaves, occurs also on *Pinus*.

Endophragmia pinicola M.B.Ellis (Fig. 771)
Colonies effuse, greyish brown, shortly hairy, inconspicuous. Conidiophores up to 100 × 3–4. Conidia straw-coloured with median, thick, dark septum,

10–14 × 5–9. On dead leaves of *P. nigra* var. *maritima* and *P. sylvestris*, Mar.–Sept., common.

Fusidium griseum, described on 'leaf litter', has been seen on leaves of *P. nigra* var. *maritima*.

Junctospora pulchra Minter & Hol.-Jech. (Fig. 772)
Colonies effuse, cottony, pale olivaceous. Conidiophores subhyaline, up to 350 × 3–5. Conidia fusiform, hyaline, 14–25 × 5–8, in chains, linked together by small, deliquescing separating cells. On rotting leaves of *P. sylvestris*, June–Nov.

Linodochium, v.s. under *Pseudohelotium pineti*.

Pseudocercospora deightonii Minter (Fig. 773)
Conidiophores arising from superficial mycelium, brown, 30–70 × 5, tapered to 1–2. Conidia hyaline, 5- to 8-septate, 30–40 × 2.5–3.5. On leaves of *P. sylvestris*, Sept.–Feb.

Pseudomicrodochium candidum (Bresad.)de Hoog (Fig. 774)
Colonies white, 1–3mm diam. Hyphae 4–5 thick. Phialides lateral, 6–8 × 3–4. Conidia adhering in packets, hyaline, 3-septate, 13–16 × 2, end cells smaller than median ones. On dead leaves and small twigs of *P. sylvestris*, June–Dec.

Sesquicillium candelabrum (Bon.)K.W.Gams
Conidiophores hyaline, smooth, verticillately branched, 2.5–3.5 thick. Phialides mostly 6–10 × 2–3. Conidia in irregular, slipped chains, hyaline, smooth, ovoid to ellipsoid, 4–5 × 2.5–3. On rotting leaves of *P. sylvestris*.

Sphaeridium candidum Fuckel
Synnemata superficial, small, spherical, compact, white. Conidia in branched chains, hyaline, straight or slightly curved, 4 × 1. On dead leaves, small twigs and cones of *P. sylvestris*, very common all the year round.

Sympodiella acicola Kendrick (Fig. 775)
Colonies effuse, dark blackish brown, hairy. Conidiophores repeatedly geniculate, dark brown, up to 300 × 2–4. Conidia cylindrical, hyaline, 7–14 × 2–2.5. On rotting leaves of *P. sylvestris*, common.

Thysanophora penicillioides, described on *Abies*, is not uncommon also on *Pinus*.

Troposporella monospora (Kendrick)M.B.Ellis (Fig. 776)
Colonies effuse or punctiform, black. Conidiophores brown, up to 50 × 1–3.5. Conidia horseshoe-shaped or helicoid, brown, end cells pale, with 4 or 5 dark, thick septa, 9–15 diam., filaments 5–9 thick.

Verticicladium, v.s. under *Desmazierella acicola*.

COELOMYCETES

Ceuthospora, v.s. under *Phacidium lacerum*.

Cytospora pinastri Fr.
Conidiomata amphigenous, multilocular, dark brown to black, about 0.25mm diam. Conidia allantoid, 4–5 × 1–1.5, issuing in milky white tendrils. On dead leaves.

Dothistroma, v.s. under *Scirrhia pini*.

Fujimyces oödes (Bayliss Elliott)Minter & Caine (Fig. 777)
Conidiomata 0.1–0.3mm diam., golden brown, sticking up on the surface like insect eggs. Conidia hyaline, 3-septate, 35–48 × 0.5–1.5. On dead leaves of *P. nigra* var. *maritima* and *P. sylvestris*, often still hanging on trees, Nov.–May.

Hendersonia acicola Münch & v. Tub.
Pycnidia 0.15mm diam. Conidia brown, 1- to 3- (mostly 2-)septate, 11–15 × 4–5. On dead leaves of *P. nigra* var. *maritima* and *P. sylvestris*.

Leptostroma states of *Lophodermium* species have orbicular or elliptical, brown or black, immersed conidiomata, often surrounded by black lines and opening by slits. In *L. conigenum* and *L. seditiosum*, conidia are 4.5–9.5 × 1; in *L. pinastri* and *L. pini-excelsae*, 4.5–6 × 1.

Phomopsis conorum (Sacc.)Died.
Conidiomata erumpent, black, up to 2mm diam., exuding yellowish conidial tendrils. A-conidia hyaline, biguttulate, 7–12 × 2.5–3.5, B-conidia mostly much curved or spiral, 20–24 × 1. On dead leaves.

Pseudocenangium, v.s. under *Phialea fumosella*.

Sclerophoma state of *Sydowia polyspora* (Bref. & Tav.)Müller (Fig. 778)
Conidiomata erumpent, black, 2–7mm diam., with 1 or several locules surrounded by thick-walled cells, without ostioles. Conidia hyaline, 4–8 × 2–3. On many species including *P. nigra*, *P. pinaster*, *P. radiata* and *P. sylvestris*, causing leaf-blight and dieback. Leaves remain hanging downwards on trees for some time.

Sirothyriella, v.s. under *Stomiopeltis pinastri*.

Sphaeropsis sapinea (Fr.)Dyko & Sutton (Fig. 779)
Pycnidia erumpent, up to 0.25mm diam., blackish brown to black, papillate. Conidia occasionally 1-septate, golden brown, with pitted walls, mostly 40–50 × 12–17. Causes browning of leaves and occurs also on twigs and cones. Very common on many pines including *P. nigra* var. *maritima*, *P. radiata* and *P. sylvestris*, Oct.–Apr.

Strasseria geniculata (Berk. & Br.)Höhnel (Fig. 780)
Conidiomata immersed, 0.1–0.35mm diam., dark brown to black. Conidia

hyaline, 8–15 × 2–3, each with a basal setula 12–15 long. On dead leaves in litter.

ON CONES

DISCOMYCETES

Cudoniella rubicunda (Rehm)Dennis
Apothecia short-stalked, up to 5mm diam., pinkish with bright reddish-purple discs. Ascospores hyaline, 6–8 × 2–3, with a small guttule at each end. On fallen cones of *P. sylvestris*.

Durella suecica (Starb.)Nannf.
Apothecia in crowded colonies on both surfaces of cone scales, up to 0.6mm diam., black with pale discs and small hyaline foot cells. Excipulum made up of dark brown, thick-walled hyphae. Ascospores fusiform, hyaline, 6.5–8.5 × 2. On *P. sylvestris*, May.

Hyaloscypha leuconica, described on *Larix*, is common also on old fallen cones of *P. nigra* var. *maritima* and *P. sylvestris*, Nov.–Mar.

Hymenoscyphus lutescens (Hedw.)Phill. (Fig. 781)
Apothecia up to 1mm diam., with short thick stalks, pale yellow, drying reddish brown. Ascospores hyaline, 10–15 × 3–4. On scales of fallen cones of *P. sylvestris*, July–Sept.

Lasiostictis fimbriata (Schw.)Bäumler (Fig. 782)
Apothecia up to 0.5mm diam., immersed, white with downy margins. Ascospores hyaline, 5- to 8-septate when mature, 50–60 × 1.5–2.5. On scales of fallen cones of *P. sylvestris*.

Lophodermium conigenum (Brunaud)Hilitzer (Fig. 783)
Apothecia up to 2mm long, all black when wet; when dry, black only in the centre, the rest grey; lips usually thick and olive green. Ascospores 90–130 × 2. Very common on fallen cones of *P. nigra* var. *maritima* and *P. sylvestris*. In best condition Feb.–June. *Leptostroma* state, with conidia 4.5–9.5 × 1, plentiful Oct.–Nov.

Micraspis strobilina Dennis (Fig. 784)
Apothecia about 1mm diam., each with a black crust which splits open to reveal a white or greyish disc. Ascospores hyaline, mostly 10–15 × 3, 0- to 3-septate. On old cones of *P. sylvestris*, June–Aug.; apparently common on the Isle of Mull but not elsewhere.

Mollisia fallax (Desm.)Gillet (Fig. 785)
Apothecia up to 3mm diam., dark brown with pale yellowish discs sometimes seated on a brown hyphal subiculum. Ascospores hyaline, 6–10 × 2–3. On old cones of *P. sylvestris*, May–Nov.

Pezizella chionea (Fr.)Dennis (Fig. 786)
Apothecia up to 1mm diam., almost white, pubescent with pale buff or yellowish discs. Ascospores hyaline, 6–9 × 1.5. On fallen cones of *P. sylvestris*; also occasionally on leaves, Apr.–Aug.

OTHER ASCOMYCETES

Anthostomella conorum (Fuckel)Sacc. (Fig. 787)
Perithecia black, up to 0.35mm diam., each with a clypeus. Ascospores dark brown, 15–17 × 7–10 with thin hyaline sheaths. On cone scales of *P. nigra* var. *maritima* and *P. sylvestris*, Mar.–Aug.

Rosellinia obliquata (Somm.)Sacc. (Fig. 788)
Perithecia clustered, up to 0.35mm diam., black. Ascospores dark brown, 10–15 × 7–9. On cones of *P. sylvestris*, Jan.–Aug.

Scopinella solani (Zukal)Malloch (Fig. 789)
Perithecia erumpent to superficial, often on a small stroma, brown, 0.2–0.3mm diam.; necks 0.3–0.7mm long, with very short terminal setae. Ascospores rectangular in one plane, elliptic–hexagonal in the other, 5–8 × 4–5, with 2 germ slits. On cones of *P. sylvestris* but also on other decaying plant material in litter, Dec.–Aug.

HYPHOMYCETES

Endophragmia boewei Crane (Fig. 790)
Colonies effuse, greyish brown, thinly hairy. Conidiophores brown, up to 200 × 2 4. Conidia pyriform, 1-septate, pale brown, smooth, 15–25 × 8–14.

Phaeostalagmus peregrinus Minter & Hol.-Jech. (Fig. 791)
Colonies effuse, brown, thinly hairy. Conidiophores dark brown, up to 200 × 3–5. Conidia hyaline, 5–8 × 2. On fallen cones of *P. sylvestris*.

Septocylindrium leucum Bayliss-Elliott & Stansfield (Fig. 792)
Colonies white, tufted, easily fragmenting. Conidiophores very short. Conidia formed in repeatedly branched chains, cylindrical, truncate at each end, hyaline, 1- to 3-septate, 10–15 × 1.5–2, covered with minute granules which soon disappear in water. On fallen cones of *P. sylvestris*.

Sporidesmium doliiforme Minter & Hol.-Jech. (Fig. 793)
Colonies effuse, black, hairy. Conidiophores brown, up to 200 × 5–8. Conidia mid to dark brown, smooth, 2- to 5-septate, 20–40 × 10–15. On rotting cones.

Trimmatostroma scutellare, described on *Larix*, is commonly found also on fallen cones of *P. nigra* var. *maritima* and *P. sylvestris*, Nov.–Apr.

Xylohypha ortmansiae Minter (Fig. 794)
Colonies effuse or more commonly pulvinate, brown to black. Conidiophores brown, up to 60 × 4–6. Conidia catenate, mostly spherical, brown, smooth or minutely verruculose, 5–6 diam. On fallen cones of *P. sylvestris*, Oct.–Mar.

COELOMYCETES

Camarosporium pini Sacc. (Fig. 795)
Pycnidia up to 0.5mm diam., erumpent, black. Conidia brown, mostly with 3 transverse septa and 1 or 2 longitudinal septa, 15–18 × 6–8. On fallen cones of *P. sylvestris*, Jan.–May.

Patellina caesia Bayliss-Elliott & Stansfield (Fig. 796)
Conidiomata patelliform, pubescent, grey, 1mm diam., superficially resembling apothecia. Conidiophores branched, hyaline. Conidia in long chains, cylindrical, hyaline, about 10 × 1.5. On fallen cones of *P. sylvestris*.

Pseudopatellina conigena Höhnel (Fig. 797)
Conidiomata erumpent, 0.2–0.4mm diam., off-white becoming brown. Conidiophores short, branched. Conidia in slimy masses, hyaline, smooth, spherical, 2.5 diam., or ellipsoid, 3 × 1.5. On fallen cones of *P. sylvestris*.

Sirococcus strobilinus Preuss (Fig. 798)
Conidiomata erumpent, dark brown to black, up to 0.5mm diam., multilocular, opening irregularly without a definite ostiole. Conidiophores branched. Conidia hyaline, 1-septate, 12–16 × 3. On *P. contorta*.

Sporonema diamandidis Minter (Fig. 799)
Conidiomata up to 1mm diam., black, opening by one or more slits. Conidia issuing in white masses, hyaline, often biguttulate, 4–8 × 1.5–2. On apophyses of fallen cones of *P. sylvestris*, Feb.–Mar.

ON WOOD AND BARK

UREDINALES

Cronartium flaccidum (Alb. & Schw.)Wint., 0,1
Two races occur, an alternating one with uredinia and telia on *Paeonia* and *Tropaeolum*, and a non-alternating one, often called *Peridermium pini*. The mycelium is perennial in the cortex of branches of *P. sylvestris* and causes large cankers. Aecia in groups, often spread over extensive areas, orange; spores mostly 25–30 × 16–23, verruculose.

Cronartium ribicola J.C.Fischer, 0,1
White pine blister rust causes stem canker of five-needled pines, especially *P. monticola* and *P. strobus*. Aecia erumpent, orange–yellow, spores 23–29 ×

18–20, walls mostly verrucose but with a small smooth area. Alternate hosts *Ribes* spp.

Melampsora populnea (Pers.)Karst., 0,1
Aecia occasionally on leaves but mostly erumpent through cortex of young shoots of *P. nigra* and its subsp. *laricio*, *P. pinaster* and *P. sylvestris*. Aecia caeomoid, spores 15–20 × 13–17, minutely verruculose. Alternate hosts *Populus alba* and *P. tremula*.

DISCOMYCETES

KEY

	Colonies effuse, pinkish or white, asci scattered over surface .... *Ascocorticium anomalum*
	Not so 1
1.	Apothecia hairy, sometimes shaggy 2
	Apothecia sometimes downy but not hairy 8
2.	Apothecia with shaggy white rims and bright orange or yellow discs 3
	Apothecia not so 4
3.	Ascospores 7–10 long, 0-septate .... *Lachnellula subtilissima*
	Ascospores 70–80 long, 7 septa .... *Lachnellula pseudofarinacea*
4.	Hairs brown mostly setiform .... *Trichodiscus pinicola*
	Hairs hyaline or very pale 5
5.	Cottony subiculum present .... *Arachnopeziza obtusipila*
	No cottony subiculum 6
6.	Hairs granulate, not tapered .... *Dasyscyphus papyraceus*
	Hairs not granulate, tapered 7
7.	Amber globules between hairs .... *Hyaloscypha velenovskii*
	No amber globules .... *Hyaloscypha stevensonii*
8.	Apothecia immersed splitting open .... *Therrya pini*
	Apothecia erumpent or superficial 9
9.	Apothecia polysporous or becoming filled with numerous ascoconidia 10
	Asci 4- or 8-spored 11
10.	Ascospores spherical, 2–3 diam., on resin .... *Sarea difformis*
	Ascoconidia 2–3 × 1–1.5 .... *Tympanis hypopodia*
11.	Ascospores dark brown, appendaged .... *Rhizina undulata*
	Ascospores hyaline or pale 12
12.	Apothecia flat, waxy 13
	Apothecia not so 14
13.	Apothecia pale cream .... *Orbilia leucostigma*
	Apothecia bright golden yellow .... *Orbilia xanthostigma*
14.	Apothecia downy, ascospores mostly 1-septate, 8–13 × 3–4 .... *Ciliolarina laricina*

	Apothecia not downy, ascospores not 1-septate 15
15.	Apothecia pale tangerine .... *Pezicula livida*
	Apothecia purple or purplish brown .... *Phaeohelotium purpureum*
	Apothecia greyish, ascospores 60–90 long .... *Gorgoniceps aridula*
	Apothecia dark brown to black, discs paler 16
16.	Ascospores mostly without septa 17
	Ascospores regularly septate 18
17.	Ascospores up to 14 long .... *Cenangium ferruginosum*
	Ascospores up to 22 long .... *Crumenulopsis sororia*
	Ascospores up to 32 long .... *Crumenulopsis pinicola*
18.	Ascospores 2- to 5-septate .... *Gremmeniella abietina*
	Ascospores with 7 transverse and some also with longitudinal septa .... *Haematomma elatina*

Arachnopeziza obtusipila, described on *Larix*, has been found also on dead twigs of *P. radiata*.

Ascocorticium anomalum (Ellis & Harkn.)Schröter (Fig. 800)
Colonies effuse, thin, pale pinkish or white. Asci forming an indefinite layer over the surface of logs of *P. sylvestris*; ascospores 4–6 × 2–2.5, Sept.–Jan., uncommon.

Cenangium ferruginosum Fr. (Fig. 801)
Apothecia up to 3mm diam., erumpent, often in groups, dark brown, pruinose, with yellow or yellowish-olive discs obscured by inrolled margins when dry. Ascospores hyaline, 11–14 × 5–6. Found all the year round on dead twigs and small branches, especially those cut off trees and lying on the ground.

Ciliolarina laricina (Raitv.)Svrček (Fig. 802)
Apothecia up to 0.5mm diam., solitary or occasionally in clusters, greyish white with brown bases, downy. Ascospores hyaline, with 1 or occasionally 2 septa, 8–13 × 3–4. On bark of fallen dead branches, sometimes also on cones, Aug.–Jan., common.

Crumenulopsis pinicola (Rebent.)Groves (Fig. 803)
Apothecia erumpent, solitary or in clusters of 3 or 4, 0.5–2mm diam., dark reddish brown with pale grey or olive discs, margins inrolled when dry. Ascospores hyaline, 18–32 × 2.5–4. On dead branches, Mar.–Apr.

Crumenulopsis sororia (P.Karsten)Groves (Fig. 804)
Apothecia 1–2mm diam., dark brown to almost black, with pale buff to olive discs, often somewhat triangular when dry, with inrolled margins. Ascospores hyaline, mostly without septa but occasionally septate, 18–22 × 4–5. Sometimes accompanied by a *Digitosporium* state with pale brown, branched conidia up to 60 long, with septate branches 3–4 thick. On small branches of *P. laricio* and *P. nigra*, Apr.–May.

Dasyscyphus papyraceus (Karst.)Sacc.
Apothecia 0.5–1mm diam., white, hairy, with yellow discs. Hairs 40–70 × 3. Ascospores hyaline, 4–6 × 1. Paraphyses with yellowish sap. On decorticated wood, Oct.

Gorgoniceps aridula (P.Karsten)P.Karsten (Fig. 805)
Apothecia shaped like little flowerpots, 0.4–0.7mm diam., grey or greyish brown, discs white or faintly greenish or yellowish, convex when moist, drying flat. Ascospores hyaline, multiseptate, 60–90 × 2–3. On dead wood and bark of *P. nigra* var. *maritima*, Oct.–Nov.

Gremmeniella abietina (Lagerb.)Morelet (Fig. 806)
Apothecia short-stalked, 1mm diam., dark brown, with creamy discs. Ascospores hyaline, 3-septate, mostly 15–20 × 3–4.5. *Brunchorstia* state has 2- to 5-septate, hyaline conidia, 25–40 × 3–3.5. On twigs of many different species including *P. nigra* var. *maritima*, *P. sylvestris* and *P. radiata*, parasitic, causes dieback.

Haematomma elatina (Ach.)A.L.Sm. (Fig. 807)
Apothecia erumpent, up to 2mm long, black with straw-coloured discs. Asci 4- to 8-spored. Ascospores hyaline or pale yellow, with up to 7 transverse septa, some with a longitudinal septum, 25–45 × 12–14. On bark of *P. sylvestris* and other conifers.

Hyaloscypha stevensonii (Berk. & Br.)Nannf. (Fig. 808)
Apothecia up to 0.3mm diam., white, drying yellowish, margins incurved. Hairs mostly 30 × 2–3. Ascospores hyaline, 6–9 × 1.5–2. On fallen branches of *P. nigra* var. *maritima* and *P. sylvestris*, Oct.–Apr., fairly common.

Hyaloscypha velenovskii, described on *Larix*, has been found also on *P. sylvestris* and *P. radiata*.

Lachnellula pseudofarinacea (Crouan & H.Crouan)Dennis (Fig. 809)
Apothecia erumpent, 0.1–0.2mm diam., in clusters, white, shaggy, with yellow to orange discs. Ascospores hyaline, mostly 7-septate, 70–80 × 1.5. Hairs smooth. On fallen branches, Mar.–May.

Lachnellula subtilissima (Cooke)Dennis (Fig. 810)
Apothecia erumpent, 1–2mm diam., often in clusters, white, shaggy, with yellow to orange discs. Ascospores hyaline, 7–10 × 2–2.5. On fallen twigs and branches of *P. nigra* var. *maritima*, *P. pinaster*, *P. radiata* and *P. sylvestris*, Jan.–July, common.

Orbilia leucostigma (Fr.)Fr. (Fig. 811)
Apothecia flat, waxy, white or very pale cream, about 1mm diam. Ascospores hyaline, allantoid, 3–4 × 1–1.5, biguttulate. On fallen branches of *P. nigra* var. *maritima*, June.

Orbilia xanthostigma, described on 'wood and bark' and found more commonly on deciduous trees, has been collected on *P. radiata*.

Pezicula livida (Berk. & Br.)Rehm (Fig. 812)
Apothecia erumpent, solitary or clustered, varying in colour with age but becoming pale tangerine, up to 2mm diam. Ascospores hyaline, becoming mostly 3- to 5-septate when mature, 22–32 × 6–8, sometimes budding off microconidia in asci. The *Cryptosporiopsis* state has hyaline conidia 25–34 × 9.5–10. On fallen branches and also on cones of many species including *P. contorta*, *P. nigra* var. *maritima*, *P. strobus* and *P. sylvestris*, common all the year round.

Phaeohelotium purpureum Dennis (Fig. 813)
Apothecia erumpent, dark purple to purplish brown, about 1.5mm diam. Ascospores hyaline, 5–8 × 1. On fallen branches of *P. sylvestris*, Jan.

Rhizina undulata Fr. (Fig. 814)
Apothecia flat or humped and undulating, up to 25cm across, reddish brown to black with pale yellow margins and with white root-like attachments below. Ascospores dark brown with a hyaline appendage at each end, mostly 25–35 × 9–10. On burnt woody pine debris, sometimes also attacking seedling roots, July–Oct.

Sarea difformis (Fr.)Fr. (Fig. 815)
Apothecia superficial, 0.5–1mm diam., dark brown to black, somewhat gelatinous when fresh, leathery when dry. Asci very thick-walled at the apex, containing large numbers of spherical, hyaline ascospores 2–3 diam. *Epithyrium* pycnidial state has subglobose, pale brown conidia 2–3.5 diam. On resin exudates from stumps of *P. sylvestris* and other conifers.

Therrya pini (Alb. & Schw.)Kujala (Fig. 816)
Apothecia up to 3mm diam., immersed, black, splitting open by lobes to expose golden brown discs. Ascospores hyaline to very pale yellow, 4- to 10-septate, 60–100 × 3–4. On small dead branches of *P. sylvestris*, either still attached to trees or cut off and lying on the ground, Feb.–July.

Trichodiscus pinicola Graddon (Fig. 817)
Apothecia up to 0.25mm diam., grey, setose. Setiform hairs dark brown, up to 100 × 4, mixed with pale brown granulate hairs 30–40 × 3–4. Ascospores hyaline, 10–12 × 1.5. On blackened areas inside bark of *P. nigra* var. *maritima*, Oct.–Nov.; found once also on cones of *P. radiata*.

Tympanis hypopodia Nyl. (Fig. 818)
Apothecia clustered, 0.5–1mm diam., black, erumpent. Ascospores hyaline, at first 8–10 × 2–4 but these soon disintegrate, forming large numbers of ascoconidia 2–3 × 1–1.5. Often accompanied by small black conidiomata with hyaline allantoid conidia 2–4 × 1.5. On bark of small cut branches, Nov.

OTHER ASCOMYCETES

KEY

	Ascomata orange or purplish red	1
	Ascomata black or blackish brown	2
1.	Ascospores 1-septate	*Nectria pinea*
	Ascospores 14- to 18-septate	*Scoleconectria cucurbitula*
2.	Ascomata star-shaped hysterothecia	*Actidium hysterioides*
	Ascomata mussel-like hysterothecia	3
	Ascomata flask-shaped	4
3.	Ascospores 150–170 long	*Lophium mytilinum*
	Ascospores not more than 50 long	*Mytilinidion*
4.	Ascospores muriform, brown	*Cucurbidothis pithyophila*
	Ascospores 1-septate, brown	*Endoxylina pini*
	Ascospores 3-septate	5
	Ascospores without septa	6
5.	Ascomata setose	*Keissleriella pinicola*
	Ascomata not setose	*Zignoella morthieri*
6.	Ascospores brown or greyish brown	7
	Ascospores hyaline	8
7.	Ascomata superficial, setose	*Coniochaeta malacotricha*
	Ascomata immersed, not setose	*Endoxyla operculata*
8.	Perithecia solitary, long-necked	9
	Perithecia clustered, necks converging on disc	10
9.	Ascospores 6–7 long	*Ceratocystis coerulescens*
	Ascospores 4–5 long	*Ceratocystis piceae*
10.	Ascospores 12–20 long	*Valsa curreyi*
	Ascospores 7–9 long	*Valsa pini*

Actidium hysterioides Fr. (Fig. 819)
Hysterothecia superficial, star-shaped, mostly with 3 to 5 arms, black, about 1mm diam. Ascospores 12–17 × 2–2.5, 1-septate, pale golden brown in mass. On bark or bare wood.

Ceratocystis coerulescens (Münch)Bakshi (Fig. 820)
Perithecia superficial, 0.1–0.2mm diam., black, with slender necks up to 0.8mm long, fimbriate at their tips. Asci with diffluent walls, spores sliming out in long tendrils, hyaline, 6–7 × 1.5–2. *Chalara* state has conidia mostly 8–10 × 3–5. On wood, causes blue stain.

Ceratocystis piceae (Münch)Bakshi (Fig. 821)
Perithecia up to 0.25mm diam., black, with slender necks up to 1mm long, fimbriate at their tips. Ascospores hyaline, allantoid, 4–5 × 1.5–2. On wood of *P. sylvestris* and other conifers, causes blue stain.

Coniochaeta malacotricha (Auersw. ex Niessl)Traverso (Fig. 822)
Perithecia superficial, 0.2–0.4mm diam., black, setose; setae 40–60 × 4–5. Ascospores dark brown, 11–14 × 9–11 × 6–8. On dead branches of *P. nigra* var. *maritima*, *P. radiata* and *P. sylvestris*, Mar.–June.

Cucurbidothis pithyophila, described on *Abies*, has been recorded also on *Pinus*.

Endoxyla operculata (Alb. & Schw.)Sacc. (Fig. 823)
Perithecia immersed, 0.5mm diam., with short, thick, slightly protuberant necks, black. Ascospores allantoid, greyish brown, 12–15 × 3–3.5. On decorticated wood of *P. sylvestris*.

Endoxylina pini Sivan. (Fig. 824)
Perithecia immersed, black, up to 0.9 × 0.5mm with stout, cylindrical, slightly protuberant necks. Ascospores brown, 1-septate, 13–16 × 2.5–3. Another ascomycete, *Scotiosphaeria endoxylinae* Sivan., with 1-septate, brown ascospores 6–9 × 2.5–3, was described by Dr Sivanesan growing between the necks of the *Endoxylina*. On blackened areas of decorticated wood.

Keissleriella pinicola D.Hawksw. & Sivan. (Fig. 825)
Pseudothecia erumpent becoming superficial, 0.1–0.15mm diam., papillate, black with short, dark brown setae around the ostiole. Ascospores 3-septate, hyaline or pale brown, 15–20 × 3.5–4.5. On decorticated wood of *P. sylvestris*, Dec.–Mar.

Lophium mytilinum (Pers.)Fr. (Fig. 826)
Hysterothecia like tiny black mussels about 1mm long. Ascospores straw-coloured, 150–170 × 2–2.5, with up to 21 septa. On twigs and logs, especially on scars of cut-off branches of *P. nigra* var. *maritima* and *P. sylvestris*, July–Aug.

Mytilinidion. Four species, all of which have tiny mussel-like hysterothecia and brown septate ascospores, occur on pines.

KEY

Ascospores with 1 to 3 septa, 16–25 × 2.5–3 .... *mytilinellum*
Ascospores with 2 to 5 septa, 30–45 × 3.5–4 .... *rhenanum*
Ascospores with 5 to 7 septa, 40–50 × 2–2.5 .... *scolecosporium*
Ascospores with 7 septa, 30–40 × 5–7 .... *gemmigenum*

Mytilinidion gemmigenum, described on *Larix*.

Mytilinidion mytilinellum (Fr.)Zogg (Fig. 762)
Found on rotting leaves and cones as well as bark, Oct.–Nov.

Mytilinidion rhenanum Fuckel (Fig. 827)

Mytilinidion scolecosporium Lohman
Mostly on wood and weathered stumps of *P. strobus* but has been found on *P. sylvestris*.

Nectria pinea Dingley (Fig. 828)
Perithecia 0.5mm diam., orange becoming reddish brown, usually in small groups on a thin stroma, erumpent through periderm, often with brown septate hairs between them. Ascospores becoming pale brown, 1-septate, 16–22 × 7–10. On dead branches of *P. laricio, P. sylvestris* and other conifers, Sept.–May.

Scoleconectria cucurbitula (Tode)Booth (Fig. 829)
Perithecia up to 0.4mm diam., pomiform, dull purplish red, grouped on erumpent stromata, sometimes covered with green scurf. Ascospores hyaline with 14 to 18 septa, 35–55 × 2.5–3.5, forming allantoid ascoconidia, 3–4.5 × 1 inside the asci which they eventually fill. *Zythiostroma* state usually precedes the perithecial state. Conidiomata multilocular, yellow to orange–brown, 0.2–0.4mm diam., conidia allantoid, 4–6 × 1. On dead twigs, branches and occasionally leaves of *P. pinea, P. radiata, P. strobus, P. sylvestris* and other conifers, Mar.–May and Sept.–Dec.

Valsa (*Leucostoma*) *curreyi*, described on *Larix*, occurs also on *Pinus*.

Valsa pini (Alb. & Schw.)Fr.
Perithecia 0.5mm diam., immersed in groups of 5 to 10, with black necks erumpent through small discs. Ascospores allantoid, 7–9 × 1.5–2. On dead twigs of *P. strobus*, Nov.–Feb.

Zignoella morthieri (Fuckel)Sacc.
Perithecia black, papillate, mostly superficial but with bases immersed in whitened areas of rotten but still firm wood. Ascospores oblong or somewhat clavate, rounded at ends, hyaline, 3-septate, 18–24 × 7–8. On dead branches of *P. sylvestris*.

HYPHOMYCETES

Antennatula state of *Strigopodia resinae* (Sacc. & Bres.)Hughes (Fig. 830)
Colonies effuse, dense, woolly or felted, dark olivaceous brown to black, composed of 5 to 11 thick, rough-walled hyphae. Conidia brown, 5- to 14-septate, 50–120 × 11–14, borne directly on the hyphae. On branches, associated with resinous exudates.

Cheiromycella microscopica (Karst.)Hughes (Fig. 831)
Sporodochia punctiform, brown. Conidia aggregated in firm slimy masses, usually branched, cheiroid, golden brown, 8–23 long, each branch 1- to 5-septate, 3–7 thick. On dead wood, especially of *P. sylvestris*, common but rather inconspicuous.

Dendrodochium citrinum Grove (Fig. 832)
Fructifications bright lemon yellow, some with a thin fringe of white hairs, about 1mm diam. Conidia hyaline, yellow in mass, 1.5–2 diam. Very common

on cut branches of *P. nigra* var. *maritima*, *P. radiata* and *P. sylvestris* and occasionally on cones and dead leaves, Aug.–Apr.

Dendrodochium pinastri Paol
Fructifications yellow or yellowish brown. Conidia oblong rounded at ends, 3.5–5 × 1. On branches of *P. nigra* and *P. pinaster*.

Hormiactella asetosa Hol.–Jech (Fig. 833)
Colonies discrete or effuse, greenish grey to greyish brown, hairy. Conidiophores brown, up to 120 × 3–4. Conidia in long branched chains, cylindrical, 1-septate, pale olivaceous brown, 14–22 × 2–3.5. On fallen twigs, branches and sometimes cones of *P. sylvestris*, Mar.–Sept., occasionally on other conifers.

Leptographium lundbergii Lagerb. & Melin (Fig. 834)
Colonies effuse, grey to black, cottony or hairy. Conidiophores brown, penicillately branched, up to 200 × 7–16. Conidia wedge-shaped, hyaline, 8–12 × 3–6, aggregated in slimy heads. On wood of *P. sylvestris*, often associated with bark beetles.

Oramasia hirsuta, described on *Salix* bark, has been found also on twigs, leaves and cones of *P. sylvestris*.

Parasympodiella clarkii Sutton (Fig. 835)
Colonies effuse, thin, brown, hairy. Conidiophores brown, up to 70 × 3–4. Conidia in short chains, pale brown, 3-septate, mostly 15–17 × 2.5–3. On fallen dead twigs, Nov.–Dec.

Polyscytalum verrucosum Sutton (Fig. 836)
Colonies effuse, brown, thin, hairy. Conidiophores brown, up to 90 × 3–5. Conidia in branched chains, hyaline, slightly verruculose, 7–9 × 2–3. On dead twigs.

Septonema fasciculare (Corda)Hughes (Fig. 837)
Colonies effuse, brown, hairy. Conidiophores reddish brown, up to 350 × 3–4. Conidia in long branched chains, brown, mostly 3-septate and 10–15 × 4–5. On rotten wood and bark of *P. sylvestris*.

Xylohypha pinicola D. Hawksw. (Fig. 838)
Colonies effuse, dark blackish brown to black. Conidiophores dark brown, up to 160 × 4–7. Conidia in simple or branched chains, brown, 4–10 × 3.5–5. On decorticated wood of *P. sylvestris*.

COELOMYCETES

Epithyrium, v.s. under *Sarea difformis*.

Oncospora pinastri (Moug.)Died. (Fig.839)
Conidiomata erumpent, caespitose, cup-shaped, black. Conidia hyaline,

curved, multiguttulate, 30–50 × 3.5–5.5. On bark of *P. sylvestris*, uncommon.

Phomopsis occulta Traverso
Conidiomata erumpent, black, up to 1mm diam. A-conidia biguttulate, 8–11 × 2–2.5, B-conidia often hamate, 20–27 × 1. On twigs of *P. patula* and *P. sylvestris*.

Strasseria geniculata (Berk. & Br.)Höhnel (Fig. 780)
Conidiomata immersed becoming erumpent, up to 0.4mm diam., black, thick-walled. Conidia hyaline, allantoid, 10–15 × 2–3.5, each with a slender basal setula. On dead twigs of *P. nigra*, *P. strobus* and *P. sylvestris*, Jan.–May.

Zythiostroma, v.s. under *Scoleconectria cucurbitula*.

PLATANUS: Plane

Dichomera mutabilis (Berk.& Br.)Sacc.
Conidiomata subepidermal, up to 2mm diam., black, plurilocular. Conidia oblong–ellipsoid, brown, with 3 to 5 transverse septa and an occasional longitudinal septum, 16–18 × 6–7. On dead twigs, Feb.–Mar.

Apiognomonia errabunda, described on leaves of *Fagus*, occurs also on *P. occidentalis* and *P. orientalis*, especially in its *Discula* state.

Hapalocystis berkeleyi Auersw. ex Fuckel (Fig. 840)
Perithecia immersed, in small groups, with necks just bursting through bark. Ascospores brown, 2-septate with blunt hyaline appendages, 25–35 × 14–16. On dead twigs.

POPULUS: Poplars

ON LEAVES AND BUD-SCALES

UREDINALES

Three species of *Melampsora* form uredinia and telia on poplars.

KEY

    Walls of uredinial paraphyses evenly 2–3 thick .... *allii-populina*
    Walls of uredinial paraphyses 3–10 thick 1
1.   Urediniospores 15–25 × 11–18 .... *populnea*
    Urediniospores 30–40 × 13–17 .... *larici-populina*

Melampsora allii-populina Kleb.,11,111
Uredinia on yellowish spots, mostly hypophyllous, up to 1mm diam., reddish orange; paraphyses usually capitate with narrow stalks, 50–60 × 14–22, wall

2–3 thick; spores 25–38 × 11–18, echinulate except at apex. Telia hypophyllous, small, blackish brown, spores 35–60 × 6–10.

Melampsora larici-populina Kleb., 11,111
Uredinia mostly hypophyllous, causing angular yellow spots on upper surface; paraphyses clavate or capitate, 40–70 × 14–18, thick-walled, up to 10 at apex; spores 30–40 × 13–17, echinulate except at apex. Telia epiphyllous, widely distributed, reddish brown to black, spores 40–50 × 7–10.

Melampsora populnea (Pers.)Karst., 11,111
Uredinia hypophyllous, 0.5mm diam., yellowish orange; paraphyses 40–60 × 8–23, clavate, wall 3–9 thick; spores 15–25 × 11–18, echinulate. Telia hypophyllous, dark brown, spores 23–60 × 7–12. Alternate hosts *Larix*, *Mercurialis* and *Pinus*.

DISCOMYCETES AND TAPHRINALES

Calycellina populina (Fuckel)Höhnel (Fig. 841)
Apothecia erumpent, up to 1mm diam., subsessile, brownish when young, disc ivory, brown basal ring. Ascospores hyaline, 10–14 × 2.5–3.5, mostly 12 × 3. On dead fallen leaves, Sept.–Mar.

Drepanopeziza. Three species are found on poplars, the apothecia on fallen decaying leaves in early spring, and acervuli of the *Marssonina* states in summer and early autumn on living leaves. The apothecia are erumpent, brown and rather difficult to find on rotting leaves. The *Marssonina* states, on the other hand, cause brown spots on green leaves and are seen easily.

KEY TO MARSSONINA STATES

    Leaf spots less than 1mm diam., conidia 12–16 long .... *punctiformis*
    Leaf spots up to 5mm diam., conidia more than 18 long 1
1.   Septum just below middle of conidium .... *populi-alba*
    Septum much nearer base of conidium .... *populorum*

Drepanopeziza populi-alba (Kleb.)Nannf. (Fig. 842)
Apothecia 0.25–0.5mm diam., erumpent, dark brown with orange–brown discs. Ascospores hyaline, 14–16 × 5–7. Conidia of *Marssonina* state hyaline, 18–20 × 5–9. On leaves of *P. alba*; apothecia Mar., acervuli Aug.–Sept.

Drepanopeziza populorum (Desm.)Höhnel (Fig. 843)
Apothecia 0.2–0.3mm diam., brown to black. Ascospores 12–16 × 5–9. Conidia of *Marssonina* state 17–25 × 6–11. On various hybrid poplars.

Drepanopeziza punctiformis Gremmen (Fig. 844)
Apothecia up to 1mm diam. Ascospores 6–17 × 4–7. Conidia of *Marssonina* state 12–16 × 5–7. On hybrid poplars, especially *P. nigra* × Serotina,

apothecia Mar., acervuli July–Sept., common.

Hymenoscyphus immutabilis (Fuckel)Dennis (Fig. 845)
Apothecia erumpent, short-stalked, 1–1.5mm diam., white, drying golden brown. Ascospores hyaline, 10–13 × 4–4.5. On fallen dead leaves, Oct.–Dec.

Hymenoscyphus phyllogenus (Rehm)O. Kuntze (Fig. 846)
Apothecia 0.3–0.5mm diam., white, drying yellowish; hyphae of excipulum staining deeply with cotton blue. Ascospores hyaline, 11–15 × 3.5–5, often with the pointed end upwards in the ascus. On mesophyll of rotting fallen leaves especially those of *P. nigra* × Serotina, Oct.–Nov.

Pezizella gemmarum (Boud.)Dennis (Fig. 847)
Apothecia 1–1.5mm diam., white or greyish white, puberulent, with short protuberant, clavate, finely roughened hairs. Ascospores hyaline, 6–8 × 2–2.5. On fallen bud-scales, Nov.–Apr.

Rutstroemia conformata, described on *Alnus*, has been found also on skeletonised leaves of *P. canescens*, May–June.

Taphrina populina Fr. (Fig. 848)
Yellow leaf blister. Asci form a yellow lining to the concave side of the blisters; they eventually become filled with bud spores 2 × 1. On living leaves, especially common on *P. nigra* × Serotina and × Regenerata, July–Sept.

OTHER ASCOMYCETES

Linospora ceuthocarpa (Fr.)Lind
Perithecia solitary or in pairs, immersed in black stromata 0.5–1mm diam., visible on both leaf surfaces. Ascospores hyaline, multiseptate, 100–120 × 2. On fallen leaves of *P. tremula*, Apr.–July.

Microthyrium fagi, described on *Fagus*, has been found also on fallen dead leaves of *Populus*.

Uncinula adunca, powdery mildew described on *Salix*, occurs also on *Populus*.

Venturia. Two species form their pseudothecia on fallen dead leaves and shoots of poplars in spring, and conidia of their *Pollaccia* states on living leaves and shoots in summer and early autumn.

KEY

Ascospores 8–14 × 4–6, conidia 18–21 × 7–8 .... *macularis*
Ascospores 20–30 × 10–15, conidia 25–35 × 9–13 .... *populina*

Venturia macularis (Fr.)Müller & v.Arx (Fig. 849)
Pseudothecia immersed, blackish brown, 0.1–0.15mm diam., some with brown setae up to 60 long. Ascospores olivaceous brown, 1-septate, 8–14 ×

4–6. *Pollaccia* state causes round or irregular, sometimes confluent, buff spots, 1–2cm diam., each with a blackish purple margin; conidia hyaline to pale olive, mostly 1-septate, 18–21 × 7–8. On *P. alba* and *P. tremula*.

Venturia populina (Vuill.)Fabric. (Fig. 850)
Pseudothecia 0.15–0.2mm diam., some with a few short setae. Ascospores pale olivaceous brown, 1-septate, 20–30 × 10–15. *Pollaccia* state causes brown to black spots; conidia mostly 2-septate, 25–35 × 9–13.

ON CATKINS

Ciboria caucus (Rebent.)Fuckel
Apothecia long-stalked, up to 8mm diam., pale brown with umber discs. Ascospores hyaline, 7.5–10.5 × 4.5–6. On overwintered, fallen male catkins of *P. alba*, *P. nigra* and *P. tremula*, Mar.–Apr.

Taphrina johanssonii Sadeb.
Asci form a yellow coating to swollen carpels, especially those of *P. tremula*; bud spores inside asci 4–10 × 1.5–4.

ON WOOD AND BARK

DISCOMYCETES

Encoelia fascicularis (Alb. & Schw.)P.Karsten (Fig. 851)
Apothecia erumpent, clustered, 1–3mm diam., brown, outer surface dark and roughened like an acorn cup when dry. Ascospores hyaline, 10–15 × 2.5–3.5. On dead twigs and branches of *P. tremula*, Nov.–May.

Hyalinia rubella and *Pachydisca ascophanoides*, described on *Salix*, occur also on *Populus*.

OTHER ASCOMYCETES

Cryptodiaporthe populea (Sacc.)Butin (Fig. 852)
Perithecia 0.5mm diam., immersed, clustered, pushing up the periderm, with necks just protruding. Ascospores hyaline, becoming 1-septate, 17–23 × 6–7. *Discosporium* state has conidiomata about 0.8mm diam. which exude greyish-brown masses of conidia in wet weather. Separate conidia appear hyaline, 9–12 × 7–9. On twigs and branches of *P. nigra* × Serotina and other poplars causing dieback and canker. Perithecia mature Feb.–Mar.

Cryptosphaeria populina (Pers.)Sacc. (Fig. 853)
Perithecia immersed, about 0.4mm diam., with tips of necks appearing as black dots on the bark surface. Asci long-stalked. Ascospores pale olivaceous brown, 8–12 × 2–2.5. On dead branches of *P. nigra* × Serotina and other poplars, June–Sept.

Cryptosporella populina (Fuckel)Sacc.
Perithecia 0.5mm diam., immersed in groups of 2 to 10, with convergent necks and no stroma. Ascospores narrowly ellipsoid, hyaline, 13–16 × 4. Conidia of *Phaeostromella* state 10–18 × 3–4.5. On dead twigs and branches of *P. nigra*, uncommon.

Dothiora sphaerioides (Pers.)Fr. (Fig. 854)
Pseudothecia forming locules in erumpent, pulvinate, brown stromata about 0.5mm diam. Ascospores hyaline with 5 to 7 transverse septa and often a longitudinal septum, 20–30 × 5–10. On dead branches of *P. tremula*, uncommon.

Leucostoma niveum (Hoffm.)Höhnel (Fig. 855)
Perithecia immersed in groups of up to 10 with tips of black necks emerging through a snow-white disc. Asci mostly 4-spored. Ascospores hyaline, curved, 12–16 × 3. On dead twigs of *P. alba*, *P. nigra* and *P. tremula*, Apr.–May. The *Cytospora* state, found in good condition Feb.–Mar., has similar white discs; allantoid conidia 6–7 × 1.5–2 form long crimson tendrils.

Massarina emergens (P.Karsten)Holm (Fig. 856)
Pseudothecia immersed, 0.3–0.4mm diam. Ascospores hyaline, 1- to 3-septate, 18–26 × 6–7. On dead twigs and branches of *P. tremula*.

Valsa sordida Nitschke
Perithecia immersed in groups of 5 to 12 with black necks emerging through a greyish white, dark-bordered disc. Ascospores hyaline, slightly curved, about 12 × 2. On branches especially those of *P. nigra* × Serotina but also on *P. alba*, *P. balsamifera* and *P. tremula*, Feb.–Apr. Found commonly, Dec.–Jan., in its *Cytospora* state; allantoid conidia 4–6 × 1 issue in conspicuous golden-yellow tendrils.

HYPHOMYCETES AND COELOMYCETES

Camarosporium propinquum, described on *Corylus*, occurs also on twigs of *Populus*.

Cytospora, v.s. under *Leucostoma niveum* and *Valsa sordida*.

Discosporium, v.s. under *Cryptodiaporthe populea*.

Excipularia fusispora, described on *Clematis*, has been recorded also on branches of *Populus*.

Phaeostromella, v.s. under *Cryptosporella populina*.

Phomopsis putator (Sacc.)Wehmeyer
Conidiomata black, erumpent through small slits in bark. Conidia fusiform, hyaline, biguttulate, 9–11 × 2.5. On dead attached twigs of *P. alba* and *P. canescens*.

Troposporella fumosa Karst. (Fig. 857)
Colonies scattered, small, pulvinate, fawn or snuff-coloured. Conidiophores up to 100×3–5. Conidia helicoid, golden brown, smooth, 12–18 diam., with filaments 3–5 thick, 7- to 15-septate, coiled $1\frac{1}{2}$ to 2 times. On bark.

PRUNUS: Almond, Apricot, Blackthorn, Bullace, Bird Cherry, Cherry Laurel, Peach and Plum

UREDINALES

Of the three rusts found on *Prunus*, one occurs on cultivated species and the other two on wild species:

On *P. armeniaca*, *P. domestica* and *P. persica*: *Tranzschelia discolor*
on *P. padus*: *Pucciniastrum areolatum*
on *P. spinosa*: *Tranzschelia pruni-spinosae*

Pucciniastrum areolatum (Fr.)Otth, 11,111
Uredinia hypophyllous on reddish spots, up to 5mm diam.; spores shortly echinulate, 15–20×10–15. Telia mostly on upper leaf surface, forming red to brown crusts up to 1cm diam; spores brown, 20–30×8–15, smooth, 2, 3 or 4 celled. On leaves of *P. padus*. Alternate host *Picea abies*.

Tranzschelia discolor (Fuckel)Tranz. & Litv., 11,111 (Fig. 858)
Uredinia hypophyllous, cinnamon brown, on small yellow or brown spots, spores 20–40×10–18, echinulate at base. Telia blackish brown; spores 30–43×20–25, upper cell brown, lower cell pale, almost smooth. On leaves of *P. armeniaca*, *P. domestica* and *P. persica*, July–Sept., uncommon. Alternate host *Anemone coronaria*.

Tranzschelia pruni-spinosae (Pers.)Diet., 11, 111 (Fig. 859)
Uredinia and spores similar to those of *T. discolor*. Teliospores with both cells dark brown and verrucose, 30–45×10–20. On leaves of *P. spinosa*, July–Sept., uncommon.

DISCOMYCETES AND TAPHRINALES

Apostemidium leptospora (Berk. & Br.)Boud. (Fig. 860)
Apothecia 1–1.5mm diam., dark blackish brown, discs greenish yellow when fresh. Ascospores hyaline, up to 280×2. On damp dead twigs of *P. padus* and sometimes other Rosaceae, Mar.–Sept.

Blumeriella jaapii (Rehm) v.Arx (Fig. 861)
Apothecia hypophyllous, 0.2–0.3mm diam., at first hemispherical, becoming urceolate, dark brown with subhyaline discs. Asci 50–60×12–14. Ascospores

oblong, often slightly curved, sometimes rounded at apex and pointed at base, hyaline, 25 × 2.5. Paraphyses filiform, thickened at apex. On rotten leaves, May. Acervuli of *Microgloeum* and *Phloeosporella* states are found on the lower surface of angular brown spots on living leaves of *P. cerasus*, *P. padus*, *P. domestica* and ornamental species. Conidia of the former are 7.5–9.5 × 1.5 and those of the latter 1-septate and 60–75 × 3. Centres of lesions may eventually fall out, leaving 'shot-holes'.

Dermea. Apothecial states of the three species which occur are very similar but the *Foveostroma* conidial states are quite different.

KEY

```
     Conidia 40–65 long .................................................................... cerasi
     Conidia 20–35 long .................................................................... 1
1.   Conidia 2.5–4 wide ..................................................................... padi
     Conidia 5–7 wide ..................................................................... prunastri
```

Dermea cerasi (Pers.)Fr. (Fig. 862)
Apothecia erumpent singly or in small groups from cracks in bark, up to 1.5mm diam., blackish brown to black, leathery. Ascospores hyaline to yellowish, occasionally 1- to 3-septate, 14–19 × 4–6. Conidiomata of *Foveostroma* state clustered, golden brown, up to 2mm diam., conidia hyaline, 0- or 1-septate, 40–65 × 3. On twigs of *P. avium*, *P. cerasus* and *P. serotina*.

Dermea padi (Alb. & Schw.)Fr. (Fig. 863)
On dead twigs of *P. domestica*, *P. padus* and *P. spinosa*.

Dermea prunastri (Pers.)Fr.
Ascospores 15–20 × 5–7.5. Conidiomata conical, black, conidia 20–30 × 5–7. On dead twigs of *P. domestica* and *P. spinosa*.

Encoelia fuckelii Dennis (Fig. 864)
Apothecia erumpent, dark brown but covered by white granules, discs pale. Ascospores 5 × 0.5. On dead branches of *P. spinosa*, Apr.

Eupropolella britannica Greenhalgh & Morgan-Jones (Fig. 865)
Apothecia amphigenous, 0.2–0.3mm diam., black, discoid, immersed below grey areas and splitting open when ripe. Mature ascospores 1- to 3-septate, pale brown, 12–18 × 5–7. Parasitic on leaves of *P. laurocerasus*, Apr.–July.

Monilinia laxa (Aderh. & Ruhl)Honey ex Whetzel
Apothecia brown, on long stalks arising from blackened mummified fruit. Ascospores hyaline 8–18 × 5–8. Paraphyses sometimes branched. *Monilia* state forms grey cushions 1–2mm diam. on flowers, fruit (causing brown rot), leaves and twigs. Conidia hyaline, 8–23 × 4–15, in long branched chains. Mostly on *P. domestica* but also on *P. armeniaca*, *P. cerasus* and *P. persica*.

Patellariopsis atrovinosa, described on *Crataegus*, has been found also on fallen dead branches of P. laurocerasus.

Pezicula houghtonii (Phill.)Groves (Fig. 866)
Apothecia up to 1mm diam., clustered, erumpent, cinnamon-yellow to brown. Ascospores hyaline to pale yellowish green, 0- to 3-septate, $22-38 \times 9-13$. On dead branches of *P. laurocerasus* and *P. lusitanica*, Sept.–Oct.

Taphrina. Four species occur, causing various forms of distortion and malformation, with infected parts often coloured bright red or yellow. Effects seen include leaf curl, witches' brooms, and 'pocket' plums. Asci form a layer over the surface of swollen fruit or swollen and convoluted areas of leaves. They contain initially 8 broadly oval or subspherical hyaline spores about $4-9 \times 3-7$ which frequently bud inside the asci to form ascoconidia.

Taphrina deformans (Berk.)Tul. (Fig. 867)
Causes leaf curl of *P. amygdalus* and *P. persica*, infected parts bright red, very conspicuous. Fairly common, June–July.

Taphrina padi (Jacz.)Mix
On flowers and fruit of *P. padus*. Fruit become elongated with persistent styles.

Taphrina pruni Tul.
Causes 'pocket' plums on *P. domestica* and both 'pocket' plums and shoot distortion on *P. spinosa*. Fruit remain pale, become elongated and often depressed on one side. Infected shoots are stunted, swollen, pale yellow tinged with red, and bear reduced strap-like leaves.

Taphrina wiesneri (Rathay)Mix
Causes witches' broom and leaf curl of *P. avium* and *P. cerasus*. Infected leaves red or pinkish and smell of new-mown hay.

Trochila laurocerasi (Desm.)Fr. (Fig. 868)
Apothecia immersed, 0.5–1mm diam., opening by teeth to expose dark olive disc. Ascospores hyaline, mostly $8-9 \times 4$. Paraphyses swollen, olivaceous. On fallen dead leaves of *P. laurocerasus*, May–Dec.

OTHER ASCOMYCETES

Amphisphaeria vibratilis (Fuckel)E. Müller
Perithecia immersed, black, thick-walled, up to 1mm diam., pustulate. Ascospores 1-septate, olivaceous brown, $18-25 \times 7-9$, with gelatinous sheaths. On dead twigs of *P. avium*, Feb.–Mar.

Calosphaeria pulchella (Pers.)Schröter (Fig. 869)
Perithecia splayed out beneath bark with the tips of their long necks protruding together through cracks. You have to strip away the bark to see them properly. Ascospores allantoid, hyaline, $5-6 \times 1$. On dead branches of

P. avium and *P. cerasus*.

Diaporthe decorticans (Lib.)Sacc. & Roum.
Perithecia 0.4–0.6mm diam., immersed in clusters of 4 to 12, with protuberant necks and a well-developed stroma dipping down between them. Ascospores hyaline, 1-septate, with 4 guttules, 14–20 × 3.5–5.5, with a small appendage at each end. On dead twigs and branches of *P. cerasus* and *P. padus*.

Diaporthe perniciosa Marchal (Fig. 870)
Perithecia immersed, with long slender black necks often protruding separately. Ascospores hyaline, 1-septate, 11–14 × 3–4. *Phomopsis* state has A-conidia 7–9 × 2–3 and B-conidia often hamate, 25–30 × 1–1.5. Parasitic on branches of *P. domestica*, *P. persica* and *P. spinosa*, Feb.–Mar.

Eutypella prunastri (Pers.)Sacc. (Fig. 871)
Perithecia 0.4–0.5mm diam., densely crowded in stromata up to 1cm long which often fissure the bark transversely and through which protrude black necks with sulcate ostioles. Ascospores allantoid, 5–7 × 1, golden brown in mass. Common on dead, often still attached branches of *P. spinosa*, Apr.–May.

Podosphaera tridactyla (Wallr.)de Bary (Fig. 872)
Powdery mildew. Mycelium tends to disappear and few conidia are formed. Cleistothecia dark brown, about 0.1mm diam., with long appendages several times dichotomously branched at tips. Ascospores 18–30 × 12–15. On living leaves and twigs of *P. domestica*, *P. laurocerasus*, *P. padus* and *P. spinosa*, July–Oct.

Polystigma fulvum DC.
Perithecia immersed in fleshy stromata which initially form yellow ochre spots up to 1cm diam., visible on both leaf surfaces. Ascospores hyaline, 14–16 × 5–6. On leaves of *P. padus*.

Polystigma rubrum (Pers.)DC. (Fig. 873)
Stromata up to almost 1mm diam., in thickened orange to yellow spots visible on both leaf surfaces. On attached leaves only the *Polystigmina* state is found, with immersed conidiomata containing large numbers of hyaline, hamate conidia, mostly 25–30 × 1. Perithecia are found in stromata which have become almost black on overwintered fallen leaves; ascospores hyaline, 11–13 × 4–5. Common on *P. spinosa* growing near the sea. Found occasionally on *P. domestica* ssp. *insititiae*.

Sphaerotheca pannosa (Wallr.)Lév. var. *persicae* Woronich.
A variety of the common powdery mildew described on *Rosa*; found occasionally on *P. persica*.

Uncinula prunastri (DC.)Sacc.
Powdery mildew. Mycelium sparse, conidia few. Cleistothecia with 20 to 30 appendages attached equatorially, coiled and slightly thickened at their tips.

Asci 10 to 15 with 6 to 8 spores each. On *P. spinosa*, especially in shady situations.

Valsa ceuthospora Cooke
Perithecia immersed, 6 to 8 in a ring, raising the epidermis which is sometimes blackened. Ascospores hyaline, almost straight, 10–12 × 2–3. On dead branches of *P. laurocerasus*.

Venturia carpophila E.E.Fisher (Fig. 874)
Pseudothecia immersed, dark brown, up to 0.15mm diam., only occasionally setose. Ascospores clavate, septate just below the middle, olivaceous, 12–16 × 3–5. On overwintered fallen dead leaves of *P. armeniaca*, *P. domestica* and *P. persica*. *Cladosporium* state on living leaves and fruit forms effuse or punctiform, dark olivaceous brown, velvety colonies responsible for disease known as 'freckle' or 'scab'. Conidia mostly in simple or branched chains of 3 or 4, pale olivaceous brown, smooth or minutely verruculose, 12–20 × 4–5.

Venturia cerasi Aderh. (Fig. 875)
Pseudothecia immersed, about 0.1mm diam., with setae around papillate ostioles. Ascospores septate below the middle, pale brown, 10–14 × 4–6. On overwintered fallen leaves of *P. cerasus*. *Fusicladium* state forms small, dark olive, velvety 'scab' spots on fruit, leaves and young twigs. Conidia olivaceous brown, 16–23 × 5–7.

HYPHOMYCETES AND COELOMYCETES

Ceuthospora lauri (Grev.)Grev. (Fig. 876)
Conidiomata immersed, orbicular, black, leathery, 0.5–1mm diam., with 1 or more locules, visible on both leaf surfaces. Conidia cylindrical, hyaline, 12–14 × 2–2.5, some with a small splayed out apical appendage. On dead leaves of *P. laurocerasus*, May–July.

Cladosporium, v.s. under *Venturia carpophila*.

Cryptocline phacidiella (Grove)v.Arx.
Acervuli epiphyllous, very small, opening by 3 or 4 teeth. Conidia hyaline, granular, cylindrical, rounded at the ends, 18–20 × 7–8. Causes whitish spots up to 1.5cm diam., with narrow brown borders, on living leaves of *P. laurocerasus*, May–June.

Cytospora. Three species are found on dead branches of *Prunus*. They all have multilocular conidiomata with black necks erumpent through small whitish discs, and allantoid conidia issuing in reddish tendrils.

KEY

Conidia 6–8 × 1.5–2 .... *prunorum* Sacc. & Syd.
Conidia 5–6 × 1 1

1. On *P. laurocerasus* .... *lauro-cerasi* Fuckel
    On *P. cerasus, P. domestica* and *P. padus* ....
    .... *Leucostoma persoonii*, described on *Sorbus*.

Foveostroma, v.s. under *Dermea*.

Fusicladium, v.s. under *Venturia cerasi*.

Microgloeum, v.s. under *Blumeriella jaapii*.

Monilia, v.s. under *Monilinia laxa*.

Phloeosporella, v.s. under *Blumeriella jaapii*.

Phomopsis, v.s. under *Diaporthe perniciosa*.

Polystigmina, v.s. under *Polystigma rubrum*.

Stigmina carpophila (Lév.)M.B.Ellis (Fig. 877)
Sporodochia punctiform, olivaceous brown or black. Conidia subhyaline or pale golden brown, mostly 30–60 × 9–18, with 3 to 7 dark transverse septa. Parasitic on leaves, dormant buds, branches, flowers and fruit of *P. amygdalus*, *P. cerasus, P. laurocerasus* and *P. persica*. On leaves small purplish spots with yellow centres appear first, later these enlarge, tissues in the middle turn brown and often drop out, leaving 'shot-holes'.

PSEUDOTSUGA: Douglas Fir

Dermea balsamea (Peck)Seaver (Fig. 878)
Apothecia erumpent, about 1mm diam., black. Ascospores hyaline, 19–24 × 4–5. Conidia of *Foveostroma* state hyaline, 3-septate, up to 95 × 3. On dead branches of *P. menziesii*, Oct.–Nov.

Lachnellula calyciformis (Batsch)Dharne (Fig. 879)
Apothecia 1–4mm diam., cream-coloured, with bright orange discs. Ascospores hyaline, 4–6.5 × 1.5–2.5. On dead twigs of *P. menziesii*, Dec.–Apr.

Phaeocryptopus gaeumannii (Rohde)Petrak (Fig. 880)
Pseudothecia superficial, less than 0.1mm diam., black, lying above stomata. Ascospores 1-septate, hyaline, 11–17 × 4–5. On the lower surface of living leaves and dead leaves of *P. menziesii*, June–July.

Phomopsis pseudotsugae Wilson
Conidiomata erumpent, black. Conidia hyaline, biguttulate, 5–9 × 2–3. On dead branches and leaves of *P. menziesii*.

Rhabdocline pseudotsugae H.Sydow (Fig. 881)
Apothecia 1–3 × 0.3mm, in rows, opening by the folding back of flaps of epidermis to expose dark brown discs. Ascospores 15–22 × 7–12, remaining hyaline and nonseptate for a long time but eventually becoming 1-septate with

one brown and one hyaline cell. Parasitic on living leaves of *P. menziesii*, May–July.

Fungi found more commonly on other conifers but which occur occasionally on *Pseudotsuga* include: *Gremmeniella abietina*, *Nectria pinea*, *Pezicula livida*, *Phomopsis occulta*, *Sirococcus strobilinus*, *Sphaeropsis sapinea* and *Strasseria geniculata*, described on *Pinus*, *Nectria fuckeliana* on *Picea*, *Sarea resinae* and *Thysanophora penicillioides* on *Abies*.

PYRACANTHA: Firethorn

Podosphaera clandestina, powdery mildew described on *Crataegus*, occurs also on *P. coccinea*.

Spilocaea pyracanthae (Otth)v.Arx (Fig. 882)
Colonies orbicular or irregular, dark olivaceous or blackish olive, velvety. Conidiophores up to 40 × 5–8. Conidia obpyriform, pale to mid olivaceous brown, 14–21 × 7–10. On leaves and fruit of *P. coccinea*.

PYRUS: Pear

UREDINALES

Gymnosporangium confusum, described on *Crataegus*, occurs occasionally on *Pyrus*.

Gymnosporangium sabinae (Dicks.)Wint., 0,1
Aecia hypophyllous, up to 2.5mm high and 1.5mm wide, pale brown. Aeciospores reddish brown, finely verruculose, up to 28 × 24. On leaves, petioles and fruit, July–Sept. Alternate host *Juniperus sabina*.

DISCOMYCETES AND TAPHRINALES

Pezicula corticola, with *Cryptosporiopsis* conidial state, described on *Malus*, is sometimes associated with bark canker of *Pyrus*.

Potebniamyces pyri (Berk. & Br.)Dennis
Apothecia up to 1mm diam., dark grey, solitary or in small groups, developing within erumpent, cushion-like stromata, their discs exposed when these split open. Ascospores hyaline, 15–23 × 8–11, often with 2 large guttules. Conidia of the *Phacidiopycnis* state ellipsoid, hyaline, 10–14 × 6–9. On dead branches, Sept.–Nov.

Taphrina bullata (Berk.)Tul.
Causes blister-like spots on living leaves.

OTHER ASCOMYCETES

Mycosphaerella pyri (Auersw.)Boerema
Pseudothecia hypophyllous, densely gregarious, immersed or in part erumpent, up to 0.14mm diam., black. Ascospores fusoid, slightly curved or sigmoid, 1-septate, 27–30 × 4. On dry leaves.

Mycosphaerella sentina (Fr.)Schröter
Perithecia amphigenous, minutely papillate, erumpent, about 0.2mm diam. Ascospores 1-septate, biconic, pale olivaceous, about 15 × 5. On leaves causing small flecks.

Venturia pirina Aderh. (Fig. 883)
Pseudothecia formed on fallen rotting overwintered leaves, 0.1–0.25mm diam., immersed, some with dark brown setae up to 50 long around ostioles. Ascospores olivaceous, with septum nearest the base, 14–20 × 4–8. Found most commonly in its *Fusicladium* state which forms dark olivaceous brown, velvety colonies on leaves, bud-scales, flowers and fruit, causing scab disease. Conidia broadly fusiform, olivaceous brown, 0- or 1-septate, 17–28 × 8–10.

HYPHOMYCETES AND COELOMYCETES

Cryptosporiopsis, v.s. under *Pezicula corticola*.

Fusicladium, v.s. under *Venturia pirina*.

Phacidiopycnis, v.s. under *Potebniamyces pyri*.

Phoma limitata (Peck)Boerema
Pycnidia epiphyllous, immersed, about 0.15mm diam., black. Conidia straight or occasionally curved, ovoid to ellipsoid, hyaline, biguttulate, 6–8.5 × 3–4.5. On living leaves causing ochraceous or greyish-brown spots 2–6mm diam., with dark brown borders, Aug.–Nov.

Phomopsis ambigua (Sacc.)Trav.
Conidiomata black, 0.25–0.35mm diam., long covered but eventually erumpent through fissures. A-conidia fusoid, biguttulate, hyaline, 8–10 × 2–3, B-conidia curved or hamate, 30–40 × 1. On dead twigs.

QUERCUS: Oaks including English Oak, Holm Oak, Red Oak, Sessile Oak, Turkey Oak
Unless specified to the contrary, all records are of fungi on either *Q. robur* or *Q. petraea*.

ON LEAVES

UREDINALES

Uredo quercus Duby, 11. (Fig. 884)
Uredinia hypophyllous, 0.25mm diam., yellow to orange, spores shortly echinulate, 15–25 × 10–18. Uncommon or seldom found because it is inconspicuous; usually on sucker shoots of *Q. ilex* and *Q. robur*.

DISCOMYCETES AND TAPHRINALES

KEY

	Asci forming palisade over leaf surface .... *Taphrina caerulescens*	
	Asci in apothecia 1	
1.	Apothecia immersed or erumpent 2	
	Apothecia superficial 5	
2.	Apothecia forming pale yellow blisters which collapse	
	.... *Laetinaevia pustulata*	
	Apothecia opening by lids .... *Stegopeziza quercea*	
	Apothecia erumpent, brown with cream discs .... *Pyrenopeziza nervicola*	
	Apothecia hysterioid, opening by slits 3	
	Apothecia opening by teeth 4	
3.	Ascospores 3–4 wide .... *Hypoderma ilicinum*	
	Ascospores 1.5 wide .... *Lophodermium petiolicolum*	
4.	Apothecia 0.2mm diam. .... *Naevala perexigua*	
	Apothecia 1mm diam., 4 teeth .... *Coccomyces dentatus*	
	Apothecia to 3mm diam., many teeth .... *Coccomyces coronatus*	
5.	Apothecia hairy 6	
	Apothecia not hairy 10	
6.	Cobwebby subiculum present .... *Arachnopeziza eriobasis*	
	No subiculum 7	
7.	Hairs of 2 kinds, mostly brown, setiform .... *Trichodiscus diversipilus*	
	Hairs of 1 kind, if setiform not brown 8	
8.	Hairs granulate or with crystals .... *Dasyscyphus*	
	Hairs smooth 9	
9.	Hairs short, cylindrical .... *Phialina parenchymatosa*	
	Hairs swollen at base, upper part very slender and curved	
	.... *Hyaloscypha pseudopuberula*	
	Hairs hyaline setiform .... *Hyaloscypha pygmaea*	
10.	Apothecia arising from small sclerotia 11	
	No sclerotia 12	
11.	Stalks of apothecia hairy .... *Ciborinia hirtella*	
	Stalks of apothecia smooth .... *Ciborinia candolleana*	
12.	Apothecia short-stalked, at first globose with small opening	
	.... *Pycnopeziza pachyderma*	

	Apothecia not so 13
13.	Apothecia rather bright yellow 14
	Apothecia not bright yellow 15
14.	Apothecia 0.5mm diam. .... *Calycellina punctiformis*
	Apothecia 2–6mm diam. .... *Hymenoscyphus epiphyllus*
15.	Ascospores filiform up to 140 × 1 .... *Pocillum cesatii*
	Ascospores not filiform, much shorter 16
16.	Apothecia stalked 17
	Apothecia sessile or subsessile 19
17.	Apothecia white to yellowish .... *Hymenoscyphus phyllophilus*
	Apothecia white with brown base .... *Allophylaria basalifusca*
	Apothecia almost black, disc grey .... *Lanzia coracina*
	Apothecia brown or hazel 18
18.	Ascospores mostly 3–4 wide .... *Rutstroemia petiolorum*
	Ascospores 5–6 wide .... *Rutstroemia sydowiana*
19.	Apothecia white or whitish 20
	Apothecia amber, yellow ochre or reddish 21
	Apothecia brown or brownish 22
	Apothecia blackish .... *Niptera muelleri-argoviensis*
20.	Ascospores 7-septate .... *Gorgoniceps charnwoodensis*
	Ascospores 1-septate .... *Calycellina rivelinensis*
21.	Apothecia translucent amber, ascospores 11–14 long ....
	.... *Pezizella roburnea*
	Apothecia yellow ochre becoming reddish, ascospores 15–19 long ....
	.... *Pezizella rubescens*
22.	Ascospores 6–9 long .... *Mollisia rabenhorstii*
	Ascospores 11–15 long .... *Mollisia spectabilis*

Allophylaria basalifusca Graddon (Fig. 885)
Apothecia up to 0.5mm diam., short-stalked, white except at the base where they are blackish brown; excipular hyphae with thick glassy walls. Ascospores hyaline, 18–25 × 3.5–4. Crowded along and adjacent to midribs on the lower side of fallen dead leaves, Dec.

Arachnopeziza eriobasis (Berk.)Korf (Fig. 886)
Apothecia 0.5–1.5mm diam., pale yellow, fringed with very slender, matted, hyaline hairs and seated on a cobwebby subiculum. Ascospores hyaline, 1-septate, 6–11 × 1.5–2. On fallen dead leaves in damp situations, July–Oct.

Calycellina punctiformis (Grev.)Höhnel (Fig. 887)
Apothecia up to 0.5mm diam., downy, bright lemon yellow, with a ring of dark brown cells at the base. Ascospores hyaline, 1-septate, 12–16 × 1.5–2. On fallen dead leaves of *Q. borealis*, *Q. petraea* and *Q. robur*, July–Oct., common.

Calycellina rivelinensis Dennis (Fig. 888)
Apothecia sessile, 0.1mm diam., white. Ascospores hyaline, 1-septate, 10–12

× 3. On fallen rotting leaves, June–Sept.

Ciborinia candolleana (Lév.)Whetzel (Fig. 889)
Apothecia 1–4mm diam., long-stalked, yellowish brown, arising from small round lens-shaped, black sclerotia. Ascospores hyaline, mostly 7–9 × 3–4. On rotting leaves, May–July. The fungus causes brown spots on living leaves in late summer.

Ciborinia hirtella (Boud.)Batra & Korf (Fig. 890)
Apothecia about 1mm diam., pale brown, arising from black, bean-shaped sclerotia 2 × 1mm, their often very long stalks covered with pale brown hairs up to 160 × 4.5. Ascospores hyaline, 6–9 × 3–4. Among buried leaf litter, Mar.–June.

Coccomyces coronatus (Schum.)de Not. (Fig. 891)
Apothecia similar to those of the following species, *C. dentatus*, but up to 3mm diam. and opening by numerous teeth. Ascospores 30–60 × 2–3.5. On bleached patches on fallen dead leaves, Aug.–Nov. *Leptothyrium* state has mostly epiphyllous conidiomata 0.15–0.2mm diam., black, shining. Conidia hyaline, 3–5 × 0.5–1.

Coccomyces dentatus (Kunze & Schm.)Sacc. (Fig. 892)
Apothecia embedded, up to 1mm diam., flat, black, shining, opening by 4 teeth to expose the pale translucent yellow disc. Ascospores hyaline, 45–55 × 1–2. On bleached patches on fallen dead leaves. Nov.–Apr.

Dasyscyphus. Seven species occur on fallen dead leaves; all have hairy apothecia.

KEY

	Apothecia brown, common on *Q. ilex* .... *fuscescens*
	Apothecia white or pale 1
1.	Hairs without crystals, cylindrical or slightly swollen at tip, ascospores 6–10 × 1.5–2.5 .... *virgineus*
	Hairs commonly with crystals, especially at tip 2
2.	Discs orange–yellow .... *patulus*
	Discs very pale or creamy 3
3.	Ascospores not more than 6 long 4
	Ascospores 9 or more long 5
4.	Hairs very thick-walled, up to 70 × 7 .... *capitatus*
	Hairs thin-walled, up to 20 × 4 .... *coruscatus*
5.	Ascospores 15–23 × 2–2.5 .... *ciliaris*
	Ascospores 9–11 × 2.5–3 .... *soppittii*

Dasyscyphus capitatus (Peck)Le Gal (Fig. 893)
Apothecia up to 0.3mm diam., short-stalked, white, covered with thick-walled, hyaline hairs up to 70 × 7, capped by groups of large crystals. Ascospores 5–6 × 1. On fallen dead leaves, June–Aug.

Dasyscyphus ciliaris (Schrader)Sacc. (Fig. 894)
Apothecia stalked, up to 0.5mm diam., white with creamy discs, covered with thin-walled, hyaline hairs, up to 100 × 6, capped by groups of small crystals. Ascospores hyaline, some 1-septate, 15–23 × 2–2.5. On fallen dead leaves, July–Oct.

Dasyscyphus coruscatus Graddon (Fig. 895)
Apothecia sessile, up to 0.5mm diam., like little round pieces of crystallised ginger, covered with thin-walled, hyaline hairs up to 20 × 4, capped by groups of sharp-pointed crystals. Ascospores hyaline, 4–6 × 1.5–2. On fallen dead leaves of *Q. borealis* and *Q. robur*, Oct.–Dec.

Dasyscyphus fuscescens (Pers.)Rehm (Fig. 896)
Apothecia stalked, up to 1mm diam., brown, hairy, with cream or pearly discs. Hairs up to 90 × 5–6, thick-walled, reddish brown except for one or two cells at the apex, which are hyaline and often capped by large crystals. Ascospores hyaline, 6–9 × 2–2.5. Very common on fallen dead leaves of *Q. ilex*, Sept.–June.

Dasyscyphus patulus (Pers.)Sacc.
Apothecia up to 0.5mm diam., white, hairy, with orange–yellow discs, very similar to and perhaps only a form of *D. bicolor* which occurs on dead twigs. Ascospores hyaline, 7–9 × 2–2.5. On fallen dead leaves Apr.–May.

Dasyscyphus soppittii Massee (Fig. 897)
Apothecia stalked, up to 2mm diam., white, hairy, with creamy discs. Hairs tapered, thin-walled, hyaline, some with crystals near the tips or along the sides. Ascospores hyaline, 9–11 × 2.5–3. Common on fallen dead leaves of *Q. borealis* and *Q. robur*, July–Feb.

Dasyscyphus virgineus, described on 'leaf litter'.

Gorgoniceps charnwoodensis, described on *Castanea* cupules, has been found on dead leaves of *Quercus*.

Hyaloscypha pseudopuberula Graddon (Fig. 898)
Apothecia subsessile, 0.3–0.5mm diam., white, downy, with cream discs. Hairs about 20 long, distally very slender and curved. Ascospores hyaline, 9–12 × 3–4. On both sides of rotten fallen leaves, Sept.–Jan.

Hyaloscypha pygmaea (Mouton)Boud. (Fig. 899)
Apothecia white or whitish, about 0.1mm diam., hairs setiform, pointed at tip, swollen at base. Ascospores hyaline, 3–4 × 1–1.5. On fallen dead leaves, Nov.–Dec.

Hymenoscyphus epiphyllus (Pers.)Rehm ex Kaufmann (Fig. 900)
Apothecia sessile or subsessile, usually about 2mm diam., but occasionally up to 6mm, bright yellow. Paraphyses often branched. Ascospores hyaline,

sometimes 1-septate, 16–19 × 3.5–5. On decaying leaves and other litter, Sept.–Nov.

Hymenoscyphus phyllophilus, described on *Fagus* leaves, occurs also occasionally on *Quercus*.

Hypoderma ilicinum de Not. (Fig. 901)
Apothecia subcuticular, elliptical, black, opening by slits. Ascospores hyaline, guttulate, 20–25 × 3–4. On large, pale spots surrounded by thin black lines, on dead leaves of *Q. borealis* and *Q. ilex*, Sept.–Nov.

Laetinaevia pustulata Graddon (Fig. 902)
Apothecia hypophyllous, embedded, up to 0.3mm diam., appearing when fresh as pale yellow blisters, when dry collapsing to form pale orange cups. Paraphyses with pale brown tips. Ascospores hyaline, fusiform, 11.5–13 × 3–4. On fallen dead leaves, Dec.

Lanzia coracina (Durieu & Lév.)Spooner (Fig. 903)
Apothecia stalked, 0.5–0.8mm diam., almost black, with greyish discs and toothed margins. Ascospores hyaline, 11–13 × 5–6. On petioles and veins of fallen leaves of *Q. ilex*, sometimes on blackened areas, Mar.

Lophodermium petiolicolum Fuckel (Fig. 904)
Apothecia elliptical, black, 1–1.5 × 0.5mm, opening by slits. Ascospores hyaline, 30–50 × 1.5. On petioles and main veins of fallen dead leaves, June.

Mollisia rabenhorstii (Auersw.)Rehm
Apothecia 0.2–0.5mm diam., reddish brown, with greyish-yellow discs; excipulum of large-celled parenchyma drawn out at the edge into blunt, septate, 30–36 × 6–9, hairs. Ascospores hyaline, spindle-shaped, 6–9 × 1.5–2. On lower surface of fallen dead leaves, June–Aug.

Mollisia spectabilis Kirschst. (Fig. 905)
Apothecia 1.5–3mm diam., greyish or yellowish brown, velvety, with pale opaque discs. Ascospores hyaline, 11–15 × 4–5. On both surfaces of fallen dead leaves, especially in wet places, Oct.–Nov.

Naevala perexigua (Rob. ex Desm.)K. Holm & Holm (Fig. 906)
Apothecia immersed, 0.2mm diam., with pale fawn discs exposed by the splitting of a dark brown covering layer into 4 or 5 lobes. Ascospores hyaline, 6–8 × 3–3.5. Scattered over the underside of fallen dead leaves, May–June.

Niptera muelleri-argoviensis Rehm (Fig. 907)
Apothecia up to 0.5mm diam., blackish, with pale to dark grey discs. Ascospores hyaline, 1-septate, 9–10 × 1.5–2. On the underside of fallen dead leaves, typically of *Q. ilex*, Aug.–Oct.

Pezizella roburnea Velen. (Fig. 908)
Apothecia 0.2–0.5mm diam., translucent amber, each with a dark border and a dark spot in the centre where attached to the leaf. Swollen tips of

paraphyses contain yellow oil drops. Ascospores hyaline, 11–14 × 3–3.5. Usually on bleached areas on dead leaves of *Q. borealis* and *Q. robur*, Sept.–Feb.

Pezizella rubescens Mouton (Fig. 909)
Apothecia up to 0.5mm diam., pale yellow-ochre or becoming slightly reddish, each with a dark brown ring forming the base. Ascospores hyaline, mostly 15–19 × 3–3.5. On fallen dead leaves of *Q. borealis* and *Q. robur*, in wet places, Aug.–Dec.

Phialina parenchymatosa (Velen.)Graddon (Fig. 910)
Apothecia sessile, 0.1–0.2mm diam., white or yellowish, downy. Hairs cylindrical, hyaline, up to 25 × 5. Paraphyses swollen at tips, often branched. Ascospores hyaline, 4–6.5 × 2. On lower surface of fallen dead leaves, Nov.

Pocillum cesatii (Mont.)de Not. (Fig. 911)
Apothecia 0.3–0.4mm diam. and about 0.5mm tall, dark brown with pale discs and margins. Ascospores hyaline, up to 140 × 1. On fallen dead leaves, Mar.–May.

Pycnopeziza pachyderma (Rehm)White & Whetzel (Fig. 912)
Apothecia at first globose with a small opening then saucer-shaped, 1–4mm diam., yellowish brown, furfuraceous, with short blackish-brown stalks. Ascospores hyaline, 7–9 × 3–4. Conidiomata of *Acarosporium* state erumpent, 0.4–1mm diam., black, with stellate dehiscence, the lobes folding back to expose a pinkish mass of conidia which are cylindrical, 1-septate, hyaline, 15–22 × 2.5–3 and formed in dichotomously branched chains. On petioles and main veins of fallen dead leaves, Apr.–June.

Pyrenopeziza nervicola, described on *Castanea*, occurs also on *Quercus*.

Rutstroemia petiolorum, described on *Fagus*, is found occasionally on petioles of old fallen leaves of *Quercus*, Oct.–Nov.

Rutstroemia sydowiana (Rehm)White (Fig. 913)
Apothecia stalked, up to 3mm diam., golden brown, striate; discs with toothed margins. Ascospores hyaline, 12–15 × 5–6. On petioles of old fallen leaves, Aug.–Dec.

Stegopeziza quercea (Fautr. & Lambotte)Spooner (Fig. 914)
Apothecia about 1mm diam., brown, with discs exposed by the pushing off of little lids; marginal hairs clavate, pale brown, up to 50 × 15, encrusted with crystals. Paraphyses lanceolate, up to 100 × 7. Ascospores hyaline, 5–7 × 1–2. On suckers, May. The type was on *Q. borealis*.

Taphrina caerulescens (Desm. & Mont.)Tul.
Causes discoloured spots, raised above, concave below, where they are quite pale. On living leaves, uncommon.

Trichodiscus diversipilus Graddon (Fig. 915)
Apothecia up to 0.15mm diam., setose, pale brown. Setae brown, up to 120 long, interspersed with much shorter, obtuse, pale, granulate hairs. Ascospores hyaline, 5–8 × 1.5–2. On fallen dead leaves, Oct.–Nov.

OTHER ASCOMYCETES

Apiognomonia errabunda, described on *Fagus*, is common also on fallen dead leaves of *Quercus*.

Epibelonium gaeumannii Müller (Fig. 916)
Ascomata hypophyllous, superficial, brown, 0.15–0.25mm diam. Asci bitunicate, 2- to 4-spored, covered by a brownish epithecium. Ascospores hyaline, mostly 3-septate, 11–14 × 3.5–4.5. On fallen leaves of *Q. ilex*, Mar.

Guignardia punctoidea (Cooke)Schröter (Fig. 917)
Perithecia epiphyllous, in small groups, 0.15–0.2mm diam., black, thick-walled. Ascospores hyaline, 8–13 × 3–5. On fallen dead leaves, Apr.–May.

Hyponectria cookeana (Auersw.)Barr (Fig. 918)
Perithecia immersed, up to 0.15mm diam., dark grey to black, visible through epidermis on lower surface. Ascospores hyaline, 8–10 × 2–4. On fallen dead leaves, Apr.–May.

Hypospilina bifrons (DC.)Traverso
Perithecia 0.25mm diam., immersed singly or in small groups in black stromata. Ascospores hyaline, with a septum near the base, 11–13 × 4. On fallen dead leaves, Mar.–May.

Microsphaera alphitoides Griffon & Maublanc (Fig. 919)
Powdery mildew. Colonies amphigenous. Conidia plentiful, ellipsoidal. Asci 5 to 12, each with 8 spores. Especially common on leaves of sucker growths. Cleistothecia commoner in some years than others, ascospores ripe Oct.–Nov.

Microthyrium ilicinum de Not. (Fig. 920)
Thyriothecia amphigenous, up to 0.27mm diam., brown, with cells around ostiole not forming a dark collar. Asci 8-spored, spores hyaline, 1-septate, 11–14 × 2.5–3.5, with 4 apical cilia. On rotting leaves, especially those of *Q. ilex*, but also on *Q. cerris*, *Q. petraea* and *Q. robur*, causing a greying of yellow–brown leaves, Apr.–Oct.

Microthyrium microscopicum Desm. (Fig. 921)
Thyriothecia mostly epiphyllous, up to 0.18mm diam., with cells around ostiole thickened and forming a dark collar. Asci 8-spored. Ascospores hyaline, 9–14 × 2–3.5, 1-septate with 2 apical cilia. Common all the year round on fallen dead leaves of *Q. borealis* and *Q. robur*.

Mycosphaerella punctiformis (Pers.)Starb. (Fig. 922)
Pseudothecia hypophyllous, immersed, 0.1–0.14mm diam., black. Ascospores hyaline, 1-septate, 8–14 × 2–3. On fallen dead leaves of *Q. cerris* and *Q. robur*.

Plagiostoma pustula (Pers.)v.Arx (Fig. 923)
Perithecia 0.25–0.35mm diam., brown, immersed but seen easily with the naked eye, solitary or in groups often close to veins. Ascospores hyaline or pale olive, 1- to 3-septate, 17–24 × 4–5. On fallen dead leaves of *Q. borealis* and *Q. robur*, Mar.–Aug.

Sphaerulina myriadea (DC.)Sacc. (Fig. 924)
Pseudothecia 0.1–0.15mm diam., forming greyish patches on upper surface of fallen dead leaves, May–June. Asci in bundles attached by their bases. Ascospores hyaline, 3-septate, 25–38 × 2.5–3.

HYPHOMYCETES

Beltrania querna Harkn. (Fig. 925)
Colonies effuse, velvety or hairy, brown to black. Setae dark brown, up to 400 long, with radially lobed basal cells. Conidiophores pale olive or brown, up to 200 × 3–7. Conidia biconic with appendage at free end, olivaceous brown, with a hyaline band well above the centre, 18–30 × 7–10, appendage 2–6 long. On fallen dead leaves of *Q. ilex*, common.

Chalara hughesii Nag Raj & Kendrick (Fig. 926)
Colonies effuse, brown, velvety. Conidiophores often consisting of a phialide only, brown up to 60 × 7.5, with long cylindrical necks 2–3 wide. Conidia catenate, hyaline, 1-septate, 12–16 × 2–2.5. On fallen dead leaves of *Q. ilex*.

Chalara kendrickii Nag Raj (Fig. 927)
Colonies effuse, brown, hairy. Conidiophores often consisting of a phialide only, brown, up to 60 × 10, with necks about 5 wide. Conidia hyaline, 1-septate, 8–12 × 3–4. On fallen dead leaves of *Q. borealis*, Dec.

Cylindrium elongatum Bon. (Fig. 928)
Colonies effuse, white. Conidia hyaline, 15–18 × 2, formed in long chains. Common on fallen decaying leaves, especially those of *Q. ilex*, July–Apr.

Dictyochaeta querna P.M. Kirk (Fig. 929)
Colonies effuse, blackish brown to black, hairy. Setae up to 400 × 5–8, dark brown, often fertile. Conidiophores pale brown, 30–150 × 4–5, polyphialidic. Conidia curved, hyaline, mostly 15–18 × 1.5–2. On dead leaves and cupules, July–Sept.

Diplocladiella scalaroides, described on *Ulex*, has been found also on dead leaves and peduncles of *Q. ilex* and *Q. robur*.

Fusidium aeruginosum Link (Fig. 930)
Colonies effuse, bright greenish yellow, velvety. Conidiophores forming a superficial palisade, 14–18 × 2. Conidia 17–23 × 2–2.5, in long chains. A very common and early coloniser of fallen leaves.

Henicospora minor, described on *Rubus*, occurs also on fallen dead leaves of *Q. ilex*.

Lobatopedis foliicola P.M.Kirk (Fig. 931)
Colonies amphigenous, effuse, dark blackish brown, hairy. Conidiophores dark brown, up to 100 × 8–15, with radially lobed basal cells. Conidia, some in chains, mid dark brown, 7- to 14-septate, 50–90 × 8–15. On fallen dead leaves of *Q. borealis* and *Q. robur*, Aug.–Oct.

Parapleurotheciopsis ilicina P.M.Kirk (Fig. 932)
Colonies effuse, brown, hairy, rather inconspicuous. Conidiophores olivaceous brown, up to 150 × 4–5. Conidia catenate, hyaline to very pale brown, 0- to 3-septate, 16–24 × 4–5, developing from denticles on solitary terminal ramoconidia. On fallen dead leaves of *Q. ilex*, Apr.–Oct.

Subramaniomyces fusisaprophyticus (Matsushima)P.M.Kirk (Fig. 933)
Colonies discrete or effuse, white to buff, velvety. Conidiophores pale brown, up to 30 × 3.5–5.5. Conidia in short, sometimes branched chains, attached to denticles, mostly fusiform, subhyaline to pale olivaceous brown, 17–19 × 2.5–3.5, terminal ones brown, 25–30 × 2.5–3. On decaying leaves of *Q. ilex*, Aug.

Subulispora britannica Sutton (Fig. 934)
Colonies effuse, greyish brown. Conidiophores brown, up to 35 × 3–4. Conidia hyaline, 2- to 8-septate, 45–85 × 2.5–3. On fallen dead leaves of *Q. ilex*.

COELOMYCETES

Camarosporium oreades (Fr. apud Duby)Sacc. (Fig. 935)
Pycnidia in concentric circles. Conidia brown, with transverse and longitudinal septa, 9–12 × 7–9. Causes greyish yellow spots on fading leaves, Aug.–Sept.

Cryptocline cinerescens (Bubák)v.Arx (Fig. 936)
Acervuli pale to dark brown. Conidia pale brown, 8–20 × 5–8. On fallen dead leaves in litter.

Leptothyrium ilicinum Sacc. (Fig. 937)
Conidiomata mostly hypophyllous, 0.2–0.35mm diam., blackish brown. Conidia hyaline, pale olivaceous in mass, 20–27 × 3. On fallen dead leaves of *Q. ilex*, Feb.–Mar.

Pilidium acerinum, described on 'leaf litter', is common on fallen leaves of *Quercus*.

Septoria quercicola (Desm.)Sacc.
Pycnidia hypophyllous, brown, about 0.1mm diam. Conidia slightly curved, becoming 3-septate, 25–30 × 3–4. On reddish-brown spots 1–2mm diam.

ON FRUIT (ACORNS AND CUPULES)

Arachnopeziza aurelia (Pers.)Fuckel (Fig. 938)
Apothecia 0.5–3mm diam., golden yellow, with long orange hairs, seated on a thin, white or yellowish, cobwebby subiculum. Ascospores hyaline, 3-septate, mostly 15–19 × 4. Common, especially on fallen cupules, but also on other debris, Oct.–June.

Ciboria batschiana (Zopf)Buchwald (Fig. 939)
Apothecia up to 1.5cm diam., brown, stalked, arising from blackened cotyledons. Ascospores hyaline, mostly 7–10 × 4–5. On old fallen mummified acorns, Sept.–Nov.

Hymenoscyphus fructigenus, described on *Corylus*, is common also on *Quercus* cupules.

Phomopsis glandicola Grove
Conidiomata black, arranged in lines. Conidia hyaline, 6–7 × 1.5–2. On fallen acorns.

Scopinella caulincola (Fuckel)Malloch (Fig. 940)
Perithecia superficial or with the base immersed, dark reddish brown, about 0.2mm diam., with necks up to 0.8mm long, fimbriate at their tips. Asci evanescent. Ascospores brown, 9–11 × 7–8. Inside fallen acorn cups, Aug.

Tricladium castaneicola, described on *Castanea*, has been found also on *Quercus* cupules.

ON WOOD AND BARK

DISCOMYCETES

KEY

    Apothecia immersed to erumpent opening by slits 1
    Apothecia superficial or becoming so 2
1. Apothecia in dense groups, blackish .... *Ascodichaena rugosa*
    Apothecia with cream discs, splitting bark transversely .... *Colpoma quercinum*
2. Apothecia hairy 3
    Apothecia not hairy 11
3. Apothecia on cobwebby subiculum .... *Eriopeziza caesia*

	No cobwebby subiculum 4
4.	Apothecia brown or blackish brown 5
	Apothecia white or pale coloured 7
5.	Apothecia 0.15–0.2mm diam. .... *Dematioscypha dematiicola*
	Apothecia 3–4mm diam. 6
6.	Ascospores up to 14 long .... *Haglundia elegantior*
	Ascospores up to 9 long .... *Haglundia perelegans*
7.	Hairs smooth 8
	Hairs wholly or partly granulate 9
8.	Hairs tapered, up to 30 long .... *Hyaloscypha hyalina*
	Hairs cylindrical, up to 70 long .... *Hyaloscypha quercina*
9.	Discs orange, hairs granulate .... *Dasyscyphus bicolor*
	Discs not orange, hairs with smooth end cell 10
10.	Crystals between hairs .... *Dasyscyphus crystallinus*
	No crystals .... *Dasyscyphus niveus*
11.	Apothecia bluish green 12
	Apothecia not bluish green 13
12.	Ascospores 6–8 long .... *Chlorociboria aeruginascens*
	Ascospores 10–14 long .... *Chlorociboria aeruginosa*
13.	Ascospores brown, kidney-shaped .... *Bulgaria inquinans*
	Ascospores muriform .... *Triblidium caliciiforme*
	Ascospores hyaline, not muriform 14
14.	Ascospores over 160 long, filiform .... *Vibrissea flavovirens*
	Ascospores less than 100 long 15
15.	Ascospores 70–80 long, multiseptate .... *Schizoxylon friabilis*
	Ascospores 30–40 long, to 7-septate .... *Strossmayeria basitricha*
	Ascospores less than 30 long 16
16.	Apothecia stalked 17
	Apothecia sessile 18
17.	Apothecia brown, up to 2cm diam. .... *Rutstroemia firma*
	Apothecia whitish, disc convex .... *Cudoniella acicularis*
18.	Apothecia clustered, erumpent 19
	Apothecia not clustered and erumpent 20
19.	Apothecia cinnamon, pruinose .... *Pezicula cinnamomea*
	Apothecia brown, discs drying yellow .... *Mollisia discolor* var. *longispora*
20.	Apothecia ochraceous with yellow discs .... *Bisporella fuscocincta*
	Apothecia black or blackish 21
21.	Ascospores 1-septate, 10–15 long .... *Niptera exsiliens*
	Ascospores 0- or 1-septate, 7–12 long .... *Durella commutata*
	Ascospores 3-septate 15–20 long .... *Durella macrospora*

Ascodichaena rugosa, described on *Fagus*, occurs also on *Quercus*.

Bisporella fuscocincta (Graddon)Dennis (Fig. 941)
Apothecia sessile, up to 1.25mm diam., dark ochraceous with yellowish discs, each with a narrow, very dark margin. Ascospores hyaline, 6–9 × 2. Paraphyses branched near base. On cut surface of sawn-off trunks, Dec.

Bulgaria inquinans (Pers.)Fr. (Fig. 942)
Apothecia erumpent, up to 4cm diam., gelatinous, blackish brown, scurfy, with smooth black discs. Ascospores dark brown, kidney-shaped, 10–13 × 6–7. On fallen trunks and branches, Oct.–Apr.

Chlorociboria aeruginascens (Nyl.)Kanouse ex Ramamurthi, Korf & Batra (Fig. 943)
Apothecia up to 5mm diam., stalked, bluish green, the discs sometimes yellowish green. Ascospores hyaline, mostly 6–8 × 1–1.5. On rotting logs and branches, the wood of which is stained blue–green over large areas, May–Nov.

Chlorociboria aeruginosa (Pers.)Seaver ex Ramamurthi, Korf & Batra (Fig. 944)
Apothecia rather similar to those of *C. aeruginascens* but dark blue–green, with thicker stipes and yellow discs. Ascospores hyaline, 10–14 × 1.5–2.5. On rotting logs and branches; this also stains the wood blue–green.

Colpoma quercinum (Pers.)Wallr. (Fig. 945)
Apothecia up to 2mm long, dark grey to black, developing below bark, which splits usually transversely or obliquely; discs cream-coloured. Ascospores hyaline, 70–90 × 1.5. *Conostroma* state has hyaline conidia 5–7 × 1.5. On dead attached or fallen twigs, Oct.–July, common.

Cudoniella acicularis (Bull.)Schröter (Fig. 946)
Apothecia 2–4mm diam., with convex discs and tall stout stalks, white becoming pale greyish brown when old. Ascospores hyaline, 0- to 3-septate, 15–23 × 4–5. Often crowded in crevices on rather hard old stumps, Sept.–Dec.

Dasyscyphus bicolor (Bull.)Fuckel (Fig. 947)
Apothecia 1–2mm diam., white, hairy, with orange discs; hairs thick-walled up to 200 × 5. Ascospores hyaline, 7–10 × 1.5–2. On fallen dead twigs, uncommon.

Dasyscyphus crystallinus (Fuckel)Sacc. (Fig. 948)
Apothecia white to honey-coloured, about 1mm diam., long-stalked, hairy. Hairs granulate except for the smooth swollen end cells, crystals present between hairs. Paraphyses long, lanceolate, up to 5 thick, often with one or more septa. Ascospores hyaline, 8–14 × 2–2.5. On dead branches and occasionally on petioles, Mar.–Apr.

Dasyscyphus niveus (Hedw.)Sacc. (Fig. 949)
Apothecia white, drying cream, about 1mm diam., stalked, hairy, the hairs often gummed together giving the appearance of water droplets on the

surface. Hairs granulate except for smooth swollen end cells. Paraphyses 2 thick. Ascospores hyaline, 6–8 × 1.5. On old stumps and decorticated wood, Sept.–May, common.

Dematioscypha dematiicola (Berk. & Br.)Svrček (Fig. 950)
Apothecia 0.15–0.2mm diam., with dark olivaceous brown discs fringed with hyaline to very pale brown hairs about 50 long. Ascospores hyaline, biguttulate, 5–8 × 2. *Haplographium* state very common all the year round, the dark brown to black conidiophores up to 250 long, forming effuse colonies with numerous raised, spherical, hyaline or cinnamon, glistening, slimy, conidial heads clearly visible under a dissecting microscope. Conidia hyaline, 2.5–4 × 1–2. Apothecia found Oct.–Apr., on fallen dead branches.

Durella commutata Fuckel (Fig. 951)
Apothecia about 0.5mm diam., black. Ascospores hyaline, 0- or 1-septate, 7–12 × 2.5–3. Tips of paraphyses embedded in an olivaceous matrix forming an epithecium. On decorticated wood, Jan.–Apr.

Durella macrospora Fuckel (Fig. 952)
Apothecia up to 0.4mm diam., black. Ascospores hyaline, 3-septate, 15–20 × 3.5–4. Paraphyses branched near their apices, which are swollen and dark brown. On decorticated wood.

Eriopeziza caesia (Pers.)Rehm (Fig. 953)
Apothecia 0.2–0.5mm diam., olivaceous brown, covered with narrow whitish hairs, discs bluish grey, seated on a white, cobwebby subiculum. Ascospores hyaline, 5–6 × 1.5–2. On decorticated wood all the year round.

Haglundia elegantior Graddon (Fig. 954)
Apothecia up to 4mm diam., dark olivaceous brown, discs bluish grey, like small birds' nests, frosted at the margin with hyaline or very pale hairs up to 70 × 5. Ascospores hyaline, 8–14 × 2–4. On very decayed stumps and the underside of old logs, Nov.–May.

Haglundia perelegans Nannf. (Fig. 955)
Apothecia up to 3mm diam., blackish brown, hairy, discs grey or greyish yellow, hairs rather pale greyish brown, up to 120 × 4. Ascospores hyaline, 7–9 × 1.5. On rotten wood, May–Oct.

Hyaloscypha hyalina (Pers.)Boud. (Fig. 956)
Apothecia in swarms, 0.2–0.5mm diam., white, fringed with hyaline, tapered hairs up to 30 × 3. Ascospores hyaline, 6–10 × 2–3. Very common on dead branches all the year round.

Hyaloscypha quercina Velen.
Apothecia gregarious, 0.3–1mm diam., lobed, white, fringed with septate, cylindrical hairs 25–70 × 2–3. Ascospores hyaline, narrowly cylindrical, rounded at ends, slightly curved, 5–8 × 2. On logs, Mar.–May.

Mollisia discolor (Mont.)Phill. var. *longispora* Le Gal (Fig. 957)
Apothecia often lobed, up to 1.5mm diam., externally dark brown, with discs which become quite yellow when dry, in clusters erumpent through cracks in bark. Ascospores hyaline, 9–16 × 2–2.5. Common on dead attached and fallen twigs, Mar.–Aug.

Niptera exsiliens Speg.
Apothecia erumpent, 0.25–1mm diam., sessile or shortly stalked, blackish with pale grey discs, white at margin. Ascospores hyaline, slightly clavate, 1-septate, 10–15 × 2–3. On fallen dead twigs, sometimes growing on *Colpoma quercinum*, Dec.–Feb.

Pezicula cinnamomea (DC.)Sacc. (Fig. 958)
Apothecia erumpent, usually in clusters, cinnamon, pruinose. Ascospores hyaline, mostly 3-septate, 17–30 × 6–9. *Cryptosporiopsis* state has hyaline conidia 19–27 × 9.5–11.5. On dead twigs and small branches, Sept.–Jan.

Rutstroemia firma (Pers.)P.Karsten (Fig. 959)
Apothecia up to 2cm diam., yellowish brown, with blackish brown stalks up to 3cm long. Ascospores hyaline, becoming 3- to 5-septate, often budding off tiny spherical conidia, mostly 15–18 × 5–6. On fallen dead branches, Sept.–Jan., quite common.

Schizoxylon friabilis (Phill. & Plowr.)Dennis (Fig. 960)
Apothecia superficial, 0.4–0.6mm diam., whitish powdery, at first hemispherical with a pore, then splitting open more widely to expose the brownish yellow discs. Balls of crystals between hyphae in excipulum. Ascospores hyaline, 70–80 × 3, multiseptate. On dead branches, June–Sept., uncommon.

Strossmayeria basitricha (Sacc.)Dennis (Fig. 961)
Apothecia about 0.5mm diam., white, drying amber. Ascospores hyaline, becoming up to 7-septate, 30–40 × 4–5. On decorticated wood, seated among dark brown conidiophores of *Pseudospiropes simplex*, uncommon.

Triblidium caliciiforme Rebent. (Fig. 962)
Apothecia erumpent, up to 3mm diam., almost black, leathery, opening by teeth to expose sunken grey discs. Asci 4-spored. Ascospores muriform, hyaline to pale yellow, 35–55 × 15–21. On bark, Sept.–Nov., rare.

Vibrissea flavovirens (Pers.)Korf & Dixon
Apothecia up to 2.5mm diam., dark blackish brown, with greenish-yellow discs. Ascospores hyaline, thread-like, 160–190 long, only partially filling asci which are 260–300 × 6. On fallen trunks near streams, May–June.

OTHER ASCOMYCETES

Amphisphaeria bufonia (Berk. & Br.)Ces. & de Not. (Fig. 963)
Perithecia immersed, 0.7–0.8mm diam., black, raising the periderm in

pimples, with necks just pushing through to the surface. Ascospores brown, 1-septate, 20–30 × 9–11, with gelatinous sheaths. On dead attached branches, Dec.–Feb.

Anthostoma dryophilum (Currey)Sacc. (Fig. 964)
Perithecia immersed in greenish-brown pustulate stromata. Ascospores mid dark brown, 10–15 × 3–5. On fallen rotten branches.

Botryosphaeria melanops (Tul.)Wint. (Fig. 965)
Pseudothecia and conidial locules often in the same stromata. Ascospores hyaline, 35–47 × 16–20. Conidia hyaline 40–50 × 9–11. On fallen branches, Mar.–Apr.

Calosphaeria cyclospora (Kirschst.)Petrak (Fig. 966)
Perithecia black, 0.7–0.9mm diam., clustered below the bark, with their necks erumpent through cracks. Ascospores hyaline, horseshoe-shaped, about 4 × 1 measured across the tips not round the curve. On fallen branches. The bark has to be cut away to see the fungus properly.

Calosphaeria dryina (Currey)Nitschke (Fig. 967)
Perithecia 0.4–0.6mm diam., brown to black, in small clusters below the bark with necks erumpent through cracks. Ascospores hyaline, curved, some more strongly than others, occasionally septate, 10–12 × 2. On dead branches, Dec.–Jan.

Calospora arausiaca (Fabr.)Sacc. (Fig. 968)
Perithecia about 0.3mm diam., in groups of 4 to 12 per stroma. Ascospores hyaline, with 1 to 5, mostly 3, septa, 43–63 × 9–11. On fallen dead branches.

Caryospora callicarpa (Currey)Nitschke ex Fuckel (Fig. 969)
Pseudothecia erumpent, remaining partly immersed in wood, 0.5–0.75mm diam., black. Ascospores 60–75 × 28–32, 4- or 5-septate, two large central cells dark brown, smaller cells at each end pale. On decorticated wood, Mar.–May, uncommon.

Caudospora taleola (Fr.)Starb. (Fig. 970)
Perithecia 0.3–0.4mm diam., in clusters, with small blackish ostiolar discs exposed through cracks in bark. Ascospores hyaline, 1-septate, mostly 20–24 × 8–9, some with hyaline appendages at each end and laterally near the septum. The appendages are seen most clearly in water mounts but they are often lacking on old spores and the fungus then looks like a *Diaporthe*, readily distinguished from *D. leiphemia* by its much wider spores. On dead attached and fallen branches, Jan.–Oct., quite common.

Ceratocystis moniliformis, described on *Fagus*, occurs also occasionally on *Quercus*.

Diaporthe leiphemia (Fr.)Sacc. (Fig. 971)
Perithecia 0.4–0.6mm diam., in groups of 5 to 20, in stromata which break

through the periderm at regular intervals giving it the appearance of a nutmeg grater. Ostiolar discs pale brown, becoming cracked and roughened when old. Ascospores hyaline, 1-septate, 15–20 × 2.5–5.5. A very common primary colonist on branches still attached or freshly fallen, and found in best condition on those still bearing withered leaves, Oct.–Mar.

Diatrypella quercina (Pers.)Cooke (Fig. 972)
Perithecia 0.6–0.7mm diam., immersed in pulvinate, dark brown to black erumpent and prominent stromata about 2mm diam.; necks slightly protuberant, with sulcate ostioles. Asci polysporous. Ascospores strongly curved, pale golden brown in mass, 8–12 × 2–3. On attached and fallen branches, Oct.–May, very common especially in dry situations.

Enchnoa infernalis (Kunze ex Fr.)Fuckel (Fig. 973)
Perithecia about 1mm diam., nestling in feltwork of dark brown hyphae beneath the periderm, which becomes raised and cracked. Ascospores curved, very pale brown, mostly 20–30 × 5–6. On small dead branches.

Hypoxylon confluens (Tode)Westend. (Fig. 974)
Perithecia partly immersed in rotten wood, large, grey to black, with black, papillate ostioles; stroma thin and often poorly developed. Ascospores dark brown, 15–22 × 8–12. On old, usually very rotten wood, Oct.–Apr.

Hypoxylon udum (Pers.)Fr. (Fig. 975)
Perithecia partly immersed in rotten wood, large, black, with papillate ostioles, in small groups, with poorly developed stroma. Ascospores brown, 27–37 × 9–13. Usually on very rotten wood, Oct.–Apr., uncommon.

Persiciospora masonii (Kirchst.)P. Cannon & D. Hawksw. (Fig. 976)
Perithecia 0.25–0.3mm diam., brown, with black necks, seated on a creamy hyphal subiculum. Asci up to 200 × 25. Ascospores becoming dark olivaceous brown when old, mostly 30–34 × 15–17, with two terminal pores, walls faintly reticulate. On wood, Aug., rare.

Pseudovalsa longipes (Tul.)Sacc. (Fig. 977)
Perithecia about 0.5mm diam., in groups of up to 15, immersed in conico-truncate stromata which push up and split the periderm; discs dark grey or blackish, about 0.8mm diam. Ascospores cylindrical to fusiform, often slightly curved, brown, up to 9-septate, 30–80 × 6–11. *Coryneum* state has brown, 5- to 7-pseudoseptate conidia mostly 50–70 × 13–19. Common on dead branches, Dec.–May.

Pseudovalsa umbonata (Tul.)Sacc. (Fig. 978)
Perithecia and stromata rather similar to those of *P. longipes*. Ascospores mostly 3- to 5-septate, golden brown with hyaline tips, 35–55 × 13–18. Conidia of *Coryneum* state 4- to 6-pseudoseptate, pale brown, 40–60 × 19–24. On dead branches, Jan.–Aug.

Rhamphoria pyriformis (Pers.)Höhnel (Fig. 979)
Perithecia immersed or superficial, about 0.4 × 0.2mm, black. Ascospores muriform, hyaline, 25–35 × 5–8, budding off minute conidia which eventually fill the asci. On wet rotten wood, Nov.–Apr.

Thuemenella britannica Rifai & Webster
Stromata 1–3mm diam., subglobose or pulvinate, flesh-coloured, with perithecia near the upper surface. Ascospores obovate to ellipsoid or subcylindrical, green, verrucose, 9–15 × 4–5. On rotten wood.

HYPHOMYCETES

Arthrobotryum stilboideum Ces. (Fig. 980)
Synnemata black, up to 2.5mm high, 15–40 thick, with slimy heads. Conidia very pale brown, mostly 3-septate, 10–16 × 3–4. On wood.

Bactrodesmium submoniliforme Hol.–Jech. (Fig. 981)
Conidiophores very pale brown, characteristically moniliform, 4–5 thick. Conidia pale brown, 5- to 10-septate, constricted at septa, 25–45 × 7–12. On rotten wood, May.

Clonostachys compactiuscula, described on *Fagus*, occurs also occasionally on *Quercus*.

Corynesporopsis quercicola (Borowska)P.M.Kirk (Fig. 982)
Colonies effuse, dark blackish brown to black, hairy. Conidiophores dark brown, up to 130 × 3–4. Conidia in short chains, arising through terminal pores, 2-septate, end cells pale, median cell dark brown, 12–18 × 6–9. On rotten wood, Sept.–May.

Dactylaria chrysosperma (Sacc.)Bhatt & Kendrick (Fig. 983)
Colonies effuse, yellowish brown. Conidiophores brown, up to 120 × 3–5. Conidia borne on pegs, 1-septate, hyaline or yellowish, 18–26 × 3–4. On rotten wood.

Dactylaria purpurella (Sacc.)Sacc. (Fig. 984)
Colonies effuse, shortly hairy, white to pale brown. Conidiophores up to 500 × 2–4, pale brown near base, otherwise hyaline. Conidia on pegs, hyaline, mostly 3-septate, 18–25 × 3–4. On rotten wood.

Gonatobotryum fuscum Sacc. (Fig. 985)
Colonies effuse, hairy, dark brown. Conidiophores dark brown, up to 700 long, 12–15 thick, nodose, with terminal and intercalary conidiogenous ampullae up to 28 diam. Conidia in chains of two, brown, 10–25 × 6–13. On rotten wood and wood blocks.

Helminthosporium microsorum D.Sacc. (Fig. 986)
Colonies punctiform, black, hairy. Conidiophores fasciculate, arising from

stromata, up to 500 × 8–14, dark brown. Conidia developing through pores at and just below the apex, with 9 to 17 pseudosepta, pale to mid golden brown, 60–160 × 12–22. On twigs of *Q. ilex*, May–Dec.

Paradendryphiopsis cambrensis M.B.Ellis (Fig. 987)
Colonies effuse, dark blackish brown to black, shortly hairy. Conidiophores branched near apex, brown, up to 150 × 6–9. Conidia mostly 3-septate, with brown median cells and a pale brown cell at each end, 25–40 × 10–14. On dead wood.

Taeniolella pulvillus (Berk. & Br.)M.B.Ellis (Fig. 988)
Colonies pulvinate, compact. Conidiophores caespitose, pale to mid brown, 3–6 thick. Conidia brown, smooth or occasionally verrucose, 2- to 11-septate, 25–90 × 7–9. On dead twigs.

COELOMYCETES

Conostroma, v.s. under *Colpoma quercinum*.

Coryneum elevatum (Riess)Sutton (Fig. 989)
Acervuli punctiform, dark brown, up to 1.5mm diam. Conidia broadly fusiform, 5- to 7-pseudoseptate, pale brown, mostly 55–65 × 20–25. On twigs of *Q. ilex*, May–Sept.

Coryneum neesii Sutton (Fig. 990)
Acervuli dark brown, up to 0.8mm diam. Conidia brown, 6- to 8-pseudoseptate, mostly 70–80 × 18–22. On dead twigs, uncommon.

Coryneum, v.s. also under *Pseudovalsa longipes* and *P. umbonata*.

Cryptosporiopsis, v.s. under *Pezicula cinnamomea*.

Cytospora intermedia Sacc.
Conidiomata erumpent, multilocular with small grey discs. Conidia allantoid, 5–6 × 1.5. On thin dead twigs. The *Cytospora* state of the plurivorous *Valsa ceratophora*, which also has conidiomata with small grey discs but conidia 4–5 × 1, is very common on oak twigs.

Fusicoccum noxium Ruhl
Conidiomata erumpent, conical, plurilocular, dark greyish. Conidia ellipsoidal, hyaline, guttulate, 12–15 × 4–5.5, in pale rather pinkish slimy masses. Causes discoloured patches on attached living twigs, May.

Phomopsis quercella (Sacc. & Roum.)Died.
Conidiomata becoming erumpent through longitudinal slits, black. A-conidia fusiform, pointed at each end, biguttulate, hyaline, 8–12 × 2–3. B-conidia curved or hooked, 20–30 × 1.5–2. On dead twigs.

RHAMNUS: Buckthorn

Puccinia coronata Corda, 0,1
Crown rust. Aecia mostly hypophyllous on yellow or yellowish-purple spots, cylindrical or cupulate, with whitish cut and revolute margins; spores orange, finely verruculose, $17–25 \times 13–20$. May–June. Alternate hosts many different grasses.

Cercospora rhamni Fuckel (Fig. 991)
Colonies hypophyllous, effuse, olivaceous brown, cottony or velvety. Conidiophores mid pale brown, $50–80 \times 6–7$. Conidia pale or mid pale golden brown, 3- to 7-septate, $60–170 \times 4–6$. On living and fading leaves.

Cucurbitaria rhamni (Nees)Ces. & de Not (Fig. 992)
Pseudothecia about 0.6mm diam., black, concentrically rugose, collapsing, at first grouped on a thin subiculum which tends to disappear. Ascospores rather pale golden brown, muriform, $18–22 \times 8–9$. On dead branches.

Diaporthe fibrosa (Pers.)Nitschke ex Fuckel (Fig. 993)
Perithecia 0.6–0.8mm diam., clustered in pustules, with thick protuberant necks. Ascospores hyaline, 1-septate, $11–14 \times 6–8$. On dead twigs and branches, Apr.–June.

Nectria punicea (Schmidt)Rabenh.
Perithecia 0.3–0.4mm diam., ovoid with pointed ostiolar papillae, reddish brown, smooth-walled, grouped on erumpent stromata 2.5–3mm diam. Ascospores broadly fusiform or ellipsoid, hyaline, smooth, 1-septate, $14–16 \times 6$. On dead branches, Feb.

RHODODENDRON: Azaleas and Rhododendrons

UREDINALES AND EXOBASIDIALES

Chrysomyxa rhododendri de Bary, 11, 111
Uredinia mostly hypophyllous, up to 2mm diam., orange; spores finely verruculose, $23–27 \times 17–22$. Telia reddish brown with a varnished look; spores in columns, hyaline, cylindrical, $20–30 \times 10–14$. On various species but telia mostly on *R. ponticum*. Alternate hosts *Picea abies* and *P. sitchensis*.

Exobasidium vaccinii (Fuckel)Woron.
Colonies white or pale pinkish, powdery, covering the surface of thickened and distorted leaves, young shoots and flowers. Spores hyaline, elongate fusiform, often curved, $10–20 \times 2.5–5$, 0- to 3-septate. Galls fairly common, May–Oct. *E. japonicum* Shirai, with oblong–reniform spores about 14.5×4, is less common.

DISCOMYCETES

Lophodermium vagulum Wilson & Robertson (Fig. 994)
Apothecia small, black, shield-shaped, opening by longitudinal slits. Ascospores hyaline, 35–55 × 1. On fallen dead leaves, uncommon.

Lophomerum ponticum Minter (Fig. 995)
Apothecia subcuticular, black, shining, about 0.5mm long, with white or grey lips surrounding a slit-like opening. Mature ascospores 3-septate, hyaline, 50–75 × 2–3. On the lower surface of dead leaves of *R. ponticum*, either attached to plants or lying on the ground, Sept.–Mar.; collections made in Feb. usually in very good condition.

Ovulinia azaleae Weiss (Fig. 996)
Large hyaline, pyriform, 1-septate conidia, 35–55 × 20–30, can be seen with a lens on the surface of brown spots on petals, May–Sept. They are formed terminally on very short conidiophores. Black sclerotia formed in autumn but apothecia not recorded with certainty in Britain. Uncommon.

OTHER ASCOMYCETES

Botryosphaeria rhodorae (Cooke)Barr (Fig. 997)
Pseudothecia immersed, 0.15–0.2mm diam., thick-walled, blackish brown. Ascospores hyaline or yellowish, 15–24 × 6–9, some with small gelatinous caps at ends. Two pycnidial states occur, one with macroconidia 12–17 × 9.5, the other with microconidia 5–8 × 1–2. Causes brown or greyish-brown spots on living and newly dead leaves, Feb.–Aug.

Chaetapiospora rhododendri (Tenwall)v.Arx (Fig. 998)
Perithecia erumpent, inconspicuous, similar in some ways to *Pseudomassaria thistletonia* but with long dark brown setae around the ostiole plainly visible under a dissecting microscope. Found on both surfaces of dead leaves of various species including *R. ponticum*, especially near the midribs, Feb.–Apr. Ascospores hyaline to pale yellow with a septum near the base, 12–18 × 3.5–6.

Dennisiella babingtonii (Berk.)Bat. & Cif. (Fig. 999)
A sooty mould which covers large areas of the upper surface of living leaves with a thin, olivaceous or blackish, filmy, mycelial mat from which arise dark brown setae about 200 long. Pseudothecia superficial, black, about 0.2mm diam. Ascospores hyaline with up to 6 septa, 15–33 × 5–7. On leaves of *Rhododendron* and various other evergreens in areas of high rainfall.

Lembosina aulographoides (Bomm., Rouss. & Sacc.)Theiss. (Fig. 1000)
Thyriothecia black, up to about 0.3 × 0.2mm, opening by slits. Ascospores pale brown, 1-septate, minutely verruculose, 18–23 × 8–11. On bark of

attached twigs, especially those of *R. ponticum*, Feb.–June.

Morenoina rhododendri J.P.Ellis (Fig. 1001)
Thyriothecia simple or branched, about 0.3–0.4 × 0.1mm, margin not fimbriate. Ascospores 1-septate, hyaline to very pale brown, smooth or minutely verruculose, 8–10 × 3–4. Pycnothyria up to 0.1mm diam., conidia 2–3 × 0.5–1. On dead twigs of *R. ponticum*, May–June.

Mycosphaerella rhododendri Lindau (Fig. 1002)
Pseudothecia mostly immersed, about 0.1mm diam., black. Ascospores 1-septate, hyaline, 8–10 × 1.5–2. On living and dead leaves of *R. ponticum*, May.

Phomatospora gelatinospora Barr
Perithecia epiphyllous, immersed, blackish brown. Ascospores uniseriate in asci, hyaline, smooth, elliptic, slightly constricted in middle, with 1 or 2 large guttules or with a pseudoseptum, 15–18 × 5.5–6.5, surrounded by gelatinous sheaths. In swarms on dead but still attached leaves, Apr.–May.

Protoventuria arxii (Müller)Barr (Fig. 1003)
Pseudothecia superficial on an olivaceous hyphal subiculum, about 0.2mm diam., black, covered with short, dark brown setae. Ascospores hyaline or pale olivaceous, 1-septate, mostly 10–14 × 4. On dead leaves and twigs, May.

Pseudomassaria thistletonia (Cooke)v.Arx (Fig. 1004)
Perithecia immersed, dark brown to black, about 0.2–0.3mm diam. Ascospores hyaline, with a septum near the base, 18–23 × 5–8. On dead leaves of *R. ponticum*, Feb.–Mar.

HYPHOMYCETES

Brachysporium dingleyae Hughes (Fig. 1005)
Colonies effuse, brown, hairy. Conidiophores up to 250 × 5–7. Conidia 3-septate, 22–25 × 11–15, basal cell small, pale, other cells larger, brown or dark brown. On rotten wood, Apr.–Sept.

Cercoseptoria handelii (Bubák)Deighton (Fig. 1006)
Colonies amphigenous. Conidiophores fasciculate, pale olivaceous, with indistinct scars, 20–50 × 2.5–4. Conidia pale straw-coloured, 6- to 14-septate, 60–100 × 3–4. On living leaves, causing brown or purple spots with broad, very dark purple margins.

Graphium smaragdinum (Alb. & Schw.)Sacc. (Fig. 1007)
Synnemata up to about 0.5mm high, dark blackish brown, with bright emerald-green, shining, slimy, conidial heads when fresh. Conidia hyaline, green in mass, 2–3.5 × 1.5–2. On dead wood, Mar.–Aug.

Pycnostysanus azaleae (Peck)Mason (Fig. 1008)
Synnemata dark olivaceous or blackish brown with a pale dusting of conidia at

the apex, up to 2mm high and 0.5mm wide. Conidia in chains, 4–6 diam. or 6–12 × 4–6, pale brown or olivaceous. The cause of bud blast and twig blight. Terminal flower buds are first affected; later, lateral buds, stems and leaves are attacked. When a twig is infected, flowers fail to develop and leafy shoots become necrotic. It is easy to spot infected buds, which turn brown and later silvery, and synnemata make them look as if they are thickly coated with black spines.

COELOMYCETES

Ceuthospora rhododendri Grove
Similar to *C. lauri* described on *Prunus*.

Coleophoma empetri (Rostr.)Petrak (Fig. 1009)
Pycnidia black, immersed, flattened at the base, 0.1–0.15mm wide. Conidia hyaline, guttulate, with guttules mostly near the ends, 13–18 × 2–3. On dead leaves of *R. ponticum*, Oct.–Apr.

Cytospora subclypeata Sacc.
Conidiomata 0.5–0.7mm diam., raising the epidermis, which is dark reddish brown and shining; discs very small, grey. Conidia allantoid, 4–5 × 1. On dead branches and leaves, Feb.–Nov.

Gloeosporium rhododendri Briosi & Cav. (Fig. 1010)
Acervuli black, wrinkled, shining, in concentric rings. Conidia cylindrical, hyaline, 15–20 × 4–5. Causes large, zonate, brown or greyish brown, purple-bordered spots on leaves, especially those of *R. ponticum*.

Macrophoma falconeri Henn.
Pycnidia erumpent, 0.25–0.3mm diam., black, thick-walled. Conidia hyaline, guttulate, rather thick-walled, 15–30 × 10–14. Causes large, reddish brown, marginal spots on leaves of *R. falconeri*, Oct.

Monochaetia karstenii (Sacc. & Syd.)Sutton (Fig. 1011)
Acervuli opening irregularly. Conidia 4-septate, median cells rather pale brown, end cells hyaline, 15–17 × 5–5.5, with basal appendage 1–3 long, and a single, often branched apical setula up to 20 long. On brown, grey or whitish, sometimes bordered spots on leaves, Oct.–Nov.

Pestalotiopsis guepini (Desm.)Steyaert (Fig. 1012)
Acervuli amphigenous, up to 0.2mm diam., oozing black spore masses. Conidia 4-septate, 21–29 × 6.5–8.5, median cells rather pale olivaceous brown, end cells hyaline, with basal appendage 4–12 long, and 2 to 5, mostly 3, apical setulae up to 33 long. On dead leaves, Oct.–Nov.

Pestalotiopsis sydowiana (Bres.)Sutton (Fig. 1013)
Acervuli mostly epiphyllous, up to 0.4mm diam., black. Conidia 4-septate

22–29 × 8–11, upper median cells dark brown, lower one pale brown, end cells hyaline, basal appendage 3–6 long, 2 to 4 apical setulae up to 30 long. On dead leaves of *R. hybridum* and *R. ponticum*, Aug.–Sept.

Seimatosporium arbuti (Bonar)Shoemaker (Fig. 1014)
Acervuli up to 0.2mm diam. Conidia 4-septate, 15–21 × 4.5–6, median cells brown, end cells hyaline, appendage at each end up to 12 long. On dead leaves.

Seimatosporium mariae (Clinton)Shoemaker (Fig. 1015)
Acervuli up to 0.25mm diam. Conidia 5-septate, 20–23 × 6–7, median cells pale brown, end cells hyaline, basal appendage up to 25 long, apical appendage up to 30 long. On dead leaves.

Septoria azaleae Vogl
Pycnidia amphigenous, immersed, black, 0.1–0.15mm diam. Conidia hyaline, 1- to 4-septate, slightly curved, 15–30 × 1.5–2.5. Causes leaf scorch, the leaves turning yellow to rusty brown from the tip backwards.

RIBES: Black Currant, Flowering Currant, Gooseberry, Red Currant

UREDINALES

Four rusts are found on different species of *Ribes*.

KEY

	With telia	1
	With aecia	2
1.	Teliospores 1-septate .... *Puccinia ribis*	
	Teliospores aseptate, in columns .... *Cronartium ribicola*	
2.	Aecia caeomoid, peridium rudimentary or none .... *Melampsora epitea*	
	Aecia aecidioid, with distinct, white, torn peridium .... *Puccinia caricina*	

Ribes species	Rusts
alpinum	M. epitea
nigrum	C. ribicola, P. caricina
rubrum	C. ribicola
sanguineum	C. ribicola, P. caricina
spicatum	P. ribis
sylvestre	C. ribicola
uva-crispa	C. ribicola, M. epitea, P. caricina

Cronartium ribicola J.C.Fischer, 11, 111 (Fig. 1016)
Uredinia hypophyllous, 0.1–0.25mm diam., bullate, yellowish, on angular brown spots; spores orange, echinulate, 20–30 × 15–18. Telia brownish yellow, arising in uredinia and forming firmly cemented columns up to 2 × 0.2mm; spores 30–60 × 10–20. Sometimes almost the whole lower leaf surface is covered, July–Oct. Alternate hosts *Pinus* spp.

Melampsora epitea Thüm. var. *epitea*, 0,1
Aecia caeomoid, orange; spores minutely verruculose, 15–25 × 10–20. Alternate host *Salix purpurea*.

Puccinia caricina DC., 0,1
Aecia sometimes on fruits and stems as well as leaves, causing distortion, orange, with white, cut and recurved peridia; spores verruculose, 16–25 × 12–20. Alternate hosts *Carex* spp.

Puccinia ribis DC., 111
Telia epiphyllous, chestnut brown; spores verrucose, brown, 20–30 × 15–20. On *R. spicatum*, rare.

PERONOSPORALES

Plasmopara ribicola Schröter (Fig. 1017)
Downy mildew. Colonies thin, white, effuse. Sporangia mostly 15–20 × 11–15, occasionally up to 23 × 18. On yellowed leaves of *R. rubrum*, June.

DISCOMYCETES

Drepanopeziza ribis (Kleb.)Höhnel (Fig. 1018)
Apothecia erumpent, 0.2–0.3mm diam., very dark brown with paler discs. Ascospores hyaline, 12–15 × 5–7. On overwintered dead leaves lying on the ground, Mar.–May. Acervuli of *Gloeosporidiella* state with hyaline, falcate conidia 17–26 × 6–7, and microconidia 4–6.5 × 2–3, cause brown lesions 1–2mm diam. on living leaves of *R. nigrum*, *R. rubrum* and *R. uva-crispa*, July–Sept.

Godronia ribis (Fr.)Seaver (Fig. 1019)
Apothecia erumpent, dark brown, pimply, often in groups. Ascospores hyaline, mostly 3-septate and 30–35 × 3–4. Conidiomata of *Fuckelia* state multilocular, convoluted; conidia hyaline, 1-septate, 8–10 × 3–4. On dead branches of *R. rubrum*.

Godronia uberiformis Groves (Fig. 1020)
Apothecia erumpent, up to 1mm high, urn-shaped, blackish brown, toothed margins creamy on inside. Ascospores up to 11-septate, hyaline, 80–95 × 2.

Pycnidia of *Topospora* state erumpent, up to 1.3 × 0.8mm; conidia hyaline, 3-septate, 17–22 × 3–3.5. On dead twigs of *R. nigrum*.

OTHER ASCOMYCETES

Botryosphaeria ribis Grossenbacher & Duggar (Fig. 1021)
Pseudothecia immersed becoming erumpent, up to 4mm diam., black. Ascospores hyaline, 18–22 × 7–10. *Fusicoccum* state has hyaline, fusoid conidia, 17–25 × 5–7, and there are also microconidia 2–3 × 1. On dead twigs of *R. uva-crispa*.

Diaporthe strumella (Fr.)Fuckel (Fig. 1022)
Perithecia 0.4–0.5mm diam., immersed in groups of 10 to 20, with necks erumpent through prominent black discs up to 2 × 1mm. Ascospores hyaline, 1-septate, 15–18 × 3–4. On attached, dead branches of *R. nigrum*, *R. rubrum* and *R. uva-crispa*, Oct.–Apr., common. *Phomopsis* state has fusoid, hyaline, biguttulate A-conidia 6–10 × 2.5–3.5 and filiform B-conidia about 30 × 1.5.

Dothiora ribesia (Pers.)Barr (Fig. 1023)
Pseudothecia numerous as locules in black erumpent stromata 2–3mm diam. Ascospores long remaining hyaline and 1-septate, some eventually becoming pale brown and 3-septate, 18–26 × 5–8. On dead twigs of *R. nigrum* and *R. rubrum*.

Microsphaera grossulariae (Wallr.)Lév. (Fig. 1024)
Powdery mildew. Conidia few, in chains, 20–28 × 10–15. Cleistothecia chestnut brown to black, about 0.1mm diam., appendages 5 to 20. Asci 3 to 10, mostly with 4 to 5 spores. Mainly on the upper surface of living leaves of *R. nigrum*, *R. rubrum*, *R. sanguineum* and *R. uva-crispa*, especially where planted close together or heavily shaded.

Mycosphaerella ribis (Fuckel)Lindau
Pseudothecia hypophyllous, immersed, about 0.1mm diam. Ascospores almost cylindrical, slightly curved, hyaline, 1-septate, 28–33 × 3. On dead leaves of *R. nigrum*, Mar.–May.

Sphaerotheca mors-uvae (Schw.)Berk. & Curt. (Fig. 1025)
American gooseberry mildew. Conidia formed in long chains, mainly on young shoots, but leaves and berries also attacked, hyaline, 25–30 × 17–20. Cleistothecia surrounded by a dense mat of brown hyphae; they each contain one 4- to 8-spored ascus. On *R. nigrum*, *R. rubrum* and *R. uva-crispa*.

Thyronectria berolinensis (Sacc.)Seaver (Fig. 1026)
Perithecia 0.3–0.35mm diam., grouped on stromata, brick red, collapsing in a cupulate manner, walls rough. Ascospores muriform, hyaline to straw-coloured, 18–21 × 6–9. Prior to the formation of perithecia, stromata subtend

sporodochia which produce hyaline conidia, 6–9 × 2–2.5. On dead branches, uncommon.

COELOMYCETES

Ascochytella grossulariae (Oudem.)Died. (Fig. 1027)
Pycnidia immersed, becoming erumpent, black, up to 0.2mm diam. Conidia pale brown, 1-septate, 7–10 × 2–2.5. On dead twigs of *R. uva-crispa*, Apr.

Fuckelia, v.s. under *Godronia ribis*.

Fusicoccum, v.s. under *Botryosphaeria ribis*.

Gloeosporidiella, v.s. under *Drepanopeziza ribis*.

Phomopsis, v.s. under *Diaporthe strumella*.

Topospora, v.s. under *Godronia uberiformis*.

Trullula melanochlora (Desm.)Höhnel (Fig. 1028)
Conidiomata becoming erumpent, discoid, black. Conidia in long, sometimes branched chains, oblong, olivaceous brown, 4–6.5 × 1.5–2.5. Abundant on prickle bases of *R. uva-crispa*, Feb.

ROBINIA: Locust Tree, False Acacia

Aglaospora profusa (Fr.)de Not. (Fig. 1029)
Pseudothecia up to 1mm diam., in clusters of 2 to 5, immersed in bark, surrounded by a stroma. Ascospores golden brown, 3-septate, 50–64 × 16–18. On dead twigs, Oct.–Jan.

Cucurbitaria elongata (Fr.)Grev. (Fig. 1030)
Pseudothecia 0.3–0.5mm diam., in clusters or lines on a brown hyphal subiculum. Ascospores golden brown, muriform, 20–30 × 9–11. On dead twigs and branches. *Camarosporium* state has immersed pycnidia up to 0.8mm diam., black. Conidia golden brown, muriform 17–27 × 8–9.

Diaporthe oncostoma (Duby)Fuckel (Fig. 1031)
Perithecia immersed, up to 0.7mm diam., in small groups, necks erumpent through black discs. Ascospores hyaline, 1-septate, 16–20 × 3–4. On dead twigs Dec.–Feb. *Phomopsis* state with erumpent, black conidiomata up to 0.5mm diam.; A-conidia 8–10 × 2–2.5, B-conidia flexuous or hooked, 18–30 × 1.

ROSA: Roses

UREDINALES

Four species of *Phragmidium* occur but only *P. mucronatum* and *P. tuberculatum* are at all common.

KEY

    Teliospores mostly with 10 or 11 septa, on *R. rubrifolia* .... *fusiforme*
    Teliospores never with more than 8 septa 1
1.   Telia brown, on *R. pimpinellifolia* .... *rosae-pimpinellifoliae*
    Telia black 2
2.   Apical papilla of teliospore gradually tapered from a broad base. On
        R. arvensis, *R. canina* and many cultivated roses .... *mucronatum*
    Apical papilla of teliospore with narrow base. On *R. rubiginosa*,
        R. rugosa and cultivars .... *tuberculatum*

Phragmidium fusiforme Schröter, 0, 1, 11, 111
Easily distinguished from all other rose rusts, by the number of septa in the teliospores. Rare.

Phragmidium mucronatum (Pers.)Schlecht. (Fig. 1032)
Aecia caeomoid, bright orange, hypophyllous and on living stems and fruits. Uredinia hypophyllous, small, pale orange, with curved paraphyses; spores finely echinulate, with small pores (about 2 diam.), 20–27 × 17–22. Telia black, spores mostly 5- to 7-septate, 70–90 × 23–30. Very common everywhere.

Phragmidium rosae-pimpinellifoliae Diet., 1, 11, 111
Aecia caeomoid, yellowish orange. Uredinia minute, orange, hypophyllous. Telia brown, spores 5- to 7-septate, 80–110 × 25–35, smooth to minutely verruculose.

Phragmidium tuberculatum J.Müller, 0, 1, 11, 111 (Fig. 1033)
Aecia caeomoid. Uredinia small, pale yellow, with curved paraphyses; spores coarsely echinulate, with large pores (4.5 diam.), 20–27 × 18–24. Telia black; spores 60–110 × 30–35, mostly 4- or 5-septate. Fairly common.

PERONOSPORALES

Peronospora sparsa Berk.
A downy mildew which occurs occasionally on cultivars in cool glasshouses causing leaf fall.

DISCOMYCETES

Diplocarpon rosae Wolf (Fig. 1034)
Known in Britain only in its *Marssonina* (often called *Actinonema*) state, which causes 'black spot' disease. Acervuli epiphyllous, blackish, seated on subcuticular fibrils which radiate from a central point. Conidia 1-septate, hyaline, 15–19 × 5–7. On purplish-brown spots on leaves of *R. canina* and cultivated roses; very common especially in wet summers.

Pezicula rubi, described on *Rubus*, is common also on *Rosa*.

Tapesia rosae (Pers.)Fuckel (Fig. 1035)
Apothecia about 1mm diam., brown with white-edged olivaceous discs when fresh, seated on an extensive greyish brown, felted, cottony subiculum. Ascospores hyaline, 8–11 × 2–3. Common on dead stems of *R. canina*, especially where tangled in matted grass, Jan.–June and Sept.

Zoellneria rosarum Velen. (Fig. 1036)
Apothecia 0.5–0.75mm diam., pinkish yellow, with pale discs surrounded by long brown hairs. Ascospores hyaline, 7–12 × 2.5–3. There is an accompanying setose *Amerosporium* state with conidia 10–11 × 2. On current year's leaves of *R. canina* which have fallen prematurely, especially under isolated bushes on calcareous soil, Aug.–Dec.

OTHER ASCOMYCETES

Botryosphaeria dothidea (Moug. ex Fr.)Ces. & de Not. (Fig. 1037)
Pseudothecia aggregated in blackish stromatal crusts 1–1.5cm diam. which are often cracked concentrically. Ascospores hyaline to pale yellow, 21–25 × 9–10. On living and dead stems of *R. canina*, fairly common, Jan.–Oct.

Chaetosphaeria bramleyi Booth (Fig. 1038)
Colonies effuse, dark blackish brown, hairy. Perithecia sparse, black, up to 0.2mm diam., surrounded by conidiophores. Ascospores hyaline, 1-septate, 7–10 × 2–2.5. *Catenularia* state with conidiophores up to 140 × 4, and wedge-shaped conidia 1.5–2.5 × 1–1.5. On dead stems.

Diaporthe incarcerata (Berk. & Br.)Nitschke (Fig. 1039)
Perithecia immersed, necks scarcely protruding, surrounded by black lines. Ascospores hyaline, 1-septate, 14–17 × 4–4.5. Conidiomata of *Phomopsis* state erumpent, black, 0.5–1mm diam.; A-conidia hyaline, biguttulate, 8–10 × 2–3, B-conidia flexuous or hamate, 25–30 × 1. On dead branches and prickles of *R. canina* and cultivated roses, Apr.–June, fairly common.

Griphosphaeria corticola, described on *Rubus*, is not uncommon also on *Rosa*.

Pseudomassaria sepincolaeformis (de Not.)v.Arx (Fig. 1040)
Perithecia about 0.3mm diam., immersed but showing through to the surface as shining, dark grey areas, with short black necks just protruding. Ascospores hyaline, with septum towards the base, 18–21 × 8–10. On dead twigs, Nov.–Jan.

Saccothecium sepincola (Fr.)Fr. (Fig. 1041)
Pseudothecia up to 0.2mm diam., immersed in rows, black, thick-walled, with asci characteristically attached firmly by their bases to a central column of tissue. Ascospores hyaline to pale yellow, with 2 to 5 transverse septa and occasionally a longitudinal septum, 15–22 × 5–7. On dead twigs, especially of *R. canina*, Feb.–Apr., fairly common.

Sphaerotheca pannosa (Wallr.)Lév. (Fig. 1042)
Powdery mildew with conspicuous persistent mycelium forming a white to grey or buff felt on leaves, young shoots, flowers and fruits. Cleistothecia, rarely formed, contain one 8-spored ascus each. Very common.

COELOMYCETES

Amerosporium, v.s. under *Zoellneria rosarum*.

Camarosporium rosae Grove (Fig. 1043)
Pycnidia just over 0.1mm diam., black. Conidia brown, mostly with 3 transverse septa, some with 1 longitudinal septum, 12–17 (20) × 6. On dead stems, Feb.–May.

Coniothyrium wernsdorffiae Laub.
Pycnidia immersed, 0.5mm diam., dark grey. Conidia yellowish brown, 5–8 × 4.5–6. Causes oval brown spots on small branches, the so-called brand canker.

Cytospora rosarum Grev.
Conidiomata erumpent, 0.25–0.3mm diam., with small greyish discs. Conidia allantoid, hyaline, 4–6 × 1–1.5. On dead branches, Oct.–Mar.

Marssonina, v.s. under *Diplocarpon rosae*.

Myxosporium rosae Fuckel
Acervuli erumpent, black, up to 0.5mm diam. Conidia hyaline, straight or slightly curved, 10–12 × 3–4. On dead branches, a cause of canker, May.

Pestalotiopsis versicolor (Speg.)Steyaert
Acervuli at first covered then opening to expose black spore masses. Conidia 19–25 × 7–9.5, 4-septate, with a hyaline cell at each end and three brown median cells, the two upper ones very dark; 3 apical setulae up to 28 long, basal appendage 2–5 long. On dead branches of *R. canina*.

Phomopsis, v.s. under *Diaporthe incarcerata.*

Seimatosporium rosarum (P.Henn.)Sutton (Fig. 1044)
Acervuli about 0.2mm diam., black. Conidia mostly 2-septate (occasionally 3-septate), 10–12 × 3–5, pale brown, the basal cell palest and sometimes bearing a filiform appendage (setula) up to 10 long. On dead leaves of *R. canina.*

Septoria rosae Desm.
Pycnidia epiphyllous, about 0.1mm diam., black. Conidia hyaline, 25–75 × 2.5–3, becoming 3- to 7-septate. On living leaves causing pale brown, grey or whitish spots with reddish-purple borders, June–Nov., very common.

RUBUS: Blackberry, Dewberry, Raspberry, Stone Bramble

UREDINALES

Five rusts are found on *Rubus* but one of them is extremely rare and recorded only from Scotland.

KEY

	On *R. idaeus* only .... *Phragmidium rubi-idaei*
	On *R. saxatilis* only, very rare .... *Phragmidium acuminatum*
	On *R. caesius, R. fruticosus* or *R. laciniatus* 1
1.	Telia whitish or yellowish, uredinia commonly found on stems and petioles .... *Kuehneola uredinis*
	Telia black, uredinia hypophyllous, conspicuous crimson and purple spots on upper leaf surface above soil 2
2.	Teliospores mostly 5-septate .... *Phragmidium bulbosum*
	Teliospores mostly 3-septate .... *Phragmidium violaceum*

Kuehneola uredinis (Link)Arth., 0, 1, 11, 111 (Fig. 1045)
Aecia uredinoid. Uredinia often forming long yellow lines on stems. Telia whitish or yellowish, spores in chains. On *R. fruticosus,* common.

Phragmidium acuminatum (Fr.)Cooke, 1, 11, 111
A very rare rust found only a few times in Scotland. Telia black, spores mostly 5- to 6-septate, 50–110 × 25–30.

Phragmidium bulbosum (Str.)Schlecht., 0, 1, 11, 111 (Fig. 1046)
Teliospores with 3 to 6, mostly 5, septa, 60–75 × 25–30. On *R. caesius* and one section of *R. fruticosus* (agg.), common.

Phragmidium rubi-idaei (DC.)Karst., 0, 1, 11, 111 (Fig. 1047)
Uredinia hypophyllous, very small, orange. Telia hypophyllous, small, black,

spores 5- to 9- (mostly 6- to 7-)septate, brown, 80–130 × 30–35. On *R. idaeus*, common, May–Oct.

Phragmidium violaceum (C.F.Schultz)Wint., 0, 1, 11, 111 (Fig. 1048)
Uredinia orange–yellow. Telia black, spores, 1- to 4- (mostly 3-)septate, 60–100 × 30–35. On *R. fruticosus* and *R. laciniatus*, very common.

PERONOSPORALES

Peronospora rubi Rabenh.
Downy mildew, forms pale grey or greyish-brown effuse colonies on lower surface of leaves of *R. caesius* and *R. fruticosus*. Sporangia about 22 × 11–13.

DISCOMYCETES

Calycellina populina, described on *Populus*, is quite common on fallen dead leaves of *R. fruticosus*, Sept.–Mar.

Dasyscyphus. Five species are found on *Rubus*; they all have hairy apothecia.

KEY

	Apothecia white hairy with orange discs .... *bicolor* var. *rubi*
	Apothecia white hairy with pale discs .... *virgineus*
	Apothecia pale brown or greyish brown 1
1.	Apothecia to 1mm diam., stalked .... *clandestinus*
	Apothecia not more than 0.4mm diam., sessile 2
2.	Hairs all pale brown, cylindrical, spinulose .... *dumorum*
	Some hairs dark brown, tapered .... *misellus*

Dasyscyphus bicolor (Bull.)Fuckel var. *rubi* (Bres.)Dennis
Apothecia subsessile, 1–2mm diam., white hairy with contrasting orange or dark yellow ochre discs; hairs up to 160 × 4, thick-walled and capped with crystals. Ascospores hyaline, 6–8 × 1–2. On dead canes of *R. idaeus*, July–Aug.

Dasyscyphus clandestinus (Bull.)Fuckel (Fig. 1049)
Apothecia up to 1mm diam., stalked, pale brown, covered by long, straw-coloured hairs tipped with masses of calcium oxalate. Ascospores hyaline, 5–7 × 1.5. On dead canes of *R. idaeus*, Apr.–Aug.

Dasyscyphus dumorum (Rob. ex Desm.)Massee (Fig. 1050)
Apothecia sessile or subsessile, up to 0.4mm diam., pale brown with whitish rims; hairs cylindrical, thin-walled, pale brown, spinulose, 20–40 × 3–4. Ascospores hyaline, 4–7 × 1–1.5. On the underside of fallen dead leaves of *R. fruticosus* all the year round but especially Mar.–Nov., very common.

Dasyscyphus misellus (Rob. ex Desm.)Höhnel (Fig. 1051)
Apothecia sessile or subsessile, 0.1–0.2mm diam., pale brown with whitish rims; hairs of two kinds, those on the edge cylindrical to clavate, pale, spinulose, 20–30 × 3–4, those lower down on the flanks tapered, darker brown and up to 70 long. Ascospores hyaline, 5–6 × 1. On damp fallen leaves in dense shade, Sept.–Oct.

Dasyscyphus virgineus, a plurivorous species described on 'wood and bark', is common on dead stems of *R. fruticosus*, Mar.–Sept.

Echinula asteriadiformis Graddon (Fig. 1052)
Apothecia up to 0.2mm diam., sessile, hyaline to pale straw-coloured, like tiny brittle-stars with 5 to 8 projecting, tapered arms made up of bundles of hyphae. Ascospores hyaline, 10–12 × 4–5. On dead leaves of *R. fruticosus*, Apr.–July, common but not at all easy to see.

Hyaloscypha. Four species have been found on *Rubus*, three of them on dead stems and one on dead leaves. They all have sessile or subsessile apothecia fringed with tapered, hyaline, smooth-walled hairs.

KEY

    Hairs up to 120 long .... *richonis*
    Hairs not more than 50 long 1
1.   Ascospores 6–9 × 1.5–2 .... *hyalina* f.
    Ascospores 8–11 × 2–3 .... *lectissima*
    Ascospores 12–18 long .... *salicina*

Hyaloscypha hyalina (Pers.)Boud., f. (Fig. 1053)
Apothecia up to 0.5mm diam., drying brown, fringed with hyaline hairs up to 40 × 3–4. Ascospores hyaline, 6–9 × 1.5–2. On dead stems of *R. fruticosus*, Sept.–Mar.

Hyaloscypha lectissima (P.Karsten)Raitv. (Fig. 1054)
Apothecia 0.2–0.35mm diam., creamy yellowish, fringed with hyaline hairs up to 40 × 3–5. Ascospores hyaline, 8–11 × 2–3. On dead stems of *R. fruticosus*, Nov.–Jan.

Hyaloscypha richonis (Boud.)Dennis
Apothecia up to 0.5mm diam., pink, drying reddish brown. Hairs up to 120 × 3–4, with reddish oil drops. Ascospores hyaline, 5–6 × 2–2.5. Near base of dead stems of *R. fruticosus*, Nov.

Hyaloscypha salicina Velen.
Apothecia 0.2–0.6mm diam., ochraceous grey, with hairs 25–50 long, tapered part solid. Ascospores 12–18 long, sometimes slightly curved. On underside of dead leaves of *R. fruticosus*, July.

Hymenoscyphus separabilis (P.Karsten)Dennis
Apothecia in swarms, very short-stalked, up to 1mm diam., pale yellow or

ivory, downy, with pale orange to cream discs. Ascospores hyaline, 12–13 × 2–2.5. Paraphyses septate, end cells containing yellowish oil drops. On dead stems of *R. fruticosus* and *R. idaeus*, Sept.–Jan.

Hypoderma rubi (Pers.)de Not. (Fig. 1055)
Apothecia subepidermal, variously shaped, often navicular, black, shining, up to 2mm long. Ascospores hyaline with 2 large guttules, 20–25 × 3–4. Sometimes accompanied by its *Leptostroma* state with conidia 4–6 × 1. On dead stems of *R. fruticosus*, July–Nov., collections made later usually effete, very common.

Incrupila melatheja (Fr.)Dennis
Apothecia 0.5mm diam., subsessile, black, clothed with yellow-encrusted, tapered hairs, up to 60 × 3. Ascospores hyaline, 4–6 × 1. On damp dead stems of *R. fruticosus*, Mar.

Mollisia clavata Gremmen (Fig. 1056)
Apothecia 0.4–0.5mm diam., dark grey, olivaceous brown or black, downy, often pale-rimmed when fresh, with clavate protruding cells which are pale brown on the flanks and hyaline around the rim. Ascospores hyaline, clavate, 7–10 × 2–2.5. On dead stems of *R. fruticosus*, Apr.–Nov.

Mollisia stromatica Graddon
Apothecia developing singly or in clusters from small embedded hyaline stromata, up to 1.5mm diam., smooth, dark brown, with discs watery grey, drying pale ochraceous. Ascospores 7–9.5 × 1.5–2. On dead branches of **R. fruticosus**, Dec.

Mollisina rubi (Rehm)Höhnel (Fig. 1057)
Apothecia sessile, 0.15–0.25mm diam., white with creamy discs. Marginal cells bear short, thread-like, often branched processes. Ascospores 4–6 × 1.5–2. Very common, especially on the lower surface of dead leaves of *R. fruticosus*, all the year round.

Mollisiopsis lanceolata, described on *Filipendula*, has been recorded on dead stems of *R. fruticosus*. The lanceolate paraphyses project well above the **asci** and the ascospores are 6–8 × 1.5–2.

Pezicula rubi (Lib.)Niessl (Fig. 1058)
Apothecia erumpent, about 1mm diam., dull orange, somewhat pruinose. Ascospores often 1- to 3-septate, 20–26 × 7–8. On rather thick, dead stems of *R. fruticosus*, July–Jan., common.

Plectania melastoma (Sow.)Fuckel
Apothecia urn-shaped, up to 2cm diam., black, stippled with red granules along the edge. Ascospores hyaline, 24–28 × 10–11. On dead stems half-buried in moss on damp ground in woods, Jan.–June.

Ploettnera exigua (Niessl)Höhnel (Fig. 1059)
Apothecia erumpent, 0.2–0.3mm diam., dark bluish-green to almost black, often staining the surrounding host tissues bluish green. Ascospores 12–16 × 6–8. The swollen tips of the paraphyses and even some of the spores occasionally are blue–green. On dead stems and leaves of *R. fruticosus*, July–Jan., fairly common.

Pyrenopeziza escharodes (Berk. & Br.)Rehm (Fig. 1060)
Apothecia erumpent, 0.3–0.5mm diam., brown, hairy, discs yellowish grey. Projecting cells up to 60 × 4, 2- or 3-septate, pale brown, often with hyaline ends. Ascospores hyaline, 6–8.5 × 2–2.5. On dead stems of *R. fruticosus*, Feb.–Oct.

Pyrenopeziza rubi (Fr.)Rehm (Fig. 1061)
Apothecia erumpent, up to 1mm diam., dark brown with olivaceous sunken discs when fresh. In mature specimens marginal cells elongate to form hairs up to 80 long, outer ones brown, inner ones hyaline. Ascospores hyaline, 7–9 × 2–3. On dead canes of *R. idaeus*, Apr.–Aug., quite common.

Rutstroemia fruticeti Rehm (Fig. 1062)
Apothecia 1.5–3mm diam., with short fat stalks, chocolate brown with brownish-orange discs when quite fresh. Ascospores hyaline, 12–16 × 4–5. On dead stems of *R. fruticosus*, Feb.–Apr.

Rutstroemia rubi Velen.
Apothecia up to 3mm diam., pale brown with darker striations and a sparse covering, especially near the margin, of tapered, septate, brown hairs up to 400 × 12–20. Ascospores slightly curved, hyaline, 14–20 × 5–6. On dead stems of *R. fruticosus*, Mar.–Apr.

Stegopeziza dumeti (Sacc. & Speg.)Spooner (Fig. 1063)
Apothecia developing under the epidermis, then raising or lifting off little lids to expose the brown discs. Paraphyses lanceolate, septate. Ascospores hyaline, 5–7 × 1–1.5. On dead stems of *R. caesius* and possibly also *R. fruticosus*, Mar.–May.

Torrendiella ciliata Boud. (Fig. 1064)
Apothecia about 1mm diam., stalked, brown, with paler discs, setose. Setae brown or burnt sienna, septate, up to 300 long. Ascospores hyaline, curved, 15–20 × 5–7. On newly dead leaves of *R. fruticosus*, often still attached to stems and confined to areas surrounded by narrow black lines, Aug.–Jan.

Velutarina rufo-olivacea, described on 'wood and bark', a plurivorous species which has powdery buff or tan apothecia with olivaceous discs about 3mm diam., is found frequently on dead stems of *R. fruticosus*.

OTHER ASCOMYCETES

Anthostomella. Four species have been found on *R. fruticosus*; two of them, *A. appendiculosa* and *A. rubicola*, are confined to this host. Of the others, *A. clypeoides* has been seen more frequently on *Epilobium*, and *A. tomicoides* on *Carex*, and they are described on these plants.

KEY

Ascospores 10–15 × 3–5	*clypeoides*
Ascospores 15–19 × 5–8	*tomicoides*
Ascospores over 22 long	1
1. Ascospores 5–6 wide	*rubicola*
Ascospores 8–10 wide	*appendiculosa*

Anthostomella appendiculosa (Berk. & Br.)Sacc. (Fig. 1065)
Perithecia immersed, 0.5–0.7mm diam., rarely solitary, more frequently in groups under a black shining clypeus which may be up to 1cm long. Small whitish spots on the clypeus with black dots in their centres show where perithecia lie. Ascospores 28–36 × 8–10, with a large upper cell, which becomes brown in the ascus, and a small cell which remains hyaline. On dead stems, Sept.–May, uncommon.

Anthostomella rubicola (Speg.)Sacc. & Trotter (Fig. 1066)
Clypeus and perithecia similar to those of *A. appendiculosa* but smaller, the blackened areas seldom more than 1–2mm long. Ascospores 23–30 × 5–6, each with a small hyaline basal cell, and a large upper cell which does not usually turn brown in the ascus, but does so after release. On dead stems, Feb.–Oct., common; most collections made in Apr.

Apioporthe vepris (de Lacr.)Wehmeyer (Fig. 1067)
Perithecia immersed, about 0.2mm diam., usually in little clusters, breaking through to the surface where their necks protrude only slightly through small black discs. Ascospores hyaline, septate below the middle, mostly 6–9 × 2–2.5. In fresh collections a filiform appendage is seen at each end of the spore. The *Phomopsis* state has conidia 5–7 × 1–1.5. On dead stems, leaves and petioles of *R. fruticosus* and *R. idaeus*, Jan.–May, common.

Appendiculella calostroma (Desm.)Höhnel (Fig. 1068)
Colonies appear as small black patches, especially near the base of green stems, but also on living leaves. Mycelium superficial, brown, hyphopodiate. Cleistothecia black, tuberculate, with a few worm-like appendages. Ascospores brown, 3-septate, 45–50 × 15. On *R. fruticosus* in damp, sheltered woods, Mar.–May. This is sometimes overgrown by *Dimerium meliolicola* (Petrak)Hansf. (Fig. 1069), which has black pseudothecia 0.1–0.15mm diam., and 1-septate, brown ascospores 13–17 × 4–6.

Chaetosphaeria callimorpha (Mont.)Sacc. (Fig. 1070)
Colonies usually black and hairy or velvety. Perithecia black, shining, about 0.2mm diam., nestling among conidiophores and setae of the *Codinaea* state. Ascospores hyaline, 1-septate, 12–17 × 3.5–4. Often, glistening white conidial heads can be seen capping short conidiophores which lie between much longer setae. Setae up to 400 long, blackish brown. Conidia hyaline, 10–15 × 2–3. Common on dead stems of *R. fruticosus*, especially near the base of plants, Oct.–May. Two plurivorous wood and bark species of *Chaetosphaeria* are found occasionally on dead *Rubus* stems, *C. inaequalis* and *C. preussii*, both of which have much smaller ascospores which often break into two parts while still in the ascus.

Clypeosphaeria notarisii Fuckel (Fig. 1071)
Perithecia immersed, about 0.5mm diam., black, seated singly or in pairs beneath a blackish clypeus, their necks showing as black dots in paler areas. Ascospores brown, with 4 large guttules and sometimes 3 thin pseudosepta, 18–27 × 5–6. On dead stems of *R. fruticosus* and *R. idaeus*, Jan.–Mar., but ascospores usually not quite mature in Jan., common.

Coleroa chaetomium (Kunze ex Fr.)Rabenh. (Fig. 1072)
Pseudothecia superficial, black, setose, solitary or in clusters scattered over the whole of the upper surface of the leaf; setae dark brown, up to 60 long. Ascospores pale olivaceous brown, 1-septate, 11–14 × 5–6. On living leaves of *R. caesius* and *R. idaeus*, Apr.–Aug.

Didymella applanata (Niessl)Sacc. (Fig. 1073)
Pseudothecia immersed, 0.2–0.25mm diam., blackish brown. Ascospores hyaline, 1-septate, mostly 14–17 × 5–6. *Phoma* state has conidia 3–5 × 1.5–2.5. Causes 'spur blight', pale grey patches on canes, especially of *R. idaeus*, and mostly near the nodes, Mar.–May.

Didymosphaeria oblitescens (Berk. & Br.)Fuckel (Fig. 1074)
Pseudothecia immersed, 0.3–0.4mm diam., with necks just emerging through the centre of small round or oval brown spots. Ascospores clear brown, verruculose, 1-septate, 10–20 × 5–7. On dead stems of *R. fruticosus*, Dec.–Jan.

Didymosphaeria rubicola Berl. (Fig. 1075)
Pseudothecia immersed, 0.25–0.3mm diam. Ascospores olivaceous brown, with striate walls, 1-septate, 16–26 × 8–11. On dead stems of *R. fruticosus*, May, uncommon.

Elsinoë veneta (Burkh.)Jenkins
Stromata immersed, about 0.35mm diam., with spherical asci formed singly within each of the scattered pseudothecial locules. Ascospores hyaline, 3-septate, 18–21 × 7–8. Acervuli of the *Sphaceloma* state contain hyaline conidia 5–7 × 2.5–3. On greyish, sunken, purple-bordered spots on canes and occasionally leaves of *R. idaeus* and *R. fruticosus*, May–Oct. The disease caused

is known as 'cane-spot' and stems sometimes become very cracked and rough.

Gloniopsis praelonga (Schw.)Zogg (Fig. 1076)
Hysterothecia up to 3mm long, dark grey to black, frequently furrowed. Ascospores with transverse and longitudinal septa, hyaline to yellow, mostly 20–26 × 7–10, occasionally up to 30 × 12, often with gelatinous sheaths. Very common on old dead stems of *R. fruticosus* all the year round.

Gnomonia rubi (Rehm)Winter (Fig. 1077)
Perithecia scattered or in small groups, immersed, up to 0.3mm diam., black, with necks projecting well above the surface. Ascospores hyaline, 1-septate, 10–17 × 2.5–3 with an appendage (setula) at each end up to 6 long. On fallen dead leaves and occasionally also on stems of *R. fruticosus*, all the year round.

Gnomoniella rubicola Pass. (Fig. 1078)
Perithecia immersed singly or in small groups, black, up to 0.25mm diam., with protruding necks. Ascospores hyaline, biguttulate, 6–7 × 2–2.5. On dead stems and leaves of *R. fruticosus*, Mar.–Dec., common but inconspicuous.

Griphosphaeria corticola (Fuckel)Höhnel (Fig. 1079)
Perithecia 0.3–0.5mm diam., black, immersed singly, in pairs or threes, with necks just showing through a small grey clypeus. Ascospores hyaline, mostly 3-septate, rarely with 5 septa, 12–17 × 5–7. *Seimatosporium* state has acervuli up to 0.35mm diam. and brown conidia, with 3 thick septa and side walls tending to collapse inwards, 13–15 × 5–6.5. On dead attached stems of *R. fruticosus* and occasionally *R. idaeus*, Mar.–Aug., very common especially in its *Seimatosporium* state.

Herpotrichia herpotrichoides (Fuckel)P. Cannon (Fig. 1080)
Pseudothecia erumpent, 0.3–0.6mm diam., overlaid by brown hyphae. Ascospores at first hyaline and 1-septate, becoming brown and 3-septate, 20–30 × 6–9, with gelatinous sheaths. On dead stems of *R. fruticosus* and *R. idaeus*, Feb.–June.

Leptosphaeria coniothyrium (Fuckel)Sacc. (Fig. 1081)
Pseudothecia immersed, 0.3mm diam., black, with small dark clypeus. Ascospores 3-septate, pale olivaceous brown, 12–17 × 3.5–4.5. *Coniothyrium* state has broadly ellipsoidal or globose, pale brown conidia 3–5 × 2–3. On *R. caesius, R. fruticosus* and especially *R. idaeus*, on fruiting stems at soil level. The disease caused is called 'cane-blight'; leaves wilt and wither and stems become very brittle and snap readily. Pseudothecia found Dec.–Feb.

Leptosphaeria praetermissa (P.Karsten)Sacc.
Pseudothecia immersed, up to 0.5mm diam. Ascospores more or less ellipsoidal, olivaceous, 3-septate, 18–24 × 6–8. On dead canes of *R. idaeus*, Apr.–July.

Lophiostoma. Three species and one variety are found on *Rubus*. They all have laterally compressed black necks to their perithecia.

KEY

Substrate stained green, ascospores striate .... *viridarium*
Not so 1
1. Ascospores 3-septate in asci, up to 3 wide .... *hysterioides*
   Ascospores 1-septate in asci, more than 3 wide 2
2. Ascospores 11–18 × 3–5 .... *fuckelii*
   Ascospores 15–21 × 4–5.5 .... *fuckelii* var. *pulveraceum*

Lophiostoma fuckelii Sacc. (Fig. 1082)
Pseudothecia partly immersed, black, 0.15–0.2mm diam., with strongly compressed necks. Ascospores hyaline or sometimes slightly coloured when old, 1-septate, occasionally with a small appendage at each end. On dead stems of *R. fruticosus*, Mar.–Oct.

Lophiostoma fuckelii Sacc. var. *pulveraceum* (Sacc)Chesters & Bell (Fig. 1083)
Similar to *L. fuckelii* except for the rather larger ascospores which tend to become 3-septate outside the asci. On dead stems of *R. fruticosus*, Feb.–Oct.

Lophiostoma hysterioides (Schw.)Sacc. (Fig. 1084)
Pseudothecia mostly immersed, with ridge-like necks. Ascospores hyaline, 3-septate, 14–20 × 3. On dead stems of *R. idaeus* and *R. saxatilis*, Mar.–May.

Lophiostoma viridarium Cooke (Fig. 1085)
Pseudothecia immersed, with only the compressed necks, 0.5–0.75 long, showing on the surface; surrounding wood stained green. Ascospores brown, 3-septate, with striate walls, 28–38 × 9–12. On dead stems of *R. fruticosus*, Apr.–May. We have seen this only from Cornwall.

Microthyrium versicolor (Desm.)Höhnel (Fig. 1086)
Thyriothecia superficial, orbicular, up to 0.3mm diam., dark brown, with scutellum made up mostly of elongated, narrow, wavy, radiating brown cells, but of square cells around ostiole. Ascospores hyaline, 1-septate, slipper-shaped, 11–13.5 × 3.5–4.5, with two apical cilia which tend to disappear. On dead stems of *R. fruticosus*, Apr.–Nov.

Morenoina clarkii J.P.Ellis (Fig. 1087)
Thyriothecia mostly bent, lobed and branched, blackish brown, up to 1.5mm long. Ascospores hyaline to very pale straw-coloured, 1-septate, 14–19 × 4.5–6.5. On dead stems of *R. fruticosus*, Mar.–Oct.

Nectria mammoidea Phill. & Plowr. var. *rubi* (Osterw.)Weese
Perithecia at first orange–red, later dark reddish brown or even blackish brown. Ascospores 1-septate, hyaline becoming pale brown and faintly rugulose, 12–14 × 5–6.5. Conidia hyaline, curved, 3- to 5-septate, 45–60 × 6–8. Usually found at the junction between stem and root of *R. fruticosus* and *R. idaeus*, July–Aug.

Paradidymella clarkii D. Hawksw. & Sivan. (Fig. 1088)
Perithecia immersed, 0.15–0.25mm diam., black, some with a clypeus. Ascospores hyaline, 1-septate, 18–23 × 2.5–3.5. On dead stems of *R. fruticosus*, Oct.

Saccothecium sepincola, described on *Rosa*, is not uncommon also on dead stems of *R. fruticosus*.

Schizothyrium speireum (Fr.)Holm & K.Holm (Fig. 1089)
Thyriothecia superficial black 'fly-specks'; scutellum of interlocking cells like a jigsaw puzzle, splitting irregularly. Ascospores hyaline, 1-septate, 7–10 × 3–4. On dead canes of *R. idaeus*, May.

Sydowiella depressula (Karst.)Barr
Perithecia immersed, about 0.5mm diam., with protuberant black necks. Ascospores hyaline, 1-septate, constricted at septum, 20–30 × 6–10, some with gelatinous appendages. On silvery grey areas on dead canes of *R. idaeus*, June–Aug.

Sphaerotheca alchemillae, powdery mildew described on *Alchemilla*, is found occasionally on *R. idaeus*.

HYPHOMYCETES

Chalara kendrickii, described on *Quercus* leaves, has been found also on dead stems of *R. fruticosus*.

Codinaea, v.s. under *Chaetosphaeria callimorpha*.

Endophragmia boothii M.B.Ellis (Fig. 1090)
Colonies effuse, blackish brown to black, hairy. Conidiophores dark brown, up to 150 × 3.5–5. Conidia rather pale brown, mostly 3-septate, 22–30 × 11–14. On decorticated dead stems of *R. fruticosus*, Oct.–Apr.

Endophragmia parva M.B.Ellis (Fig. 1091)
Colonies effuse, grey, hairy, scarcely visible to the naked eye. Conidiophores brown, up to 120 × 2–4. Conidia straw-coloured or pale brown, 1- or 2-septate, 15–18 × 3–4. On dead stems of *R. fruticosus*, Sept.

Helicoon fuscosporum Linder (Fig. 1092)
Colonies effuse, greyish brown, hairy. Conidiophores brown, up to 85 × 4. Coiled conidia pale brown, 25–30 × 20–30, filaments 3–4 thick. On wet dead stems of *R. idaeus*.

Henicospora minor Kirk & Sutton (Fig. 1093)
Colonies effuse, inconspicuous. Conidiophores 4–10 × 3–4. Conidia pale olivaceous brown, with 5 pseudosepta, 16–26 × 4–5. On dead stems of *R. fruticosus*, Sept.–Nov.

Pseudocercospora rubi (Sacc.)Deighton (Fig. 1094)
Colonies epiphyllous. Conidiophores fasciculate, mid pale brown, 15–55 × 3–4. Conidia straw-coloured, often minutely verruculose, 3- to 7-septate, 40–105 × 3–4. On leaves of *R. fruticosus* causing brown or whitish, often irregular spots.

Sporidesmium clarkii P.M.Kirk (Fig. 1095)
Colonies effuse, blackish brown, hairy. Conidiophores up to 150 × 5–7, brown. Conidia obclavate, mostly 4- or 5-septate, brown with paler apical cell, 30–50 × 9–12. Conidiophores and conidia occasionally bear small lateral ampulliform conidiogenous cells with filiform secondary conidia 10–14 × 0.5–1. On dead branches of *R. fruticosus*, Apr.

Sporidesmium rubi M.B.Ellis (Fig. 1096)
Colonies effuse, blackish brown to black, hairy. Conidiophores dark brown, 100–200 × 5–8. Conidia brown, 7- to 11-septate, 47–67 × 8–10. On dead stems of *R. fruticosus*, June.

Triposporium elegans Corda (Fig. 1097)
Colonies effuse, black, hairy. Conidiophores up to 230 × 5–8, swollen at the base and tapered towards the apex. Conidia 3- or 4-armed, arms dark brown, tapered, mostly 20–40 long, 9–12 thick at base, 3- to 9-septate. Very common on dead stems of *R. fruticosus* and occasionally on *R. idaeus*, all the year round.

COELOMYCETES

Camarosporium rubicolum (Sacc.)Sacc. (Fig. 1098)
Pycnidia immersed or erumpent, black, up to 0.5mm diam., thick-walled. Conidia mostly with 3 transverse and 1 or 2 longitudinal septa, dark brown, 14–22 × 6–8. On dead stems of *R. fruticosus*, Feb.–May.

Coniothyrium, v.s. under *Leptosphaeria coniothyrium.*

Cytospora clypeata Sacc.
Conidiomata immersed, 0.5–1mm diam., with greyish to black discs appearing through slits in the epidermis. Conidia hyaline, allantoid, 6 × 1. On dead stems of *R. fruticosus*, Feb.–June.

Diplodia rubi Fr. (Fig. 1099)
Pycnidia immersed, black, rather prominent. Conidia dark brown, 1-septate, 15–20 × 6–9. On dead branches of *R. fruticosus*, Mar.–May.

Hapalosphaeria deformans (Syd.)Syd. (Fig. 1100)
Pycnidia formed in outer walls of anthers, 80–120 diam., brown. Conidia spherical or subspherical, hyaline, 3–5 diam. Causes stamen blight of *R. fruticosus* and *R. idaeus*; the flowers become filled with conidia.

Leptostroma, v.s. under *Hypoderma rubi*.

Leptothyrina rubi (Duby)Höhnel (Fig. 1101)
Conidiomata epiphyllous, subcuticular, 0.15–0.2mm diam., black, shining. Conidia 3.5–4.5 × 1. On bleached areas on fallen leaves of *R. fruticosus*, Jan.–Feb.

Peltasterinostroma rubi Punithalingam
Pycnothyria superficial, about 0.2mm diam., in a black subiculum, with abundant superficial mycelium composed of dark brown, 8–16 thick hyphae and paler, more slender hyphopodiate hyphae. Conidia hyaline, 9–11 × 2.5–3. On dead stems of *R. fruticosus*, May.

Phoma idaei Oud.
Pycnidia immersed, 0.2–0.25mm diam., below blackened areas of epidermis 1–2mm long. Conidia hyaline, biguttulate, 7–8 × 2.5–3.5. On dead stems of *R. idaeus*.

Phoma, v.s. also under *Didymella applanata*.

Phomopsis mulleri (Cooke)Grove
Conidiomata immersed, about 0.3mm diam., black. Conidia hyaline, biguttulate, often slightly curved, 8–10 × 2–3. On dead stems of *R. fruticosus* and *R. idaeus*, Feb.–Mar.

Phomopsis, v.s. also under *Apioporthe vepris*.

Seimatosporium, v.s. under *Griphosphaeria corticola*.

Septocyta ruborum (Lib.)Petrak (Fig. 1102)
Conidiomata immersed, up to 0.3mm diam., blackish brown. Conidia hyaline, 0- to 3-septate, 22–30 × 1.5–2. On living green stems of *R. fruticosus*, staining long stretches of these blackish purple, with white flecks indicating the location of conidiomata, Apr.–May.

Septoria rubi Westend. (Fig. 1103)
Pycnidia epiphyllous, about 0.1mm diam., dark brown, usually only one or a few to each spot. Conidia hyaline, curved, mostly 3- or 4-septate, 35–60 × 1.5–2. Causes pale brown or whitish spots with reddish purple borders. On living leaves of *R. fruticosus* and *R. idaeus*, Feb.–Apr., very common.

Sphaceloma, v.s. under *Elsinoë veneta*.

Truncatella laurocerasi (Westend.)Steyaert
Acervuli few, 0.2–0.45mm diam. Conidia 3-septate with the two median cells chestnut brown, end cells hyaline, 17–22 × 6–8, with a single branched apical setula. On dead leaves of *R. idaeus*.

RUSCUS: Butcher's Broom

Guignardia istriaca Bubák
Pseudothecia immersed, about 0.15mm diam., black. Ascospores hyaline, 14–19 × 5–7. On dead cladodes and on spots on living cladodes, Apr.

Paraphaeosphaeria rusci (Wallr.)O. Eriksson (Fig. 1104)
Pseudothecia immersed, 0.15–0.2mm diam., black. Ascospores straw-coloured, minutely verruculose, 4-septate, 19–24 × 4.5–6, the cell next to the basal one inflated. Associated conidial state has straw-coloured, minutely verruculose conidia, 7–10 × 3–9. On dead cladodes, Feb.–Mar.

Phomopsis rusci (Westend.)Grove
Conidiomata immersed, up to 0.5mm diam., black. Conidia fusoid, hyaline, biguttulate, 6–9 × 2–3. On dead cladodes, Sept.

Phyllosticta hypoglossi (Mont.)Allesch.
Pycnidia up to 0.25mm diam., in a subepidermal stroma. Conidia hyaline, ovoid to subglobose, 8–16 × 6–10, each with a gelatinous coat and a caducous apical appendage. There is also a *Dothiorella* state with cylindrical or dumb-bell-shaped conidia 6–10 × 1.5–3. On living and dead cladodes, which have a silvery appearance.

Pycnofusarium rusci Hawksw. & Punithalingam
Acervuli mostly epiphyllous, subepidermal, reddish yellow or scarlet. Conidia *Fusarium*-like, with foot-cells, curved, 1- or 2-septate, 30–45 × 2.5–4. On dead cladodes, March.

SALIX: Willows

ON LEAVES

UREDINALES

Seven species of *Melampsora* are recorded on willows but only one of them, *M. amygdalina* on *S. triandra*, is autoecious; all the others have alternate aecial hosts, not always represented in Britain. The commonest and most widespread is *M. epitea*, which is found on all *Salix* species except *S. alba* var. *vitellina* and *S. pentandra*. A fairly full description is given of this species; the others are listed and differential diagnostic features indicated under the different hosts.

M. allii-fragilis Kleb., 11, 111.
M. amygdalinae Kleb., 0, 1, 11, 111.
M. capraearum Thüm., 11, 111; aecia on *Larix*.
M. epitea Thüm., 11, 111; aecia on *Euonymus*, *Larix*, *Ribes*, and various orchids.

M. larici-pentandrae Kleb., 11, 111.
M. ribesii-viminalis Kleb., 11, 111.
M. salicis-albae Kleb., 11, 111.

Melampsora epitea Thüm., 11, 111
Uredinia amphigenous, orange-yellow, 0.5–1.5mm diam., paraphyses capitate, mostly thick-walled, up to 90 long, with heads 15–25(35) diam.; spores ellipsoidal or subspherical, shortly echinulate, 12–25 × 10–18. Telia mostly hypophyllous, up to 1mm diam. or larger by confluence, yellowish brown to blackish brown; spores prismatic, pale brown, smooth, 20–50 × 7–15, wall 2 thick, only slightly thicker at apex. There are two varieties, var. *epitea* and var. *reticulatae* (A.Blytt)Jørst.

On *S. alba* and var. *vitellina*: *M. salicis-albae* forms uredinia without paraphyses, up to 5mm long on twigs, as well as the paraphysate ones on leaves; spores oblong or clavate, smooth towards apex. Telia amphigenous.
On *S. arbuscula*: *M. epitea* only.
On *S. atro-cinerea*, *S. aurita*, *S. caprea* and *S. cinerea*: *M. capraearum* forms hypophyllous uredinia with corresponding yellow spots on upper leaf surface. Telia epiphyllous; spores thickened to 10 at apex.
On *S. daphnoides*: *M. epitea* only.
On *S. fragilis*: *M. allii-fragilis* forms mostly hypophyllous uredinia with corresponding reddish-yellow spots on upper leaf surface; paraphyses thin-stalked, spores smooth towards apex. Telia mostly epiphyllous.
On *S. herbacea*, *S. lanata*, *S. lapponum*, *S. myrsinites* and *S. nigricans*: *M. epitea* only.
On *S. pentandra*: *M. allii-fragilis* has urediniospores 22–33 × 13–15 and paraphyses with thin pedicels. Telia mostly epiphyllous; *M. larici-pentandrae* with urediniospores 26–44 × 12–16. Telia hypophyllous.
On *S. phylicifolia*, *S. purpurea*, *S. repens* and *S. reticulata*: *M. epitea* only.
On *S. triandra*: *M. amygdalina* has caeomoid aecia. Uredinia mostly hypophyllous with corresponding pale yellow spots on upper leaf surface; spores smooth at apex, verrucose below.
On *S. viminalis*: *M. ribesii-viminalis* has minute hypophyllous uredinia (0.25mm diam); spores 15–19 × 14–16. Telia epiphyllous.

DISCOMYCETES

Calycellina indumenticola Graddon (Fig. 1105)
Apothecia 0.1–0.15mm diam., hyaline to white, with clavate marginal cells. Ascospores hyaline, 6–8 × 1.5–2. On hairs on lower surface of fallen dead leaves of *S. caprea*, *S. cinerea* and *S. cinerea* × *aurita*, Mar.–July.

Drepanopeziza salicis (Tul. & C. Tul.)Höhnel (Fig. 1106)
Apothecia erumpent, up to 0.2mm diam., blackish brown with grey discs.

Ascospores hyaline, 10–16 × 6–7. On both surfaces of fallen dead leaves of *S. alba*, *S. amygdalina*, *S. fragilis*, *S. triandra* and *S. viminalis*, Mar.–May. Acervuli of the *Monostichella* state with slightly curved, hyaline, biguttulate conidia, 14–17 × 4.5–5.5, can be seen on living leaves, Aug.–Sept.

Drepanopeziza sphaerioides (Pers.)Höhnel (Fig. 1107)
This is said to be similar to *D. salicis*. We have so far not seen apothecia, but the *Marssonina* state with hyaline, 1-septate conidia 15–17 × 5–7 is not uncommon on living leaves and small twigs of weeping willows (*S. alba* var. *vitellina* × *babylonica*).

Hyalotricha salicicola Graddon (Fig. 1108)
Apothecia up to 0.4mm diam., white, covered with thick-walled hairs 100–130 × 3. Ascospores hyaline, 4–5.5 × 2. On fallen dead leaves of *S. caprea* and *S. cinerea*, Oct.–Dec.

Hymenoscyphus caudatus, described on 'leaf litter', is very common on fallen willow leaves.

Pyrenopeziza fuckelii Nannf. (Fig. 1109)
Apothecia erumpent, becoming superficial, 0.2–0.4mm diam., grey or brown with whitish, irregular margins. Ascospores hyaline, 8–10 × 1.5–2. On fallen dead leaves of *S. aurita*, *S. caprea* and *S. cinerea*, May–Aug.

Rhytisma salicinum (Pers.)Fr.
Tarspot. Stromata black, wrinkled, variable in size but often 2–5mm wide. Apothecia seldom seen, straw-coloured discs exposed through slits in stromata. Ascospores hyaline, somewhat curved, 60–90 × 1.5–3. *Melasmia* state has cylindrical conidia 5–6 long. On fallen dead leaves of *S. aurita*, *S. caprea* and *S. viminalis*.

OTHER ASCOMYCETES

Capnodium salicinum Mont. (Fig. 1110)
Pseudothecia up to 0.2mm diam., black, shining, seated on thick, dark mycelial mats of olivaceous brown hyphae. Ascospores dark brown, with 3 or 4 transverse septa and 1 longitudinal septum, 18–24 × 8–12, mostly about 20 × 9. *Fumagospora* state has long-beaked, black pycnidia with fimbriate tips, and brown conidia with 3 transverse septa and 1 longitudinal septum, about 14 × 6. On living leaves and twigs.

Isothea saligna (Ehrh. ex Pers.)Berk. (Fig. 1111)
Perithecia 0.3–0.4mm diam., lodged on their sides, usually singly, in immersed black stromata 0.5–1mm diam., seen easily on the upper leaf surface but obscure on the lower surface where slender necks protrude. Ascospores hyaline, septate, up to 120 × 2. On dead leaves of *S. aurita* and *S. caprea*, Mar.–Aug.

Mycosphaerella punctiformis, a plurivorous species described on 'leaf litter', is common on leaves of *S. aurita*, *S. caprea* and *S. triandra*, Apr.–May.

Trichothyrina salicis J.P.Ellis (Fig. 1112)
Thyriothecia 90–120 diam., with entire margins, and ostiolar collars of 3 to 6 rows of dark brown, thick-walled cells. Ascospores 1-septate, hyaline, 6.5–9 × 1.5–2. On dead leaves of *S. atrocinerea*, *S. cinerea*, etc., Apr.–Oct.

Uncinula aduncta (Wallr.)Lév.
Powdery mildew. Conidia plentiful on both leaf surfaces, 25–35 × 11–20. Cleistothecia 0.1–0.15mm diam., appendages spirally coiled at tips. Asci usually 5 to 12, each with 3 to 6 spores. On *S. atrocinerea*, *S. caprea*, *S. cinerea*, *S. repens*, *S. triandra*, etc.

Venturia. Three species are found on *Salix* in Britain.

KEY

Ascospores 9–12 × 3–5 .... *minuta*
Ascospores 11–15 × 3.5–5 .... *saliciperda*
Ascospores 11–18 × 5–7 .... *chlorospora*

Venturia chlorospora (Ces.)P. Karsten (Fig. 1113)
Pseudothecia immersed, 0.1–0.15mm diam., dark brown, with setae up to 60 long around the ostiole. Ascospores pale olivaceous or brown, 1-septate well above the middle, 11–18 × 5–7. On dead leaves of *S. alba*, *S. caprea*, *S. cinerea* and *S. fragilis*, Apr.–June.

Venturia minuta Barr (Fig. 1114)
Pseudothecia immersed or erumpent, less than 0.1mm diam., with setae up to 60 long around the ostiole. Ascospores olivaceous, 1-septate just above the middle, 9–12 × 3–4. On dead leaves of *S. atrocinerea*, and *S. aurita*, Feb.–July.

Venturia saliciperda Nüesch (Fig. 1115)
Pseudothecia immersed, 0.1–0.12mm diam., with dark brown setae up to 90 long around the ostiole. Ascospores pale olive, 1-septate just above the middle, 11–15 × 3.5–5. On dead leaves, Apr.–May. The *Pollaccia* state, which causes large brown to black spots on living leaves and shoots of *S. alba*, *S. amygdalina* and *S. babylonica* in summer, has brown or olivaceous conidia mostly 1-septate above the middle, 16–22 × 7–8.

COELOMYCETES AND HYPHOMYCETES

Fumagospora, v.s. under *Capnodium salicinum*.

Marssonina state of *Drepanopeziza triandrae* Rimpau (Fig. 1116)
Acervuli grey or whitish on small, dark, often confluent spots. Conidia curved,

Fungi Specific to Trees, Shrubs and Woody Climbers

hyaline, 1-septate, 17–25 × 5–7. On living leaves of *S. cinerea* and *S. triandra*.

Marssonina, v.s. also under *Drepanopeziza sphaerioides*.

Melasmia, v.s. under *Rhytisma salicinum*.

Monostichella, v.s. under *Drepanopeziza salicis*.

Pollaccia, v.s. under *Venturia saliciperda*.

Ramularia rosea (Fuckel)Sacc. (Fig. 1117)
Colonies hypophyllous, pale reddish. Conidiophores in fascicles of 6 to 30, 25–35 × 2.5–4. Conidia hyaline, 0- to 3-septate, 15–27 × 2.5–3.5. On living leaves of *S. caprea*, *S. triandra* and *S. viminalis*, causing very small brown spots.

ON CATKINS

Crocicreas amenti (Batsch)S.Carp. (Fig. 1118)
Apothecia up to 1mm diam., creamy white, downy, stalked. Ascospores clavate, curved, hyaline, 7–11 × 2.5–4. On decaying female catkins, Mar.–Apr., common.

ON WOOD AND BARK

DISCOMYCETES

Apostemidium. Two species and one variety are quite common on wet twigs of willows, Apr.–Aug. Apothecia blackish brown or black, usually 1mm or less diam., often urn-shaped or obconical, with flat or convex, pale bluish grey, greenish or yellowish discs and, when fresh, long bundles of extruded ascospores.

KEY

Asci more than 250 long, spores up to 220 long ....
.... *fiscellum* var. *submersum*
Asci less than 250 long 1
1. Ascospores 100–140 long .... *guernisaci*
   Ascospores 160–200 long .... *fiscellum*

Apostemidium fiscellum (P.Karsten)P. Karsten (Fig. 1119)
Apothecia up to 0.4mm diam. Asci up to 240 × 5; spores 160–200 × 1. The var. *submersum* Graddon has apothecia up to 0.75mm diam., asci 250–260 × 6 and spores 210–220 long.

Apostemidium guernisaci (Crouan & H. Crouan)Boud.
Apothecia up to 1mm diam. Asci 160–190 × 5–6; spores 100–140 long.

Cryptomyces maximus (Fr.)Rehm (Fig. 1120)
Stromata seen as shining black blisters with yellow margins up to 10cm long. Discs when exposed yellowish brown. Ascospores yellowish, 30–37 × 13–17, with gelatinous coats. On thin living branches of *S. fragilis* and *S. viminalis*, uncommon.

Dasyscyphus pudibundus (Quélet)Sacc. (Fig. 1121)
Apothecia solitary or clustered, rather short-stalked, 1–2mm diam., hairy, white when fresh, turning pinkish brown with age. Ascospores hyaline, 7–9 × 1.5–2. On dead twigs of *S. atrocinerea* and *S. cinerea*, June–Sept.

Godronia fuliginosa (Fr.)Seaver (Fig. 1122)
Apothecia 1–2mm diam., numerous, clustered on a flat, erumpent, dark stroma, furrowed, incurved, blackish brown or rusty black, with grey discs. Ascospores hyaline, mostly 7-septate, 60–85 × 2.5–3.5. Associated pycnidia of *Topospora* state contain 3-septate, curved, hyaline conidia, 20–30 × 2–3. On dead branches of *S. caprea* and *S. fragilis*, July–Aug.

Hyalinia rubella (Pers.)Nannf. (Fig. 1123)
Apothecia up to 2mm diam., light red, *Orbilia*-like but with margins appearing toothed owing to the presence of groups of adhering, non-septate hairs, 30–40 × 3–5. Ascospores hyaline, 8–12 × 0.5. On bark of dead branches, all the year round.

Hymenoscyphus. Five species occur on *Salix* but two of these, *H. imberbis* and *H. vernus*, are found no more commonly on willows than on dead branches of other trees in wet places and are described as plurivorous on 'wood and bark'.

KEY

Ascospores 22–30 long	*salicellus*
Ascospores 15–21 long, fresh apothecia reddish brown	*subferrugineus*
Ascospores mostly 12–14 long, fresh apothecia yellowish	*conscriptus*
Ascospores mostly 8–11 long	1
1. Apothecia white, drying reddish brown	*imberbis*
Apothecia yellow, drying brown, stalks blackish	*vernus*

Hymenoscyphus conscriptus (P.Karsten)Korf ex Kobayasi *et al.*
Apothecia stalked, 1–5mm diam., white or yellowish, brownish or ferruginous when dry. Ascospores hyaline, with pointed ends, mostly 12–14 × 3.5–5. On dead branches, Nov.–Dec.

Hymenoscyphus salicellus (Fr.)Dennis (Fig. 1124)
Apothecia short-stalked, 1–2mm diam., pale ochraceous to reddish brown. Ascospores hyaline, 22–30 × 5–7. On dead freshly cut branches, Aug.–Oct.

Hymenoscyphus subferrugineus (Nyl.)Dennis
Apothecia short-stalked, 1–3mm diam., reddish brown. Ascospores hyaline, with rounded ends, 15–21 × 4–6. On dead branches, Oct.–Dec.

Mollisia aquosa, described on 'wood and bark', has been found on dead branches of *Salix*.

Ocellaria ocellata (Pers.)Schröter (Fig. 1125)
Apothecia immersed, 1–2mm diam., with orange to yellowish-brown discs exposed by bursting the bark. Ascospores 4 or 8 to the ascus, hyaline or yellowish, 20–35 × 10–14. On dead branches, Mar.–May.

Orbilia. Six species have been found but none of them more commonly on *Salix* than on other trees. The apothecia are sessile, translucent or waxy-looking, white to rosy orange or copper coloured, and their paraphyses are swollen into a knob at the apex. *O. comma* shows marked preference for *Ulmus*; the others are plurivorous 'wood and bark' species.

KEY

	Apothecia deep cup-shaped, much wider at top than bottom ...... *cyathea*
	Apothecia flatter, not deep cup-shaped 1
1.	Ascospores up to 15 long .... *occulta*
	Ascospores 10–11 long, curved .... *curvatispora*
	Ascospores less than 9 long 2
2.	Ascospores straight, oblong, 3–5 long .... *alnea*
	Ascospores comma-shaped, 5–7 long .... *comma*
	Ascospores narrowly clavate, 6–8 long .... *sarraziniana*

Pachydisca ascophanoides Boud.
Apothecia sessile, lenticular, white to ochraceous, 0.25–0.4mm. Asci 80–90 × 12–13. Ascospores hyaline, granular, 13–15 × 3.5–4. Paraphyses septate near base, swollen to 4–4.5 at apex, containing small granules. On rotten wood.

Pyrenopeziza salicis (Feltg.)Nannf.
Apothecia 0.3–0.6mm diam., greyish brown, developing beneath the epidermis on a thick web of yellowish-brown hyphae, eventually breaking through to the surface. Marginal cells more or less free, about 20 × 5. Ascospores hyaline, 8–15 × 2–2.5. Accompanied by a *Phacidiella* state with long, branched chains of hyaline, somewhat barrel-shaped conidia, 5–11 × 2–3. Parasitic on twigs of *S. caprea* and *S. viminalis*, Mar.–May.

Scutellinia scutellata (L.)Lambotte (Fig. 1126)
Apothecia up to 1cm diam. or occasionally even larger, with bright red discs fringed with pointed, blackish-brown hairs up to 1mm long and much shorter, paler, obtuse hairs. Ascospores 17–21 × 10–14, hyaline with slightly rough walls. Very common on wet wood, especially on branches which have been cut and laid to form paths, May–Nov.

OTHER ASCOMYCETES

Cryptodiaporthe salicella (Fr.)Petrak (Fig. 1127)
Perithecia 0.3–0.4mm diam., black, immersed, with separately erumpent ostioles. Ascospores hyaline, 1-septate, 15–20 × 2–3. Conidiomata of *Diplodina* state black, up to 1mm diam. Conidia mostly 1-septate, hyaline, 16–20 × 4–4.5. On dead twigs of *S. babylonica*, *S. caprea*, *S. fragilis*, etc., Mar.–June; *Diplodina* state Jan.–Feb.

Cryptodiaporthe salicina (Pers.)Wehmeyer (Fig. 1128)
Perithecia similar to those of *C. salicella*, but ascospores much wider, 15–22 × 4.5–7.5 and often slow in forming their septa. On dead twigs, Dec.–Mar.

Cryptosphaerella annexa (Nitschke)Höhnel
Perithecia in groups beneath the periderm, 0.8–1mm diam. Asci long-stalked, polysporous; spores curved, 7–9 × 1. On dead twigs of *S. caprea* and *S. cinerea*, Jan.–Feb.

Cucurbitaria rubefaciens Petrak (Fig. 1129)
Pseudothecia 0.35–0.6mm diam., mostly in clusters of 2 to 4, often causing reddening of surrounding tissues. Ascospores honey-coloured to mid pale brown, with 3 to 5 transverse and 1 or more longitudinal septa, 18–26 × 8–10. On dead branches of *S. caprea*, Mar.–May.

Diaporthe tesella (Pers.)Rehm
Perithecia immersed, 0.5–0.8mm diam., in groups of 3 to 10, with black necks separately erumpent around small blackish discs. Ascospores cylindric–fusoid, hyaline, 1-septate, constricted and often curved or bent at the septum, 35–55 × 7–9, some with an appendage at each end. On dead twigs, Dec.–Feb.

Diatrype bullata (Hoffm.)Fr. (Fig. 1130)
Perithecia immersed in conspicuous, pulvinate, blackish-brown stromata. Ascospores curved, straw-coloured or pale brown in mass, 5–8 × 1–1.5. On dead branches, Mar.–May; fairly common, especially on *S. cinerea*.

Fenestella salicis (Rehm)Sacc. (Fig. 1131)
Pseudothecia clustered in a stroma, erumpent through bark. Ascospores straw-coloured, muriform, 22–28 × 10–11. On attached dead branches, Feb.

Hypocreopsis lichenoides (Tode)Seaver (Fig. 1132)
Stromata superficial, up to 10cm diam., pale yellowish brown or somewhat orange, branched in a radiating manner and lobed at the edge, with ostioles showing as dots on the upper surface. Ascospores broadly fusiform, 1-septate, hyaline, 24–30 × 8–9. On dead branches, May–Sept.

Hypoxylon mammatum (Wahlenb.)Miller (Fig. 1133)
Stromata up to 2mm thick, white pruinose at first, becoming black; perithecia 5 or more to a stroma. Ascospores mid to dark brown, 20–36 × 6–14. *Nodulisporium* state forming pillars or coremia. A rare species in Britain,

recorded a few times on dead branches.

Hypoxylon serpens (Pers.)Kickx var. *effusum* (Nitschke)Miller
Stromata effuse, thin, black, shining; perithecia immersed with protruding papillate ostioles. Ascospores brown, 6–7 × 3–3.5. On dead bark of *S. alba*, Apr., rare.

Hypoxylon multiforme, common on *Betula*, *H. rubiginosum* generally on *Fraxinus*, and *H. serpens*, a plurivorous 'wood and bark' species, have been found occasionally on willows including *S. alba* and *S. caprea*.

Hysterographium elongatum (Wahlenb.)Corda
Hysterothecia superficial, about 3 × 0.6mm, black, opening by longitudinal slits bordered by swollen lips. Ascospores blackish brown, with 7 to 9 transverse septa and an occasional longitudinal septum, 42–46 × 14–17. *H. mori*, a smaller-spored plurivorous species, has been recorded also on willows.

Lophiostoma macrostomoides (de Not.)Ces. & de Not. (Fig. 1134)
Pseudothecia immersed or partly immersed, with large protruding compressed necks, black, 0.5–0.7mm diam. Ascospores golden brown, 5- to 7-septate, mostly 28–37 × 6–9. On dead branches of *S. atrocinerea*, *S. aurita* and *S. cinerea*, Oct.–May.

Nectria. Seven species have been found on *Salix* wood and bark, but none of them commonly. Only *N. citrino-aurantia* and *N. coryli* show preference for willow; the others are plurivorous, with *N. coccinea* most frequent on *Fagus*, *N. galligena* on *Malus*, and *N. episphaeria* always overgrowing or associated with *Diatrype stigma* and other pyrenomycetes.

KEY

	Ascospores 7–8 × 2.5–3 .... *citrino-aurantia*
	Ascospores 7–12 × 3–5 1
	Ascospores more than 12 long 3
1.	Large numbers of small allantoid ascoconidia formed which eventually fill the asci .... *coryli*
	No ascoconidia 2
2.	Lateral collapse of perithecia involving apical disc .... *episphaeria*
	Lateral collapse not involving disc .... *viridescens*
3.	Ascospores 12–17 × 5–7 .... *coccinea*
	Ascospores 15–20 × 6–8 .... *galligena*
	Ascospores 17–24 × 7–9 .... *veuillotiana*

Nectria citrino-aurantia Delacr. ex Desm.
Perithecia about 0.1mm diam., yellow, crowded on and partly embedded in pulvinate yellow stromata up to 1.5mm across. Ascospores 1-septate, 7–8 × 2.5–3. On twigs, few records.

Nectria coryli Fuckel
Perithecia up to 0.35mm diam., clustered on erumpent stromata, red, showing pomiform collapse on drying. Ascospores hyaline, 1-septate, 11–13 × 3–3.5, forming numerous allantoid ascoconidia 3–5 × 1–1.5. Similar but slightly larger conidia develop over the surface of young stromata. On dead twigs, Jan.–Aug., uncommon.

Physalospora miyabeana Fukushi
Perithecia erumpent, dark brown, up to 0.2mm diam. Ascospores hyaline, ellipsoid or oblong, some slightly curved, 14–17 × 5–7. Acervuli of *Colletotrichum* state 0.3–0.7mm diam., setose, sometimes arranged concentrically, spore mass pink, setae brown, 40–50 × 3–4. Conidia hyaline 13–21 × 4–7. On *S. vitellina*, causing blackening of leaves, canker and dieback of growing tips, from end of May onwards.

Rosellinia desmazieresii (Berk. & Br.)Sacc. (Fig. 1135)
Perithecia up to 1.5mm diam., black, with papillate ostioles, seated on a pale greyish-brown felt-like subiculum. Ascospores dark brown, mostly 25–30 (35) × 7–8. Conidia 5–6 × 2–2.5. Parasitic on *S. repens*, Apr.–May.

Valsa salicina (Pers.)Fr.
Perithecia immersed, 0.3–0.4mm diam., 2 to 8 per stroma, discs small, pale grey. Asci mostly 4-spored; spores hyaline, allantoid, 16–22(30) × 5–8. On dead twigs.

Valsella salicis Fuckel (Fig. 1136)
Perithecia immersed, 0.3–1mm diam., 5 to 12 per stroma, with grey discs up to 0.5mm across which are seen through small slits in bark. Asci polysporous, spores allantoid, hyaline, 5–8 × 1–1.5. On dead twigs of *S. aurita*, etc., Dec.–Feb.

HYPHOMYCETES

Bactridium flavum Kunze (Fig. 1137)
Sporodochia lemon yellow, 0.2–0.5mm diam. Conidia pale yellow, 2- to 4-septate, 100–200 × 40–50. On rotten wood, especially wet stumps, Dec.–July.

Oramasia hirsuta Urries (Fig. 1138)
Sporodochia erumpent, 0.1–0.2mm diam., brown or olivaceous brown, setose; setae brown, septate, up to 300 long. Conidia hyaline, slightly hooked at apex, 16–22 × 2–2.5. On bark, Dec.–Jan.

Taeniolella stilbospora (Corda)Hughes (Fig. 1139)
Colonies pulvinate or effuse, dark olivaceous brown to black, velvety. Conidia mid to dark brown, 3- to 24-septate, 25–140 × 7–11. Common on dead branches, frequently staining wood bright reddish purple.

Trimmatostroma salicis Corda (Fig. 1140)
Sporodochia pulvinate, black, powdery. Conidia curved or bent, often forked, with up to 13 transverse and occasionally 1 or a few longitudinal septa, olivaceous brown, smooth or verruculose, 12–38 × 4–10. On twigs and branches. *T. betulinum*, described on *Betula*, is found also on *Salix* and seems to be much commoner than *T. salicis*.

COELOMYCETES

Camarosporium salicinum (Vize)Grove (Fig. 1141)
Pycnidia immersed, 0.5mm diam., black. Conidia dark brown, 15–18 × 6–7, with 3 transverse septa and often 1 longitudinal septum. On twigs of *S. alba* and *S. fragilis*, May–June.

Colletotrichum, v.s. under *Physalospora miyabeana*.

Cytospora salicis (Corda)Rabenh.
Conidiomata erumpent, about 0.5mm diam., conical with greyish discs. Conidia allantoid, 4–7 × 1–1.5. Very common on dead, often attached, twigs of *S. alba*, *S. caprea*, *S. cinerea*, *S. viminalis* and *S. vitellina*, Jan.–Aug.

Diplodina, v.s. under *Cryptodiaporthe salicella*.

Phacidiella, v.s. under *Pyrenopeziza salicis*.

Phomopsis salicina (Westend.)Died.
Conidiomata 0.2–0.3mm diam., black. Conidia hyaline, elliptic to oblong, biguttulate, 6–8 × 2–3. On dead twigs of *S. alba*, *S. atrocinerea*, *S. babylonica*, *S. caprea*, *S. viminalis*, etc., Oct.–May.

Pleurophoma pleurospora (Sacc.)Hohnel (Fig. 1142)
Pycnidia becoming superficial, blackish brown to black, 0.15–0.3mm diam. Conidiophores hyaline, up to 50 × 2, septate. Conidia hyaline, 3.5–4.5 × 1.5. On twigs of *S. fragilis*, May–Nov.

Topospora, v.s. under *Godronia fuliginosa*.

SAMBUCUS: Elder

ASCOMYCETES

Diaporthe circumscripta (Fr. ex Mont.)Otth ex Fuckel
Perithecia immersed, 0.2–0.5mm diam., usually clustered in small groups, with necks collectively erumpent through pustulate discs. Ascospores 1-septate, hyaline, 10–15 × 2.5–4. *Phomopsis* state has conidiomata up to 0.5mm diam., with oblong–fusoid, hyaline, biguttulate A-conidia 5–9 × 2.5–3, and

curved or hooked B-conidia 16–23 × 1. On dead twigs.

Dothidea sambuci (Pers.)Fr. (Fig. 1143)
Stromata erumpent, blackish brown to black, about 1mm diam., with numerous pseudothecial locules. Ascospores mid pale brown, 1-septate below the middle, 20–26 × 8–9. On dead twigs, Mar.–Apr.

HYPHOMYCETES

Balanium stygium Wallr. (Fig. 1144)
Colonies effuse, black, velvety. Conidiophores dark brown, repeatedly dichotomously and trichotomously branched. Conidia very dark brown to black, 1-septate, 13–23 × 8–14. On dead wood, uncommon.

Cercospora depazeoides (Desm.)Sacc. (Fig. 1145)
Colonies amphigenous. Conidiophores fasciculate, olivaceous brown, 60–120 × 4–5. Conidia pale olive, smooth or minutely verruculose, 3- to 9-septate, 50–100 × 4–6. On living leaves, causing large, round or angular, often conspicuously zonate, pale greyish-brown spots, Aug.–Sept., quite common.

Ramularia sambucina Sacc. (Fig. 1146)
Colonies white, hypophyllous. Conidiophores hyaline, 25–45 × 2.5–5. Conidia hyaline, 0- to 2-septate, 17–35 × 4–5. On living leaves, causing pale brown or grey, dark-bordered spots, 3–8mm diam.

Sporidesmium. Four species are found, some of them quite commonly on wood and bark of elder.

KEY

	Conidia narrowly obclavate, pale .... *leptosporum*
	Conidia conical or obpyriform, mid or dark brown 1
1.	Conidia 5- to 8-septate, 14–25 wide .... *altum*
	Conidia 2- to 4-septate, 7–12 wide 2
2.	Conidia conico-truncate, protruding at base .... *cookei*
	Conidia rounded at base .... *aturbinatum*

Sporidesmium altum (Preuss)M.B.Ellis (Fig. 1147)
Colonies black, effuse. Conidiophores dark brown, 100–500 × 4.5–6. Conidia 5- to 8-septate, mid to dark brown, smooth, 35–65 × 14–25. It usually stains the surrounding wood green, Sept.–May.

Sporidesmium aturbinatum (Hughes)M.B.Ellis (Fig. 1148)
Colonies black, effuse. Conidiophores dark brown, 60–250 × 3–6. Conidia 2- to 4-septate, 17–28 × 8–12, lower cells usually dark brown and verrucose, upper cells pale. On rotten wood, Apr.–Aug.

Sporidesmium cookei (Hughes)M.B.Ellis (Fig. 1149)
Colonies black, effuse. Conidiophores dark brown, 70–200 × 3–4.5. Conidia 2- or 3-septate, 17–25 × 7–10, lower cells mid to dark brown, occasionally verruculose. On dead branches, Sept.–May.

Sporidesmium leptosporum (Sacc. & Roum.)Hughes (Fig. 1150)
Colonies brown, effuse, hairy. Conidiophores mid to dark brown, 35–100 × 3.5–5, often with barrel-shaped proliferations. Conidia 5- to 21-septate, 25–130 × 5–8, subhyaline or pale straw-coloured, often brown around the scar. Occurs on many different substrata but is especially common on dead branches of elder all the year round.

COELOMYCETES

Ascochytula deformis (Karst.)Grove
Pycnidia immersed, black, 0.2–0.25mm diam. Conidia hyaline, yellowish in mass, 1-septate, 8–13 × 2–3. On dead twigs, Apr.–May.

Phoma sambuciphila Oudem.
Pycnidia with base sunken in wood, black, 0.1–0.15mm diam. Conidia slightly curved, hyaline, 5 × 2. On decorticated branches, Sept.–Feb.

Phomopsis, *v.s.* under *Diaporthe circumscripta*.

SORBUS: Rowan or Mountain Ash, White Beam and Wild Service Tree. (Mostly on Mountain Ash.)

UREDINALES

Gymnosporangium cornutum Kern, 0, 1
Aecia hypophyllous on yellow or orange, thickened spots, cylindrical or horn-shaped, up to 5mm high; spores brown, minutely verruculose, 20–30 × 18–25. Telial host *Juniperus communis*. Common in Scotland, uncommon in England.

Ochropsora ariae (Fuckel)Ramsb., 11, 111
Uredinia hypophyllous, minute, often on pale spots, subepidermal, with a ring of paraphyses; spores pale brown, verruculose, 20–28 × 16–21. Telia hypophyllous on yellow or reddish spots, pale indian red, waxy, crust-like; spores up to 70 × 10–14, finally 3-septate. Aecial host *Anemone nemorosa*.

DISCOMYCETES

Dermea ariae (Pers.)Tul. ex P.Karsten (Fig. 1151)
Apothecia erumpent, 0.5–1mm diam., blackish brown. Ascospores hyaline or pale yellowish, mostly 12–15 × 3–4. Associated *Foveostroma* state has narrowly fusiform or sickle-shaped conidia 15–20 × 2–2.5. On dead branches, Feb.–May.

Hyaloscypha quercina Velen. var. *resinacea* Dennis
Apothecia 0.3–1mm diam., dark brown, covered with septate hairs up to 140 × 3, which are coated with an amorphous brown exudate. Ascospores hyaline, 5–7 × 2.5. On fallen wood.

Pezizellaster serrata (Hoffm.)Dennis (Fig. 1152)
Apothecia sessile, up to 0.8mm diam., white or pinkish becoming darker on drying, with hairs at margin grouped to form distinct teeth up to 100 long. Ascospores hyaline, 5–6 × 2. On inner surface of bark, Apr.

Tympanis conspersa, described on *Malus*, is found also on dead twigs of *Sorbus*.

OTHER ASCOMYCETES

Coronophora gregaria (Lib.)Fuckel (Fig. 1153)
Perithecia immersed, often arranged in circles under the bark, which becomes pushed up and split, black, without ostioles, 1–2mm diam., tending to collapse. Asci polysporous, with long stalks; spores curved, hyaline, 7–9 × 1–2. On dead twigs and branches.

Diaporthe impulsa (Cooke & Peck)Sacc. (Fig. 1154)
Perithecia immersed, clustered, 0.3–0.6mm diam., black, with necks emerging through grey discs. Ascospores hyaline, 1-septate, constricted at the septum, 13–18 × 2.5–5.5. *Phomopsis* state has hyaline, biguttulate conidia 9–12 × 2–2.5. On dead but often still attached branches.

Dothiora pyrenophora (Fr.)Fr. (Fig. 1155)
Stromata gregarious, erumpent, black, up to about 1mm diam., with one or a few pseudothecial locules. Ascospores with transverse and incomplete longitudinal septa, hyaline, 26–47 × 6–10. Sept.–Jan. *Dothichiza* state with conidia 6–8 × 2.5–3.5 causes canker of small branches, Mar.–Apr.

Eutypella sorbi (Alb. & Schw.)Sacc.
Perithecia 0.3–0.5mm diam., in erumpent, circular stromata up to 4mm diam., necks projecting, black, with sulcate ostioles. Ascospores allantoid, 7–9 × 1.5–2, yellowish. Conidia of *Cytospora* state form red tendrils; they are allantoid, 3.5–4 × 1. On dead twigs and branches, fairly common.

Leucostoma persoonii Höhnel
Perithecia immersed, 3 to 10 per stroma, discs pure white. Ascospores

cylindrical, curved, hyaline, 10–12 × 2.5–3. On dead twigs. *Cytospora* state has white discs and allantoid conidia 5–6 × 1, which ooze out forming red tendrils.

Podosphaera aucupariae Eriksson
Powdery mildew. Amphigenous, few conidia. Cleistothecia with appendages 4 to 6 times dichotomously branched. The single ascus contains 6 to 8 spores. Mostly on young shoots.

HYPHOMYCETES AND COELOMYCETES

Corynespora cambrensis M.B.Ellis (Fig. 1156)
Colonies effuse, dark blackish brown. Conidiophores brown, 60–100 × 6–9. Conidia pale olivaceous brown, with 2 to 8 pseudosepta, 20–85 × 5–10. On dead branches, Mar.–May.

Cytospora, v.s. under *Eutypella sorbi* and *Leucostoma persoonii*.

Dothichiza, v.s. under *Dothiora pyrenophora*.

Foveostroma, v.s. under *Dermea ariae*.

Phomopsis, v.s. under *Diaporthe impulsa*.

Rhabdospora inaequalis (Sacc. & Roum.)Sacc. (Fig. 1157)
Conidiomata up to 0.25mm diam., erumpent, black. Conidia slightly curved, narrowly fusiform, hyaline, 11–18 × 2–3. On attached dead twigs, Mar.

Septoria sorbi Lasch
Pycnidia hypophyllous, blackish brown, 0.15–0.2mm diam., gregarious. Conidia 60–70 × 3.5–4, curved, faintly yellow, some 2-septate. On small brown spots along margins of leaves, June–July.

Septosporium bulbotrichum Corda (Fig. 1158)
Colonies dark brown to black, setose. Conidia muriform, straw-coloured to dark brown, 27–80 × 12–35; often one or more cells of the conidium swell separately, darken and become transformed into short-beaked pycnidia. A rare species which has been found in Scotland on inner bark of *S. aucuparia*.

SYMPHORICARPOS: Snowberry

COELOMYCETES

Ascochytula symphoricarpi (Pass.)Died.
Pycnidia immersed, 0.1–0.15mm diam., brown or blackish brown. Conidia cylindric–fusoid, becoming 1-septate, hyaline to golden brown, 8–11 × 2–3. On dead stems, June.

Hendersonia fiedleri Westend. var. *symphoricarpi* Cooke (Fig. 1159)
Pycnidia about 0.3mm diam., immersed but raising and splitting the epidermis. Conidia 3-septate, straw-coloured, 14–18 × 4. On dead twigs.

TAMARIX: Tamarisk

COELOMYCETES

Coniothyrium tamaricis Oud.
Pycnidia immersed, black, about 0.1mm diam. Conidia becoming pale olivaceous, 5–8 × 3.5–4.5. On dead, attached twigs, May–Aug.

Coryneopsis tamaricis (Mig.)Grove
Acervuli up to 0.5mm diam. Conidia 3-septate, golden brown, 27–35 × 10–13 (mostly about 30 × 12). Conidiophores hyaline, 15–45 × 2–3. On thin, dead, attached twigs, Apr.–Sept. This species would perhaps be better accommodated in *Stilbospora*.

Cytospora tamaricis Brun.
Conidiomata up to 0.5mm diam., with black discs. Conidia allantoid, hyaline, 4–6 × 1.5–2. On dead attached twigs, Apr.–Sept.

TAXUS: Yew

DISCOMYCETES

Chaenothecopsis caespitosa (Phill.)D.Hawksw. (Fig. 1160)
Fructifications 2–4mm tall, branched, reddish brown to black; discs of apothecia at ends of branches 0.15–0.35mm diam., black. Ascospores brown, 1-septate, 9–14 × 3–4.5. On rotten wood, Oct., apparently rare.

OTHER ASCOMYCETES

Anthostomella formosa Kirschst. var. *taxi* (Grove)S. Francis (Fig. 1161)
Perithecia immersed, 0.2–0.3mm diam., without a clypeus. Ascospores 14–17 × 7–9, chestnut brown with a hyaline basal dwarf cell. On dead twigs and leaves, Sept.–May, uncommon.

Botryosphaeria foliorum (Sacc.)v.Arx & E.Müller (Fig. 1162)
Pseudothecia mostly epiphyllous, erumpent, black, up to 0.25mm diam. Ascospores hyaline, mostly 16–22 × 8–10. On dead leaves and twigs, Mar.–May. Pycnidia of *Phyllosticta* state amphigenous, sometimes in short lines, 0.2mm diam., black, splitting epidermis. Conidia obovoid, tapered

slightly at base, hyaline, with 1 large guttule or several smaller ones, 10–11 × 5–7.

Dothiora taxicola (Peck)Barr (Fig. 1163)
Stromata immersed, small, black, with a single pseudothecial locule. Ascospores 3-septate, hyaline, mostly 20–27 × 5–7. On dead leaves.

HYPHOMYCETES

Capnobotrys dingleyae Hughes (Fig. 1164)
Colonies effuse, dark blackish brown to black, spongy. Moniliform hyphae golden brown or reddish brown, up to 27 thick. Conidia 1- to 6- (mostly 2-) septate, reddish brown, verruculose, 10–45 × 7–14. A 'sooty mould' on living leaves and twigs.

Corynespora pruni (Berk. & Curtis)M.B.Ellis (Fig. 1165)
Colonies black, usually tufted. Conidiophores blackish brown, 60–280 × 7–11, with barrel-shaped proliferations. Conidia olivaceous brown, dark at scar, with 4 to 9 pseudosepta, 50–130 × 10–16. On wood and bark, uncommon.

Endophragmia coronata Sutton (Fig. 1166)
Colonies effuse, thinly hairy. Conidiophores dark brown, up to 70 × 3–4. Conidia pale brown, pseudoseptate, mostly cuneiform with short projections at the top, 25–35 × 4–7; a few are cylindrical and up to 80 long. On dead leaves, Sept.

Endophragmia taxi M.B.Ellis (Fig. 1167)
Colonies effuse, shortly hairy, blackish brown. Conidiophores brown, 30–80 × 2–3.5. Conidia brown, with 1 thick dark septum, 12–20 × 7–12. On dead leaves, Apr.–May.

Sporidesmium larvatum Cooke & Ellis (Fig. 1168)
Colonies effuse, dark brown to black. Conidiophores reddish brown, 5–9 × 3–6. Conidia dark reddish brown, 10- to 15-septate, 50–80 × 10–12. On decorticated wood, Jan., uncommon.

Sterigmatobotrys macrocarpa (Corda)Hughes (Fig. 1169)
Colonies effuse, dark blackish brown, thinly hairy. Conidiophores branched penicillately at apex, blackish brown, up to 500 × 4–7. Conidia aggregated in slimy heads, 2-septate, hyaline at first, later brown with median cell darkest, 16–25 × 6–8. On dead wood.

Thysanophora taxi (Schneider)Stolk & Hennebert (Fig. 1170)
Colonies effuse, hairy, blackish brown. Conidiophores olivaceous brown, up to 1mm high, 3–6 thick, with concolorous phialides, 17–24 × 5–7 at apex. Conidia catenate, hyaline to olivaceous, slightly verruculose, 6.5–10 × 2–3.5.

On fallen dead leaves, Sept.

COELOMYCETES

Cytospora taxi Sacc.
Conidiomata erumpent, plurilocular, black, up to 2mm diam. Conidia allantoid, hyaline, 6–10 × 1.5. On dead twigs and leaves, Nov.–Mar.

Diplodia taxi (Sow.)de Not.
Pycnidia amphigenous, immersed, black. Conidia ellipsoid–oblong, rounded at ends, remaining hyaline for a long time, eventually becoming smoky brown and 1-septate, 20–25 × 8–10. On dead leaves, Sept.–Mar.

Phoma allostoma Died.
Pycnidia immersed, mostly hypophyllous, 0.1–0.15mm diam., black, thick-walled. Conidia cylindrical, truncate and guttulate at each end, 5–8 × 1–1.5. On dead, attached leaves and twigs, Nov.–Mar.

Phomopsis occulta, described on *Pinus*, has been found on dead twigs of *Taxus*.

Phyllosticta, v.s. under *Botryosphaeria foliorum*.

THUJA: Western Red Cedar and White Cedar

Chloroscypha seaveri Seaver
Apothecia up to 0.7mm diam., short-stalked, black, subgelatinous. Ascospores pale yellowish, 25–30 × 8–12. Tips of paraphyses gummed together with olivaceous brown exudate to form epithecia. On dead leaves.

Didymascella thujina (Durand)Maire (Fig. 1171)
Apothecia immersed, about 1mm diam., dark olivaceous brown, opening by little lids which become detached. Ascospores 21–25 × 13–17, 1-septate near apex, brown, pitted, the small cell often pale. On dead attached leaves of *T. occidentalis* and *T. plicata*, June–Aug.

Dothiora thujae (Grove)Barr
Pseudothecia 0.15–0.25mm diam., black, solitary or in groups, thick-walled. Ascospores golden brown, with 3 to 7 (mostly 5) transverse septa and 1 longitudinal septum, 20–30 × 6–9, some with gelatinous sheaths. On cone scales of *T. occidentalis*.

Kabatina thujae Schneider & v.Arx
Acervuli erumpent, pulvinate, brown, up to 0.17mm diam. Conidia hyaline, 5–8 × 2.5–3.5. On branches of *T. occidentalis* and *T. plicata*.

Phoma thujana Thümen
Pycnidia mostly epiphyllous, immersed, about 0.1mm diam., black. Conidia

hyaline, 3–5 × 2–2.5. On leaves, especially leaf bases, and on twigs.

Truncatella hartigii (Tubeuf)Steyaert
Acervuli immersed, with extruded black spore masses. Conidia 3-septate, 19–25 × 7–9, median cells large, brown, end cells hyaline. Apical appendages (setulae) 1 to 4, about 20 × 1. On bark of *T. plicata*.

TILIA: Common Lime

DISCOMYCETES

Encoelia tiliacea (Fr.)P.Karsten
Apothecia erumpent, in small groups, 1–8mm diam., brown, leathery. Ascospores slightly curved, hyaline, mostly 15–18 × 3–3.5. On fallen dead branches, Aug.–Oct., uncommon.

OTHER ASCOMYCETES

Cryptodiaporthe hranicensis (Petrak)Wehmeyer (Fig. 1172)
Perithecia immersed, in clusters of 10–20, with long, converging, excentric, erumpent necks. Ascospores hyaline, mostly 1-septate, 11–15 × 3–4. Associated *Amphicytostroma* state has basal locules with long ostiolar channels, and hyaline, allantoid conidia 4–6.5 × 1–1.5. On fallen dead branches, Mar.–Apr.

Diaporthe velata (Pers.)Nitschke (Fig. 1173)
Perithecia immersed, 0.4–0.5mm diam., black, with black stromatic line deep in wood. Ascospores hyaline, 1-septate, 8–12 × 2.5–3. *Phomopsis* state has fusoid A-conidia 8–12 × 2.5 and hooked, filiform B-conidia 16–20 × 1. On dead branches, conidia June–Dec., ascospores Dec.–Mar.

Hercospora tiliae (Pers.)Fr. ex Tul. & C.Tul. (Fig. 1174)
Perithecia immersed, 0.4mm diam., black, grouped in small stromata. Ascospores hyaline, 1-septate, mostly 15–18 × 7–8. *Rabenhorstia* state has hyaline conidia 13–18 × 7–8, with large guttules. On dead twigs and branches, conidia June–Nov., asci Dec.–Mar.

Pseudomassaria chondrospora (Ces.)Jacz. (Fig. 1175)
Perithecia immersed, 0.25–0.35mm diam., inconspicuous. Ascospores hyaline or yellowish, with a septum cutting off a small basal cell, and occasionally another septum, 20–34 × 10–11. On thin twigs, Feb.–Apr., uncommon.

Splanchnonema ampullaceum (Pers.)Shoemaker & Le Clair (Fig. 1176)
Pseudothecia 0.6–0.8mm diam., immersed, inconspicuous, raising bark only slightly, with small cracks. Ascospores 1-septate, rather dark brown, 30–50 ×

15–18, with thick gelatinous sheaths. On twigs and small branches, Dec.–Feb.

HYPHOMYCETES

Cercospora microsora Sacc. (Fig. 1177)
Conidiophores arising from well-developed stromata, mid pale olivaceous brown, 20–35 × 4–5. Conidia pale olive, 3- to 8- (mostly 3-)septate, 35–90 × 3–4. Causes brown, often confluent, spots on leaves and large, shining, black lesions on twigs.

Corynespora olivacea (Wallr.)M.B.Ellis (Fig. 1178)
Colonies discrete, blackish brown, round, 0.5–1.5mm diam. Conidiophores arising from surface of large stromata, brown or dark brown, 0- to 2-septate, 12–30 × 8–11, some with oval or spherical terminal proliferations. Conidia golden brown, black at scar, with 5 to 14 pseudosepta, 50–105 × 12–19. On dead branches all the year round, fairly common.

Exosporium tiliae Link (Fig. 1179)
Colonies discrete, punctiform, dark brown. Conidiophores arising from stromata, 3- to 6-septate, 50–150 × 8–12. Conidia golden brown, with 7 to 18 pseudosepta, 70–195 × 12–21, with protruding, conico-truncate, black scar at base. On dead twigs and branches, June–Sept.

COELOMYCETES

Amphicytostroma, v.s. under *Cryptodiaporthe hranicensis*.

Camarosporium tiliae Sacc. & Penz. (Fig. 1180)
Pycnidia immersed, black. Conidia brown, with transverse and longitudinal septa, 16–20 × 7–10. On dead branches, Nov.–June.

Cytospora carphosperma Fr.
Conidiomata immersed, 1mm diam., black, with whitish discs. Conidia issuing in yellow tendrils, allantoid, 5–6.5 × 1–1.5. On dead twigs, Dec.–June.

Diplodia tiliae Fuckel
Pycnidia immersed, black. Conidia dark brown, 1-septate, 20–24 × 8–10. On dead twigs and branches, Oct.–Nov.

Gloeosporium tiliae Oud.
Acervuli hypophyllous, minute, brown. Conidia oblong–ovoid, hyaline, 10–18 × 4–7. Sometimes causing ochraceous, dark-bordered spots up to 2cm diam. on leaves, July-Oct.

Lamproconium desmazieresii (Berk. & Br.)Grove (Fig. 1181)
Acervuli immersed, 0.5–1mm diam., black. Conidia thick-walled, pale

purplish slate coloured, 30–35 × 7–10. On living and dead twigs and branches.

Phomopsis, v.s. under *Diaporthe velata.*

Rabenhorstia, v.s. under *Hercospora tiliae.*

ULEX: Furze, Gorse or Whin and Dwarf Gorse

UREDINALES

Uromyces pisi-sativi, described on *Pisum*, is just occasionally found on *Ulex*.

USTILAGINALES

Thecaphora deformans, described on *Lathyrus*, sometimes attacks *U. minor*. Infected pods become somewhat deformed and have shrivelled, infected seeds containing rusty or chocolate brown masses of spore balls.

DISCOMYCETES

Crocicreas complicatum (P. Karsten)S. Carp. (Fig. 1182)
Apothecia up to 1mm diam., cream or pale buff. Upturned hyphal tips on the surface of the receptacle stain deeply in cotton blue and bear little lumps of hyaline exudate. Ascospores hyaline, some 1-septate, 7–10 × 1.5–2. On wood of *U. europaeus*.

Dasyscyphus pygmaeus (Fr.)Sacc. (Fig. 1183)
Apothecia mostly 2–3mm diam., with long, slender stalks and yellow or yellowish discs which dry somewhat orange. Hairs 1-septate, 20–40 × 4–5. Ascospores hyaline, 7–11 × 1.5–2.5. Paraphyses lanceolate, 5 wide. On roots and branches, especially where burnt and partly buried in soil or matted grass, Apr.–Oct.

Habrostictis rubra, described on *Ulmus*, occurs occasionally also in cracks in bark of gorse.

Mollisia ventosa, with ascospores up to 21 long, described on 'wood and bark', is quite common on dead branches, and two other plurivorous species, *M. cinerea* and *M. ligni*, also occur.

Mollisiopsis dennisii Graddon (Fig. 1184)
Apothecia up to 1mm diam., buff with white margins. Lanceolate paraphyses protrude well above the tips of asci. Ascospores hyaline, 6–8 × 2. On dry dead spines and branchlets of *U. europaeus*, May–Nov.

Phaeangella ulicis (Cooke)Sacc. (Fig. 1185)
Apothecia erumpent, 1–1.5mm diam., dark brown, scurfy, with inrolled margins. Flesh turns crimson in a solution of KOH. Ascospores hyaline or faintly coloured, 7–11 × 2.5–3. On small dead twigs on ground, Feb.

Trichodiscus ulicicola Graddon (Fig. 1186)
Apothecia 0.1–0.15mm diam., covered with dark brown setae up to 120 × 4 and pale brown hairs up to 50 × 5. Ascospores hyaline, 0- to 3-septate, 14–22 × 2. On burnt branches of *U. europaeus*, Nov.

Velutarina rufo-olivacea, a plurivorous species described on 'wood and bark', is not uncommon on fallen or cut branches.

OTHER ASCOMYCETES

Coniochaeta ligniaria (Grev.)Cooke (Fig. 1187)
Perithecia superficial, 0.2–0.35mm diam., black or blackish brown, setose. Setae dark brown, 30–50 × 4–5. Ascospores blackish brown, 12–18 × 8–12 × 5–7, some with gelatinous sheaths. On dead, sometimes burnt wood, Nov.–Apr.

Cucurbitaria elongata, described on *Robinia*, has been found on dead branches of *Ulex*.

Daldinia vernicosa (Schw.)Ces. & de Not.
Pulvinate stromata erumpent from bark or superficial on wood, brown to black, mostly 4–10mm diam. but occasionally up to 3cm, showing concentric zoning when sliced through vertically. Ascospores dark brown, 7–12 × 4.5–7. Common on burnt branches.

Diaporthe inaequalis (Currey)Nitschke (Fig. 1188)
Perithecia immersed, 0.4–0.7mm diam., black, with necks erumpent through brownish, pulvinate discs. Ascospores hyaline, 1-septate, mostly 12–18 × 7–9. *Phomopsis* state has A-conidia 7–10 × 2–3 and B-conidia curved, 21–27 × 2. On dead twigs, Dec.–July.

Diaporthe nucleata (Currey)Cooke
Rather similar to *D. eres*, which occurs also, but with ascospores 15–20 × 2.5–3. On dead branches.

Dothidea puccinioides (DC.)Fr. (Fig. 1189)
Stromata erumpent, roughly hemispherical, up to 2mm diam., black, with a number of pseudothecial locules. Asci 2- to 4-spored; spores brown, 1-septate, 20–30 × 9–16. On dead branches.

Lophiostoma angustilabrum (Berk. & Br.)Cooke (Fig. 1190)
Pseudothecia either immersed with only their flattened necks erumpent or semi-immersed, about 0.5mm diam., black. Ascospores hyaline, 1-septate,

25–32 × 4–5.5, sometimes with an appendage at each end. Common on dead branches, Mar.–Oct.

Microthyrium cytisi Fuckel var. *ulicis* J.P.Ellis (Fig. 1191)
Superficial mycelium abundant. Thyriothecia 0.2–0.28mm diam., brown, with wide, almost black collar around ostiole. Ascospores hyaline, 1-septate, 12–14.5 × 3–4. On dead branchlets and spines of *U. europaeus*, often still attached to plants, but only on bleached areas, Feb.–Sept.

Microthyrium cytisi var. *ulicis-gallii* J.P.Ellis (Fig. 1192)
Ascospores 14–19 × 3–4.5. On dead leaves and branchlets of *U. gallii* and *U. minor*, Jan.–Mar.

Trichosphaerella decipiens Bomm., Rouss. & Sacc. (Fig. 1193)
Perithecia superficial, up to 0.2mm diam., but usually smaller, yellowish olive, setose, tending to collapse. Setae blackish brown, 20–40 × 4–5. Ascospores hyaline, soon dividing into two parts each about 4 × 2. On burnt branches on ground.

HYPHOMYCETES

Diplocladiella scalaroides Arnaud (Fig. 1194)
Colonies effuse, brown, shortly hairy. Conidiophores brown, geniculate, 25–45 × 3–4. Conidia triangular, 2-horned, brown, 15–30 from horn tip to horn tip. On dead wood of *U. europaeus*, uncommon.

Hansfordia pulvinata (Berk. & Curtis)Hughes (Fig. 1195)
Colonies effuse, grey or olivaceous grey. Conidiophores repent or erect, very variable in length, 2–5 thick, brown. Conidia borne on short denticles, almost spherical, hyaline to pale brown, minutely echinulate, 4–7 diam. On dead branches of *U. europaeus*, sometimes overgrowing other fungi.

Phaeostalagmus cyclosporus, a plurivorous species described on 'wood and bark', is very common on dead branches of gorse.

Phialocephala fusca Kendrick (Fig. 1196)
Colonies effuse, olivaceous brown or blackish brown, hairy or velvety. Conidiophores brown, up to 300 × 5–7. Conidia 2–4 × 1.5–2.5, forming dark brown or black slimy heads. On wood of *U. europaeus*, occasional.

Sporidesmium cambrense M.B.Ellis (Fig. 1197)
Colonies effuse, grey to black. Conidiophores dark brown, up to 300 × 7–11. Conidia 6- to 10-septate, mid to dark brown except for cells at each end, which are hyaline or very pale, 40–60 × 12–15. On dead branches of *U. europaeus*, Mar., occasional.

COELOMYCETES

Ascochytula ulicis Grove
Pycnidia immersed, about 0.2mm diam., rather thick-walled. Conidia fusoid, 1-septate, pale yellowish olive, 10–14 × 2. On dead branchlets of *U. europaeus*, May–June.

Coniothyrium sphaerospermum Fuckel
Pycnidia erumpent, punctiform, black. Conidia subglobose, yellowish, 2–3 diam. On spines of *U. europaeus* and *U. gallii*, June–Nov.

Diplodia ulicis Sacc. & Speg.
Pycnidia immersed, about 0.2mm diam., black. Conidia ovoid or ellipsoid, olivaceous brown, becoming 1-septate, 20–25 × 10–11. On rotting branches and spines, July.

Phomopsis, v.s. under *Diaporthe nucleata*.

Septoria slaptoniensis Hawksw. & Punithalingam
Pycnidia up to 0.12mm diam., brown or black. Conidia straight or curved, 1- to 3-septate, hyaline, 17–22 × 1.5–2. On dead leaves of *U. europaeus*, June.

ULMUS: Elms

ON LEAVES

Dasyscyphus deflexus Graddon (Fig. 1198)
Apothecia sessile, up to 0.3mm diam., white, hairy. Hairs up to 45 × 5, uncinate, granulate. Ascospores hyaline, 4–5 × 1.5–2. On fallen dead leaves, Nov.–Dec.

Mycosphaerella oedema (Fr. ex Duby)Schröter
Pseudothecia immersed, often clustered, 0.1–0.15mm diam. Ascospores hyaline, 1-septate, fusiform to somewhat clavate, 23–25 × 4. On dead leaves, Apr.–May.

Mycosphaerella ulmi Kleb.
Pseudothecia immersed, 0.1mm diam. Ascospores fusoid, hyaline, medianly 1-septate, 26–28 × 2.5–4. The *Phloeospora* state is common on living leaves of *U. glabra* and *U. procera*, Sept.–Oct., and causes irregular brown spots. Acervuli hypophyllous, up to 0.2mm diam., thin, brown, with whitish masses of conidia oozing out. Conidia falcate, hyaline, 3- to 5-septate, 30–60 × 4.5–6.

Platychora ulmi (Schleicher ex Duval)Petrak (Fig. 1199)
Stromata subepidermal, pulvinate, 2–3mm diam., often appearing silvery, dotted with black papillate ostioles and containing many pseudothecial locules. Ascospores with septum near base, hyaline or pale olive, 10–13 × 4–5. On

fallen dead leaves of *U. glabra* and *U. procera*, Jan.–Apr., very common. The *Piggotia* state with groups of black subcuticular acervuli up to 1mm diam. is found on the upper side of living leaves, June–Sept. Conidia cuneiform, pale olivaceous brown, 9–11 × 4.5–6.

Taphrina ulmi (Fuckel)Johansson
Causes diffuse blotching of mainly sucker leaves, Aug.–Sept. Blotches at first pale or slightly yellow, with asci on wrinkled upper surface. Necrotic areas eventually fall out.

ON WOOD AND BARK

DISCOMYCETES

Encoelia siparia (Berk. & Br.)Nannf. (Fig. 1200)
Apothecia erumpent in clusters, yellowish becoming dark brown, up to 6mm diam. Ascospores hyaline, 12–18 × 2–3. On branches, July–Nov.

Habrostictis rubra Fuckel (Fig. 1201)
Apothecia just immersed in wood where bark has cracked away, 1–2mm diam., pink with orange-tinged discs when fresh, and toothed margins. Ascospores hyaline, 11–15 × 2. Paraphyses with swollen, flask-shaped tips. On dead branches, Jan.–Aug.

Orbilia comma Graddon (Fig. 1202)
Apothecia up to 1.5mm diam., waxy, rosy peach to orange. Ascospores hyaline, comma-shaped, 5–7 × 1.5–2. On and inside bark, Sept.–Mar.

OTHER ASCOMYCETES

Amphisphaeria umbrina (Fr.)de Not. (Fig. 1203)
Perithecia erumpent from bark, about 1mm diam., papillate, black. Ascospores 1-septate, brown, 17–26 × 6–8. On dead branches.

Anthostoma melanotes (Berk. & Br.)Sacc. (Fig. 1204)
Stromata effuse. Perithecia immersed, black, with short, protuberant necks. Ascospores brown, 12–15 × 5–7. On decorticated branches.

Ceratocystis ulmi (Buisman)C.Moreau (Fig. 1205)
Perithecia erumpent through cracks in bark or superficial on damp wood, blackish brown to black, 0.1–0.14mm diam., with dark necks pale and fimbriate around ostiole. Ascospores crescent-shaped, hyaline, 4.5–6 × 1.5. Synnemata of *Pesotum* state up to over 1mm tall and 70 thick at base, dark brown; conidia on denticles, hyaline, 2.5–10 × 1–2. The cause of Dutch elm disease, which is spread by two species of elm bark beetle *Scolytus scolytus* and *S. multistriatus*. The first symptoms, wilting of shoots followed by yellowing or

browning of leaves, appear from June onwards. In transverse sections of diseased branches, dark spots, sometimes forming a ring, are seen. Dieback of affected branches begins at the tip, which may be curled over in the form of a crook. Leaves fall and large branches or even whole trees may die in a few weeks. Dead branches or trees serve as breeding grounds for the elm bark beetles, and the fungus fruits in tunnels made by their larvae. Eggs are laid and larvae overwinter in bark, emerging in enormous numbers the following spring. Before breeding, young adults feed on twigs of healthy trees, which they may infect with the fungus.

Cryptosporella hypodermia (Fr.)Sacc. (Fig. 1206)
Perithecia 0.5mm diam., immersed in groups of 1 to 10 with their necks just showing through cracks in bark. Ascospores hyaline or very pale straw-coloured, 35–65 × 8–10. On dead twigs and branches, Dec.–May.

Diaporthe eres Nitschke (Fig. 97)
Perithecia immersed, up to 0.8mm diam., often loosely grouped with necks protruding through blackened surface of wood or bark. Ascospores hyaline, constricted at the septum, 10–14 × 2.5–4. On dead branches, especially of *U. procera*, July–Feb.

Eutypella stellulata (Fr.)Sacc. (Fig. 1207)
Perithecia immersed in pustulate stromata, in groups of 4 to 15, with sulcate ostioles visible through cracks in bark, 0.5–0.6mm diam., black. Ascospores allantoid, very pale brown in mass, 8–12 × 1.5–2. On dead branches of *U. procera*, Dec.–Feb.

Hapalocystis bicaudata Fuckel (Fig. 1208)
Perithecia immersed in small groups, slightly raising the bark, with necks just protuberant. Ascospores 1- to 6- (mostly 3-)septate, brown, 40–70 × 16–23, with a thick, coiled hyaline appendage at each end. On dead attached branches, May–June. Accompanying *Stilbospora* state has dark smoky-brown conidia, 40–70 × 18–23, with 3 to 5, most commonly 3, septa.

Lopadostoma gastrinum (Fr.)Traverso (Fig. 1209)
Perithecia up to 0.8mm diam., in groups of up to 30 or sometimes even more, immersed in pulvinate stromata 2–5mm diam., which are dark brown to black on the outside but white inside. Ascospores mid to dark brown, 10–12 × 5–6. Mainly on branches which have fallen and lain on moist ground for a long time.

Nectria aurantiaca (Tul. & C.Tul.)Jacz.
Perithecia up to 0.4mm diam., red, warted, usually formed in small groups around the base of erect red or reddish-brown synnemata 1–2mm high, with creamy conidial heads. Ascospores hyaline to pale brown, 1-septate, 18–27 × 7–9. Conidia hyaline, 12–18 × 6–7. On bark, Aug.–Dec.

Quaternaria dissepta (Fr.)Tul. & C.Tul. (Fig. 1210)
Perithecia 0.6–0.8mm diam., in groups of 2 to 5, embedded in pulvinate stromata which split the bark, necks with sulcate ostioles slightly protuberant. Ascospores curved, straw-coloured to golden brown, 20–40 × 5–8. On dead twigs and branches, Feb.–May.

Splanchnonema foedans (Fr.)O.Kuntze (Fig. 1211)
Pseudothecia 0.6–0.7mm diam., immersed in bark. Ascospores mid to dark brown, with one large cell and two smaller cells, 40–55 × 18–24. On dead twigs and branches.

Thyridaria rubro-notata, described on *Acer*, but almost equally common on *Ulmus*.

HYPHOMYCETES

Bloxamia leucophthalma (Lév.)Höhnel (Fig. 1212)
Sporodochia black, scattered or gregarious and often confluent. Conidiophores forming a close palisade, brown, up to 35 × 2.5–3.5. Conidia endogenous, catenate, hyaline, 2–3 × 2. On wood of dead branches, Sept.–Apr.

Corynespora proliferata Loerakker (Fig. 1213)
Colonies effuse, dark brown, hairy. Conidiophores mid to dark brown, up to 90 × 7–11. Conidia pale to mid brown, with 3 to 17 pseudosepta, 30–300 × 9–12. On dead bark.

Nematogonium ferrugineum (Pers.)Hughes (Fig. 1214)
Colonies effuse, orange, powdery or granular. Conidiophores up to 2mm long, 8–20 thick, with terminal and sometimes also intercalary swellings, proliferating. Conidia in chains, hyaline or yellowish, 4–20 × 3–16. On bark, not uncommon.

COELOMYCETES

Aposphaeria ulmicola (Berk.)Sacc.
Pycnidia superficial, about 0.1mm diam., black, gregarious and forming oblong dark patches. Conidia hyaline, 2 × 1–1.5. On decorticated branches, Nov.

Coryneum compactum Berk. & Br. (Fig. 1215)
Acervuli erumpent through periderm, up to 0.7mm diam., dark brown. Conidiophores pale brown, up to 30 × 4. Conidia mid pale brown, 4- to 6-pseudoseptate, 40–55 × 18–22. On dead twigs, Mar.

Diplodia melaena Lév. (Fig. 1216)
Pycnidia crowded, breaking through epidermis singly or in groups, about 0.5mm diam., black. Conidia 1-septate, mid dark golden brown, 20–25 × 8–11. On attached twigs and dying branches, Mar.–Apr.

Fusicoccum ulmi Oudem.
Conidiomata immersed, dark brown to black, with few locules. Conidia fusoid, hyaline, 40–50 × 7–9. On branches of *U. procera*.

Phomopsis oblonga (Desm.)Traverso
Conidiomata up to 0.3mm diam., eventually erumpent, usually through slits. A-conidia oblong or fusoid, hyaline, biguttulate, 9–10 × 2.5–3; B-conidia uncinate, 25–30 × 1. On dead twigs of *U. minor* and *U. procera*, Jan.–June.

Seimatosporium macrospermum (Berk. & Br.)Sutton (Fig. 1217)
Acervuli erumpent, pulvinate, olivaceous brown to black, up to 0.4mm diam. Conidiophores hyaline, 10–30 × 2–3. Conidia 5-septate, 27–37 × 9–12, median cells brown, end cells hyaline or pale brown. On dead branches, Jan., uncommon.

Seiridium intermedium (Sacc.)Sutton (Fig. 1218)
Acervuli up to 0.2mm diam., black. Conidiophores hyaline, 10–40 × 2–3. Conidia 5-septate, 28–35 × 8–11, cell at each end hyaline, other cells brown, often appearing longitudinally striate. On and inside bark, Aug.–Oct.

Stilbospora, *v.s.* under *Hapalocystis bicaudata*.

VACCINIUM: Bilberry, Bog Whortleberry, Cowberry, Cranberry

UREDINALES AND EXOBASIDIALES

Pucciniastrum vaccinii (Wint.)Jørst., 11, 111
Uredinia hypophyllous, up to 0.2mm diam., long covered by epidermis, pale yellow or reddish yellow; spores minutely echinulate, 18–30 × 13–20. Telia very rarely seen. 11 fairly common on *V. myrtillus*, occasional on *V. oxycoccus*, *V. uliginosum* and *V. vitis-idaea*, May–Oct.

Exobasidium vaccinii (Fuckel)Woron.
Colonies powdery, white or pinkish, effuse. Basidiospores hyaline, fusoid, 5–8 × 1–2. On lower side of leaves and on stems causing swelling and distortion. Common on *V. myrtillus*, May–Sept.

DISCOMYCETES

Coccomyces leptideus (Fr.)B.Eriksson (Fig. 1219)
Apothecia embedded, lenticular, black, 1–2mm diam., opening by 4 or 5 teeth

to expose pale discs. Asci 4-spored; spores hyaline 65–90 × 2.5–4. On dead twigs of *V. myrtillus* and *V. vitis-idaea*.

Lophodermium maculare (Fr.)de Not.
Apothecia broadly elliptical, up to 0.7 × 0.4mm, black, opening by slits. Paraphyses not swollen at tips, protruding beyond asci to form an epithecium. Ascospores hyaline, 60–95 × 1.5–2. On pale papery spots, surrounded by black stromatic lines, on dead leaves of *V. uliginosum*.

Lophodermium melaleucum (Fr.)de Not. (Fig. 1220)
Apothecia broadly elliptical, up to 1 × 0.4mm, black, with creamy-white or yellowish lips. Paraphyses slightly swollen at tips. Ascospores hyaline, 50–60 × 1.5–2. On pale spots on leaves and twigs of *V. vitis-idaea*.

Monilinia. Four species which form apothecia on mummified fruit, and conidia mostly on leaves and shoots occur.

On *V. myrtillus*: *M. baccarum*.
On *V. oxycoccus*: *M. oxycocci* (Woronin)Honey; we have not seen this species.
On *V. uliginosum*: *M. megalospora* (Woronin)Whetzel, said to have ascospores up to 26 × 19.
On *V. vitis-idaea*: *M. urnula*.

Monilinia baccarum (Schröter)Whetzel (Fig. 1221)
Apothecia long-stalked, up to 1cm diam., pale brown, arising in groups of 2 or 3 from fallen mummified berries of *V. myrtillus*, May. Asci often with 4 mature spores in the upper part, which are 16–22 × 6–9, and 4 smaller ones below. Conidia of *Monilia* state in branched chains, spherical, hyaline, 20–30 diam., formed on young twigs, which tend to curve.

Monilinia urnula (Weinm.)Whetzel
Apothecia long-stalked, up to 1mm diam., dark brown, on fallen mummified fruits of *V. vitis-idaea*, May–June. Ascospores hyaline, 15–19 × 7.5–9. Conidia of *Monilia* state on dead shoots, pale buff, top-shaped, 25–30 × 15–17, formed in short chains.

Pezicula myrtillina P.Karsten
Apothecia erumpent from deep-seated stromata, greyish white, powdery, up to 0.6mm diam. Ascospores fusiform, hyaline, guttulate, 22–28 × 5.5–6. Paraphyses swollen at the tip up to 12. On dry dead branches of *V. myrtillus*, July–Dec.

Sporomega degenerans (Fr.)Corda
Apothecia resemble those of *Colpoma*, are elongated and open to expose yellow discs. Ascospores hyaline, filiform, rounded at apex pointed at base, 80–90 × 1.5–2. On dead twigs of *V. uliginosum*, uncommon.

Terriera cladophila (Lév)B. Eriksson
Apothecia *Lophodermium*-like, usually less than 1mm long, opening by slits to

expose dark brown to black discs. Ascospores hyaline, filiform, 50 × 1. Paraphyses not curled at tips, where they are encrusted with brown exudate. On pale spots on dead attached twigs of *V. myrtillus*, Apr.–June.

OTHER ASCOMYCETES

Gibbera. Three species with black, setose pseudothecia seated on erumpent dark brown to black crusts or stromata occur on *Vaccinium*.

KEY

Ascospores 10–15 × 3–5 .... *myrtilli*
Ascospores 12–18 × 5–8 .... *vaccinii*
Ascospores 20–30 × 7–9 .... *elegantula*

Gibbera elegantula (Rehm)Petrak
Pseudothecia 0.2–0.25mm diam.; setae up to 120 × 5. Ascospores 1-septate, dark olivaceous, 20–30 × 7–9. On fallen dead leaves of *V. myrtillus*, May.

Gibbera myrtilli (Cooke)Petrak (Fig. 1222)
Pseudothecia 0.15mm diam.; setae up to 150 × 5. Ascospores hyaline to pale olivaceous, 1-septate, 10–15 × 3–5. On dead leaves of *V. myrtillus* and *V. uliginosum*.

Gibbera vaccinii (Sow.)Fr.
Pseudothecia up to 0.3mm diam., with setae up to 60 long. Ascospores pale straw-coloured, 1-septate, 12–18 × 5–8. On dead leaves and twigs of *V. vitis-idaea*.

Leptosphaerulina myrtillina (Sacc. & Fautr.)Petrak
Pseudothecia erumpent, about 0.1mm diam. Ascospores hyaline, 3-septate, medianly constricted, 35–50 × 15–18, in very thick-walled asci. On living leaves and ends of twigs, causing brown, purple-bordered spots, Sept.–Oct.

Meliola niessliana Winter (Fig. 1223)
Colonies amphigenous, with black cleistothecia 0.3mm diam., attached to superficial, hyphopodiate mycelium which bears also erect, black or blackish-brown setae up to 1mm high. Ascospores 3-septate, dark brown, mostly 50–60 × 15–18. On leaves of *V. vitis-idaea*, uncommon.

Physalospora vitis-idaeae Rehm
Perithecia immersed, 0.25mm diam. Ascospores ellipsoid, hyaline, with granular contents, 20–25 × 10–12. On leaves of *V. vitis-idaea*, Sept.

Podosphaera myrtillina (Schubert)Kunze
Powdery mildew. Colonies mostly hypophyllous, thin, inconspicuous, few conidia. Cleistothecia with appendages 4 to 6 times dichotomously branched, each containing one 8-spored ascus. On leaves of *V. myrtillus*.

Pseudomassaria vaccinii Dennis
Perithecia 0.2mm diam., immersed just beneath the epidermis. Ascospores hyaline with a septum near the base, 22–26 × 7–8. On dead twigs of *V. myrtillus*, Jan.

Stigmatea conferta (Fr.)Fr. (Fig. 1224)
Pseudothecia subcuticular or subepidermal, setose, black, 0.1–0.2mm diam., in clusters subtended by a brown hyphal pseudostroma. Setae dark brown, 50–60 × 5–6. Ascospores greenish yellow, 1-septate, 15–18 × 6–7. On living and overwintered leaves of *V. uliginosum*.

COELOMYCETES

Seimatosporium vaccinii (Fuckel)Erikss.
Acervuli about 0.1mm diam., black. Conidiophores hyaline, up to 30 × 1.5–2. Conidia fusiform, 3-septate, mid pale brown, end cells often paler than median ones, 13–18 × 4.5–5.5. On dead twigs of *V. myrtillus*.

VIBURNUM: Guelder Rose, Laurustinus and Wayfaring Tree

Ascochyta viburni Sacc.
Pycnidia epiphyllous, immersed, brown, 0.2mm diam. Conidia oblong-ellipsoid, hyaline, 1-septate, 9–12 × 2–3. On living leaves of *V. opulus*, causing pale spots up to 1cm diam., with purplish borders, Aug.–Sept.

Cytospora lantanae Bres.
Conidiomata multilocular, 0.25–0.3mm diam., with erumpent grey discs. Conidia allantoid, hyaline, 5–7 × 1. On dead twigs of *V. lantana* and *V. opulus*, Jan.–May.

Diaporthe beckhausii Nitschke
Perithecia immersed, about 0.3mm diam., with long necks, gregarious. Ascospores hyaline, 1-septate, 10–14 × 3–4, each with a small filiform appendage at each end. On dead twigs of *V. opulus*, Jan.

Diplodia lantanae Fuckel
Pycnidia in clusters, black. Conidia oblong, dark brown, becoming 1-septate, 20–24 × 8. On dead branches of *V. lantana*.

Microsphaera viburni (Duby)Blumer
Powdery mildew. Colonies mostly hypophyllous, thin, few conidia. Cleistothecia with 3 to 8 asci, each with 6 to 8 spores. Appendages four times dichotomously branched, with recurved tips. On *V. opulus*, leaves redden and tend to fall.

Phomopsis tinea (Sacc.)Died.
Conidiomata erumpent, black. A-conidia oblong–fusoid, biguttulate, hyaline, 8–10 × 2–2.5, B-conidia curved or hooked, 18–25 × 1.5. On dead branches of *V. lantana*, *V. opulus* and *V. tinea*, Apr.–May.

Stigmina tinea (Sacc.)M.B.Ellis (Fig. 1225)
Colonies mostly hypophyllous, punctiform, dark brown. Conidiophores up to 30 × 3–7, pale brown. Conidia 3- to 11-septate, rather pale straw-coloured, 45–105 × 3–4. On leaves of *V. davidii* and *V. opulus* causing large, often angular, dark brown spots, visible on both surfaces.

Valsa opulina Sacc. & Syd.
Perithecia immersed, 0.3mm diam., 5 to 8 in a ring around a black disc. Ascospores allantoid, hyaline, 18 × 2.5. On dead branches of *V. opulus*.

VITIS: Grape Vine

Phomopsis viticola Sacc.
Conidiomata becoming erumpent, up to 0.4mm long, grey to black. Conidia elliptic–fusoid, hyaline, 7–10 × 2–2.5. On dead branches, Apr.–July.

Plasmopara viticola (Berk. & Curtis) Berl. & de Toni (Fig. 1226)
Downy mildew. Colonies on lower leaf surface show as corresponding pale green patches on the upper surface. Sporangia very variable in size, up to 35 × 19. Embedded oospores 30–38 diam.

Uncinula necator (Schw.)Burrill
Powdery mildew. Colonies white, powdery. Conidia plentiful, hyaline, 32–40 × 17–21, usually in long chains. Cleistothecia rarely formed, have appendages which coil at their tips; they contain 4 to 6 asci each with 4 to 7 spores. Most conspicuous on leaves but infects also flowers and young fruit.

PLURIVOROUS FUNGI ON HERBACEOUS PLANTS

DISCOMYCETES

KEY TO GENERA

	Ascospores purplish with anastomosing striations .... *Ascobolus*
	Ascospores hyaline 1
1.	Apothecia with very dark discs and lemon-yellow powdery margins .... *Schizoxylon*
	Apothecia not so 2
2.	Apothecia stalked, arising from black sclerotia .... *Sclerotinia*
	No sclerotia 3
3.	Apothecia immersed in host tissues 4
	Apothecia superficial or erumpent becoming so 5
4.	Apothecia with yellow or orange discs and white lobed margins .... *Stictis*
	Apothecia purplish brown with pinkish discs .... *Duebenia*
	Apothecia black with greenish-yellow discs .... *Hypoderma*
5.	Apothecia obviously hairy or downy with clumps of hairs 6
	Apothecia smooth or appearing smooth 8
6.	Apothecia pinkish or purplish grey, downy with tufts of dark brown hairs .... *Pseudombrophila*
	Apothecia brown, downy with clumps of short dark brown hairs in rows .... *Pirottaea*
	Apothecia distinctly hairy 7
7.	Hairs refractive, 'glassy' .... *Unguicularia*
	Hairs with narrow hooked tips .... *Unguiculella*
	Hairs grouped around margin and set at an angle like cogs ..... *Urceolella*
	Hairs otherwise .... *Dasyscyphus*
8.	Apothecia stalked 9
	Apothecia sessile or subsessile 10
9.	Apothecia tending to have dentate margins, flesh tough, elastic .... *Crocicreas*
	Apothecia with smooth margins, flesh soft .... *Hymenoscyphus*
10.	Apothecia lemon yellow .... *Calycellina*
	Apothecia white or very pale .... *Pezizella*
	Apothecia brown 11
11.	Apothecia superficial .... *Mollisia*
	Apothecia erumpent .... *Pyrenopeziza*

Ascobolus foliicola Berk. & Br. (Fig. 1227)
Apothecia 1–5mm diam., greenish yellow with black tips of asci protruding from surface of discs, distinct margin and the outer surface often covered by reddish-brown particles. Ascospores purplish brown with anastomosing striations, 15–22 × 8–12. On rotting stems and leaves, May–July.

Calycellina chlorinella (Ces.)Dennis (Fig. 1228)
Apothecia sessile or subsessile, 0.5mm diam., pale lemon yellow when fresh, seated on blackened areas on dead stems. Ascospores hyaline, slightly curved, 5–10 × 0.5–1, mostly 5–7 × 1. Oct.–Nov., quite common.

Crocicreas. Apothecia stalked, sometimes with dentate margins, flesh somewhat elastic and difficult to cut owing to the presence of thick-walled hyphae.

KEY

Ascospores 7–10 × 1.5–2 .... *cyathoideum*
Ascospores 12–14 × 2–2.5 .... *dolosellum*
Ascospores 16–20 × 4–4.5 .... *coronatum*

Crocicreas coronatum (Bull.)S.Carp. (Fig. 1229)
Apothecia up to 3mm diam., with toothed margins and slender stalks, white, cream or pinkish. Ascospores hyaline, 16–20 × 4–4.5. Common on dead stems, Sept.–Nov.

Crocicreas cyathoideum (Bull.)S.Carp. (Fig. 1230)
Apothecia up to 2mm diam., with entire or slightly indented margins, incurved when young, and slender stalks; rather pale cream, sometimes drying pinkish. Ascospores hyaline, mostly 7–10 × 1.5–2. On dead stems of many different plants, Mar.–Oct., very common.

Crocicreas dolosellum, with short teeth, described on *Fraxinus*, is found also sometimes on dead herbaceous stems.

Dasyscyphus. Six species, all plurivorous and with hairy apothecia, occur commonly on dead herbaceous stems.

KEY

	Apothecia stalked	1
	Apothecia sessile	2
1.	Apothecia pale brown, hairs straw-coloured	*clandestinus*
	Apothecia white with cream discs, hairs white	*virgineus*
2.	Apothecia covered with bright sulphur-yellow hairs	3
	Apothecia not so	4
3.	Ascospores 10–15 long	*mollissimus*
	Ascospores 30–37 long, often septate	*sulphureus*
4.	Apothecia pale straw-coloured, hairs short	*grevillei*
	Apothecia brown with pale discs, hairs long	*nidulus*

Dasyscyphus clandestinus, described on *Rubus idaeus*, is found also on dead stems of *Arctium, Epilobium, Filipendula*, etc.

Dasyscyphus grevillei (Berk.)Massee (Fig. 1231)
Apothecia 0.2–0.4mm diam., sessile, with incurved margins, pale straw-coloured, hairs granulate, 20–35 × 3–4. Paraphyses about 2 thick. Ascospores hyaline, 7–10 × 1.5–2. Very common, especially on dead stems of Umbelliferae, Apr.–Aug.

Dasyscyphus mollissimus(Lasch)Dennis (Fig. 1232)
Apothecia sessile, up to 2mm diam., whitish or pale brown, densely covered with long bright sulphur-yellow hairs. Ascospores hyaline, 10–15 × 1.5–2. Very common, especially on standing dead stems of Umbelliferae, Apr.–July; common also on *Lamium* and *Urtica*.

Dasyscyphus nidulus (Schm. & Kunze)Massee (Fig. 1233)
Apothecia sessile, like little birds'-nests, up to 1mm diam., brown, with whitish or very pale yellow discs. Hairs smooth, brown except for one or two cells at the tip which remain colourless, up to 160 × 4–5. Ascospores hyaline, 8–12 × 1–1.5. Common in May and June, especially on thick dead stems of marsh plants such as *Epilobium hirsutum, Eupatorium cannabinum* and *Filipendula ulmaria*.

Dasyscyphus sulphureus (Pers.)Massee (Fig. 1234)
Superficially very similar to *D. mollissimus* with long, bright-sulphur-yellow hairs, but it is found later in the year, Aug.–Oct., and is especially common on *Urtica dioica*. Ascospores hyaline, 0- to 3-septate, mostly 30–37 × 2–2.5.

Dasyscyphus virgineus, described and figured on 'wood and bark', has apothecia up to 1mm diam., white with cream discs, usually long-stalked and covered with white, granulate hairs. Ascospores hyaline, 6–10 × 1.5–2.5. One of the commonest of all discomycetes, most frequently on dead twigs, beech cupules and debris but found also on dead stems of *Angelica sylvestris, Epilobium angustifolium, E. hirsutum, Rumex crispus, Teucrium scorodonia*, etc., Feb.–Aug., especially abundant in spring.

Duebenia compta (Sacc.)Nannf. ex Hein
Apothecia immersed, up to 3 × 0.5mm, dark purplish brown, opening to expose pinkish-orange discs. Asci with 4 to 7 spores; spores hyaline, 6–10 × 3. Paraphyses slender, clavate. Uncommon.

Hymenoscyphus, with apothecia usually stalked and smooth; flesh easy to cut, made up of thin-walled hyphae.

KEY

Ascospores tapered to a point at the base, often with a slender setula at each end .... *scutula*
Ascospores not so 1

1. Apothecia rather pale brick red .... *sublateritius*
   Apothecia ivory white or cream when fresh 2
   Apothecia yellow or yellowish when fresh 3
2. Apothecia gregarious, short-stalked .... *herbarum*
   Apothecia not gregarious, long-stalked .... *pileatus*
3. Ascospores 7.5–8 × 1.5 .... *limonium*
   Ascospores 8–12 × 2–2.5 .... *repandus*
   Ascospores 16–20 × 3–3.5 .... *vitellinus*

Hymenoscyphus herbarum (Pers.)Dennis (Fig. 1235)
Apothecia gregarious, up to 3mm diam., with very short, thick stalks, ivory white to cream when fresh, drying yellow ochre when old. Ascospores hyaline, 1-septate, 12–17 × 2.5–3. On dead stems, especially those of *Urtica dioica*, Sept.–Dec., common.

Hymenoscyphus limonium (Cooke & Peck)Dennis
Apothecia lemon yellow, with slender stalks. Ascospores 7.5–8 × 1.5. On small dead stems, e.g. those of *Centaurea nigra* and *Senecio erucifolius*, on the ground under grass and other vegetation, Oct.–Nov.

Hymenoscyphus pileatus (Karst.)O.Kunze (Fig. 1236)
Apothecia 1–2mm diam., long-stalked, cream-coloured. Ascospores hyaline, 17–30 × 3–3.5. Found mostly on debris of plants growing in marshy places, Sept.–Nov.

Hymenoscyphus repandus, described on *Epilobium*, is found also on *Eupatorium*, *Filipendula*, *Iris*, *Mentha*, etc.

Hymenoscyphus scutula (Pers.)Phill. (Fig. 1237)
Apothecia up to 4mm diam., with rather flat discs, stalked, creamy or pale yellow when fresh, the disc drying dark ochraceous. Ascospores hyaline, tapered below to a point, often with a slender setula at each end, mostly 20–27 × 4–5. Very common on dead herbaceous stems, Sept.–Nov.

Hymenoscyphus sublateritius (Berk. & Br.)Dennis
Apothecia up to 2mm diam., short-stalked, with flat discs, rather pale brick red. Ascospores elliptic–fusoid, hyaline, 14–20 (25) × 4–5, not tapered to a point at the base as they are in *H. scutula*.

Hymenoscyphus vitellinus, described on *Filipendula*, has been found also on dead stems of *Epilobium*, *Silene* and *Teucrium*. Apothecia lemon yellow or deep cream.

Hypoderma commune (Fr.)Duby (Fig. 1238)
Apothecia developing beneath epidermis, 1–2.5mm long, black, splitting longitudinally to expose a greenish-yellow disc. Ascospores hyaline, 17–20 × 3–4. On dead herbaceous stems, Sept.–Nov.

Mollisia. Most species of *Mollisia* on herbaceous stems tend to be host-limited but *M. clavata* described on *Rubus* has been seen on various herbaceous plants, *M. coerulans* described on *Eupatorium* occurs also on *Artemisia* and *Centaurea*, *M. fusco-striata* described on *Filipendula* has been recorded also on *Solanum dulcamara*, and *M. pastinacae* on other Umbelliferae, although its main host is *Pastinaca*.

Pezizella discreta (P.Karsten)Dennis (Fig. 1239)
Apothecia up to 0.5mm diam., short-stalked, white. Ascospores hyaline, often very slightly curved, 5–8 × 1–1.5. On dead stems of *Artemisia vulgaris*, *Cirsium arvense* and *Epilobium angustifolium*, Oct.–Nov.

Pezizella glareosa Vel.
Apothecia 1–2mm diam., in clusters of 2 to 5, white, not changing colour, sessile, flattened, somewhat flexuous or lobed. Asci 30–50 × 5. Paraphyses filiform, simple or branched at base. Ascospores 4–5 × 1. On dead stems of *Cirsium arvense*, etc., Sept.–Oct.

Pirottaea nigrostriata, described on *Heracleum*, has been found also on *Artemisia vulgaris* and *Polygonum cuspidatum*.

Pseudombrophila deerata (P. Karsten)Seaver
Apothecia 3–9mm diam., becoming reflexed, almost sessile or with short stalks, pinkish or purplish grey, downy on the outside, with tufts of thin-walled, dark brown hairs up to 120 × 4–5. Asci up to 150 × 14. Spores elliptical, hyaline, smooth, 12–16 × 8–9. On rotting stems; recorded on *Brassica*, *Cirsium* and *Heracleum*, May–Nov.

Pyrenopeziza revincta (P.Karsten)Gremmen (Fig. 1240)
Apothecia 0.5–1mm diam., grey to yellowish fawn, with pale rims. Ascospores 8–10 × 2. Common on all sorts of dead herbaceous stems.

Schizoxylon berkeleyanum (Durieu & Lév.)Fuckel (Fig. 1241)
Apothecia erumpent, up to 1mm diam., with dark olivaceous to almost black discs and broad, pale lemon-yellow, powdery margins. Ascospores hyaline, multiseptate, up to 250 × 1.5–2.5. Paraphyses much branched at tips. On dead stems of *Epilobium*, etc., uncommon.

Sclerotinia sclerotiorum (Lib.)de Bary (Fig. 1242)
Apothecia mostly 2–8mm diam., on stalks up to 2cm long, arising from overwintered, black sclerotia which vary greatly in size but are most commonly in the range 5–30 × 3–10mm. Discs golden brown. Ascospores hyaline, 9–13 × 4–6. Sclerotia are found in decaying parts of herbaceous plants. Apothecia should be looked for in Apr.–June.

Stictis stellata Wallr. (Fig. 1243)
Apothecia sunken, with yellow to orange discs and white, lobed margins. Ascospores hyaline, multiseptate, up to 200 × 1.5–2. Common on dead, woody herbaceous stems, e.g. those of *Epilobium hirsutum* and *Eupatorium cannabinum*, Oct.–May.

Unguicularia. Apothecia more or less urn-shaped, covered with refractive, 'glassy' hairs.

KEY

Ascospores 9–11 × 3–3.5 .... *incarnatina*
Ascospores not more than 7 × 2 1
1. Hairs up to 40 × 5 .... *cirrhata*
    Hairs not more than 25 × 4 .... *millepunctata*

Unguicularia cirrhata (Crouan & H.Crouan)Le Gal (Fig. 1244)
Apothecia urn-shaped, mostly about 0.2mm diam., but may become up to 0.6mm if kept damp, pale greyish brown with white margins, covered with hyaline, highly refractive hairs up to 40 × 5. Ascospores hyaline, 5–6 × 1.5–2. Found most commonly near the base of dead standing stems of *Cirsium*, *Eupatorium*, *Filipendula* and *Rumex* in wet areas, Apr.–June, and occasionally at other times of the year.

Unguicularia incarnatina (Quélet)Nannf.
Apothecia sessile, 0.2–0.3mm diam., flesh-coloured, downy. Hairs up to 30 × 4. Ascospores hyaline, slightly curved, 9–11 × 3–3.5. On dead stems of *Centaurea nigra*, *Cirsium palustre*, *Epilobium hirsutum* and *Filipendula ulmaria*, Apr.–July.

Unguicularia millepunctata (Lib.)Dennis (Fig. 1245)
Apothecia urn-shaped, up to 0.2mm diam., white or very pale brown; refractive hairs not more than 25 × 4. Ascospores 6–7 × 1–1.5. Very common especially on dead stems of Compositae and Umbelliferae, but also on *Agrimonia*, *Epilobium* and *Ononis*, May–Oct.

Unguiculella eurotioides (P.Karsten)Nannf. (Fig. 1246)
Apothecia 0.2–0.3mm diam., pale yellow, covered with hyaline, hooked hairs up to 50 × 4. Ascospores hyaline, 6–8 × 2–2.5. Paraphyses hooked at the apex in the same way as the hairs. On dead herbaceous stems; also occasionally on effete pyrenomycetes on bark, Aug.–Sept., uncommon.

Unguiculella hamulata (Feltg.)Höhnel (Fig. 1247)
Apothecia not more than 0.2mm diam., very pale brown or whitish, covered with hooked hairs 20–30 long. Ascospores hyaline, 5–6 × 2–2.5. Paraphyses filiform. On dead stems of *Arctium*, *Heracleum*, *Urtica*, etc., May–Sept.

Urceolella crispula (P.Karsten)Boud. (Fig. 1248)
Apothecia urn-shaped, up to 0.3mm diam., white to very pale fawn, with hairs in groups around the margin set at an angle rather like little cogs. The hairs are bent at the base, very thick-walled and up to 130 × 5–6. Ascospores hyaline, 6–9 × 1.5–2. Found most commonly on dead stems of Umbelliferae, e.g. *Anthriscus*, *Heracleum* and *Smyrnium*, but also occasionally on other plants such as *Artemisia vulgaris*; May–Nov.

OTHER ASCOMYCETES

KEY TO GENERA

    Ascomata yellow to yellowish brown 1
    Ascomata blackish brown to black 2
1. Ascomata (perithecia) pale tangerine, ascospores with 3 or more septa ..
    .... *Calonectria*
    Ascomata (perithecia) ochre, honey or golden brown, ascospores
    1-septate .... *Nectria*
2. Ascomata seated on brown hyphal subiculum 3
    Ascomata not so 4
3. Ascospores without septa, brown, boat-shaped .... *Rosellinia*
    Ascospores becoming septate and brown .... *Herpotrichia*
4. Necks of ascomata laterally compressed and slotted .... *Lophiostoma*
    Necks not so 5
5. Ascomata covered with long hairs or setae .... *Lasiosphaeria*
    Ascomata more or less smooth 6
6. Ascospores hyaline, without septa .... *Glomerella*
    Ascospores hyaline, 1-septate .... *Diaporthe*
    Ascospores brown, muriform .... *Pleospora*
    Ascospores over 100 long, very narrow 7
    Ascospores less than 100 long, broader 8
7. Ascospores with 1 or 2 cells near the middle much swollen ....
    .... *Ophiobolus*
    Ascospores without swollen cells .... *Leptospora*
8. Asci very long-stalked .... *Diapleella*
    Asci not very long-stalked .... *Leptosphaeria*

Calonectria pseudopeziza, described on 'wood and bark', is found also not infrequently on woody herbaceous stems such as those of *Arctium* and *Eupatorium*.

Diapleella clivensis (Berk. & Br.)Munk (Fig. 1249)
Pseudothecia immersed, black, up to 0.35mm diam. Asci long-stalked; these often ooze out through the ostioles and sit in black heaps on the stem surface. Ascospores rather dark golden or reddish brown, 3-septate, 20–25 × 6–9. On dead stems, especially those of Compositae, e.g. *Centaurea nigra, Cirsium arvense* and *Senecio jacobaea*, May–June.

Diaporthe arctii, described on *Arctium*, is very common also on dead stems of other Compositae, Umbelliferae, etc., and often extensive areas of the surface are blackened by it.

Diaporthe pardalota, described on *Lonicera*, is common also on dead stems of *Convallaria majalis, Epilobium angustifolium, Polygonatum multiflorum, Polygonum cuspidatum, Rumex acetosa*, etc.

Glomerella cingulata (Stonem.)Spaulding & v.Schrenk (Fig. 1250)
Perithecia immersed, up to 0.3mm diam., dark brown to black, with short necks lined by hair-like periphyses. Ascospores hyaline, 10–30 × 4–7. On dead stems and leaves of many different plants. Commonly found in its *Colletotrichum* state. Acervuli about 0.5mm diam., with brown, septate setae, up to 200 × 4–7, protruding through pinkish, slimy masses of conidia. Conidia hyaline or very pale brown, guttulate, mostly 10–20 × 3–5. These sometimes germinate on the surface of the host and mycelium can be seen there bearing brown appressoria, 5–15 × 4–10, each with a small hyaline spot at its centre.

Herpotrichia macrotricha (Berk. & Br.)Sacc. (Fig. 1251)
Pseudothecia 0.3–0.5mm diam., black, in clusters seated on a dense, dark brown subiculum, and covered by brown, septate hyphae. Ascospores mostly 35–50 × 4–6, at first hyaline and 1-septate, becoming brown and 3- to 5-septate, sometimes with a colourless appendage at each end. On dead stems such as those of *Epilobium* lying on damp ground, Oct.–Apr.

Lasiosphaeria phyllophila, described on 'wood and bark', is not uncommon also on dead herbaceous stems such as those of *Epilobium angustifolium*.

Leptosphaeria. Six plurivorous species are described.

KEY

	Ascospores all or mostly 3-septate	1
	Ascospores all or mostly 4-septate	3
	Ascospores 5-septate	*ogilviensis*
1.	Walls of pseudothecia concentrically ringed	*doliolum*
	Walls of pseudothecia not so	2
2.	Ascospores 23–30 long	*purpurea*
	Ascospores 30–40 long	*macrospora*
3.	Ascospores 6–7 wide with mucilaginous coats	*gloeospora*
	Ascospores 4–4.5 wide, sometimes appendaged	*modesta*

Leptosphaeria doliolum (Pers.)Ces. & de Not. (Fig. 1252)
Pseudothecia erumpent, becoming free, 0.4–0.5mm diam., black, with characteristic concentric ridges (usually 2) on the surface. Ascospores 3-septate, rather pale golden brown, 16–28 × 4–7 (mostly 18–23 × 4–6). Very common throughout the year on all sorts of dead stems but especially those of Umbelliferae and *Urtica dioica*. Pseudothecia are sometimes parasitised by *Didymosphaeria conoidea* Niessl (Fig. 1253), which has brown, 1-septate ascospores 8–12 × 4–5.

Leptosphaeria gloeospora (Berk. & Currey)Sacc.
Pseudothecia 0.2–0.4mm diam., immersed, short-necked, ostioles lined with stiff brown hairs up to 100 × 4. Ascospores fusiform, almost hyaline, with

thin mucilaginous coats, 4-septate, the second cell somewhat swollen, 23–33 × 6–7.

Leptosphaeria macrospora (Fuckel)Thümen (Fig. 1254)
Pseudothecia immersed, 0.3–0.4mm diam., black, flattened. Ascospores straw-coloured, usually 3-septate, with second cell swollen, occasionally with 4 or 5 septa, 30–40 × 4.5–7. Mainly on dead stems of Compositae and *Rumex*, Apr.–June.

Leptosphaeria modesta, described on *Scabiosa*, is found also occasionally on dead stems of many other herbaceous plants.

Leptosphaeria ogilviensis (Berk. & Br.)Ces. & de Not.
Pseudothecia immersed, 0.2–0.3mm diam. Ascospores golden brown, 30–40 × 3.5–4.5, 5-septate, constricted at the median septum, third cell down somewhat enlarged. Mostly on dead stems of Compositae, e.g. *Senecio jacobaea*.

Leptosphaeria purpurea Rehm (Fig. 1255)
Pseudothecia 0.2–0.3mm diam., black, often flattened or collapsed and covered with dark brown hyphae, partly immersed, with the surrounding host tissue stained reddish purple. Ascospores rather pale golden brown, 3-septate, mostly 23–30 × 4.5–5. On dead stems, especially those of Compositae, June–July.

Leptospora rubella (Pers.)Rabenh. (Fig. 1256)
Pseudothecia 0.2–0.3mm diam., at first immersed, black with red-tipped necks protruding through the epidermis, sometimes becoming free on old stems, the surrounding host tissue stained reddish purple. Ascospores pale straw-coloured in mass, mostly 120–180 × 1–1.5, parallel to one another and often somewhat spirally twisted inside the asci. Very common on all sorts of herbaceous stems, both thin and thick, Apr.–Aug.

Lophiostoma, characterised by having pseudothecia with laterally compressed, slotted necks and transversely septate ascospores. Six plurivorous species and one variety occur on dead herbaceous stems.

KEY

	Spores with 1 to 3 septa, hyaline, ripening straw-coloured	1
	Spores with 5 septa or more, golden brown	4
1.	Asci clavate with tapering stalks	2
	Asci cylindrical with button-like stalks	3
2.	Spores less than 20 long	*fuckelii*
	Spores mostly 20–30 long	*angustilabrum*
	Spores more than 30 long	*semiliberum*
3.	Spores not more than 25 long, host tissues not stained red	*vagabundum*

    Most spores more than 25 long, host tissues stained red ....
.... *origani* var. *rubidum*
4.   Spores elliptical to fusiform .... *caulium*
    Spores pyriform to clavate .... *caudatum*

Lophiostoma angustilabrum (Berk. & Br.)Cooke
Pseudothecia immersed or semi-immersed, black, about 0.5mm diam. Ascospores mostly 24–30 × 4–5.5, long remaining hyaline with one median septum and 3 to 6 large guttules. Very common, especially on *Agrimonia eupatoria*, *Rumex* and *Urtica dioica*, Feb.–Oct.

Lophiostoma caudatum, described on *Phragmites*, has been found also on dead stems of *Epilobium hirsutum* and *Urtica dioica* in marshy places.

Lophiostoma caulium (Fr.)Ces. & de Not. (Fig. 1257)
Pseudothecia immersed or semi-immersed. Ascospores mostly 5-septate, 20–30 × 5–6. Found all the year round on many different herbaceous stems but especially common on species of *Rumex* and on *Urtica dioica*.

Lophiostoma fuckelii, described on *Rubus* on which it occurs most commonly, but many collections have been made also on *Agrimonia*, *Epilobium*, *Filipendula*, *Rumex*, *Teucrium* and Umbelliferae.

Lophiostoma origani (J.Kunze)Winter var. *rubidum* (Sacc., Rouss. & Bomm.)Chesters & Bell (Fig. 1258)
Pseudothecia immersed with just the necks protruding, or partially erumpent, host tissues stained red to deep magenta. Ascospores 1- to 3-septate, 25–30 × 4–5, often with colourless sheaths. Especially common on *Epilobium*, *Filipendula* and *Urtica*.

Lophiostoma semiliberum, a plurivorous species described on grasses but found occasionally on dead stems of other plants such as *Artemisia* and *Centaurea*.

Lophiostoma vagabundum (Sacc.)Chesters & Bell (Fig. 1259)
Pseudothecia usually immersed with only the necks protruding, 0.2–0.3mm diam. Ascospores hyaline, 1-septate with 4 to 6 guttules, occasionally becoming 3-septate and straw-coloured when old and outside the asci, 18–25 × 3–5. Found all the year round on dead stems of many different plants, especially common on *Centaurea*, *Epilobium* and *Rumex*.

Nectria arenula (Berk. & Br.)Berk. (Fig. 1260)
Perithecia superficial, scattered or in small groups, seated on a thin subiculum, pale yellow ochre or golden brown, up to 0.25mm diam., the upper part tending to collapse when dry. Ascospores hyaline, 1-septate, slightly curved, longitudinally striate, 15–21 × 3–4. Quite common on dead stems throughout the year. We have found it on *Armoracia rusticana*, *Artemisia vulgaris*, *Cirsium arvense*, *Foeniculum vulgare*, *Galium aparine* and *Urtica dioica*, and also on *Equisetum* and grass culms.

Nectria ellisii Booth (Fig. 1261)
Perithecia superficial, scattered or in groups, about 0.15mm diam., honey-coloured, covered with white, sometimes branched hairs, up to 75 × 2.5. Ascospores hyaline, 1-septate, 11–16 × 3. On dead stems of herbaceous plants, e.g. *Epilobium hirsutum, Eupatorium cannabinum, Filipendula ulmaria, Iris pseudacorus* and *Silene dioica*, growing in damp places, May–Dec.

Nectria inventa Pethybridge (Fig. 1262)
Perithecia seated on an erumpent stroma, up to 0.5mm diam., yellow to dark brown, the upper part covered with hyaline, septate, pointed hairs. Ascospores hyaline, 1-septate, 8–12 × 3–3.5. Only known in its perithecial state from Pethybridge's work although in its *Verticillium* state it is extremely common on dead herbaceous stems throughout the year, where it forms bright brick-red colonies with white margins. Conidiophores tree-like, the lower part unbranched and pale yellowish or burnt sienna, the upper part much branched, hyaline and bearing whorls of phialides. Conidia pale reddish brown in mass, 3.5–5 × 2–2.5.

Ophiobolus acuminatus (Sow.)Duby (Fig. 1263)
Pseudothecia immersed with short necks just protruding through the epidermis, black, up to 0.5mm diam. Ascospores straw-coloured or golden brown in mass, multiseptate, up to 160 × 3, with two cells near the centre swollen to 5; constricted at the middle and often dividing at this point to form two part-spores. Very common especially on dead stems of Compositae, e.g. *Cirsium vulgare*, Mar.–June.

Ophiobolus erythrosporus (Riess)Winter
Pseudothecia immersed, sometimes becoming superficial, up to 0.5mm diam., black, often flattened, with short cylindrical necks. Ascospores multiseptate, up to 140 × 3–4, yellowish or somewhat pinkish in mass when fresh, curved, with one short cell just above the middle distinctly swollen. Not uncommon on dead herbaceous plants, e.g. *Urtica dioica*.

Pleospora, with brown, muriform ascospores.

KEY

	Pseudothecia with brown setae around ostiole .... *ambigua*
	Pseudothecia without brown setae around ostiole 1
1.	Ascospores with 7 transverse septa .... *herbarum*
	Ascospores with not more than 5 transverse septa 2
2.	Ascospores 18–21 × 8–11 .... *phaeocomoides*
	Ascospores 20–25 × 6–8 .... *scrophulariae*

Pleospora ambigua, described on *Reseda* occurs also on *Epilobium* and various Compositae.

Pleospora herbarum (Pers.)Rabenh. ex Ces. & de Not. (Fig. 1264)
Pseudothecia subepidermal sometimes becoming superficial, up to 0.5mm

diam., papillate and often dorsiventrally flattened, black. Ascospores yellow to golden brown, normally with 7 transverse and 1 or 2 longitudinal septa, constricted in the middle, variable in shape but often somewhat pointed at the apex and broadly rounded at the base, 25–40 × 11–16 (most commonly 30–35 × 12–15). The *Stemphylium* conidial state has pale to rather dark brown or olivaceous brown, minutely verruculose or echinulate conidia with 3 transverse and 1 to 3 longitudinal septa, constricted at the median transverse septum, 27–42 × 24–30. Conidia are borne on the swollen apices of percurrently proliferating brown conidiophores. This is probably the commonest of all ascomycetes on dead herbaceous stems, Oct.–June.

Pleospora phaeocomoides (Berk. & Br.)Winter (Fig. 1265)
Pseudothecia immersed, up to 0.4mm diam., black. Ascospores straw-coloured with 5 transverse septa and 1 longitudinal septum, 18–21 × 8–11. Common and found throughout the year especially on Compositae and Umbelliferae but also on the slender stems of *Plantago* and on grasses.

Pleospora scrophulariae (Desm.)Höhnel (Fig. 1266)
Pseudothecia up to 0.3mm diam., sometimes lying beneath blackened parts of the epidermis. Ascospores pale yellow, with 3 to 5 transverse septa and 1 longitudinal septum, mostly 20–25 × 6–8. Fairly common on a variety of dead herbaceous stems, e.g. those of *Artemisia vulgaris*, *Scrophularia aquatica*, *Umbilicus rupestris* and *Veronica* spp.

Rosellinia necatrix Berl. ex Prill. (Fig. 1267)
Perithecia black, papillate, 1–2mm diam., shortly pedicellate, in groups on a thin ropy subiculum made up of brown hyphae, some of which have characteristic pear-shaped swellings. Ascospores boat-shaped, 30–50 × 5–8, dark brown, with longitudinal germ slits. *Dematophora* conidial state has dark brown synnemata splayed out at the top and pale brown conidia 3–4.5 × 2–2.5. Causes white root rot of various herbaceous plants and also some trees. Plants attacked include *Beta*, *Medicago*, *Narcissus*, *Paeonia* and *Phaseolus*.

HYPHOMYCETES

KEY TO GENERA

	Colonies discrete, sporodochial	1
	Colonies effuse	4
1.	Sporodochia setose, setae long, thick-walled	2
	Sporodochia not setose	3
2.	Sporodochia white or yellowish, glistening .... *Volutella*	
	Sporodochia pinkish to purplish brown .... *Sarcopodium*	
3.	Sporodochia black, conidia dark brown, muriform but with septa often obscure .... *Epicoccum*	
	Sporodochia pink or orange, conidia hyaline, curved, septate .. *Fusarium*	

4. Conidiophores in bundles, forming synnemata or loose coremia 5
    Conidiophores separate 8
5. Conidia septate with black bands at septa .... *Endophragmia*
    Conidia without septa 6
6. Conidia not in chains .... *Dematophora*
    Conidia in long chains 7
7. Conidia oblong or cubical .... *Coremiella*
    Conidia ellipsoid or oval .... *Doratomyces*
8. Conidia branched, brown, with septate arms .... *Triposporium*
    Conidia septate, often with long lateral appendage near base .... *Mycocentrospora*
    Conidia neither branched nor appendaged 9
9. Conidia septate 10
    Conidia without septa 22
10. Conidia 1-septate 11
    Conidia with 2 or more transverse septa 13
    Conidia with transverse and longitudinal septa 19
11. Conidia olivaceous brown, catenate .... *Cladosporium*
    Conidia hyaline 12
12. Conidia solitary, on pegs .... *Arthrobotrys*
    Conidia in long chains .... *Chalara*
    Conidia in slimy bundles .... *Cylindrotrichum*
13. Conidia hyaline or very pale 14
    Conidia greyish green, 3-septate .... *Fusariella*
    Conidia brown or olivaceous brown 15
14. Conidia on tapered pegs at apex of conidiophore .... *Pleurophragmium*
    Conidia leaving small scars at apex of conidiophore .... *Pseudospiropes*
15. Conidia not in chains, black bands at septa .... *Endophragmia*
    Conidia in chains 16
16. Conidiophores short with swollen cell at apex .... *Torula*
    Conidiophores quite long, erect 17
17. Conidia with protuberant scars at end .... *Cladosporium*
    Conidia without protuberant scars 18
18. Conidiophores tree-like with many short branches .... *Dendryphion*
    Conidiophores simple or loosely branched with nodose swellings .... *Dendryphiella*
19. Conidia usually in chains, beaked .... *Alternaria*
    Conidia not or seldom in chains, not beaked 20
20. Conidiophores short, conidia pyriform, twisted .... *Monodictys*
    Conidiophores rather short, conidia not twisted .... *Ulocladium*
    Conidiophores long, erect 21
21. Conidiophores with terminal and intercalary swellings .... *Stemphylium*
    Conidiophores without swellings .... *Dactylosporium*
22. Conidia green, conidiophores with branches almost at right angles .... *Trichoderma*

Conidia hyaline 23
Conidia olivaceous brown or black 30
23. Greyish-brown setae present, conidia spherical, thick-walled ....
.... *Botryotrichum*
No setae, conidia otherwise 24
24. Colonies pink or orange, conidiophores pencillately branched ....
.... *Gliocladium*
Not so 25
25. Conidia in chains .... *Septofusidium*
Conidia not in chains 26
26. Conidiophores with branches, or phialides in verticils 27
Conidiophores not so 28
27. Conidiophores hyaline .... *Verticillium*
Conidiophores brown below, hyaline above .... *Stachylidium*
28. Conidiophores simple with terminal and intercalary swellings ....
.... *Gonatobotrys*
Conidiophores branched 29
29. Conidiophores hyaline .... *Botryosporium*
Conidiophores brown .... *Botrytis*
30. Conidia not in chains 31
Conidia in chains 32
31. Conidia solitary at ends of tapered branches .... *Acremoniella*
Conidia aggregated in slimy black heads .... *Stachybotrys*
Conidia spherical, verruculose, formed along the sides of
dichotomously branched conidiophores .... *Chromelosporium*
32. Conidia with protuberant scars at ends .... *Cladosporium*
Conidia not so 33
33. Conidiophores simple with conidial chain at the apex .... *Gliomastix*
Conidiophores with conidia looking like black, round-headed pins ....
.... *Periconia*
Conidiophores tree-like, conidia joined by connectives like beads on a
string .... *Oidiodendron*
Conidiophores short, irregularly branched .... *Scopulariopsis*

Acremoniella atra (Corda)Sacc. (Fig. 1268)
Colonies effuse, cottony, at first white, later cinnamon brown. Conidiophores hyaline, up to 100 × 4–8, branches pointed at tips and often at right-angles to one another. Conidia golden brown, smooth or slightly wrinkled, 20–30 × 15–25. On dead stems, leaves and seeds.

Alternaria alternata (Fr.)Keissler (Fig. 1269)
Colonies effuse, grey, black or olivaceous black. Conidiophores brown, up to 50 × 3–6. Conidia very short beaked, formed in long, often branched chains, brown, smooth or verruculose, muriform, 20–60 × 9–18. A very common saprophyte found on all kinds of dead plant material.

Alternaria tenuissima (Kunze)Wiltshire (Fig. 1270)
Colonies effuse, thin, brown to black. Conidiophores brown, up to 115 × 4–6. Conidia solitary or in short chains, smooth or almost smooth, brown, muriform, 22–95 × 8–19 with beaks longer than those of *A. tenuis* and often swollen at the tip. On dead plants, common and widespread.

Arthrobotrys conoides Drechsler (Fig. 1271)
This and the next species form rather inconspicuous, white or slightly yellowish cobwebby colonies not infrequently encountered on dead herbaceous plants when looking at other fungi. Conidiophores hyaline, up to 500 × 4–8, with conidia borne on pegs which cover the swollen tips of branches. Conidia 1-septate, hyaline, smooth, mostly 25–45 × 10–13.

Arthrobotrys oligospora Fresen. (Fig. 1272)
Similar to *A. conoides* but with shorter, fatter conidia mostly 20–30 × 13–16.

Botryosporium longibrachiatum (Oudem.)Maire (Fig. 1273)
Colonies extensive, cobwebby or fluffy, white. Conidiophores very long and branched. Conidia quite smooth, 8–11 × 4–4.5. On dead stems and leaves of *Dahlia*, *Dianthus*, *Lycopersicon*, *Solanum*, etc.

Botryosporium pulchrum, described on *Urtica*, has been found occasionally on other herbaceous plants, e.g. on *Cirsium palustre* and *Mentha arvensis*.

Botryotrichum piluliferum Sacc. & March. (Fig. 1274)
Colonies effuse, white, buff or grey, setose. Setae up to 250 × 3–5, greyish brown, often verrucose or encrusted, especially near the base. Spherical, very thick-walled conidia, hyaline, mostly 9–16 diam. Found occasionally on dead herbaceous plants but not common.

Botrytis cinerea Pers. (Fig. 1275)
Colonies effuse, grey or greyish brown. Conidiophores tree-like, with stipes often 2mm or more long and 16–30 thick, brown except for the ends of the branches which are hyaline. Conidia colourless or very pale brown, smooth, mostly 8–14 × 6–9. The cosmopolitan 'grey mould' which damages flowers, leaves, stems, fruit and other parts of all sorts of plants, including many of economic importance.

Chalara urceolata Nag Raj & Kendrick (Fig. 1276)
Colonies effuse, thinly hairy, brown or greyish brown. Conidiophores brown, up to 230 long, spore sac 8–13 wide in broadest part. Conidia catenate, 1-septate, hyaline, 10–18 × 2.5–4. Fairly common on dead stems including those of *Epilobium hirsutum*, *Eupatorium cannabinum*, *Filipendula ulmaria*, *Heracleum sphondylium* and *Rumex* spp.

Chromelosporium ochraceum, described on tree 'leaf litter', is common also on dead herbaceous stems, e.g. those of *Eupatorium cannabinum*, *Heracleum sphondylium*, *Pastinaca sativa* and *Petasites hybridus*. We have found it also, in October, covering living leaves and stems of *Ranunculus repens*.

Cladosporium. Four species are found commonly on dead herbaceous plants, where they form olivaceous, rather velvety colonies.

KEY

    Conidia mostly globose, 3–4.5 diam. .... *sphaerospermum*
    Conidia not globose 1
1. Conidia mostly smooth and 3–7 × 2–4 .... *cladosporioides*
    Conidia distinctly rough-walled 2
2. Conidia mostly 8–15 × 4–6 .... *herbarum*
    Conidia mostly 15–25 × 7–10 .... *macrocarpum*

Cladosporium cladosporioides (Fresen.)de Vries (Fig. 1277)
Conidiophores up to 350 long but usually much shorter, 2–6 thick, olivaceous brown. Conidia formed in long, branched chains. An early secondary invader of damaged plants.

Cladosporium herbarum (Pers.)Link (Fig. 1278)
Conidiophores often nodose, olivaceous brown, up to 250 × 3–6, with vesicular swellings, when present, 7–9 diam. Conidia in fairly long, often branched chains, pale to mid brown or olivaceous brown, rather thick-walled and distinctly verruculose, with low warts, nearly always 0- or 1-septate, with scars at one end or both ends small but clearly protuberant. One of the commonest of all fungi, found throughout the year.

Cladosporium macrocarpum Preuss (Fig. 1279)
Conidiophores often geniculate and nodose, brown or olivaceous brown, up to 300 × 4–8 with vesicular swellings, when present, 9–11 diam. Conidia usually in rather short chains, 0- to 3-septate, pale to mid brown or olivaceous brown, thick-walled, densely verrucose. Another very common species, distinguished from *C. herbarum* by its broader, frequently 2- or 3-septate conidia.

Cladosporium sphaerospermum Penz. (Fig. 1280)
Conidiophores usually less than 300 × 3–5, olivaceous brown. Distinguished by its small, mostly globose conidia; these have little warts on them which are seen clearly in water or in air bubbles. Common on dead plants of all kinds.

Coremiella cubispora (Berk. & Curtis)M.B.Ellis (Fig. 1281)
Colonies olivaceous to dark blackish brown, effuse. Conidiophores usually forming loose, olivaceous brown coremia up to 800 × 200. Conidia in chains, oblong, pale to mid brown or olivaceous brown, smooth, 5–9 × 4–9. Fairly common on dying and dead leaves and stems of herbaceous plants, especially those growing in marshes and fens, e.g. *Filipendula ulmaria* and *Lythrum salicaria*, July–Oct.

Cylindrotrichum oligospermum (Corda)Bon. (Fig. 1282)
Colonies effuse, pale greyish brown, thinly hairy. Conidiophores brown, up to 600 × 3–7. Conidia often aggregated in slimy bundles, mostly 1-septate,

hyaline, 12–22 × 2.5–3. Common on dead herbaceous stems, e.g. those of *Angelica sylvestris*, *Arctium minus*, *Dipsacus fullonum*, *Eupatorium cannabinum*, *Filipendula ulmaria* and *Heracleum sphondylium*, Apr.–Sept.

Dactylosporium macropus, described on 'wood and bark', is found also occasionally on dead stems of *Arctium*, *Heracleum*, etc.

Dematophora, *v.s.* under *Rosellinia necatrix*.

Dendryphiella infuscans (Thümen)M.B.Ellis (Fig. 1283)
Colonies effuse, brown. Conidiophores smooth, brown, up to 500 × 4–6, swollen at nodes to 7–9. Conidia pale brown, smooth or minutely verruculose, 0- to 2-septate, 9–16 × 4–7. Not uncommon on dead stems of Umbelliferae and *Urtica dioica*, Apr.–May.

Dendryphiella vinosa (Berk. & Curtis)Reisinger (Fig. 1284)
Colonies effuse, rather velvety, rust-coloured to rusty black. Conidiophores reddish brown, smooth or verruculose, up to 450 × 4–6, swollen at nodes to 6–11. Conidia 3-septate, distinctly verruculose when mature, pale clear brown to burnt sienna, 16–39 × 4–8. Very common especially on the lower parts of dead standing stems of Umbelliferae and on old stems lying on the ground, May–Sept.

Dendryphion comosum Wallr. (Fig. 1285)
Colonies grey when young, later dark reddish brown to olive or black, velvety, with conidiophores close together, often completely encircling dead stems and extending along them for 2–7cm. Conidiophores dark reddish brown to black, 100–500 long, 9–14 thick at base, 5–8 at apex, branched in a tree-like manner, branches pale to mid brown. Conidia in chains, 10–50 × 5–8, pale to mid brown, minutely verruculose, with 1 to 7 (mostly 4) septa, sometimes branched. Very common on all sorts of herbaceous plants, and especially so on *Urtica dioica*, all the year round.

Dendryphion nanum (C.G.Nees)Hughes (Fig. 1286)
Colonies black, velvety, variable in size, sometimes encircling the base of dead stems and extending along them up to 10cm. Conidiophores black and shining by reflected light, dark brown by transmitted light, 80–300 long, 10–12 thick at base, 7–9 at apex; branches pale to mid brown, smooth or verruculose. Conidia solitary or in chains, with 5 to 11 (mostly 10 or 11) septa, smooth or verruculose, brown with subhyaline end cells, 45–90 × 10–12.5. Common, especially on *Brassica oleracea* and Umbelliferae.

Doratomyces. Erect dark synnemata bear long chains of small conidia.

KEY

Conidia verrucose .... *nanus*
Conidia smooth 1
1. Conidia 3–5 × 2–3 .... *microsporus*

Conidia mostly larger than 5 × 3.5 2
2. Heads ellipsoidal or cylindrical, conidia often pointed at apex; often with *Echinobotryum* state .... *stemonitis*
Heads spherical or subspherical, conidia mostly rounded at apex; no *Echinobotryum* state .... *purpureofuscus*

Doratomyces microsporus (Sacc.)Morton & Smith (Fig. 1287)
Colonies grey to black, hairy. Synnemata up to 600 high with long cylindrical or ellipsoidal heads. Occasionally found on dead herbaceous stems and also on dung.

Doratomyces nanus (Ehrenb.)Morton & Smith (Fig. 1288)
Colonies grey to brown or blackish brown. Synnemata up to 900 high, with subspherical or ellipsoidal heads. Conidia pale brown, 6–8.5 × 5–6. Seen occasionally on dead herbaceous stems.

Doratomyces purpureofuscus, described on *Brassica*, has been recorded occasionally on other herbaceous plants.

Doratomyces stemonitis (Pers.)Morton & Smith (Fig. 1289)
Colonies at first grey, becoming dark blackish brown to black. Synnemata up to 1200 high. Conidia pale brown, 6–8.5 × 4–4.5. *Echinobotryum* state usually present, attached to sides of stipes, with obpyriform, verrucose, brown conidia, 9–14 × 5–8. Common on dead herbaceous stems but found also on dead wood and on dung.

Endophragmia elliptica (Berk. & Br.)M.B.Ellis (Fig. 1290)
Colonies effuse, dark brown to black, up to 6 × 1cm. Conidiophores in tufts of 6 to 80, forming loose coremia, up to 240 long, lower part mid to dark brown, 4–5 thick, upper part subhyaline broadening gradually to 8–11. Conidia 27–54 (mostly 35–40) × 16–23, usually 5-septate, end cells hyaline or pale brown, central cells mid to dark brown, often with broad black bands at the septa. Common on dead stems, e.g. those of *Centaurea nigra*, *Epilobium hirsutum*, *Filipendula ulmaria*, *Lysimachia vulgaris* and *Polygonum cuspidatum*, Apr.–Oct.

Endophragmia hyalosperma (Corda)Morgan-Jones & Cole (Fig. 1291)
Colonies effuse, pale brown, thinly hairy. Conidiophores pale to mid brown, up to 250 × 3–5. Conidia subhyaline, mostly 4-septate, 20–30 × 10–13. Common on dead stems, Feb.–Oct.

Epicoccum purpurascens Ehrenb. (Fig. 1292)
Sporodochia pulvinate, black, up to 2mm diam., with surrounding host tissues often stained reddish purple. Conidiophores 5–15 × 3–6, hyaline to pale brown. Conidia spherical or pyriform, black by reflected light, dark golden brown by transmitted light, often with a pale protuberant basal stalk cell, muriform but with the septa obscured in mature conidia by the rough opaque wall, most commonly 15–25 diam., but smaller and much larger (up to 50

diam.) conidia are formed occasionally. An extremely common early secondary invader on all sorts of plants, frequently found on leaf spots with other saprophytes such as *Periconia byssoides*, all the year round.

Fusariella hughesii Chabelska-Frydman (Fig. 1293)
Colonies compact or somewhat effuse, greyish green. Conidiophores branched, hyaline. Conidia catenate, subhyaline, in mass pale greyish green, 3-septate, 14–25 × 2.5–3.5. On dead stems and leaves, not uncommon, especially on Umbelliferae, Apr.–June.

Fusarium species are found quite frequently on dead herbaceous plants but as their accurate identification necessitates their being grown in culture, they are not dealt with in any detail here. An excellent inexpensive modern book, *The Genus Fusarium*, by C. Booth, published by the Commonwealth Mycological Institute, Kew, deals with all species likely to be encountered. The curved, septate, narrowly fusiform conidia, often with a small oblique projection at the base forming a so-called foot-cell, are diagnostic (Fig. 1294). They form conspicuous slimy, pink or orange masses on cushion-shaped sporodochia; seen singly they are quite colourless. Many fusaria are conidial states of *Nectria* and other hypocreaceous fungi.

Gliocladium roseum, a plurivorous species described on 'wood and bark', is quite common also on dead herbaceous stems and leaves.

Gliomastix luzulae (Fuckel)Mason (Fig. 1295)
Colonies effuse, green or blackish green. Conidiophores subhyaline, slightly roughened with dark granules, up to 40 × 1.5–3.5. Conidia in chains which readily fragment, olive green, often with a darker median band, almost black in mass, 4–9 × 1.5–2.5. Quite common, especially on dead stems of *Gunnera tinctoria*, *Heracleum sphondylium*, *Oenanthe crocata* and *Petasites* spp.

Gonatobotrys simplex Corda (Fig. 1296)
Colonies effuse, white, cottony or cobwebby. Conidiophores up to 250 × 5–7, with nodose swellings which bear conidia at the ends of flat-topped pegs. Conidia hyaline, smooth, 10–15 × 5–8. On dead herbaceous stems and often also on rotten wood, commonly associated with other fungi.

Monodictys levis (Wiltshire)Hughes (Fig. 1297)
Colonies effuse, grey or greyish brown, mycelium superficial, conidiophores very short. Conidia clear, pale to mid brown or greyish brown, smooth, often twisted, 17–30 × 15–19. On dead herbaceous stems and sometimes also on rotten wood.

Mycocentrospora acerina (Hartig)Deighton (Fig. 1298)
Colonies effuse, at first hyaline, later green, grey or reddish purple and finally almost black. Conidiophores up to 50 × 5–7, hyaline or subhyaline, with broad, flat scars. Conidia mostly 150–200 × 8–15, smooth, hyaline or with the broader cells rather pale brown, basal appendage when present 30–150 ×

2–3. A plant pathogen able to live in soil and with a wide host range. It causes a leaf spot of *Arum maculatum*, leaf blight of pansies, and crown and stalk rot of celery, and is an agent of parsnip black canker in the East Anglian fens.

Oidiodendron tenuissimum, described on 'wood and bark', occurs also on dead herbaceous plants.

Periconia. Three very common plurivorous species occur. When viewed through a lens they look like black, round-headed pins.

KEY

    Conidiophores branched at apex, conidia 4–6 diam. .... *minutissima*
    Conidiophores not branched, conidia 10 diam. or more 1
1.   Conidiophores with short apical cell cut off by a septum,
    conidiogenous cells formed over apex and in a ring below septum........
    .... *byssoides*
    Conidiophores without short apical cell, conidiogenous cells formed
    over swollen apex .... *cookei*

Periconia byssoides Pers. (Fig. 1299)
Colonies effuse, grey to black, hairy. Conidiophores dark brown, black and shining by reflected light, 200–1400 × 10–20. Heads of conidia mostly 50–120 diam., brown. Conidia brown, verrucose, 10–15 diam., in short, sometimes branched chains. Very common on blackened areas of dead herbaceous stems, and on leaf spots where it is nearly always associated with other secondary invaders.

Periconia cookei Mason & M.B.Ellis (Fig. 1300)
Colonies effuse, grey to black, hairy. Conidiophores similar to those of *P. byssoides* except that there is no septum cutting off an apical cell and the apex is often swollen (17–32 diam.). Conidia brown, verrucose, 13–16 diam. Fairly common on blackened areas of herbaceous stems, especially those of Umbelliferae.

Periconia minutissima Corda (Fig. 1301)
Colonies effuse, grey to brown or black, hairy. Conidiophores brown, up to 550 × 5–10. Conidia straw-coloured or pale brown, verruculose, 4–6 or rarely 7 diam., in fairly long branched chains. Common on dead stems and leaves usually close to or on the ground.

Pleurophragmium parvisporum (Preuss)Holubová-Jechová (Fig. 1302)
Colonies effuse, thin, greyish brown, hairy. Conidiophores up to 400 × 4–6, brown. Conidia hyaline to very pale brown, smooth, 1- to 4- (mostly 3-) septate, tapered to a point at the base, attached to conidiophore apices by pointed denticles. Very common on partly decorticated stems of *Urtica dioica*; also on dead stems of *Arctium lappa*, *Conium maculatum*, *Epilobium angustifolium*, *Filipendula ulmaria*, *Heracleum sphondylium*, *Sisymbrium officinale*,

etc., all year round.

Pseudospiropes rousselianus (Mont.)M.B.Ellis (Fig. 1303)
Colonies effuse, dark blackish brown to black, thin, hairy. Conidiophores up to 300 × 6–9, dark brown, paler near the apex where there are a number of small dark scars. Conidia hyaline or subhyaline, smooth, 3- to 7-septate, 20–37 × 5–8. On dead stems of *Anthriscus sylvestris*, *Arctium minus*, *Cirsium arvense*, *Dipsacus fullonum*, *Epilobium angustifolium* and *Paeonia officinalis*, also occasionally on wood.

Pseudospiropes subuliferus, described on 'wood and bark', is found also quite often near the base of dead stems of such plants as *Arctium lappa*, *Dipsacus fullonum*, *Epilobium angustifolium* and *Foeniculum vulgare*.

Sarcopodium circinatum Ehrenb. (Fig. 1304)
Colonies superficial, cushion-like, setose, round, oval, irregular or elongated and up to 1cm long, at first pinkish then becoming brown or purplish brown, spongy or cottony. Setae flexuous or circinate, pale to mid golden brown, verrucose, thick-walled, up to 500 × 4–6. Conidia in slimy masses, hyaline, 7–10 × 2. Common on dead stems of herbaceous plants including *Arctium lappa*, *Epilobium hirsutum*, *Heracleum sphondylium*, *Mercurialis perennis*, *Ononis repens*, *Pastinaca sativa*, *Trifolium pratense* and *Urtica dioica*.

Scopulariopsis brevicaulis (Sacc.)Bainier (Fig. 1305)
Colonies effuse, at first whitish, later buff or nut brown with narrow white margins. Conidiophores branched, fertile branches (annellides) 10–25 long. Conidia catenate, very pale singly but brown in mass, coarsely warted when mature, 5–8 × 5–7. Not uncommon on dead herbaceous plants, e.g. rotting stalks of *Brassica* on rubbish heaps.

Septofusidium herbarum (Brown & Smith)Samson (Fig. 1306)
Colonies tufted, creamy white. Conidiophores up to 100 × 3–4, hyaline. Conidia in chains, hyaline, at first smooth, becoming minutely verruculose, 8–11 × 3–5. On dead stems and petioles of *Gunnera tinctoria*, *Heracleum sphondylium*, *Petasites hybridus*, *Urtica dioica*, etc.

Stachybotrys. Conidia aggregated in slimy black heads and produced from clusters of short phialides at the tops of the conidiophores.

KEY

Conidia 11–15 long with longitudinal striations .... *cylindrospora*
Conidia seldom more than 11 long, without longitudinal striations 1
1. Conidiophores hyaline, thick-walled, smooth .... *dichroa*
Conidiophores olivaceous brown to black, with thin walls often covered in part by granules .... *atra*

Stachybotrys atra Corda (Fig. 1307)
Colonies effuse, black. Conidiophores up to 100 × 3–5, often branched.

Conidia in black slimy heads, dark blackish brown to black, verrucose, 8–11 × 5–10. A common saprophyte found on many different substrata including decaying parts of herbaceous plants, but not so frequently found on these as is *S. dichroa*.

Stachybotrys cylindrospora, described on *Heracleum*. Quite common on dead stems of Umbelliferae, May–Sept., but rare on other substrata; found once on *Arctium minus*.

Stachybotrys dichroa Grove (Fig. 1308)
Conidiophores up to 270 long, 8–20 thick at base tapering to 3.5–7 just below the apex, then swelling again to 5–9. Conidia often obliquely attenuated at base, olivaceous brown to almost black, verrucose when mature, 8–14 × 4–6. The commonest *Stachybotrys* on dead herbaceous stems, usually about soil level when these are still standing, more widely distributed when they are lying on the ground, Apr.–Sept. Found on *Angelica sylvestris*, *Carlina vulgaris*, *Centaurea nigra*, *Cirsium arvense*, *Conium maculatum*, *Dipsacus fullonum*, *Eupatorium cannabinum*, *Filipendula ulmaria*, *Foeniculum vulgare*, *Iris pseudacorus*, *Oenanthe crocata*, *Senecio jacobaea*, *Smyrnium olusatrum* and *Urtica dioica*.

Stachylidium bicolor Link (Fig. 1309)
Colonies effuse, pale grey to olivaceous brown. Conidiophores up to 700 × 4–7, smooth to minutely verruculose, lower part brown, upper part hyaline or pale olivaceous and bearing whorls of phialides. Conidia aggregated in slimy balls at the tips of phialides, hyaline or pale olive, smooth, 4–8 × 2–3. Common on dead herbaceous stems, e.g. those of *Heracleum sphondylium*, *Oenanthe crocata*, *Petasites hybridus* and *Urtica dioica*, but found also on bracken fronds and dead wood.

Stemphylium state of *Pleospora herbarum* (*v.s.*) has conidia constricted as a rule at the median transverse septum only.

Stemphylium vesicarium (Wallr.)Simmons (Fig. 1310)
Colonies effuse, olivaceous brown to black, somewhat velvety. Conidiophores up to 70 × 3–8, pale brown to brown, smooth or minutely verruculose, with nodose swellings 8–11 diam. Conidia pale to mid brown or olivaceous brown, verrucose, with up to 6 transverse and several longitudinal septa, mostly constricted at the 3 major transverse septa, 20–50 × 15–26. Common on dead herbaceous plants including *Apium graveolens*, *Heracleum sphondylium*, *Honkenya peploides*, *Lactuca sativa*, *Lycopersicon esculentum* and *Phaseolus vulgaris*.

Torula herbarum (Pers.)Link (Fig. 1311)
Colonies very variable in size, sometimes only a few mm diam., at others completely encircling stems and extending along them for several centimetres, olive when young, black when old, velvety. Conidiophores very short, brown, each terminating in a characteristic swollen cell, with a dark, thick-walled, verrucose or echinulate base, and a pale upper part which is thin-walled and

often collapses to form a cup; distal cells of conidia are similar. Conidia in long, branched chains, pale olive or brown, verruculose or finely echinulate, 3- to 10- (mostly 4- or 5-)septate, 20–70 × 5–9 (average 30 × 7). Very common on dead herbaceous plants and found also sometimes on other substrata.

Trichoderma koningii Oudem. (Fig. 1312)
Colonies effuse, whitish at first but soon becoming dark green. Conidiophores hyaline, much branched, with branches set almost at right-angles, about 4 thick, and with whorls of phialides similarly set. Conidia in slimy balls at tips of phialides, smooth, pale green singly, dark green in mass, 3–5 × 2–3. Not uncommon on dead herbaceous stems, e.g. those of *Epilobium angustifolium*; also on grasses.

Triposporium elegans, described on *Rubus*, is common also on some dead herbaceous stems, e.g. those of *Epilobium angustifolium* and *Filipendula ulmaria*.

Ulocladium. Four plurivorous species form effuse, olivaceous brown, blackish brown or black, rather velvety colonies. They are often found with other saprophytic fungi on dead herbaceous plants and have been referred in the past to the genus *Stemphylium*. Conidia are brown, cruciately septate or muriform.

KEY

    Conidia often in chains .... *chartarum*
    Conidia solitary or rarely forming false chains 1
1.   Conidia mostly cruciately septate .... *atrum*
    Conidia mostly not cruciately septate 2
2.   Conidia smooth or inconspicuously roughened .... *consortiale*
    Conidia closely verruculose or verrucose .... *botrytis*

Ulocladium atrum Preuss (Fig. 1313)
Conidiophores brown, smooth or verruculose, up to 120 × 3–8. Conidia mostly spherical or subspherical, cruciately septate, 13–20 diam., golden brown or dark reddish brown, verrucose.

Ulocladium botrytis Preuss (Fig. 1314)
Conidiophores often dichotomously branched near the apex, pale to mid golden brown, smooth, up to 100 × 3–5. Conidia golden brown, closely verruculose or verrucose, frequently with a minute projecting hilum, mostly 14–24 × 9–15.

Ulocladium chartarum (Preuss)Simmons (Fig. 1315)
Conidiophores golden brown, up to 50 × 5–7. Conidia commonly in chains of 2 to 10, often with short, false beaks, which are really thin-walled germ tubes, brown, smooth or verruculose, 18–38 × 11–20.

Ulocladium consortiale (Thümen)Simmons (Fig. 1316)
Conidiophores pale golden brown, up to 60 × 4–5. Conidia mostly smooth, golden brown, 16–34 × 10–15.

Verticillium. Apart from conidial *Nectria inventa* (*v.s.*), species usually need cultural studies for accurate determination. The colourless or pale-coloured conidiophores bear whorls of short phialides towards their apices, and at the tips of these are seen slimy balls of small, colourless or very pale conidia (Fig. 1317). Two common plurivorous species which are often responsible for wilt diseases are *V. albo-atrum* Reinke & Berth. and *V. dahliae* Kleb.

Volutella ciliata (Alb. & Schw.)Fr. (Fig. 1318)
Colonies made up of variable numbers of glistening, white, sessile or shortly stalked sporodochia 0.1–0.25 mm diam., each with a fringe of long white setae. The sporodochia tend to become yellowish or sometimes flesh-coloured when old. Setae mostly 250–500 × 8–9, thick-walled, verruculose towards their tips. Conidia hyaline, smooth, 5–8 × 2, aggregated in slimy masses. Very common on dead herbaceous plants, Oct.–Apr., and best seen in the field after a shower of rain.

COELOMYCETES

Chaetospermum chaetosporum (Pat.)A.L.Smith & Ramsb. (Fig. 1319)
Conidiomata at first immersed, gelatinous, white, 0.4–2mm diam., becoming erumpent and rupturing irregularly and often at one side to expose a yellowish mass of conidia. Conidia hyaline, 30–50 × 10–15, with clusters of wavy setulae up to 60 × 1 at each end. Found occasionally on dead herbaceous stems, e.g. those of *Epilobium hirsutum* and *Valeriana officinalis*.

Colletotrichum dematium (Pers.)Grove (Fig. 1320)
Acervuli round or somewhat elongated, up to 0.5mm diam., black, setose. Setae dark blackish brown, paler at their sharply pointed tips, septate, up to 250 × 4–8. Conidia hyaline, smooth, mostly 20–27 × 2–3, falcate. Very common on dead stems and petioles of many different plants including *Endymion non-scriptus*, *Heracleum sphondylium*, *Iris pseudacorus*, *Lunaria annua* and *Silene maritima*.

Colletotrichum, *v.s.* also under *Glomerella cingulata*.

Phoma. Cultural studies are needed for accurate determination of most species, and reference is made here only to three very common plurivorous ones which can be recognised fairly easily. The pycnidia are immersed or semi-immersed, globose, unilocular, brown to black with one or sometimes several ostioles. Conidia are small, hyaline, smooth, often with guttules and occasionally 1-septate.

KEY

	Conidia 6–9 × 3–3.5, guttulate, some 1-septate	*complanata*
	Conidia 4–6 × 1.5–2	1
1.	Conidia without guttules	*herbarum*
	Conidia biguttulate	*nebulosa*

Phoma complanata (Tode)Desm.
Pycnidia up to 0.5mm diam., at first immersed, later often erumpent, black, with papillate ostioles, very thick-walled. Especially common on old stems of Umbelliferae, e.g. *Angelica sylvestris*, *Anthriscus sylvestris*, *Heracleum sphondylium* and *Oenanthe crocata*.

Phoma herbarum Westend.
Pycnidia at first immersed, later erumpent, black, tending to be flattened dorsiventrally and to become umbilicate, with small ostioles, thin-walled. Common on dead stems of many different plants, especially in the spring.

Phoma nebulosa (Pers.)Berk.
Pycnidia immersed, with protuberant necks, very small, gregarious, arranged more or less in lines and connected by thin subepidermal layers of brown hyphae which show on the surface as grey, cloudy patches. On dead stems of *Althaea rosea*, *Brassica oleracea*, *Lamium album*, *Petroselinum sativum*, etc.

Phomopsis hysteriola (Sacc.)Grove
Conidiomata often in elongated groups of 2 to 5, black, immersed or becoming erumpent. A-conidia hyaline with 2 to 4 guttules, 6–9 × 2.5–3. B-conidia uncinate, 28–30 × 1.5. On dead stems of Umbelliferae.

Phomopsis, *v.s.* also under *Diaporthe arctii*.

Pseudolachnea hispidula (Schrad.)Sutton (Fig. 1321)
Conidiomata subcuticular to erumpent, closed at first then opening to become cupulate, dark brown to black with over-arched, dark brown, thick-walled, sharply pointed setae. Conidia hyaline, 16–20 × 2–3, with apical and basal setulae 2–3 long. Very common on all sorts of dead herbaceous stems and also on wood and bark; most collections have been on *Urtica dioica* but we have found it also, especially during the winter months, on *Ballota nigra*, *Epilobium hirsutum*, *Foeniculum vulgare* and *Solanum dulcamara*.

Rhabdospora pleosporoides Sacc.
Pycnidia scattered, subepidermal, flattened dorsiventrally, up to 0.5mm diam., black, with short necks. Conidia straight or slightly curved, hyaline, 30–50 × 1–1.5. On old dead stems of *Heracleum sphondylium*, *Oenanthe crocata*, *Urtica dioica*, etc.

FUNGI SPECIFIC TO HERBACEOUS PLANTS OTHER THAN GRASSES, RUSHES, SEDGES, BUR-REEDS AND REEDMACES

ACHILLEA: Yarrow or Milfoil, Sneezewort

Puccinia cnici-oleracei Pers. ex Desm., 111 (Fig. 1322)
Not uncommon on leaves of *A. millefolium*. Very small dark brown sori on rather indistinct spots are found on both leaf surfaces, July–Nov. Teliospores golden brown, 1-septate, mostly 40–55 × 15–20.

Entyloma achilleae P.Magnus
Smut sori in leaves of *A. millefolium*, Aug., rare. Spores 10–12 diam.

Schizothyrioma ptarmicae (Desm.)Höhnel (Fig. 1323)
On living leaves and stems of *A. ptarmica*, June–Sept., uncommon. Apothecia black, up to 1 × 0.5mm, splitting open to expose pale golden-brown discs. Ascospores 2 to 6 but mostly 2 per ascus, hyaline, 1-septate, 8–14 × 3–5.

Schizothyrioma aterrimum (P.Karsten)Holm (Fig. 1324)
Superficially indistinguishable from *S. ptarmicae* and on the same host, but with the ascospores regularly 6 to 8 per ascus and always much narrower, 8–12 × 2–2.5.

ACORUS: Sweet Flag

Ascochyta acori Oudem.
Pycnidia immersed, up to 0.35 × 0.25mm, dark brown, on lower parts of leaves, Oct.–Dec. Conidia hyaline, mostly 1-septate with the septum often nearer one end, 30–40 × 10–12.

Ramularia aromatica (Sacc.)Höhnel (Fig. 1325)
Conidiophores in small groups on both leaf surfaces, 10–40 × 2.5–4. Conidia hyaline, with 1 to 4 septa, 20–70 × 2–3. On leaves, causing oval, reddish-brown to black spots, 5–20 × 2–5mm.

ADOXA: Moschatel or Townhall Clock

Three rusts occur but only one of them, *Puccinia adoxae*, is common.

Puccinia adoxae DC., 111 (Fig. 1326)
All parts of plants affected and often swollen and distorted, Mar.–May. Sori clustered, frequently confluent, covered at first by silvery epidermis, opening to expose masses of dark brown spores. Teliospores 25–35 × 15–20.

Puccinia albescens Plowr., 0, 1, 11, 111
Sori widely scattered. Aecia found Mar.–Apr., on stems and leaves, spores very pale yellow. Teliospores formed May–June, 30–45 × 15–25.

Puccinia argentata (C.F.Schultz)Winter, 0, 1
Aecia on stems and leaves, Apr.–June; spores golden yellow. Rare. Alternate host *Impatiens capensis*.

Ramularia adoxae (Rabenh.)Karsten (Fig. 1327)
Colonies white, conidiophores emerging through stomata on lower leaf surface, 20–65 × 3–4. Conidia hyaline, 0- to 3-septate, 20–40 × 4–6. On leaves, causing pale grey or pale brown spots 3–10mm diam.

AEGOPODIUM: Goutweed or Ground Elder

Puccinia aegopodii Röhl., 111
Sori black, shining, mostly on petioles and nerves, clustered on swollen yellowish spots, Apr.–Aug., quite common. Teliospores mostly 30–45 × 15–22.

Mycosphaerella podagrariae (Fr.)Petrak
Almost superficial pseudothecia formed on overwintered leaves; ascospores hyaline, constricted at the median septum, 18–19 × 5, with the lower cell somewhat narrower than the upper one. This fungus is much better known in its *Phloeospora* state which is found quite commonly causing angular, pale yellow or whitish spots on living leaves, July–Sept.; conidia hyaline, multiguttulate, sometimes 1- to 3-septate, 50–80 × 2.5–4.

Plasmopara nivea (Unger)Schröter (Fig. 1328)
Downy mildew. Colonies on lower surface of living leaves, white; corresponding areas on upper surface pale yellow; July. Sporangia mostly 23–27 × 17–19, hyaline.

Protomyces macrosporus Unger (Fig. 1329)
Chlamydospores pale yellow, thick-walled, 50–65 diam., embedded in small, pale brown, swollen areas on living stems and leaves, Mar.–Oct.

AETHUSA: Fool's Parsley

Puccinia nitida (Str.)Röhl., 0, 1, 11, 111
Aecia uredinoid. Teliospores mostly 30–48 × 20–25. The only rust on *A*.

cynapium, found on leaves, petioles and stems, June–Oct.

AGRIMONIA: Common Agrimony

Pucciniastrum agrimoniae (Diet.)Tranz., 11
Uredinia yellowish orange to ochraceous, tending to be confluent and sometimes covering the whole of the lower surface of living leaves; July–Sept., common. Spores echinulate, 18–20 × 14.

Only the rust is specific to *A. eupatoria*, but plurivorous fungi such as *Unguicularia millepunctata* and species of *Lophiostoma*, including *L. angustilabrum*, *L. caulium*, *L. fuckelii* and *L. vagabundum*, are found commonly on dead stems.

AJUGA: Bugle

Ramularia ajugae (Niessl)Sacc. (Fig. 1330)
Colonies white, hypophyllous. Conidiophores in fascicles of 5 to 10 emerging through stomata, 20–35 × 3–4. On leaves of *A. reptans*, causing round or oval, pale spots, 2–8mm diam., with narrow reddish-brown or purple borders.

ALCHEMILLA: Lady's Mantle

Trachyspora intrusa (Grev.)Arth., 1, 111
This systemic rust overwinters in the rhizomes of various species including *A. glabra*, *A. vestita*, and *A. xanthochlora*; infected leaves tend to be erect, small and pale with long petioles. Orange uredinoid aecia formed from Apr.–June on lower leaf surface, followed by telia. Teliospores almost spherical or oblong, 28–40 × 20–30, coarsely verrucose, brown.

Anthostomella alchemillae (A.L.Smith & Ramsb.)S.Francis
Perithecia immersed, black, neck erumpent, no clypeus. Ascospores dark brown, 15–19 × 7–9. On dead leaves of *A. alpina*, May–Sept.

Coleroa alchemillae (Grev.)Winter (Fig. 1331)
Pseudothecia superficial, black, setose, about 0.1mm diam., arranged in rows radiating from a central point. Ascospores hyaline, 1-septate, 9–11 × 3–4. On the upper surface of living leaves of *A. vulgaris*, July–Nov., often causing discoloration.

Sphaerotheca alchemillae (Grev.)Junell (Fig. 1332)
Powdery mildew. Conidia of *Oidium* state plentiful, hyaline, about 30 × 15. Cleistothecia ripening from June onwards, chestnut brown; ascospores about 25 × 16. On living and fading leaves of *A. vulgaris*.

ALISMA: Water Plantain

Doassansia alismatis (Nees)Cornu (Fig. 1333)
Spore balls of this smut are dark reddish brown, up to 0.2mm diam., embedded in yellowish-brown spots about 1cm diam. on living leaves of *A. plantago-aquatica*, July–Oct. Spores pale yellow, 10–12 diam.

Rhynchosporium alismatis (Oudem.)J.J.Davis (Fig. 1334)
Found on spots on yellowing leaves of *A. plantago-aquatica* in dried-up ponds, July–Oct. Conidia hyaline, 1-septate, 15–18 × 3–3.5.

ALLIARIA: Garlic Mustard or Jack-by-the-hedge

Leptosphaeria maculans (Desm.)Ces. & de Not. (Fig. 1335)
Pseudothecia up to 0.5mm diam., immersed, often in blackened areas of dead stems, May–Oct. Ascospores pale to mid straw-coloured, 5-septate, mostly 40–50 × 5–7; sometimes up to 60 long on other host plants. There is a *Phoma* state with conidia 3–5 × 1.5–2 which forms lesions on stems and leaves.

Peronospora niessleana Berl.
This downy mildew, found only on *A. petiolata*, has sporangia 26–29 × 17–18, longer and narrower than those of the plurivorous *P. parasitica* which occurs on many Cruciferae and is described on *Capsella*.

ALLIUM: Chives, Crow Garlic, Garlic, Leek, Onion, Ramsons

UREDINALES

Four rusts occur on *Allium* species. *A. ursinum* serves as one of the aecial hosts of *Puccinia sessilis*, which forms its uredinia and telia on *Phalaris arundinacea*. The other three species are described briefly below.

Puccinia allii Rud., II, III
On *A. cepa*, *A. schoenoprasum* and *A. vineale*. Uredinia reddish yellow, spores 24–29 × 20–24, finely echinulate. Telia blackish brown, long covered by the grey epidermis, spores 1-septate, 28–45 × 20–26; mesospores often present. Found more frequently on *A. vineale* than on other hosts.

Puccinia porri (Sow.)Wint., II (Fig. 1336)
On *A. porrum*, uredinia only in Britain; spores distinctly echinulate and with hyaline caps over the pores, 27–38 × 24–30. Not common.

Uromyces ambiguus (DC.)Fuckel, II, III (Fig. 1337)
On *A. schoenoprasum* and *A. scorodoprasum*, scarce. Uredinia covered by yellow

epidermis, spores finely verruculose, 20–28 × 18–22. Telia blackish, spores brown, smooth, 20–35 × 18–24.

USTILAGINALES

Urocystis cepulae Frost
Smut is economically important only on *A. cepa*, seedlings of which can be killed in 3–4 weeks, but it occurs also on *A. porrum*, *A. sativum* and *A. vineale*, Apr.–Nov. Sori in leaves as elongated dark streaks or isolated pustules. Spore balls 14–22 diam., each with a single dark brown spore 11–14 diam., surrounded by small, pale, sterile cells.

OTHER FUNGI

Alternaria porri (Ellis)Cif. (Fig. 1338)
On *A. cepa*, *A. porrum*, *A. sativum* and *A. schoenoprasum*, causing purple blotch disease. Leaf spots often elliptical, large and coloured some shade of purple, sometimes with a broad, pale brown or pale yellow border. Uncommon in Britain although world-wide in distribution. Conidiophores brown, up to 120 × 9–10. Conidia muriform, golden brown, 100–300 × 15–20, beak 2–4 thick.

Aspergillus niger van Tiegh. (Fig. 1339)
This common saprophyte of soil and decaying vegetable matter not infrequently causes black mould of onions, garlic and shallots. Conidiophores up to 3mm long and 15–20 thick; black conidial heads frequently splitting into columns. Conidia in chains, often verruculose or echinulate, 4–5 diam.

Botrytis. Two species of *Botrytis* and four species of *Botryotinia* with *Botrytis* states occur on *Allium*. As the apothecial states of the *Botryotinia* species are seldom seen in the field, it is convenient to treat them all together under *Botrytis*.

KEY

    Conidia spherical 1
    Conidia ellipsoidal or pyriform 2
1.   Conidia mostly 20–26 diam. .... *Botryotinia sphaerosperma*
    Conidia mostly 12–18 diam. .... *Botryotinia globosa*
2.   Conidia 7–11 × 5–6 .... *Botrytis allii*
    Conidia 10–14 × 6–9, sclerotia rare or absent .... *Botrytis byssoidea*
    Conidia 11–14 × 7–10, sclerotia large and cerebriform ....
    .... *Botryotinia porri*
    Conidia mostly 15–21 × 13–16 .... *Botryotinia squamosa*

Botryotinia globosa Buchwald (Fig. 1340)
Colonies effuse, pale, powdery, on large, dark green, collapsed, water-soaked

areas which sometimes extend over the whole leaf surface of *A. ursinum*, May–June.

Botryotinia porri (van Beyma)Whetzel
Recorded originally on *A. porrum*. We have found very large cerebriform sclerotia of this species, accompanied by the *Botrytis* state, on old stems and leaf-sheath bases of *A. vineale* in June and July.

Botryotinia sphaerosperma (Gregory)Buchwald
Recorded only on *A. triquetrum* in the Isles of Scilly.

Botryotinia squamosa Viennot-Bourgin (Fig. 1341)
In addition to the large size of its conidia, this species is characterised by the very marked concertina-like collapse of its conidiophore branches. Causes tip and leaf blight of *A. cepa* and forms numerous small black sclerotia on the bulb scales.

Botrytis allii Munn (Fig. 1342)
Distinguished from all other species by its narrowly ellipsoidal, occasionally septate conidia. On various species of *Allium*, including *A. cepa* and *A. ursinum*; causes extensive damage to onions (neck rot), and to shallots in storage.

Botrytis byssoidea Walker
Causes mycelial neck rot of onions and has been recorded also on leeks.

Cladosporium allii-cepae (Ranojević)M.B.Ellis (Fig. 1343)
Colonies effuse, grey. Conidiophores pale olive, up to 200×10–18. Conidia pale olive to golden brown, mostly 1- or 2-septate, 30–90×11–22. On scapes and leaves of *A. cepa*, uncommon in Britain.

Colletotrichum circinans (Berk.)Vogl.
Acervuli often formed in concentric circles, setae dark brown, up to 200 long, pointed at the tips. Conidia falcate, more sharply pointed at the apex than at the base, hyaline, buff in mass, 17–21×3–3.5. On dry outer bulb scales of *A. cepa*, Aug.–Sept., uncommon in Britain.

Embellisia allii (Campanile)Simmons (Fig. 1344)
Colonies dark blackish brown to black, velvety or powdery. Conidiophores brown, up to 100×5–10. Conidia mid to dark brown, with 3 to 6 very dark transverse septa and occasionally also oblique or longitudinal septa, 25–45×10–15. On bulb scales of *A. sativum*.

Peronospora destructor (Berk.)Casp.
Downy mildew. Colonies pale purplish grey, downy. Sporangiophores greyish violet, up to 150×7–15. Sporangia somewhat pyriform, 35–60×20–35. On leaves and stems of *A. cepa*, causing long pale-yellow lesions. Leaves shrivel from the tip backwards, stems twist and plants die; a whole crop may be destroyed by this pathogen.

Phytophthora porri Foister
The cause of white tip disease of *A. porrum*. The top portions of leaves die and turn white, often becoming limp. Leaves may be affected also at the edges and become twisted, with a darkening of the green tissue in the middle or towards the base, appearing water-soaked. Sporangiophores simple; sporangia obpyriform or oval, mostly about 50 × 35.

Sclerotium cepivorum Berk.
Roots and bulb bases covered by a dense coating of white mycelium, accompanied by numerous very small, black sclerotia. Leaves wilt and turn yellow. Although most important as the cause of white rot of *A. cepa*, it occurs also on other species.

ALTHAEA: Hollyhock

Puccinia malvacearum, described on *Malva*, is very common also on hollyhock leaves everywhere.

ALYSSUM: Golden Alyssum

Peronospora galligena Blumer
Downy mildew on leaves and young seedlings of *A. saxatile*, July–Nov. Causes small yellowish blisters on upper surface of leaves, which tend to be deformed. Sporangia usually ovate, occasionally spherical, 15–24 × 14–19, smooth, with granular contents.

ANAGALLIS: Scarlet Pimpernel

Alternaria anagallidis Raabe (Fig. 1345)
Conidiophores olivaceous brown, up to 100 × 5–8. Conidia with the narrow beak usually the same length as or shorter than the body, pale to mid golden brown, smooth, with up to 14 transverse septa and usually one or more longitudinal or oblique septa, 70–140 (rarely up to 200) × 14–25. On leaves of *A. arvensis*, causing reddish-brown spots, uncommon.

ANCHUSA: Field Bugloss

The orange cluster cups of *Puccinia recondita* f.sp. *recondita* are found commonly on living leaves of *A. arvensis*, Aug.–Sept., wherever rye, the alternate host, is grown in the district.

ANEMONE: Pasque Flower, Wood Anemone

UREDINALES

Ochropsora ariae (Fuckel)Ramsb., 0,1
Alternate host *Sorbus aucuparia*. Mycelium perennial in rhizomes of *A. nemorosa*, leaves of infected plants long, narrow, pale green; nearly every leaf and sepal bears aecia. Aeciospores finely verruculose, $18-26 \times 17-21$.

Tranzschelia anemones (Pers.)Nannf., 111 (Fig. 1346)
Telia almost always hypophyllous, dark brown, spores warted, 1-septate, mostly $30-40 \times 20-25$ (sometimes up to 50×30). On leaves of *A. nemorosa*.

Tranzschelia discolor (Fuckel)Tranz. & Litv., 0,1
Alternate hosts *Prunus* spp. Aecia on leaves of *A. coronaria*, Apr.–May; aeciospores 16–24 diam.

USTILAGINALES

Urocystis anemones (Pers.)Winter (Fig. 1347)
This smut, found commonly on *A. nemorosa* and occasionally on *A. pulsatilla*, forms blister-like swellings on leaves and stems; these rupture to expose black spore masses, Apr.–Sept. Spore balls irregular, with 1 to 3 mid dark brown spores 15–20 diam., partially surrounded by smaller, pale, sterile cells.

OTHER FUNGI

Dumontinia tuberosa (Hedw.)Kohn (Fig. 1348)
Apothecia 1–3cm diam., pale brown with dark brown discs and long stalks, arising singly or several together from black sclerotia in rhizomes. Ascospores hyaline, mostly $12-16 \times 6-7$. In fairly damp conditions where there is some bare ground between the *Anemone* plants, Mar.–May. Most records on *A. nemorosa*, but found occasionally also on garden species such as *A. blanda*.

Peronospora ficariae, described on *Ranunculus*, occurs also on *A. coronaria* and *A. nemorosa*.

Plasmopara pygmaea (Unger)Schröter
Downy mildew on living leaves of *A. nemorosa*, Apr. Sporangia papillate, ovoid or ellipsoid, $20-26 \times 15-19$. Oospores pale golden brown, 45–50 diam.

Septoria anemones Desm.
Pycnidia epiphyllous, 0.1mm diam., dark brown. Conidia hyaline, $20-22 \times 1-1.5$. On leaves of *A. nemorosa*, causes brownish, pale-centred spots, June–July.

Synchytrium anemones (DC.)Woronin
Resting spores spherical, 70–170 diam., epispore golden brown, thick, smooth or verruculose, with transverse ridges, endospore thin, hyaline; 1, or occasionally 2 or 3, per host cell. Causes hard, blackish galls up to 0.5mm diam. on above-ground parts of *A. nemorosa*.

ANGELICA: Wild Angelica

UREDINALES

Two rusts are recorded on *A. sylvestris* but both of them are uncommon.

Puccinia angelicae (Schum.)Fuckel, 0, 1, 11, 111
Aecia uredinoid, dark cinnamon brown. Uredinia very small, brown; spores echinulate, 25–40 × 18–28. Telia blackish brown; spores 1-septate, 30–45 × 18–24.

Puccinia bistortae DC., 0,1
Aecia yellow or orange. Alternate hosts *Polygonum bistorta* and *P. viviparum*.

OTHER FUNGI

Cercosporidium depressum (Berk. & Br.)Deighton (Fig. 1349)
On angular yellow to brown spots on leaves of *A. sylvestris*, June–Oct., common; the whole leaf eventually turns yellow. Conidiophores in dense fascicles, emerging through stomata on lower surface, olivaceous, 20–70 × 4–7. Conidia pale olive, mostly 1-septate, 20–75 × 7–11.

Diaporthopsis angelicae (Berk.)Wehmeyer
On dead stems of *A. sylvestris* and sometimes other Umbelliferae, at first inconspicuous but later causing blackening; decorticated parts often tinged violet. Perithecia up to 0.4mm wide, deeply embedded, with long, narrow necks; ascospores hyaline, 9–15 × 3–4. The *Phomopsis* state with hyaline conidia 7–8 × 2–2.5 is common.

Erysiphe heraclei, powdery mildew, described on *Heracleum*, is common also on *Angelica*.

Heterosphaeria patella (Tode)Grev. (Fig. 1350)
Apothecia erumpent, often becoming superficial, dark brown to black with margins conspicuously inrolled. Ascospores hyaline, 12–16 × 4–5. Often accompanied or preceded by its *Heteropatella* state which has 2- or 3-septate, appendaged conidia, measuring mostly 30–37 × 4. Very common on dead stems of *A. sylvestris* and other Umbelliferae, Apr.–Sept.; late summer and autumn collections usually bear both states.

Leptosphaeria libanotis (Fuckel)Niessl (Fig. 1351)
Pseudothecia large, almost black, at first immersed but often becoming superficial. Ascospores 3-septate, pale golden brown, 17–22 × 7–8. On dead stems of *A. sylvestris*, May–Sept.

Plasmopara nivea, downy mildew, described on *Aegopodium*, causes yellow blotching on upper surface of living leaves of *A. sylvestris*.

Pyrenopeziza plicata Rehm (Fig. 1352)
Apothecia erumpent, up to 0.7 × 0.5mm, almost black when dry. Ascospores hyaline, 8–12 × 2. On dead stems of *A. sylvestris*, especially around the nodes, May–Nov.

Ramularia archangelicae Lindroth
Colonies mostly epiphyllous, white. Conidiophores hyaline, 15–25 × 3. Conidia hyaline, 1- to 2-septate, 22–32 × 2–3. On leaves of *A. sylvestris*; spots whitish, limited by the veins and surrounded by brown zones.

Symphyosirinia angelicae E.A.Ellis (Fig. 1353)
Apothecia very pale brown, sometimes with slightly pinkish stipes when fresh, 2–4mm diam., up to 30mm high. Ascospores hyaline, mostly 12–17 × 3.5–4.5. Preceded and sometimes accompanied by white or rosy synnemata of its *Symphyosira* state, with hyaline, cylindrical, 1- to 3-septate conidia mostly 25–40 × 5–6. On fallen, 1-year-old mericarps of *A. sylvestris* buried under marsh vegetation, Oct.–Dec.

Also found on dead stems are many plurivorous fungi such as *Acrospermum compressum*, *Crocicreas cyathoideum*, *Dasyscyphus sulphureus*, *D. virgineus* and *Hymenoscyphus herbarum*.

ANTHRISCUS: Bur Chervil, Cow Parsley or Keck

Puccinia chaerophylli Purton, 0, 1, 11, 111 (Fig. 1354)
Aecia yellow, on leaves and petioles, May–June. Uredinia hypophyllous, small, cinnamon, spores 20–32 × 20–24. Telia found Aug.–Oct., dark brown; spores 1-septate, with reticulate walls, 25–36 × 20–25. Not uncommon in Suffolk on *A. sylvestris*.

Erysiphe heraclei, described on *Heracleum*, is common also on *Anthriscus*.

Plasmopara nivea and *Protomyces macrosporus*, described on *Aegopodium*, are found occasionally on *A. sylvestris*.

Ramularia anthrisci Höhnel (Fig. 1355)
Colonies white, mostly on lower leaf surface. Conidiophores in fascicles of 5 to 20, 20–30 × 2–3; conidia 18–60 × 2.5–3.5. Causes rather indefinite, often angular, brownish, discoloured spots on leaves of *A. sylvestris*.

Also found on dead stems are many plurivorous hyphomycetes such as *Alternaria ramulosa, Dendryphiella vinosa, Dendryphion comosum, Fusariella hughesii, Pseudospiropes rousselianus, Stachybotrys cylindrospora* and *Torula herbarum*; discomycetes including *Crocicreas cyathoideum, Dasyscyphus grevillei, D. mollissimus, Hymenoscyphus scutula, Unguicularia millepunctata* and *Urceolella crispula*; and the little cup-shaped basidiomycete *Calyptella capula*. Fuller coverage is given under *Heracleum*, which is treated as the standard umbellifer and under the general heading 'herbaceous plants'.

ANTHYLLIS: Kidney-Vetch or Ladies' Fingers

Uromyces anthyllidis Schröter, 11, 111
Sori on both leaf surfaces, scattered or sometimes in circles. Uredinia very small, cinnamon-brown, spores 18–25 diam., with sparsely echinulate walls up to 4 thick. Telia few, dark brown, spores 16–22 × 15–20, with scattered warts. Not common; found also on other Leguminosae.

Cercospora radiata Fuckel (Fig. 1356)
Colonies amphigenous; spots whitish or pale brown with dark brown borders. Conidiophores pale olivaceous brown, 30–100 × 4–5. Conidia hyaline, 4- to 8-septate, 45–90 × 4–5.

ANTIRRHINUM: Snapdragon, Weasel's Snout

Puccinia antirrhini Diet. & Holw., 11, 111
Uredinia common, mostly on lower leaf surface, reddish brown; spores echinulate, 16–24 × 20–30. Telia scarce, dark brown; spores 1-septate, 35–55 × 18–25. On *A. majus*.

Heteropatella antirrhini Buddin & Wakefield (Fig. 1357)
Acervuli up to 0.15mm diam. on leaves of *A. majus* and *A. orontium*, causing spots which later fall out giving a shot-hole effect; found also on overwintered stems. Conidia hyaline, shortly rostrate, 1- to 3-septate, 20–40 × 3.5–4.5, often with a basal appendage 3–4 long. On overwintered stems the acervuli are typical of the genus but on leaf spots they are much simpler and more open.

Myrothecium roridum, described on *Viola*, is found also not uncommonly on leaf spots and also causes shot-hole.

Peronospora antirrhini Schröter
Downy mildew, found on both *A. majus* and *A. orontium*, has sporangia 20–30 × 14–17.

FOOL APIUM: Celery, Fool's Watercress

Puccinia apii Desm., 0, 1, 11, 111
Celery rust is economically of little importance. Sori mostly on lower surface of leaves of *A. graveolens*; scarce in Britain. Uredinia cinnamon brown; spores shortly echinulate, 25–38 × 20–26. Telia very dark brown; spores smooth, 1-septate, 30–50 × 18–25.

Entyloma helosciadii Magnus
Smut sori in discoloured spots on leaves of *A. nodiflorum*, June–Oct. Spores pale to dark yellow, thin-walled, 5–12 diam.

Acremonium apii (M.A.Smith & Ramsey)W.Gams (Fig. 1358)
The cause of brown spot disease on the inner side of stems and leaves of *A. graveolens*. Conidia often hanging together in bundles at the ends of phialides, hyaline, 5–16 × 1–3; greyish-brown chlamydospores 8–15 × 5–9.

Burenia inundata (Dangeard)Reddy & Kramer (Fig. 1359)
Forms rather inconspicuous brownish blisters up to 1mm diam. on leaves of *A. nodiflorum*, July–Oct., common. Embedded chlamydospores 60–70 diam., with walls 2–3 thick, germinating almost immediately to form thin-walled asci containing numerous spores, 3.5–4.5 × 3.

Cercospora apii Fresen. (Fig. 1360)
On living leaves of *A. graveolens*, causing pale brown spots about 1cm diam., with raised borders. Conidiophores in groups of up to 30 emerging through stomata, olivaceous brown, mostly 30–70 × 4–9; conidia hyaline, 9- to 17-septate, 60–200 × 3.5–5.

Phoma apiicola Kleb.
Pycnidia immersed or erumpent, blackish brown, up to 0.25mm diam., on stems and petioles of *A. graveolens*. Spores hyaline, 3–4 × 1–2.

Septoria apiicola Speg.
Pycnidia immersed, brown to black, up to 0.2mm diam. Conidia with 1 to 5, mostly 3, septa, hyaline, 20–55 × 2–2.5. Causes lesions up to 6mm diam. on leaves of *A. graveolens*. Economically important as a disease of cultivated celery.

AQUILEGIA: Columbine

Erysiphe aquilegiae DC.
Powdery mildew. Conidia fairly plentiful. Appendages on cleistothecia straight, brown at base. Asci 3 to 8, each with 2 to 5 spores.

Haplobasidion thalictri, described on *Thalictrum*, sometimes causes buff, purple-bordered spots on leaves of *Aquilegia*.

ARABIS: Hairy Rock Cress, Tower Mustard

Albugo candida, described on *Capsella*, has been recorded on *Arabis*.

ARCTIUM: Burdock

Puccinia calcitrapae, described on *Centaurea*, is rarely found on *Arctium* but a race of it has been recorded occasionally on *A. lappa*, *A. minus* and *A. pubens*.

Arnium apiculatum (Griffiths)Lundq. (Fig. 1361)
Perithecia superficial, brown, setose, 0.5–1 × 0.2–0.5mm. Ascospores 27–40 × 18–26, dark brown with a pale septum when mature, and a conical, hyaline appendage at each end. On dead stems of *A. minus*, Feb.

Diaporthe arctii (Lasch)Nitschke (Fig. 1362)
Especially common on dead stems of *A. lappa* and *A. minus* but also on many other herbaceous plants, July–Nov. Extensive areas of stems very dark grey, with numerous black ostioles protruding. Perithecia immersed, up to 0.5mm diam. Ascospores hyaline, 1-septate, mostly 12–15 × 3–4. The *Phomopsis* state, with A-conidia fusiform, biguttulate, 7–9 × 2.5–3, and B-conidia filiform, curved, 16–22 × 0.5–1, is found earlier on the previous year's dead stems.

Erysiphe depressa (Wallr.)Schlecht.
Powdery mildew. On leaves and stems of *A. lappa* and *A. minus*. Conidia abundant, broadly ellipsoidal, often constricted near ends. Cleistothecia with many appendages. Asci usually 8 to 12, each with 2 spores.

Pyrenopeziza arctii (Phill. ex Bucknall)Nannf. (Fig. 1363)
Apothecia erumpent through the periderm, up to 1mm diam., black with pale rims. Ascospores hyaline, mostly 3-septate and 30–35 × 3 (range 25–45 × 2–3). On overwintered dead stems of *A. lappa* and *A. minus*, Mar.–May.

Pyrenopeziza depressuloides (Gremmen)Gremmen
Apothecia 0.7–0.8mm diam., black, with bluish, white-rimmed discs. Olivaceous brown, clavate cells at margin about 40 × 6. Ascospores hyaline, 7.5–11.5 × 3. On old dead stems, Sept.

Pyrenopeziza inornata Graddon (Fig. 1364)
Apothecia up to 1.5mm diam., blackish brown, with dark grey discs. Ascospores hyaline, 1- to 3-septate, 15–25 × 2.5–3. On dead stems, Apr.–Nov.

Ramularia filaris Fresen. var. *lappae* Bres. (Fig. 1365)
Conidiophores in fascicles of 10 to 30 on both leaf surfaces, hyaline, 15–80 × 2–3.5. Conidia hyaline, 0- or 1-septate, 7–20 × 2–3.5. On living leaves of *A. lappa*, causing round or angular, brown spots.

The thick dead stems of *Arctium* serve as a suitable substrate for many microfungi in addition to the host-limited species described above. These include plurivorous hyphomycetes such as *Cylindrotrichum oligospermum*, *Dactylosporium macropus* and *Volutella ciliata*; discomycetes including *Calycellina chlorinella*, *Crocicreas cyathoideum*, four species of *Dasyscyphus*, *D. clandestinus*, *D. grevillei*, *D. mollissimus* and *D. sulphureus*, *Hymenoscyphus herbarum* and *Pezizella discreta*; and other ascomycetes such as *Lophiostoma vagabundum* and *Ophiobolus acuminatus*.

ARENARIA: Thyme-leaved Sandwort

Puccinia arenariae (Schum.)Wint., 111 (Fig. 1366)
Telia on lower surface of leaves, rather pale brown, often appearing grey and powdery due to the production *in situ* of large numbers of basidiospores from the teliospores, which are 1-septate, 35–60 × 15–20. On *A. serpyllifolia* and many other Caryophyllaceae.

Peronospora campestris Gäum.
Downy mildew. On yellowed leaves of *A. serpyllifolia*. Sporangia 18–20 × 15–17.

ARMERIA: Sea Pink, Thrift

Uromyces armeriae Kickx, 0, 1, 11, 111
Aecia Mar.–June, followed by other stages. Uredinia cinnamon brown; spores minutely verruculose, 23–31 × 20–27. Telia dark brown; spores 25–35 × 20–30. Common on *A. maritima* and cultivated species.

Septoria armeriae Allesch.
On leaves of the cultivar 'Bees Ruby' it causes brown, circular spots 5–10mm diam., with purplish borders, Aug. Pycnidia mostly epiphyllous, 50–100 diam., substomatal. Conidia yellowish in mass, mostly 1-septate, 10–25 × 1.5.

ARMORACIA: Horse-radish

Ascochyta armoraciae Fuckel
Causes small, buff, dark-bordered spots on leaves, July–Oct. Pycnidia up to 0.1mm diam. Conidia mostly 1-septate, hyaline, 13–17 × 4.

Cercospora armoraciae Sacc. (Fig. 1367)
Causes round or angular, pale brown spots which sometimes coalesce and cover large areas, Sept.–Oct. Conidiophores fasciculate, pale brown, 30–85 × 5–7. Conidia hyaline, 7- to 18-septate, 50–140 × 3–5.

Ramularia armoraciae Fuckel (Fig. 1368)
Causes pale brown, sometimes brown-bordered spots which become grey or papery white, and may fall out leaving shot-holes, July–Oct. Conidiophores in small groups on lower surface of leaves, hyaline, 20–40 × 2.5–4. Conidia hyaline, 0- or 1-septate, 8–30 × 3–4.

Thielaviopsis basicola, described on *Daucus*, has been found on horse-radish, and a number of plurivorous fungi such as *Dendryphion comosum*, *Nectria arenula* and *Torula herbarum* occur also.

ARTEMISIA: Mugwort, Sea Wormwood, Wormwood

Puccinia tanaceti DC., 11, 111
Found most frequently on *A. maritima*, July–Sept., only occasionally on *A. absinthium* and *A. vulgaris*. Uredinia mostly hypophyllous, cinnamon brown, spores minutely echinulate, 25–28 × 20–22. Telia sometimes on stems, blackish brown, spores slightly verrucose at apex, otherwise smooth, 1-septate, 40–60 × 17–26.

Camarosporium aequivocum (Pass.)Sacc. (Fig. 1369)
Pycnidia dark brown, up to 0.3mm diam., sometimes formed in rows. Conidia mid pale brown, muriform, 10–15 × 7–10. On dead stems of *A. maritima* and *A. vulgaris*, Apr.–May.

Cercospora ferruginea Fuckel (Fig. 1370)
Colonies on lower surface of leaves, effuse, reddish brown, velvety, upper surface above colonies often bright yellow, Sept.–Oct. Conidiophores branched, brown, up to 350 × 4–6. Conidia subhyaline or pale brown, 2- to 5-septate, 30–90 × 5–8. On *A. absinthium* and *A. vulgaris*.

Diaporthe arctii (Lasch)Nitschke var *artemisiae* Rehm (Fig. 1371)
Perithecia immersed, 0.2–0.3mm diam., with necks 0.6–0.7 × 0.1–0.15mm. Stem surface blackened just around ostioles, no black stromatic line. Ascospores 1-septate, hyaline, 12–14 × 3–4. On dead stems of *A. vulgaris*, Mar.–July.

Erysiphe artemisiae Grev.
Powdery mildew. Cleistothecia formed abundantly on lower surface of yellowed leaves of *A. absinthium* and *A. vulgaris*, Sept.–Oct. Asci 10 to 20, each with 2 spores.

Mycosphaerella osborniae D. Hawksw. & Sivan.
Pseudothecia up to 0.065mm diam.; ascospores hyaline, 1-septate, 10–12 × 2–3. On dead stems of *A. vulgaris*, Aug.

Phomopsis oblita Sacc.
Pycnidia in large groups, dark brown, up to 0.5mm diam. Conidia fusiform,

hyaline, biguttulate, 8–12 × 2–3. On dead stems of *A. vulgaris*, May.

Pirottaea nigro-striata, described on *Heracleum*, has been found also on dead stems of *A. vulgaris*.

Pyrenopeziza artemisiae (Lasch)Rehm (Fig. 1372)
Apothecia erumpent, about 1mm diam., disc dark grey, excipulum paler. Ascospores hyaline, 9–14 × 2. On dead stems of *A. vulgaris*, Apr.–May.

ARUM: Cuckoo Pint or Lords-and-Ladies

Puccinia sessilis Schroet., 0,1
Aecia uncommon on *Arum*. Alternate host *Phalaris arundinacea*.

Ramularia ari Fautr.
Colonies epiphyllous. Conidiophores subhyaline, 20–50 × 4–6; conidia cylindrical, hyaline, 20–22 × 4. On thin papery spots with grey centres on leaves of *A. maculatum*, Apr.

ASPARAGUS: Asparagus

Puccinia asparagi DC., 0, 1, 11, 111
Aecia cause lesions on stems in May; other states also on stems found Sept.–Dec. Uredinia cinnamon brown; spores minutely echinulate, 20–30 × 18–25. Telia blackish brown; spores 1-septate, 35–45 × 18–26. Uncommon in Britain.

Botrytis tulipae, described on *Tulipa*, has been found also on *Asparagus*.

Zopfia rhizophila Rabenh. (Fig. 1373)
Cleistocarps superficial or partly immersed, black, 0.25mm diam., with walls made up of plates as in *Cephalotheca*. Ascospores dark brown, often rough-walled, mostly 60–85 × 30–45, constricted at the septum. On decaying roots.

ASTER: Michaelmas Daisy, Sea Aster

Puccinia cnici-oleracei Pers. ex Desm., 111, described on *Achillea*, has been recorded on *Aster tripolium*.

Puccinia dioicae Magn. var *extensicola* (Plowr.)Henderson, 0,1
Pycnia and aecia on *A. tripolium*, alternate host *Carex extensa*, uncommon.

Erysiphe cichoracearum DC. (Fig. 1374)
Downy mildew, common on *A. tripolium*, *A. novi-belgii* and many other

Compositae. Conidia abundant, hyaline, 25–35 × 14–18. Cleistothecia up to 0.15mm diam., with long, hyaline or brown appendages. Asci 4 to 20, each with usually 2 but occasionally 4 spores.

Phialophora asteris (Dowson)Burge & Isaac
A cause of wilt of cultivated Michaelmas daisy. Leaves become yellow with brown necrotic areas making them appear mottled; finally they wilt and become shrivelled. Cultural methods are usually needed to study the fungus but it can be grown out from cut pieces of shoot which have discoloured vascular tissue; ropes of brown hyphae bear phialides which produce large numbers of hyaline conidia 2–8 × 1–2.

Pleospora media Niessl
Pseudothecia immersed, up to 0.3mm diam., somewhat flattened, often with a few pale hyphae on the surface. Ascospores golden brown, mostly with 5 transverse septa, constricted at 3 of them and with a longitudinal septum in all central and often also the end cells, 20–25 × 10–12. On dead stems of *A. tripolium*, July.

Ploettnera solidaginis(de Not.)Hein (Fig. 1375)
Apothecia at first covered by dark brown areas, which split into 2 to 4 lobes to expose ochraceous discs about 0.5mm diam. Asci mostly 4-spored; spores hyaline 11–15 × 6–7. On dead stems of *A. tripolium*, Sept.–Nov.

Ramularia asteris (Phill. & Plowr.)Bubák (Fig. 1376)
Colonies on both leaf surfaces, white, pale pink or pale orange. Conidiophores often in fascicles of 50 or more, 30–60 × 4–6. Conidia hyaline, 0- to 3-septate, 16–50 × 4.5–6. On fading leaves of *A. tripolium*.

ASTRAGALUS: Milk-vetch, Purple Milk-vetch

Uromyces pisi-sativi (Pers.)Liro, 11, 111, described on *Pisum*, uncommon on *A. danicus*. Aecial host *Euphorbia cyparissias*.

Microsphaera astragali (DC.)Trevisan
Powdery mildew. Only a few of the cleistothecial appendages once or twice dichotomously branched, not recurved. Asci 6 to 10, each with 3 to 5 spores which measure 20–27 × 12–13. On *A. glycyphyllos*.

ATRIPLEX: Oraches

Ascochyta chenopodii, described on *Chenopodium*, causes ochraceous, dark-bordered spots up to 1cm diam. on living leaves of *A. hastata*, June–Aug.

Cercospora chenopodii, described on *Chenopodium*, causes greenish-brown spots on leaves of *A. hastata*.

Peronospora farinosa, described on *Chenopodium*, has been found on *A. hastata*.

Peronospora minor (Casp.)Gäum.
Downy mildew on *A. patula*; sporangia average 20–26 × 16–22.

Pleospora calvescens (Fr.)Tul. (Fig. 1377)
Perithecia in blackened areas, up to 0.2mm diam., with stiff, dark brown hyphae around the sides and base. Ascospores mid pale golden brown, with 3 transverse and 1 or 2 longitudinal septa, 15–21 × 6–8. On dead stems of *A. littoralis* and other Chenopodiaceae, Jan.–Mar. Pycnidial state has pale yellow, 1-septate conidia 10–13 × 4–6.

Stagonospora atriplicis (Westend.)Lind (Fig. 1378)
On living leaves of *A. hastata*, July. Causes pale spots with yellow borders. Pycnidia 0.15mm diam.; conidia hyaline, 1- to 3-septate, 20–30 × 4–5.

ATROPA: Deadly Nightshade

Diaporthe chailletii Nitschke
Perithecia up to 0.5mm diam., with narrow, elongated necks, immersed in broadly effused blackened areas of stems of *A. belladonna*; ascospores hyaline, 1-septate, 11–12 × 2–2.5.

BALLOTA: Black Horehound

Ophiobolus ulnasporus (Cooke)Sacc. (Fig. 1379)
Pseudothecia up to 0.45mm diam., short-necked, with slight pomiform collapse when dry. Curiously shaped, septate ascospores, yellowish in mass, twisted around one another in the asci, 90–120 × 3–5. On dead stems, July–Sept.

In spite of the fact that one often finds quite large, almost pure stands of *B. nigra*, few fungi seem to grow on either living or dead plants. *Erysiphe galeopsidis*, common to a number of Labiatae, is seen occasionally and we have found, from time to time, plurivorous species such as *Hymenoscyphus herbarum*, *Leptosphaeria doliolum*, *Periconia byssoides*, *Pseudolachnea hispidula* and *Pyrenopeziza adenostylidis* on dead stems.

BELLIS: Daisy

Puccinia lagenophorae Cooke, 1, 111, described on *Senecio*, is rarely seen on *B. perennis*.

Puccinia obscura Schröter, 0, 1
Spermogonia and aecia formed on *B. perennis*, Sept.–Dec., inconspicuous.

Aeciospores minutely verruculose, pale yellow, 15–22 diam. Alternate hosts *Luzula* spp.

Entyloma calendulae (Oudem.)de Bary f. *bellidis* (Kreiger)Ainsworth & Sampson
Smut sori embedded in leaves. Spores very pale yellow, smooth, 9–14 diam.

BERULA: Narrow-leaved Water Parsnip

Uromyces lineolatus (Desm.)Schröter, 0, 1
Aecia hypophyllous. Alternate host *Scirpus maritimus*.

Protomyces macrosporus, described on *Aegopodium*, has been recorded on *B. erecta*.

BETA: Beet, Mangold, Sea Beet, Sugar Beet

Uromyces betae Kickx, 0, 1, 11, 111
Sori on both leaf surfaces. Uredinia cinnamon, spores shortly echinulate, 22–30 × 16–26. Telia dark brown, spores smooth, 24–34 × 19–25. On sugar beet, beet, mangold, etc.

Aphanomyces cochlioides Drechsler
On beet and sugar beet seedlings, especially those grown on acid soils with high water-holding capacity. Hypocotyls become black and thin and cotyledons necrotic at base. Seedlings are either killed or remain stunted. Oospores in diseased tissue hyaline to yellow, 18–24 diam.

Cercospora beticola Sacc. (Fig. 1380)
Colonies amphigenous. Conidiophores fasciculate, pale brown, 40–120 × 4–6. Conidia hyaline, with up to 16 septa, 60–200 × 3.5–4.5. On living and wilting leaves of *B. vulgaris*, causes very pale brown, whitish or grey spots, each surrounded by a narrow, reddish-purple or brown border.

Cyathicula littoralis Graddon (Fig. 1381)
Apothecia up to 1.5mm diam., buff, frosted with white crystals. Ascospores hyaline, 8–13 × 2–3. On dead stems of *B. vulgaris* spp. *maritima*, Apr.–July.

Erysiphe betae (Vañha)Weltzien
Powdery mildew. Conidia plentiful, cylindrical or ellipsoidal, 30–50 × 15–20. Cleistothecia small, about 0.1mm diam., appendages numerous, brown at base. Asci 3 to 8, each with 3 to 5 spores. Ascospores 18–30 × 12–20. Powdery mildew is sometimes an important disease on sugar beet.

Gabarnaudia betae (Delacr.)Samson & W.Gams (Fig. 1382)
Colonies white, floccose. Conidiophores repeatedly branched, hyaline. Conidia in chains, sliming down with age, smooth, hyaline, 8–13 × 3.5–4.5. On dry stems of wild beet and sugar beet.

Peronospora schachtii Fuckel
Downy mildew on *B. vulgaris* ssp. *vulgaris* and ssp. *maritima*. Sporangia 20–33 × 12–26.

Pleospora bjoerlingii Byford (Fig. 1383)
Pseudothecia up to 0.4mm diam., often accompanied by smaller *Phoma* pycnidia. Ascospores muriform, pale to mid sepia, 20–25 × 7–11. Conidia hyaline, 5–8 × 3–4. On dead stems of *B. vulgaris* and *B. trigyna*. Causes black leg disease of sugar beet and mangold.

Ramularia beticola Fautrey & Lamb. (Fig. 1384)
Colonies on both leaf surfaces but more plentiful on the lower surface. Conidiophores hyaline, 30–60 × 2–4.5. Conidia hyaline, 0- or 1-septate, 16–22 × 3–5. On leaves of *B. vulgaris*, causes grey or pale brown spots 1–7mm diam., often with dark borders.

Septoria betae Westend.
Pycnidia crowded on elongated pale spots, very small (less than 0.1mm). Conidia hyaline, mostly straight, 13–18 × 1–1.5. On dead stems of *B. vulgaris*, Sept.

BETONICA: Betony

Puccinia betonicae DC., 111
Telia mostly hypophyllous, reddish brown, small, on pale irregular spots, spores 1-septate, 28–45 × 16–24. On *B. officinalis*, fairly common.

BLACKSTONIA: Yellow-wort

Peronospora chlorae de Bary
Downy mildew. Sporangia 17–20 × 12–14. Oospores pale brown, reticulate, 26–30 diam.

BRASSICA: Black Mustard, Broccoli, Brussels Sprouts, Cabbage, Cauliflower, Charlock, Colza, Kale, Kohl-rabi, Rape, Swede, Turnip

Albugo candida, white blister, described on *Capsella*, occurs occasionally on young cabbage plants.

Alternaria brassicae (Berk.)Sacc. (Fig. 1385)
Colonies amphigenous, pale olive. Conidiophores greyish olive, up to 170 × 6–11. Conidia solitary or occasionally in short chains, pale olive, with 6 to 19 transverse and several longitudinal septa, 70–350 × 20–30, beak $\frac{1}{3}$ to $\frac{1}{2}$ the length of the conidium and 5–9 thick. On leaves of *B. napa*, *B. oleracea*, *B. rapa*

and ssp. *campestris*, forming zonate, circular, light brown to greyish or dark brown spots 0.5–12mm diam.; sometimes causes black spots on cauliflower heads.

Alternaria brassicicola (Schw.)Wiltshire (Fig. 1386)
Colonies amphigenous, dark olivaceous or dark blackish brown, velvety. Conidiophores olivaceous brown, up to 70 × 5–8. Conidia mostly in chains of up to 20 or more, sometimes branched, pale to dark olivaceous brown, with 1 to 11 transverse and a few longitudinal septa, 18–130 × 8–20. On leaves of *B. oleracea* and other Cruciferae forming dark brown to almost black, circular, zonate spots 1–10mm diam. Common.

Arnium olerum (Fr.)Lundqvist & Krug (Fig. 1387)
Pale grey, felted perithecia, 0.5–0.75mm diam., with black necks, lodged in cracks in old cabbage stalks. Ascospores brown, mostly 50–60 × 30, with hyaline appendages 20–50 × 4–5.

Dendryphion nanum, described on 'herbaceous plants', is found very commonly on old *Brassica* stalks at ground level.

Doratomyces purpureofuscus (Fr.)Morton & G. Smith (Fig. 1388)
Colonies at first greenish grey, becoming dark grey to dark blackish brown. Synnemata brown, up to 900 high with spherical or subspherical heads. Conidia pale to mid brown or olivaceous brown, 5–7 × 3.5–4.5. Most collections of this fungus have been made on old cabbage stalks, although it is found occasionally on other substrata.

Erysiphe cruciferarum Opiz ex Junell
Powdery mildew of swedes (*B. napus*) and turnips (*B. rapa* ssp. *campestris*). Colonies amphigenous, white, powdery. Usually found only in *Oidium* state; conidia 30–40 × 11–15, solitary or in short chains, cylindrical. Cleistothecia when formed dark brown, about 0.1mm diam., with unbranched appendages. Asci mostly 6 to 8, 2- to 7-spored; spores 19–22 × 10–13.

Gibberella cyanogena (Desm.)Sacc. (Fig. 1389)
Perithecia scattered or in small groups, about 0.25mm diam., black, warted; wall purple by transmitted light. Ascospores hyaline, 3-septate, mostly 20–30 × 5–7. Especially common on old cabbage stalks in autumn and spring, although it is found occasionally on other substrata.

Graphium putredinis (Corda)Hughes (Fig. 1390)
Synnemata olivaceous brown or reddish brown, up to 1mm high, 40 thick at the base. Conidia pale olivaceous brown, 5–11 × 2–4. Most collections of this have been made on old cabbage stalks.

Leptosphaeria maculans, described on *Alliaria*, is commonly found also on *B. napobrassica*, *B. oleracea* and *B. rapa*, where it causes disease variously known as black leg, canker or dry rot.

Mycosphaerella brassicicola (Duby)Johanson ex Oudem. (Fig. 1391)
Pseudothecia about 0.1mm diam., dark brown; ascospores hyaline, 1-septate, 17–23 × 3–5. Mainly on outer leaves of *B. napus* and *B. oleracea*, causing ring spot disease. Lesions circular, brown or grey, with black fruit bodies zonately arranged, each lesion surrounded by a water-soaked area and a yellow zone. Pseudothecia plentiful on fallen leaves. Fruit bodies on attached leaves mostly belong to the *Asteromella* state with hyaline conidia 3–5 × 1.

Nectria. Two species have been found on old dead stems of *B. oleracea* but we have seen no fresh collections of either. *N. brassicae* Ellis & Sacc. has blood red, smooth perithecia 0.15mm diam., which collapse laterally on drying; ascospores 1-septate, pale brown when old, 10–17 × 4–5. In *N. keithii* Berk. & Br. the yellowish-brown perithecia 0.25mm diam. are formed on a small stroma; ascospores 9–13 × 3–5.

Olpidium brassicae (Woron.)Dang.
A chytrid found in roots of various *Brassica* species, especially when these are grown in wet soils; generally of little importance.

Perisporium kunzei (Fuckel)Sacc. (Fig. 1392)
Cleistothecia dark blackish brown, 1–2mm diam., lodged in furrows in old dead cabbage stalks, Apr.–May. Ascospores pale brown, 3-septate, 20–25 × 6–8, constricted at septa and occasionally breaking up into 1-celled segments.

Peronospora parasitica, downy mildew, described on *Capsella*, occurs also on *Brassica*.

Plasmodiophora brassicae Woron.
A slime mould very common in roots of *Brassica* species; it causes the well-known and very aptly named club root or finger-and-toe disease.

Pseudocercosporella capsellae (Ellis & Ev.)Deighton (Fig. 1393)
Often referred to as *Cercosporella brassicae*. Causes angular, greyish-white, sometimes brown-bordered or zonate spots on leaves of *B. nigra*, *B. oleracea*, *B. rapa* and its ssp. *campestris*. Colonies amphigenous, whitish. Conidiophores hyaline, 5–15 × 2–4. Conidia hyaline, 3- to 5-septate, 30–90 × 2–3.

Pyrenopeziza brassicae Sutton & Rawlinson
The cause of light leaf spot of broccoli, cabbage, cauliflower, kale, swede and oilseed rape. Apothecia 1mm diam., dark grey, with hazel discs; protruding cells at edge up to 30 long. Ascospores 0- or 1-septate, 13–18 × 2.5–3, paraphyses branched. Most familiar in its *Cylindrosporium* state formed on the lower surface of pale leaf spots, the small, white acervuli arranged in concentric circles. Conidia cylindrical, rounded at apex, truncate at base, hyaline, 8–16 × 3–4.5, sometimes becoming 1-septate.

Pyxidiophora petchii (Breton & Faurel)Lundqvist
Perithecia superficial, elongated flask-shaped, pale ochraceous, base 30–80 diam., neck 80–200 × 10–25. Ascospores hyaline, 1-septate, 45–54 × 5–7.

On decaying cabbage stalks, June.

BRYONIA: White Bryony

Didymella bryoniae (Auersw.)Rehm (Fig. 1394)
Pseudothecia up to 0.2mm diam., black, erumpent. Ascospores hyaline, 1-septate, 15–21 × 5–8. On dead stems.

BUNIUM

Puccinia bulbocastani Fuckel, 0, 1, 111
Aecia mostly on petioles and stems. Telia on both leaf surfaces, black; spores 26–40 × 15–24, 1-septate. On *B. bulbocastanum*, rare.

BUPLEURUM: Hare's Ear, Slender Hare's Ear

Puccinia bupleuri Rud., 11, 111
Uredinia cinnamon; spores echinulate, 20–24 × 18–22. Telia blackish brown; spores 1-septate, 26–44 × 17–30. On both leaf surfaces and on stems of *B. tenuissimum*, uncommon.

CACTACEAE: Cacti

Drechslera cactivora (Petrak)M.B.Ellis (Fig. 1395)
Colonies black, velvety, sometimes causing plants to die back from the top. Conidiophores golden brown, up to 250 × 4–6. Conidia golden brown, with 2 to 4 pseudosepta, 30–65 × 9–12.

Phomopsis cacti Grove
Conidiomata immersed, up to 0.3mm diam. Conidia biguttulate, hyaline, 6–8 × 1.5–2. On dead stems.

CALAMINTHA: Calamint

Puccinia menthae, described on *Mentha*, has been recorded on *Calamintha*.

CALENDULA: Pot Marigold

Coleosporium tussilaginis, described on *Tussilago*, occurs also on *C. officinalis*.

Entyloma calendulae (Oudem.)de Bary
Smut sori embedded in pallid, then brown, leaf spots on *C. officinalis*. Spores pale yellow, globose or polygonal, smooth, 9–14 diam.

Sphaerotheca xanthii (Castagne)Junell (Fig. 1396)
Powdery mildew. On living and fading leaves of *C. officinalis*. Conidia abundant, 25–30 × 17–19, hyaline. Cleistothecia about 0.1mm diam. Ascospores hyaline, 20–24 × 14–16.

CALLUNA: Heather or Ling

Godronia callunigera (P.Karsten)P. Karsten
Apothecia erumpent, pyriform or fig-shaped, 1–1.5mm wide, reddish brown to black, striate. Ascospores hyaline, 3-septate, tapered at ends, 30–60 × 2–4. *Topospora* pycnidial state found Apr.–May and accompanying apothecia in Aug.–Sept.; conidia sickle-shaped, hyaline, 1- to 3-septate, 15–30 × 2–3.5. On dead twigs.

Godronia cassandrae Peck f. *callunae* Groves (Fig. 1397)
Differs from *G. callunigera* in having longer ascospores, 50–85 × 2–3.5 and more or less straight, 1-septate conidia 8–16 × 1.5–3.5.

Keissleriella subalpina (Rehm)Bose (Fig. 1398)
Pseudothecia mostly superficial, black, 0.1–0.125mm diam., conical. Ascospores hyaline, 12–14 × 4.5–5, divided by a septum into a short, wide upper cell and a longer, narrower lower cell. On old stems.

Pseudophacidium callunae (P. Karsten)P. Karsten (Fig. 1399)
Apothecia erumpent, black, splitting open to show a yellow disc, 2mm diam. Ascospores cylindrical with rounded ends, sometimes slightly curved, hyaline, 9–14 × 4–5. On dead stems.

Tapesia cinerella Rehm
Apothecia about 1mm diam., pale grey, gregarious, seated on a dark blackish-brown subiculum. Ascospores straight or slightly curved, hyaline, 10–12 × 3.5. On dead stems.

Tapesia melaleucoides Rehm (Fig. 1400)
Apothecia with white to very pale yellow discs and dark grey outer walls, seated on a dark brown subiculum. Ascospores hyaline, slightly curved, 7–10 × 2.5–3. On dead stems, May–Aug.

Trichobelonium obscurum (Rehm)Rehm (Fig. 1401)
Apothecia up to 2mm diam., externally dark brown, velvety, with cream or pale grey discs, seated on a thin or dense brown subiculum. Ascospores hyaline, 3- to 7-septate, 30–45 × 2.5–3. On old stems, Apr.–Aug.

Plurivorous species commonly found on *Calluna* include the discomycetes

Dasyscyphus virgineus and *Mollisia cinerea*, and hyphomycetes *Oidiodendron tenuissimum* and *Phaeostalagmus cyclosporus*.

CALTHA: Kingcup or Marsh Marigold

Puccinia calthae Link, 0, 1, 11, 111
Telia amphigenous, brown; spores smooth-walled, 30–42 × 14–22.

Puccinia calthicola Schröter, 0, 1, 11, 111 (Fig. 1402)
Telia amphigenous but mostly hypophyllous, dark brown; spores mostly 40–60 × 24–34, with partially verruculose walls.

Botryotinia calthae Hennebert & Elliott (Fig. 1403)
Apothecia 2–4mm diam., pale brown, with stipes up to 1.5mm long, dark hairy at base, arising from dead overwintered petioles, Mar.–Apr. Ascospores ellipsoid or navicular, hyaline, 12–16.5 × 5–8.5. Paraphyses branched at base, 4–4.5 wide at apex. *Botrytis* state with conidia 8–16 × 5–9 (mostly 12 × 7).

Crocicreas starbaeckii, described on *Ranunculus*, is found also on old petioles of *Caltha*.

Erysiphe aquilegiae, described on *Aquilegia*, is found also on *Caltha*.

Pseudopeziza calthae (Phill.)Massee
Apothecia up to 0.25mm diam., erumpent, brown or greyish. Ascospores hyaline, 14–20 × 4–6, sometimes becoming 1-septate. Paraphyses branched near the apex. On brown spots on decaying leaves, Sept.–Nov.

Ramularia calthae (Erikss.)Lindr. (Fig. 1404)
Colonies white, on both surfaces of leaves but more abundant on the upper surface. Conidiophores 20–45 × 2–3. Conidia hyaline, 0- or 1-septate, 10–30 × 2–2.5. Causes white or pale brown spots, 1–6mm diam., sometimes with brown borders.

Verpatinia calthicola Whetzel (Fig. 1405)
Apothecia arise from black sclerotia with pointed ends in dead overwintered petioles, May–June. They are campanulate or turbinate, clay coloured or yellowish brown, 2–3 × 1–2mm, with pale stipes up to 15 × 0.5mm. Ascospores hyaline, somewhat flattened on one side, 6–10 × 2–3.

CALYSTEGIA: Bellbine, Large Bindweed, Sea Bindweed

Puccinia convolvuli Cast., 0, 1, 11, 111
Only one British record on *C. sepium*. Uredinia minute, brown; spores echinulate, 22–30 × 18–26. Telia dark brown; spores 40–65 × 18–30.

Thecaphora seminis-convolvuli (Desm.)Liro (Fig. 1406)
Smut sori in seeds of *C. sepium* and *C. soldanella*, Aug.–Oct. Spore mass reddish brown, granular, with spore balls 10–30 diam.; spores mostly 12–16 diam. Prior to the formation of sori, anthers of infected flowers become swollen, white or dirty yellowish, covered by oval, hyaline conidia, and the flowers often appear somewhat tattered, with torn petals.

Septoria convolvuli Desm.
Pycnidia immersed, 0.125mm diam., brown, with large ostioles. Conidia slightly curved, hyaline, guttulate or with up to 5 septa, 35–55 × 1–1.5. Causes large, round or irregular, reddish or greenish brown spots on fading leaves of *C. sepium*, July–Aug.

Stagonospora calystegiae (Westend.)Grove
Pycnidia few, immersed, on small, pale brown spots with thickened margins. Conidia slightly curved, cylindrical, rounded at the ends, hyaline, 3- to 5-septate, 25–45 × 4–5. On fading leaves of *C. sepium*, July–Sept.

CAMPANULA: Bats-in-the-Belfry, Clustered Bellflower, Creeping and Large Campanula, Harebell, Rampion

Coleosporium tussilaginis, 11, 111, described on *Tussilago*, has been recorded on *C. glomerata*, *C. latifolia*, *C. rapunculoides*, *C. rotundifolia* and *C. trachelium*.

Puccinia campanulae Berk., 111
Telia on stems, petioles and lower surface of leaves, reddish brown; spores 1-septate, 27–45 × 12–22. On *C. rotundifolia* and *C. rapunculus*, June–Aug., uncommon.

Didymella exigua (Niessl)Sacc (Fig 1407)
Pseudothecia black, immersed in dead stems of *C. latifolia*. Ascospores hyaline, 1-septate, 14–17 × 4.5–5.

Leptotrochila radians (Rob. in Desm.)P. Karsten
Apothecia 1mm diam., erumpent, arising from a dark blackish-brown, flat, branched stroma. Ascospores hyaline, 8–12 × 2.5–3. This uncommon fungus has been recorded in Britain on basal leaves of *C. rotundifolia*. In Europe it occurs with a *Sporonema* state on other species including *C. rapunculoides*, *C. rapunculus* and *C. trachelium*.

Ramularia macrospora Fresen. (Fig. 1408)
Colonies hypophyllous, white. Conidiophores 20–80 × 3–6. Conidia hyaline, 0- to 3-septate, 15–36 × 5–7. On leaves of *C. rapunculoides*, forming round or angular, brown spots 3–15mm diam.

CAPSELLA: Shepherd's Purse

Albugo candida (Pers.)O. Kuntze (Fig. 1409)
White blister or white rust of Cruciferae. Infection occurs mostly in young plants as white or whitish pustules, very variable in size, often causing distortion or gall-like swellings. All above-ground parts of plants are affected. Very common on *C. bursa-pastoris*, Oct.–Nov., but found also at other times of the year and on many other crucifers. Sporangiophores hyaline, 30–45 × 14–18. Sporangia in chains, hyaline, 11–22 diam. Oospores formed mostly in the stems, spherical, brown, warted, 35–55 diam.

Erysiphe cruciferarum, described on *Brassica*, is common also on *Capsella*.

Peronospora parasitica (Pers.)Fr. (Fig. 1410)
Downy mildew. White downy colonies on swollen stems, and on the lower surface of leaves which have corresponding yellow patches on their upper surface. Sporangia mostly 22–25 × 18–20, but sometimes up to 30 × 22. Common on Cruciferae, Jan.–Apr.

CARDAMINE: Hairy, Large and Wood Bittercress, Lady's Smock or Cuckoo-flower

Albugo candida, described on *Capsella*, is found also on *Cardamine*.

Peronospora dentariae Rabenh.
Downy mildew on *C. amara*, *C. flexuosa*, *C. hirsuta* and *C. pratensis*; sporangia 15–22 × 14–18.

Ramularia cardamines Syd. (Fig. 1411)
Colonies amphigenous, white. Conidiophores 30–50 × 3–4.5. Conidia hyaline, 0- to 3-septate, 12–35 × 3.5–4.5. On living leaves of *C. amara*, *C. flexuosa* and *C. pratensis*, causes greenish or pale brown spots 1–5mm diam., sometimes with yellow haloes.

CARDUUS: Musk Thistle, Slender Thistle, Welted Thistle

Puccinia calcitrapae, 0, 1, 11, 111, described on *Centaurea*, occurs on *Carduus crispus* and *C. nutans* but is uncommon.

Ophiobolus cirsii (P.Karsten)Sacc.
Pseudothecia scattered, immersed, black, 0.4–0.5mm diam., with erumpent necks which terminate in radiating hairs about 50 × 5. Ascospores parallel in the ascus, olivaceous, 130–170 × 4–5, somewhat curved towards the apex, with up to 20 septa, and 1 cell, just above the middle, slightly swollen. On dead stems of *C. crispus*.

Ramularia cardui Karst. (Fig. 1412)
Colonies white, amphigenous. Conidiophores often in large fascicles, 20–50 × 3–5. Conidia hyaline, 0- or 1-septate, 10–26 × 3–4. On living leaves of *C. crispus*, *C. nutans* and *C. tenuiflorus*, causing whitish or pale brown, often dark-bordered spots, 1–5mm diam.

CARLINA: Carline Thistle

Puccinia calcitrapae, described on *Centaurea*, has been recorded on *Carlina vulgaris*.

Veronaea carlinae M.B.Ellis (Fig. 1413)
Colonies effuse, mid to dark brown, woolly or velvety. Conidiophores up to 130 × 2–4. Conidia pale brown, 9–17 × 2.5–4. On dead stems, Sept.

CENTAUREA: Bluebottle or Cornflower, Hardheads or Knapweed, Greater Knapweed

UREDINALES

Puccinia calcitrapae DC., 0, 1, 11, 111
Common on *C. nigra*, uncommon on *C. scabiosa*. Aecia mostly epiphyllous, uredinoid. Uredinia mostly hypophyllous; spores echinulate, 23–30 × 17–28, with three equatorial pores. Telia mostly hypophyllous, dark brown; spores 1-septate, slightly verruculose, 25–50 × 16–26.

Puccinia cyani Pass., 0, 1, 11, 111
On *C. cyanus*, uncommon. Aecia uredinoid. Uredinia cinnamon brown; spores shortly echinulate, 23–30 × 20–24. Telia amphigenous, black; spores dark brown, minutely verruculose, 1-septate, 30–35 × 23–27.

Puccinia dioicae Magn. var. *arenariicola* (Plowr.)Henderson, 1
Aecia on *C. nigra*. Alternate host *Carex arenaria*. Rare.

Puccinia hieracii, 0, 1, 11, 111, described on *Hieracium*, occurs on *C. nigra* but is much less common than *P. calcitrapae*; its urediniospores have two supra-equatorial pores but otherwise the two rusts are rather similar.

OTHER FUNGI

Bremia lactucae, downy mildew described on *Lactuca*, is found occasionally on living leaves of *C. nigra*, Sept.–Oct.

Erysiphe cichoracearum, powdery mildew described on *Aster*, is common also on *C. nigra*.

Leptosphaeria centaureae E.Müller
Pseudothecia immersed, 0.25mm diam., with necks ending in brown hairs. Ascospores cylindrical, with the third cell swollen, 6-septate, 28–33 × 4.5–5, often with a small hyaline appendage at each end. On dead stems of *C. scabiosa*.

Leptosphaeria jaceae Holm (Fig. 1414)
Pseudothecia about 0.3mm diam., with necks terminating in radiating hairs. Ascospores olivaceous, 7- to 9-septate, 38–48 × 4.5–5, sometimes with a small appendage at each end. On dead stems of *C. jacea* and hybrids with *C. nigra*.

Pirottaea brevipila (Rob. ex Desm.)Boud. (Fig. 1415)
Apothecia erumpent, about 0.5mm diam., blackish brown, with numerous thick-walled, dark brown hairs arising from sides, discs grey. Ascospores hyaline, 1- to 3-septate, 24–32 × 3. On dead stems of *C. scabiosa*, June.

Pyrenopeziza adenostylidis (Rehm)Gremmen (Fig. 1416)
Apothecia erumpent, up to 1.5mm diam., dark brown, with greyish discs. Ascospores hyaline, 9–12 × 1.5. On dead stems of *C. nigra*, May–Nov.

Ramularia centaureae Lindr. (Fig. 1417)
Conidiophores mostly epiphyllous, 15–40 × 3–3.5. Conidia hyaline, 0- to 2-septate, 12–40 × 2.5–3.5. On leaves of *C. nigra* and *C. scabiosa*, causing greyish spots 2–5mm diam.

Fungi not specific to *Centaurea* found on dead stems include: *Diapleella clivensis*, *Leptosphaeria dolioloides*, *Lophiostoma caulium*, *L. vagabundum*, *Mollisia coerulans*, *Pyrenopeziza revincta*, *Unguicularia incarnatina* and *U. millepunctata*.

CENTAURIUM: Centaury

Taeniolina centaurii (Fuckel)M.B.Ellis (Fig. 1418)
Colonies effuse, dark brown or blackish brown, velvety. Conidia in dense clusters, much branched, olivaceous brown, up to 120 long but usually shorter, branches 3–5 thick. On leaves of *C. erythraea*, quite common.

CENTRANTHUS: Red Valerian

Ramularia centranthi Brun. (Fig. 1419)
Colonies hypophyllous, white. Conidiophores 15–50 × 3–5. Conidia hyaline, 0- or 1-septate, 15–34 × 2.5–4. On living leaves of *C. ruber* causing pale brown or grey spots 2–15mm diam., sometimes with raised purplish-brown margins, May–June.

CERASTIUM: Mouse-ear Chickweeds

Melampsorella caryophyllacearum Schröter, 11, 111
Alternate hosts *Abies* spp. Uredinia mostly hypophyllous, orange–yellow; spores sparingly echinulate, 15–30 × 12–20. Telia widespread on lower leaf surface; spores continuous or occasionally 1-septate, smooth, 12–24 diam. On *C. arvense*, *C. fontanum* ssp. *triviale*, *C. semidecandrum* and *C. tomentosum*.

Puccinia arenariae, 111, described on *Arenaria*, occurs also on *C. fontanum* ssp. *triviale*.

Ustilago violacea, anther smut described on *Silene*, occurs occasionally on *C. fontanum* ssp. *triviale*.

Didymella cerastii Gucevicz (Fig. 1420)
Pseudothecia scattered, 0.25mm diam., black. Ascospores fusiform, hyaline, 1-septate, 17–24 (27) × 5–6. On fading and dead leaves of *C. fontanum* ssp. *triviale*.

Endoconospora cerastii Gjaerum (Fig. 1421)
Conidia formed in chains over the surface of sporodochia, hyaline, 15–25 × 4–5. On *C. arvense* and *C. fontanum* ssp. *triviale*.

Leptotrochila cerastiorum (Wallr.)Schüepp (Fig. 1422)
Apothecia erumpent, yellowish, translucent with dark margins, 0.3–0.8mm diam. Ascospores hyaline, 0- or 1-septate, 8–13 × 2–4. On fading leaves and stems of *C. fontanum* ssp. *triviale* and *C. glomeratum*.

Peronospora conferta (Unger)Unger
Downy mildew on *C. fontanum* ssp. *triviale*. Sporangia 22–30 × 17–21.

Peronospora paula Gustavsson
A less common downy mildew also found on *C. fontanum* ssp. *triviale* with smaller sporangia, mostly 13–18 × 12–15. Oospores with reticulate epispore, 35–40 diam.

Ramularia alborosella (Desm.)Gjaerum (Fig. 1423)
Colonies small. Conidiophores grouped to form rosy-white synnemata. Conidia hyaline, 1-septate, 20–40 × 7–8. On leaves of *C. fontanum* spp. *triviale*.

CHAEROPHYLLUM: Rough Chervil

Puccinia chaerophylli, described on *Anthriscus*, has been rarely recorded on *C. aureum*.

Plasmopara nivea, downy mildew described on *Aegopodium* is found occasionally on *C. temulentum*.

CHEIRANTHUS: Wallflower

Albugo candida, described on *Capsella*, is found also on *Cheiranthus*.

Alternaria cheiranthi (Fr.)Bolle (Fig. 1424)
Causes yellowish spots which may occupy half the width of leaves, common. Conidiophores pale olive, up to 130 × 5–8. Conidia muriform, beaked, 20–100 × 13–32, golden brown by transmitted light but black and powdery in mass and often covering the upper surface of spots except for a marginal zone about 1mm wide. Found also on wilting stems and fruits.

Erysiphe cruciferarum, powdery mildew described on *Brassica*, is seen occasionally on wallflowers.

Peronospora parasitica, downy mildew described on *Capsella*, is common also on *Cheiranthus*.

CHELIDONIUM: Greater Celandine

Septoria chelidonii Desm.
Causes round or angular, olivaceous or pale brown spots on leaves, Aug.–Oct. Pycnidia about 0.1mm diam.; conidia straight or curved, pale yellowish in mass, 20–30 × 1.5.

CHENOPODIUM: All-seed, Fat Hen, Good King Henry, Red Goosefoot

Ascochyta chenopodii Rostrup
Pycnidia mostly epiphyllous, golden brown to black, about 0.15mm diam. Conidia straight or curved, cylindrical, rounded at each end, very pale yellow in mass, 1-septate, 12–20 × 3–4.5. On living leaves of *C. album*, causing pale buff spots with dark borders, up to 1cm diam., June–Aug.

Cercospora chenopodii Fresen. (Fig. 1425)
Colonies amphigenous. Conidiophores olivaceous brown, 30–60 × 3–6. Conidia hyaline or subhyaline, 1- to 4-septate, 35–55 × 5–7. On living leaves of *C. album* causing spots which are at first greenish brown, later pallid.

Chaetodiplodia caulina Karsten (Fig. 1426)
Pycnidia superficial, gregarious, black, up to 0.3mm diam., covered with brown setae. Conidia pale brown, 1-septate, 12–16 × 4–5.5. On dead stems of *C. album*.

Peronospora chenopodii Schlecht.
Downy mildew. Sporangia 28–32 × 17–22. On *C. polyspermum* and *C. rubrum*.

Peronospora farinosa (Fr.)Fr. (Fig. 1427)
Sporangiophores purplish brown, closely packed together. Sporangia tinged violet, ellipsoid, pedicellate, 22–30 × 16–23. Oospores 26–35 diam. On *C. album*.

Pleospora calvescens, described on *Atriplex*, is found occasionally also on dead stems of *C. album*.

Stagonospora atriplicis, described on *Atriplex*, occurs also on living leaves and stems of *C. bonus-henricus*, causing pale brown spots with yellow borders.

CHIONODOXA

Ustilago vaillantii Tul.
Smut sori in anthers, less often in ovaries of *C. sardensis*. Spore mass blackish brown, powdery. Spores olivaceous, 6–12 × 6–9.

CHRYSANTHEMUM: Corn Marigold, Marguerite or Ox-eye Daisy, Tansy

UREDINALES

Five rusts are found on wild and cultivated species of *Chrysanthemum* in Britain, the plurivorous *Coleosporium tussilaginis*, described on *Tussilago*, and four species of *Puccinia*.

On wild species:

Puccinia cnici-oleracei, 111, described on *Achillea*, has been recorded occasionally on *C. leucanthemum* and *C. segetum*.

Puccinia tanaceti, 11, 111, described on *Artemisia*, is seen only occasionally on *C. vulgare*.

On cultivated species:

Puccinia chrysanthemi Roze, 11, 111
In greenhouses all the year round and found occasionally on outdoor plants. Causes premature defoliation and stunting of plants, which form few flowers. Uredinia mostly hypophyllous, on pale yellow or brown spots; spores echinulate 25–50 × 18–26. Telia hypophyllous, dark cinnamon brown; spores 1-septate, 35–55 × 20–25.

Puccinia horiana P. Henn., 111
So-called white rust, first recorded in Essex in 1967, which has now spread quite widely over the country. Leaves spotted yellow on upper surface, telia on lower surface at first pinkish buff, later white; spores pale yellow, 1-septate, 33–45 × 12–18.

OTHER FUNGI

Bremia lactucae, described on *Lactuca*, is found occasionally on *C. segetum*.

Didymella ligulicola (Baker, Dimock & Davis)v. Arx
The cause of ray blight of cultivated varieties, mostly in the *Ascochyta* state, but pseudothecia are found on dead stems. Leaves, stems, buds and flowers are attacked and the growing points often killed. Pseudothecia erumpent on greyish spots, 0.1–0.2mm diam., papillate. Ascospores hyaline, broadly fusiform, constricted at septum, 12–16 × 4–6, upper cell abruptly swollen above septum, lower cell narrow, pointed. *Ascochyta* conidia 1- to 3-septate, 8–19 × 3–4.

Leptosphaeria dolioloides (Auersw.)P. Karsten (Fig. 1428)
Pseudothecia immersed, blackish brown, up to 0.35mm diam. Mature ascospores 40–56 × 4–6, pale straw-coloured, mostly 7- to 9-septate, the fourth cell from the top swollen. Common on dead stems of *C. vulgare*.

Oidium chrysanthemi Rabenh.
Powdery mildew. Occurs on flowers and leaves of cultivated species and on *C. segetum*. Conidia hyaline, 40–50 × 20–25, formed in long chains.

Peronospora radii de Bary
Downy mildew found on cultivated species and on *C. leucanthemum* and *C. segetum*, sometimes only on the flowers but often on the leaves as well. Sporangia in mass appearing dark violet, 30–34 × 18–22.

Protomycopsis leucanthemi Magnus
Chlamydospores terminal on hyphal branches, hyaline or very pale yellow, finely verruculose, 40–50 × 35–45, embedded in small, slightly swollen, pale yellow spots on leaves of *C. leucanthemum*, July–Sept., uncommon.

Ramularia bellunensis Speg. (Fig. 1429)
Colonies white, amphigenous. Conidiophores 40–80 × 4–5. Conidia hyaline, 0- to 3-septate, 17–35 × 4–6. On leaves of *C. frutescens* causing irregular, greyish brown, sometimes dark-bordered spots.

Rhabdospora tanaceticola Bubák & Kab.
Pycnidia subepidermal, black, 0.1–0.15mm diam. Conidia mostly curved, narrowed towards the rounded ends, hyaline to pale olive, finally 3-septate, 20–40 × 2.5–3. On dead stems of *C. vulgare*, Apr.–Aug.

Septoria. Six species occur in Britain.

KEY

Conidia 20–33 × 1 (1 to 3 septa) on *C. leucanthemum* .... *socia*
Conidia 20–30 × 1.5–2 (0 to 3 septa) on *C. vulgare* .... *tanaceti*
Conidia 35–65 × 1.5–2.5 (4 to 9 septa) on cult. spp .... *chrysanthemella*

Conidia 40–60 × 2–2.5 (4 or 5 septa) on *C. leucanthemum* and cult. spp. .... *chrysanthemi*
Conidia 70–100 × 2.5–3 (6 to 12 septa) on *C. leucanthemum*, *C. segetum* and cult. spp. .... *leucanthemi*
Conidia 50–90 × 2.5–3.5 (5 to 9 septa) on cult. spp. .... *obesa*

Septoria chrysanthemella Sacc.
Pycnidia mostly epiphyllous, up to 0.15mm diam. Conidia hyaline, tapering gradually towards the apex and more abruptly towards the base. Leaf spots brown to black, 5–10mm diam.

Septoria chrysanthemi Allesch.
Pycnidia epiphyllous, 0.1–0.12mm diam. Conidia hyaline, guttulate or faintly septate. Leaf spots buff or reddish brown with brown borders, 10–15mm diam.

Septoria leucanthemi Sacc. & Speg.
Pycnidia mostly epiphyllous, up to 0.25mm diam. Conidia hyaline, tapering gradually to the apex and more abruptly to the base. Leaf spots round or irregular, up to 2cm across, often confluent, sometimes zonate, brown becoming pale in the centre, which sometimes falls out leaving a shot-hole.

Septoria obesa Syd.
Pycnidia epiphyllous, up to 0.2mm diam. Conidia narrowly obclavate, hyaline. Causes large brown blotches, which often coalesce and may occupy most of the leaf surface.

Septoria socia Pass.
Pycnidia epiphyllous, less than 0.1mm diam. Conidia hyaline. Leaf spots chocolate brown to black, often with reddish-purple borders, up to 5mm diam.

Septoria tanaceti Niessl
Pycnidia epiphyllous, up to 0.1mm diam. Conidia straw-coloured. Leaf spots blackish brown, irregular, often confluent.

CHRYSOSPLENIUM: Golden Saxifrage

Puccinia chrysosplenii Grev., 111 (Fig. 1430)
On *C. alternifolium* and *C. oppositifolium*, uncommon. Telia of two kinds, on the upper surface of leaves mostly pulverulent, on the lower surface crustose, brown; spores 1-septate, 32–45 × 10–15.

Entyloma chrysosplenii (Berk. & Br.)Schröter
A rare smut on leaves of *C. oppositifolium*. Sori 2–5mm diam., in swollen, whitish spots. Spores hyaline, 10–12 diam.

CICHORIUM: Chicory, Endive

Puccinia hieracii, 11, 111, described on *Hieracium*, has been found occasionally on *C. endivia* and *C. intybus*.

CICUTA: Cowbane

Puccinia cicutae Lasch, 0, 1, 11, 111
Aecia on yellowish spots on leaves, petioles and stems. Uredinia mostly hypophyllous, cinnamon brown; spores verruculose to echinulate, 18–27 × 14–22. Telia hypophyllous, dark brown; spores verruculose, 1-septate, 30–45 × 19–30. Uncommon.

Ramularia cicutae Karsten (Fig. 1431)
Colonies mostly hypophyllous. Conidiophores hyaline, 18–60 × 3–4. Conidia hyaline, 0- to 3-septate, 20–45 × 3–4.5. On leaves, causing angular, brown or reddish-brown spots 1–8mm diam.

CIRCAEA: Enchanter's Nightshade

Puccinia circaeae Pers., 111
On *C. alpina*, *C. intermedia* and *C. lutetiana*, common. Telia on stems and lower leaf surfaces, often on purple or yellow spots, cinnamon brown, or grey where basidiospores are being formed; spores 1-septate, 25–40 × 10–13, smooth.

Pucciniastrum circaeae (Wint.)de Toni, 11 common, 111 rare
On *C. alpina*, *C. intermedia* and *C. lutetiana*. Uredinia small, orange or yellow, often covering the lower leaf surface; spores finely echinulate, 17–24 × 12–16. Telia hypophyllous, minute; spores 18–24 × 20–28, divided longitudinally into 2 to 4 cells.

Erysiphe circaeae Junell
Powdery mildew. Conidia few, 30–35 × 15–16. Perithecia 0.1mm diam., appendages few. Asci 3 to 5, each with 3 to 6 spores. Ascospores 20–23 × 11–12. On leaves of *C. lutetiana*, Sept.–Oct.

Ramularia circaeae Allesch. (Fig. 1432)
Groups of conidiophores scattered unevenly and rather sparsely over the lower surface of leaf spots, 20–40 × 2–3.5. Conidia hyaline, 0- or 1-septate, 11–25 × 3–4. On leaves of *C. lutetiana*, causing round or angular, pale brown spots, 0.5–4mm diam.

CIRSIUM: Creeping, Marsh, Meadow, Melancholy, Spear, Stemless and Woolly Thistles

UREDINALES

Five rusts occur. In one of these, *Puccinia dioicae* var *dioicae*, pycnia and aecia are found on *C. dissectum* and *C. palustre* only in the presence of the alternate host *Carex dioica*. Most of the others are seen on just one or two thistles.

On *C. acaule*: *Puccinia calcitrapae*
On *C. arvense*: *P. punctiformis*
On *C. dissectum*: *P. calcitrapae*
On *C. eriophorum*: *P. cnici*
On *C. heterophyllum* and *C. palustre* two species occur, *P. calcitrapae* and *P. cnici-oleracei*, neither of which is common. These two rusts are distinguished from one another quite easily. *P. calcitrapae* has uredinia and telia and the teliospores are chestnut brown, verruculose, 25–50 × 16–26; *P. cnici-oleracei* has telia only and the teliospores are yellowish brown, smooth, 40–55 × 15–20.
On *C. vulgare*: *P. cnici*

Puccinia cnici Mart., 0, 1, 11, 111
Uredinia amphigenous, reddish brown; spores echinulate, 25–35 × 20–26. Telia amphigenous, dark brown; spores golden brown, smooth to minutely verruculose, 1-septate, 30–40 × 23–25, common.

Puccinia cnici-oleracei, 111, described on *Achillea*.

Puccinia punctiformis (Str.) Röhl., 0, 1, 11, 111
Honey-coloured pycnia cover the lower surface of leaves of creeping thistles towards the end of April and give off a strong honey-like smell. Rusty uredinoid aecia appear on the same leaves in May, again covering the whole lower surface. Affected plants are pale and spindly and do not flower. Aeciospores affect other plants. Uredinia appear in July, are scattered, brown; spores minutely echinulate, 22–28 × 21–27. Telia dark brown; spores 1-septate, finely verruculose, 26–40 × 18–25. Very common.

USTILAGINALES

Thecaphora trailii Cooke
A rare smut which has been recorded in the inflorescences of *C. dissectum* and *C. heterophyllum*, replacing the florets in July. Spore balls 18–35 diam., composed of 2 to 8, pale yellow spores with verrucose outer walls. Spore masses purplish brown.

OTHER FUNGI

Albugo tragopogonis, white blister described on *Tragopogon*, is seen occasionally on *C. arvense*.

Cryptodiscus rhopaloides, described on 'wood and bark', and *Diapleella clivensis*, described on 'herbaceous plants', have been found on dead stems of *C. arvense* in June.

Erysiphe cichoracearum, powdery mildew described on *Aster*, is not uncommon on *C. arvense* and *C. vulgare*.

Leptosphaeria. Three species occur on dead stems of *C. arvense* and *C. vulgare* but none of them shows preference for these plants; two are described on 'herbaceous plants' and the other one on *Chrysanthemum*.

KEY

	Ascospores 7- to 9-septate	*dolioloides*
	Ascospores 3-septate	1
1.	Ascospores 25–30 long, substrate reddened	*purpurea*
	Ascospores 30–40 long	*macrospora*

Mollisia clavata, described on *Rubus*, has been found on dead stems of *C. arvense*.

Ophiobolus acuminatus, described on 'herbaceous plants', is very common on dead stems of *C. arvense*, and *O. cirsii*, described on *Carduus*, occurs also on dead stems of *Cirsium arvense*, *C. heterophyllum* and *C. palustre*.

Pezizella discreta and *P. glareosa*, described on 'herbaceous plants', have been found on dead stems of *C. arvense*.

Phoma rubella Grove
Pycnidia immersed, 0.25–0.3mm diam. Conidia hyaline, biguttulate, 4–7 × 1.5–2.5. On dead stems of *C. arvense* and *C. vulgare*, staining host epidermis red, Apr.–May.

Phomopsis cirsii Grove
Conidiomata 0.5mm diam., black, sometimes in rows. A-spores few, biguttulate, 12 × 3–4, B-spores numerous, curved, hamate, 20–30 × 0.5–1. On dead stems and leaves of *C. arvense* and *C. eriophorum*, Oct.–May.

Pirottaea brevipila, described on *Centaurea*, occurs also on dead stems of *Cirsium arvense*, June.

Psilachnum rubrotinctum, described on *Filipendula*, has been found on dead stems of *C. palustre*, June.

Pyrenopeziza carduorum (Rehm)Gremmen (Fig. 1433)
Apothecia erumpent, just over 1mm diam., disc grey, sides dark brown with

protruding brown hairs. Ascospores hyaline, 14–19 × 2–2.5. On dead stems of *C. arvense* and *C. palustre*, May–Aug. Two other species of *Pyrenopeziza* occur also: *P. adenostylidis*, with ascospores 9–12 × 1.5, described on *Centaurea*; and *P. revincta*, with ascospores 8–10 × 2–2.5, described on 'herbaceous plants'.

Ramularia cirsii Allesch.
Colonies small, white. Conidiophores 30–40 × 3. Conidia more or less cylindrical, hyaline, 0- to 3-septate, 30–35 × 2.5–3.5. On fading leaves of *C. arvense*.

Rhabdospora cirsii Karsten
Pycnidia 0.4mm diam., dark blackish brown, with prominent ostioles. Conidia straight or slightly curved, tapered at each end, hyaline, guttulate, 35–50 × 1–1.5. On dead stems of *C. arvense* and *C. palustre*, Feb.–May.

Unguicularia. Two plurivorous species described on 'herbaceous plants' have been found: *U. incarnatina* on *C. palustre*, and *U. millepunctata* on *C. arvense*, Apr.–Nov.

CLINOPODIUM: Wild Basil

Puccinia menthae, described on *Mentha*, occurs on *C. vulgare*.

COCHLEARIA: Common, English and Danish Scurvy-grass

Puccinia eutremae Lindr., 111
Telia on stems, petioles and both leaf surfaces, blackish brown; spores constricted and sometimes dividing into two at the septum, smooth or striate, 35–50 × 14–17. On *C. alpina* and *C. danica*, uncommon.

Ramularia cochleariae Cooke
Colonies epiphyllous, round, white. Conidia cylindrical, rounded at the ends, 25–28 × 3.5. On leaves of *C. officinalis*, causing pale spots.

COLCHICUM: Autumn Crocus, Meadow Saffron or Naked Ladies

Uromyces colchici Massee, 111
Telia amphigenous, brown; spores smooth, 30–40 × 20–30. On leaves, very rare and no recent records.
Urocystis colchici (Schlecht.)Rabenh.
Smut sori blister-like, up to 10 × 1mm, parallel with veins. Spore mass dark brown. Spore balls 15–35 × 15–22, each composed of one or two reddish

brown spores 12–16 diam., surrounded by yellowish sterile cells. On leaves, uncommon.

CONIUM: Hemlock

Puccinia conii Lagh., 11, 111
Uredinia hypophyllous, cinnamon brown; spores with top half echinulate and lower half smooth, 25–35 × 18–26. Telia blackish brown; spores smooth, 1-septate, 30–48 × 20–28.

Several plurivorous fungi, including *Crocicreas cyathoideum*, *Dasyscyphus grevillei*, *D. mollisimus*, *Diaporthe arctii* and *Dendryphiella vinosa*, have been found on dead stems.

CONOPODIUM: Pignut

Puccinia bistortae DC., 1
Aecia on leaves and petioles of *C. majus*, uncommon. Alternate host *Polygonum bistorta*.

Puccinia tumida Grev., 11, 111
Uredinia very few. Telia mostly on petioles and veins of leaves, in swollen, elongated groups, blackish brown; spores 1-septate, brown, 25–35 × 15–25. Common, Apr.–May.

CONVALLARIA: Lily-of-the-Valley

Puccinia sessilis Schröter, 0, 1
Alternate host *Phalaris arundinacea*. Race with aecia on *C. majalis* rare.

Mycosphaerella brunneola (Fr.)Allesch. & Schnabel
Pseudothecia epiphyllous, gregarious, with connecting brown hyphae, forming spots of various sizes. Ascospores hyaline, septum not quite central, 17–20 × 4. Pycnidia of '*Asteroma*' state mixed with pseudothecia; conidia hyaline, 75–105 × 2. On leaves of *C. majalis*.

CONVOLVULUS: Bindweed, Cornbine

Thecaphora seminis-convolvuli and *Septoria convolvuli*, described on *Calystegia* both occur also on *Convolvulus arvensis*.

CORONOPUS: Swine-cress, Wart-cress

Peronospora lepidii (Mc.Alp.)G.W.Wils.
Downy mildew on *C. squamatus*. Sporangia 34–40 × 20–23.

CREPIS: Hawksbeards

Puccinia crepidicola Syd., 11, 111
Common on *C. biennis*, *C. capillaris* and *C. vesicaria* ssp. *haenseleri*. Uredinia amphigenous and on stems, pale brown; spores minutely echinulate, 22–26 × 18–21. Telia blackish brown; spores 1-septate, dark brown, minutely verruculose, 30–35 × 21–25.

Puccinia major Diet., 0, 1, 11, 111
Fairly common on *C. paludosa* in some areas. Aecia hypophyllous. Uredinia amphigenous, cinnamon; spores finely echinulate, 25–30 × 21–26. Telia scattered over most of the lower leaf surface; spores 1-septate, brown, minutely verruculose, 35–48 × 23–30.

Albugo tragopogonis, white blister described on *Tragopogon*, is seen occasionally on *C. capillaris*.

CUCUBALUS: Berry Catchfly

Puccinia arenariae, described on *Arenaria*, and anther smut, *Ustilago violacea*, described on *Silene*, both occur on *C. baccifer*.

CUCUMIS: Cucumber

Alternaria cucumerina (Ellis & Everh.)Elliott (Fig. 1434)
Colonies amphigenous. Conidiophores brown, up to 110 × 6–10. Conidia pale to mid golden brown, muriform, beaked, smooth or minutely verruculose, 130–220 × 15–24; beak 1–2.5 thick. Causes leaf blight. Spots at first small, circular, water-soaked, whitish or tan, later expanding and often zonate, with a clear brown margin on the upper surface of the leaf. Not common in Britain but there is occasionally a severe outbreak on cucumbers grown in beds under glass.

Cladosporium cucumerinum Ellis & Arth. (Fig. 1435)
Colonies rather pale greyish olive, velvety or felted. Conidiophores up to 400 × 3–5. Conidia in long, branched chains, 4–25 × 2–6 (mostly 4–9 × 3–5), pale olivaceous brown, smooth, 0- to 2-septate, or sometimes minutely verruculose. On leaves, stems and fruits causing gummosis or scab. Deep

lesions up to 1cm diam., with gummy exudate, formed on young fruits where disease is most severe.

Colletotrichum orbiculare (Berk. & Mont.)v. Arx
Acervuli with or without setae, spore mass pinkish orange. Setae when present few, olivaceous, 1- or 2-septate, $60–70 \times 5–7$. Conidia hyaline, $14–15 \times 4.5–6$. On leaves, causing large ochraceous yellow zonate spots; also on fruits where deep lesions up to 10cm long may be formed. Not common in Britain but is sometimes very damaging.

Corynespora cassiicola (Berk. & Curt.)Wei (Fig. 1436)
Colonies effuse, grey or brown, thinly hairy. Conidiophores brown, up to $850 \times 4–11$, with successive apical proliferations. Conidia very variable in shape, subhyaline to pale olivaceous brown, with 4 to 20 pseudosepta, $40–220 \times 9–22$. Found occasionally on leaves, causing irregular, sometimes zonate spots.

Didymella bryoniae, described on *Bryonia*, sometimes attacks stems and fruits of cucumber.

Phomopsis cucurbitae McKeen
Conidiomata in longitudinal rows on stem internodes and on rind of fruit, up to 1mm diam., blackish brown. A-conidia ellipsoid–fusiform, hyaline, with 2 or 3 guttules, $8–12 \times 2.5–3$. B-conidia curved, $18–26 \times 1$.

Phomopsis sclerotioides van Kesteren
Conidiomata 0.3mm diam. No B-conidia. A-conidia ellipsoid–fusiform, $7–10 \times 2.5–3.5$, with two guttules. Mainly a root pathogen.

Phyllosticta cucurbitacearum Sacc.
Pycnidia about 0.1mm diam. Conidia slightly curved, hyaline with two guttules, $5–6 \times 2.5$. On leaves, causing whitish spots. Known under the above name but not a true *Phyllosticta*.

Sphaerotheca fuliginea, powdery mildew described on *Veronica*, is recorded occasionally on cucumber.

CYMBALARIA: Ivy-leaved Toadflax

Peronospora linariae Fuckel
Downy mildew found occasionally on *C. muralis*. Sporangia $24–26 \times 16–18$.

CYNARA: Artichoke

Bremia lactucae, downy mildew described on *Lactuca*, is not uncommon on *C. scolymus*.

Leveillula taurica, a mainly tropical powdery mildew described on *Helianthemum*, has been recorded in Britain also on *C. scolymus*.

Ramularia cynarae Sacc.
Conidiophores hyaline, 40–50 × 3. Conidia cylindrical, rounded at the ends, hyaline, 1- or 2-septate, 20–25 × 3–4. On leaves, causing round or irregular grey spots with dark brown margins.

CYNOGLOSSUM: Hound's-tongue

Erysiphe asperifoliorum Grev.
Conidia plentiful. Asci 5 to 15 with 2 to 4 spores each. This powdery mildew is common on a number of the Boraginaceae.

DACTYLORHIZA: Common Spotted and Marsh Orchids

Melampsora epitea Thüm. var *epitea*, 0, 1
Alternate host *Salix repens*. On leaves of *D. incarnata*, *D. maculata* and *D. praetermissa*, May–June, uncommon. Aecia hypophyllous, orange, without or with a very rudimentary peridium, often confluent; spores verruculose, 15–25 × 10–20. Formed on large, pale yellow spots.

Puccinia sessilis Schröter, 0, 1
Alternate host *Phalaris arundinacea*. On leaves of *D. fuchsii*, *D. incarnata* and *D. praetermissa*, fairly common. Aecia hypophyllous, often circinate, orange, with a well-defined, white, laciniate, recurved peridium, the cells of which have outer walls up to 10 thick; spores minutely verruculose, 20–27 diam. Formed on large yellow spots.

Cladosporium orchidis E.A. & M.B.Ellis (Fig. 1437)
Colonies parasitic, usually epiphyllous, olivaceous brown or dark brown, velvety. Conidiophores up to 100 × 3–8. Conidia pale olivaceous brown, 0- or 1-septate, 5–18 × 2–3.5. On living leaves of *D. praetermissa*, causing large, dark, blackish brown, oval or irregular spots.

DAHLIA: Dahlia

Entyloma calendulae (Oudem.)de Bary f. *dahliae* (Syd.)Viégas
Sori in living leaves causing round or elliptical, pallid to brown spots, Aug.–Oct., common. Spores spherical or polygonal, very pale yellow, smooth, 9–14 diam.

Ascochyta dahliicola (Brun.)Petrak
Spots on leaves irregular, 1–4cm diam., grey or whitish. Pycnidia up to 0.15mm diam. Conidia 1-septate, hyaline, 8–15 × 3–4.5.

Itersonilia perplexans, described on *Cynara*, has been recorded causing flower scorch of *Dahlia*.

DAUCUS: Carrot

Alternaria dauci (Kühn)Groves & Skolko (Fig. 1438)
Conidiophores pale olivaceous brown, up to 80 × 6–10. Conidia pale olivaceous brown to brown, smooth, with 7 to 11 transverse and one or a few longitudinal septa, 100–450 × 16–25, the beak up to three times the length of the body of the conidium and often once-branched. Causes leaf blight. Affected leaves and petioles of cultivated plants turn yellow and then brown to black, and when infection is severe the whole top may be killed.

Alternaria radicina Meir, Drechsler & Eddy (Fig. 1439)
Colonies effuse, dark blackish brown to black. Conidiophores brown, up to 200 × 3–9. Conidia mid to dark brown, muriform, 27–57 × 9–27. Causes black rot, common in storage, a progressive softening and blackening of the tissue of the root, infection frequently starting at the crown and extending downwards.

Chalaropsis thielavioides Peyr. (Fig. 1440)
Conidiophores hyaline, very variable in length, 4–9 thick. Conidia of two kinds: (1) cylindrical, hyaline, 8–15 × 2.5–4.5, issuing from tapered phialides, and (2) spherical or subspherical, olivaceous brown, 14–19 diam., formed at the ends of branches. Not uncommon on the surface of prepared carrots which become blackened.

Heterosphaeria patella, described on *Angelica*, is common also on dead stems of wild carrot.

Mycocentrospora acerina, described on 'herbaceous plants', has been recorded causing 'liquorice rot' of carrots in spring.

Plasmopara nivea, described on *Aegopodium*, is found occasionally on *D. carota*.

Thielaviopsis basicola (Berk. & Br.)Ferraris (Fig. 1441)
Colonies grey or olivaceous, somewhat velvety. Conidiophores up to 50 × 6–9. Conidia of two kinds: (1) golden brown ones 7–12 long, 10–17 wide, in chains of 4 to 8 which remain together for a long time resembling large, multiseptate spores; (2) hyaline phialoconidia, 7–17 × 2.5–4.5. Sometimes causes a black root rot.

DELPHINIUM: Larkspur

Erysiphe ranunculi, powdery mildew described on *Ranunculus*, is found commonly also on *Delphinium*.

DIANTHUS: Carnation, Clove Pink, Maiden Pink, Sweet William

Puccinia arenariae, 111, described on *Arenaria*, occurs on *D. barbatus* and *D. deltoides*.

Uromyces dianthi (Pers.)Niessl, 11, 111
Uredinia on stems and both leaf surfaces, small, cinnamon; spores echinulate, 20–35 × 19–25. Telia in elongated or circinate swollen groups on leaves and stems; spores brown, 20–30 × 19–23. On *D. barbatus*, *D. caryophyllus* and *D. chinensis*, common.

Ustilago violacea, anther smut described on *Silene*, is found occasionally on *D. caryophyllus*.

Alternaria dianthi Stevens & Hall (Fig. 1442)
Conidiophores olivaceous brown, up to 120 × 5–8. Conidia brown or olivaceous brown, often rather dark, smooth, with up to 9 transverse and usually several longitudinal septa, 30–120 × 10–25. Causes carnation blight. On the leaves very small purple spots are seen at first and these soon develop a broad yellowish-green border. The spots expand, the centre becomes light brown or grey, and adjacent spots tend to coalesce. Healthy tissue between the spots often turns yellow. Stem lesions at first on one side of the stem and more or less superficial often eventually extend into the pith, girdle the stem and may kill the whole plant. Common on *D. caryophyllus* and occurs also on *D. barbatus*.

Alternaria dianthicola Neergaard (Fig. 1443)
Conidiophores pale olivaceous brown, up to 150 × 4–6. Conidia pale olivaceous brown with up to 14 transverse and occasionally 1 or 2 longitudinal septa, 55–130 × 10–16. Causes leaf spot and flower-bud rot. Recorded occasionally but apparently not common in Britain.

Heteropatella valtellinensis (Trav.)Wollenw. (Fig. 1444)
Acervuli superficial, dark brown to black, about 0.2mm diam. Conidia 1- or 2-septate, hyaline, smooth, 22–30 × 4–5.5. Causes leaf rot of carnations.

Mycosphaerella dianthi (Burt.)Jørstad (Fig. 1445)
Common and widespread on leaves and sometimes inflorescences of *D. barbatus* and *D. caryophyllus* and sometimes other species in its *Cladosporium* state; pseudothecia seldom seen. Plants are sometimes completely destroyed. Colonies effuse, olivaceous grey. Conidiophores brown, up to 200 × 8–13. Conidia pale or mid pale brown or olivaceous brown, mostly 2- to 4-septate, 25–55 × 8–17.

Phialophora cinerescens (Wollenw.)Beyma (Fig. 1446)
Parasitic on *D. caryophyllus*, one of the causes of carnation wilt disease. Vascular tissue of stem discoloured and showing up in transverse section as a brown ring, top leaves of older plants irregularly mottled red. Conidia at first

hyaline, later brown, smooth, 3–6 × 1.5–2.5.

Phomopsis caryophylli Grove
Conidiomata 0.3mm long, black, each surrounded by a brown halo. Conidia fusoid, hyaline, biguttulate, 7–9 × 2–2.5. Often on bleached areas on stems and peduncles of *D. barbatus* and *D. caryophyllus*; parts of stems sometimes blackened.

Septoria dianthi Desm.
Pycnidia epiphyllous, numerous, black, 0.125mm diam. Conidia hyaline, flexuous, 30–40 × 3–4. On leaves of *D. barbatus* and *D. caryophyllus*, causing ochraceous spots 5mm diam., with wide purplish borders, common, July–Sept.

Zygophiala jamaicensis Mason (Fig. 1447)
Conidiophores mid to dark brown, up to 35 × 4–8. Conidia hyaline, 1-septate, 13–20 × 5–6. Causes greasy blotch disease with loss of bloom on leaves of commercially grown carnations.

DIGITALIS: Foxglove

Ascochyta mollieriana Wint.
Pycnidia few, epiphyllous, immersed, 0.1–0.15mm diam., pale brown. Conidia hyaline, eventually 1-septate, 9.5–12 × 3.5. On living leaves of *D. purpurea*, causing round or irregular greyish spots up to 1cm diam., with broad, dark purple borders.

Colletotrichum fuscum Laub.
Acervuli setose, on living leaves of *D. lanata* and *D. purpurea*, Oct.–Nov. Conidia hyaline, straight or very slightly curved, 14–17 × 3.5–4.

Peronospora digitalis Gäum.
Downy mildew on *D. purpurea*. Sporangia 21–37 × 16–29.

Phomopsis digitalis Hawksw. & Punithalingam
Conidiomata erumpent, black, up to 0.25mm diam. Conidia ellipsoid–fusiform, biguttulate, 6–10 × 2–3. On dead stems of *D. purpurea*, Mar.

Pyrenopeziza digitalina (Phill.)Sacc. (Fig. 1448)
Apothecia up to 0.8mm diam., dark brown, discs dark grey with pale rims. Ascospores hyaline, 5–7 × 1.5–2. Near the base of dead stems of *D. purpurea*, Apr.–July, common.

Ramularia variabilis Fuckel
Colonies thin, white. Conidiophores very short. Conidia hyaline, cylindrical, 1-septate, 15–22 × 3–4. On blackish-brown or greenish spots on leaves of *D. purpurea*.

Septoria digitalis Pass.
Pycnidia very small. Conidia hyaline, multiguttulate, 20–30 × 1.5. On blackish-brown spots on dying leaves of *D. purpurea*.

DIPLOTAXIS: Perennial Wall Rocket, Stinkweed or Wall Rocket

Albugo candida, described on *Capsella*, occurs on *D. tenuifolia*.

Peronospora parasitica, described on *Capsella*, has been recorded on *D. muralis*.

DIPSACUS: Teasel

Peronospora dipsaci Tul.
Downy mildew. Sporangia mostly 26–32 × 20–24.

Ramularia sylvestris Sacc.
Colonies hypophyllous, white, small. Conidiophores 15–20 × 3. Conidia cylindric–fusoid, hyaline, 1-septate, 20–30 × 2–2.5. On leaves of *D. fullonum* ssp. *sylvestris*.

Sphaerotheca dipsacearum (Tul. & C.Tul.)Junell
Powdery mildew. Cleistothecia about 0.1mm diam., wall small-celled, appendages straight, brown, mostly unbranched.

DORONICUM: Leopard's-bane

Ramularia doronici (Sacc.)Lindau
Colonies hypophyllous, white. Conidiophores hyaline, 30–40 × 3. Conidia oblong–fusoid, hyaline, 12–15 × 4–5. On *D. pardalianches*.

Sphaerotheca fusca (Fr.)Blumer (Fig. 1449)
Powdery mildew on *D. plantagineum*. Mycelium turning brown. Conidia abundant, becoming brown, mostly 28–30 × 15–17. Cleistothecia up to 0.1mm diam., appendages brown. Asci 8-spored, ascospores 18–25 × 14–17.

ECHIUM: Viper's Bugloss

Leptosphaeria cesatiana (Mont. ex Ces. & de Not.)Holm
Pseudothecia immersed, about 0.5mm diam. with necks up to 0.15mm long. Ascospores yellowish, 110–120 × 3.5–4, 15-septate, strongly constricted at or just above the centre. Growing on blackened areas on dead stems of *E. vulgare*.

ENDYMION: Bluebell

Uromyces muscari (Duby)Graves, 111
Telia amphigenous, dark brown, on yellowish spots; spores brown, 19–32 × 14–22. Very common on *E. non-scriptus* and not uncommon also on *E. hispanicus*, Apr.–June.

Colletotrichum liliacearum (Westend.)Duke
Acervuli black, setose, about 0.25mm diam. Conidia oblong fusoid, hyaline, slightly curved, more tapered towards one end than the other, 16–20 × 3. On dead scapes of *E. non-scriptus*, Sept.–Nov.

EPILOBIUM: Willow-herbs

UREDINALES

Puccinia epilobii DC., 111
A rare rust recorded occasionally on *E. anagallidifolium*, *E. hirsutum*, *E. montanum* and *E. obscurum*. Telia hypophyllous, reddish brown; spores 1-septate, minutely verruculose, 28–48 × 16–25.

Puccinia pulverulenta Grev., 0, 1, 11, 111 (Fig. 1450)
Common on *E. adnatum*, *E. hirsutum*, *E. montanum* and *E. parviflorum*; plants affected tend to be pale and yellowish. Aecia orange with white peridia. Uredinia hypophyllous, sometimes circinate, reddish brown; spores echinulate, 20–27 × 16–25. Telia hypophyllous, often circinate, dark brown; spores smooth, 25–35 × 15–20, 1-septate.

Pucciniastrum epilobii Otth, 11, 111
On *E. anagallidifolium*, *E. angustifolium*, *E. montanum* and *E. palustre*, Aug.–Oct., common. Alternate host *Abies grandis*. Uredinia on stems and leaves, often causing yellow or red spots; spores minutely echinulate, 15–22 × 10–15. Telia very small, pale brown; spores intercellular, divided by vertical septa into 2 to 4 cells.

PERONOSPORALES

Plasmopara epilobii (Rabenh.)Schröter
White downy mildew on leaves of *E. parviflorum*, July–Sept. Sporangia with flat papillae, 13–15 × 11–13.

DISCOMYCETES

Allophylaria macrospora (Kirscht. ex Jaap)Nannf. (Fig. 1451)
Apothecia 0.5mm diam., pale buff with whitish discs. Ascus pore blue in iodine. Ascospores 18–28 × 4–6, hyaline. On dead stems of *E. angustifolium*, Sept.–Oct.

Ciboriopsis tenuistipes (Schröter)Palmer (Fig. 1452)
Apothecia usually about 0.75mm diam., pale brown with ochraceous discs, stalks 4–5mm long and often dark at the base. Ascospores hyaline, 6–8 × 2–3. Most commonly found on fallen rotting leaves of *E. angustifolium*, May–Aug., only occasionally on other plants such as *Filipendula ulmaria*.

Dasyscyphus castaneus Graddon (Fig. 1453)
Apothecia scattered, subsessile, up to 0.6mm diam., pale brown with chestnut-coloured discs, fringed with short (up to 45 × 3–3.5) hyaline, minutely granulate hairs. Ascospores hyaline, 3–4 × 1.5. On dead stems of *E. hirsutum*, Nov.

Dasyscyphus nudipes (Fuckel)Sacc. var *minor* Dennis (Fig. 1454)
Apothecia up to 0.5mm diam., cream or white, discs becoming yellowish. Hairs up to 50 × 3, sometimes swollen at the tip and with a few rhomboidal crystals. Ascospores hyaline, 6–10 × 1–1.5. On dead stems of *E. hirsutum*. *D. clavigerus* Svrček on *E. angustifolium* is similar but has balls of small crystals at the ends of the hairs.

Hymenoscyphus repandus (Phill.)Dennis (Fig. 1455)
Apothecia pale yellow when fresh, becoming ochraceous, 1.5–2mm diam., with stalk up to 2.5mm. Ascospores hyaline, 8–12 × 2–2.5. On dead stems of *E. angustifolium*, *E. hirsutum* and *E. palustre*, May–Oct., common Almost equally common on *Filipendula ulmaria* and found also on *Eupatorium cannabinum*, *Iris pseudacorus* and *Mentha aquatica*.

Mollisia dilutella Gill. (Fig. 1456)
Apothecia solitary or in clusters, pale greyish, drying pale yellow or creamy, up to 0.75mm diam. Ascospores hyaline, 7–12 × 1–1.5. On dead stems of *E. angustifolium* and *E. hirsutum*, Sept.–Oct., common.

Pezizella punctoidea (P. Karsten)Rehm (Fig. 1457)
Apothecia shaped like minute egg-cups, yellowish white, up to 0.4mm diam., downy. Ascospores hyaline, 5–7 × 1.5–2. Rather thinly scattered, mostly on dead leaves of *E. angustifolium* and *E. hirsutum*, July–Dec., not uncommon but difficult to see.

Pyrenopeziza chamaenerii Nannf. (Fig. 1458)
Apothecia erumpent, blackish brown, discs cream to brown with white rims, up to 0.5mm diam. Ascospores hyaline, 7–10 × 2. On dead stems of *E. angustifolium*, June–Aug.

Rutstroemia hercynica (Kirscht.)Dennis (Fig. 1459)
Apothecia when fresh up to 3mm diam., buff to hazel, edged with brown teeth, turning reddish brown when dry, the teeth becoming obscured, solitary or in small groups in areas delimited by black lines. Ascus pore dark blue in iodine. Ascospores hyaline, 13–16 × 5–6. On dead stems of *E. angustifolium*, July–Oct.

Unguicularia dilatopilosa Graddon (Fig. 1460)
Apothecia up to 0.6mm diam., golden brown, clothed with densely packed, glassy, hyaline hairs, discs blackish brown. Hairs up to 70 long, tapered, solid except at the base where they are thin-walled and swollen to 8. Ascospores hyaline, 5–7 × 1–1.5. On dead stems of *E. hirsutum*, Nov.–Apr.

Other discomycetes found mainly on dead stems of *Epilobium*, which are either plurivorous or occur more commonly on other substrata, include: *Betulina fuscostipitata* (*Betula*), *Crocicreas cyathoideum*, *C. dolosellum* (*Fraxinus*), *Dasyscyphus clandestinus*, *D. nidulus*, *D. sulphureus*, *D. virgineus* (wood and bark), *Echinula asteriadiformis* (*Rubus*), *Hyaloscypha hyalina* (wood and bark), *Hyalotricha niveocincta* (*Euphorbia*), *Hymenoscyphus scutula*, *H. vitellinus* (*Filipendula*), *Hypoderma commune*, *Mollisina rubi* (*Rubus*), *Pezizella discreta*, *P. oenotherae* usually in *Hainesia* state only (*Fragaria*), *Pyrenopeziza revincta*, *Stictis stellata*, *Uncinia foliicola* (*Alnus*), *Unguicularia cirrhata*, *U. millepunctata* and *Velutarina rufo-olivacea* (wood and bark).

OTHER ASCOMYCETES

Anthostomella clypeoides Rehm (Fig. 1461)
Perithecia about 0.2mm diam., with black clypeus, solitary or in groups. Ascospores 10–15 × 3–5, with large brown cell and very small hyaline one. On dead stems of *E. angustifolium* and *E. hirsutum*, Oct.–May.

Morenoina epilobii (Lib.)v. Arx (Fig. 1462)
Thyriothecia simple, branched or bent, 150–450 × 40–60. Ascospores hyaline or pale straw-coloured, smooth, 5–7 × 2–3, 1-septate. On dead stems of *E. angustifolium*, July.

Paradidymella tosta (Berk. & Br.)Petrak (Fig. 1463)
Perithecia 0.25–0.3mm diam., blackish brown, immersed, each with a rather ill-defined clypeus, grouped in greyish areas. Ascospores hyaline, 1-septate, 10–14 × 3–4. Very common on dead stems of *E. angustifolium*, also on *E. hirsutum*, Mar.–July.

Pleospora epilobii E. Müller (Fig. 1464)
Pseudothecia immersed, up to 0.15mm diam., clothed with long brown hairs. Ascospores yellow, muriform, 18–20 × 7–9. On dead stems of *E. angustifolium*, Mar.

Sphaerotheca epilobii (Link)de Bary (Fig. 1465)
Powdery mildew, seen mostly on upper parts of plants of *E. hirsutum*, *E. montanum*, *E. palustre* and *E. parviflorum*, Aug.–Oct. Conidia plentiful. Ascospores hyaline, 25–27 × 13–14.

Sydowiella fenestrans (Duby)Petrak (Fig. 1466)
Perithecia immersed, sometimes close together in rows, black, 0.5–0.7mm diam., with thick, black, protuberant necks. Ascospores hyaline, 1-septate, mostly 15–20 × 7–9. Very common on dead stems of *E. angustifolium* often in company with *Paradidymella tosta*, Apr.–July.

Venturia maculiformis (Desm.)Winter (Fig. 1467)
Pseudothecia epiphyllous, black, mostly about 0.1mm diam., clustered in the centre of small (1–1.5mm diam.) dark brown spots on living leaves of *E. angustifolium*, *E. hirsutum*, *E. montanum* and *E. palustre*, May–Sept. Ascospores greenish or olivaceous, 1-septate, 9–14 × 3–5.

Other species found on *Epilobium* include: *Chaetosphaeria callimorpha* (*Rubus*); *C. innumera* (wood and bark); *Clypeosphaeria notarisii* (*Rubus*); *Diaporthe arctii* (*Arctium*); *D. eres* (*Ulmus*); *D. pardalota* (*Lonicera*); *Herpotrichia herpotrichoides* (*Rubus*); *H. macrotricha*, *Lasiosphaeria caudata*, *L. phyllophila*, *L. strigosa*, *Lophiostoma angustilabrum*, *L. caudatum* (*Phragmites*); *L. fuckelii* (*Rubus*); *L. origani* var. *rubidum*, *L. vagabundum*, *Nectria ellisii* and *Niesslia exosporioides* (*Carex*).

HYPHOMYCETES

Ramularia epilobiana (Sacc. & Fautr.)Sutton & Pirozynski
Colonies hypophyllous, white. Conidiophores hyaline, geniculate, up to 120 × 2.5–4. Conidia oval or ovoid, hyaline, 11–24 × 7–15. On living leaves of *E. hirsutum*, causing pale greyish-brown spots up to 8mm diam., sometimes with purplish-brown margins and surrounded by yellow haloes, Aug.

Ramularia montana Speg. (Fig. 1468)
Colonies hypophyllous, olive or greyish. Conidiophores arising from olive or brown pseudostromata, pale olive near the base, hyaline above, 30–60 × 3.5–6. Conidia hyaline, 0- to 3-septate, 16–46 × 4–6. On leaves of *E. adnatum*, *E. angustifolium*, *E. montanum*, *E. palustre*, *E. parviflorum* and *E. tetragonum*, causing pale brown, often purple-bordered spots, 2–10mm diam., Aug.–Sept.

Plurivorous hyphomycetes found mainly on dead stems include: *Ceratosporium fuscescens* (*Ilex*), *Dendryphion comosum* (*Urtica*), *Endophragmia atra*, *Menispora ciliata* (wood and bark), *Myrothecium carmichaelii* (*Thalictrum*), *Periconia byssoides*, *P. minutissima*, *Pseudospiropes obclavatus* (wood and bark), and *Torula herbarum*.

COELOMYCETES

Phomopsis epilobii (Preuss)Grove
Conidiomata black, immersed. Conidia hyaline, pointed at each end, 8–10 × 2–2.5. On dead stems of *E. angustifolium* and *E. hirsutum*, May.

Pilidium concavum (Desm.)Höhnel (Fig. 1469)
Conidiomata up to 0.15mm diam., dark brown, concave, without an ostiole. Conidia hyaline, mostly 6–8 × 1.5–2. On dead stems of *E. montanum*.

Septoria alpicola Sacc.
Pycnidia 0.1–0.15mm diam. Conidia up to 50 × 1.5. On seedling leaves of *E. montanum*, not forming spots, July–Sept.

Septoria epilobii Westend.
Pycnidia less than 0.1mm diam. Conidia up to 40 × 1.5. On living leaves of *E. hirsutum*, *E. montanum* and *E. tetragonum* causing small, olivaceous, pale-centred spots, July–Aug.

Plurivorous coelomycetes include: *Pseudolachnea hispidula* on dead stems of *E. hirsutum*; the *Hainesia* state of *Pezizella oenotherae* (*Fragaria*) on leaves of *E. angustifolium*; and *Rhabdospora pleosporoides* on dead stems of the same host.

EPIPACTIS: Marsh Helleborine

Melampsora epitea, 0, 1, on *E. palustris*. Alternate hosts *Salix* spp.

ERANTHIS: Winter Aconite

Urocystis eranthidis (Pass.)Ainsworth & Sampson
This smut forms blisters on leaves and petioles of winter aconite which eventually rupture and expose a black powdery spore mass. Spore balls 20–40 diam., composed usually of 1 dark brown spore 13–18 diam., surrounded by yellowish sterile cells, Apr.–May.

ERICA: Bell-heather, Cross-leaved Heath

Gibbera salisburgensis Niessl (Fig. 1470)
Pseudothecia superficial, black, about 0.15mm diam., covered with pointed, thick-walled setae up to 100 × 5. Asci 4-spored. Ascospores olivaceous, 1-septate, 18–24 × 5–8. On twigs and fading leaves of *E. tetralix*, Aug.

Protoventuria straussii (Sacc. & Roum.)Dennis (Fig. 1471)
Pseudothecia superficial on a brown hyphal mat, black, up to 0.35mm diam.,

sometimes with short brown setae. Ascospores pale brown, 1-septate, 14–20 × 7–10. On newly dead stems of *E. cinerea*, Apr.–May.

ERIGERON: Canadian Fleabane

Sphaerotheca erigerontis-canadensis (Lév.)Junell (Fig. 1472)
Powdery mildew. Conidia plentiful, 25–30 × 15–18. Cleistothecia 70–90 diam., 8 ascospores 16–20 × 13–16. Common on *E. canadensis*.

ERODIUM: Storksbill

Peronospora erodii Fuckel
Downy mildew. Colonies hypophyllous, greyish brown. Sporangiophores 100–300 × 4–6. Sporangia globose 16–23 diam. or ovoid 23–35 × 18–24. Oospores smooth. On living leaves of *E. cicutarium*, June–Sept.

ERYNGIUM: Sea Holly

Entyloma eryngii (Corda)de Bary
Smut sori in leaves as pale ochraceous, raised spots, 1–3mm diam. Spores thin-walled, smooth, hyaline to golden brown, 8–10 diam. On *E. maritimum*, Sept., rare.

Diaporthopsis angelicae, described on *Angelica*, has been recorded on *Eryngium*.

ERYTHRONIUM: Dog's-tooth Violet, Trout Lily

Uromyces erythronii Pass., 0, 1, 111
Aecia on petioles and lower leaf surface, very small, spores 20–30 × 15–24. Telia amphigenous, chocolate brown; spores 22–40 × 17–25, brown, veined on the surface. On *E. dens-canis*, rare.

Ustilago heufleri Fuckel
Smut sori on leaves of *E. oreganum*, conspicuous, round or elongated, covered by a thin whitish membrane. Spore mass black, dusty. Spores spherical or ovoid, dark reddish brown, 13–22 diam., thick-walled, smooth but with distinct projections from the inner into the outer wall.

EUPATORIUM: Hemp Agrimony

Allophylaria sublicoides (P. Karsten)Nannf. (Fig. 1473)
Apothecia when fresh up to 0.7mm diam., pale ochraceous. Mature

ascospores 3-septate, hyaline, 17–29 × 5–7. On dead stems of *E. cannabinum*.

Erysiphe cichoracearum, powdery mildew described on *Aster*, is quite common also on *Eupatorium*.

Fusariella sarniensis M.B.Ellis (Fig. 1474)
Colonies effuse, black, powdery. Conidiophores hyaline, up to 60 × 3–4. Conidia pale to mid grey, smooth, 3-septate, 15–18 × 5–7. On dead stems.

Leptosphaeria agnita (Desm.)Ces. & de Not. (Fig. 1475)
Pseudothecia immersed, black, 0.35–0.4mm diam. Ascospores rather pale straw-coloured, 6-septate, 27–36 × 4.5–5.5. On dead stems, May–July, common.

Leptostroma eupatorii Allesch.
Conidiomata in clusters or lines, subcuticular, black, 0.1–0.2mm diam., each opening by a slit. Conidia hyaline, 7–9 × 2. On dead stems, Apr.–May.

Mollisia coerulans Quél. (Fig. 1476)
Apothecia 0.5–1.5mm diam., pale olivaceous or bluish grey with white rims. Ascospores hyaline, 10–17 × 1.5. On dead stems, most commonly of *E. cannabinum*, but we have seen it also on *Artemisia vulgaris* and *Centaurea nigra*, Apr.–June.

Phomopsis eupatoriicola Petrak
Conidiomata sometimes causing stripes or blackish flecks, 1–6 × 0.5–3mm. Conidia spindle-shaped, hyaline, often with two guttules, 6–8 × 2.5–3.5. On old stems, June.

Sporidesmium eupatoriicola M.B.Ellis (Fig. 1477)
Colonies effuse, blackish brown or black, hairy. Conidiophores blackish brown, 70–150 × 5–7. Conidia dark brown, 60–195 × 8–11, with 14 to 31 septa. On dead stems, Mar.–Oct.

Other fungi found mostly on dead stems include: *Anthostomella tomicoides* (*Carex*), *Chromelosporium ochraceum*, *Crocicreas cyathoideum*, *Hymenoscyphus repandus* (*Epilobium*), *Leptosphaeria haematites* (*Clematis*), *Leptospora rubella*, *Melomastia mastoidea* (wood and bark), *Myrothecium carmichaelii* (*Thalictrum*), *Nectria ellisii*, *Sporidesmium cookei* (*Sambucus*), *Unguicularia cirrhata* and *U. millepunctata*.

EUPHORBIA: Spurges

UREDINALES

Five rusts are recorded on *Euphorbia* species in Britain but only one, *Melampsora euphorbiae*, is common and widespread. The others are very seldom seen and are host limited.

Melampsora euphorbiae Cast., 11, 111
Uredinia mostly hypophyllous, small, yellow, with numerous capitate paraphyses; spores echinulate, 16–22 × 13–20. Telia on stems and both leaf surfaces, reddish brown to black; spores 20–60 × 8–15. Common on *E. helioscopia* and *E. peplus*; recorded occasionally also on *E. amygdaloides*, *E. cyparissias*, *E. hiberna* and *E. paralias*.

On *E. amygdaloides*:

Endophyllum euphorbiae-sylvaticae (DC.)Wint., 0, 111
Telia aecidioid, mostly hypophyllous; spores in chains, verruculose, 17–24 × 12–22. Infected plants tend to have long shoots and narrow, pale leaves, Apr.–June.

On *E. cyparissias*:

Uromyces pisi-sativi (Pers.)Liro, 0, 1
Aecia evenly scattered on lower leaf surface, orange with white peridia; spores verruculose, 18–22 diam., May–June. Alternate hosts various Leguminosae.

Uromyces scutellatus (Pers.)Lév., 11, 111
Telia on lower surface of deformed leaves, swollen, dark brown, often with white margins; spores 22–29 × 24–27, with striately arranged warts on walls.

On *E. exigua*:

Uromyces tuberculatus Fuckel, 0, 1, 11, 111
Aecia hypophyllous. Uredinia hypophyllous, cinnamon; spores 20–25 diam. Telia on stems and both leaf surfaces, black or blackish brown; spores 20–30 × 20–24, verrucose, brown with colourless apical papilla.

OTHER FUNGI

Hyalotricha niveocincta Graddon (Fig. 1478)
Apothecia up to 0.3mm diam., golden brown with dark discs surrounded by incurved glassy hairs. Ascospores hyaline, 9–12 × 2.5–3. On dead stems of *E. amygdaloides*, Sept.–Feb.

Mycosphaerella euphorbiae Niessl ex Schröter (Fig. 1479)
Pseudothecia immersed, black, about 0.1mm diam. Ascospores hyaline, 1-septate, 8–14 × 2–3.5. On dead stems of *E. cyparissias* and *E. paralias*, May.

Naeviopsis tithymalina (Kunze)Hein (Fig. 1480)
Apothecia immersed, up to 0.75mm diam., overlying tissue becoming split into 4 pale brown lobes to expose the dark flesh pink disc. Ascospores hyaline, 9–12 × 5–7. On dead stems of *E. amygdaloides* and *E. cyparissias*.

Peronospora cyparissiae de Bary
Downy mildew on *E. amygdaloides*. Sporangia ellipsoid, pale violaceous, 17–24 × 13–16.

Peronospora euphorbiae Fuckel
On various species including *E. paralias*. Sporangia hyaline, 13–17 × 12–15; oospores brown or yellowish, thick-walled, 23–33 diam.

Peronospora valesiaca Gäum.
On *E. paralias*. Shoots turn pale yellow and are very conspicuous. Sporangia pale smoky brown, mostly globose or broadly oval, 21–28 × 19–23.

Phomopsis euphorbiae (Sacc.)Trav.
Conidiomata elongated, black, subepidermal; surrounding tissues stained blackish brown. Conidia broadly fusiform, biguttulate, 7–8 × 2.5–3.5. On dead stems of *E. amygdaloides*, *E. cyparissias* and *E. paralias*, Apr.–June.

Sphaerotheca euphorbiae (Castagne)Salmon
On *E. peplus*. The only powdery mildew recorded on *Euphorbia*. Cleistothecia surrounded by a thick felt of brown hyphae.

EUPHRASIA: Eyebright

Coleosporium tussilaginis, 11, 111, described on *Tussilago*; this orange-coloured rust is common on *E. officinalis*.

FILIPENDULA: Dropwort, Meadow-sweet

UREDINALES

Triphragmium filipendulae Pass., 1, 11, 111
A rare rust recorded only a few times on *F. vulgaris*. Similar to *T. ulmariae* but with thinner-walled aeciospores and smaller warts on the teliospores.

Triphragmium ulmariae (DC.)Wint., 0, 1, 11, 111 (Fig. 1481)
Aecia uredinoid, bright orange, conspicuous on veins and petioles in May and June; spores echinulate, 25–27 × 18–20. Uredinia hypophyllous, pale yellow. Telia hypophyllous, small, blackish brown; spores triquestrously 3-celled, mostly 40–45 diam. On *F. ulmaria*, common.

USTILAGINALES

Urocystis filipendulae (Tul.)Schröter
A rare smut recorded on *F. vulgaris* in May. Sori in petioles and midribs of radical leaves. Spore balls composed of 1 to 7 brown spores 15–25 × 10–15, surrounded by smaller sterile cells.

DISCOMYCETES

Allophylaria clavuliformis (P.Karsten)P.Karsten (Fig. 1482)
Apothecia at first yellow, becoming vermilion and then drying pale yellow, 0.3–0.5mm diam. Ascospores hyaline, with large guttules, 13–20 × 4–5. On dead stems of *F. ulmaria*, Oct.–Mar.

Calycellina spiraeae (Rob. ex Desm.)Dennis (Fig. 1483)
Apothecia greyish brown, 0.2–0.27mm diam. Ascospores hyaline, 8–10 × 4. Abundant on fallen dead leaves of *F. ulmaria*, May–Sept.

Dasyscyphus aeruginellus (P.Karsten)Korf & Dixon (Fig. 1484)
Apothecia erumpent, bluish green, up to 2mm diam. Hairs smooth, about 20 × 3. Ascospores hyaline, 7–10 × 2–3. On rotting stems of *F. ulmaria*, Oct., uncommon.

Dasyscyphus nudipes (Fuckel)Sacc. (Fig. 1485)
Apothecia 0.5–1.5mm diam., white with yellowish discs which dry pale brown. Hairs hyaline, 55–80 × 3, swollen at tips to 4–5, sometimes with crystals. Ascospores hyaline, 10–12 × 1.5–2. Very common on dead stems of *F. ulmaria*, May–Aug.

Hyaloscypha herbarum Velen.
Apothecia sessile, 0.3–0.7mm, white. Hairs 15–25 long, constricted at ends into a solid point. Asci 25–35 × 6–8. Paraphyses filiform, forked at base. Ascospores hyaline, cylindrical, obtuse, mostly arcuate, 6–16 long. On dead stems of *F. ulmaria*, Nov.–Feb.

Hymenoscyphus vitellinus (Rehm)O.Kuntze (Fig. 1486)
Apothecia 1.5–3mm diam., lemon yellow or deep cream, with pale, often rather long stalks. Ascospores hyaline, 16–20 × 3–3.5. On dead stems of *F. ulmaria*, July–Oct.

Mollisia fuscostriata Graddon (Fig. 1487)
Apothecia up to 0.5mm diam., pale brown, each with a submarginal ring of evenly spaced tufts of dark brown hyphae. Ascospores hyaline, 7–8 × 1–1.5. On dead stems of *F. ulmaria*, May–Aug.

Mollisiopsis lanceolata (Gremmen)D. Hawksw. (Fig. 1488)
Apothecia up to 0.5mm diam., grey or buff with white margins. Paraphyses projecting, lanceolate. Ascospores hyaline, 6–8 × 1.5–2. On dead stems of *F. ulmaria* (the type host); also on *Lamium*, *Lycopus* and *Rubus*, May–Aug.

Phialina ulmariae (Lasch)Dennis (Fig. 1489)
Apothecia up to 0.5mm diam., pale ochraceous with orange discs when fresh; hairs tapered, with filiform ends, often curved. Paraphyses with yellow oil drops in their tips. Ascospores hyaline, 10–13 × 1.5. Common on dead stems of *F. ulmaria*, June–Oct.

Psilachnum rubrotinctum Graddon (Fig. 1490)
Apothecia subsessile, up to 0.5mm diam., greyish white, turning pale yellow when dry, brick-red immediately on exposure to ammonia vapour. Hairs hyaline, smooth, with 3 or 4 septa, up to 60 × 5. Paraphyses pointed. Ascospores fusiform, hyaline, 6–7 × 1.5–2. On dead stems of *F. ulmaria*, June.

Pyrenopeziza millegrana Boud. (Fig. 1491)
Apothecia 0.25–0.4mm diam., reddish brown with broad white rims. Ascospores hyaline, 17–20 × 3.5–4. On dead stems of *F. ulmaria*, Apr.–Aug.

Pyrenopeziza pulveracea (Fuckel)Gremmen (Fig. 1492)
Apothecia 0.1–0.4mm diam., dark greyish brown to black with in-turned white rims. Ascospores hyaline, 5–8 × 2. On dead stems of *F. ulmaria*, May–June.

Unguicularia ulmariae (Velen.)Dennis (Fig. 1493)
Apothecia 0.15mm diam., white, with bands of glassy hairs which are mostly less than 20 × 4–5. Ascospores hyaline, 4–5 × 1. On dead stems of *F. ulmaria*, May–Sept.

Verpatinia spiraeicola Dennis (Fig. 1494)
Apothecia bell-shaped, ridged, brown, 2 × 1mm, on long stalks attached to small black sclerotia. Ascospores hyaline, 5–6.5 × 1.5–2. On rotting leaves of *F. ulmaria*, May–July.

Other discomycetes found on dead stems of *F. ulmaria* include: *Ciboriopsis tenuistipes, Dasyscyphus clandestinus, D. grevillei, D. nidulus, Helotium consobrinum* (*Rumex*), *Hymenoscyphus repandus* (*Epilobium*), *H. scutula, Sclerotinia fuckeliana* (*Castanea*), and *Unguicularia cirrhata*.

OTHER ASCOMYCETES

Diaporthe lirella (Moug. & Nestl. ex Fr.)Nitschke ex Fuckel (Fig. 1495)
Perithecia immersed beneath blackened areas of epidermis, up to 0.25mm diam., necks not protuberant. Ascospores hyaline, 4-guttulate, septum faint, mostly 10 × 2. On stems of *F. ulmaria*, Sept.–Oct.

Erysiphe ulmariae Desm.
Powdery mildew. Conidia few. Cleistothecia with many narrow appendages. Asci 6 to 15, each with 6 to 8 spores. On *F. ulmaria* and *F. vulgaris*.

Sphaerotheca alchemillae, described on *Alchemilla*, occurs also on *F. ulmaria*, the cleistothecia appearing on living leaves as early as June.

Wentiomyces sp. (Fig. 1496)
Pseudothecia superficial, dark brown to black, mostly 50–80 diam. Setae dark

brown, 30–45 × 4–6. Ascospores pale brown, 8–9 × 2–3. On dead stems of *F. ulmaria*, July.

Other species found on *Filipendula* but not confined to it include: *Actinopeltis palustris* (*Phalaris*), *Lophiostoma fuckelii* var. *pulveraceum* (*Rubus*), *Nectria arenula*, *N. ellisii* (*Phragmites*), *Ophiobolus acuminatus* and *Peroneutypa heteracantha*.

HYPHOMYCETES

Codinaea britannica M.B.Ellis (Fig. 281)
Colonies effuse, black, hairy. Setae blackish brown, up to 400 long. Conidiophores pale brown, up to 120 × 3.5. Conidia hyaline, 15–20 × 2–3, setulae 5–9 long. Common on dead stems of *F. ulmaria*, Sept.–Nov.

Endophragmia prolifera (Sacc., Rouss. & Bomm.)M.B.Ellis (Fig. 1497)
Colonies effuse, grey to black, substrate stained reddish purple. Conidiophores brown, up to 100 × 5–7. Primary conidia 28–34 × 16–20, pale to mid brown with broad black bands at the septa; secondary conidia spherical, 11–22 diam. On dead stems of *F. ulmaria*, Sept.

Ramularia ulmariae Cooke
Colonies hypophyllous. Conidiophores hyaline, 30–50 × 4. Conidia 0- or 1-septate, hyaline, 15–25 × 4–5. On living leaves of *F. ulmaria* causing whitish spots with reddish borders.

Plurivorous hyphomycetes found on dead stems of *F. ulmaria* include: *Coremiella cubispora*, *Endophragmia elliptica*, *Menispora ciliata*, *Periconia britannica* (*Brachypodium*) and *Sporidesmium eupatoriicola* (*Eupatorium*).

COELOMYCETES

Leptostroma spiraeinum Vestergr. (Fig. 1498)
Conidiomata immersed, brown to black, split longitudinally at maturity. Conidia 6–8 × 1.5–2. On dead stems of *F. ulmaria*, Mar.–June.

Thyriostroma spiraeae (Fr.)Died. (Fig. 1499)
Conidiomata subcuticular, thin, very irregular in shape, grey to black. Conidia hyaline, 5–7 × 1. On dead stems of *F. ulmaria*, common, Jan.–June.

FOENICULUM: Fennel

No fungi appear to have special preference for fennel but a number of species common generally on dead herbaceous plants or on other Umbelliferae are found on the dead stems. These include: *Dasyscyphus grevillei, D. mollissimus, Diaporthopsis angelicae* (*Angelica*), *Heterosphaeria patella* (*Angelica*), *Hyalotricha crispula, Leptosphaeria libanotidis* (*Angelica*), *Nectria arenula, Pseudospiropes subuliferus, Pyrenopeziza revincta, Stachybotrys dichroa* and *Volutella ciliata*.

FRAGARIA: Strawberry

Coniella fragariae (Oud.)Sutton (Fig. 1500)
Pycnidia immersed, pale brown. Conidia pale brown, 8–11 × 5.5–7.5, thick-walled. On receptacles.

Diplocarpon earlianum (Ellis & Ev.)Wolf (Fig. 1501)
Acervuli of the *Marssonina* state dark brown to black, 70–100 diam., crowded. Conidia hyaline, 1-septate, 17–30 × 6–8, guttulate. Causes irregular, dark purplish to brown spots 2–3mm diam., which sometimes coalesce to form large dark brown blotches. Infected leaves often turn yellow. On wild and cultivated strawberries.

Gnomonia fragariae Kleb. (Fig. 1502)
Perithecia immersed, black, up to 0.4mm diam., with protuberant necks up to 1.5mm long. Ascospores hyaline, 1-septate, 14–20 × 3–4, with a setula at each end. On dead leaves, May–Aug.

Gnomonia fructicola (Arnaud)Fall (Fig. 1503)
Perithecia immersed, dark brown to black, 0.4–0.6mm diam., with necks up to 1mm long. Ascospores hyaline, 1-septate, 10–12 × 2–3, without setulae. On dead petioles. Conidia of the so-called '*Zythia*' state hyaline, biguttulate, 5–7 × 1.5–2. The small, pale brown conidiomata are found on living leaves, where the fungus causes brown blotches with purplish borders, each surrounded by a yellow halo.

Mycosphaerella fragariae (Tul.)Lindau
Pseudothecia epiphyllous, 0.1–0.16mm diam., usually few and peripheral. Asci 50–70 × 10–13. Ascospores hyaline, fusiform or somewhat clavate, 1-septate, 11–15 × 3. *Ramularia* state has hyaline, 1- to 3-septate conidia 15–45 × 2–3.5. Mostly on leaves causing white or greyish spots 1–3mm diam., with dark reddish or purplish brown margins.

Peronospora fragariae Roze & Cornu
Downy mildew on berries, sepals and lower surface of leaves. Sporangia 20–40 × 17–36.

Sphaerotheca alchemillae, powdery mildew described on *Alchemilla*, forms cleistothecia on peduncles, pedicels, sepals and stolons. When infection is severe, margins of leaves curl upwards.

Stagonospora fragariae Briard & Har.
Pycnidia epiphyllous, pale brown, about 0.1mm diam. Conidia hyaline, 3-septate, 30–45 × 4–8. On living leaves causing brown spots about 1mm diam., with reddish-purple borders.

FUMARIA: Fumitory

Peronospora affinis Rossm.
Downy mildew of *F. officinalis*. Sporangia 22–26 × 15–18.

GAGEA: Yellow Star-of-Bethlehem

Uromyces gageae Beck, 111
Telia amphigenous, dark brown; spores smooth, 26–40 × 20–28. A rare rust on leaves of *G. lutea*, Apr.–May.

Ustilago ornithogali (Schmidt & Kunze)Magn.
A rare smut which forms elongated blisters up to 1cm long on leaves and pedicels of *G. lutea*, Apr. Spores in mass purplish brown, smooth, 13–19 diam.

GALANTHUS: Snowdrop

Botrytis galanthina (Berk. & Br.)Sacc. (Fig. 1504)
Colonies grey or greyish brown, effuse. Sclerotia black, 1–2mm diam. Conidiophores up to 1.5mm long, 16–30 thick, lower part brown, upper part hyaline. Conidia smooth, colourless or pale brown, 10–20 × 8–11. On flowers, shoots, young leaves and bulbs, more frequent in the north of England than in the south. Shoots sometimes decay down to ground level. Sclerotia formed in large numbers on rotting shoots, leaves and occasionally bulbs.

GALEOPSIS: Hemp-nettle

Erysiphe galeopsidis DC.
Powdery mildew. Conidia plentiful. Cleistothecia often arranged in radial rows with the oldest in the centre. Appendages narrow, brownish. Asci 6 to 16, each with 3 to 6 spores which apparently develop the following spring. On *G. speciosa* and *G. tetrahit*.

Pyxidiophora caulicola (D. Hawksw. & Webster)Lundq.
Perithecia superficial, pale ochraceous, up to 0.12mm diam., with necks up to

0.25mm long. Asci 2-spored; spores hyaline, 1-septate, 35–50 × 4–7, with gelatinous sheaths. On dead stems of *G. tetrahit*, Sept.

Septoria galeopsidis Westend.
Pycnidia immersed, brown, hypophyllous, less than 0.1mm diam. Conidia 30–40 × 1–1.5. On leaves of *G. tetrahit*, not common.

GALIUM: Bedstraws, Cleavers or Goosegrass, Crosswort, Sweet Woodruff

UREDINALES

Five rusts occur on *Galium*: (1) *Puccinia difformis*, (2) *Puccinia galii-cruciatae*, (3) *Puccinia galii-verni*, (4) *Puccinia punctata*, (5) *Pucciniastrum guttatum*. These are distributed as follows:

On *G. aparine*, (1), (4), both uncommon on this host
On *G. cruciata*, (2) uncommon, (3), (4)
On *G. mollugo*, (4)
On *G. odoratum*, (4), (5), both uncommon
On *G. palustre*, (4), (5) rare
On *G. saxatile*, (3) common, (4) common, (5) rare
On *G. uliginosum*, (3) common, (4) common, (5) rare
On *G. verum*, (3) uncommon, (4) common, (5) rare

KEY

Teliospores with 1 to 3 vertical septa .... *Pucciniastrum guttatum*
Teliospores with 1 transverse septum 1
1. Telia only, teliospores broadly fusiform .... *Puccinia galii-verni*
   Aecia and telia, teliospores clavate or ellipsoid 2
2. Aecia uredinoid, dark brown .... *Puccinia galii-cruciatae*
   Aecia yellow or orange cluster cups 3
3. Cinnamon-brown uredinia present .... *Puccinia punctata*
   No uredinia formed .... *Puccinia difformis*

Puccinia difformis Kunze, 0, 1, 111
Aecia hypophyllous, pale yellow with reflexed peridia; spores verruculose, 15–25 diam. Telia hypophyllous, compact, black; spores 35–55 × 15–25.

Puccinia galii-cruciatae Duby, 0, 1, 11, 111
Aecia hypophyllous, uredinoid, circinate, dark brown; spores brown, echinulate, 22–32 × 18–25. Uredinia amphigenous, very small, pale brown; spores similar to aeciospores. Telia hypophyllous and on stems, small, blackish brown; spores 25–66 × 18–25.

Puccinia galii-verni Ces., 111
Telia hypophyllous and on stems, chestnut or greyish brown; spores tapered at each end, 35–60 × 12–17.

Puccinia punctata Link, 0, 1, 11, 111 (Fig. 1505)
Aecia hypophyllous, orange with white recurved peridia; spores finely verruculose, 16–20 diam. Uredinia amphigenous, cinnamon brown; spores finely echinulate, 22–30 × 18–22. Telia amphigenous, dark brown to black; spores 30–55 × 15–24.

Pucciniastrum guttatum (Schröter)Hyl., Jørst. & Nannf., 11, 111
Uredinia less than 0.25 mm diam., yellowish orange, opening by a pore; spores minutely echinulate, 14–23 × 11–17. Telia inconspicuous on dark brown areas which may cover the whole leaf; spores golden brown, with 1 to 3 vertical septa, 20–30 diam.

USTILAGINALES

Melanotaenium endogenum (Unger)de Bary (Fig. 1506)
This smut forms sori in stems and leaf bases causing thick, black, gall-like swellings, often at the nodes of stunted shoots. Spores dark brown, mostly 17–22 diam. Usually found on *G. verum* but has been collected also on *G. mollugo*, June–July.

OTHER FUNGI

Acrospermum pallidulum Kirschst. (Fig. 1507)
Ascocarps erect, laterally compressed, opening by an apical pore, pale cream, less than 1mm tall. Ascospores hyaline, multiguttulate, up to 300 × 1. On dead stem bases of *G. cruciata*, Apr.–July.

Diplodina galii Sacc. (Fig. 1508)
Pycnidia black, 0.1–0.2mm diam. Conidia hyaline, 1-septate, 9–13 × 3. On dead stems of *G. mollugo*, Apr.

Erysiphe galii Blumer
Powdery mildew on *G. aparine*. Conidia fairly abundant. Perithecial appendages numerous, hyaline. Asci 5 to 10, each with 2 spores.

Leptosphaeria galiorum (Rob. ex Desm.)Ces. & de Not.
Pseudothecia immersed, up to 0.35mm diam., black. Ascospores pale golden brown, 3-septate, mostly 35–45 × 7–9. On dead stems of *G. aparine*.

Leptosphaeria scitula Sydow (Fig. 1509)
Pseudothecia immersed, about 0.2mm diam., black, tending to collapse and flatten when dry. Ascospores golden brown, 3-septate, mostly 25–30 × 3–3.5. On dead stems of *G. aparine*, Mar.

Leptotrochila verrucosa (Wallr.)Schüepp. (Fig. 1510)
Apothecia 0.3–1mm diam., yellowish brown with pale rims. Ascospores

hyaline, mostly 8–14 × 3–3.5. *Sporonema* state with conidiomata about 0.4mm diam. and hyaline, rod-shaped conidia 4–7 × 1–2. On living and fading leaves and stems of *G. mollugo*, *G. odoratum* and *G. saxatile*, Apr.–Oct.

Peronospora aparines Gäum.
Downy mildew on *G. aparine*, especially on seedlings, spring and autumn. Sporangia 28–34 × 19–22.

Peronospora calotheca de Bary
Downy mildew on brown areas on leaves of *G. odorata*. Sporangia 21–27 × 17–19.

Phomopsis elliptica (Peck)Grove
Conidiomata erumpent, black, about 0.1mm diam. A-conidia oblong-lanceolate, 9–10 × 2.5–3, B-conidia mostly hooked, 18–20 × 0.5. On dead stems of *G. aparine* and *G. mollugo*, Jan.–Apr.

Symphyosirinia galii E.A. Ellis (Fig. 1511)
Apothecia stalked, pale greyish brown, 1–1.5mm diam. Ascospores hyaline, 10–15 × 3–4. *Symphyosira* state with white or pale yellow synnemata up to 3mm high. Conidia white or pale rose in mass, 5- to 11-septate, 35–60 × 5–7, with 1 to 3 setulae up to 50 long. On 1-year-old fallen fruits of *G. palustre*, Sept.–Oct.

GENTIANA: Gentians

Puccinia gentianae Röhl., 1, 11, 111
Aecia on stems, peduncles and lower leaf surfaces, orange with white peridia; spores 16–22 × 14–16. Uredinia pale brown; spores finely echinulate, 20–30 × 18–24. Telia dark blackish brown; spores 1-septate, 30–38 × 24–30. On *G. acaulis* and *G. verna*, rare.

GENTIANELLA: Felwort

Uromyces gentianae Arth., 11, 111
Uredinia amphigenous, pale brown; spores echinulate, 20–25 × 16–22. Telia amphigenous, cinnamon brown; spores verruculose, 15–18 × 18–21. On *G. amarella*, rare.

GERANIUM: Cranesbills, Herb Robert

Puccinia polygoni-amphibii Pers. var *convolvuli* Arth., 0, 1
Aecia on *G. dissectum*, rare. Alternate host *Polygonum convolvulus*.

Uromyces geranii (DC.)Fr., 0, 1, 11, 111
Aecia orange. Uredinia cinnamon brown; spores minutely echinulate, 20–30 × 20–26. Telia blackish brown; spores 22–40 × 15–25, smooth with a hyaline papilla. Common on the lower surface of leaves of many species including *G. dissectum*, *G. molle*, *G. pratense*, *G. pusillum*, *G. pyrenaicum*, *G. robertianum*, *G. rotundifolium* and *G. sylvaticum*.

Coleroa circinans (Fr.)Winter (Fig. 1512)
Pseudothecia becoming superficial, about 0.1mm diam., dark brown to black, setose. Setae about 30 long. Ascospores very pale brown, 1-septate, 10–12 × 4–5. Mostly on the upper surface of living leaves of *G. dissectum*, *G. molle*, *G. pusillum* and *G. rotundifolium*, Oct.–Apr.

Hormotheca robertiani (Fr.)Höhnel (Fig. 1513)
Pseudothecia superficial, hemispherical, up to 0.15mm diam., black, shining. Ascospores pale olivaceous, 1-septate, 10–13 × 4–5. Common on the upper surface of living leaves of *G. robertianum*, especially those which have overwintered, May–Sept.

Peronospora conglomerata Fuckel
Downy mildew on *G. molle*, *G. pusillum* and occasionally other species. Sporangia 23–25 × 22–24.

Plasmopara pusilla (de Bary)Schröter
Downy mildew on *G. pratense*. Sporangiophores short with branches about 10–11 long. Sporangia ovoid or obovoid, papillate, 20–26 long or occasionally longer. Oospores golden brown, 40 diam.

Ramularia geranii (Westend.)Fuckel
Colonies hypophyllous, white. Conidiophores about 30 × 2. Conidia hyaline, 1-septate, 18–20 × 2.5–3. On round, sometimes confluent, brown spots on living leaves of *G. dissectum*, *G. pratense*, *G. pusillum* and *G. sanguineum*.

Sphaerotheca fugax Penzig & Sacc.
Powdery mildew. Conidia abundant. Perithecia surrounded by brown hyphae. On several species including *G. dissectum*, *G. molle* and *G. pusillum*.

Venturia geranii (Fr.)Winter (Fig. 1514)
Pseudothecia aggregated, immersed, black, with or without brown setae up to 25 long around the papillate ostioles. Ascospores olivaceous, 1-septate, 9–12 × 3.5–4. On dark brown spots on living leaves of *G. dissectum*, *G. molle* and *G. sylvaticum*, Nov.–May.

GEUM: Herb Bennet or Wood Avens, Water Avens

Peronospora gei H.Syd.
Downy mildew. Occurs on *G. rivale* and *G. urbanum* but British collections

mostly on cultivated species. Sporangia mostly 15–19 × 13–18, pale brown. Causes pale yellowish patches on upper surface of leaves, corresponding to colonies on the lower surface.

Ramularia gei (Eliass.)Lindroth (Fig. 1515)
Colonies mostly hypophyllous. Conidiophores in fascicles of 5 to 30, hyaline, 12–30 × 2.5–4. Conidia hyaline, 0- or 1-septate, 10–20 × 3–4. Causes pale brown spots 1–6mm diam. on leaves of *G. urbanum*, sometimes with brown or purple borders.

Sphaerotheca alchemillae, powdery mildew described on *Alchemilla*, occurs on *G. urbanum*.

GLADIOLUS: Gladiolus

Two rusts are recorded from Britain on cultivated species: *Puccinia gladioli* (Duby)Cast., one collection only, and *Uromyces transversalis* (Thümen)Wint., found on material imported from South Africa.

Urocystis gladiolicola Ainsworth
Smut sori in corms and forming dark-brown blisters on leaves parallel with the veins. Spore balls 15–28 diam., each with 1 or 2 reddish-brown spores surrounded by colourless sterile cells. On cultivated plants, June–Dec., uncommon.

Botryotinia draytonii (Buddin & Wakefield)Seaver
Apothecia 2.5–5mm diam., yellowish brown, stalks up to 12mm long arising from lens-shaped black sclerotia embedded in stems. Ascospores hyaline, biguttulate, 12–17 × 8–10. *Botrytis* state with conidia 10–22 × 8–13 on corms, leaf spots and water-soaked oval lesions on flowers. There are two types of leaf spot, one oblong, brown with a red margin, the other small, round and rust coloured.

Curvularia trifolii (Kauffm.)Boedijn f.sp. *gladioli* Parmelee & Luttrell (Fig. 1516)
Colonies greyish brown to black, hairy. Conidiophores brown, up to 150 × 5–7. Conidia 3-septate, 23–38 × 12–16, with dark-brown median cells and pale end ones. On leaves, stems and flowers, causing brown spots; can cause rotting of corms.

Penicillium corymbiferum Westl. and *P. gladioli* McCulloch & Thom both occur on corms. These green and blue–green moulds need cultures for accurate determination.

Septoria gladioli Passer.
Pycnidia black, about 0.1mm diam., clustered. Conidia cylindrical or slightly broader at one end, hyaline, with many guttules or 3-septate, 20–60 × 2.5–4.

On leaves, where it causes whitish spots 3–4mm diam. with brown borders, and on corms and stems.

Stromatinia gladioli (Drayton)Whetzel
Apothecia brown, 3–7mm diam., with stalks 6–8mm long which arise from a thin, black, subcuticular stroma. Ascospores uniseriate in asci, elliptical, hyaline, 10–17 × 6–9.5. On roots and sometimes corms, causing 'dry rot'.

GLAUCIUM: Yellow Horned Poppy

Ascochyta glaucii Died.
Pycnidia mostly 0.15–0.2mm diam., brown, darker around the ostiole. Conidia narrowly ellipsoidal, 1-septate, not constricted at the septum, 10–13 × 2.5–3. On dead stems of *G. flavum*, Apr.–May.

GLAUX: Sea Milkwort

Uromyces lineolatus (Desm.)Schröter, 0, 1
Aecia on yellowish spots on petioles and lower leaf surfaces of *G. maritima*, May–June. Alternate host *Scirpus maritimus*.

Septoria glaucis Syd.
Pycnidia amphigenous immersed, 0.1mm diam.; conidia hyaline, 40–60 × 2.5. On fading and dead basal leaves, Aug.–Sept.

GLECHOMA: Ground Ivy

Puccinia glechomatis DC., 111
Telia hypophyllous and on petioles, sometimes on brown spots; spores 30–48 × 16–24, common.

Erysiphe biocellata, powdery mildew described on *Mentha*, is found occasionally on *G. hederacea*. Few conidia formed, asci mostly with 2 spores.

Erysiphe galeopsidis, powdery mildew described on *Galeopsis*, is common also on *Glechoma*. Conidia formed abundantly. Asci mostly with 3 to 6 spores, which develop the following spring.

Ramularia calcea (Desm.)Ces. (Fig. 1517)
Colonies hypophyllous, white. Conidiophores 15–38 × 2–3. Conidia hyaline, 0- or 1-septate, 8–25 × 2–3.5. Causes white spots 2–5mm diam., with narrow brown borders, on living leaves of *G. hederacea*, very common, July–Oct.

GOODYERA: Creeping Ladies' Tresses

Uredo goodyerae Tranz., 11
Uredinia amphigenous, very small, orange to pale yellow; spores finely echinulate, 24–34 × 16–20. On *G. repens*, rare.

GYMNADENIA: Fragrant Orchid

Two rusts form aecia on *G. conopsea*: *Melampsora epitea*, with as alternate host *Salix*, and *Puccinia sessilis*, with as alternate host *Phalaris arundinacea*. In the *Melampsora* the aecia are caeomoid with at the most a very rudimentary and inconspicuous peridium. In the *Puccinia*, on the other hand, they are well-defined cluster cups each with a well-developed, white, recurved peridium.

GYSOPHILA: Maidens' Breath

Puccinia arenariae, described on *Arenaria*, is seen occasionally on *G. elegans*.

Stromatinia serica (Keay)Kohn
Apothecia 2–5mm diam., pale reddish brown with stalks tapering towards the base, arising from black sclerotia in stems. Ascospores hyaline, 14–26 × 7–11. Uncommon.

HALIMIONE: Sea Purslane

Ascochytula obiones (Jaap)Died. (Fig. 1518)
Pycnidia immersed becoming erumpent, 0.15–0.2mm diam. Conidia pale yellow or yellowish brown in mass, 1-septate, 9–12 × 2.5–4. On dead leaves, stems and propagules, Apr.–June.

Camarosporium obiones Jaap (Fig. 1519)
Pycnidia immersed, up to 0.2mm diam. Conidia smooth, pale to mid golden brown, muriform, 14–23 × 8–13. On dead stems, Apr.–Nov.

Coniothyrium obiones Jaap
Pycnidia gregarious, immersed becoming erumpent, 0.15–0.2mm diam. Conidia pale olivaceous, mostly 2.5–4 diam. On dead stems and propagules, May–Nov.

Phomopsis piceae (Fr.)Höhnel f. *obiones* Grove
Conidiomata gregarious, within blackened areas. Conidia ellipsoid to fusiform, hyaline, 7–8 × 2.5–3. On dead stems, Sept.–Oct.

Trematosphaeria circinans, described on rootstocks of *Medicago*, occurs also on *Halimione*.

HELIANTHEMUM: Rockrose

Leveillula taurica (Lév.)Arnaud (Fig. 1520)
A white or yellowish, mainly tropical mildew with a wide host range there, recorded on *H. nummularium* and *Cynara scolymus* in Britain. Cleistothecia rarely seen, up to 0.25mm diam. with up to 30 asci, 2-spored; spores 25–40 × 14–22. *Oidiopsis* state with variously shaped hyaline conidia 50–80 × 15–20, formed singly on short conidiophores.

Peronospora leptoclada Sacc.
Downy mildew. Sporangiophores 300–350 × 18, 5 or 6 times dichotomously branched above; sporangia hyaline, 25–28 × 20–22. Immersed oospores pale yellowish, 25–28 diam. On the lower surface of fading leaves.

Septoria chamaecysti Vestergr.
Pycnidia less than 0.1mm diam., black. Conidia straight or slightly curved, hyaline, occasionally guttulate, 20–40 × 1. On living leaves causing somewhat swollen, brown or whitish spots with dark brown borders, about 1mm diam., which are sometimes confluent.

HELLEBORUS: Bear's-foot, Green and Stinking Hellebores

Urocystis floccosa (Wallr.)Henderson
Smut sori mostly on leaves and petioles of *H. viridis*. Spore mass black, powdery. Spore balls 20–30 diam., usually composed of 1 to 3 smooth reddish-brown spores 13–19 diam., surrounded by 1 to 3 yellowish-brown, smooth, sterile cells, which are detached easily.

Coniothyrium hellebori Cooke & Massee
Pycnidia mostly epiphyllous in the centre of spots. Conidia pale olivaceous brown, 4–5 × 2–3. On leaves of *H. viridis*, causing dark blackish-brown, often concentrically zoned spots with paler centres, 1–3cm diam., Jan.–Sept.

Ramularia hellebori Fuckel
Colonies white. Conidiophores hyaline, 20–30 long. Conidia hyaline, 0- or 1-septate, 24–30 × 4–5. Causes pale spots with dark margins on leaves of *H. foetidus* and *H. viridis*.

HERACLEUM: Hogweed

Puccinia heraclei Grev., 0, 1, 11, 111
Aecia on petioles and lower leaf surfaces, yellow with little peridium. Uredinia amphigenous, brown; spores echinulate, 25–30 × 20–27. Telia small, dark brown to black; spores 1-septate, blackish brown with finely reticulate walls, 27–37 × 19–27. Uncommon.

Dactylella arnaudii Yadav (Fig. 1521)
Colonies effuse, white, cottony, thin. Conidiophores hyaline, up to 100 × 3–4.5. Conidia hyaline, fusiform, some rostrate, 3- to 12-septate, 30–90 × 3.5–10. On dead stems, Aug.–Sept.

Erysiphe heraclei Schleich. ex DC. (Fig. 1522)
Powdery mildew. Conidia abundant, solitary or in chains, hyaline, 35–45 × 15–20. Cleistothecia about 0.1mm diam., with numerous branched appendages. Asci 3 to 8 with 3 to 6 spores each. Ascospores 20–27 × 11–15. Very common, July–Oct.

Phloeospora heraclei (Lib.)Petrak
Conidiomata scattered or in small groups, mainly on the lower surface, immersed, blackish brown. Conidia cylindrical, arcuate, guttulate becoming 1- to 4-septate, 45–60 × 3.5–4, emerging in tendrils and forming white patches on the surface. On living and fading leaves sometimes causing slight yellowing but no distinct spots, July–Oct.

Phoma complanata (Tode)Desm.
Pycnidia black, shining, thick-walled, up to 0.5mm diam., erumpent, sometimes becoming superficial. Conidia hyaline, cylindrical or ellipsoid, guttulate, 5–9 × 2–3.5. On dead stems.

Phoma longissima (Pers.)Westend.
Pycnidia immersed, densely crowded, very small, black, linked by thick wefts of dark-brown hyphae. Conidia ovoid, biguttulate, 4–6 × 1.5–2. On dead stems, Sept.–Jan., causing parallel black stripes 0.5–2mm broad, often several cm long, and covering large areas.

Pirottaea nigrostriata Graddon (Fig. 1523)
Apothecia erumpent, up to 0.5mm diam., pale brown, with tufts of very dark brown hairs which give the outside a striate appearance. Ascospores hyaline, 9–12 × 2–2.5. Mainly on the residual cortex of dead stems, Mar.–June.

Pyrenopeziza chailletii Fuckel (Fig. 1524)
Apothecia up to 0.6mm diam., dark brown to black, with pale discs and hyaline hairs at the margin. Ascospores hyaline, 15–21 × 3–4. On dead stems, Feb.–Mar.

Ramularia heraclei (Oudem.)Sacc. (Fig. 1525)
Conidiophores in small, scattered groups, hyaline, 60–110 × 2–4. Conidia hyaline, 0- to 2-septate, 10–34 × 4–6. On lower surface of leaves, causing round or angular whitish spots 1–3mm diam., each with a narrow brown border.

Stachybotrys cylindrospora Jensen (Fig. 1526)
Conidiophores with lower part hyaline and smooth, upper part smoky grey, verrucose, up to 100 × 3–5. Conidia grey with dark longitudinal striations, 11–15 × 4–5. On dead stems, May–Sept.

Symphyosirinia heraclei E.A.Ellis (Fig. 1527)
Apothecia 3–5mm diam., pale brown, long-stalked. Ascospores becoming 1-septate, hyaline, 20–24 × 6–7. Synnemata of *Symphyosira* state white to flesh-coloured, clavate, up to 15mm high. Conidia hyaline to pale pink, mostly 5- to 7-septate and 35–50 × 6–8, rarely up to 70 long. On fallen fruits of *H. sphondylium*, Aug.

Taphridium umbelliferarum (Rostr.)Lagerheim & Juel (Fig. 1528)
Chlamydospores forming a compact layer beneath the upper epidermis, 50–75 × 30–60. On leaves, Apr.–June; affected areas often extensive, pale greyish green.

Fungi commonly found on dead stems of *H. sphondylium* but not restricted to or especially favouring this host:

Discomycetes: *Crocicreas cyathoideum, Dasyscyphus grevillei, D. mollissimus, D. sulphureus, Hymenoscyphus herbarum, Pyrenopeziza adenostylidis, P. revincta* and *Unguiculella hamulata*. **Other ascomycetes**: *Acrospermum compressum* (*Urtica*) *Arnium apiculatum* and *Diaporthe arctii* (*Arctium*), *Diaporthopsis angelicae* (*Angelica*), *Lasiosphaeria caudata, Leptospora rubella, Lophiostoma caulium, Ophiobolus erythrosporus, Pleospora herbarum*. **Hyphomycetes**: *Alternaria alternata, Botrytis cinerea, Camposporium pellucidum, Chromelosporium ochraceum, Cladosporium herbarum, Dactylosporium macropus, Dendryphiella infuscans, D. vinosa, Dendryphion comosum, D. nanum, Dictyosporium toruloides, Endophragmia hyalosperma, Epicoccum purpurascens, Fusariella hughesii, Gliomastix luzulae, Periconia byssoides, P. cookei, Pleurophragmium parvisporum, Septofusidium herbarum, Stachybotrys dichroa, Torula herbarum, Trichocladium opacum, Trichoderma koningii, Verticillium* state of *Nectria inventa*.

HESPERIS: Dame's Violet

Erysiphe cruciferarum, described on *Brassica*, and *Leptosphaeria maculans*, described on *Alliaria*, have been found on *H. matronalis*.

HIERACIUM: Hawkweeds

Puccinia hieracii Mart., 0, 1, 11, 111
Aecia uredinoid, cinnamon brown, on reddish yellow spots. Uredinia on stems and leaves, very small; spores echinulate, 20–30 × 17–25. Telia often on stems, very dark brown; spores minutely verruculose, 1-septate, mostly 30–40

× 20–28. The typical form is found on a number of species including *H. umbellatum* and *H. vulgatum*. The variety which occurs on *H. pilosella* is *P. hieracii* var. *piloselloidarum* (Probst)Jørst.

Entyloma calendulae (Oudem.)de Bary f. *hieracii* Schröter
Sori embedded in leaves of *H. murorum* and *H. vulgatum*. Spores very pale yellow, smooth, 9–14 diam.

Phomopsis hieracii H.C.Greene
Conidiomata erumpent, black, up to 0.25mm diam. A-conidia fusiform or ellipsoid, mostly 6–10 × 2–3, B-conidia usually curved, 14–20 × 0.5–1. On dead stems of *H. pilosella*, Aug.

Septoria mougeotii Sacc. & Roum.
Pycnidia epiphyllous, immersed, blackish brown, about 0.1mm diam. Conidia 15–40 × 1. On leaves of *H. umbellatum*, causing large, yellowish spots.

HIPPOCREPIS: Horse-shoe Vetch

Uromyces anthyllidis, described on *Anthyllis*, has been recorded on *Hippocrepis*.

HONKENYA: Sea Sandwort

Peronospora honckenyae H.Syd.
Downy mildew on dwarfed, yellowish plants of *H. peploides*. Sporangia 20–23 × 15–18.

HUMULUS: Hops

Ascochyta humuli Kab. & Bubák
Pycnidia epiphyllous, immersed, brown, up to 0.15mm diam. Conidia hyaline, mostly 1-septate, 8–15 × 3–5. Causes rather irregular, greyish leaf spots, with narrow dark-brown borders.

Dasyscyphus humuli (Phill.)Dennis
Apothecia 0.2–0.4mm diam., sessile, with incurved margins, pale straw-coloured; hairs cylindrical, granulate, up to 30 × 5. Paraphyses cylindrical, narrow. Ascospores narrowly clavate, hyaline, 6–10 × 1–1.5. On dead stems.

Diaporthe sarmenticia Sacc.
Perithecia gregarious, up to 0.35mm diam., scarcely visible on the surface but circumscribed in the wood by a wavy black line. Ascospores hyaline, 1-septate, constricted at the septum, tapered towards the ends, about 10 × 4–5. On dead stems.

Mycocentrospora cantuariensis (Salmon & Wormald)Deighton (Fig. 1529)
Colonies amphigenous. Conidiophores in small groups above stomata and sometimes arising from external mycelium, up to 115 × 14, pale olivaceous brown. Conidia pale olivaceous, up to 500 × 10–21, with 5 to 19 septa. On leaves, causing suborbicular, pale greyish brown, purple-bordered spots 1–5mm diam., each surrounded by a yellowish zone. Uncommon.

Pseudoperonospora humuli (Miyabe & Tak.)C.G.Wilson (Fig. 1530)
Downy mildew on underside of leaves and on cones; colonies dark grey, velvety. Sporangia hyaline to pale brown, smooth 22–29 × 16–19. Causes irregular yellowish spots limited by veins, and sometimes stunting and leaf curl, Aug.–Sept. and May.

Sphaerotheca macularis (Wallr.)Lind (Fig. 1531)
Powdery mildew. Conidia plentiful, barrel-shaped, hyaline, in chains, mostly 25–35 × 15–22. Cleistothecia up to 0.12mm diam., with brown appendages, containing one 8-spored ascus. Ascospores hyaline, 20–25 × 14–17. Aug.–Sept.

Verticillium. Progressive and fluctuating wilts are caused by *V. albo-atrum* and *V. dahliae*, and black root rot by *Phytophthora citricola*; cultural studies are needed for determination of these. Other fungi found on hops include grey mould, *Botrytis cinerea*; *Gibberella pulicaris*, a canker organism described on *Lupinus*; *Pezizella discreta*; *Pezizellaster tami*; and *Sclerotinia sclerotiorum*.

HYDROCOTYLE: Pennywort

Puccinia hydrocotyles Cooke, 11
Uredinia amphigenous but mostly on the lower leaf surface, small, cinnamon brown, often circinate; spores echinulate, 25–33 × 20–26. Rare.

Septoria hydrocotyles Desm.
Pycnidia epiphyllous, less than 0.1mm diam., brown. Conidia hyaline, guttulate, sometimes with 3 septa, 16–25 × 1–2. Causes irregular, pale brown or whitish spots on leaves.

HYPERICUM: St. John's Worts, Tutsan

Melampsora hypericorum Wint., 1, 111
Aecia hypophyllous, caeomoid, orange; spores mostly 20–28 × 10–18. Telia hypophyllous, reddish brown; spores 30–40 × 10–17. Common on *H. androsaemum*, less so on other species.

Erysiphe hyperici (Wallr.)Blumer
Powdery mildew producing abundant conidia. Cleistothecia contain 3 to 7 asci

each with 3 to 5 spores. On *H. hirsutum*, *H. maculatum*, *H. perforatum* and *H. tetrapterum*.

Keissleriella ocellata (Niessl)Bose
Pseudothecia subepidermal, without setae, up to 0.3mm diam. Ascospores 2-septate, with large oil drops, mostly hyaline but sometimes becoming brownish and verruculose when old, 16–21 × 6–7. On blackened areas on dead stems of *H. maculatum* and *H. perforatum*.

Mycosphaerella elodis (A.L.Sm. & Ramsb.)Tomilin
Pseudothecia erumpent, black, 0.1–0.15mm diam. Ascospores hyaline, with almost median septum, 14–16 × 2–3.5. On leaves of *H. elodes*, Aug.

Seimatosporium hypericinum (Ces.)Sutton (Fig. 1532)
Acervuli brown, 0.2mm diam. Conidia curved, 3-septate, 15–19 × 4.5–5.5, median cells pale brown, cell at each end hyaline with two unbranched setulae. On dead stems of *H. perforatum* and *H. tetrapterum*.

Septoria hyperici Desm.
Pycnidia epiphyllous, brown, less than 0.1mm diam. Conidia 30–50 long, guttulate, sometimes with 3 or 4 septa. Causes reddish-brown, yellow-bordered spots on leaves of *H. elodes*, *H. hirsutum*, *H. perforatum*, *H. pulchrum* and *H. tetrapterum*, Aug.–Oct.

HYPOCHAERIS: Cat's Ear, Smooth Cat's Ear

Puccinia hieracii Mart. var. *hypochaeridis* (Oud.)Jørst., 0, 1, 11, 111
Aecia uredinoid. Uredinia amphigenous, cinnamon brown; spores 20–30 × 17–25, echinulate. Telia mostly on stems, blackish brown; spores 1-septate, minutely verruculose, mostly 26–40 × 18–28. Rare on *H. glabra*, common on *H. radicata*.

Phomopsis albicans Rob. & Desm.
Conidiomata crowded, blackish brown, up to 0.2mm diam. Conidia elliptic to fusiform, biguttulate, hyaline, 8–11 × 2–2.5. On bleached spots on dead peduncles of *H. radicata*.

IBERIS: Wild Candytuft

Peronospora parasitica, downy mildew described on *Capsella*, has been recorded on *I. amara*.

IMPATIENS: Balsam

Puccinia argentata (C.F.Schulz)Wint., 11, 111
Uredinia hypophyllous, ochraceous; spores finely reticulate, 16–22 × 14–20. Telia brown; spores smooth, 1-septate, mostly 25–35 × 12–20. On *I. capensis*, rare. Aecial host *Adoxa moschatellina*.

IRIS: Gladdon or Stinking Iris, Yellow Flag, cultivated species

Puccinia iridis Rabenh., 11, 111
Uredinia amphigenous, reddish brown; spores echinulate, 25–30 × 20–24. Telia hypophyllous, black; spores smooth, 1-septate, 30–50 × 15–22. On *I. foetidissima* and various cultivated species.

Ascochyta pseudacori Allesch. (Fig. 1533)
Pycnidia immersed, black, 0.2–0.25mm diam. Conidia hyaline, 1-septate, 17–30 × 4.5–6. On dead leaves, Dec.

Belonium nigromaculatum Graddon (Fig. 1534)
Apothecia up to 0.4mm diam., black with grey discs and dirty-white margins; excipulum studded with small, dark projecting hairs about 15 long. Ascospores hyaline, 8–10 × 2. On 1-year-old dead leaves of *I. pseudacorus*, June–July.

Belonopsis iridis (Crouan & H.Crouan)Graddon (Fig. 1535)
Apothecia mollisioid, 1–5mm diam. Ascospores 45–65 × 2.5–3, hyaline, with 3 or occasionally 4 septa. On dead stems of *I. pseudacorus*, June.

Botrytis state of *Botryotinia convoluta* (Drayton)Whetzel
Sclerotia characteristically convoluted, often confluent, black, shiny. Conidia 7–18 × 5–13, mostly 10–12 × 8–10. Parasitic on rhizomes.

Drechslera iridis (Oudem.)M.B.Ellis (Fig. 1536)
Conidiophores mid to dark brown, up to 150 × 10–14. Conidia dark golden brown, 4- to 9-pseudoseptate, end cells often very pale, 45–90 × 16–29. On leaves and especially leaf bases of cultivated species, the cause of ink disease; oval or elongated colonies brown at first becoming black on senescent leaves.

Ectostroma iridis Fr. (Fig. 1537)
Black stromata on green parts as well as on large, elongated, brown, dead areas surrounded by yellow haloes. No spores. A very striking disease on living leaves of *I. pseudacorus*, June–Aug.

Leptosphaeria vectis (Berk. & Br.)Ces. & de Not. (Fig. 1538)
Pseudothecia immersed, black. Ascospores 4-septate, becoming pale brown, 20–22 × 6–7. On dead leaves of *I. foetidissima*.

Mycosphaerella iridis (Desm.)Schröter (Fig. 1539)
Pseudothecia amphigenous, immersed, black, about 0.1mm diam. Ascospores hyaline, 1-septate, 16–21 × 5–7. On brown patches on living leaves of *I. pseudacorus*, July–Oct., quite common.

Mycosphaerella macrospora (Kleb.)Jørstad (Fig. 1540)
Mostly found only in its *Cladosporium* state. Colonies effuse or punctiform, dark olivaceous brown, hairy. Conidiophores up to 150 × 8–20, brown. Conidia pale to mid olivaceous brown, echinulate, 35–70 × 13–25. Causes oval, brown lesions with grey centres and dark-brown margins, especially at the ends of living and fading leaves of *I. pseudacorus* and cultivated species, common.

Nectriella dacrymycella (Nyl.)Rehm (Fig. 1541)
Perithecia at first subepidermal, becoming erumpent, 0.3–0.5mm diam., yellow to pale orange. Ascospores hyaline, 1-septate, 16–23 × 4–5. On fibrous dead leaves and stems of *I. pseudacorus*, June–Dec., fairly common.

Phoma pseudacori Brun.
Pycnidia up to 0.15mm diam. Conidia ellipsoid, hyaline, often becoming 1-septate, 5 × 2.5. On elongated grey spots, especially at tips of fading leaves of *I. pseudacorus*, Aug.

Phomopsis iridis (Cooke)Hawksw. & Punith.
Conidiomata black, becoming erumpent, 0.4mm diam. Conidia fusiform, hyaline, with two guttules, 6–8 × 2–3. On old dead flowering shoots of *I. foetidissima* and cultivated species.

Trematosphaeria heterospora (de Not.)Winter (Fig. 1542)
Pseudothecia immersed or erumpent, black, 0.25–0.35mm diam. Ascospores ellipsoid or broadly fusiform, 3-septate, 35–45 × 10–12, hyaline to dark brown, the end cells often somewhat paler than the median ones. On rhizomes of *I. germanica*.

Other fungi found on dead leaves and stems of *I. pseudacorus* include *Anthostomella limitata* (*Carex*), *Ceriophora palustris* (*Carex*), *Myrothecium carmichaelii* (*Thalictrum*), *Neottiospora caricina* and *Niesslia exosporioides* (*Carex*), *Orbilia curvatispora*, *Stachybotrys dichroa*, *Volutella arundinis* and *V. melaloma* (*Carex*).

JASIONE: Sheep's Bit

Puccinia campanulae, 111, described on *Campanula* is seen occasionally on *J. montana*.

KICKXIA: Fluellen

Melanotaenium hypogeum (Tul.)Schellenb.
Smut sori in swollen rootstocks of *K. spuria*. Spores dark brown, smooth, 20–24 × 14–20. Rare.

KNAUTIA: Field Scabious

Two smuts, one of which is fairly common, form their sori in the anthers of *K. arvensis*. Both have almost round spores with reticulate walls.

Ustilago flosculorum (DC.)Fr.
Spore mass brownish purple, spores 12–16 diam., uncommon.

Ustilago scabiosae (Sow.)Winter
Spore mass honey-coloured, flower heads appearing dusty. Spores 8–11 diam., fairly common and widespread, July–Aug.

Erysiphe knautiae Duby
Powdery mildew. Conidia hyaline, 27–34 × 15–20. Cleistothecia about 0.1mm diam., containing usually 3 to 6 asci each with 2 to 5 spores.

Peronospora violacea Berk.
Downy mildew. Colonies very small. Sporangia ellipsoid to ovate, violaceous, 30–39 × 17–19. Oospores 22–24 diam., brown, epispore irregularly plicate and 5–8 thick. On petals.

Ramularia knautiae (Massal.)Bubák (Fig. 1543)
Colonies mostly hypophyllous. Conidiophores hyaline, 30–90 × 2.5–3. Conidia hyaline, 0- or 1-septate, 10–23 × 2 3.5. On leaves, causing round or angular, dark purple spots 0.5–2mm diam.

Septoria scabiosicola Desm.
Pycnidia few, minute, black. Conidia hyaline, guttulate or indistinctly septate, 40–60 × 1–2. Causes purple spots with white centres on leaves, Sept.–Oct., common.

LACTUCA: Lettuce

Puccinia opizii Bubák, 0, 1
Cup-shaped aecia on leaves of *L. sativa* and *L. virosa*; spores finely verruculose, 16–20 diam. Alternate hosts *Carex appropinquata* and *C. paniculata*. Occasional outbreaks.

Botrytis cinerea, the plurivorous grey mould described on 'herbaceous plants', is the most common and widespread cause of disease in lettuces. Plants often collapse completely.

Bremia lactucae Regel (Fig. 1544)
Downy mildew. Delicate, felted, white and somewhat mealy colonies mostly on lower surface of leaves, with corresponding areas on upper surface yellowish or very pale green. Usually only the outer leaves are attacked. Sporangia hyaline, 16–24 × 14–20. Chiefly on lettuces grown under glass but may be found outside where plants are overcrowded.

Marssonina panattoniana (Berl.)P. Magn.
Acervuli 0.1–0.5mm diam., brown. Conidia hyaline, 1-septate, obclavate, slightly curved, occasionally beaked, 12–17 × 3–4. On living leaves, causing pale brown spots with dark margins the centres of which often fall out to give a shot-hole effect. The common name for the disease is ring spot.

Pleospora herbarum, a plurivorous ascomycete with brown, muriform spores, is found sometimes on leaf spots.

Sclerotinia minor Jagger
Apothecia brown with pale buff-coloured discs up to 8mm diam., on stalks about 1cm long, arise singly from small black sclerotia. Ascospores elliptical or slightly flattened on one side, 10–20 × 5–10. On rotting plants, June–Dec. Although the sclerotia are common enough, apothecia are found very seldom.

Septoria lactucae Passer.
Pycnidia mostly epiphyllous, 0.1–0.2mm diam., yellowish brown, with broad ostioles. Conidia hyaline, often slightly curved, 1- to 3-septate, 25–50 × 1.5–2. Causes brown or greyish spots, sometimes with yellow haloes, on living leaves.

LAMIUM: Henbit, Red Dead-nettle, Spotted Dead-nettle, White Dead-nettle

Melanotaenium lamii Beer (Fig. 1545)
Smut sori forming swellings 0.8–2cm diam. in underground stems of *L. album*, Feb.–June. Spore mass black, firm; spores smooth, dark brown, thick-walled, 17–22 diam. Uncommon.

Erysiphe galeopsidis, powdery mildew described on *Galeopsis*, is common on *L. album* and *L. purpureum* and is found also on yellow dead-nettle, *Lamiastrum galeobdolon*.

Peronospora lamii A. Braun
Downy mildew on *L. amplexicaule*, *L. maculatum* and *L. purpureum*. Sporangia becoming dingy violaceous, mostly 20–24 × 17–18.

Pirottaea veneta Sacc. & Speg. (Fig. 1546)
Apothecia 0.2–0.7mm diam., almost black. Much of the surface of the excipulum is covered by clumps of dark brown setae and groups of dark brown cells. Ascospores hyaline, 15–19 × 2–3. On thin, dry, dead stems of *L. album*, Feb.–June.

Ramularia lamiicola Massal. (Fig. 1547)
Colonies amphigenous, white. Conidiophores hyaline, 30–55 × 3–5. Conidia hyaline, 0- to 2-septate, 6–24 × 3–4.5. On leaves of *L. album* and *L. purpureum* causing round or angular, pale brown spots, 1–6mm diam., sometimes with dark brown borders.

Septoria lamii Passer.
Pycnidia epiphyllous, brown, less than 0.1mm diam. Conidia hyaline, 3- to 5-septate, 40–50 × 1–1.5. On living leaves of *L. album* and *L. purpureum* causing small brown or white spots with reddish borders.

LAPSANA: Nipplewort

Puccinia lapsanae Fuckel, 0, 1, 11, 111 (Fig. 1548)
Aecia on purple spots. Uredinia amphigenous, very small, brown; spores finely echinulate, 18–22 × 15–18. Telia amphigenous and on stems, blackish brown, very small; spores finely echinulate, 1-septate, 23–33 × 17–26. Very common.

Ramularia lapsanae (Desm.)Sacc. (Fig. 1549)
Colonies white, mostly hypophyllous. Conidiophores hyaline, 30–70 × 3–4. Conidia hyaline, 0- or 1-septate, 7–20 × 2.5–4.5. On leaves, causing pale, irregular spots which are at first inconspicuous but later spread over a large area.

Sphaerotheca erigerontis-canadensis, powdery mildew described on *Erigeron*, has been recorded on *Lapsana*.

LATHYRUS: Bitter Vetch, Everlasting Pea, Marsh Pea, Meadow Vetchling, Narrow-leaved Everlasting Pea, Sea Pea

UREDINALES

Two rust species and one variety occur: *Uromyces pisi-sativi*, described on *Pisum*, *U. viciae-fabae*, described on *Vicia*, and its var. *orobi* (Schum.)Jørst., which is uncommon and confined to *L. montanus*. *U. viciae-fabae* is recorded on *L. palustris* and both this species and *U. pisi-sativi* on *L. pratensis*. On *L. pratensis*, if aecia are found, the rust is *U. viciae-fabae* and this species has smooth teliospores; it is found more frequently than *U. pisi-sativi*, which has verrucose teliospores with warts arranged in lines, and which forms its aecia on *Euphorbia cyparissias*.

USTILAGINALES

Thecaphora deformans Dur. & Mont.
Smut sori in seeds of *L. pratensis*, Aug.–Sept., few records. Spore mass reddish brown, spore balls 20–60 diam., each composed of up to 20 or even more spores with verrucose outer walls.

OTHER FUNGI

Ascochyta lathyri Trail
Pycnidia amphigenous, up to 0.1mm diam., brown. Conidia hyaline, 1-septate, 6–10 × 2–2.5. On fading leaves of *L. japonicus* ssp. *maritimus* and *L. sylvestris*, causing large, irregular, greyish-brown spots, Apr.–Oct.

Erysiphe trifolii, powdery mildew described on *Trifolium*, occurs on *L. pratensis*.

Isariopsis carnea Oudem. (Fig. 1550)
Colonies at first white then flesh-coloured. Conidiophores 2–3 thick, often grouped to form loose synnemata. Conidia 10–16 × 7–8, hyaline, sometimes with large guttules and appearing septate. On narrow, elongated black spots on living and fading leaves of *L. pratensis*.

Leptosphaeria niessliana Rabenh. ex Niessl (Fig. 1551)
Pseudothecia black, about 0.25mm diam., with protruding necks. Ascospores pale straw-coloured, 4-septate, 24–33 × 3.5–5. On dead stems of *L. latifolius* and *L. sylvestris*.

Peronospora viciae (Berk.)Casp., downy mildew described on *Vicia*, is found occasionally on *L. pratensis*.

Ramularia alba (Dowson)Nannf.
On *L. odoratus*. Colonies amphigenous and on stems, white, mealy, indefinite. Conidiophores short, hyaline. Conidia hyaline, smooth, occasionally 1-septate, 3.5–12 × 2.5–5.5, mostly limoniform, sometimes subspherical or cylindrical.

Ramularia deusta (Fuckel)Karakulin
Colonies small, pinkish. Conidiophores narrow, hyaline. Conidia mostly about 12 × 4. On small blackish-brown spots on leaves of *L. montanus*, *L. palustris* and *L. pratensis*.

LAVANDULA: Lavender

Phomopsis lavandulae (Gabotto)Ciferri & Vegni
Conidiomata black, 0.15–0.2mm diam. Conidia oval or fusiform, hyaline, guttulate, 4–5 × 2. On living and dead stems causing 'shab' disease, June–Aug.

Septoria lavandulae Rob.
Pycnidia few, epiphyllous, black, less than 0.1mm diam. Conidia straight or curved, hyaline, 25–35 × 1–2. Causes whitish spots, with raised purple borders, on fading leaves, July–Sept.

LAVATERA: Tree Mallow

Puccinia malvacearum and *Colletotrichum malvarum* described on *Malva* occur also on *Lavatera*.

LEONTODON: Hawkbits

Puccinia hieracii, described on *Hieracium*, is common on *L. autumnalis* and *L. hispidus* but found seldom on *L. taraxacoides*.

LILIUM: Lilies

Uromyces aecidiiformis (Str.)Rees, 0, 1, 111
Aecia on leaves, stems and petioles; spores yellowish, minutely verruculose. Telia mostly hypophyllous on yellow spots, dark brown; spores brown with anastomosing ridges on walls, 30–44 × 22–30. On *L. candidum*, rare.

Botrytis elliptica (Berk.)Cooke (Fig. 1552)
Characterised by its large, ellipsoidal conidia, 16–35 × 10–24. On many lilies including *L. candidum*, at first causing brown spots on leaves, later attacking flowering shoots.

LIMONIUM: Sea Lavender

Uromyces limonii (DC.)Berk., 0, 1, 11, 111
Aecia in clusters on red or reddish-purple spots on leaves. Uredinia cinnamon brown; spores verruculose, 22–30 × 20–27. Telia often circinate, black; spores smooth, 25–50 × 15–25. Common on *L. vulgare* and recorded also on cultivated species.

Erysiphe limonii Junell (Fig. 1553)
Few conidia formed. Cleistothecia 0.1–0.125mm diam., containing 3 to 6 asci each with 3 to 5 spores. Ascospores hyaline, 24–26 × 12–14. On living leaves of *L. vulgare*, June–Aug.

LIMOSELLA: Mudwort

Doassansia limosellae (Kunze)Schröter
Sori amphigenous and on petioles, brown or black, swollen. Spore balls brown, up to 150 diam.; spores pale brown, smooth, 9–11 diam. A rare smut found a few times only, Oct.

LINARIA: Yellow Toadflax

Entyloma linariae Schröter
Smut sori mostly hypophyllous, pale yellow, up to 3.5mm long. Spores thick-walled, golden brown, 9 to 16 diam. Conidia hyaline, sickle-shaped, 9–11 × 2–3. On leaves of *L. vulgaris*, Oct., seldom recorded.

Melanotaenium cingens (Beck)Magnus
Smut sori in stems and leaves of *L. vulgaris*, July–Aug., at first covered by host tissue. Spore mass firm, black. Spores dark brown, smooth, 13–18 × 10–16. Few records.

Didymaria linariae Pass. (Fig. 1554)
Colonies amphigenous, white. Conidiophores hyaline, up to 50 × 5. Conidia hyaline, 1-septate, 19–32 × 6–7. On leaves of *L. vulgaris*.

LINNAEA: Linnaea

Metacoleroa dickiei (Berk. & Br.)Petrak (Fig. 1555)
Setose brown pseudothecia about 0.1mm diam., in a mat of superficial, dark hyphae, are seated on small, black, shining stromata on the upper surface of leaves. Ascospores brown, 1-septate, 10–15 × 4–5.

LINUM: Flax

Melampsora lini (Ehrenb.)Desm., 11, 111
On stems and both leaf surfaces of *L. catharticum*, May–Oct., common. Uredinia orange, with capitate paraphyses; spores shortly echinulate, 17–25 × 14–20. Telia reddish brown to black; spores 35–55 × 10–20.

Melampsora lini var. *liniperda* Körnicke, 0, 1, 11, 111
On *L. usitatissimum*. This differs in having aecia on the leaves, and teliospores up to 80 long.

Alternaria linicola Groves & Skolko (Fig. 1556)
Conidiophores olivaceous brown, up to 80 × 6–8. Conidia with beak usually about the same length as the body, golden brown, smooth, with up to 16 transverse and one or more longitudinal septa, up to 300 long, body 17–25 and beak 1–3 thick. On *L. usitatissimum*, causing damping-off of seedlings, and, on mature plants, brown, rectangular leaf spots.

Colletotrichum lini (Westerdijk)Tochinai (Fig. 1557)
Acervuli small, with 2 to 10 dark brown, septate setae up to 150 × 4. Conidia hyaline, guttulate, slightly curved, 15–20 × 3–4.5. On *L. catharticum* and *L. usitatissimum*; causes seedling blight.

Guignardia fulvida Sanderson
Found usually only in its conidial state when it is commonly called *Polyspora lini* by plant pathologists. Sporodochia very small and inconspicuous, appearing as gelatinous, milky masses of conidia overlying stomata. Conidia hyaline, variable in shape, 8–20 × 4. Causes stem-break and browning of cultivated flax, *L. usitatissimum*.

Phoma exigua Desm. var. *linicola* (Naum. & Vass.)Maas
Pycnidia immersed, dark brown, 0.1–0.3mm diam. Conidia elliptical, hyaline, biguttulate, occasionally 1-septate, 5–8 × 2.5–4. A cause of foot rot of cultivated flax.

LISTERA: Twayblade

The aecia of two rusts, *Melampsora epitea* and *Puccinia sessilis*, are found on leaves of *L. ovata*. For comparison between these species see *Dactylorhiza*.

LOTUS: Bird's Foot Trefoils

Uromyces anthyllidis, described on *Anthyllis*, has been recorded on *L. hispidus*.

Uromyces pisi-sativi, described on *Pisum*, is widespread on *L. corniculatus* and has been found occasionally on *L. angustissimus* and *L. tenuis*.

Cercospora loti Hollós (Fig. 1558)
Colonies amphigenous. Conidiophores pale olivaceous brown, 50–110 × 4 5. Conidia hyaline, 6- to 17-septate, 50–110 × 3–4. On leaves of *L. corniculatus*, causing reddish brown or grey spots.

Peronospora lotorum H.Sydow
Downy mildew on *L. corniculatus*. Sporangia 24–25 × 19–21.

Peronospora trifoliorum de Bary
Downy mildew on *L. uliginosus*. Sporangia pale violaceous, 19–26 × 15–19. Oospores brown, smooth, 25–34 diam.

Ramularia schulzeri Bäumler (Fig. 1559)
Conidiophores in small fascicles, emerging through stomata, hyaline, 20–40 × 2.5–4. Conidia hyaline, 0- or 1-septate, 9–22 × 3.5–4.5. On leaves of *L. corniculatus*, causing rather ill-defined yellowish-brown spots.

Ramularia sphaeroidea Sacc. (Fig. 1560)
Colonies hypophyllous, white, velvety. Conidiophores twisted, hyaline, 40–120 × 3. Conidia mostly spherical, 8–12 diam. On living and fading leaves of *L. corniculatus* and *L. uliginosus*, June–Oct.

LUNARIA: Honesty

Albugo candida and *Peronospora parasitica*, described on *Capsella*, occur also on *Lunaria*.

Septoria lunariae Ellis & Dearness
Pycnidia black, 0.1–0.15mm diam. Conidia somewhat curved, hyaline to pale straw, 0- to 5-septate, 25–30 × 3. On bleached spots on fruits of *L. annua*.

LYCHNIS: Ragged Robin

Ustilago violacea, anther smut described on *Silene*, occurs on *L. flos-cuculi*.

LYCOPERSICON: Tomato

Alternaria solani, described on *Solanum*, also causes zonate brown spots on leaves of tomato.

Alternaria tomato (Cooke)Jones (Fig. 1561)
Conidiophores mid to dark brown, mostly about 100 × 4–7. Conidia with the beak usually as long as or slightly longer than the body, pale to mid golden brown, smooth, with 5 to 9 transverse septa and 1 or more longitudinal septa, 50–140 long, body 10–20(14) and beak 1.5–2 thick. On leaves, small sunken blackish-brown spots are formed, which tend to become pale in the centre; on fruits the spots resemble nail heads.

Botrytis cinerea, grey mould described on 'herbaceous plants', is often the cause of stem, leaf and fruit rot of tomato plants in cold greenhouses.

Colletotrichum coccodes (Wallr.)Hughes (Fig. 1562)
Acervuli on roots and stems up to 0.3mm diam., setose; setae brown, about 100 long, tapered towards the paler tips. Conidia hyaline, honey-coloured in mass, 16–24 × 3–4. The cause of 'black dot' disease.

Didymella lycopersici Kleb. (Fig. 1563)
Pseudothecia immersed, papillate, pale brown, up to 0.3mm diam. Ascospores hyaline, 1-septate, 13–18 × 4.5–5.5. Pycnidia immersed or erumpent, 0.2–0.25mm diam.; conidia ellipsoid or ovoid, hyaline, pink in mass, biguttulate and often 1-septate, 5–10 × 2–3.5. On leaves, stems and fruits. A cause of stem and fruit rot, found mostly in the pycnidial state. Dark brown cankers girdle stems just above soil level and cause plants to collapse.

Fulvia fulva (Cooke)Ciferri (Fig. 1564)
Colonies hypophyllous, effuse, buff to brown or purplish brown, velvety. Conidiophores up to 200 long, 2–4 thick near base, 7–8 thick at nodes. Conidia 0- to 3-septate, 12–47 × 4–10. Leaf mould disease, especially of plants grown under glass; the upper leaf surface above infected areas is at first yellowish, later reddish brown.

Phoma destructiva Plowr.
On green and ripe fruits. Pycnidia erumpent, brown or blackish brown, 0.1–0.2mm diam. Conidia hyaline, biguttulate, 3.5–5 × 2.

Phytophthora infestans, described on *Solanum*, also causes tomato late blight, common on outdoor plants in wet seasons. Dark-brown or blackish-purple areas appear on leaves and stems, and brown or blackish-brown blotches on fruits, which later rot.

Pyrenochaeta lycopersici Schneider & Gerlach, which causes brown root rot, is found only as mycelium in corky roots of tomatoes. Cultural studies are needed for determination of this and also of the various species of *Verticillium* which are responsible for wilt diseases.

LYCOPUS: Gipsywort

Mollisia lycopi Rehm (Fig. 1565)
Apothecia erumpent from blackened areas, up to 1.3mm diam., almost black, with greyish-olive discs. Ascospores pyriform, hyaline, 8–10 × 4–5. On dead stems, Mar.

Pyrenopeziza lycopincola (Rehm)Boud. (Fig. 1566)
Apothecia blackish with thin white margins, 0.5–1mm diam. Ascospores oblong, rarely slightly clavate, straight or curved, hyaline, with a guttule at each end, 7–10 × 2–2.5; paraphyses filiform. On blackened areas on rotten stems, July–Aug.

Other fungi found on dead stems include *Ciboriopsis tenuistipes* (*Epilobium*), *Lophiostoma angustilabrum*, *Mollisiopsis lanceolata* (*Filipendula*).

LYSIMACHIA: Creeping Jenny, Yellow Loosestrife, Yellow Pimpernel

Ramularia lysimachiarum Lindroth (Fig. 1567)
Colonies mostly hypophyllous, white. Conidiophores hyaline, 20–35 × 3–4. Conidia 0- to 2-septate, hyaline, 9–26 × 3–4. On leaves of *L. nummularia*, causing round or irregular, brown or yellowish spots 5–10mm diam., sometimes spreading over the whole leaf.

Septoria lysimachiae Westend.
Pycnidia epiphyllous, blackish brown, about 0.1mm diam. Conidia straight or slightly curved, hyaline, guttulate, 0- to 3-septate, 60–75 × 2.5–3. On leaves of *L. nemorum*, *L. nummularia* and *L. vulgaris*.

LYTHRUM: Purple Loosestrife

Dasyscyphus salicariae Rehm (Fig. 1568)
Apothecia pale yellowish, with white hairs, discs drying yellow ochre. Hairs 40–50 × 3–4. Ascospores hyaline with small guttules, 6–11 × 1.5–2. On and inside the bases of dead stems, Aug.

Stenella lythri (Westend.)Mulder (Fig. 1569)
Colonies brown or reddish brown, effuse, sometimes covering almost the whole lower surface of the leaf. Conidiophores pale brown, 80–140 × 3–4.5. Conidia solitary or in chains, pinkish in mass, verruculose, 0- to 3-septate, 20–45 × 4–5. On dying leaves, causing a reddening of the upper surface.

MALVA: Mallows

Puccinia malvacearum Mont., 111 (Fig. 1570)
Telia on stems, petioles and the lower surface of leaves, pale reddish or purplish brown, pulvinate; spores pale yellowish brown, 1-septate, mostly 50–75 × 15–25. On *M. moschata*, *M. neglecta*, *M. pusilla*, *M. sylvestris* and other Malvaceae, causing yellow spots, Apr.–Nov., very common.

Colletotrichum malvarum Southworth
Acervuli setose, yellowish brown, flesh-coloured when sporulating freely; setae stiff, dark brown, septate, about 100 × 3–4. Conidia cylindrical, rounded at ends, hyaline, 12–28 × 4–5. Parasitic on stems.

Ramularia keithii Massee
Colonies white. Conidia cylindrical, rounded at ends, 0- or 1-septate, hyaline, about 30 × 5. On leaves of *M. moschata*.

MATRICARIA: Pineapple-weed

Peronospora leptosperma de Bary (Fig. 1571)
White downy mildew on *M. matricarioides*. Sporangia 30–50 × 16–26. Oospores pale brown, 38–45 diam.

Sphaerotheca erigerontis-canadensis, powdery mildew described on *Erigeron*, occurs also on *M. matricarioides*.

MATTHIOLA: Stock or Gilliflower

White rust, *Albugo candida*, and downy mildew, *Peronospora parasitica*, described on *Capsella*, and *Alternaria raphani*, described on *Raphanus*, have been recorded also on *Matthiola*.

MECONOPSIS: Welsh Poppy

Two species of downy mildew are recorded, *Peronospora arborescens* (Berk.)Caspary, with sporangia 15–22 × 13–18; and *P. cristata* Tranz., with sporangia 21–26 × 19–21.

MEDICAGO: Alfalfa or Lucerne, Medicks

Uromyces pisi-sativi 11, 111, described on *Pisum*, is found occasionally on *M. arabica*, *M. lupulina*, *M. polymorpha* and *M. sativa*.

Ascochyta pisi, described on *Pisum*, has been recorded on *M. arabica*.

Cercospora zebrina Pass. (Fig. 1572)
Colonies amphigenous. Conidiophores fasciculate, mid pale brown, 20–80 × 3–5. Conidia hyaline, 8- to 20-septate, 30–180 × 3–4.5. Causes dark brown spots on living leaves of *M. arabica*, July–Oct.

Colletotrichum trifolii Bain & Essary
Acervuli setose; setae often flexuous, brown, 40–60 × 4–7. Conidia cylindrical, rounded at both ends, hyaline, 11–13 × 3–4. Causes dark blackish-brown spots on stems of *M. sativa*.

Leptotrochila medicaginis (Fuckel)Schüepp
Apothecia erumpent, yellowish brown with blackish rims, 0.2–0.7mm diam., on pale brown spots on living leaves of *M. lupulina*, *M. polymorpha* and *M. sativa*. Ascospores hyaline, 7.5–10 × 3–5. Conidiomata of the *Sporonema* state unilocular, up to 0.15mm diam.; conidia hyaline, 4–5.5 × 2–2.5.

Peronospora trifoliorum, downy mildew described on *Trifolium*, has been found on *M. lupulina* and *M. sativa*.

Phoma medicaginis Malbr. & Roum. var. *medicaginis*
Pycnidia on stems of *M. sativa* causing black stem disease. Conidia cylindrical, hyaline, 6–12 × 2–3.5, often becoming 1-septate.

Physoderma alfalfae (Lagerh.)Karling
Crown wart disease of *M. sativa*. Conspicuous large warts on rudimentary leaves and buds, infected plants aborted. Resting sporangia 40–60 diam., smooth, golden brown, with a ring of blunt haustoria, up to 40 in each host cell.

Pseudopeziza medicaginis (Lib.)Sacc. (Fig. 1573)
Apothecia 0.5–1mm diam., yellowish brown, jelly-like, surrounded by torn parts of epidermis but without the surrounding cellular excipulum which is present in the otherwise rather similar *Leptotrochila*. Ascospores hyaline, 8–11 × 4–6. On leaves of *M. lupulina* and *M. sativa*, causing brown or yellowish spots, June–Dec.

Stagonospora meliloti (Lasch)Petrak (Fig. 1574)
Pycnidia yellowish to blackish brown, 0.2–0.3mm diam. Conidia straight or curved, cylindrical, rounded at the ends, at first guttulate, then 1- to 4-septate, 15–27 × 3–4.5. On leaves of *M. arabica*, causing whitish or pale buff, brown-bordered spots.

Trematosphaeria circinans (Sacc.)Winter (Fig. 1575)
Pseudothecia pyriform, with base immersed, black, 0.3–0.4mm diam. Ascospores broadly fusiform, 3-septate, 27–35 × 10–11, median cells dark brown, end cells almost hyaline, verruculose when old. On rootstocks of *M. sativa*.

MELAMPYRUM: Cow-wheat

Coleosporium tussilaginis, 11, 111, described on *Tussilago*, is common on *M. arvense* and *M. pratense*.

Puccinia nemoralis Juel, 0, 1
Aecia on *M. pratense* white-margined, epiphyllous, very conspicuous on reddish-purple swollen areas; infected plants prematurely blackened, with number and size of flowers reduced. Aeciospores in chains, finely verruculose, 13–20 × 10–14. Alternate host *Molinia caerulea*.

MELILOTUS: Melilots

Erysiphe trifolii, described on *Trifolium*, occurs on *M. officinalis*.

Ramularia meliloti Ellis & Ev.
Colonies effuse, white. Conidiophores fasciculate, 15–25 × 3. Conidia 0- or 1-septate, hyaline, 6–10 × 3.

MENTHA: Mints

Puccinia menthae Pers., 0, 1, 11, 111
A very common rust found on a number of mints, both wild and garden, including *M. aquatica, M. arvensis, M. longifolia, M. rotundifolia* and *M. spicata*. Systemically infected plants have elongated, swollen shoots with abundant aecia. Uredinia buff to cinnamon; spores mostly 20–26 × 16–22. Telia dark brown; spores mostly 25–30 × 18–24.

Erysiphe biocellata Ehrenb.
Powdery mildew. Few conidia formed. Cleistothecia contain 4 to 15 asci, each usually with 2 or occasionally 4 spores. On *M. aquatica* and *M. arvensis*.

Hymenoscyphus repandus, described on *Epilobium*, and *H. scutula* var. *solani*, described on *Solanum*, both occur also on dead stems of *M. aquatica*.

Ramularia menthicola Sacc. (Fig. 1576)
Colonies amphigenous. Conidiophores in groups of 5 to 15, hyaline, 40–60 × 3–5. Conidia 0- to 2-septate, hyaline, 15–35 × 3–5. On living leaves of *M. aquatica*, causing round, white to grey or pale brown spots 1–3mm diam., with dark brown borders.

MERCURIALIS: Annual Mercury, Dog's Mercury

Melampsora populnea (Pers.)Karst., 0, 1
Orange aecia conspicuous on pale yellow spots on leaves, petioles and stems of *M. perennis*; spores 15–24 × 12–16. Common, Apr.–June. Alternate hosts *Populus alba* and *P. tremula*.

Ascochyta mercurialis Bres.
Pycnidia epiphyllous, 0.1–0.15mm diam., yellowish brown. Conidia hyaline, 0- or 1-septate, 8–10 × 2.5–3.5. Causes pale spots with dark brown borders on living leaves of *M. perennis*, May–Sept.

Cercospora mercurialis Pass. (Fig. 1577)
Colonies amphigenous. Conidiophores mid pale olivaceous brown, 10–90 × 4–5. Conidia hyaline, 3- to 12-septate, 40–120 × 4–5. On leaves of *M. annua* and *M. perennis*, causing whitish or pale brown, often dark-bordered spots, common, Aug.–Sept.

Phoma macrocapsa Trail (Fig. 1578)
Pycnidia at first immersed, becoming superficial, 0.6–0.7mm diam., blackish brown. Conidia hyaline, 4–5 × 1–1.5. On dead stems of *M. perennis*, May.

Pyrenopeziza mercurialis (Fuckel)Boud. (Fig. 1579)
Apothecia up to 1mm diam., dark blackish brown to black, with white fimbriate margins and grey discs. Ascospores hyaline, 9–11 × 2. On dead stems of *M. perennis*, Feb.–June, common.

Spilopodia melanogramma Boud. (Fig. 1580)
Apothecia 0.25–0.5mm diam., black, rugose, with indented margins and grey or olivaceous discs. Ascospores hyaline, 10–12 × 3–4. The apothecia develop on elongated black stromata; the effect on the host stems is very striking, making them look like 'liquorice allsorts' sideways on. On dead stems of *M. perennis*, Feb.–May.

Synchytrium mercurialis (Lib.)Fuckel
Causes pale yellowish blisters on leaves and stems of *M. perennis*, Apr.–June. Resting spores usually solitary, spherical or broadly ellipsoidal, mostly 70–180 diam.; exospore golden brown, thick, smooth or transversely or spirally ridged. Infected cells surrounded by one or two layers of hypertrophied cells.

MEUM: Meu or Spignel

Nyssopsora echinata (Lév.)Arth., 111
A rare rust found only in Scotland on leaves, stems and fruits of *M. athamanticum*. Telia black; spores 25–35 × 25–30, 3-celled, as in *Triphragmium*, but ornamented with curved spines up to 18 long.

MOEHRINGIA: Three-nerved Sandwort

Puccinia arenariae, 111, described on *Arenaria*, is found on *M. trinervia*.

Peronospora arenariae (Berk.)Tul.
Downy mildew. Sporangia 20–22 × 13–15.

Ramularia arenariae A.L.Sm. & Ramsb. (Fig. 1581)
Colonies on leaves effuse, white. Conidiophores hyaline, 20–30 × 3–4. Conidia hyaline, 1- to 3-septate, 7–30 × 2–4.

MUSCARI: Grape Hyacinth

Uromyces muscari, 111, described on *Endymion*, has been recorded on *M. polyanthum*.

Ustilago vaillantii, anther smut described on *Chionodoxa*, occurs also on *M. botryoides*, Apr.

Sclerotinia bulborum (Wakker)Rehm
Causes black slime disease of bulbs. Apothecia not yet found in Britain.

MYCELIS: Wall Lettuce

Puccinia maculosa (Str.)Röhl, 1, 11, 111
Aecia on large yellow and purple spots. Uredinia hypophyllous, very small, pale brown; spores echinulate, 16–24 diam. Telia blackish brown; spores minutely roughened, 27–35 × 17–24. An uncommon rust.

MYOSOTIS: Forget-me-nots

Entyloma fergussoni (Berk. & Br.)Plowr.
Smut sori immersed. Spores pale brown, smooth, 10–12 diam. Conidia often seen as a whitish coating on the surface, cylindrical, 8–30 × 2.5–4. Causes pale, circular spots on living leaves of *M. arvensis*, *M. laxa* ssp. *caespitosa* and *M. scorpioides*, May–July.

Erysiphe asperifoliorum, described on *Cynoglossum*, a powdery mildew found also on *M. arvensis* and *M. sylvatica*.

Peronospora myosotidis de Bary
Downy mildew on *M. arvensis* and *M. ramosissima*. Sporangia 20–23 × 13–18. Oospores golden brown, reticulate, 24–30 diam.

MYOSOTON: Water Chickweed

Ustilago violacea, anther smut described on *Silene*, is found occasionally on *Myosoton*.

Ascochyta anisomera Kab. & Bubák
Pycnidia up to 0.15mm diam. Conidia hyaline, straight or slightly curved, with a septum near one end, 20–25 × 8–9. On large yellowish areas on fading leaves, Aug.

MYRRHIS: Sweet Cicely

Puccinia chaerophylli, 0, 1, 11, 111, described on *Anthriscus*, and *Erysiphe heraclei*, described on *Heracleum*, have been recorded on *Myrrhis*.

NARCISSUS: Daffodil, Jonquil, Pheasant's Eye

Puccinia schroeteri Pass., 111
Telia epiphyllous, black, surrounded by purplish-brown zones; spores 40–60 × 25–30, with reticulate walls. On *N. jonquilla*, *N. majalis* and *N. pseudonarcissus*, few records.

Botryotinia narcissicola (Gregory)Buchwald
On *N. pseudonarcissus*, sometimes rotting the leaves at ground level, also bulbs; disease referred to as smoulder or grey mould. Sclerotia black, smooth, spherical, 1–1.5mm diam. Usually found only as *Botrytis* state with pale brown conidia 8–16 × 7–12. Apothecia may develop on sclerotia in dead overwintered leaves in Feb. or Mar.; they are brown, up to 2.5mm diam., on long stalks with blackish bases. Ascospores hyaline, 10–20 × 5–9.

Botryotinia polyblastis (Gregory)Buchwald
On living leaves of *N. pseudonarcissus* causing a yellow blotch disease often called 'fire'; it also causes brown spots on flowers. Sclerotia immersed, elongated, flattened, 3–8 × 1.5–3mm. *Botrytis* conidia when young pyriform, when mature almost spherical with a protuberant hilum, colourless to pale brown, 30–50 diam.; on germination they form numerous germ tubes. Apothecia may develop on sclerotia in overwintered leaves during Feb.–Mar.; they are 3–4mm diam., brown, with long stalks. Ascospores hyaline, 14–20 × 7–10.

Ramularia vallisumbrosae Cav.
Colonies amphigenous, white, powdery. Conidiophores erumpent in bundles, hyaline, sometimes branched. Conidia hyaline, granular, 0- to 3-septate, 14–44 × 4. On living leaves of *N. pseudonarcissus*, causing oblong, yellow or ochraceous spots.

Stagonospora curtisii (Berk.)Sacc.
Pycnidia amphigenous, immersed, up to 0.2mm diam. Conidia hyaline, 1- to 3-septate, 10–25 × 4–7. On leaves of various cultivated species causing large, irregular, brown spots, especially at the tips; sometimes also on the flowers. The disease is referred to as leaf scorch.

Stromatinia narcissi Drayton & Groves
This causes a scale speck disease on bulbs. Numerous black sclerotules 0.5–2mm diam. are formed on the papery outer scales. Apothecia have not been seen in Britain.

NARTHECIUM: Bog Asphodel

Cladosporium magnusianum (Jaap)M.B.Ellis (Fig. 1582)
Colonies tufted, olivaceous brown. Conidiophores pale olivaceous or golden brown, up to 250 × 4–7. Conidia catenate, pale to mid golden brown, closely

verruculose or echinulate, 3- to 6-septate, 16–35 × 7–11. On bleached areas at tips of leaves.

Entyloma ossifragi Rostrup
Sori immersed; spores golden brown, 10–16 diam. In leaves causing blackish-brown lesions up to 1cm long, which sometimes coalesce and occupy large areas.

NASTURTIUM: Watercress

Albugo candida and *Peronospora parasitica*, described on *Capsella*, occur also on *Nasturtium*.

Spongospora subterranea (Wallr.)Lagerh. f.sp. *nasturtii* Tomlinson
Causes crook root disease of *N. officinale* and *N. officinale* × *N. microphyllum*.

NIGELLA: Love-in-the-Mist

Cercospora nigellae Hollós (Fig. 1583)
Colonies on both surfaces of leaves and on capsules. Conidiophores fasciculate, pale olive or brown, 20–60 × 3–5. Conidia hyaline, 1- to 5-septate, 30–80 × 4–5.

ODONTITES: Red Rattle

Coleosporium tussilaginis, 11, 111, described on *Tussilago*, is found also on *Odontites*.

Peronospora sordida, downy mildew described on *Scrophularia*, with sporangia 20–25 × 16–18, occurs also on *Odontites*.

Plasmopara densa (Rabenh.)Schröter
Downy mildew with sporangiophores about 200 long and sporangia 13–15 × 11–15.

Sphaerotheca melampyri Junell
Powdery mildew. Conidia few, hyaline, 25–30 × 15–17. Cleistothecia 70–90 diam., with branched, brown appendages. Ascospores 18–21 × 14–16.

OENANTHE: Water Dropworts

Uromyces lineolatus (Desm.)Schröter, 0, 1
Aecia clustered on yellowish spots on leaves and petioles of *O. crocata*, *O. fistulosa* and *O. lachenalii*; spores 20–27 × 17–26. Alternate host *Scirpus maritimus*; uncommon.

Entyloma helosciadii, smut described on *Apium*, has been found also on leaves of *O. crocata*.

Nectria boothii D. Hawksw. (Fig. 1584)
Perithecia superficial, 0.1–0.17mm diam., orange, with pale, spine-like projections when young, darkening with age. Ascospores hyaline to pale brown, 12–17 × 4–5, 1-septate. On dead stems of *O. crocata*, Sept.

Plasmopara nivea, downy mildew described on *Aegopodium*, occurs also on *Oenanthe*.

Pyrenopeziza subplicata Rehm (Fig. 1585)
Apothecia erumpent, solitary or in groups, up to 0.8 × 0.5mm. Marginal hairs 15–30 × 4–5. Ascospores hyaline, straight or slightly curved, 7–15 × 2–3. On dead stems of *O. crocata*.

Stachybotrys oenanthes M.B.Ellis (Fig. 1586)
Colonies effuse, grey to black, shortly hairy. Conidiophores smoke grey to black, up to 180 × 6–9, swollen at the base to 10–16 and at the apex to 9–12. Conidia pale to dark grey, smooth or verrucose, 9–12 × 4.5–8. On dead stems of *O. crocata*, June.

Other fungi found on dead stems of *O. crocata* include *Anthostomella limitata* (*Carex*), *Dasyscyphus grevillei*, *D. nidulus* (*Filipendula*), *Dendryphiella vinosa*, *Gliomastix luzulae* and *Unguicularia millepunctata*.

OENOTHERA: Evening Primrose

Septoria oenotherae Westend.
Pycnidia up to 0.1mm diam., blackish brown. Conidia slightly curved, guttulate or 3-septate, 30–55 × 1.5–2. On leaves of *O. biennis*, causing brown spots 2–3mm diam., with wine-red borders.

ONOBRYCHIS: Sainfoin

Uromyces pisi-sativi, 11, 111, described on *Pisum*, is found occasionally on *Onobrychis*.

Ascochyta orobi Sacc.
Pycnidia immersed, pale brown, about 0.1mm diam. Conidia cylindrical with rounded ends, or ellipsoid, hyaline, 1-septate, guttulate, 13–18 × 4–5.5. On leaves, causing pale spots with dark brown borders.

Ramularia onobrychidis Allesch. (Fig. 1587)
Colonies amphigenous, white. Conidiophores hyaline, 30–60 × 2.5–4. Conidia hyaline, 0- to 3-septate, 7–40 × 4–6. On leaves, causing round, brown spots 1–5mm diam.; attacked leaflets are killed and fall readily.

Sclerotinia trifoliorum, described on *Trifolium*, is found occasionally on *Onobrychis*.

Septoria orobina Sacc.
Pycnidia scattered, punctiform. Conidia hyaline, about 30 × 1. On living leaves causing irregular, ochraceous, dark-bordered spots.

ONONIS: Restharrow

Crocicreas cyathoideum (Bull.)S. Carp. var. *cacaliae* (Pers.)S. Carp. (Fig. 1588)
Apothecia 0.5–0.6mm diam., pale brown, with thin-walled reddish-brown terminal cells, sometimes in clumps. Ascospores hyaline, 8–10 × 1.5. On dead stems, July.

Erysiphe cruchetiana Blumer
Powdery mildew. Conidia abundant. Cleistothecial appendages coralloid, irregularly branched. Asci 4 to 8, each with 3 to 5 spores.

Ophiobolus fruticum (Rob. ex Desm.)Sacc. (Fig. 1589)
Pseudothecia black, up to 0.5mm diam. Ascospores twisted inside asci, hyaline but yellowish in mass, 11- to 16-septate, up to 130 × 4–5. On dead stems, July.

Pyrenopeziza compressula Rehm.
Apothecia erumpent, 0.1–0.45mm diam., dark brown with grey discs. Ascospores hyaline, 8–12 × 1.5–3. On dead stems.

Ramularia winteri Thüm. (Fig. 1590)
Colonies predominantly hypophyllous, white. Conidiophores hyaline, 25–40 × 2–3. Conidia hyaline, 1- to 4-septate, 20–45 × 3–7, mostly 20–30 × 5–7 On living and fading leaves, not forming definite spots, Aug.

Other fungi found on dead stems include *Dinemasporium strigosum, Lasiosphaeria caudata, Lophiostoma caulium, Pleospora herbarum, Sarcopodium circinatum* and *Unguicularia millepunctata*.

ORIGANUM: Wild Marjoram

Puccinia menthae, 0, 1, 11, 111, described on *Mentha*, occurs on the leaves, and *P. thymi*, 111, described on *Thymus*, on stems and petioles, where it causes witches' brooms.

ORNITHOGALUM: Star of Bethlehem

Puccinia hordei Otth, 0, 1
Aecia have been found on *O. pyrenaicum*; alternate hosts *Hordeum* spp.

Puccinia liliacearum Duby, 0, 111 (Fig. 1591)
Telia amphigenous, forming silvery-grey blisters which eventually open to expose the dark reddish-brown spore mass; spores mostly 40–60 × 25–30. Infected areas are yellowish, and the plants seldom flower. On *O. pyrenaicum* and *O. umbellatum*, Mar.–May.

ORTHILIA

Pucciniastrum pyrolae, 11, described on *Pyrola*; a rare rust which has been recorded also on *Orthilia*.

OXALIS: Pink Bulbous Oxalis, Wood Sorrel

Puccinia oxalidis Diet. & Ellis, 11, 111
Uredinia hypophyllous, pale yellow; spores minutely echinulate, 17–24 × 16–20. Telia hypophyllous, brownish yellow; spores smooth, 20–28 × 14–20. On *O. corymbosa* and *O. latifolia*.

Mycosphaerella depazeiformis (Auersw.)Lindau
Pseudothecia less than 0.1mm diam., immersed. Ascospores clavate or cylindrical, hyaline, 1-septate, 10–11 × 2.5–3. Causes circular brown spots 2–4mm diam., on living leaves of *O. acetosella*, Aug.

OXYRIA: Mountain Sorrel

Puccinia oxyriae Fuckel, 11, 111
Uredinia cinnamon brown, on small purple spots; spores echinulate, 24–30 × 20–25. Telia brown; spores 30–45 × 15–25. Uncommon.

Ustilago vinosa Tul.
Smut sori inside inflated perianth. Spore mass pinkish purple; spores finely reticulate, violet, 7–9 diam., July, rare.

Cercoseptoria oxyriae (Trail)Gjaerum (Fig. 1592)
Conidiophores in small groups, scattered sparsely over the surface of leaf spots. Conidiophores hyaline, 20–35 × 2–3. Conidia hyaline, 2- to 4-septate, 60–140 × 2–2.5. On living leaves, causing white or pale brown spots 1–3mm diam., visible on both surfaces.

PAEONIA: Peony

Cronartium flaccidum (Alb. & Schw.)Wint., 11, 111
Uredinia hypophyllous, yellowish; spores echinulate, 20–30 × 15–20. Telia hypophyllous; spores smooth, 20–60 × 10–15, extruded as golden or reddish-brown waxy masses. On *P. mascula*, few records; alternate host *Pinus sylvestris*.

Botrytis paeoniae Oudem. (Fig. 1593)
Sclerotia small, usually about 1mm diam. Conidiophores brown, often swollen at base. Conidia 10–22 × 7–11, mostly 12–18 × 8–10. Causes wilt of young shoots of peony plants in spring; mature leaves later also become infected.

Cladosporium chlorocephalum (Fresen.)Mason & M.B.Ellis (Fig. 1594)
Colonies effuse, olivaceous brown to black. Conidiophores up to 700 × 8–15, blackish brown or black, bearing spherical or oval, olive-green heads of conidia, in chains, on short branches. Conidia olive or pale brown, smooth or verruculose, 6–14 × 4–9, terminal ones spherical, 3.5–7 diam. On dead stems and leaves, Dec.–May.

Septoria paeoniae Westend. var *berolinensis* Allesch.
Pycnidia epiphyllous, immersed, brown, usually few. Conidia curved, tapered at the ends, hyaline, guttulate, 20–30 × 1.5–2. On leaves, causing irregular, zonate, yellowish brown, purple-bordered spots, sometimes with pale grey centres, May–Sept.

PAPAVER: Poppies

Entyloma fuscum Schröter
Smut sori in leaves. Spores hyaline to pale brown, 10–15 diam. Conidia forming white tufts on surface, curved, hyaline, 10–20 × 3. Causes round yellowish spots on basal leaves, May–July, very few records.

Peronospora arborescens (Berk.)Caspary
Downy mildew on leaves of *P. dubium*, *P. rhoeas* and *P. somniferum*. Sporangia pale violaceous, 15–22 × 13–18.

Phomopsis morphaea (Sacc.)Grove
Conidiomata erumpent, black, about 0.2mm diam. Conidia hyaline, biguttulate, 7–8 × 2.5–3. On dead stems and capsules of *P. orientale* and other garden poppies.

Pleospora papaveracea (de Not.)Sacc. (Fig. 1595)
Pseudothecia on dead stems of *P. rhoeas* and *P. somniferum*. Ascospores pale yellow, with 3 or 4 transverse septa and often a longitudinal septum, 22–27 × 7–10. The *Dendryphion* state is found on leaves, stems and capsules, forming black, hairy colonies. Conidiophores up to 600 × 5–10, sometimes up to 20 thick at base. Conidia solitary or in chains, pale olive, smooth, on stems mostly

3-septate and 17–28 × 5–9 but on leaves sometimes up to 8-septate and 60 long.

PARENTUCELLIA: Yellow Bartsia

Coleosporium tussilaginis, 11, 111, described on *Tussilago*, occurs also on *Parentucellia*.

PARIETARIA: Pellitory-of-the-wall

Ramularia parietariae Passer. (Fig. 1596)
Colonies white, amphigenous. Conidiophores hyaline, 50–100 × 2.5–4. Conidia hyaline, 0- to 3-septate, 5–25 × 3–4.5. On living leaves, causing round or oval, pale brown or whitish spots 2–6mm diam.

PARIS: Herb Paris

Puccinia sessilis Schröter, 0, 1
Aecia rarely found on *P. quadrifolia*. Alternate host *Phalaris arundinacea*.

PARNASSIA: Grass of Parnassus

Puccinia caricina DC. var *uliginosa* (Juel)Jørstad, 1
Aecia recorded in Scotland. Alternate host *Carex nigra*.

PASTINACA: Parsnip

Erysiphe heraclei, powdery mildew described on *Heracleum*, is found also on *Pastinaca*.

Itersonilia pastinacae Channon (Fig. 1597)
Mycelium with clamp connections. Ballistospores reniform or pyriform, hyaline, 11–20 × 7–11, borne singly on sporophores. Chlamydospores thick-walled, 9–12 diam. On roots, causing canker; also on leaves.

Phloeospora heraclei, described on *Heracleum*, has been found also on *Pastinaca*.

Phomopsis diachenii Sacc.
Conidiomata multilocular, up to 0.4mm diam. A-conidia hyaline, 8–16 × 2.5–4.5, B-conidia curved, sometimes hamate, 15–20 × 0.5. On stems and fruits.

Plasmopara nivea, downy mildew described on *Aegopodium*, occurs also on *Pastinaca*.

Pyrenopeziza pastinacae (Nannf.)Gremmen (Fig. 1598)
Apothecia erumpent, up to 1mm diam., with watery white discs and thick, darker rims. Ascospores hyaline, 6–7 × 1.5–2. On greyish areas on dead, decorticated, overwintered stems, May–July.

Ramularia pastinacae Bubák (Fig. 1599)
Colonies mostly hypophyllous, white. Conidiophores hyaline, 20–45 × 3–5. Conidia hyaline, 0- to 4-septate, 14–42 × 3–5. On leaves, causing whitish or pale brown, often dark-bordered spots 1–6mm diam.

PEDICULARIS: Lousewort

Puccinia caricina DC. var. *paludosa* (Plowr.)Henderson, 0, 1
Aecia on *P. palustris*. Alternate hosts *Carex bigelowii*, *C. elata*, *C. nigra* and *C. panicea*.

Puccinia clintoniae Peck var. *sylvaticae* Savile, 111
Telia on leaves, stems and calyces of *P. palustris* and *P. sylvatica*, reddish brown, uncommon; spores smooth or striate, 25–35 × 15–17.

Ramularia obducens Thümen
Colonies hypophyllous, thin, white. Conidia hyaline, 12–17 × 3.5–4.5. On living leaves of *P. palustris*.

PELARGONIUM: Geranium

Puccinia pelargonii-zonalis Doidge, 11, rarely 111
Uredinia cinnamon brown, on small pale spots; spores finely echinulate, 22–30 × 20–22. Teliospores found occasionally in same sori, 35–70 × 17–28. On leaves of *P.* × *hybridum*.

PETASITES: Butterbur, Winter Heliotrope

Coleosporium tussilaginis, 11, 111, described on *Tussilago*, occurs on *P. hybridus*.

Ramularia purpurascens Wint. (Fig. 1600)
Colonies amphigenous. Conidiophores in groups of 3 to 10, hyaline; conidia 0- to 1-septate, 8–25 × 2.5–3. On living leaves of *P. fragrans*, causing round or angular, grey or whitish spots 1–5mm diam., each with a dark-brown border surrounded by a purplish zone, Sept.

Plurivorous fungi found on dead stems and petioles of *P. hybridus* include *Chromelosporium ochraceum* and *Gliomastix luzulae*.

PETROSELINUM: Parsley

Puccinia nitida, 0, 1, 11, 111, described on *Aethusa*, has been recorded also on *Petroselinum*.

Alternaria petroselini (Neergaard)Simmons (Fig. 1601)
Colonies effuse, dark blackish brown to black, often covering large areas of leaf surface. Conidiophores rather pale olivaceous brown, up to 80×6–8. Conidia mostly solitary, golden brown, usually smooth, muriform, 35–105×18–34.

Septoria petroselini Desm.
Pycnidia amphigenous, 0.1mm diam. Conidia hyaline, guttulate or with a few thin septa, 30–40×1–2. On leaves and stems, causing whitish or pale brown, sometimes dark-bordered, spots, July–Aug.

PEUCEDANUM: Milk Parsley, Sulphur-weed

Puccinia angelicae, 0, 1, 11, 111, described on *Angelica*, occurs on *P. palustre*.

Puccinia rugulosa Tranz., 0, 1, 11, 111
Aecia uredinoid. Uredinia hypophyllous, very small, brown; spores minutely echinulate, 27–33×19–25. Telia very dark brown; spores 1-septate, striate, 35–50×20–28. A rare rust on *P. officinale*.

Diaporthopsis angelicae, described on *Angelica*, is found sometimes on dead stems of *P. palustre*.

Symphyosirinia angelicae, described on *Angelica*, has been recorded on fruits of *P. palustre*.

PHASEOLUS: French Beans, Runner Beans

Uromyces appendiculatus (Pers.)Unger, 0, 1, 11, 111
Aecia seldom found. Uredinia cinnamon; spores echinulate, 20–30×20–25. Telia amphigenous, blackish brown; spores smooth or slightly warted, 26–36×20–30. Not uncommon on leaves of *P. coccineus* and *P. vulgaris*.

Ascochyta boltshauseri Sacc.
Pycnidia immersed, brown, 0.1–0.15mm diam. Conidia oblong rounded at ends, with 1, 2 or sometimes more septa, hyaline, 22–28×7–8. Causes yellowish-brown spots on leaves and sometimes on pods.

Ascochyta phaseolorum Sacc.
Pycnidia amphigenous but mostly epiphyllous, immersed, 0.1–0.2mm diam., honey-coloured, darker around the ostiole. Conidia straight or slightly bent, hyaline, 1-septate, 10–16 × 3–5. On leaves, causing grey to brown sometimes zonate spots.

Colletotrichum lindemuthianum (Sacc. & Magn.)Briosi & Cav. (Fig. 1602)
Acervuli up to 0.3mm diam., setose. Setae brown, up to 100 × 4–8. Conidia hyaline, pale salmon-pink in mass, 10–20 × 3–5.5. Leaf, stem and pod anthracnose, grey to brown, sunken lesions, with raised reddish-brown borders, June–Aug., widely distributed.

Diaporthe phaseolorum Cooke & Ellis
Perithecia black, up to 0.5mm diam., with necks up to 0.5mm long. Ascospores hyaline, 1-septate, 9–11 × 2.5–3.5. Conidiomata of *Phomopsis* state about 0.1mm diam.; A-conidia hyaline, 6–8 × 2–2.5, B-conidia curved or hamate, 20–30 × 0.5–1. On dead stems and pods.

Macrophomina phaseolina (Tassi)Goid. (Fig. 1603)
Pycnidia immersed, becoming erumpent, dark brown, 0.1–0.2mm diam.; conidia hyaline, 15–30 × 5–10. It also forms numerous immersed, smooth, hard, black sclerotia up to 1mm diam. On beans and many other plants; few records from Britain.

PHYSOSPERMUM: Bladder-seed

Puccinia physospermi Pass., 0, 111
Telia hypophyllous, brown; spores with wavy walls, 38–45 × 25–27. A very rare rust.

Cercospora physospermi Deighton (Fig. 1604)
Colonies hypophyllous. Conidiophores pale olivaceous to brown, 15–35 × 4–7. Conidia hyaline, mostly 5-septate, 30–115 × 4–6. On leaves causing inconspicuous yellowish spots.

PICRIS: Hawkweed Ox-tongue, Bristly Ox-tongue

Puccinia hieracii, 0, 1, 11, 111, described on *Hieracium*, occurs but is uncommon on *P. hieracioides*.

Ramularia filaris Fresen. (Fig. 1605)
Colonies amphigenous. Conidiophores hyaline, 15–45 × 2–4. Conidia hyaline, 0- to 3-septate, 10–40 × 2.5–3.5. On leaves of *P. echioides*, causing round or angular, greyish-brown spots up to 1.5cm diam.

Ramularia picridis Fautr. & Roum. (Fig. 1606)
Colonies hypophyllous. Conidiophores hyaline, 20–50 × 2.5–3.5. Conidia hyaline, 0- to 3-septate, 15–52 × 3–4.5. On leaves of *P. echioides* and *P. hieracioides*, causing brown spots 2–10mm diam., often with purple borders.

PIMPINELLA: Burnet Saxifrage, Greater Burnet Saxifrage

Puccinia pimpinellae (Str.)Röhl., 0, 1, 11, 111
Aecia hypophyllous, often on the nerves. Uredinia small, cinnamon; spores echinulate, 22–30 × 20–25. Telia blackish brown; spores with reticulate walls, 1-septate, 30–37 × 20–25. On *P. major* and *P. saxifraga*, common.

Symphyosira rosea Keissler (Fig. 1607)
On fallen fruits of *P. major*. Synnemata with white stalks and pale pink heads. Conidia hyaline, 3-septate, mostly 30–45 × 4–6, occasionally up to 60 long.

PISUM: Field Pea, Garden Pea

Uromyces pisi-sativi (Pers.)Liro, 11, 111
Uredinia mostly hypophyllous, up to 1mm diam., cinnamon; spores finely verruculose, 20–30 × 15–22. Telia dark brown; spores 22–28 × 16–24, finely verruculose, sometimes appearing somewhat striate, pedicels short. Aecial host *Euphorbia cyparissias*. Uncommon.

Uromyces viciae-fabae, 0, 1, 11, 111, described on *Vicia*, has smooth teliospores with long pedicels.

Ascochyta pisi Lib.
Pycnidia at first immersed, then erumpent, up to 0.25mm diam., brown. Conidia hyaline, mostly 1-septate, 11–16 × 3–4.5. On living leaves, stems and pods, causing brown spots with pale centres and dark, somewhat prominent, margins.

Didymella pinodes (Berk. & Bloxam)Petrak
Pseudothecia on stems and pods, immersed, dark brown, 0.1–0.15mm diam. Ascospores 1-septate, hyaline with guttules, 12–18 × 4–8. Pycnidia of *Ascochyta* state 0.1–0.2mm diam., dark brown to black; conidia hyaline with 1 or sometimes 2 or 3 septa, 8–16 × 3–5. Causing leaf, stem and pod spot, and foot rot. Spots at first very small and purple, later enlarging and becoming dark brown, somewhat zonate but without well-defined margins.

Erysiphe pisi DC.
Powdery mildew. Conidia plentiful, 30–38 × 17–21. Cleistothecia about 0.1mm diam., appendages short, usually about same as diam. of cleistothecia. Asci 3 to 10, each with 3 to 5 spores.

Peronospora viciae, downy mildew described on *Vicia*, occurs also on *Pisum*.

Phoma medicaginis Malbr. & Roum. var *pinodella* (L.K.Jones)Boerema
Pycnidia small. Conidia hyaline, with minute guttules, 0- or 1-septate, 5–10 × 2.5–4. At base of stems, causing foot rot.

PLANTAGO: Plantains

Ascochyta plantaginis Sacc. & Speg. (Fig. 1608)
Pycnidia dark brown, 0.1–0.2mm diam. Conidia mostly 1-septate, 8–12 × 2–3. Causes brown spots with pale centres, up to 6mm diam., especially on young, living leaves of *P. major*, June–Nov.

Cercospora plantaginis Sacc. (Fig. 1609)
Colonies amphigenous. Conidiophores pale olivaceous brown, mostly 40–70 × 3.5–5 but sometimes longer. Conidia hyaline, multiseptate, 50–100 × 2–3.5. On living and fading leaves of *P. lanceolata* causing pale spots with brown margins, Mar., uncommon.

Erysiphe sordida Junell (Fig. 1610)
Powdery mildew. Conidia abundant. Cleistothecia 0.1–0.13mm diam., containing 6 to 20 asci, each usually with 2 spores. On *P. coronopus*, *P. major*, *P. maritima* and *P. media*, Sept.–Oct.

Peronospora alta Fuckel (Fig. 1611)
Downy mildew. Colonies effuse, greyish. Sporangia becoming pale brown or violaceous, 25–35 × 18–25. Common on *P. major*, occasional on *P. media*, July–Aug.

Phoma polygramma (Fr.)Sacc.
Pycnidia erumpent, arranged in rows, about 0.1mm diam., dark brown, often connected with one another by brown hyphae. Conidia hyaline, 5–7 × 2.5–3. On dead peduncles and occasionally leaves and petioles of *P. lanceolata*, Feb.–Mar.

Phomopsis subordinaria (Desm.)Trav.
Conidiomata immersed, 0.25–0.5mm diam., covered by blackish-brown epidermis. A-conidia oblong–fusoid or subclavate, often slightly curved, hyaline, 7–12 × 2–2.5; B-conidia straight, curved or hamate, 20–23 × 0.5–1. On the upper part of living and dead stems of *P. lanceolata* and *P. media*, causing them to bend over, June–Nov.

Pyrenopeziza plantaginis Fuckel (Fig. 1612)
Apothecia erumpent, up to 0.25mm diam., seldom larger, dark greyish brown, with white fimbriate margins when fresh. Ascospores hyaline, 10–15 × 3. On fading and dead attached leaves of *P. lanceolata*, and sometimes on brown spots on the living leaves, Oct.–July, and always in good condition Feb.–Mar., common.

Ramularia rhabdospora (Berk. & Br.)Nannf. (Fig. 1613)
Colonies amphigenous, white. Conidiophores hyaline, 40–70 × 3–4. Conidia hyaline, 0- to 3-septate, 20–50 × 3.5–5. On living leaves causing brown or greyish-brown, sometimes bordered, spots on *P. lanceolata*, *P. major* and *P. media*, Mar.–Oct., very common.

Sphaerotheca plantaginis (Castagne)Junell (Fig. 1614)
Powdery mildew on *P. lanceolata*, common, Aug.–Oct. Conidia plentiful, 27–31 × 15–17. Cleistothecia 80–90 diam., each containing 1 ascus.

Spilopodia nervisequia (Fr.)Boud. (Fig. 1615)
Apothecia 0.5–1mm diam., dark blackish brown to black, with indented margins, discs grey, seated on stromatic strands along the veins of dead and often frayed leaves of *P. lanceolata*, Feb.–Apr. Ascospores hyaline, 10–12 × 3.5–4.

Taeniolella plantaginis (Corda)Hughes (Fig. 1616)
Colonies effuse, black. Mycelium partly superficial. Conidiophores brown, 2–9 thick. Conidia flexuous, vermiform, smooth to verrucose, dark brown, 30–150 × 8–10, with 6 to 28 septa. On senescent leaves of *P. media*.

Common plurivorous fungi found on dead stems of *P. lanceolata* include *Leptospora rubella* and *Pleospora herbarum*.

POLEMONIUM: Jacob's Ladder

Puccinia polemonii Diet. & Holw., 111
Telia mostly on petioles of *P. caeruleum*, dark reddish brown; spores 1-septate, 35–45 × 14–15, pale. Very few records.

POLYGONATUM: Solomon's Seal

Kabatiella microsticta Bubák (Fig. 1617)
Conidiophores emerging through stomata in tufts, yellowish, 20–50 × 3–8. Conidia hyaline, 5–12 × 2–4. On living leaves of *P. multiflorum*, causing large brownish spots spreading back from tips.

POLYGONUM: Amphibious Bistort, Bistort or Snake-root, Black Bindweed, Knotgrass, Persicaria, Water-pepper, etc.

UREDINALES

Five rusts including one variety occur on different species of *Polygonum* but only *P. viviparum* has more than one.

On *P. amphibium*: *Puccinia polygoni-amphibii*
On *P. aviculare*: *Uromyces polygoni-aviculare*
On *P. bistorta*: *Puccinia bistortae*
On *P. convolvulus*: *Puccinia polygoni-amphibii* var. *convolvuli*
On *P. viviparum*: *Puccinia bistortae*, which has teliospores often finely verruculose, without an apical papilla; and *Puccinia septentrionalis*, teliospores quite smooth, with a rather large apical papilla.

Puccinia bistortae DC., 11, 111
On *P. bistorta* and *P. viviparum*, uncommon. Uredinia dull orange; spores shortly echinulate, 20–25 × 19–20. Telia hypophyllous, dark brown; spores broadly ellipsoid, often finely verruculose, 1-septate, 25–40 × 17–25. Aecial hosts *Angelica sylvestris* and *Conopodium majus*.

Puccinia polygoni-amphibii Pers., 11, 111
On *P. amphibium*, common. Uredinia mostly hypophyllous, brown; spores faintly echinulate, 18–30 × 15–21. Telia amphigenous, blackish brown; spores smooth, 1-septate, 35–50 × 17–22.

Puccinia polygoni-amphibii var. *convolvuli* Arth., 11, 111
On *P. convolvulus*. Aecial host *Geranium dissectum*. Few records.

Puccinia septentrionalis Juel, 11, 111
On *P. viviparum* in Scotland. Uredinia hypophyllous, yellowish; spores echinulate, 20–22 diam. Telia dark brown, spores 1-septate, with rather large hyaline papilla, 30–48 × 14–23.

Uromyces polygoni-aviculare (Pers.)Karst., 0, 1, 11, 111
On *P. aviculare*. Aecia scarce, other states common. Uredinia cinnamon; spores 18–25 × 18–24, minutely verruculose. Telia dark brown; spores smooth, 23–38 × 15–22.

USTILAGINALES

Five smuts are found on different species, three in flowers, one on leaves and the other in bulbils.

On *P. bistorta*: in leaves, *Ustilago bistortarum*
On *P. convolvulus*, *P. lapathifolium* and *P. persicaria*: in flowers (1) *Ustilago anomala*, spores with delicately reticulate walls, mesh never more than 2 wide; (2) *Ustilago utriculosa*, spores with more coarsely reticulate walls, mesh up to 4 wide
On *P. hydropiper*: in flowers (1) *Ustilago anomala*, spores with reticulate walls; (2) *Sphacelotheca hydropiperis*, spores verruculose
On *P. viviparum*: in leaves, *Ustilago bistortarum*; in bulbils in inflorescences, *Sphacelotheca inflorescentiae*

Sphacelotheca hydropiperis (Schum.)de Bary
Sori in flowers, each enveloped in a pale grey false membrane which projects from the perianth. Spore mass blackish purple, with central columella. Spores minutely and closely verruculose, 10–14 diam. On *P. hydropiper*, Sept.–Oct., fairly common.

Sphacelotheca inflorescentiae (Trel.)Jaap
Sori in bulbils in inflorescences of *P. viviparum*, Scotland, June, few records. Spores minutely verruculose, 11–16 diam.

Ustilago anomala Kunze
Sori in flowers. Spores violet to purplish brown, 10–13 diam., walls reticulate, mesh not more than 2 wide. On *P. convolvulus*, *P. hydropiper*, *P. lapathifolium* and *P. persicaria*, Aug.–Oct., fairly common.

Ustilago bistortarum (DC.)Körn.
Sori around the margins of leaves or scattered over the surface. Spore mass blackish purple. Spores minutely verruculose, 10–16 diam. On *P. bistorta* and *P. viviparum*, July–Aug., few records.

Ustilago utriculosa (Nees)Tul. & C. Tul.
Sori in flowers. Spore mass brownish purple. Spores 11–14 diam., walls reticulate, mesh 2–4 wide. On *P. convolvulus*, *P. lapathifolium* and *P. persicaria*, Aug.–Sept., uncommon.

OTHER FUNGI

Bostrichonema alpestre Ces. (Fig. 1618)
Colonies hypophyllous, white. Conidiophores emerging through stomata in groups of 3 to 8, spirally twisted, hyaline, 80–160 × 2–7. Conidia on short pegs, hyaline, smooth or verruculose, becoming 1-septate, 20–33 × 13–16. Causes dark reddish brown, round or angular spots up to 3mm diam. on leaves of *P. viviparum*.

Ceriospora polygonacearum (Petrak)Pirozynski & Morgan-Jones (Fig. 1619)
Pseudothecia 0.15–0.25mm diam., immersed in black, flattened stromata in stems, with necks protruding through a dark brown clypeus, and just showing. Ascospores hyaline, 1-septate, mostly 20–25 × 4–6, with appendages usually 5–10 but occasionally up to 40 long. *Chaetoconis* conidial state with black, sclerotium-like pycnidia 0.25mm diam., conidia hyaline, 1- or 2-septate, 30–50 × 3–4, including apical appendage. On dead stems of *P. cuspidatum* and *P. sachalinense*, Mar.–Apr.

Endophragmia cesatii (Mont.)M.B.Ellis (Fig. 1620)
Colonies effuse, hairy, blackish brown. Conidiophores up to 160 × 3–7, pale brown. Conidia 3-septate, 22–44 × 9–13, median cells dark brown, end cells hyaline or pale brown. On dead stems of *P. cuspidatum*, May.

Erysiphe polygoni DC.
Powdery mildew. Conidia abundant. Cleistothecia up to 0.1mm diam., yellow to blackish brown, containing 3 to 12 asci, each with 3 or 4 spores. On *P. amphibium*, *P. avenastrum* and *P. aviculare*, very common.

Hymenoscyphus scutula (Pers.)Phill. var *fucatus* Phill.
Apothecia slender-stalked, brownish yellow, with incurved margins. Ascospores hyaline, 24–34 × 5–7. On dead stems lying in water.

Mollisia polygoni (Lasch)Gillet (Fig. 1621)
Apothecia erumpent, 0.5–1mm diam., reddish brown, disc sometimes pale. Ascospores hyaline, 7–8 × 2. On dead stems of *P. amphibium*, July.

Peronospora polygoni (Halsted)Fisher
Downy mildew on *P. convolvulus*. Sporangia 28–32 × 16–20.

Pezizella effugiens (Desm.)Rehm (Fig. 1622)
Apothecia erumpent, convex or pulvinate, white with a slight yellowish tinge. Ascospores hyaline, 5–8 × 1. On dead stems of *P. cuspidatum*, Apr.

Phomopsis polygonorum (Cooke)Grove
Conidiomata immersed, black, 0.3–0.5mm long, epidermis blackened. Conidia ellipsoid to fusiform, biguttulate, 6–9.5 × 2.5–3. On dead stems of *P. cuspidatum* and *P. sachalinense*.

Plagiostoma devexa (Desm.)Fuckel (Fig. 1623)
Perithecia immersed, black, 1–4mm diam., somewhat flattened and elongated, with neck at one end which protrudes through the epidermis. Rather inconspicuous. Ascospores hyaline, 1-septate, 8–11 × 2.5–3. On dead stems of *P. amphibium*.

Ramularia bistortae Fuckel
Colonies thin, white. Conidia oblong–ovate, hyaline, about 12 × 6. On leaves of *P. bistorta*, June.

Septoria polygonorum Desm.
Pycnidia epiphyllous, immersed, 0.1–0.15mm diam., brown. Conidia flexuous, hyaline, guttulate or with a few indistinct septa, 25–70 × 1–1.5. On living leaves of *P. persicaria*, causing fawn spots 2–6mm diam., with purplish borders, July–Oct., common. Found occasionally also on *P. bistorta*, *P. convolvulus* and *P. hydropiper*.

Plurivorous microfungi found on dead stems of *P. cuspidatum* and similar large species include *Dendryphion comosum*, *Dictyosporium toruloides*, *Endophragmia atra*, *E. boewei*, *E. elliptica*, *E. nannfeldtii*, *Gibberella cyanogena*, *Helminthosporium velutinum*, *Lophiostoma vagabundum*, *Melanomma pulvis-pyrius*, *Mcnispora glauca*, *Periconia byssoides*, *Pleurophragmium parvisporum*, *Pseudospiropes nodosus*, *Pyrenopeziza revincta* and *Torula herbarum*.

POTENTILLA: Barren Strawberry, Cinquefoils, Silverweed, Tormentils

UREDINALES

Frommea obtusa (Strauss)Arth. (Fig. 1624)
Aecia epiphyllous, uredinoid, orange. Uredinia hypophyllous, yellow; spores echinulate, 20–25 × 15–20. Telia pale brown; spores 1- to 6-septate, smooth, 55–140 × 20–30. On *P. erecta* and *P. reptans*, uncommon.

Phragmidium fragariae (DC.)Rabenh., 0, 1, 11, 111 (Fig. 1625)
Aecia mostly hypophyllous, caeomoid, orange. Uredinia hypophyllous, yellowish orange; spores 20–25 × 18–22, verrucose. Telia blackish brown; spores 1- to 4-septate, 45–70 × 23–28. On *P. sterilis*, very common.

Phragmidium potentillae (Pers.)Karst., 0, 1, 11, 111
Aecia amphigenous, orange. Uredinia hypophyllous, small, with paraphyses; spores echinulate, 21–24 × 17–20. Telia black; spores 2- to 5-septate, smooth, 40–80 × 20–30. On *P. anglica* and *P. tabernaemontani*, few records.

OTHER FUNGI

Cercospora comari Peck (Fig. 1626)
Colonies mostly hypophyllous. Conidiophores golden brown, 150–300 × 5–7. Conidia straw-coloured to golden brown, with 1 to 6, mostly 3, septa, 45–75 × 6–8. On living leaves of *P. palustris* causing irregular, reddish-brown spots.

Coleroa potentillae (Wallr. ex Fr.)Winter (Fig. 1627)
Pseudothecia epiphyllous, superficial, in small clusters along the veins, black, setose, up to 0.1mm diam.; setae dark brown, 15–35 × 4–5. Ascospores pale greyish olive, 1-septate, 10–14 × 4–5. On living leaves of *P. anserina*, Aug.–Oct.

Diplocarpon earlianum, described on *Fragaria*. The *Marssonina* state causes roundish, red or reddish-brown spots on small, yellowish but still living leaves of *P. anserina* and *P. reptans*, often scattered over the whole of the upper surface, July–Oct.

Gnomonia comari P. Karsten
Perithecia immersed, 0.3–0.35mm diam., with thick necks. Ascospores hyaline, clavate, straight or slightly curved, 1-septate, 7–8 × 2. On dead stems of *P. palustris*, June.

Leptotrochila repanda (Fr.)P. Karsten
Apothecia erumpent, 0.3–1mm diam., at first pale greenish then almost black, with greyish discs. Ascospores straight or slightly curved, spindle-shaped, with oil drops, often becoming 1-septate, 11–19 × 2.5–3.5. On living and dying

stems and leaves of *P. norvegica*, especially along the main nerve on the underside, June–July.

Peronospora potentillae de Bary
Downy mildew on *P. reptans*. Sporangia pale violaceous, 20–26 × 15–19. Oospores yellow, smooth-walled, 22–24 diam.

Plagiostoma tormentillae (Lind)Bolay
Perithecia immersed, valsoid, black, 0.2–0.25 × 0.1–0.15mm, with lateral necks up to 0.1mm long. Asci mostly 20–30 × 3–5, with apical ring. Ascospores hyaline, 1-septate below the middle, 6–10 × 1.5–2.5, with a filiform appendage at each end. On dead stems of *P. erecta*, Apr.–May.

Ramularia arvensis Sacc. (Fig. 1628)
Colonies hypophyllous, white. Conidiophores 20–50 × 2–3.5. Conidia hyaline, 0- or 1-septate, 20–35 × 2.5–3.5. On leaves of *P. anserina* and *P. reptans*, causing pale brown or white spots 1–2mm diam. with narrow reddish-brown or purple borders.

Septoria tormentillae Desm.
Pycnidia numerous, dark brown, up to 0.1mm diam. Conidia curved or flexuous, pale yellowish, occasionally 7- or 8- septate, 40–55 × 1–1.5. On leaves and stipules of *P. erecta*, causing ochraceous or dingy white spots 1–4mm long.

Sphaerotheca alchemillae, powdery mildew described on *Alchemilla*, occurs on *P. reptans* and *P. sterilis*.

Taphrina potentillae (Farlow)Johansson
On *P. erecta*, causes yellow thickened spots on leaves and yellow thickening of stems. Asci 20–50 × 7–15, each with up to 8 spores 2–4 diam. Common in wet boggy places.

Venturia palustris Sacc., Bomm. & Rouss. (Fig. 1629)
Pseudothecia amphigenous, immersed, black, up to 0.15mm diam., sometimes setose, with brown setae up to 40 long. Ascospores pale olivaceous, 1-septate, 10–14 × 2–4.5. On leaves of *P. palustris*, causing irregular, brown or purplish spots.

POTERIUM: Salad Burnet

Phragmidium sanguisorbae (DC.)Schröter, 0, 1, 11, 111
Aecia amphigenous, caeomoid, orange. Uredinia hypophyllous, small, yellowish orange, with curved paraphyses; spores 18–24 × 17–20. Telia hypophyllous; spores with 1 to 4, mostly 3, septa, 40–70 × 20–25. On *P. polygamum* and *P. sanguisorba*, common.

PRIMULA: Auricula, Cowslip, Oxlip, Primrose

Puccinia primulae Duby, 1, 11, 111
Aecia hypophyllous, on yellow spots. Uredinia hypophyllous, brown; spores echinulate, 20–23 × 17–20. Telia blackish brown; spores smooth, 1-septate, 24–30 × 15–18. On *P. vulgaris*, common.

Urocystis primulicola Magn.
The conidial state of this smut can be found in March attacking the anthers and ovaries of *P. vulgaris*, and filling the corolla tubes with a white powdery mass of smooth, hyaline conidia 4–12 × 4–6. Sori in ovaries July–Sept.; spore mass blackish brown; spores 10–15 diam., smooth, forming balls 30–60 × 20–50. Not common.

Ascochyta primulae Trail
Pycnidia epiphyllous, pale brown, 0.1mm diam. Conidia cylindrical, 0- or 1-septate, 5–6 × 2–2.5. On leaves of *P. vulgaris*, causing large, whitish, yellow-bordered spots, Aug.

Cercosporella primulae Allesch. (Fig. 1630)
Conidia hyaline, 20–120 × 3–5, with up to 12 septa. On leaves of *P. vulgaris*, causing chocolate brown spots, sometimes with grey centres which eventually become thin and papery.

Mycosphaerella primulae (Auersw. & Heufl.)Schröter
On leaves of *P. auricula*. Pseudothecia black, erumpent. Ascospores hyaline, 1-septate, 22–28 × 4–6.

Peronospora oerteliana Kühn
Downy mildew on *P. veris* and *P. vulgaris*, May, uncommon.

Ramularia primulae Thüm. (Fig. 1631)
Colonies mostly hypophyllous. Conidiophores in groups of 6–15 emerging through stomata, hyaline, 25–60 × 3–5. Conidia hyaline, mostly 0- or 1-septate, 6–40 × 3–5. On leaves of *P. elatior*, *P. veris* and *P. vulgaris*, May–June, causing round or angular, pale brown spots, each surrounded by a bright yellow or pale orange border; the centres sometimes drop out, leaving shot-holes.

PRUNELLA: Self-heal

Puccinia brunellarum-moliniae Cruchet, 0, 1
Aecia hypophyllous on purplish or yellowish-brown spots, June–July; spores echinulate, 17–22 × 14–17. Alternate host *Molinia caerulea*.

Leptotrochila brunellae (Lind)Dennis (Fig. 1632)
Apothecia 1–2mm diam., dark brown to black, in small groups seated on mats of radiating black hyphae which envelop stem bases and cover some of the lower leaves of plants growing on paths and rides in woodlands, Jan.–May. Ascospores hyaline, becoming 1-septate, 12–20 × 4–5.

Rosenschoeldia abundans (Dobrozr.)Petrak (Fig. 1633)
Pseudothecia black, 0.3–0.35mm diam., in clusters on erumpent, elongated, pulvinate, black stromata on living stems, Sept.–Oct., uncommon. Ascospores subhyaline, 1- to 6-septate, 30–40 × 1.5–3.

PULICARIA: Fleabane

Uromyces junci (Desm.)Tul., 0, 1
Aecia hypophyllous, on spots with purple and yellow borders, May–July; spores finely verruculose, 18–21 diam. Alternate hosts *Juncus effusus, J. inflexus* and *J. subnodulosus*.

Ramularia cupulariae Passer. (Fig. 1634)
Colonies amphigenous, white. Conidiophores hyaline, 20–70 × 3–4.5. Conidia hyaline, 0- to 2-septate, 10–25 × 3–5. On leaves, causing yellow or pale brown spots 3–8mm diam., uncommon.

Sphaerotheca xanthii, powdery mildew described on *Calendula*, occurs on *Pulicaria*.

PYROLA: Wintergreen

Chrysomyxa pirolata Wint., II, III
Uredinia covering lower leaf surface uniformly; spores verruculose, 20–35 × 14–24. On *P. minor* and *P. rotundifolia*, rare. Few records of telia.

Pucciniastrum pyrolae Diet. ex. Arth., II
Uredinia hypophyllous, scattered or in small groups, causing small reddish spots on the upper leaf surface; spores 24–40 × 10–20, finely echinulate. On *P. media* and *P. minor*, rare.

RANUNCULUS: Buttercups, Crowfoots, Goldilocks, Lesser Celandine, Spearworts

UREDINALES

Five rusts are found on different species, four of them in rather similar aecial states but only one, *Uromyces ficariae* on *R. ficaria*, in the telial state.

On *R. acris*: (1) *Uromyces dactylidis*. Aecia shortly cup-shaped, with slightly torn recurved margins; spores pale yellow, round or angular, 12–25 diam., common. (2) *Puccinia recondita* f.sp. *perplexans*. Aecia often somewhat cylindrical with white, incised margins; spores orange, tending to be ellipsoid, 20–29 × 14–26, uncommon.

On *R. auricomus*: *U. dactylidis* only.

On *R. bulbosus* and *R. repens*: (1) *U. dactylidis*. Aecia with slightly torn recurved margins. Alternate hosts *Dactylis*, *Festuca* and *Poa*. (2) *Puccinia magnusiana*. Aecia with distinct, white, incised margins. Alternate host *Phragmites*, very common.

On *R. ficaria*: (1) *U. ficariae*. Dark chocolate-brown telia only. (2) *U. dactylidis*. Aeciospores without refractive granules in their walls. Very common. (3) *U. rumicis*. Aeciospores with refractive granules in their walls. Few records.

On *R. flammula* and *R. lingua*: *P. magnusiana*.

On *R. sceleratus*: *U. dactylidis*.

Puccinia magnusiana Körn., 0, 1
Aecia on *R. bulbosus*, *R. flammula*, *R. lingua* and *R. repens*, cup-shaped with white incised margins, in small groups on round yellowish spots on lower surface of leaves, elongated on petioles and stems; spores yellow, minutely verruculose, 15–25 diam. Alternate host *Phragmites australis*.

Puccinia recondita Rob. & Desm. f.sp. *perplexans* Plowr., 0, 1
Aecia on *R. acris*, hypophyllous, in crowded groups, often somewhat cylindrical, with white incised margins; spores orange, tending to be ellipsoid, minutely verruculose, 20–29 × 14–26. Alternate host *Alopecurus pratensis*, uncommon.

Uromyces dactylidis Otth., 0, 1
Aecia on *R. acris*, *R. auricomus*, *R. bulbosus*, *R. ficaria*, *R. repens* and *R. sceleratus*, causing yellow spots on leaves, elongated on petioles, shortly cup-shaped with slightly torn recurved margins; spores pale yellow, round or angular, minutely verruculose, 17–25 diam., without refractive granules in their walls. Alternate hosts *Dactylis*, *Festuca rubra*, *Poa nemoralis*, *P. pratensis*, *P. trivialis*, very common.

Uromyces ficariae Tul., 111 (Fig. 1635)
Telia amphigenous and on petioles, dark chocolate brown; spores smooth, mostly 25–35 × 20–25. On *R. ficaria*, Mar.–May, very common.

Uromyces rumicis (Schum.)Wint., 0, 1
A few records of aecia on *R. ficaria*. Similar to *U. dactylidis* but spores have refractive granules in their walls. Alternate hosts *Rumex* spp.

USTILAGINALES

Four smuts occur on *Ranunculus*: two species of *Entyloma*, which cause round, rather indistinct, pale yellowish or yellowish-brown spots on leaves, and may produce powdered white masses of conidia on the surface; and two species of *Urocystis*, which form swollen, silvery blisters that rupture to expose dark brown to black masses of spore balls.

Entyloma ficariae (Cornu & Roze)Fischer v. Waldh.
Spores embedded in leaves, spherical, smooth, pale straw, thin-walled, 10–14 diam. Conidia formed on both leaf surfaces, hyaline, fusiform or ellipsoidal, 30–45 × 2. On *R. ficaria* and *R. sceleratus*, very common, Apr.–May.

Entyloma microsporum (Unger)Schröter (Fig. 1636)
Sori slightly swollen, pale yellowish brown. Spores embedded, pale golden brown, thick-walled, smooth, 12–21 × 10–18. Conidia fusiform, hyaline, 12–18 × 2–3. On *R. acris* and *R. repens*. Most records Sept.–Oct., but has been collected in May.

Urocystis ficariae (Liro)Moesz
Sori in leaves, petioles and flowers of *R. ficaria*. Spore mass powdery, black. Spore balls 25–65 diam., with usually 1 to 3 spores and 8 sterile cells. Spores dark brown, 11–21 diam.

Urocystis ranunculi (Lib.)Moesz (Fig. 1637)
Sori in leaves and stems of *R. repens* as silvery blisters which burst to expose black masses of spore balls. Spores dark brown, smooth, mostly 16–18 × 14–16, each surrounded by a ring of pale sterile cells. Apr.–Sept., fairly common.

OTHER FUNGI

Crocicreas starbaeckii (Rehm)S. Carp. (Fig. 1638)
Apothecia goblet-shaped, up to 0.5mm diam., but usually smaller, whitish at margin, grey or smoky brown below. Ascospores hyaline, 7–10 × 1–1.5. On fibrous remains of previous year's leaves and runners of *R. repens*, in damp places, common but has to be searched for, Apr.–Sept.

Erysiphe ranunculi Grev. (Fig. 1639)
Powdery mildew on living leaves of *R. acris*, *R. auricomus*, *R. bulbosus*, *R. ophioglossifolius*, *R. repens* and *R. sardous*, common. Conidia fairly abundant. Cleistothecia containing 2 to 8 asci, each with 3 to 6 spores.

Leptotrochila ranunculi (Fr.)Schüepp (Fig. 1640)
Apothecia erumpent, mostly hypophyllous, up to 0.5mm diam.; pale smoky-grey or yellowish discs with brown, slightly toothed margins. Ascospores hyaline, 12–15 × 3.5–5. On large, brown, blotched areas on living leaves of *R. acris* and *R. repens*, Oct.–July, common.

Niptera sp. (Fig. 1641)
Apothecia up to 0.45mm diam., black with pale grey discs. Ascospores hyaline, 1-septate, 9–11 × 2–2.5. Quite common on fibrous remains of *R. repens*.

Peronospora ficariae (Nees von Esenbeck)Tul.
Downy mildew on *R. ficaria*. Sporangia 25–28 × 21–22.

Peronospora ranunculi Gaüm.
Downy mildew on *R. acris*, *R. bulbosus*, *R. flammula* and *R. repens*. Sporangia 26–28 × 20–24.

Ramularia aequivoca (Ces.)Sacc.
Conidiophores hyaline, about 20–24 × 2–3. Conidia cylindric–fusoid, hyaline, 15–20 × 2–2.5. Causes pale spots with brown margins on leaves of *R. auricomus* and *R. repens*.

Ramularia didyma Unger (Fig. 1642)
Colonies small, white, hypophyllous. Conidiophores hyaline, 50–90 × 3–5. Conidia hyaline, 1-septate, 20–29 × 7–12. On brown spots 2–4mm diam., which may coalesce and form larger brown areas, on living leaves of *R. acris*, *R. parviflorus* and, very commonly, *R. repens*, July–Oct.

Ramularia scelerata Cooke
Colonies hypophyllous. Conidiophores short, hyaline. Conidia cylindric-ellipsoid, hyaline, 18–21 × 3–4. Causes large brown spots on radical leaves of *R. sceleratus*.

Septoria ficariae Desm.
Pycnidia amphigenous, immersed, less than 0.1mm diam., black. Conidia straight or slightly curved, hyaline, 25–35 × 1–1.5. Causes small, brown-bordered spots, with pale grey centres on leaves of *R. ficaria*, May–July.

RAPHANUS: Radish

Alternaria brassicicola, described on *Brassica*, sometimes causes leaf spots on *R. sativus*.

Alternaria raphani Groves & Skolko (Fig. 1643)
Conidiophores olivaceous brown, up to 150 × 3–7. Conidia commonly in chains of 2 or 3, mid to dark golden brown, smooth or minutely verruculose, muriform, 50–130 × 14–30. Forms circular black spots up to 4mm diam. on seed pods of *R. sativus* ('black pod blotch').

Erysiphe cruciferarum, powdery mildew described on *Brassica*, occurs on *R. maritimus*.

Leptosphaeria maculans, described on *Alliaria*, has been recorded on *R. maritimus*.

Leptosphaeria raphani D. Hawksw. & Sivan. (Fig. 1644)
Pseudothecia erumpent, becoming superficial, black, about 0.3mm diam. Ascospores golden brown, 3-septate, 20–26 × 4–5. On dead stems of *R. maritimus*, Aug.

Peronospora parasitica, downy mildew described on *Capsella*, has been found on *R. raphanistrum*.

RESEDA: Dyer's Rocket or Weld

Cercospora resedae Fuckel (Fig. 1645)
Colonies amphigenous. Conidiophores fasciculate, golden brown, 25–70 × 4–5. Conidia hyaline or subhyaline, 3- to 11-septate, 40–140 × 3–4. On living leaves of *R. luteola*, causing greyish or pale brown spots.

Pleospora ambigua (Berl. & Bres.)Wehmeyer
Pseudothecia erumpent, dark brown to black, 0.2–0.3mm diam., with pale brown hyphae around the sides and base, and dark brown setae, up to 80 long, around the ostiole. Ascospores golden brown, with 5 to 7 transverse septa and 1 longitudinal septum, 20–26 × 10–12. On dead stems of *R. luteola*, June.

RHEUM: Rhubarb

Puccinia phragmitis (Schum.)Körn., 0, 1
Aecia with incised and recurved white margins, clustered on reddish-purple leaf spots. Alternate host *Phragmites australis*.

Ramularia rhei Allesch. (Fig. 1646)
Colonies mostly hypophyllous. Conidiophores in fascicles of 3 to 20 emerging through stomata, hyaline, 20–50 × 3–4.5. Conidia hyaline, 0- or 1-septate, 10–30 × 3–4. On leaves, causing round, pale brown spots, 2–7mm diam., with broad crimson borders.

RHINANTHUS: Yellow Rattle

Coleosporium tussilaginis, 11, 111, described on *Tussilago*, occurs on *Rhinanthus*.

Ephelina lugubris (de Not.)Höhnel
Apothecia erumpent, about 1mm diam., dark brown to black with grey discs, formed in clusters on black subepidermal stromata 1–3cm long. Ascospores

hyaline, 8–11 × 3–6. On overwintered stems at soil level.

Plasmopara densa, downy mildew described on *Odontites*, is quite common also on *Rhinanthus*.

Sphaerotheca fuliginea, powdery mildew described on *Veronica*, has been recorded also on *Rhinanthus*.

RUBIA: Wild Madder

Mycosphaerella peregrina (Cooke)Lindau
Pseudothecia scattered or gregarious, immersed, punctiform. Ascospores 1-septate, hyaline, 15–18 × 4. On stems and leaves.

RUMEX: Docks, Sheep's Sorrel, Sorrel

UREDINALES

Five rusts are found on different species of *Rumex*.

On *R. acetosa*: *Puccinia acetosae*, 11, common; urediniospores with 2 supra-equatorial pores and only a few spaced-out short spines. *P. phragmitis*, 0, 1, common; aeciospores white in mass, verruculose. *Uromyces acetosae*, 0, 1, 11, 111; aeciospores creamy in mass, almost smooth, urediniospores with 3 equatorial pores, minutely and densely verruculose.

On *R. acetosella*: *P. acetosae*, 11, 111, rare; teliospores 1-septate. *U. acetosae*, 0, 1, 11, 111, common; teliospores without septa, verruculose, with thin, hyaline, deciduous pedicels. *U. polygoni-avicolariae*, 1, 11, 111, rare; teliospores without septa, smooth, with long, thick, persistent pedicels.

On *R. conglomeratus*, *R. crispus*, *R. hydrolapathum* and *R. obtusifolius*: *P. phragmitis*, 0, 1, common; aeciospores white in mass. *U. rumicis*, 11, 111; cinnamon to brown uredinia and telia.

On *R. maritimus* and *R. sanguineus*: *U. rumicis*, 11, 111.

Puccinia acetosae Körn., 11, 111
Uredinia amphigenous, cinnamon brown; spores 25–30 × 20–23. Telia amphigenous, brown; spores 30–45 × 20–26, minutely verruculose.

Puccinia phragmitis (Schum)Körn., 0, 1
Aecia white with cut and recurved margins, clustered on reddish-purple spots; spores 17–25 diam. Alternate host *Phragmites australis*.

Uromyces acetosae Schröter, 0, 1, 11, 111
Aecia creamy, on reddish-purple spots; spores 19–22 × 13–19. Uredinia amphigenous, cinnamon brown; spores 19–26 × 18–23. Telia dark brown; spores 20–26 × 20–24.

Uromyces polygoni-aviculariae, 0, 1, 11, 111, described on *Polygonum*.

Uromyces rumicis (Schum.)Wint., 11, 111
Uredinia amphigenous, cinnamon; spores with short, scattered spines, 22–28 × 20–24. Telia brown; spores smooth, 25–35 × 20–24.

USTILAGINALES

Ustilago kuehneana Wolff
Smut sori in ovaries and anthers, as blisters on upper part of stems, and occasionally on leaves. Spore mass pinkish purple. Spores spherical, with reticulate walls, mostly 14–16 diam. On *R. acetosa*, *R. acetosella* and *R. crispus*, June–Sept., not uncommon.

OTHER FUNGI

Crocicreas cyathoideum var. *cacaliae*, described on *Ononis*, has been found on dead stems of *R. acetosa*, July–Sept.

Helotium consobrinum Boud. (Fig. 1647)
Apothecia up to 3mm diam., bright yellow with a paler stalk which is white-hairy at the base. Ascospores hyaline, 16–25 × 3–3.5, sometimes with 1 or 2 septa. Most commonly on dead stems of *R. acetosella* and *R. conglomeratus* but also occasionally on other herbaceous stems, June–Oct.

Hymenoscyphus rumicis (Velen.)Dennis (Fig. 1648)
Apothecia up to 2mm diam., with very long, narrow stalks, cream becoming pale brown, with yellowish discs. Ascospores broadly cylindrical with rounded ends, hyaline, mostly 1-septate, 9–13 × 3–4. On fallen fruits of *R. acetosa*, June–Aug.

Hymenoscyphus scutula (Pers.)Phillips f. *alba* (Le Gal)Dennis
Apothecia ivory white about 2mm diam., otherwise as *H. scutula*. On dead stems.

Keissleriella gallica (E.Müller)Bose (Fig. 1649)
Pseudothecia immersed, 0.25mm diam., dark brown to black, with short, dark brown, thick-walled setae around the ostiole. Ascospores hyaline, 3-septate, 18–22 × 4–5. On blackened areas on dead stems of *R. acetosa*, July–Aug.

Mollisia polygoni (Lasch)Gillet var. *rumicis* (Sacc.)Sacc.
Apothecia pinkish brown, 0.5–1mm diam. Ascospores hyaline with 2 guttules, 6–8 × 1.5–2. On dead stems of *R. acetosella*, Mar.–Apr.

Peronospora rumicis Corda
Downy mildew on *R. acetosa* and *R. acetosella*. Colonies greyish violet. Sporangia 25–32 × 16–20.

Phomopsis durandiana (Sacc. & Roum.)Died.
Conidiomata immersed, 0.5mm diam., black. Conidia oblong–fusiform, hyaline, with 2 guttules, 7–10 × 2–3. On dead stems of *R. acetosa* and *R. obtusifolius*, May–Aug.

Plagiostoma devexa, described on *Polygonum*, has been found on dead stems of *R. obtusifolius*.

Ramularia pratensis Sacc. (Fig. 1650)
Colonies white, hypophyllous. Conidiophores 20–50 × 3–4. Conidia hyaline, 0- or 1-septate, 12–30 × 2.5–3.5. On living leaves of *R. acetosa*, *R. acetosella* and *R. hydrolapathum*, causing round or oval, pale brown, purple-bordered spots 2–6mm diam., May–Oct., common.

Ramularia rubella (Bon.)Nannf. (Fig. 1651)
Colonies white. Conidiophores hyaline, up to 90 × 4–5. Conidia hyaline, without septa, 25–40 × 9–11. On living leaves of *R. acetosa*, *R. conglomeratus*, *R. crispus* and *R. obtusifolius*, Mar.–Nov., very common.

Venturia rumicis (Desm.)Winter (Fig. 1652)
Pseudothecia immersed, black, sometimes with, but often without, setae around the ostiole. Ascospores olivaceous, 1-septate, 14–19 × 5–8. On small, dark purplish-brown spots, sometimes with reddish-purple borders, on living and fading leaves of *R. acetosa*, *R. acetosella*, *R. crispus* and *R. obtusifolius*, Nov.–July, very common.

Plurivorous species found on dead stems include *Chalara urceolata*, *Dasyscyphus sulphureus*, *Dendryphiella vinosa*, *Diaporthe pardalota*, *Hymenoscyphus scutula*, *Leptosphaeria macrospora*, *L. ogilviensis*, *Sclerotinia fuckeliana* and *Torula herbarum*.

SAGINA: Pearlworts

Puccinia arenariae, 111, described on *Arenaria*, occurs on *S. intermedia*, *S. maritima*, *S. nodosa*, *S. procumbens* and *S. saginoides*.

SAGITTARIA: Arrowhead

Doassansia sagittariae (Westend.)C.Fisch. (Fig. 1653)
Smut spore balls up to 80 diam., pale golden brown, each composed of numerous spores 10–12 diam., surrounded by a cortical layer of larger irregular cells. On leaves of *S. sagittifolia* causing yellow or pale brown, minutely pimpled spots up to 1cm diam., June–Aug., uncommon.

SALICORNIA: Glassworts

Uromyces salicorniae de Bary, 1, 11, 111
Aecia mostly on young plants in May. Uredinia cinnamon; spores minutely echinulate, 25–35 × 20–25. Telia dark brown; spores smooth, 25–35 × 20–28 with long pedicels. On leaves and stems of *S. europaea*, *S. perennis* and *S. ramosissima*, uncommon.

Stagonosporopsis salicorniae (P.Magnus)Died.
Pycnidia scattered, up to 0.15mm diam., brown, darker around the ostiole. Conidia cylindrical, rounded at the ends, hyaline to pale olivaceous, with 1 or 2 septa and often 2 guttules, 10–18 × 3.5–5. On lower part of stems of *S. europaea*, July.

SAMOLUS: Brookweed

Entyloma henningsianum Syd.
Smut sori embedded, spores very pale yellow, 9–15 diam. In leaves, causing pale yellow to brown spots 4–8mm diam., few records.

SANGUISORBA: Great Burnet

Xenodochus carbonarius Schlecht., 1, 111 (Fig. 1654)
Aecia hypophyllous, orange, on purple or yellow spots; spores verruculose, 18–26 × 17–22. Telia amphigenous, black; spores smooth, up to 300 × 25–28, with 3 to 21 septa.

Sphaerotheca ferruginea (Schlecht.)Junell (Fig. 1655)
Conidia plentiful. Cleistothecia dark brown with pale appendages. Ascospores 22–25 × 14–15.

SANICULA: Sanicle

Puccinia saniculae Grev., 0, 1, 11, 111
Aecia clustered on purplish-brown spots. Uredinia small, pale; spores echinulate, 26–39 × 19–28. Telia brown, spores smooth, 1-septate, 25–45 × 20–25. On lower surface of leaves and on petioles, common.

SAPONARIA: Soapwort

Septoria saponariae (DC.)Savi & Becc.
Pycnidia crowded, mostly epiphyllous, dark brown, 0.1–0.15mm diam.

Conidia cylindrical, somewhat curved, hyaline, 1- or 3-septate, 40–50 × 3.5–4.5. Causes round or irregular fawn-coloured spots on leaves.

SAXIFRAGA: Dovedale Moss, Saxifrages

Six rust species and varieties occur but most of them are very uncommon.

On *S. aizoides*: *Melampsora epitea* var. *reticulatae* with orange caeomoid aecia. *Puccinia pazschkei* var. *jueliana* with dark brown telia.

On *S. granulata*: *Melampsora vernalis* with orange aecia, and brown subepidermal telia with 1-celled spores adhering laterally. *Puccinia saxifragae* with dark brown telia, spores 2-celled.

On *S. hypnoides*: *M. epitea* var. *epitea*.

On *S. oppositifolia*: *M. epitea* var. *epitea* with orange aecia. *P. pazschkei* var. *jueliana* with dark brown telia.

On *S. spathularis*, *S. stellaris* and *S. umbrosa*: *P. saxifragae*.

On cultivated species such as *S. aizoon* and *S. cotyledon*: *P. pazschkei* var. *pazschkei*.

Melampsora epitea Thüm. var *epitea*, 0, 1
Aecia caeomoid, orange; spores minutely verruculose, 15–25 × 11–21. Alternate host *Salix herbacea*.

Melampsora epitea var *reticulatae* (A.Blytt)Jørst., 0,1
Alternate host *Salix reticulata*.

Melampsora vernalis Wint., 0, 1, 111
Aecia orange; spores minutely verruculose, 17–25 × 12–22. Telia subepidermal, brown or blackish brown; spores 1-celled, adhering laterally, 35–55 × 14–18.

Puccinia pazschkei var. *jueliana* (Diet.)Savile, 111
Teliospores with thick apical wall forming a cap.

Puccinia pazschkei Diet. var *pazschkei*, 111
Telia mostly epiphyllous, dark reddish-brown; spores 1-septate, rugose or verruculose, 25–40 × 14–18.

Puccinia saxifragae Schlecht., 111
Telia mostly hypophyllous, dark brown; spores conical, 1-septate, longitudinally striate, 25–45 × 15–20.

SCABIOSA: Small Scabious

Ustilago intermedia Schröter
In anthers, Aug. Spore mass very dark purple. Spores globose or subglobose, mostly 10–13 diam., walls reticulate with mesh 1–3 across.

Leptosphaeria modesta (Desm.)Auersw. (Fig. 1656)
Pseudothecia subepidermal, pyriform, black, 0.3mm diam., ostiole lined with brown hairs about 30 long. Ascospores pale straw-coloured, 4-septate, mostly 28–32 × 4–4.5, some with a small hyaline appendage at each end. On dead stems, May–Sept.

SCILLA: Squill

Uromyces muscari, leaf rust described on *Endymion*, and *Ustilago vaillantii*, anther smut described on *Chionodoxa*, have both been recorded on *S. verna*.

SCLERANTHUS: Knawel

Peronospora scleranthi Rabenh.
Downy mildew. Colonies white. Sporangia 18–20 × 13–15.

SCROPHULARIA: Figwort, Water Betony

Uromyces scrophulariae Fuckel, 0, 1, 111
On *S. auriculata*, *S. nodosa*, *S. scorodonia* and *S. umbrosa*, uncommon; it should be looked for on stems just above soil level, as well as on leaves, July–Sept. Aecia in round clusters on yellowish spots. Telia small, dark brown; spores smooth, 20–35 × 12–18.

Didymella commanipula (Berk. & Br.)Sacc. (Fig. 1657)
Pseudothecia at first immersed, becoming erumpent, black, collapsing. Ascospores hyaline, 1-septate, 15–18 × 3.5–4.5. On dead stems and capsules of *S. auriculata*.

Peronospora sordida Berk.
Downy mildew on *S. auriculata* and *S. nodosa*, July–Aug. Sporangia pale olivaceous, 20–25 × 16–18.

Ramularia scrophulariae Fautr. & Roum. (Fig. 1658)
Colonies hypophyllous. Conidiophores hyaline, 20–40 × 2–3. Conidia hyaline, 0- or 1-septate, 8–23 × 2.5–3. On leaves of *S. nodosa*, causing round, brown, grey or white spots with reddish-purple borders, Sept.–Oct.

SEDUM: Rose-root, Stonecrops

Puccinia umbilici, described on *Umbilicus*, has been recorded on *S. rosea*.

SELINUM

Puccinia angelicae, described on *Angelica*, has been recorded on *S. carvifolia*.

SEMPERVIVUM: Houseleek

Endophyllum sempervivi de Bary, 0, III
Telia aecioid, amphigenous with well-developed peridia; spores pale golden brown, minutely verruculose, 20–35 × 20–28. On various species including *S. tectorum*, uncommon.

SENECIO: Cineraria, Groundsels, Ragworts

UREDINALES

Coleosporium tussilaginis, described on *Tussilago*, and three species of *Puccinia* are found on *Senecio*.

On *S. aquaticus*: *Puccinia glomerata*, III.
On *S. cambrensis*: *P. lagenophorae*, I, III.
On *S. cruentus*, *S. squalidus* and *S. vulgaria*: *C. tussilaginis*, II, III, uredinia orange, powdery, spores 20–40 × 17–25; telia subepidermal, reddish, waxy, spores 3-septate; common. *P. lagenophorae*, I, III, aecia cupulate, in large conspicuous groups on leaves and stems, spores 10–16 diam.; telia dark brown, spores 0- or 1-septate, common.
On *S. jacobaea*: *C. tussilaginis*, II, III. *P. dioicae* var. *schoeleriana*, 0, 1, aecia orange with torn white peridia, spores mostly 20–25 diam.; common. *P. glomerata*, III, telia brown, spores 1-septate.
On *S. sylvaticus* and *S. viscosus*: *C. tussilaginis*.

Puccinia dioicae Magn. var. *schoeleriana* (Plowr. & Magn.)Henderson, 0, 1
Alternate host *Carex arenaria*.

Puccinia glomerata Grev., III
Teliospores 25–47 × 20–30.

Puccinia lagenophorae Cooke, 0, 1, III
Telia usually surrounded by groups of aecia; spores 20–40 × 13–18.

OTHER FUNGI

Albugo tragopogonis, white blister described on *Tragopogon*, occurs on *S. cruentus* and *S. squalidis* and is common on the latter.

Alternaria cinerariae Hori & Enjoji (Fig. 1659)
Colonies effuse, mid to dark olivaceous brown. Conidiophores olivaceous brown, up to 100 × 5–8. Conidia golden brown, smooth, muriform, body 50–140 × 15–40, beak up to 80 × 6–9. Parasitic on leaves of *S. cruentus* (cineraria), causing reddish or olivaceous brown spots up to 1cm diam., often with grey centres.

Alternaria dennisii M.B.Ellis (Fig. 1660)
Colonies olivaceous brown, velvety. Conidiophores brown, up to 100 × 4–8. Conidia pale to mid brown, smooth, with 3 to 12 transverse and 0 to 3 longitudinal septa, 20–70 × 7–11. On living leaves and stems of *S. jacobaea*, Sept.

Ascochyta senecionicola Petrak
Pycnidia immersed, about 0.1mm diam. Conidia cylindrical, rounded at the ends, 1-septate, with 1 or 2 oil drops in each cell, 6–12 × 3–4. On living leaves of *S. squalidus*, causing greyish-brown spots 0.5–2cm diam.

Bremia lactucae, downy mildew described on *Lactuca*, occurs on living plants of *S. vulgaris*.

Erysiphe fischeri Blumer
Powdery mildew on *S. viscosus* and *S. vulgaris*. Conidia abundant. Cleistothecia with numerous 2-spored asci.

Leptosphaeria derasa (Berk. & Br.)Auersw.
Pseudothecia immersed, conical or pyriform, black, sometimes hairy, 0.25–0.3mm diam. Asci 90–120 × 11–14; spores subcylindrical, olivaceous, mostly 8-septate, fourth cell down slightly swollen, 40–50 × 4–4.5, with a short, curved, hyaline appendage at each end. On dead stems of *S. jacobaea*.

Ramularia pruinosa Speg. (Fig. 1661)
Colonies mostly hypophyllous. Conidiophores hyaline, 30–80 × 2.5–4.5. Conidia hyaline, 0- to 3-septate, 10–40 × 3.5–5. On leaves of *S. jacobaea*, causing angular, pale brown spots 2–4mm diam.

Ramularia senecionis (Berk. & Br.)Sacc. (Fig. 1662)
Colonies mostly hypophyllous. Conidiophores hyaline, 30–70 × 3.5–4.5. Conidia hyaline, 0- or 1-septate, 15–34 × 3–5. On living leaves of *S. aquaticus* and *S. viscosus*, causing round or angular, yellowish or pale brown spots 2–10mm diam.

Septoria senecionis-silvatici Syd.
Pycnidia epiphyllous, immersed, pale brown, less than 0.1mm diam. Conidia hyaline with few septa, 30–50 × 1.5. On leaves of *S. aquaticus* and *S. jacobaea* causing pale brown, sometimes dark-bordered, spots, July–Oct.

Sphaerotheca fusca, powdery mildew described on *Doronicum*, has been recorded on *S. cruentus*.

Sphaerotheca xanthii, downy mildew on *Calendula*, occurs also on *S. jacobaea*.

Plurivorous fungi found mainly on dead stems include *Diapleela clivensis, Leptosphaeria dolioloides, L. doliolum, L. ogilviensis, Pyrenopeziza adenostylidis* and *Sarcopodium circinatum*.

SERRATULA: Saw-wort

Puccinia hieracii, 11, 111, described on *Hieracium*, has been recorded a few times on *Serratula*.

SESELI

Puccinia libanotis Lindr., 0, 1, 11, 111
Aecia uredinoid, large, reddish brown. Uredinia very small; spores echinulate, 27–34 × 22–28. Telia small, dark brown; spores smooth, 1-septate, 34–50 × 15–24. On *S. libanotis*, rare.

Leptosphaeria libanotis, described on *Angelica*, occurs on dead stems of *S. libanotis*.

SHERARDIA: Field Madder

Leptotrochila verrucosa, described on *Galium*, has been recorded also on *Sherardia*.

SILAUM: Pepper Saxifrage

Puccinia angelicae, 0, 1, 11, 111, described on *Angelica*, has been recorded on *S. silaus*.

SILENE: Campions, Catchflies

UREDINALES

Four rusts are recorded on different species:

On *S. alba* and *S. dioica*: *Puccinia arenariae*, 111, common; teliospores 35–60 × 15–20, often germinating in situ. *Puccinia behenis*, 11, 111, uncommon; telia seldom seen, spores 1-septate, 25–40 × 15–26; uredinia very small, spores echinulate, 20–25 × 18–22.

On *S. maritima* and *S. vulgaris*: *Puccinia behenis*, 11, 111. *Uromyces behenis*, 1, 111, uncommon; aecia very conspicuous on purplish spots; teliospores smooth, 25–35 × 20–26, with persistent pedicels.

On *S. nutans*: *Uromyces inaequialtus*, 11, 111, rare.

Puccinia arenariae, described on *Arenaria*.

Puccinia behenis Otth, 11, 111; *v.s.*

Uromyces behenis (DC.)Unger, 1, 111
Aecia hypophyllous, cup-shaped, pale yellow; spores minutely verruculose, 15–20 diam. Telia hypophyllous and on stems, blackish brown to black.

Uromyces inaequialtus Lasch, 11, 111
Uredinia amphigenous on purplish spots; spores 21–25 × 19–23, verruculose. Telia with uredinia, black; spores smooth, 25–30 × 20–25, with long pedicels.

USTILAGINALES

Ustilago violacea (Pers.)Roussel (Fig. 1663)
Anther smut on *S. acaulis*, *S. alba*, *S. dioica*, *S. maritima*, *S. otites* and *S. vulgaris*, common everywhere, May–Oct. Spore mass purple, infected flowers easily detected. Spores finely reticulate, pale purple, 6–11 (mostly 8) diam.

OTHER FUNGI

Didymaria kriegeriana Bres. (Fig. 1664)
Colonies white. Conidiophores hyaline, up to 120 × 3–4. Conidia hyaline, 1-septate, minutely verruculose, 18–35 × 6–9. Causes pale brown spots on living leaves of *S. dioica*, Oct.–Nov.

Diplosporonema delastrei (Delacr.)Petrak (Fig. 1665)
Acervuli amphigenous, yellowish, about 0.15mm diam. Conidiophores hyaline, 20–30 × 2–4. Conidia hyaline, with 1 or 2 septa, 21–36 × 5–6. On living leaves and occasionally stems of *S. dioica*, causing pale yellow to pale brown, sometimes purple-bordered, spots, July–Oct., common; recorded also on *S. maritima* and *S. vulgaris*.

Leptosphaeria silenes-acaulis de Not.
Pseudothecia immersed, less than 0.1mm diam., conical, black. Ascospores golden brown, 3-septate, 30–40 × 5–7. On leaves and capsules of *S. acaulis*.

Phomopsis silenes D. Hawksw. & Punithalingam
Conidiomata immersed, black, 0.15–0.3mm diam. Conidia fusiform, hyaline, biguttulate, 6–9 × 2–3. On dead stems of *S. dioica*, Aug.

Pleospora androsaces Fuckel (Fig. 1666)
Pseudothecia immersed, black, 0.15–0.2mm diam., with dark brown divergent setae up to 200 × 8 on upper part and around ostioles. Ascospores dark reddish-brown, muriform, 45–60 × 20–24. On dead leaves of *S. acaulis*, Apr.

Pyrenopeziza lychnidis (Sacc.)Rehm (Fig. 1667)
Apothecia erumpent, sometimes becoming superficial, 0.5–1mm diam., dark brown to black. Excipulum with characteristic groups of dark cells. Ascospores hyaline, occasionally 1-septate, 14–21 × 2. On dead stems of *S. dioica*, Apr.–June, common.

Ramularia lychnicola Cooke
Colonies hypophyllous, white, small. Conidia hyaline, 12–15 × 4. On living leaves of *S. alba* and *S. dioica*, causing round, pale brown spots.

Septoria lychnidis Desm.
Pycnidia mostly epiphyllous, less than 0.1mm diam., brown. Conidia hyaline, 1- to 7-septate, 30–50 × 2–3. On living leaves of *S. dioica*, causing reddish-brown or pale brown spots with reddish-brown borders, May–Sept.

SINAPIS: Charlock

Peronospora parasitica, downy mildew described on *Capsella*, is common on *S. arvensis*.

SISYMBRIUM: Hedge Mustard, London Rocket

Albugo candida, white blister described on *Capsella*, is common on *S. officinale*.

Erysiphe cruciferarum, powdery mildew described on *Brassica*, occurs on *S. officinale*.

Peronospora parasitica, downy mildew described on *Capsella*, has been recorded on *S. irio*.

SMYRNIUM: Alexanders

Puccinia smyrnii Biv.-Bernh., 0, 1, 111
Aecia amphigenous, yellowish orange. Telia hypophyllous on pale yellow spots, blackish brown; spores coarsely tuberculate, 30–50 × 17–27. Common and can be found all the year round.

Alternaria ramulosa (Sacc.)Joly (Fig. 1668)
Colonies effuse, dark blackish brown to black, hairy or velvety. Conidiophores thick-walled, dark brown, up to 650 × 8–15, with short, or sometimes long,

branches at the apex. Conidia brown, muriform, mostly 30–60 × 15–25, smooth. On dead stems Apr.–May, common.

Plurivorous species found on dead stems include *Calyptella capula* and *Dasyscyphus mollissimus*.

SOLANUM: Bittersweet or Woody Nightshade, Black Nightshade, Potato

Alternaria solani Sorauer (Fig. 1669)
Conidiophores solitary or in small groups, pale brown or olivaceous brown, up to 110 × 6–10. Conidia pale to mid golden or olivaceous brown, with 9 to 11 transverse and 0 or a few longitudinal septa, 150–300 long including the beak, 15–19 thick. On *S. nigrum* and *S. tuberosum*. The cause of early blight of potatoes, affecting all parts above ground. On leaves it causes round, oval or irregular, brown or dark brown, often concentrically ridged, target spots. Under favourable conditions, leaf spots enlarge rapidly and may involve as much as half a leaf. Thin-leaved early varieties appear to be more susceptible than thick-leaved later ones.

Colletotrichum coccodes, described on *Lycopersicon*, is found also on dead stems of *S. tuberosum*.

Cucurbitaria dulcamarae (Schm.)Fuckel
Pseudothecia 0.3mm diam., black, closely packed together in small groups which break through the epidermis. Ascospores golden brown, mostly with 5 transverse septa and 1 longitudinal septum, 21–30 × 7–9. Pale brown subiculum with 3-septate brown conidia 17–19 × 5. On dead stems of *S. dulcamara*, Oct.

Diaporthe sarothamni Auersw. ex Nitschke var. *dulcamarae* (Nitschke)Wehmeyer (Fig. 1670)
Perithecia in groups of 2 or 3, 0.4–0.6mm diam., with short necks erumpent through the blackened surface. Ascospores hyaline, 1-septate, 13–17 × 3–5. On dead stems of *S. dulcamara*, Apr.–June.

Helminthosporium solani Dur. & Mont. (Fig. 1671)
Colonies dark brown to black, hairy. Conidiophores mid to dark brown, paler near the apex, smooth or occasionally verruculose, with small pores at the apex and laterally in whorls beneath the septa, up to 600 × 6–9 (15 at base). Conidia subhyaline to brown with 2 to 8 pseudosepta, 24–85 × 7–11. On dead stems of *S. dulcamara*, Sept.–Feb., and on stems and tubers of *S. tuberosum*. The cause of 'silver scurf' disease; the skin becomes discoloured brown or silvery in patches, the effect being most striking when tubers are washed.

Hymenoscyphus scutula (Pers.)Phill. var. *solani* P.Karsten (Fig. 1672)
Apothecia up to 1.5mm diam., yellow ochre. Ascospores hyaline, 14–17 ×

3–4. On dead stems of *S. tuberosum*.

Nectria solani Reinke & Berth.
Perithecia golden brown with rough walls, 0.3–0.4mm diam., formed in groups on pulvinate, erumpent, yellow stromata. Ascospores hyaline, minutely verruculose when mature, 1-septate, 10–14 × 3.5–5. *Gliocladium* state, associated with perithecia, produces hyaline conidia 5–7 × 2.5–3. On old tubers of *S. tuberosum*, Sept.–May, rare.

Nectriopsis solani (Reinke & Berth.)Booth
Perithecia dark red, smooth-walled, 0.3–0.4mm diam., sometimes showing pinched collapse when dry. Ascospores pale brown and verrucose when mature, 1-septate, 14–17 × 7–10. *Fusarium* state with hyaline, fusiform, 3-septate conidia 25–35 × 5–8. On rotting tubers of *S. tuberosum*, Oct.–July, rare.

Phoma exigua Desm. var. *foveata* (Foister)Boerema
Pycnidia immersed, brown to black, 0.2–0.25mm diam. Conidia ellipsoid or cylindrical, straight or occasionally slightly curved, hyaline, biguttulate, sometimes 1-septate, 5–10 × 2–3.5. Causes lesions on tubers of *S. tuberosum* known to plant pathologists as 'gangrene'.

Phytophthora erythroseptica Pethybridge
The cause of pink rot of potatoes. Tubers cut open show a pink colour in a few minutes, which deepens and eventually becomes purplish brown or black. Sporangia obpyriform or ovoid, about 32 × 20. Oospores spherical with thick dark epispore, 29–30 diam.

Phytophthora infestans (Mont.)de Bary
Sporangiophores branched, emerging through stomata, hyaline, bearing ovoid to elliptical, subapiculate, hyaline sporangia 27–30 × 15–20. The cause of potato late blight. The first signs of blight appear on potato leaves as brown spots or blotches, especially on the margins and at the tips, and the dead areas tend to curl up. These brown areas increase rapidly in warm wet weather, and the under surface is often seen to be silvery towards the edge. Infected leaves turn brown and rot, stems become blackened, and the plants putrefy. Infected tubers show a watery, reddish discoloration in and under the skin, which extends inwards. The time when blight is likely to appear in England varies, but is generally about May or June in the west, July in the south-east and July or August in the north. The fungus occurs in most countries where potatoes are grown.

Polyscytalum pustulans (Owen & Wakefield)M.B.Ellis (Fig. 1673)
Colonies effuse, grey, powdery. Conidiophores simple or branched, lower part pale brown, upper part hyaline or subhyaline, up to 140 × 2–4. Conidia dry, in long, often branched chains, which fragment readily, 0- or 1-septate, hyaline, smooth, 6–18 × 2–3. Parasitic on tubers of *S. tuberosum*, causing 'skin spot' disease; it occurs also on roots, stolons and stems. Skin spot is most

in evidence in tubers after storage and may weaken or destroy the eyes.

Spongospora subterranea (Wallr.)Lagerh.
The cause of powdery or corky scab of potatoes. Raised, more or less circular brown spots are formed on the tubers; these burst open and release a brown, snuff-like, powdery mass of spore balls, each 20–80 diam., which are rather like little sponges and contain smooth, golden brown cysts 3.5–4.5 diam.

Synchytrium endobioticum (Schilb)Perc.
This fungus causes a very striking wart disease of potatoes. Outgrowths on tubers and also sometimes on stems are cauliflower-like, dark brown and sometimes quite large. Dark brown, thick-walled, spherical resting sporangia are found inside the host cells.

SOLIDAGO: Golden-rod

Puccinia eriophori Thuem., 0, 1
Aecia hypophyllous with white torn margins. Alternate host *Trichophorum cespitosum*, rare.

Puccinia virgae-aureae (DC.)Lib., III
Telia hypophyllous, black, seated on yellowish or purple spots; spores 1-septate, smooth, 30–55 × 13–20. Uncommon.

Erysiphe cichoracearum, powdery mildew described on *Aster*, occurs on *S. virgaurea*.

Leptosphaeria planiuscula (Riess ex Rabenh.)Ces. & de Not.
Pseudothecia in rows, immersed, black, 0.3–0.35mm diam., with short necks. Ascospores fusiform, pale golden brown, 35–55 × 6–8, 5-septate, constricted at the middle septum and with the two median cells larger than the others, occasionally shortly appendaged. On dead stems, Apr.–Aug.

Septoria virgaureae Desm.
Pycnidia epiphyllous, immersed, blackish brown, about 0.1mm diam. Conidia multiguttulate, 80–100 × 1.5. On round, whitish or brown spots on fading leaves, Aug.

SONCHUS: Sow-thistles

UREDINALES

Two rusts are found on *S. arvensis*, *S. asper*, *S. oleraceus* and *S. palustris*. *Coleosporium tussilaginis*, described on *Tussilago*, has orange–yellow uredinia without paraphyses, and waxy reddish–orange telia with 3-septate colourless spores in a single layer; this is common. *Miyagia pseudosphaeria* has yellowish uredinia each surrounded by a ring of dark brown clavate paraphyses, often compound telia divided up by paraphyses, and 1-septate pale brown spores; uncommon.

Miyagia pseudosphaeria (Mont.)Jørst., 0, 1, 11, 111 (Fig. 1674)
Aecia uredinoid, amphigenous. Uredinia hypophyllous; spores verruculose, 25–38 × 15–21. Telia hypophyllous and on stems, paraphyses clavate, reddish brown; spores smooth, 30–60 × 20–30. Mesospores often present.

OTHER FUNGI

Alternaria sonchi J.J.Davis (Fig. 1675)
Colonies amphigenous. Conidiophores olivaceous brown, up to 80 × 5–9. Conidia golden brown, with 4 to 8 transverse and 0 or a few longitudinal septa, 60–130 × 15–26. On leaves of *S. asper* and *S. oleraceus*, causing greyish-brown spots with narrow purple margins.

Ascochyta sonchi (P.Henn.)Syd.
Pycnidia immersed, 0.1mm diam. Conidia hyaline, becoming 1-septate, 8–10 × 2–3. On leaves of *S. asper* and *S. oleraceus*, causing brown, dark-bordered spots.

Bremia lactucae, downy mildew described on *Lactuca*, occurs on *S. arvensis* and *S. oleraceus*.

Erysiphe cichoracearum, powdery mildew described on *Aster*, is not uncommon on *S. arvensis*, *S. oleraceus* and *S. palustris*.

SPERGULA: Corn Spurrey

Puccinia arenariae, described on *Arenaria*, occurs on *S. arvensis*.

SPERGULARIA: Sand Spurrey, Sea Spurrey

Uromyces sparsus (Schm. & Kunze)Cooke, 0, 1, 11, 111
Aecia in rings, orange–red. Uredinia brownish cinnamon; spores smooth or slightly echinulate, 22–30 × 15–20. Telia dark brown; spores smooth, 20–32 × 16–22. On *S. marina*, *S. media* and *S. rubra*, rare.

Albugo lepigoni (de Bary)O.Kuntze
Sori on most parts of plants above ground, yellowish or pale buff. Sporangia cylindrical, spherical or subspherical, hyaline, 15–25 (mostly 18–20) diam. Oospores spherical, brown; epispore tuberculate to spinulose or papillate. On *S. marina*.

Peronospora lepigoni Fuckel
Downy mildew on *S. marina* and *S. rubra*. Sporangia 28–31 × 17–18.

SPINACIA: Spinach

Cladosporium variabile (Cooke)de Vries (Fig. 1676)
Colonies effuse, blackish olive, sometimes with reddish-purple pigment, velvety or felted. Aerial hyphae tortuous and spirally coiled. Conidiophores mostly up to 150 × 6–8, brown, smooth. Conidia brown or olivaceous brown, densely verrucose, 0- to 3-septate, 5–30 × 3–13 (commonly 15–25 × 7–10). Parasitic on living leaves, causing pale yellow spots each with a dark fruiting area in the middle.

Peronospora farinosa, downy mildew described on *Chenopodium*, is not uncommon also on *Spinacia*.

SPIRANTHES: Lady's Tresses

Uredo oncidii P. Henn.
Urediniospores 20–30 × 15–22. A very rare rust which we have seen only once on *S. spiralis*.

STACHYS: Woundworts

Erysiphe galeopsidis, powdery mildew described on *Galeopsis*, occurs on *S. palustris* and *S. sylvatica*.

Ramularia stachydis (Pass.)Massal. (Fig. 1677)
Colonies amphigenous, white. Conidiophores 20–60 × 2.5–3.5. Conidia hyaline, 0- to 3-septate, 10–36 × 3–4. On living leaves of *S. palustris*, causing oval or angular, pale brown spots 3–12mm long.

Septoria stachydis Rob. & Desm.
Pycnidia epiphyllous, 0.1mm diam., blackish brown. Conidia slightly curved, 0- to 3-septate, 25–40 × 1–2. On living leaves of *S. palustris* and *S. sylvatica*, causing dark brown spots 5–10mm diam., June–Nov.

STELLARIA: Chickweed, Stitchworts

UREDINALES

Melampsorella caryophyllacearum, 11, 111, described on *Cerastium*, is recorded occasionally on *S. graminea*, *S. holostea* and *S. media*. Spores ellipsoid or subspherical and mostly without septa.

Puccinia arenariae, 111, described on *Arenaria*, is found commonly on *S. alsine*, *S. graminea*, *S. holostea*, *S. media*, *S. nemorum* and *S. palustris*. Spores 1-septate, oblong fusiform to clavate.

OTHER FUNGI

Ascochyta stellariae Fautr.
Pycnidia 0.15–0.2mm diam., honey-coloured, dark brown around the ostiole. Conidia straight or slightly curved, oblong, rounded at the ends, multi-guttulate, 23–30 × 6–7. On fading and dead leaves of *S. alsine*, May.

Didymella holosteae Sydow (Fig. 1678)
Pseudothecia up to 0.15mm diam., black. Ascospores hyaline, 1-septate, 14–21 × 5–7. On dying and dead leaves of *S. holostea*, Feb.

Lanzia stellariae (Velen.)Spooner (Fig. 1679)
Apothecia 1–2mm diam., solitary or in little groups, pale yellow or whitish when fresh, becoming honey-brown when old, base of stalk blackish brown. Ascospores hyaline, biguttulate, 15–19 × 5–7. On the lower part of stems of *S. alsine*, May–Oct., quite common.

Mycosphaerella isariophora (Desm.)Johanson (Fig. 1680)
Pseudothecia hypophyllous, immersed, 0.1–0.13mm diam. Ascospores oblong–cylindrical or somewhat clavate, hyaline, 1-septate, 10–14 × 3–4. On dead leaves of *S. holostea*, Apr.–May.

Peronospora alsinearum Casp.
Downy mildew on *S. media*, especially on the upper leaves of seedlings in May; plants pale and stunted. Sporangia ellipsoid, 24–27 × 18–20.

Septoria stellariae Rob. ex Desm. (Fig. 1681)
Pycnidia epiphyllous, numerous, brown, up to 0.1mm diam. Conidia slightly curved, hyaline, 3- to 6-septate, 50–80 × 1.5–2. On leaves and stems of *S. graminea* and *S. media*, May–Oct., causing whitish or pale brown spots.

SUAEDA: Herbaceous Seablite, Shrubby Seablite

Uromyces chenopodii (Duby)Schröter, 1, 11, 111
Aecia in circles. Uredinia cinnamon brown; spores minutely echinulate, 20–25 × 18–21. Telia dark brown; spores smooth, 25–35 × 18–20. On *S. fruticosa* and *S. maritima*, most records from East Anglia.

Cryptovalsa suaedicola Spooner
Perithecia 0.3–0.35mm diam., immersed, black, scattered, with stromatic layer dipping between them. Asci polysporous. Ascospores allantoid, hyaline, olivaceous in mass, about 5–7 × 1. On small dead branches of *S. fruticosa*, July.

Phoma suaedae Jaap
Pycnidia erumpent, black, 0.15–0.2mm diam. Conidia hyaline, with 2 to 4 guttules, 5–10 × 3–4. On branches of *S. maritima*, Aug.

Stagonospora suaedae Syd.
Pycnidia immersed, up to 0.1mm diam. Conidia hyaline, 0- to 3-septate, 12–25 × 3–5. On dead leaves of *S. maritima*, July.

SUCCISA: Devil's-bit Scabious

Ustilago succisae P. Magn. (Fig. 1682)
Smut sori in anthers, spore mass creamy. Spores hyaline with reticulate walls, 11–14 diam. Common and widespread, Aug.–Oct.

Erysiphe knautiae, powdery mildew described on *Knautia*, occurs also on *Succisa*.

Ramularia succisae Sacc.
Colonies hypophyllous, white. Conidiophores hyaline, 15–20 × 3–3.5. Conidia hyaline, 0- to 3-septate, 18–25 × 2.5–4. On living leaves, causing pale reddish spots with dark red margins.

Synchytrium succisae de Bary & Woron.
Resting spores solitary or in compound galls, numerous, spherical or ellipsoidal, 50–80 diam., containing red granules; walls brown, 4–6 thick. Galls on leaves, stems and petioles, often large, reddish orange or purplish red, July–Sept. Sporangial sori, formed in spring and early summer, contain 100 to 150 sporangia.

SYMPHYTUM: Comfrey

Melampsorella symphyti Bubák, 11, 111
Uredinia hypophyllous, often over whole surface, small, yellow; spores echinulate, 23–33 × 19–25. Telia hypophyllous; spores 10–20 diam. On *S. asperum*, *S. officinale* and *S. tuberosum*; infected leaves usually small and pale.

Entyloma serotinum Schröter
Smut sori in leaves of *S. officinale*, Sept., circular, bordered. Spores 10–15 diam.

Erysiphe asperifoliorum, powdery mildew described on *Cynoglossum*, has been recorded on *S. peregrinum*.

Peronospora symphyti Gäum.
Downy mildew on *S. officinale*.

TAMUS: Black Bryony

Cercospora scandens Sacc. & Wint. (Fig. 1683)
Colonies amphigenous. Conidiophores rather pale olivaceous brown, 40–90 × 4–5. Conidia hyaline, 2- to 16-septate, 80–250 × 4–5. On leaves, causing round, olive to brown or almost black spots.

Cryptosporium tami Grove (Fig. 1684)
Acervuli honey brown, 0.15–0.25mm diam. Conidia hyaline, mostly 20–27 × 2–3. On dead stems, Mar.–May.

Phomopsis tamicola (Desm.)Trav.
Conidiomata black, 0.3–0.5mm diam., covered by discoloured epidermis. A-spores biguttulate, hyaline, 8–9 × 2.5–3; B-spores arcuate, 20–25 × 1–1.5. On dry dead stems.

Pirottaea nigro-striata, described on *Heracleum*, has been found on dead stems of *Tamus*.

TARAXACUM: Dandelions

UREDINALES

Puccinia hieracii, 0, 1, 11, 111, described on *Hieracium*, differs from *P. variabilis* in having spermogonia and large, cinnamon-brown, uredinoid aecia, found usually in Apr.–May, on the upper surface of leaves of *T. officinale*.

Puccinia variabilis Grev., 1, 11, 111
Aecia amphigenous, in small clusters spread uniformly over the whole leaf, cup-shaped, orange, with white, cut margins; spores verruculose, 20–25 × 15–20. Uredinia minute, brown, on very small yellow or purplish spots; spores 0-septate, echinulate, 22–30 × 20–26. Telia dark brown; spores 1-septate, minutely verruculose, 30–40 × 20–25. Common on *T. officinale* and *T. palustre*.

OTHER FUNGI

Protomyces pachydermis Thümen
Thick-walled chlamydospores 30–45 diam., formed inside small purplish swellings on leaves and peduncles of *T. officinale*, common.

Ramularia taraxaci Karsten
Colonies hypophyllous, white. Conidiophores hyaline, 35–45 × 2–3. Conidia hyaline, 18–30 × 2–3. Causes round, pale spots with purple margins on leaves of *T. officinale*, Sept.–Oct.

Sphaerotheca erigerontis-canadensis, powdery mildew described on *Erigeron*, occurs on *T. officinale*.

Synchytrium taraxaci de Bary & Woron.
Resting spores usually solitary, 50–80 diam., exospore brown, contents orange. Sporangia mostly 20 to 50 per sorus, angular, 50–80 diam., with orange granular contents. Causes yellow, orange or red galls on leaves, petioles and peduncles of *T. officinale*.

TEUCRIUM: Wood Sage

Puccinia annularis (Str.)Röhl., 111
Telia hypophyllous, pulvinate, reddish or purplish brown, seated on pale yellow or brown spots; spores smooth, 1-septate, mostly 30–50 × 15–20. On living leaves of *T. scorodonia*, July–Oct., common.

Calycellina chlorinella, a little, pale yellow discomycete described on 'herbaceous plants', we find to be common on blackened areas on dead stems of *T. scorodonia*, Oct. Nov.

THALICTRUM: Meadow Rue

UREDINALES

On *T. alpinum*: *Puccinia recondita* f.sp. *borealis*, 0, 1, not causing marked discoloration of host; uncommon. *P. septentrionalis*, 1, causing extensive, dark purple, swollen spots on leaves, petioles and stems.

On *T. flavum* and *T. minus*: *P. recondita* f.sp. *persistens*, 0, 1. *Tranzschelia anemones*, 111, dark brown telia with warted, 1-septate spores, rare on these hosts.

Puccinia recondita Rob. & Desm. f.sp. *borealis* Juel, 0, 1
Aecia rarely found. No British collections on its alternate grass hosts known.

Puccinia recondita f.sp. *persistens* Plowr., 0, 1

Alternate hosts *Agropyron junceiforme* and *A. repens*.

Puccinia septentrionalis Juel, 1
Alternate host *Polygonum viviparum*.

Tranzschelia anemones, 111, described on *Anemone*.

USTILAGINALES

Urocystis sorosporioides Körn.
Sori in leaves and occasionally in petioles and stems of *T. minus*, causing blisters which finally rupture to expose the black spore mass. Spore balls 27–40 diam., each composed of 3 to 7 smooth brown spores 10–14 diam., surrounded by paler sterile cells. Recorded in June, seldom collected.

OTHER FUNGI

Haplobasidion thalictri Erikss. (Fig. 1684A)
Colonies amphigenous, olivaceous brown, effuse. Conidiophores 25–60 × 4–9, with ampullae 10–18 diam. Conidia pale golden brown, 5–10 diam. On leaves, causing buff spots with broad purple borders.

Myrothecium carmichaelii Grev. (Fig. 1685)
Sporodochia sessile, up to 1.5mm diam., often confluent, at first green, later black with a white margin. Conidia hyaline to pale olive, green or black in mass, 10–13 × 1–1.3. On dead stems and leaves of *T. flavum* and occasionally other marsh and fen plants.

Phoma jacquiniana Cooke & Massee
Pycnidia immersed, 0.2mm diam., black. Conidia hyaline, ellipsoid, 9–10 × 4–5. On dead stems of *T. flavum* and *T. minus*, Jan.–July.

Pyrenopeziza thalictri (Peck)Sacc. (Fig. 1686)
Apothecia erumpent, about 0.5mm diam., blackish brown with pale fimbriate margins. Marginal septate hyphae 25–40 × 2.5–4. Asci with 4 or 8 spores. Ascospores hyaline, 0- to 3-septate, 18–30 × 2.5–3.5. Found near the bases of dead stems of *T. flavum*, June.

THESIUM: Bastard Toadflax

Puccinia thesii Duby, 0, 1, 11, 111
Aecia orange with white recurved margins. Uredinia small, brown; spores 27–30 × 20–24. Telia elongated, brown; spores 1-septate, mostly 35–50 × 17–24. Generally rare but in places in Suffolk abundant.

THLASPI: Field Penny-cress

Peronospora parasitica, downy mildew described on *Capsella*, has been recorded on *T. arvense*; sporangia mostly 22–25 × 18–20.

Peronospora thlaspeos-arvensis Gäum.
Downy mildew specific to *T. arvense*; sporangia 23–32 × 18–21.

THYMUS: Thyme

Puccinia thymi (Fuckel)Karsten, III
Telia on petioles and stems of *T. praecox* ssp. *arctii* (common wild thyme) and *T. pulegioides*, rare; spores 1-septate, 25–33 × 15–24.

TRAGOPOGON: Goat's-beard or Jack-go-to-bed-at-noon, Salsify

Puccinia hysterium (Str.)Röhl., 0, I, III
Aecia orange with white laciniate margins. Telia dark brown; spores 1-septate, tuberculate, 30–45 × 20–30. On *T. pratensis*, Apr.–Sept.

Ustilago tragopogonis-pratensis (Pers.)Roussel
Sori in inflorescences of *T. porrifolius* and *T. pratensis*, destroying the florets, powdery, blackish purple. Smut spores pale purplish, with reticulate walls, 12–14 diam., May–June, not uncommon.

Albugo tragopogonis (DC.)S.F.Gray
White rust or blister. Sori mostly hypophyllous and on stems, white or yellowish. Sporangiophores clavate, 40–55 × 11–15. Sporangia in chains, often cylindrical or cuboid, hyaline or pale yellow, 18–25 × 12–20. Oospores immersed, spherical, very dark brown, with reticulate, spiny or tuberculate walls, 40–65 diam. On *T. pratensis*, June–Sept.

Bremia lactucae, downy mildew described on *Lactuca*, is found occasionally on *T. pratensis*.

TRIENTALIS: Chickweed Wintergreen

Urocystis trientalis (Berk. & Br.)Lindeberg
Smut sori forming blisters on leaves and stems. Spore mass black. Spores dark golden brown, smooth, 11–18 diam., united in large numbers to form spore balls up to 90 diam. Conidia hyaline, mostly obpyriform, 7–15 × 4–5, forming white patches on the surface.

Septocylindrium magnusianum Sacc.
Colonies amphigenous, velvety, dirty white. Conidia catenate, cylindrical, 1-septate, 20–25 × 4. On dead leaves.

TRIFOLIUM: Clovers, Hare's Foot, Trefoils

UREDINALES

On *T. campestre*: *Uromyces anthyllidis*, uncommon.
On *T. dubium*: *U. anthyllidis* with rough-walled urediniospores and teliospores, both more than 16 diam.; uncommon. *U. minor* with aeciospores 14–16 diam. and smooth-walled teliospores; rare.
On *T. fragiferum*: *U. trifolii*.
On *T. hybridum*: *U. trifolii-repentis*, uncommon.
On *T. incarnatum*, *T. medium* and *T. pratense*: *U. fallens*, uncommon.
On *T. molinieri*: *U. minor*, rare.
On *T. repens*: *U. trifolii* with no aecia or uredinia and with large, often confluent telia; spores almost smooth; common. *U. trifolii-repentis* with echinulate urediniospores and small, scattered telia; uncommon.

Uromyces anthyllidis, described on *Anthyllis*.

Uromyces fallens (Arth.)Barth., 0, 1, 11, 111
Spermogonia and aecia amphigenous. Uredinia hypophyllous, pale brown; spores finely echinulate with 4 to 7 pores, 20–28 × 20–24. Telia hypophyllous, blackish brown; spores smooth or with a line of small warts, 20–24 × 16–18.

Uromyces minor Schröter, 1, 111
Aecia and telia formed together; teliospores 20–24 × 14–18.

Uromyces trifolii (R.A.Hedw. ex DC.)Fuckel, 111
Telia up to 2mm diam., often confluent and appearing larger, on nerves and petioles, causing distortion, dark brown; spores smooth or with a line of small warts, 20–30 × 17–25.

Uromyces trifolii-repentis Liro, 0, 1, 11, 111
Aecia seldom found. Uredinia amphigenous and on stems, cinnamon brown; spores echinulate with 2 to 4 pores, 22–24 × 18–24. Telia hypophyllous, small, dark brown; spores smooth or with a few warts, 20–28 × 18–24.

OTHER FUNGI

Ascochyta trifolii Bond. & Truss.
Pycnidia epiphyllous, brown, 0.1–0.15mm diam. Conidia straight or slightly

curved, cylindrical, rounded at ends, hyaline, 1- to 3-septate, 15–22 × 4–6. On leaves of *T. pratense* causing brown, zonate spots.

Botrytis anthophila Bondartzev (Fig. 1687)
Grows systemically in shoots of *T. pratense* and sporulates in the anthers making them appear grey. Conidia hyaline, mostly 11–16 × 4–5.

Cercospora zebrina Pass. (Fig. 1572)
Colonies amphigenous. Conidiophores fasciculate, mid pale brown, 20–80 × 3–5. Conidia hyaline, 8- to 20-septate, 30–180 × 3–4.5. On leaves of *T. repens* causing dark brown spots, July–Sept., common.

Cymadothea trifolii (Pers.)Wolf (Fig. 1688)
Pseudothecia formed as locules in and around the edge of overwintered black stromata. Ascospores hyaline to pale yellowish brown, 1-septate, 20–25 × 7–8. Found most frequently in its striking *Polythrincium* state, July–Oct., where the same stromata bear olivaceous-brown, velvety groups of undulate and often twisted conidiophores. These are up to 100 × 6–9 with large, flat scars. Conidia hyaline to pale brown, smooth or verruculose, 1-septate, 17–24 × 13–24. On *T. fragiferum*, *T. molinieri* and *T. repens*, causing 'black blotch' or 'sooty blotch' disease.

Erysiphe trifolii Grev.
Powdery mildew. Conidia plentiful, mostly 30–40 × 16–20. Cleistothecia about 0.1mm diam.; appendages long and rather straight, occasionally branched near tips. Asci 3 to 10, each with 3 or 4 spores. Ascospores 20–25 × 10–15. On *T. arvense*, *T. campestre*, *T. dubium*, *T. hybridum*, *T. medium*, *T. pratense* and *T. repens*.

Kabatiella caulivora (Kirchner)Karakulin (Fig. 1689)
Conidia curved, hyaline, 10–25 × 3–4.5, formed in groups on the swollen ends of conidiophores where these break through the epidermis. Causes brown, dark-margined spots on stems, petioles and leaves of *T. pratense*, Apr.–July. The leaves hang down and wilt, and stems become broken and discoloured.

Leptosphaerulina trifolii (Rostrup)Petrak (Fig. 1690)
Pseudothecia immersed, 0.2mm diam., pale brown. Ascospores hyaline or very pale straw-coloured, with 3 or 4 transverse septa and often a longitudinal septum, 25–45 × 10–20 (mostly 30–35 × 12–15). Causes small brown spots, which sometimes coalesce and form larger pale areas, on leaves of *T. pratense* and *T. repens*.

Mycosphaerella carinthiaca Jaap
Pseudothecia amphigenous, dark brown, less than 0.1mm diam. Asci 30–40 × 8–10; spores fusiform, often slightly curved, hyaline, 1-septate, 10–14 × 2.5–3. On living leaves of *T. medium* and *T. pratense* causing angular, brown spots which begin at the edge and are enclosed by the veins, Apr.–May.

Peronospora trifoliorum de Bary (Fig. 1691)
Downy mildew on *T. dubium*, *T. pratense* and *T. repens*. Sporangia 23–30 × 16–21, pale brown.

Phoma medicaginis var. *medicaginis*, described on *Medicago*, has been recorded on *T. subterraneum*.

Pseudopeziza trifolii (Biv.-Bernh.)Fuckel (Fig. 1692)
Apothecia erumpent, about 0.5mm diam., pale yellowish brown, jelly-like, each surrounded by a fringe of torn epidermis. Ascospores hyaline, 10–15 × 4.5–5.5. On living leaves of *T. fragiferum*, *T. pratense*, *T. repens* and *T. striatum*, often but not always on brown dead spots, Apr.–Jan.

Sclerotinia spermophila Noble
Apothecia up to 1.5mm diam., cinnamon brown, with long, shortly hairy stipes, black at the base. Ascospores hyaline, ellipsoidal, 12–19 × 11–12. *Botrytis* state with conidia 12–19 × 7–12. Arising from black sclerotia in mummified seeds of *T. repens*.

Sclerotinia trifoliorum Jakob & Eriksson (Fig. 1693)
Apothecia up to 8mm diam., reddish brown with slightly paler discs, and stipes up to 3cm long. Ascospores hyaline, 13–17 × 7–9. Parasitic on *T. pratense*, *T. repens* and other Leguminosae. The apothecia arise from buried, irregular, black sclerotia about 2 × 1cm, which rot away. They should be looked for Sept.–Nov. where round bare patches are seen among sizeable colonies of clover plants.

Stemphylium sarciniforme (Cav.)Wiltshire (Fig. 1694)
Conidiophores golden brown, up to 50 × 6–10, with vesicular swellings 11–14 diam. Conidia smooth, golden brown, muriform, 30–50 × 22–33. On living leaves of *T. campestre* and *T. pratense* causing dark brown, zonate spots 1–8mm diam.

TRIGLOCHIN: Sea Arrow-grass

Pleospora triglochinicola Webster (Fig. 1695)
Pseudothecia scattered, subepidermal, becoming superficial, black, up to 0.4mm diam. Ascospores golden brown, muriform, 45–65 × 17–24. *Stemphylium* state with conidia characterised by an oblique conical projection which is formed frequently to one side of the basal scar, pale golden brown, smooth, muriform, 45–80 × 20–30. On dead leaves and inflorescences of *T. maritima*; pseudothecia Jan.–Feb., conidia June–Aug.

TRIPLEUROSPERMUM: Scentless Mayweed

Entyloma matricariae Rostrup
Smut sori appearing as brown spots on leaves and sometimes stems. Spores immersed, hyaline to straw-coloured, 10–12 diam. Conidia on surface, hyaline, 0- to 3-septate, 6–25 × 2–3. Aug.–Sept.

Peronospora leptosperma de Bary (Fig. 1571)
Downy mildew on leaves, stems and involucres. Colonies white. Sporangia mostly 35–45 × 16–22. Oospores pale brown, 40–45 diam.

Septoria matricariae Sydow
Pycnidia dark brown, 50–70 diam. Conidia hyaline, septate, mostly bent, 30–60 × 1–1.5.

TROLLIUS: Globe Flower

Urocystis anemones, smut described on *Anemone*, has been recorded on *Trollius*.

TROPAEOLUM: Nasturtium

Cronartium flaccidum (Alb. & Schw.)Wint., 11, 111
Uredinia hypophyllous, pustular, pale; spores echinulate, 20–30 × 15–20. Telia hypophyllous, waxy or horny, golden brown; spores 30–55 × 10–15, smooth, in long chains. Alternate host *Pinus sylvestris*. Scarce.

Acroconidiella tropaeoli (Bond)Lindquist & Alippi (Fig. 1696)
Colonies mostly hypophyllous, greyish brown, cottony. Conidiophores olivaceous brown, up to 180 × 5–10. Conidia developing through pores in conidiophore wall, almost always 2-septate, olivaceous brown, echinulate, 30–50 × 15–27. On leaves and occasionally stems, causing brownish or purple spots up to 1cm diam., often confluent and sometimes with broad yellow margins. Rare in Britain.

TULIPA: Tulip

Puccinia prostii Duby 0, 111
Telia amphigenous, blackish brown; spores with unusually large spines, 55–60 × 17–20. Very rare.

Botrytis tulipae Lind (Fig. 1697)
Similar to *B. cinerea* but with larger conidia, mostly 16–20 × 10–13. The cause of 'fire' disease.

Various fungi, such as *Fusarium oxysporum* f. *tulipae*, *Penicillium* species and *Sclerotium tuliparum*, cause bulb rots; cultural studies are needed for their determination.

TUSSILAGO: Coltsfoot

Coleosporium tussilaginis (Pers.)Berk., 11, 111
Uredinia hypophyllous, orange–yellow, powdery; spores densely verruculose, 20–40 × 17–25. Telia waxy, reddish orange; spores hyaline, smooth, 0- to 3-septate, 60–100 × 16–24. Alternate hosts *Pinus nigra* and *P. sylvestris*. Very common.

Puccinia poarum Niels., 0, 1
Aecia hypophyllous, on round, yellow or purplish red, thickened spots, 1–2cm diam., cup-shaped, orange with white, cut, revolute margins; spores verruculose, 20–25 × 17–20. Alternate hosts *Poa pratensis* and *P. trivialis*. Very common, May–June.

Asteroma impressum Fuckel
Pycnidia black, 0.2mm diam.; conidia not seen. They are surrounded by radiating, blackish brown, branched, subcuticular mycelial strands and cause reddish-brown spots on leaves. Quite common.

UMBILICUS: Wall Pennywort

Puccinia umbilici Duby, 111
Telia on stems, petioles and both leaf surfaces, often in rings on yellow spots, dark reddish brown; spores 20–35 × 18–25. Fairly common.

URTICA: Stinging Nettles

Nearly all records refer to collections made on *U. dioica* but *Peronospora debaryi* is confined to *U. urens* and *Pseudoperonospora urticae*, *Septoria urticae* and the aecia of *Puccinia caricina* have been found on both *U. dioica* and *U. urens*. More than 50 different microfungi are found on nettles. The little creamy-white cup-shaped basidiomycete *Calyptella capula*, which can at first sight be mistaken for a discomycete, occurs frequently on stem bases.

UREDINALES

Puccinia caricina DC., 0, 1
Aecia mostly hypophyllous, yellowish orange with white, cut, recurved

margins; spores verruculose 15–25 × 13–20. Very common in early spring on living leaves of plants growing in damp places near sedges.

PERONOSPORALES

Peronospora debaryi Salmon & Ware
Downy mildew. Colonies white to pale greyish violet. Sporangia 23–32 × 19–25. Oospores spherical, 30–40 diam., on *U. urens*.

Pseudoperonospora urticae (Lib.)Salmon & Ware (Fig. 1698)
Downy mildew. Colonies hypophyllous. Sporangia 20–40 × 14–22. Forms greyish-brown or greyish-lilac patches on yellowed leaves of *U. dioica* and *U. urens*, Sept. and May.

DISCOMYCETES

Calloria neglecta (Lib.)Hein (Fig. 1699)
Apothecia up to 1mm diam., pinkish orange to tangerine. Ascospores 1- to 3-septate, 8–14 × 3–3.5. Paraphyses swollen at their tips. Very common on dead stems, Mar.–May. The *Cylindrocolla* conidial state forms round or irregular, thin, clear, orange patches on the dead stems during the winter and is in very good sporulation Jan.–Feb. Conidia hyaline, 5–18 × 1–1.5, in long, branched chains.

Erinella discolor Mouton (Fig. 1700)
Apothecia sessile, 1–1.5mm diam., excipulum pale yellowish brown, strigose with numerous cylindrical, septate, more or less fasciculate hairs up to 500 × 4–5, hyaline above, pale straw at least in part below; discs fleshy, ochraceous. Asci 100–120 × 8, minute pore blued with iodine. Paraphyses slightly tapered, 2–3 thick. Ascospores commonly curved, attenuated below, multi-guttulate, becoming 6- to 10-septate, 50–76 × 2–2.5. On dead stems, Nov.

Laetinaevia carneoflavida (Rehm)Nannf. ex Hein (Fig. 1701)
Apothecia up to 0.25mm diam., pink with a darker edge. Ascospores hyaline, 0- or 1-septate, 10–15 × 3–4. Paraphyses filiform, branched. On damp dead stems, June–July.

Naemacyclus caulium Höhnel (Fig. 1702)
Apothecia immersed, up to 1mm diam., the greyish discs becoming exposed when the overlying greyish-green host tissue splits into 4 lobes. Ascospores hyaline, mostly 7-septate, 40–50 × 1.5. Near the base of dead stems, Apr., rare.

Pyrenopeziza urticicola (Phill.)Boud. (Fig. 1703)
Apothecia up to 0.5mm diam., reddish brown becoming blackish brown, with white fimbriate margins; discs pale greyish brown. Ascospores hyaline, 6–7 × 1–1.5. On dead stems, Apr.–Aug., common.

Plurivorous species most of which are common on dead stems of *Urtica* include *Allophylaria macrospora* (*Epilobium*), *Crocicreas coronatum*, *C. cyathoideum*, *Dasyscyphus grevillei*, *D. mollissimus*, *D. sulphureus*, *Hymenoscyphus herbarum* and *Unguiculella hamulata*.

OTHER ASCOMYCETES

Acrospermum compressum Tode (Fig. 1704)
Ascomata club-shaped, somewhat compressed laterally, with an apical opening, up to 3mm tall, dark blackish brown. Ascospores thread-like, up to 400 × 1, in narrow cylindrical asci. Common especially near the base of dead stems in damp situations, Feb.–June.

Erysiphe urticae (Wallr.)Klotzsch
Powdery mildew. Conidia abundant. Cleistothecia containing 6 to 10 asci, each with 3 to 5 spores; appendages numerous, narrow. On living leaves, Sept.–Oct.

Leptosphaeria acuta (Hoffm)P.Karsten (Fig. 1705)
Pseudothecia mostly superficial, conical, black, 0.3–0.5mm diam. Ascospores yellow, 8- to 12- septate, 40–60 × 5–7.5. Very common especially near the base of dead stems, Feb.–May. At other times of the year often represented by its superficially very similar *Phoma* conidial state with hyaline, biguttulate conidia 4–5 × 1.5–2. *L. acuta* is sometimes parasitised by *Nectria leptosphaeriae* Niessl (Fig. 1706), with orange to red perithecia, and hyaline to pale brown, 1-septate ascospores 17–25 × 6–8; also by a hyphomycete *Pleurophragmium acutum* (Grove)M.B.Ellis (Fig. 1707) with brown conidiophores up to 90 × 2–4 bearing hyaline conidia 9–12 × 2–3, on tapered denticles.

Mycosphaerella superflua (Auersw.)Petrak (Fig. 1708)
Pseudothecia immersed, crowded, about 0.15mm diam., black. Ascospores hyaline, 1-septate, 12–17 × 3–4.5. Common on dead stems and usually in good condition in March. The *Ramularia* conidial state forms white colonies 1–3mm diam. on the lower side of living leaves, Apr.–Sept. Conidiophores in fascicles emerging through stomata, hyaline, 30–40 × 2–4. Conidia in chains, 0- to 1-septate, 15–30 × 3–5.5.

Plagiosphaera immersa (Trail)Petrak
Perithecia grouped below grey patches of epidermis, flattened, with short, often lateral necks. Asci with refractive apical ring. Ascospores parallel,

cylindrical, slightly bent, hyaline, multiguttulate, 50–60 × 2–3. On dead stems, May–Aug.

Common plurivorous species found on dead stems include *Diaporthe arctii*, *Gibberella cyanogena* (*Brassica*), *G. pulicaris* (*Lupinus*), *Leptosphaeria doliolum*, *Leptospora rubella*, *Lophiostoma angustilabrum*, *L. caudatum*, *L. caulium*, *L. origani* var. *rubidum*, *Nectria arenula* and *Ophiobolus erythrosporus*.

HYPHOMYCETES

Arthrinium urticae M.B.Ellis (Fig. 1709)
Colonies pulvinate, small, round or elongated, up to 1mm long, blackish brown to black. Conidiophores 40–75 × 1.5–2, hyaline with dark brown bands at septa. Conidia brown, 5–6 × 3–4. On dead stems, Nov., uncommon.

Botryosporium pulchrum Corda (Fig. 1710)
This forms extensive, white, cobwebby or fluffy colonies on living nettle plants, and may be seen any time from May to Nov. The conidiophores are hyaline, very long and branched, with the main trunks about 12–15 thick. Conidia minutely verruculose, hyaline, 6–9 × 3–4.

Endophragmia atra (Berk. & Br.)M.B.Ellis (Fig. 1711)
Colonies tufted, dark brown to black, up to 6 × 1cm. Synnemata black, up to 400 × 25–60; individual conidiophores dark brown, 4–4.5 thick, expanding at their subhyaline apices to 10–15. Conidia 4-septate, brown with pale or hyaline end cells, and black bands at septa, 28–43 × 15–23. On dead stems, Oct.–Jan., fairly common.

Gyrothrix verticillata Pirozynski (Fig. 1712)
Colonies effuse, thin, up to 5mm diam., pale greyish brown. Setae pale brown, up to 300 × 3–4, with whorls of 3 or 4 horizontal branches. Conidia in a thick, whitish layer around the bases of setae, hyaline, 10–15 × 1.5–2. On dead stems, Sept., uncommon.

Polyscytalum berkeleyi M.B.Ellis (Fig. 1713)
Colonies effuse, pale grey or olivaceous grey, powdery. Conidiophores up to 80 × 3–5, lower part olivaceous brown, upper part subhyaline. Conidia dry, in long, often branched chains, hyaline to pale olive, smooth, 0- to 3-septate, 10–30 × 2–3.5. On dead stems just above soil level, Mar.–May, common.

Ramularia on living leaves, *v.s.* under *Mycosphaerella superflua*.

Many plurivorous hyphomycetes are found on the dead stems all the year round. In any fair-sized patch of nettles one can always find *Dendryphion comosum*, looking like forests of tiny greyish-brown trees; blackened areas with upright conidiophores of *Periconia byssoides*, which resemble little round-headed pins and velvety, olivaceous to black colonies of *Torula herbarum*. Rather less conspicuous but frequently found are *Pleurophragmium parvisporum* and *Periconia minutissima*. Less frequent but still commonly recorded are *Alternaria alternata, Botrytis cinerea, Camposporium pellucidum, Cladosporium herbarum, Dendryphiella infuscans, D. vinosa, Dendryphion nanum, Dictyosporium toruloides, Epicoccum purpurascens, Periconia cookei, Trichocladium opacum* and *Volutella ciliata*.

COELOMYCETES

Apomelasmia state of *Aporhytisma urticae* (Wallr.)Höhnel (Fig. 1714)
Conidiomata up to 0.5mm long, black and shining, sometimes confluent, immersed in an almost black stroma which is always extensive and sometimes nearly encircles the dead stem. Conidia hyaline, mostly 18–26 × 3–4.5, Feb.–Mar., not common.

Phoma, v.s. under *Leptosphaeria acuta.*

Pyrenochaeta fallax Bres. (Fig. 1715)
Pycnidia immersed in grey areas on dead stems, 0.1–0.18mm diam., dark brown with brown, septate setae 100–200 × 5–7 around the ostiole. Conidia hyaline, 4–6.5 × 1.5–2.

Septoria urticae Rob. & Desm.
Pycnidia epiphyllous, immersed, less than 0.1mm diam. Conidia hyaline, curved or flexuous, 30–50 × 1–2. On leaves of *U. dioica* and *U. urens*, causing greyish-brown spots which eventually fall out leaving shot-holes, May–Aug.

The plurivorous coelomycete *Pseudolachnea hispidula* is not uncommon on dead stems.

VALERIANA: Marsh Valerian, Valerian

Puccinia commutata P. & H. Sydow, 0, 1, 111
Telia dark brown, tending to be confluent and arranged in long rows; spores smooth, 1-septate, 45–65 × 20–30. On *V. officinalis*, very rare.

Uromyces valerianae Fuckel, 0, 1, 11, 111
Aecia hypophyllous. Uredinia amphigenous, small, brown, on yellowish spots; spores echinulate, 20–30 diam. Telia dark brown; spores smooth, 20–30 × 14–20. On *V. dioica* and *V. officinalis*, common.

Peronospora valerianae Trail
Downy mildew on *V. officinalis*.

Ramularia valerianae (Speg.)Sacc.
Colonies hypophyllous. Conidiophores hyaline, up to 50 × 3–4. Conidia hyaline, 1- to 3-septate, 15–50 × 5–8. Causes large greyish spots on living leaves of *V. dioica* and *V. officinalis*.

VALERIANELLA: Corn Salad or Lamb's Lettuce

Peronospora valerianellae Fuckel
Downy mildew. Colonies white. Sporangia 17–20 × 15–17.

VERBASCUM: Mulleins

Erysiphe verbasci (Jacz.)Blumer
Powdery mildew. Conidia plentiful. Cleistothecia, with numerous hyaline or pale brown appendages, contain 10 to 20 asci each with usually 2 but occasionally 4 spores. On *V. nigrum* and *V. thapsus*, Oct.–Nov.

Peronospora verbasci Gäum.
Downy mildew on *V. nigrum* and *V. thapsus*. Sporangia 19–20 × 17–18.

Phomopsis verbasci Hawksw. & Punithalingam
Conidiomata in lines, immersed, becoming erumpent, conical, black, up to 1mm diam. A-conidia fusiform to ellipsoid, hyaline, biguttulate, 5–8 × 2–3; B-conidia curved, 15–25 × 0.5–1. On dead stems of *V. thapsus*, Mar.

Septocylindrium bellocense Massal. & Sacc.
Colonies effuse, white. Conidiophores hyaline, up to 140 × 4–5. Conidia catenate, hyaline, multiseptate, very variable in length, up to 200 × 4–6. On fading leaves of *V. nigrum*, Oct.

VERONICA: Brooklime, Speedwells

UREDINALES

Three rusts occur but only one is at all common.

On *V. alpina*: *Puccinia albulensis*, rare.
On *V. montana*: *P. veronicae*, common.
On *V. spicata*: *P. veronicae-longifoliae*, rare.

Puccinia albulensis Magn., 111
Telia hypophyllous and on stems; spores smooth, 25–45 × 12–18.

Puccinia veronicae Schröter, 111 (Fig. 1716)
Telia hypophyllous, on brown spots; spores 30–40 × 14–17.

Puccinia veronicae-longifoliae Savile, 111
Telia hypophyllous; spores 40–50 × 13–15.

USTILAGINALES

Schroeteria delastrina (Tul.)Winter (Fig. 1717)
This smut forms grey or greyish-brown spore masses inside the seed capsules of *V. arvensis*, May–June. Spores mostly in pairs, 10–13 diam., pale greyish brown, verrucose.

OTHER FUNGI

Asterina veronicae (Lib.)Cooke (Fig. 1718)
Thyriothecia 0.1–0.15mm diam., black, round, flat, linked together by a superficial network of dark brown, hyphopodiate mycelium. Ascospores brown, 1-septate, 14–17 × 6–7. Pycnothyria of conidial state similar to thyriothecia but smaller; pycnospores reddish brown, each with a hyaline band at the centre, 18–21 × 7–9. On living leaves and stems of *V. officinalis*, June–July, fairly common.

Discogloeum veronicae (Lib.)Petrak (Fig. 1719)
Acervuli up to 0.2mm long. Conidia hyaline, 12–19 × 3–4. Causes brown spots, less than 1mm diam., with white or pale grey centres, on living leaves and stems of *V. persica*, May–Sept.

Peronospora agrestis Gäum.
Downy mildew, with sporangia 18–23 × 14–18, on *V. agrestis*, *V. arvensis*, *V. filiformis*, *V. persica* and *V. polita*, Feb.–Apr.

Peronospora grisea (Unger)Unger
Downy mildew, with dingy violaceous sporangia 22–30 × 15–22, and brownish oospores 30–38 diam., on *V. agrestis*, *V. beccabunga*, *V. hederifolia*, *V. persica* and *V. serpyllifolia*, Apr.–June.

Sphaerotheca fuliginea (Schlecht.)Pollacci
Powdery mildew. Mycelium dense, brown when old. Conidia few, ellipsoid to barrel-shaped, 25–35 × 14–20. Single 8-spored ascus; spores 18–28 × 12–18. On *V. chamaedrys*.

VICIA: Broad Bean, Tares and Vetches

UREDINALES

The common rust on all *Vicia* species is *Uromyces viciae-fabae*. A second very much less commonly reported species, *U. ervi*, is found occasionally on *V. hirsuta*; the urediniospores in this have 2 germ pores (3 to 5 in *U. viciae-fabae*) and the teliospores are smaller, 20–27 × 15–20 instead of 26–38 × 18–28.

Uromyces ervi West., 0, 1, 11, 111
Aecia amphigenous. Uredinia rarely formed. Telia amphigenous and on stems and petioles, blackish brown. On *V. hirsuta* only.

Uromyces viciae-fabae (Pers.)Schröter, 0, 1, 11, 111
Aecia mostly hypophyllous. Uredinia amphigenous, pale brown; spores echinulate, 20–30 × 20–26. Telia blackish brown; spores smooth. Very common on *V. angustifolia*, *V. bithynica*, *V. cracca*, *V. fabae*, *V. hirsuta*, *V. lutea*, *V. sativa* and *V. sepium*.

OTHER FUNGI

Ascochyta fabae Speg.
Pycnidia immersed, 0.1–0.25mm diam., golden or reddish brown. Conidia hyaline, 1- to 3-septate, 16–24 × 4–6. On leaves, pods and stems of *V. faba*, causing slightly sunken, pale brown spots up to 1cm diam., with thick, dark brown borders, Feb.–Mar.

Botrytis fabae Sardiña (Fig. 1720)
Conidiophores not normally found in the field, but develop and form conidia when spotted leaves are kept in a damp chamber. Conidia always much larger than those of the plurivorous grey mould *B. cinerea*, 14–30 × 11–20 (mostly 16–25 × 13–16). The cause of 'chocolate-spot' disease; reddish or chocolate brown spots on leaves in winter and spring, may coalesce later and cause a damaging black blight.

Cercospora zonata Wint. (Fig. 1721)
Colonies amphigenous but mostly epiphyllous. Conidiophores fasciculate, mid pale olivaceous brown, 30–60 × 4–7. Conidia hyaline, 3- to 15-septate, 40–140 × 4.5–6.5. On leaves of *V. faba*, causing brown or grey, often zonate spots.

Diaporthe phaseolorum (Cooke & Ellis)Sacc. var. *sojae* (Lehman)Wehmeyer
Morphologically similar to *D. phaseolorum*, described on *Phaseolus*, has been recorded in Britain on *V. faba*.

Erysiphe pisi, powdery mildew described on *Pisum*, occurs on *V. sepium*.

Microsphaera baeumleri Magnus
Powdery mildew. Conidia abundant. Cleistothecia, with only some of the appendages dichotomously branched, contain 7 to 12 asci each with 3 to 5 spores. On *V. sylvatica*, occasional.

Peronospora viciae (Berk.)Casp. (Fig. 1722)
Sporangiophores up to 700 × 7–13. Sporangia hyaline to very pale violet or brown, 20–30 × 15–20. Oospores pale brown, 25–40 diam., with raised reticulate walls. On the underside of yellowed leaflets and on stems of *V. angustifolia*, *V. faba*, *V. hirsuta*, *V. sativa*, *V. sepium* and *V. tetrasperma*.

Sclerotinia trifoliorum Jakob & Eriksson var. *fabae* Keay
On *V. faba*, similar to *S. trifoliorum* described on *Trifolium*, but with ascospores sometimes up to 28 × 15.

VINCA: Greater and Lesser Periwinkles

Puccinia vincae Berk, 0, 1, 11, 111
Aecia uredinoid, hypophyllous; spores echinulate, some hyaline, some brown, 35–40 × 22–25. Teliospores brown, 1-septate, verrucose, 37–40 × 20–23. On *V. major*, uncommon.

Ascochyta vincae (Thümen)Grove
Pycnidia few, mostly epiphyllous, black. Conidia hyaline, narrowly fusiform, sometimes slightly curved, eventually 1-septate, 10–14 × 2. On living leaves of *V. major* and *V. minor*, causing large, pale brown spots, with narrow dark brown margins, Mar.–May.

Ceuthospora feurichii Bubák (Fig. 1723)
Conidiomata immersed, 0.15–0.2mm diam., olivaceous grey to black. Conidia hyaline, 10–12 × 1.5–2.5. On stems and dead or dying leaves of *V. minor*, Feb.–June.

Diaporthe eumorpha (Durieu & Mont.)Maire
Perithecia few, scattered, immersed, 0.2–0.3mm diam.; stem surface blackened. Ascospores hyaline, 1-septate, with four guttules, 9–15 × 2.5–4. On stems, causing spots 0.2–0.5mm diam.

VIOLA: Pansy, Violets

UREDINALES

On *V. palustris*: *Puccinia fergussonii*, rare.
On *V. canina*, *V. cornuta*, *V. hirta*, *V. lutea*, *V. odorata*, *V. reichenbachiana*, *V. riviniana* and *V. tricolor*: *Puccinia violae*, very common.

Puccinia fergussonii Berk. & Br., 111
Telia hypophyllous, in groups, on large yellowish spots, chocolate brown; spores somewhat fusiform, smooth, 1-septate, 25–45 × 13–17.

Puccinia violae DC., 0, 1, 11, 111
Aecia on stems, leaves and other parts. Uredinia amphigenous; spores echinulate, 20–28 × 18–23. Telia hypophyllous, small, brown; spores minutely verruculose, 1-septate, 20–40 × 16–23.

USTILAGINALES

Urocystis violae (Sow.)Fisch. von Waldh.
On *V. odorata*, *V. reichenbachiana*, *V. riviniana* and cultivated violets. Smut sori cause swelling and distortion of petioles, veins and upper parts of rootstocks. Spore mass dark brown. Spore balls up to 70 long, each with 4 to 8 smooth-walled, reddish-brown spores 10–16 diam., surrounded by yellowish sterile cells. Common, Nov.–July.

OTHER FUNGI

Alternaria violae Galloway & Dorsett (Fig. 1724)
Conidiophores in fascicles, pale brown, up to 40 × 4–8. Conidia pale to mid golden brown, muriform, 60–90 × 11–17, with beaks 2.5–4 thick. On leaves of cultivated species, causing pale brown, sometimes zonate spots.

Ascochyta violae Sacc. & Speg.
Pycnidia immersed, brown, darker around the ostiole, 0.1–0.2mm diam. Spores hyaline, with 2 to 4 guttules, finally 1-septate, 10–16 × 2.5–3.5. On living leaves of *V. odorata* and cultivated species, causing brownish spots which whiten and drop out when dry, May–Sept.

Cercospora murina Ell. & Kell. (Fig. 1725)
Colonies hypophyllous. Conidiophores branched, pale olivaceous brown, 50–150 × 3–5. Conidia pale olivaceous, 3- to 9-septate, 25–105 × 4–5. On leaves of *V. palustris*.

Cercospora violae Sacc. (Fig. 1726)
Colonies amphigenous. Conidiophores fasciculate, mid pale olivaceous brown, 30–55 × 4–6. Conidia hyaline, 7- to 17-septate, 70–180 × 4–5. On leaves of *V. canina* and *V. riviniana*, causing pale brown or greyish spots, June–July.

Myrothecium roridum Tode ex Fr. (Fig. 1727)
Sporodochia sessile, up to 1.5mm diam., often confluent, at first green, later black, with white margins. Phialides 10–12 × 1–2. Conidia colourless to pale olive, green to black in mass, mostly 6–8 × 1.5–2.5. On leaves of *V. riviniana*, *V. tricolor* and cultivated species, causing pale brown, sometimes zonate, spots which eventually drop out leaving shot-holes, June–July.

Peronospora violae de Bary
Downy mildew on *V. arvensis*. Sporangia pale dingy violet, 20–26 × 15–19.

Ramularia agrestis Sacc. (Fig. 1728)
Colonies white, amphigenous. Conidiophores in fascicles of 5 to 20, hyaline, 30–60 × 3–4.5. Conidia hyaline, 0- to 3-septate, 12–32 × 4.5–7. On living leaves of *V. tricolor*, causing pale brown or whitish spots 2–5mm diam., with narrow brown borders.

Ramularia lactea (Desm.)Sacc. (Fig. 1729)
Colonies amphigenous. Conidiophores hyaline, 10–50 × 2–3.5. Conidia hyaline, 0- or 1-septate, 10–30 × 2–3.5. On living leaves of *V. canina*, *V. odorata* and *V. riviniana*, causing white or pale brown spots 2–8mm diam., Sept.–Oct.

ZINNIA: Zinnia

Alternaria zinniae M.B.Ellis (Fig. 1730)
Colonies effuse, greyish brown to dark blackish brown, shortly hairy. Conidiophores brown, up to 150 × 5–10. Conidia pale to dark brown, muriform, smooth or minutely verruculose, 55–105 × 19–28, with pale beaks, 55–185 × 1.5–2.5, which are often swollen at their tips to 3–4. On leaves, flowers and stems, causing pale grey or brown, purple-margined spots.

PLURIVOROUS FUNGI ON GRASSES

UREDINALES

Six rust species and varieties are plurivorous: *Puccinia brachypodii* var. *poae-nemoralis*, *P. coronata*, *P. graminis*, *P. recondita*, *P. striiformis* and *Uromyces dactylidis*. No aecial state is known for *P. striiformis*; the cluster cups of *P. graminis* are found on *Berberis* and *Mahonia*; and those of *P. brachypodii* var. *poae-nemoralis* on *Berberis* only; in *P. coronata* they occur on *Frangula alnus* and *Rhamnus catharticus*; and in *Uromyces dactylidis* on various species of *Ranunculus*. The aecial hosts of *P. recondita*, which is split into a number of formae speciales, mostly belong to the Boraginaceae and Ranunculaceae. Separation of the six species is sometimes difficult when only uredinia are present.

KEY

	Teliospores 1-celled .... *Uromyces dactylidis*
	Teliospores with more than 1 cell 1
1.	Teliospores with long pedicels .... *Puccinia graminis*
	Teliospores with short pedicels 2
2.	Clavate or capitate paraphyses abundant in uredinia ....
	.... *P. brachypodii* var. *poae-nemoralis*
	Paraphyses in uredinia few or absent 3
3.	Teliospores crowned with tooth-like projections .... *P. coronata*
	Teliospores not crowned with tooth like projections 4
4.	Walls of urediniospores quite colourless .... *P. striiformis*
	Walls of urediniospores brownish .... *P. recondita*

Puccinia brachypodii Otth var. *poae-nemoralis* (Otth)Cummins & Greene, 11, 111
Uredinia mostly on small yellow spots on upper surface of leaves, scattered, minute; spores yellow with almost colourless, minutely verruculose walls, 22–27 × 18–23, with 5 or 6 spores; paraphyses numerous, clavate or capitate, curved, thick-walled, 50–80 long with heads up to 20 wide. Telia on lower surface of leaves, oblong, black, long covered by epidermis; spores chestnut brown, paler below, 35–50 × 16–23, 1-septate; pedicels short, brownish. On *Anthoxanthum*, *Glyceria*, *Poa* and *Puccinellia*.

Puccinia coronata Corda, 11, 111 (Fig. 1731)
Crown rust. Uredinia minute, orange, scattered or in patches on both leaf surfaces in summer; spores yellow, 15–38 × 12–35, with finely echinulate

walls; peripheral, thin-walled clavate paraphyses few or none. Telia black, on lower leaf surface from mid Aug., sometimes arranged in circles around uredinia; spores brown, crowned with long, tooth-like projections, 1-septate, 30–60 × 12–20, pedicels short. On *Agropyron*, *Agrostis*, *Alopecurus*, *Arrhenatherum*, *Avena*, *Calamagrostis*, *Dactylis*, *Deschampsia*, *Festuca*, *Glyceria*, *Holcus*, *Lolium*, *Phalaris* and *Poa*.

Puccinia graminis Pers., 11, 111 (Fig. 1732)
Black stem rust. Uredinia on both leaf surfaces and commonly also on sheaths and stems, yellowish brown, scattered or in rows, 2–3mm long, often confluent, the joined sori extending to 1cm or more; spores yellowish brown, walls echinulate, pores 4; in the subspecies *graminis*, recorded mostly on cereals, 26–40 × 16–22, in the subspecies *graminicola*, on wild grasses, 20–30 × 14–20. Telia mostly on sheaths and stems, forming narrow, black lines 0.5–1cm long; spores chestnut brown, 35–60 × 12–22, 1-septate; pedicels up to 60 long, persistent, often brownish. Very common on *Agropyron* and *Agrostis*, and frequently found also on many other wild grasses and on cereals.

Puccinia recondita Rob. & Desm., 11, 111
Brown leaf rust. Uredinia mostly on upper surface of leaves, scattered, cinnamon brown, 1–2mm long, without paraphyses; spores 15–24 × 17–32, walls brownish, minutely echinulate, contents orange, pores 6 to 8. Telia on lower surface, small, black, oblong; paraphyses dark brown, somewhat curved, subepidermal, peripheral and sometimes dividing up larger sori; spores 35–60 × 12–24, brown, 1-septate, pedicels short, often dark just below the spore. Occasionally spores without a septum or with 2 or more septa are formed. Formae speciales are found on *Agropyron*, *Agrostis*, *Alopecurus*, *Bromus*, *Holcus*, *Secale*, *Trisetum* and *Triticum*.

Puccinia striiformis Westend., 11, 111 (Fig. 1733)
Yellow rust or stripe rust. Sometimes also called spring rust because it appears very early in the year. Uredinia arranged in long lines, up to 8cm, on yellowed areas, most commonly on the upper surface of leaves; spores 25–30 × 14–26, walls colourless, shortly echinulate, contents yellow or orange, pores 8 to 10. Telia dark brown to black, formed in long, thin lines on leaves, stems and inflorescences, long remaining covered by epidermis; paraphyses numerous, brown, thick-walled, curved, surrounding little groups of spores; spores 30–70 × 11–24, 1-septate, chocolate brown, with very short pedicels. On *Agropyron*, *Brachypodium*, *Bromus*, *Elymus*, *Hordeum*, *Secale* and *Triticum*, common. Very common on wheat.

Uromyces dactylidis Otth, 11, 111
Uredinia yellowish brown, up to 1mm long, scattered or in rows, without paraphyses; spores yellow or yellowish brown, finely echinulate, 21–32 × 18–25, with 4 to 9 pores. Telia mostly on lower surface of leaves, small, black, shining; paraphyses numerous, brown, agglutinated, often dividing up sori;

spores without septa, golden brown, 18–30 × 14–22, with pedicels about same length as spores. On *Dactylis, Festuca* and *Poa*.

USTILAGINALES

There are four common plurivorous smuts on grasses. Three of these, *Urocystis agropyri, Ustilago serpens* and *U. striiformis*, commonly called stripe smuts, are found on leaves and seem very similar in the field. The sori form long blisters parallel to the veins; when ripe, these rupture exposing black masses of spores. Eventually the leaves split into ribbons. Infected plants are stunted and rarely flower. Occasionally, sori are formed on the glumes. In *Ustilago hypodytes*, the stem smut, dark brown to black sori encircle and often cover the whole surface of one or several internodes, becoming exposed when leaf sheaths split. Infected plants are usually sterile.

Urocystis agropyri (Preuss)Schröter (Fig. 1734)
Spore balls usually composed of one or two golden brown, smooth-walled spores 12–16 diam., surrounded by pale sterile cells 7–10 diam. On *Agropyron, Agrostis, Arrhenatherum* and *Lolium*.

Ustilago hypodytes (Schlecht.)Fr. (Fig. 1735)
Spores spherical, 4–5 diam. or oval up to 7 long, pale yellowish brown, smooth. On *Agropyron, Ammophila, Bromus, Elymus, Festuca* and *Trisetum*.

Ustilago serpens (Karsten)B.Lindeb.
Spores coarsely echinulate or verrucose, golden brown, 11–18, most commonly about 14 diam. On *Agropyron, Bromus* and *Calamagrostis*.

Ustilago striiformis (Westend.)Niessl (Fig. 1736)
Spores dark golden brown, echinulate, 9–14, mostly 11 diam. On *Arrhenatherum, Dactylis, Deschampsia, Festuca, Holcus, Lolium, Phalaris, Phleum, Poa* and *Sesleria*.

DISCOMYCETES

KEY TO GENERA

	Apothecia black, hysterioid .... *Lophodermium*
	Apothecia not so 1
1.	Apothecia sessile 2
	Apothecia stalked 5
2.	Apothecia lemon or primrose-yellow .... *Calycella*
	Apothecia brown or honey-coloured 3
3.	Apothecia developing through little shields .... *Micropeziza*
	Apothecia not developing through shields 4
4.	Outer surface dotted with raised clusters of dark brown cells . *Belonium*

        Outer surface not so .... *Mollisia*
5.   Apothecia distinctly hairy .... *Dasyscyphus*
      Apothecia sometimes downy but not distinctly hairy 6
6.   Apothecia 0.2–0.4mm diam., white, downy .... *Pezizella*
      Apothecia larger 7
7.   Apothecia tough-fleshed, difficult to cut, often with toothed margins
        or encrusted with shining crystals .... *Crocicreas*
      Apothecia soft-fleshed, easy to cut, margins not toothed, no crystals ....
      .... *Hymenoscyphus*

Belonium incurvatum Graddon
Apothecia scattered, cupulate, up to 0.75mm diam., mostly superficial but with base immersed and anchored by dark brown hyphae, discs pale grey or honey-coloured, outer surface dotted with raised clusters of dark brown, globose cells, margin strongly inrolled when dry, composed of long, hyaline, slender hyphae. Ascospores hyaline, narrowly fusiform, with several small guttules at each end, rarely 1-septate, 19–22 × 2.5–3. Paraphyses up to 3 wide at tip. On dead stems.

Calycella scolochloae (de Not.)Dennis (Fig. 1737)
Apothecia sessile, 0.2–0.8mm diam., lemon or primrose-yellow. Ascospores hyaline, 3-septate, 10–16 × 2–3. On dead stems of *Agropyron repens*, *Arrhenatherum elatius*, etc., lying on the ground, July–Aug.

Crocicreas culmicolum (Desm.)S.Carp. (Fig. 1738)
Apothecia with dentate margins, usually about 1mm but sometimes up to 1.5mm across, white or pale yellow, occasionally with yellow discs and slightly pinkish stalks. Ascospores hyaline, 3-septate, mostly 21–28 × 3–3.5. On dead stems of *Agropyron*, *Calamagrostis*, *Deschampsia* and *Phalaris*, July–Sept.

Crocicreas stramineum (Berk. & Br.)S. Carp.
Apothecia erumpent, subsessile, up to 0.5mm diam., whitish, glistening, with pale yellowish or flesh-coloured discs. Outer surface covered with hyaline hairs up to 15 × 3 with masses of calcium oxalate crystals lying between them. Paraphyses cylindrical, 1.5 thick. Ascospores hyaline, becoming 1-septate, 6–10 × 1.5–2. On dead grass stems, Apr.–May.

Crocicreas tomentosum (Dennis)S.Carp. (Fig. 1739)
Apothecia about 1mm diam., ivory to pale yellow or grey, with protruding white shining hairs on the outside; stalk dark grey to almost black. Hairs 2–3 thick, encrusted with small crystals. Ascospores hyaline, 9–12 × 2–3.5. On half-buried bases of *Phalaris* and other grass stems, Aug.–Feb.

Dasyscyphus. Eight species and one variety are found on grasses and several of them are very common. *Phragmites australis* is a particularly good host plant, and examination of dead and especially fallen stems in summer seldom fails to yield a plentiful supply of apothecia. The apothecia are distinctly hairy, ascospores hyaline without septa and paraphyses mostly lanceolate.

KEY

    Apothecia yellowish brown with brown hairs .... *palearum*
    Apothecia white or cream, never pink 1
    Apothecia becoming flesh-coloured or brownish pink 2
1.   Hairs 100 or more long, tapered, spores 10–16 × 1–1.5 .... *acutipilus*
    Hairs less than 50 long, cylindrical, spores 5–6 × 1–1.5 .... *acutus*
2.   Hairs tapered to a point, almost smooth, pale yellow, 150–200 long
    .... *albotestaceus*
    Hairs never tapered to a point, hyaline, granulate, less than 80 long
    3
3.   Hairs often swollen at the apex 4
    Hairs cylindrical or very slightly tapered 5
4.   Spores 7–11 × 1–1.5 .... *tenuissimus*
    Spores 10–14 × 1.5–2.5 .... *rhodoleucus*
5.   Spores 5–6 × 1 .... *carneolus*
    Spores 8 or more long 6
6.   Spores pointed at each end, paraphyses hyaline, not becoming septate
    .... *carneolus* var. *longisporus*
    Spores blunt-ended, paraphyses often becoming 1-septate and
    olivaceous with age .... *controversus*

Dasyscyphus acutipilus (P. Karsten)Sacc. (Fig. 1740)
Apothecia 0.2–0.4mm diam., short-stalked, white or cream, fringed with white hairs about 3 thick and 100 or more long. Ascospores 10–16 × 1–1.5. Common, especially on dead stems of *Phalaris* and *Phragmites*, but found also on *Agrostis*, *Arrhenatherum*, *Brachypodium*, *Deschampsia* and *Glyceria*, Apr.–Aug.

Dasyscyphus acutus (Velen.)Dennis (Fig. 1741)
Apothecia 0.2–0.35mm diam., short-stalked; hairs mostly 20–40 × 2. Paraphyses thin-walled, tending to collapse. Ascospores 5–6 × 1–1.5. On dead stems and leaves of *Arrhenatherum*, *Deschampsia*, etc., not common, Aug.–Oct.

Dasyscyphus albotestaceus (Desm.)Massee (Fig. 1742)
Apothecia 0.5–1mm diam., short-stalked, flesh-pink, fringed with tapered, pale yellow hairs 150–200 × 3–4. Spores 7–11 × 1.5–2. On dead leaves and stems of *Agropyron*, *Arrhenatherum*, *Brachypodium*, *Bromus*, *Dactylis*, *Festuca*, *Holcus*, *Phalaris* and *Phragmites*, common, Feb.–Aug.

Dasyscyphus carneolus (Sacc.)Sacc. (Fig. 1743)
Apothecia up to 1mm diam., short-stalked, white at first, then flesh-coloured and finally brownish pink, fringed with white hairs up to 70 × 4. Ascospores 5–6 × 1. On dead stems of *Molinia* and *Glyceria*, and found once also on *Scirpus maritimus*; Jan.–July.

Dasyscyphus carneolus var. *longisporus* Dennis (Fig. 1744)
Apothecia up to 0.5mm diam., white at first then flesh-coloured and finally brownish pink, fringed with white hairs up to 75 × 3–4. Ascospores pointed at each end, 8–12 × 1–1.5. Mostly on dead leaves of *Agropyron, Ammophila, Arrhenatherum, Bromus, Dactylis, Deschampsia, Glyceria, Holcus, Molinia, Phalaris* and *Phragmites* but found occasionally also on *Carex acutiformis* and *Scirpus maritimus*, common, Feb.–Oct., especially abundant June–Aug.

Dasyscyphus controversus (Cooke)Rehm (Fig. 1745)
Apothecia up to 1.5mm diam., pale pinkish buff with the disc drying reddish brown, fringed with white or pale yellow hairs 50–60 × 3–4. Ascospores 8–10 × 1–1.5, bluntly rounded at ends. Paraphyses becoming olivaceous and often 1-septate. Very common on dead stems of *Phragmites* and recorded on other grasses such as *Calamagrostis, Glyceria, Holcus* and *Phalaris* and occasionally on other plants, including *Cladium mariscus* and *Typha latifolia*, May–Oct.

Dasyscyphus palearum (Desm.)Massee (Fig. 1746)
Apothecia up to 1mm diam., stalked, yellowish brown, with granulate, brown hairs up to 100 × 3–5, the ends of which remain hyaline and tend to stain deeply in cotton blue. Ascospores 8–15 × 1.5–2. Fairly common and found sometimes in great abundance on dead stems of *Agropyron;* occurs also on *Ammophila, Arrhenatherum, Phragmites* and *Triticum*, Mar.–Aug.

Dasyscyphus rhodoleucus (Sacc.)Sacc.
Apothecia 0.2–0.4mm diam., short-stalked, pinkish, fringed with granulate, white hairs up to 50 × 3–4, distinctly swollen at their tips. Ascospores 10–14 × 1.5–2.5. On unidentified grass stems, uncommon.

Dasyscyphus tenuissimus (Quélet)Dennis (Fig. 1747)
Apothecia 0.3–0.5mm diam., stalked, white at first, becoming pink or pinkish brown, covered with hairs 40–60 × 3–4, swollen to 5–6 at the apex. Ascospores slightly curved, 7–11 × 1–1.5. On dead stems of *Arrhenatherum, Glyceria, Holcus* and *Phragmites*, May–Aug.

Hymenoscyphus. Two plurivorous species and one variety with pale, smooth, stalked apothecia, and hyaline spores without septa, occur on grasses.

KEY

Spores less than 15 long	*robustior*
Spores more than 15 long	1
1. Spores 2.5–3.5 wide	*pileatus*
Spores 4–5.5 wide	*scutula* var. *suspecta*

Hymenoscyphus pileatus (P.Karsten)O.Kuntze (Fig. 1236)
Apothecia 1–2mm diam., cream, with long white stalks. Ascospores 15–30 × 2.5–3.5. On debris of grasses and other plants in marshy places, Oct.–Nov.

Hymenoscyphus robustior (P.Karsten)Dennis (Fig. 1748)
Apothecia up to 2mm diam., when fresh, yellow with pink stalks, brownish when old. Ascospores mostly 7–12 × 3–3.5. Most commonly found on dead stems of *Agropyron* and *Phragmites*, but has been collected also on *Ammophila* in a damp spot, and on various marsh and fen plants, including *Cladium*, *Iris*, *Scirpus* and *Typha*, June–July.

Hymenoscyphus scutula (Pers.)Phill. var. *suspecta* (Nyl.)P. Karsten (Fig. 1749)
Apothecia up to 2mm diam., ivory white, with long stalks. Ascospores mostly 20–28 × 4–5.5. On dead stems of *Arrhenatherum*, *Dactylis*, *Festuca* and *Glyceria*, Oct.–Jan.

Lophodermium. The erumpent, flat, elliptical apothecia, with a slit down the centre which gapes widely to expose a disc when mature and damp, are very common on dead stems and leaves of grasses. *L. arundinaceum* and *L. gramineum* are widespread; *L. culmigenum* is collected less frequently. Best collecting months Mar.–Aug.

KEY

Asci up to 120 × 10–12, spores 60–100 × 1.5–2 .... *arundinaceum*
Asci not more than 100 × 10, spores 55–85 × 1 1
1. Asci cylindrical, spore-sheath thick, paraphyses up to 3 wide ....
.... *culmigenum*
Asci clavate, spore-sheath very thin, paraphyses 1–2 wide ..... *gramineum*

Lophodermium arundinaceum (Schrader)Chev. (Fig. 1750)
On dead stems, sheaths and leaves. Very common on *Phragmites*; found also on *Agropyron*, *Ammophila*, *Festuca* and *Glyceria*. Conidiomata of the conidial state resemble the apothecia, but are filled with a pale olivaceous mass of slightly curved conidia, which measure 14–25 × 2; they are found Nov.–Mar., and sometimes later accompanying apothecia.

Lophodermium culmigenum (Fr.)de Not. (Fig. 1751)
Most collections of this species have been made on dead stems of *Agropyron* and *Arrhenatherum*; it occurs also on *Bromus*, *Dactylis*, *Deschampsia* and *Elymus*.

Lophodermium gramineum (Fr.)Chev. (Fig. 1752)
On dead leaves of *Agropyron*, *Ammophila*, *Brachypodium*, *Deschampsia*, *Glyceria*, *Molinia*, etc., very common everywhere.

Micropeziza. Apothecia sessile, about 0.5mm diam., pale brown with still paler discs. They develop through little mats or shields of pale or very dark brown radiating hyphae, and retain small bits of these around their rims. Spores hyaline.

KEY

Spores 1-septate, 2.5–3 wide .... *karstenii*
Spores without septa, multiguttulate, 1.5–2 wide .... *poae*

Micropeziza karstenii Nannf. (Fig. 1753)
Hyphae of shields very dark brown. Ascospores 11–15 × 2.5–3. On dead stems of *Brachypodium, Bromus, Deschampsia, Phalaris*, etc., and collected once also on *Juncus articulatus*; July–Dec.

Micropeziza poae Fuckel (Fig. 1754)
Hyphae of shields pale brown, thin-walled. Ascospores 11–13 × 1.5–2. On dead stems of *Molinia* and *Poa*, Aug.

Mollisia. Apothecia saucer-shaped, brown or grey. Ascospores hyaline.

KEY

Spores 5–7 × 1, curved .... *caricina*
Spores 8–13 × 1–2, without guttules, or with one at each end .. *palustris*
Spores 7–10 × 2–2.5, with many small guttules .... *poaeoides*

Mollisia caricina, described on *Carex*, is found also on grasses in wet places.

Mollisia palustris, described on *Juncus*, is common on dead stems of grasses, including *Arrhenatherum, Deschampsia, Glyceria, Molinia, Phalaris* and *Phragmites*, in damp situations, Mar.–Sept.

Mollisia poaeoides Rehm (Fig. 1755)
Apothecia 0.2–0.4mm diam., pale grey, with dark brown margins. On dead culms of *Ammophila, Molinia* and *Poa*, Nov.–Feb.

Pezizella eburnea (Rob. ex Desm.)Dennis (Fig. 1756)
Apothecia 0.2mm diam., short-stalked, white, with a downy fringe. Ascospores hyaline, 6–8 × 1–1.5. Common on dead stems and leaves of *Agropyron, Ammophila, Arrhenatherum, Dactylis, Deschampsia, Festuca, Glyceria* and *Holcus*, and seen occasionally on *Carex* and *Scirpus*, Apr.–Sept. *P. turgidella* (P. Karsten)Sacc. has apothecia up to 0.4mm diam. and ascospores 8–12 × 1.5–2.

OTHER ASCOMYCETES

KEY TO GENERA

Black banana-shaped sclerotia from inflorescences .... *Claviceps*
Not so 1
1. Stromata present 2
   No stromata 3
2. Stromata immersed, black, appearing as flecks on both leaf surfaces

	.... *Phyllachora*
	Stromata white or yellow, encircling living stems and forming thick collars, embedded perithecia orange .... *Epichloe*
	Stromata pulvinate, white, dotted with olive-green blobs .... *Creopus*
3.	Ascomata superficial 4
	Ascomata all or mostly immersed in substratum 9
4.	Ascomata erect, brown, club-shaped or navicular with small pore at apex .... *Acrospermum*
	Ascomata more or less spherical, barrel-shaped or flattened 5
5.	Colonies white or buff, powdery, ascocarps (cleistothecia) appendaged .... *Erysiphe*
	Colonies not so 6
6.	Ascomata setose 7
	Ascomata not setose 8
7.	Ascomata (perithecia) barrel-shaped, yellowish brown with a ring of simple, pale brown setae at top .... *Tubeufia*
	Ascomata (perithecia) spherical or flask-shaped, almost black with dark brown, often branched or coiled setae .... *Chaetomium*
8.	Ascomata flat, dark, radiate thyriothecia with ostiolar collars .... *Trichothyrina*
	Ascomata spherical, walls violet or bluish grey by transmitted light .... *Gibberella*
9.	Necks of ascomata (pseudothecia) laterally compressed .... *Lophiostoma*
	Necks not laterally compressed 10
10.	Ascomata hairy or setose 11
	Ascomata smooth 13
11.	Ascospores 70–115 × 3–4, pseudothecia hairy .... *Ophiosphaerella*
	Ascospores shorter and broader, pseudothecia with setae around ostiole 12
12.	Ascospores with transverse and longitudinal septa .... *Pyrenophora*
	Ascospores with transverse septa only .... *Keissleriella*
13.	Dark brown clypeus present .... *Paradidymella*
	No clypeus 14
14.	Ascospores with transverse and longitudinal septa .... *Pleospora*
	Ascospores with transverse septa only 15
	Ascospores without septa 17
15.	Ascospores 1-septate, hyaline .... *Mycosphaerella*
	Ascospores 1- to 3-septate, hyaline .... *Monographella*
	Ascospores 2-septate, golden brown .... *Paraphaeosphaeria*
	Ascospores with 3 or more septa, mostly straw-coloured or pale golden brown 16
16.	Ascospores not more than 50 long .... *Leptosphaeria*
	Ascospores 70–150 long, hyaline or straw-coloured .... *Gaeumannomyces*
17.	Ascospores 26–30 × 8–10 .... *Botryosphaeria*
	Ascospores 7–16 × 2–4 .... *Phomatospora*

Acrospermum graminum Lib. (Fig. 1757)
Ascocarps erect, brown, with small pore at top, less than 1mm high. Asci very long and narrow; spores hyaline, thread-like. On dead stems of *Brachypodium, Festuca, Poa*, etc.

Botryosphaeria festucae (Lib.)v.Arx & E.Müller (Fig. 1758)
Pseudothecia up to 0.35mm diam., immersed, with just the rounded necks showing. Ascospores hyaline to pale yellow, mostly 26–30 × 8–10. On dead leaves and stems of *Agrostis, Festuca, Molinia, Phragmites* and *Poa*, June–Aug.

Chaetomium. Species belonging to this genus have dark brown to black perithecia clothed, especially in the upper part, by dark brown setae. These setae may be simple or branched, straight, wavy or spirally coiled, smooth or ornamented in various ways. The lemon-shaped, brown ascospores are freed from their asci inside the perithecia, and emerge through the ostiole to be held by the terminal setae in large, very dark masses.

Chaetomium elatum Kunze ex Grev. (Fig. 1759)
This is the commonest species in piles of damp hay and straw; its terminal setae are dichotomously branched, and the ascospores measure about 12 × 8–9. *C. globosum* Kunze and *C. cochliodes* Palliser are found sometimes in the same habitat; in the former the setae are straight, slightly curved or wavy, and in the latter coiled or contorted setae are mixed with other types.

Claviceps purpurea (Fr.)Tul. (Fig. 1760)
Ergot is the name given to the sclerotia of this fungus, small, black, banana-shaped bodies found projecting from inflorescences of *Agropyron, Agrostis, Alopecurus, Ammophila, Anthoxanthum, Avena, Brachypodium, Bromus, Dactylis, Deschampsia, Elymus, Festuca, Holcus, Lolium, Molinia, Nardus* and *Secale*, Sept.–Nov. These fall to the ground, and in late spring or early summer germinate to produce one or a number of creamy to purplish drumstick-like stromata, in the heads of which are embedded little flask-shaped perithecia, containing numerous long, cylindrical asci, each with 8 slender, hyaline spores. Germinated sclerotia are found sometimes loose among wet vegetation. They can also be grown quite easily. In autumn collect some infected grass seed (*Lolium* is excellent), and scatter it over the surface of soil in a flower-pot. Place outside in the garden for the winter. Examine after a shower of rain towards the end of June, when the ergots develop stalked stromata with mature perithecia. If these are brought indoors, and placed under a dissecting microscope in sunlight, you can watch the needle-like ascospores being shot up into the air, where they float around. When similar spore liberation takes place in a field where grasses are in flower, ascospores, in the same way as pollen grains, are caught by their projecting feathery stigmas; they germinate and germ tubes penetrate to the ovaries. Numerous colourless conidia, about 5–7 × 3–4, belonging to the *Sphacelia* state are budded off from the mycelium. In response to infection the grass produces

sugary honeydew in which the conidia become immersed. Flies and other insects attracted by the honeydew help to spread the fungus. The *Sphacelia* state should be looked for in July, when parts of inflorescences are covered in orange slime. Sometimes other fungi such as *Fusarium* and *Cladosporium* are present also. Eventually new ergots develop in the infected ovaries.

Creopus spinulosus (Fuckel)Moravec (Fig. 1761)
Stromata pulvinate, white, dotted with small olive-green blobs which show the position of immersed perithecia. Ascospores break into two parts at the septum, and are green in mass; part-spores measure 4–6 × 2–3. On decaying leaves and stems of *Ammophila*, *Dactylis*, *Deschampsia*, *Glyceria* and other grasses, also occasionally on *Juncus*, Aug.–Nov.

Epichloe typhina (Pers.)Tul. (Fig. 1762)
Choke disease. The cylindrical, at first white, then yellow, stromata are found in summer encircling living stems of grasses and forming thick collars up to 5cm long. Numerous orange or rusty-orange perithecia develop in and project above the surface of the stromata. Good material with mature asci can be collected in August. Ascospores hyaline, multiseptate, about 150 × 2, sometimes break into pieces. White collars seen earlier in the year bear only small hyaline conidia. Infected plants rarely flower. On *Agrostis*, *Bromus*, *Dactylis*, *Festuca*, *Holcus*, *Phleum* and *Poa*.

Erysiphe graminis DC. (Fig. 1763)
Powdery mildew. Colonies effuse, white or buff, mostly on upper surface of living leaves, surrounded by yellow haloes. In the *Oidium* state, found from Feb. onwards, the hyaline, barrel-shaped conidia, 20–30 × 13–17, are formed in long chains on erect conidiophores; their ability to germinate in conditions of low humidity enables the fungus to spread rapidly in dry weather. Cleistothecia formed mostly on basal leaves and sheaths, July–Oct, 0.15–0.3mm diam., contain 10 to 20 asci, each with 8 hyaline spores 20–24 × 10–13. Different physiologic races occur on *Agropyron*, *Avena*, *Bromus*, *Dactylis*, *Festuca*, *Hordeum*, *Lolium*, *Secale* and *Triticum*.

Gaeumannomyces graminis (Sacc.)v.Arx & Olivier
In one or more of its varietal forms this species has been found on a number of different grasses including *Arrhenatherum*, *Avena*, *Dactylis*, *Deschampsia*, *Hordeum* and *Triticum*; also on *Carex acutiformis* and *C. pendula*, Mar.–Oct. Perithecia black, up to 0.5mm diam., immersed in lower parts of leaf sheaths and stem bases, with their necks protruding. Ascospores hyaline to pale yellowish brown, septate. Brown, often lobed hyphopodia are borne on hyphae which form superficial dark mycelial mats surrounding perithecia.

G. graminis var. *avenae*, described on *Avena*, has ascospores mostly 120–150 × 3.

G. graminis var. *graminis* (Fig. 1764)
Ascospores 70–100 × 2.5.

G. graminis var. *tritici* Walker (Fig. 1765)
Ascospores 80–120 × 2.

Gibberella zeae (Schw.)Petch (Fig. 1766)
Perithecia clustered around lower nodes and stem bases, up to 0.25mm diam., black, with walls violet or bluish grey by transmitted light. Ascospores hyaline or pale straw-coloured, 3-septate, 18–25 × 3–4.5. *Fusarium* state has hyaline, curved, 3- to 7-septate conidia mostly 25–55 × 4–5, with foot cells. On *Arrhenatherum, Avena, Glyceria, Hordeum, Phragmites, Triticum*, etc.

Keissleriella culmifida (P.Karsten)Bose (Fig. 1767)
Pseudothecia immersed, 0.2–0.4mm diam., black, with very short necks which protrude through small pale areas; short, thick-walled setae surround the ostioles. Ascospores hyaline or pale yellow, 3-septate, 21–26 × 5–6. On dead leaves and stems of various grasses including *Agropyron, Agrostis, Dactylis, Elymus* and *Phleum*; occasionally also on *Typha*, May–Oct.

Leptosphaeria. Nine plurivorous species occur on grasses, and there are many more which are host-limited. They should be looked for in spring and summer. The pseudothecia are usually scattered, black, at first immersed, with necks just showing at the surface. Ascospores are septate and mostly pale to dark straw-coloured.

KEY TO PLURIVOROUS SPECIES

	Spores with 3 septa 1
	Spores all or mostly with 5 septa 2
	Spores with 6 to 9 septa 3
	Spores mostly with 10 septa, 35–45 × 5–6 .... *graminis*
1.	Spores 18–20 × 4–5 .... *tritici*
	Spores 19–25 × 4 .... *nodorum*
	Spores 20–30 × 7–8 .... *eustoma*
2.	Spores 17–24 × 3–4, second cell enlarged .... *nigrans*
	Spores 20–30 × 4–5, fourth cell short and fat .... *fuckelii*
	Spores 26–30 × 5.5–7 .... *luctuosa*
3.	Spores 25–33 × 4–6 .... *herpotrichoides*
	Spores 33–45 × 5–7 .... *culmifraga*

Leptosphaeria culmifraga (Fr.)Ces. & de Not. (Fig. 1768)
On dead stems of *Agrostis, Calamagrostis, Deschampsia, Festuca, Poa*, etc.

Leptosphaeria eustoma (Fuckel)Sacc. (Fig. 1769)
Very common, not only on grasses including *Dactylis, Deschampsia, Festuca, Glyceria* and *Triticum*, but also on other monocotyledons such as *Schoenoplectus* and *Typha*.

Leptosphaeria fuckelii Niessl ex Voss (Fig. 1770)
Especially common on stems of *Phalaris arundinacea*, but not uncommon also on *Arrhenatherum, Calamagrostis, Deschampsia* and *Phragmites*.

Leptosphaeria graminis (Fuckel)Sacc. (Fig. 1771)
Mostly on *Phragmites*; found occasionally on *Elymus, Phalaris* and *Secale*.

Leptosphaeria herpotrichoides de Not. (Fig. 1772)
On dead leaves and stems of *Deschampsia, Phalaris* and *Phragmites*; also on *Carex* and *Luzula*.

Leptosphaeria luctuosa Niessl ex Sacc. (Fig. 1773)
We have found this only on dead stems of *Deschampsia caespitosa*, but there are British records also on *Ammophila, Calamagrostis* and *Dactylis*.

Leptosphaeria nigrans (Rob. ex Desm.)Ces. & de Not. (Fig. 1774)
Common on dead stems and leaves of *Dactylis, Deschampsia, Elymus, Festuca, Glyceria*, etc.

Leptosphaeria nodorum E.Müller (Fig. 1775)
On dead stems, especially at the nodes, often with its *Septoria* state which has hyaline, 3-septate conidia 25–28 × 2–2.5. Hosts include *Agropyron, Ammophila, Arrhenatherum, Dactylis, Deschampsia, Festuca, Hordeum, Poa* and *Triticum*.

Leptosphaeria tritici (Garovaglio)Pass.
Recorded on *Dactylis* and *Lolium*.

Lophiostoma semiliberum (Desm.)Ces. & de Not. (Fig. 1776)
Pseudothecia black, 0.7–1mm diam., at first mostly immersed, but soon becoming partly erumpent and sometimes quite free on the surface. The large, laterally compressed necks lie parallel to the long axis of the stem and have little ridges on them. Ascospores mostly 30–40 × 6–8, long remaining hyaline and constricted at the median septum, but eventually a few of them become straw-coloured and 3-septate. On *Agropyron* (9 collections), *Brachypodium, Dactylis, Deschampsia, Phalaris* and *Phragmites* (41 collections), Dec.–Apr.

Monographella nivalis (Schaffnit)E. Müller & v.Arx (Fig. 1777)
Snow mould. Perithecia immersed, black, usually less than 0.2mm diam., most commonly located in the leaf sheath close to the base of the stem. Ascospores hyaline, 1- to 3-septate, 12–18 × 3–4. *Fusarium* state has 1- to 3-septate conidia 10–30 × 3–5. On *Agropyron, Avena, Hordeum, Triticum* and various turf grasses (see also under *Fusarium*).

Mycosphaerella. Three plurivorous species are found on grasses. They all have immersed black pseudothecia which contain asci arranged in fascicles, and 1-septate, hyaline ascospores.

KEY

Spores 11–14 × 3–4 .... *recutita*
Spores 15–18 × 3–4.5 .... *lineolata*
Spores 18–25 × 5–7 .... *allicina*

Mycosphaerella allicina (Fr.)Vestergren
Recorded on various grasses, including *Avena* and *Triticum*; also on *Carex* and on other herbaceous plants.

Mycosphaerella lineolata (Rob. ex Desm.)Schröter (Fig. 1778)
Pseudothecia 0.1mm diam., arranged in long rows. On dead stems and leaves of grasses including *Ammophila*, *Deschampsia* and *Phragmites*; also occasionally on *Carex*, *Eriophorum* and *Typha*.

Mycosphaerella recutita (Fr.)Johanson
On *Dactylis*, *Deschampsia*, etc.; recorded also on *Juncus*.

Ophiosphaerella herpotricha (Fr.)Walker
Pseudothecia immersed, with protuberant necks, black, up to 0.5mm diam., covered with dark brown, thick-walled hairs. Ascospores filiform, straw-coloured or pale brown, multiguttulate, with up to 20 septa, 120–180 × 2–3. Pycnidia sometimes accompany pseudothecia and are about the same size; conidia golden brown in mass, obclavate–rostrate, 5- to 6-septate, 70–115 × 3–4. On basal internodes of *Agropyron*, *Arrhenatherum*, *Bromus*, *Festuca*, *Holcus*, *Hordeum* and *Triticum*, Mar.–July.

Paradidymella holci D.Hawksw. & Sivan. (Fig. 1779)
Perithecia immersed, singly or in pairs, with dark brown clypeus. Ascospores 1-septate, hyaline or subhyaline, smooth or slightly rough-walled, 10–12 × 2.5–3.5. On dead stems; recorded originally on *Holcus lanatus*. Stems of the part of the type that we examined, with ripe perithecia embedded in them, were quite hairless and belonged to some other grass.

Paraphaeosphaeria michotii, with brown, mostly 2-septate, ascospores 14–22 × 4–5, described on *Cladium* is found quite commonly also on grasses, including *Agropyron*, *Dactylis*, *Deschampsia*, *Festuca* and *Nardus*.

Phomatospora. Perithecia small, dark brown to black, spherical, immersed, with just the tips of short necks showing on the surface where the surrounding tissue is often pale and greyish. Ascospores hyaline, without septa.

KEY

Spores 7–11 × 2–3 .... *berkeleyi*
Spores 9–16 × 3–4 .... *dinemasporium*

Phomatospora berkeleyi Sacc. (Fig. 1780)
Perithecia 0.2mm diam., inconspicuous. On dead stems of *Dactylis*, *Deschampsia*, *Festuca*, *Phragmites*, etc., Feb.–Sept.

Phomatospora dinemasporium Webster (Fig. 1781)
Perithecia 0.3mm diam., collected mostly in June and July. Asci with 4 or 8 spores. Best known in its *Dinemasporium* conidial state which is found all the year round on dead stems, sheaths and leaves of grasses including *Agropyron, Agrostis, Arrhenatherum, Brachypodium, Bromus, Festuca, Molinia, Phalaris* and *Poa*; also on rushes, sedges and many other plants. The conidiomata are superficial, black, setose, at first closed and more or less spherical, but opening out and becoming cup-shaped, about 0.2mm diam. Setae dark brown, up to 300 × 4–7. Conidia slightly curved, 8–19 × 1.5–2.5, with a setula at each end.

Phyllachora graminis (Pers.)Fuckel (Fig. 1782)
Visible on both surfaces of living and dead leaves as black flecks about 2–3 × 1mm, often surrounded by yellow haloes. These flecks are the outer crusts of stromata in which are embedded a number of small perithecia, sometimes accompanied or replaced by pycnidia. Ascospores hyaline, 10–15 × 5–6. Conidia curved, needle-like, hyaline, 12–18 × 0.5–0.7. Very common on *Agropyron, Agrostis, Brachypodium, Bromus* and occasionally other grasses, but not *Dactylis* and *Festuca*, which have their own *Phyllachora* species.

Pleospora. Seven plurivorous species occur on grasses. Pseudothecia mostly immersed, small, black, often with very short papillate ostioles; ascospores usually pale straw-coloured to golden brown, with transverse and longitudinal septa.

KEY

	Spores with 5 transverse septa	1
	Spores with 7 transverse septa, 25–40 × 7–10	*herbarum*
	Spores mostly with more than 7 transverse septa, substrate often stained reddish purple	3
1.	Spores often more than 25 long (20–32 × 7–10)	*vagans*
	Spores not more than 24 long	2
2.	End cells of spores without longitudinal septa	*infectoria*
	End cells often with a longitudinal septum	*phaeocomoides*
3.	Spores with 6 to 9 transverse septa, 20–26 × 6–8	*rubelloides*
	Spores with 9 to 11 transverse septa, 26–34 × 10–12	*straminis*
	Spores with 9 to 13 transverse septa, 30–40 × 8–12	*rubicunda*

Pleospora herbarum, one of the commonest of all microfungi, not only on many different grasses but on herbaceous plants of all kinds, and sometimes on trees and shrubs. It is described and figured under the general heading 'herbaceous plants'. There is a *Stemphylium* conidial state which is equally common. Pseudothecia with ripe ascospores are most plentiful in spring and early summer.

Pleospora infectoria Fuckel (Fig. 1783)
Ascospores yellowish, 18–23 × 6–8. On dead stems of *Dactylis, Deschampsia, Hordeum, Triticum* and *Zea*; also on *Typha*, Feb.–Oct. The *Alternaria* conidial state is found on leaves as well as stems.

Pleospora phaeocomoides, a plurivorous species described on 'herbaceous plants', occurs not infrequently on dead stems of grasses including *Arrhenatherum, Bromus, Dactylis, Elymus, Phleum* and *Triticum*, Feb.–Oct.

Pleospora rubelloides (Plowr. ex Cooke)Webster (Fig. 1784)
Ascospores yellow to brown, flattened on one side, curved on the other, tapered at ends. On dead stems of *Agropyron, Ammophila, Dactylis, Elymus* and *Festuca*, also on *Scirpus*; Apr.–Aug.

Pleospora rubicunda Niessl (Fig. 1785)
Ascospores mid pale golden brown, tapered to the rounded ends. On dead stems of grasses, including *Elymus*, Mar.–May.

Pleospora straminis Sacc. & Speg. (Fig. 1786)
Ascospores golden brown, oblong rounded at ends. On dead stems of *Dactylis, Hordeum, Lolium* and *Triticum*, Apr.–Sept.

Pleospora vagans Niessl (Fig. 1787)
Ascospores always rather pointed at the ends, pale lemon yellow or straw-coloured. It has a *Hendersonia* pycnidial state with 3- to 7-septate, pale straw-coloured conidia 25–45 × 4–6. Common on dead stems of many different grasses, including *Agropyron, Agrostis, Alopecurus, Ammophila, Anthoxanthum, Arrhenatherum, Avena, Brachypodium, Bromus, Calamagrostis, Dactylis, Deschampsia, Elymus, Festuca, Holcus, Hordeum, Lolium, Phalaris, Phragmites* and *Triticum*; and occasionally on other monocotyledons, e.g. *Typha*, Apr.–Aug.

Pyrenophora trichostoma (Fr.)Fuckel (Fig. 1788)
Pseudothecia scattered or in small groups, immersed, up to 0.5mm diam., with brown, septate setae around the ostioles. Ascospores pale golden brown, smooth, with 3 transverse and 1 or 2 longitudinal septa, 40–55 × 17–25. On dead stems and sheaths of *Briza, Bromus, Dactylis, Secale*, etc.

Trichothyrina. Two species, *T. alpestris* and *T. nigroannulata*, occur mainly on various grasses but also occasionally on other substrata; *T. ammophilae* grows only on *Ammophila*. The flattened, radiate thyriothecia have well-defined ostiolar collars and the ascospores are hyaline, 1-septate.

KEY

Ostiolar collar of thyriothecium with a ring of holes at the base, often surmounted by a number of acute convergent setae .... *alpestris*
Ostiolar collar composed of several rings of small, dark, thick-walled cells without holes or setae 1

1. Ascospores fusiform, without cilia .... *nigroannulata*
   Ascospores cylindrical, often with cilia .... *ammophilae*

Trichothyrina alpestris (Sacc.)Petrak (Fig. 1789)
Thyriothecia up to 0.2mm diam. Ascospores hyaline, 11–15 × 2.5–3.5, cilia in a cluster of 6, attached just above the septum. On dead stems of *Arrhenatherum*, *Dactylis*, *Deschampsia*, *Molinia* and *Phalaris*, also on *Carex*; Apr.–Nov.

Trichothyrina ammophilae, described on *Ammophila*.

Trichothyrina nigroannulata (Webster)J.P.Ellis (Fig. 1790)
Thryiothecia 60–90 diam., ostiolar collar of 1 to 3 rings of dark brown, thick-walled cells. Ascospores 7.5–10.5 × 1.5–3. Mainly on dead stems and leaves of grasses, including *Ammophila*, *Dactylis* and *Phragmites*, but also on *Carex*, *Juncus*, *Pteridium* and leaves of deciduous trees.

Tubeufia helicomyces Höhnel (Fig. 1791)
Colonies effuse, raised, cottony, pale grey to pale brown, tinged with pink when conidia of the *Helicosporium* state are being produced freely. Perithecia like little yellowish-brown barrels up to 0.35mm high, each with a ring of pale brown setae 50–100 long at the top. Ascospores hyaline, 7- to 9-septate, 70–140 × 2–3. Conidia hyaline, coiled, flat, septate, 14–21 diam., with filaments 2–2.5 thick. On dead leaves and stems of grasses such as *Arrhenatherum*, *Dactylis*, *Deschampsia*, *Festuca*, *Phalaris* and *Phragmites* in wet places, Mar.–Nov.; especially common on decaying culms of *Glyceria maxima* lying in the very damp basal leaf mat, where both states can be found together in good condition, June–Aug.

HYPHOMYCETES

Alternaria. Colonies effuse, grey, blackish brown or black, superficial. Conidia in chains, rostrate, brown, with transverse and usually also oblique or longitudinal septa. *Alternaria* species on grasses are not as a rule parasitic, are found mostly in company with other fungi, and seldom form pure colonies. *A. alternata*, a plurivorous species described on 'herbaceous plants', with long chains of short-beaked, rough-walled conidia mostly 30–50 × 10–17, and *A. tenuissima*, another plurivorous species with nearly smooth conidia mostly 50–90 × 8–16, in shorter chains and with longer beaks, are found occasionally. Much more frequently found than either of these is the *Alternaria* state of *Pleospora infectoria* (*v.s.*), which forms long chains of rough-walled conidia 20–70 × 8–14, some with short, others with long, beaks.

Arthrinium. Two species, *A. phaeospermum* (Corda)M.B.Ellis (Fig. 1792) and the *Arthrinium* state of *Apiospora montagnei* Sacc. (Fig. 1793), are extremely common on dead stems and are less frequently found on dead leaves of many different grasses throughout the year, sporulation being perhaps most abundant in summer and early autumn. Young colonies appear as small black heaps of conidia, but later join up, may completely encircle stems and extend along them for up to 10cm. The conidia are a very characteristic shape, resembling two watch-glasses placed face to face, the rim hyaline and the sides brown; in *A. phaeospermum* they measure 8–12 in face view and are 5–7 thick; corresponding measurements for the other species are 5.5–8 and 3–4.5. Genera of grasses on which we have found these two species include *Agropyron, Ammophila, Arrhenatherum, Arundinaria, Bambusa, Bromus, Calamagrostis, Dactylis, Deschampsia, Elymus, Glyceria, Phalaris, Phragmites, Spartina* and *Triticum*. They grow also on sedges and occasionally other plants.

Cercosporidium graminis (Fuckel)Deighton (Fig. 1794)
Conidiophores in fascicles emerging through stomata. Conidia 1- to 2-septate, pale olive or brown, 21–44 × 10–12. Causes chocolate-coloured spots with pale centres on leaves and sheaths. On various grasses including *Alopecurus geniculatus, Dactylis glomerata, Glyceria declinata, G. fluitans, Poa compressa* and *P. trivialis*.

Cladosporium. Three species of *Cladosporium* described more fully on 'herbaceous plants' are found commonly on grasses. They grow mostly on dead leaves and stems which have been rain soaked, and form extensive olivaceous, rather velvety colonies. Tips of leaves which may have been nipped by frost or damaged in some other way often become infected, and colonies sometimes spread a little way on to living parts. Conidia in chains, olivaceous, 0- to 3-septate, with scars at ends. *C. cladosporioides* has ramoconidia 0- to 1-septate, up to 30 × 2–5; conidia in long, branched chains, mostly smooth and without septa, 3–7 × 2–4. *C. herbarum* and *C. macrocarpum* have nodose conidiophores. In *C. herbarum*, conidia are rather thick-walled, distinctly verruculose with low warts, nearly always 0- or 1-septate, 8–15 × 4–6. In *C. macrocarpum*, conidia are densely verrucose, thick-walled, 0- to 3-septate, mostly 15–25 × 7–10.

Dictyosporium elegans Corda (Fig. 1795)
This curious fungus with brown conidia shaped rather like a hand with the fingers held close together, and measuring 50–80 × 24–31, has been found on several occasions on stubble of *Avena* and *Hordeum*.

Drechslera biseptata (Sacc. & Roum.)Richardson & Fraser (Fig. 1796)
Conidiophores of two kinds, flexuous or geniculate ones which are thin-walled and up to 80 × 3–8, and subulate, dark brown, thick-walled ones up to 800 × 8–14. Conidia pale to mid brown, smooth or verrucose, with 2 or 3 pseudosepta, mostly 23–33 × 14–17. Not uncommon on dead stems, leaves and inflorescences of *Avena, Dactylis, Triticum*, etc.

Drechslera dematioidea (Bubák & Wróblewski)Subram. & Jain (Fig. 1797)
Conidiophores brown, up to 350 × 5–9. Conidia mostly with 3 or 4

pseudosepta, thick-walled, rather dark brown, 20–70 × 10–16. On dead leaves and inflorescences of *Anthoxanthum, Avena, Holcus* and *Lolium*.

Epicoccum purpurascens, described on 'herbaceous plants', commonly accompanies other fungi on dead parts of grasses. The small, black, pulvinate sporodochia are often seated on host tissues stained reddish purple.

Fusarium species are found quite frequently on grasses, but as their accurate identification often necessitates their being grown in culture, they are not dealt with in any detail here. They all have hyaline, curved, multiseptate conidia which in mass appear pinkish or pale orange. The *Fusarium* state of *Gibberella zeae* (Fr.)Sacc. (Fig. 1766) occurs quite commonly in the inflorescences of grasses such as *Arrhenatherum elatius*, often accompanying the *Sphacelia* state of *Claviceps purpurea*. Its conidia measure mostly 25–50 × 3–5. One of the commonest turf diseases is caused by the *Fusarium* state of *Monographella nivalis* (*v.s.*), the so-called snow mould. The effect of this fungus on turf is very striking: circular bleached or brown patches appear, which later may become quite bare. The main grasses affected are *Agrostis stolonifera, Festuca ovina, F. rubra* and *Poa annua*.

Helicosporium, *v.s.* under *Tubeufia helicomyces*.

Myrothecium. Sporodochia sessile, stalked or cupulate; conidia pale olive, sometimes with striate walls, form large, dark green to black masses, which are slimy when wet, and horny when dry.

KEY

Sporodochia cupulate .... *atroviride*
Sporodochia stalked or sessile, not cupulate 1
1. Conidia with striate walls .... *cinctum*
   Conidial walls not striate .... *masonii*

Myrothecium atroviride (Berk. & Br.)Tulloch (Fig. 1798)
Sporodochia up to 0.8mm diam., olivaceous black. Conidia 9–12 × 2–3. On *Deschampsia caespitosa* and unnamed grasses, uncommon.

Myrothecium cinctum (Corda)Sacc. (Fig. 1799)
Sporodochia sometimes sessile, with slimy black spore masses, each surrounded by a broad white hyphal fringe, but more often stalked. Conidia 7–12 × 2.5–3.5, with longitudinally striate walls. A marsh and fen species, very common on dead leaves and stems of *Glyceria* and *Phragmites*; also on *Carex* and *Typha*, Mar.–May.

Myrothecium masonii Tulloch (Fig. 1800)
Sporodochia stalked, up to 1mm high and 20–70 thick, white to dark grey, with slimy black heads. Conidia pale olive, 4.5–9 × 1.5–2.5. On *Glyceria* and *Phragmites*; also on rushes and sedges.

Periconia. Brown, rough-walled, spherical conidia in chains, which are often branched, typify the genus, and in many species these form round heads on top of long brown or black stalks. Eleven species have been recorded on grasses in Britain. Most of these are found much more commonly on other substrata and are described on these: e.g. *P. atra* and *P. digitata* are rush and sedge species, and *P. typhicola* grows mainly on *Sparganium* and *Typha*; *P. byssoides* is typically on larger herbaceous stems and very rarely occurs on grasses. *P. circinata* and *P. macrospinosa* have been found only on *Triticum*. There are two very common plurivorous species, *P. hispidula* and *P. minutissima*; two much less common ones, *P. britannica* and *P. glyceriicola*; *P. igniaria* is not uncommon where burning has taken place.

KEY TO PLURIVOROUS SPECIES

    Conidiophores with well-defined conidial heads 1
    Conidiophores not so 3
1. Heads unilateral, usually without branches .... *britannica*
    Heads terminal, with branches inside 2
2. Conidia verruculose, 4–6 diam. .... *minutissima*
    Conidia echinulate, 7–10 diam. .... *igniaria*
3. Conidiophores forming small, black, bristly mats .... *hispidula*
    Conidiophores widely scattered, forming thin, pale colonies ... *glyceriicola*

Periconia britannica M.B.Ellis (Fig. 1801)
Colonies effuse, black, hairy. Conidiophores up to 1200 × 8–15. Conidia brown, becoming verruculose, 4–7 diam. On dead stems of *Brachypodium sylvaticum* and undetermined grasses, May–July.

Periconia glyceriicola Mason & M.B.Ellis (Fig. 1802)
Conidiophores up to 350 × 5–8, pale brown, with 3 to 5 whorls of branches, or single branches, arising at different levels below the apex. Conidia pale olive to brown, verruculose, 6–9 diam. On *Festuca*, *Glyceria* and *Phragmites*, Dec.–Apr.

Periconia hispidula (Pers.)Mason & M.B.Ellis (Fig. 1803)
Conidiophores subulate, up to 900 × 5–8 (12 at base), dark brown. Conidia thick-walled, shortly echinulate, 10–16 diam., formed in branched chains in exactly the same way as in all other species of *Periconia*, but from the middle of the conidiophore, not the apex, which is sterile and setose. Found most frequently on dry dead leaves of *Glyceria maxima* but commonly also on other grasses, including *Calamagrostis*, *Deschampsia*, *Festuca*, *Phalaris*, *Phragmites* and *Sesleria*, and on sedges, throughout the year.

Periconia igniaria Mason & M.B.Ellis (Fig. 1804)
Conidiophores up to 550 × 6–10, branched inside the head, with the branches mostly rough-walled. Conidia brown, echinulate, 7–10 diam. Found

usually on plants which have been scorched or prematurely killed by burning. Grass hosts include *Ammophila, Dactylis, Phalaris* and *Phragmites*, Aug.–Dec.

Periconia minutissima, described on 'herbaceous plants', a typical *Periconia* with conidiophores like tiny black round-headed pins. On dead stems and leaves of *Agropyron, Arrhenatherum, Arundinaria, Bambusa, Glyceria, Holcus, Phalaris* and *Phragmites* throughout the year.

Pithomyces chartarum (Berk. & Curt.)M.B.Ellis (Fig. 1805)
Colonies black, at first round, up to 0.5mm diam., later sometimes becoming confluent. Conidia mid to dark brown, echinulate or verruculose, mostly with 3 transverse and 1 or 2 longitudinal septa, 18–29 × 10–17, each with a protruding fractured denticle at its base. On dead leaves of *Holcus* and other grasses, not common generally in Britain, but where conditions are right it occurs sometimes in great abundance.

Pseudocercosporella herpotrichoides (Fron)Deighton (Fig. 1806)
Conidiophores simple or occasionally branched, up to 40 × 2–3.5, colourless to pale olive, arising from immersed, pale brown stromata and emerging in large tufts through stomata. Conidia hyaline, 1- to 7-septate, 20–70 × 1–3. Lesions pale, with dark borders, on leaf blades, bases, sheaths and stems of *Agropyron, Bromus, Hordeum, Poa, Secale, Triticum*, etc. It is considered to be one of the important causes of foot rot in wheat.

Rhynchosporium orthosporum, described on *Dactylis*, has been found occasionally also on *Agrostis* and *Lolium*.

Rhynchosporium secalis (Oud.)Davis (Fig. 1807)
Leaf blotch or scald. The blotches are irregular in shape, up to 2.5cm long, at first water-soaked, later pale grey or greyish brown, sometimes zonate, and often with dark reddish-brown margins. Conidia beaked, hyaline, 1-septate, mostly 10–20 × 3–4, formed singly, or in small groups, on very short protrusions from cells of thin subcuticular stromatic plates. On leaves and sheaths of *Agropyron, Alopecurus, Hordeum, Lolium* and *Secale*, Apr.–Sept.

Rotula graminis (Desm.)Crane & Schoknecht (Fig. 1808)
Colonies dark brown, round or oval, up to 1.5 × 0.5mm. Conidia in very long, sometimes branched chains, which break up into segments with 0, 1, 2 or many septa, brown, minutely verruculose; individual cells broader than long, 4–5 × 4–6. Found in Suffolk on *Deschampsia caespitosa*, and recorded on other grasses, but uncommon.

Tetraploa aristata, described on *Cladium*, is very common also on grasses, including *Agropyron, Ammophila, Arrhenatherum, Arundinaria, Dactylis, Deschampsia, Phalaris* and *Phragmites*. Its curious conidia with four long arms are recognised easily.

Torula herbarum, described on 'herbaceous plants', is found frequently on dead stems and leaves of grasses, forming velvety, olive to black colonies varying in size from a few mm to several cm.

Ulocladium atrum and *U. chartarum*, described on 'herbaceous plants', are found frequently on grasses. They have brown, verruculose, muriform or cruciately septate conidia.

COELOMYCETES

Actinothyrium graminis, described on *Molinia*, is found also occasionally on leaves and stems of other grasses, including *Deschampsia*, *Holcus* and *Phalaris*.

Ascochyta. Three plurivorous species are found on grasses, easily distinguished by the size of their conidia, which are hyaline or pale straw-coloured and usually 1-septate.

KEY

Conidia 14–17 × 2.5–3 .... *leptospora*
Conidia 17–20 × 6–7 .... *avenae*
Conidia 24–32 × 8–10 .... *rhodesii*

Ascochyta avenae (Petrak)Sprague & Johnson
Leaf spots pale ochraceous with purplish-brown borders. Pycnidia immersed, golden brown, up to 0.2mm diam. Conidia ellipsoid, pale straw-coloured, occasionally 2-septate but mostly medianly 1-septate, smooth or minutely verruculose, 17–20 × 6–7. On *Avena*, *Hordeum*, *Lolium* and *Triticum*.

Ascochyta leptospora (Trail)Hara
Leaf spots fawn with reddish-purple borders. Pycnidia brown, about 0.2mm diam. Conidia more or less cylindrical, rounded at ends, some slightly curved, hyaline, 14–17 × 2.5–3, medianly 1-septate, rarely 2-septate. On *Agropyron*, *Agrostis*, *Ammophila*, *Avena*, *Festuca* and *Poa*.

Ascochyta rhodesii Punithalingam (Fig. 1809)
Leaf spots pale straw-coloured, inconspicuous. Pycnidia dark brown, up to 0.25mm diam. Conidia broadly ellipsoid, hyaline, 1-septate, 24–32 × 8–10. On *Agropyron*, *Deschampsia* and *Lolium*.

Colletotrichum graminicola (Ces.)Wilson (Fig. 1810)
Acervuli black, setose, up to 0.4 × 0.2mm, formed on pale yellow, purplish or grey leaf spots, and sometimes also on dead stems. Setae usually few, brown, up to 150 long, 5–7 thick near base. Conidia falcate, hyaline, 20–28 × 3.5–4.5. Has been collected on a few grasses, including *Agrostis*, *Glyceria* and *Poa*, but appears to be uncommon in Britain.

Dilophospora state of *Lidophia graminis* (Sacc.)Walker & B. Sutton (Fig. 1811)
Twist. Pycnidia 0.15–0.25mm diam., immersed, frequently aggregated in rows between veins, with stromatic tissue between them. Conidia hyaline, 3-septate

at maturity, 10–15 × 1.5–2, with branched appendages at each end. Often found on leaves which die back from the tips, but also on inflorescences which become much distorted and twisted. On *Agrostis, Alopecurus, Arrhenatherum, Dactylis, Holcus, Phalaris, Poa, Trisetum* and *Triticum*, May–Oct., sporulates abundantly in July.

Dinemasporium, v.s. under *Phomatospora dinemasporium*.

Hendersonia, v.s. under *Pleospora vagans*.

Pseudoseptoria donacis (Pass.)Sutton (Fig. 1812)
Pycnidia less than 0.1mm diam., dark brown, just piercing the epidermis, crowded or in rows. Leaf spots mostly near edges and tips, oval, 2–5mm long, very pale brown or grey, each with a narrow purplish-brown border. Conidia falcate, hyaline, 18–30 × 2–3.5. On leaves, sheaths and stems of many grasses, including *Agropyron, Agrostis, Alopecurus, Arrhenatherum, Avena, Elymus, Lolium, Phalaris, Phragmites, Poa, Secale* and *Triticum*, May–July.

Pseudoseptoria stomaticola (Bauml.)Sutton
Recorded on *Dactylis, Molinia* and *Phleum*, is similar to *P. donacis*, but has conidia 15–18 × 1.5–2.

Septoria, v.s. under *Leptosphaeria nodorum*.

Stagonospora subseriata (Desm.)Sacc. (Fig. 1813)
Pycnidia numerous, scattered or in short rows, brown to black, less than 0.2mm diam., just piercing the epidermis. Conidia hyaline with 5 to 8 guttules and, when mature, 3 to 6 (most commonly 4) septa, 30–40 × 5–6.5. On dead leaves and stems of *Deschampsia, Molinia* (common on this), *Nardus* and *Sieglingia*, Apr.–May.

FUNGI SPECIFIC TO GRASSES

AGROPYRON: Couch

Very few fungi specific to *Agropyron* are common but a number of plurivorous ones are found very frequently.

UREDINALES

Rusts include several formae speciales of plurivorous species and these are listed below under their hosts.

On *A. caninum*: *Puccinia striiformis*, 11, 111.
On *A. junceiforme*: *P. recondita* f.sp. *persistens*, 11, 111, rare.
On *A. repens*: *P. coronata*, 11, 111. *P. graminis*, 11, 111. *P. recondita* f.sp. *agropyrina* Eriks., 11, 111, for which no aecial state is known: the most common rust on this plant; small, cinnamon-brown uredinia are scattered over the upper surface of leaves, July–Aug. *P. recondita* f.sp. *persistens* Plowr., 11, 111, with aecia on *Thalictrum flavum* and *T. minus*, has been recorded occasionally.

USTILAGINALES

Stem smut, *Ustilago hypodytes*, is widespread and common on *A. caninum, A. junceiforme, A. pungens* and *A. repens*. The large-spored leaf stripe smut, *Ustilago serpens*, has been collected on *A. junceiforme* and *A. repens*, and leaves of *A. pungens* and *A. repens* are found commonly split into ribbons by *Urocystis agropyri*.

DISCOMYCETES

Belonopsis graminea (P.Karsten)Sacc. & Syd. (Fig. 1814)
Apothecia like pale brown or straw-coloured saucers less than 1mm diam., some with a white fringe, not easy to see as they are very much the colour of dead grass. Ascospores hyaline, 3-septate, 20–30 × 3–4. On dead stems of *A. repens*.

Crocicreas furvum (Graddon)S. Carp. (Fig. 1815)
Apothecia up to 1.5mm diam., long-stalked, dark brown, minutely scurfy, with ochraceous discs which have crenulate margins when dry. Ascospores hyaline, 14–20 × 2. On dead stems of *A. repens*, Aug.

Plurivorous species include *Calycella scolochloae* on *A. repens*; *Crocicreas culmicolum* on dead stems of *A. pungens*; *Dasyscyphus albotestaceus* on *A. pungens*; *D. carneolus* var. *longisporus* and *D. palearum* on *A. pungens* and *A. repens*; *Hymenoscyphus robustior* on *A. repens*; *Lophodermium arundinaceum*, *L. culmigenum* and *L. gramineum* on *A. repens*; *L. culmigenum* also on *A. pungens*; and *Pezizella eburnea* on *A.* × *obtusiusculum*.

OTHER ASCOMYCETES

Anthostomella chionostoma (Dur. & Mont.)Sacc. (Fig. 1816)
Perithecia raising epidermis, 0.4–0.6mm diam., with black clypeus the same width. Ascospores brown with wavy germ slits, 18–28 × 7–13. On dead leaves of *A. pungens*, Sept., uncommon.

Leptosphaeria pontiformis (Fuckel)Sacc. (Fig. 1817)
Perithecia 0.2–0.25mm diam., black, immersed, usually in short rows, raising the epidermis. Ascospores pale straw-coloured, 8- to 11-septate, 35–43 long, with the third or occasionally the fourth cell swollen and 4–5 wide. On basal internodes of *A. caninum* and *A. repens*, Jan.–Apr.

Micronectriella agropyri Apinis & Chesters
Perithecia immersed, golden brown, up to 0.15mm diam., with cylindrical necks up to 0.2mm long. Ascospores hyaline to pale straw-coloured, mostly 3-septate or occasionally with 5 septa, 16–22 × 4–6. On dead leaves of *A. pungens*.

Plurivorous species include *Claviceps purpurea* (ergot), often in great abundance on *A. pungens* and *A. repens*; *Erysiphe graminis* (powdery mildew), *Keissleriella culmifida*, *Leptosphaeria nodorum*, *Lophiostoma semiliberum*, *Monographella nivalis* and *Paraphaeosphaeria michotii* on *A. repens*; *Phyllachora graminis* on *A. caninum* and *A. repens*; *Pleospora rubelloides* on *A. junceiforme* and *A. repens*; and *P. vagans* on *A. repens*.

HYPHOMYCETES

Species most frequently encountered are all plurivorous ones: *Arthrinium phaeospermum* and the *Arthrinium* state of *Apiospora montagnei* on *A. junceiforme* and *A. repens*; *Cladosporium cladosporioides* and *C. herbarum* on *A. pungens* and *A. repens*; *Periconia minutissima* and *Tetraploa aristata* on *A. repens*. Less commonly found are the *Drechslera* state of *Pyrenophora tritici-repentis* described below, *Pseudocercosporella herpotrichoides* and *Rhynchosporium secalis*, all recorded on *A. repens*.

Drechslera state of *Pyrenophora tritici-repentis* (Died.)Drechsler (Fig. 1818)
Conidiophores solitary or in groups of 2 or 3, emerging through stomata, brown, up to 400 × 6–12. Conidia cylindrical with the basal segment

characteristically conical or the shape of a snake's head, pale straw-coloured, usually with 5 to 7 pseudosepta, 80–250 × 14–20. Attacked leaves gradually lose their colour and wither from the tip backwards, becoming at first yellow and finally grey.

COELOMYCETES

Two plurivorous species of *Ascochyta*, *A. leptospora* and *A. rhodesii* are recorded on *A. junceiforme* and *A. repens*; and *Ascochyta psammae* (described on *Ammophila*), the *Dinemasporium* state of *Phomatospora dinemasporium* and *Pseudoseptoria donacis* on *A. repens*.

AGROSTIS: Bents

UREDINALES

Plurivorous species and formae speciales only.

On *A. canina*: *Puccinia coronata*, 11, 111. *P. graminis*, 11, 111.
On *A. stolonifera* and *A. tenuis*: *P. coronata*, 11, 111. *P. graminis*, 11, 111. *P. recondita* f.sp. *agrostidis* Oud., 11, 111.

USTILAGINALES

Tilletia sphaerococca (Rabenh.)Fischer v. Waldh.
Bunt is widespread on *A. canina*, *A. stolonifera* and *A. tenuis*, Sept.–Oct. Sori about 1mm long in the ovaries remain partly hidden by the glumes; spores dark brown in mass, singly paler, with reticulate walls, mostly 25–30 diam.

The plurivorous stripe smut *Urocystis agropyri* occurs on *A. stolonifera*, and *Entyloma crastophilum*, described on *Holcus*, has been recorded.

DISCOMYCETES

The plurivorous *Dasyscyphus acutipilus* has been recorded on dead stems.

OTHER ASCOMYCETES

Didymella hebridensis Dennis
Pseudothecia immersed, 0.2mm diam. Asci 40–70 × 9–12. Ascospores hyaline, 1-septate, 17–20 × 4–6. Causes grey, oblong or irregular spots on dying leaves of *A. stolonifera*, Aug.

Plurivorous species found on *Agrostis* include *Botryosphaeria festucae*, *Claviceps purpurea* with rather short ergots, *Epichloe typhina, Keissleriella culmifida, Leptosphaeria culmifraga, Phomatospora dinemasporium, Phyllachora graminis* and *Pleospora vagans.*

HYPHOMYCETES

Drechslera fugax (Wallr.)Shoemaker (Fig. 1819)
Conidiophores dark brown, up to 250 × 7–12. Conidia subhyaline to golden brown, with 4 to 8 pseudosepta, mostly 60–90 × 18–20. On withered leaves of *A. stolonifera.*

Drechslera state of *Pyrenophora erythrospila* Paul (Fig. 1820)
Conidiophores rather dark brown, up to 340 × 6–8. Conidia yellowish to mid pale olivaceous brown, with 4 to 6 pseudosepta, mostly 40–70 × 11–13, with basal cell often longer than the others and sometimes cut off by a dark septum. On living leaves of *A. tenuis*, causing straw-coloured spots up to 2.5mm long, each surrounded by a reddish border.

Hadrotrichum virescens Sacc. & Roum. (Fig. 1821)
Sporodochia small, pulvinate, dark brown or olivaceous brown, resembling rust pustules. Conidiophores pale olivaceous brown, closely packed together forming a palisade, up to 60 high. Conidia pale to mid olivaceous brown, verruculose, 10–15 × 10–12. On living leaves of *A. tenuis*, causing dark olive-green to brown spots.

Mastigosporium rubricosum (Dearn. & Barth.)Nannf. (Fig. 1822)
Conidia solitary or in little white heaps on the surface, hyaline, 3-septate, 35–50 × 11–15, each with a projecting frill around the hilum. On leaves of *A. stolonifera* causing elliptical, purplish spots with grey centres.

Rhynchosporium orthosporum, described on *Dactylis*, has been found occasionally on *A. stolonifera.*

COELOMYCETES

Ascochyta agrostis Polosova
Pycnidia 0.1–0.15mm diam., reddish brown. Conidia cylindrical, hyaline, 1-septate, 10–13 × 2.5. Causes inconspicuous pale yellow spots on leaves of *A. alba* and *A. tenuis. Ascochyta leptospora*, a plurivorous species, has been recorded also on *A. tenuis.*

Other plurivorous species found include *Colletotrichum graminicola*, the *Dilophospora* state of *Lidophia graminis* on *A. tenuis*, and *Pseudoseptoria donacis.*

ALOPECURUS: Fox-tails

UREDINALES

Only plurivorous species and a forma specialis occur infrequently, all on *A. pratensis*: *Puccinia coronata*, 11, 111; *P. graminis*, 11, 111; *P. recondita* f.sp. *perplexans* Plowr., 11, 111.

OTHER FUNGI

Mastigosporium album Riess (Fig. 1823)
Conidia hyaline, 4- or 5-septate, up to 70 × 16 but mostly 50–55 × 15, each with 3 slender appendages. On leaves of *A. pratensis* causing small, very dark and often almost black spots, the conidia showing up as little white heaps in the centre.

Plurivorous species include: ascomycetes, *Claviceps purpurea* and *Pleospora vagans*; hyphomycetes, *Passalora graminis* on *A. geniculatus* and *Rhynchosporium secalis*; coelomycetes, the *Dilophospora* state of *Lidophia graminis* on *A. geniculatus* and *A. myosuroides*, and *Pseudoseptoria donacis*.

AMMOPHILA: Marram

UREDINALES

Puccinia elymi, described on *Elymus*, is recorded also on *Ammophila*. Capitate paraphyses few or absent from uredinia; teliospores often 2- or 3-septate.

Puccinia pygmaea Eriks. var. *ammophilina* (Mains)Cummins & Greene, 11, 111 (Fig. 1824)
Uredinia bright yellow, powdery, on the upper surface between the nerves, found quite easily when leaves are fresh green and ribbon-like, but difficult to see when they are bluish green and incurled; capitate paraphyses numerous, spores 28–35 × 27–30. Telia on both surfaces, black; spores 1-septate, 30–60 × 15–20.

USTILAGINALES

Ustilago hypodytes, the plurivorous stem smut, is seen occasionally on *Ammophila*.

DISCOMYCETES

Belonium psammicola (Rostrup)Nannf. (Fig. 1825)

Apothecia 0.5mm diam., blackish brown with honey-coloured discs. Ascospores hyaline, becoming 1-septate, 6–8 × 2. On dead leaves, June–Aug.

Belonopsis graminea, described on *Agropyron*, has been found also on *Ammophila*.

Hysteropezizella valvata (Mont.)Nannf. (Fig. 1826)
Apothecia about 1mm long, dark grey, each lodged in a cavity in the leaf and exposed when an overlying greyish part of the epidermis, which forms a lid, is pushed off. Ascospores hyaline, 6–7 × 1.5. Paraphyses lanceolate, 5 wide. On dead leaves, Sept.–Nov.

Pyrenopeziza arenivaga (Desm.)Boud. (Fig. 1827)
Apothecia often erumpent, up to 0.5mm diam., externally dark brown to black, with translucent raw-sienna discs. Ascospores hyaline, 15–18 × 5–6. On dead stems and leaves, July–Aug. Apothecia on the inrolled surface of the leaf tend to become quite superficial, whereas others formed in a space beneath the epidermis remain immersed.

Rutstroemia maritima (Rob. ex Desm.)Dennis (Fig. 1828)
Apothecia erumpent, usually solitary, subsessile, cinnamon. Ascospores hyaline, 14–17 × 5.5–7. On dead leaves.

Plurivorous species found on *Ammophila* include *Dasyscyphus carneolus* var. *longisporus*, *D. palearum*, *Hymenoscyphus robustior*, *Lophodermium arundinaceum*, *L. gramineum*, *Mollisia poaeoides* and *Pezizella eburnea*.

OTHER ASCOMYCETES

Anthostomella lugubris (Rob. ex Desm.)Sacc. (Fig. 1829)
Perithecia scattered, immersed, about 0.3mm diam., each with a black clypeus up to 0.7mm long. Ascospores brown, 20–25 × 8–11. On bleached dead leaves, June–Sept.

Anthostomella phaeosticta (Berk.)Sacc. (Fig. 1830)
Perithecia up to 0.4mm diam., usually in little groups below dark brownish-purple areas. Ascospores each with a large brown cell 10–14 × 6–7.5, and a small hyaline cell about 2 diam. On dead leaves, May–Aug.

Chitonospora ammophilae Bomm., Rouss. & Sacc. (Fig. 1831)
Pseudothecia immersed, black, about 0.35mm diam., sometimes in groups with surrounding epidermis stained brown or grey. Ascospores thick-walled, dark brown, sometimes with a pale area at each end, 3-septate, mostly 26–30 × 11–15. On dead leaves, July–Sept.

Didymella scotica Dennis
Perithecia immersed, black, 0.2–0.3mm diam. Asci 100–120 × 15. Ascospores broadly fusiform, hyaline, medianly 1-septate, 18–22 × 6–7. On dead leaves, July.

Leptosphaeria ammophilae (Lasch)Ces. & de Not. (Fig. 1832)
Pseudothecia about 0.35mm diam., in groups, immersed. Ascospores straw-coloured, 6- to 8-septate, 40–56 × 14–15. On dead stems and leaves, July–Aug.

Leptosphaeria marram Cooke (Fig. 1833)
Pseudothecia immersed, with small black necks just piercing the epidermis which is stained brown around the ostioles. Ascospores pale straw-coloured singly, golden brown in mass, 3-septate, 27–33 × 5–7. On dead stems and leaves, May–June.

Microthyrium gramineum Bomm., Rouss. & Sacc. (Fig. 1834)
Thyriothecia about 0.1mm diam.; cells of scutellum 3–4 × 4–8, pale brown, not darker around ostiole, not forming a fringe at the margin. Ascospores hyaline, 1-septate, 8–11 × 2.5–3.5; 2-guttulate larger upper cell with 2 tufts of 3 cilia attached laterally. On dead, very grey leaves lying on the sand between clumps of living plants. *M. ilicinum*, with thyriothecia up to 0.27mm diam. slightly fringed at margin, and ascospores 11–14 long with 4 apical cilia, described on *Quercus*, has been found also on *Ammophila*.

Phomatospora arenariae Sacc., Bomm. & Rouss. (Fig. 1835)
Perithecia up to 0.3mm diam., black, becoming erumpent. Ascospores hyaline, 13–17 × 4–5. On dead basal leaf-sheaths and rhizomes, July–Aug.

Plejobolus arenarius (Bomm., Rouss. & Sacc.)O. Eriksson
Pseudothecia immersed, ellipsoidal, up to 0.7mm long, with necks off-centre and a grey clypeus. Asci up to 400 × 10. Long, multiseptate, hyaline ascospores break up inside asci to form large numbers of part-spores 4–6 × 2–2.5. On dead basal parts.

Trichothyrina ammophilae J.P.Ellis (Fig. 1836)
Thyriothecia 60–80 diam.; ostiolar collars of 3 or 4 rings of dark brown, thick-walled cells. Ascospores cylindrical, hyaline, 1-septate, 8–9.5 × 1.5–2.5; 2 to 4 cilia attached to middle of spore or sometimes absent. On dead stems and leaves, most commonly on old greying stems lying on the sand; seen with a hand lens as very small, black, shining fructifications with well-defined margins and papillate ostioles, May–Nov. A plurivorous species, *T. nigro-annulata*, is found also on marram; it has fusiform ascospores without cilia.

Tubeufia parvula Dennis (Fig. 1837)
Pseudothecia superficial, reddish brown, less than 0.1mm diam. Ascospores hyaline, 3-septate, 14–15 × 3.5–4. On dead inflorescences, May–June.

Plurivorous species include *Claviceps purpurea, Creopus spinulosus, Leptosphaeria nodorum, Pleospora rubelloides* and *P. vagans*.

HYPHOMYCETES

Thyrostromella myriana (Desm.)Höhnel (Fig. 1838)
Sporodochia punctiform, black, usually less than 0.1mm diam. Conidia brown or olivaceous brown, smooth or verruculose, becoming muriform, 20–30 × 12–19. On dead leaves and stems.

Plurivorous species include *Arthrinium phaeospermum* and the *Arthrinium* state of *Apiospora montagnei*, both very common, *Cladosporium herbarum*, *Epicoccum purpurascens*, *Periconia igniaria*, *Tetraploa aristata* and *Torula herbarum*.

COELOMYCETES

Amarenographium metableticum (J.W.H.Trail)O. Eriksson (Fig. 1839)
Pycnidia dark brown, 0.2–0.3mm diam., with short necks just piercing the epidermis, which becomes stained brown around the ostioles. Conidia brown, muriform, 38–46 × 13–18. Locally very common on dead attached leaves, Jan.–July.

Ascochyta psammae Oud.
Pycnidia becoming erumpent, about 0.3mm diam., greyish brown to black. Conidia broadly fusiform to ellipsoid, pointed at apex, flat or rounded at base, hyaline or pale yellow, 1-septate with septum usually below centre, 13–16 × 4–5. Causes dark brown spots on leaves, Feb.–Sept. The plurivorous *A. leptospora* also occurs on marram.

Coniothyrium psammae Oud. (Fig. 1840)
Pycnidia black, up to 0.15mm diam. Conidia pale brown, mostly 7–10 × 3–4. On dead leaves, Aug.–Jan.

Phoma ammophilae Dur. & Mont. (Fig. 1841)
Pycnidia immersed, black, rather shallow, sometimes flattened, about 0.2mm diam. Conidia hyaline, 4–6 × 1.5–2. On dead leaves, Jan.–Apr.

Psammina bommeriae Rouss. & Sacc. (Fig. 1842)
Acervuli blackish olive, up to 1mm diam., usually with a central pale greenish spore mass. Conidiophores branched. Conidia complex with up to 30 septate arms, each 10–30 × 1.5–2.5, almost hyaline singly, pale olive or pale brown in mass. On dead leaves, Apr.–Nov., with good sporulation Oct.–Nov.

Rhodesia subtecta (Rob.)Grove
Acervuli at first covered by epidermis, then opening widely, 0.2–0.6mm diam. Conidia rose-coloured (pinkish red) in mass, mostly 5–7 × 2–3. On the outer surface of inrolled leaves, July–Aug.

Tiarospora perforans (Rob.)Höhnel (Fig. 1843)
Numerous black protuberant necks seen easily with the naked eye; an incision reveals black, bean-shaped pycnidia up to 0.5mm long, the bases of which show up clearly on the inrolled leaf surface. Conidia hyaline to very pale

olivaceous brown, 1-septate, 25–35 × 12–15. Helmet-shaped appendages at each end of the conidium are best seen in water mounts made from fresh material collected Jan.–Mar., but collections can be made up to July.

ANISANTHA: Barren Brome

Few fungi are found on *A. sterilis*. The sori of brome ear smut, *Ustilago bullata*, are seen sometimes in the spikelets, and three plurivorous rusts occur: *Puccinia graminis*, 11, 111, *P. recondita* f.sp. *bromina*, 11, 111 (see under *Bromus*) and *P. striiformis*, 11, 111.

ANTHOXANTHUM: Sweet Vernal-grass

Two plurivorous rusts, *Puccinia graminis*, 11, 111, and *P. brachypodii* var. *poaenemoralis*, 11, 111, are seen occasionally on *A. odoratum*, and *Claviceps purpurea*, *Pleospora vagans* and *Drechslera dematioidea* have been recorded.

ARRHENATHERUM: False Oat

Arrhenatherum elatius, a coarse strongly growing perennial grass, up to 1.2m tall, is very common and often dominant on waste land and roadside verges which may be cut but are not grazed. Apart from one rust it has no fungi specific to it but, in common with other perennial grasses which form extensive stands, it is colonised by many plurivorous species, most of which are to be found on dead stems.

UREDINALES

Puccinia brachypodii Otth var. *arrhenatheri* (Kleb.)Commins & Greene, 11, 111 (Fig. 1844)
The commonest rust on this host. The minute uredinia on small yellow spots on the upper surface of leaves are distinguished from those of the two plurivorous species *P. coronata* and *P. graminis* which occur also on this grass by the presence of abundant clavate or capitate paraphyses; spores yellow, with almost colourless, minutely verruculose walls, 22–27 × 18–23. Oblong black telia, long covered by the epidermis, are found from early summer on the lower surface of the same leaves; spores on short pedicels, chestnut brown, paler below, 35–50 × 16–23.

USTILAGINALES

Ustilago avenae, described on *Avena*, appears to be almost always present in the inflorescences of *Arrhenatherum* from early June onwards, and two plurivorous smuts, *Urocystis agropyri* and *Ustilago striiformis*, frequently ribbon the leaves.

OTHER FUNGI

Plurivorous species include the following: discomycetes, *Calycella scolochloae, Dasyscyphus acutipilus, D. acutus, D. albotestaceus, D. carneolus* var. *longisporus, D. palearum, D. tenuissimus, Hymenoscyphus scutula* var. *suspecta, Lophodermium culmigenum, Mollisia palustris, Pezizella eburnea*; other ascomycetes, *Gaeumannomyces graminis, Gibberella zeae, Leptosphaeria fuckelii, L. nodorum, Phomatospora dinemasporium, Pleospora phaeocomoides, P. vagans, Tubeufia helicomyces*; hyphomycetes, *Arthrinium phaeospermum, Periconia minutissima, Tetraploa aristata*; coelomycetes, *Dilophospora* state of *Lidophia graminis* and *Pseudoseptoria donacis*.

ARUNDINARIA: Bamboo (fungi on all bamboos included here)

UREDINALES

Two rusts have been found on living leaves of *Arundinaria* in Britain but there are few records.

Puccinia kusanoi Diet., 11, 111
Uredinia hypophyllous, inconspicuous, pale cinnamon brown; spores pale brown, echinulate, 25–33 × 22–26, clavate paraphyses sometimes present, up to 12 wide at tip. Telia also hypophyllous, dark brown; spores brown, smooth, 35–82 × 16–23, pedicels up to 200 long.

Puccinia longicornis Pat. & Har., 11, 111
Uredinia similar to those of *P. kusanoi*; spores pale brown, echinulate, 30–36 × 27–31, capitate paraphyses plentiful, up to 23 wide at tip. Telia sometimes purplish brown; spores brown, smooth or minutely verruculose, 1-septate, mostly 50–90 × 14–20, often rostrate.

ASCOMYCETES

Eupropolella arundinariae (Cash)Dennis (Fig. 1845)
Apothecia about 1mm long, at first covered by blackened epidermis, splitting to expose flesh-coloured discs. Ascospores hyaline, 3-septate, 15–25 × 3–5. On dead stems lying on the ground in the centre of large clumps, May.

Morenoina arundinariae J.P.Ellis (Fig. 1846)
Thyriothecia simple or branched, black, up to 1 × 0.1mm. Ascospores hyaline or straw-coloured, finely echinulate, 1-septate, 15–18 × 4–6. On rotting stems, May–June.

Morenoina websteri J.P.Ellis (Fig. 1847)
Thyriothecia star-shaped, 0.4–0.8 × 0.15–0.2mm. Ascospores hyaline to pale

straw-coloured, minutely and distantly verruculose, 1-septate, 10–14 × 4–5. On dead stems.

Pezizella nigrocorticata, described on *Carex*, has been found also on *Arundinaria*.

HYPHOMYCETES

Chaetendophragmia britannica P.M.Kirk (Fig. 1848)
Colonies effuse, inconspicuous. Conidiophores pale golden brown, up to 150 × 4–6, with lobed bases. Conidia 45–55 × 7.5–9, 5-septate, 2 cells golden brown, others subhyaline, with 1 to 3 hyaline lateral appendages. On dead stems, Sept.

Corynespora foveolata (Pat.)Hughes (Fig. 1849)
Colonies effuse, hairy, dark chocolate brown. Conidiophores pale to mid brown, 25–420 × 4–6, with up to 7 successive, cylindrical, percurrent proliferations. Conidia rather pale brown, smooth or verruculose, 4- to 11-septate, 28–100 × 7–9. Very common on dead stems and on bamboo stakes where these are in contact with soil.

Doratomyces microsporus, described on 'herbaceous plants', is quite common on old bamboo stakes, the synnemata forming effuse, grey to almost black colonies.

Gliomastix state of *Wallrothiella subiculosa* Höhnel (Fig. 1850)
Colonies effuse, loose, brown, with irregularly branched conidiophores bearing phialides up to 50 × 3.5. Conidia pale brown singly, dark brown in mass, smooth, 3–8 × 2–4. On dead stems, not common.

Gonytrichum macrocladum (Sacc.)Hughes (Fig. 1851)
Colonies olivaceous, effuse. Conidiophores brown, up to 350 × 4–6, tapering to a pale setiform apex, with short encircling collar branches below and setiform ones above the middle. Conidia colourless or pale brown, smooth, 3–4.5 × 2–3. On old canes lying on the ground or half-buried.

Nigrospora sphaerica (Sacc.)Mason (Fig. 1852)
Colonies at first white with discrete, shining, black conidia visible under a hand lens, later brown or black, when sporulation is abundant. Conidia 14–20 (mostly 16–18) diam. On dead stems. The *Nigrospora* state of *Khuskia oryzae* Hudson with conidia 10–15 (mostly 12–14) diam. has been recorded also on this host.

Pteroconium intermedium M.B.Ellis (Fig. 1853)
Sporodochia small, erumpent, cleft-like, dark blackish brown to black. Conidiophores arising from stromata, up to 50 × 2.5–3.5, hyaline with thick, dark brown transverse septa. Conidia lenticular, 15–20 wide, 8–12 thick. On dead stems.

Sporidesmium pseudobambusae P.M.Kirk
Colonies shortly hairy, effuse, pale olivaceous brown. Conidiophores dark brown, 20–70 × 4–5. Conidia pale straw-coloured, 9- to 20-pseudoseptate, 60–100 × 7.5–9.5, 3.5–4 wide at base, some with hyaline gelatinous cap at the tip. On dead stems, Sept.

Plurivorous species include *Arthrinium phaeospermum* and the *Arthrinium* state of *Apiospora montagnei*, which are almost always present, both on dead stems inside standing clumps and on separate bamboos used as plant stakes, often forming extensive sooty black patches; *Periconia minutissima* and *Tetraploa aristata* are also very common.

AVENA: Oats

UREDINALES

Most records of fungi on *Avena* have been made on the cultivated oat *A. sativa*, but the two plurivorous rusts *Puccinia coronata*, 11, 111, and *P. graminis*, 11, 111, found on this host occur also on *A. fatua*, the wild oat, and *A. strigosa*.

USTILAGINALES

Ustilago avenae (Pers.)Rostr. (Fig. 1854)
Loose smut: attacks the spikelets, often completely destroying the ovaries and sometimes leaving the rachis quite bare. Spore mass becoming powdery. Spores dull olivaceous brown, paler on one side, minutely echinulate, 5–8 diam.

Ustilago hordei, covered smut described on *Hordeum*, common also on *Avena*, has smooth spores 7–11 diam.

ASCOMYCETES

Gaeumannomyces graminis (Sacc.)v.Arx & Olivier var. *avenae* (E.M.Turner) Dennis (Fig. 1855)
Perithecia erumpent, black, 0.3–0.5mm diam. Ascospores hyaline to pale brown, mostly 120–150 × 3. On the lower part of dead stems.

Leptosphaeria avenaria Weber
Pseudothecia immersed, up to 0.15mm diam., black. Ascospores pale yellow, 3-septate, 23–28 × 4.5–6. Pycnidia of *Septoria* state up to 0.15mm diam., brown to black; conidia 3-septate 25–45 × 3–4. On leaves, causing spots which are either purplish brown with orange borders or pale brown with dark borders.

Plurivorous species include *Claviceps purpurea*, *Erysiphe graminis*, *Gibberella zeae*, *Monographella nivalis*, *Mycosphaerella alliaria* and *Pleospora vagans*.

HYPHOMYCETES

Drechslera avenacea (Curtis ex Cooke)Shoemaker (Fig. 1856)
Conidiophores up to 1mm long, 14–16 thick just above the basal swelling, dark brown to blackish brown. Conidia straw-coloured to dark brown, mostly with 4 to 6 pseudosepta, 50–95 × 13–17.5. On dead stems.

Drechslera state of *Pyrenophora avenae* Ito & Kurib. (Fig. 1857)
Conidiophores up to 350 × 8–11, brown. Conidia pale to mid yellowish or olivaceous brown, mostly with 2 to 6 pseudosepta, 50–110 × 15–19. Causes eye-spot lesions on living leaves, at first small, with white centre surrounded by a reddish brown halo, with red border; later these spots coalesce and elongate to form short longitudinal stripes.

Two plurivorous species, *Drechslera biseptata* and *D. dematioidea*, have been recorded on dead stems and leaves.

COELOMYCETES

Ascochyta. Two species and one variety with hyaline, 1-septate conidia occur on *Avena*: (1) *A. avenae* (Petrak)Sprague & Johnson, which causes pale ochraceous spots with purplish borders, especially on seedling leaves, and has conidia 17–20 × 6; (2) *A. avenae* var. *confusa* Punithalingam, which causes buff, brown-bordered spots on leaves and has narrower conidia, 18–24 × 4–5; and (3) the plurivorous *A. leptospora* with still narrower conidia, 14–17 × 2.5–3.

Septoria, v.s. under *Leptosphaeria avenaria*.

Plurivorous species include *Pseudoseptoria donacis*.

BAMBUSA, see under **ARUNDINARIA**

BRACHYPODIUM: False-bromes

UREDINALES

Puccinia brachypodii Otth var. *brachypodii*, 11, 111
Uredinia mostly on upper surface of leaves, small, reddish brown, often in lines on elongated brown spots, capitate paraphyses abundant; spores yellow,

finely verruculose, 20–23 × 18–20. Telia blackish brown; spores 25–40 × 15–25, pedicels very short. Common on living and fading leaves of *B. sylvaticum*, July–Nov.; recorded also on *B. pinnatum*. The plurivorous *P. striiformis* has been found on *B. sylvaticum*.

USTILAGINALES

Tilletia olida (Riess)Schröter
Sori in leaves of *B. pinnatum*, forming long, grey striae. Spores globose to ellipsoidal, reddish brown, with reticulate walls, 17–26 diam. Uncommon.

DISCOMYCETES

Belonopsis filispora (Cooke)Nannf. (Fig. 1858)
Apothecia just under 1mm diam., almost the colour of dead stems, sometimes with a white fringe, anchored by brown, radiating hyphae. Ascospores hyaline, mostly 3-septate and 38–43 × 3. On dead stems of *B. sylvaticum* lying on the ground, May–Nov., common.

Phragmonaevia hysterioides, described on *Carex*, has been recorded on *B. pinnatum*.

Pirottaea exilispora Graddon (Fig. 1859)
Apothecia subglobose, up to 0.4mm diam., pale grey with very dark setae up to 60 × 6. Ascospores hyaline, guttulate, becoming 1-septate, 16–23 × 2–2.5. On dead stems of *B. sylvaticum*, Apr.

Scutomollisia fimbriomarginata Graddon (Fig. 1860)
Apothecia up to 1mm diam., pale golden brown, with greyish brown discs and white fimbriate margins, young ones emerging through dark brown shields 0.15mm diam., or lifting these off. Ascospores hyaline, guttulate, 11–12 × 2–2.5. On dead stems of *B. pinnatum*, Apr.

Scutomollisia integromarginata Graddon (Fig. 1861)
Apothecia up to 1mm diam., creamy when young, becoming pale grey, margin white; emerging through centre of brown shields. Ascospores hyaline, 1-septate, 12–16 × 2–2.5. On dead stems of *B. pinnatum*, Feb.–Apr.

Plurivorous discomycetes include *Dasyscyphus acutipilus* and *D. albotestaceus* on *B. sylvaticum*, and *Lophodermium gramineum* and *Micropeziza karstenii* on *B. pinnatum*.

OTHER ASCOMYCETES

Anthostomella tomicum, described on *Deschampsia*, has been recorded on *B. pinnatum*. Plurivorous species include *Acrospermum graminum*, *Claviceps purpurea* and *Phomatospora dinemasporium* on *B. pinnatum*; and *Lophiostoma semiliberum*, *Phyllachora graminis* and *Pleospora vagans* on *B. sylvaticum*.

HYPHOMYCETES

The plurivorous *Periconia britannica* has been recorded several times on *B. sylvaticum*.

BRIZA: Quaking-grass or Totter-grass

Only plurivorous species including the rust *Puccinia graminis* and the ascomycete *Pyrenophora trichostoma* have been found on *B. media*.

BROMUS: Bromes

UREDINALES

A forma specialis of the plurivorous brown rust, *Puccinia recondita* f.sp. *bromina* Eriks., 11, 111, occurs on *B. commutatus*, *B. mollis*, *B. ramosus* and *B. secalinus*, and *P. graminis* on *B. ramosus*.

USTILAGINALES

Ustilago bullata Berk.
The ear smut of brome grasses is not uncommon in spikelets of *B. mollis* and has been found also on *B. secalinus*. Sori up to 1cm long, each covered by a thin membrane of host tissue. Spores golden brown, usually minutely verruculose, 9–10 (12) diam. Two plurivorous smuts attack *B. erectus*: *Ustilago hypodytes* and *U. serpens*.

DISCOMYCETES

Plurivorous species include *Dasyscyphus carneolus* var. *longisporus*, *Lophodermium culmigenum* and *Micropeziza karstenii* on *B. ramosus*.

OTHER ASCOMYCETES

Plurivorous species include *Claviceps purpurea, Epichloe typhina, Erysiphe graminis* and *Phomatospora dinemasporium* on various species, and *Phyllachora graminis, Pleospora phaeocomoides* and *P. vagans* on *B. ramosus*.

HYPHOMYCETES

Drechslera state of *Pyrenophora bromi* (Died.)Drechsler (Fig. 1862)
Conidiophores yellowish brown, 100–150 × 7–11. Conidia very pale straw-coloured, with 2–10 pseudosepta, 100–250 × 14–26. Leaf spots at first small, dark brown with yellow haloes, later elongating and coalescing; finally large areas turn yellow and leaves wither.

Plurivorous species include *Arthrinium phaeospermum* and *Pseudocercosporella herpotrichoides*.

COELOMYCETES

Ascochyta gracilispora, described on *Dactylis*, has been recorded on *B. erectus* and *B. ramosus*.

CALAMAGROSTIS: Purple Small-reed, Wood Small-reed or Bush-grass

UREDINALES

Puccinia pygmaea Eriks. var. *pygmaea*, 11, 111
On leaves of *C. epigejos*. Uredinia on upper surface, often in rows, orange yellow; capitate paraphyses plentiful; spores finely echinulate, 28–35 × 27–30. Telia on both surfaces, small, black; spores brown, 1-septate, 30–60 × 15–20, pedicels short.

The plurivorous *P. coronata*, 11, 111, occurs on both *C. canescens* and *C. epigejos*.

USTILAGINALES

The plurivorous smut *Ustilago serpens* has been found on leaves of *C. canescens*.

OTHER FUNGI

The following plurivorous species have been recorded, mostly on *C. epigejos*: discomycetes, *Crocicreas culmicolum* and *Dasyscyphus controversus*; other ascomycetes, *Leptosphaeria culmifraga*, *L. fuckelii*, *L. luctuosa* and *Pleospora vagans*; hyphomycetes, *Arthrinium phaeospermum* and *Periconia hispidula*.

CYNODON: Bermuda-grass

Ustilago cynodontis (Pass.)Henn.
Smut sori mostly in inflorescences, dark brown, dusty. Spores reddish brown, smooth, 5–8 diam. Uncommon.

CYNOSURUS: Crested Dog's-tail

UREDINALES

Only the plurivorous *Puccinia graminis*, 11, 111, has been recorded on *C. cristatus*.

COELOMYCETES

Ascochyta cynosuricola Punithalingam
Similar to the plurivorous *A. leptospora* but with very pale yellow conidia 15–18 × 2.5–3, and thickened septa. On dry leaves and sheaths of *C. cristatus*, Aug.

Ascochyta subalpina Sprague & Johnson
Pycnidia up to 0.15mm diam. Conidia hyaline to pale straw-coloured, mostly medianly 1-septate, 12–14 × 2. On dead leaves of *C. cristatus*, June–Aug.

DACTYLIS: Cocksfoot

This densely tufted perennial grass is grown for hay and pasture, and is also widespread along the edges of woods and on roadside verges. It is one of the best known and most easily recognised species, and many plurivorous microfungi have been recorded on its dead stems and leaves.

UREDINALES

Four rusts in stages 11 and 111 occur, the plurivorous species *Puccinia coronata*, *P. graminis* and *Uromyces dactylidis*, and the variety *P. striiformis* var. *dactylidis* Manners, which has small urediniospores, 18–25 × 15–20.

USTILAGINALES

Only the plurivorous smut *Ustilago striiformis* has been found on *Dactylis*.

DISCOMYCETES

Plurivorous species include *Dasyscyphus albotestaceus*, *D. carneolus* var. *longisporus*, *Hymenoscyphus scutula* var. *suspecta*, *Lophodermium culmigenum* and *Pezizella eburnea*.

OTHER ASCOMYCETES

Phyllachora dactylidis Delacr. (Fig. 1863)
Numerous small black flecks, sometimes with yellow haloes, are found very commonly on both sides of living and dead leaves, Oct.–Apr. These are the outer crusts of stromata in which perithecia are embedded. Ascospores hyaline to pale brown, 12–15 × 8–9, when mature mostly 15 × 9.

Plurivorous species include *Claviceps purpurea*, *Epichloe typhina*, the thick orange collars of which are especially abundant and striking on this host, *Creopus spinulosus*, *Erysiphe graminis*, *Gaeumannomyces graminis*, *Keissleriella culmifida*, *Leptosphaeria eustoma*, *L. luctuosa*, *L. nigrans*, *L. nodorum*, *L. tritici*, *Lophiostoma semiliberum*, *Mycosphaerella recutita*, *Paraphaeosphaeria michotii*, *Phomatospora berkeleyi*, *P. dinemasporium*, *Pleospora infectoria*, *P. phaeocomoides*, *P. rubelloides*, *P. straminis*, *P. vagans*, *Pyrenophora trichostoma*, *Trichothyrina alpestris*, *T. nigroannulata* and *Tubeufia helicomyces*.

HYPHOMYCETES

Mastigosporium muticum (Sacc.)Gunnerbeck (Fig. 1864)
Leaf spots 1–3mm long, purple with pale grey centres. Conidia in little white heaps on the surface, 3-septate, mostly 40–50 × 12–13; hilum without protruding frill. Common, Sept.–Oct.

Rhynchosporium orthosporum Caldwell (Fig. 1865)
Leaf blotch, with symptoms similar to those of the plurivorous *R. secalis*, but the conidia are cylindrical without beaks, hyaline, 1-septate, 14–22 × 2.5–4.5. Leaf tipping is sometimes severe.

Plurivorous species include *Arthrinium phaeospermum* and the *Arthrinium* state of *Apiospora montagnei*, *Cercosporidium graminis*, which causes leaf spots, *Drechslera biseptata*, *Periconia igniaria* and *Tetraploa aristata*.

COELOMYCETES

Ascochyta gracilispora Punithalingam
Pycnidia immersed, brown, up to 0.2mm diam. Conidia hyaline, cylindrical, medianly 1-septate, 13–16 × 1.5. Causes yellowish-brown leaf spots up to 1cm diam., Feb.–May.

Microdiscula phragmitis, described on *Phragmites*, has been found on dead stems of *Dactylis*.

Plurivorous species include the *Dilophospora* state of *Lidophia graminis* and *Pseudoseptoria stomaticola*, which is sometimes widespread in wet seasons.

DESCHAMPSIA: Tufted Hair-grass, Wavy Hair-grass

Deschampsia caespitosa and *D. flexuosa* are perennial grasses which are often densely tufted and the former can form very large tussocks; the leaves and stems when they die provide suitable substrata for many interesting microfungi.

UREDINALES

Four rusts, two specific and two plurivorous, are found on *Deschampsia*: *Puccinia coronata*, *P. deschampsiae* and *P. graminis* on *D. caespitosa*, and *Uromyces airae-flexuosae* on *D. flexuosa*.

Puccinia deschampsiae Arth., 11
Only the uredinia are found in Britain, on the upper surface of leaves, scattered or in short rows, yellow; paraphyses abundant, clavate or capitate, 10–18 thick, bent but not abruptly narrowed to form a neck; spores 25–32 × 19–26, shortly echinulate. Quite common on *D. caespitosa*.

Uromyces airae-flexuosae Ferd. & Winge, 11, 111
Uredinia mostly on upper surface of leaves, small, yellow; spores remotely verruculose, 20–26 × 18–22; no paraphyses. Telia on both surfaces, dark brown; spores golden brown, smooth, 26–36 × 16–24, with pedicels up to 35 long. Found only occasionally on *D. flexuosa*.

USTILAGINALES

The plurivorous stripe smut *Ustilago striiformis* is found occasionally on leaves of *D. caespitosa*.

DISCOMYCETES

Crocicreas megalosporum (Rea)S.Carp. var. *gramineum* (Rehm)S. Carp. (Fig. 1866)
Apothecia stalked, with discs 1–2mm diam., very pale flesh-coloured or buff. Ascospores hyaline, 16–21 × 4–6. On dead leaves and stems of *D. caespitosa*, Oct.–Dec.

Hymenoscyphus nitidulus (Berk. & Br.)Phill.
Apothecia up to 0.4mm diam., with short fat stalks, pale pinkish buff. Ascospores hyaline, fusiform, some slightly curved, occasionally 1-septate, 8–13 × 2.5. On dead leaves of *D. caespitosa*, Jan.

Mollisia mutabilis (Berk. & Br.)Massee (Fig. 1867)
Apothecia sessile, often about 0.2–0.3mm diam. but sometimes larger, with whitish or pale grey discs. Ascospores hyaline, often 1-septate, 13–18 × 2–4. On dead leaves and stems of *D. caespitosa*, Feb.–Mar.

Pezizella nigro-corticata and *Phragmonaevia hysterioides*, described on *Carex*, have been recorded on *D. caespitosa*.

Pseudohelotium alaunae Graddon (Fig. 1868)
Apothecia up to 0.5mm diam., white at first, becoming pale yellow or yellow–ochre. Ascospores often shaped like hockey sticks, hyaline, becoming 3- to 5-septate, 35–42 × 2.5–3. On dead leaf sheaths at the base of stems, deep inside tussocks of *D. caespitosa*, Oct.–May.

Trichodiscus heterotrichus Graddon (Fig. 1869)
Apothecia up to 0.5mm diam., short-stalked, dark brown, hairy, with yellowish grey discs. Hairs of two kinds: (1) straight, acutely pointed, dark brown, up to 160 × 4–5; (2) flexuous or spirally twisted, hyaline to pale brown. Ascospores at first hyaline, becoming olive brown, 1- to 3-septate, 19–24 × 6. On dead leaves of *D. caespitosa*, Sept.–Feb.

Plurivorous species recorded on *D. caespitosa* include *Crocicreas culmicolum*, *Dasyscyphus acutipilus*, *D. acutus*, *D. carneolus* var. *longisporus*, *Lophodermium culmigenum*, *L. gramineum*, *Micropeziza karstenii*, *Mollisia palustris* and *Pezizella eburnea*.

OTHER ASCOMYCETES

Anthostomella tomicum (Lév.)Sacc. (Fig. 1870)
Perithecia up to 0.8mm diam., clypeus well developed. Ascospores brown, mostly 12–15 × 6–7. On dead leaves and stems of *D. caespitosa* and *D. flexuosa*, Feb.–Sept.

Cephalotheca clarkii Dennis (Fig. 1871)
Perithecia superficial, 0.2mm diam., black, setose, with walls made up of loosely attached polygonal plates. Setae black, up to 500 × 5. Ascus walls diffluent, but free spores sometimes found in clusters of 8, brown, verruculose, 8–10 × 7–8. On dead stems and leaves of *D. caespitosa*.

Oomyces carneo-albus (Lib.)Berk. & Br. (Fig. 1872)
Stromata erumpent, up to 0.7mm high, 0.5mm wide at base, pale flesh-coloured when fresh, each containing 3 to 7 locules. Ascospores thread-like, 1–1.5 thick, hyaline, multiseptate. On dead leaves of *D. caespitosa*, Jan.–June, uncommon.

Trematosphaeria clarkii Sivan. (Fig. 1873)
Pseudothecia immersed, black, about 0.3mm diam. Ascospores 7- to 10-septate, golden brown with pale or hyaline end cells, mostly 95–120 × 25–30. On dead leaf sheaths of *D. caespitosa*, Oct.

Plurivorous species found on *D. caespitosa* include *Claviceps purpurea, Creopus spinulosus, Gaeumannomyces graminis, Leptosphaeria culmifraga, L. eustoma, L. fuckelii, L. nigrans, L. nodorum, Lophiostoma semiliberum, Mycosphaerella lineolata, M. recutita, Paraphaeosphaeria michotii, Phomatospora berkeleyi, Pleospora infectoria, P. vagans, Trichothyrina alpestris* and *Tubeufia helicomyces*.

HYPHOMYCETES

Curvularia protuberata Nelson & Hodges (Fig. 1874)
Colonies effuse, greyish brown, hairy. Conidiophores up to 500 × 3–5. Conidia brown, with paler end cells, 4-septate, 27–38 × 10–14. On dead stems and leaves of *D. flexuosa*.

Mastigosporium deschampsiae Jørstad (Fig. 1875)
Conidia hyaline, mostly 5- to 7-septate, 40–85 × 15–30, found singly or in little white heaps on the surface. Causes small, purplish-brown spots, often with fawn centres and yellow haloes, on living leaves of *D. caespitosa*.

Plurivorous species on dead leaves and stems of *D. caespitosa* include *Arthrinium phaeospermum, Myrothecium atroviride, Periconia hispidula, Rotula graminis* and *Tetraploa aristata*.

COELOMYCETES

Plurivorous species on *D. caespitosa* include *Actinothyrium graminis*, *Ascochyta rhodesii* and *Stagonospora subseriata*.

ELYMUS: Lyme-grass

UREDINALES

Puccinia elymi West.
The uredinia of this rust are difficult to distinguish from those of *P. striiformis* which also occurs on this host; they are yellow, up to 3mm long, with spores 22–32 × 20–28. Telia hypophyllous, dark grey; spores brown, mostly 2- or 3-septate, 45–90 × 10–17, with short pedicels.

USTILAGINALES

The plurivorous stem smut, *Ustilago hypodytes*, infects plants systemically, the mycelium perennating in the rhizomes. On some dunes lyme-grass produces no flowering spikes, and inspection of the stems shows that these almost always are infected by the smut.

ASCOMYCETES

Anthostomella arenaria O.Eriksson (Fig. 1876)
Perithecia about 0.3mm diam. Ascospores with large brown cell and small hyaline cell, 10–12 × 5. On dead leaves, Aug.–Sept.

Anthostomella chionostoma, described on *Agropyron*, has been recorded on *Elymus*.

Leptosphaeria marram, described on *Ammophila*, frequently occurs also on dead leaves of *Elymus*.

Niesslia exosporioides, described on *Carex*, has been found occasionally on *Elymus*.

Plurivorous species recorded include *Claviceps purpurea*, *Keissleriella culmifida*, *Leptosphaeria nigrans*, *Lophodermium culmigenum*, *Pleospora phaeocomoides*, *P. rubelloides* and *P. vagans*.

HYPHOMYCETES

Black colonies of the plurivorous species *Arthrinium phaeospermum* and the *Arthrinium* state of *Apiospora montagnei* are common on dead stems.

COELOMYCETES

Coniothyrium psammae, described on *Ammophila*, occurs also on *Elymus*.

Stagonospora arenaria Sacc. var. *minor* Trail (Fig. 1877)
Pycnidia dark brown, 0.2–0.25mm diam., with short necks. Conidia hyaline to pale straw-coloured, 3-septate, 20–28 × 3–4. On dead leaves and stems, May–Oct.

FESTUCA: Fescues

UREDINALES

Four rusts, three of them plurivorous species, are found on *Festuca*.

Puccinia coronata, 11, 111, on *F. altissima*, *F. arundinacea*, *F. gigantea*, *F. ovina* and *F. pratensis*.

Puccinia festucae Plowr., 11, 111
Aecia on *Lonicera periclymenum*. Uredinia epiphyllous, very small, yellow; spores minutely echinulate, 21–30 diam. Telia hypophyllous, also very small, dark brown; spores pale brown, 1-septate, 50–80 × 15–18, usually crowned with 1 to 6 tooth-like projections. On *F. longifolia*, *F. ovina*, *F. rubra* and its var. *arenaria*.

Puccinia graminis, 11, 111, on *F. arundinacea*, *F. gigantea*, *F. pratensis* and *F. rubra*.

Uromyces dactylidis, 11, 111, on *F. ovina* and *F. rubra*.

USTILAGINALES

Plurivorous smuts include *Urocystis agropyri* on *F. arundinacea*, *Ustilago hypodytes* on *F. gigantea*, and *U. striiformis* on *F. ovina* and *F. rubra*.

DISCOMYCETES

Dasyscyphus palearum (Desm.)Massee var. *niger* Dennis

A variety with black hairs recorded on *F. rubra*, May–June.

Gloeotinia granigena, described on *Lolium*, has been recorded on seeds of *Festuca*.

Plurivorous species recorded include *Dasyscyphus albotestaceus* and *Hymenoscyphus scutula* var. *suspecta* on *F. rubra*, *Lophodermium arundinaceum* on *F. arundinacea*, and *Pezizella eburnea* on *F. gigantea*.

OTHER ASCOMYCETES

Phyllachora sylvatica Sacc. & Speg. (Fig. 1878)
Black stromata in lines, on leaves which sometimes turn bright yellow. Ascospores 15–22 × 7–8. On leaves of *F. ovina* and *F. rubra*.

Plurivorous species include *Acrospermum graminum* and *Botryosphaeria festucae* on *F. rubra*; *Claviceps purpurea* on *F. arundinacea* and *F. longifolia*; *Epichloe typhina* on *F. gigantea* and *F. rubra*; *Erysiphe graminis* on *F. pratensis*; *Leptosphaeria culmifraga* on *F. rubra*; *L. eustoma*, *L. nigrans* and *L. nodorum* on *F. arundinacea*; *Paraphaeosphaeria michotii* and *Phomatospora dinemasporium* on *F. rubra*; *Pleospora rubelloides*, *P. vagans* and *Tubeufia helicomyces* on *F. arundinacea*.

HYPHOMYCETES

Drechslera state of *Pyrenophora dictyoides* Paul & Parbery (Fig. 1879)
Conidiophores solitary or in groups, brown, up to 250 × 6–10. Conidia sometimes in chains, pale to mid straw-coloured, mostly with 4 to 7 pseudosepta, 50–90 × 14–20. Fairly common on *F. pratensis* causing net-blotch lesions.

Spermospora lolii, described on *Lolium*, has been recorded also on *F. arundinacea* and *F. pratensis*.

Plurivorous species include *Periconia glyceriicola* on *F. rubra*, and *P. hispidula* on *F. arundinacea*, *F. pratensis* and *F. rubra*.

COELOMYCETES

Ascochyta festucae Punithalingam
Pycnidia immersed, golden brown or reddish brown, 0.15mm diam., with smaller secondary pycnidia inside them. Conidia hyaline, 1-septate, 9–14 × 3.5. On fading and dead leaves of *F. ovina* and *F. rubra*. The plurivorous *A. leptospora* also occurs on *F. rubra*.

Hendersonia culmicola Sacc. var. *minor* (Sacc.)Sacc.
Pycnidia up to 0.5mm diam., becoming erumpent through a slit. Conidia slightly curved, hyaline to pale yellow, 1- to 3-septate, 15–25 × 2–3. On dead stems of *F. arundinacea*.

GLYCERIA: Sweet-grasses

Unless otherwise stated, records are on *G. maxima*.

UREDINALES

Two plurivorous rusts are recorded on *Glyceria*, *Puccinia brachypodii* var. *nemoralis*, 11, 111, on *G. fluitans*, and *P. coronata*, 11, 111, on *G. maxima*.

USTILAGINALES

Ustilago longissima (Schw.)Meyen (Fig. 1880)
Parallel, elongated, olivaceous brown sori on leaves vary greatly in length. Spores pale olivaceous brown, minutely verruculose, 4–6 diam. This fungus is sometimes present in such quantity that great clouds of spores are liberated when plants are disturbed. Shoots of smutted plants seldom flower. Very common on *G. maxima* but found also on *G. fluitans*, May–Oct.

DISCOMYCETES

Dennisiodiscus prasinus (Quélet)Svrček (Fig. 1881)
Apothecia up to 2mm diam., sessile, discs olivaceous, fringed with orange or reddish hairs up to 150 long. Ascospores hyaline, 10–19 × 2. On dead leaves and stems lying in the damp basal leaf mat, Apr.–Aug.

Niptera pilosa, described on *Carex*, has been found on *G. maxima*.

Ombrophila ambigua Höhnel (Fig. 1882)
Apothecia 1–2mm diam., short-stalked, white to hyaline or pale grey, and gelatinous at first, often turning pale yellowish brown when old and dry. Ascospores hyaline, 10–16 × 2–2.5. On rotting stems, June–Oct.

Rutstroemia calopus (Fr.)Rehm (Fig. 1883)
Apothecia 2–4mm diam., with fawn to ochraceous discs, stipes up to 3mm long. Ascospores hyaline, 12–13 × 5. On dead, often rotting stems and leaves, Apr.–May. There is also a variety of this with reddish-brown discs, and ascospores up to 18 × 6.

Rutstroemia lindaviana, described on *Phragmites*, has been recorded on *Glyceria*.

Plurivorous species on dead leaves and stems include *Dasyscyphus acutipilus*, *D. carneolus* and its var. *longisporus*, *D. controversus*, *D. tenuissimus*, *Hymenoscyphus scutula* var. *suspecta*, *Lophodermium arundinaceum*, *L. gramineum*, *Mollisia caricina*, *M. palustris* and *Pezizella eburnea*.

OTHER ASCOMYCETES

Acanthophiobolus helicosporus and *Anthostomella caricis*, described on *Carex*, are recorded also on *Glyceria*.

Leptosphaeria culmifraga (Fr.)Ces. & de Not. var. *propinqua* Sacc. (Fig. 1884) Pseudothecia scattered. Ascospores dark straw-coloured, 7- to 9-septate, 32–45 × 7, fourth cell swollen. On dead leaves and stems, July–Oct. Two plurivorous species of *Leptosphaeria*, *L. eustoma* and *L. nigrans*, occur also quite frequently.

Niesslia exosporioides, described on *Carex*, has been found also quite often on *Glyceria*.

Plurivorous species include *Creopus spinulosus*, *Gibberella zeae* and *Tubeufia helicomyces*, the last-named being very common in the wet basal leaf mat.

HYPHOMYCETES

Cylindrotrichum ellisii Morgan-Jones (Fig. 1885)
Colonies effuse, greyish brown, shortly hairy. Conidiophores mid to dark brown, up to 200 × 5–8. Conidia hyaline, 3-septate, 14–19 × 3–5. On dead leaves, May.

Volutella arundinis and *V. melaloma*, described on *Carex*, are found commonly also on *Glyceria*.

Plurivorous species include *Arthrinium sphaerospermum*, *Cercosporidium graminis* on *G. declinata* and *G. fluitans*, *Myrothecium cinctum* and *M. masonii*, *Periconia glyceriicola*, *P. hispidula* and *P. minutissima*.

HELICOTRICHON: Meadow Oat-grass

UREDINALES

The commoner of the two rusts found on *H. pratense* is the plurivorous *Puccinia graminis*; *P. pratensis* is rare.

Puccinia pratensis Blytt, 11, 111
Uredinia amphigenous, small, reddish brown; spores 27–30 × 25–27, echinulate, with 5 to 7 pores. Telia amphigenous; spores reddish brown, thick-walled, verrucose, 35–45 × 25–28.

HOLCUS: Creeping Soft-grass, Yorkshire Fog

UREDINALES

Two rusts are found on *H. lanatus* and *H. mollis*: crown rust, *Puccinia coronata*, and *P. holcina* Eriks., 11, 111, which morphologically resembles the plurivorous brown rust *P. recondita*. Telia of the latter are uncommon.

USTILAGINALES

Entyloma crastophilum Sacc.
This smut forms numerous black, rather rust-like sori about 0.5mm long on leaves of *H. lanatus*, Aug.–Sept.; spores round or angular, brown, 9–13 diam.

Tilletia holci (Westend.)Schröter
Sori in swollen ovaries, partly obscured by glumes. Spore mass blackish brown, spores brown with reticulate walls, 22–28 diam. On *H. lanatus* and *H. mollis*, June–Sept., widespread. The plurivorous stripe smut *Ustilago striiformis* also occurs on these hosts.

ASCOMYCETES

Plurivorous species include discomycetes, *Dasyscyphus albotestaceus*, *D. carneolus* var. *longisporus*, *D. controversus*, *D. tenuissimus* and *Pezizella eburnea*; and other ascomycetes, *Claviceps purpurea*, *Epichloe typhina*, *Leptosphaeria culmifraga* and *Pleospora vagans*.

HYPHOMYCETES

Ramularia holci-lanati (Cav.)Deighton
Conidiophores up to 170 × 2, in fascicles. Conidia hyaline, 0- or 1-septate, minutely verruculose, 13–27 × 5–11. Causes dark brown spots with yellow margins on living leaves of *H. lanatus* in spring and autumn, mostly on plants growing in rather damp places.

Plurivorous species include *Drechslera dematioidea*, *Periconia minutissima* and *Pithomyces chartarum*.

COELOMYCETES

Colletotrichum holci (Syd.)Grove (Fig. 1886)
Acervuli on both surfaces of leaves, black, about 100 × 50. Setae curved, up to 100 long, brown, tapered to a paler apex. Conidia curved, sometimes subacute at one or both ends, hyaline, with many small guttules, 20–30 × 3–5. On oblong brown spots on living and fading leaves of *H. lanatus* and *H. mollis*, July–Aug.

Plurivorous species include the *Dilophospora* state of *Lidophia graminis*.

HORDEUM: Barleys

UREDINALES

Three rusts occur: *Puccinia hordei* and two plurivorous ones, *P. graminis*, 11, 111, on *H. murinum* and *H. vulgare*, and *P. striiformis* 11, 111, on *H. marinum*, *H. murinum* and *H. vulgare*.

Puccinia hordei Otth, 11, 111
This species, which is found on the cultivated barleys *H. distichon* and *H. vulgare* as well as occasionally on common wall barley *H. murinum*, has as its aecial host *Ornithogalum pyrenaicum*. It is distinguished easily from other brown leaf rusts by the presence of large numbers of unicellular teliospores (mesospores) mixed with the 1-septate ones. Uredinia on both leaf surfaces very small, cinnamon brown; spores yellow, with echinulate walls, 21–27 × 15–20. Telia on both leaf surfaces and sometimes on stems; spores with short pedicels, smooth, brown, 1-septate ones 40–55 × 15–25, non-septate ones 25–45 × 16–25.

USTILAGINALES

Two smuts are found in the spikelets of cultivated barleys, replacing the ovaries.

Ustilago hordei (Pers.)Lagerh.
Covered smut, so called because the firm, blackish-brown spore mass remains partly or completely covered by the glumes. Spores smooth, pale olivaceous brown, almost hyaline on one side, 7–11 (mostly 8–9) diam.

Ustilago nuda (Jensen)Rostr.
Loose smut, so called because the spore mass becomes powdery and exposed and finally blows away leaving the rachis bare. Spores minutely echinulate, pale olivaceous or golden brown, almost hyaline on one side, 5–9 (mostly 6–7) diam.

ASCOMYCETES

Didymella exitialis (Morini)E. Müller (Fig. 1887)
Pseudothecia immersed, about 0.1mm diam. Ascospores hyaline, 1-septate, 16–21 × 3.5–4. Pycnidia of *Ascochyta* state erumpent, golden brown, 0.1mm diam.; conidia hyaline, 1-septate, 15–18 × 3.5–4. On leaves, causing scorch, also on stems of *H. vulgare*, Mar.–Aug.

Plurivorous species include *Erysiphe graminis*, *Gaeumannomyces graminis*, *Gibberella zeae*, *Leptosphaeria nodorum*, *Monographella nivalis*, *Pleospora infectoria*, *P. straminis* and *P. vagans*.

HYPHOMYCETES

Drechslera state of *Pyrenophora graminea* Ito & Kuribay. (Fig. 1888)
Conidiophores in groups of 2 to 6, brown, up to 250 × 6–9. Conidia subhyaline to golden brown, with 1 to 7 pseudosepta, mostly 50–80 × 18–20; short secondary conidiophores formed from apical and often also basal cells, and bearing secondary conidia, are produced regularly. Causes leaf stripe disease of *H. vulgare*.

Drechslera state of *Pyrenophora japonica* Ito & Kuribay. (Fig. 1889)
Conidiophores solitary or in groups, dark brown, up to 250 × 7–10. Mature conidia dark golden brown, mostly with 4 to 6 pseudosepta, 50–150 × 15–21; basal cell longer than broad; secondary conidiophores and conidia often produced. Causes eye-spot lesions on leaves.

Plurivorous species include *Pseudocercosporella herpotrichoides* and *Rhynchosporium secalis*.

COELOMYCETES

Ascochyta. Four species occur, including pycnidial *Didymella exitialis* (*v.s.*) and the plurivorous *A. avenae*.

Ascochyta hordei Hara
Pycnidia erumpent, golden brown, up to 0.15mm diam. Conidia pale straw-coloured, minutely verruculose, 1-septate, 17–20 × 3.5–4. On leaves of *H. vulgare*, causing pale buff spots 4–10mm diam., with reddish-brown margins.

Ascochyta hordeicola Punithalingam
Pycnidia up to 0.2mm diam. Conidia hyaline, 1-septate, 14–16 × 3.5–4. On living leaves of *H. vulgare*, causing fawn spots up to 1cm diam., with purplish-brown margins.

Septoria passerinii Sacc.
Pycnidia immersed, 0.1–0.15mm diam., dark brown. Conidia hyaline, mostly 30–40 × 1.5–2. Microconidia 3–7 × 1. On leaves of *H. distichon* and *H. vulgare*, causing speckled blotch disease.

KOELERIA: Crested Hair-grass

Puccinia longissima Schröter, 11, 111
A rare rust recorded a few times on *K. cristata*. Uredinia on upper leaf surface, orange–brown, 0.5mm long; spores straw-coloured or pale brown, minutely echinulate, 25–30 diam. Telia dark brown; spores golden brown, 1-septate, 60–120 × 12–20, lower cell larger and paler than upper one.

Crocicreas gramineum (Fr.)Fr. var. *incertellum* (Rehm)S.Carp.
Apothecia about 0.3mm diam., erumpent, subsessile, with inrolled margins, greenish grey. Ascospores hyaline, 6–8 × 1.5–2. Paraphyses much longer than asci, lanceolate, 4 thick. On dead leaves, July, rare.

LOLIUM: Darnel, Italian and Perennial Rye-grass

UREDINALES

Two plurivorous rusts occur, *Puccinia coronata*, 11, 111, on *L. multiflorum* and *L. perenne*, and *P. graminis*, 11, 111, recorded on *L. perenne* only.

USTILAGINALES

Three smuts are found, the plurivorous stripe smuts *Urocystis agropyri* and *Ustilago striiformis* on leaves of *L. perenne*, and *Tilletia lolii* in ovaries of *L. multiflorum*, *L. perenne* and *L. temulentum*.

Tilletia lolii Auersw.
Sori partly hidden by glumes, 5–7mm long. Spores pale brown with reticulate walls, 18–22 diam. Few records.

ASCOMYCETES

Gloeotinia granigena (Quélet)T.Schumacher (Fig. 1890)
The cause of blind seed disease of rye-grass, *L. perenne*. Apothecia 2–3.5mm diam., with stalks up to 8mm long, pinkish cinnamon when fresh. Ascospores hyaline, mostly 8–11 × 2.5–4. On fallen seed grains, May–July.

Plurivorous species recorded, mostly on *L. perenne*, include *Claviceps purpurea*, *Erysiphe graminis*, *Leptosphaeria tritici*, *Pleospora straminis* and *P. vagans*.

HYPHOMYCETES

Drechslera siccans (Drechsler)Shoemaker (Fig. 1891)
Conidiophores up to 400 × 7–11, brown. Conidia usually rather pale straw-coloured and 60–100 × 16–18, with 4 to 6 pseudosepta. On living leaves of *L. multiflorum* and *L. perenne*, causing dark brown, elongated spots, which coalesce to form brown blotches.

Drechslera state of *Pyrenophora dictyoides*, described on *Festuca*, also causes net-blotch lesions on *Lolium* leaves.

Spermospora lolii MacGarvie & O'Rourke
Conidia hyaline, 2- to 6-septate, 40–70 × 3.5–5, occasionally with a short lateral appendage near the base. Causes greyish or reddish-brown spots on living leaves of *L. perenne*.

Plurivorous species on *L. perenne* include *Drechslera dematioidea* and *Rhynchosporium secalis*.

COELOMYCETES

Ascochyta desmazieresii Cav.
Pycnidia golden brown, up to 0.2mm diam. Conidia hyaline, mostly 1-septate, rarely 2- or 3-septate, 14–20 × 2–2.5. Causes reddish-brown spots with brick-red margins, on leaves of *L. multiflorum* and *L. perenne*. Punithalingam (1979) records four other species of *Ascochyta* and two varieties on *Lollum*: *A. anthoxanthi* Kalymbetov, Bysova *et al.*, with conidia 30–32 × 6–7; *A. missouriensis* Sprague & Johnson, with conidia 15–16 × 4.5–5.5; the plurivorous species *A. avenae* and *A. rhodesii*, and two varieties of the plurivorous *A. leptospora*, var. *acuta*, with conidia 14–16 × 2.5, and var. *major* with conidia 15–18 × 3.5.

The plurivorous *Pseudoseptoria donacis* occurs on *L. perenne*.

MOLINIA: Purple Moor-grass

UREDINALES

Two rather similar rusts with different aecial hosts occur on purple moor-grass; both have small brown uredinia and black pulvinate telia.

Puccinia brunellarum-moliniae Cruchet, 11, 111
Aecia on *Prunella vulgaris*. Urediniospores golden brown, spiny, 20–28 × 20–24. Teliospores brown, smooth, 1-septate, 32–50 × 20–30; pedicels hyaline or yellowish, often brown at the top, up to 120 long.

Puccinia nemoralis Juel, 11, 111
Aecia on *Melampyrum pratense*. Urediniospores 20–27 × 18–26, conspicuously spiny and with 3 capped pores. Teliospores 30–46 × 20–27; pedicels colourless.

DISCOMYCETES

Belonium hystrix (de Not.)Höhnel (Fig. 1892)
Apothecia black with greyish discs, up to 0.5mm diam., emerging through narrow slits in the epidermis; excipulum setose, setae dark brown, 20–40 × 4–6. Ascospores hyaline, 0- to 3-septate, mostly 15–16 × 3. Inconspicuous but quite common on dead stems, and in good sporulation July–Oct.

Hysteropezizella melatephroides (Rehm)Dennis
Apothecia erumpent, 0.2–1mm diam., rusty or blackish brown with pale yellow discs. Paraphyses forked, swollen to 4–6 at tip. Ascospores fusiform, hyaline, 1- to 3-septate, 18–22 × 4–6, surrounded by thick gelatinous sheaths. Mostly on dead stems.

Plurivorous species include *Dasyscyphus carneolus* var. *longisporus*, *Lophodermium gramineum*, *Micropeziza poae*, *Mollisia palustris* and *M. poaeoides*.

OTHER ASCOMYCETES

Gloniella moliniae (de Not.)Sacc. (Fig. 1893)
Hysterothecia dark brown to black, about 1mm long. Ascospores hyaline, 3-septate, 16–25 × 3–4. On thin dead stems, Oct.–Nov.

Plurivorous species include *Botryosphaeria festucae*, *Claviceps purpurea*, *Phomatospora dinemasporium* and *Trichothyrina alpestris*.

COELOMYCETES

Actinothyrium graminis Kunze (Fig. 1894)
Pycnothyria numerous, black, up to 0.5mm diam. but often confluent and then appearing bigger, subcuticular, with fimbriate margins. Conidia often curved, hyaline, 35–68 × 1. The most common and conspicuous fungus on dead leaves and stems. Good sporulating material can be found Mar.–Aug.; at other times of the year pycnothyria are often effete.

Plurivorous species include *Pseudoseptoria stomaticola* and *Stagonospora subseriata*, the latter being common on this host.

NARDUS: Mat-grass

No rusts or smuts are recorded on mat-grass and only a few other fungi.

Leptosphaeria nardi (Fr.)Ces. & de Not. (Fig. 1895)
Pseudothecia scattered, immersed, black, up to 0.2mm diam. Ascospores pale straw-coloured, 5-septate, the fourth cell from the top swollen and longer than the third cell, 23–29 × 4–5. On dead leaves and stems, July–Sept.

Pyrenopeziza maculans (Rehm)Boud.
Apothecia sessile, about 0.3mm diam., dark brown, with greyish-white discs. Asci 30–40 × 6–7. Ascospores straight or slightly curved, hyaline, 9–11 × 2–3. On dead stems.

Scutomollisia puncta (Rehm)Nannf. (Fig. 1896)
Apothecia developing beneath shields, up to 0.2mm diam., dark brown. Ascospores hyaline, each with a thick, refractive septum, 15–22 × 3.5–4.5.

Plurivorous species recorded include ascomycetes, *Claviceps purpurea* and *Paraphaeosphaeria michotii*, and a coelomycete, *Stagonospora subseriata*.

PHALARIS: Reed Canary-grass

UREDINALES

The plurivorous *Puccinia coronata* and one other rust occur on *P. arundinacea*.

Puccinia sessilis Schröter (Fig. 1897)
Various races of this species are recorded with different aecial hosts; those with aecia on *Allium ursinum* and various marsh orchids are fairly common, those with aecia on *Arum*, *Convallaria* and *Paris* much less so. Uredinia on both leaf surfaces very small, golden brown; spores minutely echinulate, 20–27 × 18–24. Telia black; spores brown, 1-septate, 35–50 × 15–22, with short, brown pedicels.

USTILAGINALES

The plurivorous stripe smut *Ustilago striiformis* is found on the leaves, and *Tilletia menieri* in the ovaries.

Tilletia menieri Hariot & Pat. (Fig. 1898)
Spore mass very dark chocolate, filling the infected grains, July–Aug. Spores pale brown with reticulate walls, 20–25 diam.

DISCOMYCETES

Calycellina phalaridis (Lib. ex P.Karsten)Höhnel (Fig. 1899)
Apothecia up to 1.5mm diam., whitish when moist, pale yellow ochre when dry, each one seated on a small brown pad which remains behind when the apothecium has dropped off; the flesh sometimes contains crystals. Ascospores hyaline, 18–30 × 2–3. On dead stems, May–July, quite common.

Dennisiodiscus prasinus, described on *Glyceria*, and *Niptera pulla*, described on *Carex*, are found occasionally on *Phalaris*.

Plurivorous species include *Crocicreas culmicolum*, *C. tomentosum*, *Dasyscyphus acutipilus*, *D. albotestaceus*, *D. carneolus* var. *longisporus*, *D. controversus*, *Micropeziza karstenii*, *Mollisia caricina* and *M. palustris*.

OTHER ASCOMYCETES

Acanthostigmella genuflexa Höhnel (Fig. 1900)
Pseudothecia superficial, dark brown to black, setose, less than 0.1mm diam.; setae dark brown, up to 100 × 4–6. Ascospores very pale olivaceous brown, mostly 2-septate, 10–12 × 2–3. On dead stems, June–July.

Actinopeltis palustris J.P.Ellis (Fig. 1901)
Thyriothecia crowded, 0.1–0.15mm diam., dark reddish brown, usually with 3 to 10 very dark setae, 16–20 × 3–4, splayed out horizontally from a dark collar surrounding the ostiole. Ascospores hyaline, 1-septate, 11–14.5 × 2.5–3, each with a cluster of 6 cilia just above the septum. On dead stems, Feb.–May.

Leptosphaeria mosana Mouton (Fig. 1902)
Pseudothecia gregarious, immersed, 0.3–0.5mm diam., dark blackish brown. Ascospores golden brown, sometimes minutely verruculose, 7- to 9- (mostly 8-)septate, 35–45 × 7–9. On dead stems, July–Aug. Several plurivorous species of *Leptosphaeria* occur, most commonly *L. fuckelii*, but also *L. graminis* and *L. herpotrichoides*.

Morenoina phragmitis, described on *Phragmites*, has been collected several times on *Phalaris*.

Plurivorous species include *Lophiostoma semiliberum*, *Phomatospora dinemasporium*, *Pleospora vagans*, *Trichothyrina alpestris* and *Tubeufia helicomyces*.

HYPHOMYCETES

Codinaea britannica, described on *Filipendula*, *Endophragmia elliptica*, described on 'herbaceous plants', and *Periconia atra*, described on *Carex*, have been recorded on *Phalaris*.

Plurivorous species include *Arthrinium phaeospermum* and the *Arthrinium* state of *Apiospora montagnei*, *Periconia hispidula*, *P. igniaria*, *P. minutissima* and *Tetraploa aristata*.

COELOMYCETES

Microdiscula phragmitis, described on *Phragmites*, occurs also *Phalaris*.

Plurivorous species include *Actinothyrium graminis*, the *Dilophospora* state of *Lidophia graminis*, and *Pseudoseptoria donacis*.

PHLEUM: Timothy-grass or Cat's-tail (All records on *P. pratense*)

UREDINALES AND USTILAGINALES

The plurivorous *Puccinia graminis*, 11, 111, is the only rust recorded, and *Ustilago striiformis* the only smut.

ASCOMYCETES

Leptosphaeria linearis (Sacc.)E. Müller (Fig. 1903)
Pseudothecia up to 0.3mm diam., often in small groups or rows, surrounded by brown hyphae. Ascospores pale straw-coloured, 5- to 7- (mostly 6-)septate, with the second or third cell slightly swollen, mostly 25–35 × 4.5–5. On old stems.

Plurivorous species include *Epichloe typhina*, *Keissleriella culmifida* and *Pleospora phaeocomoides*.

HYPHOMYCETES

Cladosporium phlei (C.T.Gregory)de Vries (Fig. 1904)
Colonies effuse, thin, hairy, pale olivaceous. Conidiophores pale to mid olivaceous brown, up to 200 × 5–9. Conidia solitary or in chains, pale olive or olivaceous brown, echinulate, 1- to 3-septate, 15–36 × 7–14. On leaves causing small, grey or fawn, purple-bordered eye-spot lesions.

Drechslera phlei (Graham)Shoemaker (Fig. 1905)
Conidiophores mostly 60–150 × 6–8, pale to mid golden brown. Conidia sometimes in chains, pale to mid straw-coloured, with 4 to 6 pseudosepta, mostly 50–90 × 14–16, occasionally up to 240 long. Causes pale brown necrotic streaks and blotches, with chlorotic borders, on leaves.

COELOMYCETES

Plurivorous species include *Pseudoseptoria stomaticola*.

PHRAGMITES: Reed

UREDINALES

Puccinia magnusiana Körn., 11, 111 (Fig. 1906)
Aecial hosts *Ranunculus bulbosus*, *R. flammula*, *R. lingua* and *R. repens*. Uredinia on both leaf surfaces, scattered, very small (1–2mm long), pale golden brown, with numerous clavate paraphyses; spores straw-coloured, finely echinulate, 20–35 × 12–20. Telia on leaves very small and numerous, often scattered over the whole surface, also on stems and leaf sheaths where they are long and narrow, dark blackish brown; spores 1-septate, 30–55 × 15–25. Common in East Anglia, July–May.

Puccinia phragmitis (Schum.)Körn., 11, 111 (Fig. 1907)
Aecial hosts *Rheum raponticum* (occasional), *Rumex* species including *R. acetosa*, *R. crispus* and *R. hydrolapathum*. Uredinia much larger than those of *P. magnusiana* and without paraphyses; spores 25–35 diam. Telia large and thick, black; spores 1-septate, up to 65 × 25, with very long pedicels (up to 200). Very common, July–May.

USTILAGINALES

Ustilago grandis Fr.
A very striking smut but one which is seldom found and seems to be confined to East Anglia. Sori forming streaks or completely surrounding the stem and occupying whole internodes. Spore mass blackish brown, spores 10–12 × 7–10, pale brown, smooth.

DISCOMYCETES

Mollisia hydrophila (P.Karsten)Sacc. (Fig. 1908)
Apothecia 0.5–1.5mm diam., white becoming pale yellow–ochre on drying,

brown at base, often seen on sectioning to be packed with crystals; seated on small, brown, hyphal mats which remain behind after the apothecia have dropped off. Ascospores hyaline, 8–14 × 1.5–2.5. On damp bases of dead stems, June–Aug.

Niptera excelsior (P. Karsten)Dennis (Fig. 1909)
Apothecia 1–2mm diam., greyish brown, with hyaline to pale yellowish-green discs when fresh. Ascospores hyaline, with up to 8 septa, 50–80 × 3–4. On dead wet stems, Oct.–May.

Niptera lacustris, described on *Juncus*, and *N. pulla*, described on *Carex*, have been found also on wet stems of *Phragmites*.

Perrotia phragmiticola (P. Henn. & Ploett.)Dennis (Fig. 1910)
Apothecia erumpent, subsessile, 0.5–0.8mm diam., hairy, reddish when fresh, brown on drying. Hairs pale brown, up to 90 × 4.5, encrusted with reddish-brown granules. Ascospores hyaline, mostly 1-septate, rarely 2- or 3-septate, 18–20 × 2–2.5. On dead standing stems, Oct.–Nov.

Rutstroemia lindaviana (Kirscht.)Dennis (Fig. 1911)
Apothecia stalked, up to 2mm diam., mid pale brown, with very pale brown discs. Ascospores hyaline, 4–5 × 1.5–2. Erumpent from blackened areas on dead, often very rotten leaves and stems lying in mud, May–Sept.

Tapesia evilescens (P.Karsten)Sacc. (Fig. 1912)
Subiculum thin, effuse, brown. Apothecia 0.5–1mm diam., with whitish to grey dics and darker excipulum. Ascospores 6–11 × 1–1.5. On dead stems, Apr.–Aug.

Tapesia knieffii (Wallr.)J.Kunze (Fig. 1913)
Subiculum extensive, dark blackish brown to black. Apothecia 1–2mm diam., bright yellow or greyish yellow. Ascospores hyaline, mostly 15–20 × 2. Usually near the bases of dead stems in wet places, May–Aug., fairly common.

Plurivorous species on dead leaves and stems include *Dasyscyphus acutipilus*, common; *D. albotestaceus*, *D. carneolus* var. *longisporus*, *D. controversus*, very common; *D. palearum*, *D. tenuissimus*, *Hymenoscyphus robustior*, *Lophodermium arundinaceum*, common; and *Mollisia palustris*.

OTHER ASCOMYCETES

Acanthophiobolus helicosporus, *Anthostomella punctulata* and *A. tomicoides*, all described on *Carex*, have been recorded on *Phragmites*.

Buergenerula typhae (Fabre)von Arx (Fig. 1914)
Pseudothecia erumpent, dark grey, 0.5mm diam. Ascospores golden brown, mostly 5- or 6-septate, 40–47 × 8–10, each with a hyaline basal appendage and sometimes also a minute hyaline apical one. On dead stems.

Claviceps microcephala (Wallr.)Tul.
Ergots very small, germinate to produce spherical white stromata about 1mm diam. on long, very slender, bright crimson stalks. Otherwise similar to the plurivorous *C. purpurea*.

Keissleriella linearis E. Müller ex Dennis
Pseudothecia immersed, sometimes in short rows, up to 0.5mm diam., black, with brown setae up to 40 × 4–5 around the ostiole. Ascospores hyaline, 3-septate, with mucilaginous envelopes, 40–45 × 7–9. On dark areas on dead stems.

Leptosphaeria arundinacea (Sow.)Sacc. (Fig. 1915)
Pseudothecia solitary on dead sheaths but tending to be confluent on stems, black, erumpent, when solitary about 0.25mm diam. Ascospores 21–30 × 3–5, hyaline and 1-septate at first, later straw-coloured and 3-septate. Common, July–Aug. The plurivorous species *L. fuckelii*, *L. graminis* and *L. herpotrichoides* occur, and *L. graminis*, with mostly 10-septate ascospores 35–45 × 5–6, is probably commoner on reed than on any other grass. Several other species of *Leptosphaeria* have been recorded but are not at all common. They include *L. albo-punctata* (Westend.)Sacc. with 5- to 7-septate, golden-brown ascospores 30–55 × 10–15; *L. baldingerae* Fautr. & Lamb. with 6- to 9-septate, brown ascospores mostly 40–45 × 8–11, and *L. grandispora* Sacc. with straw-coloured, 7- to 8-septate ascospores 29–32 × 6–8.

Lophiostoma arundinis (Pers.)Ces. & de Not. (Fig. 1916)
Pseudothecia black, with laterally compressed slot-like necks, mostly found in linear groups, partly immersed. Ascospores golden brown, paler at the ends, 5-septate, usually 35–40 × 7–8. Common on dead stems, Oct.–May.

Lophiostoma caudatum Fabr. (Fig. 1917)
Pseudothecia black, immersed, usually with just the compressed necks showing. Ascospores when mature 5- or 6-septate, golden brown, paler and tapered towards the base, mostly 26–30 × 7–8. On dead stems, Jan.–Apr., uncommon.

The plurivorous *Lophiostoma semiliberum* is found very commonly on *Phragmites*, Dec.–Apr.

Morenoina phragmitis J.P.Ellis (Fig. 1918)
Thyriothecia simple or branched, often confluent and then appearing irregularly lobed, up to 0.7mm diam., black. Ascospores hyaline or straw-coloured, 1-septate, minutely and distantly verruculose, 8–10 × 3–4. Orbicular pycnothyria up to 0.12mm diam., with hyaline conidia 3–4 × 0.5–1 present. On dead stems, Apr.–Oct.

Nectria ellisii, a plurivorous species described on 'herbaceous plants', has been collected several times on dead stems of *Phragmites*.

Scirrhia rimosa (Alb. & Schw.)Nitschke ex Fuckel (Fig. 1919)
Stromata up to 3 × 1mm, black, subepidermal but splitting the epidermis longitudinally; when sectioned seen to contain rows of pseudothecial cavities. Ascospores hyaline, 1-septate, 17–24 × 5–6. Forming cracked grey spots and stripes on dead leaf sheaths.

Wettsteinina niesslii E.Müller (Fig. 1920)
Pseudothecia immersed, black, up to 0.4mm diam. Ascospores hyaline to yellowish, 50–65 × 7–13, with gelatinous sheaths, mostly 1- to 3-septate but may become up to 8-septate when old. On dead wet stems, Feb.

Plurivorous species include *Botryosphaeria festucae*, *Gibberella zeae*, *Mycosphaerella lineolata*, *Phomatospora berkeleyi*, *Pleospora vagans*, *Trichothyrina nigroannulata* and *Tubeufia helicomyces*.

HYPHOMYCETES

Acremonium alternatum Link (Fig. 1921)
A fluffy, white species found occasionally on dead stems and leaves. Phialides arranged alternately on the hyphae. Conidia mostly in chains although occasionally in heads, 3–5 × 1–2.

Deightoniella arundinacea (Corda)Hughes (Fig. 1922)
Conidiophores pale brown, 40–80 × 4–9, with darker basal cell 10–12 wide. Conidia pale to mid brown, 2-septate, 20–50 × 11–18. Found on reeds where these grow on hard trodden ground along the sides of paths; neighbouring plants growing on softer ground do not seem to be attacked. Infection is systemic and sporulation takes place Apr.–Oct. When the fungus starts to form conidia, infected plants can be detected easily, not only by their stunted growth but also by the dark grey colour of the leaves; this colour is due to the fungus itself as there is, in the early stages, no discoloration of the host tissues.

Gyrothrix podosperma, *Periconia atra* and *P. digitata*, described on *Carex*, have been found on *Phragmites*.

Plurivorous species include *Arthrinium phaeospermum* and the *Arthrinium* state of *Apiospora montagnei*, both very common and often forming extensive sooty colonies, especially on dead stems, *Myrothecium cinctum*, *M. masonii*, *Periconia glyceriicola*, *P. hispidula* (the small, black, bristly colonies of which can be found all the year round on rather dry dead leaves), *P. igniaria*, *P. minutissima* and *Tetraploa aristata*.

COELOMYCETES

Camarosporium feurichii Henn. (Fig. 1923)
Pycnidia erumpent, black, up to 0.15mm diam. Conidia brown, smooth, with

3 transverse and 1 or 2 longitudinal septa, 15–17 × 5–6. On dead stems, May–Oct., uncommon.

Cytoplacosphaeria rimosa (Oud.)Petrak (Fig. 1924)
Conidiomata subepidermal, dark brown to black, up to 1.5mm diam., aggregated and forming linear stromata which split the epidermis. Conidia hyaline, 1- or 2-septate, 15–21 × 3–3.5. On dead stems. Possibly conidial *Scirrhia*.

Hendersonia culmiseda Sacc. (Fig. 1925)
Pycnidia immersed, up to 0.25mm diam. Conidia straw-coloured, 1- to 3-septate, 14–20 × 4–6. On dead stems and leaves, Feb.–Aug.

Microdiscula phragmitis (Westend.)Höhnel
The yellow conidiomata of this fungus, which measure 0.5mm diam., are easily mistaken in the field for discomycetes. When sectioned they are seen to consist of small-celled stromata, with locules lined by penicillately branched conidiophores which produce very large numbers of hyaline, rod-shaped conidia 2–4 × 0.5. On dead stems and rhizomes, June–Nov.

Phoma arundinacea (Berk)Sacc.
Pycnidia black, about 0.2mm diam., splitting the epidermis into longitudinal fissures. Conidia biguttulate, hyaline, 8–9 × 2.5–3. On dead stems, Feb.–Oct.

Pseudorobillarda phragmitis (Cunnell)Morelet (Fig. 1926)
Pycnidia scattered, immersed, yellowish brown, up to 0.3mm diam. Conidia hyaline to pale olivaceous, 1-septate, 16–23 × 2.5–4; basal appendage with 3 divergent branches, 15–22 long. On wet dead stems, July, uncommon.

Pseudoseptoria donacis, a plurivorous species recorded on *Phragmites*.

Stagonospora cylindrica Cunnell
Pycnidia immersed, black, up to 1 × 0.4mm. Conidia hyaline, cylindrical, mostly 3-septate, 50–80 × 8–11.5. On dead stems, Sept.

Stagonospora elegans (Berk.)Sacc. & Trav.
Pycnidia boat-shaped, erumpent, black, up to 1.3 × 0.4mm, gregarious. Conidia fusiform, hyaline, 4- to 6-septate, guttulate, 50–85 × 8.5–14. Mostly on submerged parts of dead standing stems.

POA: Meadow-grasses

UREDINALES

Five rusts, four of them plurivorous species, occur on *Poa*:

Puccinia poarum Nielsen, 11, 111
Aecial host *Tussilago farfara*. Few uredinia are formed and these have no

hyaline capitate paraphyses. Telia plentiful, black, in short rows; spores 1-septate, chestnut brown, paler towards the base, 30–45 × 16–23, with short brownish pedicels; capitate paraphyses present. On leaves and sometimes also stems of *P. pratensis* and *P. trivialis*.

Of the plurivorous species *Puccinia graminis* is found on *Poa annua, P. pratensis* and *P. trivialis*; *Puccinia brachypodii* var. *poae-nemoralis* on *P. annua* and *P. nemoralis*; *Puccinia coronata* on *P. pratensis*; and *Uromyces dactylidis* on *P. annua, P. nemoralis, P. pratensis* and *P. trivialis*.

USTILAGINALES

The plurivorous stripe smut *Ustilago striiformis* is found occasionally on *P. pratensis*.

ASCOMYCETES

Plurivorous species include discomycetes *Micropeziza poae*, on *P. trivialis* and *Mollisia poaeoides*; other ascomycetes *Botryosphaeria festucae, Epichloe typhina, Leptosphaeria culmifraga, L. nodorum, Monographella nivalis* and *Phomatospora dinemasporium*.

HYPHOMYCETES

Drechslera poae (Baudys)Shoemaker (Fig. 1927)
Conidiophores mid to dark brown, up to 250 × 8–12, often geniculate. Conidia straw-coloured to golden brown, mostly with 5 to 8 pseudosepta and 60–100 × 17–23. On leaves of *P. pratensis*, causing dark, often reddish-brown, elongated spots the centres of which become bleached.

Spermospora poagena (Sprague)MacGarvie & O'Rourke
Conidia fusiform, hyaline, 1- to 9-septate, 40–105 × 1.5–3. Causes a leaf spot of *P. pratensis*.

Plurivorous species include *Cercosporidium graminis* on *P. compressa* and *P. trivialis*, and *Pseudocercosporella herpotrichoides*.

COELOMYCETES

Plurivorous species include *Ascochyta leptospora, Colletotrichum graminicola, Dilophospora* state of *Lidophia graminis* and *Pseudoseptoria donacis*.

PUCCINELLIA: Saltmarsh-grasses

Uredinia of the plurivorous rust *Puccinia brachypodii* var. *poae-nemoralis* are seen occasionally on leaves of *P. maritima* and *P. rupestris*. Stripe smut, *Ustilago striiformis*, has been recorded on *P. maritima*, and *Micronectriella agropyri*, described on *Agropyron*, has been found on dead leaves of the same species.

SECALE: Rye

UREDINALES

Three plurivorous species occur: *Puccinia graminis*, *P. recondita* and *P. striiformis*. The forma specialis *P. recondita* f.sp. *recondita* on rye has as its aecial host *Anchusa arvensis*.

USTILAGINALES

Urocystis occulta (Wallr.)Rabenh.
This stripe smut is uncommon in Britain. The very long, dark blisters are formed on leaves, stems and inflorescences in June. Spores reddish brown, 12–16 diam., in spore balls each composed of 1 or 2 (rarely 3 or 4) spores surrounded by pale sterile cells 5–10 diam.

OTHER FUNGI

Gloeotinia granigena, described on *Lolium*, has been recorded on seeds.

Plurivorous species include ascomycetes *Erysiphe graminis*, *Claviceps purpurea*, *Leptosphaeria graminis* and *Pyrenophora trichostoma*; hyphomycetes *Pseudocercosporella herpotrichoides* and *Rhynchosporium secalis*; and the coelomycete *Pseudoseptoria donacis*.

SESLERIA: Blue Sesleria

Glomerella montana (Sacc.)v.Arx & E. Müller (Fig. 1928)
Perithecia solitary or in twos or threes, each group surrounded by a dark brown rudimentary clypeus. Ascospores hyaline, mostly 14–16 × 5 7. On dead leaves, May–June.

Plurivorous species recorded on *Sesleria* include the stripe smut *Ustilago striiformis* and the hyphomycete *Periconia hispidula*.

SPARTINA: Cord-grasses

No rusts or smuts have been found on *S. anglica* and *S.* × *townsendii*, which are abundant on tidal mud flats and have been planted frequently as mud binders, nor have we seen ergot or mildew on them. There are, however, a number of interesting ascomycetes and a few hyphomycetes, some of which seem to be quite common and others found only occasionally.

ASCOMYCETES

Amphisphaeria melanommoides Sacc. (Fig. 1929)
Perithecia superficial or partly immersed, black, about 1mm diam. Ascospores brown, 1-septate, 23–29 × 9–15. On dead roots and rhizomes.

Gnomonia salina Gareth Jones
Perithecia scattered, superficial or partly immersed, up to 0.6mm diam., with necks up to 0.2mm long, dark brown to black. Ascospores cylindrical to ellipsoid, hyaline, 1-septate, 35–70 × 20–30, with a guttule in each cell and a small gelatinous appendage at each end. On dead stems, Nov.–Dec.

Haligena spartinae Gareth Jones (Fig. 1930)
Perithecia solitary, immersed, short-necked, thick-walled, dark brown to black, up to 0.35mm diam. Asci deliquescent. Ascospores ellipsoid, hyaline, 2- to 9- (mostly 5-)septate, 40–90 × 15–21, with a tapered appendage up to 18 × 1.5 at each end. On dead fallen stems.

Leptosphaeria. Five species have been found, mostly on dead stems.

KEY

	Ascospores 5- to 7-septate .... *albopunctata*
	Ascospores 3- to 5-septate .... *maritima*
	Ascospores not more than 3-septate 1
1.	Ascospores mostly more than 50 long .... *marina*
	Ascospores not more than 40 long 2
2.	Ascospores more than 25 long .... *pelagica*
	Ascospores less than 25 long .... *oraemaris*

Leptosphaeria albopunctata (Westend.)Sacc.
Pseudothecia immersed, up to 0.35mm diam. Ascospores golden brown, 30–55 × 10–15.

Leptosphaeria marina Ell. & Ev. (Fig. 1931)
Pseudothecia erumpent, up to 0.5mm diam. Ascospores hyaline, 48–72 × 10–14.

Leptosphaeria maritima Sacc.
Pseudothecia erumpent, up to 0.25mm diam. Ascospores golden brown, 32–40 × 8–12.

Leptosphaeria oraemaris Linder
Pseudothecia erumpent, up to 0.25mm diam. Ascospores brown, 18–25 × 5–8.

Leptosphaeria pelagica Gareth Jones
Pseudothecia erumpent, up to 0.25mm diam. Ascospores hyaline, 28–40 × 8–12.

Micronectriella agropyri, described on *Agropyron*, has been recorded on *Spartina*.

Passeriniella discors (Sacc. & Ellis)Apinis & Chesters (Fig. 1932)
Pseudothecia erumpent, up to 0.4mm diam. Ascospores 3-septate, with brown central cells and hyaline end ones, 28–33 × 10–12. On dead stems, May–Aug.

Pleospora spartinae (Webster & Lucas)Apinis & Chesters
Pseudothecia immersed, up to 0.4mm diam., black, with inconspicuous conical necks; surrounding host tissue stained black. Ascospores boat-shaped, golden brown, thick-walled, usually constricted at the 5 transverse septa, and often with a longitudinal septum in the third or fourth cell, 26–38 × 10–13. *Stagonospora* pycnidia, which may be found on their own or accompanying the pseudothecia, are dark brown, up to 0.2mm diam., and contain yellowish to pale brown, 5- to 7-septate conidia 45–70 × 4–6. On dead stems, Aug.

Sphaerulina pedicellata T.W. Johnson (Fig. 1933)
Pseudothecia immersed, dark blackish brown. Ascospores hyaline, 3-septate, tapered towards the base, 35–50 × 6–9. On dead stems, in good condition May–June.

HYPHOMYCETES

Embellisia phragmospora (van Emden)Simmons (Fig. 1934)
This species, which was described originally from soil with high salinity in The Netherlands, has been found in Britain on rotten leaves of *Spartina*. Conidiophores less than 60 long, 2.5–5 thick. Conidia pale to mid brown, mostly 3- to 7-septate, 20–50 × 6–13.

The plurivorous *Arthrinium phaeospermum* has been recorded on dead stems.

TRISETUM: Golden Oat-grass

Black stem rust *Puccinia graminis* and brown rust *P. recondita* f.sp. *triseti* have been recorded on *T. flavescens*. The plurivorous stem smut *Ustilago hypodytes* and the *Dilophospora* state of *Lidophia graminis* occur also.

TRITICUM: Wheat

UREDINALES

Three plurivorous rusts occur on wheat: *Puccinia graminis*, *P. recondita* and *P. striiformis*. No aecial host is known for the forma specialis *P. recondita* f.sp. *tritici* Eriks. & Henn.

USTILAGINALES

Tilletia tritici (Bjerk.)R. Wolff (Fig. 1935)
Bunt of wheat is found seldom nowadays owing to the widespread use of seed treatments. Sori in ovaries partly hidden by glumes. Spores pale brown with reticulate walls, 14–20 diam., mixed with hyaline sterile cells 12–17 diam., July–Aug.

Ustilago nuda (Jens.)Rostr.
Loose smut is common and attacks spikelets, replacing ovaries, June–Aug. The black spore mass becomes powdery and exposed and finally blows away leaving the rachis bare. Spores minutely echinulate, pale olivaceous or golden brown, almost hyaline on one side, 5–9(mostly 6–7) diam.

ASCOMYCETES

Didymella exitialis, described on *Hordeum*, has been found also on *Triticum*.

Gelasinospora cerealis Dowding (Fig. 1936)
On wheat grains, seldom seen but figured here because its curious ascospores, if found, can be recognised easily; they are dark reddish brown with hyaline pits in their walls, and measure 25–37 × 23–27. The short-necked black perithecia are 0.3–0.4mm diam.

Gibellina cerealis (Pass.)Pass. ex Roum. (Fig. 1937)
Perithecia immersed, surrounded by a greyish-brown mycelial mat, up to 0.5mm diam., pale brown with short black necks which protrude above the surface. Ascospores golden brown, 1-septate, 25–35 × 8–11. On leaf sheaths at the base of stems, causing dark-edged lesions, June–Oct. The only specimens of this we have seen came from Rothamsted Experimental Station, where it was found regularly for many years.

Plurivorous species include only one discomycete, *Dasyscyphus palearum*, but a number of other ascomycetes: *Erysiphe graminis*, mildew (less important on wheat than on barley and oats but may be severe when a mild spring, which is dry, follows a mild winter), *Gaeumannomyces graminis*, *Gibberella zeae*, *Leptosphaeria eustoma*, *L. nodorum* (common with its *Septoria* state which accompanies the pseudothecia on glumes, leaves and stems and causes brown lesions) and *L. tritici*, *Monographella nivalis*, *Mycosphaerella allicina*, *Pleospora infectoria*, *P. straminis* and *P. vagans*.

HYPHOMYCETES

Curvularia inaequalis (Shear)Boedijn (Fig. 1938)
Found occasionally on wheat grains, roots and dead stems forming small grey or black velvety patches, or just loose conidia may be detected. Conidiophores brown, up to 250 × 3–7. Conidia 4-septate, 24–45 × 9–16, sometimes with all cells brown or dark brown but frequently with one or both end cells paler than the others.

Drechslera state of *Cochliobolus sativus* (Ito & Kuribay.)Drechsler ex Dastur (Fig. 1939)
Conidiophores brown, up to 220 × 6–10. Conidia dark olivaceous brown, usually with 6 to 10 pseudosepta, 60–100 × 18–23. Causes dark brown elongated spots on leaves, also seedling blight and foot rot. Uncommon in Britain.

Periconia circinata (Mangin)Sacc. (Fig. 1940)
Conidiophores brown, curved and often circinate at the tip, up to 200 × 9–10. Conidia brown, echinulate, 15–22 diam. Found occasionally on stubble of wheat plants affected by foot rot.

Periconia macrospinosa Letebvre & A.G. Johnson (Fig. 1941)
Conidiophores very dark brown, 100–400 × 6–10. Conidia dark brown to black, very coarsely echinulate, 18–35 diam. including the spines, which are 2–7 long and 1.5–5 wide at the base. Mainly on roots.

Plurivorous species include *Drechslera biseptata* and *Pseudocercosporella herpotrichoides*.

COELOMYCETES

Ascochyta hordeicola, described on *Hordeum*, and the plurivorous *A. avenae* have been recorded on wheat.

Septoria tritici Roberge
Pycnidia in rows, immersed, brown to black, up to 0.15mm diam. Conidia hyaline, slightly curved and tapered towards the apex, 2- or 3-septate, 45–70 × 1.5–2. On leaves between the veins, causing yellowish, sometimes purple-

bordered spots up to 1cm long, May–Aug.

Plurivorous species include the *Dilophospora* state of *Lidophia graminis* and *Pseudoseptoria donacis*.

ZEA: Maize

Ascochyta straminea Punithalingam
Pycnidia immersed, up to 0.18mm diam. Conidia hyaline or pale straw-coloured, 1- to 2-septate, 18–24 × 5–7. On living leaves causing grey or yellowish-brown spots.

Ustilago zeae (Beckm.)Unger
Sori mainly in the inflorescence, up to 10cm long and causing distortion. Spore mass powdery, very dark brown; spores brown, echinulate, 8–12 diam.

FUNGI ON RUSHES, SEDGES, BUR-REEDS AND REEDMACES

CAREX: Sedges

UREDINALES

All rusts on *Carex* belong to the genus *Puccinia*. Revision of British material was carried out by D.M.Henderson in *Notes R. B. G. Edinburgh*, **23**, 1961. He recognises four species: *P. caricina* with 15 varieties, *P. dioicae* with 5 varieties, *P. microsora* and *P. opizii*. *P. microsora*, a rare rust found only on *C. vesicaria*, is distinguished by the formation of amphispores (thick- and dark-walled urediniospores) as well as urediniospores and teliospores in the same sorus. In *P. caricina* the urediniospores always have three or more pores. In *P. opizii* the urediniospores are less than 20 long and have two equatorial pores; in *P. dioicae* they are more than 20 long with two pores above the centre. British species and varieties are listed below with their aecial hosts in Britain where known, and an index is provided to show which of them can be expected to be found on any particular sedge.

Puccinia caricina DC.
    var. *magnusii* (Kleb.)Henderson
    var. *paludosa* (Plowr.)Henderson, aecia on *Pedicularis palustris*
    var. *pringsheimiana* (Kleb.)Henderson, aecia on *Ribes* spp., especially *R. uva-crispa*
    var. *ribesii-pendulae* (Hasler)Henderson
    var. *ribis-nigri-lasiocarpae* (Hasler)Henderson
    var. *ribis-nigri-paniculatae* (Kleb.)Henderson
    var. *uliginosa* (Juel)Jørst., aecia on *Parnassia palustris*
    var. *urticae-acutae* (Kleb.)Henderson, aecia on *Urtica dioica*
    var. *urticae-acutiformis* (Kleb.)Henderson
    var. *urticae-flaccae* (Hasler)Henderson
    var. *urticae-hirtae* (Kleb.)Henderson
    var. *urticae-inflatae* (Hasler)Henderson
    var. *urticae-ripariae* (Hasler)Henderson
    var. *urticae-vesicariae* (Kleb.)Henderson
Puccinia dioicae Magn., aecia on *Cirsium palustre* and *C. dissectum*
    var. *arenariicola* (Plowr.)Henderson, aecia on *Centaurea nigra*
    var. *extensicola* (Plowr.)Henderson, aecia on *Aster tripolium*
    var. *schoeleriana* (Plowr. & Magn.)Henderson, aecia on *Senecio jacobaea*
    var. *silvatica* (Schroet.)Henderson, aecia on *Taraxacum officinale*

Puccinia microsora Körn. ex Fuckel
Puccinia opizii Bubák, aecia on *Lactuca sativa* and *L. virosa*

INDEX

Carex acuta: *Puccinia caricina* var. *pringsheimiana*
C. *acutiformis*: *P. caricina* var. *urticae-acutiformis*
C. *appropinquata*: *P. opizii*
C. *aquatilis*: *P. caricina*
C. *arenaria*: *P. dioicae* var. *arenariicola*, very rare. *P. dioicae* var. *schoeleriana*, common
C. *bigelowii*: *P. caricina* var. *paludosa*
C. *binervis*: *P. caricina*
C. *capillaris*: *P. dioicae* var. *silvatica*
C. *demissa*: *P. dioicae*
C. *dioica*: *P. dioicae*
C. *elata*: *P. caricina* var. *paludosa*, urediniospores with 3 pores, not more than 30 × 22. *P. caricina* var. *urticae-acutae*, urediniospores with 4 pores, more than 34 × 26
C. *extensa*: *P. dioicae* var. *extensicola*
C. *flacca*: *P. caricina* var. *urticae-flaccae*
C. *hirta*: *P. caricina* var. *urticae-hirtae*
C. *hostiana*: *P. caricina*
C. *laevigata*: *P. caricina*
C. *lasiocarpa*: *P. caricina* var. *ribis-nigri-lasiocarpae*
C. *lepidocarpa*: *P. dioicae*
C. *maritima*: *P. dioicae* var. *schoeleriana*
C. *nigra*: *P. caricina* var. *paludosa*. *P. caricina* var. *pringsheimiana*. *P. caricina* var. *uliginosa*. *P. caricina* var. *urticae-acutae*. Since their spore measurements overlap, it is difficult to separate these four varieties, all of which are fairly common on this sedge. They all have different aecial hosts, as indicated above, and these should be sought in the immediate neighbourhood of infected plants.
C. *paniculata*: *P. caricina* var. *ribis-nigri-paniculatae*, urediniospores with 3 equatorial pores, common. *P. opizii*, urediniospores with 2 equatorial pores, uncommon.
C. *pendula*: *P. caricina* var. *ribesii-pendulae*
C. *pseudocyperus*: *P. caricina* (var. *caricina*), urediniospores with 3 pores, less than 30 diam., uncommon. *P. caricina* var. *urticae-acutiformis*, urediniospores with 3 or 4 pores, more than 30 diam.
C. *riparia*: *P. caricina* var. *magnusii*, teliospores mostly less than 60 long. *P. caricina* var. *urticae-ripariae*, teliospores mostly more than 60 long
C. *rostrata*: *P. caricina* var. *urticae-inflatae*
C. *vesicaria*: *P. caricina* var. *urticae-vesicariae*, no amphispores. *P. microsora*, abundant amphispores, rare

USTILAGINALES

Anthracoidea (*Cintractia*): three species have sori in the spikelets of *Carex*, replacing the ovaries and partly covered by the glumes.

Anthracoidea arenariae (H. Syd.)Nannf.
Similar to *A. caricis* described below but with spores regularly round or ellipsoidal, not polygonal, 13–23 × 11–19. On *C. arenaria*.

Anthracoidea caricis (Pers.)Bref. (Fig. 1942)
Spore mass black, at first with spores firmly agglutinated and covered by a thin, grey false membrane, later powdery. Spores round or polygonal or broadly navicular, dark brown, smooth or minutely verruculose, 15–28 diam. On *C. bigelowii*, *C. capillaris*, *C. caryophyllea*, *C. dioica*, *C. flacca*, *C. hirta*, *C. nigra*, *C. panicea*, *C. pilulifera* and *C. pseudocyperus*, June–Sept.

Anthracoidea subinclusa (Körn.)Bref. (Fig. 1943)
Distinguished by its coarsely warted, dark brown spores, mostly 15–20 diam. On *C. lasiocarpa*, *C. riparia* and *C. rostrata*.

Farysia thuemenii (Fischer v. Waldh.)Nannf. (Fig. 1944)
Sori in spikelets, replacing ovaries. Spore mass olivaceous brown, becoming powdery, with long, pale, straw-coloured elaters which can be seen projecting. Spores mostly 6–11 × 4–6, pale olive, minutely verruculose. On *C. riparia*, June–July.

Schizonella melanogramma (DC.)Schröter (Fig. 1945)
Sori in leaves of *C. ericetorum* and *C. flacca*, forming black striae, spore mass agglutinated, black. Spores in pairs, 9–15· diam., pale brown on the inner side shading outwards to dark brown.

Urocystis fischeri Körn.
Sori in leaves of *C. flacca*, causing elongated blisters which eventually rupture to expose a black, powdery spore mass, June–July. Spore balls 20–45 diam. Spores dark reddish brown, 14–18 × 14–15, in groups of 1 to 4 surrounded by pale yellowish sterile cells.

DISCOMYCETES

KEY TO GENERA

	Apothecia immersed	1
	Apothecia superficial or erumpent	6
1.	Apothecia elliptical or hysterioid, blackish with paler discs	2
	Apothecia not so	3
2.	Ascospores without septa, 20–25 × 2.5 .... *Hypoderma*	
	Ascospores without septa, 30–50 × 1 .... *Lophodermium*	

	Ascospores 3-septate, 20–26 × 4–5 .... *Phragmonaevia*
3.	Ascospores long and narrow, thread-like .... *Stictis*
	Ascospores not so 4
4.	Apothecia raising and splitting epidermis to expose orange discs .... *Merostictis*
	Discs exposed by pushing up or shedding operculum-like pieces of host epidermis 5
5.	Apothecia not more than 0.15mm diam. .... *Europolella*
	Apothecia up to 0.3mm diam. .... *Hysteropezizella*
6.	Apothecia arising from blackish brown or black sclerotia, stalked .... *Myriosclerotinia*
	Apothecia on old blackened fruits, stalked .... *Ciboria*
	Apothecia erumpent from blackened areas of rotting leaves and stems, stalked .... *Rutstroemia*
	Apothecia otherwise 7
7.	Apothecia hairy 8
	Apothecia smooth or slightly downy 12
8.	Hairs long, orange or reddish, surrounding olive discs ..... *Dennisiodiscus*
	Hairs and discs not so 9
9.	Paraphyses lanceolate 10
	Paraphyses not lanceolate 11
10.	Hairs long, granulate .... *Dasyscyphus*
	Hairs short, smooth .... *Psilachnum*
11.	Hairs clavate, granulate .... *Clavidisculum*
	Hairs cylindrical, smooth, hyaline .... *Microscypha*
	Hairs scintillating, with hooked tips .... *Hyaloscypha*
12.	Apothecia sessile, mostly brown or brownish, more or less saucer-shaped, 'mollisioid' 13
	Apothecia not so 16
13.	Ascospores mostly without septa 14
	Ascospores mostly with septa 15
14.	Apothecia superficial .... *Mollisia*
	Apothecia erumpent .... *Pyrenopeziza*
15.	Hyphal subiculum present .... *Trichobelonium*
	Without subiculum .... *Niptera*
	Apothecia developing beneath plates of brown hyphae .... *Micropeziza*
16.	Flesh of apothecia tough, formed of hyphae with thick glassy walls .... *Crocicreas*
	Flesh not so 17
17.	Apothecia mostly about 1mm diam., stalked or subsessile .... *Hymenoscyphus*
	Apothecia mostly less than 1mm diam., sessile or very short-stalked 18
18.	Apothecia greenish yellow, not more than 0.3mm diam., ascospores 1- to 3-septate, 25–28 × 6–7 .... *Actinoscypha*

Apothecia pale yellowish, ascospores 4–7 × 1 .... *Calycellina*
Apothecia and ascospores otherwise .... *Pezizella*

Actinoscypha muelleri Graddon (Fig. 1946)
Apothecia up to 0.3mm diam., greenish yellow. Ascospores hyaline, becoming 3-septate, 25–28 × 6–7. On dead attached leaves of *C. flacca*.

Calycellina caricina Dennis
Apothecia with short, fat stalks, pale yellowish, 0.5mm diam. Asci pointed at tips, with pore blued in iodine, 25–35 × 4. Ascospores hyaline, 4–7 × 1. Paraphyses filiform. On dry dead leaves of *C. riparia*.

Ciboria aschersoniana (Henn. & Ploettn.)Whetzel (Fig. 1947)
Apothecia rather pale brown, up to 2.5mm diam., with stalks up to 8mm long. Ascospores hyaline, mostly 10–12 × 4–4.5. On old blackened fruits of *C. acutiformis*, *C. appropinquata* and *C. elata* in wet, peaty places, under cover of marsh vegetation, end of Mar. to end of May. White or yellowish sporodochia, with small hyaline conidia formed inside phialides, which may represent a conidial state of this species, are found Oct.–Feb.

Clavidisculum caricis Raitv. (Fig. 1948)
Apothecia sessile, 0.15–0.2mm diam., white, with mostly clavate, granulate hairs up to 40 × 3–7. Ascospores hyaline, 12–15 × 2–2.5. On dead leaves and leaf bases of *C. hirta* and *C. pendula*, May–Aug.

Crocicreas megalosporum (Rea)S. Carp. (Fig. 1949)
Apothecia 1–2.5mm diam., short-stalked, off-white, drying cream and finally dark reddish brown. Excipulum tough, formed of hyphae with glassy walls. Ascospores hyaline, 26–30 × 5–6. On dead leaves of *C. rostrata*, Aug.

Dasyscyphus callimorphus (P.Karsten)Sacc.
Apothecia sessile or subsessile, 0.3–0.5mm diam., white, with yellow to orange discs. Hairs finely granulate, tapered, often 2-septate, up to 50 × 3–4. Paraphyses lanceolate, about 3 thick. Ascospores hyaline, 16–20 × 1–2. On *C. ericetorum*, May.

Dasyscyphus caricis (Desm.)Sacc. (Fig. 1950)
Apothecia up to 0.6mm diam., short-stalked, with yellowish discs, and yellowish granulate hairs up to 50 × 4. Ascospores hyaline, 7–8 × 1.5. On dead leaves of *C. paniculata* and *C. riparia*, May–Aug.

Dasyscyphus carneolus var. *longisporus*, described on grasses, has been found on *C. acutiformis*.

Dennisiodiscus prasinus, described on *Glyceria*, has been found on wet, dead leaves of *C. riparia*.

Eupropolella celata Graddon (Fig. 1951)
Apothecia immersed, up to 0.15mm diam., pale brown, each exposed by the

pushing up of a sort of operculum or lid made up of host tissue. Ascospores hyaline, guttulate, 17 × 5. On dry leaves of *C. flacca*, Sept.

Hyaloscypha scintillans (Graddon)Dennis (Fig. 1952)
Apothecia superficial, sessile, up to 0.4mm diam., ochraceous, covered on the outside with scintillating hairs. Hairs smooth, up to 80 × 4, pale brown with hyaline, hooked tips. Ascospores hyaline, 1- to 3-septate, 15–25 × 3–5. On dead leaves of *C. pendula*, growing among black setae belonging to *Chaetochalara cladii*, Dec.

Hymenoscyphus citrinulus (P.Karsten)Schröter
Apothecia subsessile, 1mm diam., pale yellow, drying ochraceous, with a few protruding hairs. Ascospores hyaline, slightly curved, sometimes 1-septate, 9–10 × 2. On dead stems and leaves of *C. otrubae* and *C. sylvatica*, Feb.–May.

Hymenoscyphus magnificus (Velen.)Dennis
Apothecia erumpent from a stromatic base in the mesophyll, up to 1mm diam., reddish brown with pale stalks. Ascospores hyaline, fusiform, 25–32 × 4.5–6. Paraphyses contain brown granules. On dead leaves of *C. rostrata*, Aug.

Hypoderma alpinum Spooner (Fig. 1953)
Apothecia elliptical, 0.3–0.5mm long, black, at first opening by a slit, then more widely. Ascospores hyaline, 20–25 × 2.5. On leaves of *C. aquatilis* and *C. bigelowii*, July.

Hysteropezizella. Apothecia immersed, with discs exposed by the pushing up or shedding of operculum-like pieces of epidermis.

KEY

	Paraphyses cylindric–clavate .... *olivacea*
	Paraphyses lanceolate 1
1.	Ascospores 10–17 × 3–4 .... *diminuens*
	Ascospores 5–7 × 1–1.5 .... *dowardensis*

Hysteropezizella diminuens (P. Karsten)Nannf.
Apothecia hypophyllous, immersed, up to 0.3mm diam., discs brown, exposed by the throwing-off of an operculum-like piece of epidermis. Ascospores hyaline, 10–17 × 3–4. Paraphyses lanceolate, up to 6 wide. On dead leaves of *C. atrata*, *C. panicea* and *C. pilulifera*, June–Aug.

Hysteropezizella dowardensis Graddon (Fig. 1954)
Apothecia up to 0.3mm diam., with ochraceous discs exposed by pushing up and to one side an operculum, which remains attached. Ascospores 5–7 × 1–1.5. Paraphyses lanceolate, septate, some with yellow contents, up to 6.5 wide. On dry leaves of *C. flacca*, Sept.

Hysteropezizella olivacea (Mouton)Nannf.
Apothecia up to 0.3mm diam., discs brown, exposed by shedding of operculum. Ascospores 15–20 × 4–5, becoming brown and sometimes septate. On dead leaves of *C. bigelowii* and *C. limosa*, June–July.

Lophodermium caricinum (Rob. ex Desm.)Duby (Fig. 1955)
Almost black, immersed apothecia about 1mm long, opening by slits. Ascospores hyaline, 35–50 × 1. Mostly near the base of dead leaves of *C. elata*, *C. flacca*, *C. paniculata* and *C. silvatica*, May–June.

Merostictis seriata (Lib.)Défago (Fig. 1956)
Apothecia 0.3–0.4mm diam., immersed, raising and splitting the epidermis to expose the orange disc. Ascospores 7–9 × 2–2.5, hyaline with 2 guttules. On dead leaves of *C. riparia* and *C. rostrata*.

Micropeziza cornea (Berk. & Br.)Nannf. (Fig. 1957)
Apothecia superficial, sessile, up to 0.5mm diam., yellowish to pale pinkish brown, developing beneath plates of dark brown hyphae and bearing pieces of these around the edge making this appear minutely dentate. Ascospores hyaline with 4 guttules, and often becoming 1-septate, 12–19 × 3–3.5. On dead stems and leaves of *C. acutiformis* and *C. paniculata*, Mar.–July.

Microscypha ellisii Dennis (Fig. 1958)
Apothecia up to 0.5mm diam., shortly stipitate, white. Hairs at margin smooth, 20–40 × 3. Ascospores hyaline, 10–12 × 2. On dead leaves of *C. acutiformis*, *C. otrubae* and *C. riparia*, Apr.–Aug., fairly common.

Mollisia: five species are found on *Carex*.

KEY

	Ascospores 5–7 × 1, curved .... *caricina*
	Ascospores more than 7 long 1
1.	Ascospores never more than 10 long 2
	Ascospores often more than 10 long 3
2.	Apothecia creamy white to ochraceous .... *chionea*
	Apothecia grey or greyish olive, on glumes .... *dactyligluma*
3.	Apothecia with fimbriate margins, ascospores often 1-septate .... *humidicola*
	Apothecia without fimbriate margins, ascospores without septa .... *palustris*

Mollisia caricina Fautr. (Fig. 1959)
Apothecia pale greyish brown, up to 1.5mm diam., flat with slightly upturned rims. Ascospores hyaline, curved, 5–7 × 1. On dead stems and leaves of sedges and other plants in wet places; type host *C. hirta*, Aug.–Dec.

Mollisia chionea Massee & Crossl. (Fig. 1960)
Apothecia up to 1mm diam., creamy white to ochraceous. Ascospores hyaline,

sometimes slightly clavate, 7–9 × 1.5–2. On dead stem bases of *C. pendula*, Oct.–Jan.

Mollisia dactyligluma Cooke
Apothecia 0.25–0.3mm diam., grey to greyish olive. Ascospores slightly clavate, 7–10 × 1.5. On glumes of *C. acutiformis*, June–July.

Mollisia humidicola Graddon (Fig. 1961)
Apothecia up to 0.75mm diam., pale brown, with grey discs and hyaline, fimbriate margins. Protruding marginal hyphae up to 40 × 6, 3- or 4-septate. Ascospores hyaline, often slightly curved, guttulate, frequently becoming 1-septate, 12–17 × 2. On dead leaves and sheaths of *C. flacca* deeply buried in damp moss, also on *C. paniculata*, Jan.–Sept.

Mollisia palustris, described on *Juncus*, has been found on *C. acutiformis* and *C. otrubae*.

Myriosclerotinia duriaeana (Tul. & C. Tul.)Buchwald
Apothecia 5–10mm diam., fawn or brownish, on short, stout stalks formed singly on narrow, fusiform, slightly curved black sclerotia inside stems. Ascospores hyaline, 9–17 × 5–6.5. Apart from the sclerotia and slightly narrower spores, this is distinguished from the commoner *M. sulcata* by the microconidial sporodochia, which form round black spots in paired groups at regular intervals of 5–15mm along stems in late summer. On *C. chordorrhiza*, *C. disticha* and *C. paniculata*, May–June.

Myriosclerotinia sulcata (Whetzel)Buchwald (Fig. 1962)
Apothecia 5–10mm diam., fawn, with long slender stalks, formed on fusiform, deeply furrowed sclerotia. Ascospores 11–17 × 5–8. Microconidial sporodochia scattered over upper parts of stems in late summer, sclerotia formed later, and apothecia Apr.–June, on *C. acuta*, *C. disticha*, *C. elata*, *C. nigra*, *C. paniculata*, *C. riparia*, *C. rostrata* and *C. vulpina*.

Niptera: three species found on *Carex*, one of which is quite common.

KEY

Ascospores not more than 2.5 wide .... *pilosa*
Ascospores more than 2.5 wide 1
1. Ascospores 16–22 long .... *phaea*
Ascospores 30–45 long .... *pulla*

Niptera phaea (Rehm)Sacc.
Apothecia sessile, pale reddish brown, up to 0.5mm diam. Ascus pore stains deep blue in Melzer; spores hyaline, 1-septate, 16–22 × 4–5. Paraphyses sometimes forked, swollen at tips to 3–5. On dead leaves and stems of *C. binervis*, *C. echinata* and *C. lasiocarpa*, June–Oct.

Niptera pilosa (Crossl.)Boud. (Fig. 1963)
Apothecia up to 1.25mm diam., greyish brown with pale rims. Ascospores hyaline, 1- to 3-septate, 15–27 × 2–2.5. Mainly on leaf bases of *C. acutiformis*, *C. pendula* and *C. riparia*, Feb.–Sept., quite common.

Niptera pulla (Phill. & Keith)Boud.
Apothecia up to 1mm diam., blackish with grey discs. Ascospores 3-septate, 30–45 × 3.5–5. Mostly below water level, Mar.–May.

Pezizella: five species have been recorded on *Carex*.

KEY

Ascospores 3-septate .... *nigro-corticata*
Ascospores without septa 1
1. Apothecia very bright yellow .... *eriophori*
   Apothecia dingy yellow .... *turgidella*
   Apothecia greyish white, spores 9–10 × 1.5–2 .... *incerta*
   Apothecia white, spores 6–8 × 1–1.5 .... *eburnea*

Pezizella eburnea, described on grasses, has been found on *C. acutiformis*.

Pezizella eriophori, described on *Eriophorum*, has been recorded also on *Carex*.

Pezizella incerta (P.Karsten)Rehm
Apothecia in swarms, subsessile, cup-shaped, up to 0.5mm diam., greyish white. Ascospores slightly clavate, hyaline, 9–10 × 1.5–2; paraphyses filiform. On dead leaves of *C. arenaria*, uncommon, July–Nov.; type host *C. pallescens*.

Pezizella nigro-corticata Graddon (Fig. 1964)
Apothecia sessile, up to 0.3mm diam., blackish, with ochraceous discs. Ascospores hyaline, 3-septate, 14–18 × 3–4.5. On dead leaves of *C. pendula*.

Pezizella turgidella (P.Karsten)Sacc.
Apothecia very short-stalked, up to 0.2mm diam., dingy yellow. Ascus pore blue in Melzer. Ascospores 7–13 × 1.5–2. On dead leaves of *C. riparia*, May.

Phragmonaevia hysterioides (Desm.)Rehm (Fig. 1965)
Apothecia immersed, minute, hysterioid with blackish-brown rims and reddish-brown or tawny discs. Ascospores sometimes curved, hyaline, at first 4-guttulate then 3-septate, 20–26 × 4–5. On dry, dead leaves of *C. acutiformis*, June–Sept.

Psilachnum: three species are found on *Carex*.

KEY

Ascospores 10–15 long .... *asemum*
Ascospores 5–8 long 1
1. Hairs broad, clavate .... *granulosellum*
   Hairs very slender .... *lateritio-album*

Psilachnum asemum (Phill.)Dennis (Fig. 1966)
Apothecia minute, very short-stalked, whitish, drying dark brown, hairy. Marginal hairs hyaline, 1- to 3-septate, smooth, up to 50 × 3–4. Paraphyses lanceolate, up to 60 × 4–5. Ascospores 10–15 × 2–3 (4). On dead leaves, July.

Psilachnum granulosellum Höhnel (Fig. 1967)
Apothecia 0.1–0.2mm diam., white or pale yellow. Hairs clavate, 20–30 long, swollen to 5–12 at tip. Paraphyses lanceolate. Ascospores hyaline, 5–8 × 1.5–2. On dead leaves of *C. flacca* and *C. pendula*, Mar.–May.

Psilachnum lateritio-album (P. Karsten)Höhnel (Fig. 1968)
Apothecia 0.1–0.4mm diam., clay-coloured, with white rims. Hairs very slender. Ascospores hyaline, 6–8 × 1.5. Paraphyses lanceolate, sometimes branched. On dead leaves of *C. acutiformis* and *C. riparia*, July–Aug.

Pyrenopeziza fuscescens (Rehm)Défago
Apothecia erumpent, 0.2–0.45mm diam., cupulate, pale brown, with hyaline, elongated marginal cells protruding well above the disc and often incurved. Asci with conical tips, and small pore blued by Melzer. Ascospores hyaline, 16–18 × 3–3.5. Paraphyses cylindrical. On dead leaves of *C. bigelowii*, July: type host *C. pendula*.

Rutstroemia lindaviana, described on *Phragmites*, has been found on *Carex*.

Stictis elongatispora Graddon (Fig. 1969)
Apothecia immersed, with very dark discs and white rims. Ascospores hyaline, up to 350 × 1.5, spirally coiled inside asci. On dead leaves of *C. flacca* and *C. hirta*, May–Aug.

Stictis pusilla Speg.
Apothecia 0.2–0.4mm diam., immersed, with sunken yellow discs and white pruinose margins. Ascospores hyaline, 80–100 × 1–1.5, coiled inside asci. Paraphyses slender with knob-like swelling at apex. On *C. pendula*.

Trichobelonium asteroma (Fuckel)Rehm (Fig. 1970)
Apothecia up to 1.5mm diam., grey with white rims, seated on a thin, brown subiculum. Ascospores hyaline, 3-septate, 35–55 × 3–4. On dead, often reddened leaf bases of *C. acutiformis* and *C. riparia*, Apr.–Nov.

OTHER ASCOMYCETES

KEY TO GENERA

	Ascomata superficial	1
	Ascomata immersed or partly so	4
1.	Ascomata spherical, setose	2
	Ascomata flattened, radiate (thyriothecia)	3

2. Ascospores 1-septate, 10–12 × 2–2.5 .... *Niesslia*
   Ascospores multiseptate, 160 × 1.5–2 .... *Acanthophiobolus*
3. Thyriothecia elongated, often branched, without ostiolar collars
   .... *Morenoina*
   Thyriothecia round, with ostiolar collars .... *Trichothyrina*
4. Clypeus usually present, sometimes small 5
   No clypeus 6
5. Ascospores brown, 1-celled or with a large brown cell and a small
      hyaline one .... *Anthostomella*
   Ascospores hyaline, 2-septate .... *Buergenerula*
6. Ascospores hyaline to slightly yellowish 7
   Ascospores brown, straw-coloured or yellow 11
7. Ascospores without septa 8
   Ascospores 1-septate 9
   Ascospores mostly with more than 1 septum 10
8. Ascospores 9–16 × 3–4 .... *Phomatospora*
   Ascospores 24–32 × 8–9 .... *Physalospora*
9. Ascospores not more than 7 wide .... *Mycosphaerella*
   Ascospores 9–10 wide .... *Didymella*
10. Ascospores 35–45 × 6–8 .... *Metasphaeria*
    Ascospores 50–65 × 7–13, with gelatinous sheaths .... *Wettsteinina*
    Ascospores 70–150 × 2–3 .... *Gaeumannomyces*
11. Ascospores brown, 1-septate, often with a hyaline appendage at each
       end .... *Ceriophora*
    Ascospores 2-septate .... *Paraphaeosphaeria*
    Ascospores with 3 or more septa .... *Leptosphaeria*

Acanthophiobolus helicosporus (Berk. & Br.)J. Walker (Fig. 1971)
Pseudothecia superficial, 0.2–0.3mm diam., dark brown to black, setose; setae up to 300 long. Ascospores hyaline, multiseptate, twisted inside asci, up to 160 × 1.5–2. Common on dead leaves of *C. acutiformis*, *C. hirta*, *C. otrubae*, *C. paniculata*, *C. pendula* and *C. riparia*, May–Oct.

Anthostomella: five species, mostly on dead leaves, perithecia immersed, usually beneath a well-defined dark clypeus.

KEY

   Ascospores 2-celled, 14–19 × 5–8, one cell brown, one hyaline ....
   .... *tomicoides*
   Ascospores 1-celled, brown 1
1. Ascospores 6–9 × 3–4, often reniform .... *punctulata*
   Ascospores 8–12 × 4–5, not reniform .... *limitata*
   Ascospores 12–15 × 6–7 .... *caricis*
   Ascospores 15–23 × 7–9 .... *tumulosa*

Anthostomella caricis S.Francis (Fig. 1972)
Perithecia up to 0.3mm diam. On *C. paniculata* and *C. pendula*, Mar.–Oct.

Anthostomella limitata Sacc. (Fig. 1973)
Perithecia up to 0.2mm diam., with little or sometimes no clypeus. On *C. acutiformis*, *C. paniculata* and *C. riparia*, Apr.–Dec.

Anthostomella punctulata (Rob. ex Desm.)Sacc. (Fig. 1974)
Perithecia up to 0.2mm diam. Found most commonly on *C. pendula* but occasionally also on *C. acutiformis*, Feb.–Oct.

Anthostomella tomicoides Sacc. (Fig. 1975)
Perithecia up to 0.4mm diam. On *C. paniculata* and *C. riparia*.

Anthostomella tumulosa (Rob. ex Desm.)Sacc. (Fig. 1976)
Perithecia up to 0.3mm diam., clypeus black, conspicuous. On *C. pendula*, Jan.–July.

Buergenerula biseptata (E.Rostrup)H. Syd. (Fig.1977)
Pseudothecia immersed, up to 0.25mm diam., each with a small clypeus, on bleached areas within brown leaf spots. Ascospores hyaline, 2-septate, 22–32 × 8–10. On *C. acutiformis* and *C. vesicaria*, Dec.–Jan.

Ceriophora palustris (Berk. & Br.)Höhnel (Fig. 1978)
Perithecia immersed, globose, black, up to 0.5mm diam., with tip of neck showing on the surface. Ascospores brown, 1-septate, 18–26 × 6–8, often, but not always, with a hyaline appendage at each end. On dead leaves of *C. acutiformis*, *C. paniculata*, *C. pendula* and *C. riparia*, common, Mar.–Aug.

Didymella proximella (P. Karsten)Sacc. (Fig. 1979)
Pseudothecia immersed, about 0.2mm diam., with small protuberant necks. Ascospores hyaline, 1-septate, with 2 guttules in each cell, mostly 20–25 × 9–10. On dead leaves of *C. acutiformis*, *C. extensa*, etc., Mar.–July.

Gaeumannomyces graminis, described on grasses, is fairly common also on the lower parts of leaf sheaths and stems of *C. acutiformis* and *C. pendula*.

Leptosphaeria: five species are recorded on *Carex*.

KEY

```
     Ascospores 3-septate ............................................................. 1
     Ascospores 5-septate ............................................................. 2
     Ascospores 6- to 9-septate ................................. herpotrichoides
1.   Ascospores 18–21 × 4 ..................................................... caricicola
     Ascospores up to 30 × 10 .............................................. epicarecta
2.   Ascospores 17–24 × 3–4, with second cell enlarged .................. nigrans
     Ascospores 25–30 × 4–5 ................................................... caricis
```

Leptosphaeria caricicola Fautr.
Pseudothecia minute, immersed. Ascospores olivaceous. On dead leaves of *C. hirta* and *C. riparia*.

Leptosphaeria caricis Schröter (Fig. 1980)
Pseudothecia scattered, immersed, 0.3–0.5mm diam. Ascospores straw-coloured. On dead leaves of *C. otrubae*, etc.

Leptosphaeria epicarecta (Cooke)Sacc.
Pseudothecia scattered, covered by grey-spotted epidermis. Ascospores yellow, with second cell slightly swollen. On dead leaves of *C. riparia* and *C. sylvatica*, Oct.–Nov.

Leptosphaeria herpotrichoides, described on grasses, occurs also on *Carex*.

Leptosphaeria nigrans, described on grasses, has been found on *C. pendula*.

Metasphaeria cumana (Sacc. & Speg.)Sacc. f. *macrospora* Fautr. (Fig. 1981)
Pseudothecia immersed, black. Ascospores hyaline, at first 1-septate but eventually with 3 to 5 septa, 35–45 × 6–8. On dead leaves of *C. acutiformis* and *C. riparia*, July–Aug.

Morenoina. Three species described on *Juncus* have been recorded also on *Carex*: *M. fibriata* on *C. binervis*, *M. minuta* on *C. flacca*, and *M. paludosa* on *C. elata*.

Mycosphaerella allicina and *M. lineolata*, described on grasses, are found also occasionally on sedges.

Mycosphaerella perexigua (P. Karsten)Johanson
Pseudothecia immersed, 0.05mm diam., black, linked together by brown hyphae. Ascospores hyaline, 1-septate, 14–16 × 3–5. On *C. rupestris*, July.

Niesslia exosporioides (Desm.)Winter (Fig. 1982)
Perithecia superficial, black, setose, 0.1–0.15mm diam.; setae 25–50 × 4. Ascospores hyaline, 1-septate, 10–12 × 2–2.5. Common on dry dead leaves of *C. acutiformis*, *C. elata*, *C. paniculata*, *C. pendula* and especially *C. riparia*, Apr.–Aug.

Paraphaeosphaeria michotii, described on *Cladium*, is quite common also on *Carex*.

Phomatospora dinemasporium, described on grasses, occurs occasionally on sedges.

Physalospora alpestris Niessl
Perithecia immersed, about 0.1mm diam., pale brown, darker around the ostiole. Ascospores hyaline, guttulate, 24–32 × 8–9. On dead leaves of *C. lepidocarpa* and *C. serotina*, Aug.–Oct.

Trichothyrina alpestris and *T. nigroannulata*, described on grasses, have been found on dead leaves of *Carex*.

Trichothyrina norfolciana J.P.Ellis (Fig. 1983)
Thyriothecia 0.2–0.27mm diam., dark brown, with fimbriate margins. Ascospores hyaline, 1- to 6-septate, 23–27 × 3.5–5. On dying and dead leaves of *C. riparia*, May–Sept.

Wettsteinina niesslii, described on *Phragmites*, has been recorded on *C. paniculata*.

HYPHOMYCETES

Arthrinium. The small, dark brown to black, pulvinate sporodochia are very common on dead, often dry, bleached leaves of sedges. Eight species and one variety occur, the two commonest and most characteristic ones being *A. puccinioides* and *A. sporophleum*. They all have brown, smooth-walled conidia.

KEY

	Conidia polygonal in face view .... *puccinioides*	
	Conidia round in face view 1	
	Conidia neither polygonal nor round 2	
1.	Conidia 8–12 diam. in face view, lenticular .... *phaeospermum*	
	Conidia 5.5–8 diam. in face view, lenticular .... conidial *Apiospora montagnei*	
2.	Conidia curved 3	
	Conidia not curved 4	
3.	Conidia 8–11 × 5–6 .... *curvatum* var. *minus*	
	Conidia 11–15 × 6–8 .... conidial *Pseudoguignardia scirpi*	
4.	Conidia oblong or irregular in outline 5	
	Conidia fusiform, navicular or limoniform 6	
5.	Sterile cells spherical .... *morthieri*	
	Sterile cells bottle-shaped, often 1-septate .... *fuckelii*	
6.	Conidia over 35 long .... *caricicola*	
	Conidia under 16 long .... *sporophleum*	

Arthrinium caricicola Kunze (Fig. 1984)
Conidia 30–53 × 7.5–13. On *C. acutiformis*, *C. digitata*, *C. echinata*, *C. ericetorum*, *C. riparia* and *C. vaginata*, Apr.–July, uncommon.

Arthrinium curvatum var. *minus*, described on *Schoenoplectus*, has been found on *C. hirta* and *C. riparia*.

Arthrinium fuckelii Gjaerum (Fig. 1985)
Conidia 13–20 × 5.5–9.5 × 5.5–7. On dead leaf tips of *C. caryophyllea*, June, rare.

Arthrinium morthieri Fuckel (Fig. 1986)
Conidia 12–16 × 7–9 × 4–8. On *C. binervis, C. flacca, C. panicea* and *C. praecox*, June–Sept.

Arthrinium phaeospermum, described on grasses, has been found on *C. acutiformis* and *C. riparia*.

Arthrinium puccinioides (DC.)Kunze (Fig. 1987)
Colonies small, under 0.4mm diam., usually round, raised, very dark brown. Conidia 9–14 diam. On dead, often bleached leaves, seen most frequently in summer. This appears to be the commonest and most widely distributed species in Britain on *C. acutiformis, C. appropinquata, C. arenaria, C. elata, C. flacca, C. riparia, C. saxatilis, C. sylvatica* and *C. vesicaria*.

Arthrinium sporophleum Kunze (Fig. 1988)
Sporodochia oval, pulvinate, up to 1 × 0.6mm, dark blackish brown. Conidia 11–15 × 5–7.5. Very common especially on newly dead leaves of *C. riparia* but also on *C. acutiformis, C. appropinquata, C. flava, C. hirta, C. paniculata* and *C. pendula*. The period of maximum growth is Mar.–Apr.; by May colonies have a weathered appearance and there are only a few conidiophores left, although heaps of conidia may remain.

Arthrinium state of *Apiospora montagnei*, described on grasses, has been found on *C. riparia*.

Arthrinium state of *Pseudoguignardia scirpi*, described on *Scirpus*, has been recorded on *C. acutiformis*.

Cladosporium cladosporioides and *C. herbarum*, described on 'herbaceous plants', occur on sedges, but not commonly.

Clasterosporium caricinum Schw. (Fig. 1989)
Colonies black, velvety, up to 24cm long, often covering the whole width of the abaxial surface of a leaf base. Found all the year round. Initially parasitic on green leaves which turn brown prematurely. Mycelium superficial, hyphopodiate; old colonies sometimes with setae. Conidiophores brown, up to 130 × 4–7. Conidia brown, 7- to 12-septate, 70–360 × 11–20. On *C. acutiformis, C. riparia* and occasionally *C. elata*, mostly on plants growing on marshes subjected to periodic flooding.

Cordella clarkii M.B.Ellis (Fig. 1990)
Colonies black, granular, shiny. Setae olivaceous, brown, up to 38 × 4–7. Conidia lenticular, blackish brown, 13–15 diam., 6–8 thick. On dead leaves of *C. pilulifera*, Feb., uncommon.

Epicoccum purpurascens, described on 'herbaceous plants', is found occasionally on sedges as a secondary invader.

Gyrothrix podosperma (Corda)Rabenh. (Fig. 1991)
Colonies superficial, velvety, dark reddish brown, up to 5mm diam.

Subdichotomously branched brown setae up to 250 × 3.5–4.5; branches verrucose. Conidia hyaline, 8–16 × 1.5–2. On dead leaves of *C. riparia*, uncommon.

Myrothecium cinctum, described on grasses, is common also Mar.–May on dead leaves of *C. acutiformis*, *C. elata*, *C. paniculata*, *C. pendula*, *C. pseudocyperus* and *C. riparia*.

Myrothecium masonii, also described on grasses, is found much less frequently on sedges.

Periconia. Eight species grow on dead leaves and stems of *Carex* in Britain. Three of these, *P. hispidula*, *P. igniaria* and *P. minutissima*, are described on grasses, on which they are more commonly found. The other five, *P. atra*, *P. curta*, *P. digitata*, *P. funerea* and *P. laminella*, are found mainly on *Carex*, although other marsh and fen plants, such as *Cladium*, *Eriophorum* and *Juncus*, may be colonised by them.

KEY

	Conidiophores with well-defined conidial heads	1
	Heads not well defined	5
1.	Branches in head short, in verticils of 4 to 7	*atra*
	Branches long, often widely spaced and irregular	2
2.	Distal branches smooth	3
	Distal branches with rough projections	4
3.	Conidia 4–6 diam.	*minutissima*
	Conidia 7–11 diam.	*digitata*
4.	Stipe under 200 long, conidia shortly echinulate	*curta*
	Stipe over 300 long, conidial spines 1 long	*igniaria*
5.	Apex of conidiophores sterile, setiform	*hispidula*
	Apex of condiophore and branches fertile	6
6.	Conidiophore stipes spirally twisted	*funerea*
	Conidiophore stipes straight	*laminella*

Periconia atra Corda (Fig. 1992)
Characterised by its rather large heads of tightly compacted conidia and by rings of short, fat branches which arise close together beneath the apex of the long, stout conidiophores. Conidia 5–9 diam. On *C. acutiformis*, *C. appropinquata*, *C. binervis*, *C. elata*, *C. paniculata*, *C. pendula* and *C. riparia*, Apr.–Sept.

Periconia curta (Berk.)Mason & M.B.Ellis (Fig. 1993)
Characterised by the very large (up to 450 diam.) heads of conidia and the complex system of rough-walled branches inside the heads. Conidia 6–8 diam. On *C. acutiformis*, *C. appropinquata*, *C. elata*, *C. paniculata*, *C. pseudocyperus* and *C. riparia*, Jan.–Dec.

Periconia digitata (Cooke)Sacc. (Fig. 1994)
Branches seen clearly in mature heads, where the conidia are relatively loosely compacted. Mostly a winter species although a few collections have been made in the summer. On *C. acutiformis, C. elata, C. otrubae, C. paniculata, C. pseudocyperus* and *C. riparia*.

Periconia funerea (Ces.)Mason & M.B.Ellis (Fig. 1995)
Characterised by its spirally twisted conidiophores. Conidia 8–12 diam. On *C. acutiformis, C. hirta* and *C. pseudocyperus*.

Periconia hispidula has been found on *C. acutiformis, C. appropinquata, C. elata, C. paniculata* and *C. riparia*.

Periconia igniaria occurs on *C. acutiformis, C. appropinquata* and *C. riparia*.

Periconia laminella Mason & M.B.Ellis (Fig. 1996)
Conidiophores narrow and loosely branched. Conidia 9–11 diam. On *C. acutiformis* and *C. riparia*.

Periconia minutissima has been found on *C. acutiformis* and *C. riparia*.

Tetraploa aristata, described on *Cladium*, has been recorded on *C. paniculata* and *C. riparia*.

Veronaea caricis M.B.Ellis (Fig. 1997)
Colonies effuse, greyish brown, hairy. Conidiophores brown, up to 550 × 4–6, with numerous scars towards the apex. Conidia 1-septate, hyaline or subhyaline, verruculose, 20–25 × 6–7. On dead leaves of *C. pendula*, Sept.

Volutella arundinis Desm. ex Fr. (Fig. 1998)
Sporodochia oblong, pale rose-coloured, with septate, thick-walled setae which are hyaline and up to 500 × 5. Conidia hyaline, 4–6 × 1–1.5. On dead leaves of *C. acutiformis, C. paniculata, C. pseudocyperus* and *C. riparia*.

Volutella melaloma Berk. & Br. (Fig. 1999)
Sporodochia about 1mm diam., orange or pinkish orange, surrounded by brown setae up to 350 × 5–9. Conidia hyaline, 14–18 × 3–4, often with a small delicate setula at one end. On dead leaves of *C. acutiformis, C. elata, C. pseudocyperus* and *C. riparia*.

COELOMYCETES

Ascochyta sodalis Grove
Pycnidia up to 0.12mm diam. Conidia hyaline, 1-septate, 10–15 × 3–5. On dead leaves of *C. arenaria, C. nigra* and *C. ovalis*, Dec.–July.

Eriospora leucostoma Berk. & Br. (Fig. 2000)
Spots pitchy brown or black, thin at margin. Pycnidia about 0.5mm diam.

Conidia hyaline, 50–100 × 1, up to 8-septate. On dead leaves of *C. paniculata* and *C. riparia*, Feb.–Apr.

Neottiospora caricina (Desm.)Höhnel (Fig. 2001)
Pycnidia immersed, 0.2–0.4mm diam., olivaceous, opening by a broad operculum. Conidia issuing in slimy orange masses, 10–15 × 2–4, each with a funnel-shaped appendage at one end, which is best seen in water mounts. On dead leaves of *C. acutiformis*, *C. appropinquata*, *C. elata*, *C. pendula* and *C. riparia*, Mar.–Aug., common.

Septoria caricicola Sacc.
Pycnidia immersed, about 0.15mm diam. Conidia very pale yellow, with 6 to 8 (mostly 7) septa, 35–55 × 4. Causes round or almost round, whitish, brown-bordered spots on leaves of *C. acutiformis* and *C. riparia*, Apr.–Sept.

Septoria caricis Passer.
Conidia hyaline, guttulate and finally with a thin septum, 20–35 × 2.5–3. Mostly on dead tips of living leaves of *C. arenaria*, *C. paniculata* and *C. pendula*, July–Aug.

Stagonospora. Pycnidial fungi with two or more septa in their colourless or very pale yellow, fusiform or cylindrical conidia. Seven species are recorded on *Carex* in Britain.

KEY

    Conidia more or less cylindrical 1
    Conidia more or less fusiform 2
1. Pycnidia large, conidia 25–70 × 2.5–5 .... *macropycnidia*
    Pycnidia small, conidia 18–32 × 4–6 .... *vitensis*
2. Conidia 4–6 wide, mostly 2-septate .... *caricinella*
    Conidia 5–7 wide, mostly 5-septate .... *caricis*
    Conidia 8–11 wide, mostly 6- to 8-septate .... *paludosa*
    Conidia 10–14 wide, mostly 6- to 9-septate .... *gigaspora*
    Conidia 14–18 wide, mostly 5- to 8-septate .... *anglica*

Stagonospora anglica Cunnell
Pycnidia up to 0.5 × 0.35mm. Conidia hyaline with many small guttules in each cell, 68–109 × 14–18. On dead leaves of *C. acutiformis*, Nov.

Stagonospora caricinella Brun.
Pycnidia up to 0.2mm diam. Conidia hyaline, often acutely pointed at ends, 17–27 × 4–6, mostly with 2, rarely with 3 or 4, septa. On dead leaves of *C. acutiformis*, *C. pseudocyperus* and *C. riparia*.

Stagonospora caricis (Oud.)Sacc. (Fig. 2002)
Pycnidia up to 0.2mm diam. Conidia hyaline, 25–45 × 5–7, mostly with 5, rarely 4, 6 or 7, septa. On dead leaves of *C. acutiformis*, *C. arenaria*, *C. muricata*, *C. otrubae* and *C. pseudocyperus* all the year round.

Stagonospora gigaspora (Niessl)Sacc.
Pycnidia up to 0.2mm diam. Conidia 58–84 × 10–14. On dead upper parts of leaves of *C. riparia*.

Stagonospora macropycnidia Cunnell
Characterised by its very large pycnidia (up to 1mm diam.) and long, narrow conidia with 3 or rarely 5 septa. On dead leaves of *C. acutiformis, C. pseudocyperus* and *C. riparia*, Nov.–May.

Stagonospora paludosa (Sacc. & Speg.)Cunnell (Fig. 2003)
Pycnidia up to 0.2mm diam. Conidia hyaline, 41–67 × 8–11, mostly 6- to 8-septate. Common on dead leaves and stems of *C. acutiformis, C. pseudocyperus* and *C. riparia*, Sept.–Apr.

Stagonospora vitensis Unam. (Fig. 2004)
Pycnidia up to 0.35mm diam. Conidia with 2, 3 or occasionally 4 septa, 18–32 × 4–6, hyaline. On dead leaves and sheaths of *C. disticha, C. hirta, C. otrubae, C. ovalis* and *C. riparia*, May–Aug.

CLADIUM: Saw-sedge

Few fungi are found on living leaves of the great saw-sedge; in Britain there is only one rust and there are no smuts or downy or powdery mildews. When they eventually die, the tough leaves are for a long time resistant to decay and provide a suitable substratum for a number of fungi, ensuring at the same time close cover and insulation.

UREDINALES

Puccinia cladii Ell. & Tr., 11
Uredinia amphigenous, dark cinnamon brown; spores 28–40 × 20–30, minutely echinulate, July–Sept., uncommon. Telia not so far recorded in Britain.

DISCOMYCETES

Dasyscyphus controversus, described on grasses, has been found on *Cladium*.

Dasyscyphus imbecillis (P. Karsten)Sacc. var. *cladii* Dennis (Fig. 2005)
The variety on *Cladium* differs from *D. imbecillis* on *Eriophorum* in having multiguttulate ascospores, and septate paraphyses. The short-stalked apothecia have orange discs when freshly collected in July.

Hyaloscypha cladii Nag Raj & Kendrick
Found associated with *Chaetochalara cladii*, Apr. Apothecia up to 0.2mm diam., brown, with pale discs and pale brown hairs 40–75 × 2.5–3.5. Asci 45–60 × 10–12; spores hyaline, guttulate, 1-septate, 12–16 × 4–5. Paraphyses 1.5–2 thick.

Hymenoscyphus robustior, described on grasses, *Hysterostegiella fenestrata*, described on *Schoenoplectus*, and *Mollisia caricina*, described on *Carex*, occur also on *Cladium*.

OTHER ASCOMYCETES

Acanthophiobolus helicosporus, described on *Carex*, is found on *Cladium*.

Anthostomella. Three species are found on *Cladium*.

KEY

Ascospores 1-celled, 12–16 × 5–6 .... *leptospora*
Ascospores with large brown cell and small hyaline one, 18–28 × 5–7 ..
.... *fuegiana*
Ascospores with a brown central cell, a hyaline cell at one end and an appendage at the other .... *scotina*

Anthostomella fuegiana Speg. (Fig. 2006)
Perithecia up to 0.3mm diam., clypeus black, Jan–Sept.

Anthostomella leptospora (Sacc.) S. Francis (Fig. 2007)
Perithecia up to 0.25mm diam., clypeus small, Apr.–Dec.

Anthostomella scotina (Dur. & Mont.)Sacc. (Fig. 2008)
Perithecia up to 0.25mm diam., clypeus black, up to 0.4mm. Ascospores 20–22 × 3–5. Apr.–Dec.

Didymopleella cladii (Larsen & Munk)Munk (Fig. 2009)
Pseudothecia immersed, up to 0.25mm diam. Ascospores 13–16 × 5–6, each composed of a large, olivaceous brown cell and a small colourless one. Common on dead leaves and stems, Dec.–June.

Leptosphaeria cladii Cruchet (Fig. 2010)
Pseudothecia immersed or erumpent, black, up to 0.3mm diam. Ascospores mostly with 5 to 7, occasionally 8 septa; dark reddish brown, 27–34 × 7–8. On dead stems and leaves, Feb.–May.

Morenoina minuta and *M. paludosa*, described on *Juncus*, both occur also on *Cladium*.

Paraphaeosphaeria michotii (Westend.)O.Eriksson ex Shoemaker & Eriksson (Fig. 2011)

Pseudothecia immersed, up to 0.2mm diam. Ascospores 2-septate, golden brown, sometimes minutely verruculose, 14–22 × 4–6. On dead stems and leaves, Jan.–July, common.

Schizothyrium pomi (Fr. ex Mont.)v.Arx (Fig. 2012)
Thyriothecia superficial black 'fly-specks'; scutellum of interlocking cells like a jig-saw puzzle, splitting longitudinally or irregularly. Ascospores hyaline, 1-septate, 10–14 × 3–5. On dead leaves and stems.

HYPHOMYCETES

Chaetochalara cladii Sutton & Pirozynski (Fig. 2013)
Colonies effuse, setose; setae brown, 80–350 × 5–8. Conidiophores brown, 30–50 × 4.5–9.5. Conidia catenate, hyaline, 0- or 1-septate, 8–25 × 4–6.

Chalara cladii M.B.Ellis (Fig. 2014)
Colonies effuse, velvety, reddish brown to black. Conidiophore stipes brown, up to 860 × 3.5–8, terminal phialides 40–70 × 6.5–10. Conidia colourless to brown, end walls verruculose, 1-septate, 11–17 × 5–7. On dead leaves throughout the year.

Periconia curta, *P. digitata* and *P. laminella*, described on *Carex*, have been found also on *Cladium*.

Sporidesmium cladii M.B.Ellis (Fig. 2015)
Colonies effuse, grey. Conidiophores 5–68 × 2.5–5, brown, arising from a superficial mycelial network. Conidia pale brown, 4- to 13-septate, 50–180 × 7–8.5. On dead leaves, Feb.–Dec.

Sporidesmium paludosum M.B.Ellis (Fig. 2016)
Colonies effuse, dark blackish brown. Conidiophores brown, 20–30 × 5–6. Conidia with up to 41 septa, 60–800 × 9–12. On dead leaves and stems, May, uncommon.

Tetraploa aristata Berk. & Br. (Fig. 2017)
Colonies effuse, brown or dark greyish brown. Conidia brown, mostly 25–40 × 14–29, with septate appendages 12–80 long. A second smaller type of conidium with much longer appendages, up to 330, is sometimes formed. This curious fungus can be found all the year round and is especially common on saw-sedge. Spring gatherings show large numbers of developing conidia whereas those made in late summer, autumn and winter have mostly mature conidia.

Volutella arundinis, described on *Carex*, is found commonly also on *Cladium*.

COELOMYCETES

Pestalotiopsis disseminata (Thüm)Steyaert (Fig. 2018)
Acervuli becoming black and granular. Conidia 24–28 × 7–8, end cells hyaline, intermediate three cells pale brown, apical setulae up to 20 long, basal appendage 5–10 long. On dead leaves, sometimes found in great abundance.

ELEOCHARIS: Spike-rushes

USTILAGINALES

Entorrhiza scirpicola (Correns)Sacc. & Syd.
Spores ellipsoid, 13–16.5 × 9–13, yellowish or purplish brown when mature, with irregular spiral markings. In roots of *E. quinqueflora*, June–July, causing cylindrical or fusiform, whitish galls.

Ustilago marina Durieu
Sori in swollen rhizomes and roots of *E. parvula*. Spores pale brown or yellow, 9–13 diam. or 14–16 × 8–10. Rare.

DISCOMYCETES

Niptera phaea, *N. pulla* and *Psilachnum lateritio-album*, described on *Carex*, have been found on *E. palustris*.

Rutstroemia plana Henderson
Apothecia pale reddish brown, with flat yellowish-brown discs 2–4mm diam., and stalks 2–4 × 0.5–1mm, dark and tomentose at the base. Ascospores biguttulate, 14–16 × 4.5–5.5. On faintly bleached areas, surrounded by rather indistinct black lines, on dead stems of *E. palustris*, June.

Sclerotinia eleocharidis Henderson
Apothecia cupulate, fawn to brown, with discs 2–5mm diam., stalks 3–12 × 1mm, dark reddish brown and tomentose at base, arising from black sclerotia 5–10 × 1–1.5mm embedded in or attached to the previous season's dead stems of *E. palustris*, June. Ascospores hyaline, 12–15 × 5–7.

Scutomollisia operculata Nannf.
Apothecia superficial, at first covered by blackish brown shields up to 0.25mm diam., made up of radiating rows of cells, which are pushed off or over to one side as the apothecia develop. Excipulum mollisioid, marginal cells clavate. Ascospores hyaline, 8–11 × 2–2.5. On *E. multicaulis*, June.

OTHER ASCOMYCETES

Claviceps nigricans Tul.
Ergots black, up to 12mm long, often curved, projecting from inflorescences of *E. palustris* and *E. uniglumis*, Aug.–Oct., uncommon generally but locally abundant in East Anglia. Stromata formed Apr.–May; stalks at first bluish violet with cinnamon–buff heads, darkening to almost black with age. Perithecia markedly protuberant in old heads. Ascospores 80–90 × 1.

Loramyces juncicola Weston (Fig. 2019)
Perithecia superficial, 0.5–2mm diam., very dark brown, covered with short clavate hairs. Ascospores hyaline, 1-septate, 15–22 × 3.5–6, each with a gelatinous sheath and a 50–60 long slender basal appendage. On dead, mostly submerged stems of *E. palustris*, June–Aug.

Pleospora palustris Berl. (Fig. 2020)
Pseudothecia immersed, black, up to 0.25mm diam. Ascospores muriform, straw-coloured, mostly 40–50 × 14–17 when mature. On dead stems of *E. palustris*, June.

Pyrenophora scirpicola, described on *Schoenoplectus lacustris*, has been recorded on *E. palustris*.

HYPHOMYCETES AND COELOMYCETES

Pseudocercosporella scirpi, described on *Schoenoplectus*, has been found also on flowering stems of *E. palustris*.

Stagonospora heleocharidis Trail (Fig. 2021)
Pycnidia scattered, immersed, up to 0.2mm diam. Conidia pale yellowish, mostly 5- or 6-septate, 30–40 × 5–7. On dead leaves and stems of *E. palustris* and *E. uniglumis*, Jan.–June.

ERIOPHORUM: Cotton-grasses

DISCOMYCETES

Dasyscyphus fugiens, described on *Juncus*, has been found on *E. angustifolium*.

Dasyscyphus imbecillis (P.Karsten)Sacc. (Fig. 2022)
Apothecia up to 0.5mm diam., orange or flesh-coloured, with hyaline hairs up to 80 × 3–4. Ascospores hyaline, 15–23 × 2–2.5. On dead leaves of *E. angustifolium*.

Dasyscyphus sydowii Dennis
Apothecia 0.2–0.3mm diam., short-stalked, white then flesh-coloured. Hairs 45–50 × 3–4, minutely granulate. Asci 30–40 × 4; spores fusiform, hyaline, 9–10 × 1.5–2. Paraphyses lanceolate, 4 thick. On dead leaves of *E. angustifolium*, June.

Hyaloscypha paludosa Dennis (Fig. 2023)
Apothecia up to 1mm diam., white or off-white, fringed with smooth, tapered hairs, 40–110 × 3–4. Ascospores hyaline, 8–14 × 2–2.5. On dead leaves and stems of *E. angustifolium*, June–Oct.

Micropeziza cornea, described on *Carex*, has been recorded on *E. vaginatum*.

Microscypha ellisii Dennis var. *eriophori* Dennis
Apothecia up to 0.5mm diam., short-stalked, white; marginal hairs 30–35 × 2.5. Ascospores hyaline, 14–18 × 2. Paraphyses filiform. On dead leaves of *E. angustifolium*, Apr.

Mollisia palustris, described on *Juncus*, occurs on *E. angustifolium*.

Myriosclerotinia dennisii (Svrček)Schwegler
Apothecia up to 9mm diam., dark reddish brown, furfuraceous, with blackish hairy stalks up to nearly 2cm arising from narrow, cylindrical, black sclerotia about 1.5cm long. Ascospores hyaline, 11–17 × 4–5.5. On dead stems of *E. angustifolium* and *E. vaginatum*, Apr.–May.

Niptera pilosa, described on *Carex*, has been found on *E. angustifolium*.

Pezizella eriophori Dennis
Apothecia up to 1mm diam., sessile or short-stalked, yellow; paraphyses full of yellowish droplets. Ascospores hyaline, 8–12 × 2.5–3. On rotting leaves of *E. angustifolium*, June–July.

Rutstroemia henningsiana (Ploettn.)Dennis (Fig. 2024)
Apothecia 1–3.5mm diam., buff or yellowish brown, with stalks up to 10 × 1mm, dark brown at base. Ascospores hyaline, 13–16 × 4–6. On dead leaves of *E. angustifolium*, surrounded by thin black lines, Apr.–May.

OTHER FUNGI

Mycosphaerella lineolata, described on grasses, has been recorded on *Eriophorum*.

Periconia digitata and *Stagonospora vitensis*, described on *Carex*, have been found on *E. angustifolium*.

JUNCUS: Rushes

UREDINALES

Puccinia cancellata Sacc., 11, 111
A rare rust found only on *J. acutus*. Urediniospores verruculose, 22–23 diam. Teliospores 1-septate, smooth, 35–45 × 20–28.

Uromyces junci (Desm.)Tul., 11, 111
Aecial host *Pulicaria dysenterica*. Urediniospores golden brown, finely echinulate, 20–28 × 16–22. Teliospores dark brown, smooth, 25–40 × 12–18. On *J. effusus*, *J. inflexus* and *J. subnodulosus*, from July on.

USTILAGINALES

Entorrhiza aschersoniana (Magnus)Lagerh.
Sori in swellings in roots of *J. bufonius* up to 10 × 3mm, July. Spores yellowish brown, verrucose, 17–20 × 15–17.

Entorrhiza casparyana (Magnus)de Toni
Spores 14–21 diam., golden brown, coarsely verrucose. In roots of *J. articulatus*, causing swellings up to 1cm long, Aug.

Urocystis junci Lagerh.
Sori in lower parts of stems of *J. acutiflorus*, Aug. Spore mass powdery, blackish brown. Spore balls 20–50 diam., composed of 1 to 4 spores, each 12–16 diam., surrounded by yellowish sterile cells 7–10 × 3–5.

DISCOMYCETES

Belonopsis iridis, described on *Iris*, has been recorded on *Juncus*.

Ciboria juncorum Velen.
Apothecia up to 2mm diam., yellowish, with slender stalks. Ascospores ovoid, hyaline, uniseriate in asci, 8–11 × 4–5. Paraphyses 2 thick. On fallen fruits, June.

Crocicreas culmicolum, described on grasses, has been found on *J. subnodulosus*.

Cudoniella junciseda (Velen.)Dennis
Apothecia up to 2mm diam., short-stalked, white, drying pale brown. Ascospores slightly clavate, hyaline, 9–14 × 2.5–4. On dead fallen stems and fruits of *J. effusus*, July.

Dasyscyphus species are found in great profusion on dead parts of rushes, usually just above water level.

KEY

    Ascospores more than 30 long .... *apalus*
    Ascospores 15–20 long, clavate .... *clavisporus*
    Ascospores not more than 15 long 1
1.   Hairs colourless, about 20 × 4 ....*fugiens*
    Hairs hyaline becoming yellowish brown .... *diminutus*
    Hairs dark brown to black .... *rehmii*

Dasyscyphus apalus (Berk. & Br.)Dennis (Fig. 2025)
The commonest species, especially on dead *J. effusus* and *J. inflexus* during the winter months. Apothecia up to 1mm diam., pale reddish brown, with long white hairs. Ascospores hyaline, up to 45 × 1.5, becoming 1- to 3-septate.

Dasyscyphus clavisporus Mouton
Apothecia 0.5mm diam., discs yellow, stalks buff, hairs white or pale brown, finely granulate, 45–70 × 4–5. Ascospores clavate, hyaline, 15–20 × 3–4. Paraphyses not more than 2 wide. On dead stems of *J. effusus*, July.

Dasyscyphus diminutus (Rob. ex Desm.)Sacc. (Fig. 2026)
Apothecia up to 0.5mm diam., pale brown with yellowish discs, very short-stalked. Hairs 30–40 × 3–4, hyaline becoming yellowish brown. Ascospores hyaline, 10–15 × 1.5. On dead stems of *J. effusus*, *J. inflexus* and *J. subnodulosus*, May–Oct.

Dasyscyphus fugiens (Phill. ex Bucknall)Massee (Fig. 2027)
Minute whitish apothecia not more than 0.2mm diam. Hairs colourless, granulate, about 20 × 4. Ascospores hyaline, 6–8 × 1.5. Very common on dead leaves, stems and inflorescences of *J. acutiflorus*, *J. articulatus*, *J. effusus*, *J. inflexus*, *J. maritimus* and *J. subnodulosus*, Dec.–Oct.

Dasyscyphus rehmii (Staritz)Sacc. (Fig. 2028)
Apothecia up to 1.5mm diam., yellowish, covered with septate, dark brown hairs 60–80 × 4–5, coarsely granulate but often with bare zones below the septa. Ascospores hyaline, 10–15 × 2–2.5. On *J. conglomeratus* and *J. squarrosus*, July–Aug.

Gorgoniceps micrometra (Berk. & Br.)Sacc.
Apothecia 0.1–0.5mm diam., blackish, clothed with very dark brown hairs up to 50 × 2–3, attached to substrate by white fibrils. Ascospores hyaline, multiseptate, about 80 × 2. On dead leaves, Apr., rare.

Hyaloscypha paludosa, described on *Eriophorum*, has been found also on *J. effusus* and *J. inflexus*.

Hysteropezizella species usually have pale greyish brown apothecia which are difficult to find; they are very small, usually less than 0.3mm diam., and are sunken in the host tissues, becoming revealed only when a disc of epidermis comes away. Ascospores hyaline.

KEY

Ascospores 10–12 long .... *pusilla*
Ascospores 15–18 long .... *exigua*
Ascospores 18–25 long .... *rehmii*

Hysteropezizella exigua (Desm.)Nannf. (Fig. 2029)
Ascospores 15–18 × 3–4, eventually often 1-septate. On *J. articulatus*, *J. inflexus* and *J. subnodulosus*, Apr.–Aug.

Hysteropezizella pusilla (Lib.)Nannf. (Fig. 2030)
Ascospores 10–12 × 3–3.5. On dead stems of *J. conglomeratus* and *J. effusus*, May–Sept.

Hysteropezizella rehmii (Jaap)Nannf. (Fig. 2031)
Ascospores 18–25 × 5–6.

Micropeziza cornea, described on *Carex*, occurs also on *J. effusus*, and *M. karstenii*, described on grasses, has been found on *J. articulatus*.

Mollisia species with brownish saucer-shaped apothecia are quite common on rushes.

KEY

Ascospores 5–7 × 1, often curved .... *caricina*
Ascospores mostly 9–10 × 1.5–2 1
Ascospores 20–28 × 2–3 .... *junciseda*
1. Ascospores mostly tapered towards base .... *palustris*
   Ascospores more or less cylindrical .... *juncina*

Mollisia caricina, described on *Carex*, has been found on *J. acutiflorus* × *articulatus* and *J. subnodulosus*.

Mollisia juncina (Pers.)Rehm (Fig. 2032)
Apothecia up to 1mm diam., at first yellowish, becoming dark brown or greyish brown. Ascospores hyaline, more or less cylindrical, often with small guttules, 9–10 × 1.5–2. On *J. articulatus*, *J. effusus* and *J. inflexus*, May–Sept.

Mollisia junciseda Karst.
Apothecia about 0.4mm diam. Characterised by its very large ascospores. In Britain recorded only on an unidentified *Juncus* species in Mull. It has been found in Europe on *J. conglomeratus*, *J. tenuis* and *J. trifidus*.

Mollisia palustris (Rob. ex Desm.)P. Karsten (Fig. 2033)
Apothecia less than 1mm diam., pale brown, sometimes darker in the centre

where they are attached, very pale at the rim, like little saucers. Ascospores hyaline, usually tapered towards the base, 9–10 (14) × 1.5–2. On rotting stems of *J. effusus*, *J. inflexus* and *J. subnodulosus*, Mar.–Oct.

Myriosclerotinia curreyana (Berk. ex Currey)Buchw. (Fig. 2034)
Apothecia 3–5mm diam., brown, often in small groups, bursting out from black sclerotia embedded in dead stems of *J. effusus* and, less commonly, *J. conglomeratus*. They are found Apr.–May, especially where a thick mat of stems is bent over and lying on wet ground. Ascospores hyaline, 7–15 × 1.5–2.5. The sclerotia are pink inside. A microconidial state with black condiomata up to 1 × 0.5mm opening by little slits, and spherical hyaline conidia 2–2.5 diam., is seen July–Sept.

Niptera apothecia resemble those of *Mollisia* but have less distinct margins and the ascospores are mostly septate. Four species occur on *Juncus*.

KEY

Ascospores 12–20 × 2, becoming 1-septate .... *melatephra*
Ascospores mostly 19–20 × 4–5, 1-septate .... *phaea*
Ascospores mostly 20–22 × 5–7, 1- to 3-septate .... *lacustris*
Ascospores 30–45 × 3.5–5, 3-septate .... *pulla*

Niptera lacustris (Fr.)Fr.
Apothecia erumpent, about 1mm diam., dark reddish brown. On dead, greyish standing stems of *J. effusus* in swampy areas, Oct.

Niptera melatephra (Lasch)Rehm (Fig. 2035)
Apothecia erumpent, cup-shaped, 0.2–0.6mm diam., grey, becoming blackish brown, with pallid discs. Ascospores tapered at each end, hyaline, packed with small guttules, and becoming 1-septate when mature, 12–20 × 2. On dead stems of *J. conglomeratus*, *J. effusus*, *J. inflexus* and *J. subnodulosus*, June–July.

Niptera phaea and *N. pulla*, described on *Carex*, have been found also on *Juncus*.

Pezizella nigrocorticata, described on *Carex*, has been recorded on *J. effusus*.

Rutstroemia lindaviana, described on *Phragmites*, has been found on *J. subnodulosus*.

Scutomollisia stenospora Nannf. (Fig. 2036)
Apothecia mollisioid, up to 0.3mm diam., developing beneath small shields of brown radiating hyphae. Ascospores hyaline, 16–22 × 2–2.5. On dead stems of *J. subnodulosus*, June–Oct.

Stictis elevata (P.Karsten)P. Karsten
Apothecia up to 0.3mm diam., with lobed, split, white sterile margins and deeply sunken yellow discs. Ascospores about 180 × 1, hyaline, multiseptate. On dead stems of *J. conglomeratus*, Aug.

Unguicularia costata (Boud.)Dennis (Fig. 2037)
Apothecia like little grey urns, 0.2–0.6mm diam., with refractive hairs 30–40 × 4–5. Ascospores hyaline, 9–10 × 1–2. On dead stems of *J. effusus*, Oct.–July.

OTHER ASCOMYCETES

Acanthophiobolus helicosporus, described on *Carex*, has been found on *J. subnodulosus*.

Anthostomella tomicoides, described on *Carex*, occurs also on *J. effusus*.

Creopus spinulosus, described on grasses, is found sometimes on *Juncus*.

Hypocrea pilulifera Webster & Rifai
Stromata pulvinate or subglobose, whitish or dirty yellow, up to 3.5mm diam., containing numerous sunken perithecia. Ascospores hyaline, minutely echinulate, unequally divided into two parts; upper part-spore subglobose, 4–6 diam., lower part-spore 4.5–6 × 3–4.5. On rotten stems of *J. effusus*, July.

Leptosphaeria juncicola Rehm ex Winter
Pseudothecia immersed, up to 0.2mm diam. Ascospores yellowish, 35–45 × 4–4.5, 3-septate, cells at each end longer than intermediate ones. On *J. trifidus*.

Leptosphaeria juncina (Auersw.)Sacc. (Fig. 2038)
Pseudothecia immersed, up to 0.1mm diam. Ascospores pale straw-coloured, 35–45 × 4–6, 3-septate, cells at each end shorter than intermediate ones. On dead stems of *J. conglomeratus* and *J. effusus*, July–Aug.

Lophiostoma alpigenum Fuckel var. *juncinum* Mouton (Fig. 2039)
Pseudothecia immersed, black, 0.2–0.5mm diam., with protuberant, laterally compressed necks. Ascospores hyaline, 8- to 10-septate, 35–45 × 7–8. On dead stems of *J. conglomeratus* and *J. subnodulosus*, June.

Loramyces juncicola, described on *Eleocharis*, has been recorded on *Juncus*.

Monascostroma innumerosa (Desm.)Höhnel (Fig. 2040)
Pseudothecia gregarious, lying beneath the epidermis, about 0.1mm diam., each containing one or a few asci. Ascospores 1-septate, becoming brown and minutely verruculose, 15–24 × 4–8. On dead stems of *J. effusus*, July–Sept.

Morenoina, with elongated, black, often branched thyriothecia usually less than 1mm long. Ascospores 1-septate, hyaline to pale straw-coloured, often minutely verruculose. Three species are found on dead leaves, sheaths and stems. Most collections have been made in May but there are records for nearly every month of the year on *J. effusus*.

Fungi on Rushes, Sedges, Bur-reeds and Reedmaces

KEY

Ascospores not more than 10 long .... *minuta*
Ascospores 9–15 long 1
1. Pycnothyria present, with conidia 5–7 × 1.5–2 .... *paludosa*
   Pycnothyria absent .... *fimbriata*

Morenoina fimbriata J.P.Ellis (Fig. 2041)
Thyriothecia sometimes branched, 0.3–1.1 × 0.08–0.12mm. Ascospores 12–15 × 4–5.5.

Morenoina minuta J.P.Ellis (Fig. 2042)
Thyriothecia unbranched, 0.1–0.25 × 0.04–0.06mm. Ascospores 7–9 × 2.5–3.5.

Morenoina paludosa J.P.Ellis (Fig. 2043)
Thyriothecia usually much branched, 0.16–0.8 × 0.06–0.1mm. Ascospores 9–11.5 × 3–4.

Mycosphaerella recutita, described on grasses, occurs also on *Juncus*.

Niesslia exosporioides, described on *Carex*, has been found on *J. subnodulosus*.

Paraphaeosphaeria michotii, described on *Cladium*, occurs on *J. inflexus*.

Phomatospora dinemasporium, described on grasses, has been recorded on *Juncus*.

Phomatospora therophila (Desm.)Sacc. (Fig. 2044)
Perithecia immersed, up to 0.25mm diam. Ascospores hyaline, 7–10 × 2.5–3. Recorded on *J. effusus*, *J. inflexus* and *J. subnodulosus*.

Phyllachora junci (Alb. & Schw.)Fuckel (Fig. 2045)
Stromata dark grey or black, 1–4 × 0.5–1mm, sometimes confluent, enclosing numerous small perithecia. Ascospores hyaline, mostly 12–15 × 3–4, becoming 1-septate. Probably the commonest fungus on dead stems and leaves of *J. effusus* and *J. conglomeratus*, Sept.–July.

Pleospora aquatica Griffiths (Fig. 2046)
Pseudothecia 0.1–0.2mm diam., black. Ascospores straw-coloured, with 5 transverse septa, and 1 longitudinal septum in all except the end cells, 27–33 × 11–13. On *J. articulatus*, Oct.

Pleospora valesiaca (Niessl)E. Müller (Fig. 2047)
Pseudothecia subepidermal, black, about 0.2mm diam. Ascospores muriform, golden brown, 35–50 (55) × 15–20. *Alternaria* conidial state has rostrate, pale straw-coloured conidia, with up to 13 septa or pseudosepta, 80–150 × 9–12. On dead stems of *J. maritimus*, Apr.–May.

Pyrenophora scirpicola, described on *Schoenoplectus*, has been found on *J. effusus*.

Tarbertia juncina Dennis
Ascomata subcuticular, up to 0.15mm diam., at first black and closed, the upper walls later flaking off irregularly to leave teeth, and then resembling a *Coccomyces* with a pale yellow disc. Ascospores hyaline, with 3 transverse septa and 1 longitudinal septum, 11–12 × 6. On stems of *J. conglomeratus*, Oct.

Trichothyrina nigroannulata, described on grasses, has been found on *J. conglomeratus*.

HYPHOMYCETES

Arthrinium curvatum var. *minus*, described on *Schoenoplectus*, occurs also on *J. conglomeratus*.

Arthrinium cuspidatum (Cooke & Harkn.)Tranzschel (Fig. 2048)
Colonies black, pulvinate or forming bands half-encircling stems and leaves of *J. gerardi*, conspicuous but uncommon. Conidiophores hyaline with thick, dark brown transverse septa. Conidia with outwardly curved horn-like processes, mid to dark brown, 15–32 × 7–11.

Cercospora juncicola (Hori & Kasai)Vassiljevsky (Fig. 2049)
Conidiophores pale to mid brown, 50–90 × 4–6. Conidia pale straw-coloured, often minutely verruculose, 2- to 7-septate, 40–140 × 3–5. On dead leaves of *J. effusus*, uncommon.

Cercosporella junci MacGarvie & O'Rourke (Fig. 2050)
Conidiophores very short. Conidia hyaline, 3- to 8-septate, 45–75 × 5–7.5. On living and dead leaves of *J. effusus*.

Chaetochalara cladii, described on *Cladium*, has been found a number of times on dead *Juncus effusus*.

Conoplea fusca Pers. (Fig. 2051)
Colonies brown, powdery. Conidiophores brown, minutely echinulate except at base, spirally twisted, up to 400 × 3–5. Conidia brown, smooth or minutely echinulate, with germ pore near base, mostly 6–8 × 4–5. On dead stems of *J. effusus*, uncommon.

Dactylaria junci M.B.Ellis (Fig. 2052)
Colonies effuse, thin, pale greyish brown, scarcely visible to the naked eye. Conidiophores brown, up to 120 × 3–5. Conidia hyaline to pale brown, 1-septate, 15–30 × 4–5. On dead leaves of *J. effusus*.

Diplorhinotrichum juncicola MacGarvie (Fig. 2053)
Conidiophores hyaline, 20–30 long with up to 8 denticles. Conidia hyaline, 1-septate, 21–28 × 4–5. On leaves of *J. effusus* causing eye-shaped or circular, pale brown or grey lesions, each surrounded by a dark purple border, Apr.–May.

Myrothecium masonii, described on grasses, has been recorded also on *Juncus*.

Periconia atra, *P. curta*, *P. digitata* and *P. funerea*, described on *Carex*, and *P. hispidula*, described on grasses, have been found on *Juncus*. All five species occur on *J. effusus*, and *P. curta* and *P. digitata* are recorded also on *J. subnodulosus*.

Selenosporella curvispora MacGarvie (Fig. 2054)
Colonies effuse, inconspicuous. Conidiophores brown, up to 250 × 5–6. Conidia hyaline or pale olive, 5–7 × 0.5. On dead leaves of *J. effusus*.

Tetraploa aristata, described on *Cladium*, has been found on *J. subnodulosus*.

Volutella arundinis, described on *Carex*, occurs on *J. effusus*.

COELOMYCETES

Actinothyrium graminis, described on *Molinia*, is found occasionally on *Juncus*.

Dinemasporium state of *Phomatospora dinemasporium*, described on grasses, occurs also on *Juncus*.

Discula, v.s. under *Myriosclerotinia curreyana* (conidial state).

Leptostroma juncacearum Sacc. (Fig. 2055)
Conidiomata subcuticular, flat, shield-shaped, black, opening by a slit. Conidia hyaline, 4–5 × 0.5–1. On dead stems of *J. conglomeratus*, *J. effusus*, *J. maritimus* and *J. subnodulosus*, Feb.–Sept., common.

Psammina bommeriae, described on *Ammophila*, is found occasionally on *J. effusus*.

Septoriella junci (Desm.)Sutton (Fig. 2056)
Conidiomata in lines, subepidermal, blackish brown. Conidia smooth, hyaline to very pale straw-coloured, multiguttulate or 1- to 7-septate, 50–80 × 2.5–3, often with a small gelatinous cap or appendage at one end or both ends. On dead stems of *J. conglomeratus*, *J. effusus* and *J. maritimus*, Feb.–Oct.

Stagonospora. Four species are found on *Juncus*:

KEY

    Conidia not more than 3.5 wide .... *junciseda*
    Conidia 4 or more wide 1
1. Conidia 9–10 wide .... *socia*
    Conidia not more than 7 wide 2
2. Conidia mostly with 2 or 3, rarely 4, septa .... *vitensis*
    Conidia with 2 to 5, mostly 4 septa .... *innumerosa*

Stagonospora innumerosa (Desm.)Sacc. (Fig. 2057)
Pycnidia numerous, up to 0.1mm diam., black, often arranged in parallel lines. Conidia smooth, hyaline, with 2 to 5, mostly 4, septa, 19–23 × 5–7. On dead stems and leaves of *J. effusus* and *J. maritimus*, July–Sept., common.

Stagonospora junciseda Sacc. (Fig. 2058)
Pycnidia up to 0.2mm diam. Conidia 3-septate, 21–30 × 3–3.5. On *J. subnodulosus*, June.

Stagonospora socia Grove
Pycnidia small, black. Conidia oblong–cylindrical, sometimes tapered at one or both ends, 4-septate, 30–35 × 9–10. On dead stems of *J. conglomeratus*, Sept.

Stagonospora vitensis, described on *Carex*, has been recorded on *J. effusus*.

LUZULA: Woodrushes

UREDINALES

Puccinia luzulae Lib., 11, 111
Uredinia amphigenous, reddish brown; spores almost or quite smooth, 30–45 × 12–15. Telia blackish brown; spores 1-septate, 45–80 × 16–24. On *L. pilosa*, uncommon.

Puccinia obscura Schröter, 11, 111
Aecial host *Bellis perennis*. Uredinia mostly hypophyllous, rusty gold, on purplish-brown spots; spores with echinulate walls, 18–25 × 15–21. Telia blackish brown; spores 1-septate, 30–50 × 14–20. Common on *L. campestris, L. forsteri, L. multiflora, L. pilosa* and *L. sylvatica*.

DISCOMYCETES

Dasyscyphus luzulinus (Phill.)Massee
Apothecia minute, short-stalked, white, pubescent, with pale yellow discs. Ascospores hyaline, fusiform, 12–20 × 5–6. At the base of dead leaves of *L. sylvatica*.

Laetinaevia luzulae Spooner (Fig. 2059)
Apothecia about 0.3mm diam., immersed, becoming erumpent, pinkish grey when fresh, drying reddish orange. Ascospores hyaline, straight or slightly curved, 30–33 × 4. On fading leaves of *L. arcuata*, July.

Orbilia luzularum Velen.
Apothecia 0.3–0.5mm diam., pellucid, colourless. Asci 25 × 5; spores narrowly acicular, straight, 5–8 long. On leaf bases of *L. sylvatica*, Mar.–May.

OTHER ASCOMYCETES

Anthostomella fuegiana, described on *Cladium*, is not uncommon on dead leaves of *L. sylvatica*, Jan.–Aug.

Anthostomella punctulata, described on *Carex*, has been found on *L. pilosa*, Mar.

Lembosia luzulae (Lib.)Höhnel (Fig. 2060)
Thyriothecia black, 0.15–0.25 × 0.06–0.08mm. Ascospores pale brown, 1-septate, 10–12 × 3–4. On dead leaves of *L. sylvatica*, May.

Leptosphaeria epicalamia (Riess)Ces. & de Not. (Fig. 2061)
Pseudothecia immersed, 0.3–0.4mm diam. Ascospores pale straw-coloured, 5-septate, 20–25 × 5–6. On dead stems and leaves of *L. sylvatica*, Mar.–June.

Leptosphaeria herpotrichoides, described on grasses, has been recorded on *Luzula*.

Stioclettia luzulina Dennis
Perithecia gregarious, immersed in a greyish stroma, necks excentric. Asci diaporthaceous; spores uniseriate, hyaline, with 6 large guttules, finally 5-septate, 25–34 × 5–6. On dry leaves of *L. sylvatica*, Sept.

COELOMYCETES

Septoria minuta Schröter
Pycnidia crowded, 0.04–0.06mm diam., black. Conidia falcate, hyaline, 10–20 × 1–2. On living and dying leaves of *L. sylvatica*, July–Sept.

Stagonospora aquatica Sacc. var. *luzulicola* Sacc. & Scalia
Pycnidia gregarious, immersed, 0.15mm diam. Conidia oblong, mostly rounded at ends, rarely acute, often truncate at base, greenish hyaline, 3-septate, 23–28 × 3.5–4. On dead leaves of *L. arcuata*.

Stagonospora luzulae (Westend.)Sacc.
Pycnidia amphigenous, scattered or in rows, immersed, black, up to 0.15mm diam. Conidia hyaline to pale straw-coloured, becoming 3-septate, 12–15 × 3–4. On withered leaves of *L. pilosa* and *L. sylvatica*, Mar.–May.

RHYNCHOSPORA: White Beak-sedge

Anthracoidea caricis, described on *Carex*. Sori of this smut have been found occasionally in spikelets of *R. alba*, replacing ovaries.

SCHOENOPLECTUS: Bulrushes

UREDINALES

Puccinia scirpi DC., 11, 111
Aecial host *Nymphoides peltata*. Uredinia reddish brown, often confluent, long covered by epidermis; spores 19–32 × 12–24. Telia dark brown; spores 1-septate, 30–60 × 12–24. A rare rust on *S. lacustris*.

DISCOMYCETES

Coleosperma lacustre Ingold (Fig. 2062)
Apothecia superficial, sessile, up to 0.3mm diam., white, soft-fleshed. Ascospores hyaline, packed with small oil drops and often surrounded by colourless sheaths, mostly 22–32 × 7–8. On submerged parts of dead stems of *S. lacustris*, May–Sept.

Hypoderma scirpinum DC.
Apothecia black, shining, up to 2 × 0.6–0.7mm, opening by a slit to expose the disc. Ascospores pale straw-coloured, guttulate, 40–56 × 5–6. It has a *Leptostroma* state with black conidiomata 0.2mm diam., and hyaline conidia 3–3.5 × 1–1.5. On dead stems of *S. lacustris*.

Hysterostegiella fenestrata (Desm.)Höhnel
Apothecia 0.1–0.4mm diam., immersed, with honey-coloured discs exposed by the lifting up of small lids. Ascospores hyaline, oblong to ellipsoid, 6–8 × 1.5–2, biguttulate. Paraphyses septate, narrowly lanceolate, 15–20 longer than asci. On dead stems of *S. lacustris* and *S. tabernaemontani*, Aug.

Myriosclerotinia scirpicola (Rehm)Buchwald (Fig. 2063)
Apothecia mostly 6–10mm diam., pale fawn to brown, downy outside; stipes rather short with dark hyphae at base. Ascospores hyaline, usually 11–16 × 5–7. Apothecia are formed in May and arise singly, or in groups, from lobed sclerotia which had become detached from dead stems of *S. lacustris* in winter and been blown ashore. Black pustules 0.5–1mm long belonging to the *Myrioconium* state are seen on stems in summer a little way below the inflorescences; they produce large numbers of hyaline, spherical conidia 2–3 diam. Sclerotia are formed lower down.

OTHER ASCOMYCETES

Didymella proximella, described on *Carex*, and *Leptosphaeria eustoma*, described on grasses, occur on *S. tabernaemontani*.

Leptosphaeria scirpina Winter (Fig. 2064)
Pseudothecia immersed, 0.2–0.25mm diam. Ascospores becoming 5-septate, hyaline, 28–38 × 5–8, enclosed in thick mucilaginous sheaths. On dead stems of *S. lacustris* and *S. tabernaemontani*.
Puccinia eriophori Thuem., 11, 111

Leptosphaeria sowerbyi (Fuckel)Sacc. (Fig. 2065)
Pseudothecia gregarious, immersed, blackish brown, up to 0.1mm diam. Ascospores pale straw-coloured to mid pale golden brown, 6-septate, 45–60 × 5–7. On dead stems of *S. lacustris*.

Metasphaeria mosana Mouton (Fig. 2066)
Pseudothecia immersed, 0.2mm diam. Ascospores hyaline, 3-septate, 40–48 × 9–12. On rotten stems of *S. lacustris*.

Pleospora rubelloides, described on grasses, has been found on *S. tabernaemontani*.

Pyrenophora scirpicola (DC.)E.Müller (Fig. 2067)
Pseudothecia immersed, up to 0.5mm diam., black, flattened. Ascospores pale yellow, muriform, mostly 50–60 × 16–20, sometimes with mucilaginous sheaths. *Alternaria* state with straw-coloured to golden-brown conidia up to 300 × 9–16. On dead standing stems of *S. lacustris* and *S. tabernaemontani*, Sept.–May.

HYPHOMYCETES AND COELOMYCETES

Alternaria, v.s. under *Pyrenophora scirpicola*.

Arthrinium curvatum Kunze var. *minus* M.B.Ellis (Fig. 2068)
Sporodochia compact, round, up to 0.3mm diam., very dark brown to black. Conidia curved, brown, 8–11 × 5–6. On dead stems of *S. tabernaemontani*, Feb.–Nov.

Myrioconium, v.s. under *Myriosclerotinia scirpicola*.

Myrothecium cinctum, described on grasses, has been found on *S. lacustris*.

Pseudocercosporella scirpi (Moesz)Deighton (Fig. 2069)
Conidia hyaline, mostly 35–90 × 3. On stems, causing irregular greyish brown spots with yellow haloes, Sept.–Oct.

Stagonospora aquatica Sacc.
Pycnidia immersed, blackish brown, up to 0.15mm diam. Conidia cylindric-fusoid, slightly curved, 3- to 6-septate, 25–40 × 5–6. On dead stems of *S. lacustris* and *S. tabernaemontani*.

SCHOENUS: Bog-rush

Drepanopeziza schoenicola Graddon (Fig. 2070)
Apothecia blackish brown, up to 0.25mm diam., erumpent from a pale stroma. Ascospores 4 per ascus, hyaline, 17–19 × 7. Paraphyses curved, upper part brown, somewhat swollen. On dry leaves of *S. nigricans*, Aug.–Sept.

SCIRPUS: Club-rushes

UREDINALES

Uromyces lineolatus (Desm.)Schröter, 11, 111
Aecial hosts include *Berula erecta*, *Glaux maritima* and three species of *Oenanthe*. Uredinia hypophyllous, cinnamon, up to 1mm long; spores minutely echinulate, 23–35 × 17–25. Telia blackish brown; spores smooth, 25–45 × 15–25. On *S. maritimus*, uncommon.

OTHER FUNGI

Arthrinium state of *Pseudoguignardia scirpi* Gutner (Fig. 2071)
Colonies amphigenous, oval to round, often confluent, dark brown, up to 0.6mm long. Conidia brown, curved, 11–15 × 6–8.

Dasyscyphus carneolus var. *longisporus* and *Hymenoscyphus robustior*, described on grasses, have been found on *S. maritimus*.

Micropeziza cornea, described on *Carex*, occurs on *S. sylvaticus*.

Mollisia juncina and *M. palustris*, described on *Juncus*, have been found on *S. maritimus*.

Niptera pilosa, described on *Carex*, and *Pezizella eburnea*, described on grasses, have been found on *S. maritimus*.

Psilachnum asemum, described on *Carex*, occurs on *S. sylvaticus*.

SPARGANIUM: Bur-reeds

Many of the fungi found on *Typha* have been recorded also on *S. erectum* and *S. simplex*; the one most commonly seen is *Ascochyta typhoidearum*.

TRICHOPHORUM: Deer-grass

UREDINALES

Puccinia eriophori Thuem., 11, 111
Aecial host *Solidago virgaurea*. Uredinia cinnamon brown; spores 20–24 × 18–22. Telia black; spores 1-septate, 40–60 × 17–24. Rare, reported only from Scotland.

USTILAGINALES

Anthracoidea scirpi (Kühn)Kukkonen
Sori in ovaries, 2–3mm diam., the whitish covering membrane flaking away to expose a hard, black, agglutinated spore mass. Spores dark reddish brown, mostly 16–21 diam., smooth to minutely verruculose.

ASCOMYCETES

Dibeloniella trichophoricola Graddon (Fig. 2072)
Apothecia erumpent, up to 0.2mm diam., with pale discs and smooth, dark rims. Ascospores hyaline, 1-septate, 10–14 × 3. On dead stems, Aug.–Oct.

Didymella refracta (Cooke)Sacc.
Pseudothecia scattered, covered by bleached epidermis. Ascospores hyaline, 1-septate, with 4 guttules, about 35 × 15. On dead stems.

Hysteropezizella foecunda (Phill.)Nannf.
Apothecia up to 0.3mm diam., sunken, blackish brown with paler fringe. Ascospores hyaline, 14–19 × 3–4. Paraphyses about 3 wide at tip. On dead stems and leaves.

Hysteropezizella hebridensis Graddon
Apothecia pale as they emerge, splitting host tissue lengthwise without forming any operculum, up to 0.3mm wide. Ascospores fusiform, slightly curved, hyaline, guttulate, eventually 3- or 4-septate, 30–36 × 4–5. Paraphyses 3–4 broad at apex. On dry leaves, Sept.

Mollisia fusco-paraphysata Graddon
Apothecia superficial, blackish brown, up to 0.35mm diam. Ascospores fusiform, hyaline, 1-septate, 13–16 × 3–3.5. Paraphyses often branched, swollen to 5 and brown at their tips. On dead stems and leaves.

Niptera phaea, described on *Carex*, has been recorded on *Trichophorum*.

Sclerotinia gregoriana Palmer
Apothecia 1–3.5mm diam., pale yellowish to orange brown, stalks up to 14 ×

5mm, lower part covered with brown hyphae, arising from cylindrical sclerotia inside rotting stems. Ascospores hyaline, 8–18 × 2.5–5.5, ellipsoid to fusiform, often broader at one end and flattened on one side, sometimes becoming 1-septate. Paraphyses simple or branched, up to 3 thick. Recorded June–Oct.

COELOMYCETES

Tiarosporella paludosa (Sacc. & Fiori)Höhnel (Fig. 2073)
Pycnidia black, up to 0.25mm diam., immersed in old, decomposing stems and leaves, with small ostioles surrounded by a blackish brown clypeus. Conidia hyaline, 25–45 × 4.5–7, with tentacle-like apical appendages.

TYPHA: Reedmaces

Most records have been made on dead stems and leaves of *T. latifolia* but many of the fungi found on this host grow equally well on *T. angustifolia* and also on *Sparganium*.

DISCOMYCETES

Dasyscyphus controversus, described on grasses, *D. fugiens*, described on *Juncus*, *Hyaloscypha paludosa*, described on *Eriophorum*, and *Hymenoscyphus robustior*, described on grasses, have all been found on *Typha*.

Lophodermium typhinum (Fr.)Lambotte (Fig. 2074)
Apothecia 0.5–0.75mm long, black, each opening by a slit. Ascospores hyaline, mostly 45–55 × 1. Recorded Apr.–Sept.

Micropeziza cornea, described on *Carex*, occurs also on *Typha*.

Mollisia typhae (Cooke)Sacc.
Apothecia minute, erumpent, black with smoky discs. Ascospores fusiform, hyaline, 12–14 × 2.5–3.

Niptera pilosa, described on *Carex*, has been recorded on *Typha*.

Orbilia arundinacea Velen. (Fig. 2075)
Apothecia 1–5mm diam., sessile, often slightly lobed, white, rose or yellowish wine-coloured. Ascospores hyaline, 6–10 × 0.5–1. Paraphyses filiform, not thickened at tip. Recorded July–Aug.

OTHER ASCOMYCETES

Anthostomella limitata and *Didymella proximella*, described on *Carex*, and *Keissleriella culmifida* and *Leptosphaeria eustoma*, described on grasses, are all found occasionally on *Typha*.

Leptosphaeria typhae (P.Karsten)Sacc. (Fig. 2076)
Pseudothecia scattered, immersed, up to 0.2mm diam. Ascospores pale yellow, 3-septate, 13–25 × 3–5 (mostly 15–20 × 3–4). Recorded May–June.

Leptosphaeria typharum (Desm.)P. Karsten (Fig. 2077)
Pseudothecia commonly found in very large numbers, about 0.15mm diam., black, immersed, with necks just protruding, often in lines between the veins of dead leaves, June–July. Ascospores mid golden brown, 3-septate, 25–35 × 9–13 (mostly 25–30 × 10–12). This species has a *Scolecosporiella* state with yellowish or pale brown, 3- to 8-septate conidia 50–70 × 7–8.5 which are tapered to a fine point at the apex.

Mycosphaerella lineolata, described on grasses, *Nectriella dacrymycella*, described on *Iris*, and *Paraphaeosphaeria michotii*, described on *Cladium*, have been recorded on *Typha*.

Ophiobolus typhae Feltg.
Pseudothecia scattered. Ascospores hyaline, curved, 7- to 9-septate, multi-guttulate, 75–100 × 3.

Pleospora infectoria and *P. vagans*, described on grasses, have been found also on *Typha*.

Pyrenophora typhicola (Cooke)E. Müller (Fig. 2078)
Pseudothecia up to 0.5mm diam., black, immersed, on leaf sheaths, sometimes erumpent on dead stems. Ascospores usually very pale yellow, muriform, with gelatinous sheaths, 50–70 × 18–24 × 12–14. *Phoma* state with pycnidia 0.15–0.2mm diam., and conidia 5–6 × 2–2.5, pinkish in mass.

HYPHOMYCETES

Arthrinium sporophleum, described on *Carex*, *Myrothecium cinctum*, described on grasses, *Periconia atra*, described on *Carex*, and *P. hispidula*, described on grasses, have all been found on *Typha*.

Periconia typhicola Mason & M.B.Ellis (Fig. 2079)
Colonies effuse, pale brown. Conidiophores brown, up to 300 long, with branches inside the heads. Conidia brown, verrucose, 11–17 diam. On dead leaves, Nov.–Apr.

Volutella arundinis, described on *Carex*, is common also on *Typha*.

COELOMYCETES

Ascochyta typhoidearum (Desm.)Cunnell (Fig. 2080)
Pycnidia often numerous, immersed sometimes beneath discoloured areas, black, up to 0.25mm diam., with large ostioles. Conidia hyaline, 1-septate, guttulate, 25–35 × 4.5–6, mostly 28–30 × 5.

Colletotrichum typhae H.C.Greene
Lesions on living leaves of *T. latifolia* orange brown, becoming darker, with purplish-brown margins and pale centres when old. Brown setae 75–140 × 4–6. Conidia cylindrical, clavate or slightly constricted at the middle, hyaline, 11–21 × 3.5–5.

Hendersonia culmicola var. *minor*, described on *Festuca*, has been found on *Typha*.

Hymenopsis typhae (Peck)Sutton (Fig. 2081)
Conidiomata immersed, dark olivaceous to almost black, up to 0.4 × 0.15mm, closed at first, later opening to form acervuli. Conidia olivaceous, mostly 15–16 × 3–4, some with gelatinous appendages at ends. Found May–June.

Neottiospora caricina, described on *Carex*, occurs also on *Typha*.

Phoma, v.s. under *Pyrenophora typhicola*.

Scolecosporiella, v.s. under *Leptosphaeria typharum*.

FUNGI ON FERNS, HORSETAILS AND CLUBMOSSES

ASPLENIUM: Spleenworts, Wall-rue

Milesina murariae P. & H.Sydow, 11
Uredinia hypophyllous and on petioles, whitish, inconspicuous; spores hyaline, sparingly echinulate, 24–37 × 14–23. On *A. ruta-muraria*, rare.

ATHYRIUM: Lady-fern

Cercosporella filicis-feminae (Bres.)Höhnel (Fig. 2082)
Colonies hypophyllous, white. Conidiophores in large fascicles, tapered, hyaline, 15–30 × 2–3. Conidia hyaline, 0- to 6-septate, 30–90 × 1.5–2. On fronds of *A. filix-femina*.

Pezizella campanulaeformis (Fuckel)Dennis (Fig. 2083)
Apothecia 0.2–0.25mm diam., thick-stalked, off-white or amber, drying yellowish buff. Ascospores broadly fusiform, hyaline, guttulate, 10–12 (15) × 2.5–3. In dense clusters all along petioles of *A. filix-femina*.

Taphrina athyrii Siemaszko
Asci amphigenous, 20–30 × 7–8; spores 3–6 × 2–3. Causes small, but sometimes very abundant, angular, brown, unthickened spots covered by a whitish bloom, on pinnules which sometimes turn black and curl, Aug.–Sept.

BLECHNUM: Hard-fern

Milesina blechni Syd., 11, 111 (Fig. 2084)
Uredinia hypophyllous, yellowish, 0.3–0.4mm diam.; spores hyaline, echinulate, 25–45 × 15–22, occur all the year round. Telia on brown areas on overwintered green fronds; spores with 1 or more cells, hyaline, smooth, 8–16 × 6–10.

CYSTOPTERIS: Brittle Bladder-fern

Hyalopsora polypodii (Diet.)Morgan, 11 (Fig. 2085)
Uredinia hypophyllous, scattered, minute, yellow; spores with yellowish contents, mostly 25–35 × 15–30, some thick-walled (amphispores). On *C. fragilis*, uncommon.

DRYOPTERIS: Buckler-ferns, Male-fern

UREDINALES

Milesina carpatorum Hyl., Jørst. & Nannf., 11, 111
A very rare rust distinguished from the commoner *M. kriegeriana* by its smaller urediniospores, averaging 20 × 14. On *D. filix-mas*.

Milesina kriegeriana (Magn.)Magn., 11, 111 (Fig. 2086)
Aecia on *Abies*. Uredinia hypophyllous, sunken and opening by a pore; spores whitish, echinulate, mostly 30–40 × 16–20, found usually on wet, dead leaves, throughout the year. Telia on brown spots on fronds of current year; spores hyaline, with 1–40 cells. On *D. dilatata*, *D. filix-mas*, *D. carthusiana* and *D. pseudomas*.

ASCOMYCETES

Leptopeltis filicina (Lib.)Höhnel (Fig. 2087)
Thyriothecia subcuticular, elliptical, dark brown. Ascospores fusiform, hyaline, 1- to 3-septate, 17–20 × 3–4. On petioles and veins of *D. filix-mas*, *D. dilatata* and *D. spinulosa*.

Pezizella campanulaeformis, described on *Athyrium*, occurs also on *D. filix-mas*.

Pezizella chrysostigma (Fr.)Sacc. (Fig. 2088)
Apothecia up to 0.75mm diam., short-stalked, white or off-white, becoming yellowish when old, downy with short hairs at margin. Ascospores hyaline, 5–7 × 1. On dead petioles of *D. filix-mas*, Nov.–May, common.

Taphrina athyrii, described on *Athyrium*, is found also on *D. dilatata* and *D. filixmas*.

Taphrina filicina Rostrup ex Johansson
Asci up to 45 × 10, 8-spored; spores 3.5–6.5 × 2–3.5. Causes small, thick, yellowish spots on pinnules of *D. spinulosa*, July, uncommon.

Taphrina vestergrenii Giesenhagen
Asci up to 50 × 10, 8-spored; spores 3.5–6 × 2.5–4, budding in asci. On small, yellow blotches on pinnules of *D. borreri* and *D. filix-mas*.

EQUISETUM: Dutch Rush, Horsetails

DISCOMYCETES

Gorgoniceps boltonii (Phill.)Dennis
Apothecia scattered, cylindrical to turbinate, 0.3–0.4mm high, discs

0.1–0.2mm diam., horn-coloured, soft, watery. Asci sometimes with 2 or 4 spores. Ascospores hyaline, 40–90 × 3–4, with several large vacuoles. On dead stems of *E. fluviatile*, Apr.–May.

Hymenoscyphus. Three species are found on *Equisetum*:

KEY

Apothecia cream .... *pileatus*
Apothecia pink 1
1. Ascospores 1-septate, 2.5–3.5 wide .... *equisetinus*
    Ascospores non-septate, 3–4 wide .... *rhodoleucus*

Hymenoscyphus equisetinus (Velen.)Dennis
Apothecia about 1mm diam., short-stalked, pale pink. Ascospores hyaline, becoming 1-septate, 8–13 × 2.5–3. On dead stems of *E. fluviatile*.

Hymenoscyphus pileatus, a plurivorous species described on grasses, has been found also on dead stems of *E. fluviatile*.

Hymenoscyphus rhodoleucus (Fr.)Phill. (Fig. 2089)
Apothecia up to 1mm diam., pink when fresh. Ascospores hyaline, mostly 10–12 × 3–4. On dead stems of *E. arvense, E. fluviatile, E. palustre*, Apr.–Sept.

Psilachnum inquilinum (P.Karsten)Dennis (Fig. 2090)
Apothecia up to 0.75mm diam., short-stalked, white, downy, with yellowish discs; marginal hairs up to 50 × 2–3. Ascospores hyaline, 7–9 × 1.5–2. On black dead stem bases of *E. arvense* and *E. palustre*, Mar.–Oct., very common.

Stamnaria persoonii (Pers.)Fuckel (Fig. 2091)
Apothecia like little flesh-pink goblets, up to 1mm diam., with discs deeply sunken and surrounded by high, delicate white or pale collars. Ascospores hyaline, 15–19 × 5–7. On stems of *E. palustre*, May–June.

OTHER ASCOMYCETES

Mycosphaerella equiseti (Fuckel)Schröter (Fig. 2092)
Pseudothecia immersed, black, with short protuberant necks. Ascospores hyaline, 1-septate, 15–20 × 4–5. On dead stems of *E. palustre*, May–June.

COELOMYCETES

Ascochyta equiseti (Desm.)Grove
Pycnidia immersed, up to 0.8mm diam., blackish brown. Conidia hyaline, becoming 1-septate, 7–12 × 3–3.5. On bleached areas on dead stems of *E. arvense, E. fluviatile* and *E. hyemale*.

Phoma epitricha Sacc.
Pycnidia immersed, seated on a subiculum composed of brown, dichotomously branched hyphae. Conidia hyaline, 6–7 × 2. On dead stems of *E. palustre*, Jan.–June.

Stagonospora equiseti Fautr. (Fig. 2093)
Pycnidia immersed, dark brown, up to 0.3mm diam. Conidia hyaline, 3-septate, 15–27 × 4–5. On dead stems of *E. palustre* and *E. fluviatile*, Aug.

Titaeospora equiseti (Desm.)Vassiljevsky (Fig. 2094)
Acervuli yellowish brown, up to 0.2mm diam., sometimes in rows, staining the surrounding epidermis reddish brown. Conidia hyaline, curved, usually 1-septate, 25–40 × 3–3.5, often anastomosing in the acervuli by germ tubes formed just above the base. On dying and dead stems of *E. arvense*, *E. fluviatile* and *E. hyemale*, Mar.–Apr.

LYCOPODIUM: Clubmosses, Stag's-horn Moss

Leptosphaeria lycopodina (Mont.)Sacc.
Pseudothecia immersed, 0.2–0.3mm diam., in blackened areas, couched in a weft of olivaceous brown hyphae. Ascospores yellow or olivaceous, 3-septate, 18–26 × 6–10, the second cell slightly swollen. On the outer side of leaves and bracts of *L. alpinum*, *L. annotatum* and *L. clavatum*, summer.

OPHIOGLOSSUM: Adder's Tongue

Curvularia crepinii (Westend.)Boedijn (Fig. 2095)
Colonies amphigenous, effuse, hairy, at first grey, later black. Conidiophores up to 160 × 6–8, brown. Conidia 3-septate, with middle cells brown or dark brown and end cells much paler, 20–32 × 7–16. On living leaves of *O. vulgatum*, July.

OSMUNDA: Royal Fern

Leptopeltis nebulosa (Petrak)Holm & K.Holm (Fig. 2096)
Thyriothecia dark brown, about 0.3 × 0.1mm. Ascospores hyaline, 1- to 3-septate, 10–11 × 2.5–3.5. On dead petioles, June.

Unguicularia aspera (Fr.)Nannf. (Fig. 2097)
Apothecia urn-shaped, 0.3–0.5mm diam., golden brown, covered by refractive hairs up to 50 × 4 each with a small lumen at its base containing brown sap. Ascospores hyaline, 11–15 × 2.5–3.5. On dead stems, June.

PHYLLITIS: Hart's-tongue Fern

Milesina scolopendrii (Faull)Henderson, 11, 111 (Fig. 2098)
Uredinia mostly hypophyllous on extensive brown areas on sterile leaves of young plants throughout the year; spores hyaline, echinulate, 30–55 × 15–23. Telia found May–June. Not uncommon.

POLYPODIUM: Polypody

Milesina dieteliana (Syd.)Magn., 11, 111
Uredinia hypophyllous, 0.1–0.2mm diam., on indefinite greenish-brown spots; spores hyaline, echinulate, 24–48 × 16–26, found throughout the year. Telia hypophyllous on brown areas on overwintered fronds of *P. vulgare*, Apr.–June; spores with one or many cells. Aecial hosts *Abies alba* and *A. concolor*.

POLYSTICHUM: Shield-ferns

Two rare species of *Milesina* are recorded on *P. aculeatum* and *P. setiferum*. In *M. vogesiaca* Syd., the urediniospores are smooth and mostly about 36 × 18; in *M. whitei* (Faull) Hirats., they are finely echinulate and average 30 × 20.

PTERIDIUM: Bracken

DISCOMYCETES

Crocicreas cyathoideum (Bull.)S. Carp. var. *pteridicola* (Crouan & H. Crouan)S. Carp. (Fig. 2099)
Apothecia up to 1mm diam., short-stalked, whitish to pale grey with dark brown stripes on the outside. Ascospores hyaline, 8–10 × 1.5–2. On dead stems, especially in the axils of branches, Apr.–July. The plurivorous *C. cyathoideum* occurs on bracken also but is larger, cream-coloured, without brown stripes and has spores up to 12 × 2.5.

Cryptomycina pteridis (Rebent.)Höhnel (Fig. 2100)
Apothecia hypophyllous, subepidermal, developing in black, often parallel stromata, up to 3 × 0.5mm, overwintering before opening by longitudinal slits. Ascospores hyaline, 8–10 × 5–6. On dead fronds, May–June. Conidiomata of *Cryptomycella* state up to 0.45mm diam., opening by irregular longitudinal fissures; conidia hyaline, 17–20 × 3.5–5.

Dasyscyphus pteridialis Graddon (Fig. 2101)
Apothecia up to 0.75mm diam., dark brown near the base, pale above and covered with hyaline hairs which are swollen to 5 and granulate towards their

tips. Ascospores hyaline, 1-septate, 9–12 × 2–2.5. On dead fronds, Oct.

Dasyscyphus pteridis (Alb. & Schw.)Massee (Fig. 2102)
Apothecia up to 0.4mm diam., dark brown, hairy, with dull yellowish discs. Hairs brown, granulate. Ascospores hyaline, 7–10 × 1.5–2.5. On dead stems, especially in the axils of branches, Aug.–May.

Hyaloscypha flaveola (Cooke)Nannf. (Fig. 2103)
Apothecia up to 0.3mm diam., bright lemon-yellow, fringed with short, tapered hairs. Asci 4-spored; spores hyaline, 10–15 × 1.5–2, each with a central spot which stains deeply in lactic cotton blue. On the lower surface of damp, dead fronds, June–July.

Melittosporium pteridinum (Phill. & Bucknall)Sacc. (Fig. 2104)
Apothecia immersed, about 1mm diam., creamy buff, with pale brown discs exposed by irregular splitting of the epidermis. Ascospores hyaline, mostly 7- or 8-septate, 40–47 × 5–6. On dead stems, Mar.–Apr.

Micropodia pteridina (Nyl.)Boud. (Fig. 2105)
Apothecia up to 0.2mm diam., white to cream, downy. Ascospores hyaline, 4–6 × 1–1.5. In swarms on blackened bases of old stems, May–Aug., common.

Microscypha grisella (Rehm)H.Syd. & Syd. (Fig. 2106)
Apothecia up to 0.5mm diam., short-stalked, pale grey or greyish brown with a fringe of hyaline hairs up to 60 long. Ascospores hyaline, 7–11 × 1.5–2. On the lower side of damp dead fronds, tending to be concealed by the hairs, very common, May–Aug.

Mollisia pteridina (Nyl.)P.Karsten (Fig. 2107)
Apothecia 0.2–0.6mm diam., pale brown with greyish discs and almost white rims. Ascospores sometimes 1-septate when mature, hyaline, 6–10 × 1.5–2. On dead stems, especially in the axils, Mar.–Sept., common.

Mollisia pteridis Gillet (Fig. 2108)
Apothecia up to 1.25mm diam., dark olivaceous, with yellowish discs surrounded by a fringe of pale hairs. Ascospores hyaline, 6–8 × 1.5–2.5. On blackened areas near the base of standing dead stems, June–July, not common.

Pezizella chrysostigma, described on *Dryopteris*, has been found on *Pteridium*.

Psilachnum pteridigenum Graddon (Fig. 2109)
Apothecia up to 0.5mm diam., sessile, with hairs up to 35 × 5, white. Ascospores hyaline, 8–12 × 2.5. On dead fronds, May–Sept.

OTHER ASCOMYCETES

Diaporthopsis pantherina (Berk.)Wehmeyer (Fig. 2110)
Perithecia immersed, 0.25–0.4mm diam., lying beneath brown discoloured areas delimited by narrow black lines, with necks showing on the surface as black dots. Ascospores hyaline, guttulate, 11–12 × 4–5. On dead petioles, June–Aug. *Phomopsis* state found from Feb. onwards has narrow, black, erumpent conidiomata 1–3mm long; A-conidia hyaline, 8–10 × 2, B-conidia curved or flexuous, 20–25 × 0.5.

Didymella lophospora (Sacc. & Speg.)Sacc.
Pseudothecia epiphyllous, 0.1–0.15mm diam., densely gregarious, blackish brown. Ascospores hyaline, cylindric–fusoid, straight or curved, 16–17 × 5–6, constricted at the median septum, with splayed-out appendages at the ends. On dead fronds, July.

Didymella prominula (Speg.)Pirozynski & Morgan-Jones
Pseudothecia immersed, up to 0.18mm diam., raising the upper epidermis. Ascospores hyaline, slightly curved, constricted at the median septum, with upper cell usually wider than lower one, 13–19 × 5–9. On dead stems and fronds, June.

Leptopeltis litigiosa (Desm.)Holm & K. Holm (Fig. 2111)
Thyriothecia superficial, about 0.1mm diam., often confluent, forming dark brown to black crusts up to 1mm across. Ascospores hyaline, 1- to 3-septate, 15–18 × 2–3. *Pycnothyrium* state with allantoid conidia 4–6 × 1. On dead petioles, June–Sept.

Leptopeltis pteridis (Mouton)Höhnel
Thyriothecia subcuticular, 0.3mm diam., usually confluent and forming crusts 2–3mm diam. Ascospores hyaline, 1- to 3-septate, 12–15 × 3–4. On dead petioles and veins.

Monographos fuckelii Holm & K.Holm (Fig. 2112)
Ascocarps orbicular about 0.1mm diam. or elongated, flattened, blackish brown, plurilocular. Ascospores hyaline, 1- to 3-septate, 25–30 × 3–4. On the upper surface of dead fronds and petioles, July–Aug.

Mycosphaerella pteridis (Desm.)Schröter (Fig. 2113)
Pseudothecia epiphyllous, often in groups, immersed, 0.1–0.13mm diam. Ascospores hyaline, 1-septate, mostly 30–38 × 4–5. On dead fronds.

Phomatospora endopteris (Plowr. ex Bucknall)Phill. & Plowr. (Fig. 2114)
Perithecia deeply immersed, with long necks up to the surface. Ascospores hyaline, 10–11 × 2.5–3. On dead fronds.

Rhopographus filicinus (Fr.)Nitschke ex Fuckel (Fig. 2115)
Stromata subepidermal, black, 0.5–2mm wide and 3mm long, often confluent and forming crusts several cm long, conspicuous. Ascospores golden brown,

3- to 7-septate, mostly 30–35 × 7–8. Very common on dead petioles, Feb.–June.

Scirrhia aspidiorum (Lib.)Bubák (Fig. 2116)
Stromata subepidermal, up to 5 × 1mm, grey, multilocular, splitting the epidermis longitudinally. Ascospores hyaline, 1-septate, 10–13 × 3. On dead petioles, May–July.

COELOMYCETES AND HYPHOMYCETES

Ascochyta pteridis Bres.
Pycnidia epiphyllous, immersed to erumpent, black, up to 0.1mm diam. Conidia cylindrical, rounded at ends, often bent, 1-septate, hyaline, guttulate, 15–30 × 4–6. On very small, pale ochraceous spots with broad purplish-brown borders, on dead fronds, July.

Camarographium stephensii (Berk. & Br.)Bubák (Fig. 2117)
Conidiomata about 0.3mm diam., arranged in longitudinal series, at first covered by the epidermis which turns brown and opens by a slit. Conidia muriform, golden brown, 40–58 × 20–32. On dead petioles, May–July.

Chalara pteridina Syd. (Fig. 2118)
Colonies widely effused, brown, thinly hairy. Conidiophore stipes up to 80 × 5–8, brown; terminal phialides lageniform, 50–70 × 8–9 with necks 3–4 wide. Conidia hyaline, 1- to 3-septate, 9–17 × 2–3. On dead petioles, Apr.–Nov., very common.

Coniothyrium pteridis A.L.Smith & Ramsbottom
Pycnidia erumpent, up to 0.2mm diam., black. Conidia smoky brown, 2.5 × 1.5–2. On dead rachides and pinnae.

Cryptomycella, v.s. under *Cryptomycina pteridis.*

Phomopsis, v.s. under *Diaporthopsis pantherina.*

Pycnothyrium, v.s. under *Leptopeltis litigiosa.*

THELYPTERIS: Beech Fern, Oak Fern

Two very rare rusts are recorded: *Hyalopsora aspidiotis* on *T. dryopteris*, and *Uredinopsis filicina* on *T. phegopteris*.

Hyalopsora aspidiotus (Magn.)Magn., 11
Uredinia amphigenous, 0.2–0.5mm diam., yellow; spores 30–40 × 16–26. Amphispores with thicker walls, 35–70 × 30–40 formed also.

Uredinopsis filicina Magn., 11, 111
Uredinia hypophyllous, yellowish brown; spores hyaline, 25–45 × 8–13, with beaks up to 12 long. Amphispores with thicker walls, 15–30 × 8–20. Telia mostly hypophyllous; spores hyaline, mostly 1-septate, 18–24 × 15–16.

FUNGI PARASITIC ON RUSTS AND POWDERY MILDEWS

Ampelomyces quisqualis Ces. ex Schlecht. (Fig. 2119)
Pycnidia 0.05–0.1 × 0.03–0.05mm, often at ends of short hyphal branches, pale golden brown. Conidia hyaline to very pale brown, 5–9 × 2–4. Fairly common on powdery mildews, growing on the mycelial mat. Most of the older records are found under the name *Cicinnobolus cesatii*.

Cladosporium aecidiicola Thuemen (Fig. 2120)
Colonies effuse, dark olive or olivaceous brown, velvety. Conidiophores olivaceous brown, up to 100 × 4–6. Conidia in chains, mostly 0- or 1-septate, verrucose, olivaceous brown, 6–18 × 5–9. On old aecia of rusts including those of *Puccinia phragmitis* and *Uromyces limonii*.

Cladosporium uredinicola Speg. (Fig. 2121)
Colonies olivaceous, velvety. Conidiophores pale brown, up to 300 × 3–5. Ramoconidia 25–30 long. Conidia very pale olive, 0- to 3-septate, 3–5 diam. or 7–25 × 3–6. On uredinia of rusts including those of *Puccinia recondita* and *Triphragmium ulmariae*.

Eudarluca caricis (Fr.)O.Eriksson (Fig. 2122)
Conidiomata of *Sphaerellopsis* state in which this fungus is usually found erumpent, with 1 or more locules, blackish brown, often in rows. Conidia hyaline to very pale brown, 1-septate, 15–20 × 2.5–5, with small gelatinous caps. Commonly parasitic in sori of rusts, especially those of *Puccinia* species. Most of the older records are found under the name *Darluca filum*.

Hainesia rubi (Westend.)Sacc. (Fig. 2123)
Conidiomata 0.2–0.8mm diam., open cups with the spore mass jelly-like, yellow, turning brown. Conidia hyaline, mostly 5–8 × 2–3. Parasitic on uredinia and telia of *Phragmidium mucronatum*, *P. rubi* and *P. violaceum*.

Micropodia oedema (Desm.)Boud. (Fig. 2124)
Apothecia gregarious, arising from a stroma, erumpent, 0.15–0.17mm diam., very pale brown. Ascospores hyaline, 8–10 × 1.5–2. On uredinia and telia of *Phragmidium mucronatum*, May.

Tuberculina maxima Rostr.
Sporodochia effuse, dark purplish. Conidia pale violaceous, 10–13 diam. On aecia of *Cronartium flaccidum* and *C. ribicola*.

Tuberculina persicina (Ditm.)Sacc. (Fig. 2125)
Sporodochia globose or somewhat flattened above, roughly arranged in

circles, dark violaceous. Conidia spherical, pinkish violet, smooth, 7–8 (10) diam. On aecia of many rusts.

Tuberculina sbrozzii Cav. & Sacc.
Sporodochia effuse, orange to reddish brown, 0.5–0.7mm diam. Conidia 8–10 diam. In sori of *Puccinia vincae*.

SOME RECOMMENDED BOOKS

Introduction to Fungi by J. Webster (Cambridge University Press, 2nd edition, 1980, paperback and hardback). A good general introduction to the structure and classification of fungi based mainly on species which can be found in Britain.

British Fungi by E.A.Ellis (Jarrold Colour Publications, Book 2, 1976). This illustrates beautifully in colour aecia and other states of certain rust fungi, purple anther smut, tar spot of sycamore, ergots, *Epichloe typhina*, *Taphrina tosquinetii*, *Hymenoscyphus fructigenus* and a number of other microfungi.

British Ascomycetes by R.W.G.Dennis (J.Cramer, 3rd edition, 1978). Especially valuable for its section on discomycetes. Dr Dennis has specialised in this group for many years and provides here an excellent and authoritative account of them.

British Rust Fungi by M. Wilson and D.M. Henderson (Cambridge University Press, 1966). A standard and scholarly work with full descriptions of all species.

British Stem- and Leaf-fungi by W.B.Grove (Cambridge University Press: Vol. 1, 1935; Vol. 2, 1937). The taxonomy and nomenclature are somewhat out of date but there is much useful information in these two volumes. A more modern treatment on a world basis is provided in *The Coelomycetes* by B.C.Sutton (Commonwealth Mycological Institute, 1980).

'Guide to the literature for the identification of British fungi', 4th edition, by Margaret Holden (reprinted from *Bulletin of the British Mycological Society*, Vol. 16. Cambridge University Press, 1982) should be consulted for other books and papers.

GLOSSARY

acervuli conidiomata formed of a flat layer of pseudoparenchyma on which conidia are formed while still covered by host tissues
acicular having the form of a needle
acropetal produced in succession towards the apex
aecia 'cluster-cups' of rust fungi which contain 1-celled spores in chains
aecidioid the typical form of an aecidium with a well-defined, short peridium
allantoid sausage-shaped and slightly curved
amphigenous occurring on both surfaces of a leaf
ampullae small, flask-like structures
annellations ring-like markings on a wall
annellides closely annellate, holoblastic, conidiogenous cells
anthracnose a plant disease causing limited necrotic lesions
apothecia open ascomata bearing asci in an exposed hymenium
appressoria swellings on hyphae especially for attachment to a host plant
arthroconidia conidia formed as part of a thallus, often cylindrical, truncate at each end or one end
ascigerous having asci
ascoconidia conidia formed by ascospores often inside asci
ascomata structures bearing or containing asci
ascospores spores formed inside an ascus
ascus (plural asci) a sac-like cell containing usually 8 ascospores
ballistospores forcibly propelled spores of basidiomycetes
basipetal produced in succession towards the base
biconic cone-shaped at each end
bitunicate 2-walled
bullate blistered or puckered
byssoid made up of slender threads
caducous falling off a plant easily
caeomoid with no well-defined peridium
caespitose in groups or tufts
capitate having a well-formed head
catenate developing in chains
cerebriform convoluted like a brain
cheiroid roughly the shape of a hand
chlamydospores thick-walled non-deciduous spores
cicatrised bearing scars
cilia hair-like outgrowths
circinate coiled

clamp connections looping hyphal outgrowths which connect two adjacent cells in a hypha
clathrate latticed or net-like
clavate club-shaped
cleistothecia usually spherical ascomata without ostioles
clypeus a shield-like stroma overlying one or more perithecia
columella a sterile central axis inside a fruit body
concolorous having the same colour
confluent running together
conico-truncate having the shape of a truncated cone
conidia asexual spores which when mature are liberated from a conidiophore
conidiogenous cells cells which produce one conidium or several conidia
conidiomata structures bearing or containing conidia
conidiophore a hypha which bears one conidium or several conidia
coremia loosely bound-together fascicles
crenate having the edge indented, with rounded teeth
crenulate delicately crenate
cruciate in the form of a cross
crustose crust-like
cuneiform wedge-shaped
cupulate cup-shaped
cyathiform like a cup which is wider at the top than at the bottom
dentate toothed
denticles cylindrical or tapered, tooth-like projections
diaphanous transparent or nearly so
diaporthaceous formed as in the genus *Diaporthe*
dichotomous branching, often successively, into two more-or-less equal arms
digitate with finger-like processes
disc the ascus-producing part of a discomycete
discoid flat and circular, disc-like
discrete separate
disjunctor a connecting piece between spores in a chain
doliiform barrel-shaped
echinulate spiny
ectostroma a fungus stroma on the surface of a host plant
effete dead, no longer sporulating
effuse spread out
elaters narrow elastic threads which aid spore dispersal
endospore inner spore wall
epiphyllous borne on the upper surface of a leaf
epispore outer spore wall
epithecium a layer formed by fusion of tips of paraphyses or branches of paraphyses and covering asci in some discomycetes
erumpent bursting through to the surface
excipulum outer tissue forming flanks of apothecia

falcate sickle-shaped
fasciculate arranged in bundles
filiform slender and thread-like
fimbriate fringed
flexuous bent alternately in opposite directions
floccose cottony
foot-cell a *Fusarium* spore cell obliquely attenuated at the base
fulvous tawny
furfuraceous scurfy
fusiform spindle-shaped, tapering at each end
fusoid somewhat fusiform
geniculate bent like a knee
globose spherical or almost so
granulate covered with very small granules
guttules oil-like droplets
hamate hooked
haustoria hyphal branches which penetrate host cells
helicoid in the form of a spiral
hilum a scar formed at the point of attachment
hyaline transparent or nearly so
hymenium the spore-bearing layer of a fungus
hypha a fungus thread or filament
hyphopodia short, lateral, sometimes lobed branches of hyphae which adhere firmly to the surface of a host plant; a pale spot in the centre often indicates the point of origin of a narrow filament which penetrates the cuticle
hypocreaceous as in the Hypocreaceae
hypophyllous borne on the lower surface of a leaf
hypostroma hyphal tissue immersed in the substrate by which conidiomata are attached
hysterioid elongated and cleft down the centre
hysterothecia elongated stromata containing asci, with cleft-like openings parallel to the long axis
intercalary between apex and base
isarioid synnematous as in the genus *Isaria*
laciniate as if torn into strips
lageniform shaped like a long-necked flask
lanceolate spear-shaped
lectotype a type chosen from original material, but not by the author
lenticular shaped like a biconvex lens
limoniform lemon-shaped
locules cavities
Melzer's iodine a reagent containing iodine, potassium iodide, distilled water and chloral hydrate
mesospores resting urediniospores
mollisioid saucer-shaped as in the genus *Mollisia*

moniliform swollen at intervals like a string of beads
mononematous composed of a single thread or filament
muriform longitudinally and transversely septate
mycelium a mass or group of hyphae making up the thallus of a fungus
navicular boat-shaped
nodose having node-like swellings
obclavate the shape of a club upside down, thickened towards the base
obconical inversely conical
obovate inversely ovate
obovoid egg-shaped with the broad end at the base
obtuse rounded or blunt
ochraceous yellowish brown
oospores resting spores in Oomycetes
operculate opening by a little lid
orbicular spherical
ostiole an opening in the wall of a fruit body through which spores are extruded
papillate with a nipple-like protuberance
paraphyses sterile threads growing between asci or conidiophores
part-spores 1-celled spores resulting from the breaking up of a 2- or more-celled ascospore
patelliform like a plate with a well-defined edge
pedicel a small stalk
penicillate arranged like a paint brush
percurrent growing straight on through the open end left when the first conidium becomes detached, or through a terminal pore
periderm the membrane surrounding a sorus
peridia walls surrounding fruiting structures
perithecia subglobose or flask shaped ascomata with ostioles
phialides cells that usually produce many conidia which ooze out through one or more openings and often hang together in chains or in slimy clumps
plumose feathery
plurilocular having several locules
polyphialides phialides with more than one opening
polysporous many-spored
pomiform apple-shaped
pruinose with a frost-like or floury covering
pseudoseptate with the appearance of having septa which do not reach right across a spore from one side wall to the other
pseudothecia ascomata of loculoascomycetes with asci formed in one or more locules in a stroma which is sometimes very much reduced and may resemble a perithecium
puberulent minutely downy
pubescent softly hairy or downy
pulverulent powdery

pulvinate cushion-shaped
punctiform dot-like
pycnidia globose or flask-like conidiomata containing conidia
pycnothyria superficial shield-shaped conidiomata
pyriform pear-shaped
ramoconidia apical branches of conidiophores which resemble conidia
recurved curved backwards
reniform kidney-shaped
repent creeping
reticulate net-like
revolute having the edge rolled over
rostrate beaked or strongly attenuated at the apex
rugose coarsely wrinkled
rugulose finely wrinkled
sclerotia masses of aggregated hyphae, often with a rind of thick-walled cells
sclerotules small sclerotia
scutellum the upper wall of a thyriothecium
septum a dividing cell wall
setae bristle-like, often erect and thick-walled hyphae
setiform having the form of a seta
setose bearing setae
setulae delicate appendages arising from the surface of a spore
spermogonia pycnidium-like structures containing spermatia in rusts
spicules needle-like outgrowths
spinulose delicately spiny
sporodochia stromata bearing closely packed, short conidiophores
stipe a stalk
stipitate stalked
striae delicate lines, grooves or ridges
striate marked with striae
strigose covered with bristle-like hairs
stromata cushion-like masses of fungal cells or closely interwoven hyphae
subiculum a mycelial mat below fruiting bodies
subulate rather slender and tapered to a point like an awl
sulcate grooved
sympodial proliferating, the main axis elongating by growth of a succession of apices, each of which develops behind and to one side of the previous apex
synnematous composed of several or many tightly adpressed threads or filaments
telia sori producing teliospores
teliospores resting or overwintering spores of a rust fungus
thyriothecia more or less flattened ascomata, often with radiate upper walls
tomentose downy
tremelloid jelly-like

trigonous having three angles with plane faces between them
triradiate three-armed
truncate ending abruptly as if cut straight across
tuberculate covered with wart-like projections
turbinate top-shaped
umbilicate with a small hollow
uncinate hooked
uniseriate in one row
unitunicate 1-walled
urceolate pitcher-shaped
uredinia sori producing urediniospores
urediniospores summer spores in rust fungi
uredinoid a term applied to aecia which look like uredinia
valsoid having groups of perithecia with their necks convergent as in the genus *Valsa*
venter the expanded basal part of a fruiting body
ventricose inflated, very broadly fusiform
vermiform worm-shaped
verrucose coarsely warted
verruculose finely warted
verticillate arranged in whorls
zonate zoned or banded

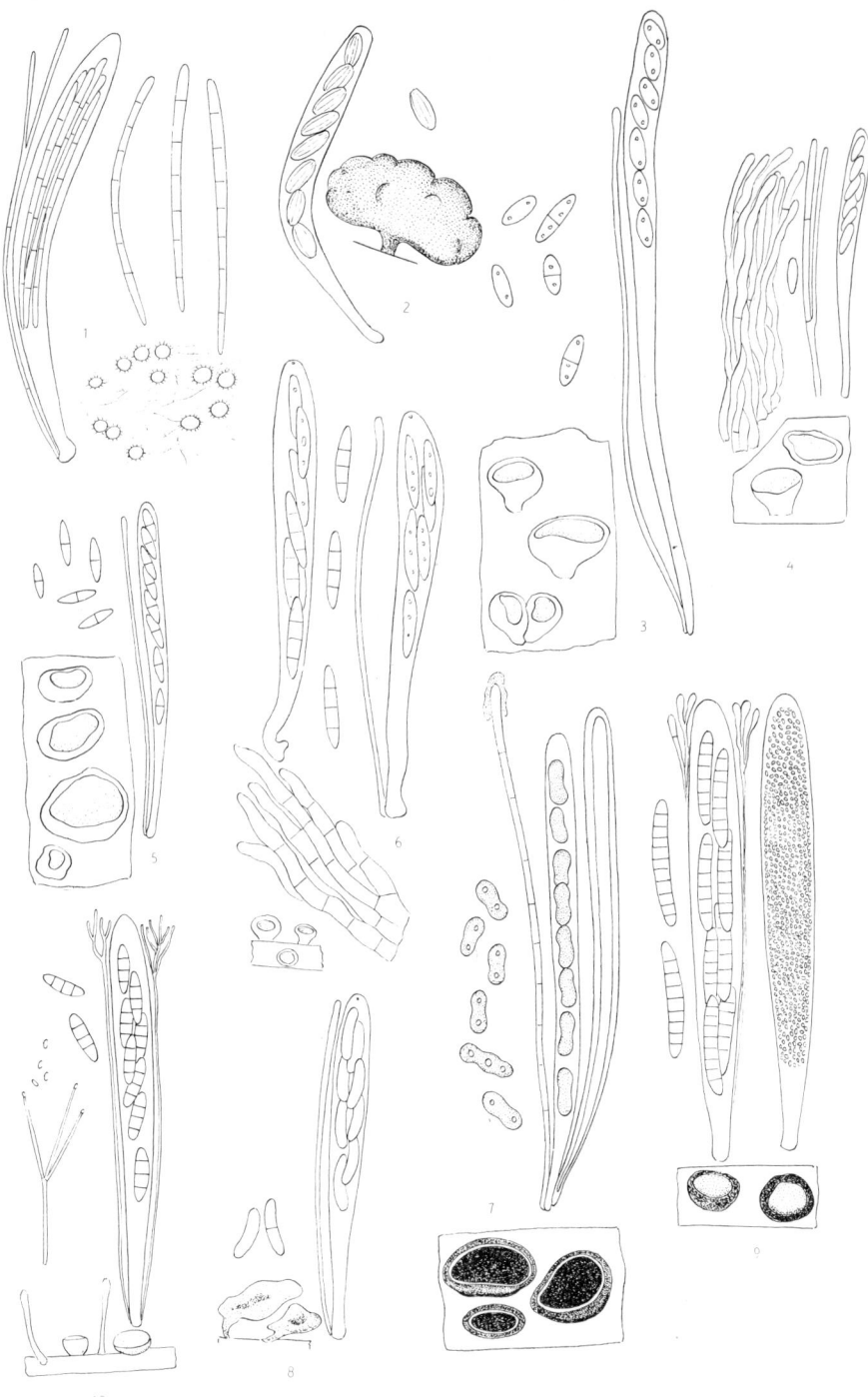

Plate 1. Discomycetes on wood and bark. 1, *Arachnopeziza aurata*; 2, *Ascotremella faginea*; 3, *Bisporella citrina*; 4, *B. subpallida*; 5, *B. sulfurina*; 6, *Calycellina ochracea*; 7, *Catinella olivacea*; 8, *Chlorencoelia versiformis*; 9, *Claussenomyces atrovirens*; 10, *C. prasinulus*.

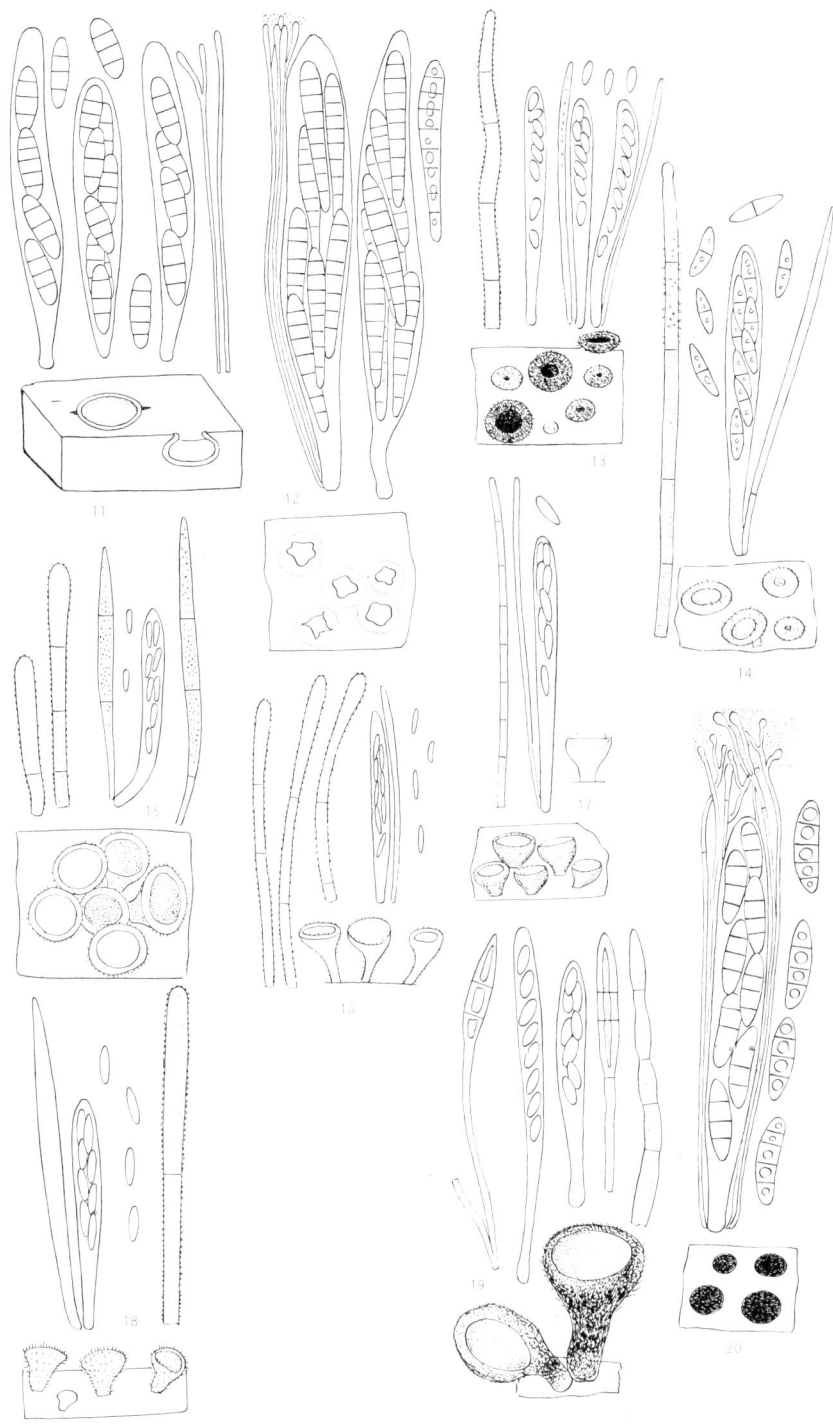

Plate 2. Discomycetes on wood and bark. 11, *Cryptodiscus pallidus*; 12, *C. rhopaloides*; 13, *Dasyscyphus cerinus*; 14, *D. corticalis*; 15, *D. fascicularis*; 16, *D. papyraceus*; 17, *D. pulveraceus*; 18, *D. virgineus*; 19, *Diplocarpa bloxamii*; 20, *Durella atrocyanea*.

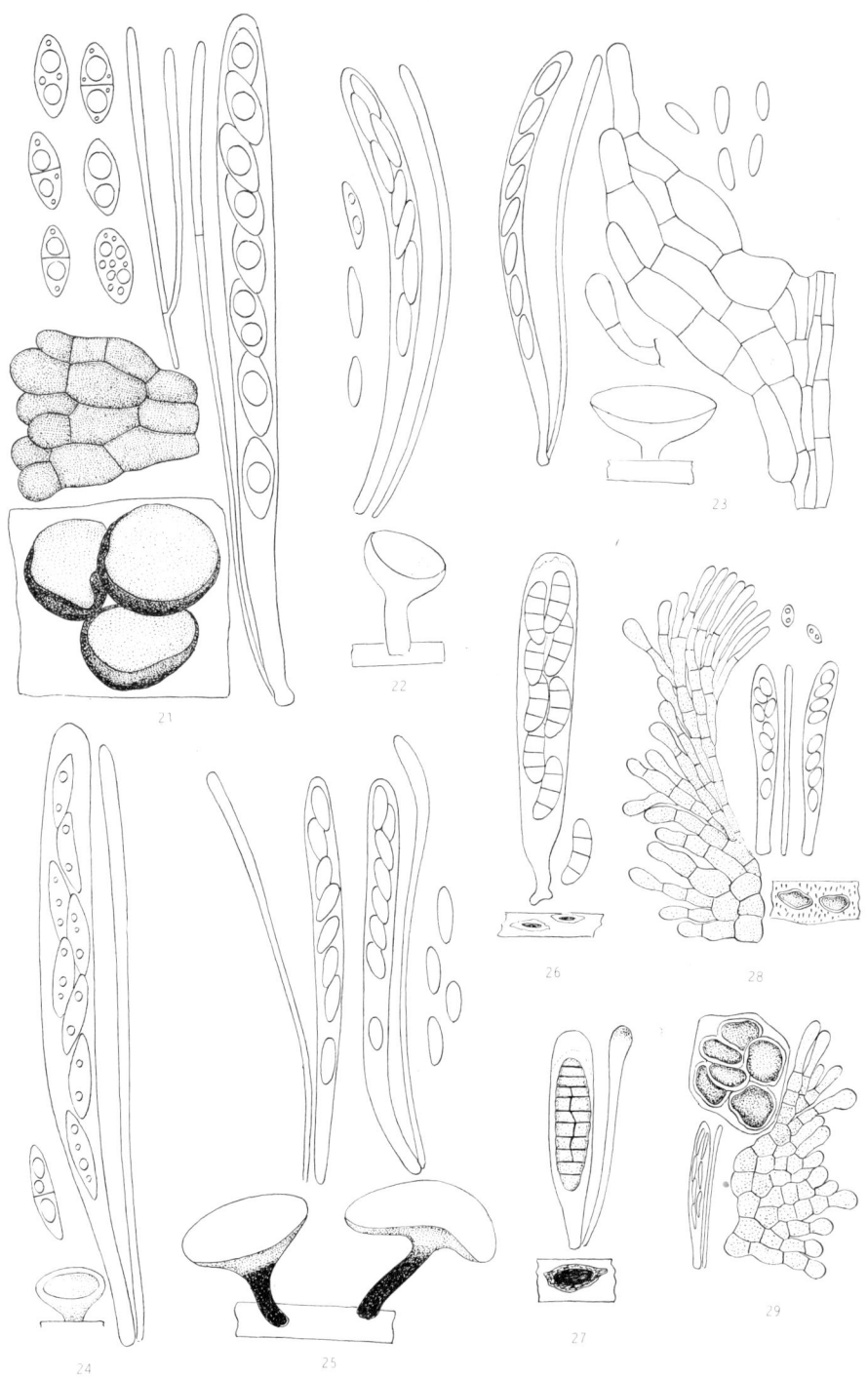

Plate 3. Discomycetes on wood and bark. 21, *Graddonia coracina*; 22, *Hymenoscyphus calyculus*; 23, *H. imberbis*; 24, *H. laetus*; 25, *H. vernus*; 26, *Melittosporiella pulchella*; 27, *Melittosporium propolidoides*; 28, *Mollisia aquosa*; 29, *M. caespiticea*.

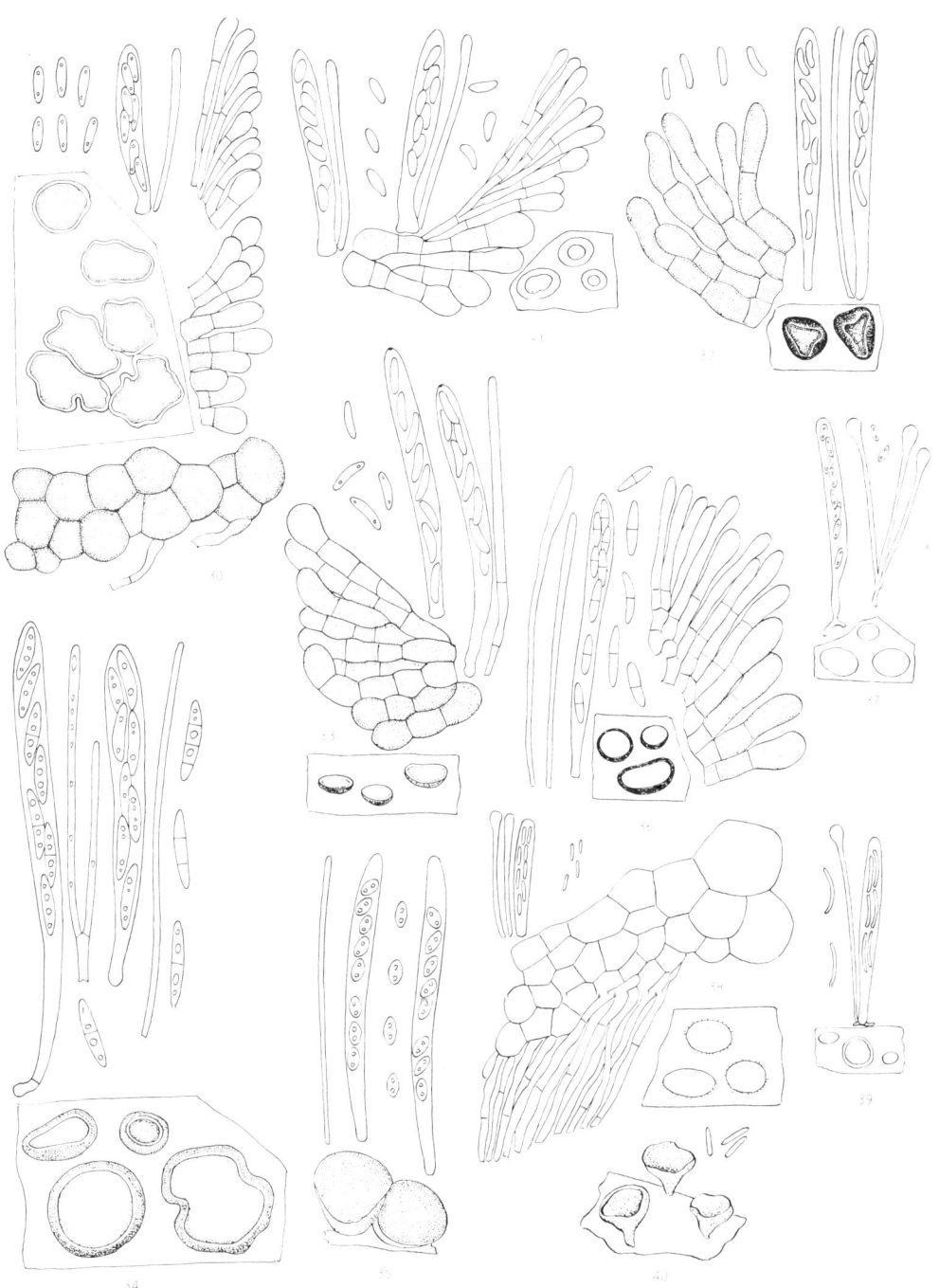

Plate 4. Discomycetes on wood. 30, *Mollisia cinerea*; 31, *M. cinerella*; 32, *M. ligni*; 33, *M. melaleuca*; 34, *M. ventosa*; 35, *Neobulgaria lilacina*; 36, *Niptera ramincola*; 37, *Orbilia alnea*; 38, *O. auricolor*; 39, *O. curvatispora*; 40, *O. cyathea*.

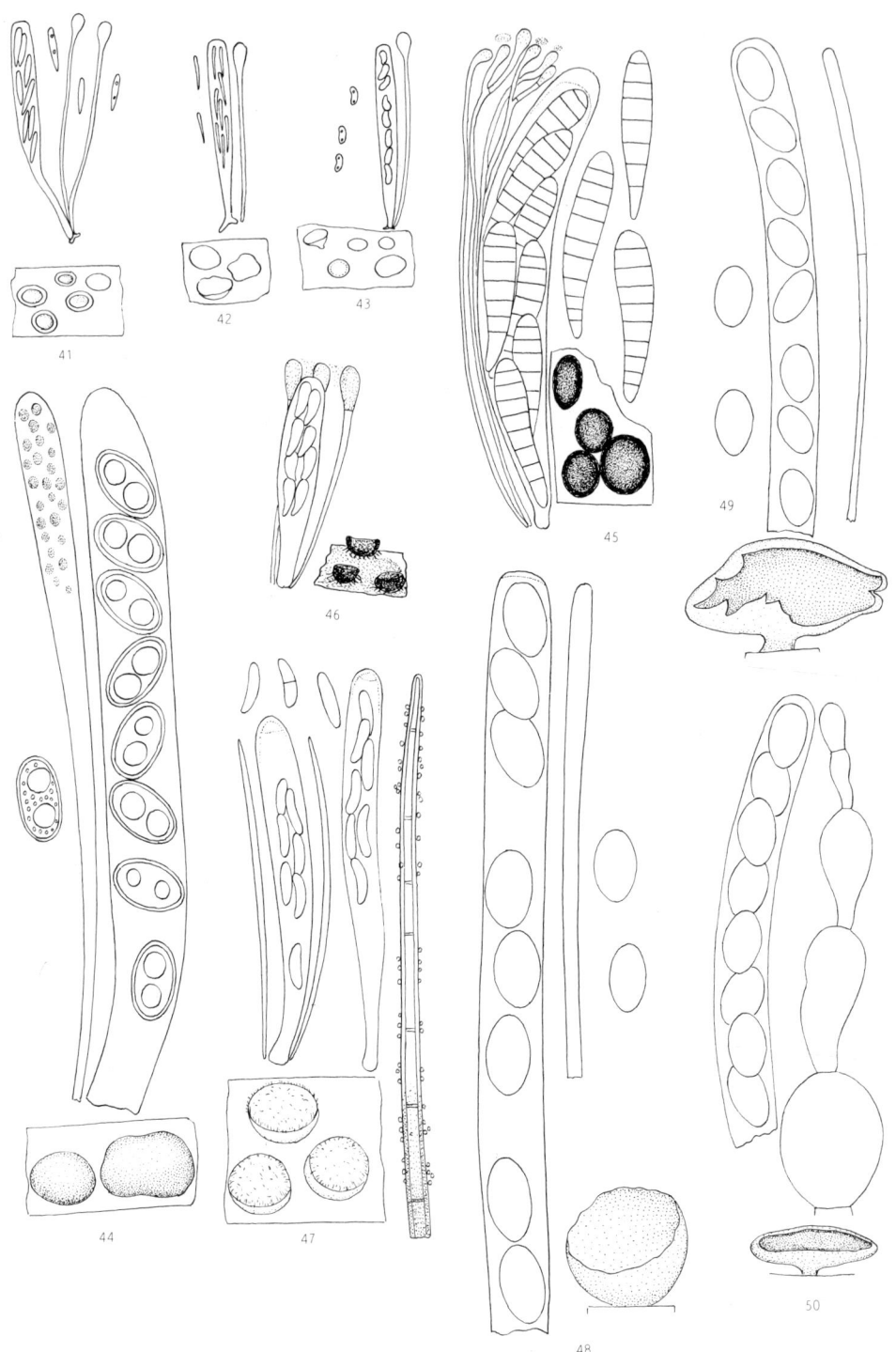

Plate 5. Discomycetes on wood and bark. 41, *Orbilia luteorubella*; 42, *O. sarraziniana*; 43, *O. xanthostigma*; 44, *Pachyella babingtonii*; 45, *Patellaria atrata*; 46, *Patinellaria sanguinea*; 47, *Perrotia flammea*; 48, *Peziza micropus*; 49, *P. repanda*; 50, *P. varia*.

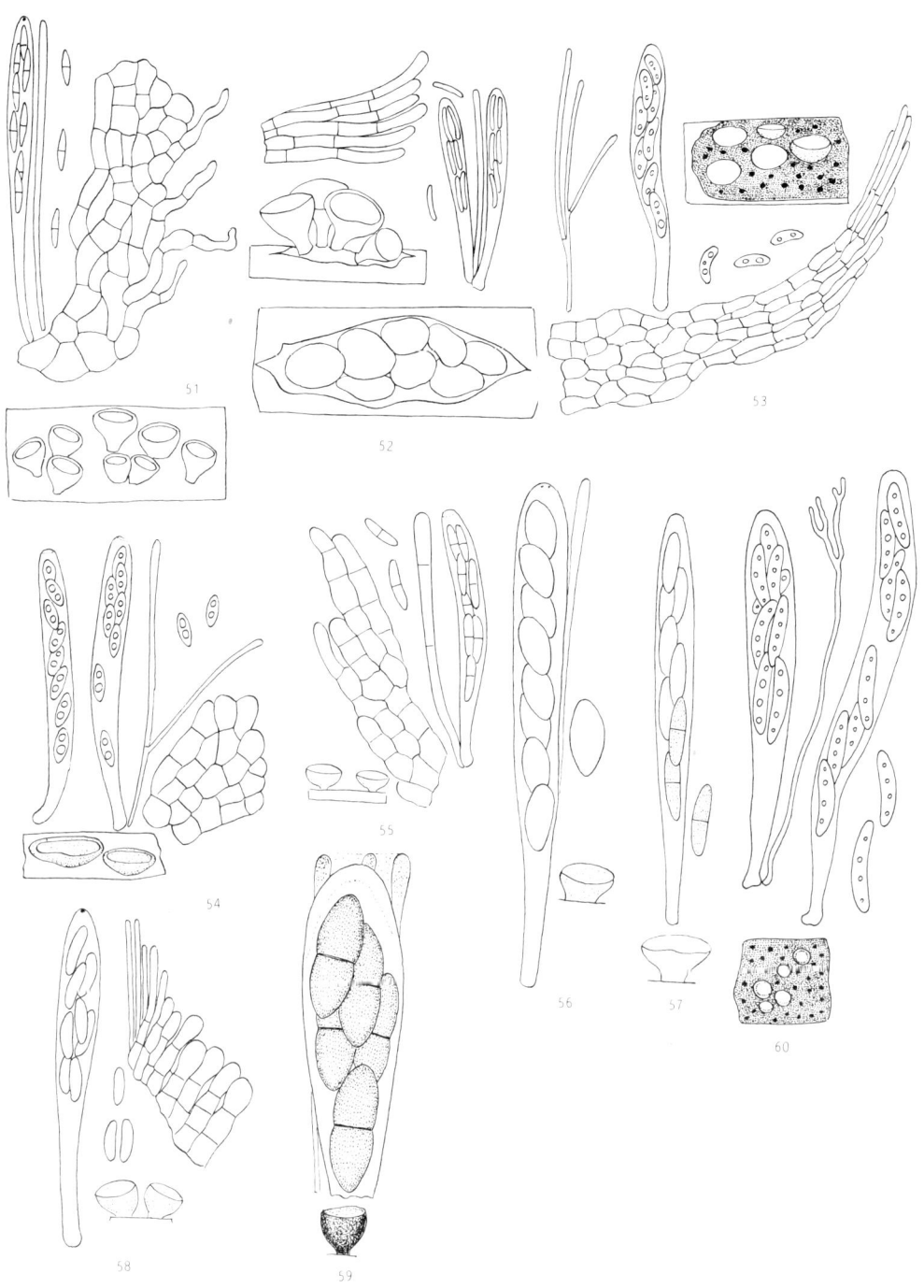

Plate 6. Discomycetes on wood and bark. 51, *Pezizella parilis*; 52, *P. vulgaris*; 53, *Phaeohelotium extumescens*; 54, *P. flexuosum*; 55, *P. italicum*; 56, *P. lilacinum*; 57, *P. monticola*; 58, *P. trabinellum*; 59, *Poetschia cratincola*; 60, *Polydesmia pruinosa*.

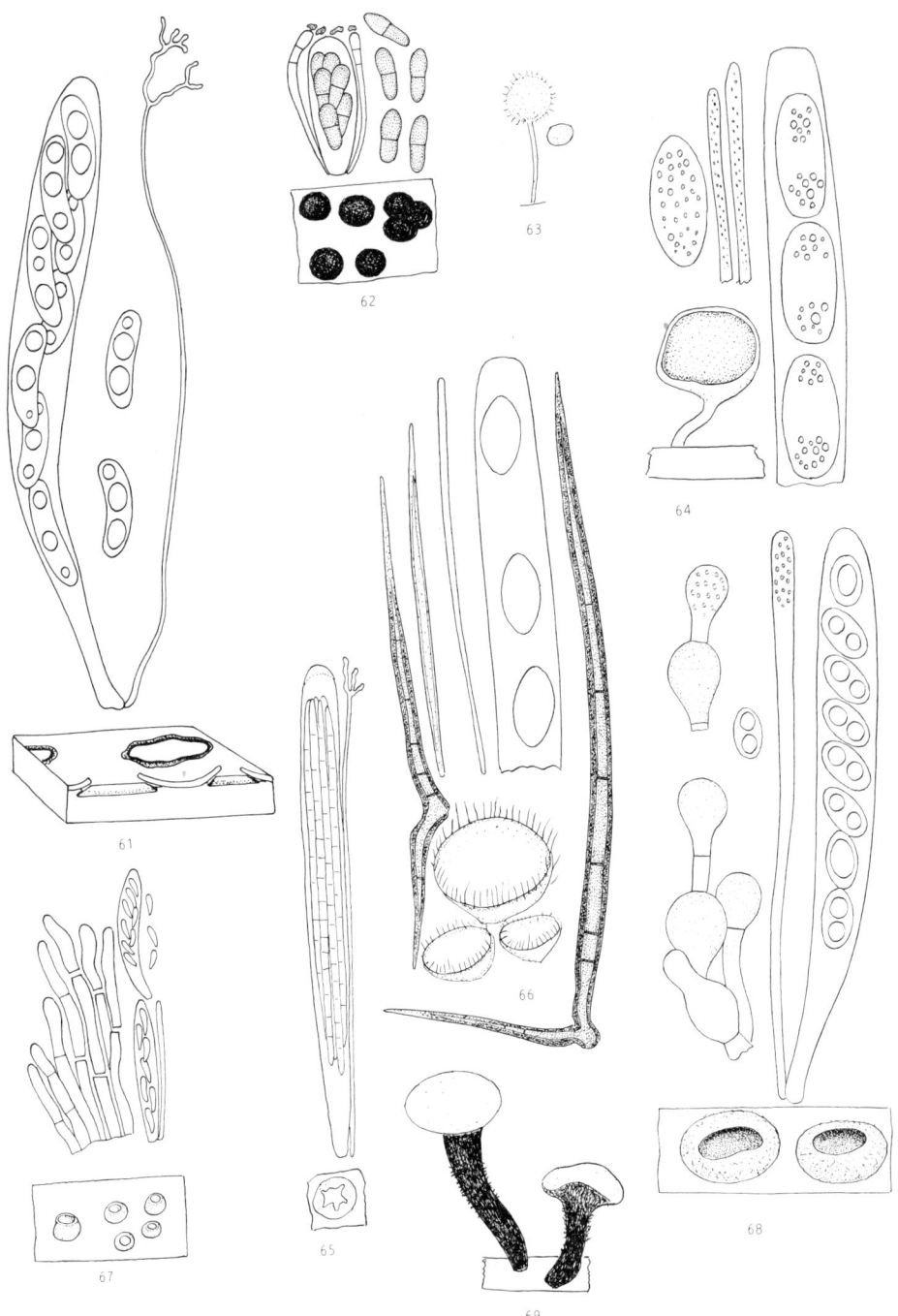

Plate 7. Discomycetes on wood and bark. 61, *Propolomyces versicolor*; 62, *Rhizodiscina lignyota*; 63, *Roesleria pallida*; 64, *Sarcosypha coccinea*; 65, *Stictis friabilis*; 66, *Trichophaeopsis bicuspis*; 67, *Unguicularia scrupulosa*; 68, *Velutarina rufo-olivacea*; 69, *Vibrissea truncorum*.

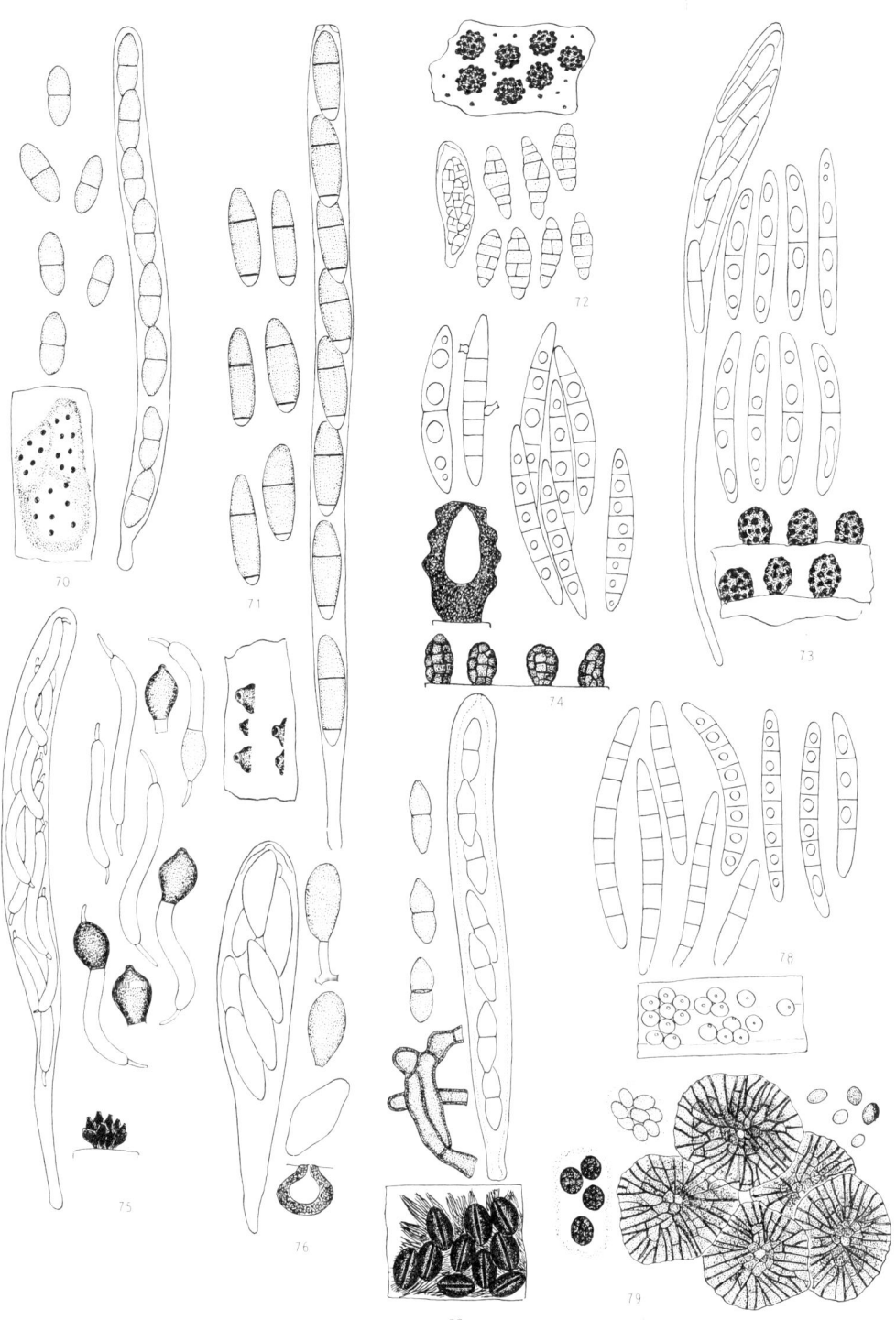

Plate 8. Ascomycetes on wood and bark. 70, *Amphisphaeria millepunctata*; 71, *Apiorhynchostoma curreyi*; 72, *Berlesiella nigerrima*; 73, *Bertia moriformis*; 74, *B. moriformis* var. *multiseptata*; 75, *Bombardia bombarda*; 76, *Botryosphaeria obtusa*; 77, *Byssolophis sphaerioides*; 78, *Calonectria pseudopeziza*; 79, *Cephalotheca sulfurea*.

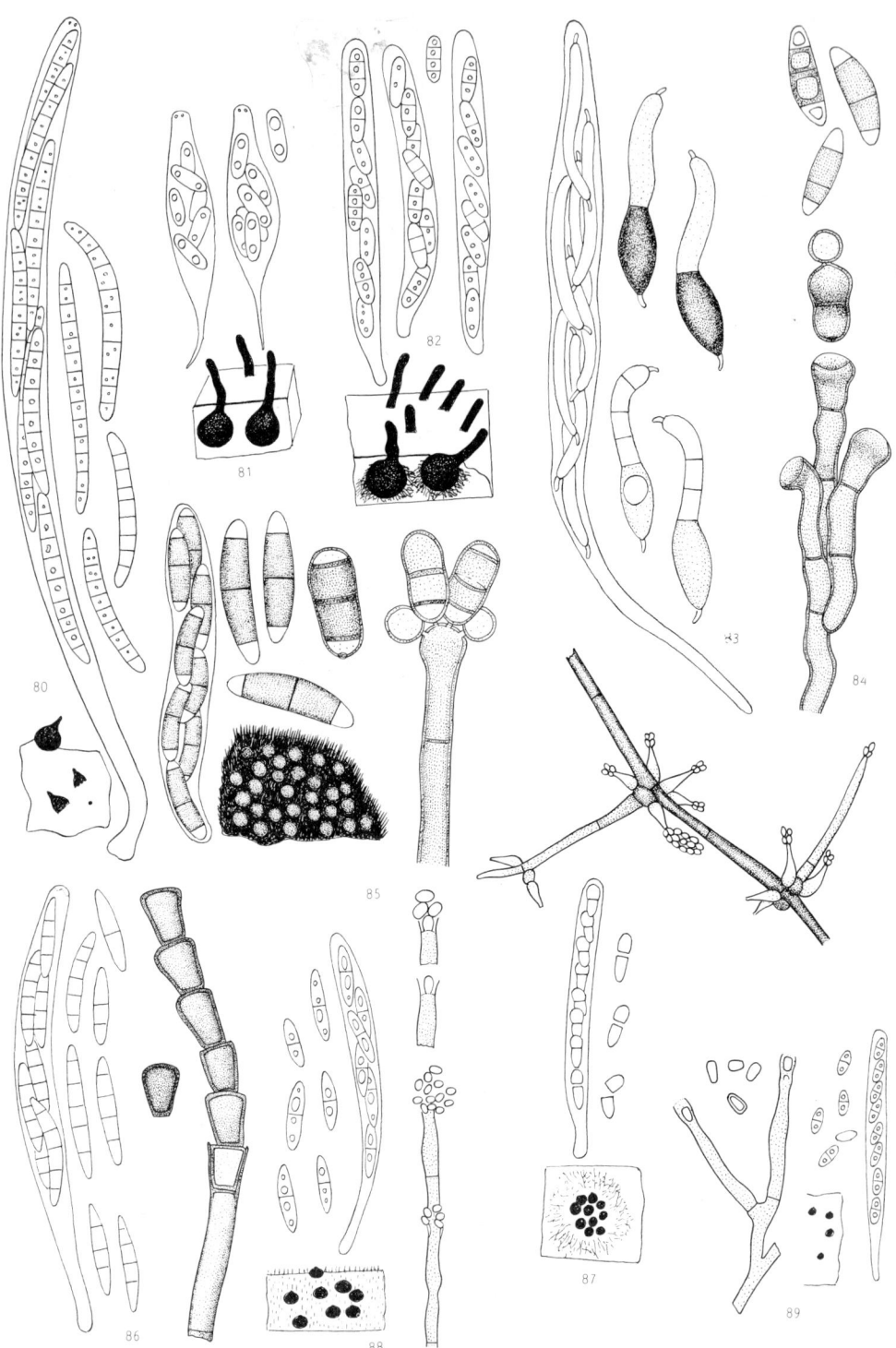

Plate 9. Ascomycetes on wood and bark. 80, *Ceratosphaeria lampadophora*; 81, *Ceratostomella ampullasca*; 82, *C. cirrhosa*; 83, *Cercophora caudata*; 84, *Chaetosphaerella fusca*; 85, *C. phaeostroma*; 86, *Chaetosphaeria cupulifera*; 87, *C. inaequalis*; 88, *C. innumera*; 89, *C. myriocarpa*.

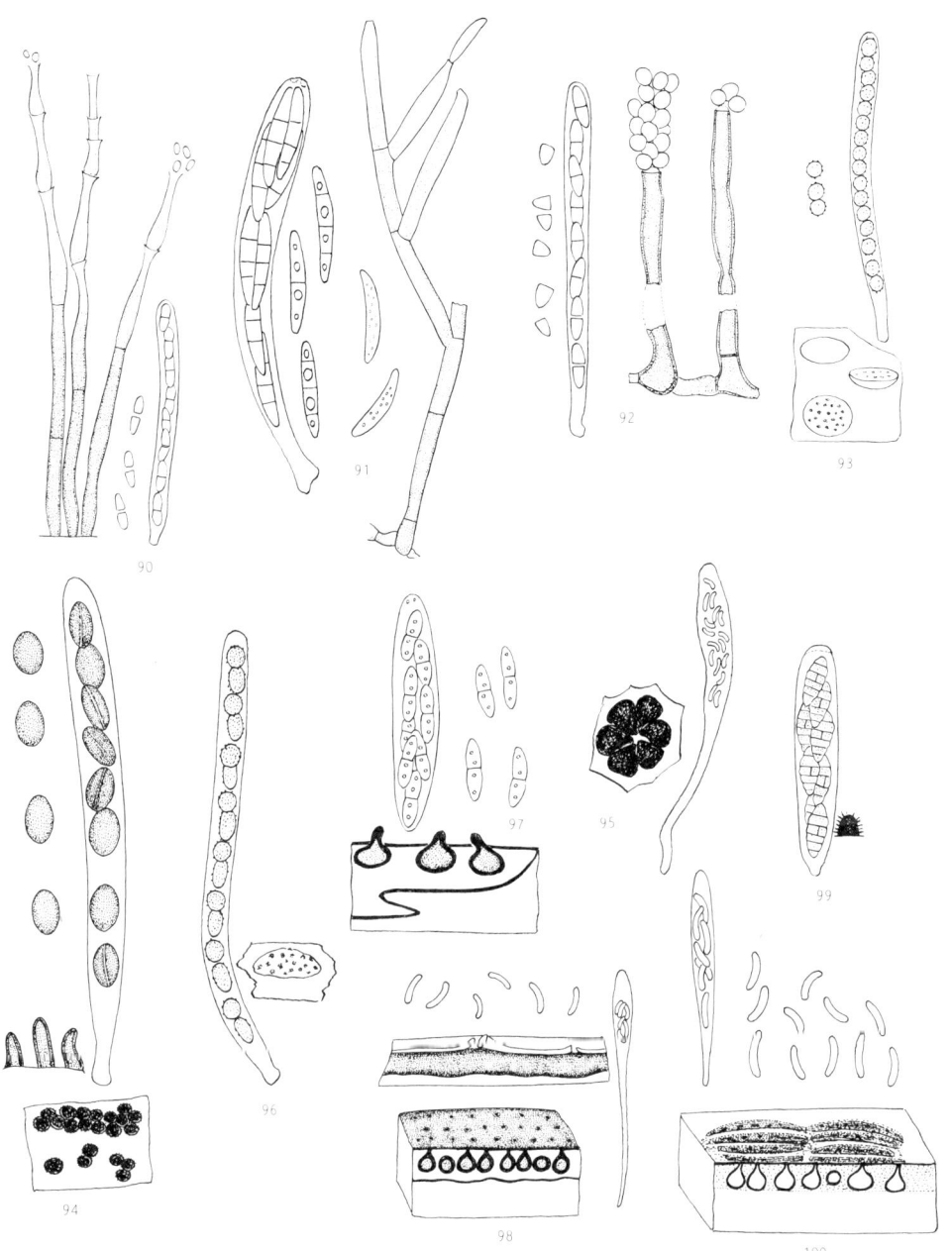

Plate 10. Ascomycetes on wood and bark. 90, *Chaetosphaeria preussii*; 91, *C. pulviscula*; 92, *C. vermicularioides*; 93, *Chromocrea aureoviridis*; 94, *Coniochaeta pulveracea*; 95, *Coronophora angustata*; 96, *Creopus gelatinosus*; 97, *Diaporthe eres*; 98, *Diatrype stigma*; 99, *Dictyotrichiella pulcherrima*; 100, *Eutypa flavovirens*.

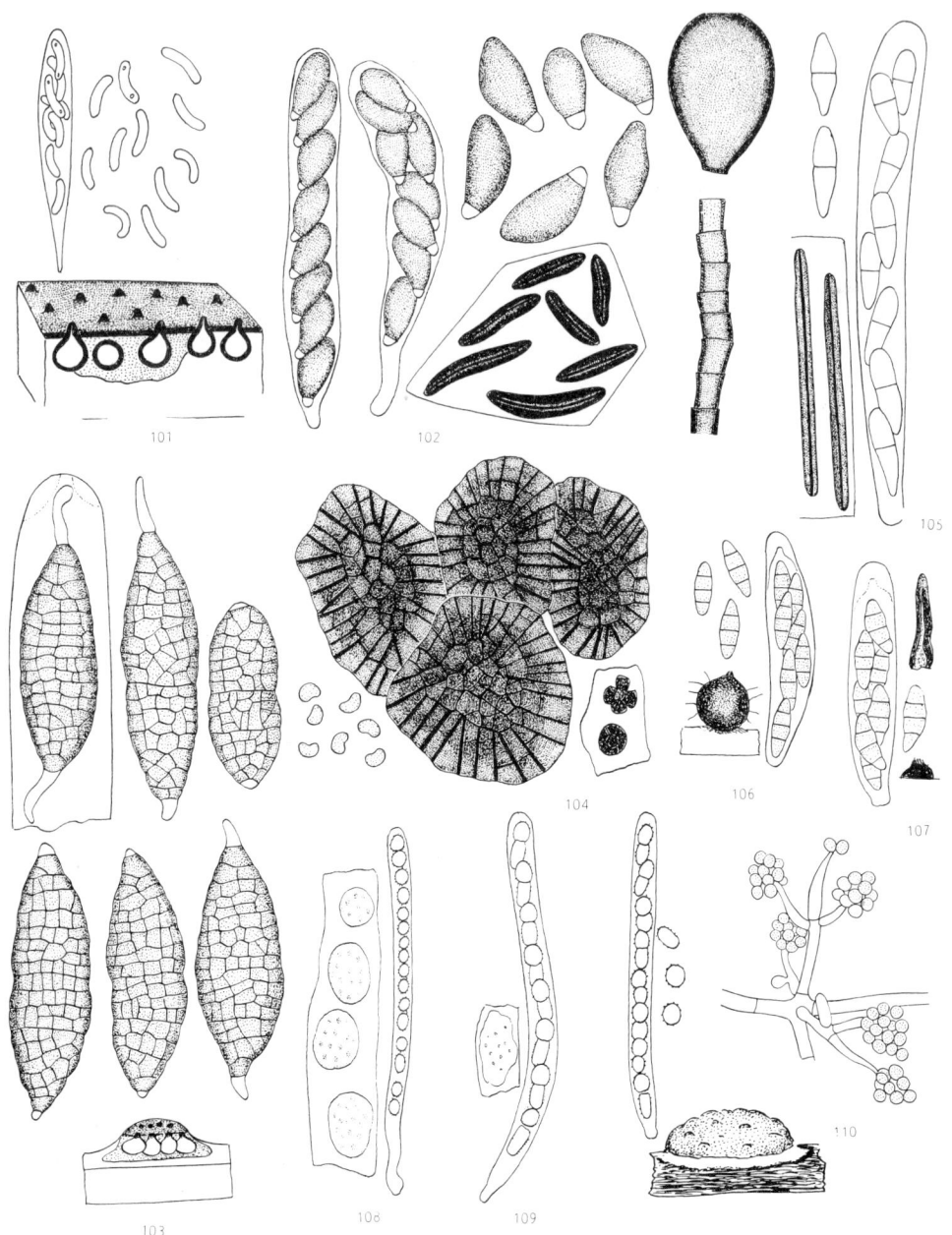

Plate 11. Ascomycetes on wood and bark. 101, *Eutypa lata*; 102, *Farlowi carmichaeliana*; 103, *Fenestella fenestrata*; 104, *Fragosphaeria reniformis*; 105, *Glon. lineare*; 106, *Herpotrichiella moravica*; 107, *H. pilosella*; 108, *Hypocrea argillacea*; 109, *citrina*; 110, *H. rufa*.

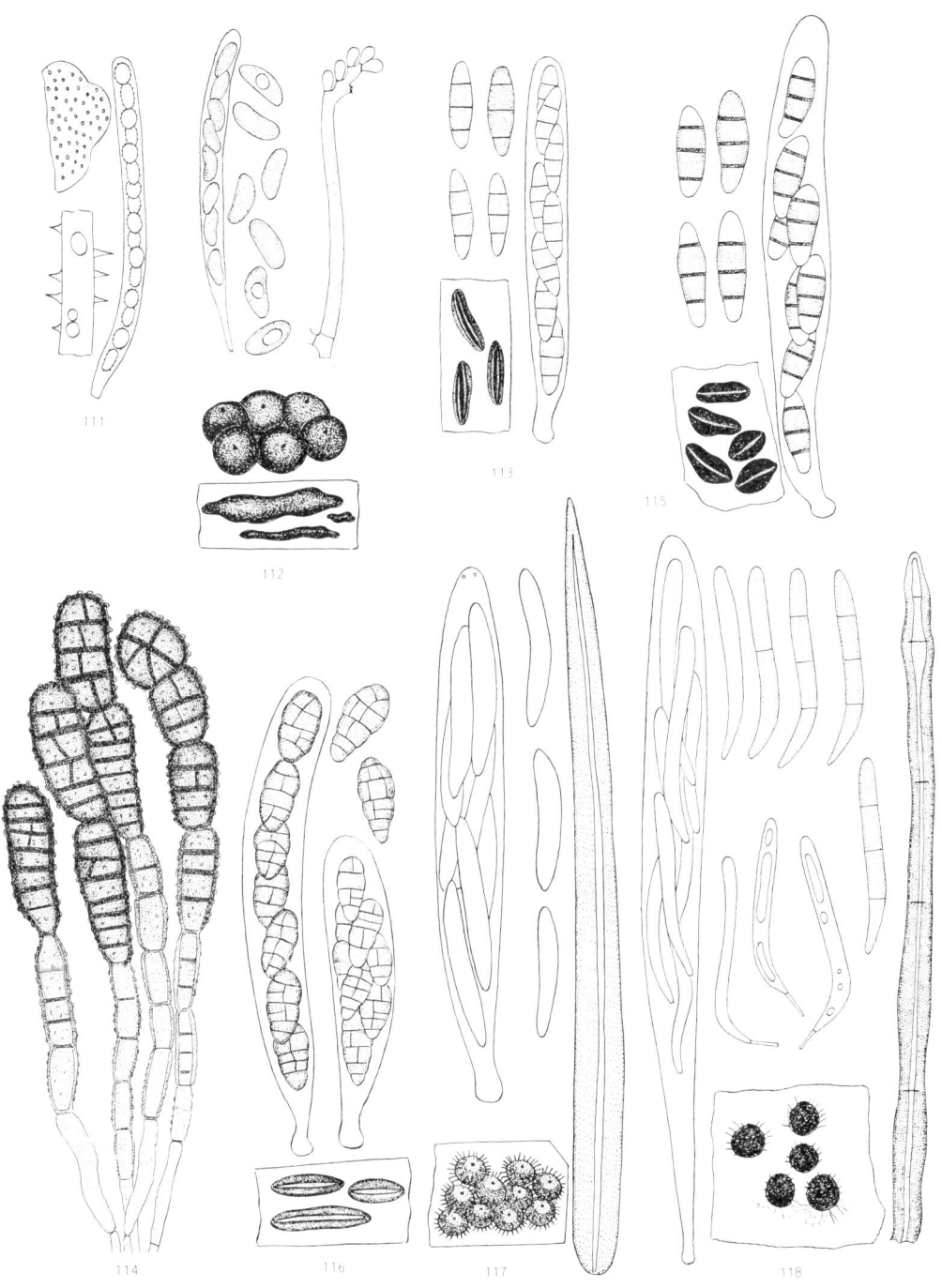

Plate 12. Ascomycetes on wood and bark. 111, *Hypocrea splendens*; 112, *Hypoxylon serpens*; 113, *Hysterium angustatum*; 114, *H. insidens*; 115, *H. pulicare*; 116, *Hysterographium mori*; 117, *Lasiosphaeria canescens*; 118, *L. caudata*.

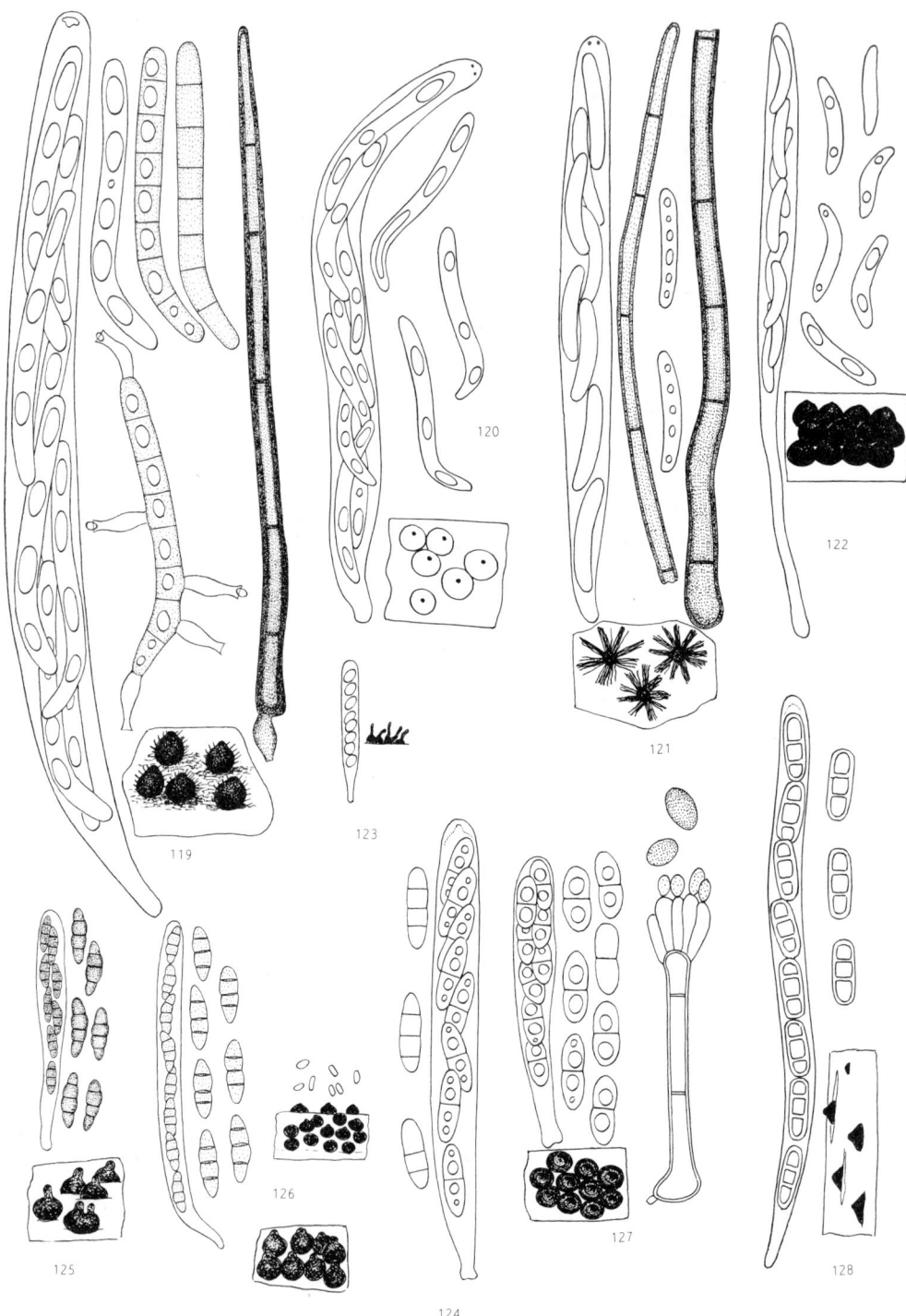

Plate 13. Ascomycetes on wood and bark. 119, *Lasiosphaeria hirsuta*; 120, *L. ovina*; 121, *L. phyllophila*; 122, *L. spermoides*; 123, *Linostomella sphaerosperma*; 124, *Lophiostoma nucula*; 125, *Melanomma fuscidulum*; 126, *M. pulvis-pyrius*; 127, *Melanopsamma pomiformis*; 128, *Melomastia mastoidea*.

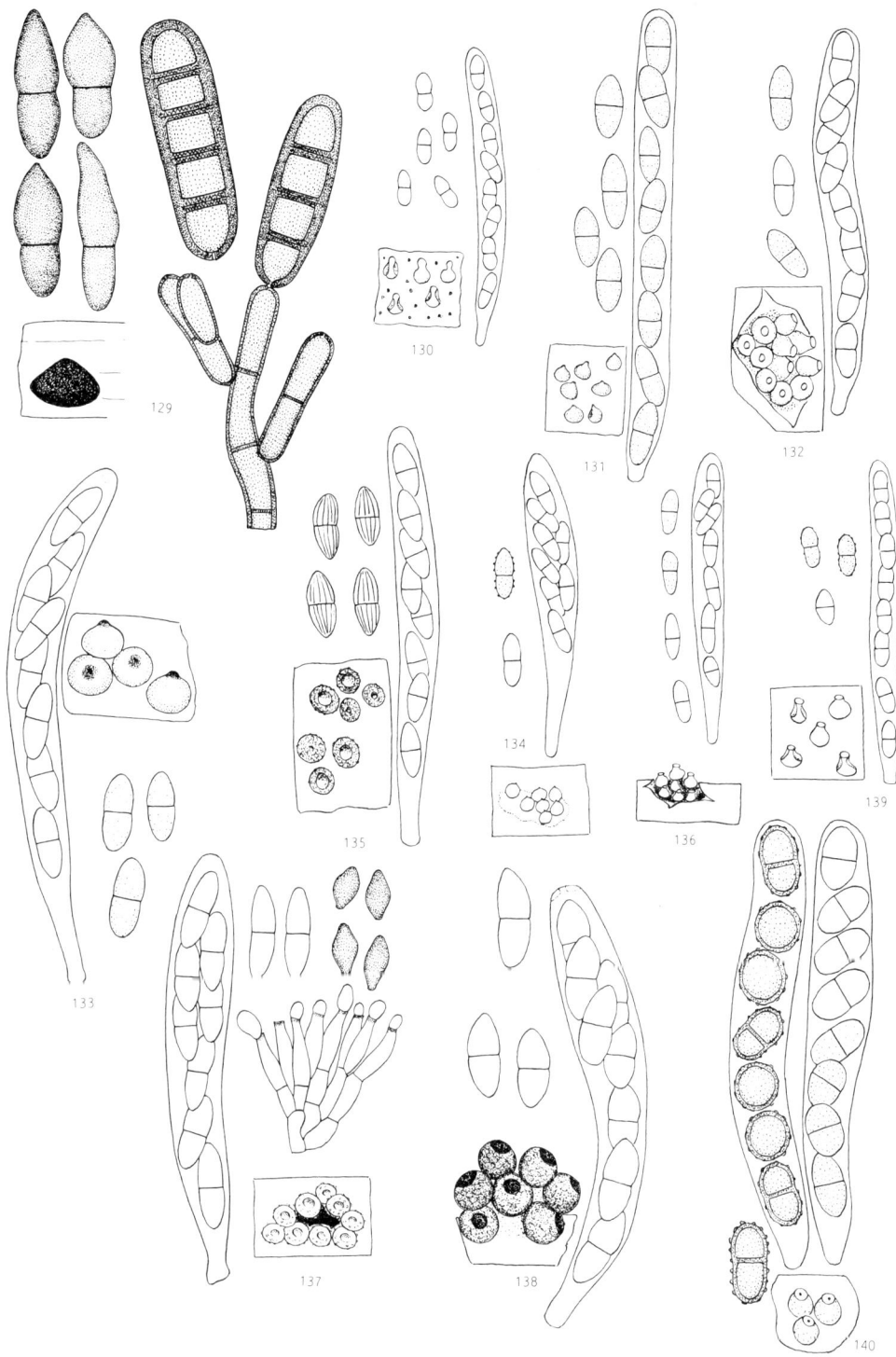

Plate 14. Ascomycetes on wood and bark. 129, *Microthelia incrustans*; 130, *Nectria episphaeria*; 131, *N. flavo-viridis*; 132, *N. magnusiana*; 133, *N. mammoidea*; 134, *N. pallidula*; 135, *N. peziza*; 136, *N. purtonii*; 137, *N. ralfsii*; 138, *N. veuillotiana*; 139, *N. viridescens*; 140, *N. wegeliniana*.

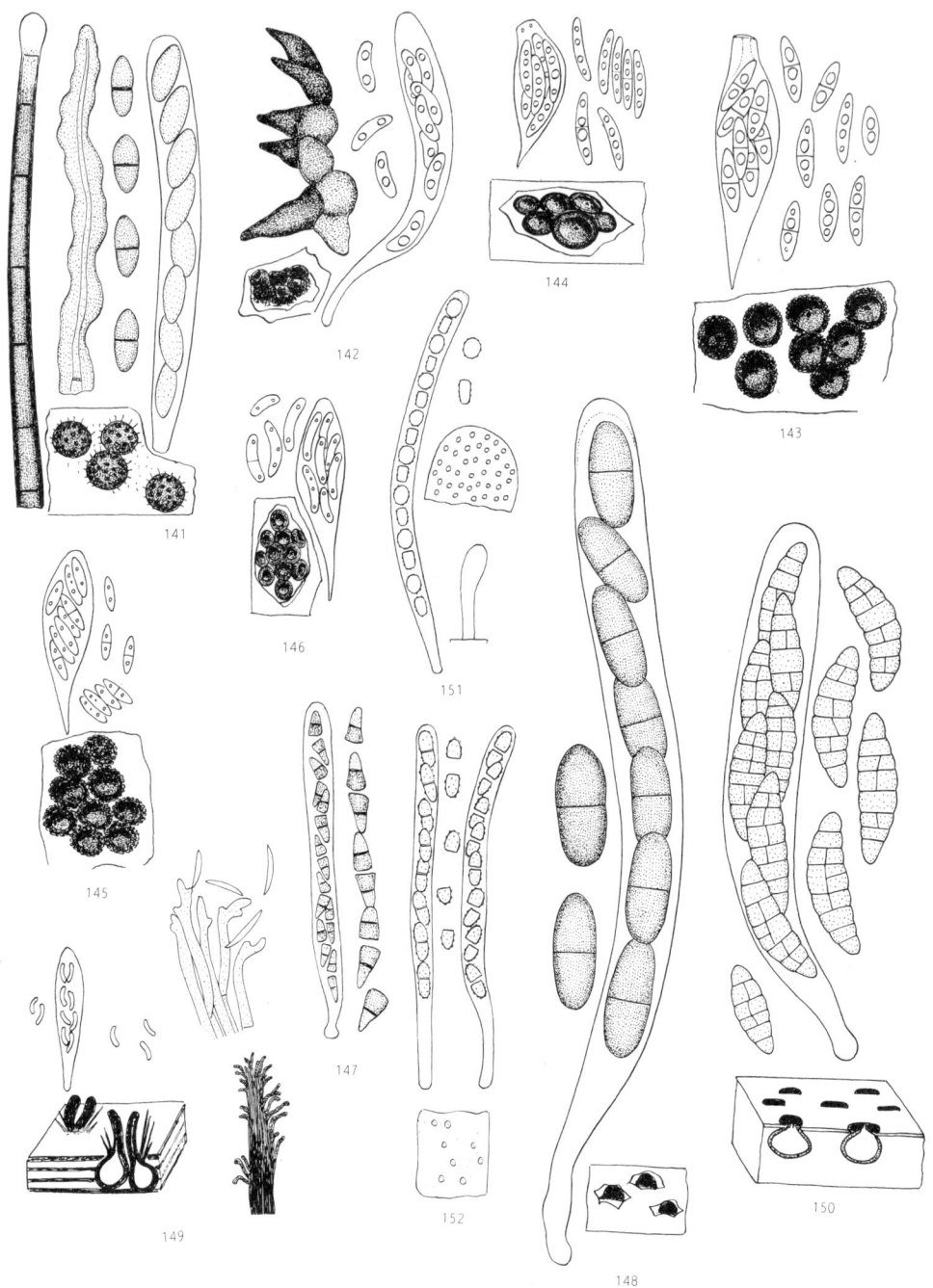

Plate 15. Ascomycetes on wood and bark. 141, *Neopeckia fulcita*; 142, *Nitschkia brevispina*; 143, *N. collapsa*; 144, *N. cupularis*; 145, *N. grevillei*; 146, *N. parasitans*; 147, *Ohleria rugulosa*; 148, *Otthia spiraeae*; 149, *Peroneutypa heteracantha*; 150, *Platystomum compressum*; 151, *Podostroma alutaceum*; 152, *Protocrea farinosa*.

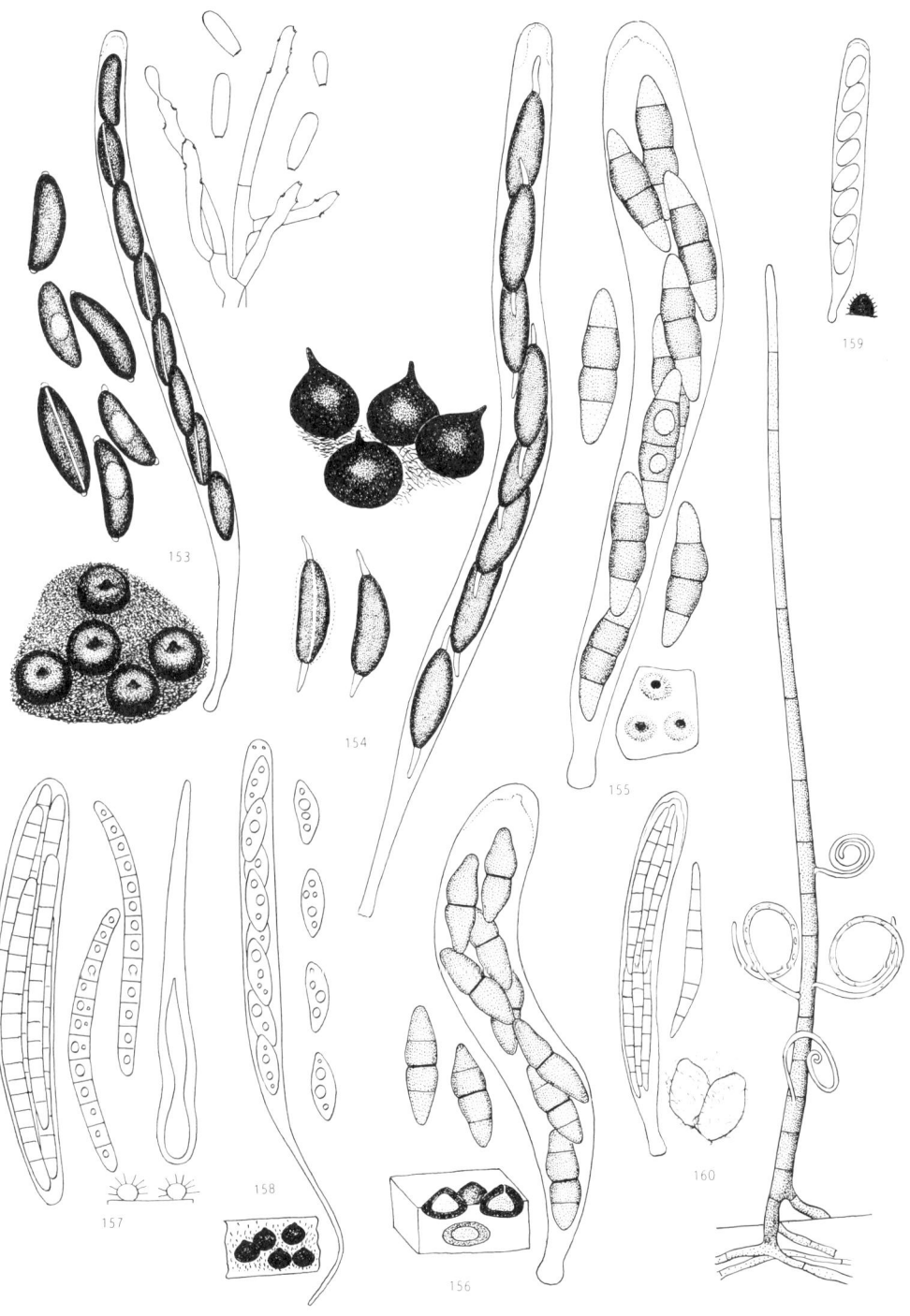

Plate 16. Ascomycetes on wood and bark. 153, *Rosellinia aquila*; 154, *R. thelena*; 155, *Trematosphaeria anglica*; 156, *T. pertusa*; 157, *Trichonectria hirta*; 158, *Trichosphaeria notabilis*; 159, *T. pilosa*; 160, *Tubeufia cerea*.

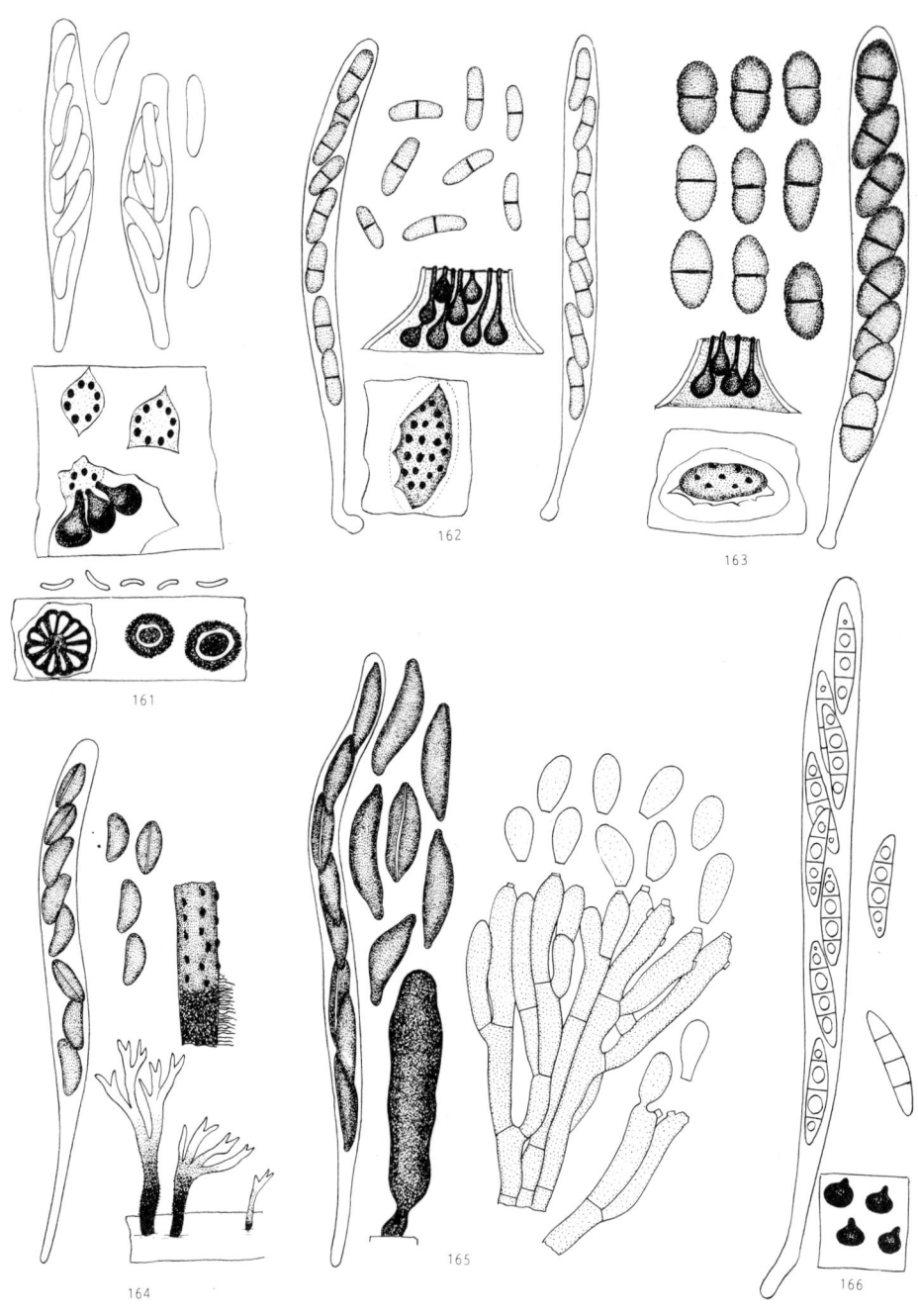

Plate 17. Ascomycetes on wood and bark. 161, *Valsa ambiens*; 162, *Valsaria foedans*; 163, *V. insitiva*; 164, *Xylaria hypoxylon*; 165, *X. polymorpha*; 166, *Zignoella ovoidea*.

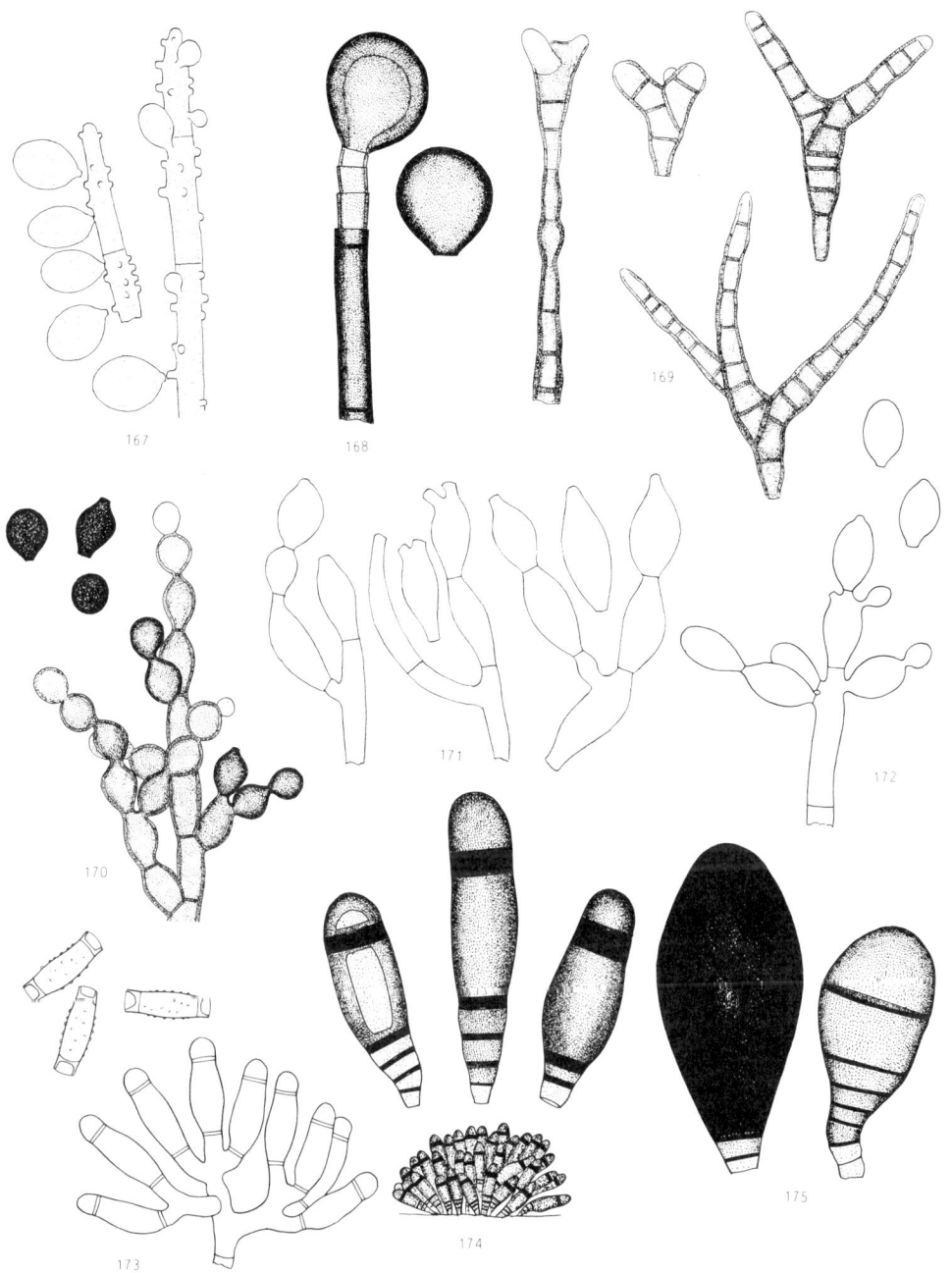

Plate 18. Hyphomycetes on wood and bark. 167, *Acladium* state of *Botryobasidium conspersum*; 168, *Acrogenospora sphaerocephala*; 169, *Actinocladium rhodosporum*; 170, *Alysidium resinae*; 171, *Alysidium* state of *Botryobasidium aureum*; 172, *Alysidium* state of *B. candicans*; 173, *Amblyosporium botrytis*; 174, *Bactrodesmium abruptum*; 175, *B. atrum*.

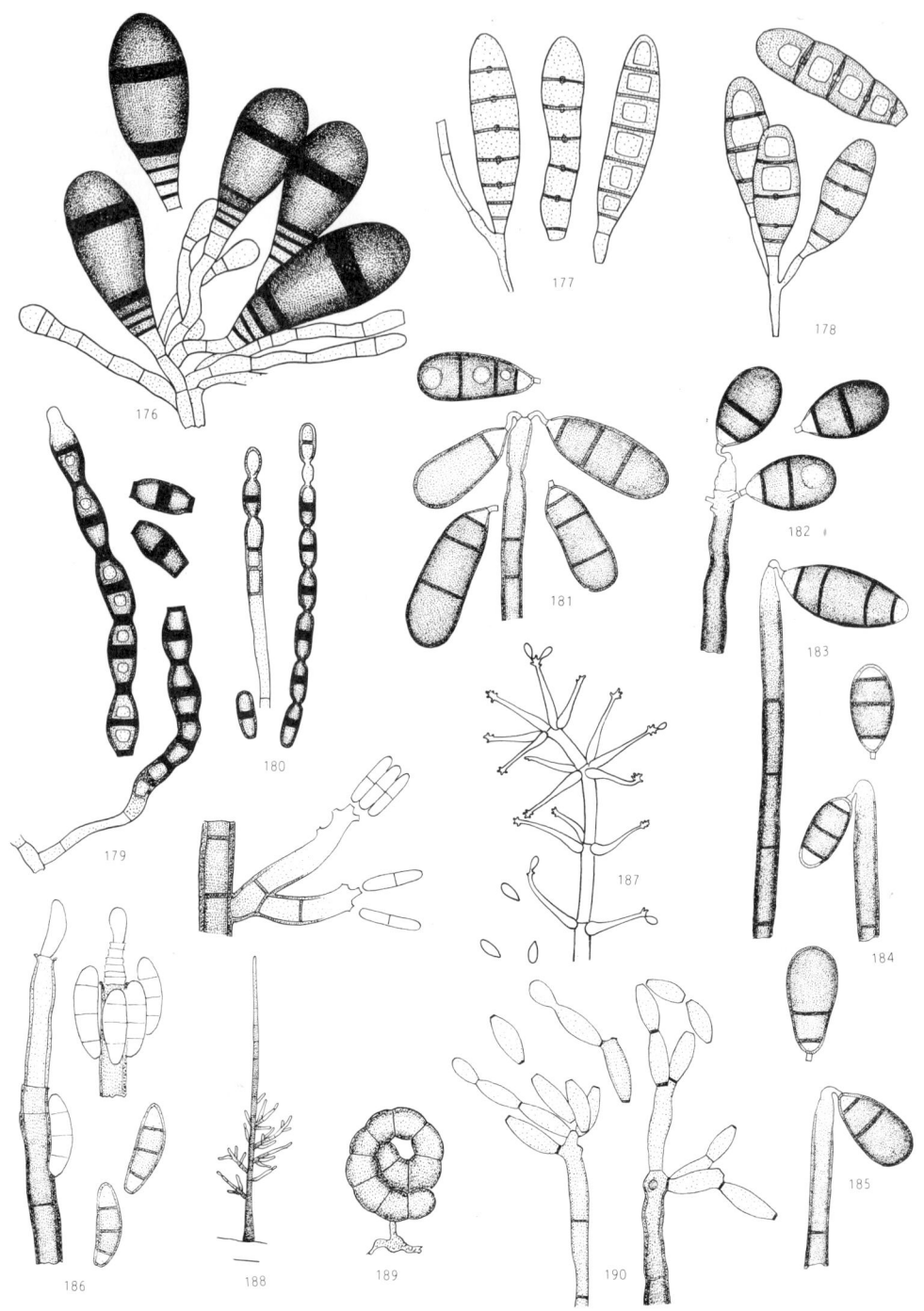

Plate 19. Hyphomycetes on wood and bark. 176, *Bactrodesmium obovatum*; 177, *B. pallidum*; 178, *B. spilomeum*; 179, *Bispora antennata*; 180, *B. betulina*; 181, *Brachysporium bloxami*; 182, *B. britannicum*; 183, *B. masonii*; 184, *B. nigrum*; 185, *B. obovatum*; 186, *Cacumisporium capitulatum*; 187, *Calcarisporium arbuscula*; 188, *Chaetopsis grisea*; 189, *Cirrenalia lignicola*; 190, *Cladosporium britannicum*.

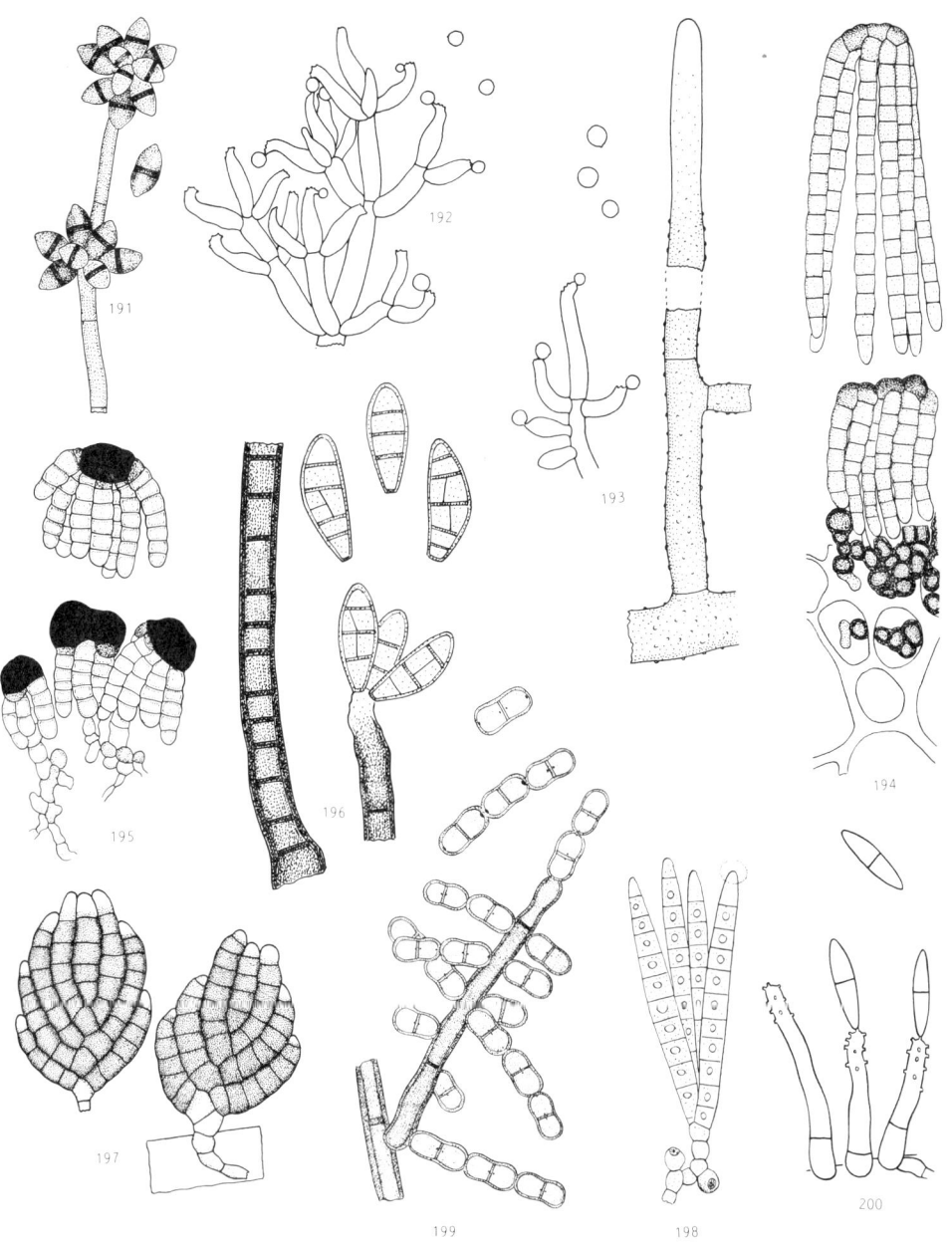

Plate 20. Hyphomycetes on wood and bark. 191, *Cordana pauciseptata*; 192, *Costantinella micheneri*; 193, *C. terrestris*; 194, *Cryptocoryneum condensatum*; 195, *C. rilstonii*; 196, *Dactylosporium macropus*; 197, *Dictyosporium toruloides*; 198, *Digitodesmium elegans*; 199, *Diplococcium spicatum*; 200, *Diplorhinotrichum candidulum*.

Plate 21. Hyphomycetes on wood and bark. 201, *Endophragmia biseptata*; 202, *E. nannfeldtii*; 203, *E. uniseptata*; 204, *Endophragmiella cambrensis*; 205, *E. corticola*; 206, *E. fallacia*; 207, *E. ovoidea*; 208, *E. pallescens*; 209, *E. subolivacea*; 210, *Gliocladium luteolum*; 211, *G. roseum*; 212, *Graphium calicioides*; 213, *Helicoma* state of *Thaxteriella pezicula*; 214, *Helicomyces roseus*; 215, *Helicoon ellipticum*.

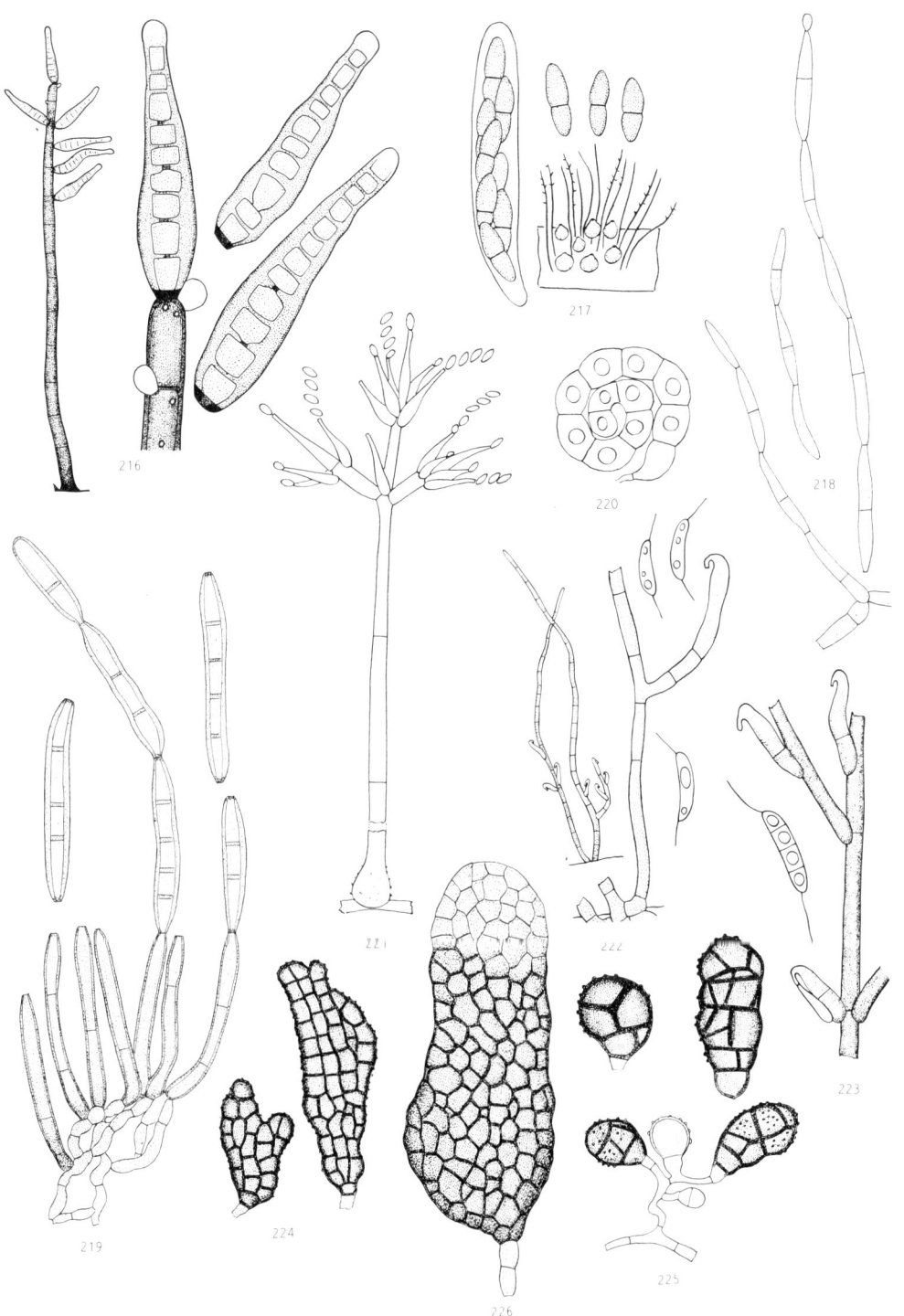

Plate 22. Hyphomycetes on wood and bark. 216, *Helminthosporium velutinum*; 217, *Letendraea helminthicola*; 218, *Heteroconium chaetospira*; 219, *H. tetracoilum*; 220, *Hobsonia mirabilis*; 221, *Mariannaea elegans*; 222, *Menispora ciliata*; 223, *M. glauca*; 224, *Monodictys antiqua*; 225, *M. castaneae*; 226, *M. lepraria*.

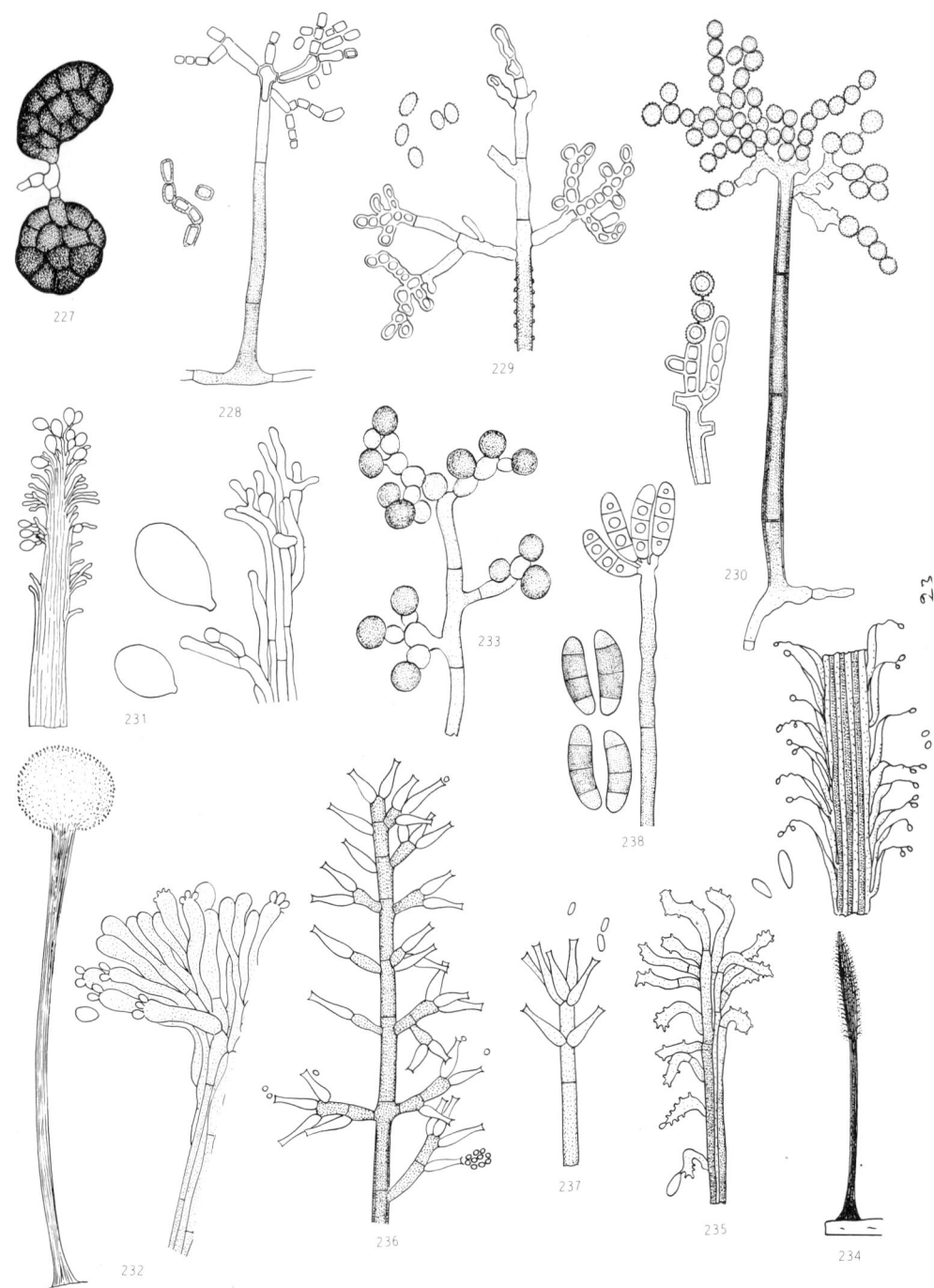

Plate 23. Hyphomycetes on wood and bark. 227, *Monodictys putredinis*; 228, *Oidiodendron griseum*; 229, *O. rhodogenum*; 230, *O. tenuissimum*; 231, *Pachnocybe albida*; 232, *P. ferruginea*; 233, *Periconia cambrensis*; 234, *Phaeoisaria clavulata*; 235, *P. clematidis*; 236, *Phaeostalagmus cyclosporus*; 237, *P. tenuissimus*; 238, *Pleurothecium recurvatum*.

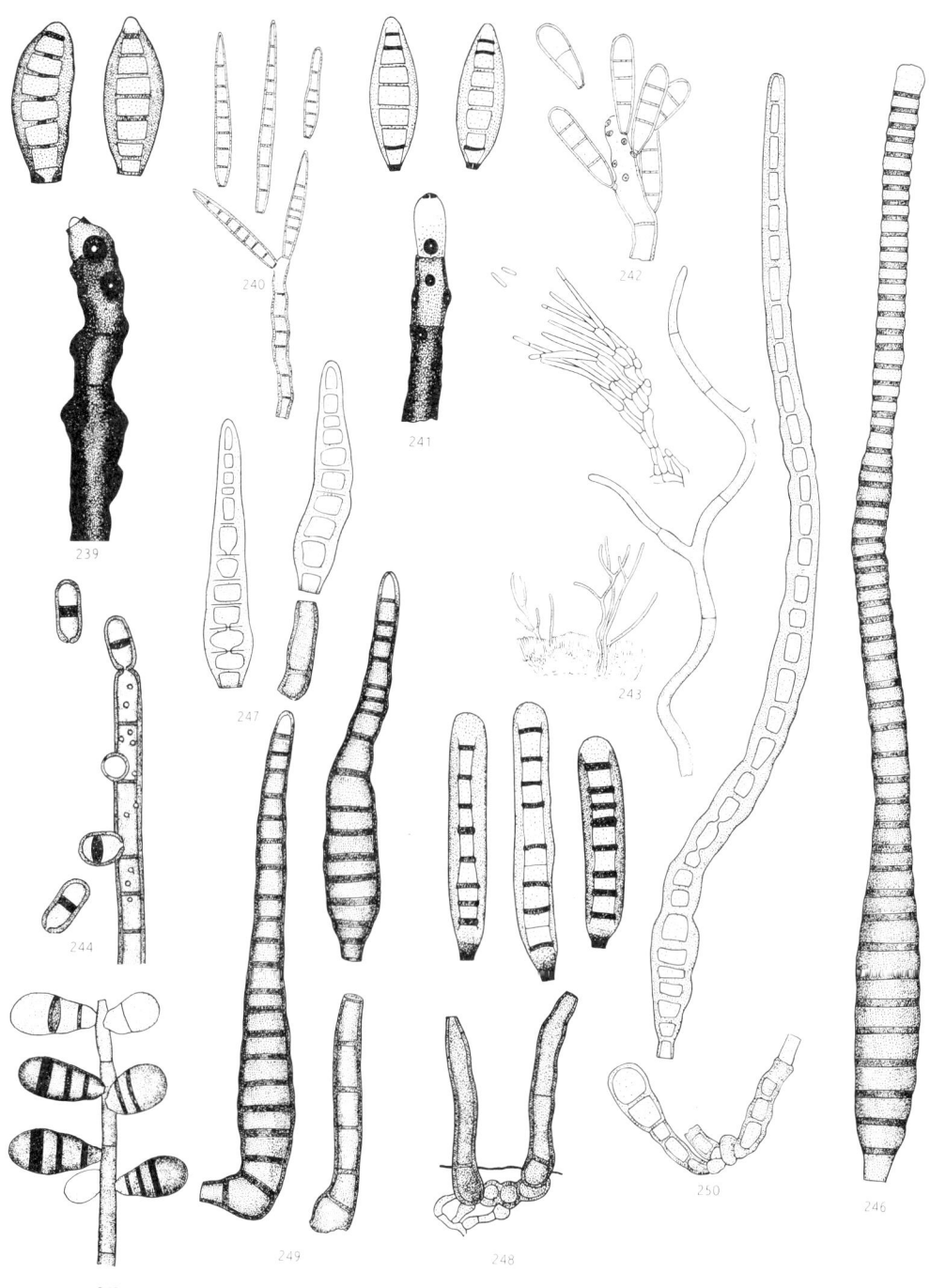

Plate 24. Hyphomycetes on wood and bark. 239, *Pseudospiropes nodosus*; 240, *P. obclavatus*; 241, *P. simplex*; 242, *P. subuliferus*; 243, *Sarcopodium tortuosum*; 244, *Spadicoides bina*; 245, *S. grovei*; 246, *Sporidesmium anglicum*; 247, *S. coronatum*; 248, *S. folliculatum*; 249, *S. pedunculatum*; 250, *S. vagum*.

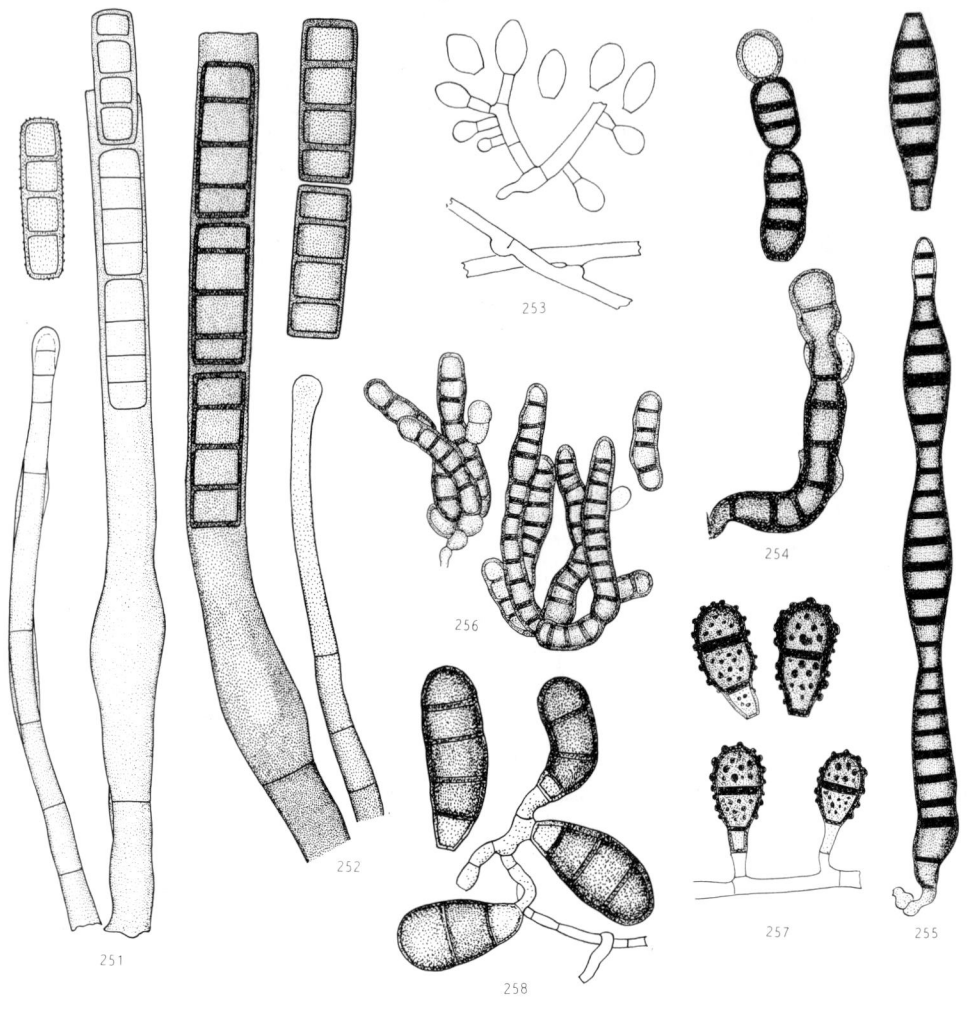

Plate 25. Hyphomycetes on wood and bark. 251, *Sporoschisma juvenile*; 252, *S. mirabile*; 253, *Sporotrichum* state of *Ceriporiopsis aneirina*; 254, *Taeniolella breviuscula*; 255, *T. rudis*; 256, *Taeniolina scripta*; 257, *Trichocladium asperum*; 258, *T. opacum*.

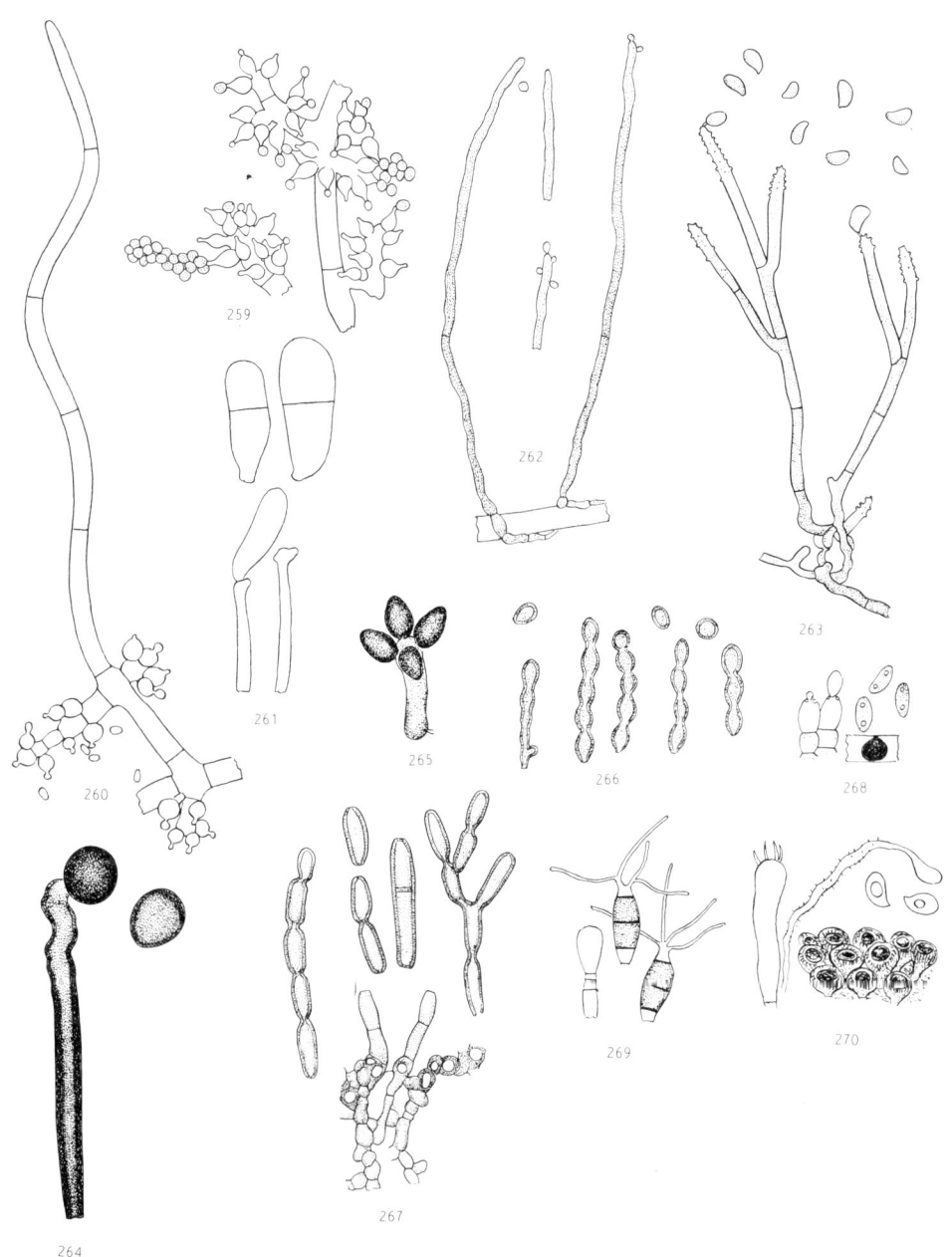

Plate 26. Various groups on wood and bark. 259, *Trichoderma hamatum*; 260, *T. polysporum*; 261, *Trichothecium roseum*; 262, *Veronaea parvispora*; 263, *Virgaria nigra*; 264, *Virgariella atra*; 265, *V. ovoidea*; 266, *Xylohypha ferruginosa*; 267, *X. nigrescens*; 268, *Microsphaeropsis olivacea*; 269, *Truncatella angustata*; 270, *Cyphellopsis anomala*.

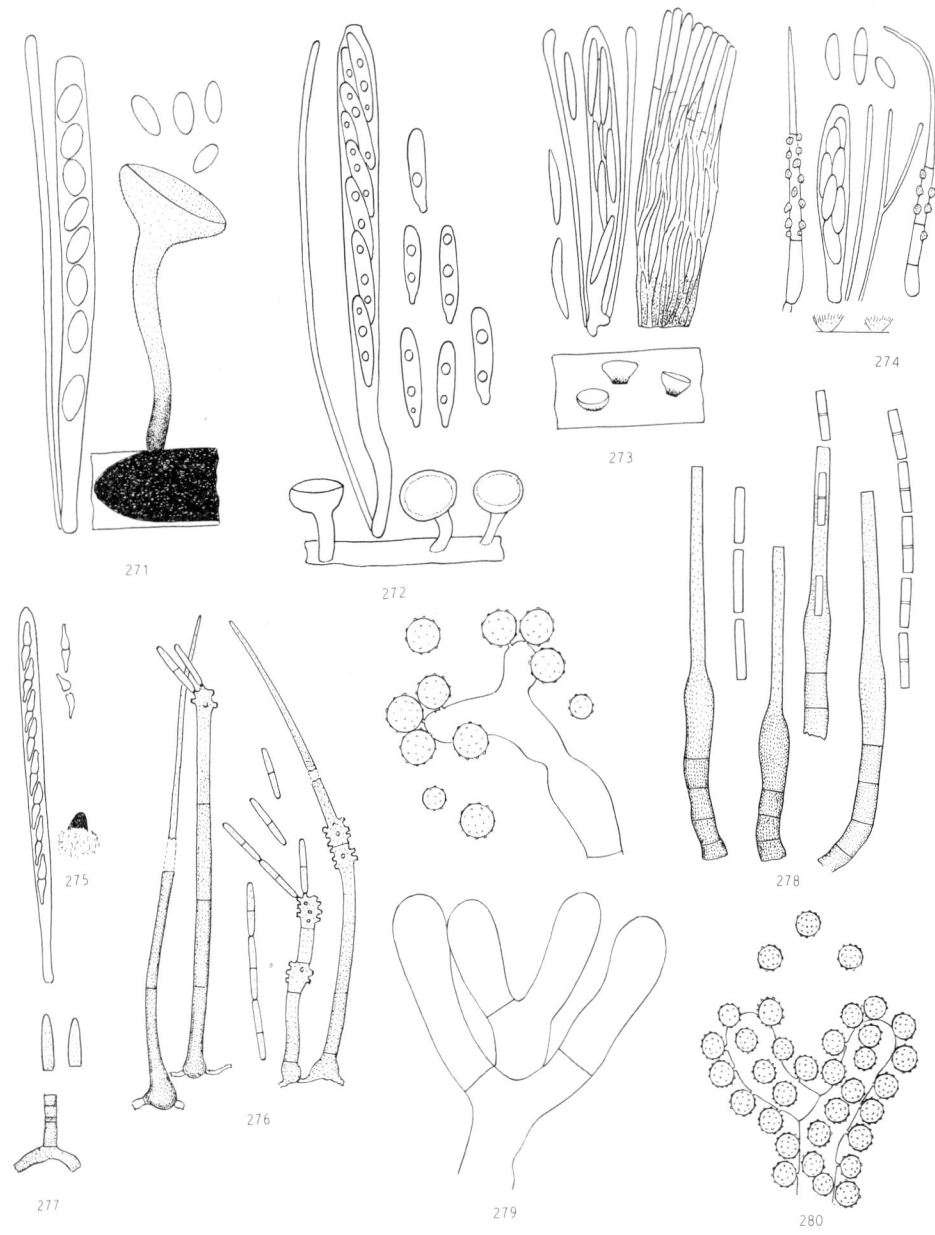

Plate 27. Leaf-litter fungi. 271, *Botryotinia fuckeliana*; 272, *Hymenoscyphus caudatus*; 273, *Phialea* sp. 1; 274, *Trichodiscus virescentulus*; 275, *Arachnocrea stipata*; 276, *Anungitea fragilis*; 277, *Belemnospora epiphylla*; 278, *Chalara aurea*; 279, *Chromelosporium carneum*; 280, *C. ochraceum*.

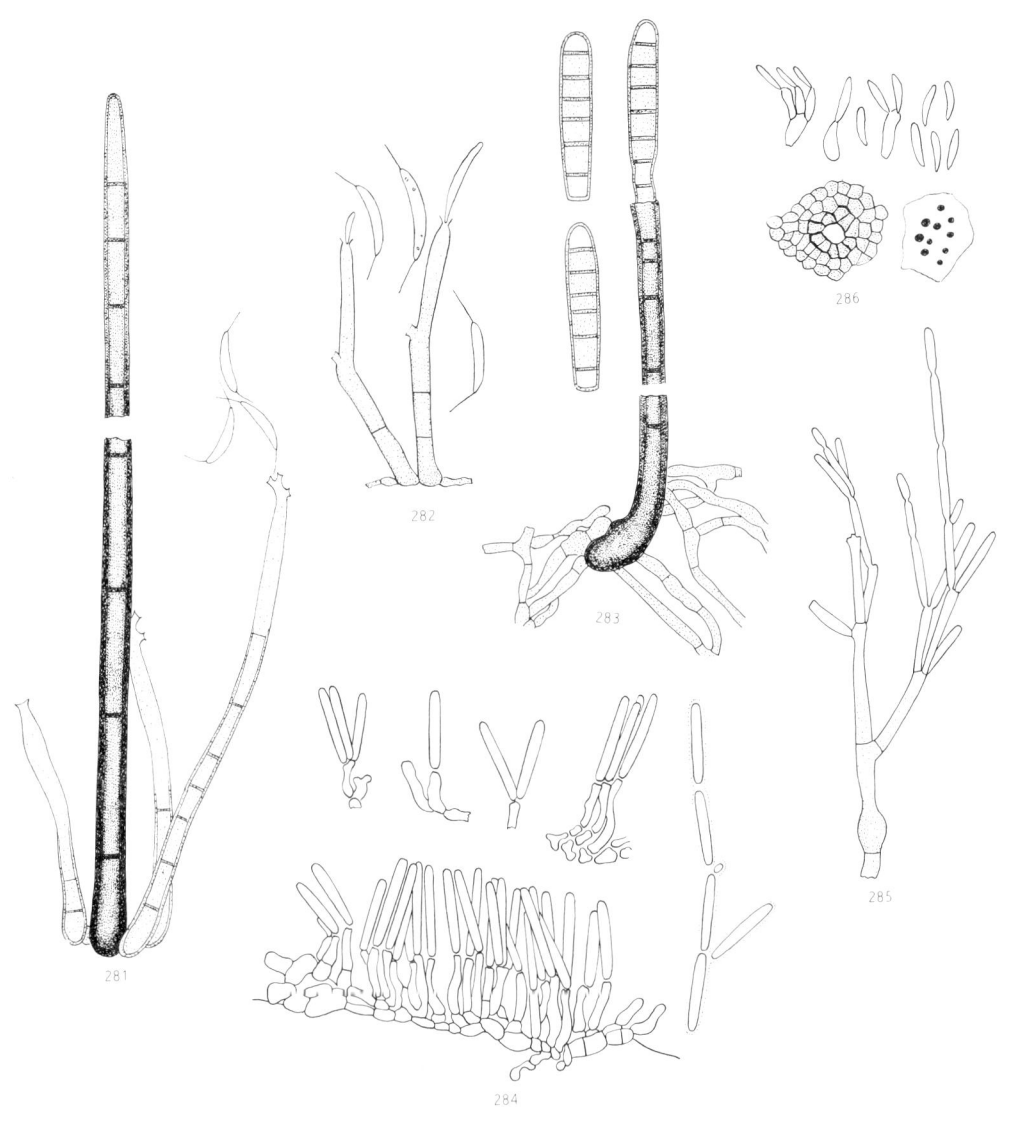

Plate 28. Leaf-litter fungi. 281, *Codinaea britannica*; 282, *C. simplex*; 283, *Endophragmia alternata*; 284, *Fusidium griseum*; 285, *Polyscytalum fecundissimum*; 286, *Pilidium acerinum*.

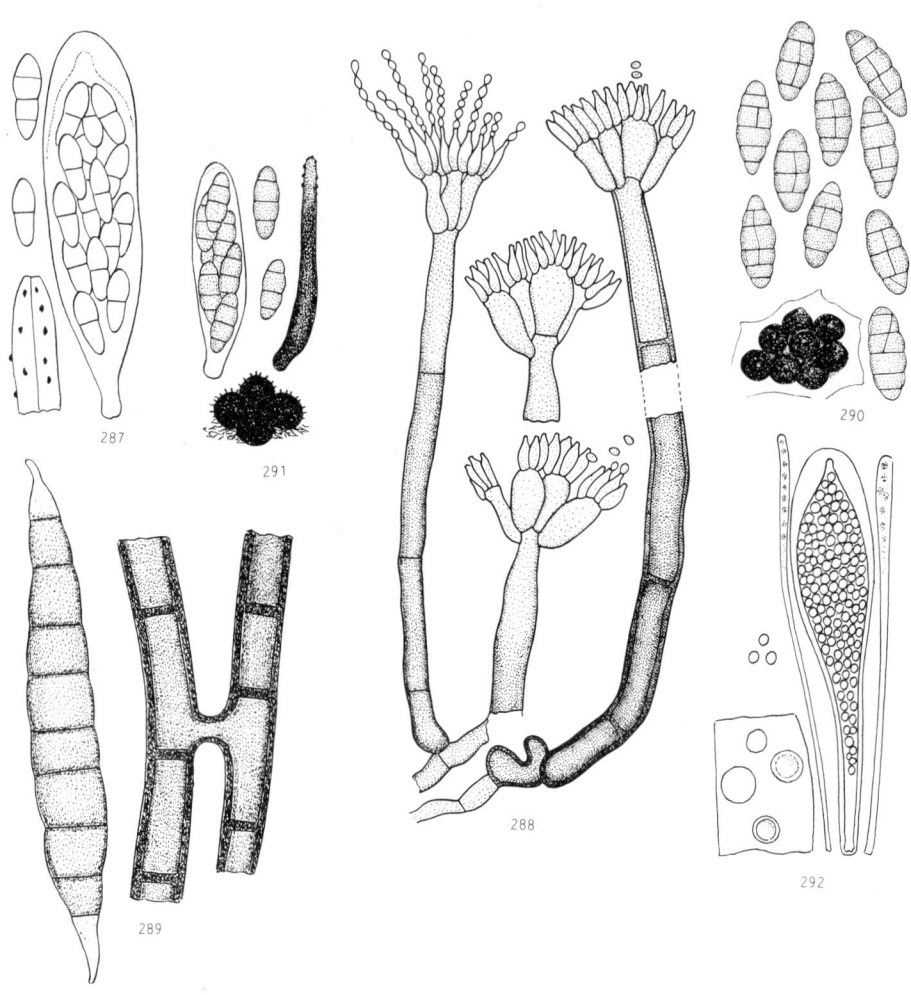

Plate 29. On *Abies*. 287, *Delphinella abietis*; 288, *Thysanophora penicillioides*; 289, *Antennatula pinophylla*; 290, *Cucurbidothis pithyophila*; 291, *Herpotrichia parasitica*; 292, *Sarea resinae*.

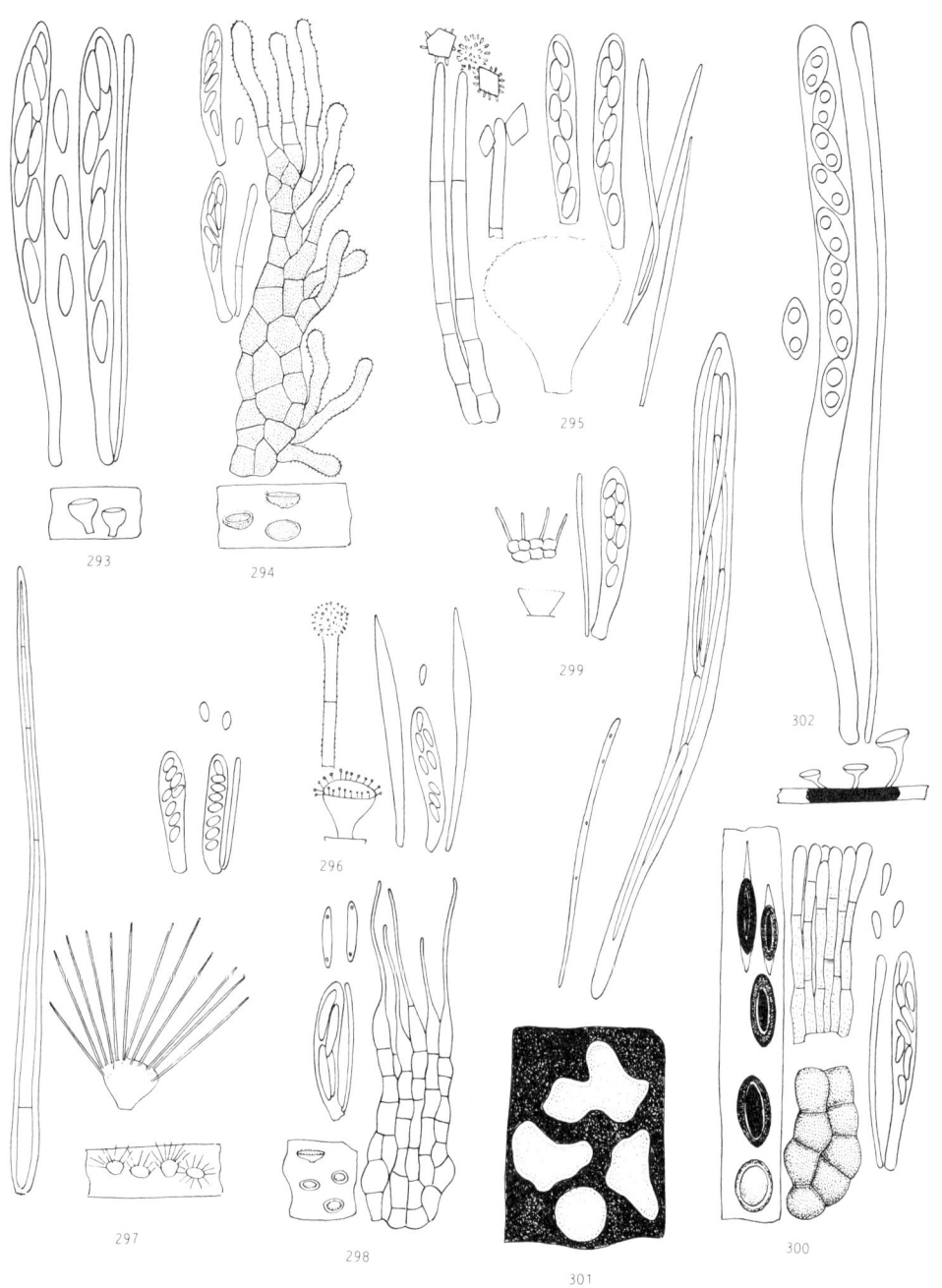

Plate 30. Discomycetes on leaves of *Acer*. 293, *Crocicreas subhyalinum*; 294, *Dasyscyphus acerinus*; 295, *D. radotinensis*; 296, *D. rhytismatis*; 297, *Hyalopeziza ciliata*; 298, *Hyaloscypha lachnobrachya*; 299, *Mollisina acerina*; 300, *Pyrenopeziza petiolaris*; 301, *Rhytisma acerinum*; 302, *Rutstroemia luteovirescens*.

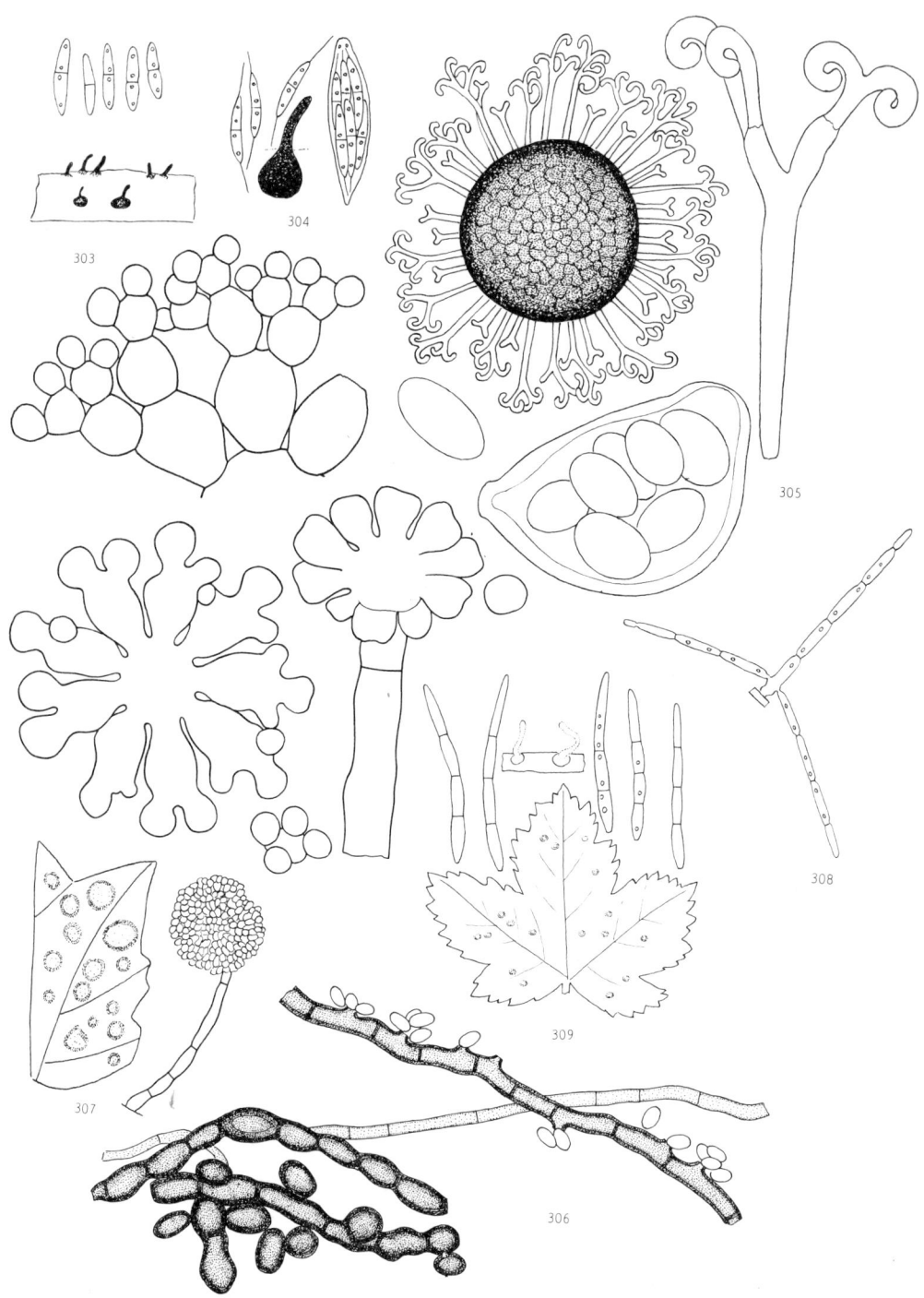

Plate 31. Various groups on leaves of *Acer*. 303, *Gnomonia cerastris*; 304, *Plagiostoma inclinata*; 305, *Uncinula bicornis*; 306, *Aureobasidium pullulans*; 307, *Cristulariella depraedens*; 308, *Trisulcosporium acerinum*; 309, *Phloeospora aceris*.

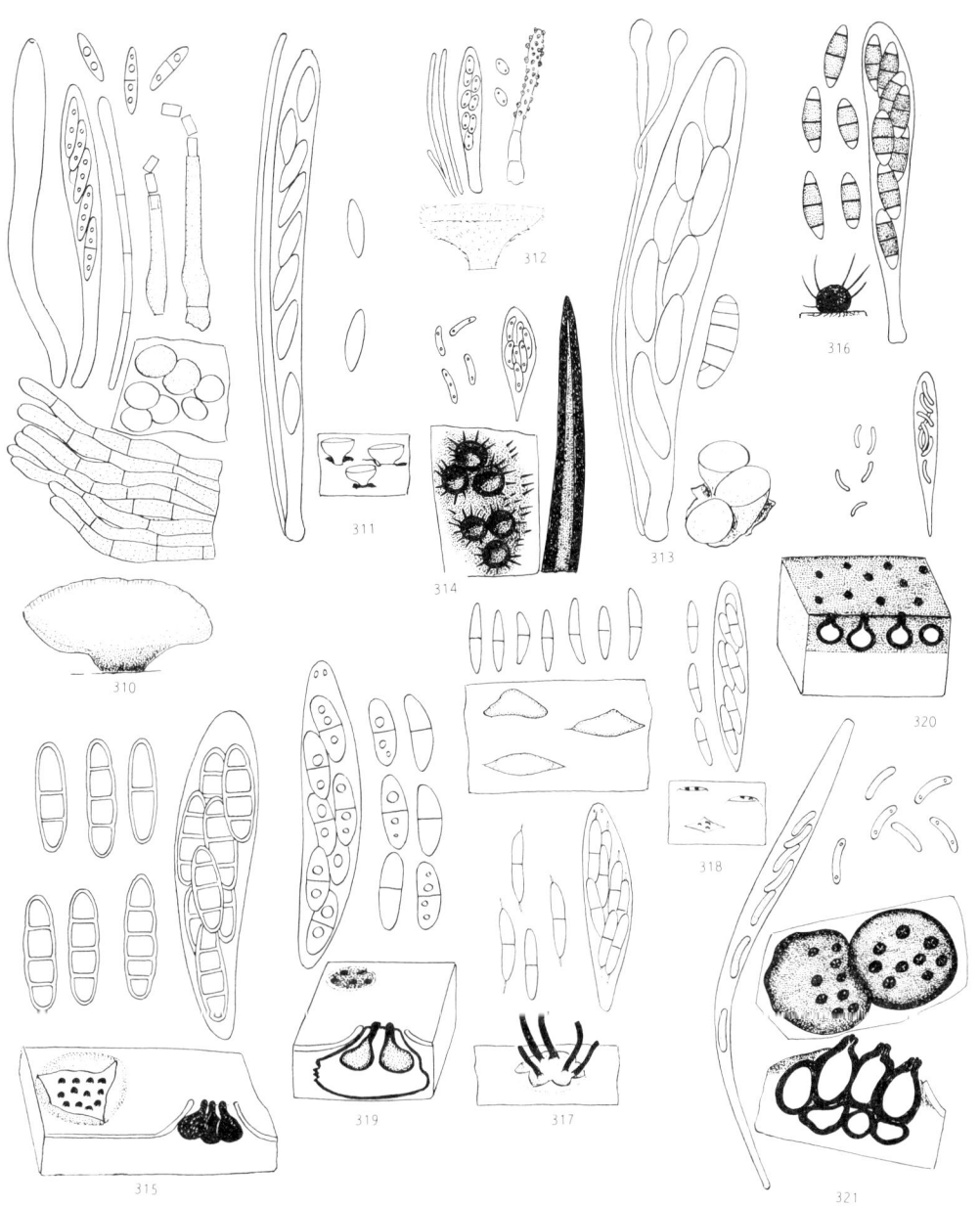

Plate 32. Discomycetes and other ascomycetes on *Acer* wood and bark. 310, *Allophylaria crystallifera*; 311, *Hymenoscyphus subpallescens*; 312, *Incrupila viridipilosa*; 313, *Pezicula acericola*; 314, *Acanthonitschkea tristis*; 315, *Calospora platanoides*; 316, *Chaetosphaerella fusispora*; 317, *Cryptodiaporthe hystrix*; 318, *C. lebiseyi*; 319, *Diaporthe varians*; 320, *Eutypa acharii*; 321, *Eutypella acericola*.

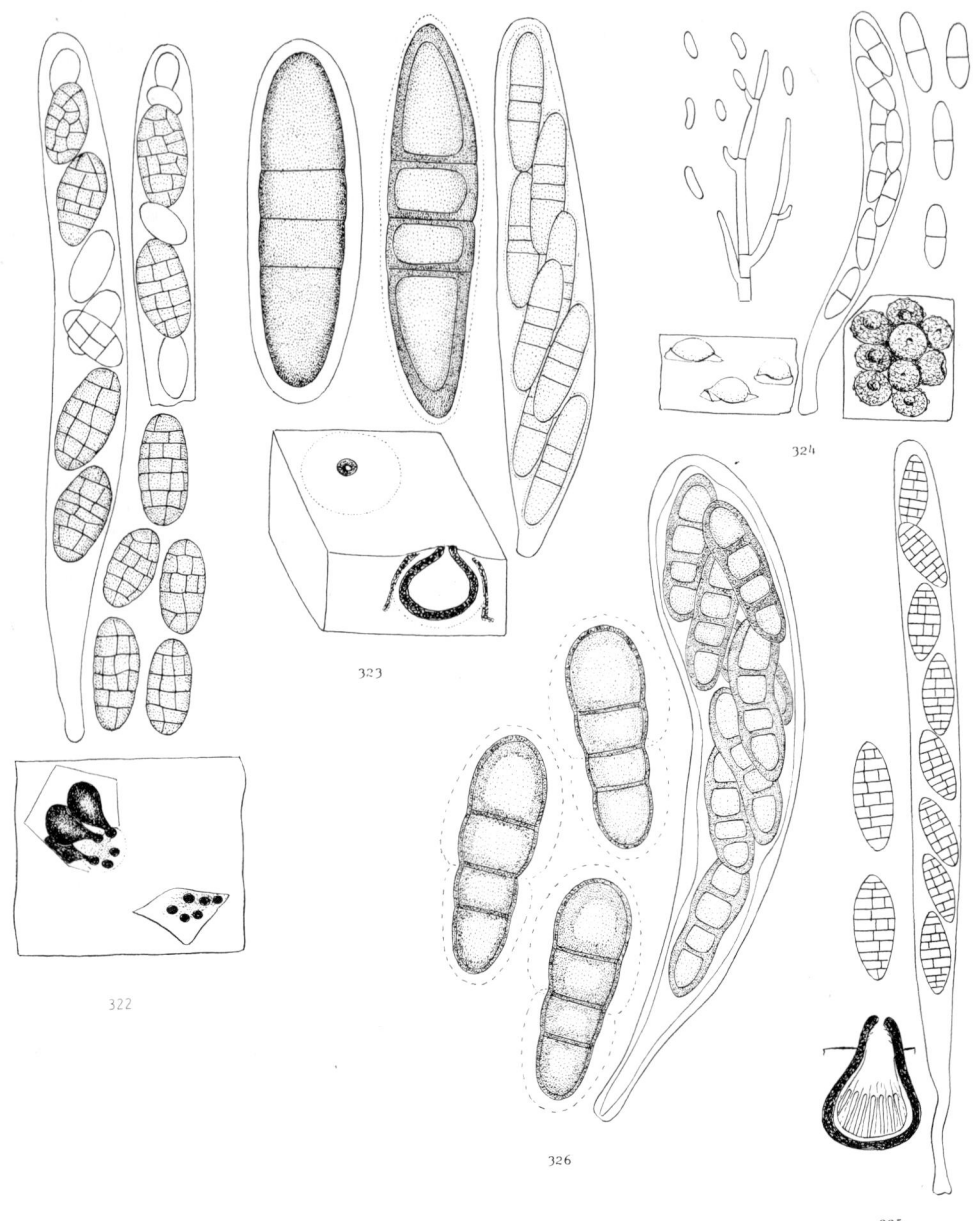

Plate 33. Ascomycetes on *Acer* wood and bark. 322, *Fenestella vestita*; 323, *Massaria inquinans*; 324, *Nectria cinnabarina*; 325, *Rhamphoria bevanii*; 326, *Splanchnonema pupula*.

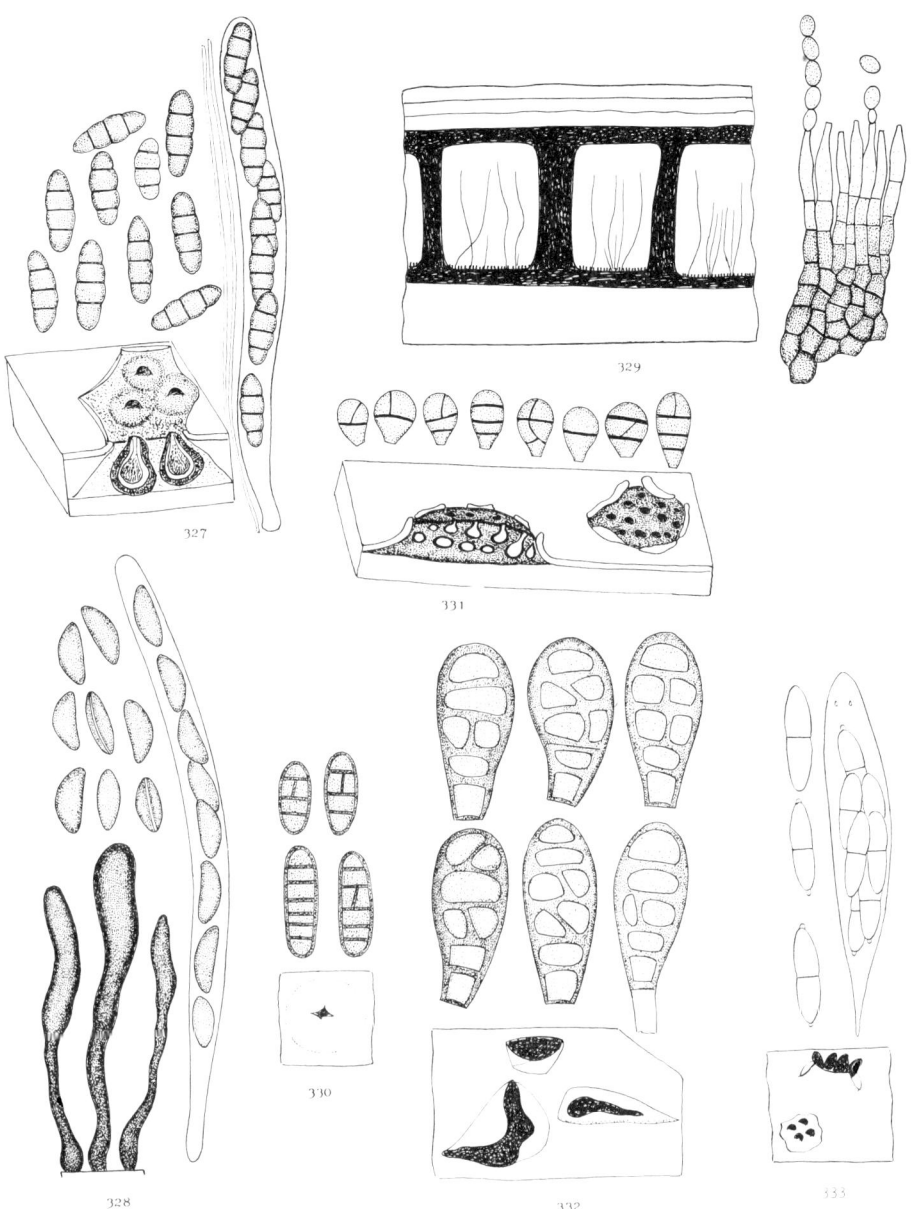

Plate 34. On wood and bark of *Acer* and of *Aesculus*. 327, *Thyridaria rubronotata*; 328, *Xylaria longipes*; 329, *Cryptostroma corticale*; 330, *Camarosporium ambiens*; 331, *Dichomera saubinetii*; 332, *Stegonosporium pyriforme*; 333, *Cryptodiaporthe aesculi*.

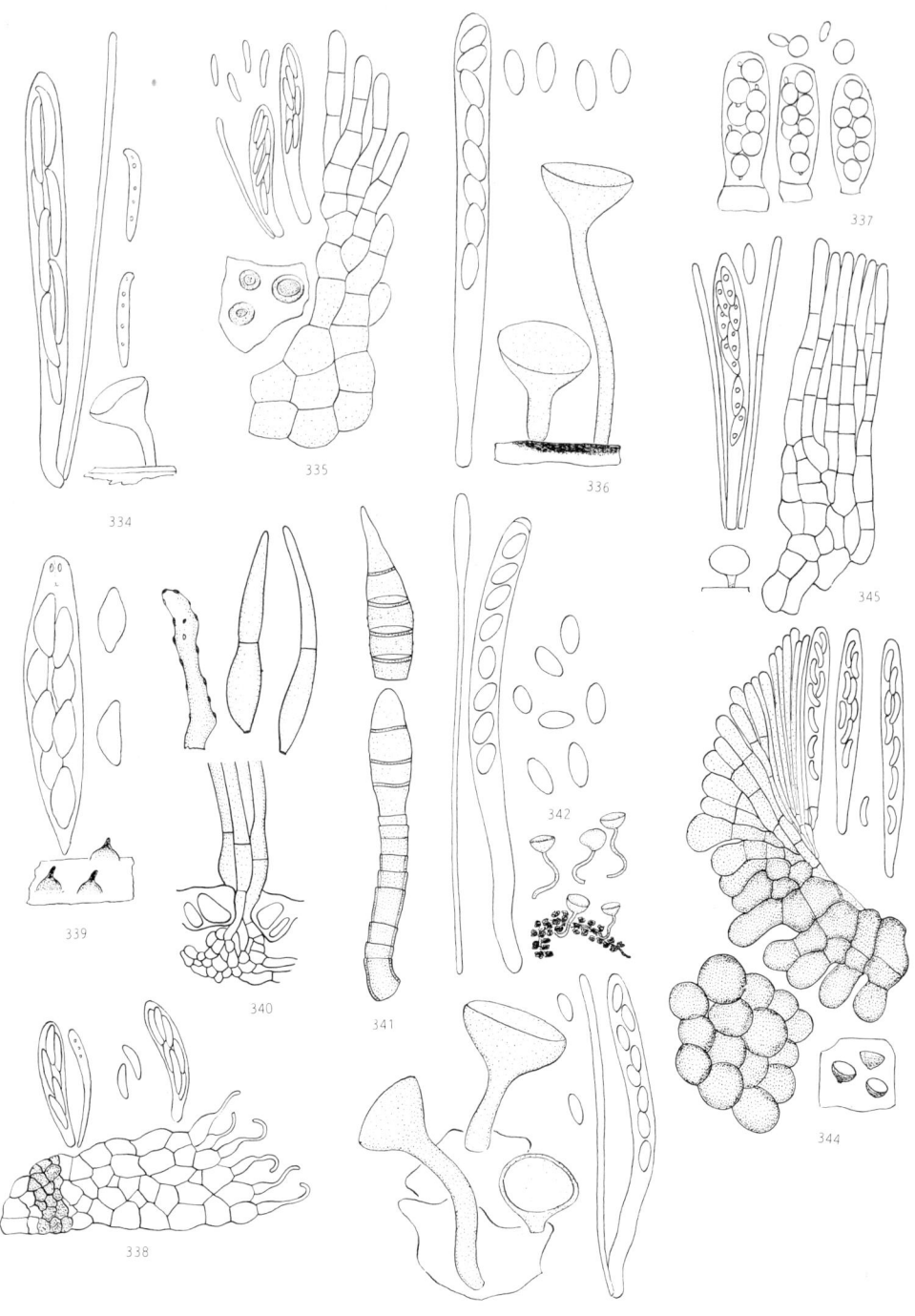

Plate 35. On *Alnus* leaves and catkins. 334, *Hymenoscyphus albopunctus*; 335, *Pyrenopeziza foliicola*; 336, *Rutstroemia conformata*; 337, *Taphrina tosquinetii*; 338, *Uncinia foliicola*; 339, *Gnomoniella tubiformis*; 340, *Passalora bacilligera*; 341, *Sporidesmium wroblewskii*; 342, *Ciboria amentacea*; 343, *C. viridifusca*; 344, *Mollisia amenticola*; 345, *Pezizella alniella*.

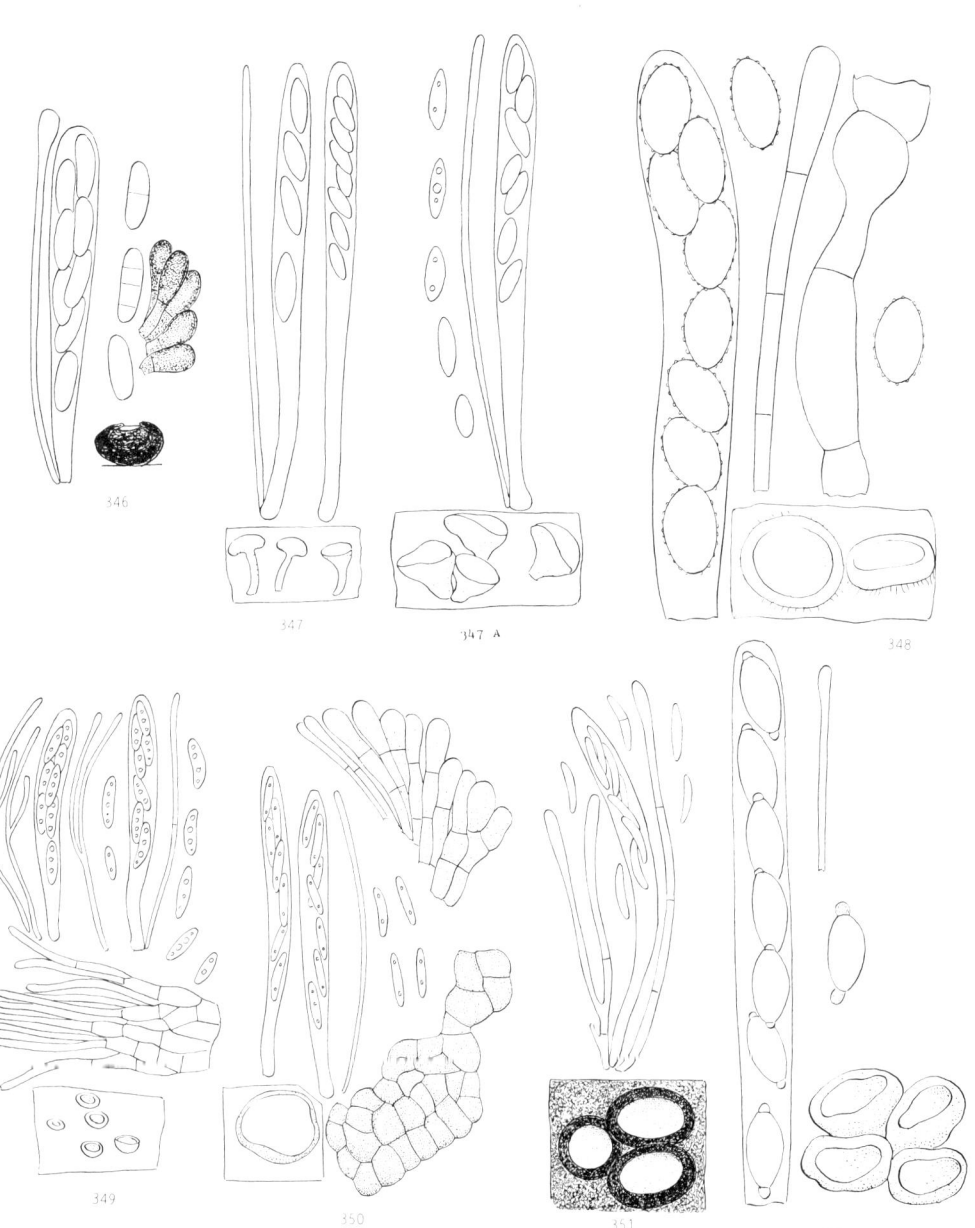

Plate 36. Discomycetes on *Alnus* wood and bark. 346, *Cenangium graddonii*; 347, *Cudoniella clavus*; 347A, *C. clavus* var. *grandis*; 348, *Miladina lecithina*; 349, *Pezizella albohyalina*; 350, *Pyrenopeziza benesuada*; 351, *Tapesia fusca*; 352, *Thecotheus rivicola*.

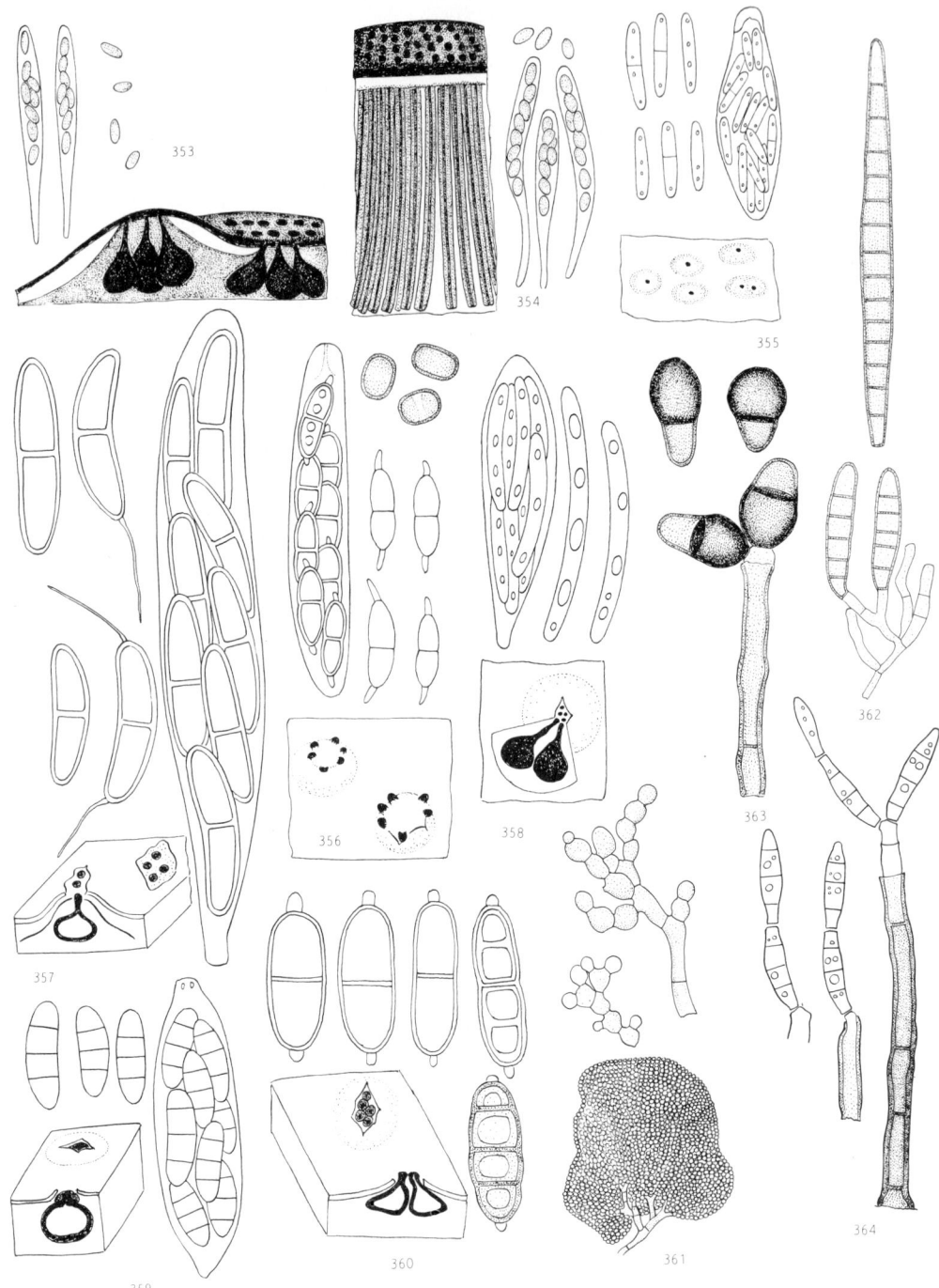

Plate 37. On *Alnus* wood and bark. 353, *Camarops microspora*; 354, *C. polysperma*; 355, *Ditopella ditopa*; 356, *Melanconis alni*; 357, *M. thelebola*; 358, *Ophiovalsa suffusa*; 359, *Phragmoporthe conformis*; 360, *Prosthecium auctum*; 361, *Pseudaegerita viridis*; 362, *Bactrodesmium longisporum*; 363, *Cordana crassa*; 364, *Pleurotheciopsis bramleyi*.

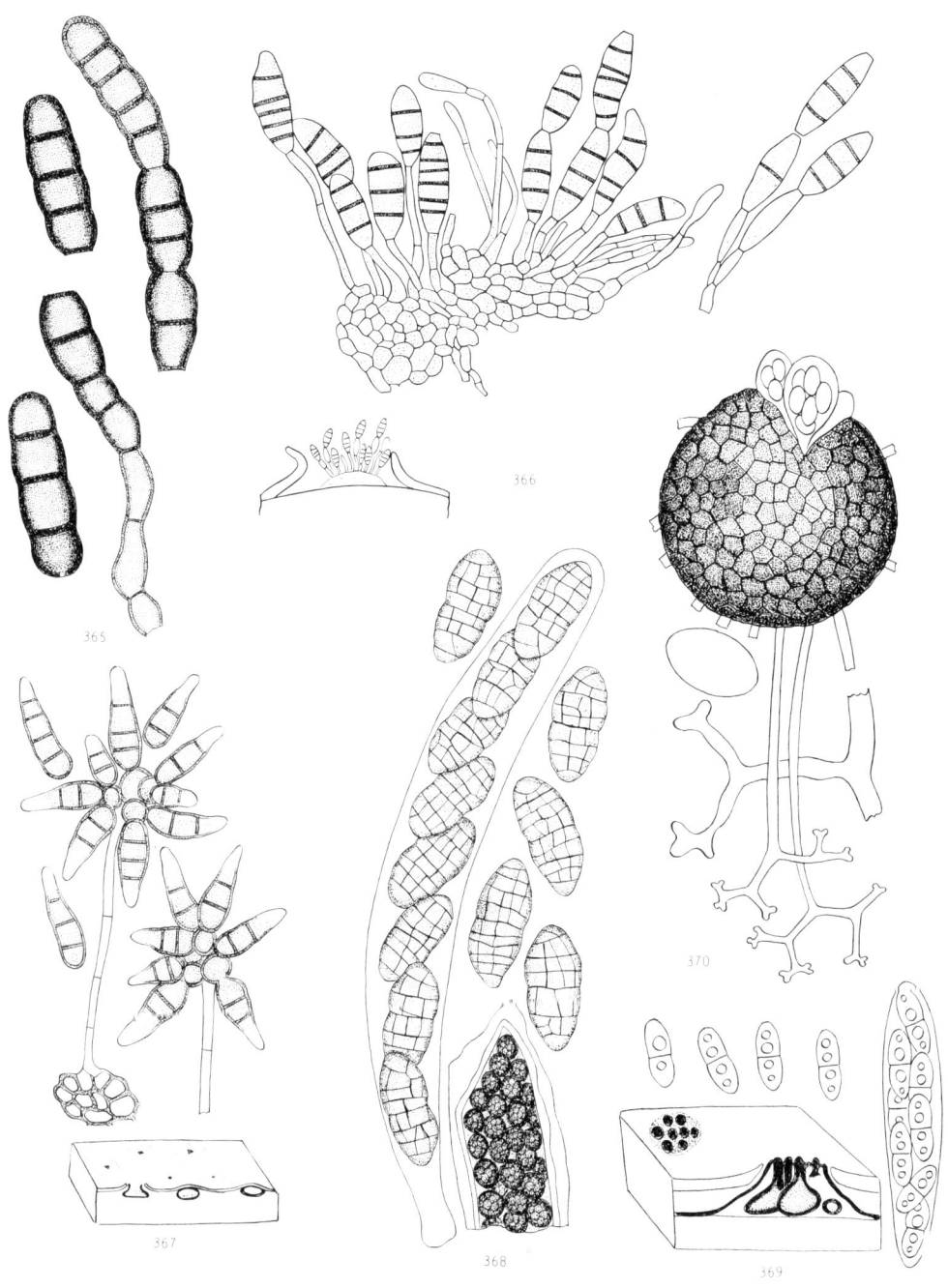

Plate 38. On *Alnus* and *Berberis*. 365, *Taeniolella alta*; 366, *Phragmotrichum rivoclarinum*; 367, *Prosthemium stellare*; 368, *Cucurbitaria berberidis*; 369, *Diaporthe detrusa*; 370, *Microsphaera berberidis*.

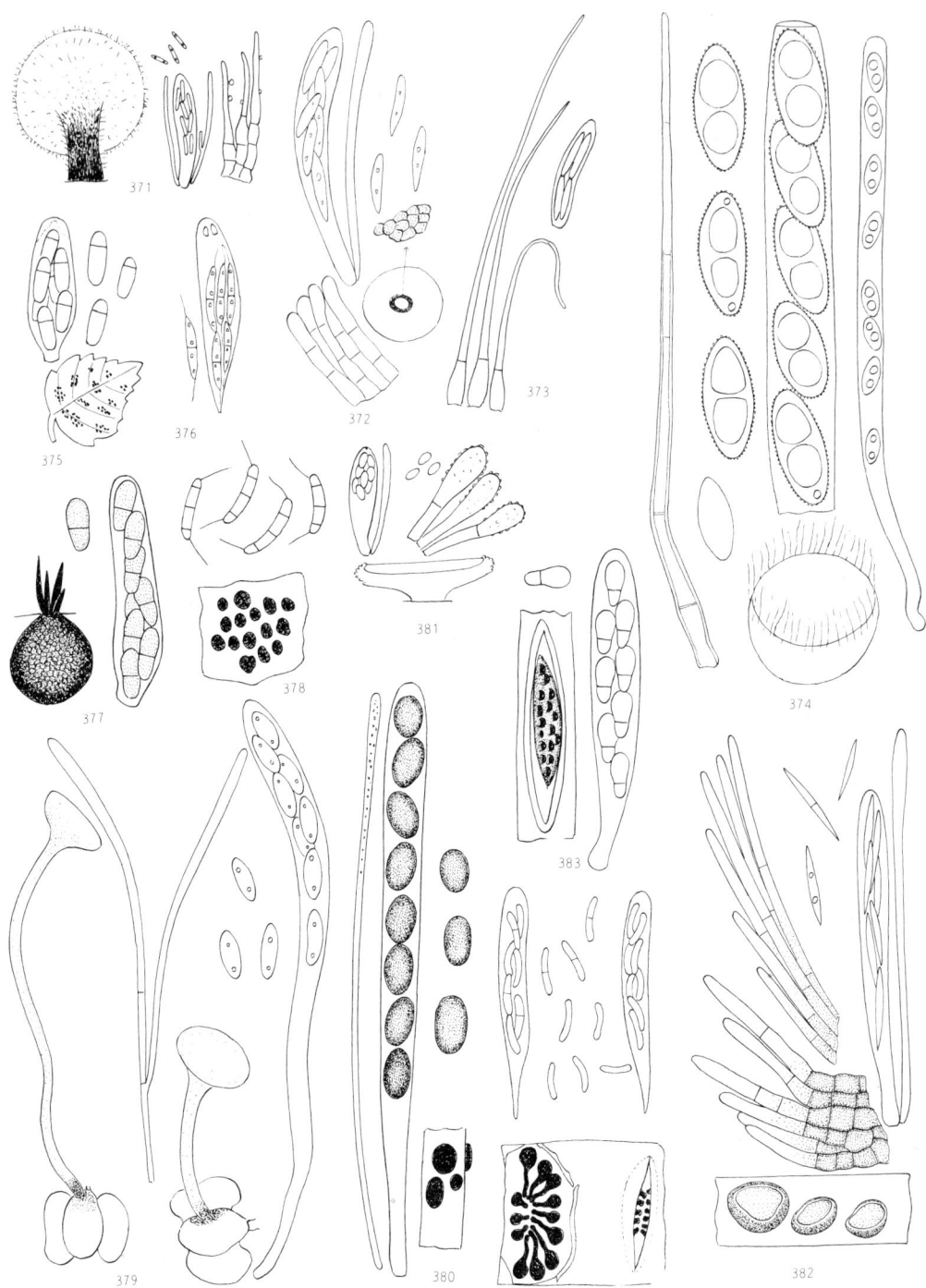

Plate 39. On *Betula* leaves, seeds, and wood and bark. 371, *Betulina fuscostipitata*; 372, *Calycellina leucella*; 373, *Hyaloscypha lachnobrachya* var. *araneocincta*; 374, *Leucoscypha erminea*; 375, *Atopospora betulina*; 376, *Gnomonia setacea*; 377, *Venturia ditricha*; 378, *Discosia artocreas*; 379, *Ciboria betulae*; 380, *Bulgariella pulla*; 381, *Clavidisculum microsporum*; 382, *Mollisia ramealis*; 383, *Anisogramma virgultorum*; 384, *Calosphaeria wahlenbergii*.

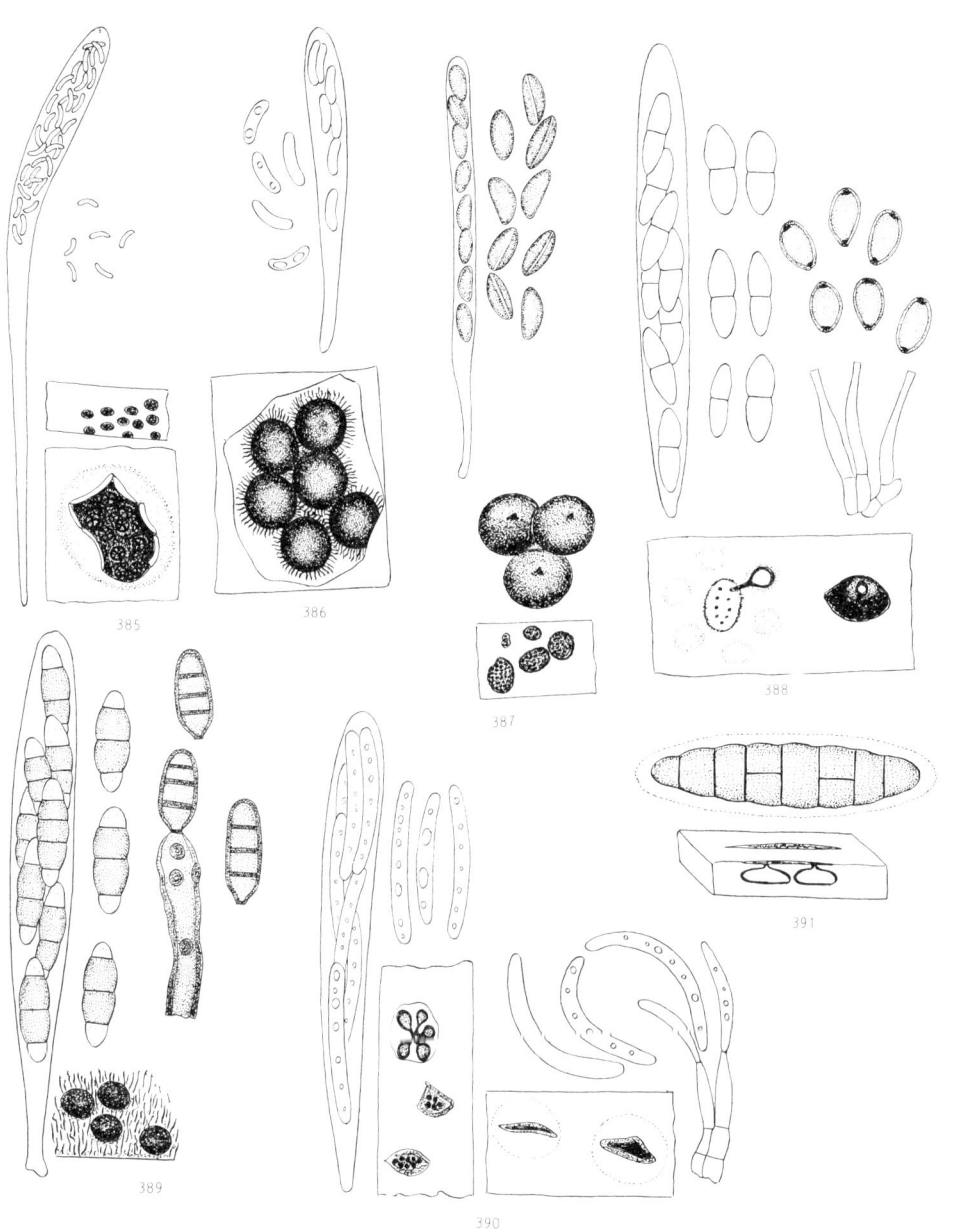

Plate 40. On *Betula* wood and bark. 385, *Diatrypella favacea*; 386, *Enchnoa lanata*; 387, *Hypoxylon multiforme*; 388, *Melanconis stilbostoma*; 389, *Melanomma subdispersum*; 390, *Ophiovalsa betulae*; 391, *Pleomassaria siparia*.

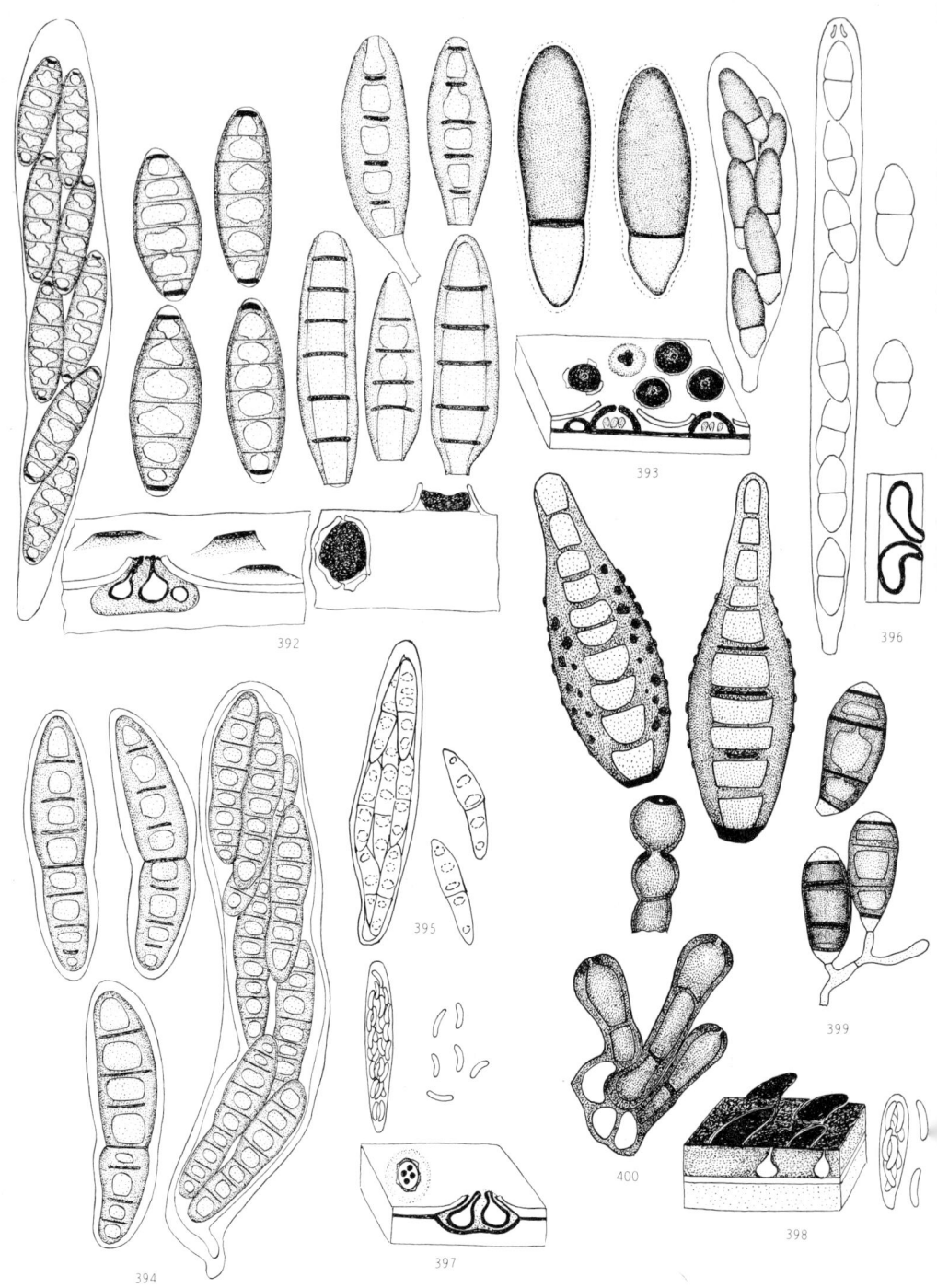

Plate 41. On *Betula* wood and bark. 392, *Pseudovalsa lanciformis*; 393, *Pteridiospora scoriadea*; 394, *Splanchnonema argus*; 395, *Stomiopeltis betulae*; 396, *Sydowiella ambigua*; 397, *Valsella adhaerens*; 398, *Xenotypa aterrima*; 399, *Bactrodesmium betulicola*; 400, *Corynespora cespitosa*.

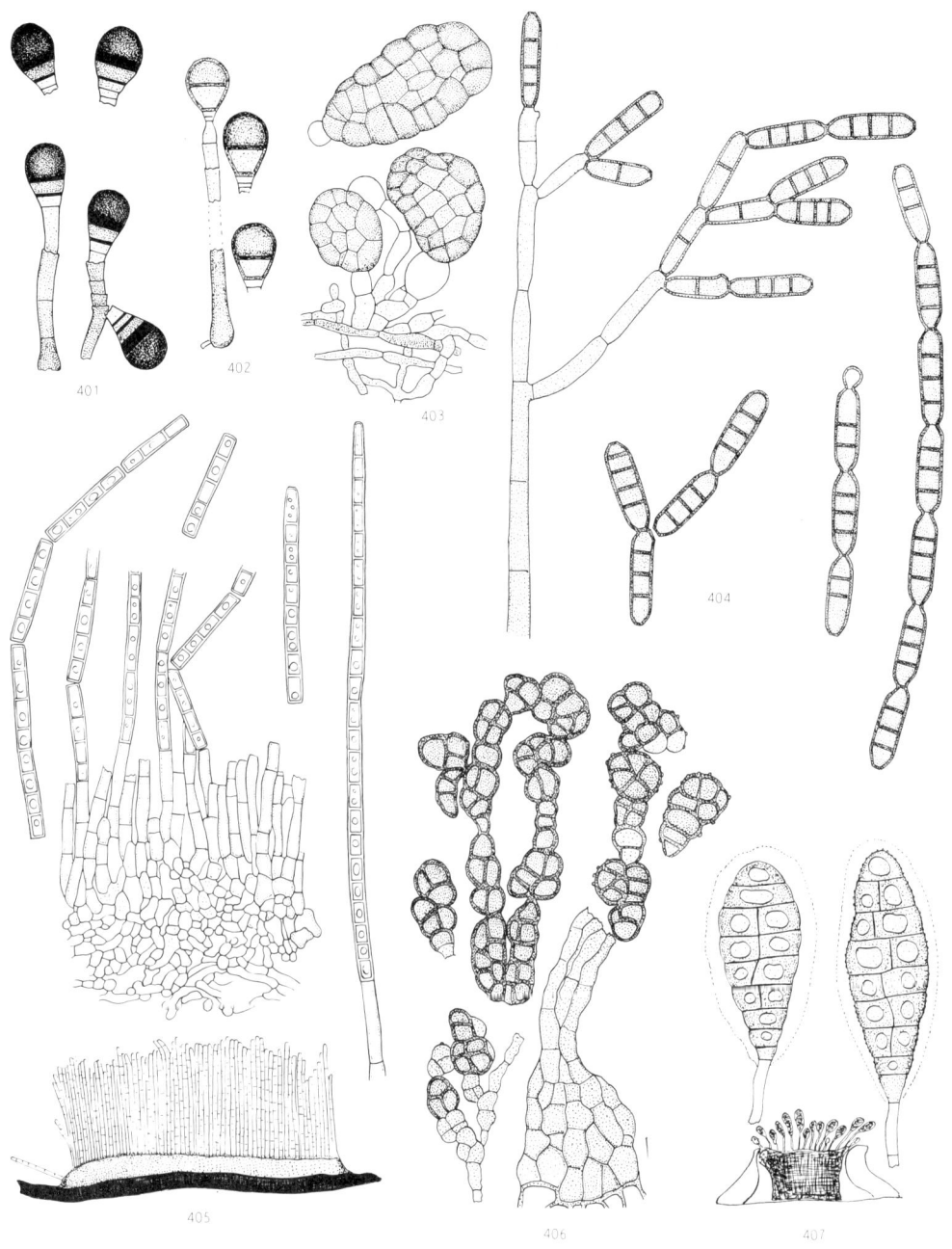

Plate 42. On *Betula* wood and bark. 401, *Endophragmia glanduliformis*; 402, *Endophragmiella suttonii*; 403, *Monodictys paradoxa*; 404, *Septonema secedens*; 405, *Septotrullula bacilligera*; 406, *Trimmatostroma betulinum*; 407, *Myxocyclus polycistis*.

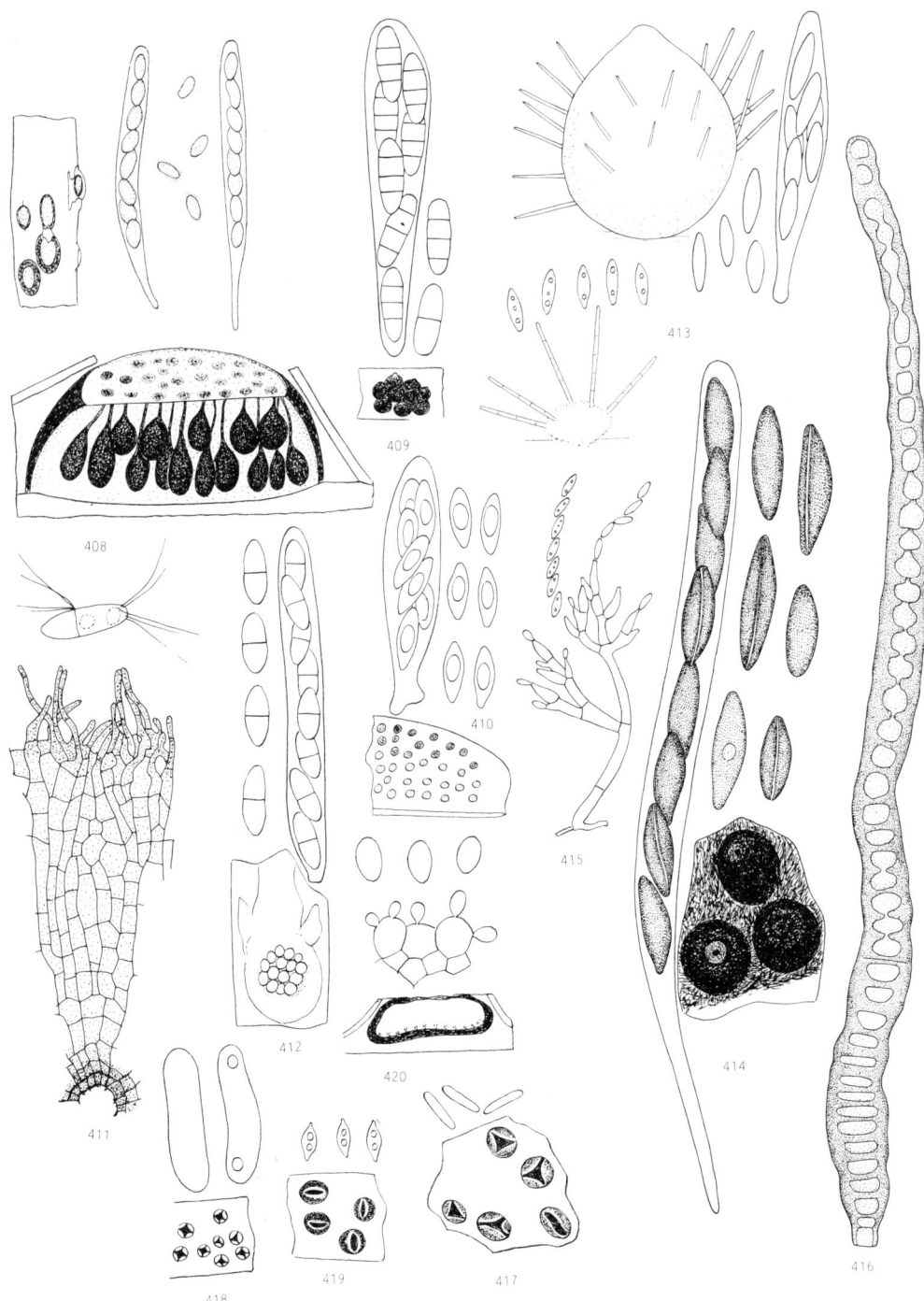

Plate 43. On *Buxus*. 408, *Camarops lutea*; 409, *Gibberella buxi*; 410, *Hyponectria buxi*; 411, *Microthyrium macrosporum*; 412, *Nectria desmazieresii*; 413, *Pseudonectria rousseliana*; 414, *Rosellinia buxi*; 415, *Sesquicillium buxi*; 416, *Sporidesmium adscendens*; 417, *Blennoria buxi*; 418, *Dothiorella candollei*; 419, *Phomopsis stictica*; 420, *Sarcophoma* state of *Guignardia miribelii*.

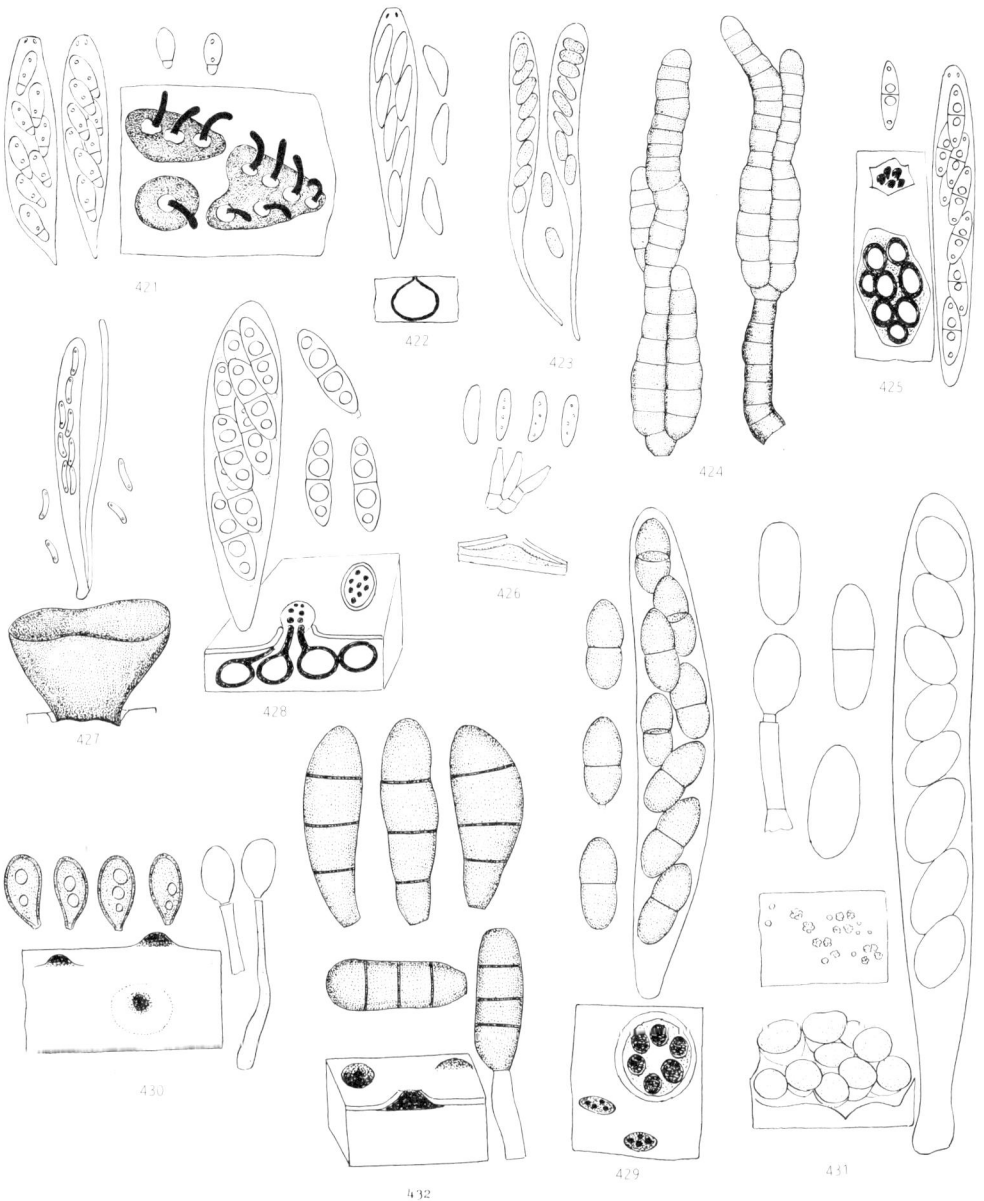

Plate 44. On *Carpinus*. 421, *Mamiania fimbriata*; 422, *Sphaerognomonia carpinea*; 423, *Anthostoma decipiens*; 424, *Ceratosporella stipitata*; 425, *Diaporthe carpini*; 426, *Discosporina deplanata*; 427, *Encoelia glaberrima*; 428, *Melanconis chrysostroma*; 429, *M. spodiae*; 430, *Melanconium stromaticum*; 431, *Pezicula carpinea*; 432, *Stilbospora macrosperma*.

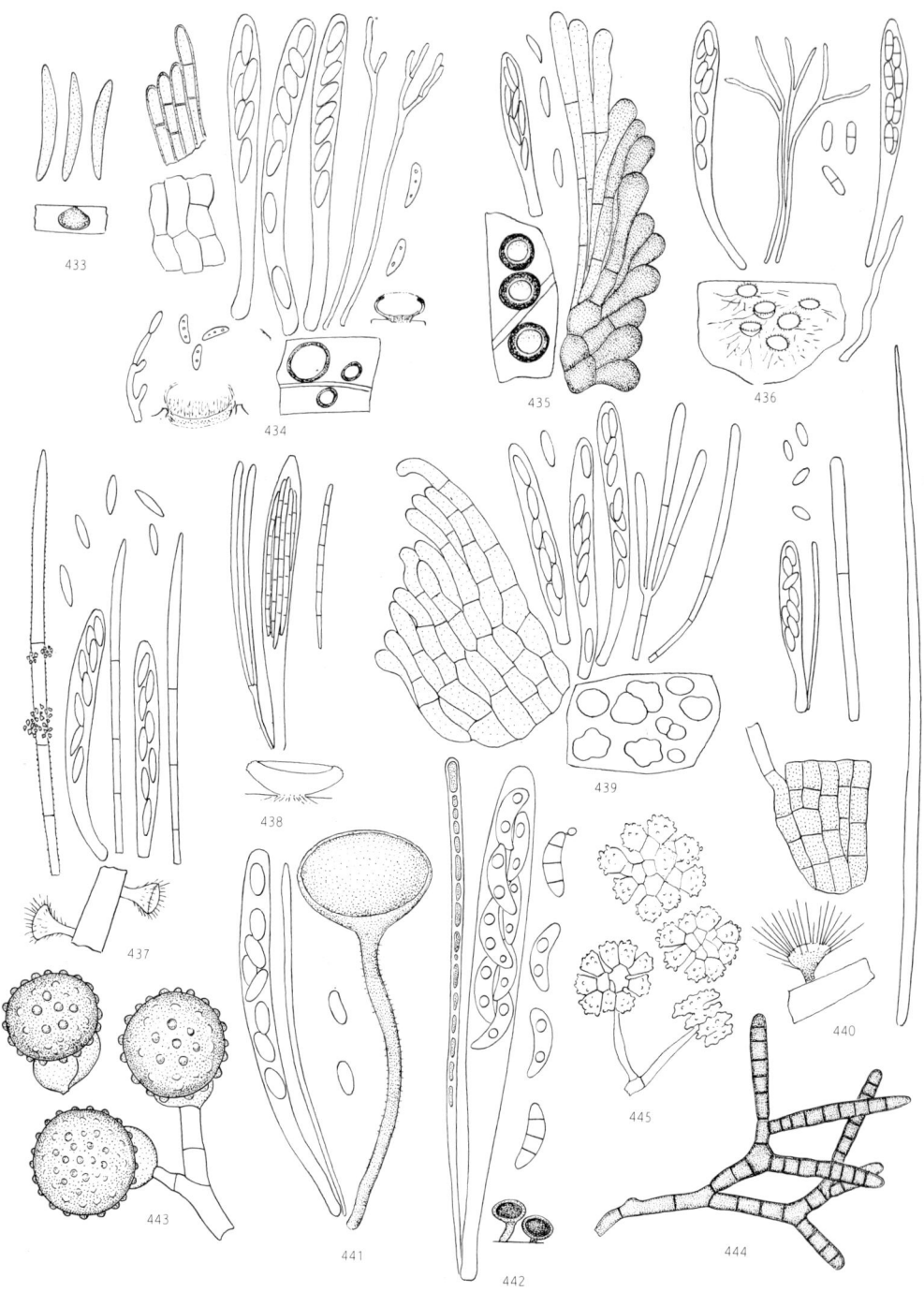

Plate 45. On *Castanea* leaves and fruit. 433, *Coniella castaneicola*; 434, *Discohainesia oenotherae*; 435, *Pyrenopeziza nervicola*; 436, *Arachnoscypha aranea*; 437, *Dasyscyphus castaneicola*; 438, *Gorgoniceps charnwoodensis*; 439, *Helotium humile*; 440, *Hyalopeziza spinicola*; 441, *Rutstroemia americana*; 442, *R. echinophila*; 443, *Acrospeira mirabilis*; 444, *Anavirga laxa*; 445, *Candelabrum spinulosum*.

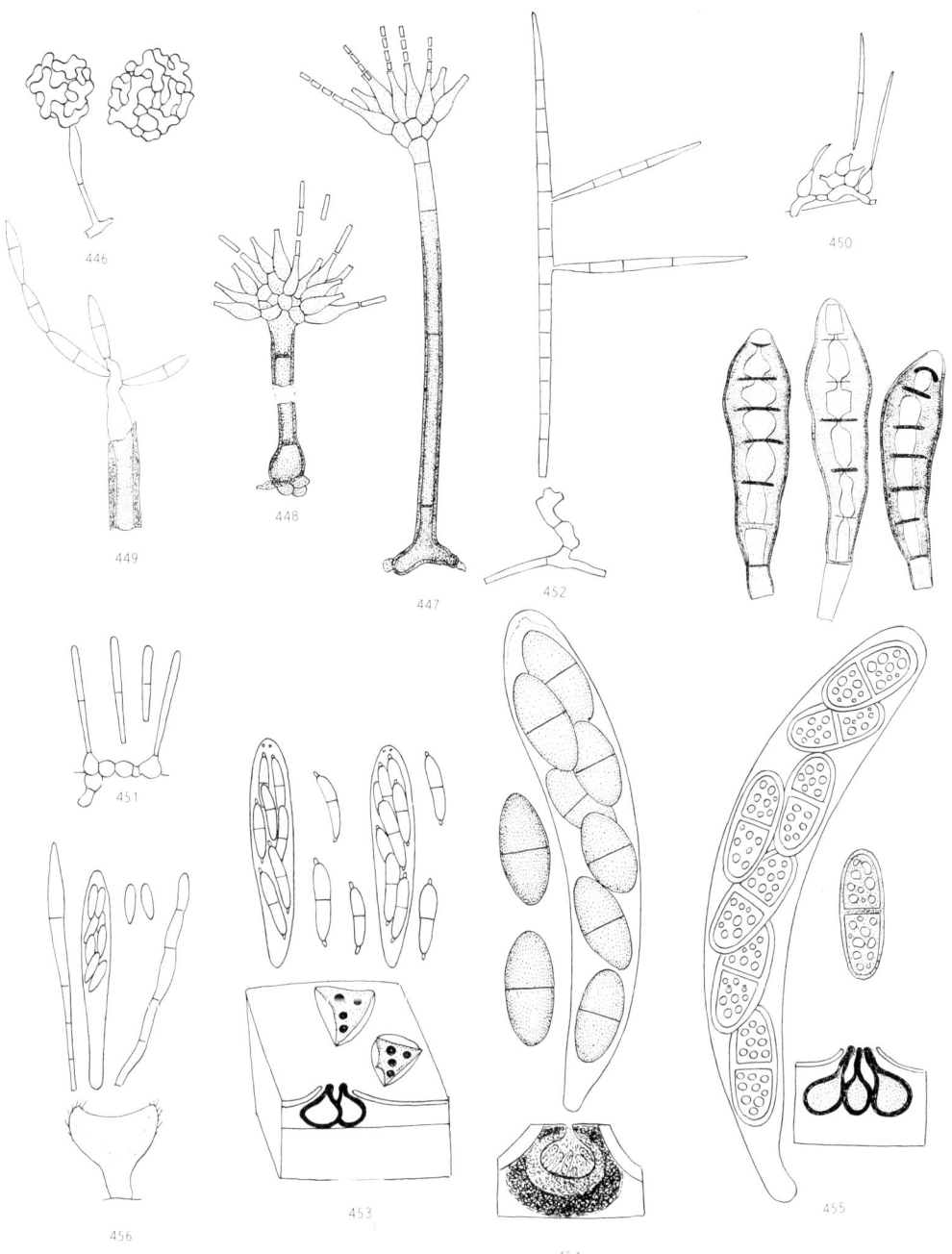

Plate 46. On *Castanea* fruit, and wood and bark. 446, *Clathrosphaerina* state of *Hyaloscypha zalewskii*; 447, *Phialocephala fumosa*; 448, *P. truncata*; 449, *Pleurotheciopsis pusilla*; 450, *Pseudomicrodochium aciculare*; 451, *P. cylindricum*; 452, *Tricladium castaneicola*; 453, *Cryptodiaporthe castanea*; 454, *Didymosphaeria superapplanata*; 455, *Melanconis modonia*; 456, *Psilachnum auranticolor*.

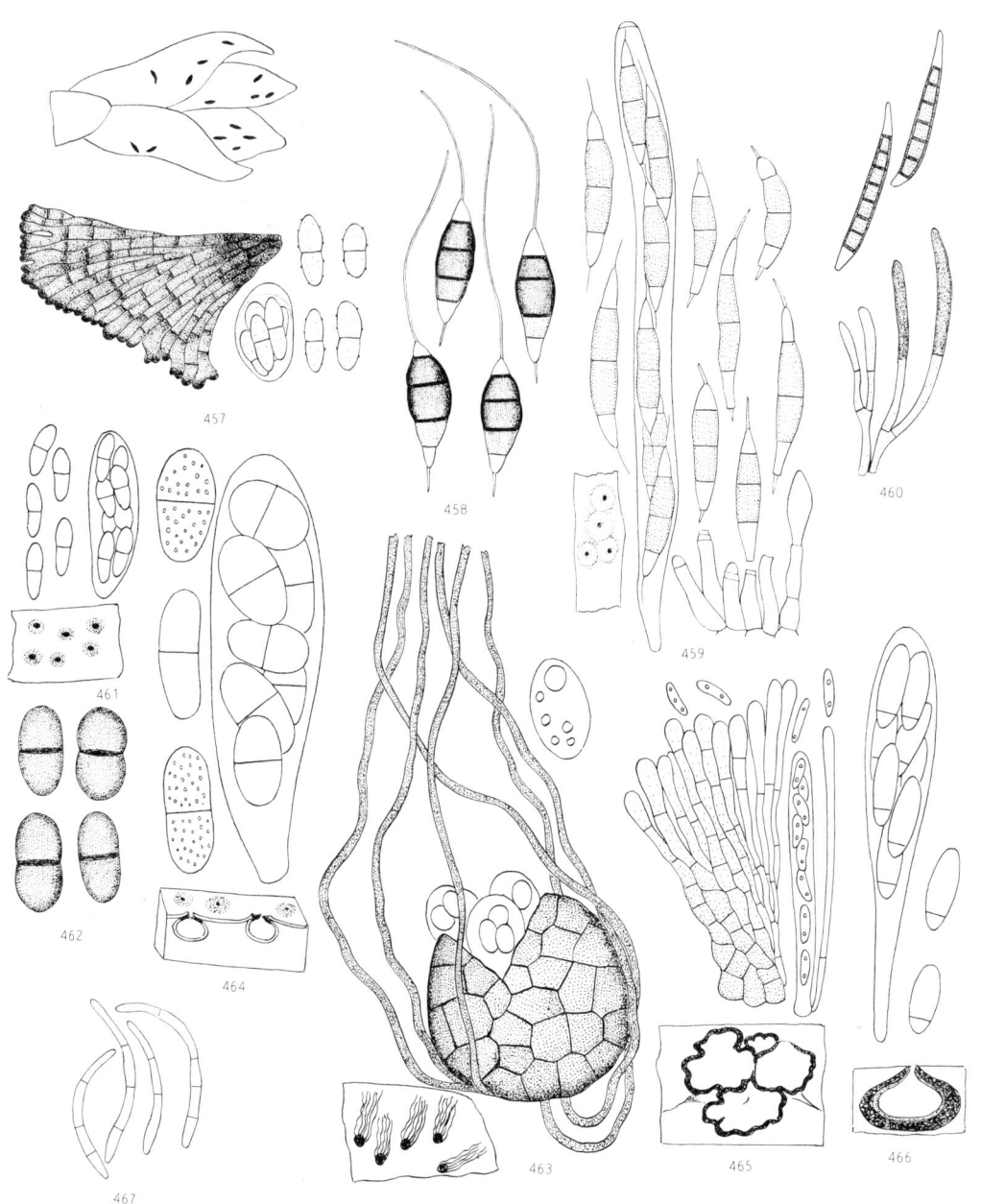

Plate 47. On *Chamaecyparis*, *Clematis* and *Cornus*. 457, *Morenoina chamaecyparidis*; 458, *Pestalotiopsis monochaetioides*; 459, *Broomella vitalbae*; 460, *Excipularia fusispora*; 461, *Didymella corni*; 462, *Diplodia mamillana*; 463, *Erysiphe tortilis*; 464, *Leiosphaerella vexata*; 465, *Mollisia discolor*; 466, *Pseudomassaria corni*; 467, *Septoria cornicola*.

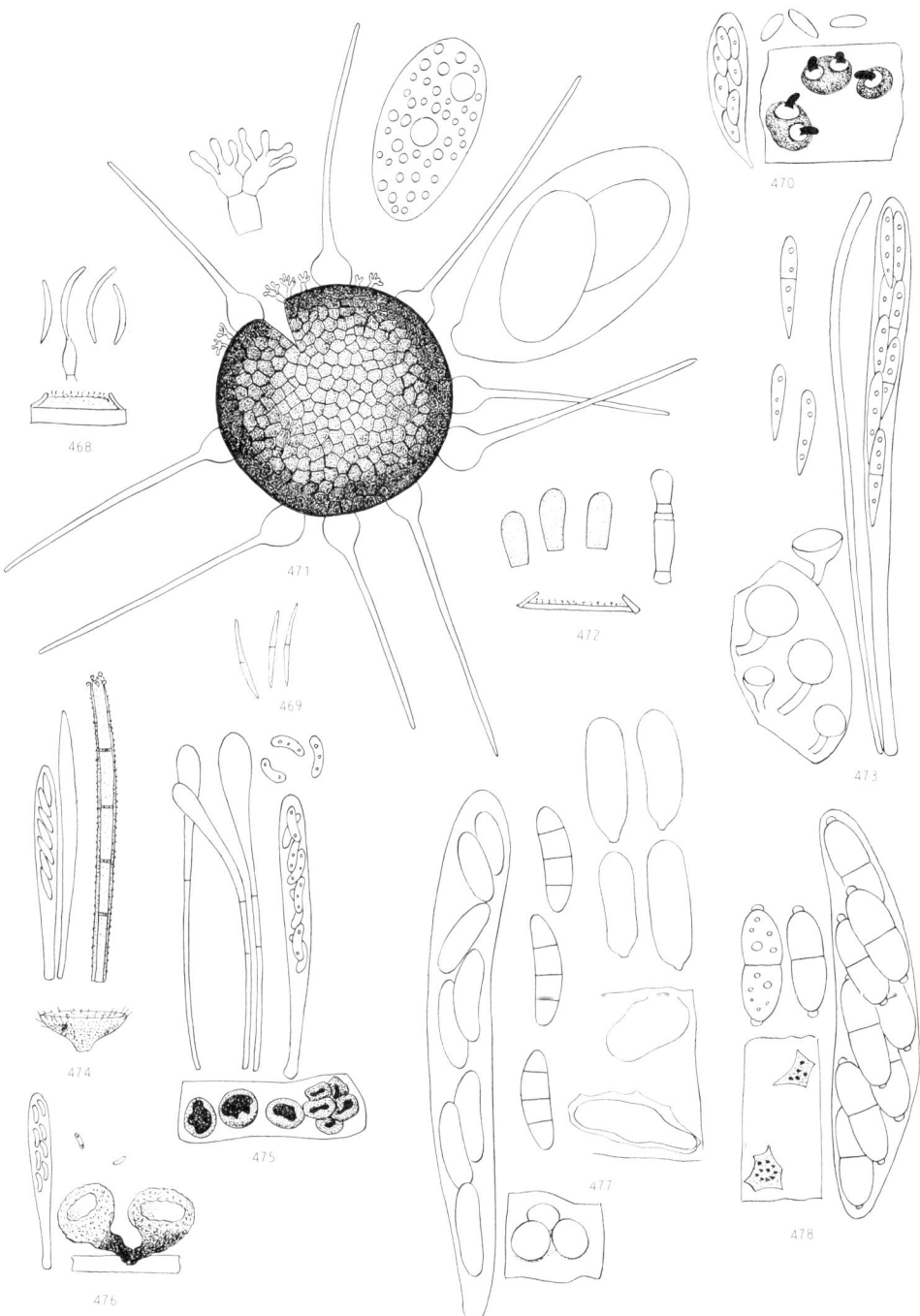

Plate 48. On *Corylus*. 468, *Asteroma coryli*; 469, *Gnomonia gnomon*; 470, *Mamiania coryli*; 471, *Phyllactinia guttata*; 472, *Piggotia coryli*; 473, *Hymenoscyphus fructigenus*; 474, *Dasyscyphus calyculiformis*; 475, *Encoelia furfuracea*; 476, *E. glauca*; 477, *Pezicula coryli*; 478, *Cryptodiaporthe pyrrhocystis*.

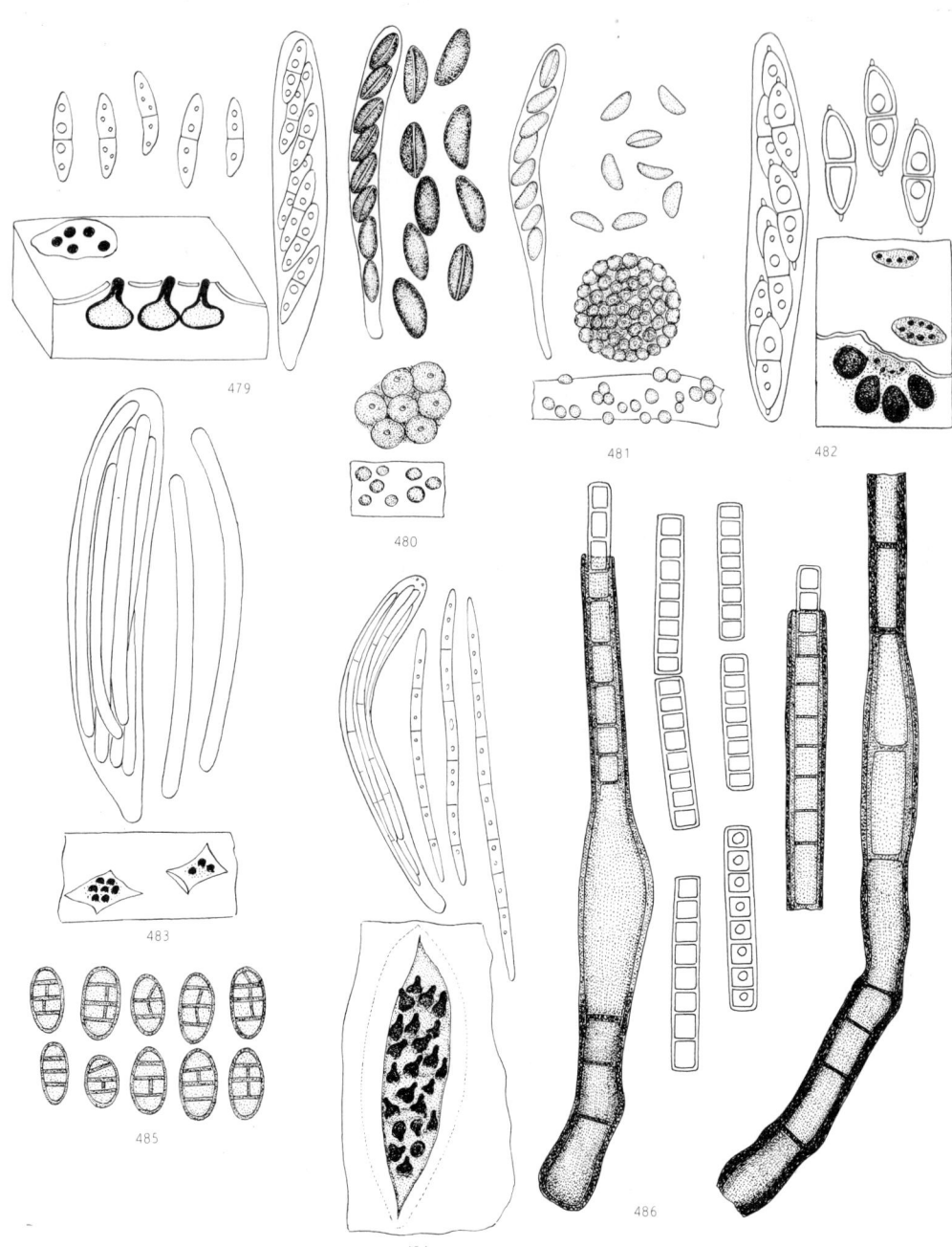

Plate 49. On *Corylus.* 479, *Diaporthe decedens*; 480, *Hypoxylon fuscum*; 481, *H. howeianum*; 482, *Melanconis flavovirens*; 483, *Ophiovalsa corylina*; 484, *Sillia ferruginea*; 485, *Camarosporium propinquum*; 486, *Chalara inflatipes.*

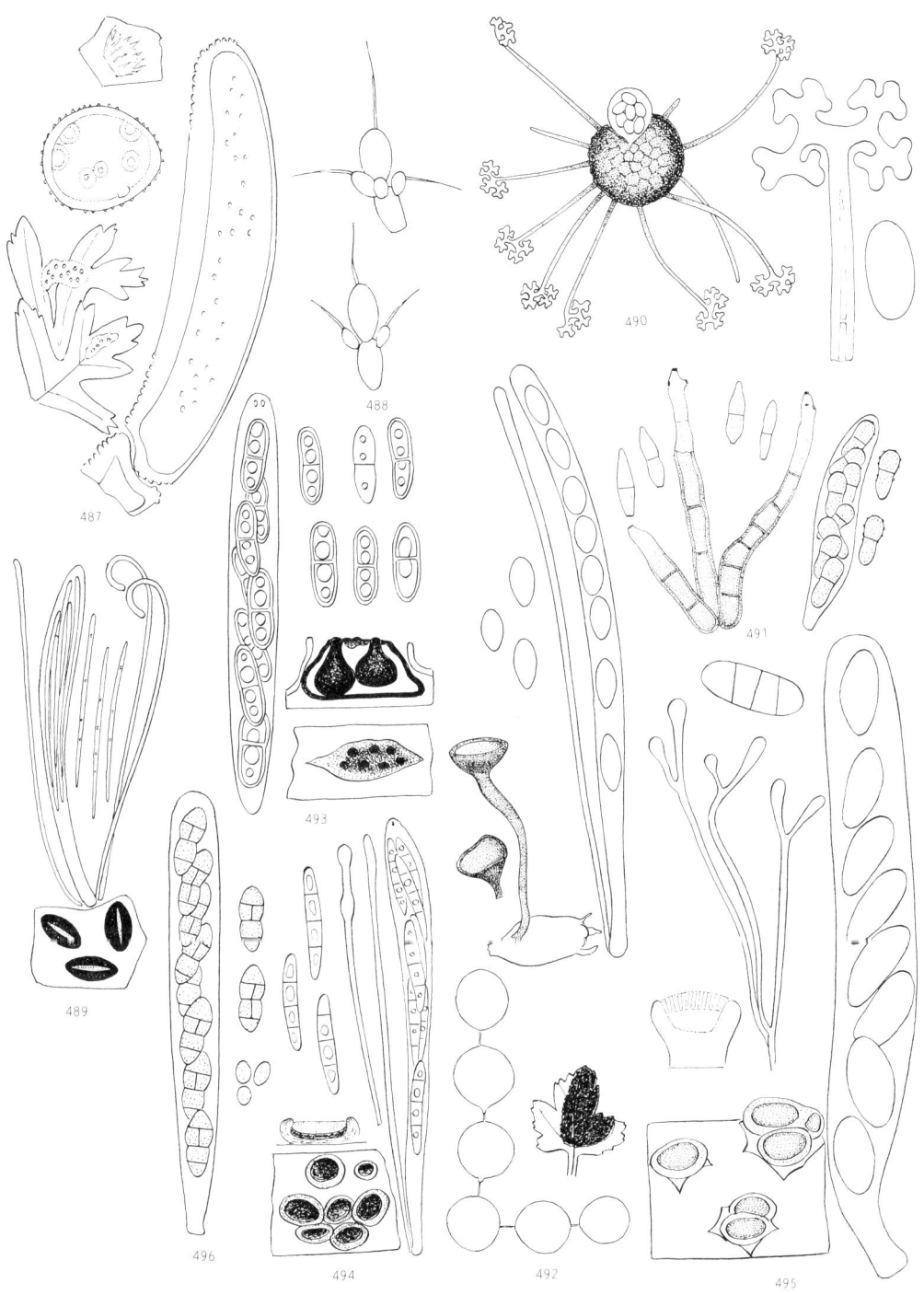

Plate 50. On *Crataegus*. 487, *Gymnosporangium clavariiforme*; 488, *Entomosporium* state of *Diplocarpon mespili*; 489, *Lophodermium hysterioides*; 490, *Podosphaera clandestina*; 491, *Venturia crataegi*; 492, *Monilinia johnsonii*; 493, *Diaporthe crataegi*; 494, *Patellariopsis atrovinosa*; 495, *Pezicula sepium*; 496, *Pleospora shepherdiae*.

Plate 51. On *Cupressus* and *Cytisus*. 497, *Pestalotiopsis funerea*; 498, *Seiridium cardinale*; 499, *Stomiopeltis cupressicola*; 500, *Trichothyrina fimbriata*; 501, *Cucurbitaria spartii*; 502, *Kalmusia sarothamni*; 503, *Pleospora cytisi*; 504, *Selenophoma* state of *Guignardia cytisi*.

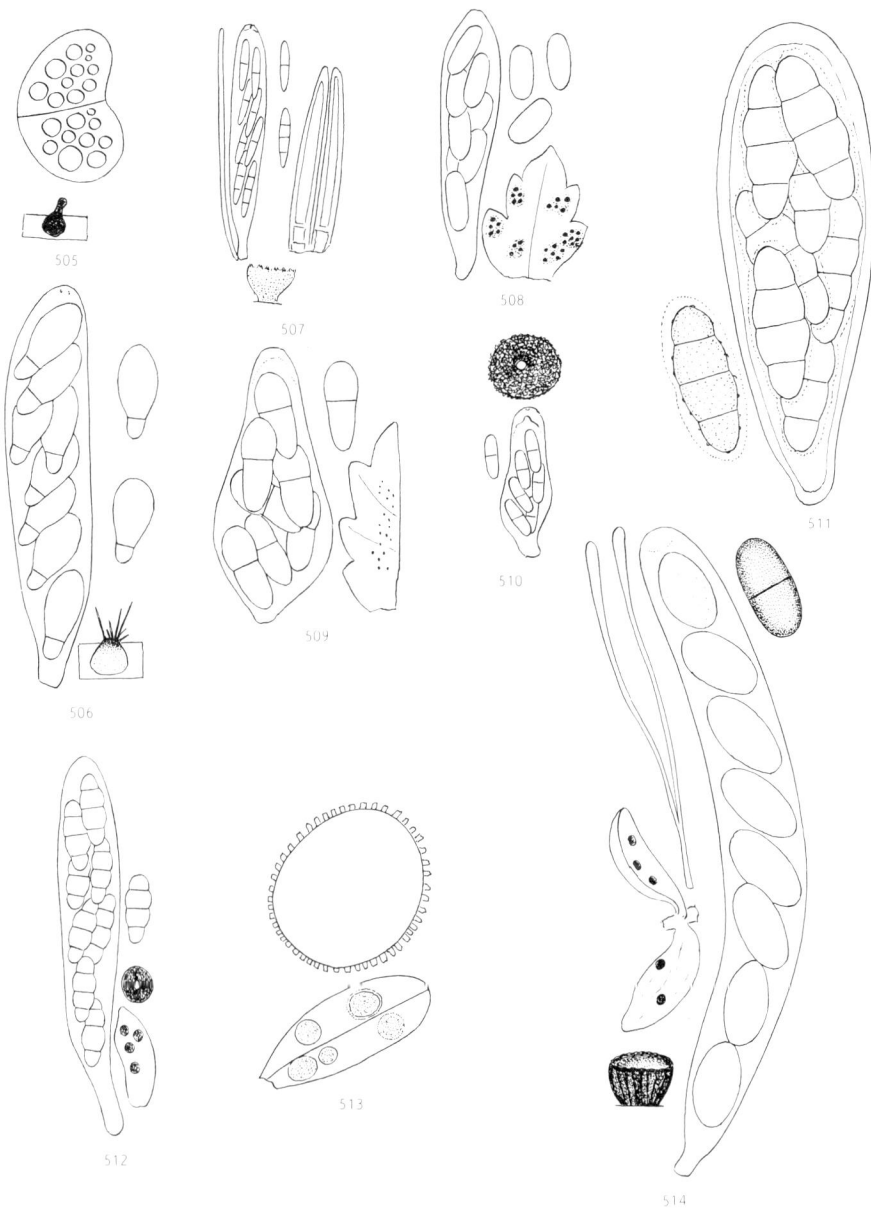

Plate 52. On *Dryas* and *Empetrum*. 505, *Cainiella johansonii*; 506, *Chaetapiospora islandica*; 507, *Grahamiella dryadis*; 508, *Isothea rhytismoides*; 509, *Mycosphaerella octopetalae*; 510, *Stomiopeltis dryadis*; 511, *Wettsteinina dryadis*; 512, *Arwidssonia empetri*; 513, *Chrysomyxa empetri*; 514, *Phaeangellina empetri*.

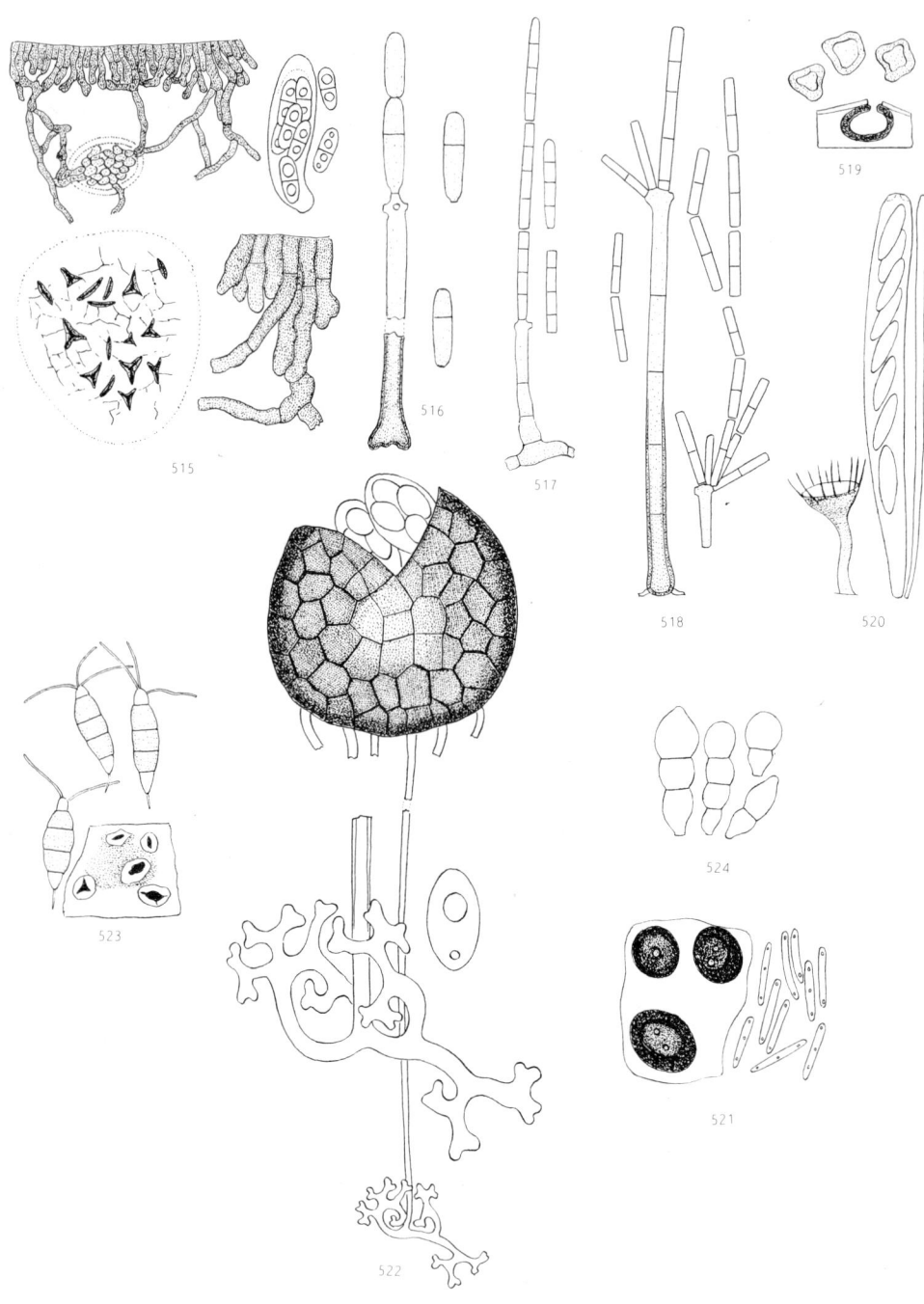

Plate 53. On *Eucalyptus* and *Euonymus*. 515, *Aulographina eucalypti*; 516, *Parapleurotheciopsis inaequiseptata*; 517, *Polyscytalum hareae*; 518, *P. truncatum*; 519, *Readeriella mirabilis*; 520, *Zoellneria eucalypti*; 521, *Ceuthospora euonymi*; 522, *Microsphaera euonymi*; 523, *Pestalotiopsis neglecta*; 524, *Septogloeum carthusianum*.

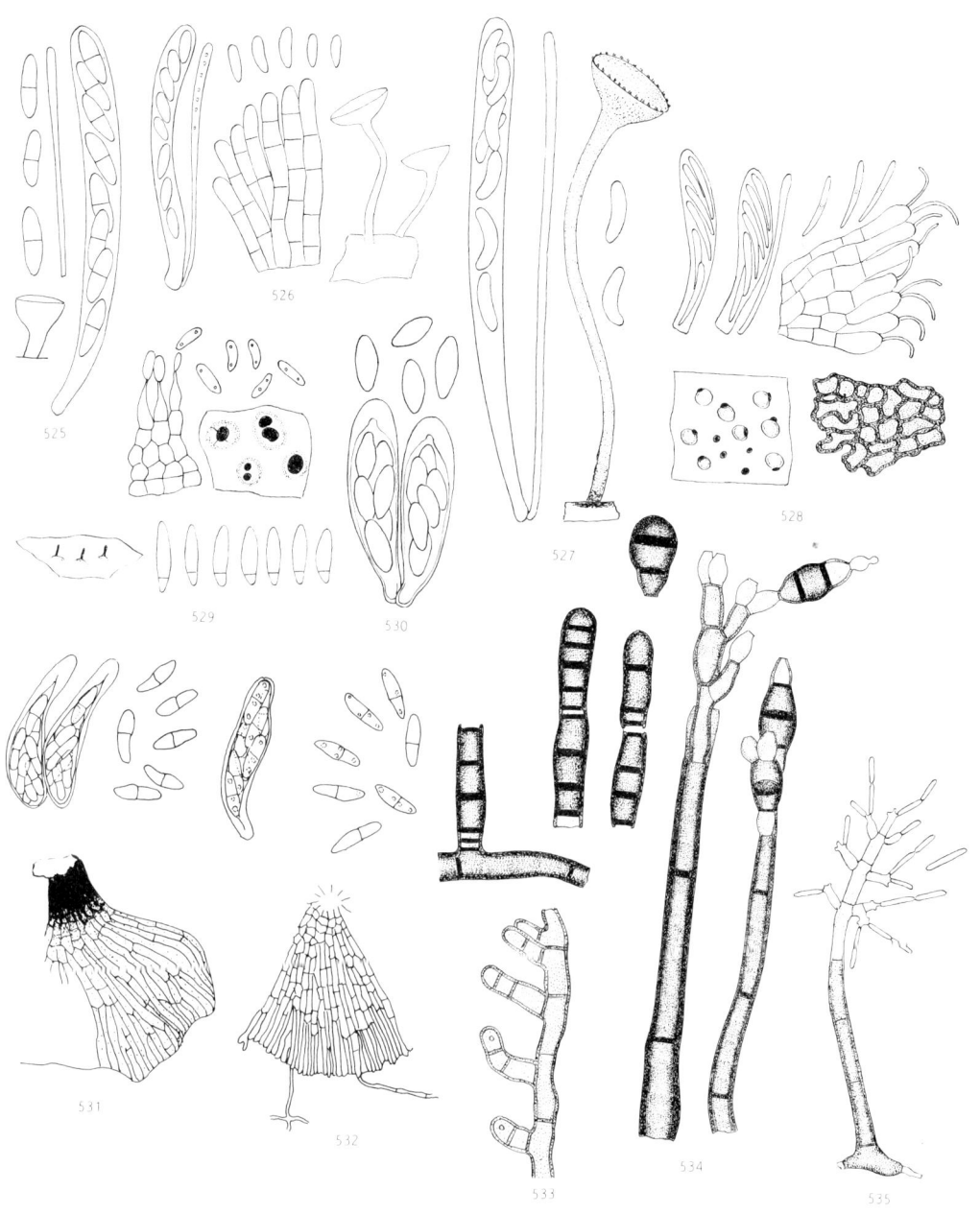

Plate 54. On *Fagus* leaves. 525, *Hymenoscyphus phyllophilus*; 526, *Pezizella fagi*; 527, *Rutstroemia petiolorum*; 528, *Scutoscypha fagi*; 529, *Apiognomonia errabunda*; 530, *Discosphaerina fagi*; 531, *Microthyrium fagi*; 532, *M. inconspicuum*; 533, *Ampulliferina fagi*; 534, *Brachysporiella laxa*; 535, *Polyscytalum fagicola*.

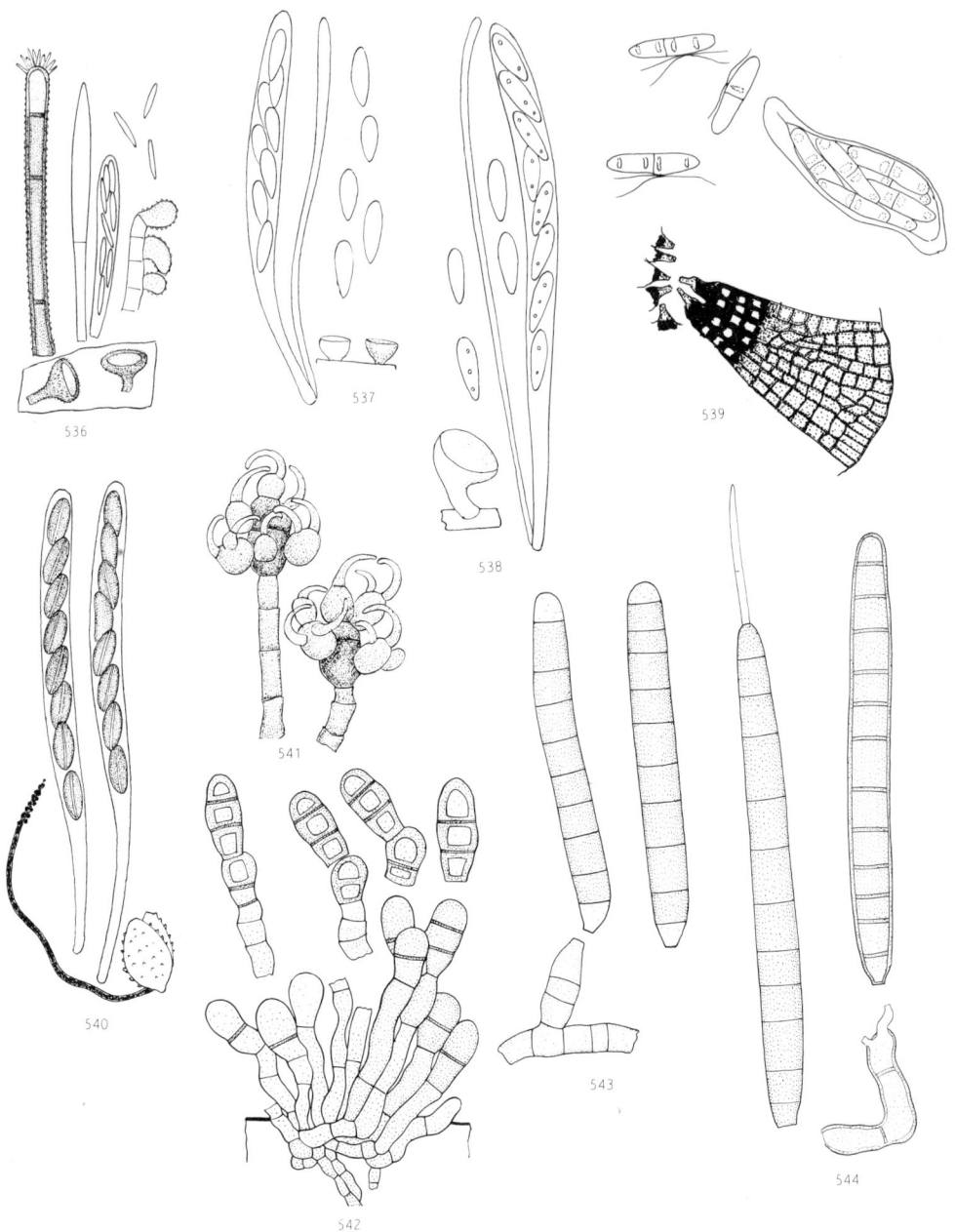

Plate 55. On *Fagus* fruit, including cupules. 536, *Dasyscyphus fuscescens* var. *fagicola*; 537, *Hymenoscyphus fagineus*; 538, *H. rokebyensis*; 539, *Trichothyrina cupularum*; 540, *Xylaria carpophila*; 541, *Arachnophora fagicola*; 542, *Bactrodesmiella masonii*; 543, *Camposporium cambrense*; 544, *C. pellucidum*.

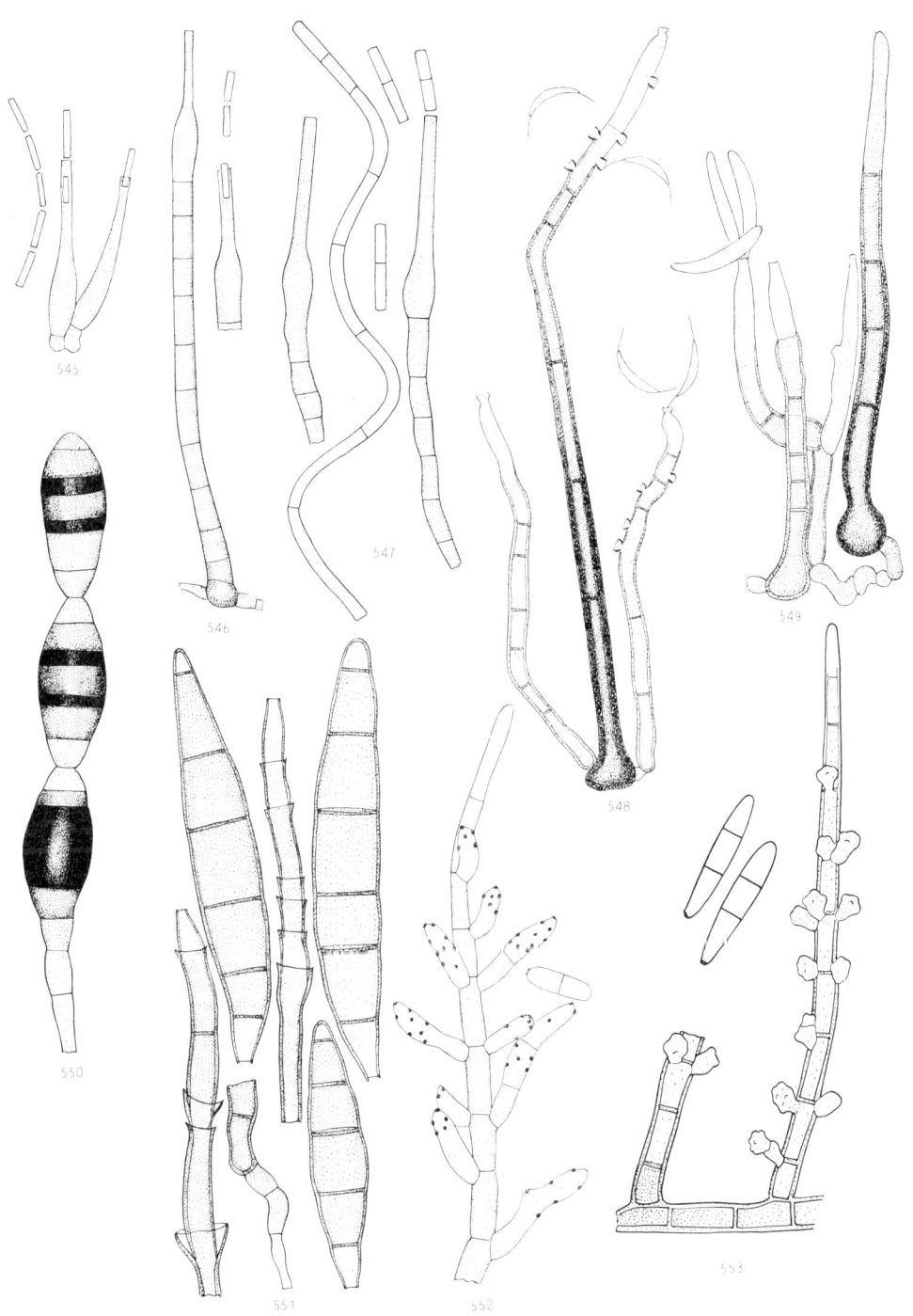

Plate 56. On *Fagus* fruit, including cupules. 545, *Chalara affinis*; 546, *C. cylindrosperma*; 547, *C. spiralis*; 548, *Codinaea fertilis*; 549, *C. hughesii*; 550, *Endophragmia catenulata*; 551, *Endophragmiella fagicola*; 552, *Haplariopsis fagicola*; 553, *Spondylocladiopsis cupulicola*.

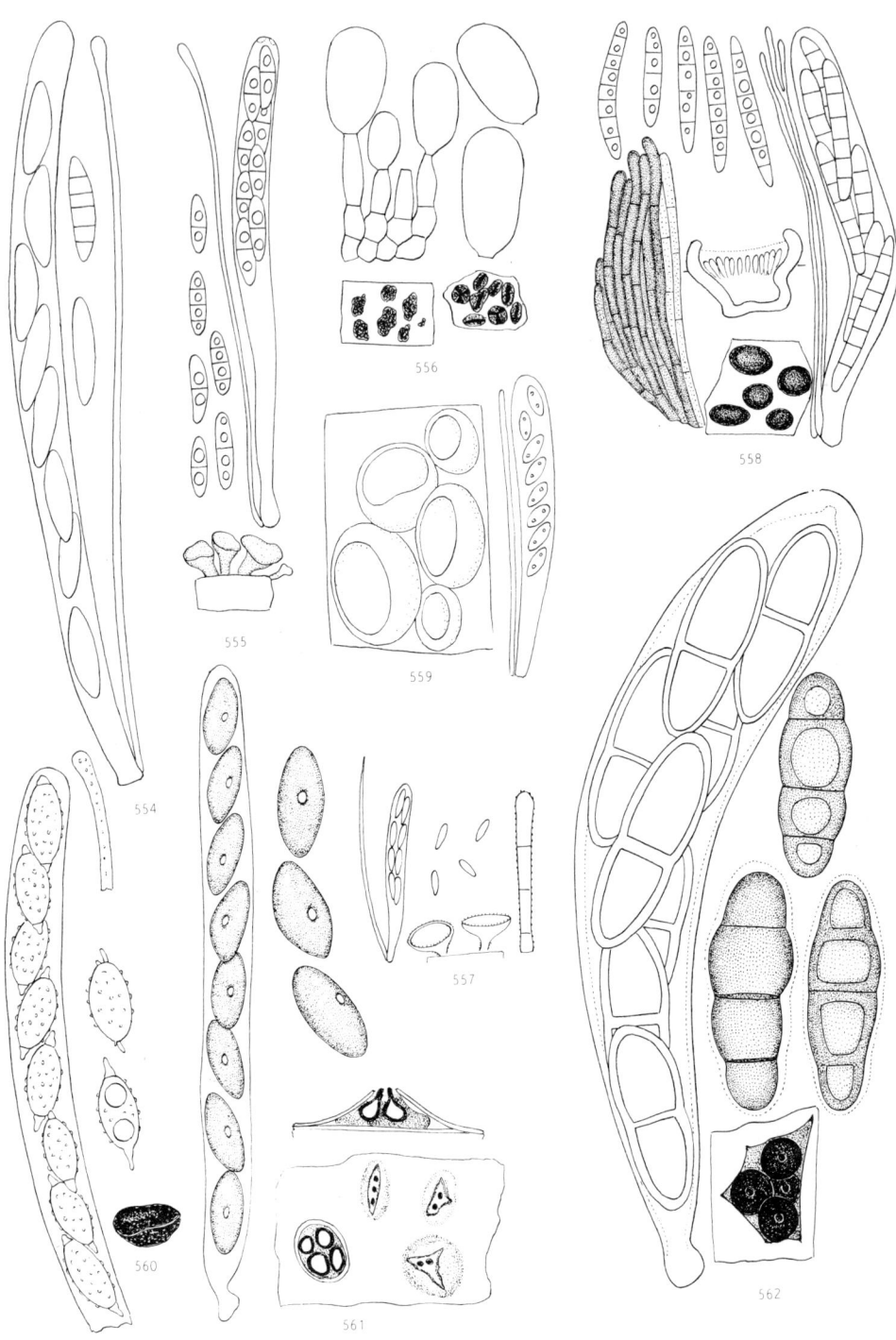

Plate 57. Discomycetes and other ascomycetes on *Fagus* wood and bark. 554, *Ascocoryne cylichnium*; 555, *A. sarcoides*; 556, *Ascodichaena rugosa*, *Polymorphum* state; 557, *Dasyscyphus brevipilus*; 558, *Durella connivens*; 559, *Neobulgaria pura*; 560, *Peziza apiculata*; 561, *Anthostoma amoenum*; 562, *Asteromassaria macrospora*.

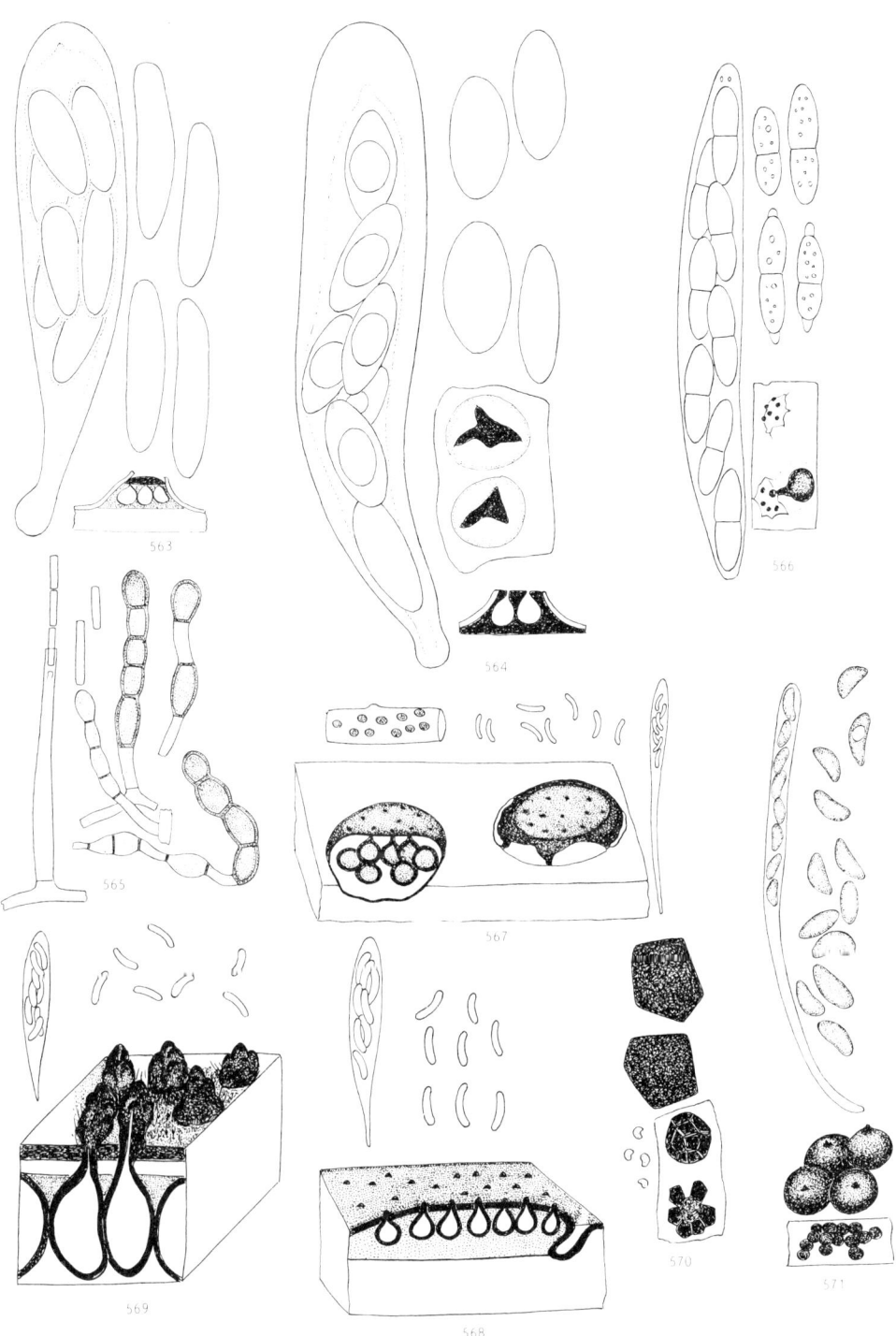

Plate 58. Ascomycetes on *Fagus* wood and bark. 563, *Botryosphaeria hoffmanni*; 564, *B. quercuum*; 565, *Ceratocystis moniliformis*; 566, *Cryptodiaporthe galericulata*; 567, *Diatrype disciformis*; 568, *Eutypa leioplaca*; 569, *E. spinosa*; 570, *Fragosphaeria purpurea*; 571, *Hypoxylon cohaerens*.

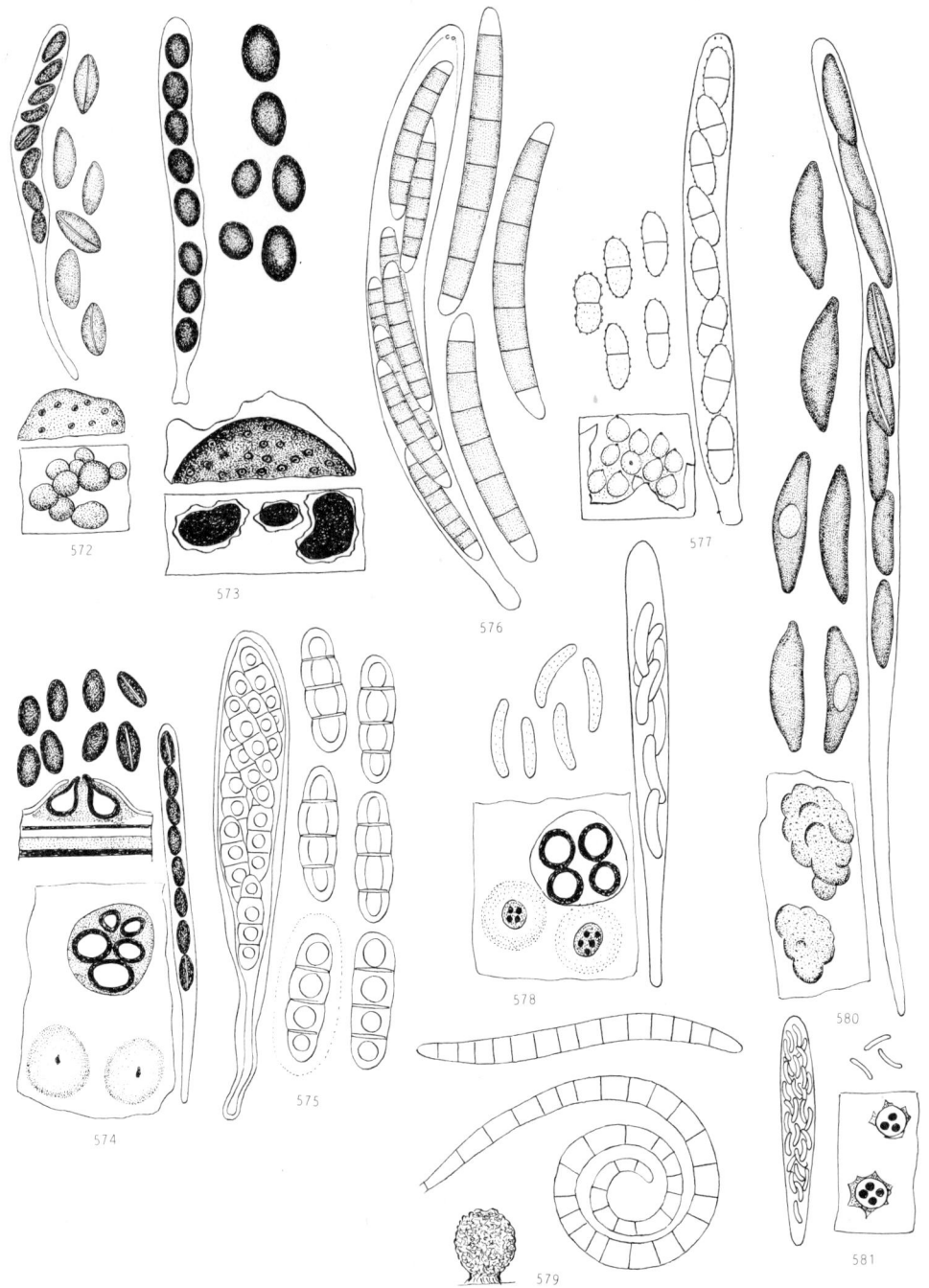

Plate 59. Ascomycetes on *Fagus* wood and bark. 572, *Hypoxylon fragiforme*; 573, *H. nummularium*; 574, *Lopadostoma turgidum*; 575, *Massarina eburnea*; 576, *Melogramma spiniferum*; 577, *Nectria coccinea*; 578, *Quaternaria quaternata*; 579, *Tubeufia helicoma*: 580, *Ustulina deusta*; 581, *Valsella amphoria*.

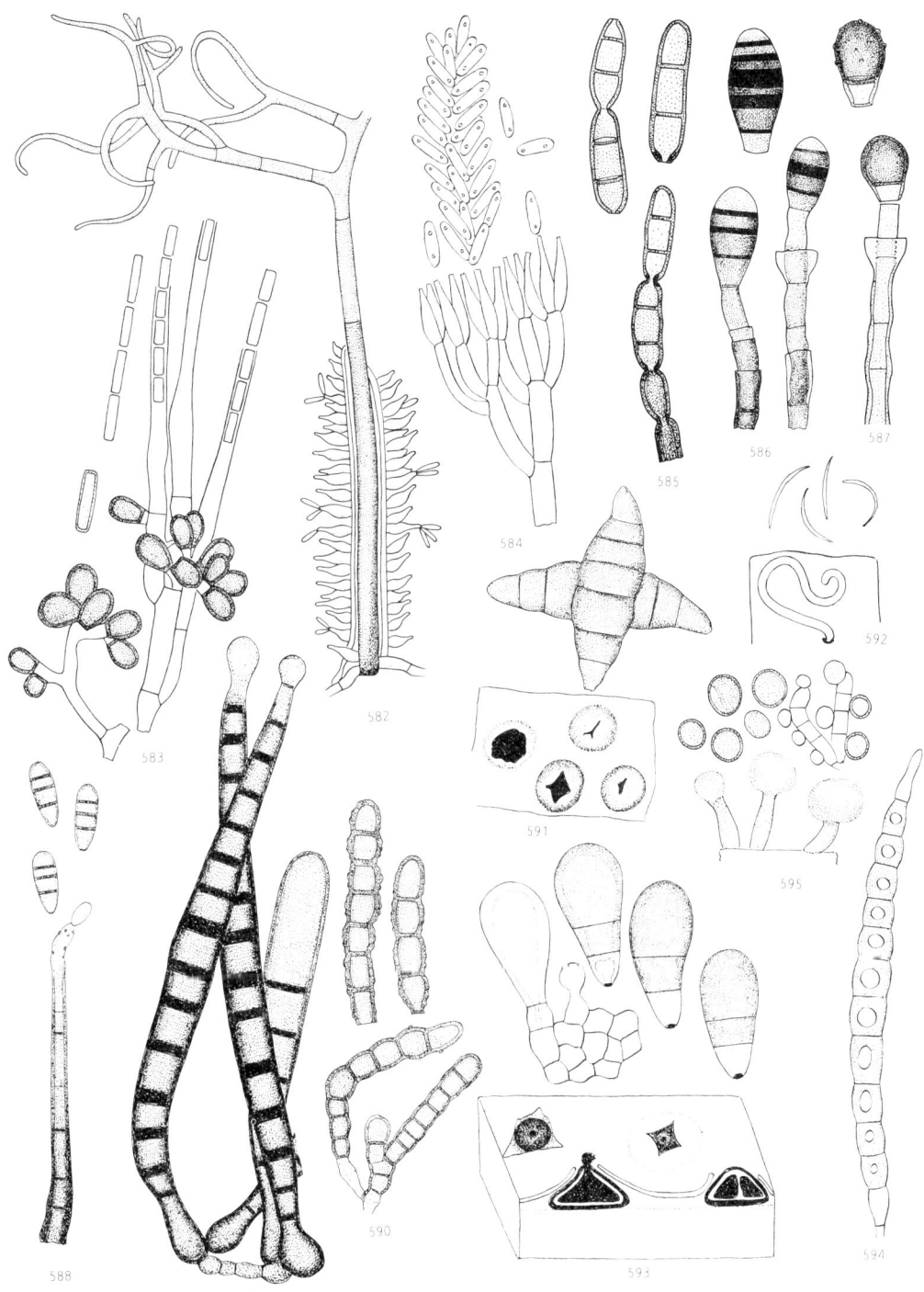

Plate 60. Hyphomycetes, coelomycetes, etc. on *Fagus* wood and bark. 582, *Ceratocladium microspermum*; 583, *Chalara ovoidea*; 584, *Clonostachys compactiuscula*; 585, *Corynespora biseptata*; 586, *Endophragmia stemphylioides*; 587, *E. verruculosa*; 588, *Pseudospiropes hughesii*; 589, *Sporidesmium hormiscioides*; 590, *Taeniolella faginea*; 591, *Asterosporium asterospermum*; 592, *Libertella faginea*; 593, *Neohendersonia kickxii*; 594, *Scolicosporium macrosporium*; 595, *Phleogena faginea*.

Plate 61. On *Frangula* and *Fraxinus*. 596, *Diaporthe syngenesia*; 597, *Karstenula rhodostoma*; 598, *Pezicula frangulae*; 599, *Crocicreas dolosellum*; 600, *Hymenoscyphus albidus*; 601, *Venturia fraxini*; 602, *Dactylospora stygia*; 603, *Laquearia sphaeralis*; 604, *Triblidium octosporum*; 605, *Cryptosphaeria eunomia*; 606, *Cryptovalsa protracta*; 607, *Daldinia concentrica*; 608, *Hypoxylon fraxinophilum*; 609, *H. rubiginosum*.

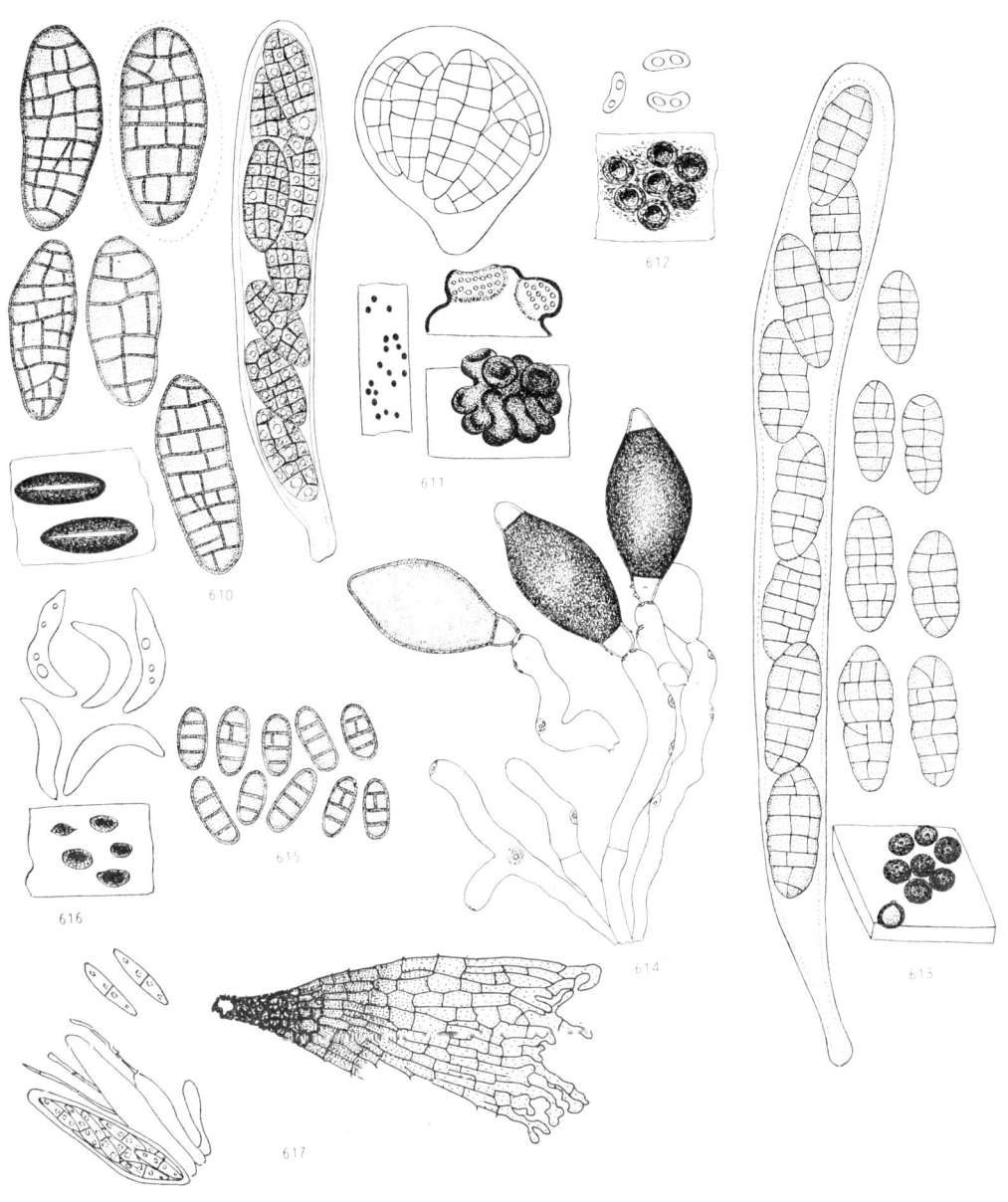

Plate 62. On *Fraxinus* and *Genista*. 610, *Hysterographium fraxini*; 611, *Myriangium duriaei*; 612, *Nitschkia confertula*; 613, *Teichospora obducens*; 614, *Brachydesmiella biseptata*; 615, *Camarosporium orni*; 616, *Gloeosporidiella turgida*; 617, *Microthyrium cytisi* var. *cytisi*.

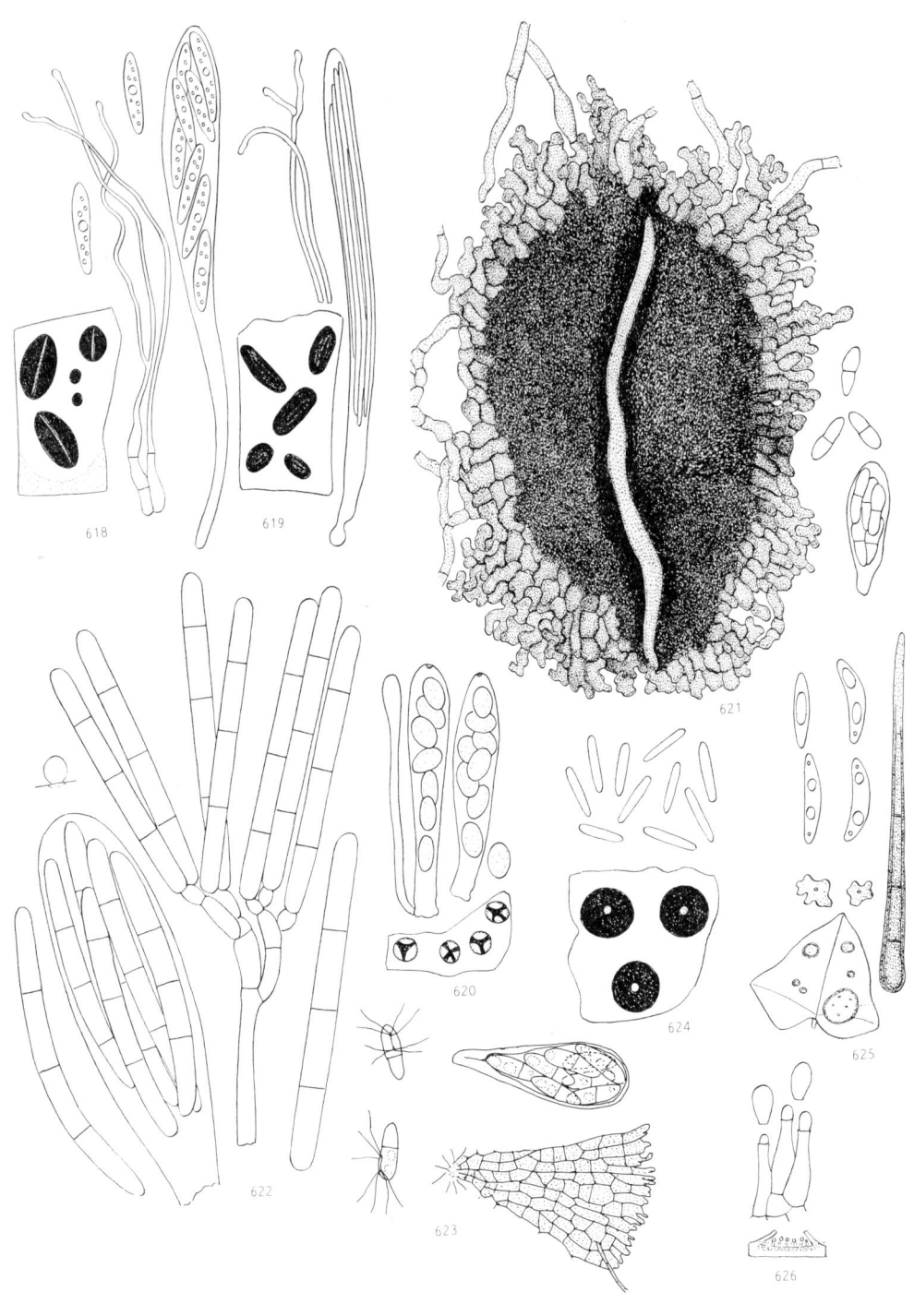

Plate 63. On leaves of *Hedera*. 618, *Hypoderma hederae*; 619, *Lophodermium hedericola*; 620, *Trochila craterium*; 621, *Aulographum hederae*; 622, *Calonectria hederae*; 623, *Microthyrium ciliatum* var. *hederae*; 624, *Ceuthospora hederae*; 625, *Colletotrichum trichellum*; 626, *Cryptocline paradoxa*.

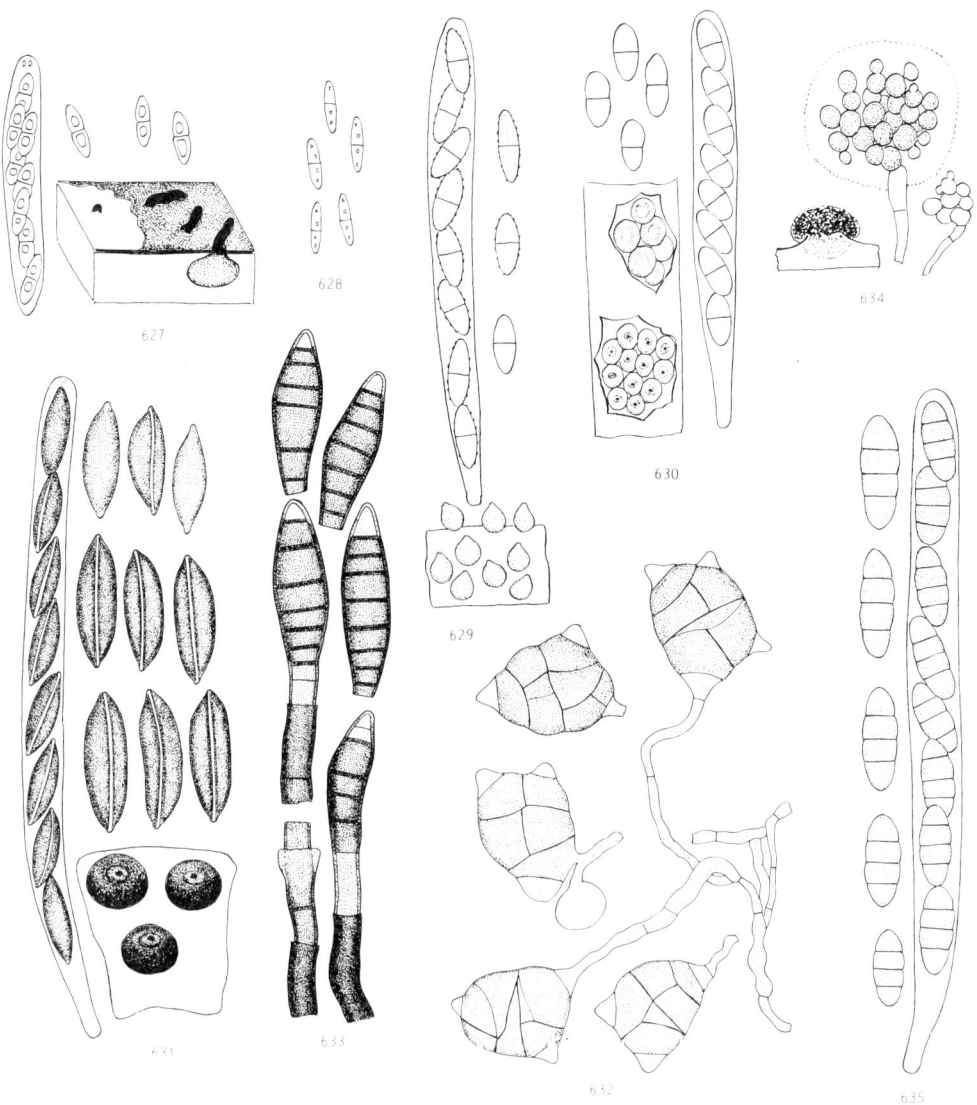

Plate 64. On wood and bark of *Hedera* and *Hippophaë*. 627, *Diaporthe hederae*; 628, *D. pulla*; 629, *Nectria hederae*; 630, *N. sinopica*; 631, *Rosellinia mammiformis*; 632, *Oncopodiella trigonella*; 633, *Sporidesmium socium*; 634, *Cheirospora botryospora*; 635, *Lepteutypa hippophaes*.

Plate 65. On leaves of *Ilex*. 636, *Microscypha enrhiza*; 637, *Phacidiostroma multivalve*; 638, *Trochila ilicina*; 639, *Calonectria erubescens*; 640, *Guignardia philoprina*; 641, *Microthyrium ciliatum*; 642, *Chaetochalara bulbosa*; 643, *Codinaea setosa*; 644, *Oidiodendron flavum*; 645, *Coleophoma cylindrospora*; 646, *Diplodia ilicicola*; 647, *Phomopsis crustosa*; 648, *Pyrenochaeta ilicis*.

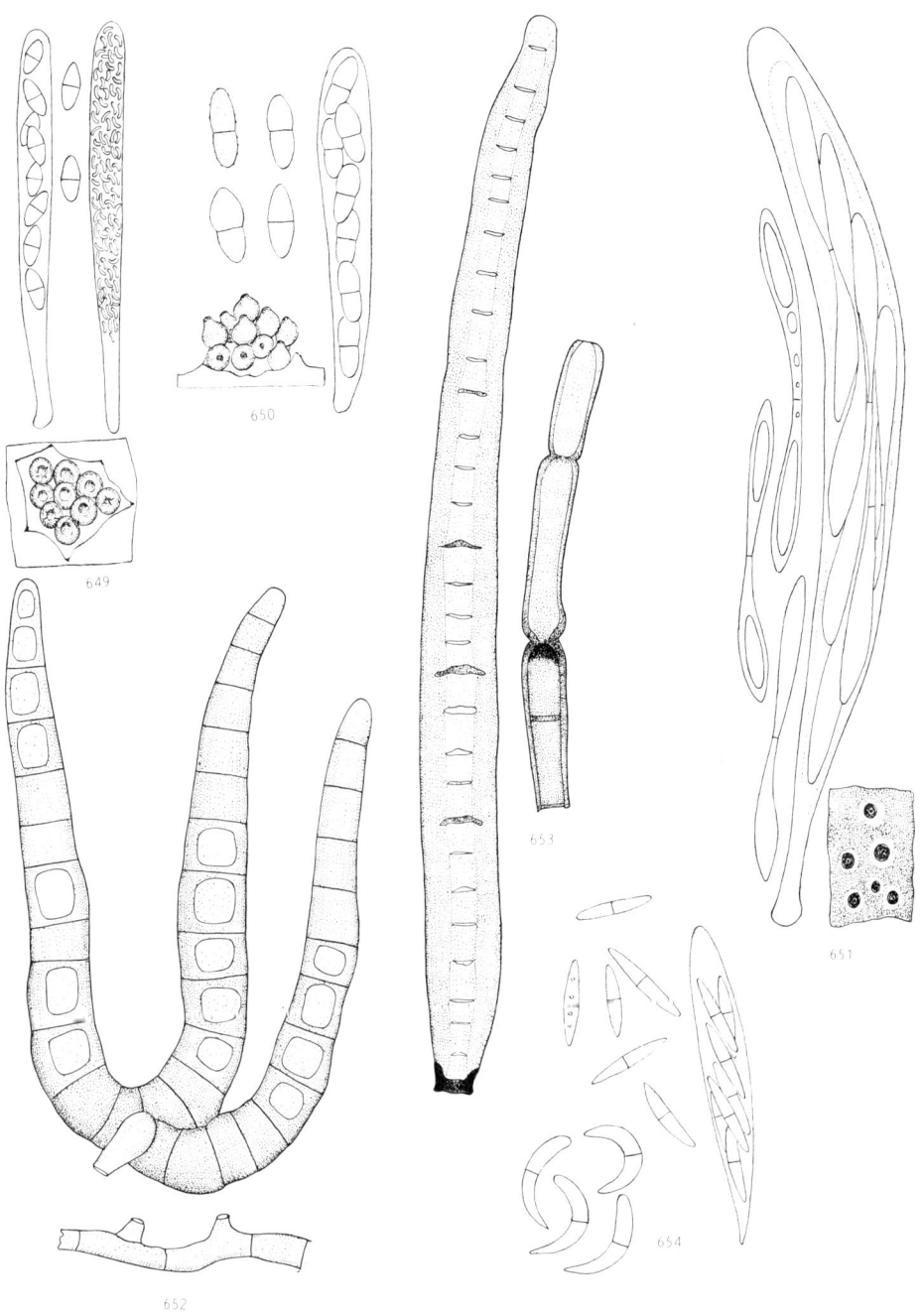

Plate 66. On wood and bark of *Ilex* and *Juglans*. 649, *Nectria aquifolii*; 650, *N. punicea* var. *ilicis*; 651, *Vialaea insculpta*; 652, *Ceratosporium fuscescens*; 653, *Corynespora smithii*; 654, *Gnomonia leptostyla*.

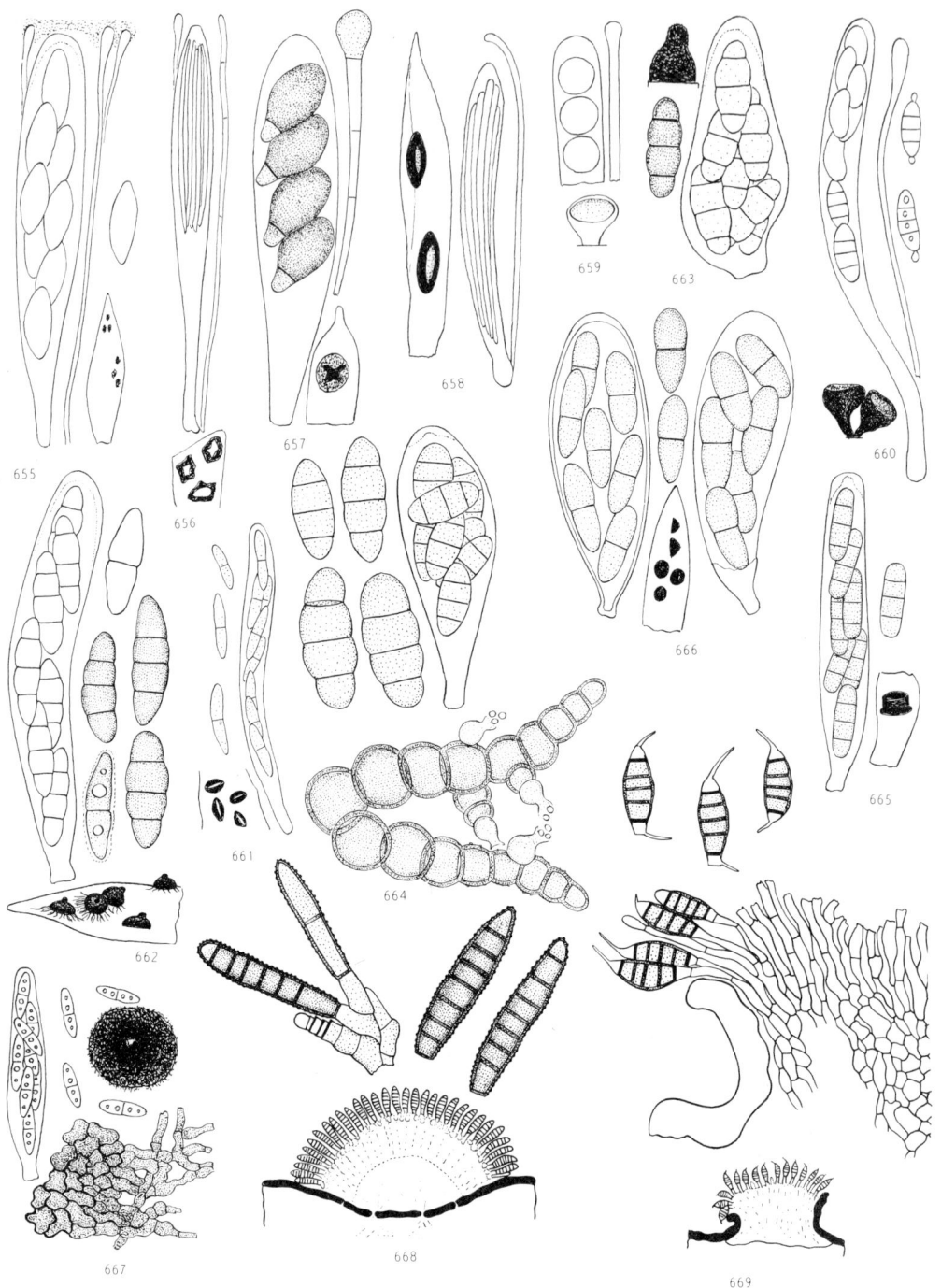

Plate 67. On *Juniperus*. 655, *Chloroscypha sabinae*; 656, *Coccomyces juniperi*; 657, *Didymascella tetraspora*; 658, *Lophodermium juniperinum*; 659, *Pithya cupressina*; 660, *Rutstroemia juniperi*; 661, *Actidium nitidum*; 662, *Herpotrichia juniperi*; 663, *Kriegeriella minuta*; 664, *Metacapnodium juniperi*; 665, *Mytilinidion acicola*; 666, *Seynesiella juniperi*; 667, *Stomiopeltis juniperina*; 668, *Stigmina glomerulosa*; 669, *Seimatosporium foliicola*.

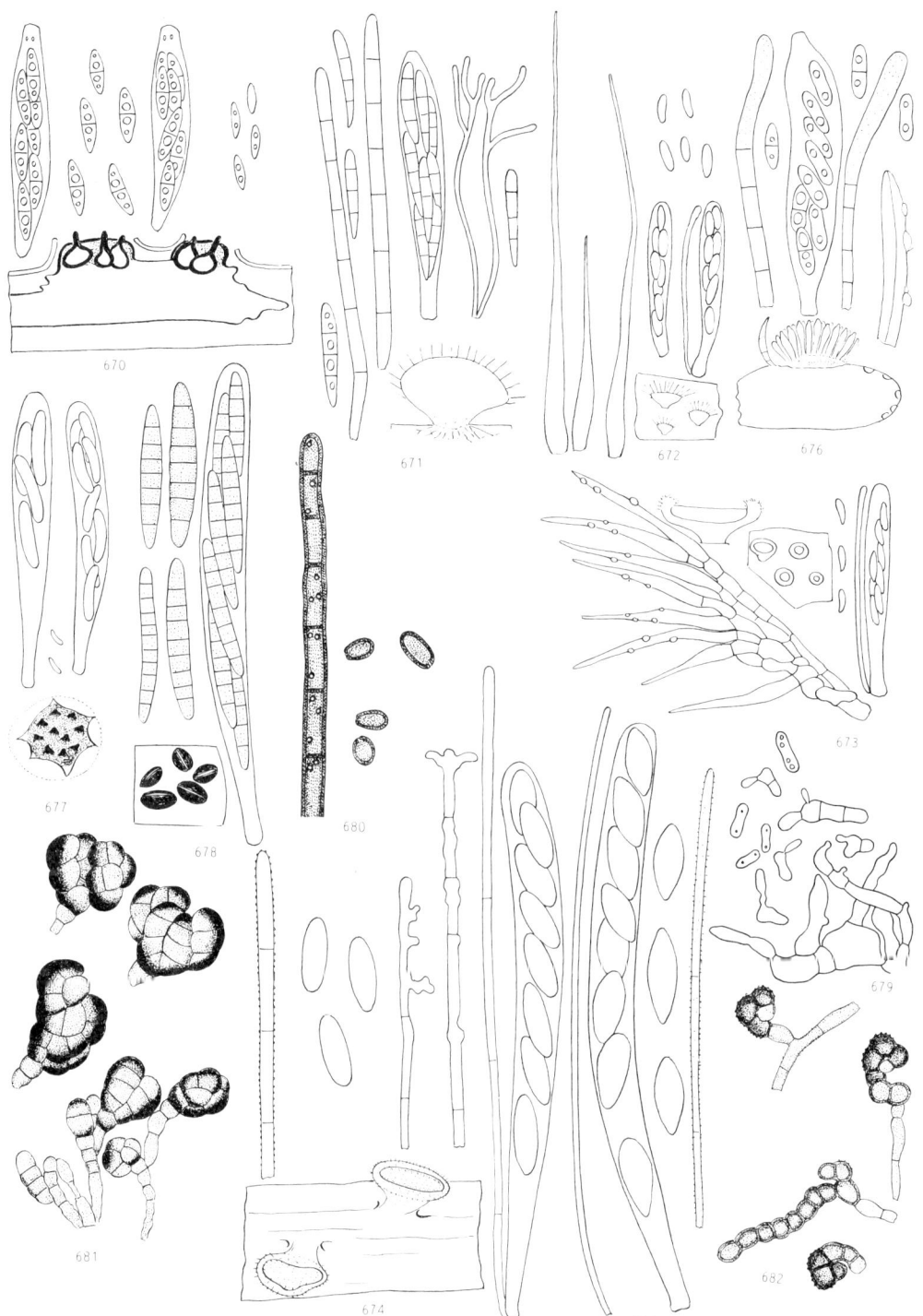

Plate 68. On *Laburnum* and *Larix*. 670, *Diaporthe rudis*; 671, *Arachnopeziza obtusipila*; 672, *Hyaloscypha leuconica*; 673, *H. velenovskii*; 674, *Lachnellula occidentalis*; 675, *L. willkommii*; 676, *Sarcotrochila alpina*; 677, *Leucostoma curreyi*; 678, *Mytilinidion gemmigenum*; 679, *Meria laricis*; 680, *Spadicoides atra*; 681, *Trimmatostroma scutellare*; 682, *Zalerion arboricola*.

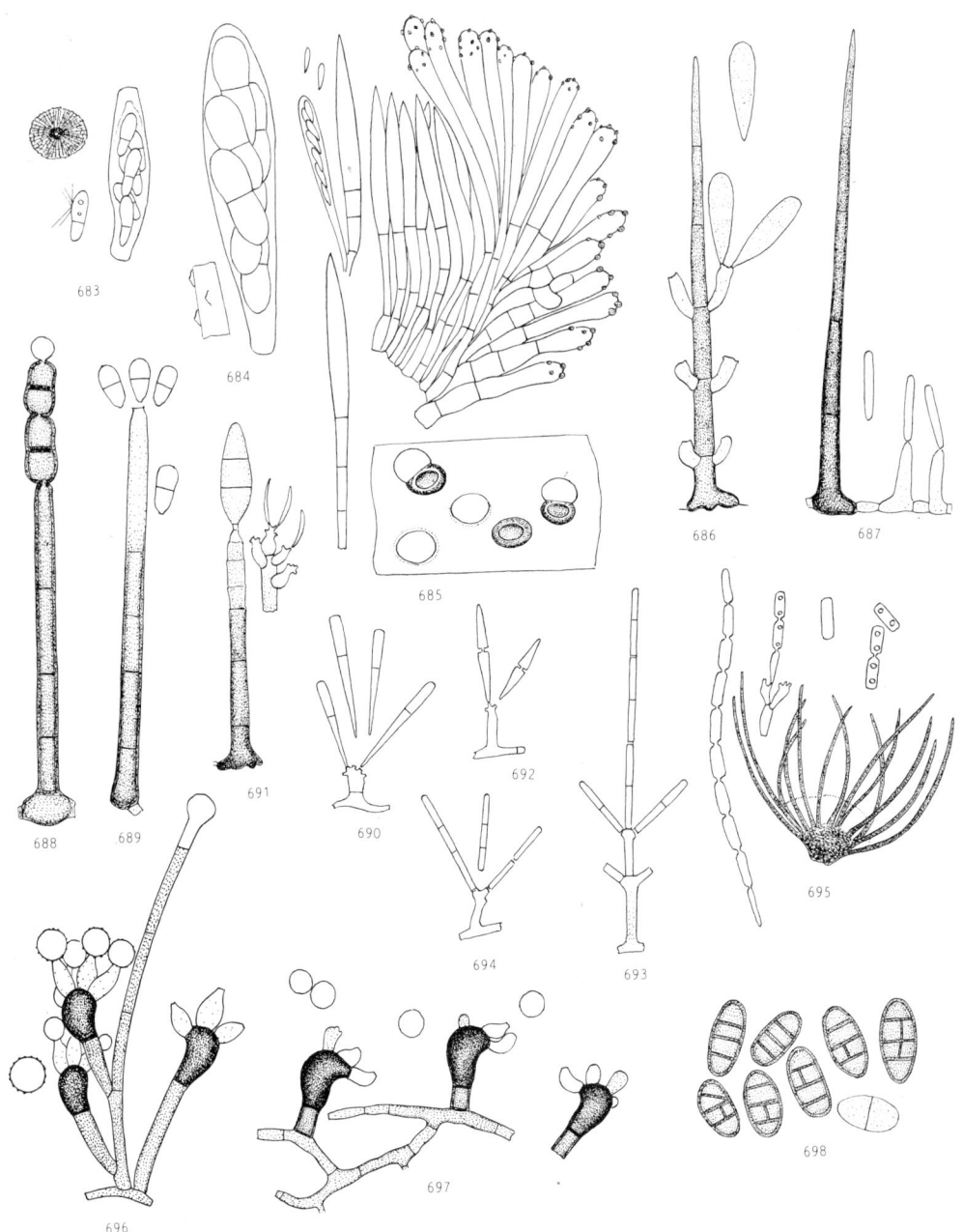

Plate 69. On *Laurus*. 683, *Microthyrium lauri*; 684, *Nectriella consolationis*; 685, *Stegopeziza lauri*; 686, *Beltraniella pirozynskii*; 687, *Circinotrichum britannicum*; 688, *Corynesporopsis uniseptata*; 689, *Cylindrotrichum clavatum*; 690, *Dactylaria obtriangularia*; 691, *Endophragmiella lauri*; 692, *Isthmolongispora minima*; 693, *Polyscytalum gracilisporum*; 694, *Subulispora minima*; 695, *Wiesneriomyces javanicus*; 696, *Zygosporium echinosporum*; 697, *Z. gibbum*; 698, *Camarosporium lauri*.

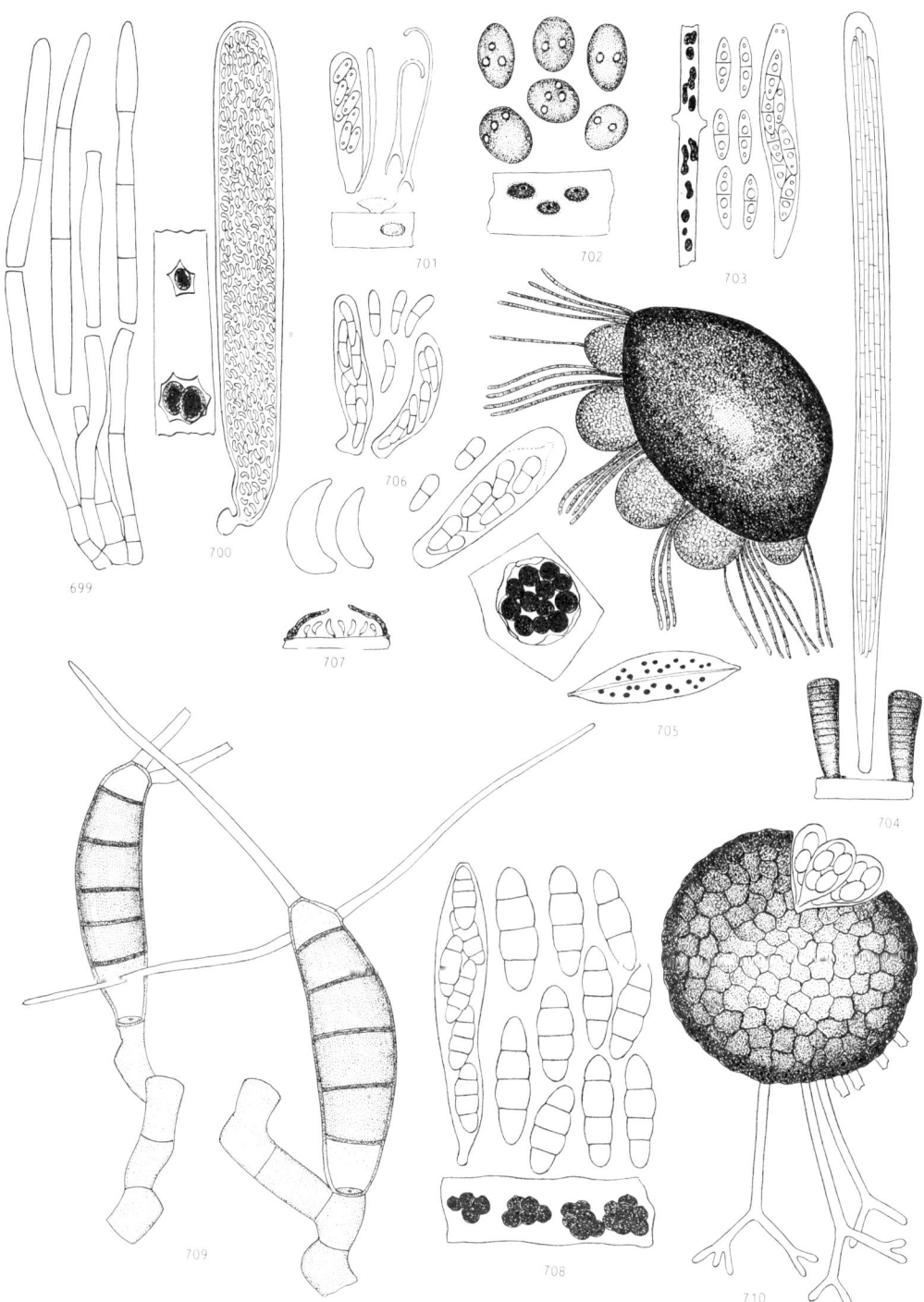

Plate 70. On *Ligustrum, Lonicera, Lupinus* and *Lycium*. 699, *Thedgonia ligustri*; 700, *Tympanis ligustri*; 701, *Unguiculella robergei*; 702, *Amphisphaerella xylostei*; 703, *Diaporthe pardalota*; 704, *Glyphium elatum*; 705, *Lasiobotrys lonicerae*; 706, *Mycosphaerella clymenia*; 707, *Kabatia periclymeni*; 708, *Gibberella pulicaris*; 709, *Pleiochaeta setosa*; 710, *Microsphaera mougeotii*.

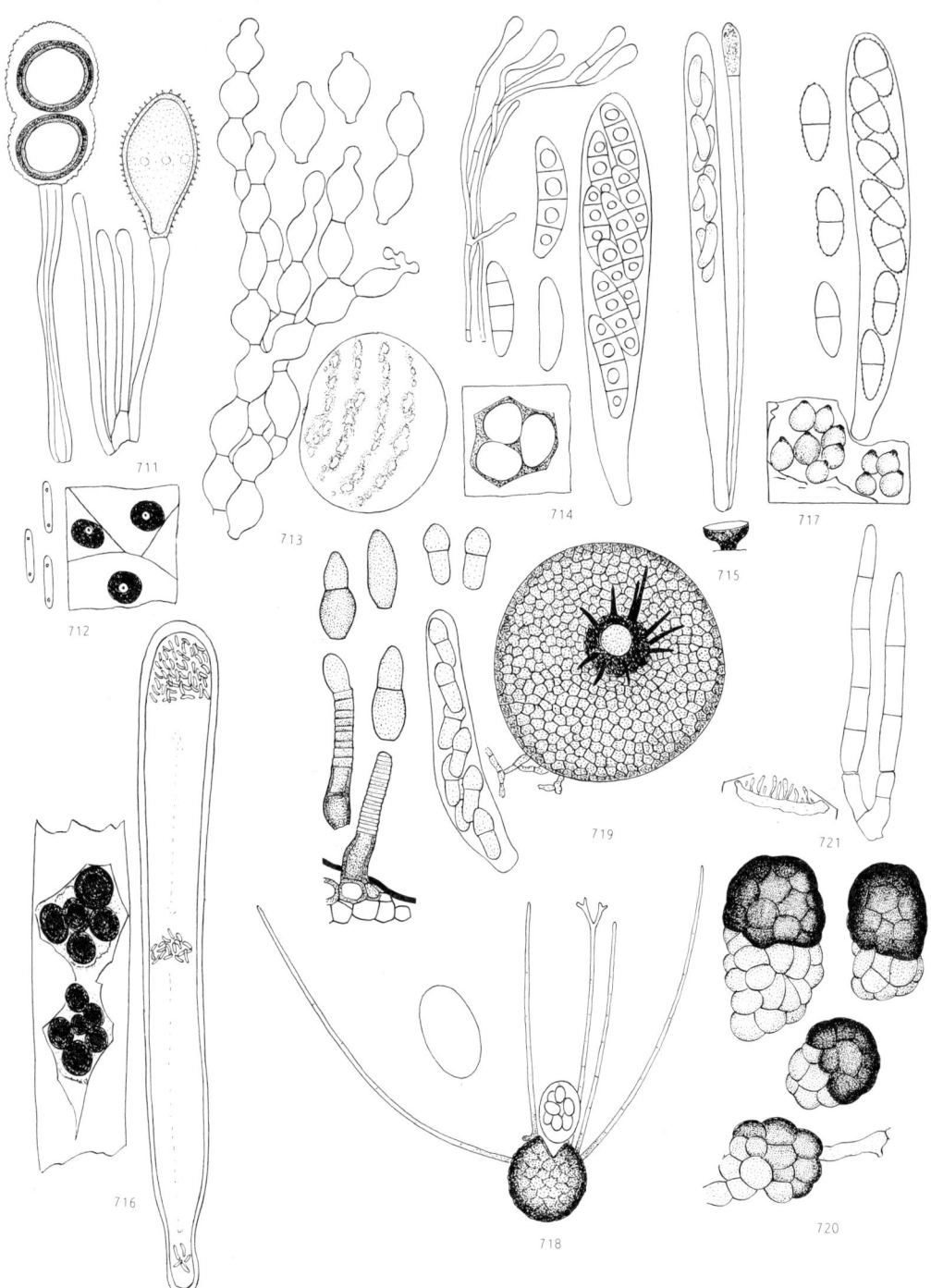

Plate 71. On *Mahonia*, *Malus* and *Morus*. 711, *Cumminsiella mirabilissima*; 712, *Ceuthospora mahoniae*; 713, *Monilinia fructigena*; 714, *Pezicula corticola*; 715, *Rutstroemia rhenana*; 716, *Tympanis conspersa*; 717, *Nectria galligena*; 718, *Podosphaera leucotricha*; 719, *Venturia inaequalis*; 720, *Monodictys melanopa*; 721, *Mycosphaerella mori*.

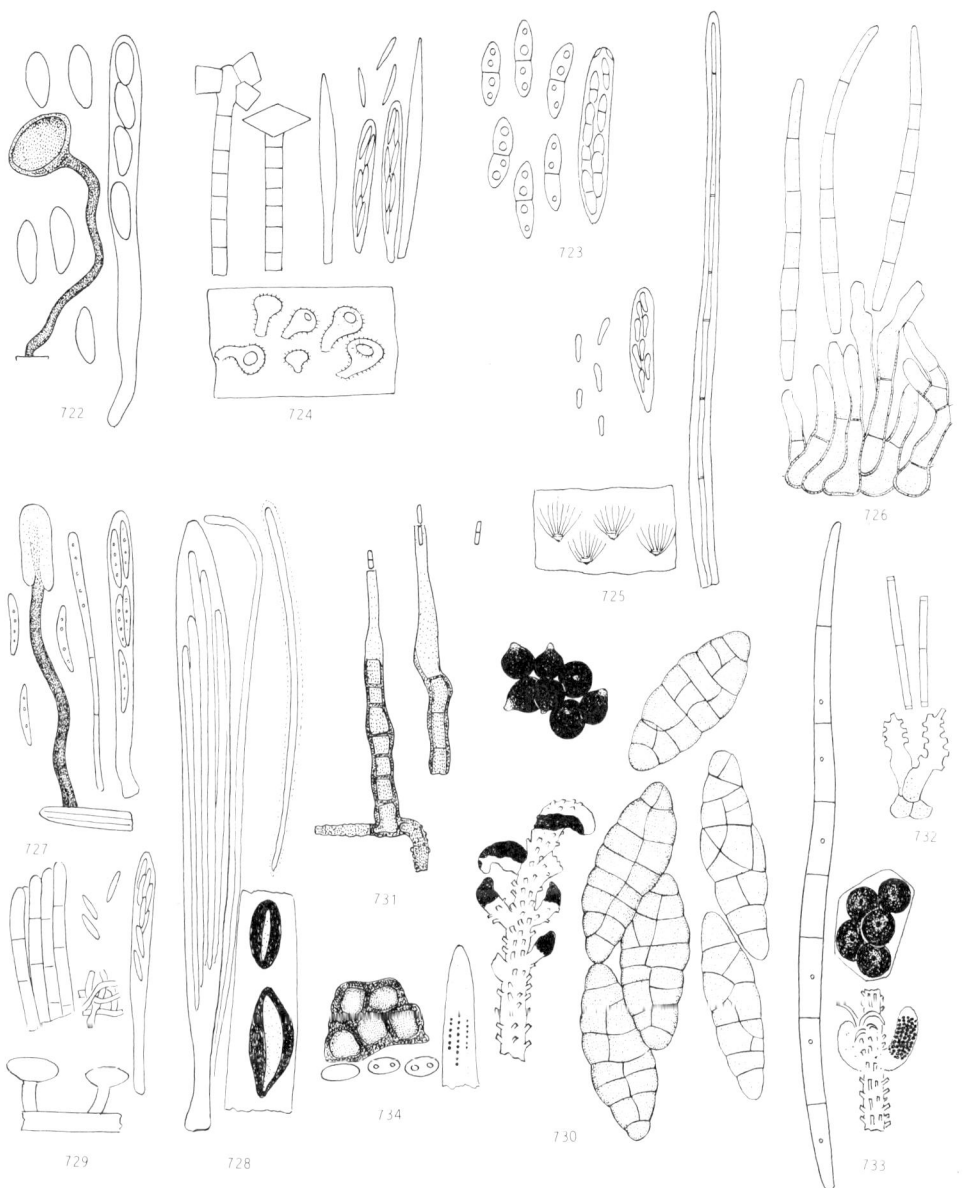

Plate 72. On *Myrica*, *Myrtus* and *Picea*. 722, *Ciboria acerina*; 723, *Cryptodiaporthe aubertii*; 724, *Dasyscyphus sulphurellus*; 725, *Hyalotricha corticicola*; 726, *Pseudocercospora myrticola*; 727, *Heyderia abietis*; 728, *Lophodermium piceae*; 729, *Pezizella subtilis*; 730, *Gemmamyces piceae*; 731, *Chalara cylindrica*; 732, *Dactylaria lepida*; 733, *Megaloseptoria mirabilis*; 734, *Rhizosphaera kalkoffii*.

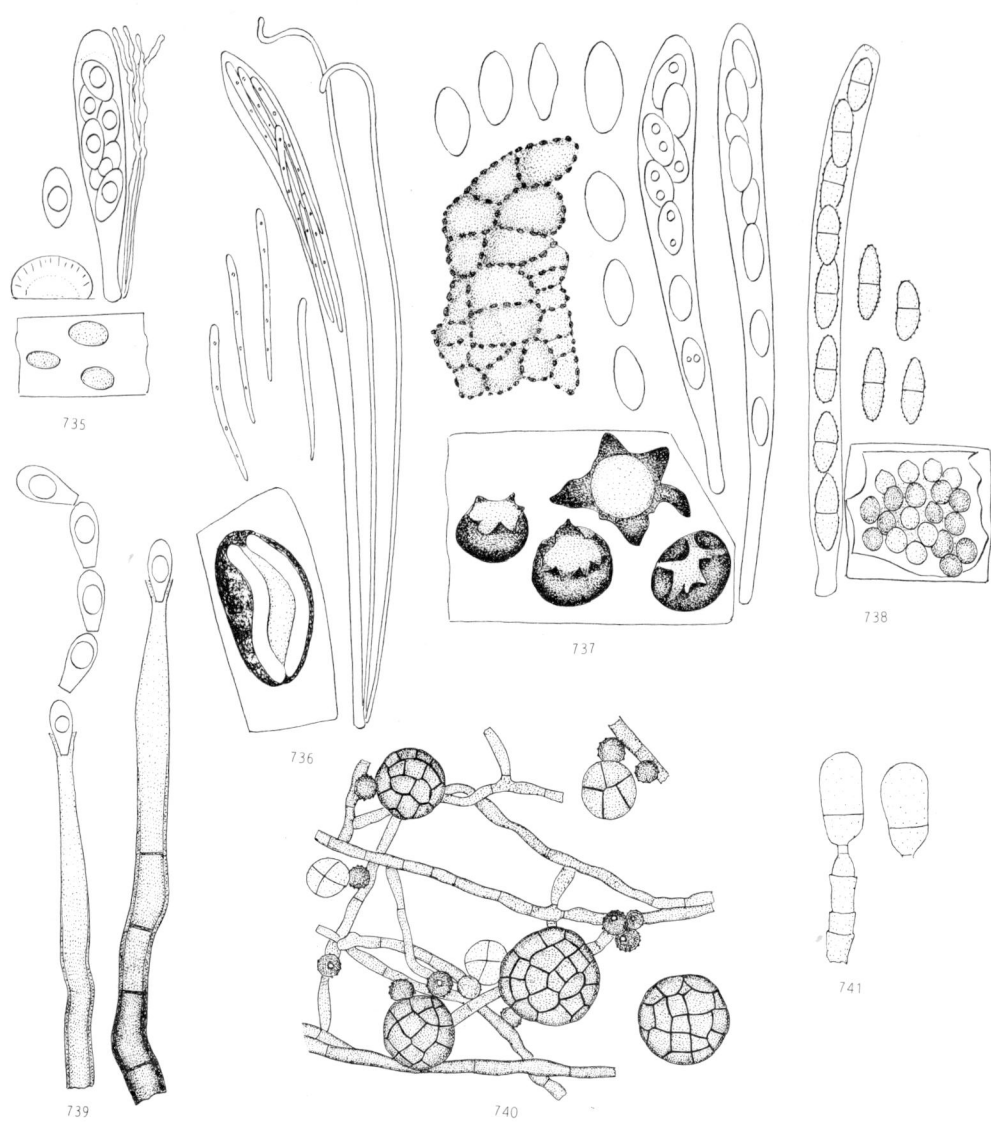

Plate 73. On *Picea*. 735, *Agyrium rufum*; 736, *Colpoma crispum*; 737, *Pseudophacidium piceae*; 738, *Nectria fuckeliana*; 739, *Catenularia piceae*; 740, *Dictyopolyschema pirozynskii*; 741, *Endophragmiella resinae*.

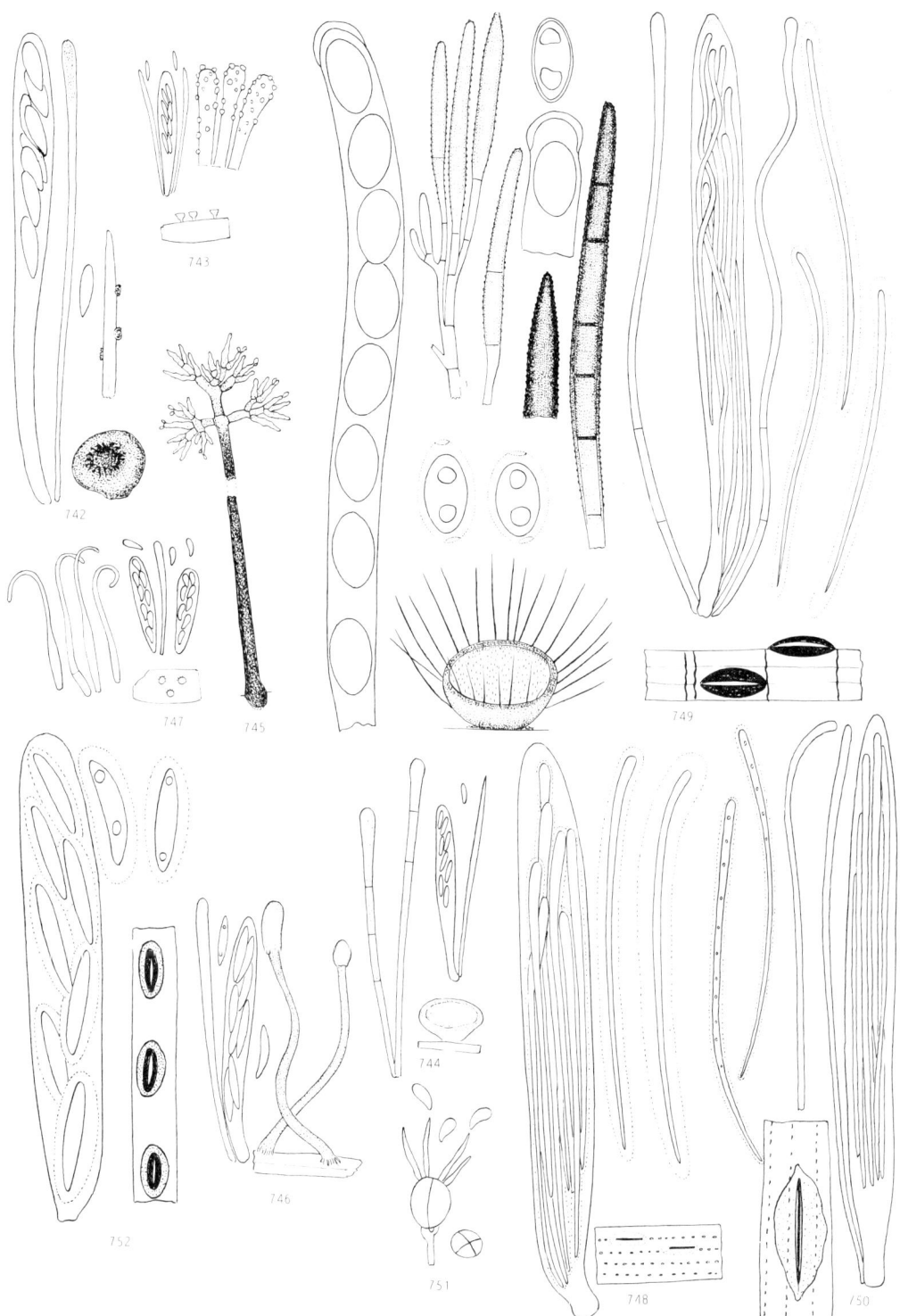

Plate 74. Discomycetes on leaves of *Pinus*. 742, *Cenangium acuum*; 743, *Dasyscyphus acuum*; 744, *D. pulverulentus*; 745, *Desmazierella acicola*; 746, *Heyderia pusilla*; 747, *Hyaloscypha laricionis*; 748, *Lophodermella conjuncta*; 749, *Lophodermium pinastri*; 750, *L. seditiosum*; 751, *Pseudostypella translucens*; 752, *Meloderma desmazieresii*.

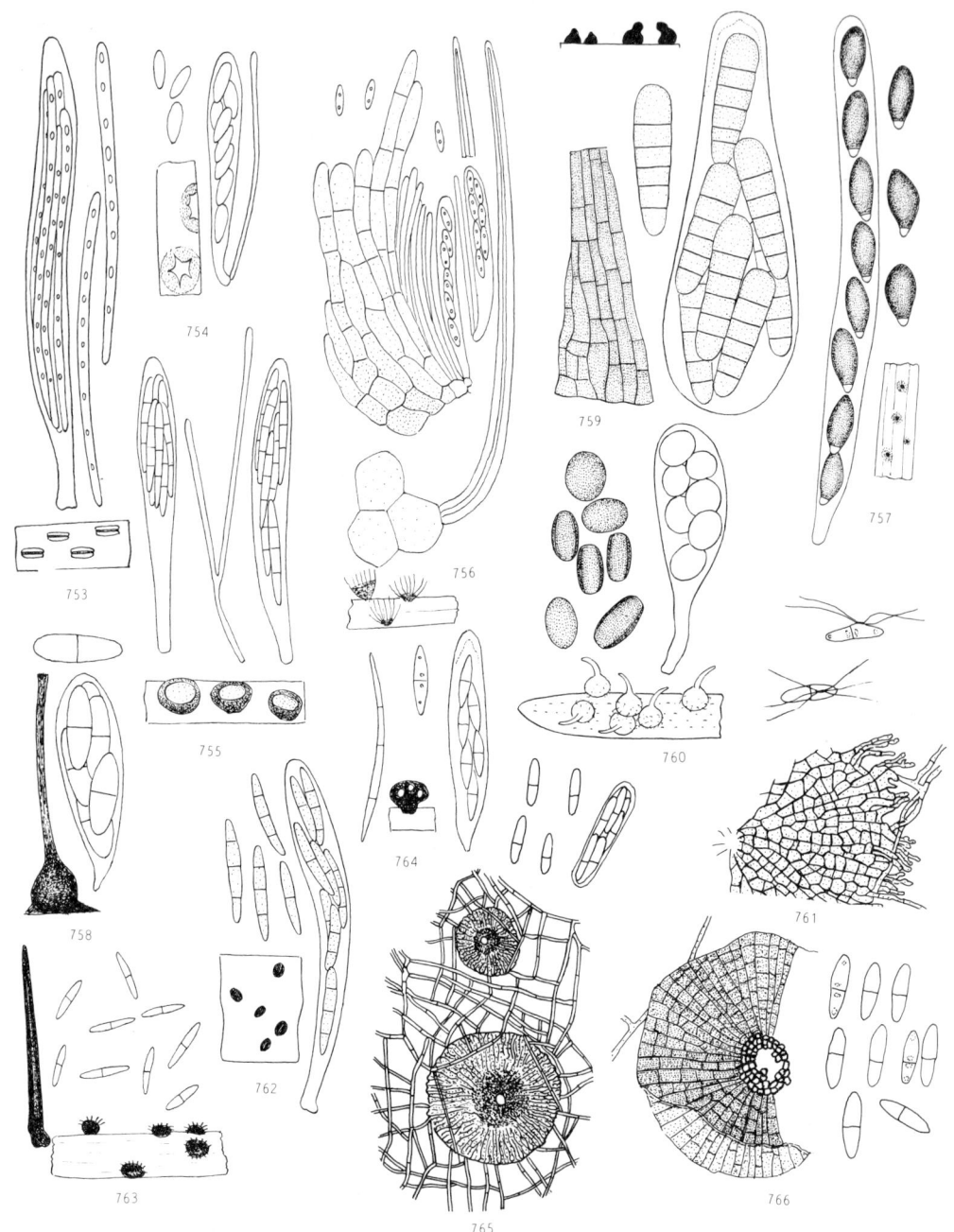

Plate 75. Discomycetes and other ascomycetes on *Pinus* leaves. 753, *Naemacyclus minor*; 754, *Phacidium lacerum*; 755, *Pseudohelotium pineti*; 756, *Urceolella trichodea*; 757, *Anthostomella formosa*; 758, *Klasterskya acuum*; 759, *Kriegeriella mirabilis*; 760, *Melanospora chionea*; 761, *Microthyrium pinophyllum*; 762, *Mytilidinidion mytilinellum*; 763, *Niesslia exilis*; 764, *Scirrhia pini*; 765, *Stomiopeltis pinastri*; 766, *Trichothyrina pinophylla*.

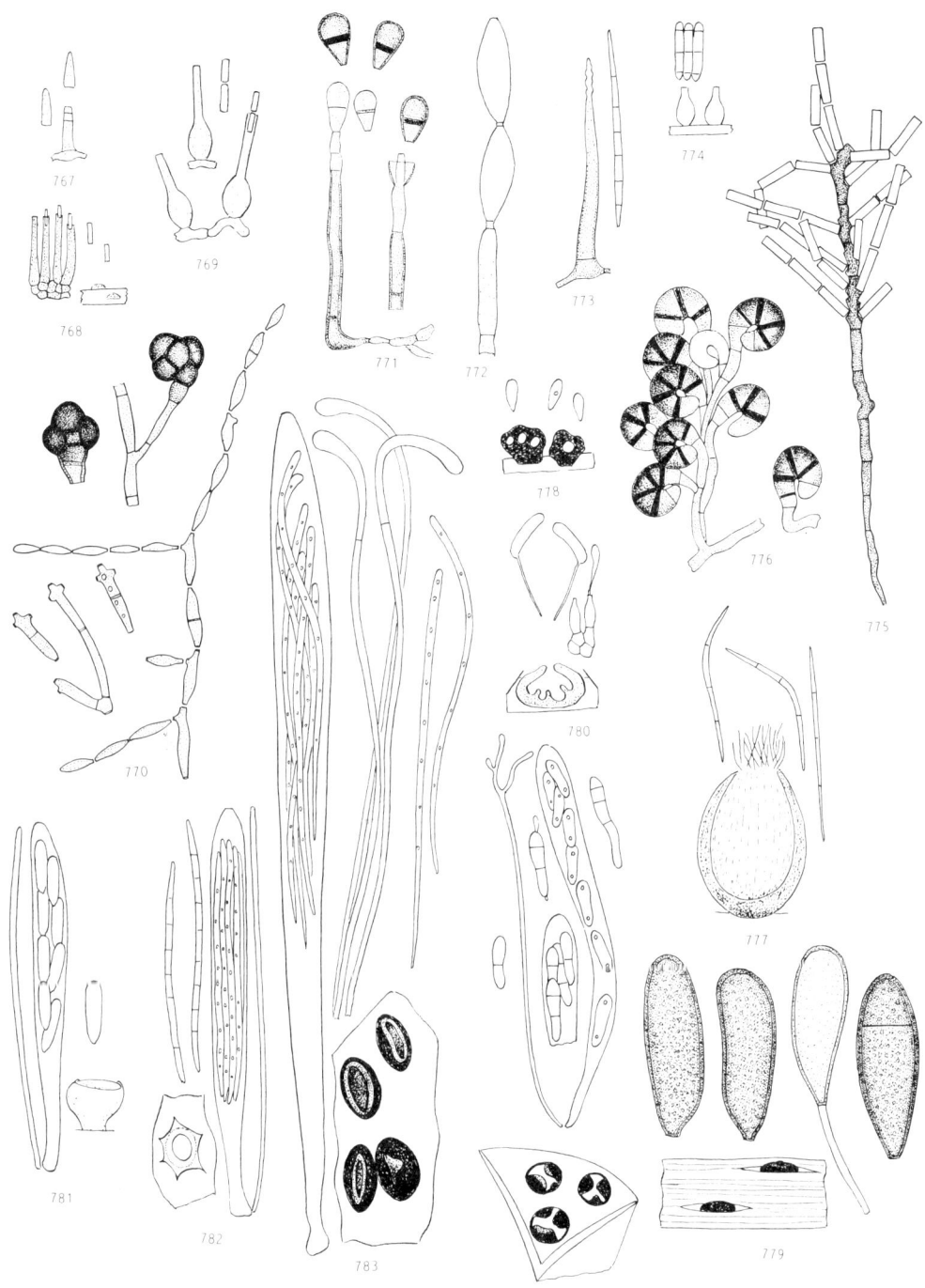

Plate 76. On *Pinus* leaves and cones. 767, *Belemnospora pinicola*; 768, *Bloxamia bohemica*; 769, *Chalara fusidioides*; 770, *Cladosporium staurophorum*; 771, *Endophragmia pinicola*; 772, *Junctospora pulchra*; 773, *Pseudocercospora deightonii*; 774, *Pseudomicrodochium candidum*; 775, *Sympodiella acicola*; 776, *Troposporella monospora*; 777, *Fujimyces oödes*; 778, *Sclerophoma* state of *Sydowia polyspora*; 779, *Sphaeropsis sapinea*; 780, *Strasseria geniculata*; 781, *Hymenoscyphus lutescens*; 782, *Lasiostictis fimbriata*; 783, *Lophodermium conigenum*; 784, *Micraspis strobilina*.

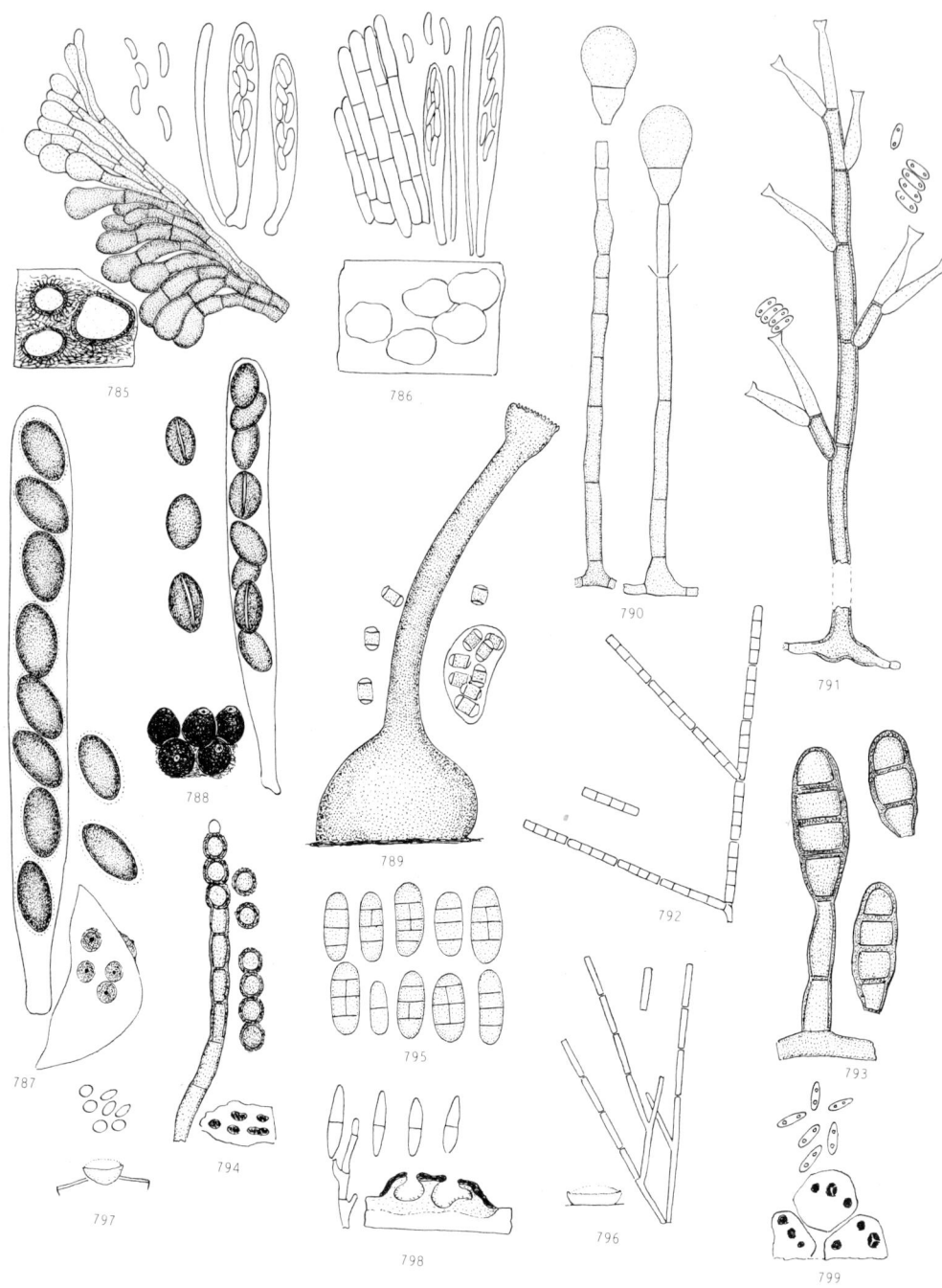

Plate 77. On *Pinus* cones. 785, *Mollisia fallax*; 786, *Pezizella chionea*; 787, *Anthostomella conorum*; 788, *Rosellinia obliquata*; 789, *Scopinella solani*; 790, *Endophragmia boewei*; 791, *Phaeostalagmus peregrinus*; 792, *Septocylindrium leucum*; 793, *Sporidesmium doliiforme*; 794, *Xylohypha ortmansiae*; 795, *Camarosporium pini*; 796, *Patellina caesia*; 797, *Pseudopatellina conigena*; 798, *Sirococcus strobilinus*; 799, *Sporonema diamandidis*.

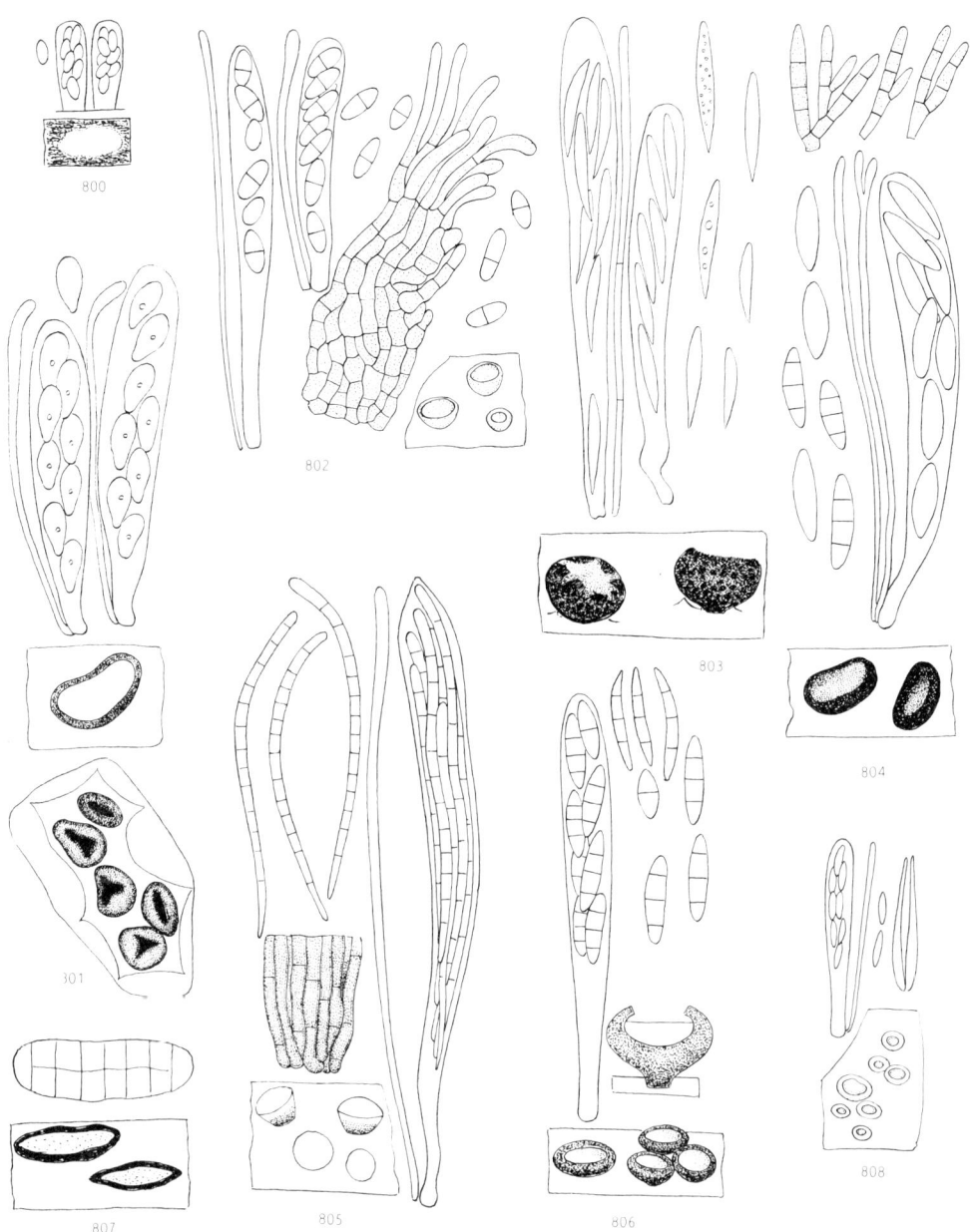

Plate 78. Discomycetes on *Pinus* wood and bark. 800, *Ascocorticium anomalum*; 801, *Cenangium ferruginosum*; 802, *Ciliolarina laricina*; 803, *Crumenulopsis pinicola*; 804, *C. sororia*; 805, *Gorgoniceps aridula*; 806, *Gremmeniella abietina*; 807, *Haematomma elatina*; 808, *Hyaloscypha stevensonii*.

Plate 79. Discomycetes on *Pinus* wood and bark. 809, *Lachnellula pseudofarinacea*; 810, *L. subtilissima*; 811, *Orbilia leucostigma*; 812, *Pezicula livida*; 813, *Phaeohelotium purpureum*; 814, *Rhizina undulata*; 815, *Sarea difformis*; 816, *Therrya pini*; 817, *Trichodiscus pinicola*; 818, *Tympanis hypopodia*.

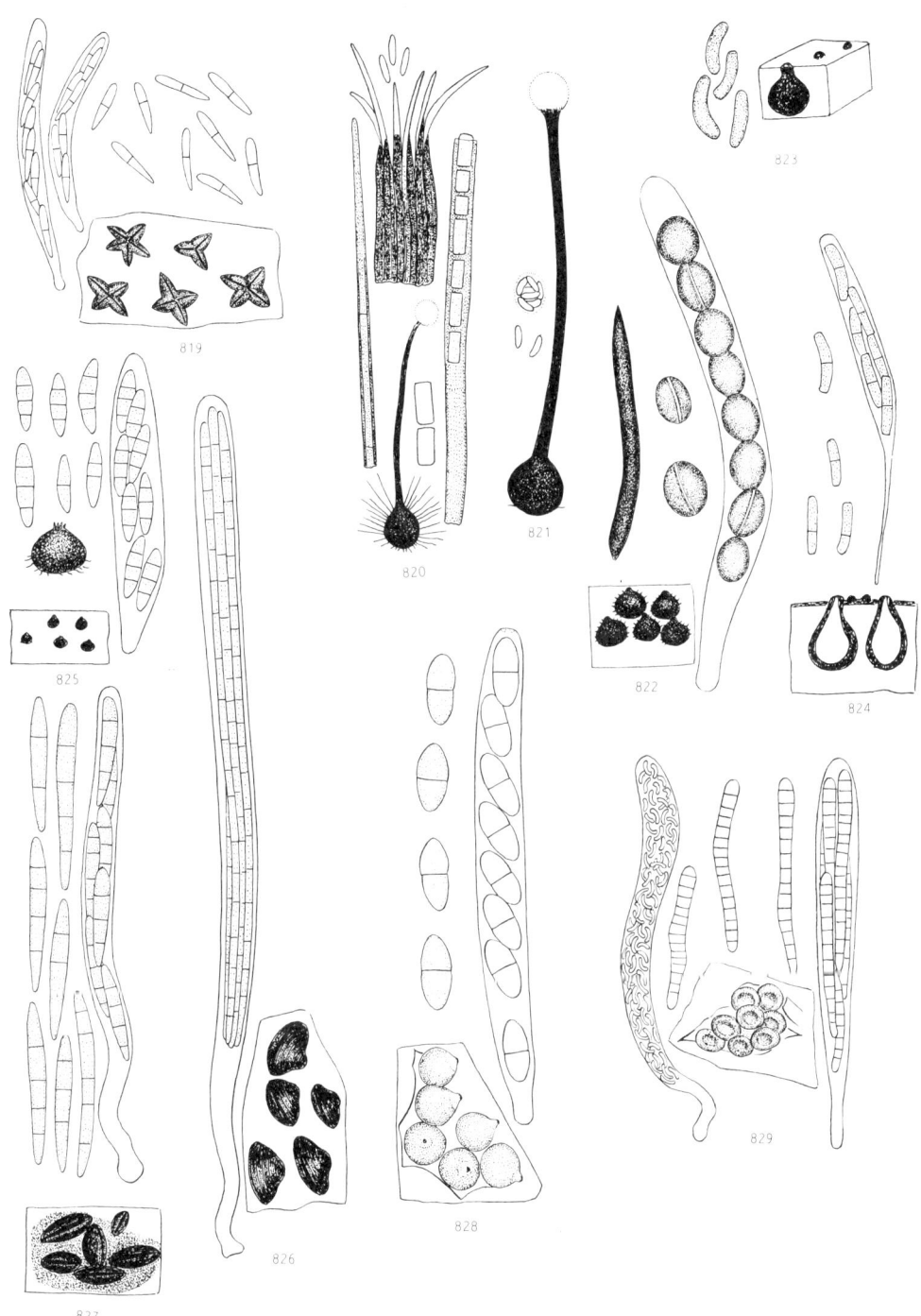

Plate 80. Ascomycetes on *Pinus* wood and bark. 819, *Actidium hysterioides*; 820, *Ceratocystis coerulescens*; 821, *C. piceae*; 822, *Coniochaeta malacotricha*; 823, *Endoxyla operculata*; 824, *Endoxylina pini*; 825, *Keissleriella pinicola*; 826, *Lophium mytilinum*; 827, *Mytilinidion rhenanum*; 828, *Nectria pinea*; 829, *Scoleconectria cucurbitula*.

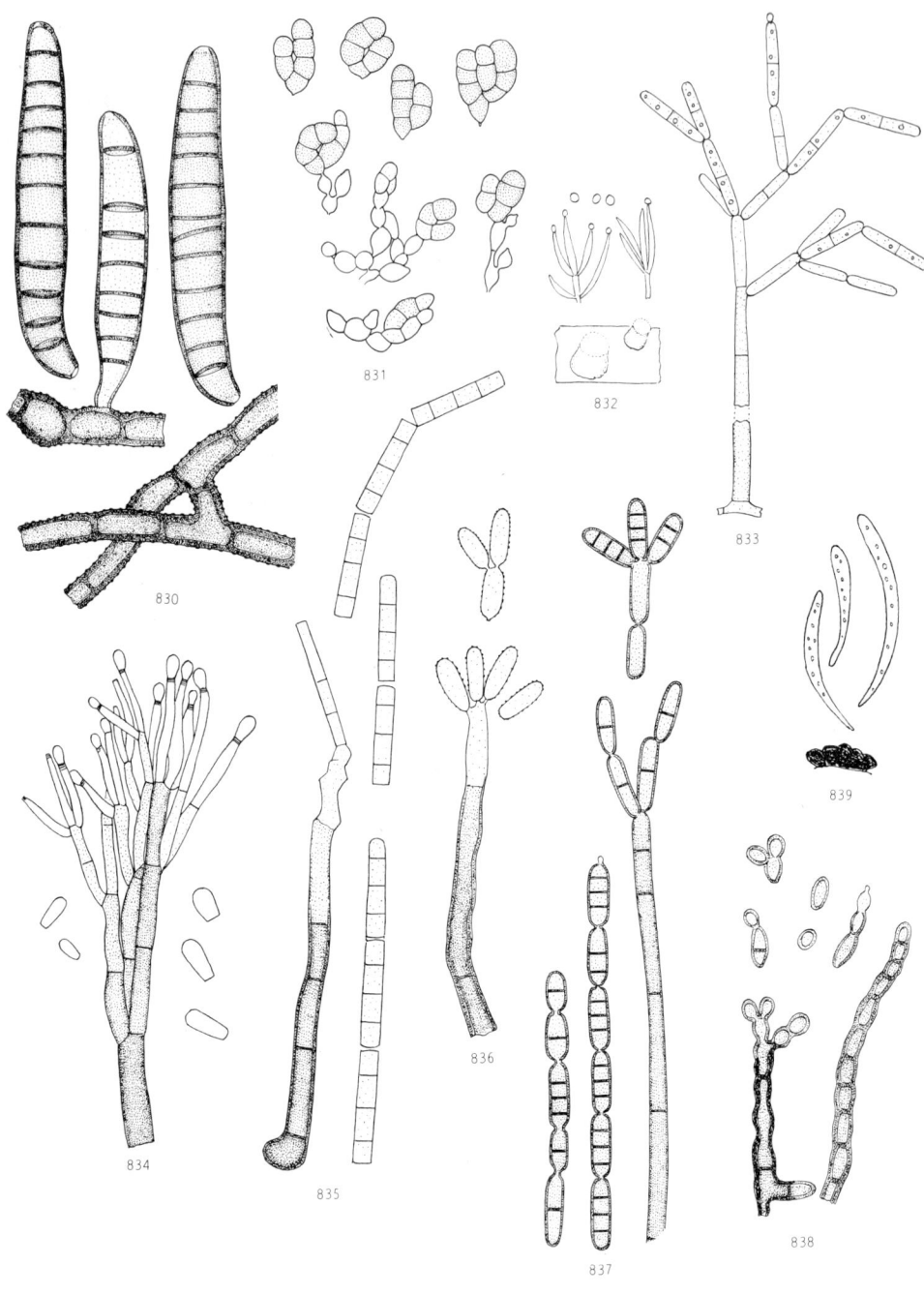

Plate 81. Hyphomycetes and coelomycetes on *Pinus* wood and bark. 830, *Antennatula* state of *Strigopodia resinae*; 831, *Cheiromycella microscopica*; 832, *Dendrodochium citrinum*; 833, *Hormiactella asetosa*; 834, *Leptographium lundbergii*; 835, *Parasympodiella clarkii*; 836, *Polyscytalum verrucosum*; 837, *Septonema fasciculare*; 838, *Xylohypha pinicola*; 839, *Oncospora pinastri*.

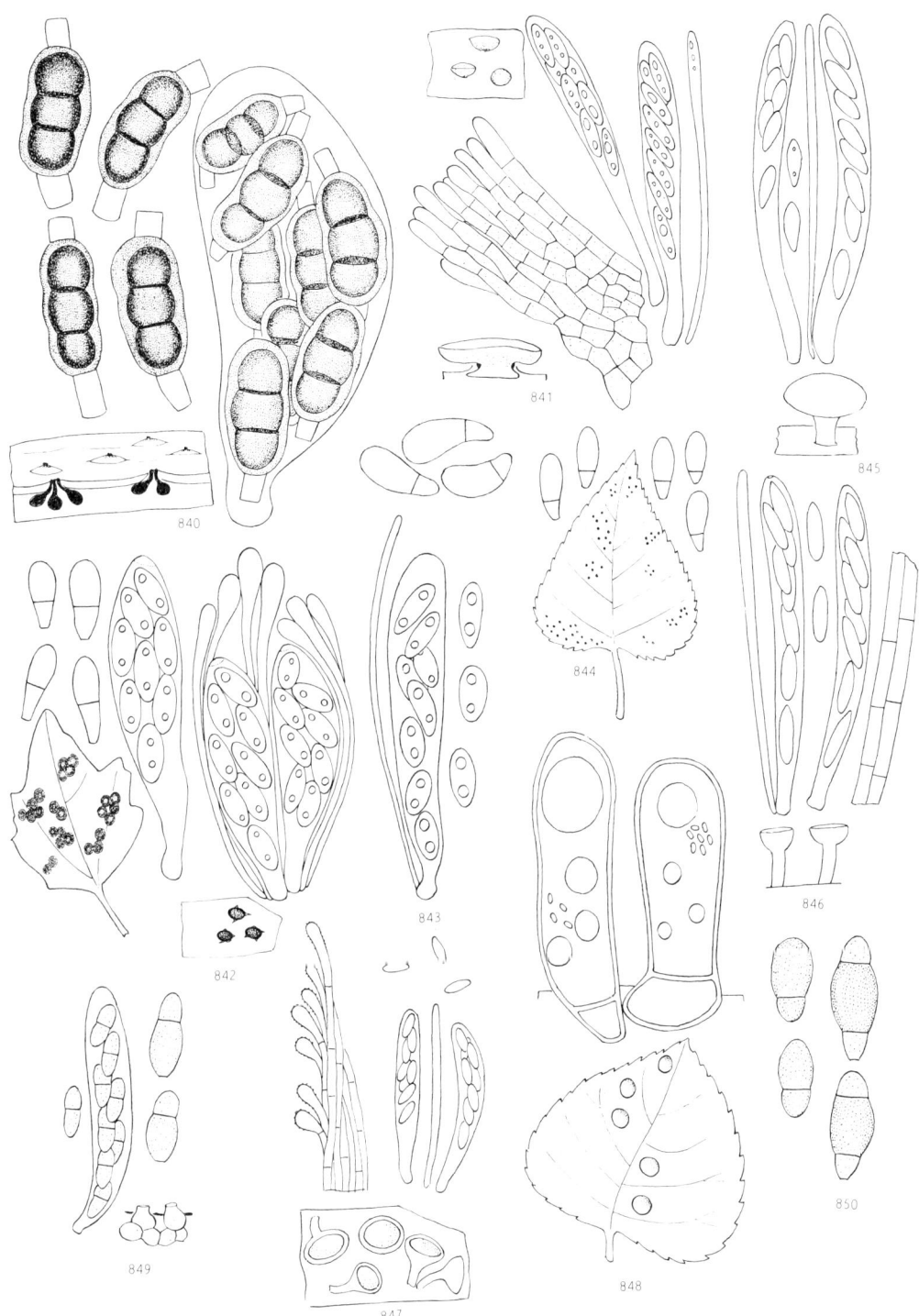

Plate 82. On *Platanus* and *Populus*. 840, *Hapalocystis berkeleyi*; 841, *Calycellina populina*; 842, *Drepanopeziza populi-alba*; 843, *D. populorum*; 844, *D. punctiformis*; 845, *Hymenoscyphus immutabilis*; 846, *H. phyllogenus*; 847, *Pezizella gemmarum*; 848, *Taphrina populina*; 849, *Venturia macularis*; 850, *V. populina*.

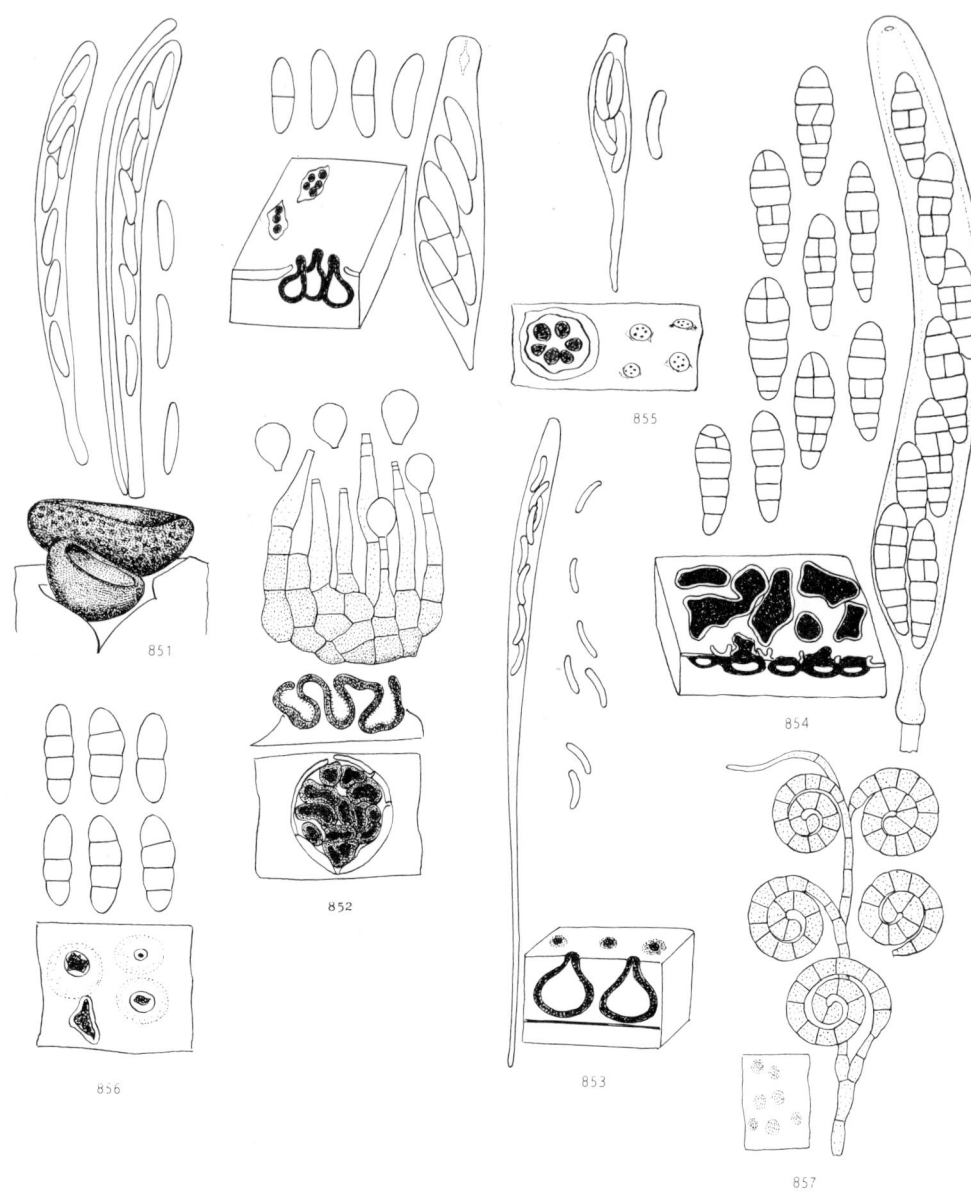

Plate 83. On *Populus*. 851, *Encoelia fascicularis*; 852, *Cryptodiaporthe populea*; 853, *Cryptosphaeria populina*; 854, *Dothiora sphaerioides*; 855, *Leucostoma niveum*; 856, *Massarina emergens*; 857, *Troposporella fumosa*.

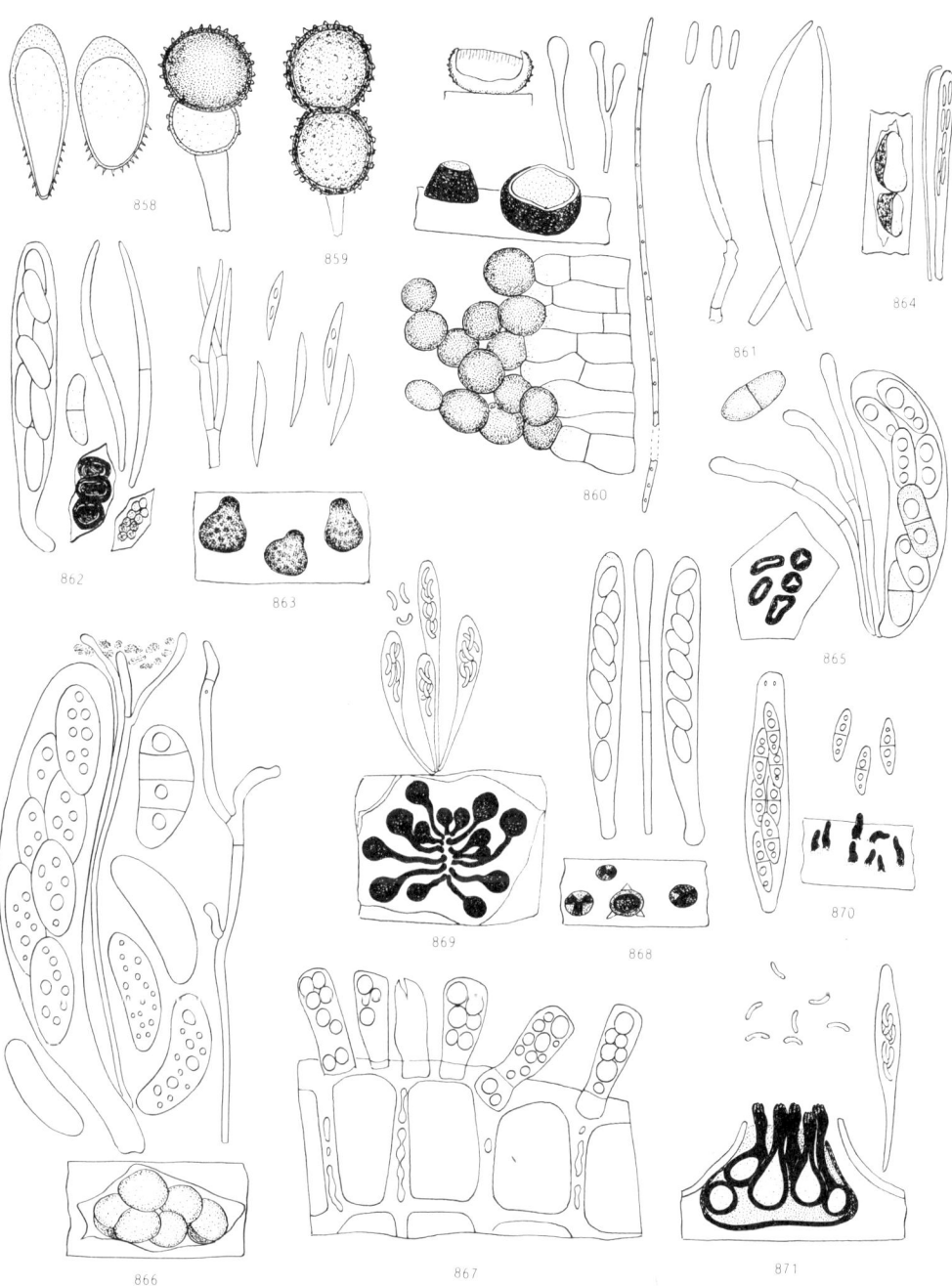

Plate 84. On *Prunus*. 858, *Tranzschelia discolor*; 859, *T. pruni-spinosae*; 860, *Apostemidium leptospora*; 861, *Blumeriella jaapii*; 862, *Dermea cerasi*; 863, *D. padi*; 864, *Encoelia fuckelii*; 865, *Eupropolella britannica*; 866, *Pezicula houghtonii*; 867, *Taphrina deformans*; 868, *Trochila laurocerasi*; 869, *Calosphaeria pulchella*; 870, *Diaporthe perniciosa*; 871, *Eutypella prunastri*.

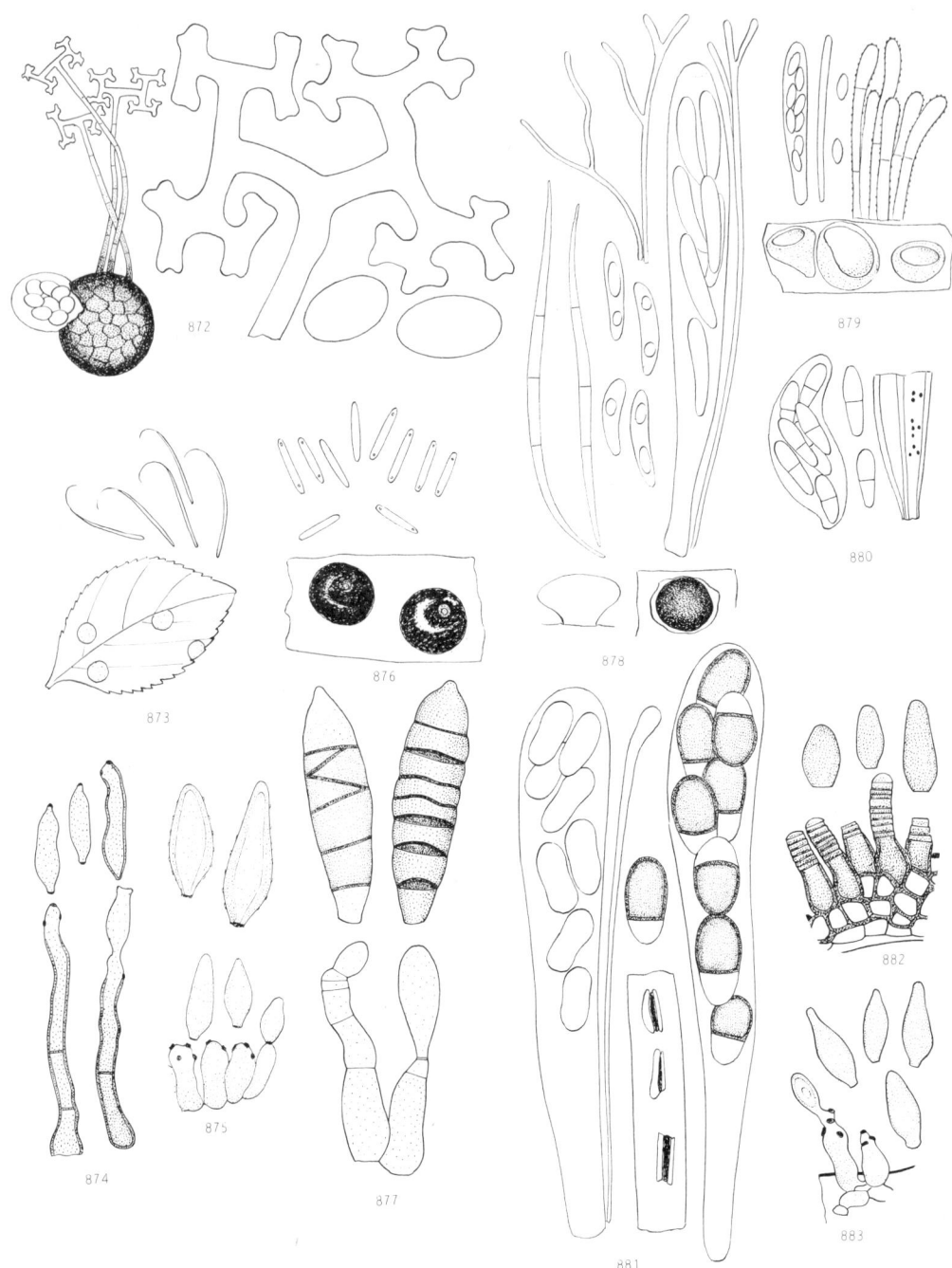

Plate 85. On *Prunus, Pseudotsuga, Pyracantha* and *Pyrus.* 872, *Podosphaera tridactyla*; 873, *Polystigma rubrum*; 874, *Venturia carpophila*; 875, *V. cerasi*; 876, *Ceuthospora lauri*; 877, *Stigmina carpophila*; 878, *Dermea balsamea*; 879, *Lachnellula calyciformis*; 880, *Phaeocryptopus gaeumannii*; 881, *Rhabdocline pseudotsugae*; 882, *Spilocaea pyracanthae*; 883, *Venturia pirina*.

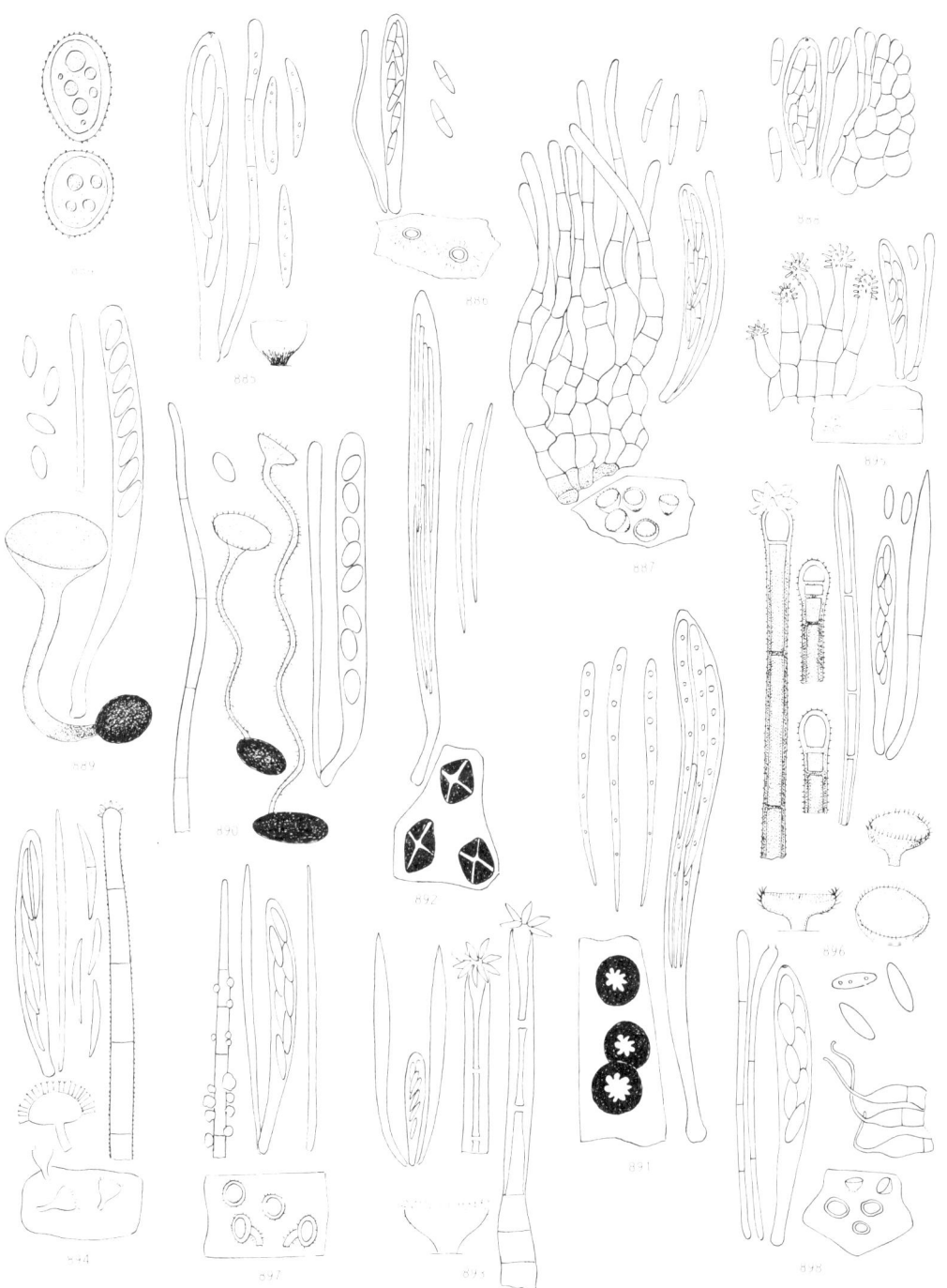

Plate 86. Discomycetes on *Quercus* leaves; also one rust. 884, *Uredo quercus*; 885, *Allophylaria basalifusca*; 886, *Arachnopeziza eriobasis*; 887, *Calycellina punctiformis*; 888, *C. rivelinensis*; 889, *Ciborinia candolleana*; 890, *C. hirtella*; 891, *Coccomyces coronatus*; 892, *C. dentatus*; 893, *Dasyscyphus capitatus*; 894, *D. ciliaris*; 895, *D. coruscatus*; 896, *D. fuscescens*; 897, *D. soppittii*; 898, *Hyaloscypha pseudopuberula*.

Plate 87. Discomycetes on *Quercus* leaves. 899, *Hyaloscypha pygmaea*; 900, *Hymenoscyphus epiphyllus*; 901, *Hypoderma ilicinum*; 902, *Laetinaevia pustulata*; 903, *Lanzia coracina*; 904, *Lophodermium petiolicolum*; 905, *Mollisia spectabilis*; 906, *Naevala perexigua*; 907, *Niptera muelleri-argoviensis*; 908, *Pezizella roburnea*; 909, *P. rubescens*; 910, *Phialina parenchymatosa*; 911, *Pocillum cesatii*; 912, *Pycnopeziza pachyderma*; 913, *Rutstroemia sydowiana*; 914, *Stegopeziza quercea*.

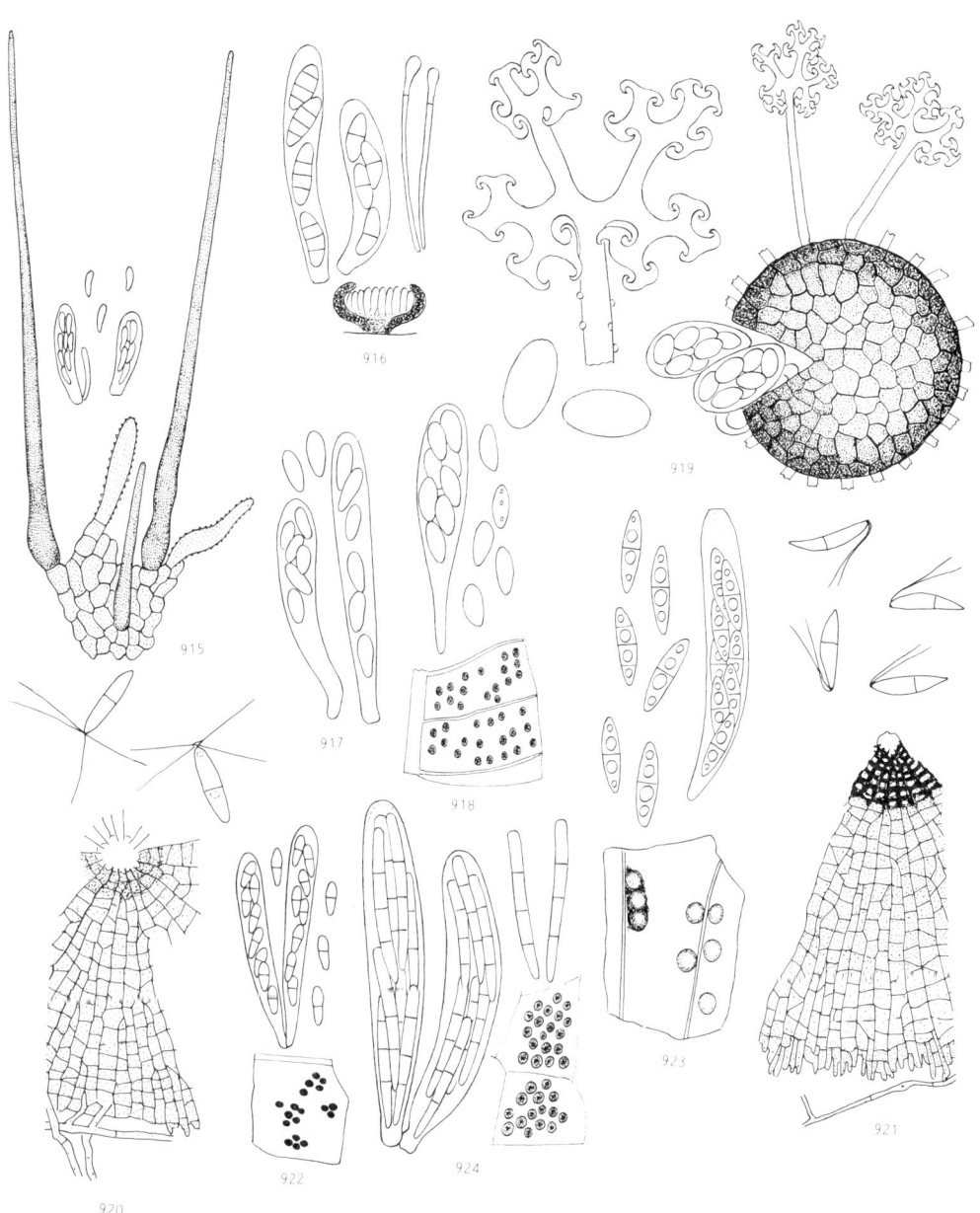

Plate 88. Discomycetes and other ascomycetes on *Quercus* leaves. 915, *Trichodiscus diversipilus*; 916, *Epibelonium gaeumannii*; 917, *Guignardia punctoidea*; 918, *Hyponectria cookeana*; 919, *Microsphaera alphitoides*; 920, *Microthyrium ilicinum*; 921, *M. microscopicum*; 922, *Mycosphaerella punctiformis*; 923, *Plagiostoma pustula*; 924, *Sphaerulina myriadea*.

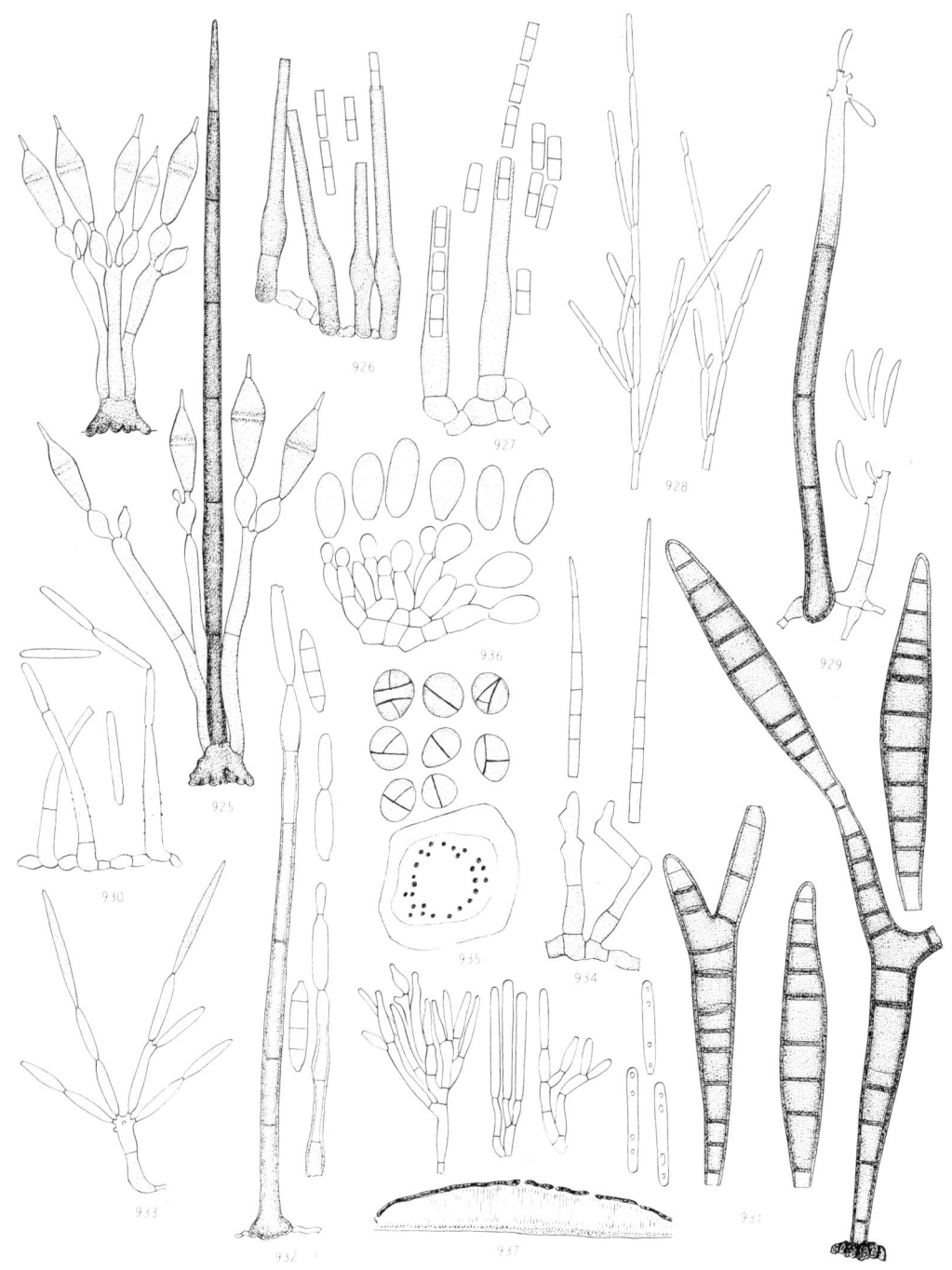

Plate 89. Hyphomycetes and coelomycetes on *Quercus* leaves. 925, *Beltrania querna*; 926, *Chalara hughesii*; 927, *C. kendrickii*; 928, *Cylindrium elongatum*; 929, *Dictyochaeta querna*; 930, *Fusidium aeruginosum*; 931, *Lobatopedis foliicola*; 932, *Parapleurotheciopsis ilicina*; 933, *Subramaniomyces fusisaprophyticus*; 934, *Subulispora britannica*; 935, *Camarosporium oreades*; 936, *Cryptocline cinerescens*; 937, *Leptothyrium ilicinum*.

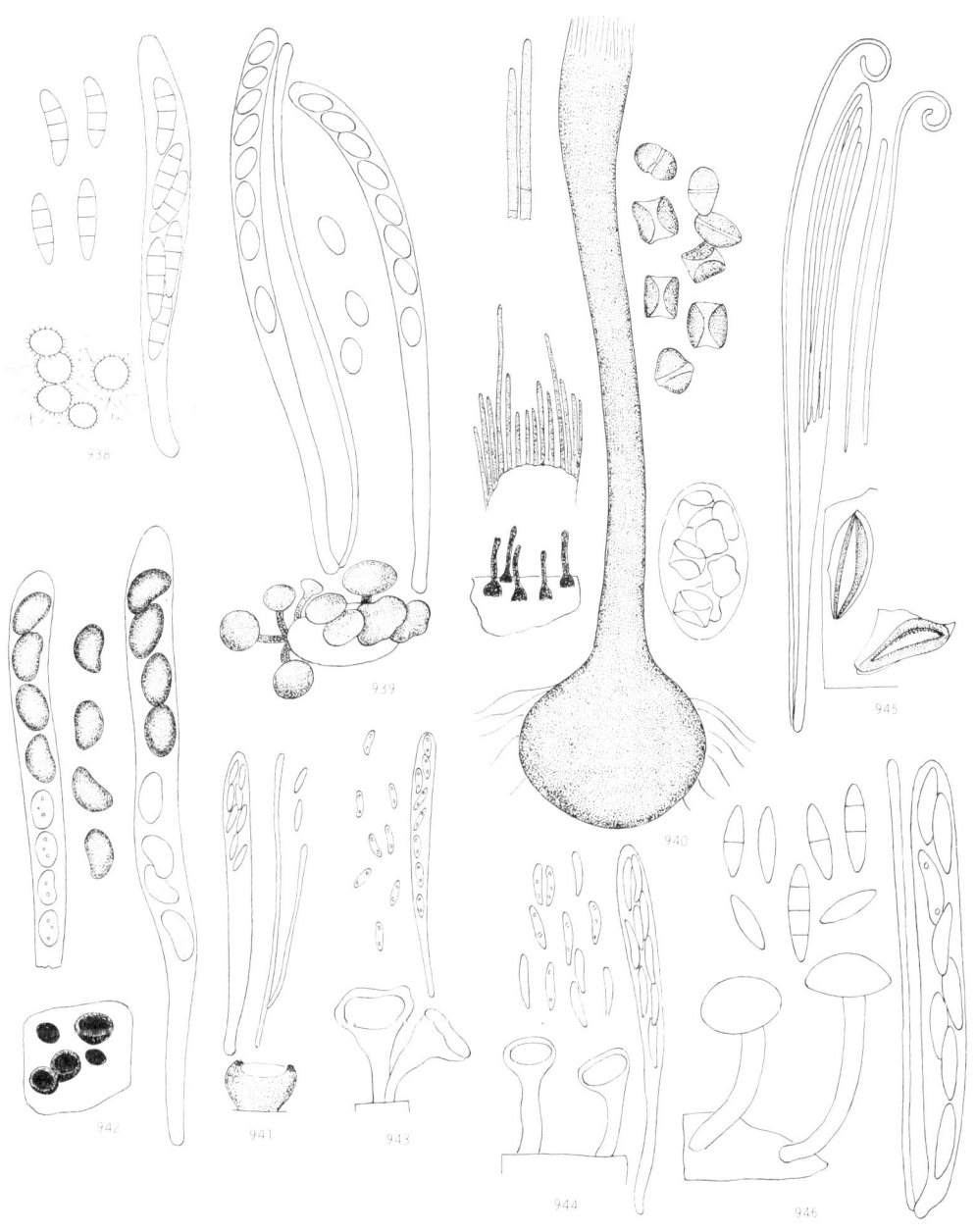

Plate 90. On *Quercus* fruit, and wood and bark. 938, *Arachnopeziza aurelia*; 939, *Ciboria batschiana*; 940, *Scopinella caulincola*; 941, *Bisporella fuscocincta*; 942, *Bulgaria inquinans*; 943, *Chlorociboria aeruginascens*; 944, *C. aeruginosa*; 945, *Colpoma quercinum*; 946, *Cudoniella acicularis*.

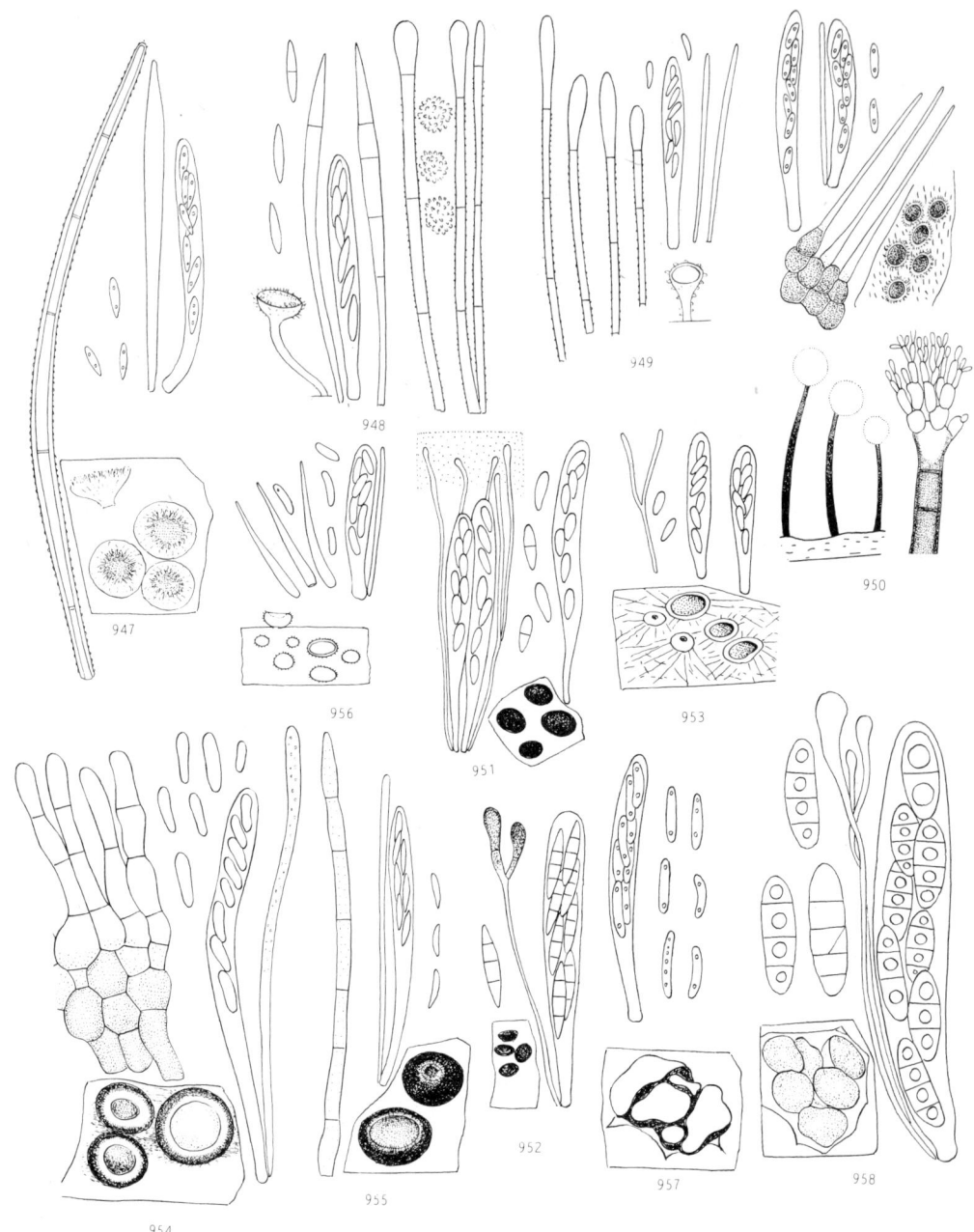

Plate 91. Discomycetes on *Quercus* wood and bark. 947, *Dasyscyphus bicolor*; 948, *D. crystallinus*; 949, *D. niveus*; 950, *Dematioscypha dematiicola*; 951, *Durella commutata*; 952, *D. macrospora*; 953, *Eriopeziza caesia*; 954, *Haglundia elegantior*; 955, *H. perelegans*; 956, *Hyaloscypha hyalina*; 957, *Mollisia discolor* var. *longispora*; 958, *Pezicula cinnamomea*.

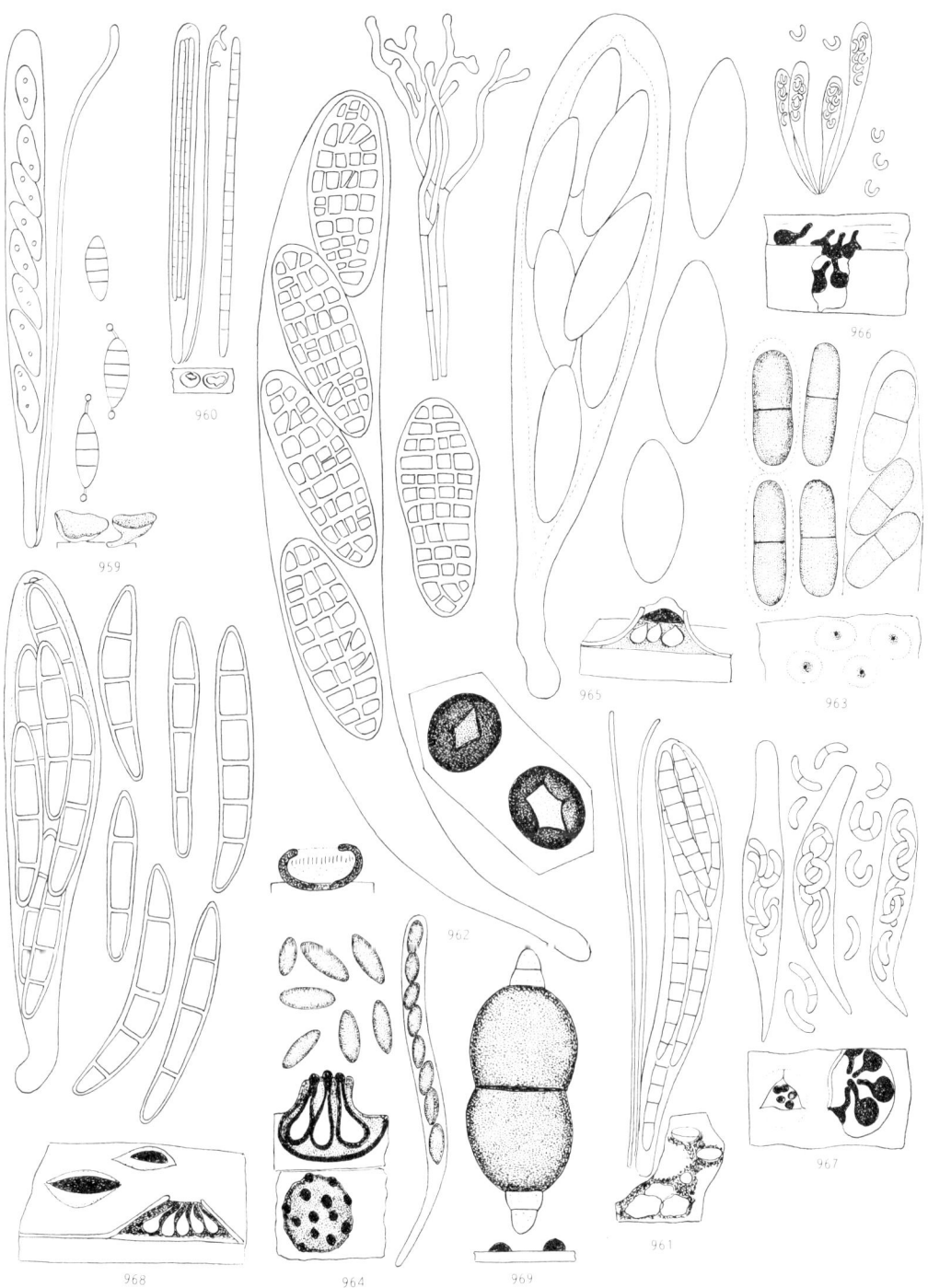

Plate 92. Discomycetes and other ascomycetes on *Quercus* wood and bark. 959, *Rutstroemia firma*; 960, *Schizoxylon friabilis*; 961, *Strossmayeria basitricha*; 962, *Triblidium caliciiforme*; 963, *Amphisphaeria bufonia*; 964, *Anthostoma dryophilum*; 965, *Botryosphaeria melanops*; 966, *Calosphaeria cyclospora*; 967, *C. dryina*; 968, *Calospora arausiaca*; 969, *Caryospora callicarpa*.

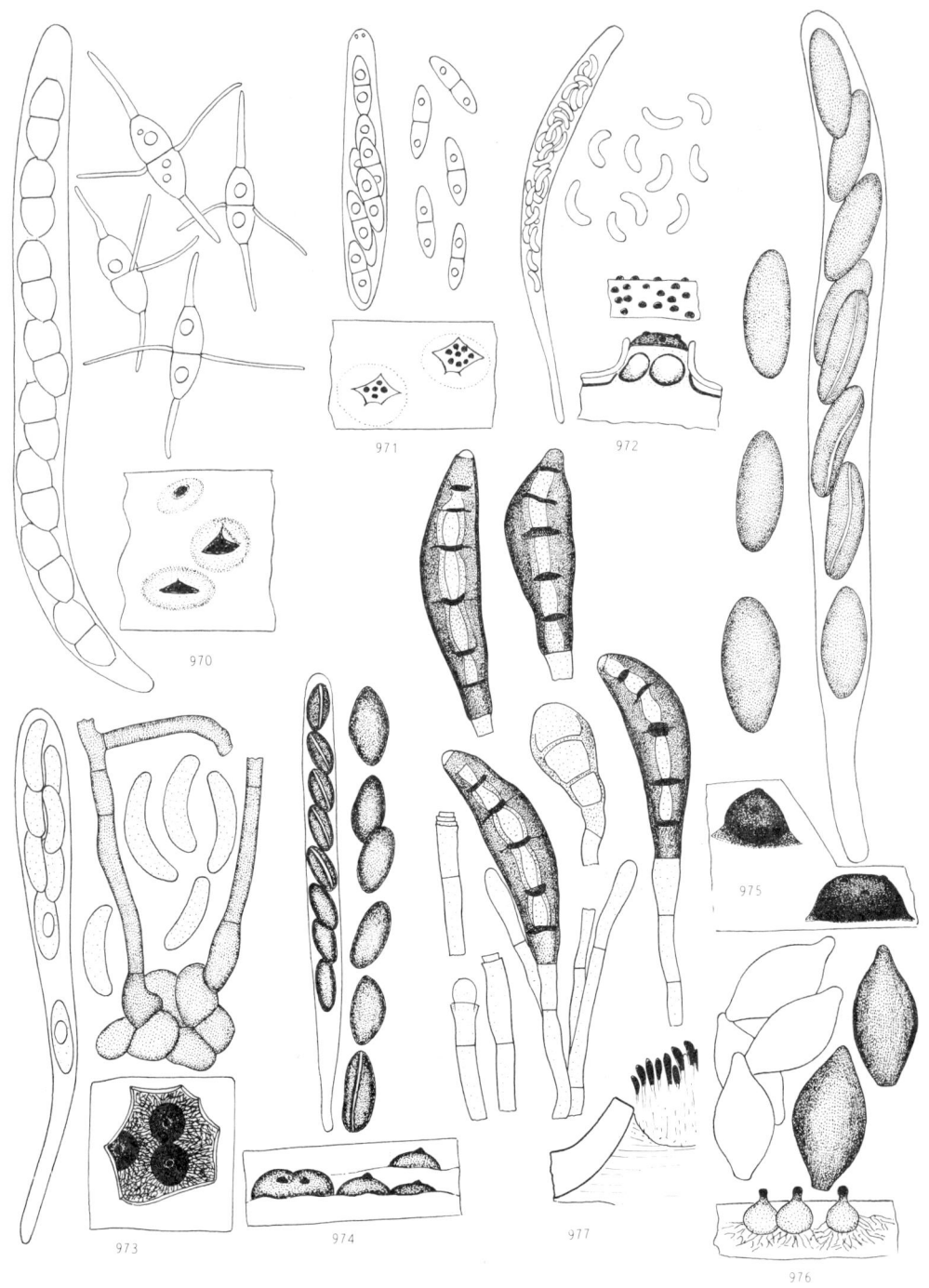

Plate 93. Ascomycetes on *Quercus* wood and bark. 970, *Caudospora taleola*; 971, *Diaporthe leiphemia*; 972, *Diatrypella quercina*; 973, *Enchnoa infernalis*; 974, *Hypoxylon confluens*; 975, *H. udum*; 976, *Persiciospora masonii*; 977, *Pseudovalsa longipes*.

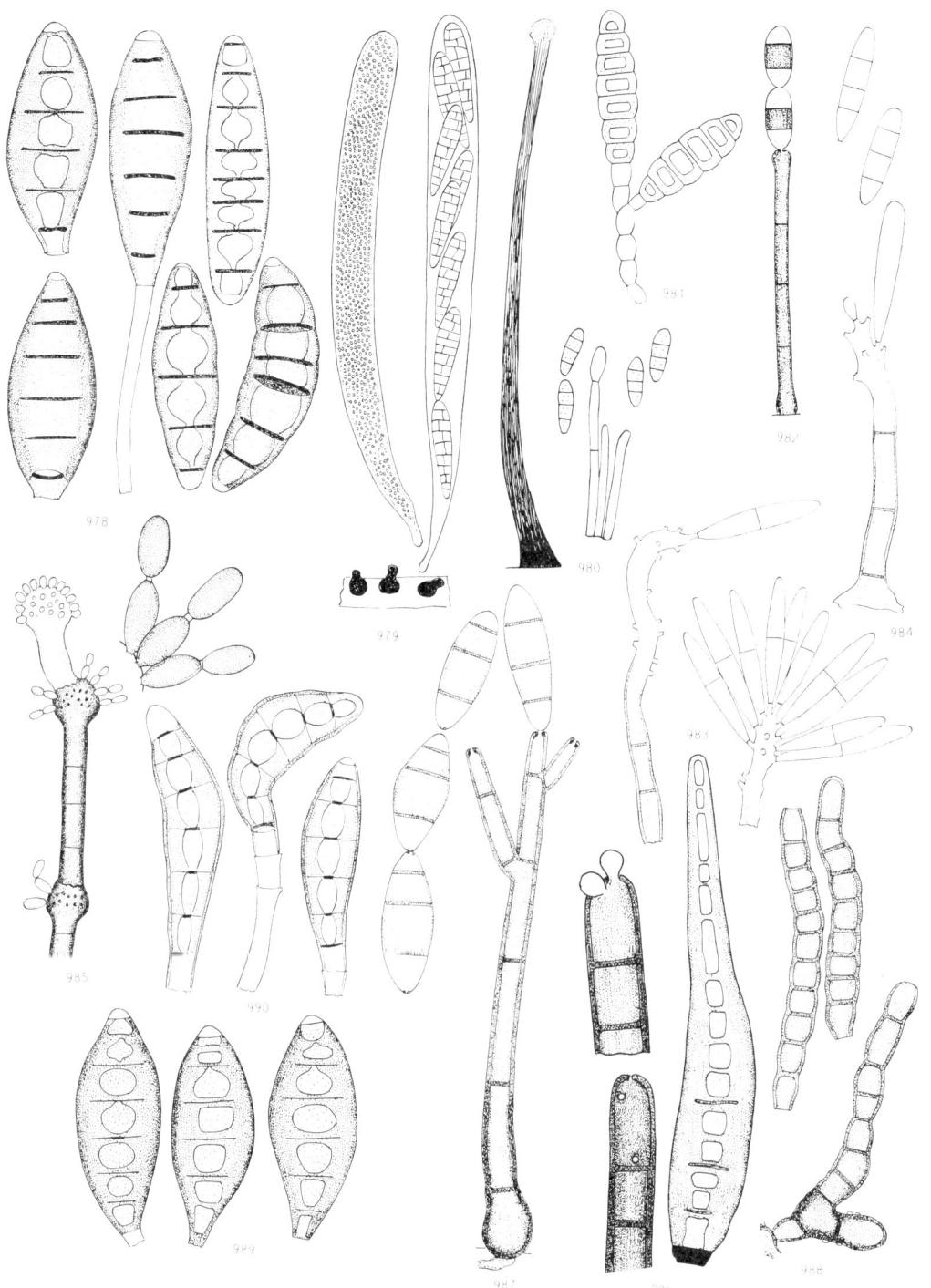

Plate 94. Various groups on *Quercus* wood and bark. 978, *Pseudovalsa umbonata*; 979, *Rhamphoria pyriformis*; 980, *Arthrobotryum stilboideum*; 981, *Bactrodesmium submoniliforme*; 982, *Corynesporopsis quercicola*; 983, *Dactylaria chrysosperma*; 984, *D. purpurella*; 985, *Gonatobotryum fuscum*; 986, *Helminthosporium microsorum*; 987, *Paradendryphiopsis cambrensis*; 988, *Taeniolella pulvillus*; 989, *Coryneum elevatum*; 990, *C. neesii*.

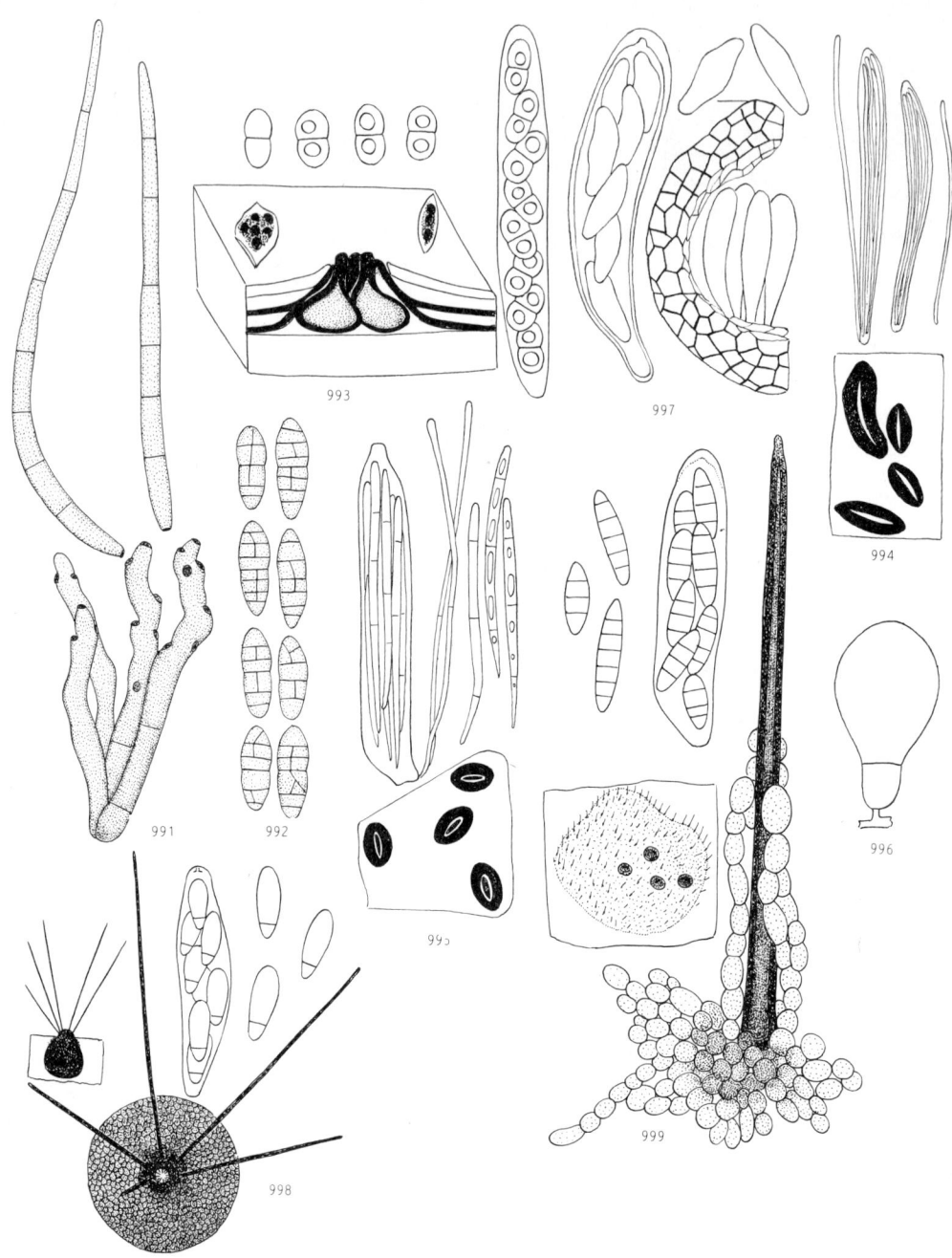

Plate 95. On *Rhamnus* and *Rhododendron*. 991, *Cercospora rhamni*; 992, *Cucurbitaria rhamni*; 993, *Diaporthe fibrosa*; 994, *Lophodermium vagulum*; 995, *Lophomerum ponticum*; 996, *Ovulinia azaleae*; 997, *Botryosphaeria rhodorae*; 998, *Chaetapiospora rhododendri*; 999, *Dennisiella babingtonii*.

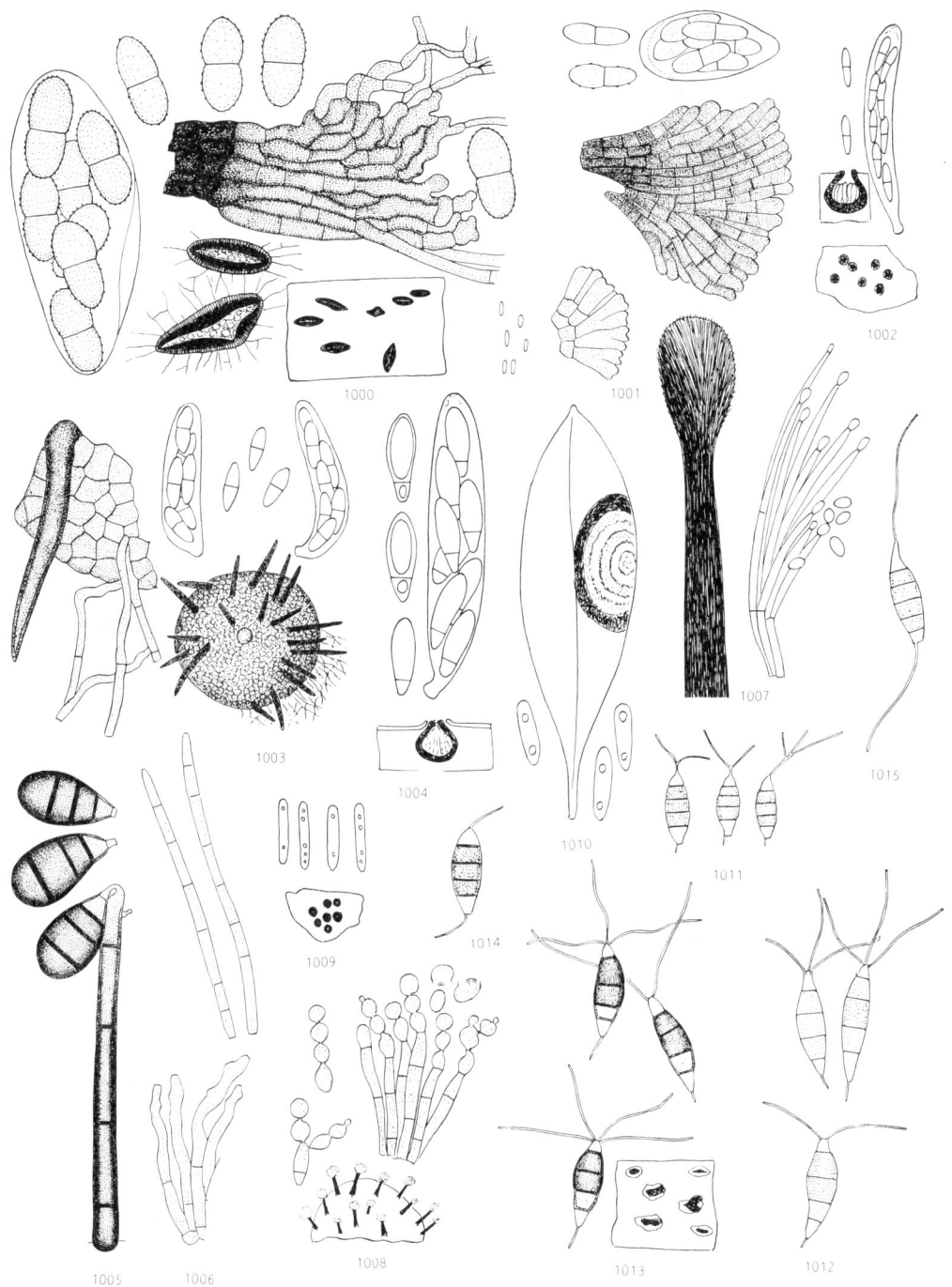

Plate 96. On *Rhododendron*. 1000, *Lembosina aulographoides*; 1001, *Morenoina rhododendri*; 1002, *Mycosphaerella rhododendri*; 1003, *Protoventuria arxii*; 1004, *Pseudomassaria thistletonia*; 1005, *Brachysporium dingleyae*; 1006, *Cercoseptoria handelii*; 1007, *Graphium smaragdinum*; 1008, *Pycnostysanus azaleae*; 1009, *Coleophoma empetri*; 1010, *Gloeosporium rhododendri*; 1011, *Monochaetia karstenii*; 1012, *Pestalotiopsis guepini*; 1013, *P. sydowiana*; 1014, *Seimatosporium arbuti*; 1015, *S. mariae*.

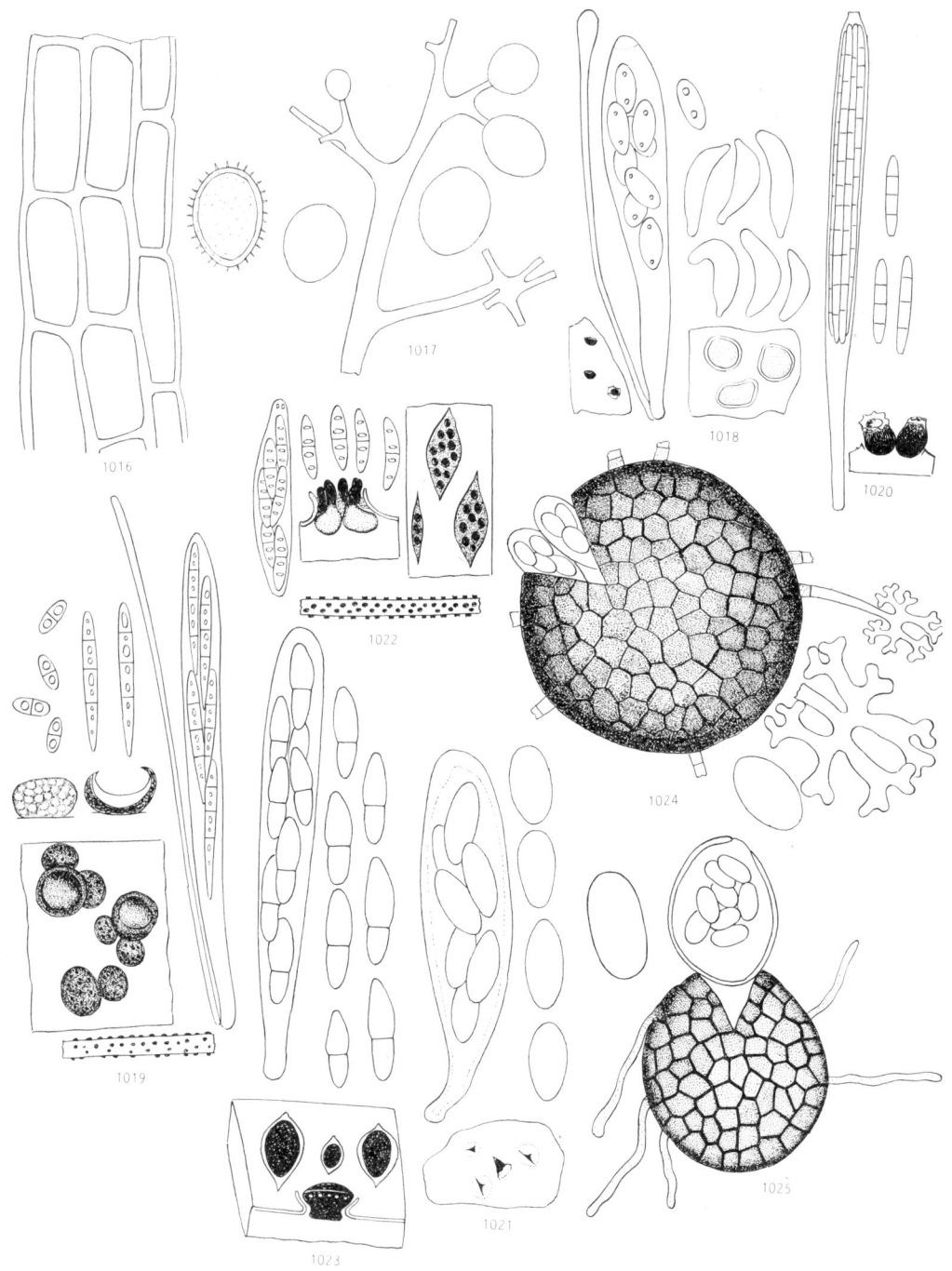

Plate 97. On *Ribes*. 1016, *Cronartium ribicola*; 1017, *Plasmopara ribicola*; 1018, *Drepanopeziza ribis*; 1019, *Godronia ribis*; 1020, *G. uberiformis*; 1021, *Botryosphaeria ribis*; 1022, *Diaporthe strumella*; 1023, *Dothiora ribesia*; 1024, *Microsphaera grossulariae*; 1025, *Sphaerotheca mors-uvae*.

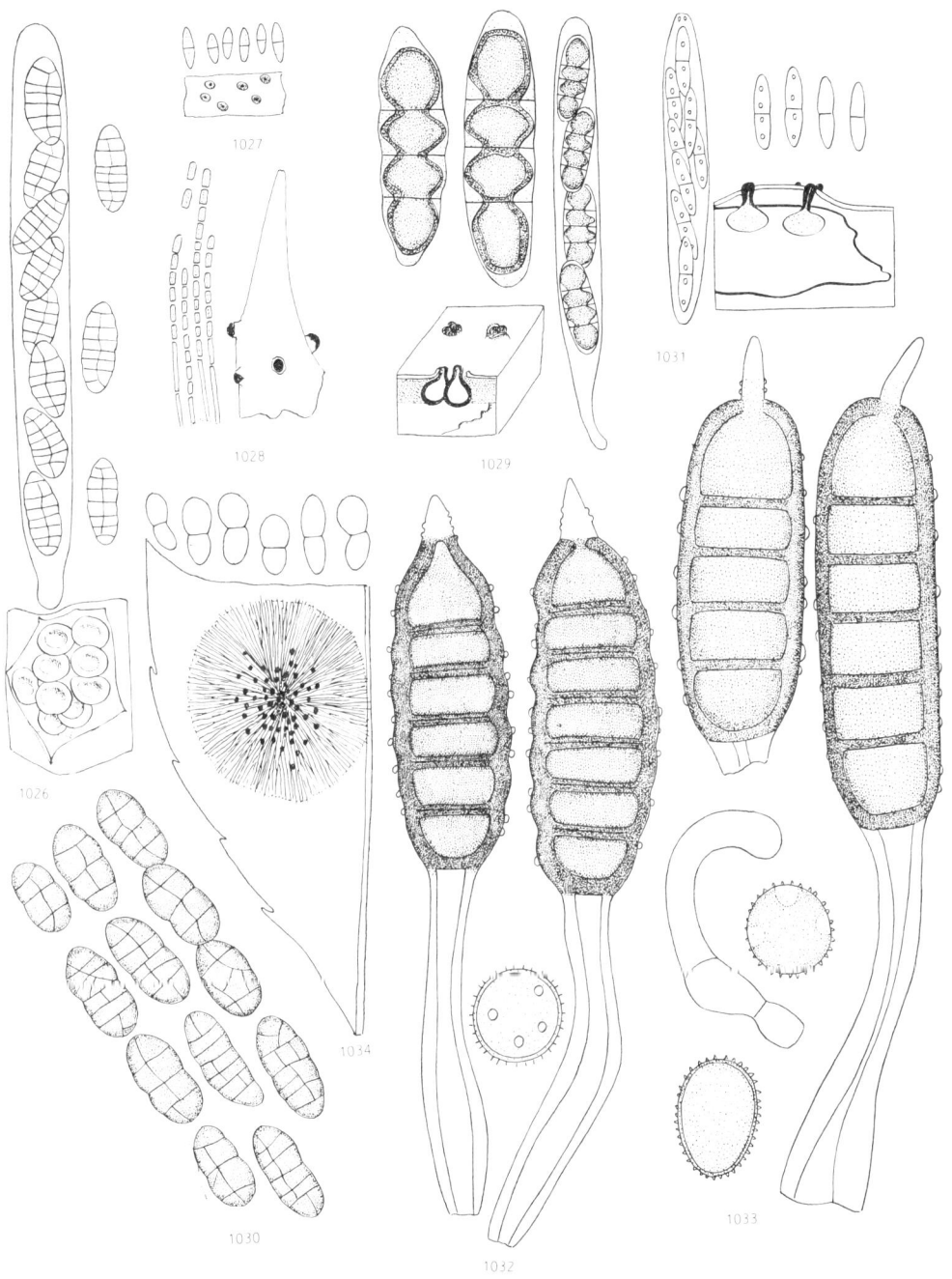

Plate 98. On *Ribes*, *Robinia* and *Rosa*. 1026, *Thyronectria berolinensis*; 1027, *Ascochytella grossulariae*; 1028, *Trullula melanochlora*; 1029, *Aglaospora profusa*; 1030, *Cucurbitaria elongata*; 1031, *Diaporthe oncostoma*; 1032, *Phragmidium mucronatum*; 1033, *P. tuberculatum*; 1034, *Diplocarpon rosae*.

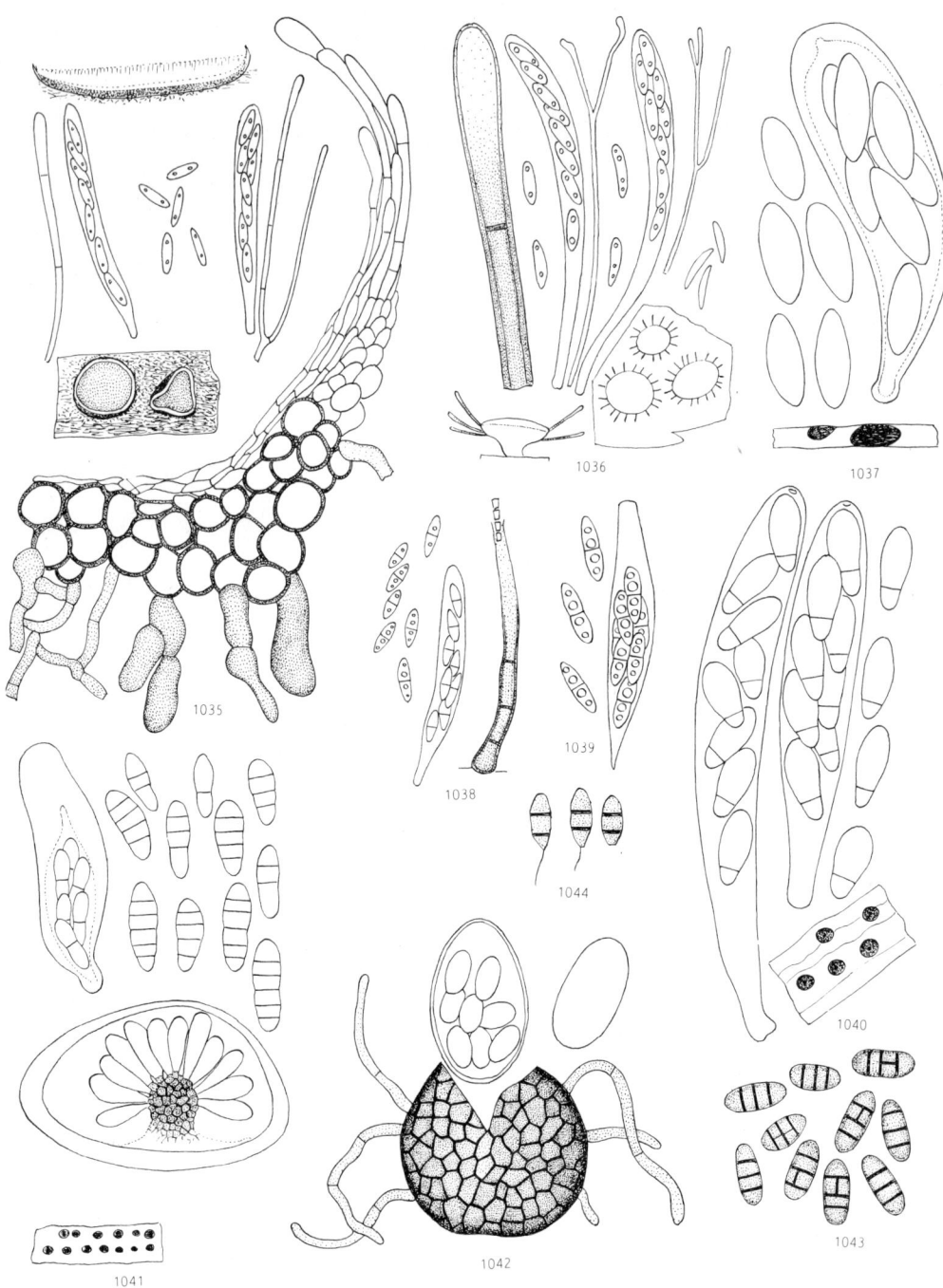

Plate 99. On *Rosa*. 1035, *Tapesia rosae*; 1036, *Zoellneria rosarum*; 1037, *Botryosphaeria dothidea*; 1038, *Chaetosphaeria bramleyi*; 1039, *Diaporthe incarcerata*; 1040, *Pseudomassaria sepincolaeformis*; 1041, *Saccothecium sepincola*; 1042, *Sphaerotheca pannosa*; 1043, *Camarosporium rosae*; 1044, *Seimatosporium rosarum*.

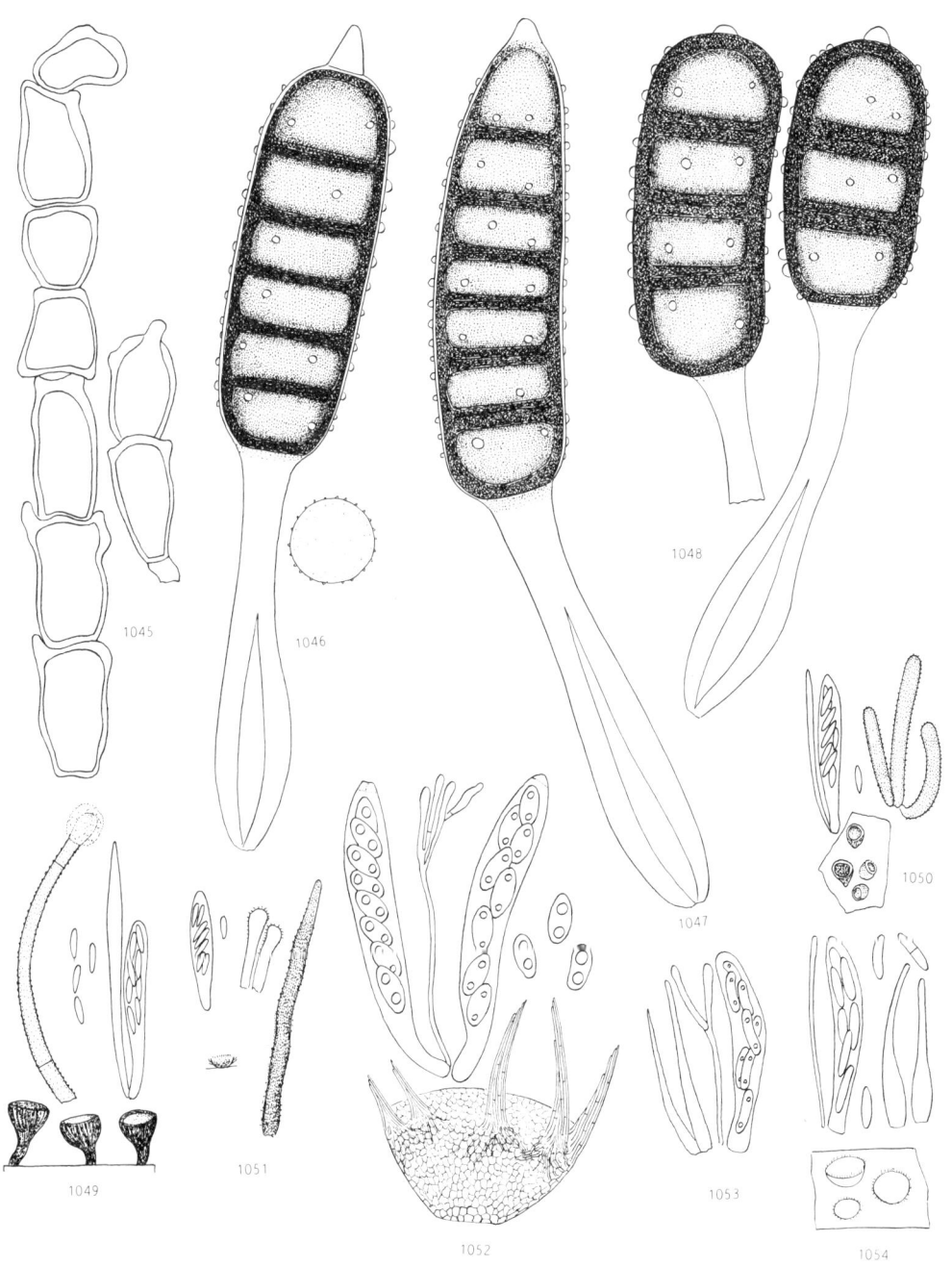

Plate 100. Rusts and discomycetes on *Rubus*. 1045, *Kuehneola uredinis*; 1046, *Phragmidium bulbosum*; 1047, *P. rubi-idaei*; 1048, *P. violaceum*; 1049, *Dasyscyphus clandestinus*; 1050, *D. dumorum*; 1051, *D. misellus*; 1052, *Echinula asteriadiformis*; 1053, *Hyaloscypha hyalina* f.; 1, 1054, *H. lectissima*.

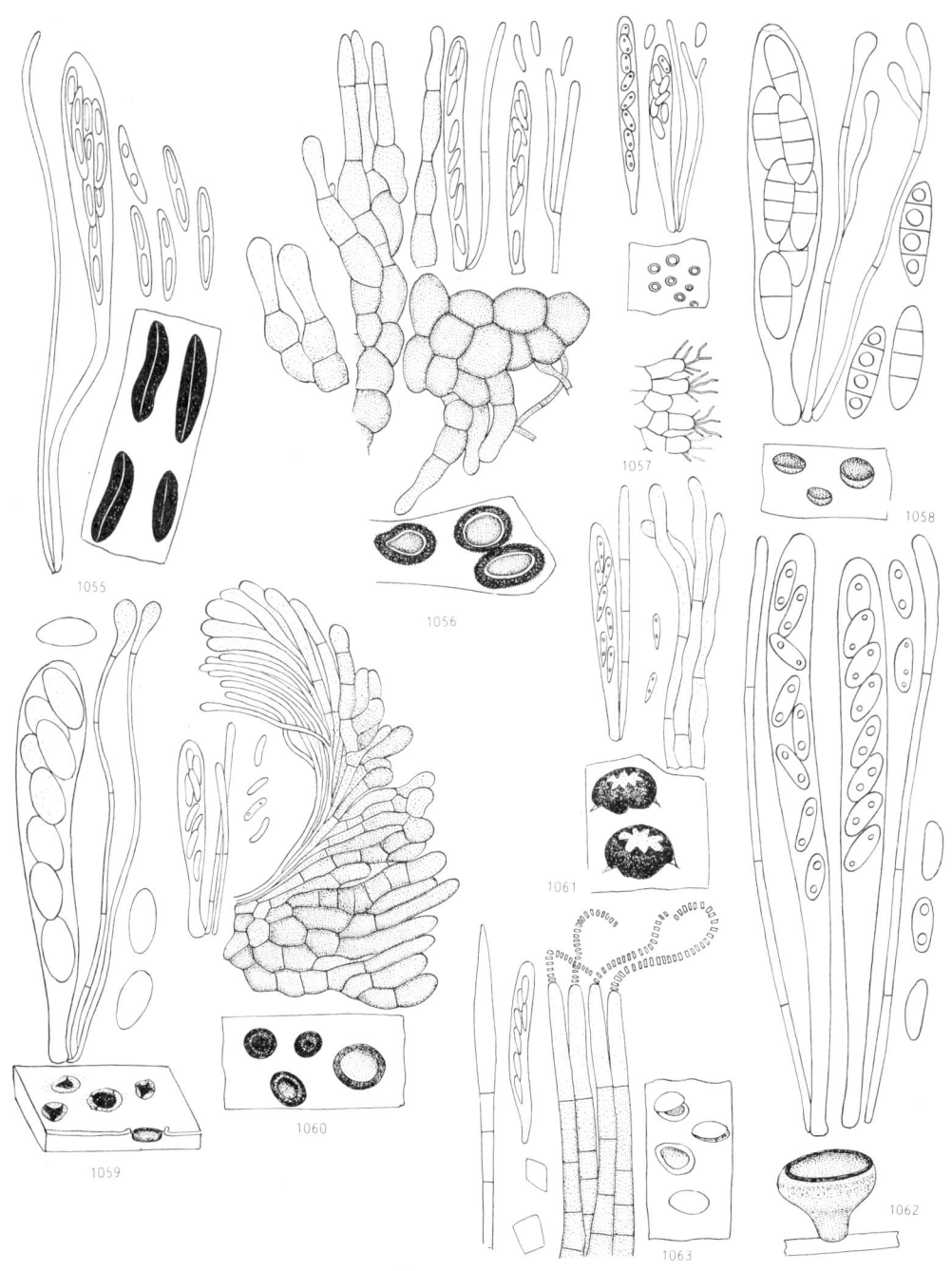

Plate 101. Discomycetes on *Rubus*. 1055, *Hypoderma rubi*; 1056, *Mollisia clavata*; 1057, *Mollisina rubi*; 1058, *Pezicula rubi*; 1059, *Ploettnera exigua*; 1060, *Pyrenopeziza escharodes*; 1061, *P. rubi*; 1062, *Rutstroemia fruticeti*; 1063, *Stegopeziza dumeti*.

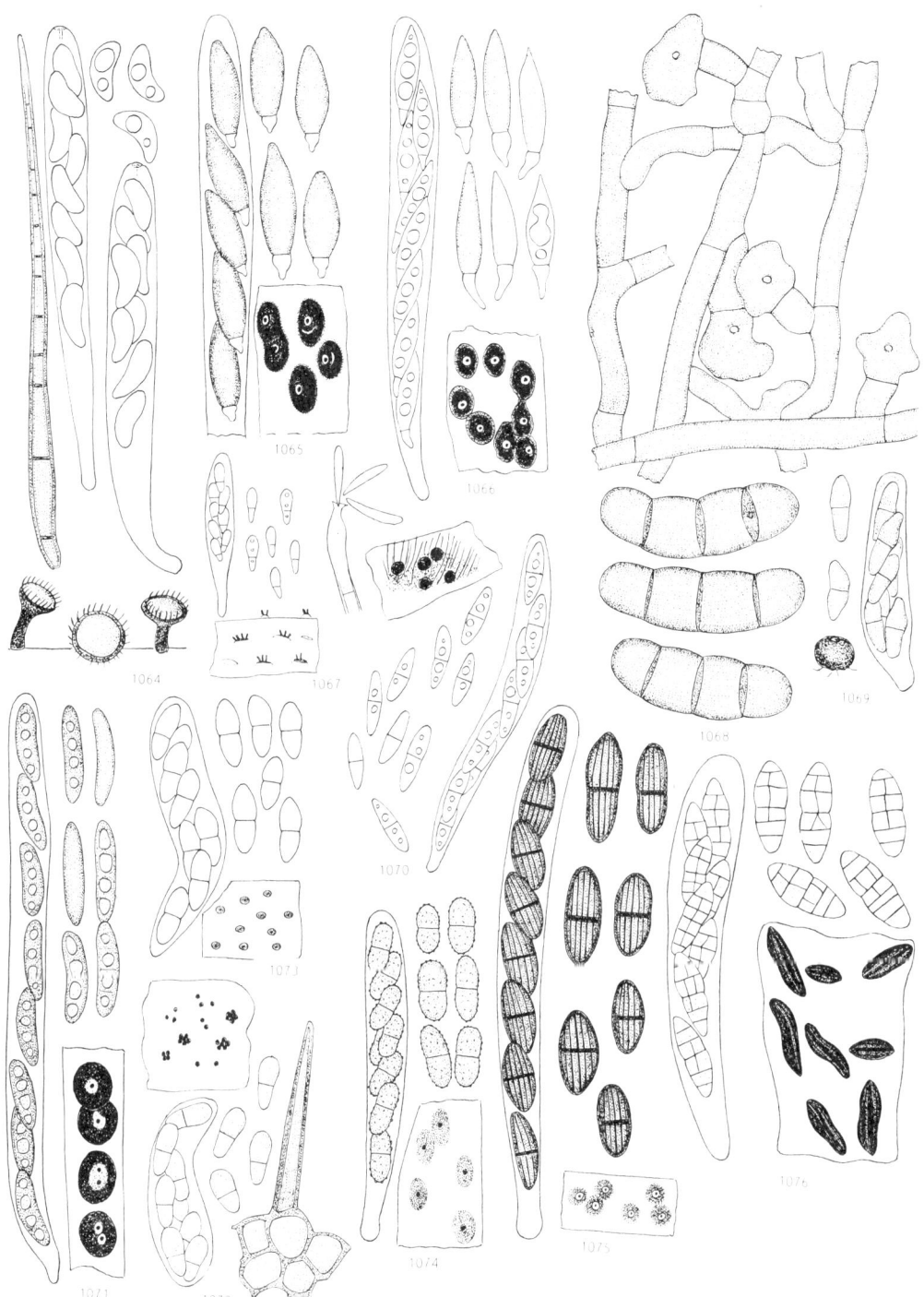

Plate 102. Discomycete and other ascomycetes on *Rubus*. 1064, *Torrendiella ciliata*; 1065, *Anthostomella appendiculosa*; 1066, *A. rubicola*; 1067, *Apioporthe vepris*; 1068, *Appendiculella calostroma*; 1069, *Dimerium meliolicola*; 1070, *Chaetosphaeria callimorpha*; 1071, *Clypeosphaeria notarisii*; 1072, *Coleroa chaetomium*; 1073, *Didymella applanata*; 1074, *Didymosphaeria oblitescens*; 1075, *D. rubicola*; 1076, *Gloniopsis praelonga*.

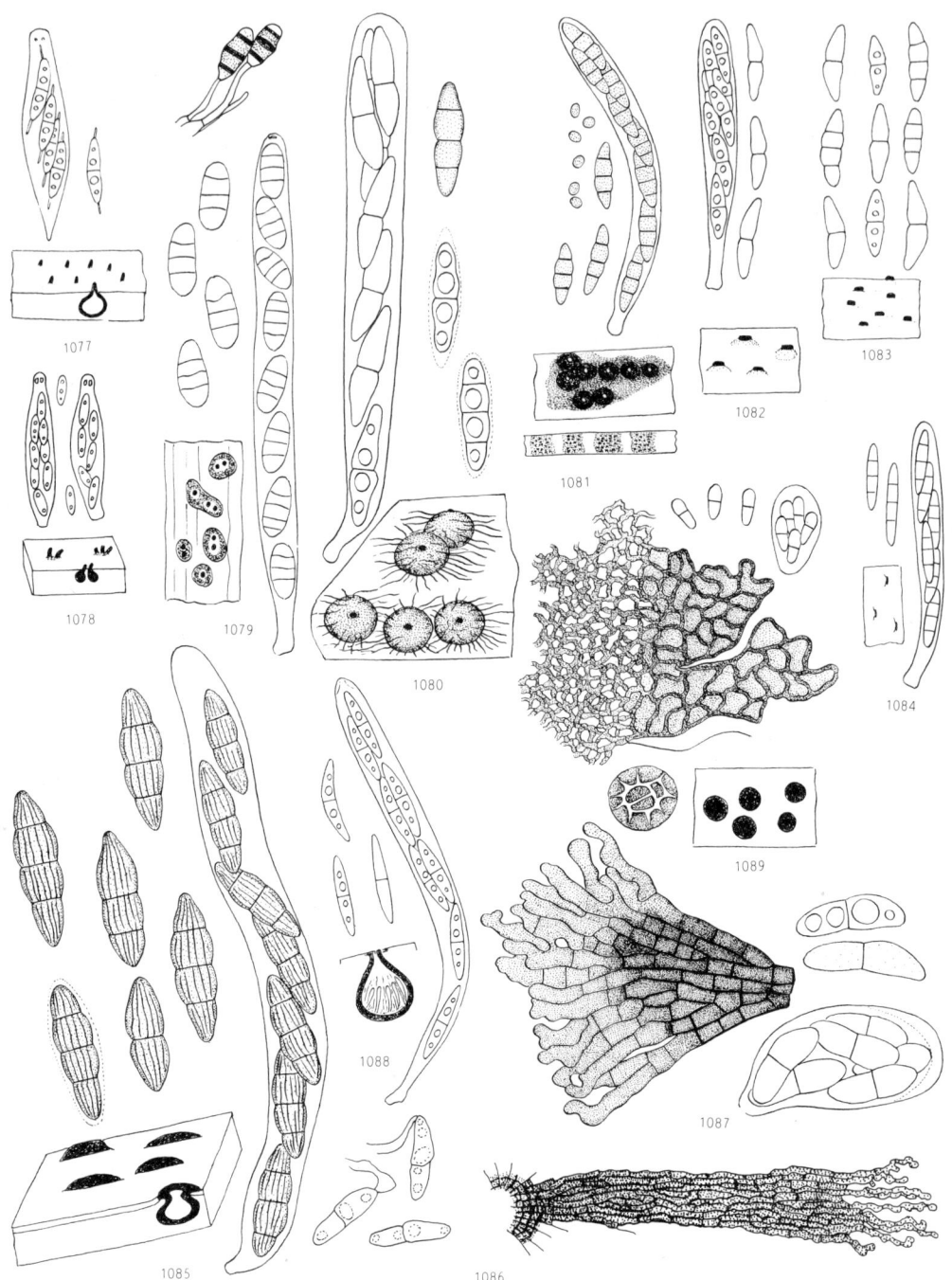

Plate 103. Ascomycetes on *Rubus*. 1077, *Gnomonia rubi*; 1078, *Gnomoniella rubicola*; 1079, *Griphosphaeria corticola*; 1080, *Herpotrichia herpotrichoides*; 1081, *Leptosphaeria coniothyrium*; 1082, *Lophiostoma fuckelii*; 1083, *L. fuckelii* var. *pulveraceum*; 1084, *L. hysterioides*; 1085, *L. viridarium*; 1086, *Microthyrium versicolor*; 1087, *Morenoina clarkii*; 1088, *Paradidymella clarkii*; 1089, *Schizothyrium speireum*.

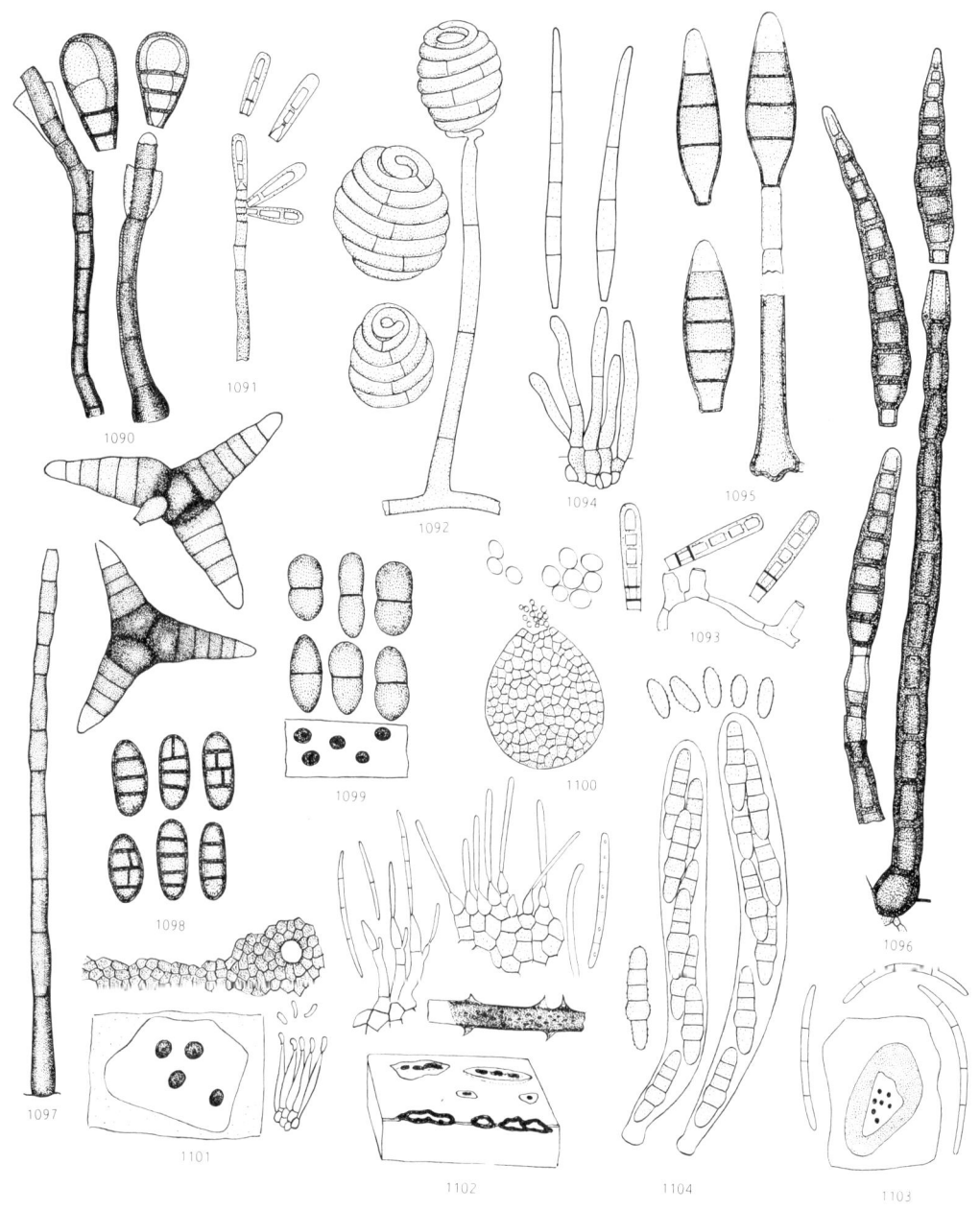

Plate 104. On *Rubus* and *Ruscus*. 1090, *Endophragmia boothii*; 1091, *E. parva*; 1092, *Helicoon fuscosporum*; 1093, *Henicospora minor*; 1094, *Pseudocercospora rubi*; 1095, *Sporidesmium clarkii*; 1096, *S. rubi*; 1097, *Triposporium elegans*; 1098, *Camarosporium rubicolum*; 1099, *Diplodia rubi*; 1100, *Hapalosphaeria deformans*; 1101, *Leptothyrina rubi*; 1102, *Septocyta ruborum*; 1103, *Septoria rubi*; 1104, *Paraphaeosphaeria rusci*.

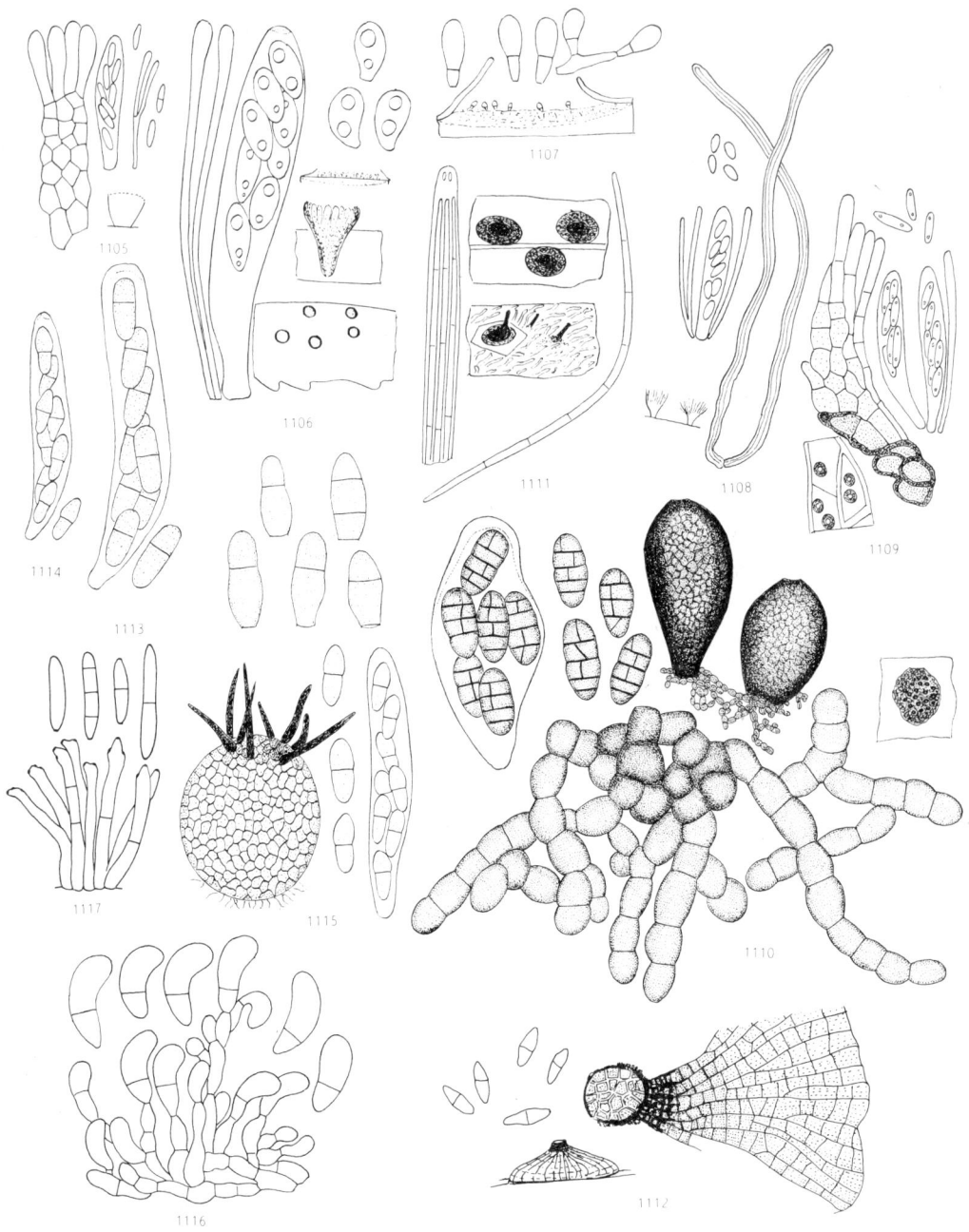

Plate 105. On *Salix* leaves. 1105, *Calycellina indumenticola*; 1106, *Drepanopeziza salicis*; 1107, *D. sphaerioides*; 1108, *Hyalotricha salicicola*; 1109, *Pyrenopeziza fuckelii*; 1110, *Capnodium salicinum*; 1111, *Isothea saligna*; 1112, *Trichothyrina salicis*; 1113, *Venturia chlorospora*; 1114, *V. minuta*; 1115, *V. saliciperda*; 1116, *Marssonina* state of *Drepanopeziza triandrae*; 1117, *Ramularia rosea*.

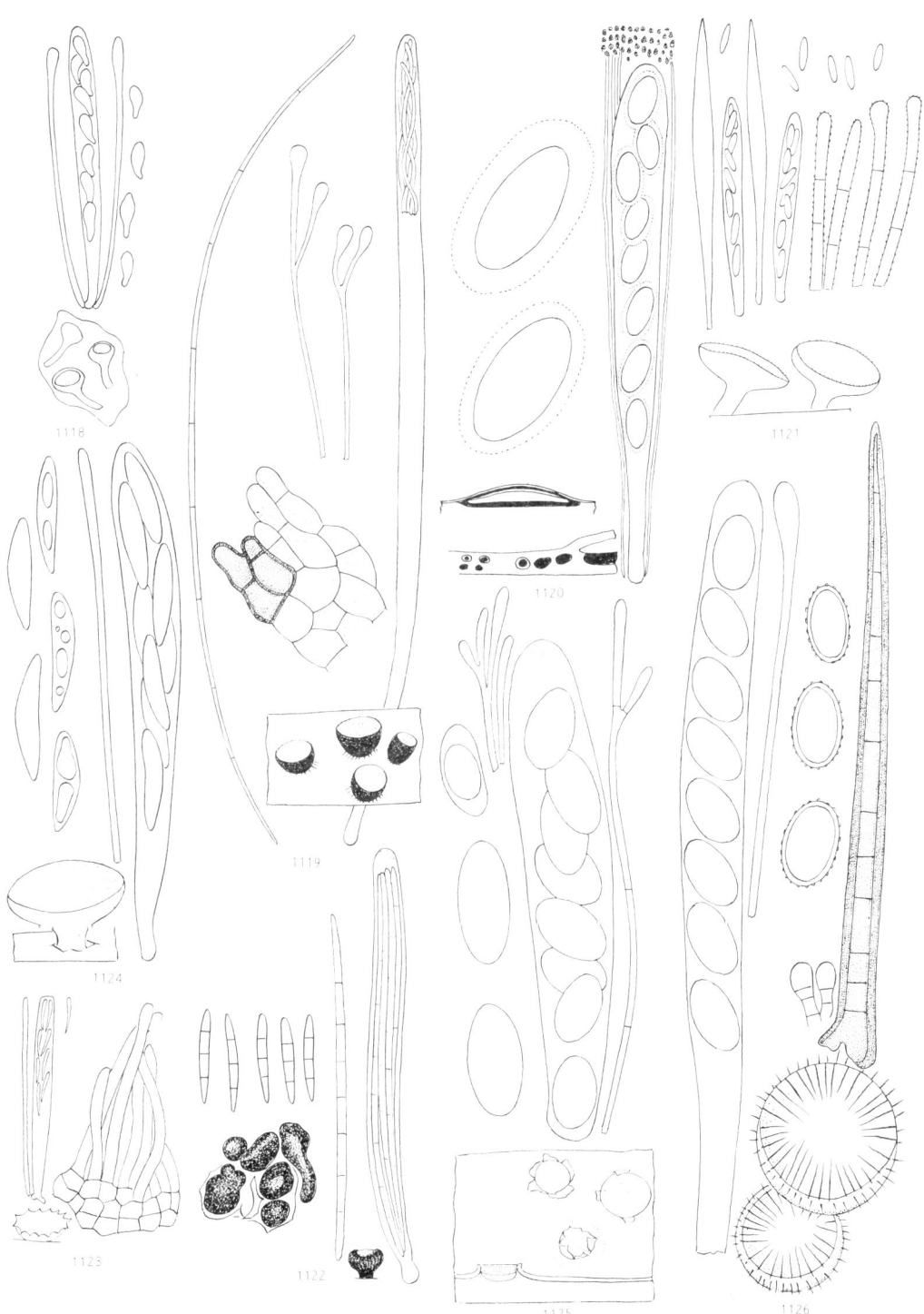

Plate 106. Discomycetes on *Salix* catkins and wood and bark. 1118, *Crocicreas amenti*; 1119, *Apostemidium fiscellum*; 1120, *Cryptomyces maximus*; 1121, *Dasyscyphus pudibundus*; 1122, *Godronia fuliginosa*; 1123, *Hyalinia rubella*; 1124, *Hymenoscyphus salicellus*; 1125, *Ocellaria ocellata*; 1126, *Scutellinia scutellata*.

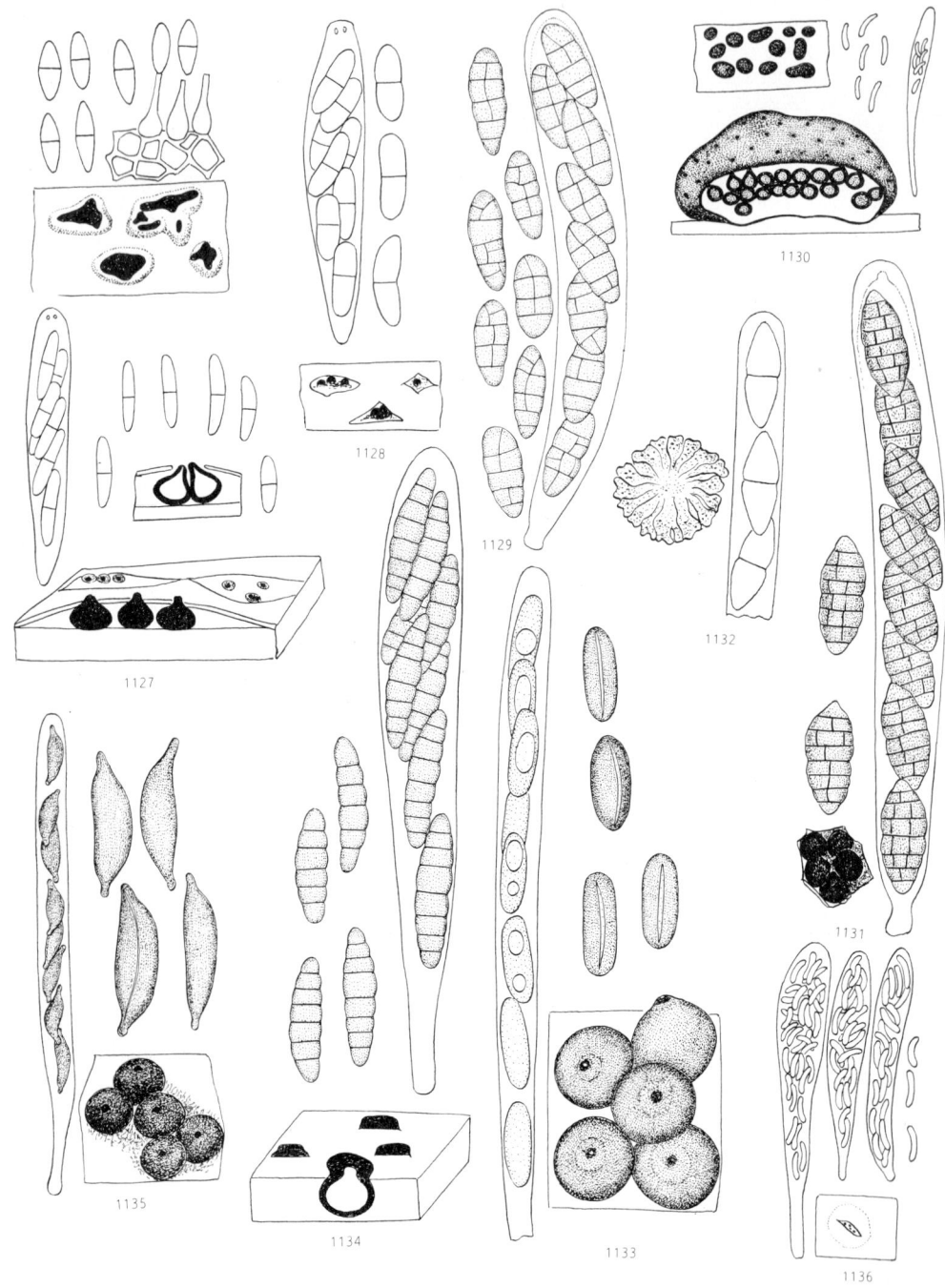

Plate 107. Ascomycetes on *Salix* wood and bark. 1127, *Cryptodiaporthe salicella*; 1128, *C. salicina*; 1129, *Cucurbitaria rubefaciens*; 1130, *Diatrype bullata*; 1131, *Fenestella salicis*; 1132, *Hypocreopsis lichenoides*; 1133, *Hypoxylon mammatum*; 1134, *Lophiostoma macrostomoides*; 1135, *Rosellinia desmazieresii*; 1136, *Valsella salicis*.

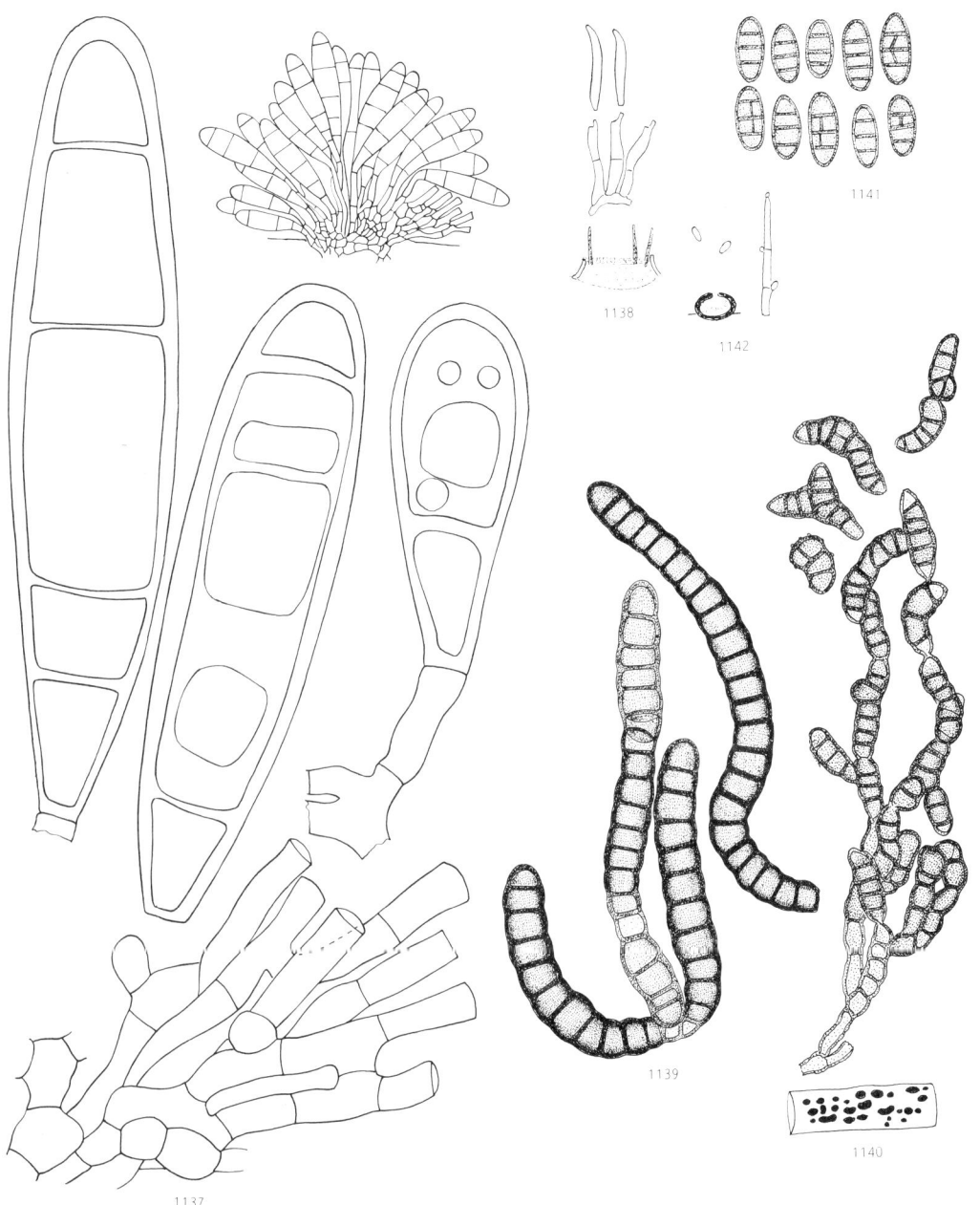

Plate 108. Deuteromycetes on *Salix* wood and bark. 1137, *Bactridium flavum*; 1138, *Oramasia hirsuta*; 1139, *Taeniolella stilbospora*; 1140, *Trimmatostroma salicis*; 1141, *Camarosporium salicinum*; 1142, *Pleurophoma pleurospora*.

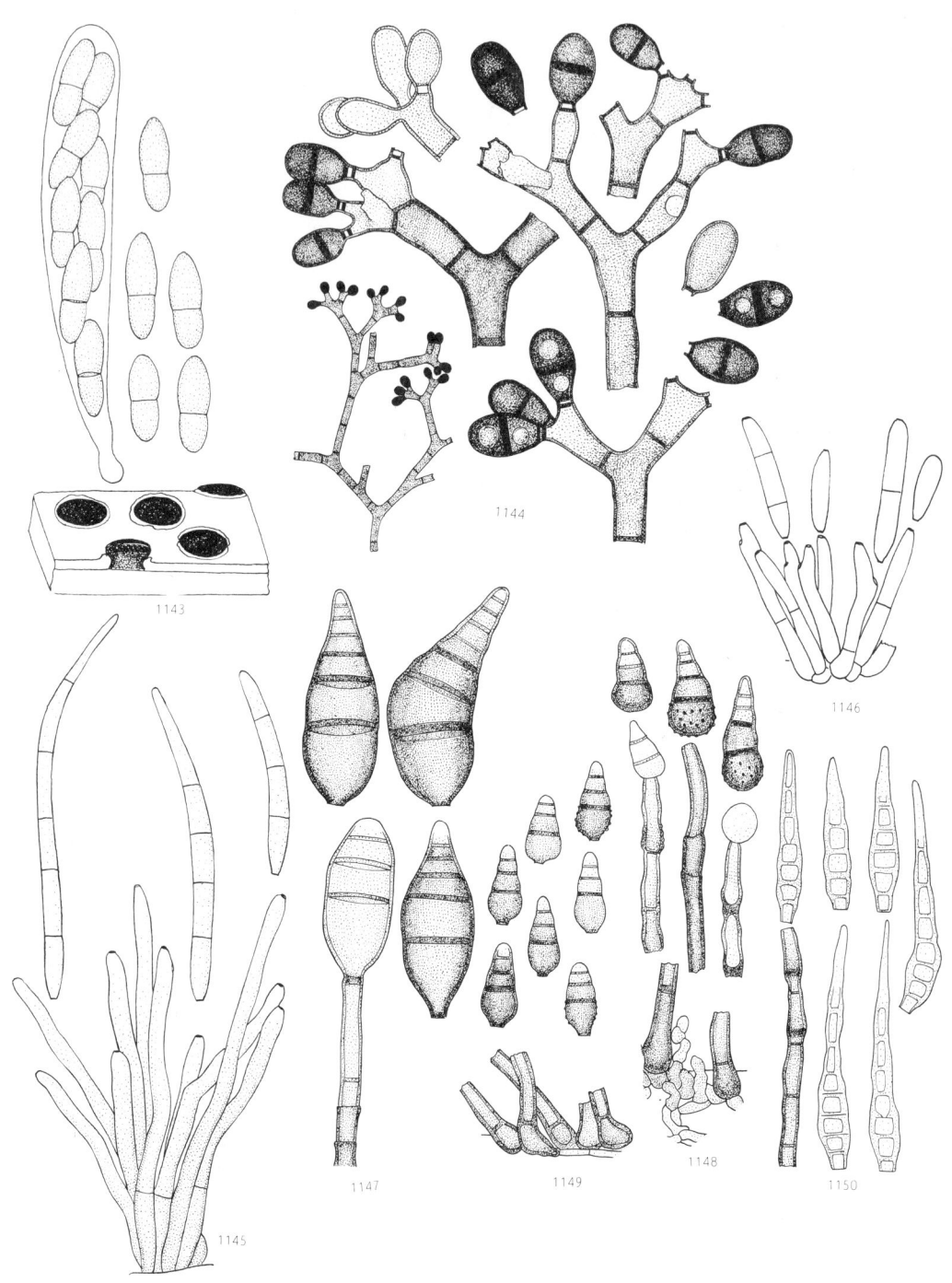

Plate 109. On *Sambucus*. 1143, *Dothidea sambuci*; 1144, *Balanium stygium*; 1145, *Cercospora depazeoides*; 1146, *Ramularia sambucina*; 1147, *Sporidesmium altum*; 1148, *S. aturbinatum*; 1149, *S. cookei*; 1150, *S. leptosporum*.

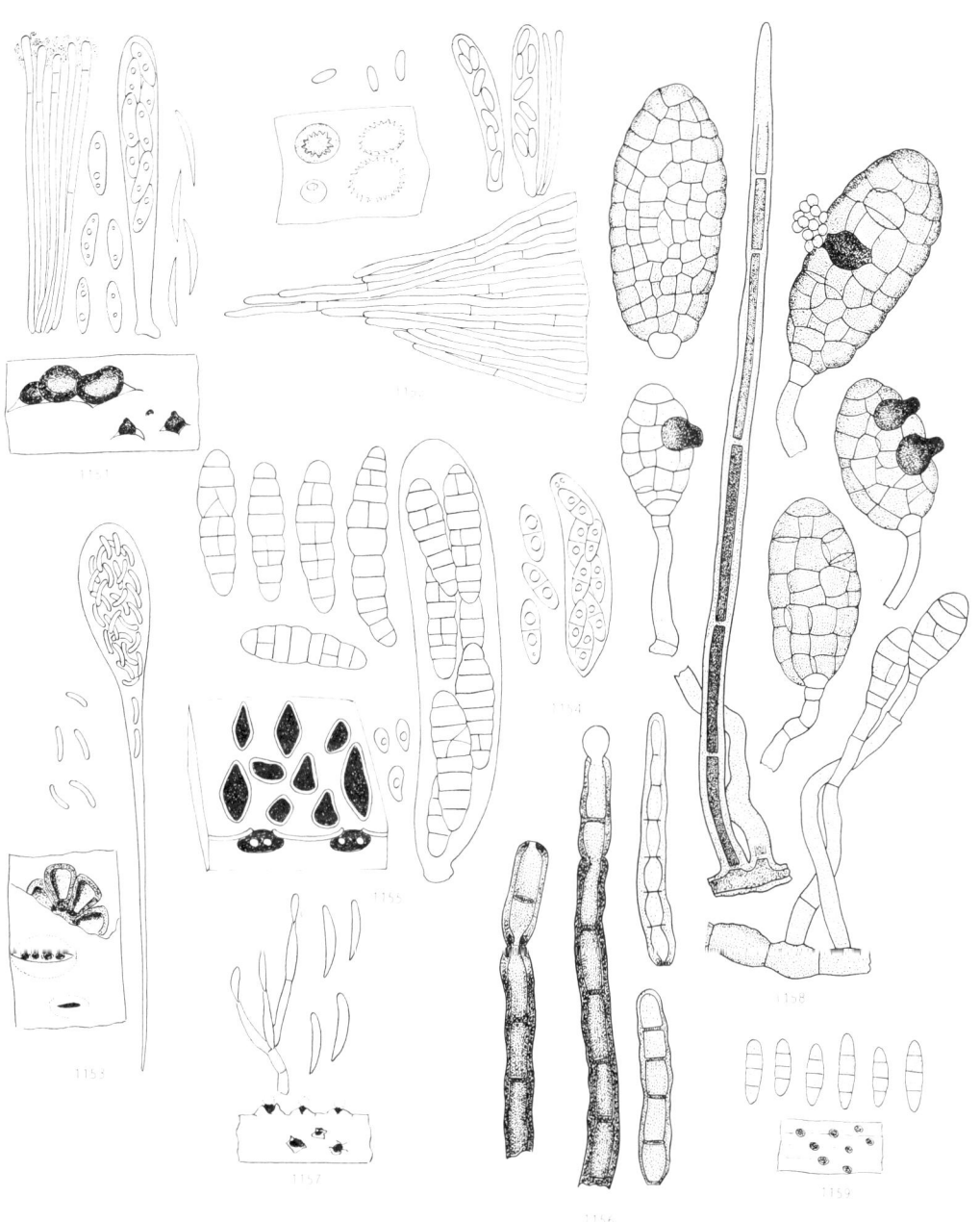

Plate 110. On *Sorbus* and *Symphoricarpos*. 1151, *Dermea ariae*; 1152, *Pezizellaster serrata*; 1153, *Coronophora gregaria*; 1154, *Diaporthe impulsa*; 1155, *Dothiora pyrenophora*; 1156, *Corynespora cambrensis*; 1157, *Rhabdospora inaequalis*; 1158, *Septosporium bulbotrichum*; 1159, *Hendersonia fiedleri* var. *symphoricarpi*.

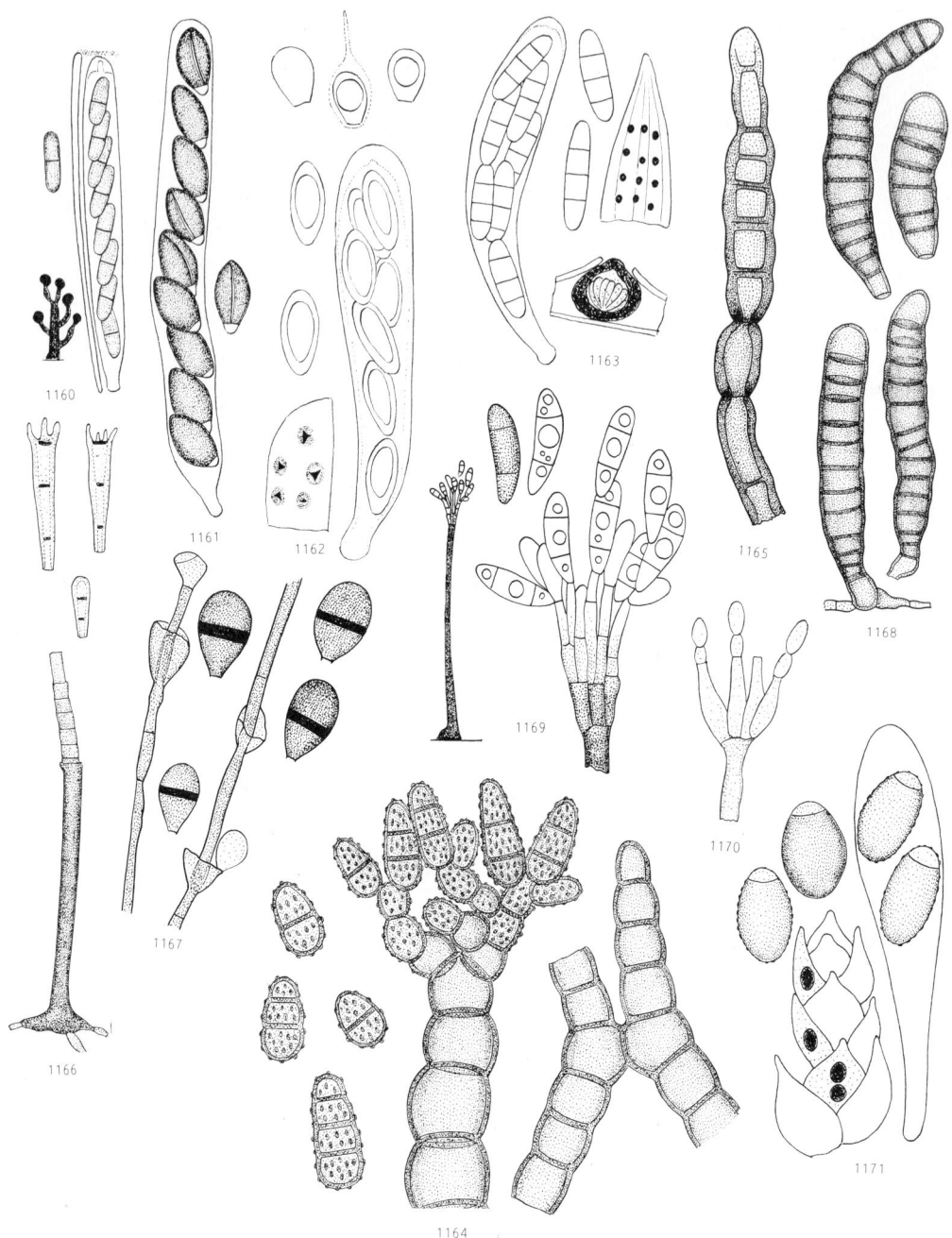

Plate 111. On *Taxus* and *Thuja*. 1160, *Chaenothecopsis caespitosa*; 1161, *Anthostomella formosa* var. *taxi*; 1162, *Botryosphaeria foliorum*; 1163, *Dothiora taxicola*; 1164, *Capnobotrys dingleyae*; 1165, *Corynespora pruni*; 1166, *Endophragmia coronata*; 1167, *E. taxi*; 1168, *Sporidesmium larvatum*; 1169, *Sterigmatobotrys macrocarpa*; 1170, *Thysanophora taxi*; 1171, *Didymascella thujina*.

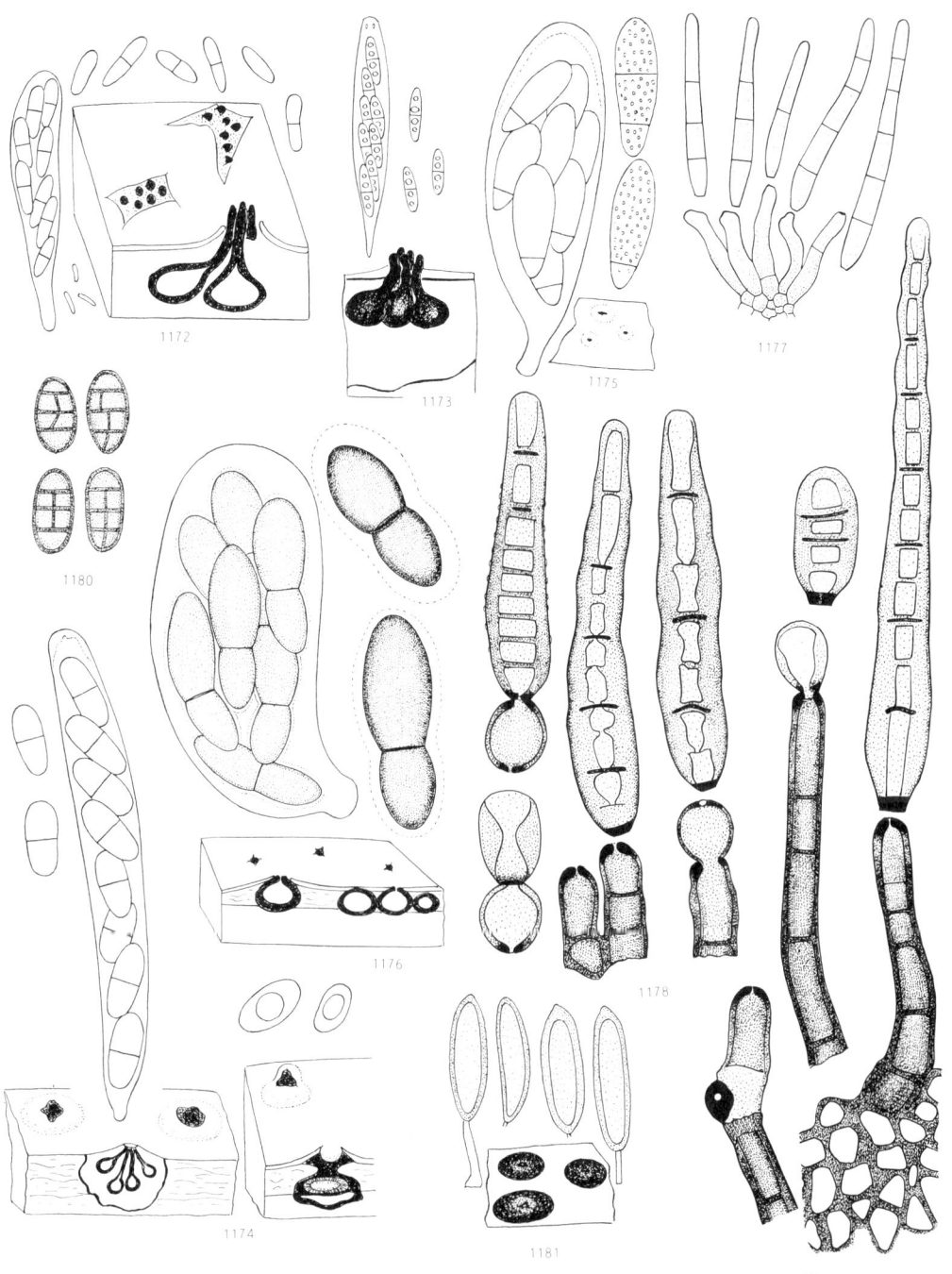

Plate 112. On *Tilia*. 1172, *Cryptodiaporthe hranicensis*; 1173, *Diaporthe velata*; 1174, *Hercospora tiliae*; 1175, *Pseudomassaria chondrospora*; 1176, *Splanchnonema ampullaceum*; 1177, *Cercospora microsora*; 1178, *Corynespora olivacea*; 1179, *Exosporium tiliae*; 1180, *Camarosporium tiliae*; 1181, *Lamproconium desmazieresii*.

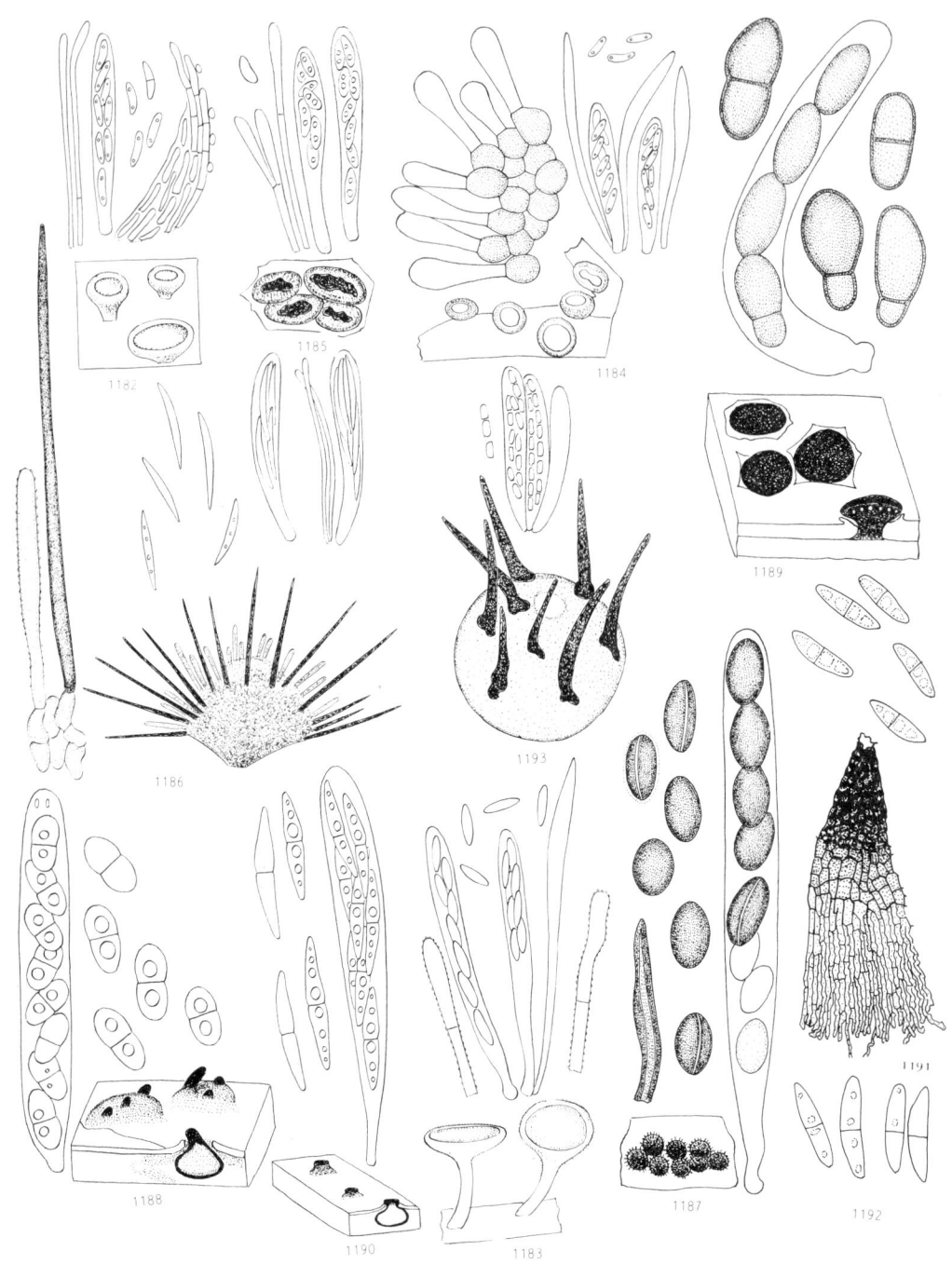

Plate 113. On *Ulex*. 1182, *Crocicreas complicatum*; 1183, *Dasyscyphus pygmaeus*; 1184, *Mollisiopsis dennisii*; 1185, *Phaeangella ulicis*; 1186, *Trichodiscus ulicicola*; 1187, *Coniochaeta ligniaria*; 1188, *Diaporthe inaequalis*; 1189, *Dothidea puccinioides*; 1190, *Lophiostoma angustilabrum*; 1191, *Microthyrium cytisi* var. *ulicis*; 1192, *M. cytisi* var. *ulicisgallii*; 1193, *Trichosphaerella decipiens*.

Plate 114. On *Ulex* and *Ulmus*. 1194, *Diplocladiella scalaroides*; 1195, *Hansfordia pulvinata*; 1196, *Phialocephala fusca*; 1197, *Sporidesmium cambrense*; 1198, *Dasyscyphus deflexus*; 1199, *Platychora ulmi*; 1200, *Encoelia siparia*; 1201, *Habrostictis rubra*; 1202, *Orbilia comma*; 1203, *Amphisphaeria umbrina*; 1204, *Anthostoma melanotes*; 1205, *Ceratocystis ulmi*; 1206, *Cryptosporella hypodermia*; 1207, *Eutypella stellulata*.

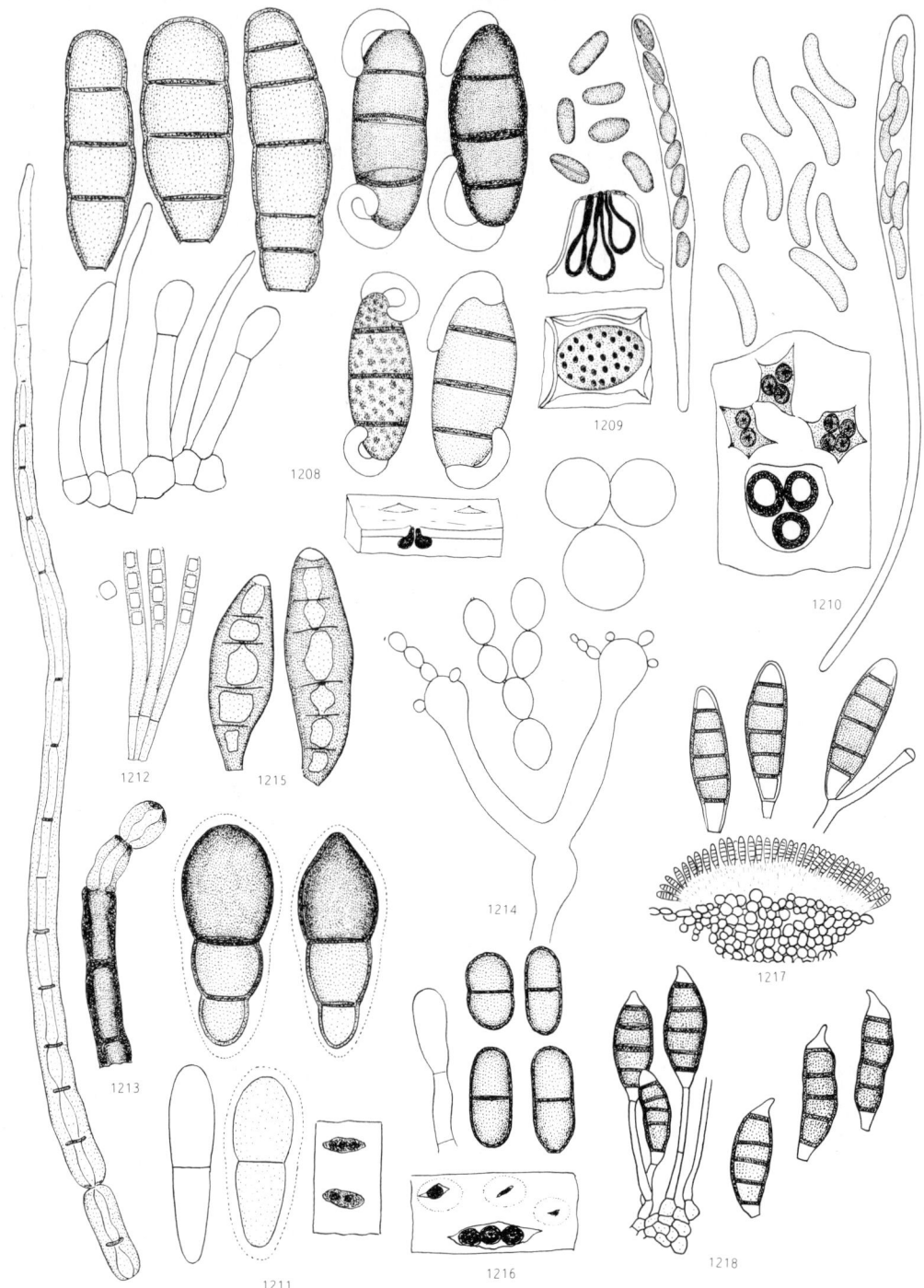

Plate 115. On *Ulmus*. 1208, *Hapalocystis bicaudata*; 1209, *Lopadostoma gastrinum*; 1210, *Quaternaria dissepta*; 1211, *Splanchnonema foedans*; 1212, *Bloxamia leucophthalma*; 1213, *Corynespora proliferata*; 1214, *Nematogonium ferrugineum*; 1215, *Coryneum compactum*; 1216, *Diplodia melaena*; 1217, *Seimatosporium macrospermum*; 1218, *Seiridium intermedium*.

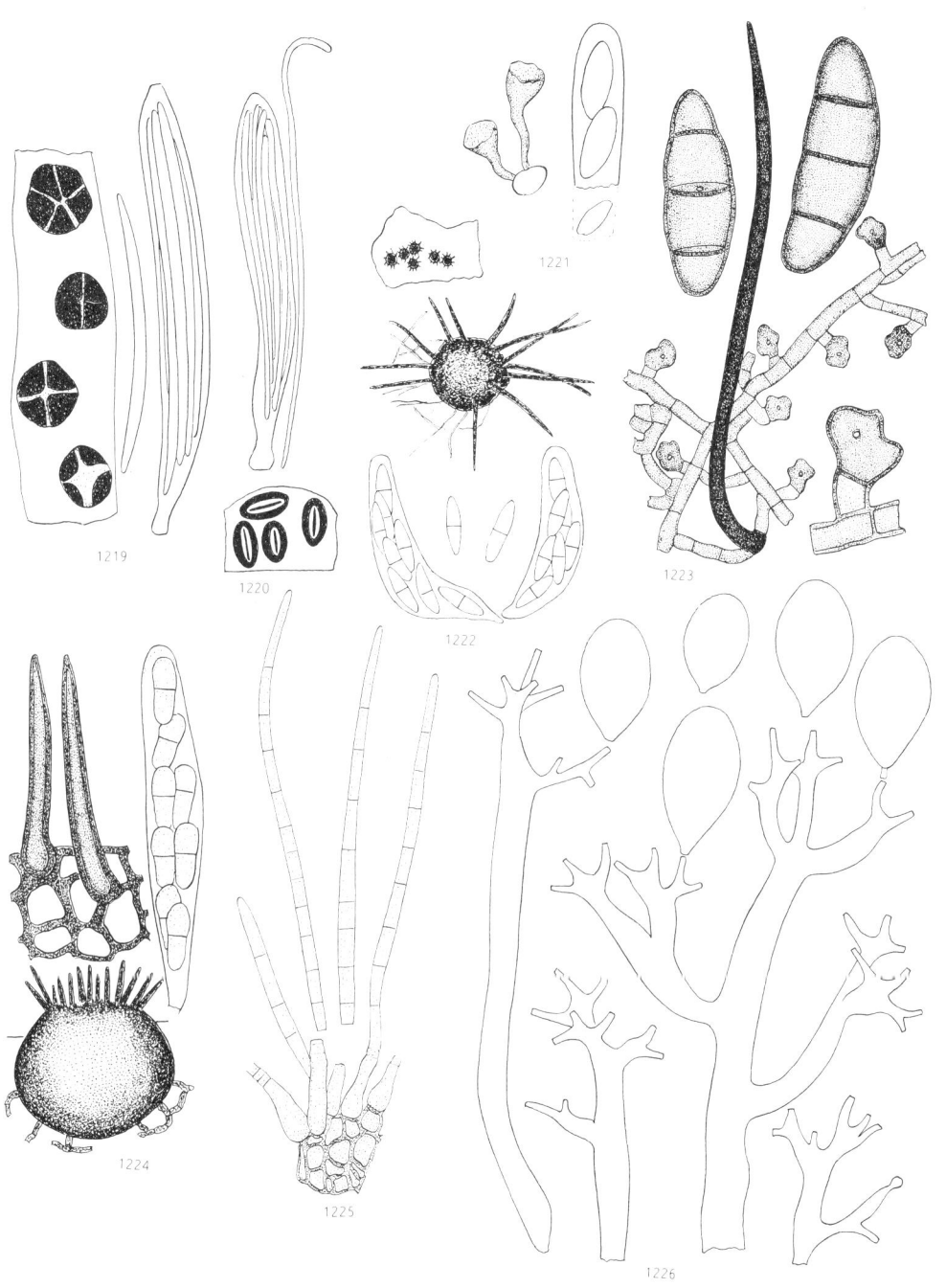

Plate 116. On *Vaccinium*, *Viburnum* and *Vitis*. 1219, *Coccomyces leptideus*; 1220, *Lophodermium melaleucum*; 1221, *Monilinia baccarum*; 1222, *Gibbera myrtilli*; 1223, *Meliola niessliana*; 1224, *Stigmatea conferta*; 1225, *Stigmina tinea*; 1226, *Plasmopara viticola*.

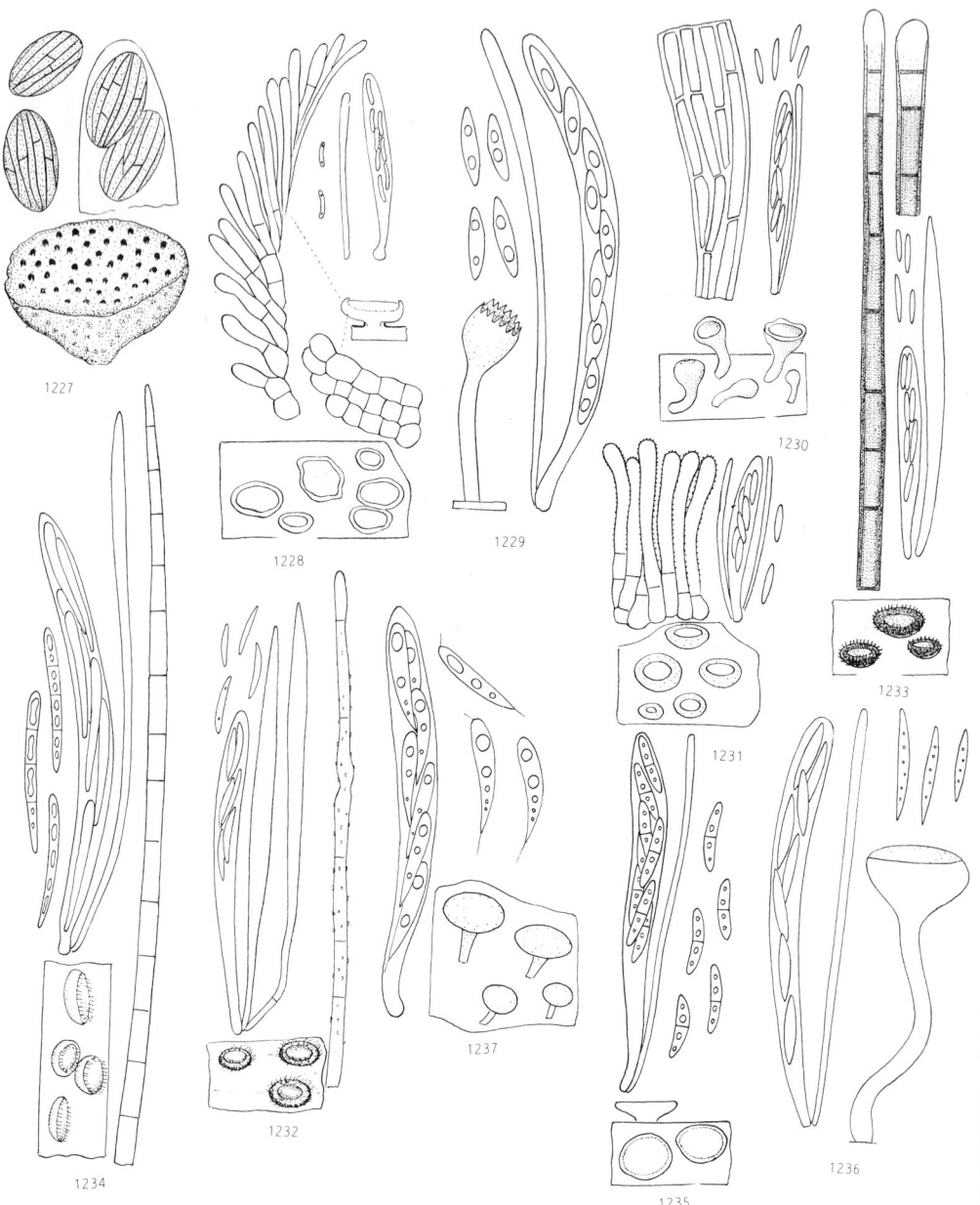

Plate 117. Discomycetes on herbaceous plants. 1227, *Ascobolus foliicola*; 1228, *Calycellina chlorinella*; 1229, *Crocicreas coronatum*; 1230, *C. cyathoideum*; 1231, *Dasyscyphus grevillei*; 1232, *D. mollissimus*; 1233, *D. nidulus*; 1234, *D. sulphureus*; 1235, *Hymenoscyphus herbarum*; 1236, *H. pileatus*; 1237, *H. scutula*.

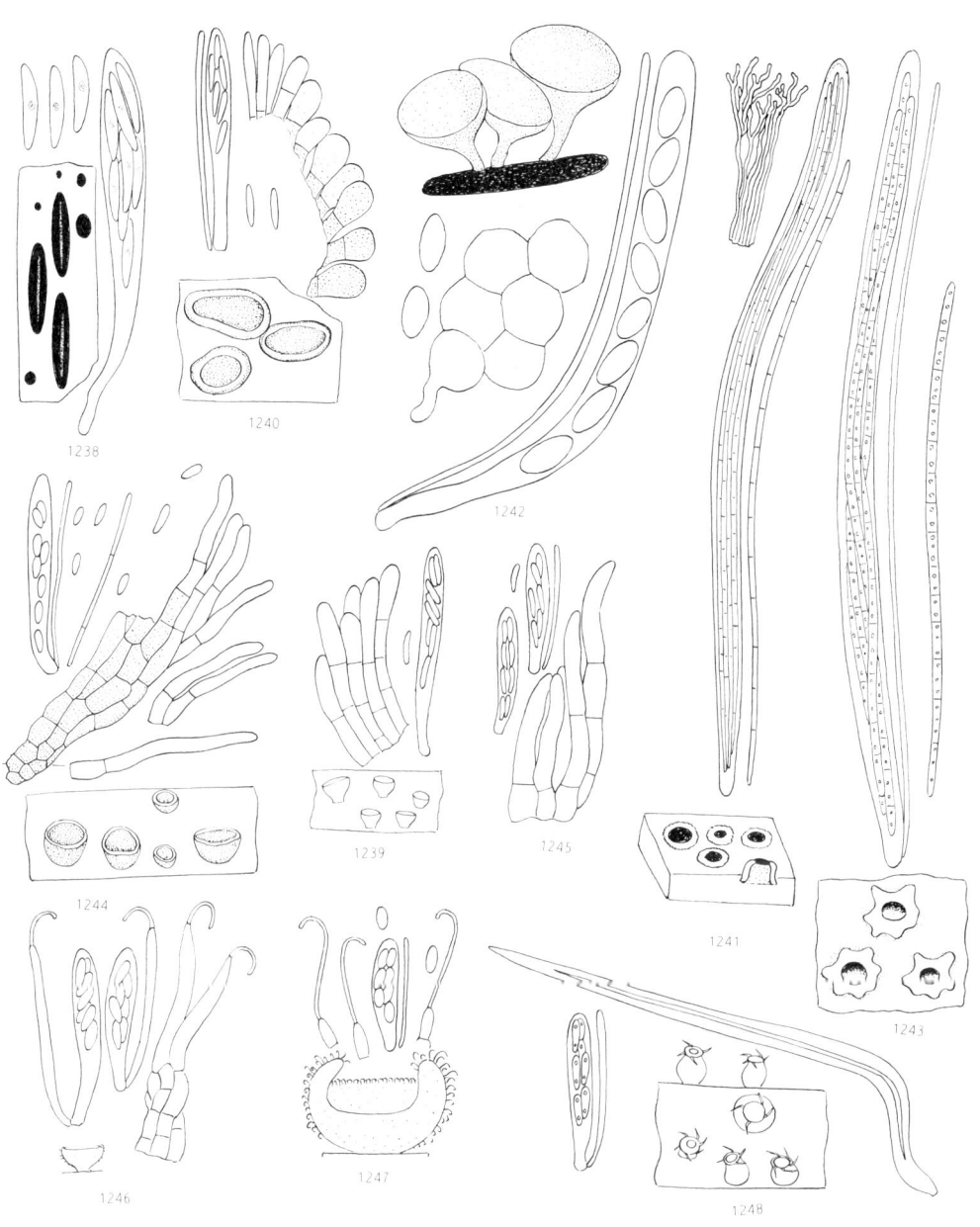

Plate 118. Discomycetes on herbaceous plants. 1238, *Hypoderma commune*; 1239, *Pezizella discreta*; 1240, *Pyrenopeziza revincta*; 1241, *Schizoxylon berkeleyanum*; 1242, *Sclerotinia sclerotiorum*; 1243, *Stictis stellata*; 1244, *Unguicularia cirrhata*; 1245, *U. millepunctata*; 1246, *Unguiculella eurotioides*; 1247, *U. hamulata*; 1248, *Urceolella crispula*.

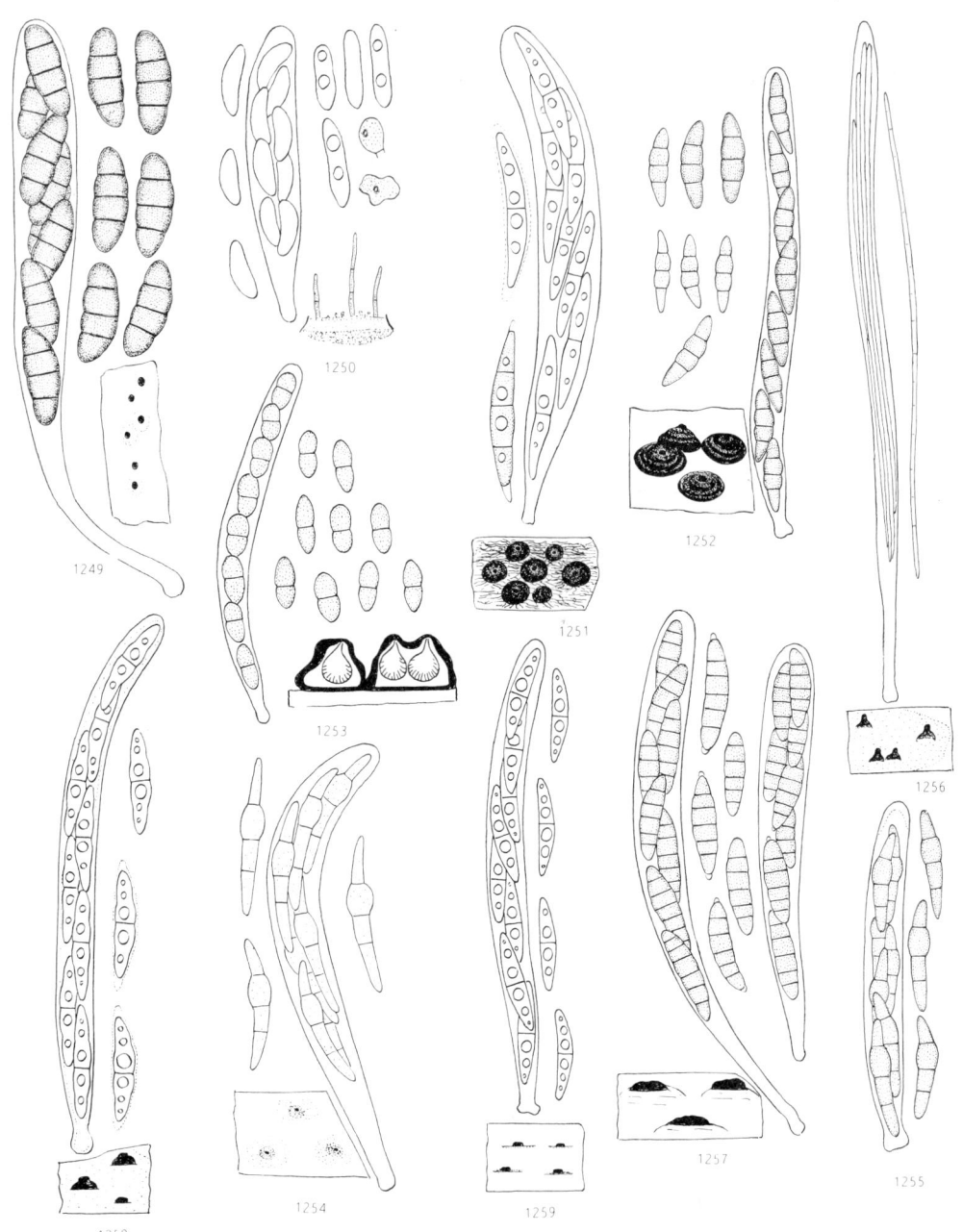

Plate 119. Ascomycetes on herbaceous plants. 1249, *Diapleella clivensis*; 1250, *Glomerella cingulata*; 1251, *Herpotrichia macrotricha*; 1252, *Leptosphaeria doliolum*; 1253, *Didymosphaeria conoidea*; 1254, *Leptosphaeria macrospora*; 1255, *L. purpurea*; 1256, *Leptospora rubella*; 1257, *Lophiostoma caulium*; 1258, *L. origani* var. *rubidum*; 1259, *L. vagabundum*.

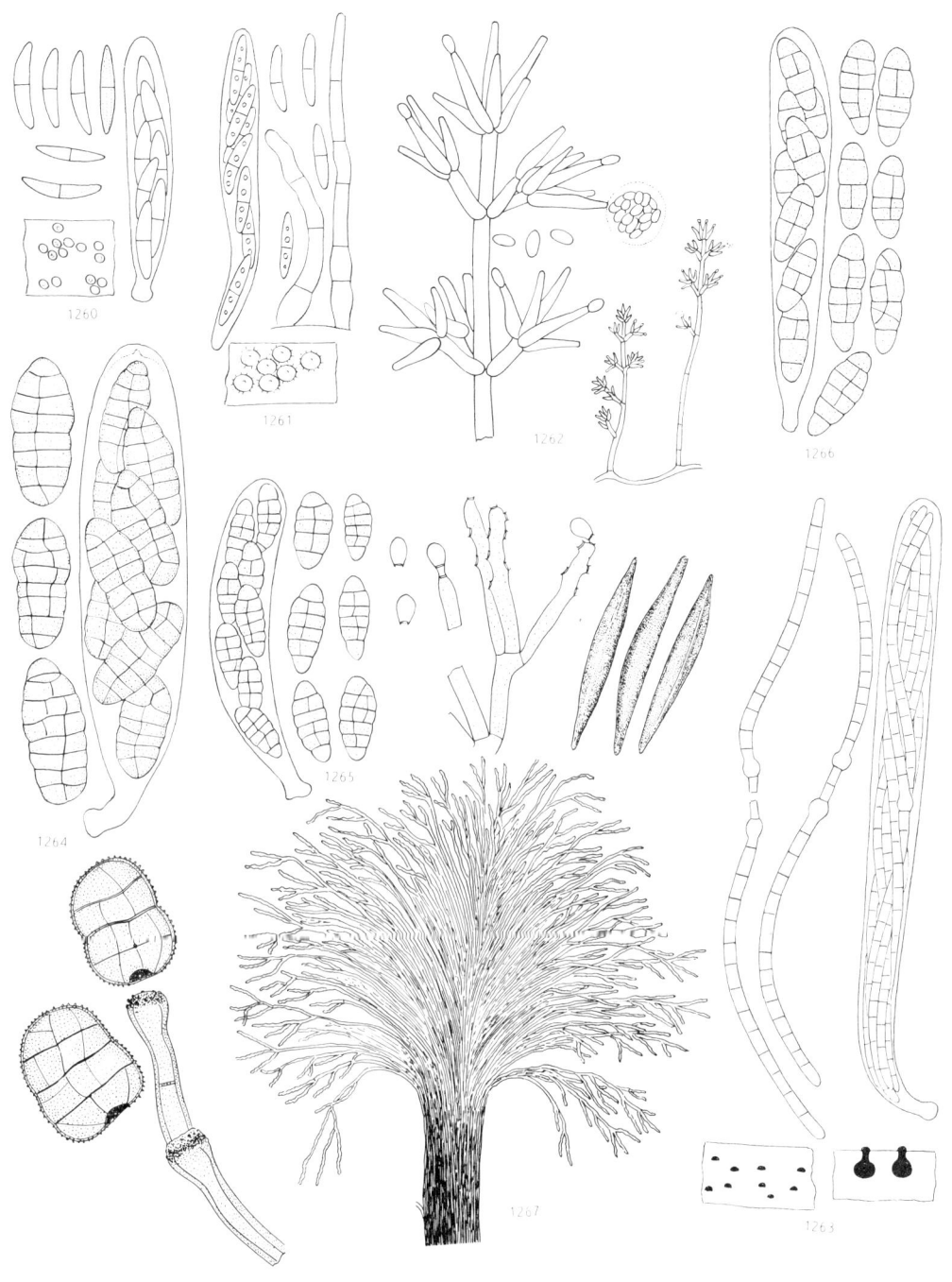

Plate 120. Ascomycetes on herbaceous plants. 1260, *Nectria arenula*; 1261, *N. ellisii*; 1262, *N. inventa*; 1263, *Ophiobolus acuminatus*; 1264, *Pleospora herbarum*; 1265, *P. phaeocomoides*; 1266, *P. scrophulariae*; 1267, *Rosellinia necatrix*.

Plate 121. Hyphomycetes on herbaceous plants. 1268, *Acremoniella atra*; 1269, *Alternaria alternata*; 1270, *A. tenuissima*; 1271, *Arthrobotrys conoides*; 1272, *A. oligospora*; 1273, *Botryosporium longibrachiatum*; 1274, *Botryotrichum piluliferum*.

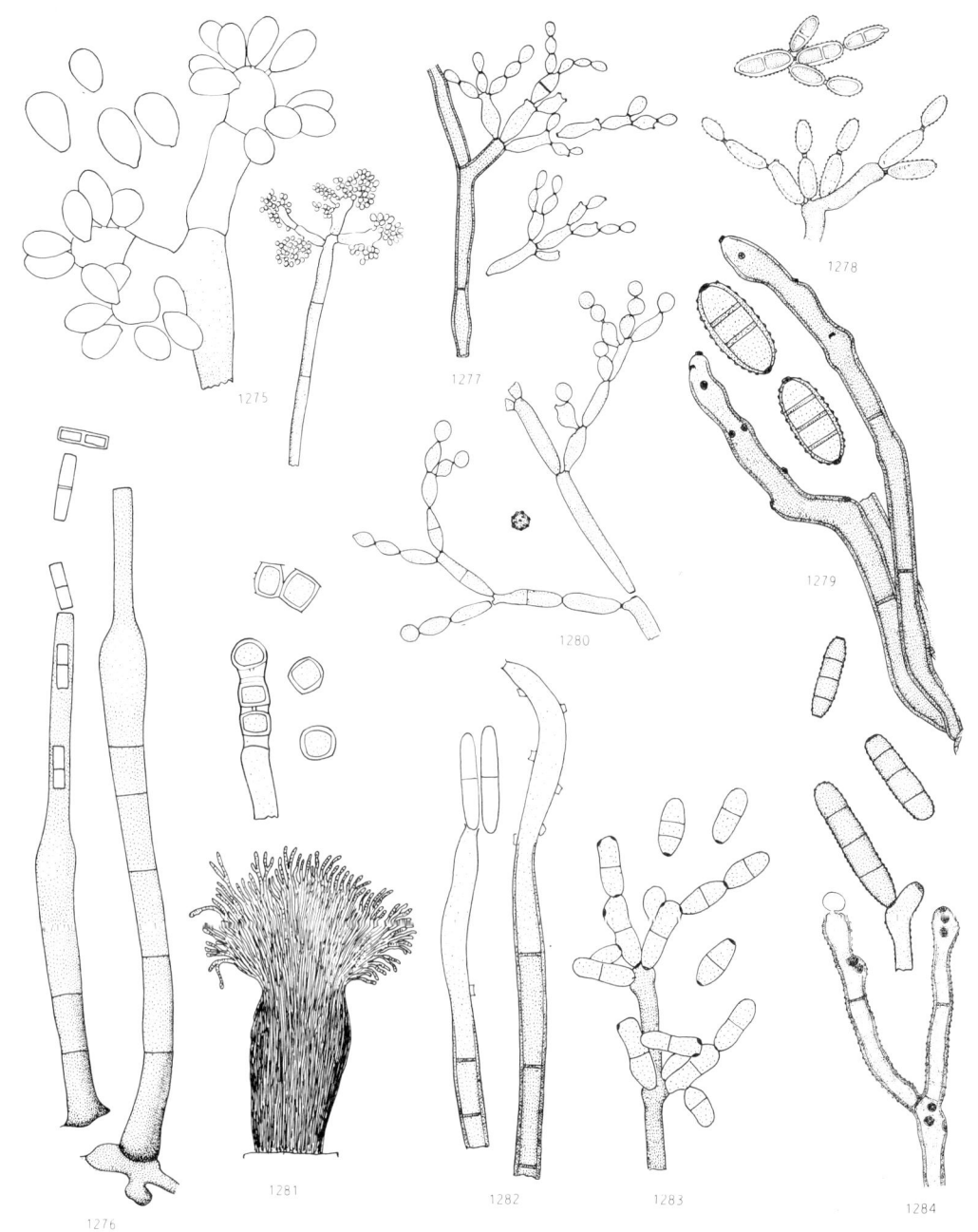

Plate 122. Hyphomycetes on herbaceous plants. 1275, *Botrytis cinerea*; 1276, *Chalara urceolata*; 1277, *Cladosporium cladosporioides*; 1278, *C. herbarum*; 1279, *C. macrocarpum*; 1280, *C. sphaerospermum*; 1281, *Coremiella cubispora*; 1282, *Cylindrotrichum oligospermum*; 1283, *Dendryphiella infuscans*; 1284, *D. vinosa*.

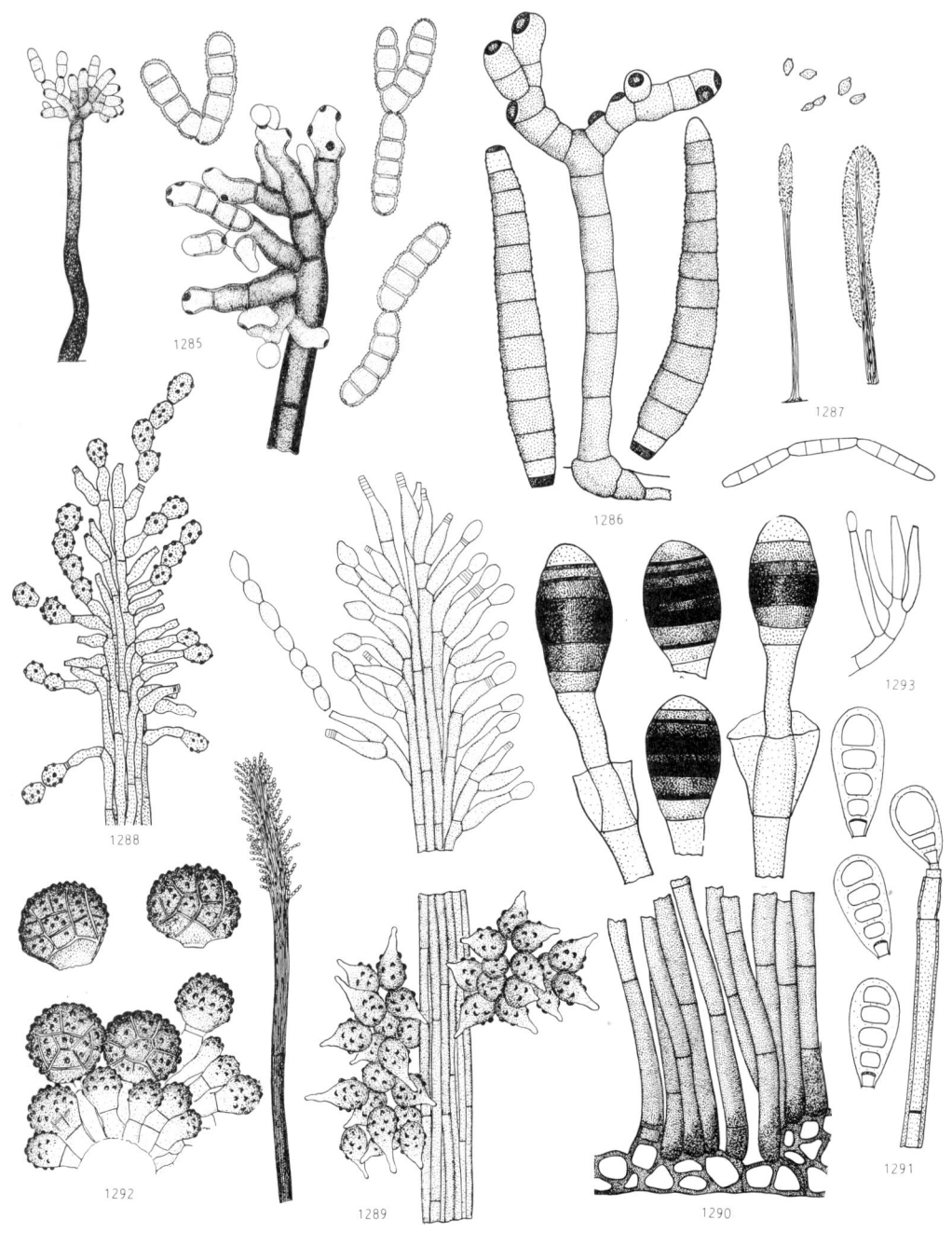

Plate 123. Hyphomycetes on herbaceous plants. 1285, *Dendryphion comosum*; 1286, *D. nanum*; 1287, *Doratomyces microsporus*; 1288, *D. nanus*; 1289, *D. stemonitis*; 1290, *Endophragmia elliptica*; 1291, *E. hyalosperma*; 1292, *Epicoccum purpurascens*; 1293, *Fusariella hughesii*.

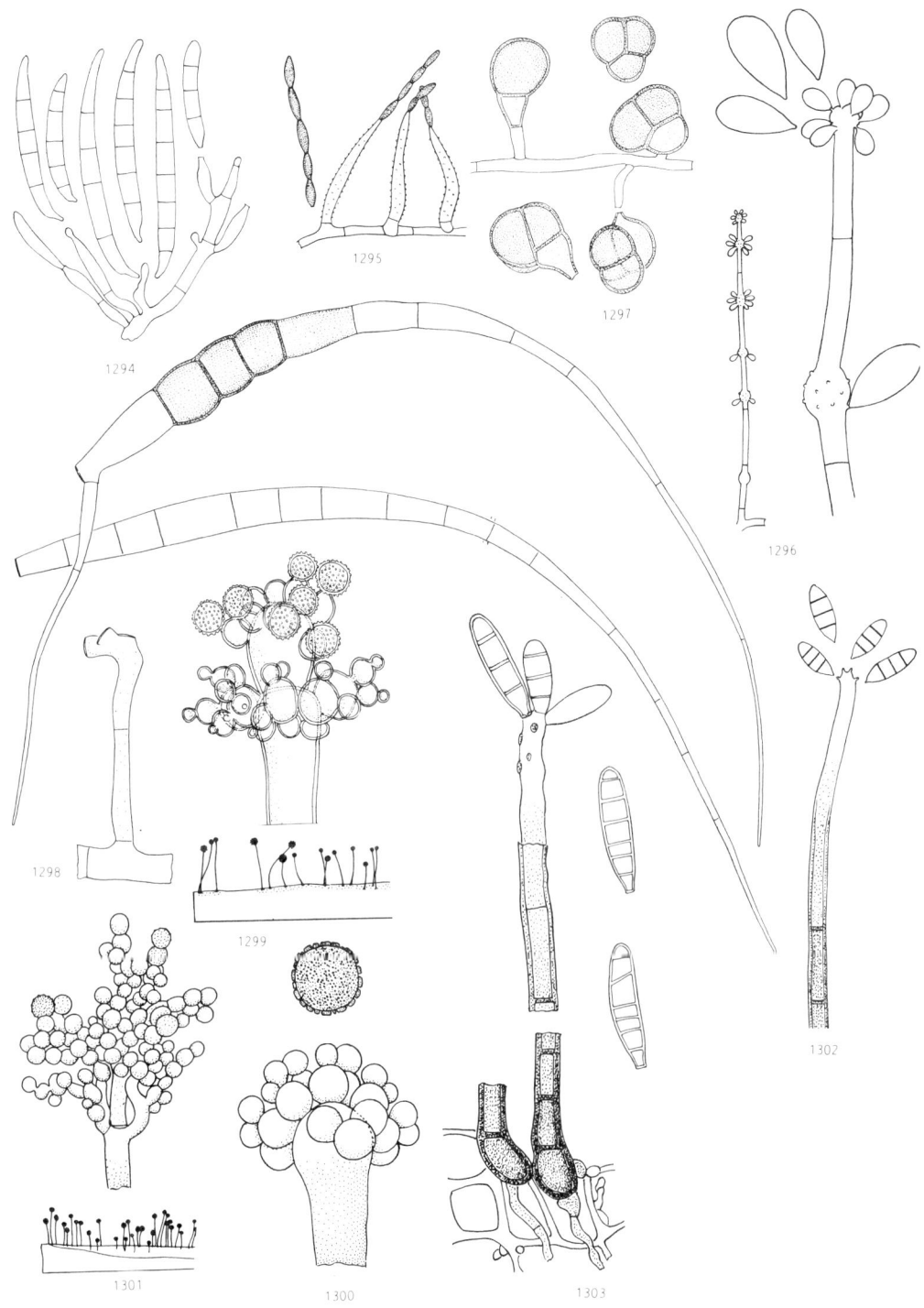

Plate 124. Hyphomycetes on herbaceous plants. 1294, *Fusarium* sp.; 1295, *Gliomastix luzulae*; 1296, *Gonatobotrys simplex*; 1297, *Monodictys levis*; 1298, *Mycocentrospora acerina*; 1299, *Periconia byssoides*; 1300, *P. cookei*; 1301, *P. minutissima*; 1302, *Pleurophragmium parvisporum*; 1303, *Pseudospiropes rousselianus*.

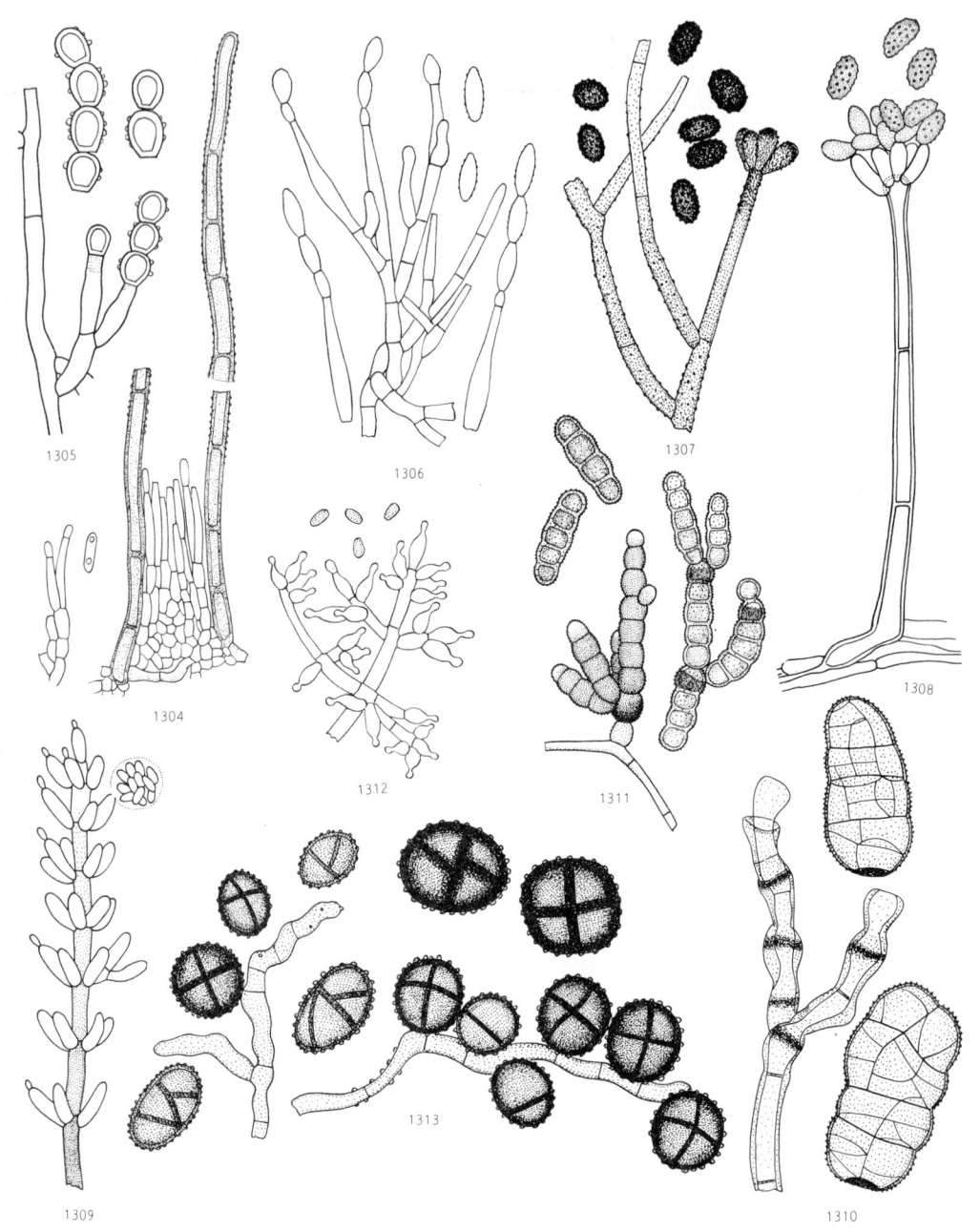

Plate 125. Hyphomycetes on herbaceous plants. 1304, *Sarcopodium circinatum*; 1305, *Scopulariopsis brevicaulis*; 1306, *Septofusidium herbarum*; 1307, *Stachybotrys atra*; 1308, *S. dichroa*; 1309, *Stachylidium bicolor*; 1310, *Stemphylium vesicarium*; 1311, *Torula herbarum*; 1312, *Trichoderma koningii*; 1313, *Ulocladium atrum*.

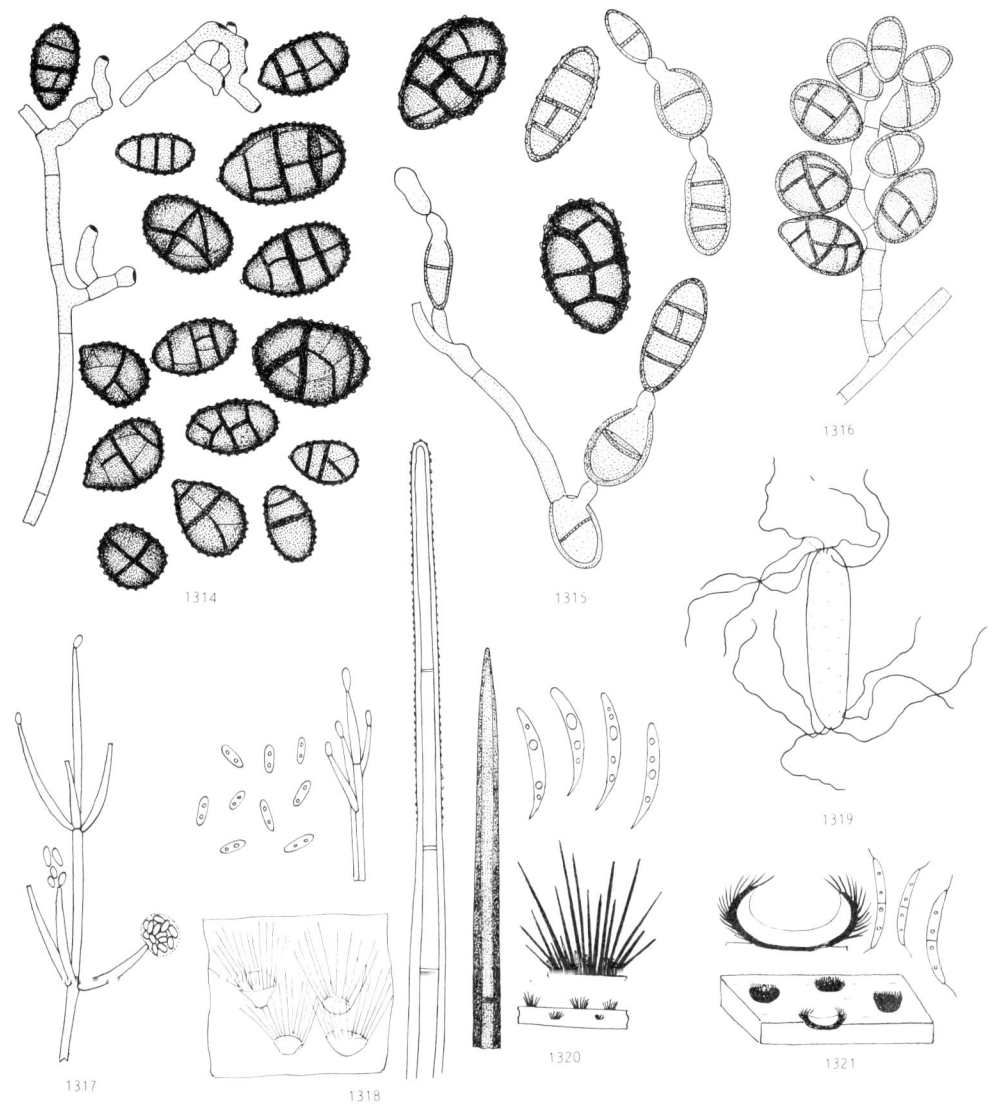

Plate 126. Hyphomycetes and coelomycetes on herbaceous plants. 1314, *Ulocladium botrytis*; 1315, *U. chartarum*; 1316, *U. consortiale*; 1317, *Verticillium* sp.; 1318, *Volutella ciliata*; 1319, *Chaetospermum chaetosporum*; 1320, *Colletotrichum dematium*; 1321, *Pseudolachnea hispidula*.

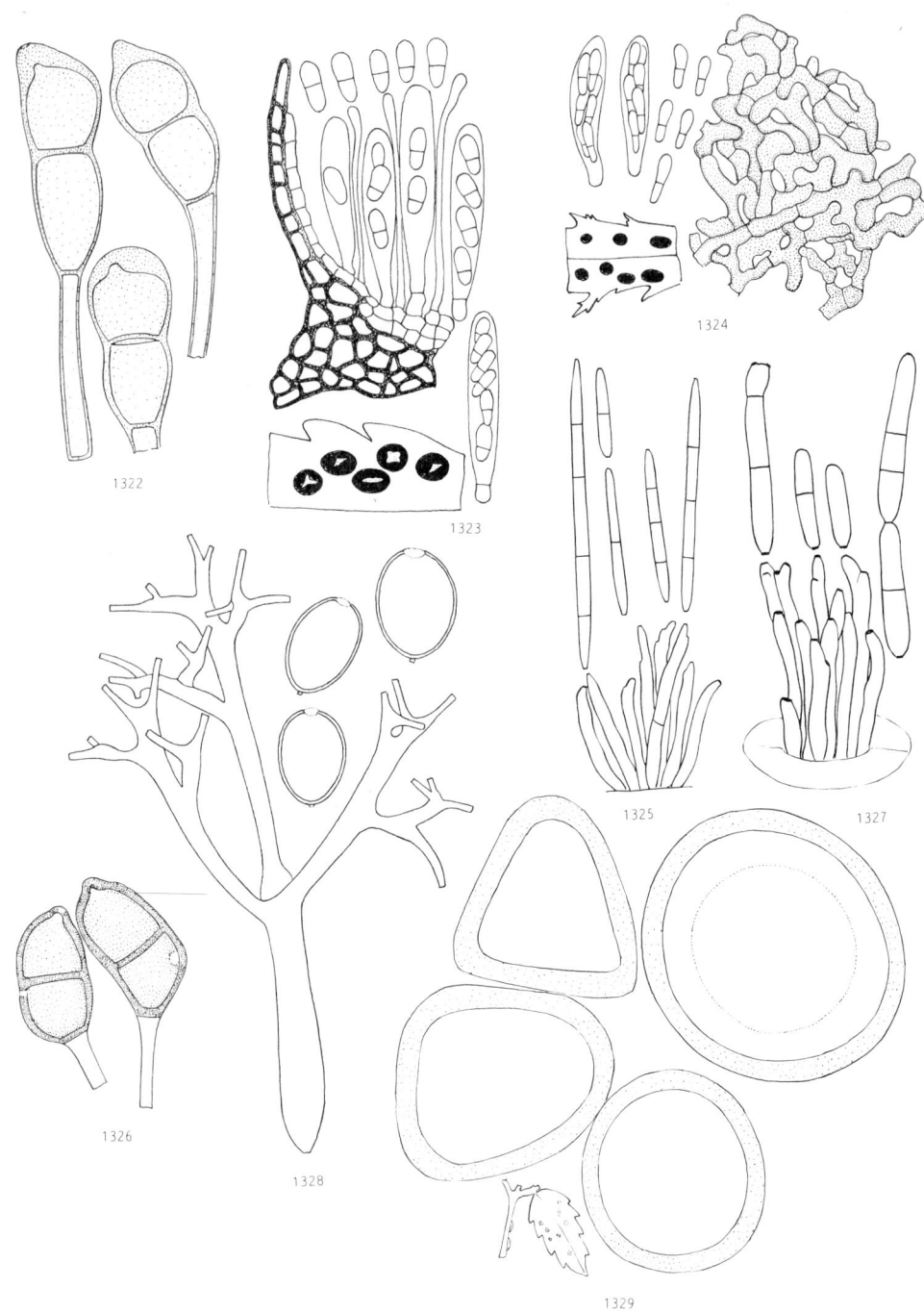

Plate 127. On *Achillea*, *Adoxa* and *Aegopodium*. 1322, *Puccinia cnici-oleracei*; 1323, *Schizothyrioma ptarmicae*; 1324, *S. aterrimum*; 1325, *Ramularia aromatica*; 1326, *Puccinia adoxae*; 1327, *Ramularia adoxae*; 1328, *Plasmopara nivea*; 1329, *Protomyces macrosporus*.

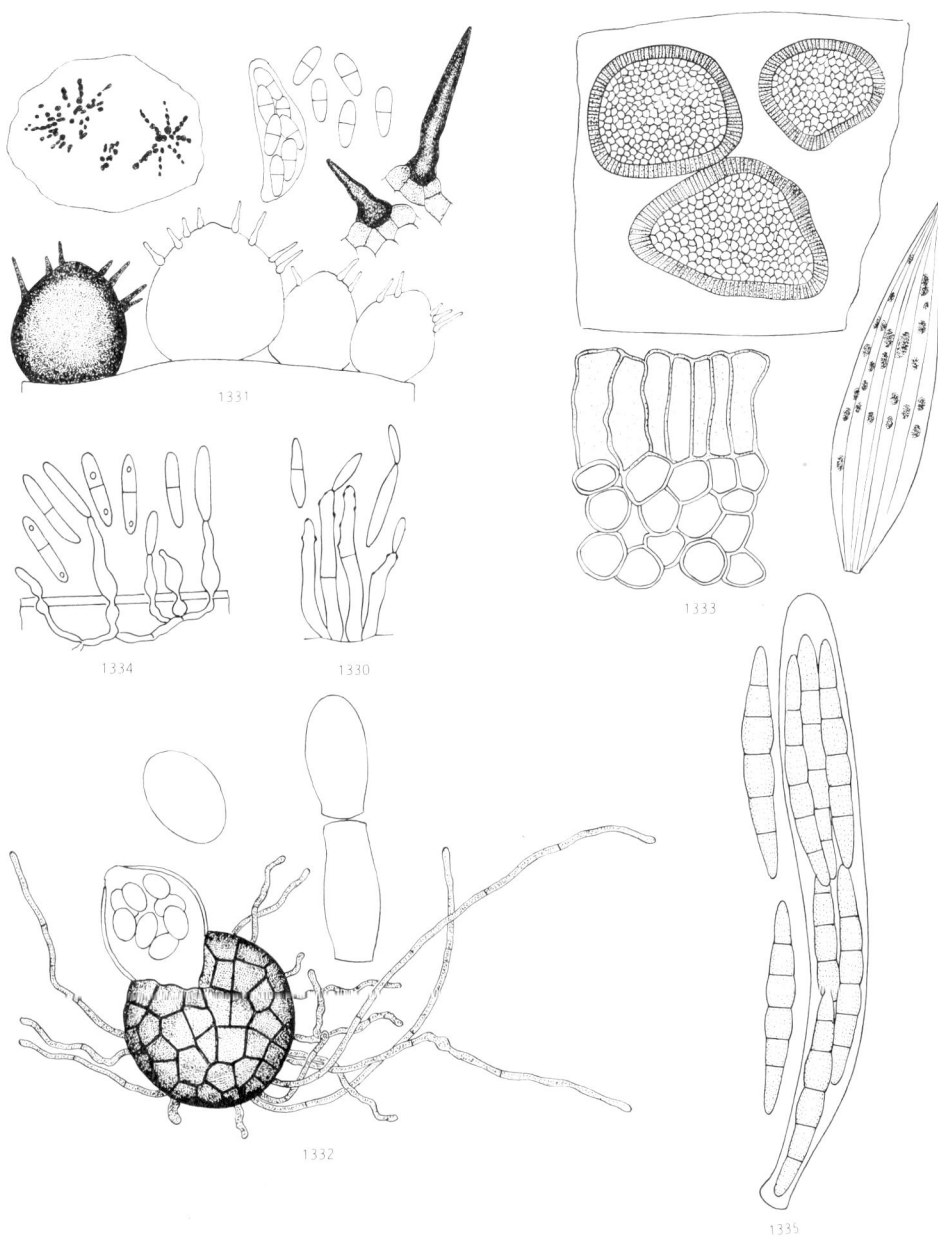

Plate 128. On *Ajuga, Alchemilla, Alisma* and *Alliaria*. 1330, *Ramularia ajugae*; 1331, *Coleroa alchemillae*; 1332, *Sphaerotheca alchemillae*; 1333, *Doassansia alismatis*; 1334, *Rhynchosporium alismatis*; 1335, *Leptosphaeria maculans*.

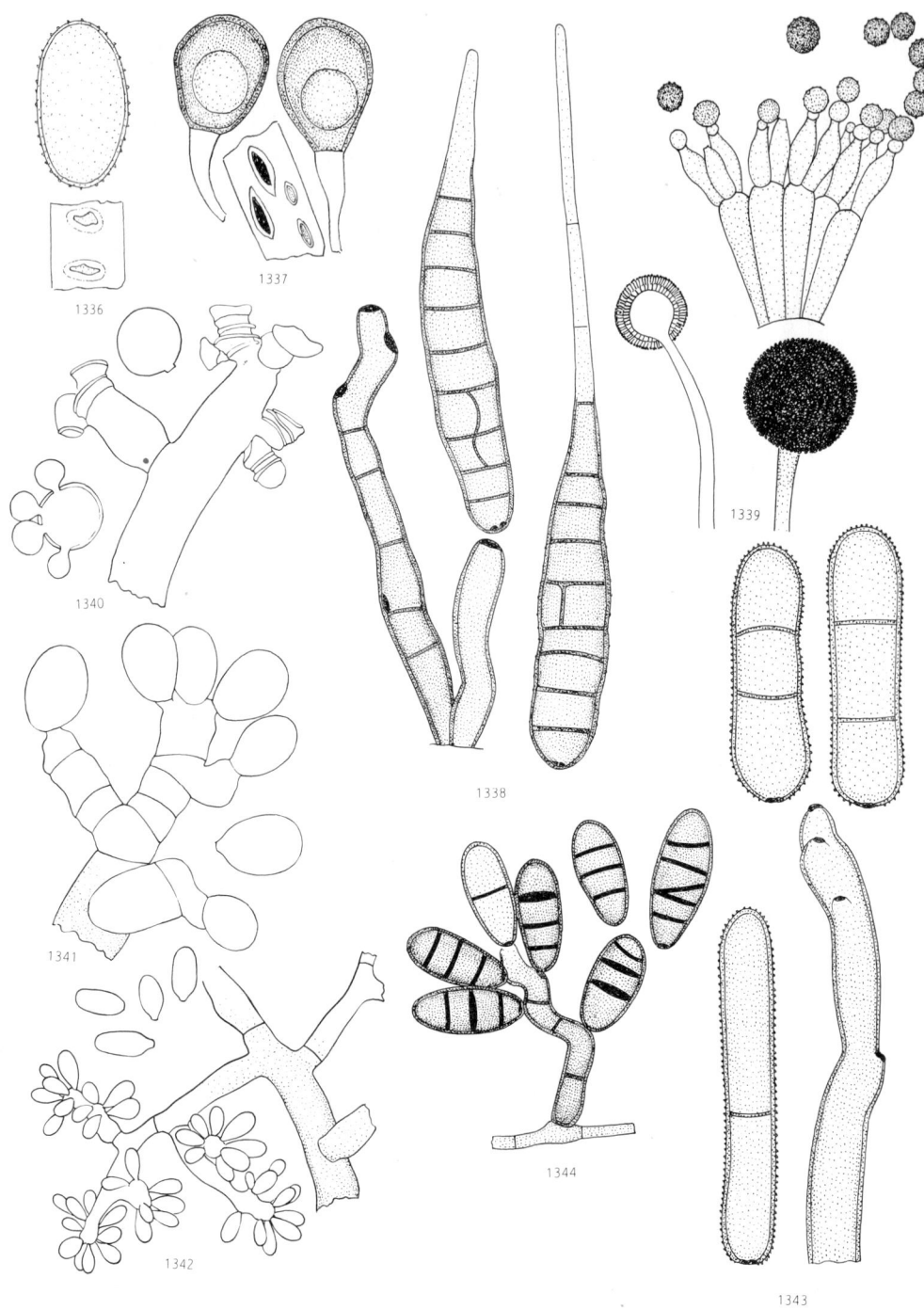

Plate 129. On *Allium*. 1336, *Puccinia porri*; 1337, *Uromyces ambiguus*; 1338, *Alternaria porri*; 1339, *Aspergillus niger*; 1340, *Botryotinia globosa*; 1341, *B. squamosa*; 1342, *Botrytis allii*; 1343, *Cladosporium allii-cepae*; 1344, *Embellisia allii*.

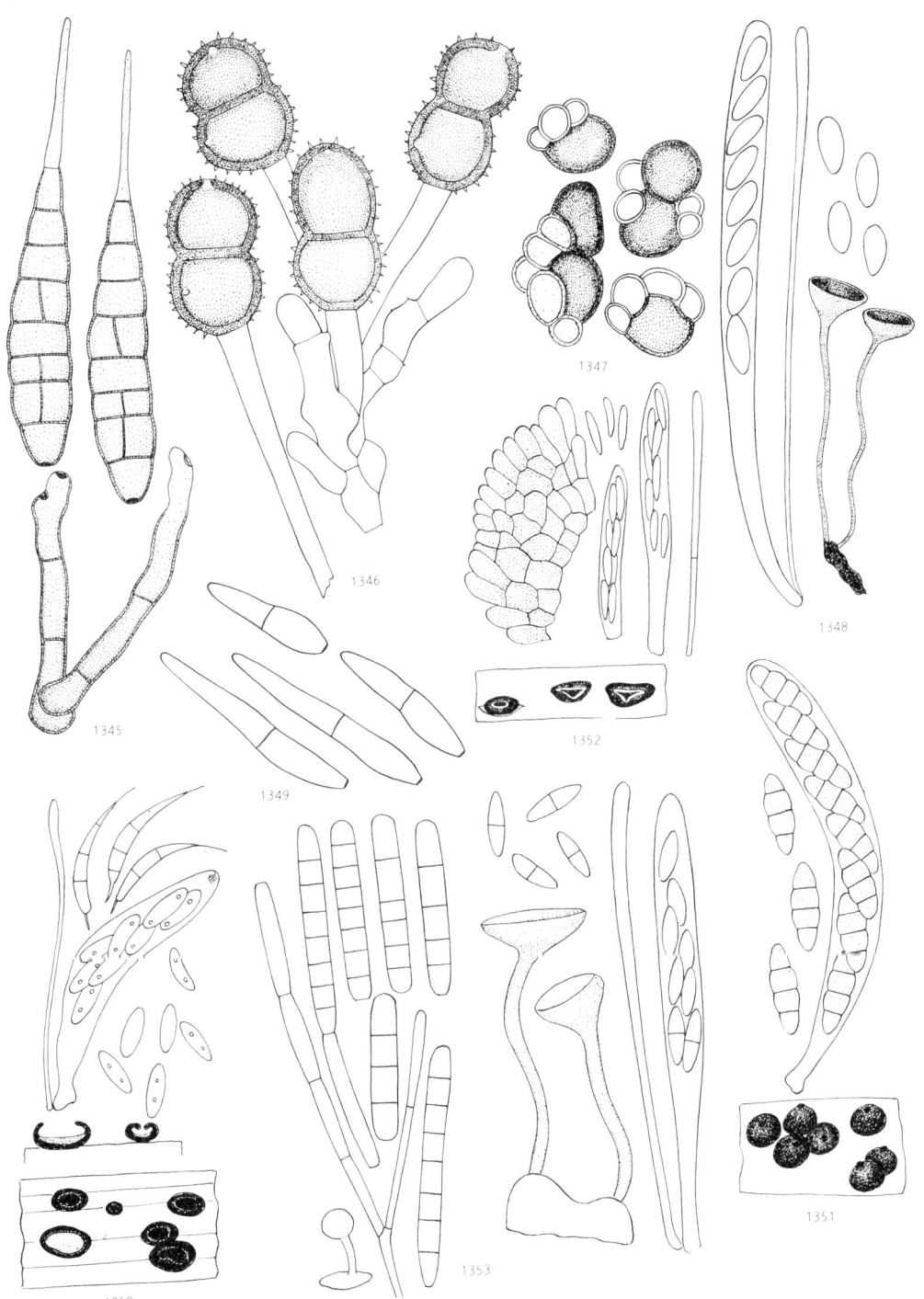

Plate 130. On *Anagallis*, *Anemone* and *Angelica*. 1345, *Alternaria anagallidis*; 1346, *Tranzschelia anemones*; 1347, *Urocystis anemones*; 1348, *Dumontinia tuberosa*; 1349, *Cercosporidium depressum*; 1350, *Heterosphaeria patella*; 1351, *Leptosphaeria libanotis*; 1352, *Pyrenopeziza plicata*; 1353, *Symphyosirinia angelicae*.

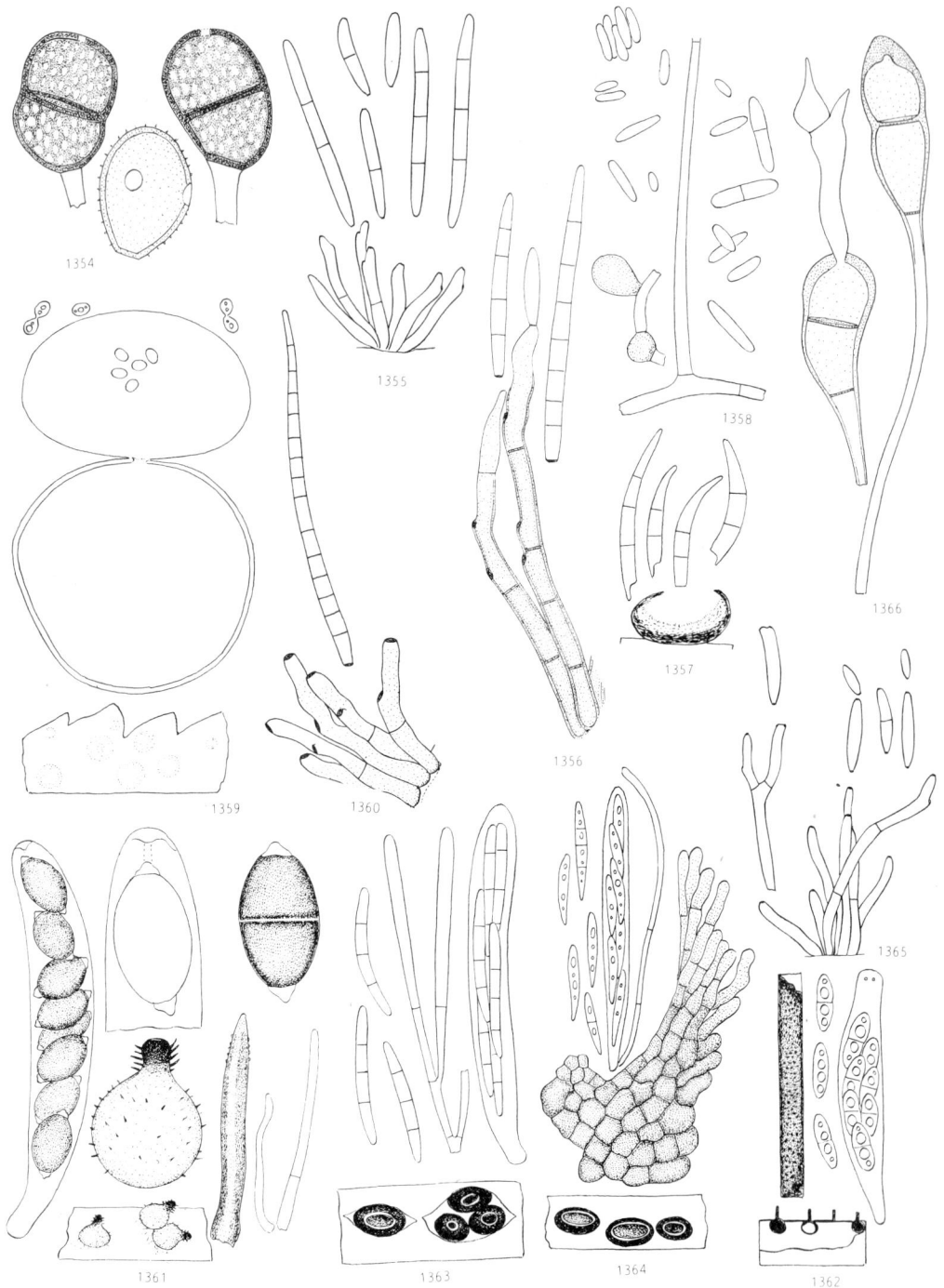

Plate 131. On *Anthriscus, Anthyllis, Antirrhinum, Apium, Arctium* and *Arenaria*. 1354, *Puccinia chaerophylli*; 1355, *Ramularia anthrisci*; 1356, *Cercospora radiata*; 1357, *Heteropatella antirrhini*; 1358, *Acremonium apii*; 1359, *Burenia inundata*; 1360, *Cercospora apii*; 1361, *Arnium apiculatum*; 1362, *Diaporthe arctii*; 1363, *Pyrenopeziza arctii*; 1364, *P. inornata*; 1365, *Ramularia filaris* var. *lappae*; 1366, *Puccinia arenariae*.

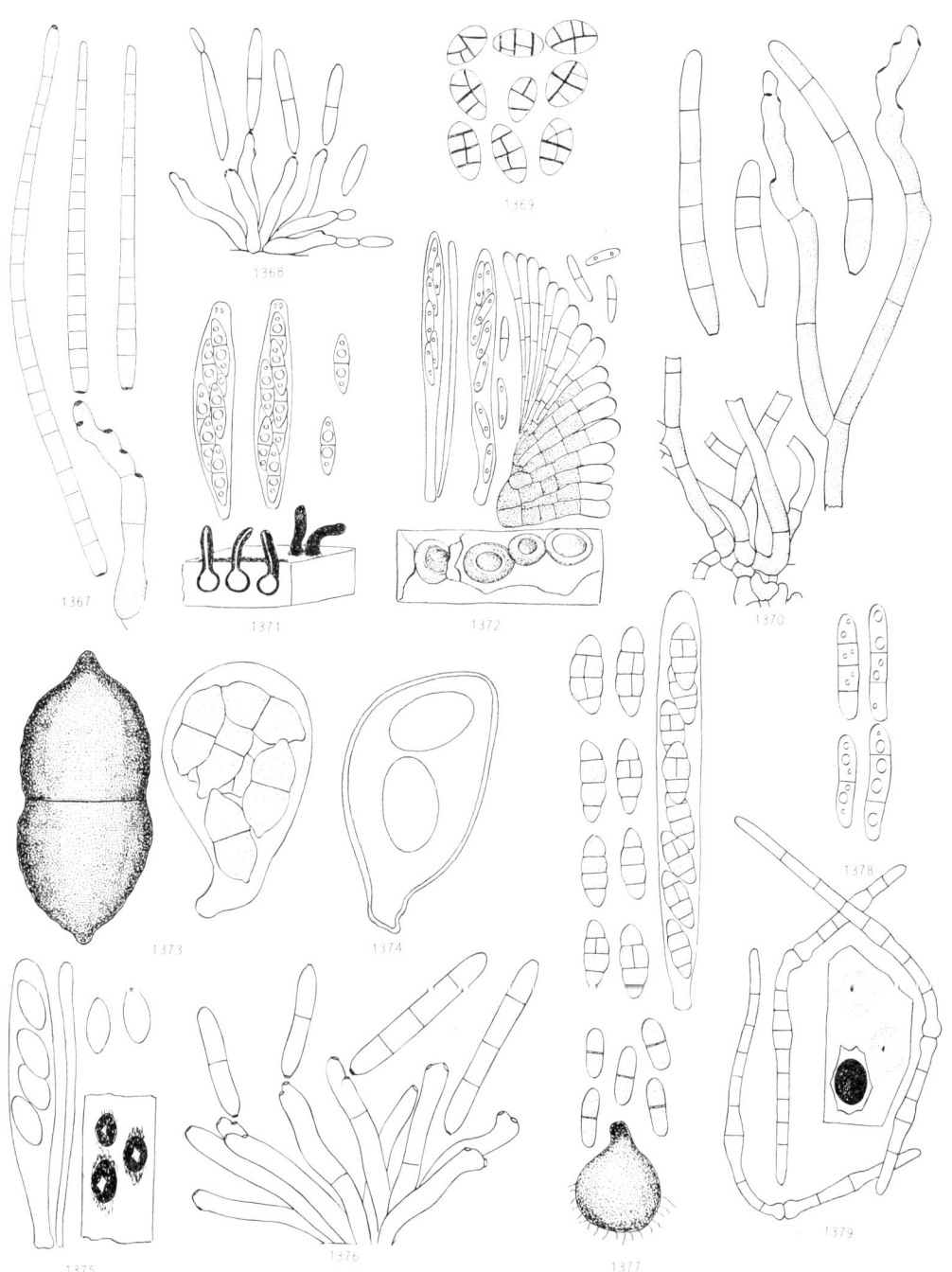

Plate 132. On *Armoracia, Artemisia, Asparagus, Aster, Atriplex* and *Ballota*. 1367, *Cercospora armoraciae*; 1368, *Ramularia armoraciae*; 1369, *Camarosporium aequivocum*; 1370, *Cercospora ferruginea*; 1371, *Diaporthe arctii* var. *artemisiae*; 1372, *Pyrenopeziza artemisiae*; 1373, *Zopfia rhizophila*; 1374, *Erysiphe cichoracearum*; 1375, *Ploettnera solidaginis*; 1376, *Ramularia asteris*; 1377, *Pleospora calvescens*; 1378, *Stagonospora atriplicis*; 1379, *Ophiobolus ulnasporus*.

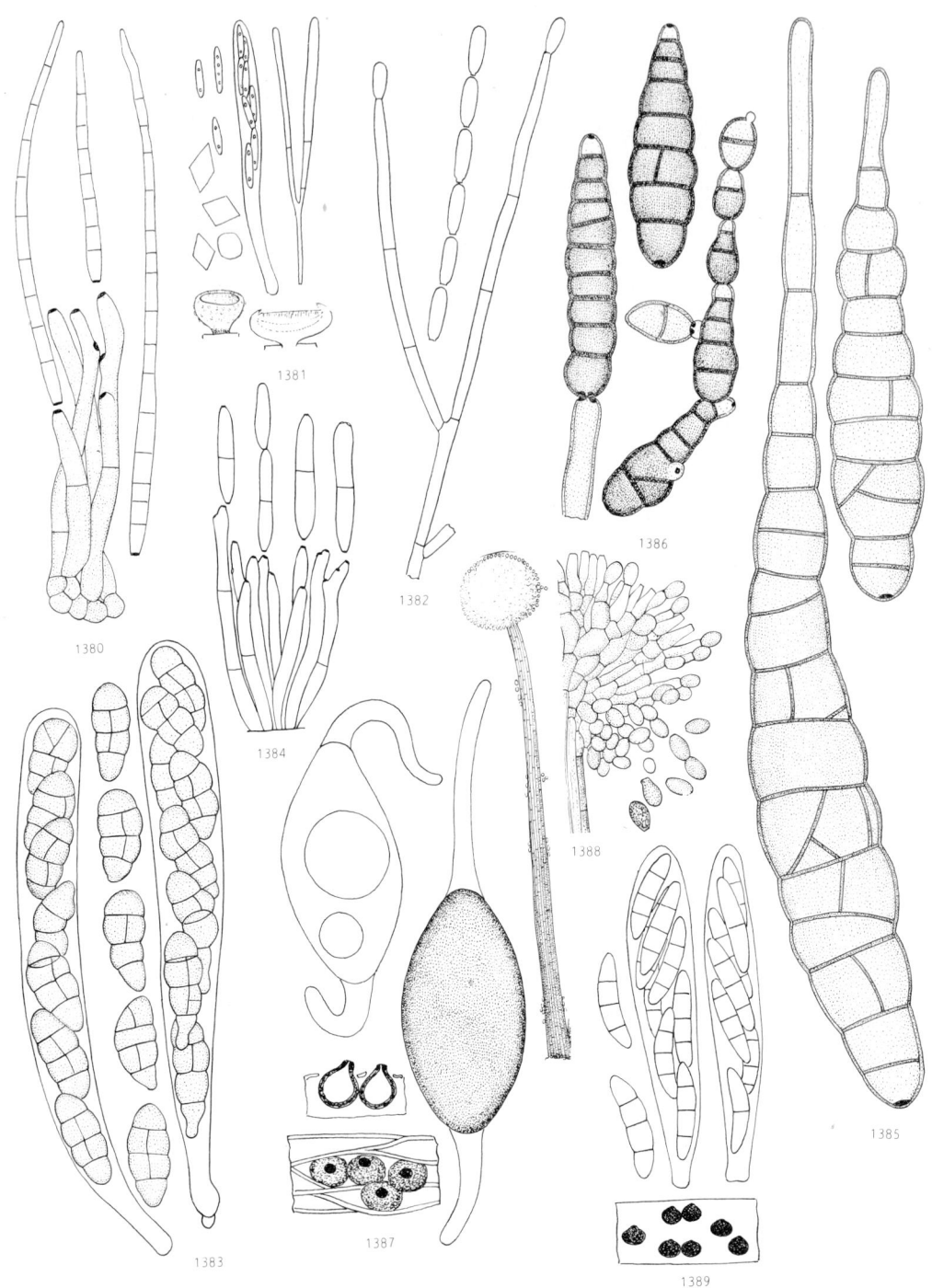

Plate 133. On *Beta* and *Brassica*. 1380, *Cercospora beticola*; 1381, *Cyathicula littoralis*; 1382, *Gabarnaudia betae*; 1383, *Pleospora bjoerlingii*; 1384, *Ramularia beticola*; 1385, *Alternaria brassicae*; 1386, *A. brassicicola*; 1387, *Arnium olerum*; 1388, *Doratomyces purpureofuscus*; 1389, *Gibberella cyanogena*.

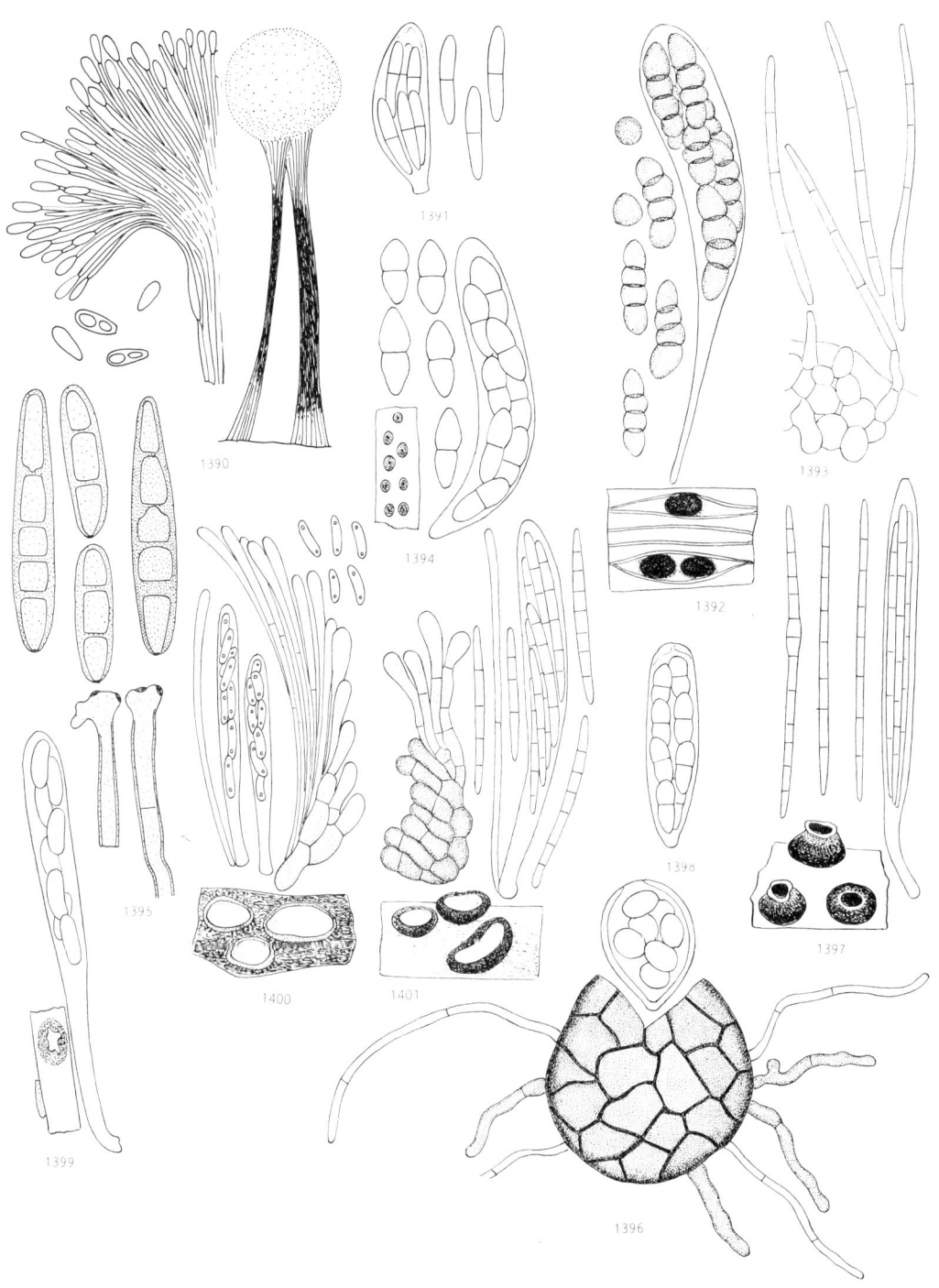

Plate 134. On *Brassica, Bryonia, Cactaceae, Calendula* and *Calluna*. 1390, *Graphium putredinis*; 1391, *Mycosphaerella brassicicola*; 1392, *Perisporium kunzei*; 1393, *Pseudocercosporella capsellae*; 1394, *Didymella bryoniae*; 1395, *Drechslera cactivora*; 1396, *Sphaerotheca xanthii*; 1397, *Godronia cassandrae* f. *callunae*; 1398, *Keissleriella subalpina*; 1399, *Pseudophacidium callunae*; 1400, *Tapesia melaleucoides*; 1401, *Trichobelonium obscurum*.

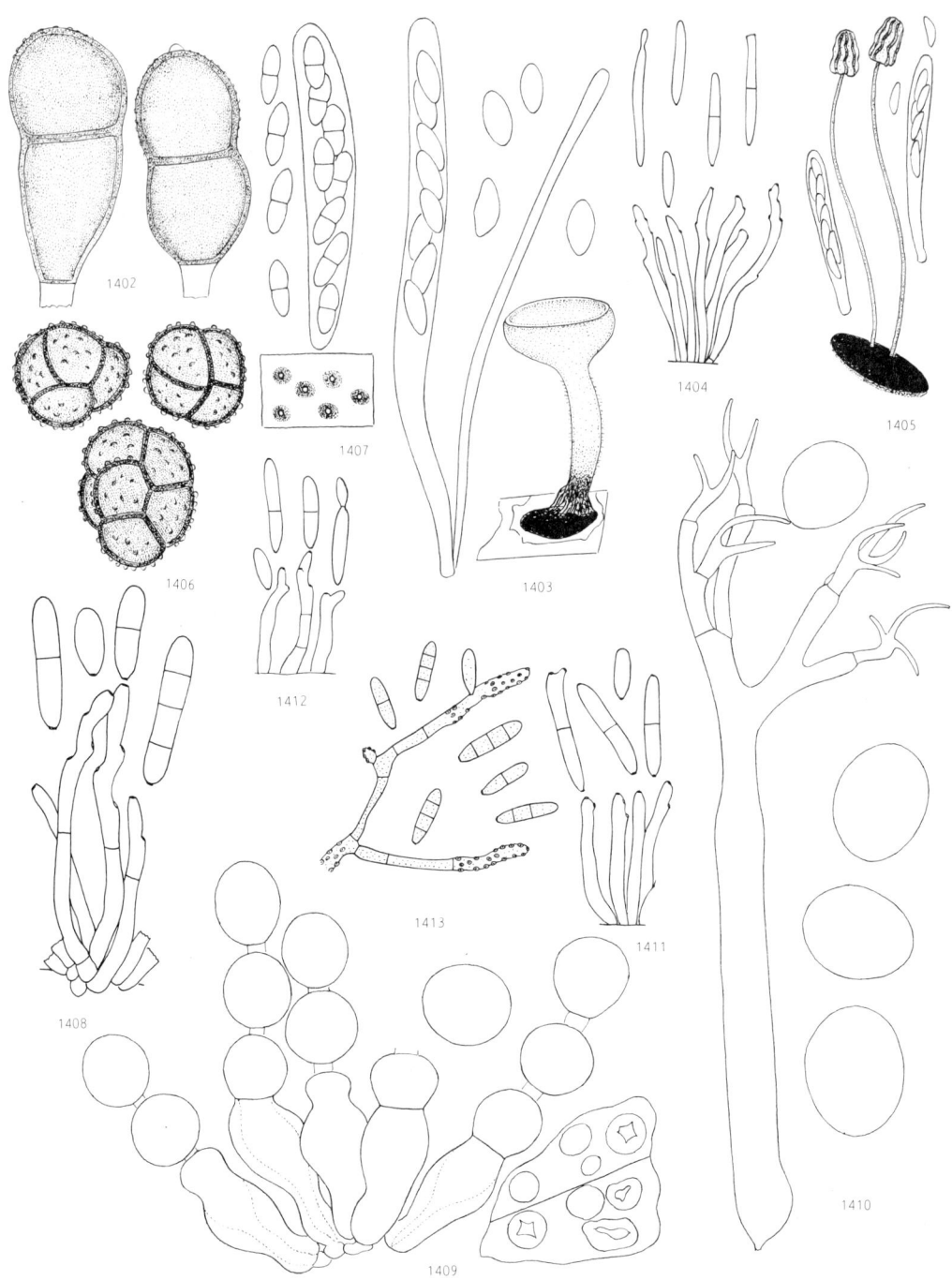

Plate 135. On *Caltha, Calystegia, Campanula, Capsella, Cardamine, Carduus* and *Carlina*. 1402, *Puccinia calthicola*; 1403, *Botryotinia calthae*; 1404, *Ramularia calthae*; 1405, *Verpatinia calthicola*; 1406, *Thecaphora seminis-convolvuli*; 1407, *Didymella exigua*; 1408, *Ramularia macrospora*; 1409, *Albugo candida*; 1410, *Peronospora parasitica*; 1411, *Ramularia cardamines*; 1412, *R. cardui*; 1413, *Veronaea carlinae*.

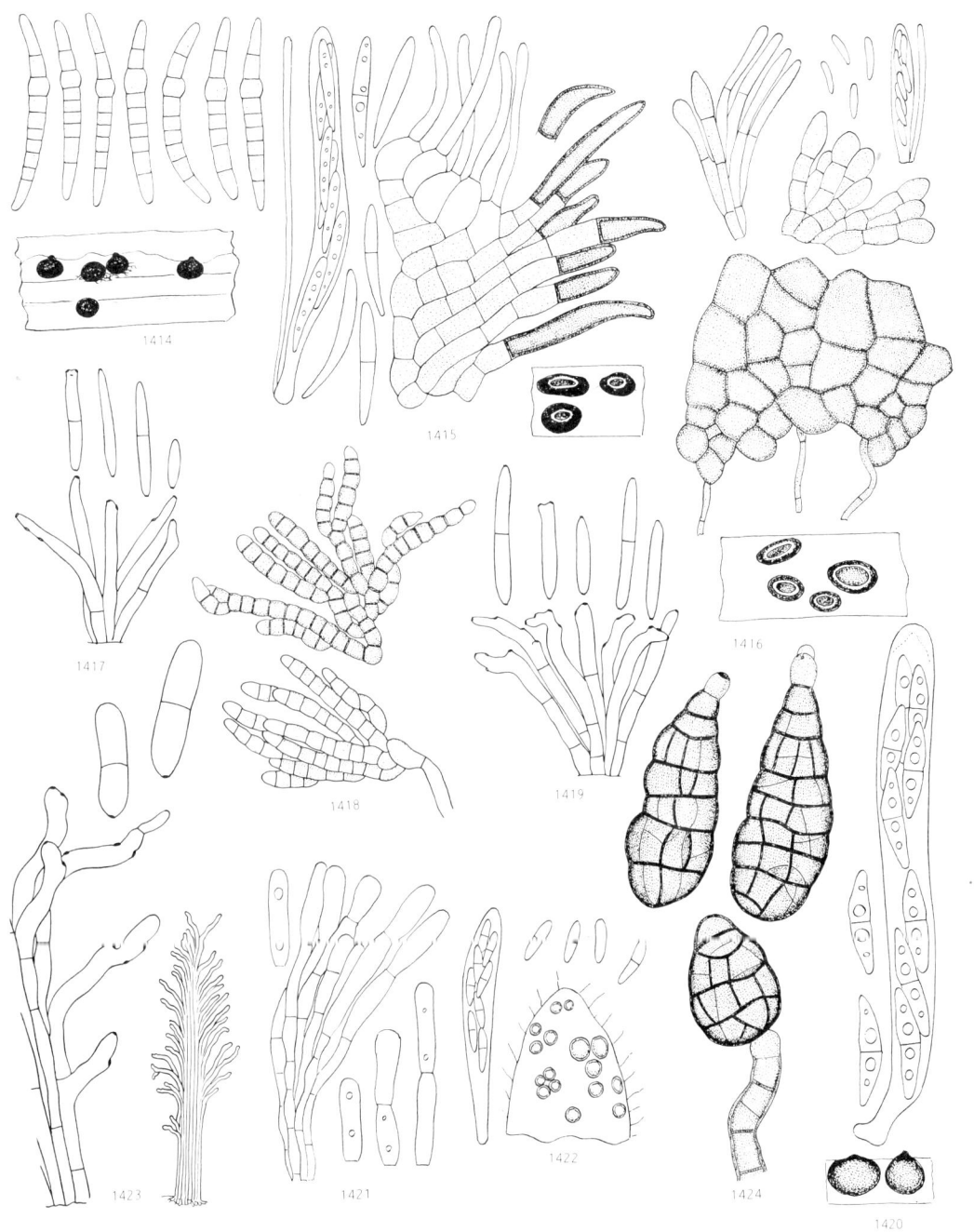

Plate 136. On *Centaurea, Centaurium, Centranthus, Cerastium* and *Cheiranthus*. 1414, *Leptosphaeria jaceae*; 1415, *Pirottaea brevipila*; 1416, *Pyrenopeziza adenostylidis*; 1417, *Ramularia centaureae*; 1418, *Taeniolina centaurii*; 1419, *Ramularia centranthi*; 1420, *Didymella cerastii*; 1421, *Endoconospora cerastii*; 1422, *Leptotrochila cerastiorum*; 1423, *Ramularia alborosella*; 1424, *Alternaria cheiranthi*.

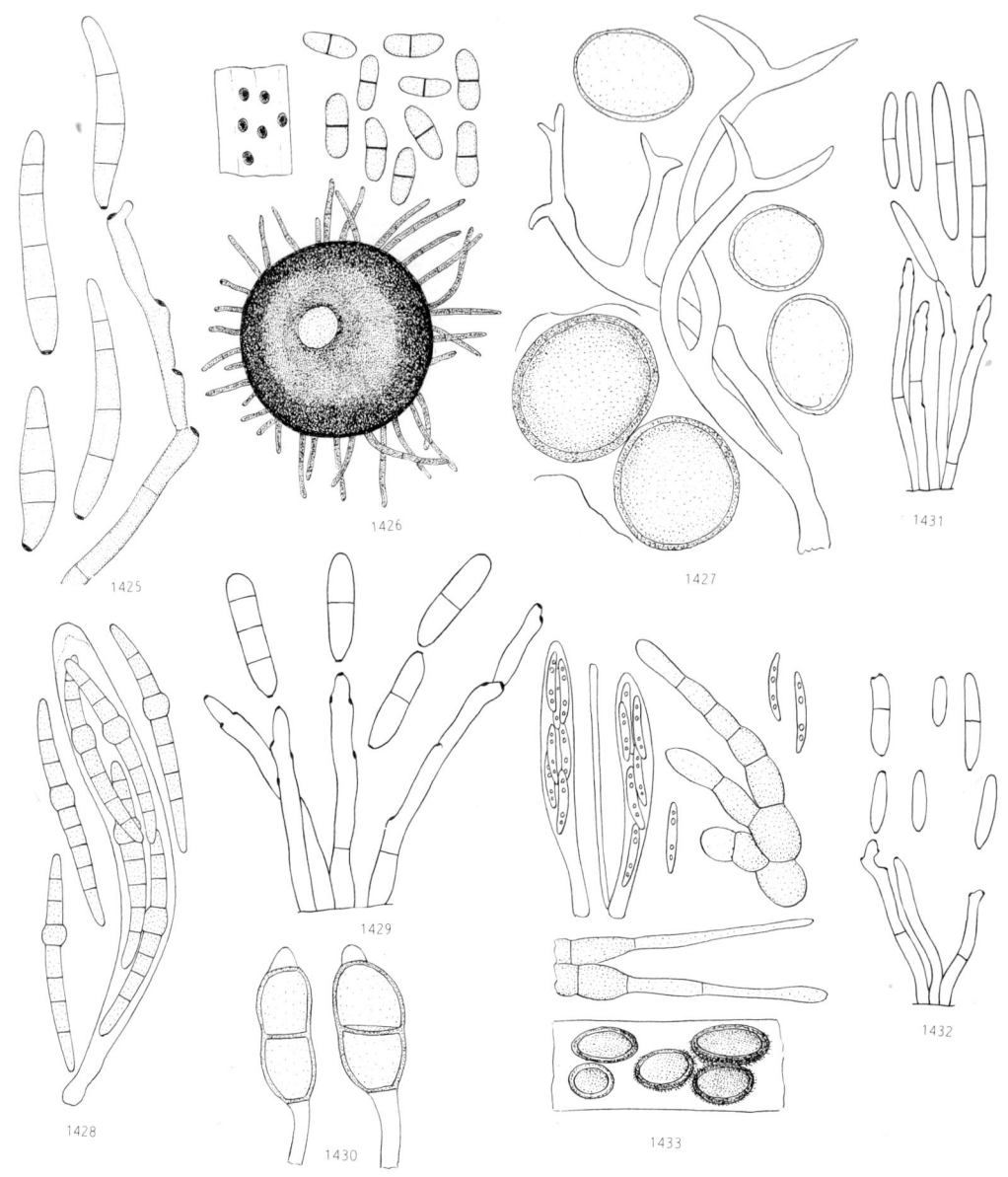

Plate 137. On *Chenopodium*, *Chrysanthemum*, *Chrysosplenium*, *Cicuta*, *Circaea* and *Cirsium*. 1425, *Cercospora chenopodii*; 1426, *Chaetodiplodia caulina*; 1427, *Peronospora farinosa*; 1428, *Leptosphaeria dolioloides*; 1429, *Ramularia bellunensis*; 1430, *Puccinia chrysosplenii*; 1431, *Ramularia cicutae*; 1432, *Ramularia circaeae*; 1433, *Pyrenopeziza carduorum*.

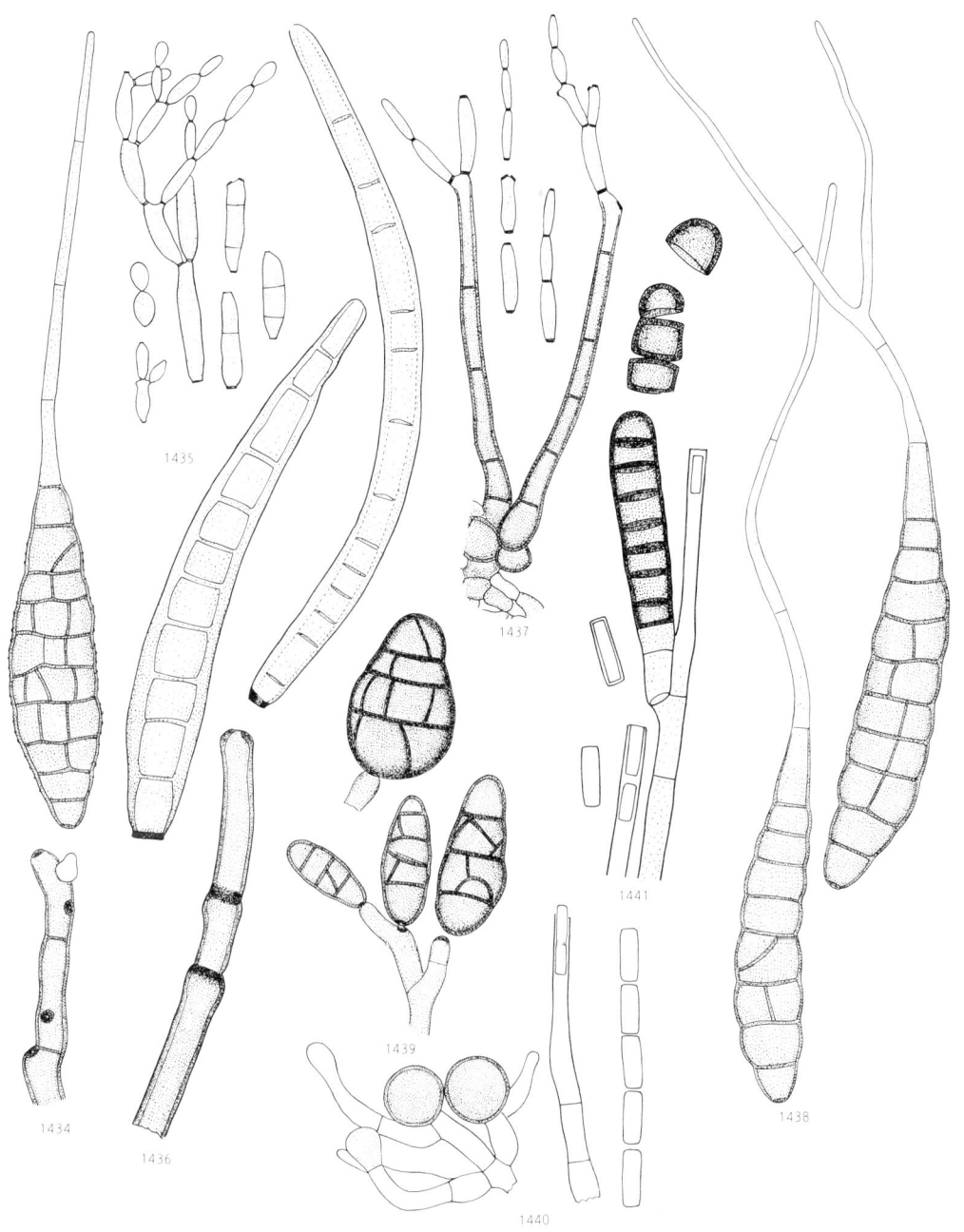

Plate 138. On *Cucumis*, *Dactylorhiza* and *Daucus*. 1434, *Alternaria cucumerina*; 1435, *Cladosporium cucumerinum*; 1436, *Corynespora cassiicola*; 1437, *Cladosporium orchidis*; 1438, *Alternaria dauci*; 1439, *A. radicina*; 1440, *Chalaropsis thielavioides*; 1441, *Thielaviopsis basicola*.

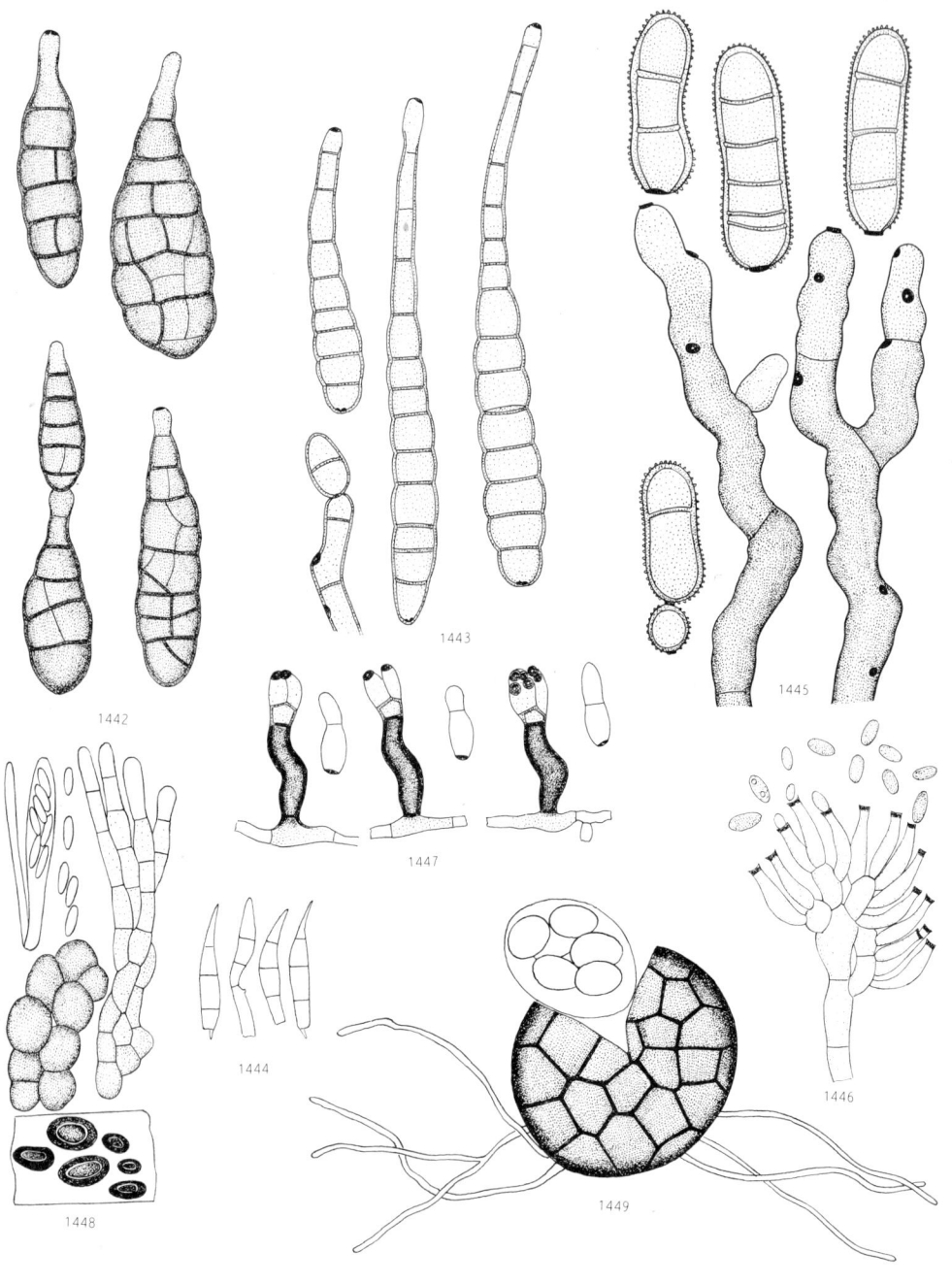

Plate 139. On *Dianthus*, *Digitalis* and *Doronicum*. 1442, *Alternaria dianthi*; 1443, *A. dianthicola*; 1444, *Heteropatella valtellinensis*; 1445, *Mycosphaerella dianthi*; 1446, *Phialophora cinerescens*; 1447, *Zygophiala jamaicensis*; 1448, *Pyrenopeziza digitalina*; 1449, *Sphaerotheca fusca*.

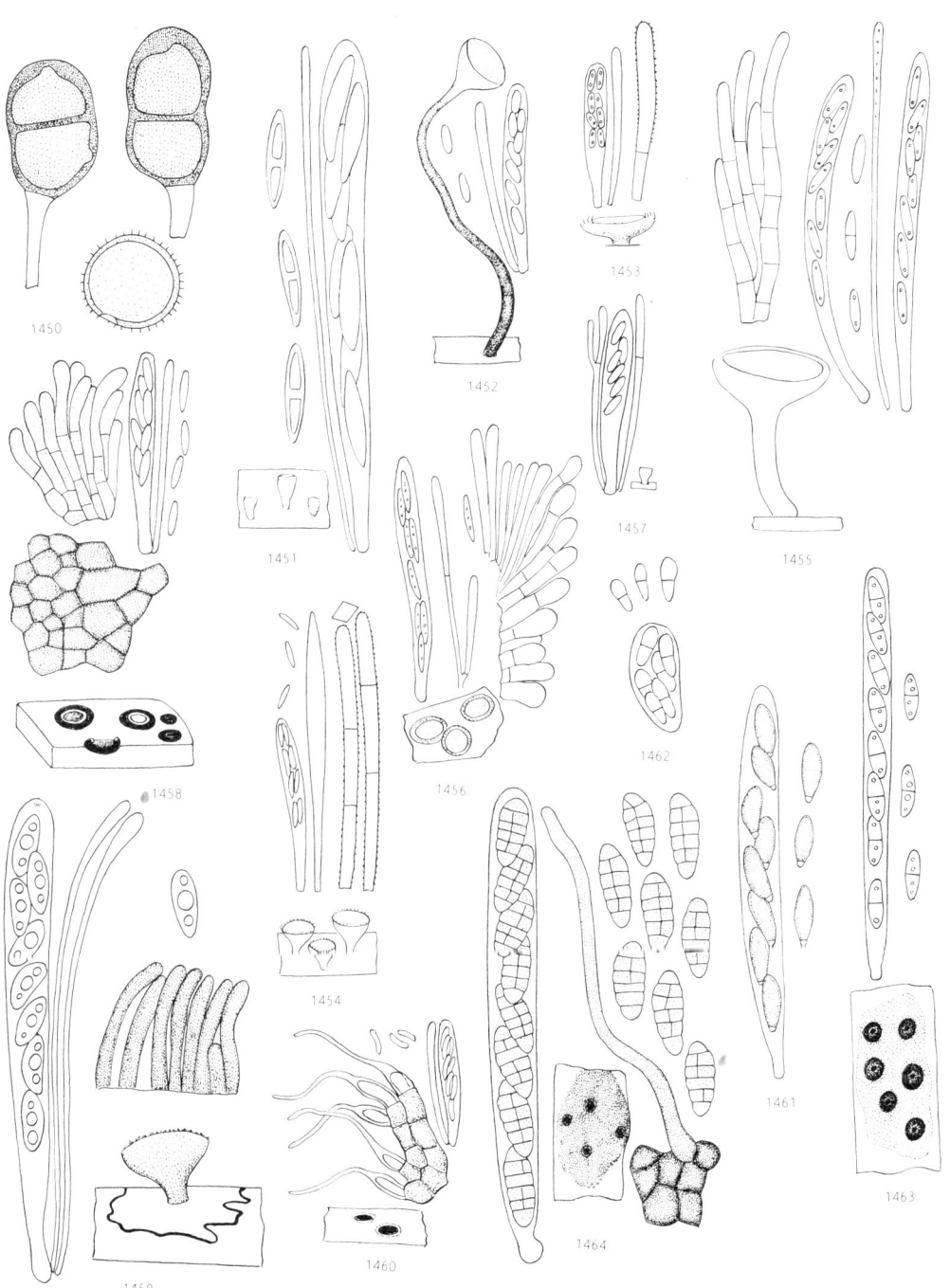

Plate 140. On *Epilobium*. 1450, *Puccinia pulverulenta*; 1451, *Allophylaria macrospora*; 1452, *Ciboriopsis tenuistipes*; 1453, *Dasyscyphus castaneus*; 1454, *D. nudipes* var. *minor*; 1455, *Hymenoscyphus repandus*; 1456, *Mollisia dilutella*; 1457, *Pezizella punctoidea*; 1458, *Pyrenopeziza chamaenerii*; 1459, *Rutstroemia hercynica*; 1460, *Unguicularia dilatopilosa*; 1461, *Anthostomella clypeoides*; 1462, *Morenoina epilobii*; 1463, *Paradidymella tosta*; 1464, *Pleospora epilobii*.

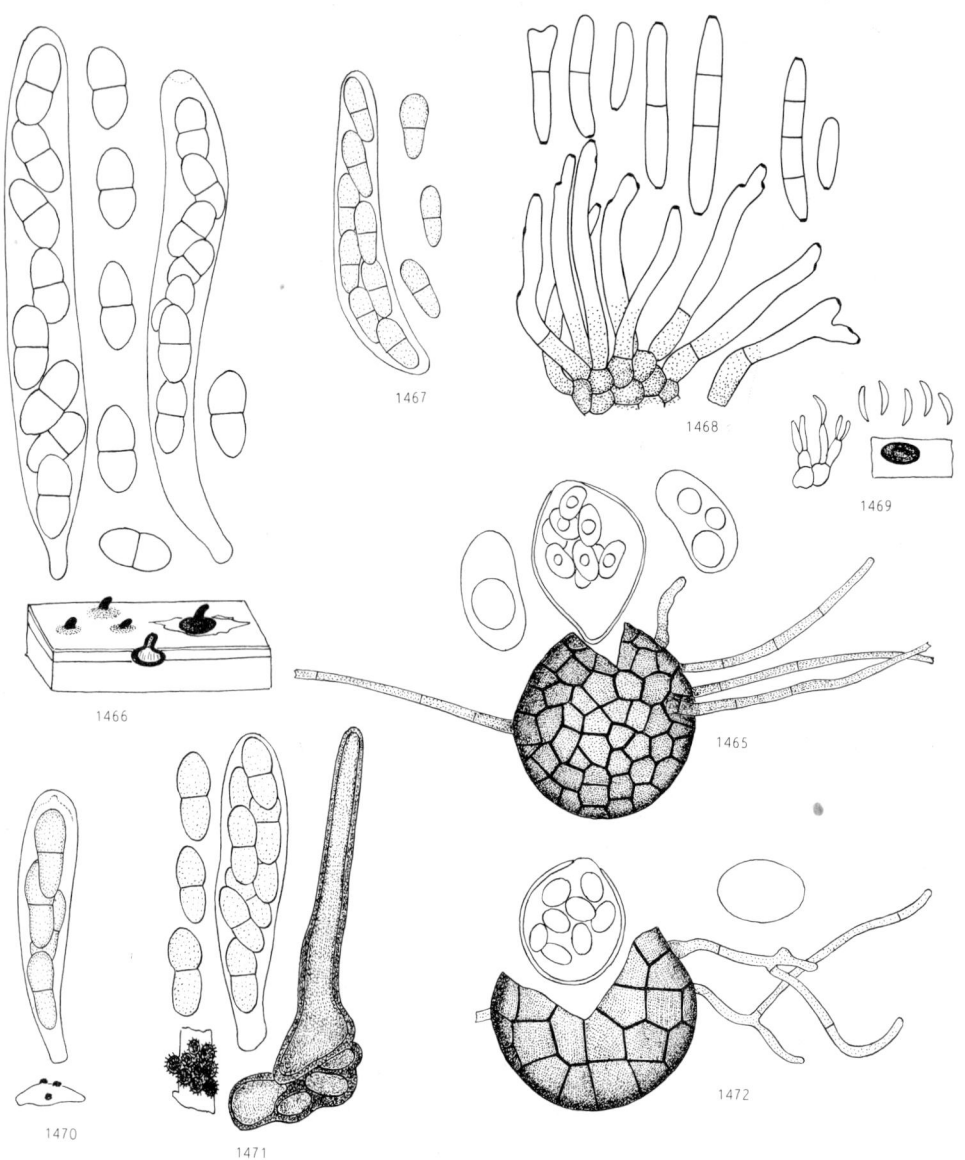

Plate 141. On *Epilobium*, *Erica* and *Erigeron*. 1465, *Sphaerotheca epilobii*; 1466, *Sydowiella fenestrans*; 1467, *Venturia maculiformis*; 1468, *Ramularia montana*; 1469, *Pilidium concavum*; 1470, *Gibbera salisburgensis*; 1471, *Protoventuria straussii*; 1472, *Sphaerotheca erigerontis-canadensis*.

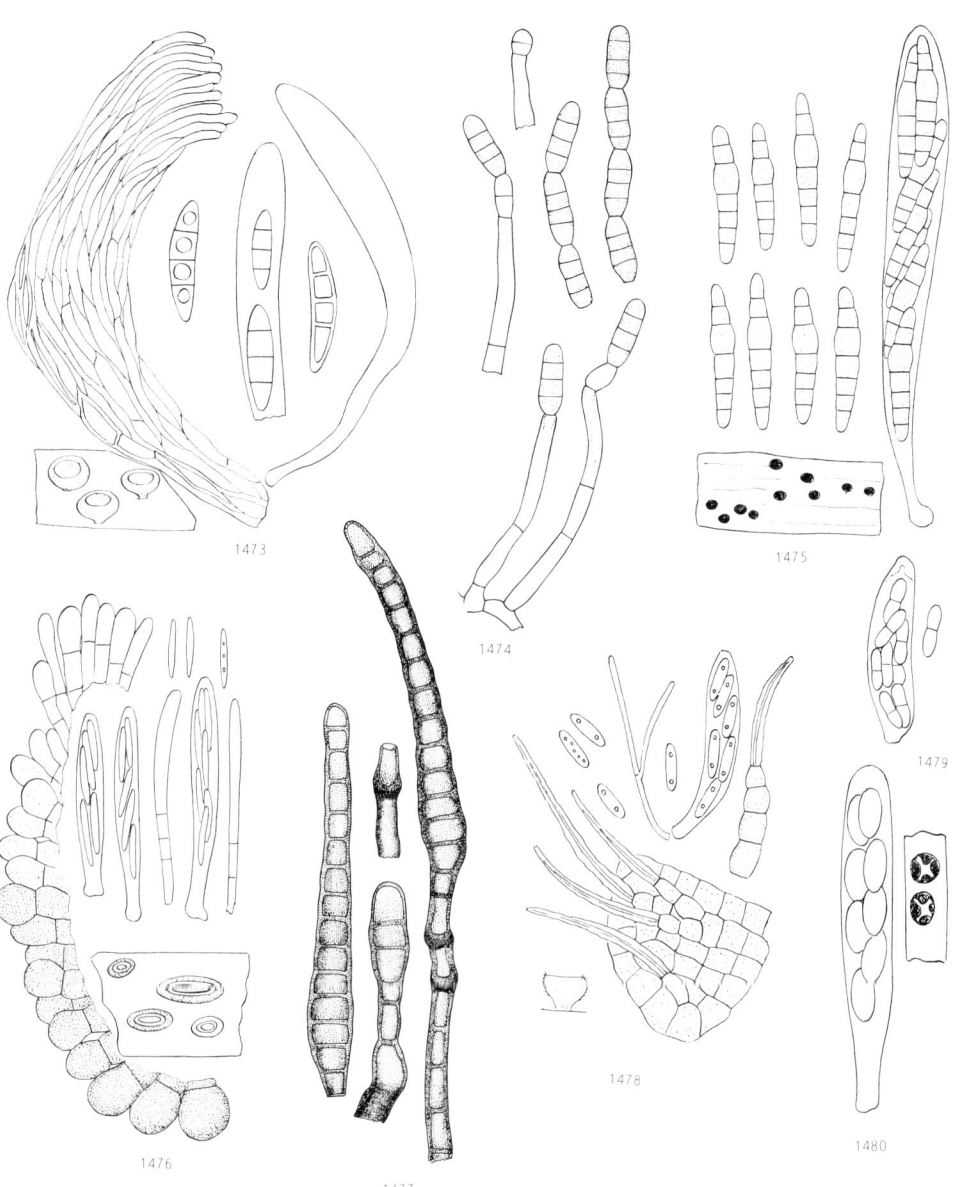

Plate 142. On *Eupatorium* and *Euphorbia*. 1473, *Allophylaria sublicoides*; 1474, *Fusariella sarniensis*; 1475, *Leptosphaeria agnita*; 1476, *Mollisia coerulans*; 1477, *Sporidesmium eupatoriicola*; 1478, *Hyalotricha niveocincta*; 1479, *Mycosphaerella euphorbiae*; 1480, *Naeviopsis tithymalina*.

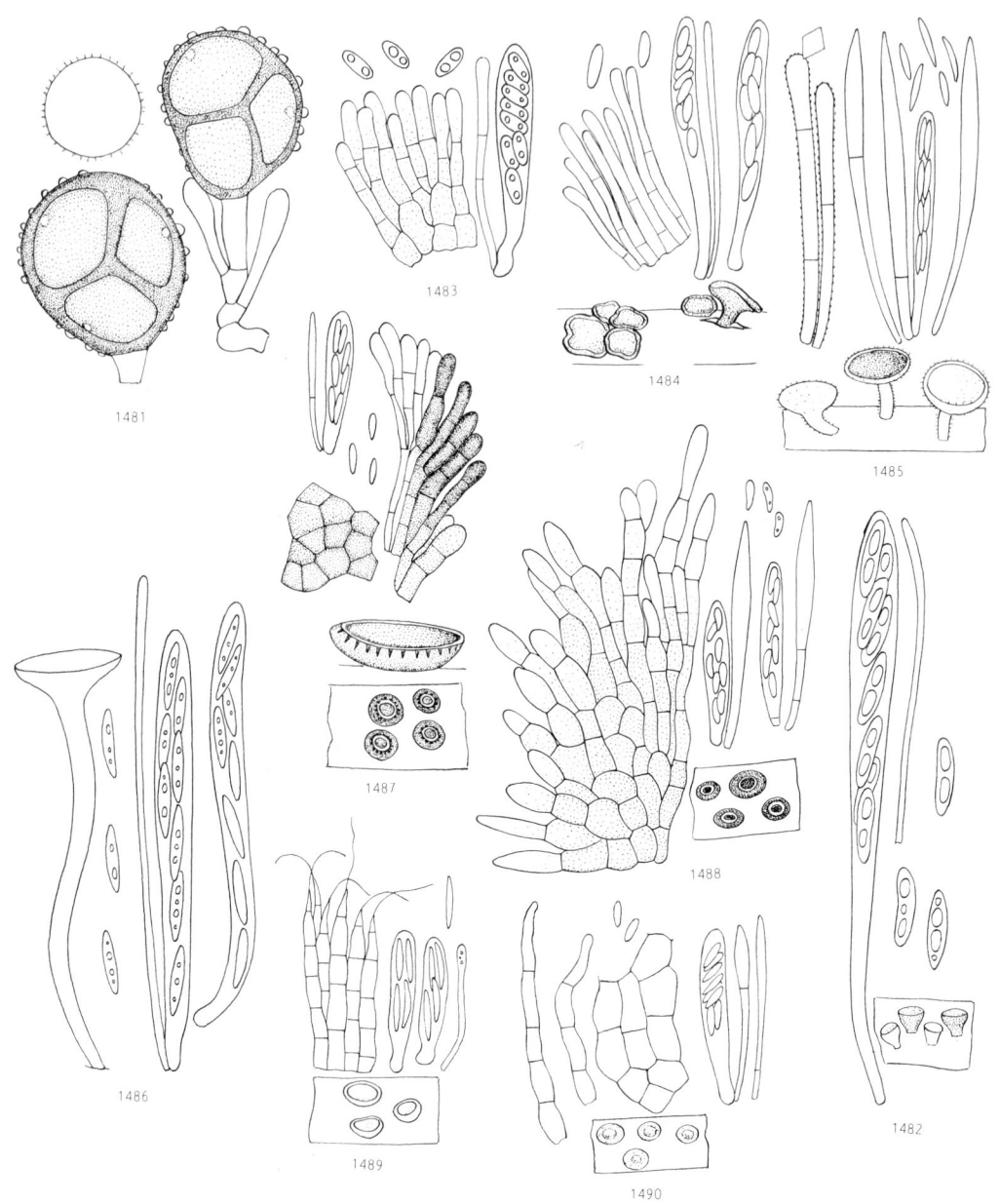

Plate 143. On *Filipendula*. 1481, *Triphragmium ulmariae*; 1482, *Allophylaria clavuliformis*; 1483, *Calycellina spiraeae*; 1484, *Dasyscyphus aeruginellus*; 1485, *D. nudipes*; 1486, *Hymenoscyphus vitellinus*; 1487, *Mollisia fuscostriata*; 1488, *Mollisiopsis lanceolata*; 1489, *Phialina ulmariae*; 1490, *Psilachnum rubrotinctum*.

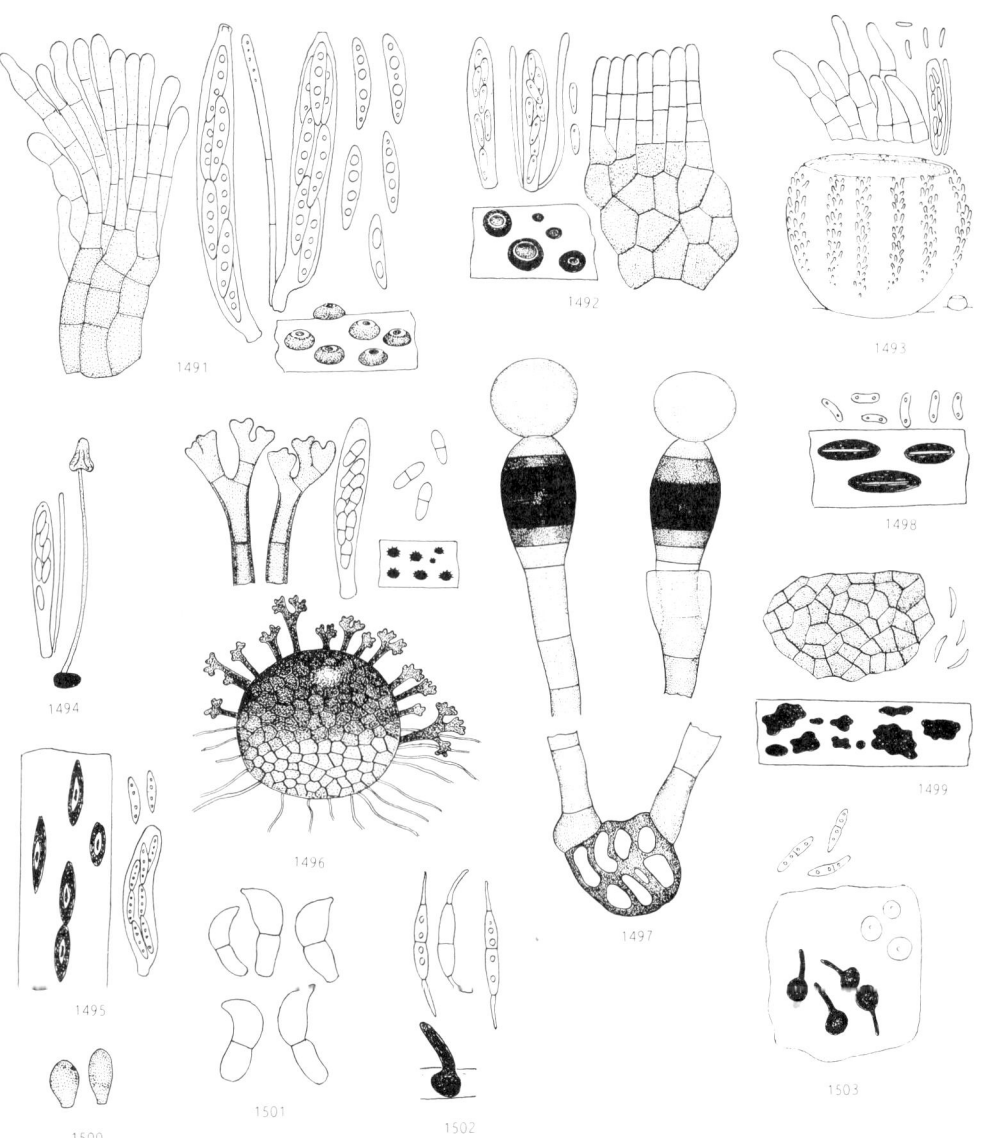

Plate 144. On *Filipendula* and *Fragaria*. 1491, *Pyrenopeziza millegrana*; 1492, *P. pulveracea*; 1493, *Unguicularia ulmariae*; 1494, *Verpatinia spiraeicola*; 1495, *Diaporthe lirella*; 1496, *Wentiomyces* sp.; 1497, *Endophragmia prolifera*; 1498, *Leptostroma spiraeinum*; 1499, *Thyriostroma spiraeae*; 1500, *Coniella fragariae*; 1501, *Diplocarpon earlianum*; 1502, *Gnomonia fragariae*; 1503, *G. fructicola*.

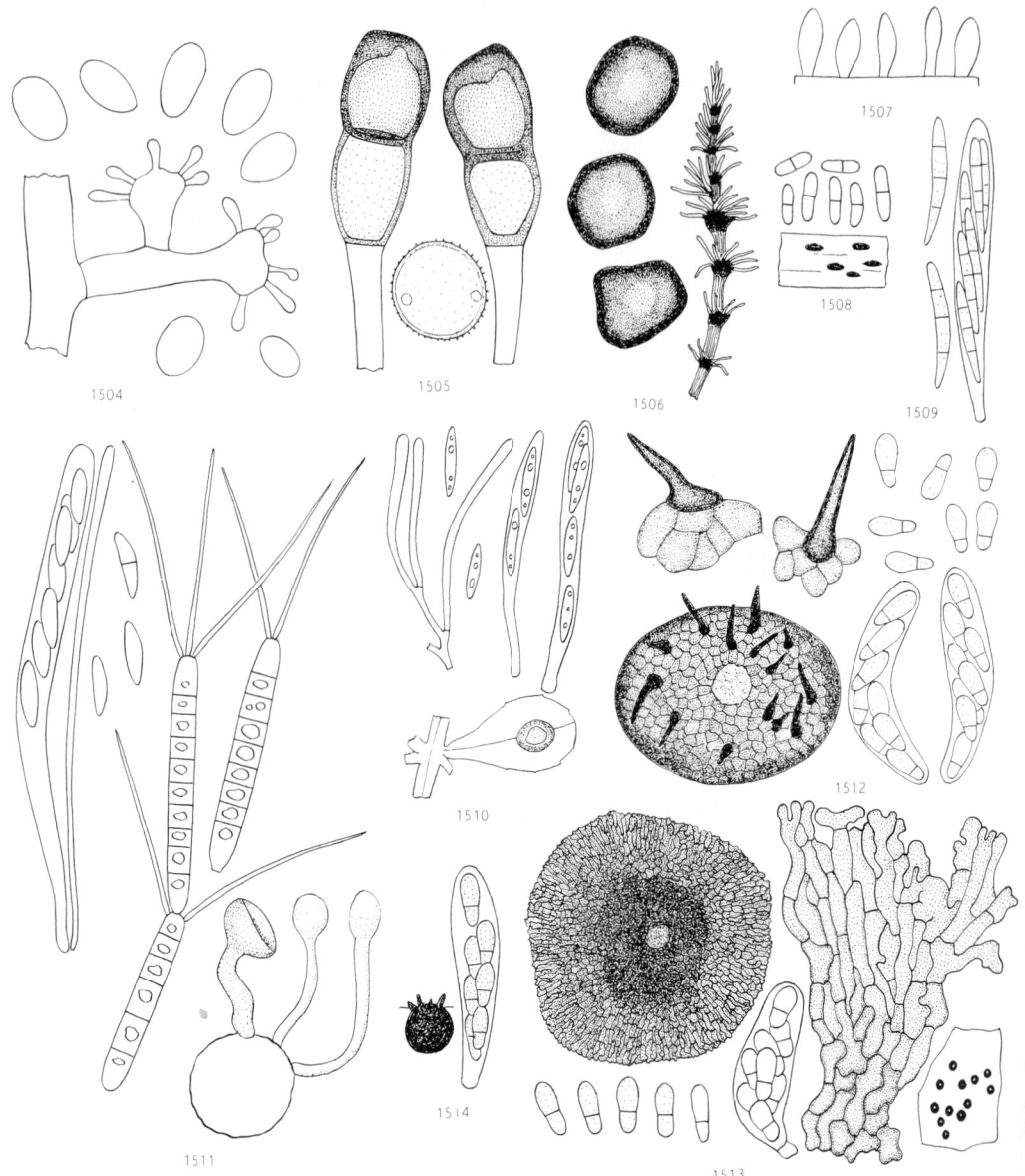

Plate 145. On *Galanthus*, *Galium* and *Geranium*. 1504, *Botrytis galanthina*; 1505, *Puccinia punctata*; 1506, *Melanotaenium endogenum*; 1507, *Acrospermum pallidulum*; 1508, *Diplodina galii*; 1509, *Leptosphaeria scitula*; 1510, *Leptotrochila verrucosa*; 1511, *Symphyosirinia galii*; 1512, *Coleroa circinans*; 1513, *Hormotheca robertiani*; 1514, *Venturia geranii*.

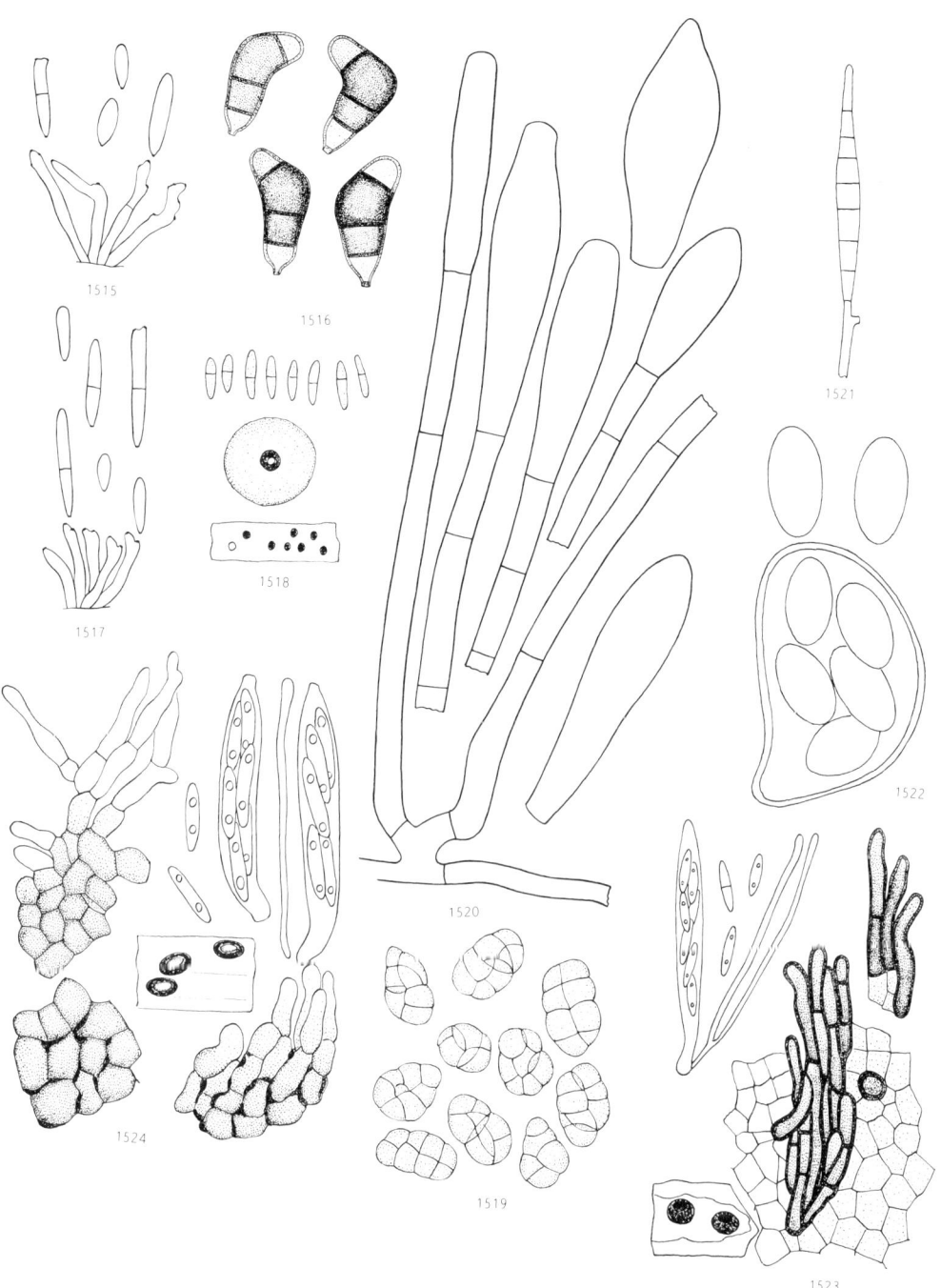

Plate 146. On *Geum, Gladiolus, Glechoma, Halimione, Helianthemum* and *Heracleum*. 1515, *Ramularia gei*; 1516, *Curvularia trifolii* f.sp. *gladioli*; 1517, *Ramularia calcea*; 1518, *Ascochytula obiones*; 1519, *Camarosporium obiones*; 1520, *Leveillula taurica*; 1521, *Dactylella arnaudii*; 1522, *Erysiphe heraclei*; 1523, *Pirottaea nigrostriata*; 1524, *Pyrenopeziza chailletii*.

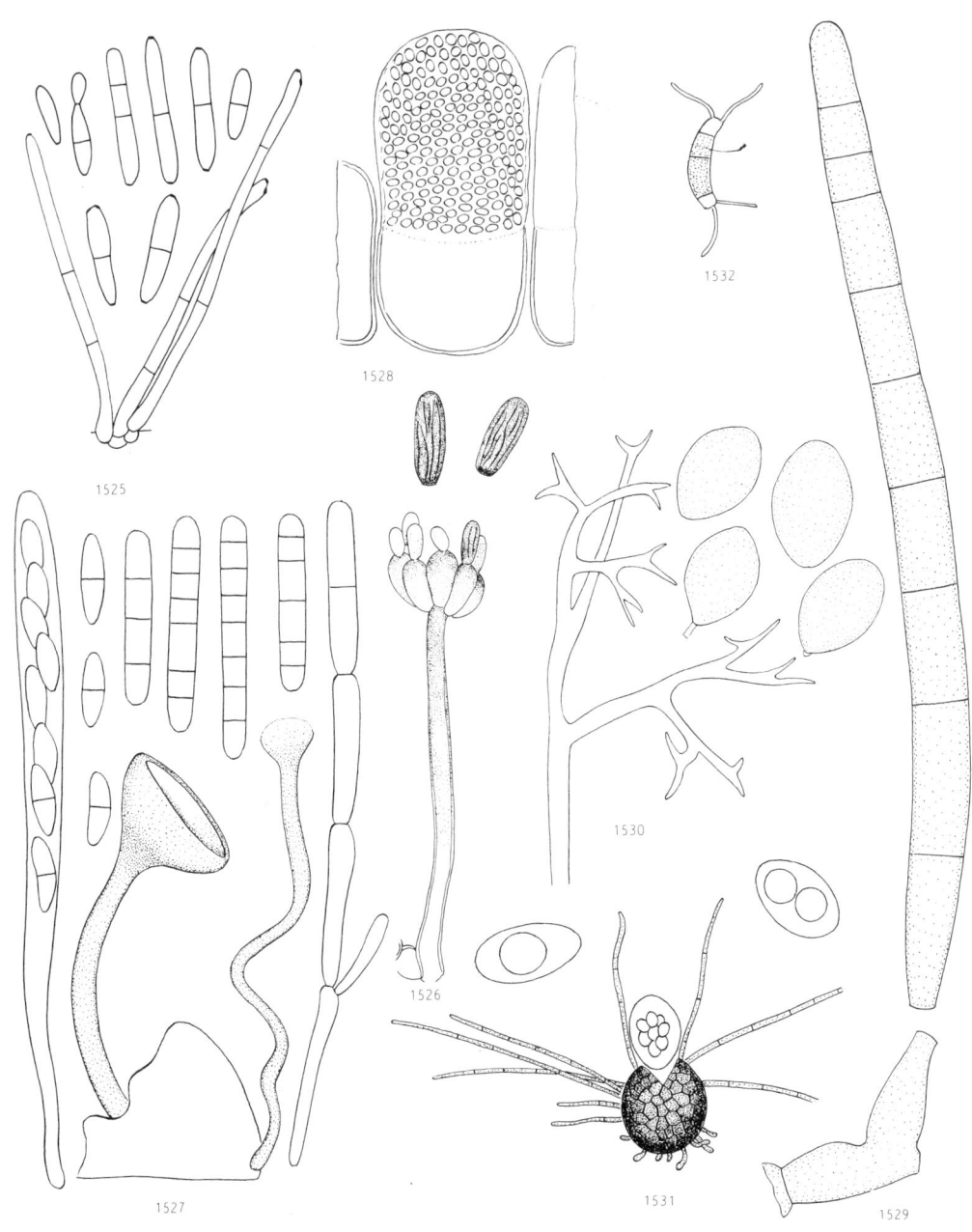

Plate 147. On *Heracleum*, *Humulus* and *Hypericum*. 1525, *Ramularia heraclei*; 1526, *Stachybotrys cylindrospora*; 1527, *Symphyosirinia heraclei*; 1528, *Taphridium umbelliferarum*; 1529, *Mycocentrospora cantuariensis*; 1530, *Pseudoperonospora humuli*; 1531, *Sphaerotheca macularis*; 1532, *Seimatosporium hypericinum*.

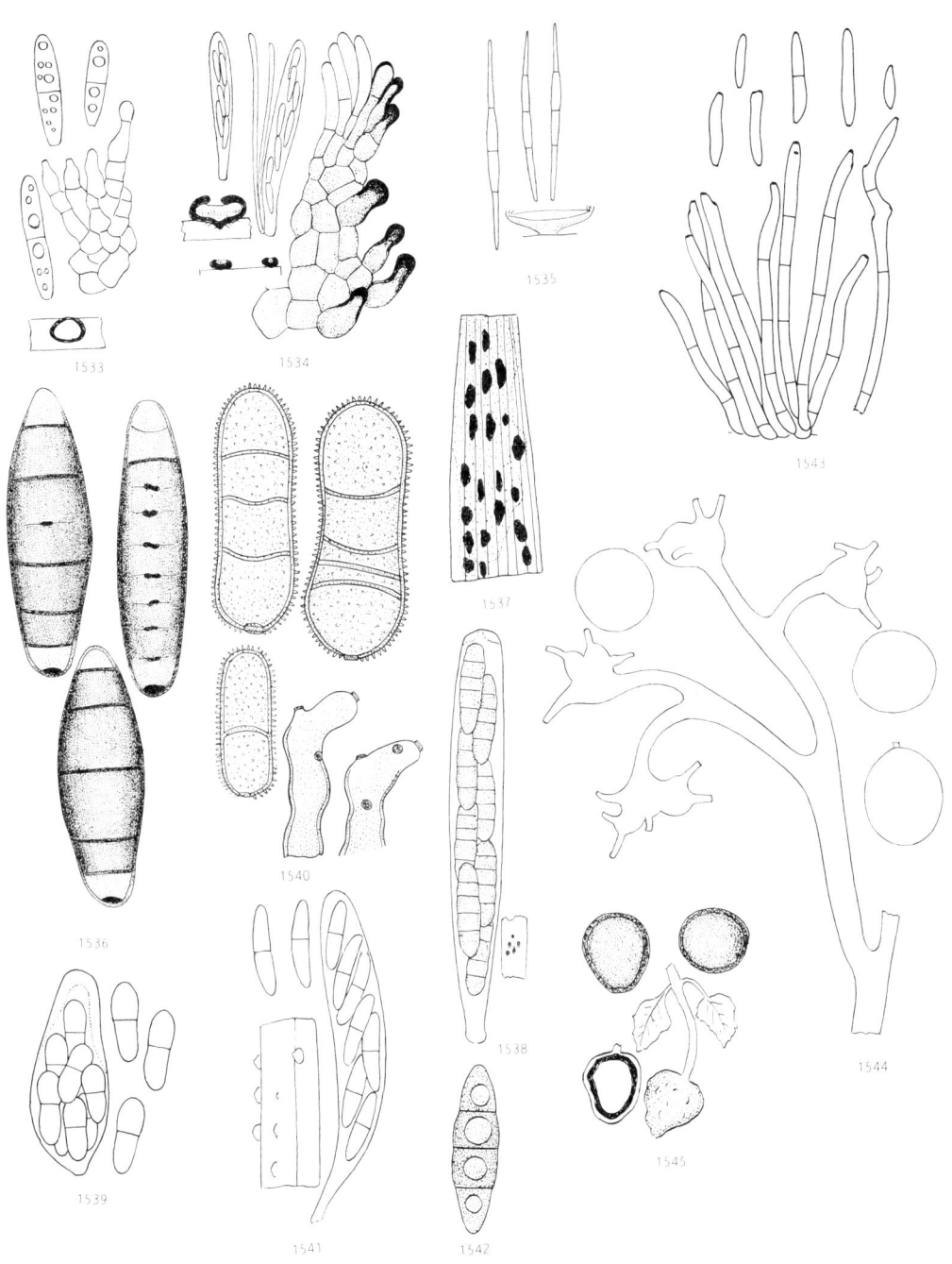

Plate 148. On *Iris, Knautia, Lactuca* and *Lamium*. 1533, *Ascochyta pseudacori*; 1534, *Belonium nigromaculatum*; 1535, *Belonopsis iridis*; 1536, *Drechslera iridis*; 1537, *Ectostroma iridis*; 1538, *Leptosphaeria vectis*; 1539, *Mycosphaerella iridis*; 1540, *M. macrospora*; 1541, *Nectriella dacrymycella*; 1542, *Trematosphaeria heterospora*; 1543, *Ramularia knautiae*; 1544, *Bremia lactucae*; 1545, *Melanotaenium lamii*.

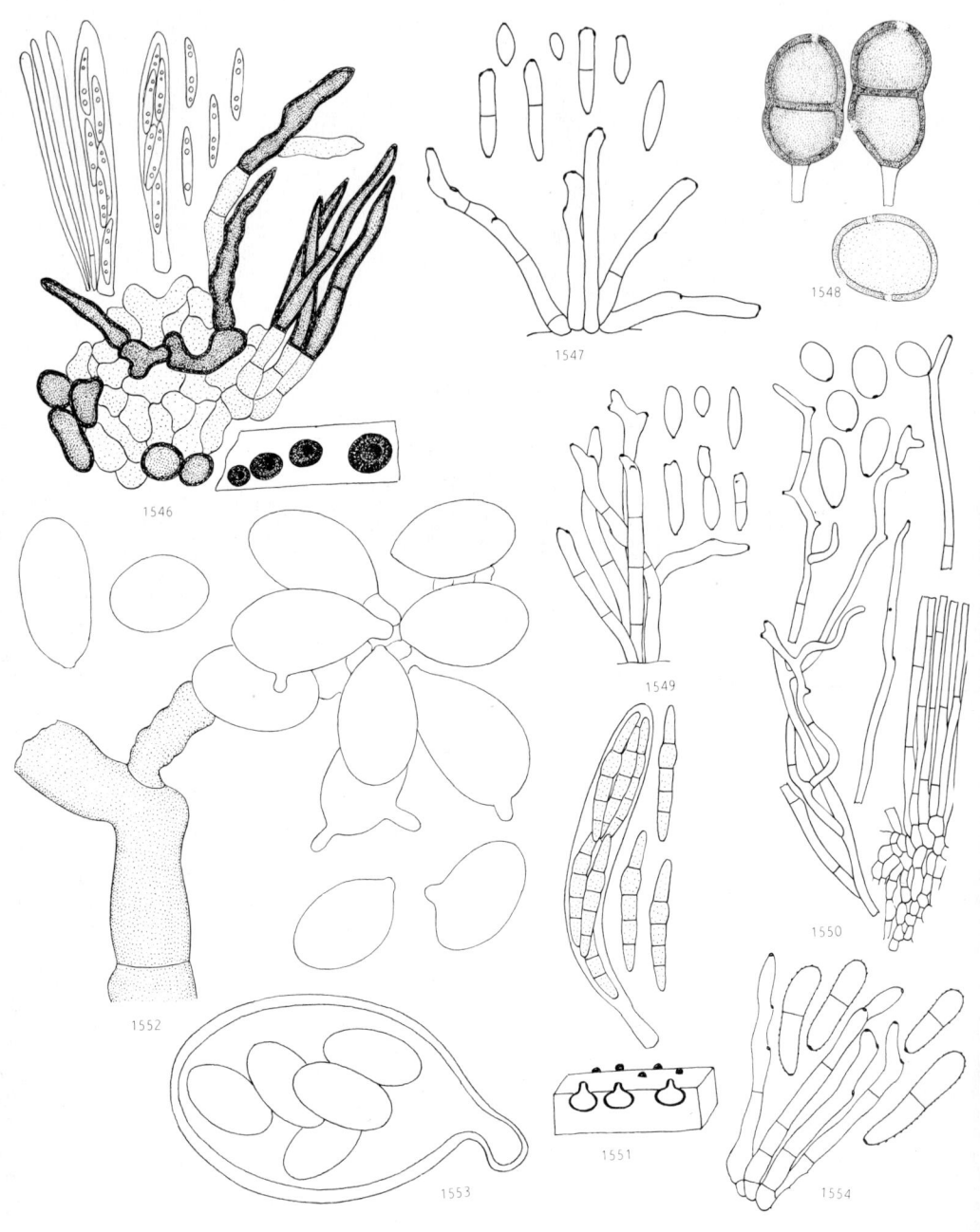

Plate 149. On *Lamium, Lapsana, Lathyrus, Lilium, Limonium* and *Linaria*. 1546, *Pirottaea veneta*; 1547, *Ramularia lamiicola*; 1548, *Puccinia lapsanae*; 1549, *Ramularia lapsanae*; 1550, *Isariopsis carnea*; 1551, *Leptosphaeria niessliana*; 1552, *Botrytis elliptica*; 1553, *Erysiphe limonii*; 1554, *Didymaria linariae*.

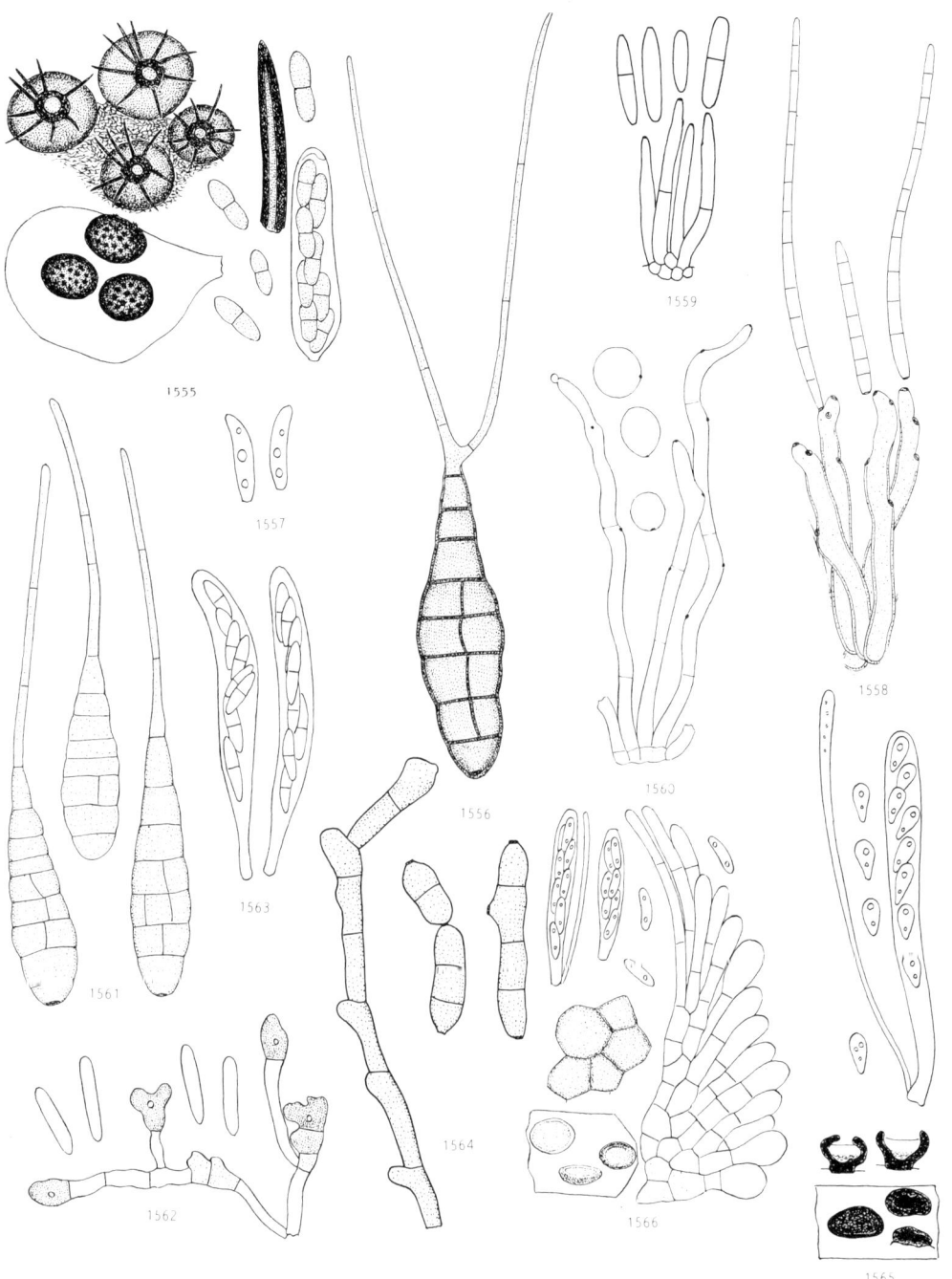

Plate 150. On *Linnaea, Linum, Lotus, Lycopersicon* and *Lycopus*. 1555, *Metacoleroa dickiei*; 1556, *Alternaria linicola*; 1557, *Colletotrichum lini*; 1558, *Cercospora loti*; 1559, *Ramularia schulzeri*; 1560, *R. sphaeroidea*; 1561, *Alternaria tomato*; 1562, *Colletotrichum coccodes*; 1563, *Didymella lycopersici*; 1564, *Fulvia fulva*; 1565, *Mollisia lycopi*; 1566, *Pyrenopeziza lycopincola*.

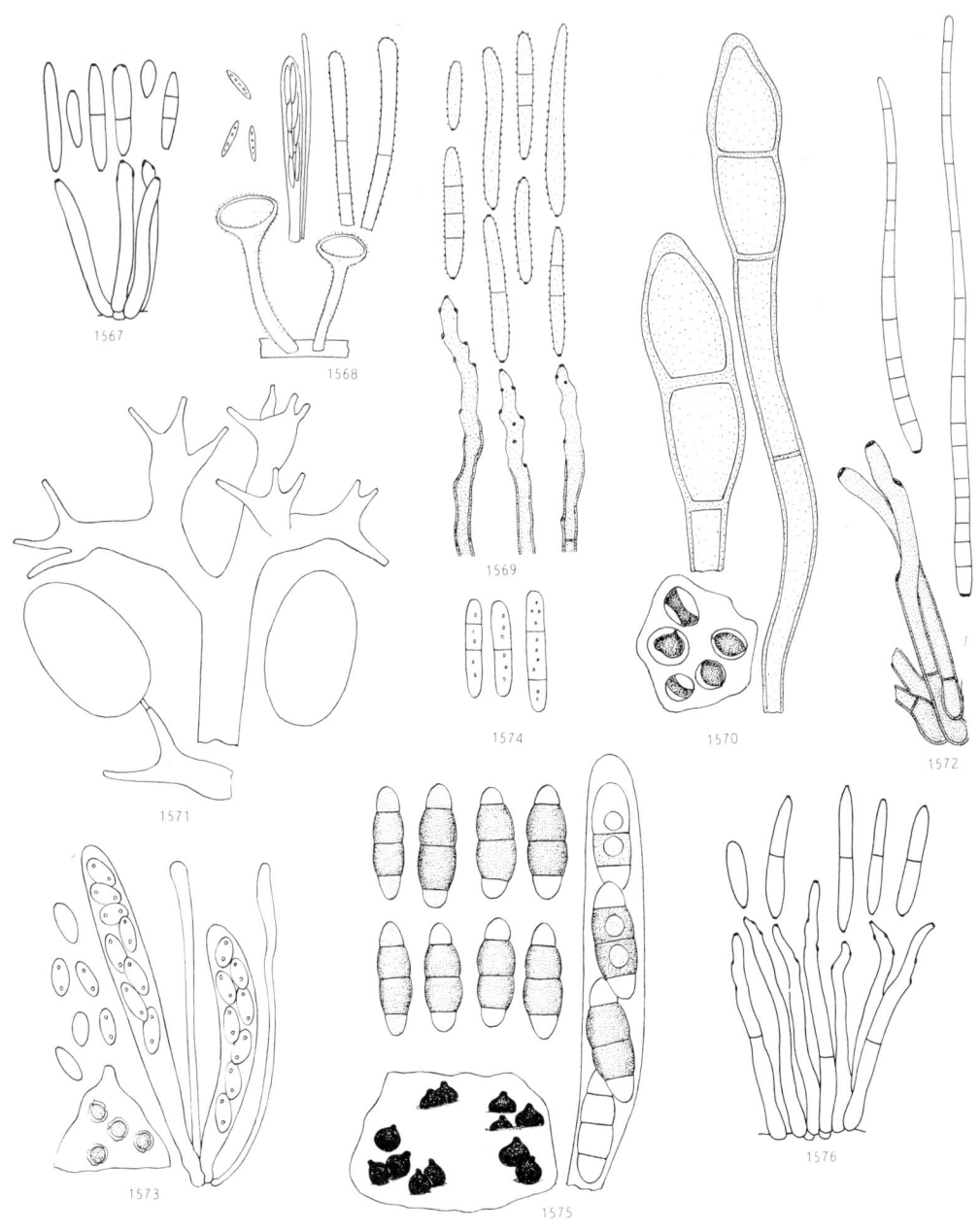

Plate 151. On *Lysimachia, Lythrum, Malva, Matricaria, Medicago* and *Mentha*. 1567, *Ramularia lysimachiarum*; 1568, *Dasyscyphus salicariae*; 1569, *Stenella lythri*; 1570, *Puccinia malvacearum*; 1571, *Peronospora leptosperma*; 1572, *Cercospora zebrina*; 1573, *Pseudopeziza medicaginis*; 1574, *Stagonospora meliloti*; 1575, *Trematosphaeria circinans*; 1576, *Ramularia menthicola*.

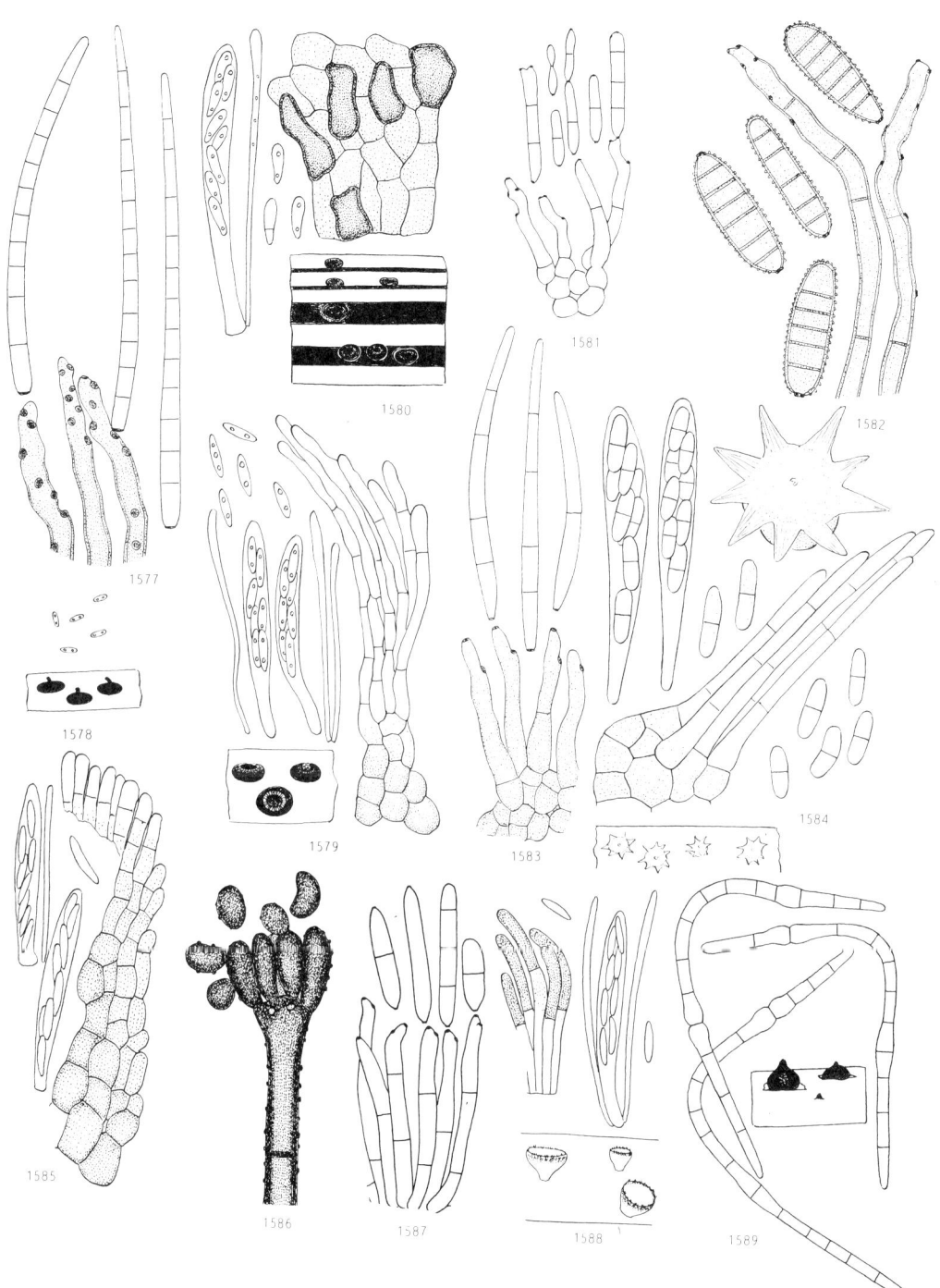

Plate 152. On *Mercurialis, Moehringia, Narthecium, Nigella, Oenanthe, Onobrychis* and *Ononis*. 1577, *Cercospora mercurialis*; 1578, *Phoma macrocapsa*; 1579, *Pyrenopeziza mercurialis*; 1580, *Spilopodia melanogramma*; 1581, *Ramularia arenariae*; 1582, *Cladosporium magnusianum*; 1583, *Cercospora nigellae*; 1584, *Nectria boothii*; 1585, *Pyrenopeziza subplicata*; 1586, *Stachybotrys oenanthes*; 1587, *Ramularia onobrychidis*; 1588, *Crocicreas cyathoideum* var. *cacaliae*; 1589, *Ophiobolus fruticum*.

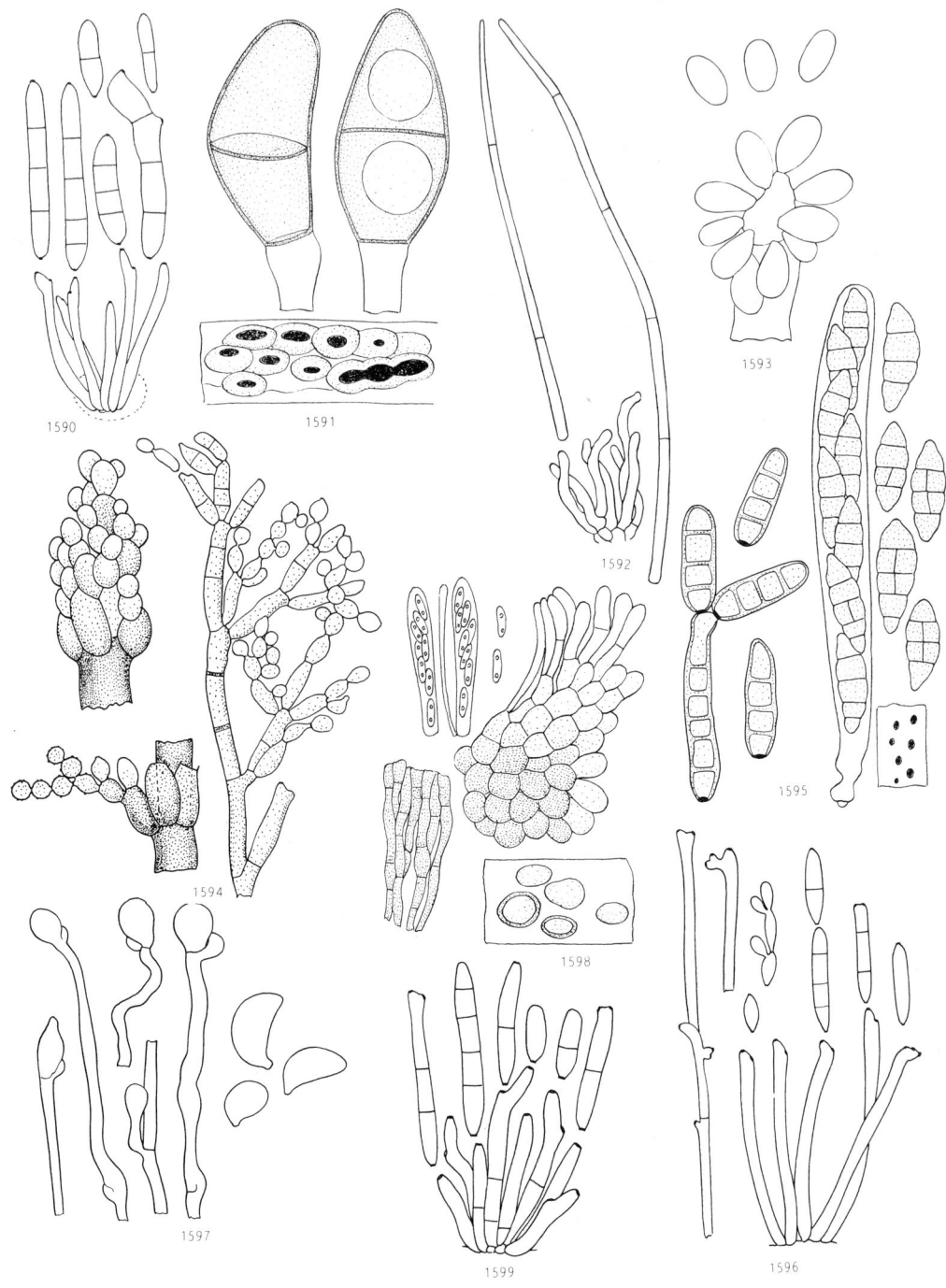

Plate 153. On *Ononis, Ornithogalum, Oxyria, Paeonia, Papaver, Parietaria* and *Pastinaca*. 1590, *Ramularia winteri*; 1591, *Puccinia liliacearum*; 1592, *Cerceseptoria oxyriae*; 1593, *Botrytis paeoniae*; 1594, *Cladosporium chlorocephalum*; 1595, *Pleospora papaveracea*; 1596, *Ramularia parietariae*; 1597, *Itersonilia pastinacae*; 1598, *Pyrenopeziza pastinacae*; 1599, *Ramularia pastinacae*.

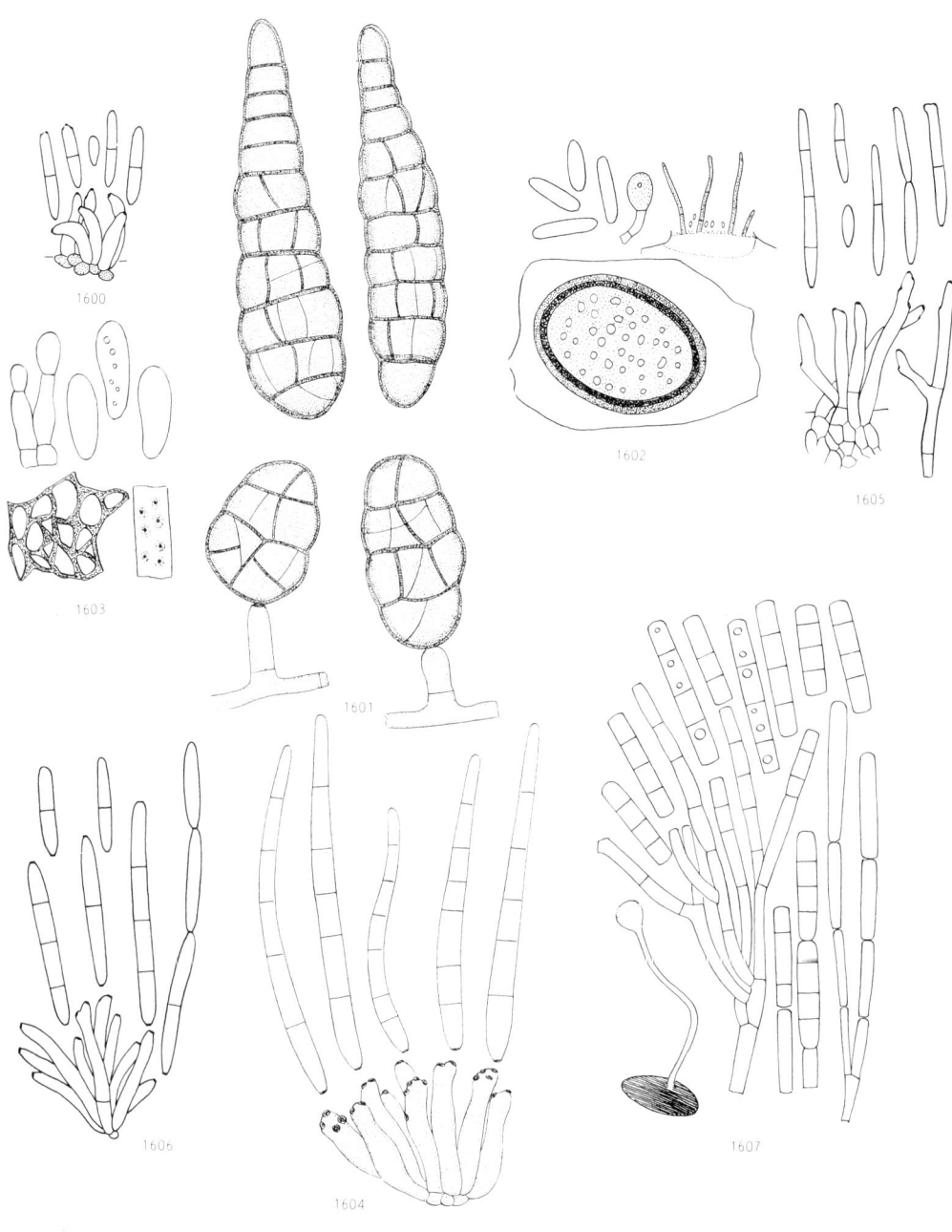

Plate 154. On *Petasites, Petroselinum, Phaseolus, Physospermum, Picris* and *Pimpinella*. 1600, *Ramularia purpurascens*; 1601, *Alternaria petroselini*; 1602, *Colletotrichum lindemuthianum*; 1603, *Macrophomina phaseolina*; 1604, *Cercospora physospermi*; 1605, *Ramularia filaris*; 1606, *R. picridis*; 1607, *Symphyosira rosea*.

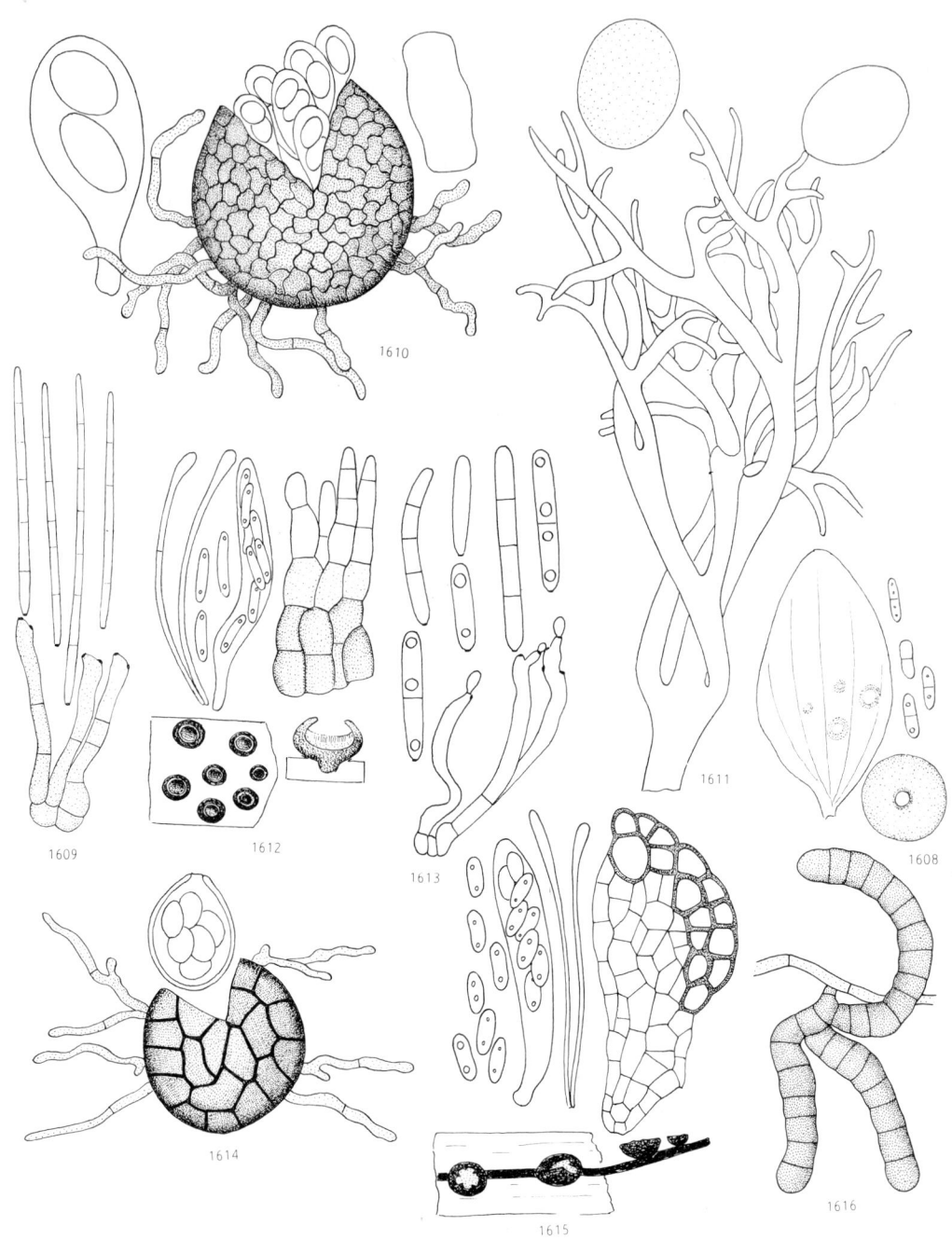

Plate 155. On *Plantago*. 1608, *Ascochyta plantaginis*; 1609, *Cercospora plantaginis*; 1610, *Erysiphe sordida*; 1611, *Peronospora alta*; 1612, *Pyrenopeziza plantaginis*; 1613, *Ramularia rhabdospora*; 1614, *Sphaerotheca plantaginis*; 1615, *Spilopodia nervisequia*; 1616, *Taeniolella plantaginis*.

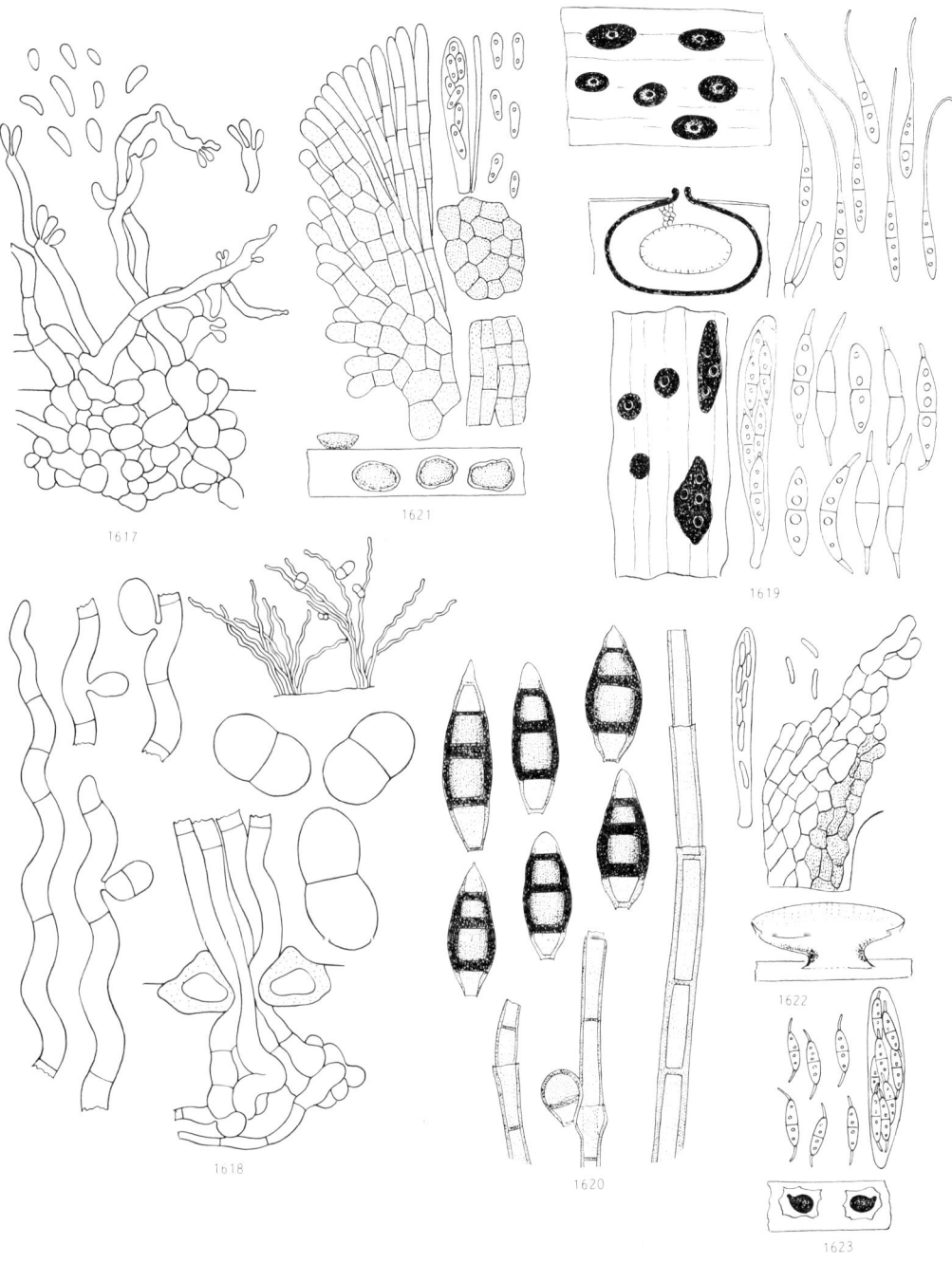

Plate 156. On *Polygonatum* and *Polygonum*. 1617, *Kabatiella microsticta*; 1618, *Bostrichonema alpestre*; 1619, *Ceriospora polygonacearum*; 1620, *Endophragmia cesatii*; 1621, *Mollisia polygoni*; 1622, *Pezizella effugiens*; 1623, *Plagiostoma devexa*.

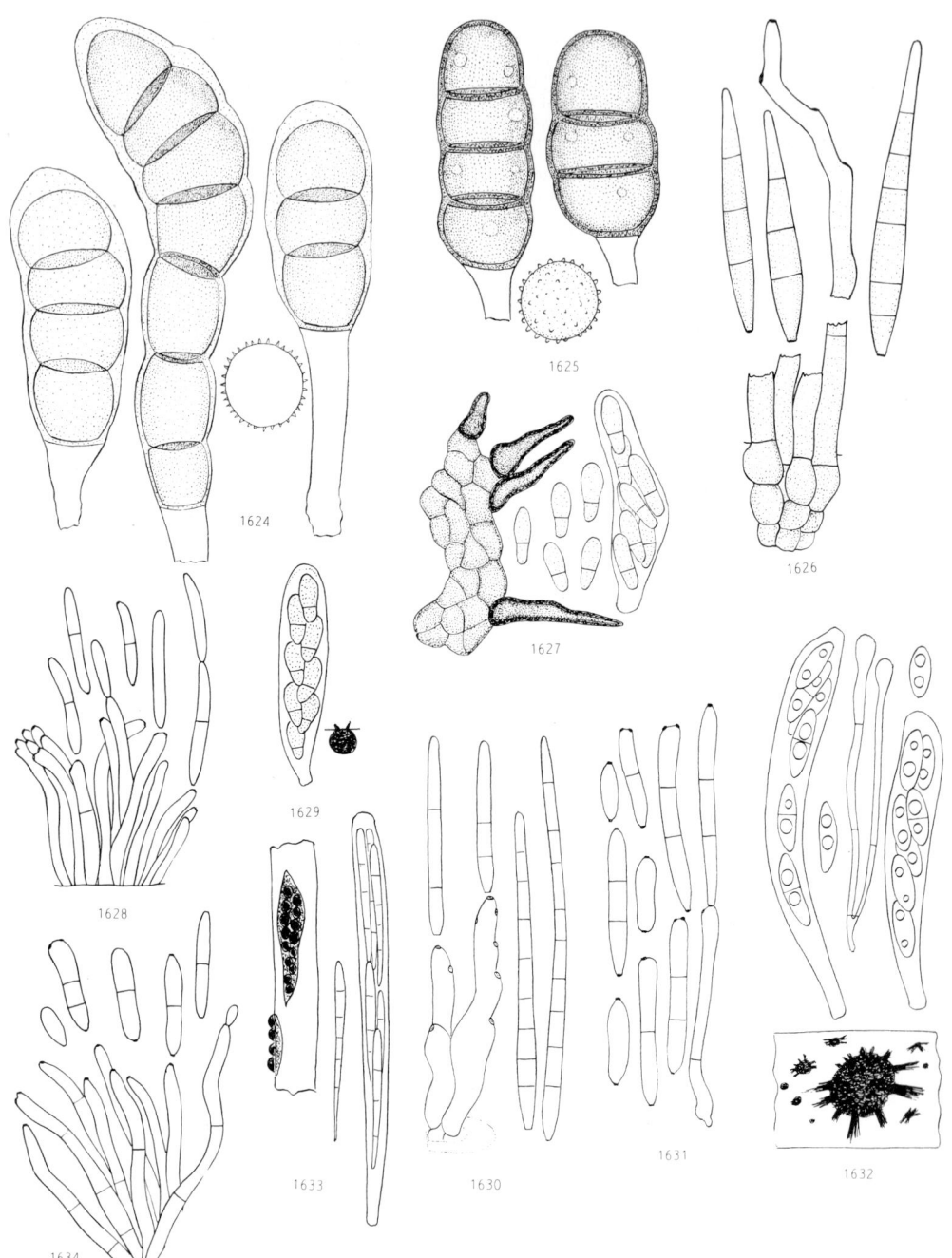

Plate 157. On *Potentilla, Primula, Prunella* and *Pulicaria*. 1624, *Frommea obtusa*; 1625, *Phragmidium fragariae*; 1626, *Cercospora comari*; 1627, *Coleroa potentillae*; 1628, *Ramularia arvensis*; 1629, *Venturia palustris*; 1630, *Cercosporella primulae*; 1631, *Ramularia primulae*; 1632, *Leptotrochila brunellae*; 1633, *Rosenschoeldia abundans*; 1634, *Ramularia cupulariae*.

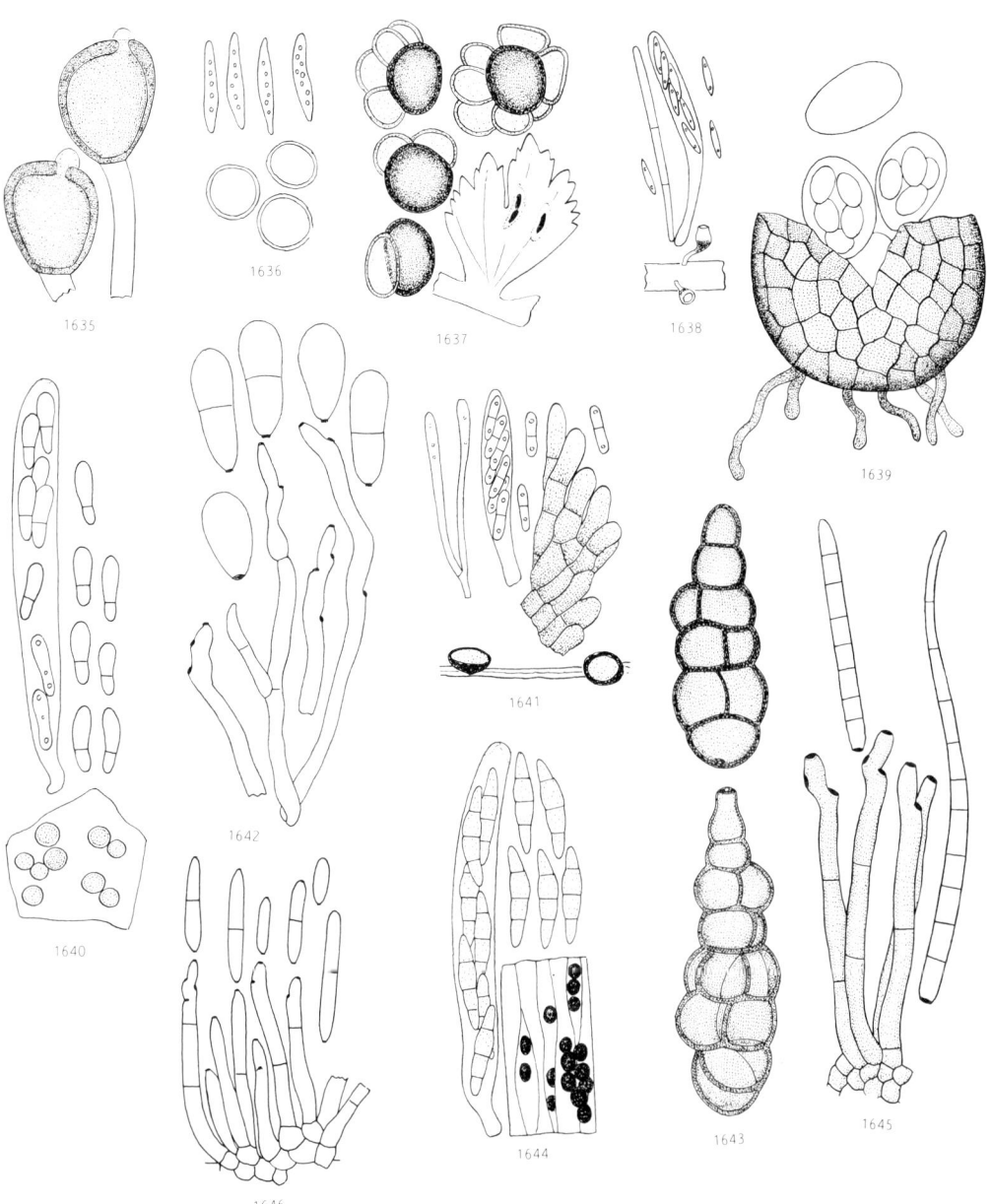

Plate 158. On *Ranunculus, Raphanus, Reseda* and *Rheum*. 1635, *Uromyces ficariae*; 1636, *Entyloma microsporum*; 1637, *Urocystis ranunculi*; 1638, *Crocicreas starbaeckii*; 1639, *Erysiphe ranunculi*; 1640, *Leptotrochila ranunculi*; 1614, *Niptera* sp.; 1642, *Ramularia didyma*; 1643, *Alternaria raphani*; 1644, *Leptosphaeria raphani*; 1645, *Cercospora resedae*; 1646, *Ramularia rhei*.

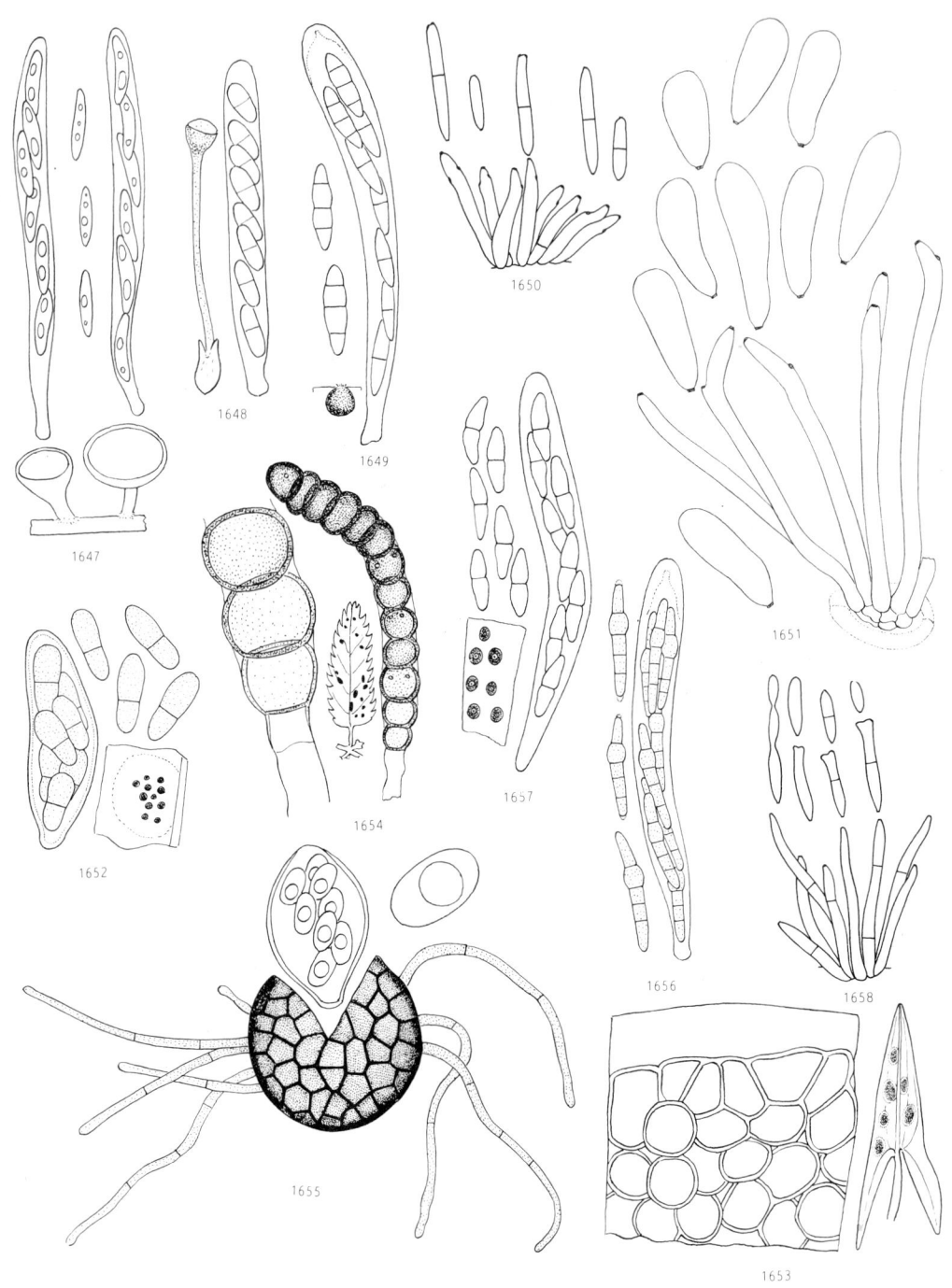

Plate 159. On *Rumex, Sagittaria, Sanguisorba, Scabiosa* and *Scrophularia*. 1647, *Helotium consobrinum*; 1648, *Hymenoscyphus rumicis*; 1649, *Keissleriella gallica*; 1650, *Ramularia pratensis*; 1651, *Ramularia rubella*; 1652, *Venturia rumicis*; 1653, *Doassansia sagittariae*; 1654, *Xenodochus carbonarius*; 1655, *Sphaerotheca ferruginea*; 1656, *Leptosphaeria modesta*; 1657, *Didymella commanipula*; 1658, *Ramularia scrophulariae*.

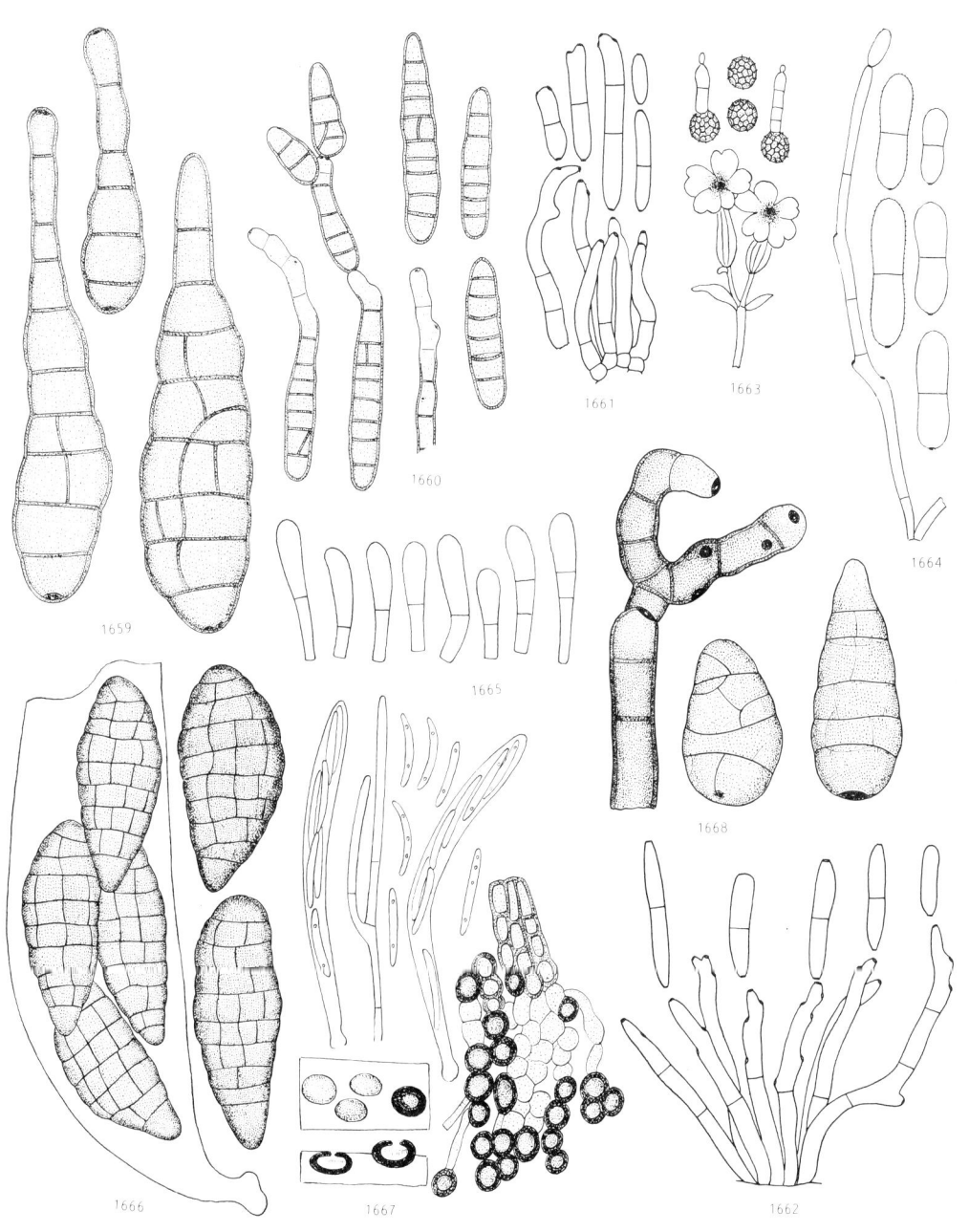

Plate 160. On *Senecio*, *Silene* and *Smyrnium*. 1659, *Alternaria cinerariae*; 1660, *A. dennisii*; 1661, *Ramularia pruinosa*; 1662, *R. senecionis*; 1663, *Ustilago violacea*; 1664, *Didymaria kriegeriana*; 1665, *Diplosporonema delastrei*; 1666, *Pleospora androsaces*; 1667, *Pyrenopeziza lychnidis*; 1668, *Alternaria ramulosa*.

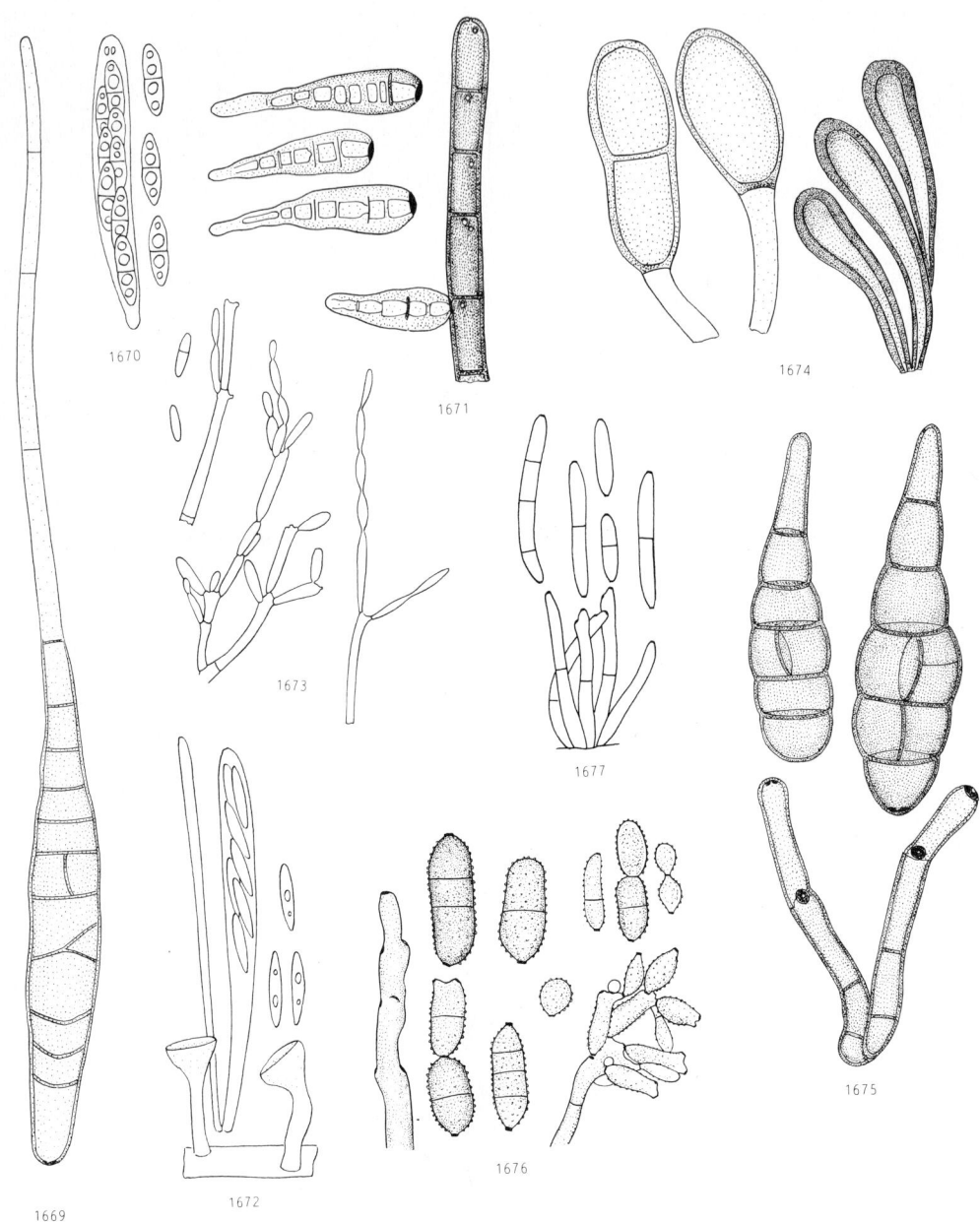

Plate 161. On *Solanum, Sonchus, Spinacea* and *Stachys*. 1669, *Alternaria solani*; 1670, *Diaporthe sarothamni* var. *dulcamarae*; 1671, *Helminthosporium solani*; 1672, *Hymenoscyphus scutula* var. *solani*; 1673, *Polyscytalum pustulans*; 1674, *Miyagia pseudosphaeria*; 1675, *Alternaria sonchi*; 1676, *Cladosporium variabile*; 1677, *Ramularia stachydis*.

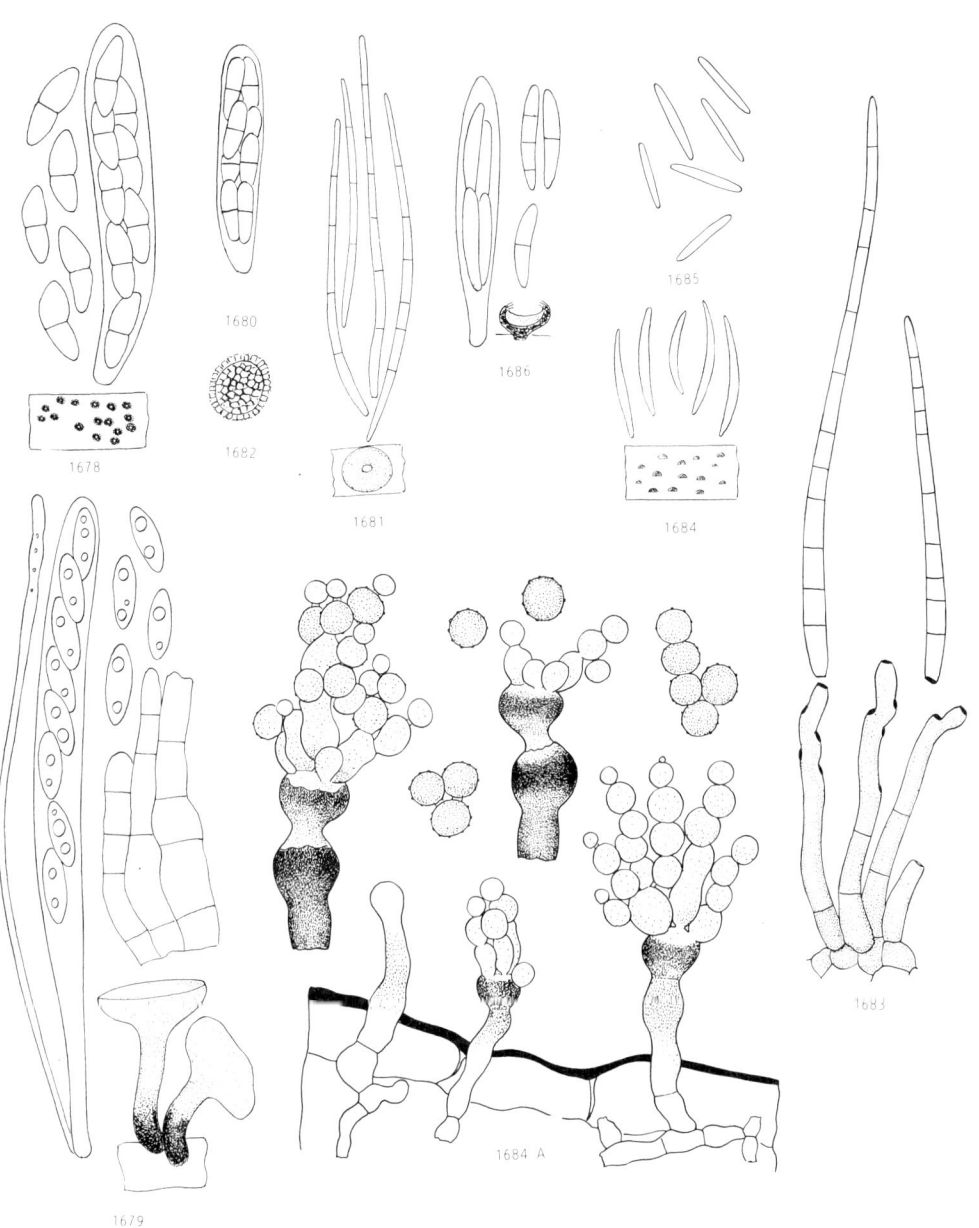

Plate 162. On *Stellaria, Succisa, Tamus* and *Thalictrum*. 1678, *Didymella holosteae*; 1679, *Lanzia stellariae*; 1680, *Mycosphaerella isariophora*; 1681, *Septoria stellariae*; 1682, *Ustilago succisae*; 1683, *Cercospora scandens*; 1684, *Cryptosporium tami*; 1684A, *Haplobasidion thalictri*; 1685, *Myrothecium carmichaelii*; 1686, *Pyrenopeziza thalictri*.

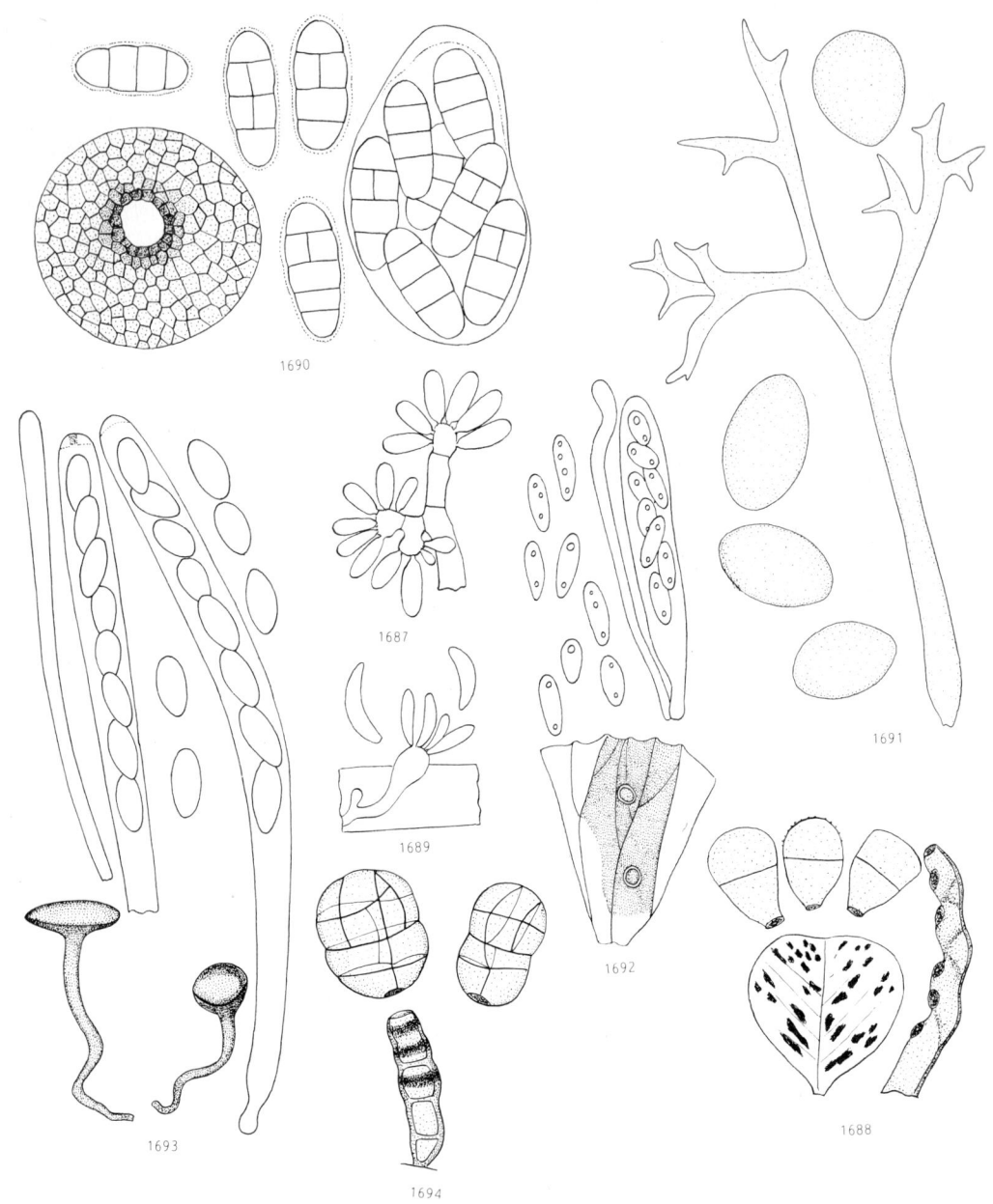

Plate 163. On *Trifolium*. 1687, *Botrytis anthophila*; 1688, *Cymadothea trifolii*; 1689, *Kabatiella caulivora*; 1690, *Leptosphaerulina trifolii*; 1691, *Peronospora trifoliorum*; 1692, *Pseudopeziza trifolii*; 1693, *Sclerotinia trifoliorum*; 1694, *Stemphylium sarciniforme*.

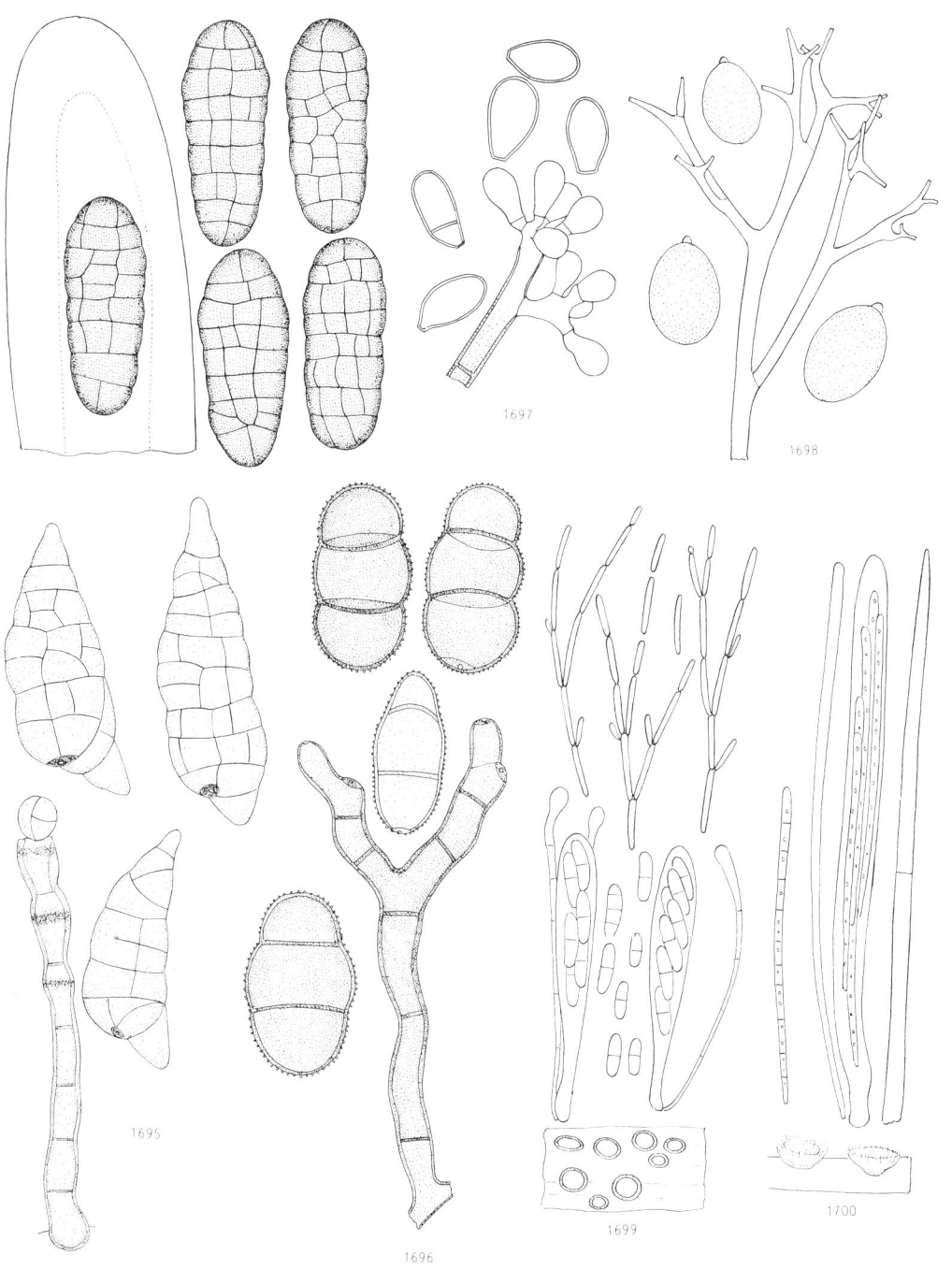

Plate 164. On *Triglochin, Tripleurospermum, Tropaeolum, Tulipa* and *Urtica*. 1695, *Pleospora triglochinicola*; 1696, *Acroconidiella tropaeoli*; 1697, *Botrytis tulipae*; 1698, *Pseudoperonospora urticae*; 1699, *Calloria neglecta*; 1700, *Erinella discolor*.

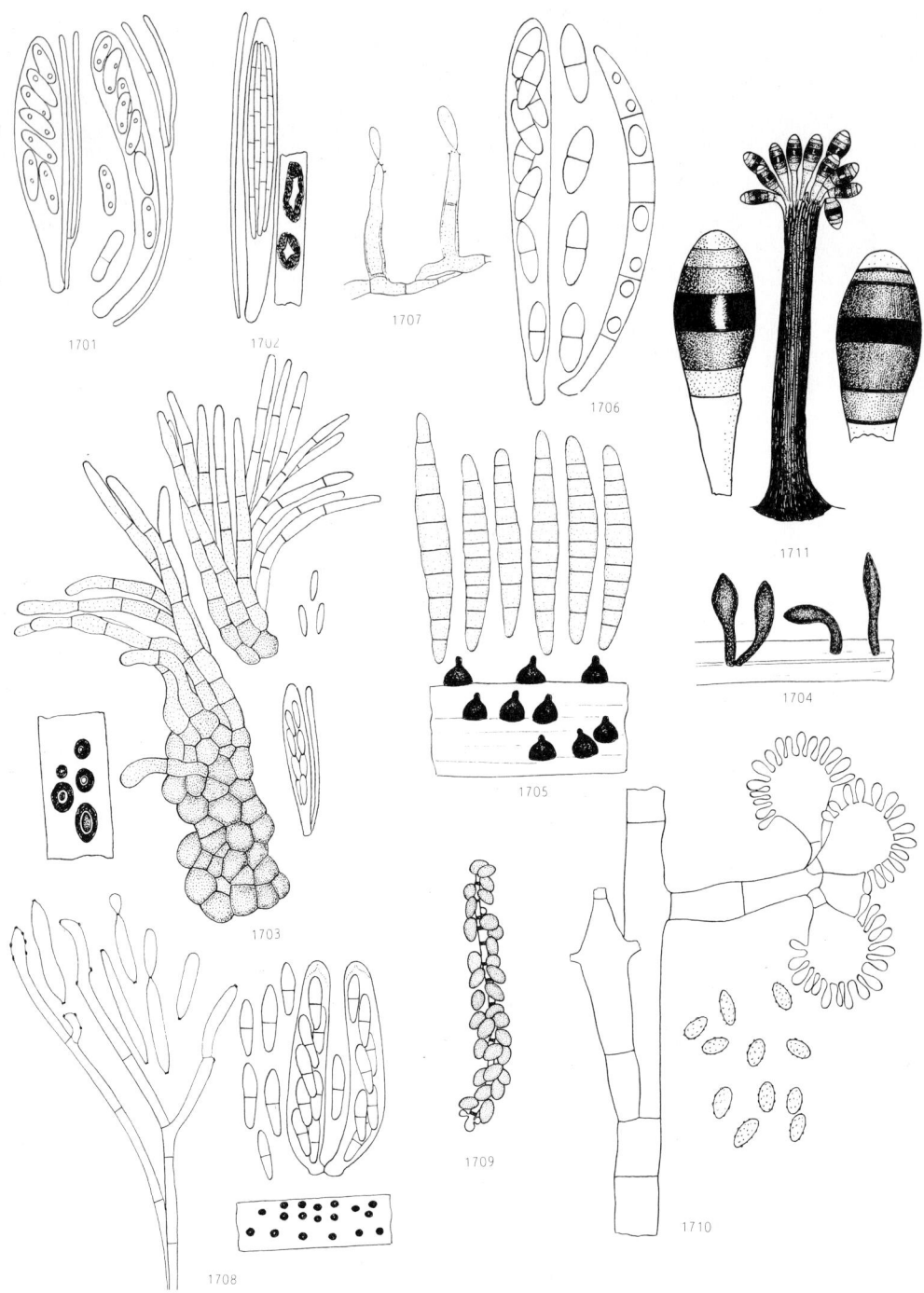

Plate 165. On *Urtica*. 1701, *Laetinaevia carneoflavida*; 1702, *Naemacyclus caulium*; 1703, *Pyrenopeziza urticicola*; 1704, *Acrospermum compressum*; 1705, *Leptosphaeria acuta*; 1706, *Nectria leptosphaeriae*; 1707, *Pleurophragmium acutum*; 1708, *Mycosphaerella superflua*; 1709, *Arthrinium urticae*; 1710, *Botryosporium pulchrum*; 1711, *Endophragmia atra*.

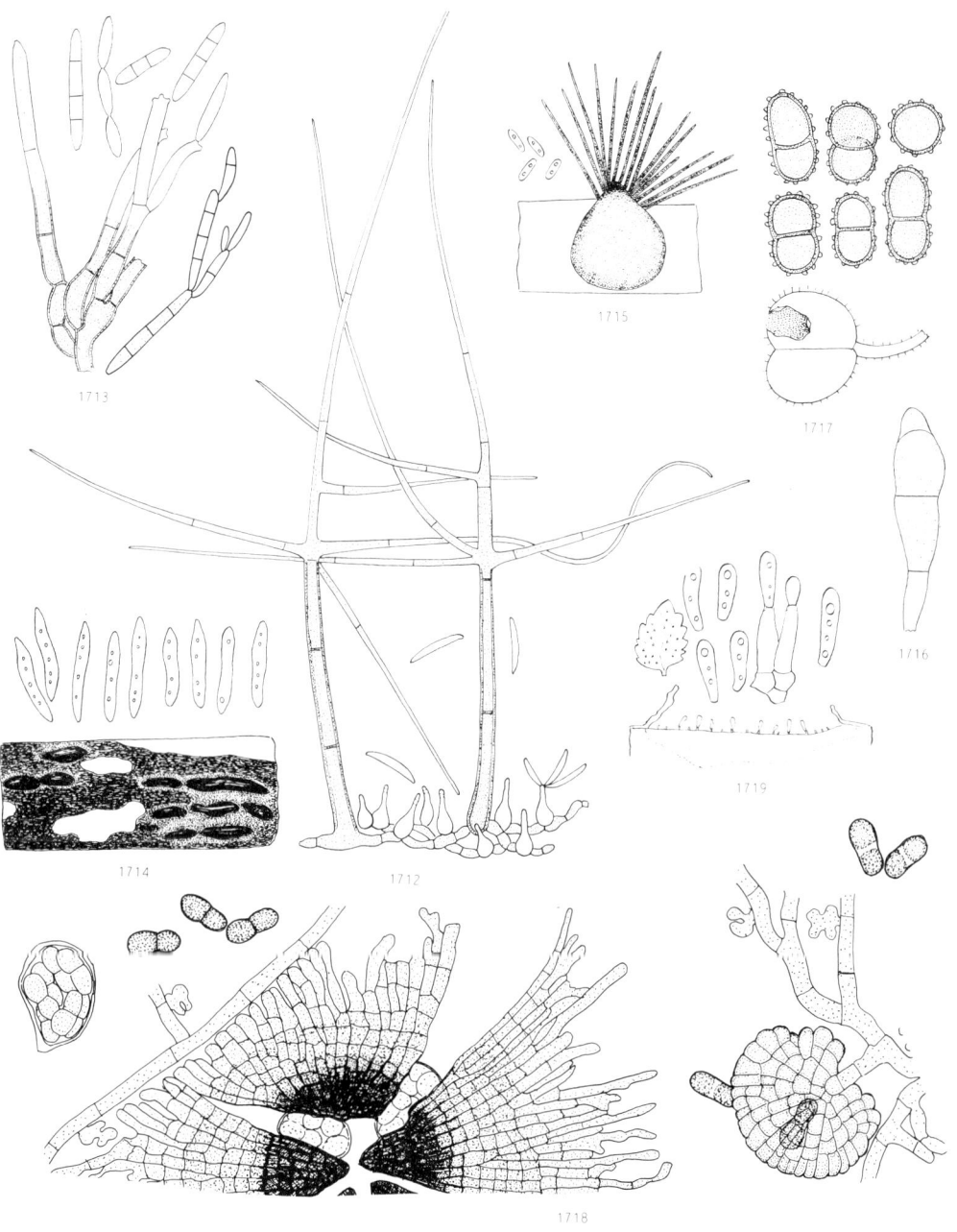

Plate 166. On *Urtica* and *Veronica*. 1712, *Gyrothrix verticillata*; 1713, *Polyscytalum berkeleyi*; 1714, *Apomelasmia* state of *Aporhytisma urticae*; 1715, *Pyrenochaeta fallax*; 1716, *Puccinia veronicae*; 1717, *Schroeteria delastrina*; 1718, *Asterina veronicae*; 1719, *Discogloeum veronicae*.

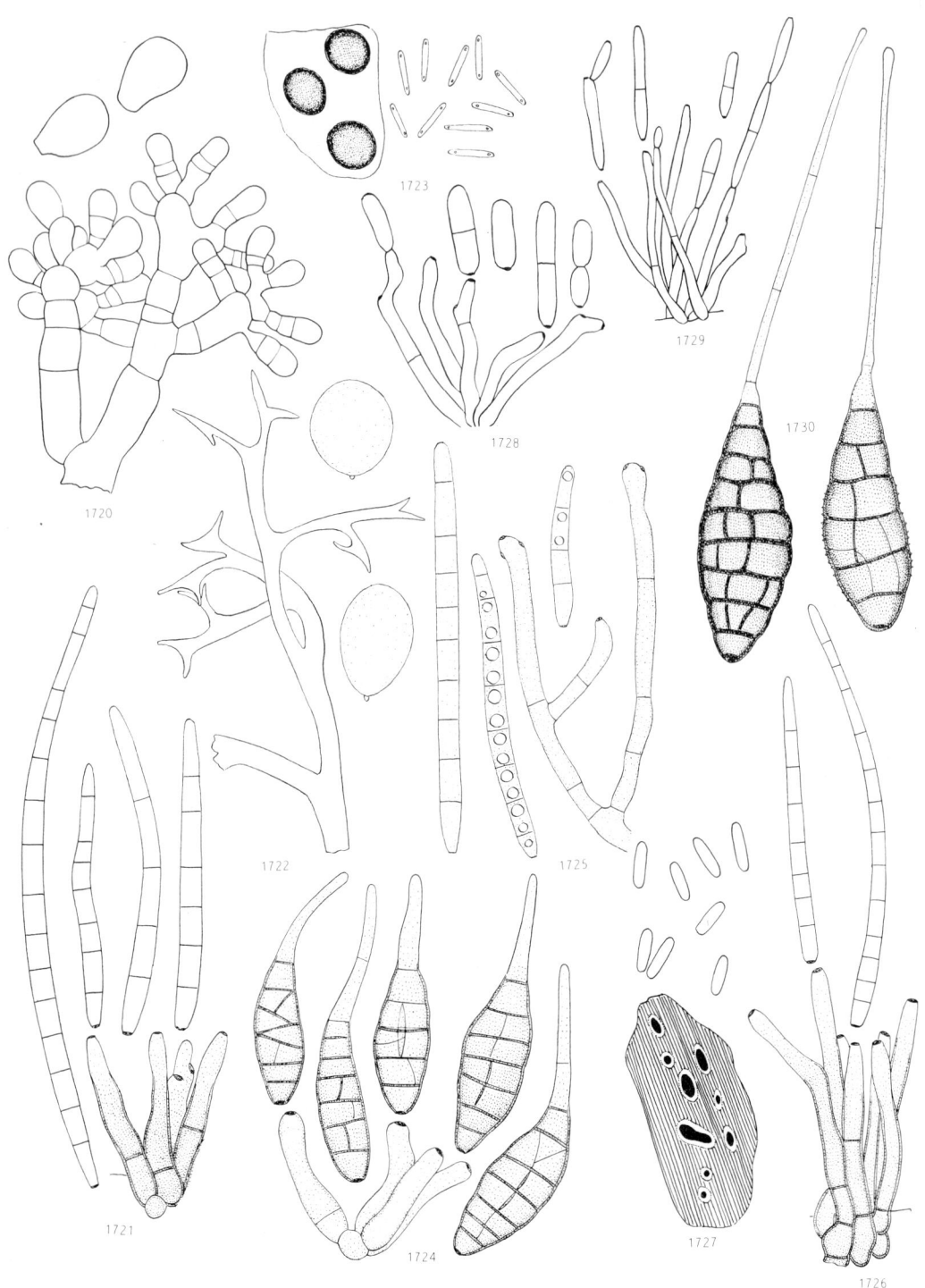

Plate 167. On *Vicia, Vinca, Viola* and *Zinnia*. 1720, *Botrytis fabae*; 1721, *Cercospora zonata*; 1722, *Peronospora viciae*; 1723, *Ceuthospora feurichii*; 1724, *Alternaria violae*; 1725, *Cercospora murina*; 1726, *C. violae*; 1727, *Myrothecium roridum*; 1728, *Ramularia agrestis*; 1729, *R. lactea*; 1730, *Alternaria zinniae*.

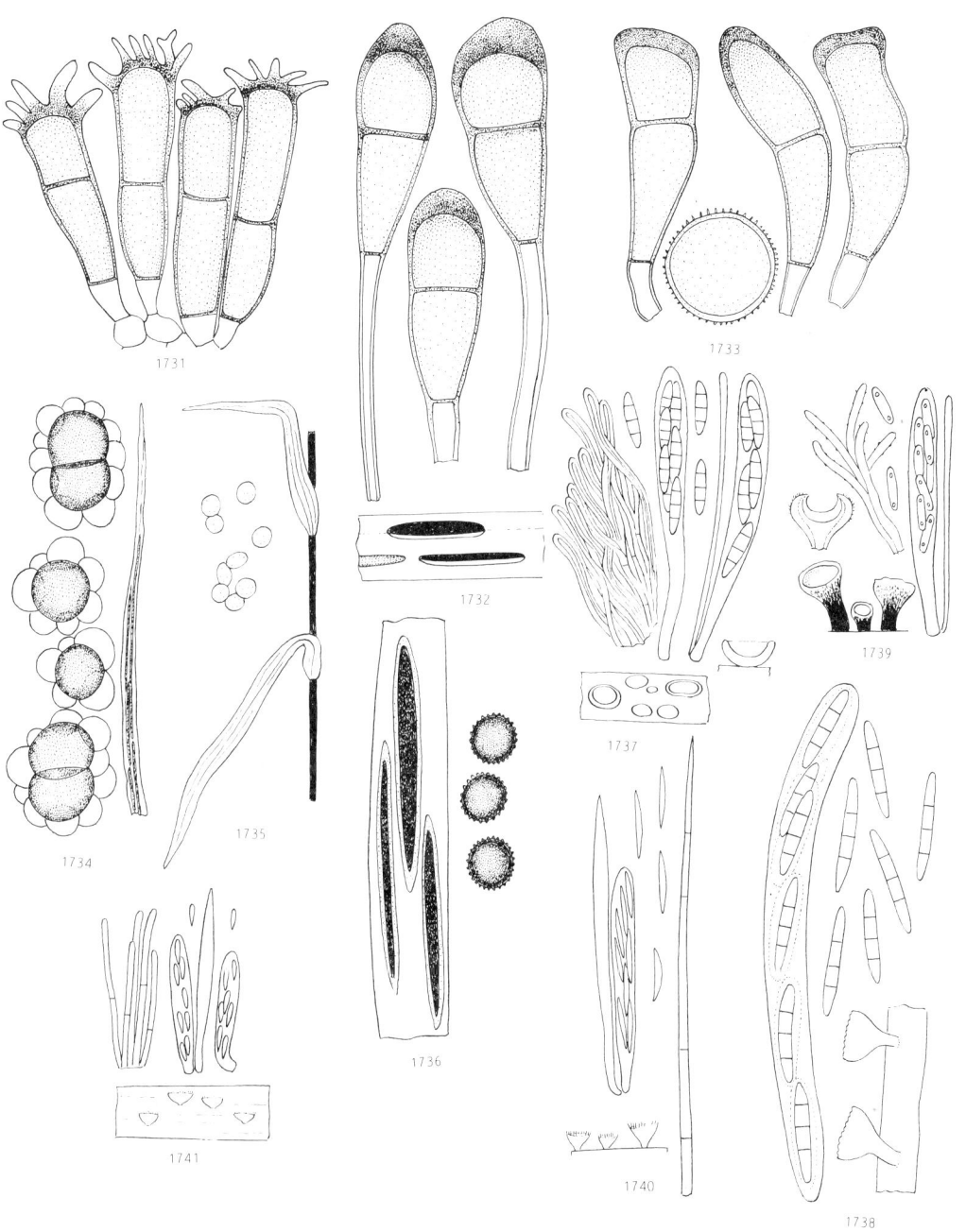

Plate 168. Rusts, smuts and discomycetes on grasses. 1731, *Puccinia coronata*; 1732, *P. graminis*; 1733, *P. striiformis*; 1734, *Urocystis agropyri*; 1735, *Ustilago hypodytes*; 1736, *U. striiformis*; 1737, *Calycella scolochloae*; 1738, *Crocicreas culmicolum*; 1739, *C. tomentosum*; 1740, *Dasyscyphus acutipilus*; 1741, *D. acutus*.

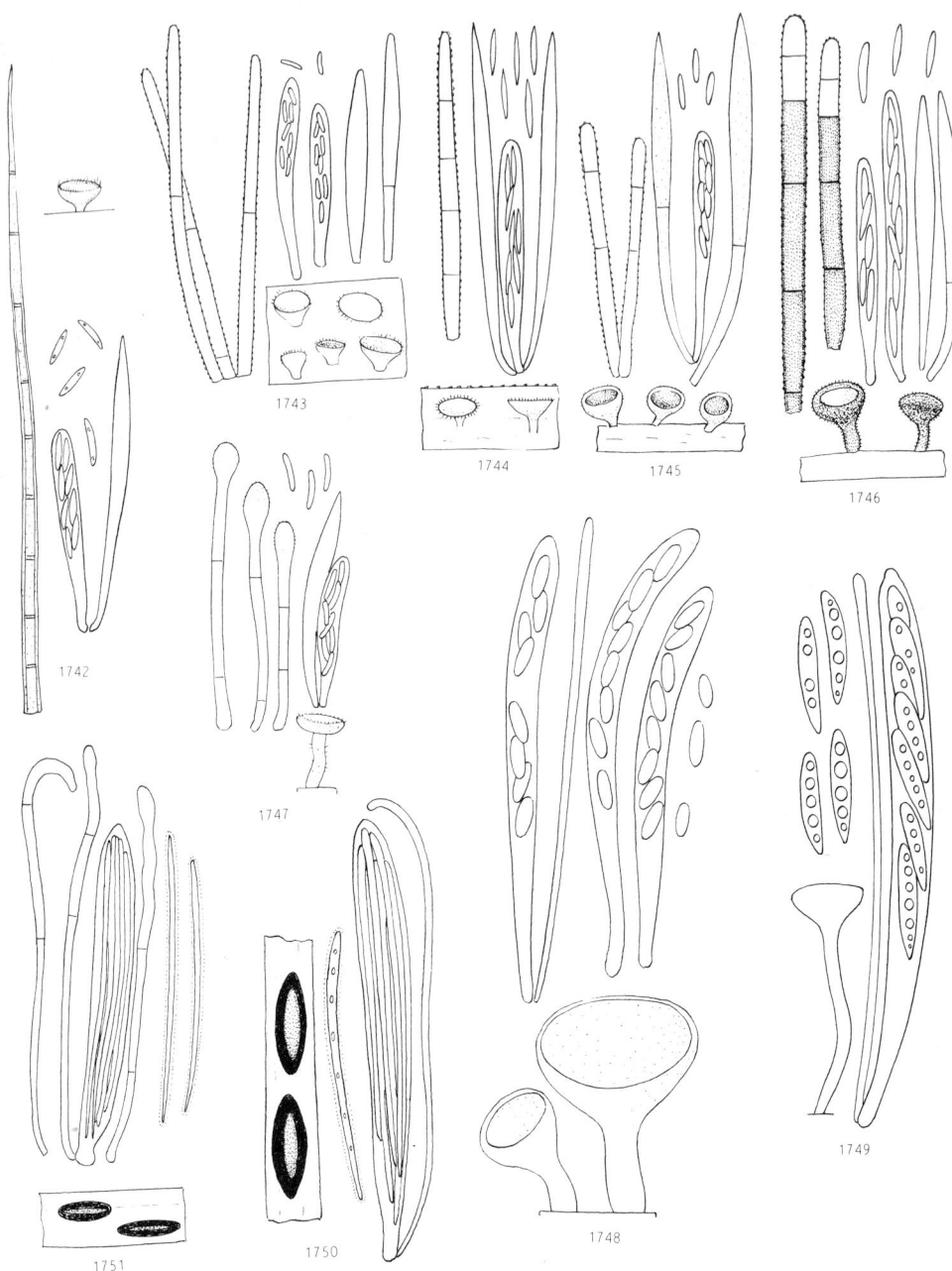

Plate 169. Discomycetes on grasses. 1742, *Dasyscyphus albotestaceus*; 1743, *D. carneolus*; 1744, *D. carneolus* var. *longisporus*; 1745, *D. controversus*; 1746, *D. palearum*; 1747, *D. tenuissimus*; 1748, *Hymenoscyphus robustior*; 1749, *H. scutula* var. *suspecta*; 1750, *Lophodermium arundinaceum*; 1751, *L. culmigenum*.

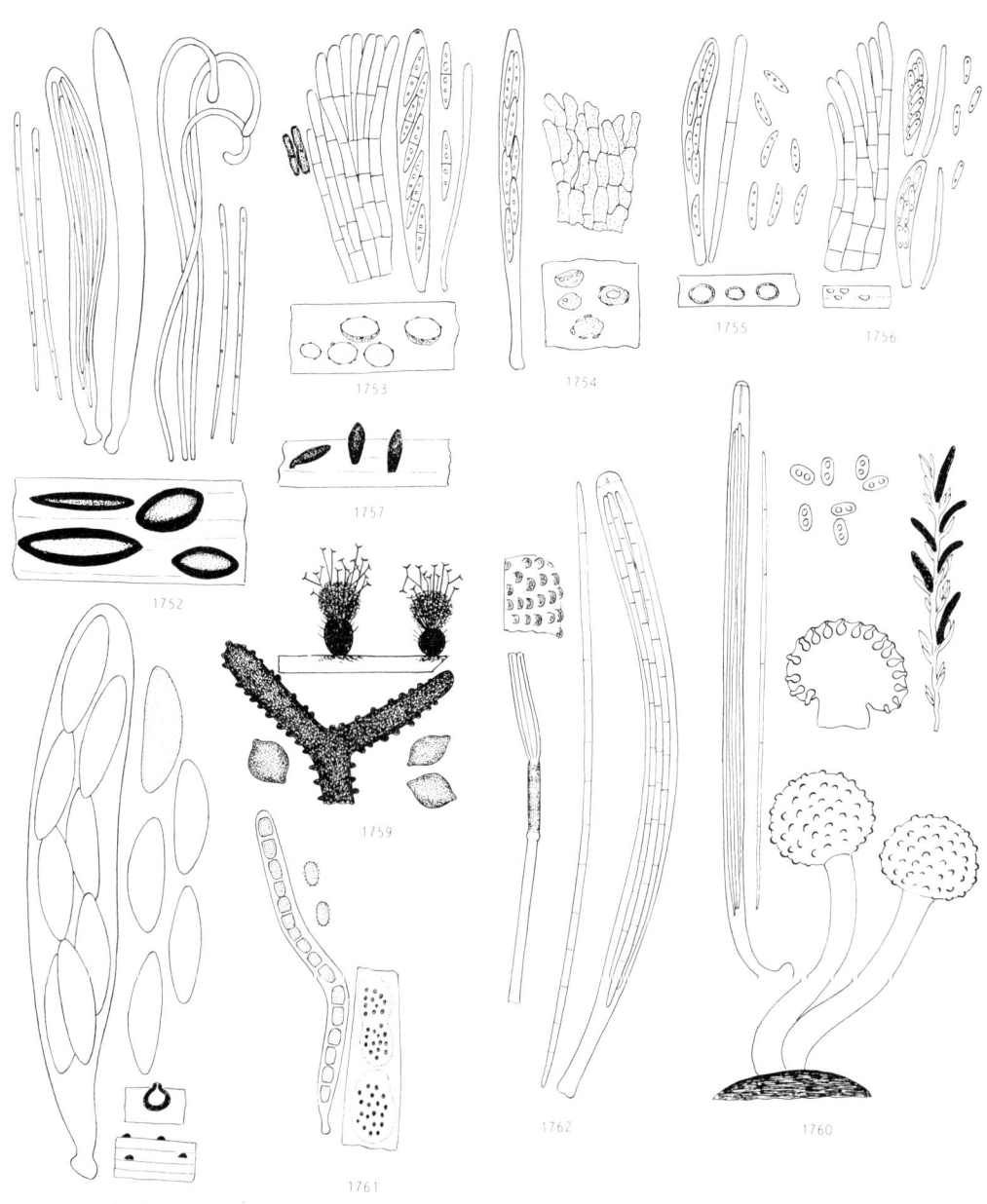

Plate 170. Discomycetes and other ascomycetes on grasses. 1752, *Lophodermium gramineum*; 1753, *Micropeziza karstenii*; 1754, *M. poae*; 1755, *Mollisia poaeoides*; 1756, *Pezizella eburnea*; 1757, *Acrospermum graminum*; 1758, *Botryosphaeria festucae*; 1759, *Chaetomium elatum*; 1760, *Claviceps purpurea*; 1761, *Creopus spinulosus*; 1762, *Epichloe typhina*.

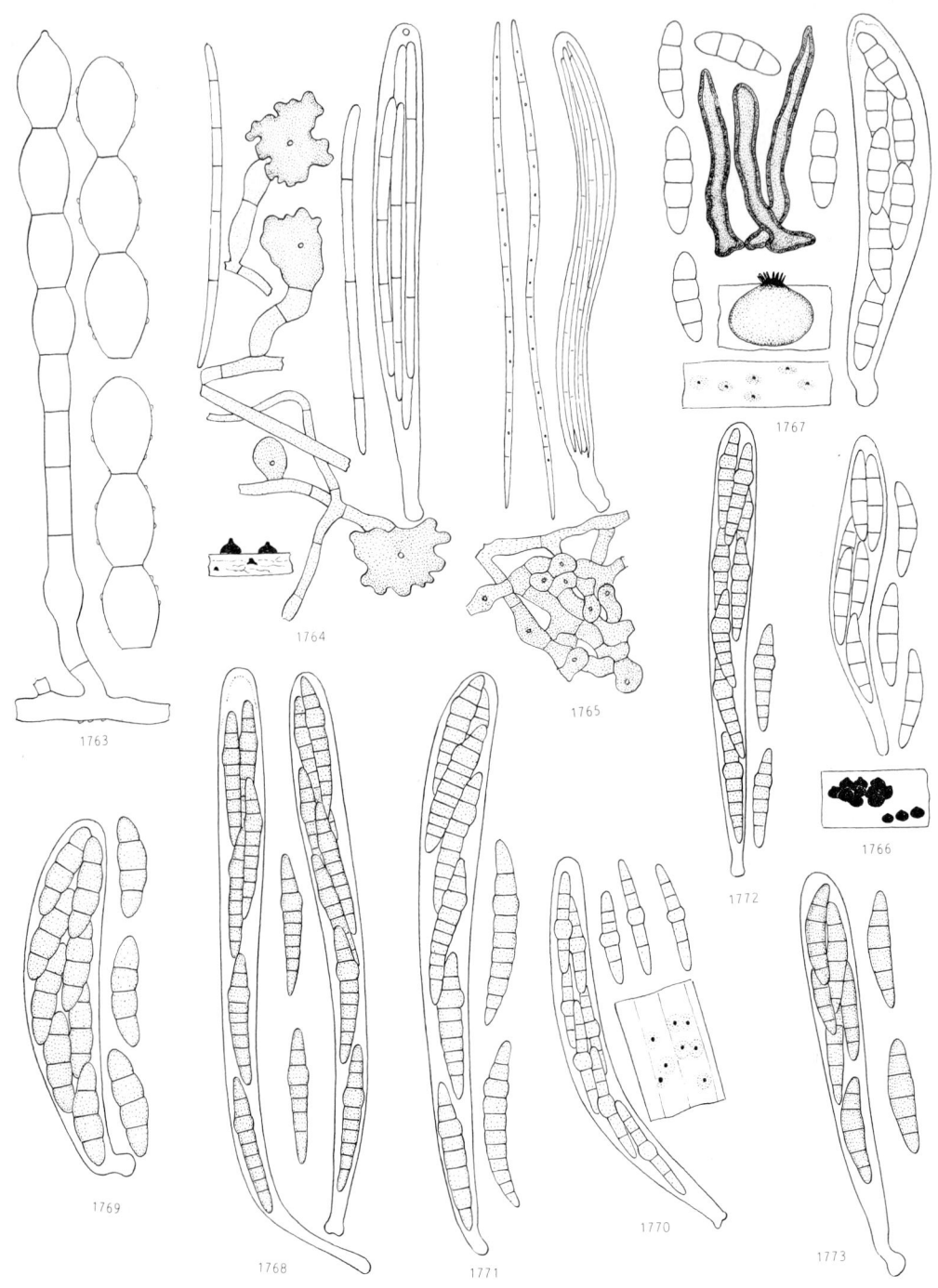

Plate 171. Ascomycetes on grasses. 1763, *Erysiphe graminis*; 1764, *Gaeumannomyces graminis* var. *graminis*; 1765, *G. graminis* var. *tritici*; 1766, *Gibberella zeae*; 1767, *Keissleriella culmifida*; 1768, *Leptosphaeria culmifraga*; 1769, *L. eustoma*; 1770, *L. fuckelii*; 1771, *L. graminis*; 1772, *L. herpotrichoides*; 1773, *L. luctuosa*.

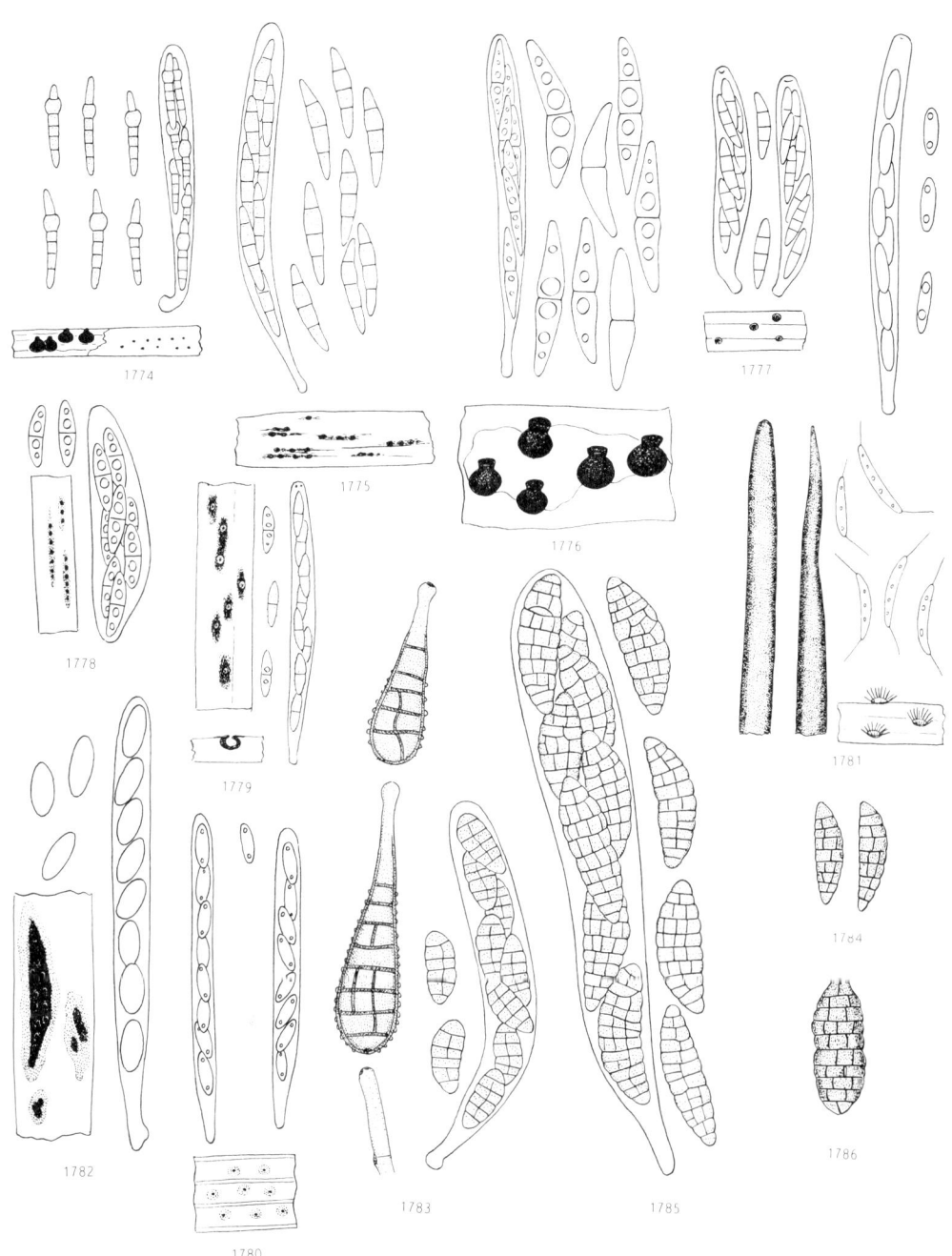

Plate 172. Ascomycetes on grasses. 1774, *Leptosphaeria nigrans*; 1775, *L. nodorum*; 1776, *Lophiostoma semiliberum*; 1777, *Monographella nivalis*; 1778, *Mycosphaerella lineolata*; 1779, *Paradidymella holci*; 1780, *Phomatospora berkeleyi*; 1781, *P. dinemasporium*; 1782, *Phyllachora graminis*; 1783, *Pleospora infectoria*; 1784, *P. rubelloides*; 1785, *P. rubicunda*; 1786, *P. straminis*.

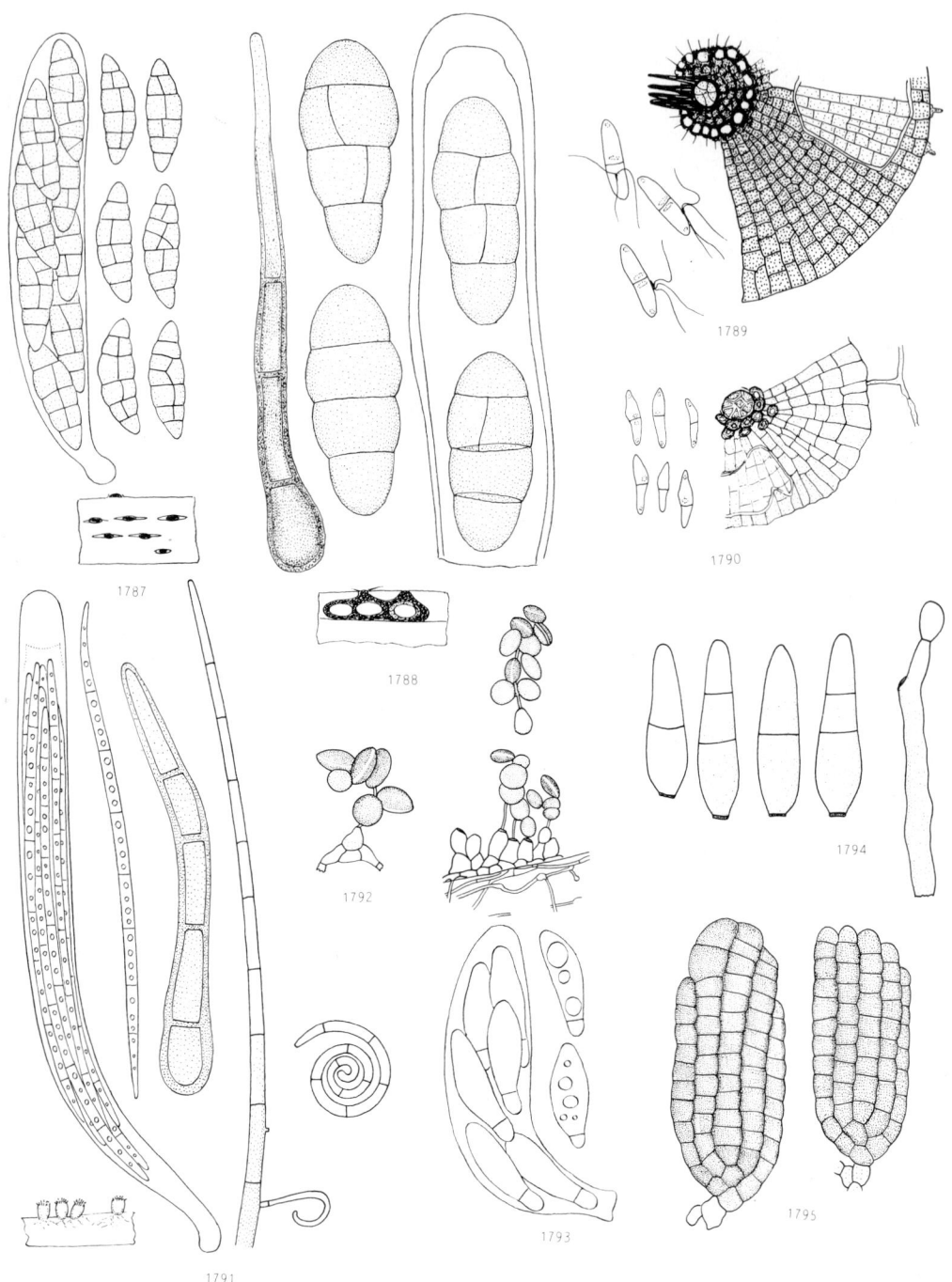

Plate 173. Ascomycetes and hyphomycetes on grasses. 1787, *Pleospora vagans*; 1788, *Pyrenophora trichostoma*; 1789, *Trichothyrina alpestris*; 1790, *T. nigroannulata*; 1791, *Tubeufia helicomyces*; 1792, *Arthrinium phaeospermum*; 1793, *Arthrinium* state of *Apiospora montagnei*; 1794, *Cercosporidium graminis*; 1795, *Dictyosporium elegans*.

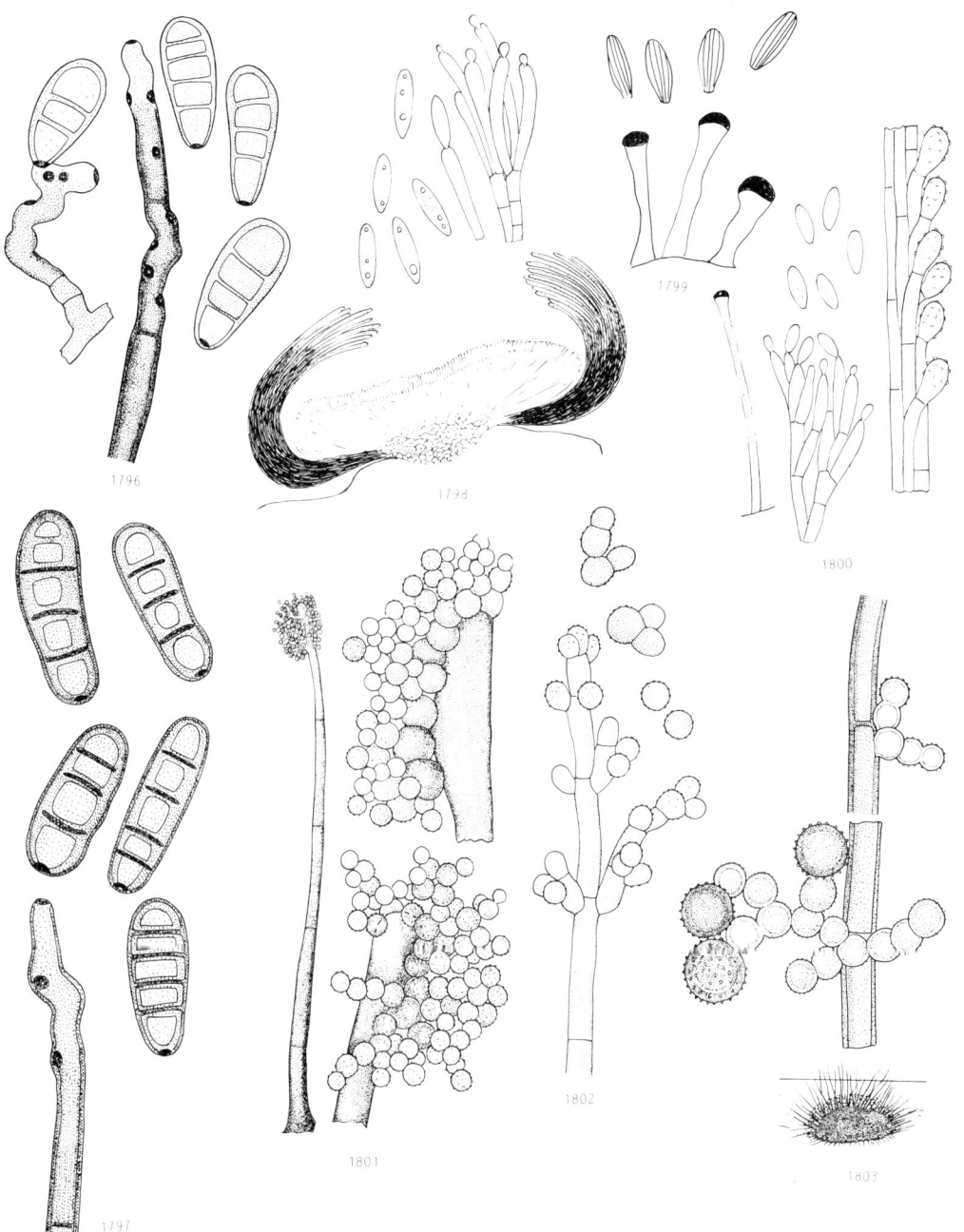

Plate 174. Hyphomycetes on grasses. 1796, *Drechslera biseptata*; 1797, *D. dematioidea*; 1798, *Myrothecium atroviride*; 1799, *M. cinctum*; 1800, *M. masonii*; 1801, *Periconia britannica*; 1802, *P. glyceriicola*; 1803, *P. hispidula*.

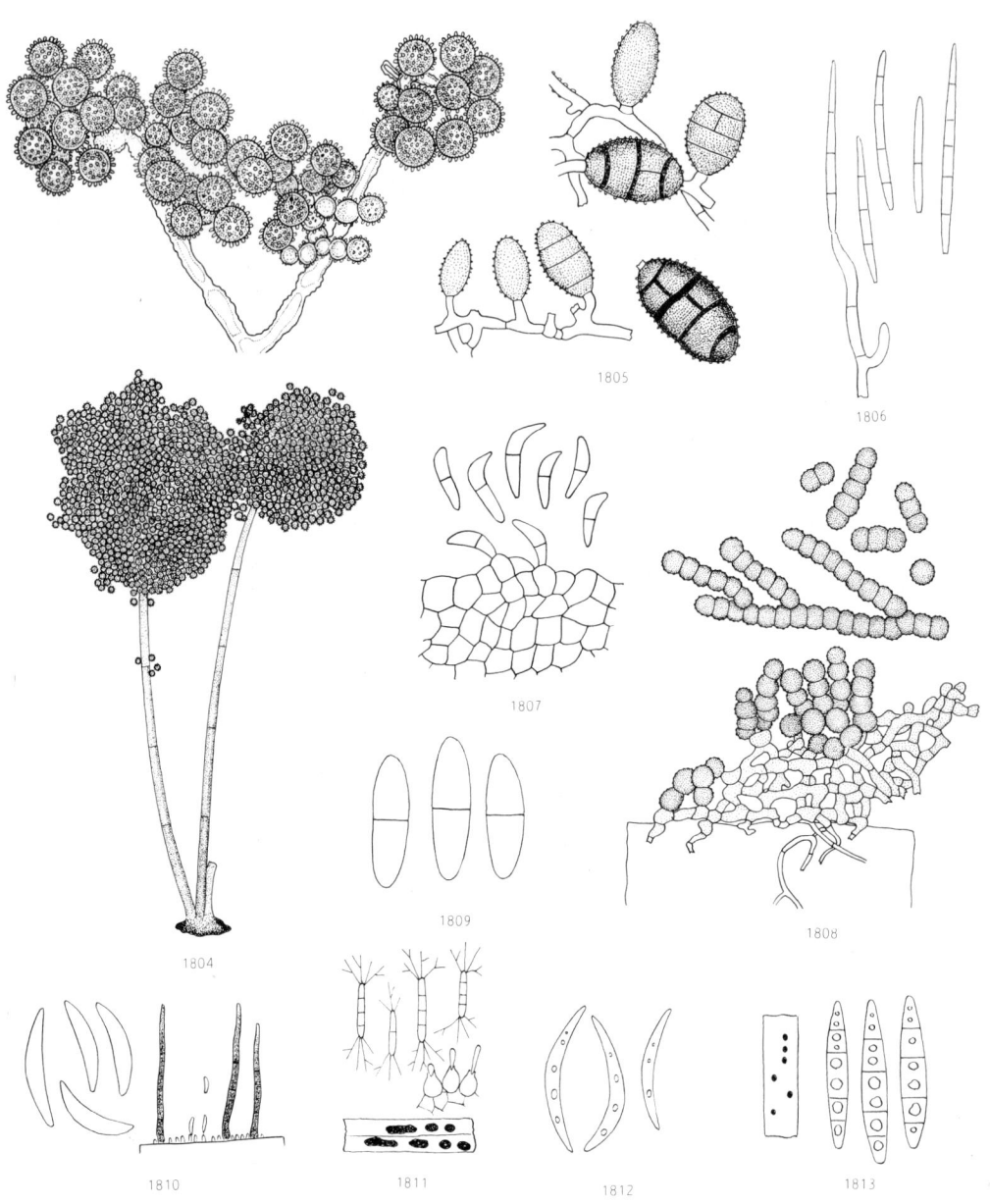

Plate 175. Hyphomycetes and coelomycetes on grasses. 1804, *Periconia igniaria*; 1805, *Pithomyces chartarum*; 1806, *Pseudocercosporella herpotrichoides*; 1807, *Rhynchosporium secalis*; 1808, *Rotula graminis*; 1809, *Ascochyta rhodesii*; 1810, *Colletotrichum graminicola*; 1811, *Dilophospora* state of *Lidophia graminis*; 1812, *Pseudoseptoria donacis*; 1813, *Stagonospora subseriata*.

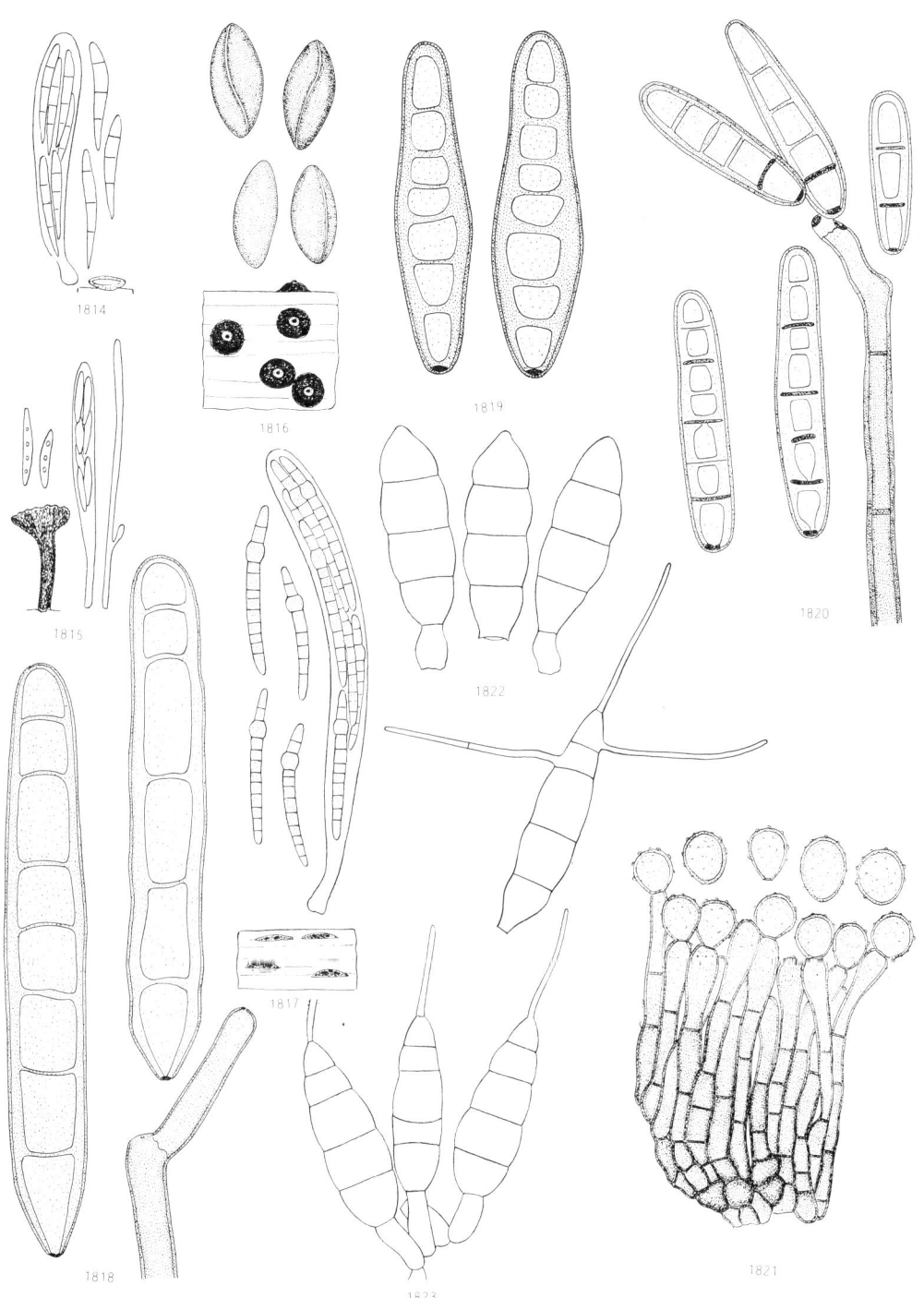

Plate 176. On *Agropyron*, *Agrostis* and *Alopecurus*. 1814, *Belonopsis graminea*; 1815, *Crocicreas furvum*; 1816, *Anthostomella chionostoma*; 1817, *Leptosphaeria pontiformis*; 1818, *Drechslera* state of *Pyrenophora tritici-repentis*; 1819, *D. fugax*; 1820, *D.* state of *P. erythrospila*; 1821, *Hadrotrichum virescens*; 1822, *Mastigosporium rubricosum*; 1823, *M. album*.

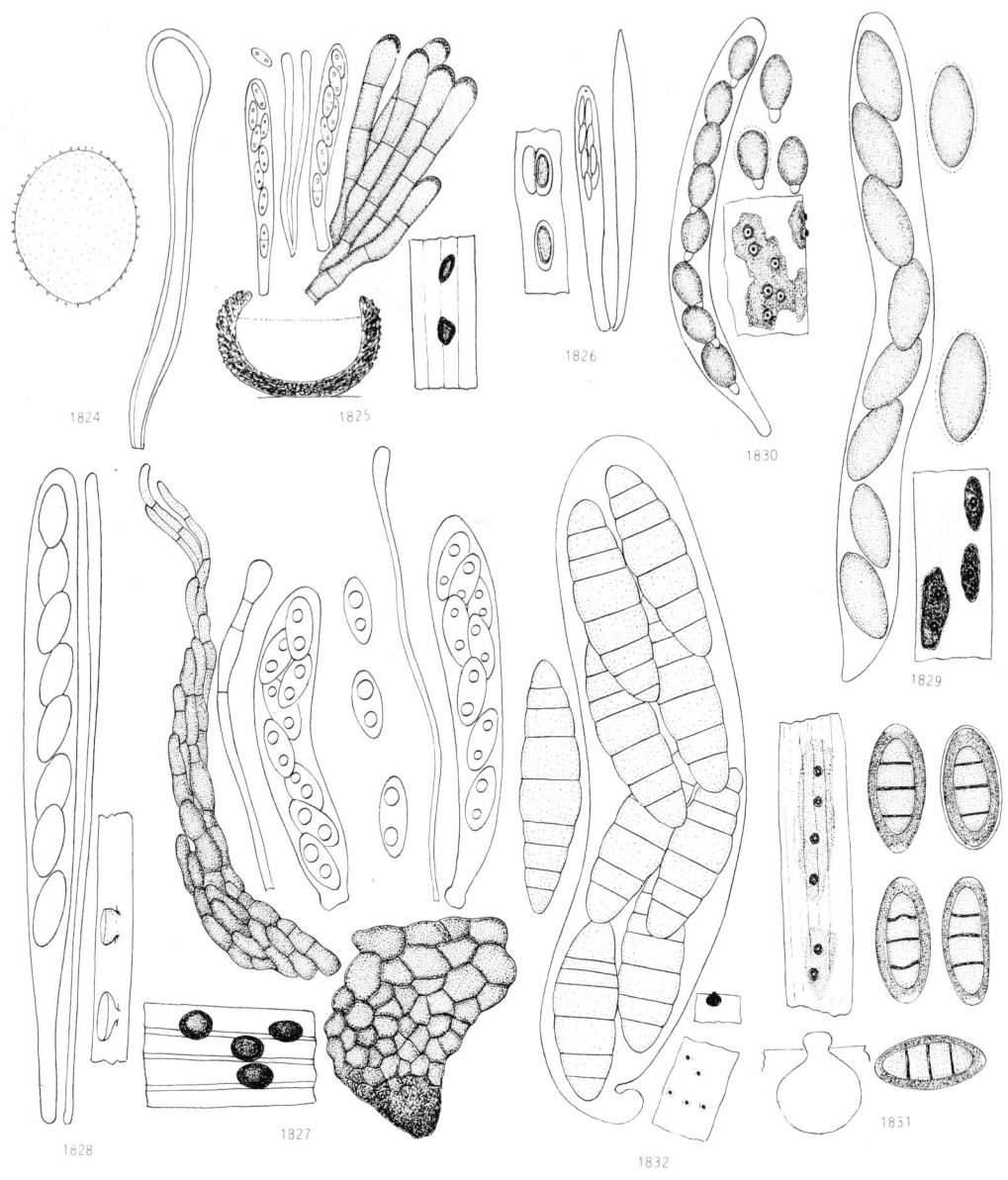

Plate 177. On *Ammophila*. 1824, *Puccinia pygmaea* var. *ammophilina*; 1825, *Belonium psammicola*; 1826, *Hysteropezizella valvata*; 1827, *Pyrenopeziza arenivaga*; 1828, *Rutstroemia maritima*; 1829, *Anthostomella lugubris*; 1830, *A. phaeosticta*; 1831, *Chitonospora ammophilae*; 1832, *Leptosphaeria ammophilae*.

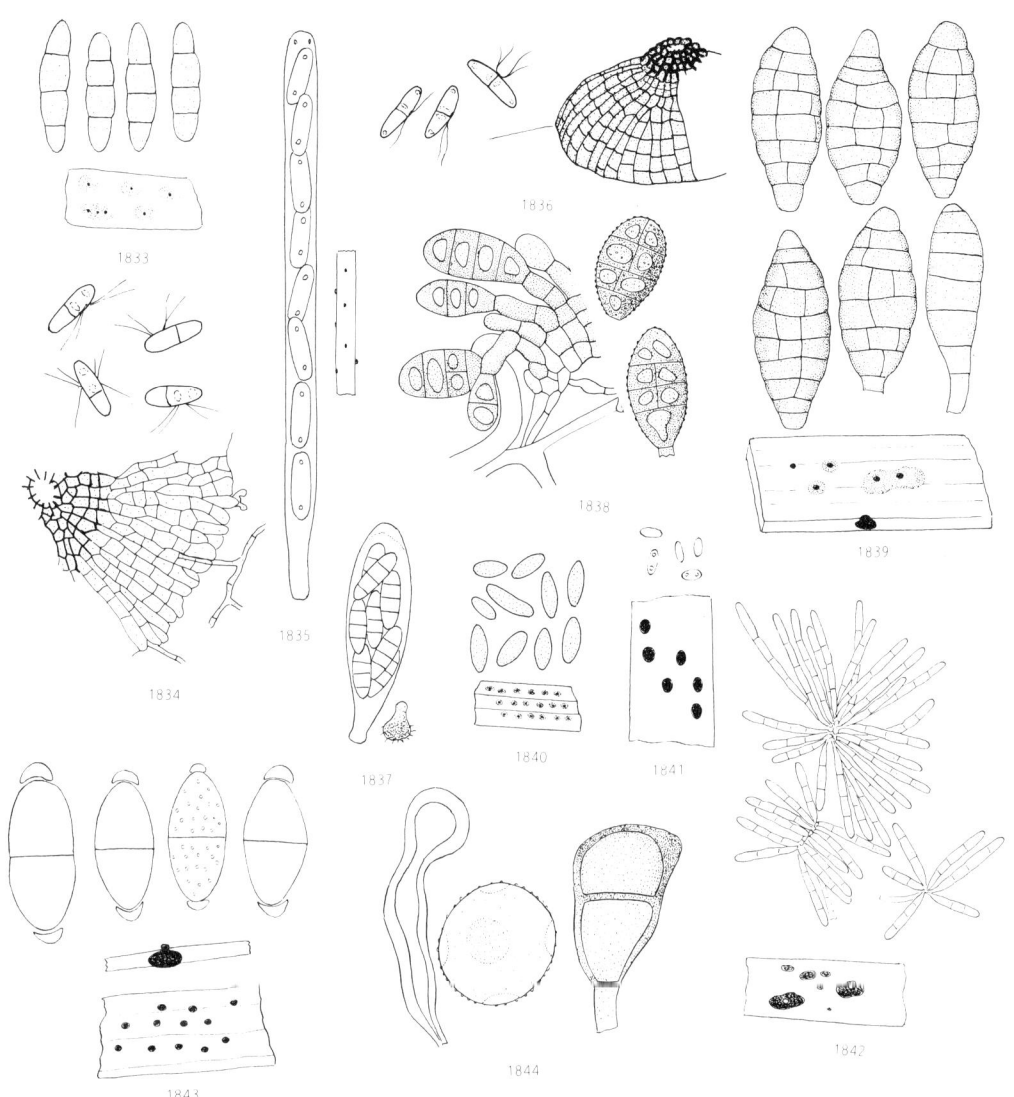

Plate 178. On *Ammophila* and *Arrhenatherum*. 1833, *Leptosphaeria marram*; 1834, *Microthyrium gramineum*; 1835, *Phomatospora arenariae*; 1836, *Trichothyrina ammophilae*; 1837, *Tubeufia parvula*; 1838, *Thyrostromella myriana*; 1839, *Amarenographium metableticum*; 1840, *Coniothyrium psammae*; 1841, *Phoma ammophilae*; 1842, *Psammina bommeriae*; 1843, *Tiarospora perforans*; 1844, *Puccinia brachypodii* var. *arrhenatheri*.

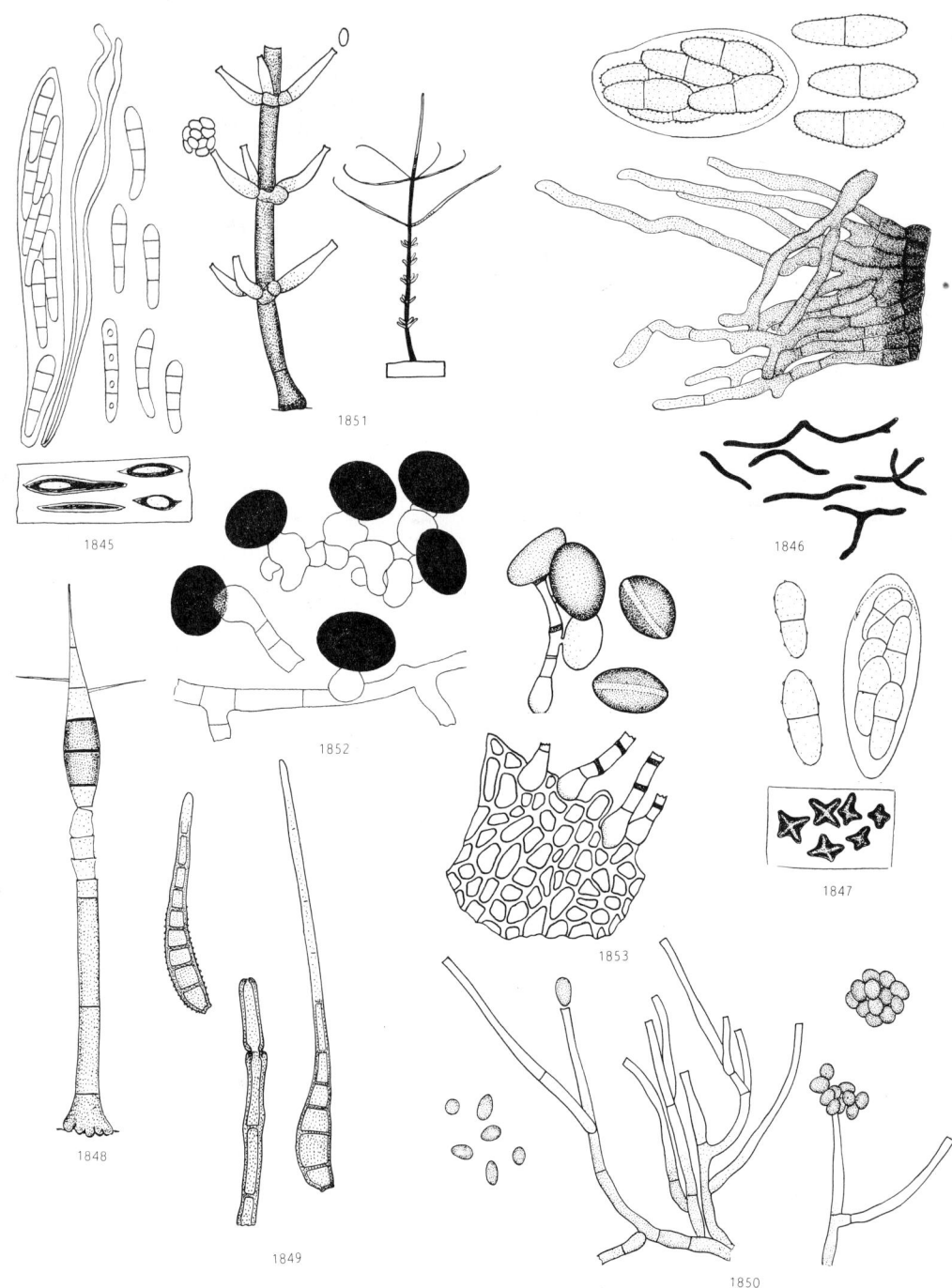

Plate 179. On *Arundinaria*. 1845, *Eupropolella arundinariae*; 1846, *Morenoina arundinariae*; 1847, *M. websteri*; 1848, *Chaetendophragmia britannica*; 1849, *Corynespora foveolata*; 1850, *Gliomastix* state of *Wallrothiella subiculosa*; 1851, *Gonytrichum macrocladum*; 1852, *Nigrospora sphaerica*; 1853, *Pteroconium intermedium*.

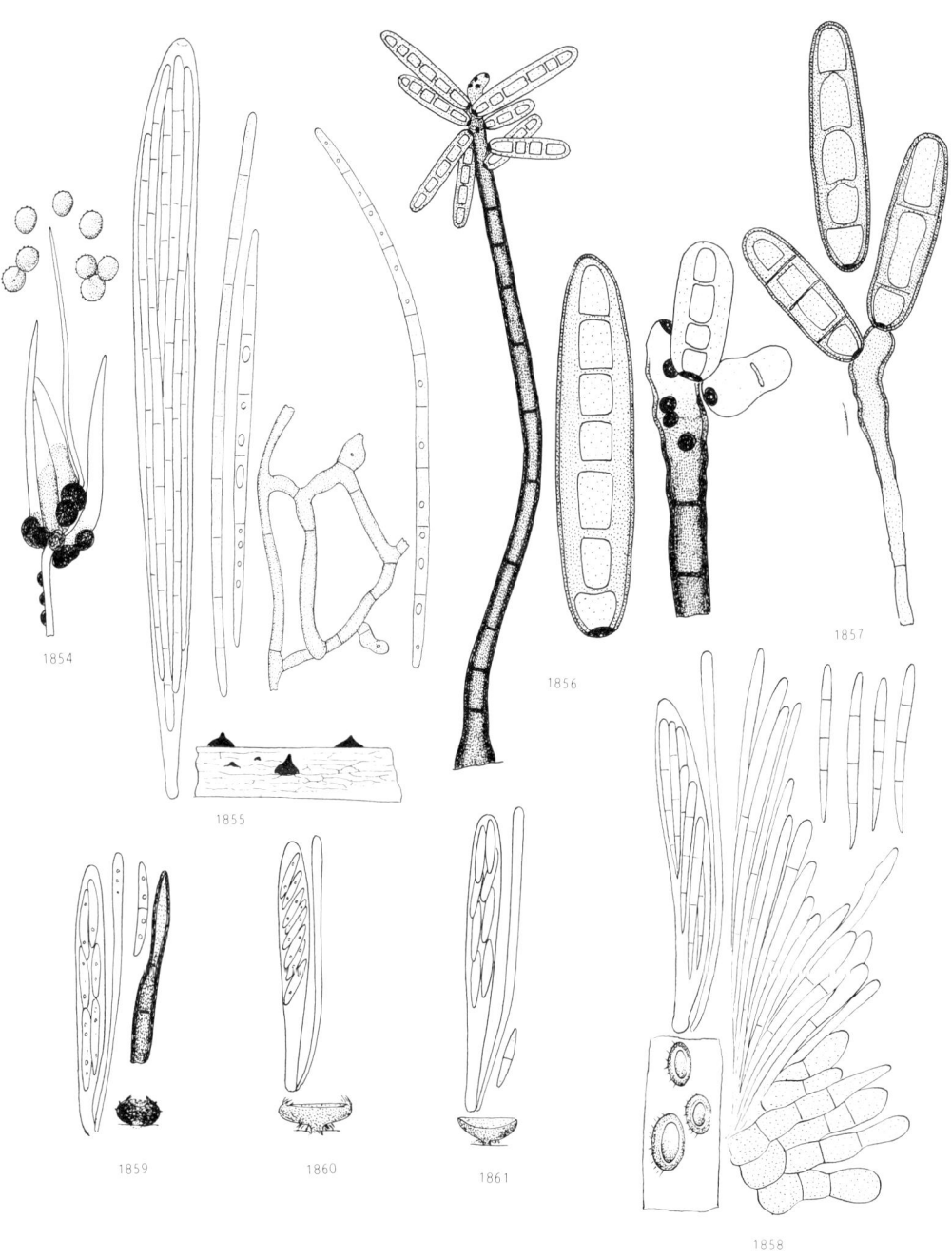

Plate 180. On *Avena* and *Brachypodium*. 1854, *Ustilago avenae*; 1855, *Gaeumannomyces graminis* var. *avenae*; 1856, *Drechslera avenacea*; 1857, *D.* state of *Pyrenophora avenae*; 1858, *Belonopsis filispora*; 1859, *Pirottaea exilispora*; 1860, *Scutomollisia fimbriomarginata*; 1861, *S. integromarginata*.

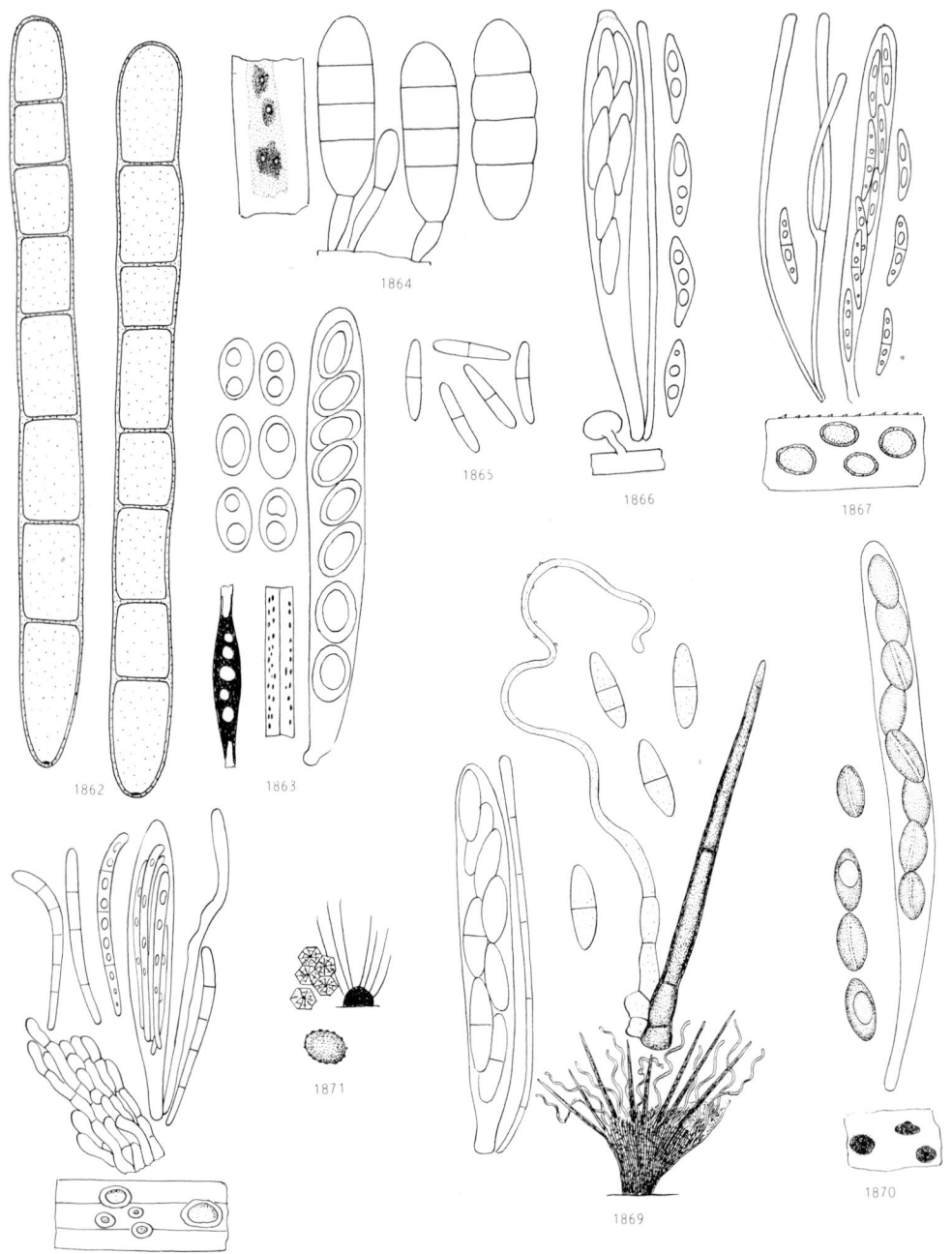

Plate 181. On *Bromus*, *Dactylis* and *Deschampsia*. 1862, *Drechslera* state of *Pyrenophora bromi*; 1863, *Phyllachora dactylidis*; 1864, *Mastigosporium muticum*; 1865, *Rhynchosporium orthosporum*; 1866, *Crocicreas megalosporum* var. *gramineum*; 1867, *Mollisia mutabilis*; 1868, *Pseudohelotium alaunae*; 1869, *Trichodiscus heterotrichus*; 1870, *Anthostomella tomicum*; 1871, *Cephalotheca clarkii*.

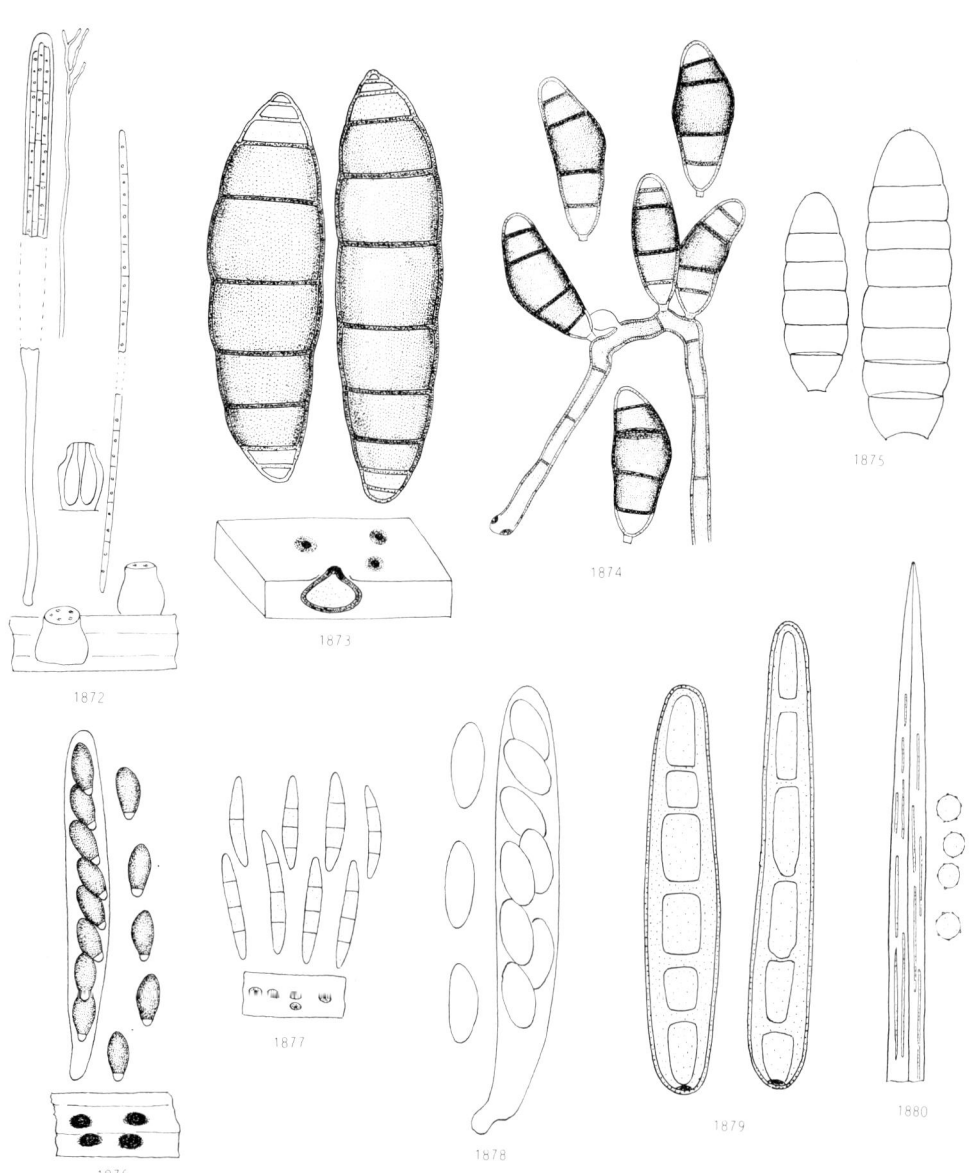

Plate 182. On *Deschampsia, Elymus, Festuca* and *Glyceria*. 1872, *Oomyces carneo-albus*; 1873, *Trematosphaeria clarkii*; 1874, *Curvularia protuberata*; 1875, *Mastigosporium deschampsiae*; 1876, *Anthostomella arenaria*; 1877, *Stagonospora arenaria* var. *minor*; 1878, *Phyllachora sylvatica*; 1879, *Drechslera* state of *Pyrenophora dictyoides*; 1880, *Ustilago longissima*.

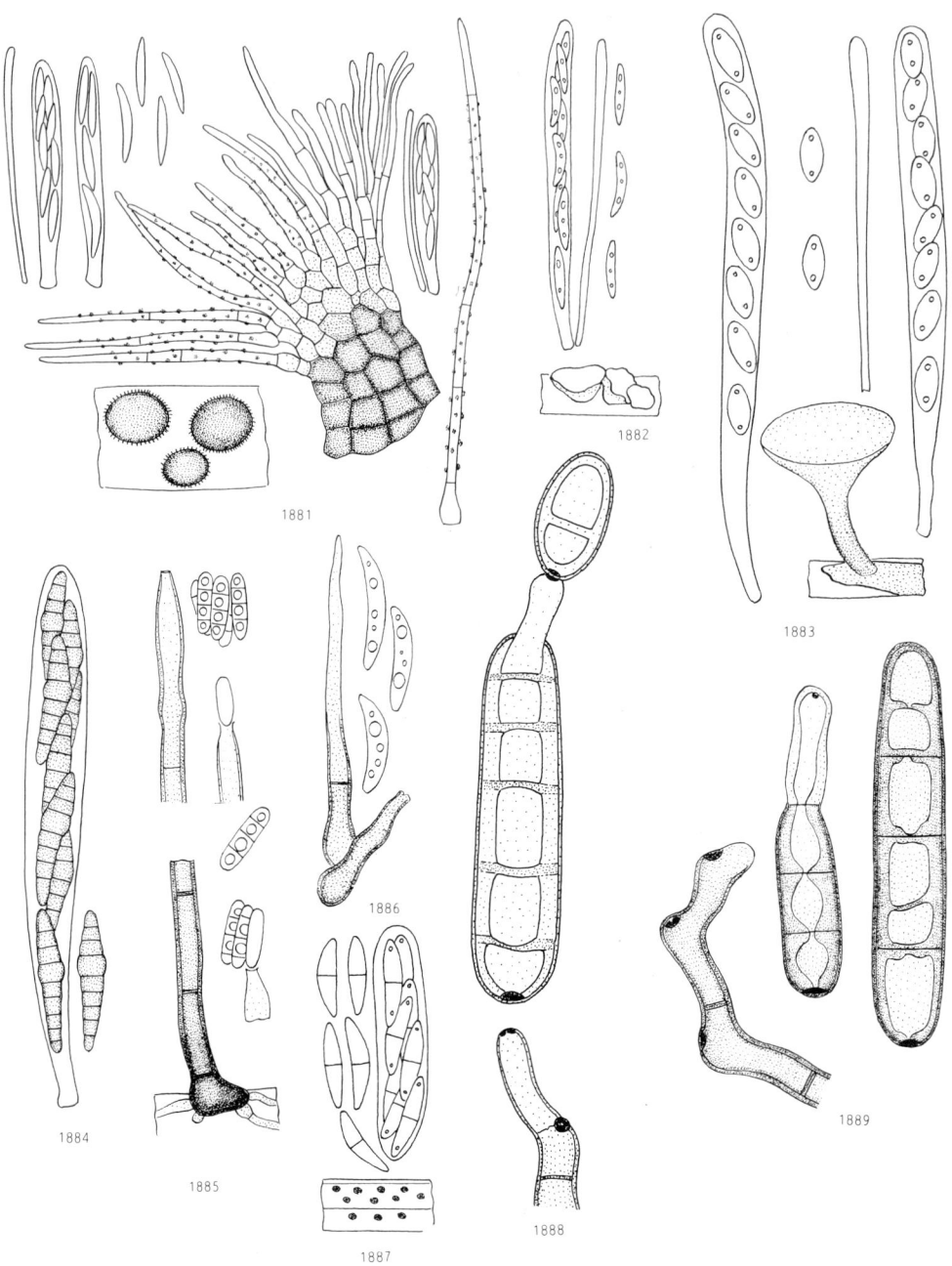

Plate 183. On *Glyceria*, *Holcus* and *Hordeum*. 1881, *Dennisiodiscus prasinus*; 1882, *Ombrophila ambigua*; 1883, *Rutstroemia calopus*; 1884, *Leptosphaeria culmifraga* var. *propinqua*; 1885, *Cylindrotrichum ellisii*; 1886, *Colletotrichum holci*; 1887, *Didymella exitialis*; 1888, *Drechslera* state of *Pyrenophora graminea*; 1889, *D.* state of *P. japonica*.

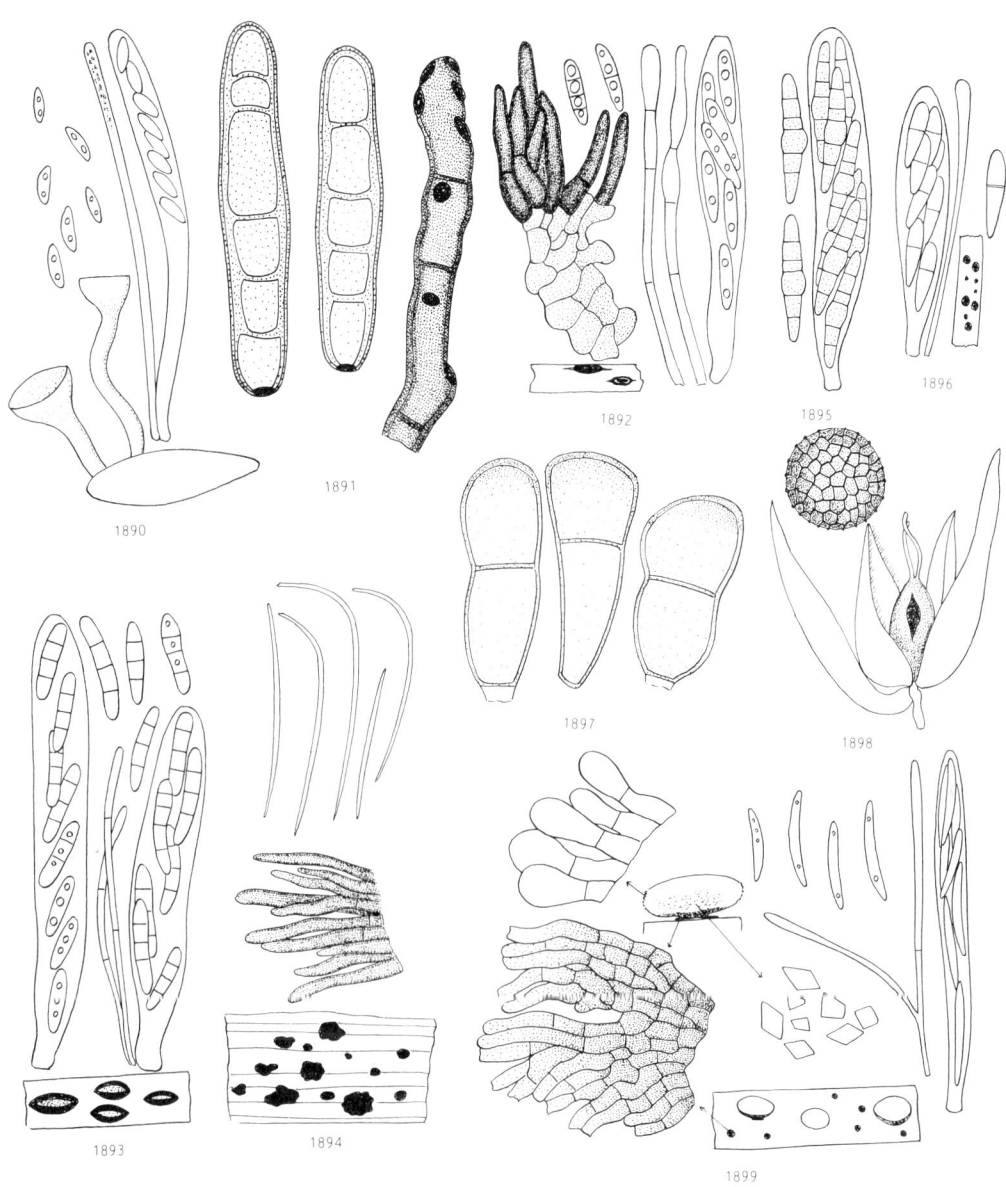

Plate 184. On *Lolium, Molinia, Nardus* and *Phalaris*. 1890, *Gloeotinia granigena*; 1891, *Drechslera siccans*; 1892, *Belonium hystrix*; 1893, *Gloniella moliniae*; 1894, *Actinothyrium graminis*; 1895, *Leptosphaeria nardi*; 1896, *Scutomollisia puncta*; 1897, *Puccinia sessilis*; 1898, *Tilletia menieri*; 1899, *Calycellina phalaridis*.

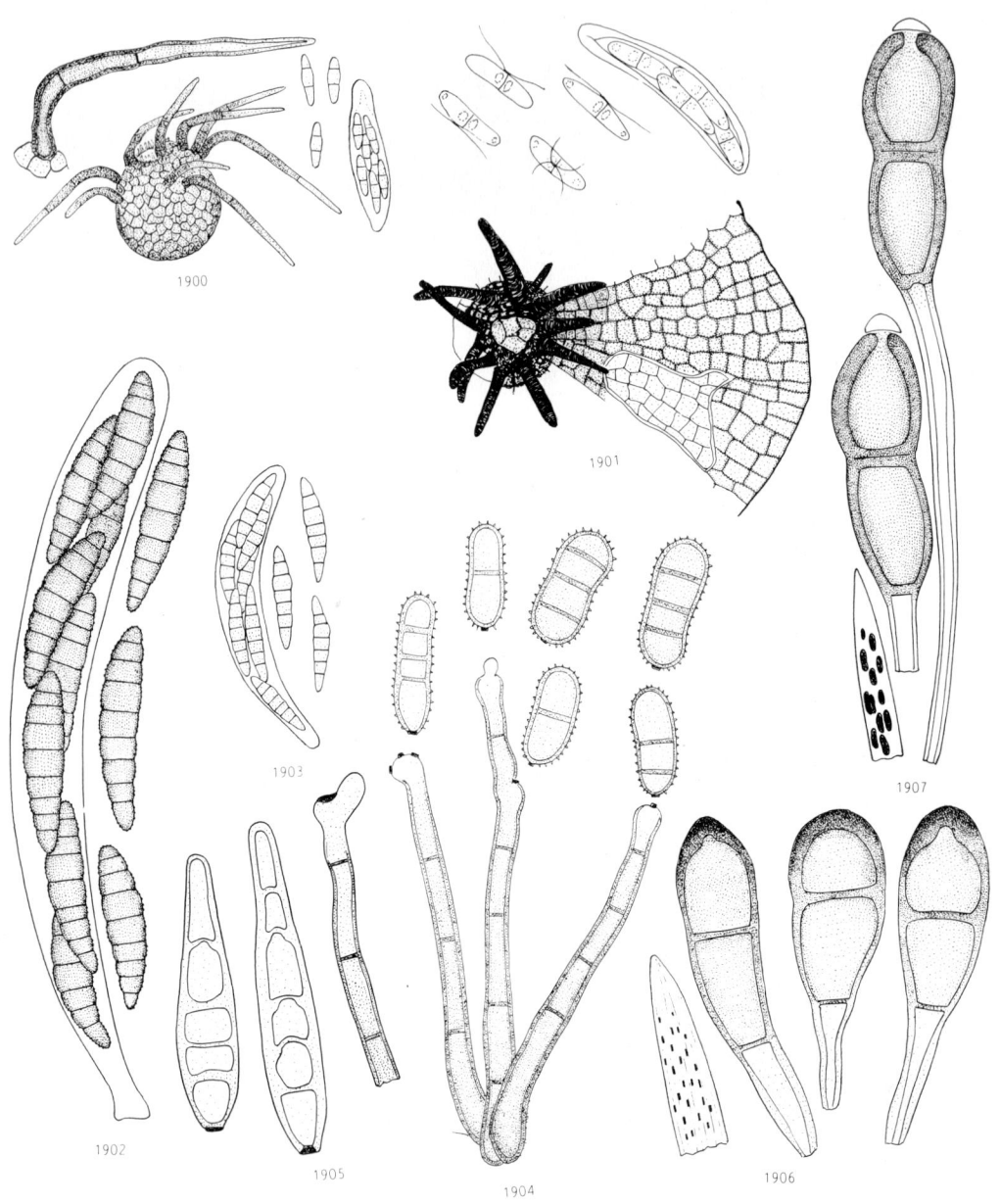

Plate 185. On *Phalaris*, *Phleum* and *Phragmites*. 1900, *Acanthostigmella genuflexa*; 1901, *Actinopeltis palustris*; 1902, *Leptosphaeria mosana*; 1903, *L. linearis*; 1904, *Cladosporium phlei*; 1905, *Drechslera phlei*; 1906, *Puccinia magnusiana*; 1907, *P. phragmitis*.

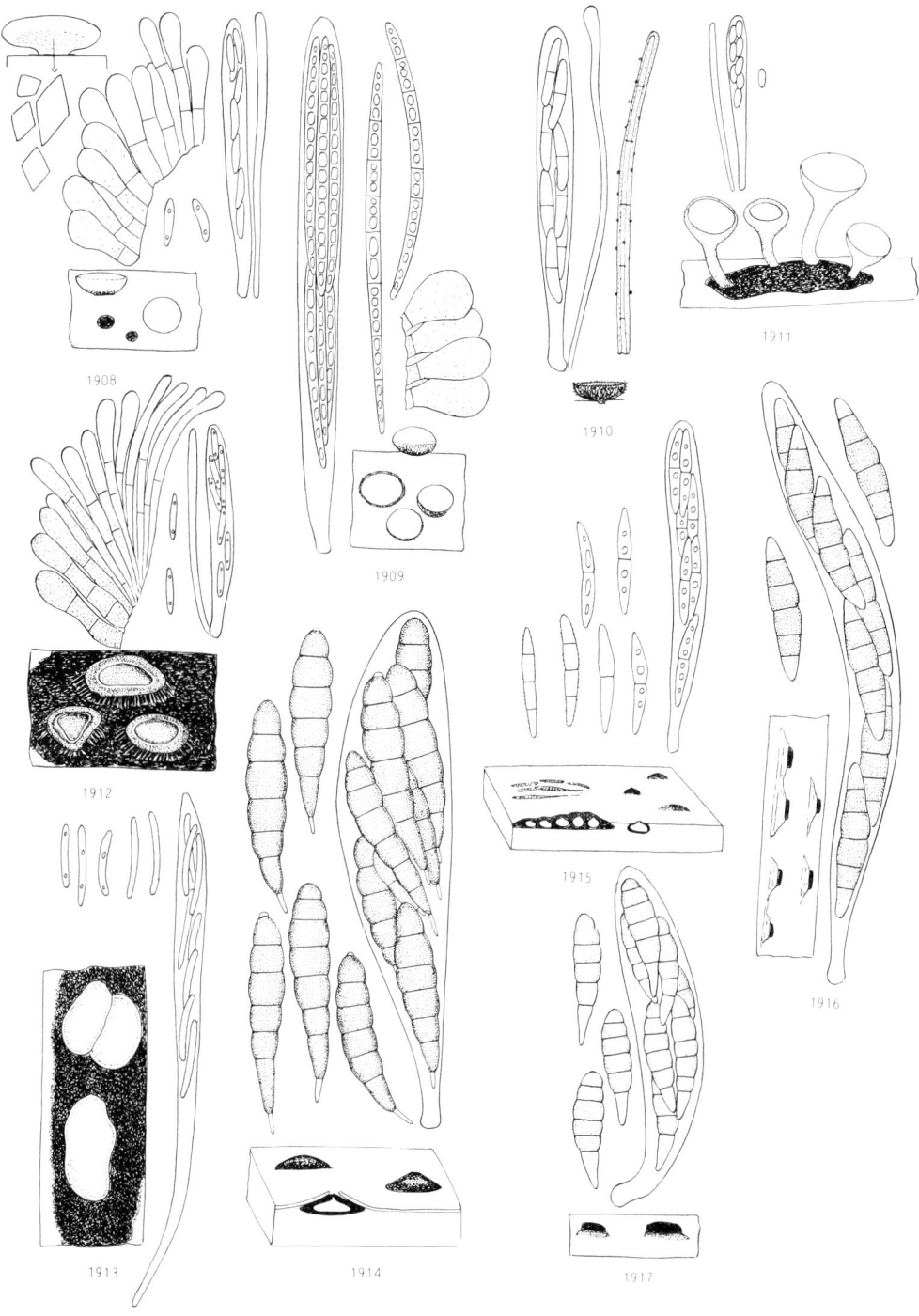

Plate 186. On *Phragmites*. 1908, *Mollisia hydrophila*; 1909, *Niptera excelsior*; 1910, *Perrotia phragmiticola*; 1911, *Rutstroemia lindaviana*; 1912, *Tapesia evilescens*; 1913, *T. knieffii*; 1914, *Buergenerula typhae*; 1915, *Leptosphaeria arundinacea*; 1916, *Lophiostoma arundinis*; 1917, *L. caudatum*.

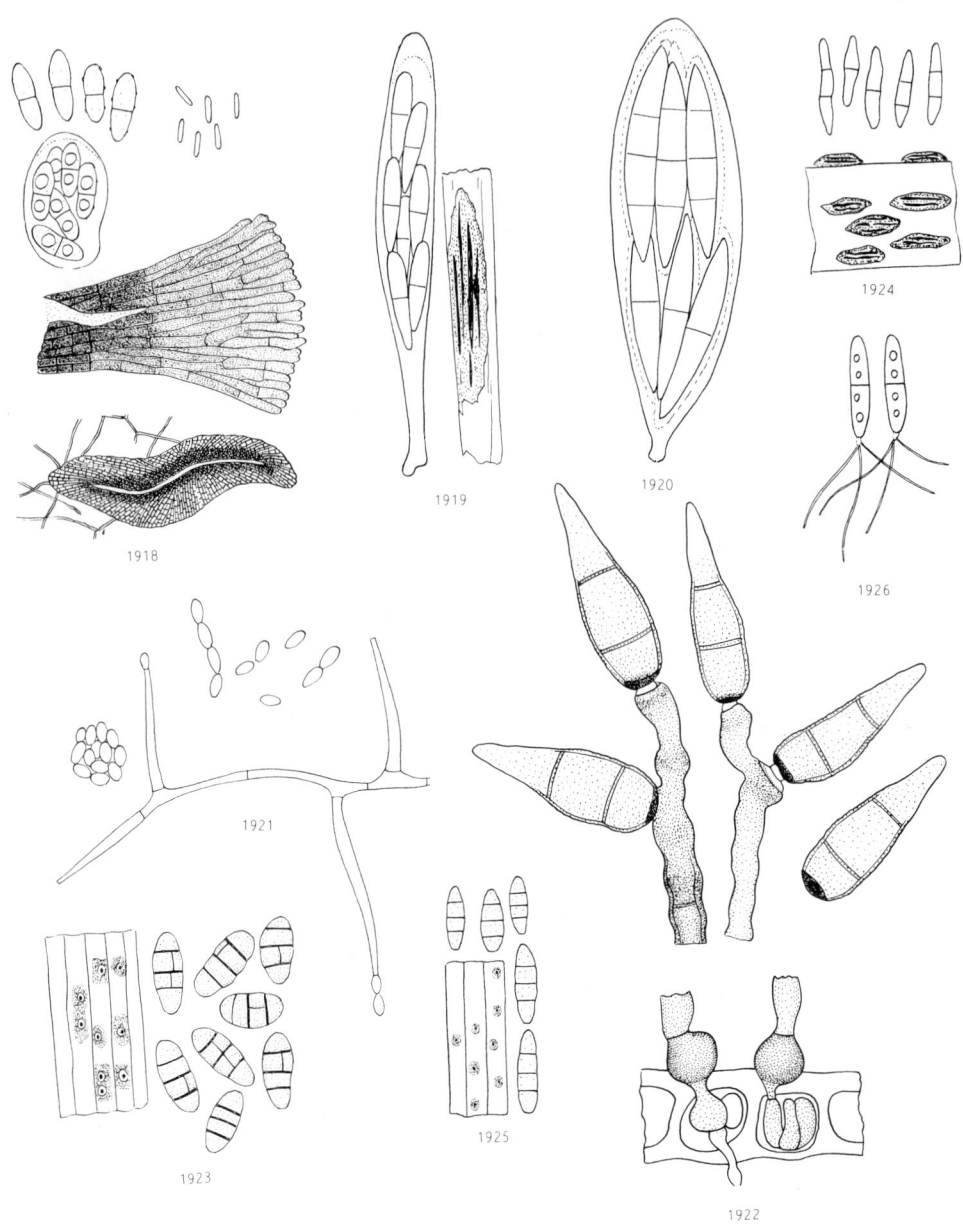

Plate 187. On *Phragmites*. 1918, *Morenoina phragmitis*; 1919, *Scirrhia rimosa*; 1920, *Wettsteinina niesslii*; 1921, *Acremonium alternatum*; 1922, *Deightoniella arundinacea*; 1923, *Camarosporium feurichii*; 1924, *Cytoplacosphaeria rimosa*; 1925, *Hendersonia culmiseda*; 1926, *Pseudorobillarda phragmitis*.

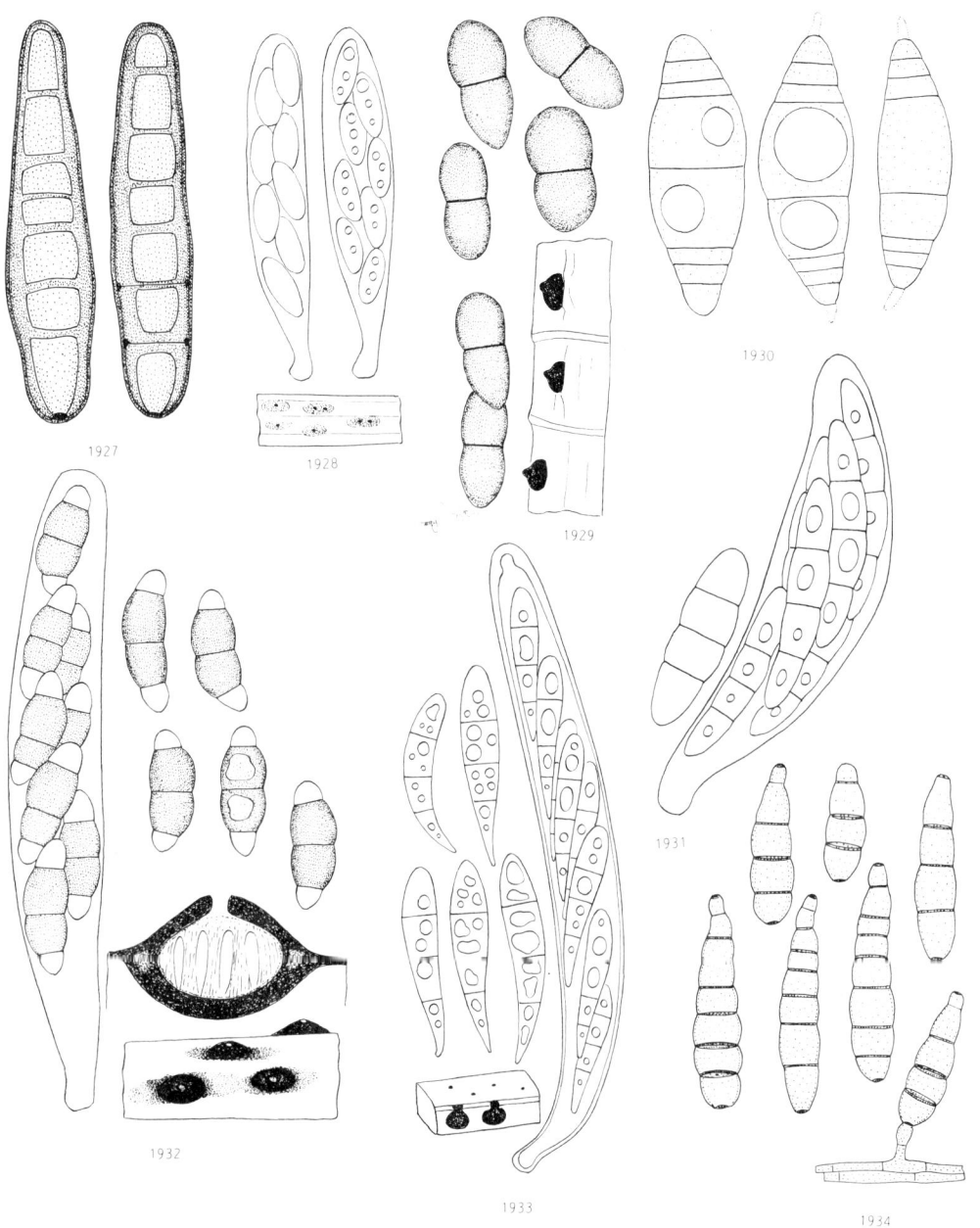

Plate 188. On *Poa*, *Sesleria* and *Spartina*. 1927, *Drechslera poae*; 1928, *Glomerella montana*; 1929, *Amphisphaeria melanommoides*; 1930, *Haligena spartinae*; 1931, *Leptosphaeria marina*; 1932, *Passeriniella discors*; 1933, *Sphaerulina pedicellata*; 1934, *Embellisia phragmospora*.

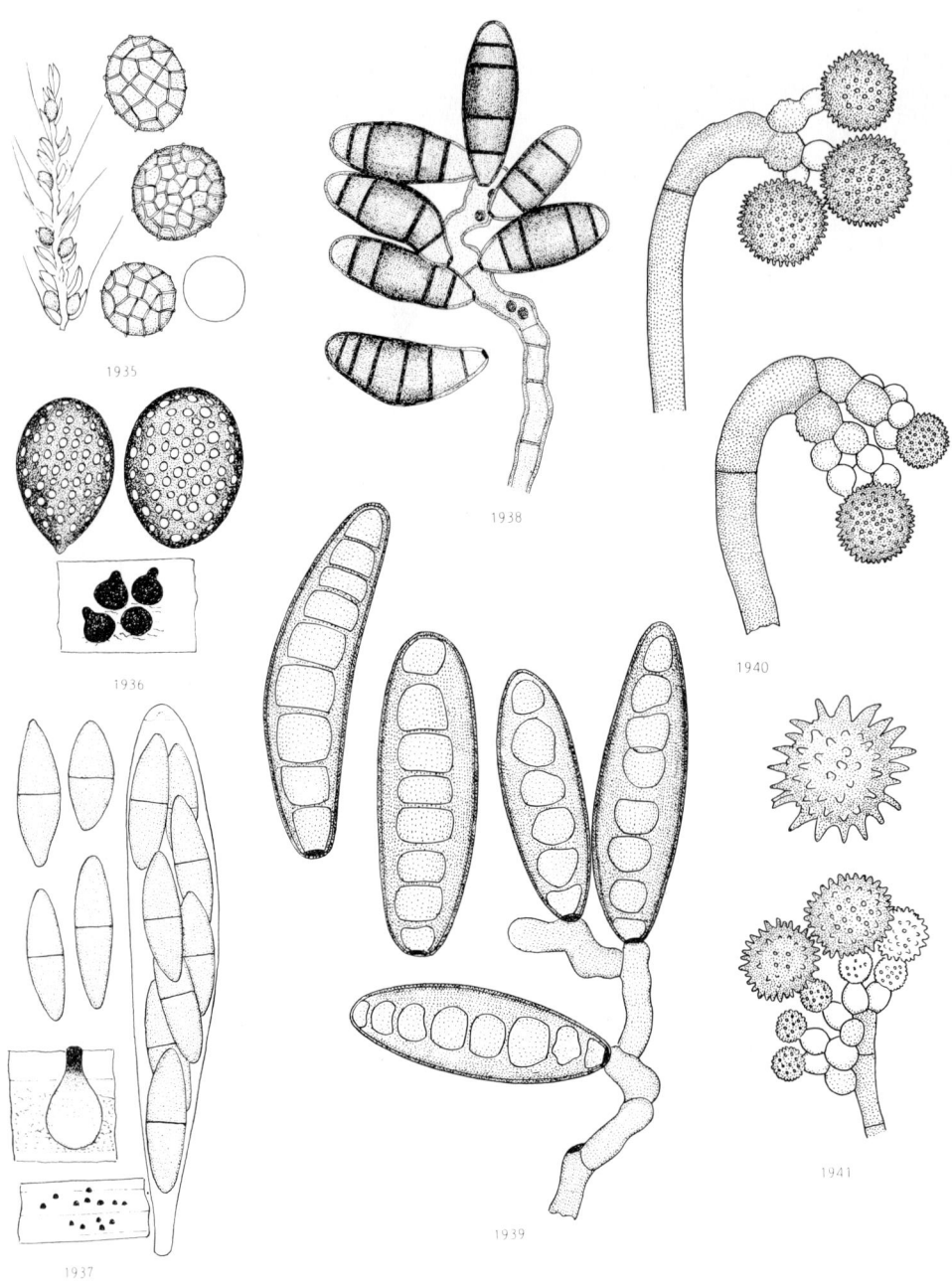

Plate 189. On *Triticum*. 1935, *Tilletia tritici*; 1936, *Gelasinospora cerealis*; 1937, *Gibellina cerealis*; 1938, *Curvularia inaequalis*; 1939, *Drechslera* state of *Cochliobolus sativus*; 1940, *Periconia circinata*; 1941, *P. macrospinosa*.

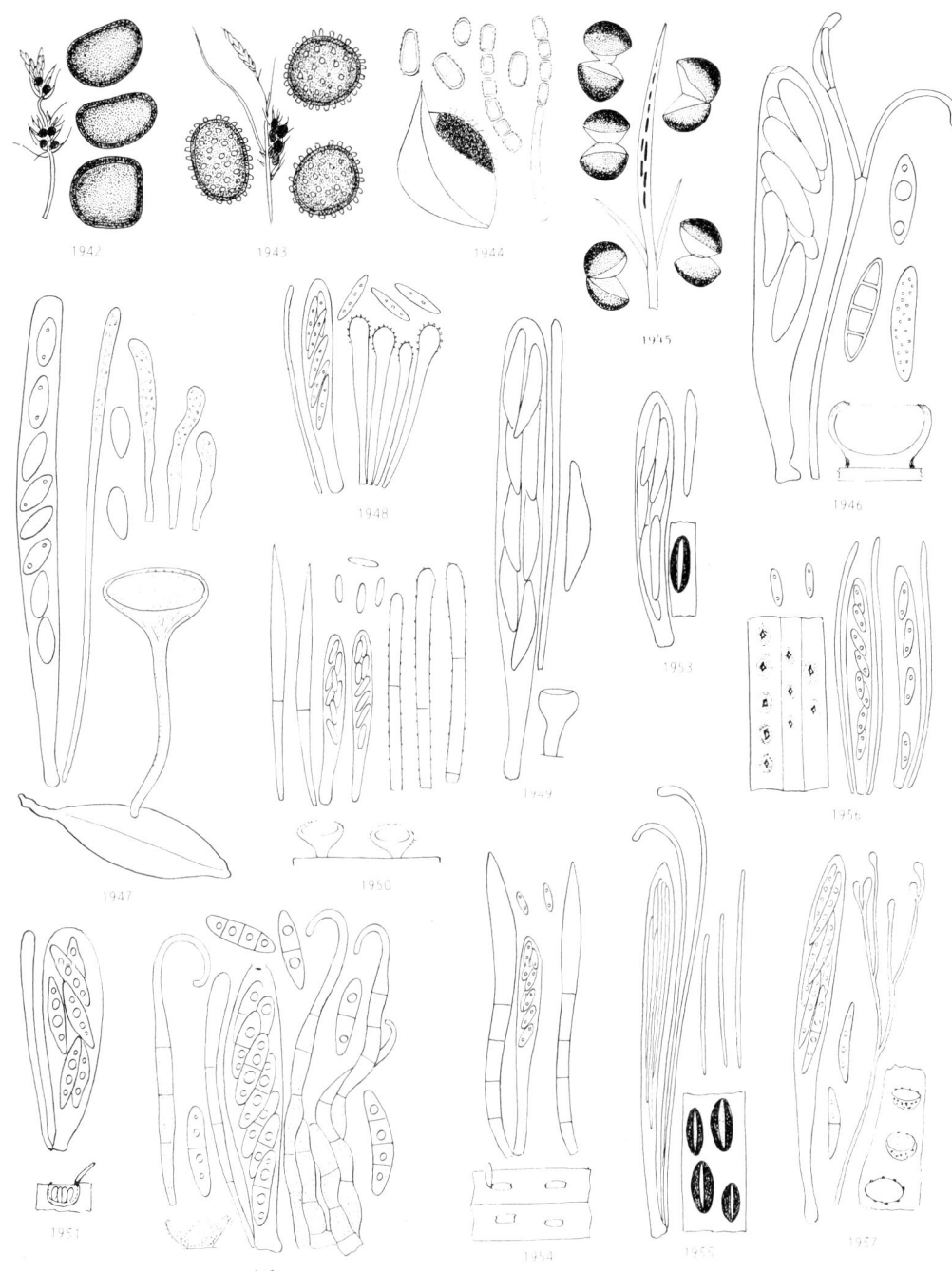

Plate 190. Smuts and discomycetes on *Carex*. 1942, *Anthracoidea caricis*; 1943, *A. subinclusa*; 1944, *Farysia thuemenii*; 1945, *Schizonella melanogramma*; 1946, *Actinoscypha muelleri*; 1947, *Ciboria aschersoniana*; 1948, *Clavidisculum caricis*; 1949, *Crocicreas megalosporum*; 1950, *Dasyscyphus caricis*; 1951, *Eupropolella celata*; 1952, *Hyaloscypha scintillans*; 1953, *Hypoderma alpinum*; 1954, *Hysteropezizella dowardensis*; 1955, *Lophodermium caricinum*; 1956, *Merostictis seriata*; 1957, *Micropeziza cornea*.

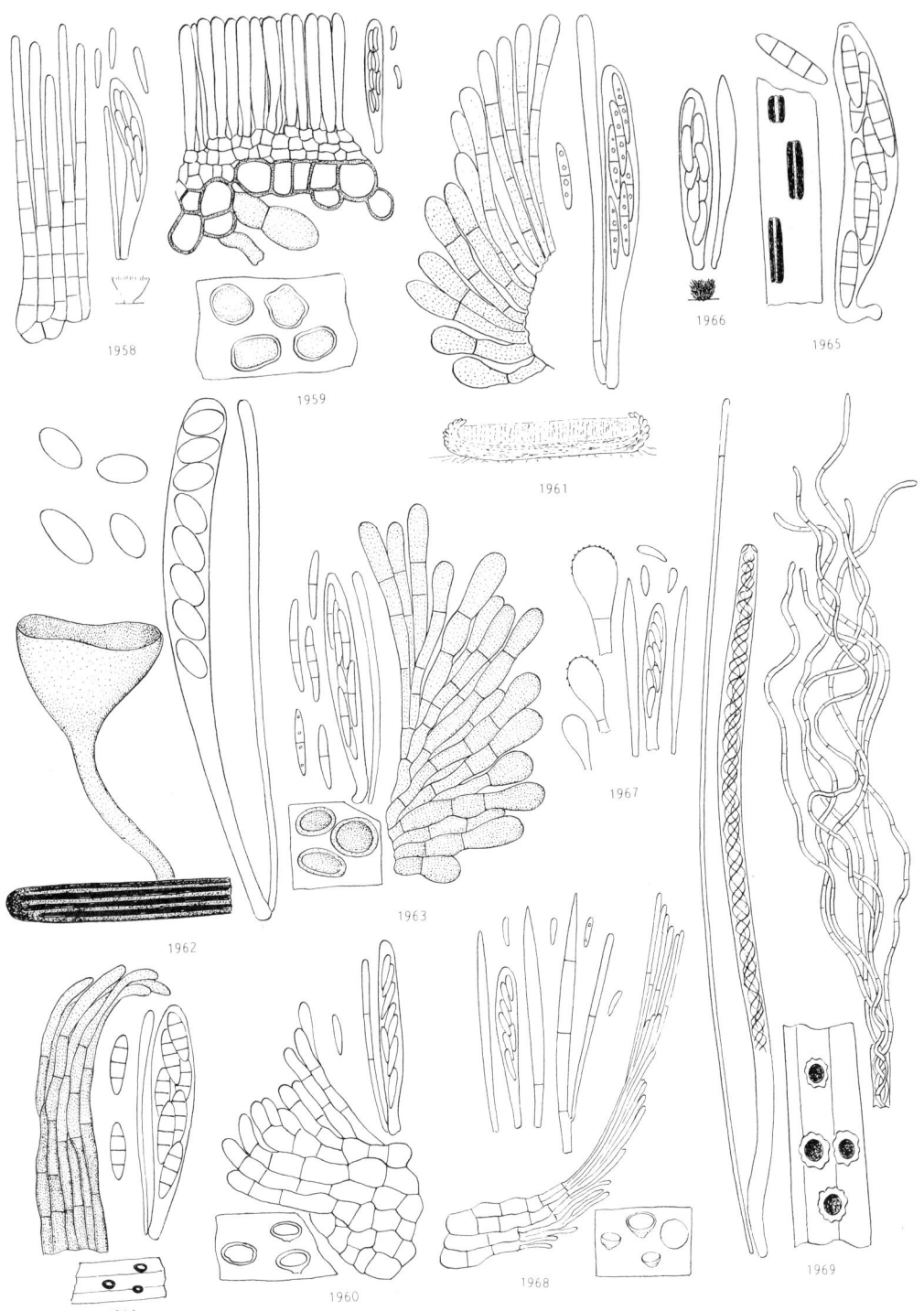

Plate 191. Discomycetes on *Carex*. 1958, *Microscypha ellisii*; 1959, *Mollisia caricina*; 1960, *M. chionea*; 1961, *M. humidicola*; 1962, *Myriosclerotinia sulcata*; 1963, *Niptera pilosa*; 1964, *Pezizella nigro-corticata*; 1965, *Phragmonaevia hysterioides*; 1966, *Psilachnum asemum*; 1967, *P. granulosellum*; 1968, *P. lateritio-album*; 1969, *Stictis elongatispora*.

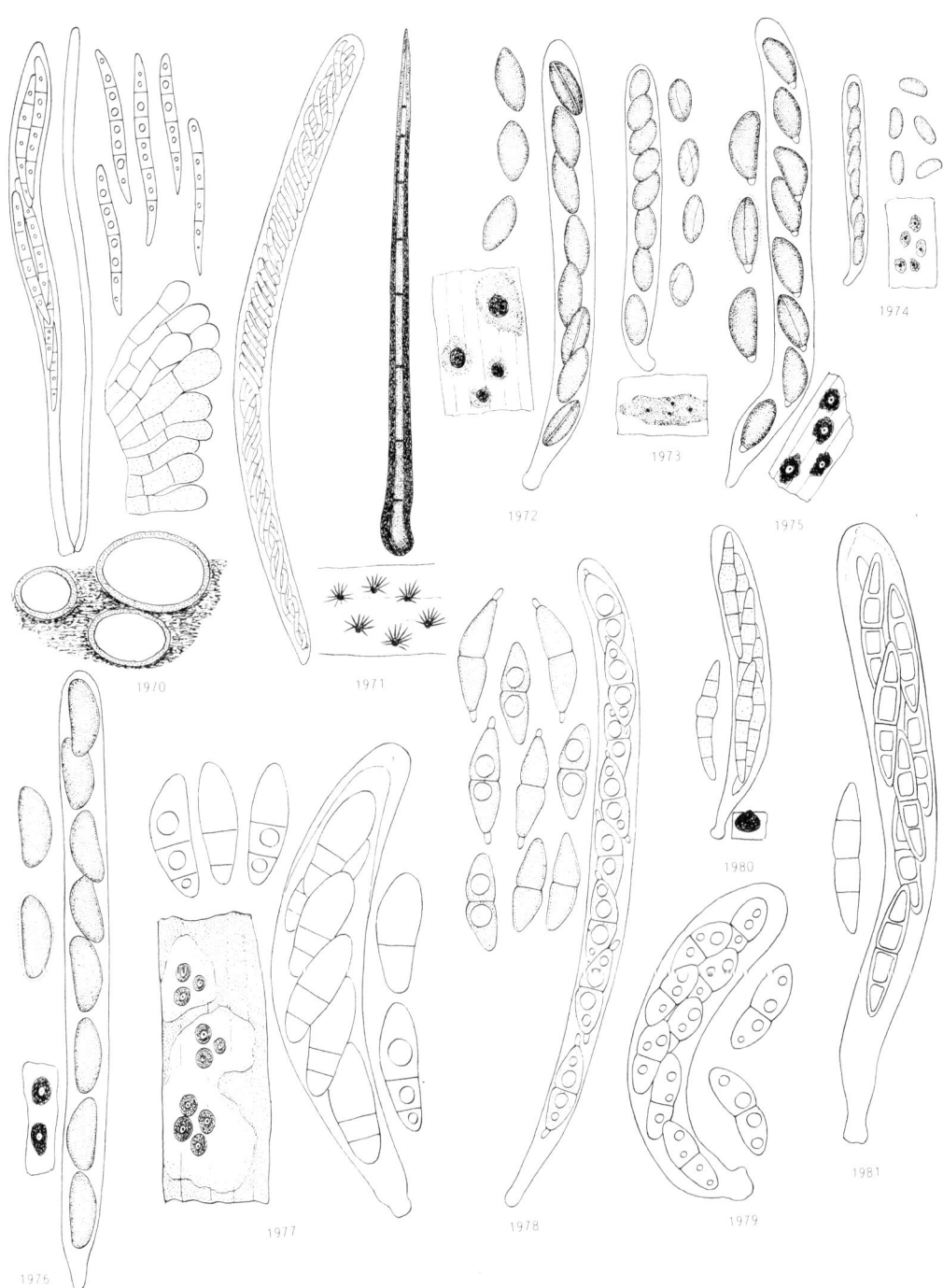

Plate 192. Discomycetes and other ascomycetes on *Carex*. 1970, *Trichobelonium asteroma*; 1971, *Acanthophiobolus helicosporus*; 1972, *Anthostomella caricis*; 1973, *A. limitata*; 1974, *A. punctulata*; 1975, *A. tomicoides*; 1976, *A. tumulosa*; 1977, *Buergenerula biseptata*; 1978, *Ceriophora palustris*; 1979, *Didymella proximella*; 1980, *Leptosphaeria caricis*; 1981, *Metasphaeria cumana* f. *macrospora*.

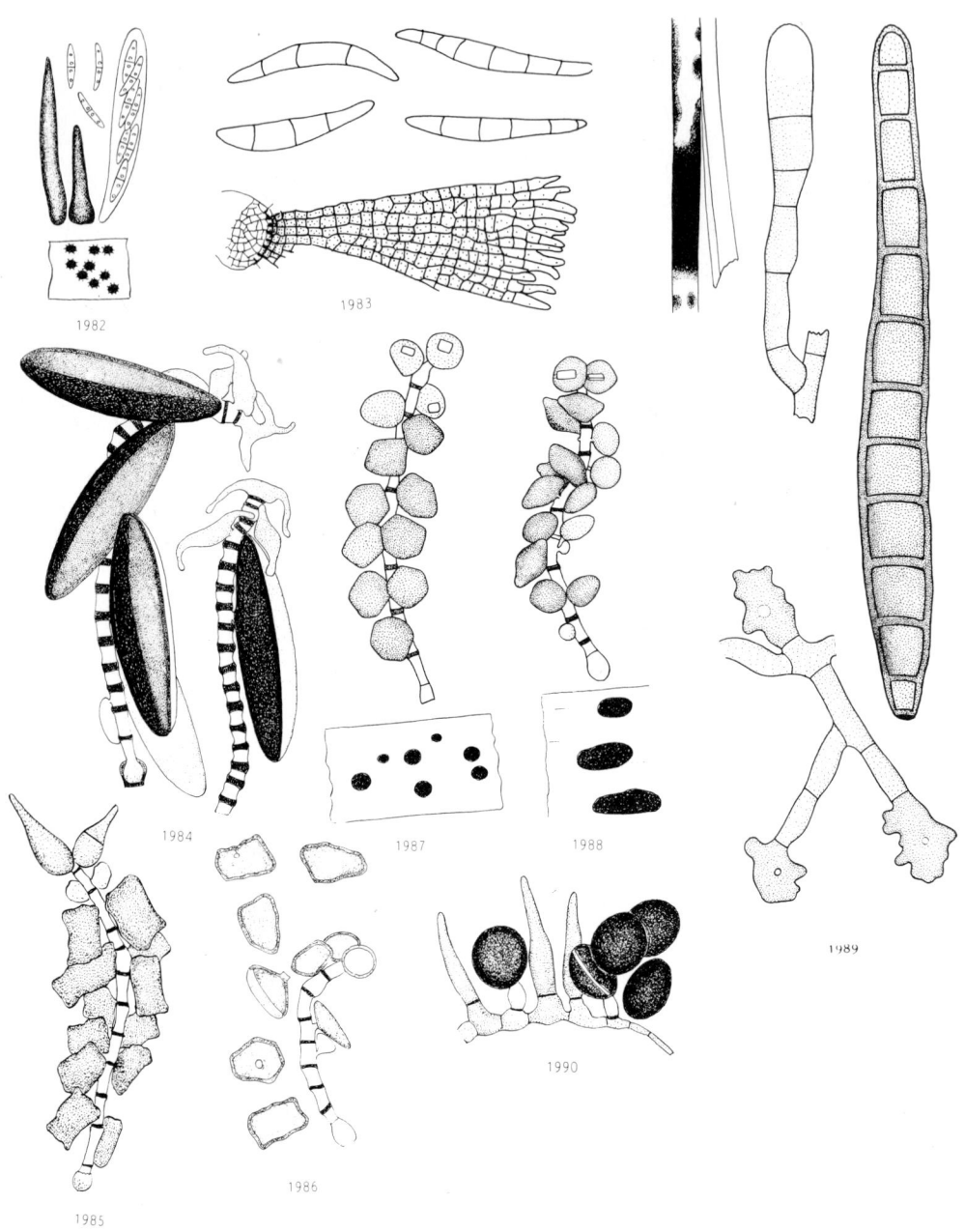

Plate 193. Ascomycetes and hyphomycetes on *Carex*. 1982, *Niesslia exosporioides*; 1983, *Trichothyrina norfolciana*; 1984, *Arthrinium caricicola*; 1985, *A. fuckelii*; 1986, *A. morthieri*; 1987, *A. puccinioides*; 1988, *A. sporophleum*; 1989, *Clasterosporium caricinum*; 1990, *Cordella clarkii*.

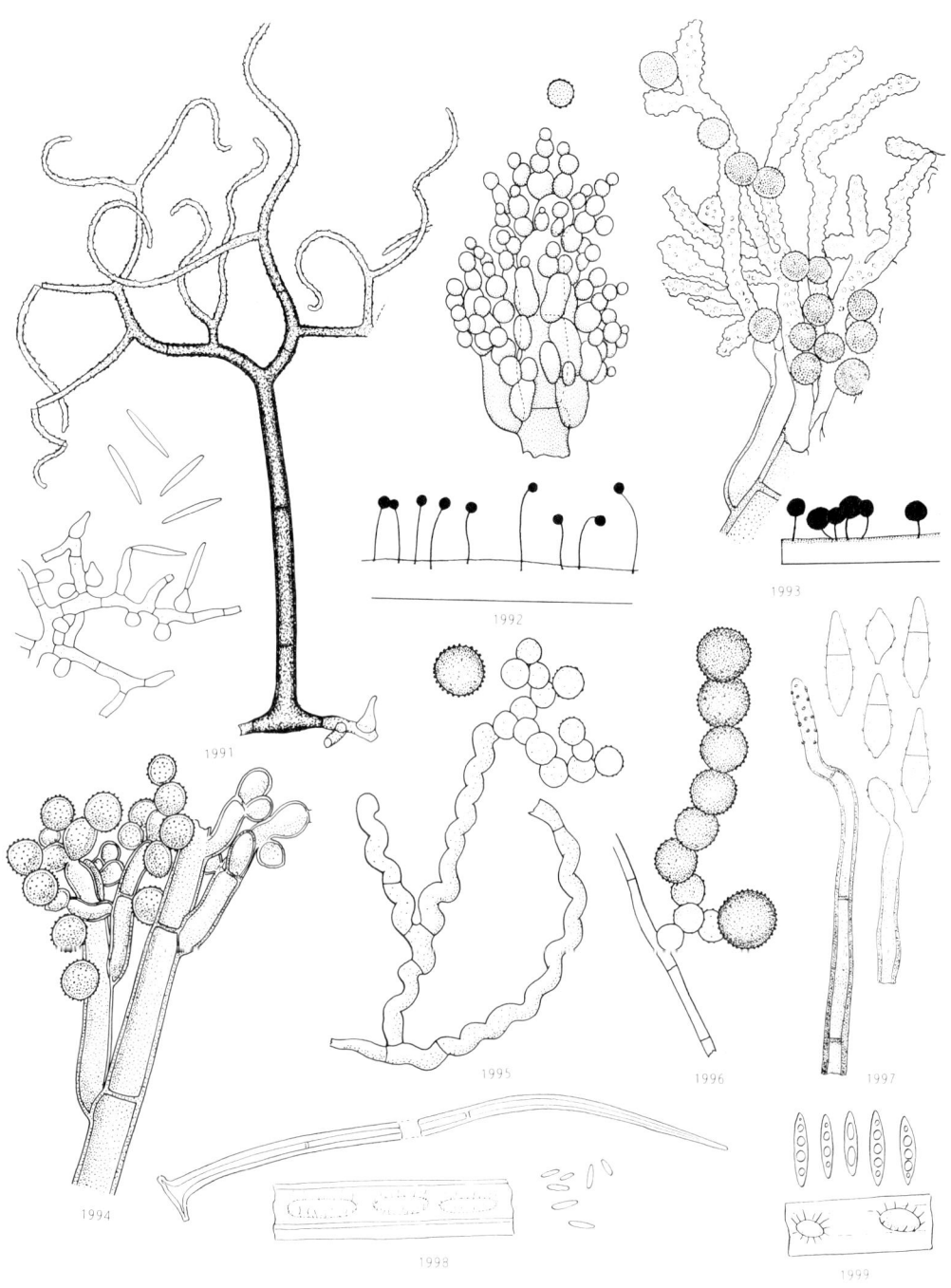

Plate 194. Hyphomycetes on *Carex*. 1991, *Gyrothrix podosperma*; 1992, *Periconia atra*; 1993, *P. curta*; 1994, *P. digitata*; 1995, *P. funerea*; 1996, *P. laminella*; 1997, *Veronaea caricis*; 1998, *Volutella arundinis*; 1999, *V. melaloma*.

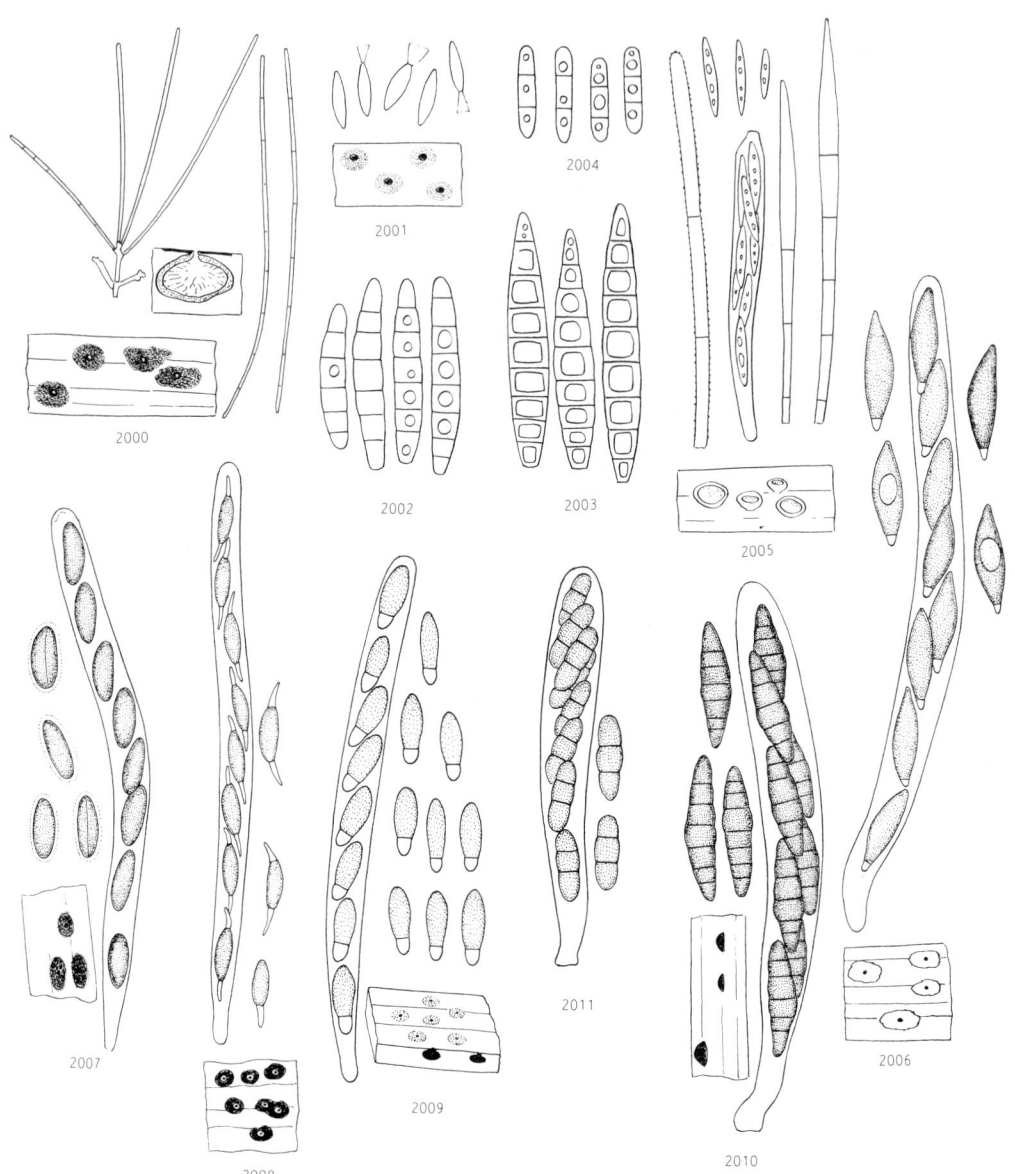

Plate 195. On *Carex* and *Cladium*. 2000, *Eriospora leucostoma*; 2001, *Neottiospora caricina*; 2002, *Stagonospora caricis*; 2003, *S. paludosa*; 2004, *S. vitensis*; 2005, *Dasyscyphus imbecillis* var. *cladii*; 2006, *Anthostomella fuegiana*; 2007, *A. leptospora*; 2008, *A. scotina*; 2009, *Didymopleela cladii*; 2010, *Leptosphaeria cladii*; 2011, *Paraphaeosphaeria michotii*.

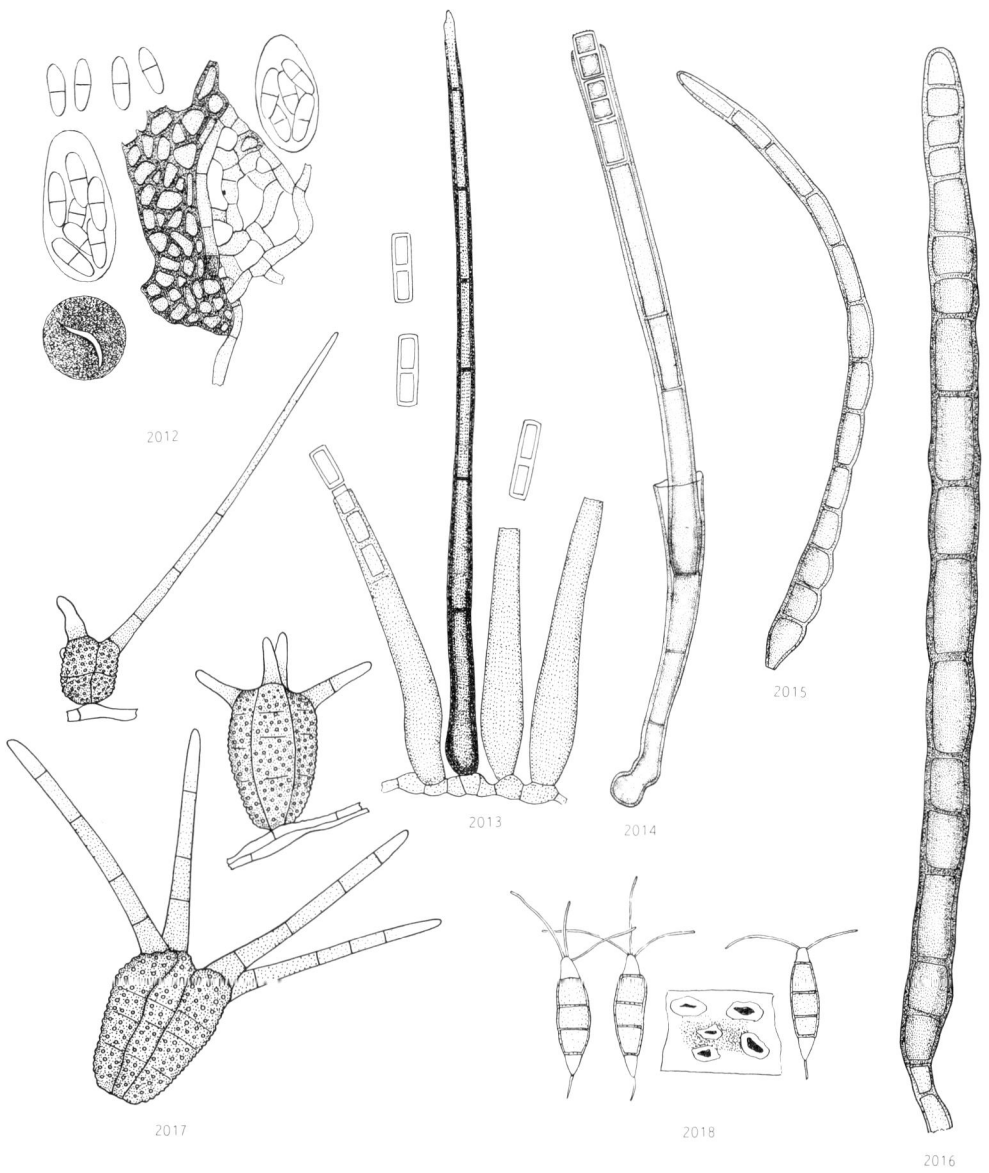

Plate 196. On *Cladium*. 2012, *Schizothyrium pomi*; 2013, *Chaetochalara cladii*; 2014, *Chalara cladii*; 2015, *Sporidesmium cladii*; 2016, *S. paludosum*; 2017, *Tetraploa aristata*; 2018, *Pestalotiopsis disseminata*.

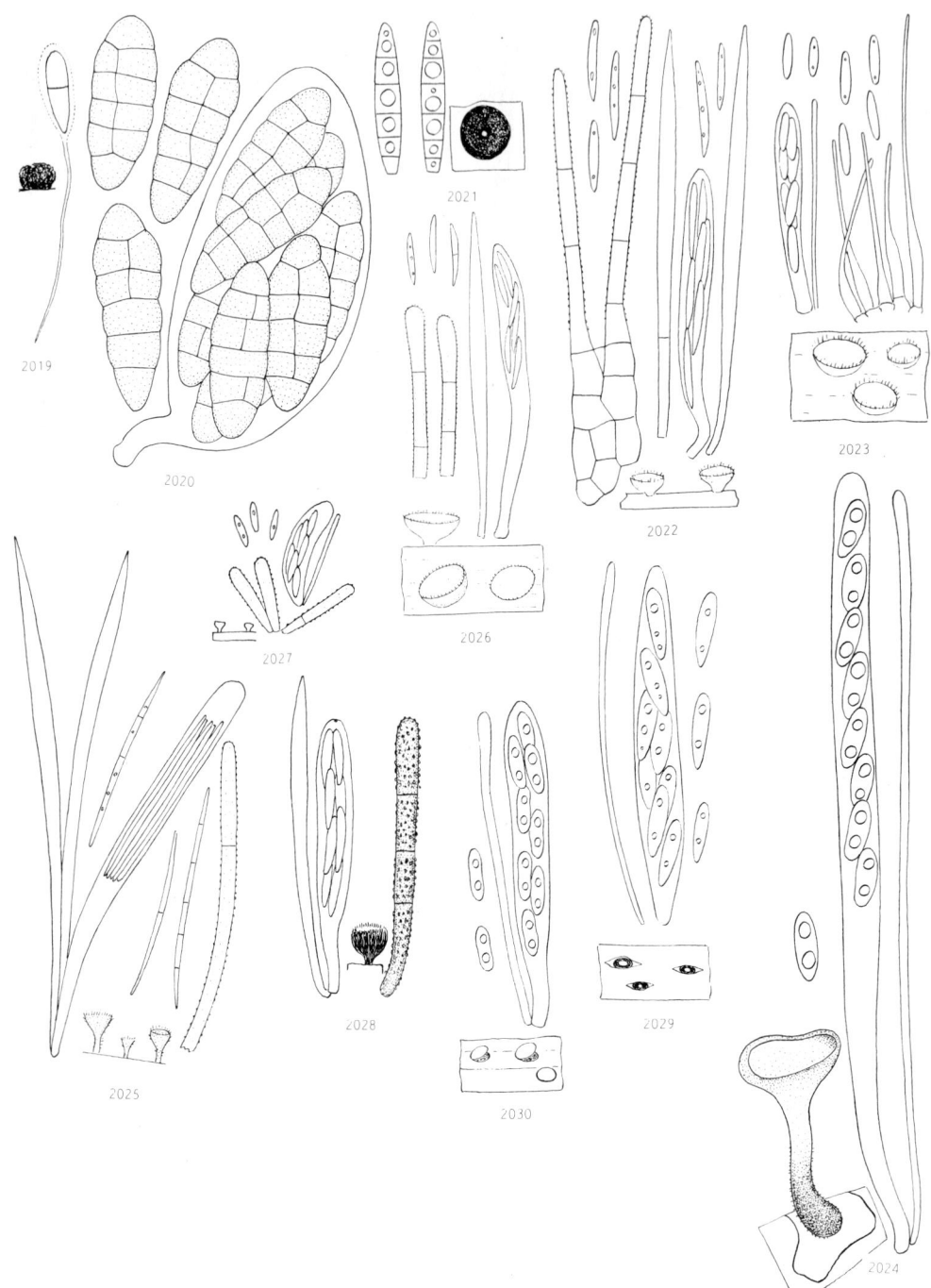

Plate 197. On *Eleocharis*, *Eriophorum* and *Juncus*. 2019, *Loramyces juncicola*; 2020, *Pleospora palustris*; 2021, *Stagonospora heleocharidis*; 2022, *Dasyscyphus imbecillis*; 2023, *Hyaloscypha paludosa*; 2024, *Rutstroemia henningsiana*; 2025, *Dasyscyphus apalus*; 2026, *D. diminutus*; 2027, *D. fugiens*; 2028, *D. rehmii*; 2029, *Hysteropezizella exigua*; 2030, *H. pusilla*.

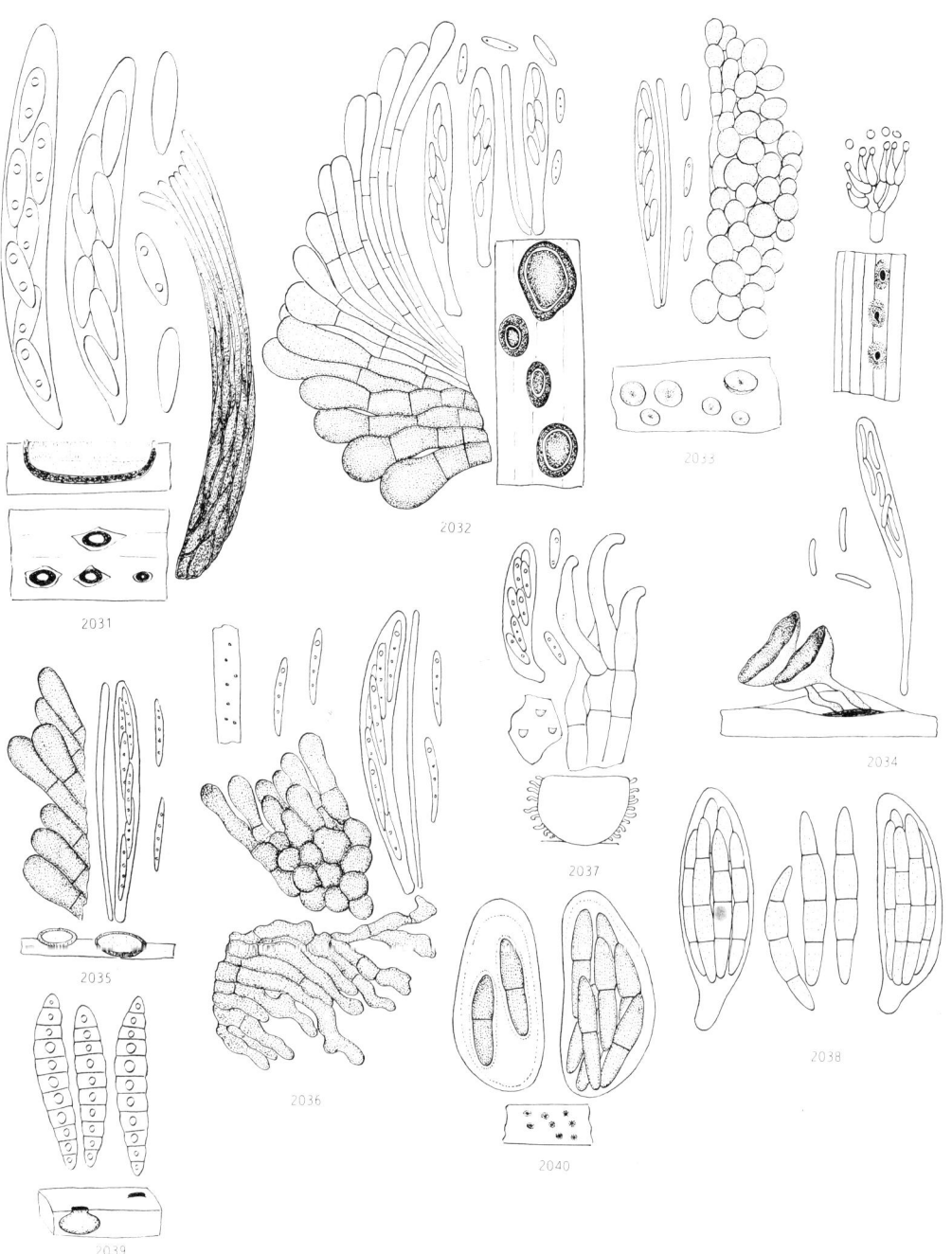

Plate 198. Discomycetes and other ascomycetes on *Juncus*. 2031, *Hysteropezizella rehmii*; 2032, *Mollisia juncina*; 2033, *M. palustris*; 2034, *Myriosclerotinia curreyana*; 2035, *Niptera melatephra*; 2036, *Scutomollisia stenospora*; 2037, *Unguicularia costata*; 2038, *Leptosphaeria juncina*; 2039, *Lophiostoma alpigenum* var. *juncinum*; 2040, *Monascostroma innumerosa*.

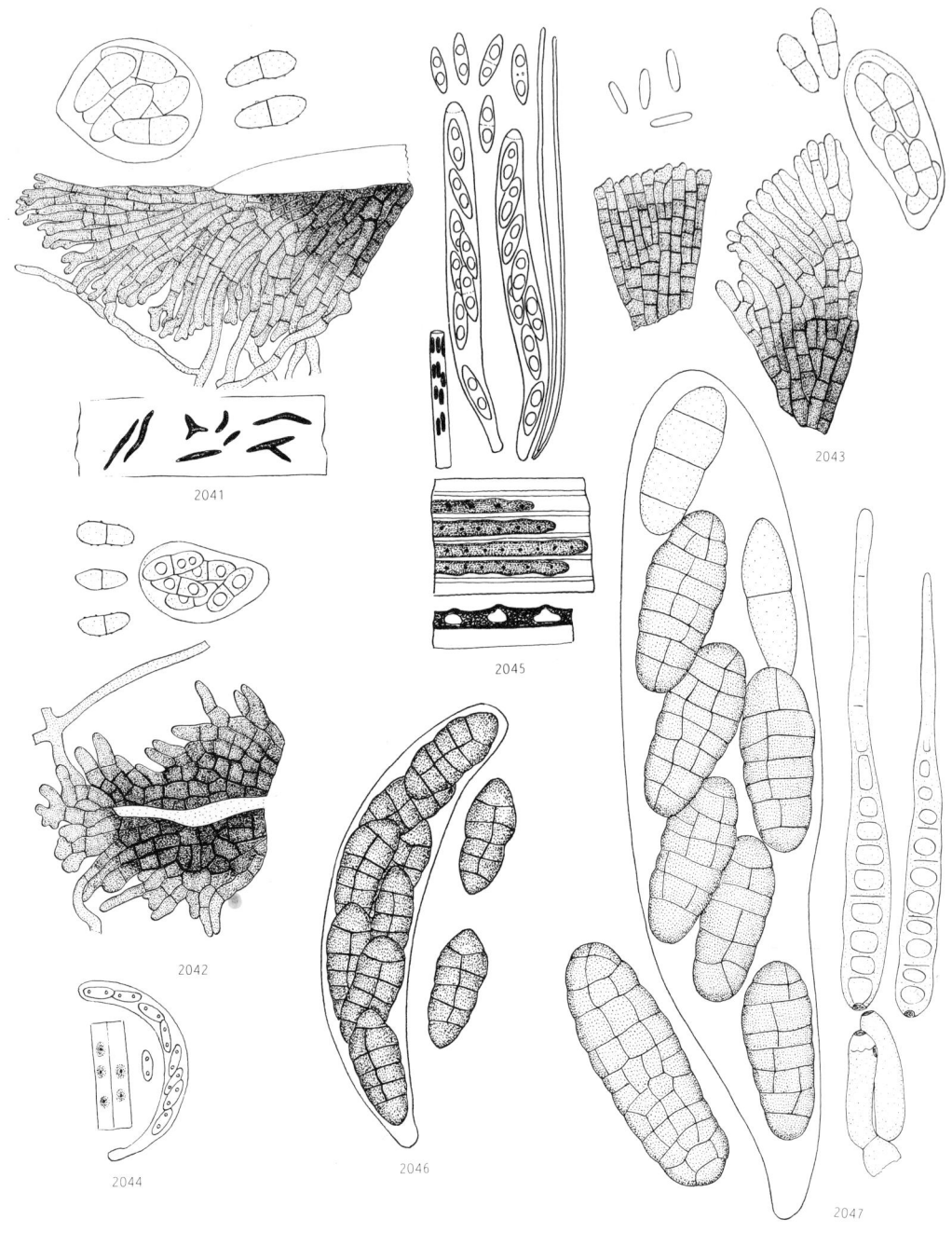

Plate 199. Ascomycetes on *Juncus*. 2041, *Morenoina fimbriata*; 2042, *M. minuta*; 2043, *M. paludosa*; 2044, *Phomatospora therophila*; 2045, *Phyllachora junci*; 2046, *Pleospora aquatica*; 2047, *P. valesiaca*.

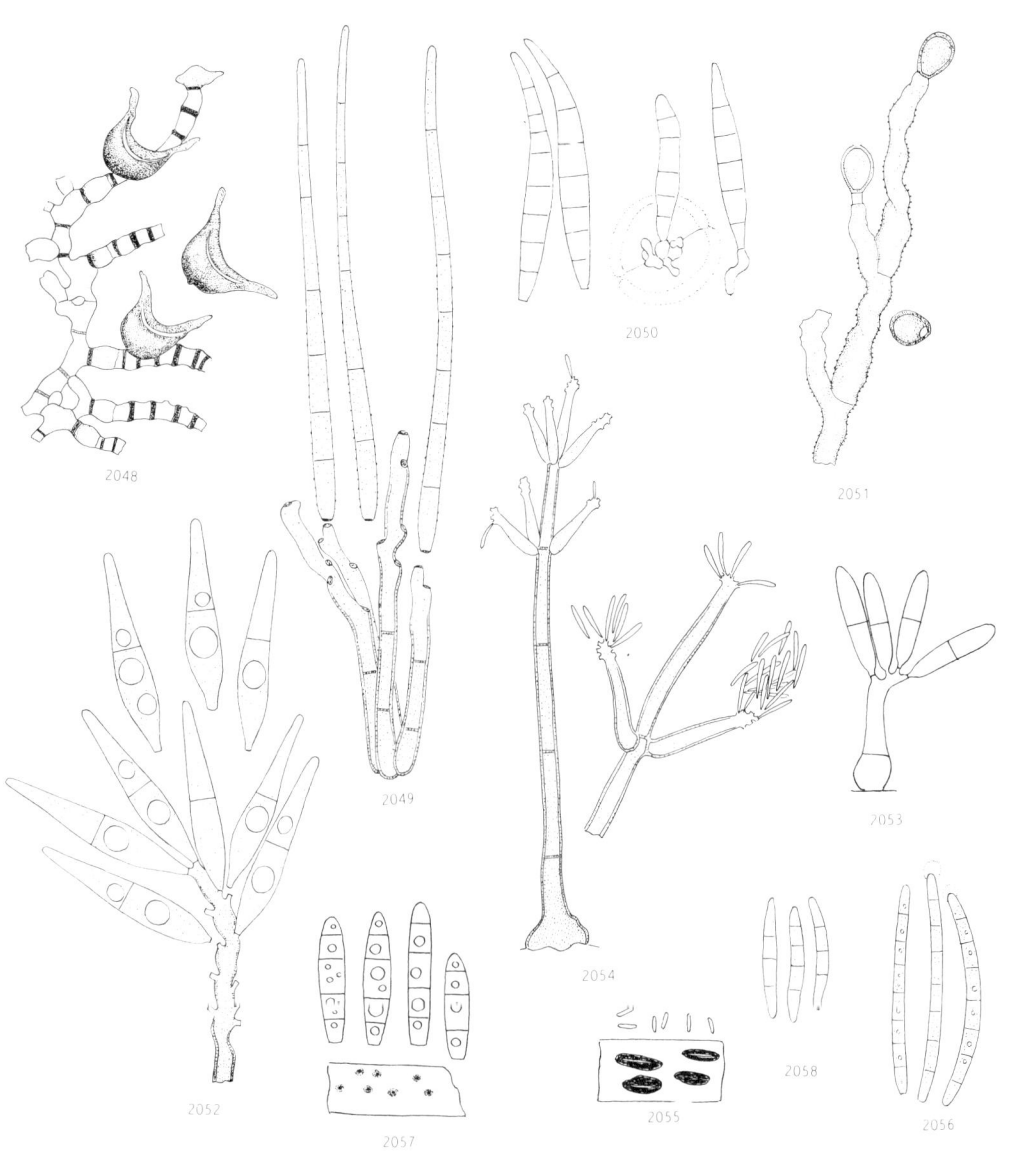

Plate 200. Hyphomycetes and coelomycetes on *Juncus*. 2048, *Arthrinium cuspidatum*; 2049, *Cercospora juncicola*; 2050, *Cercosporella junci*; 2051, *Conoplea fusca*; 2052, *Dactylaria junci*; 2053, *Diplorhinotrichum juncicola*; 2054, *Selenosporella curvispora*; 2055, *Leptostroma juncacearum*; 2056, *Septoriella junci*; 2057, *Stagonospora innumerosa*; 2058, *S. junciseda*.

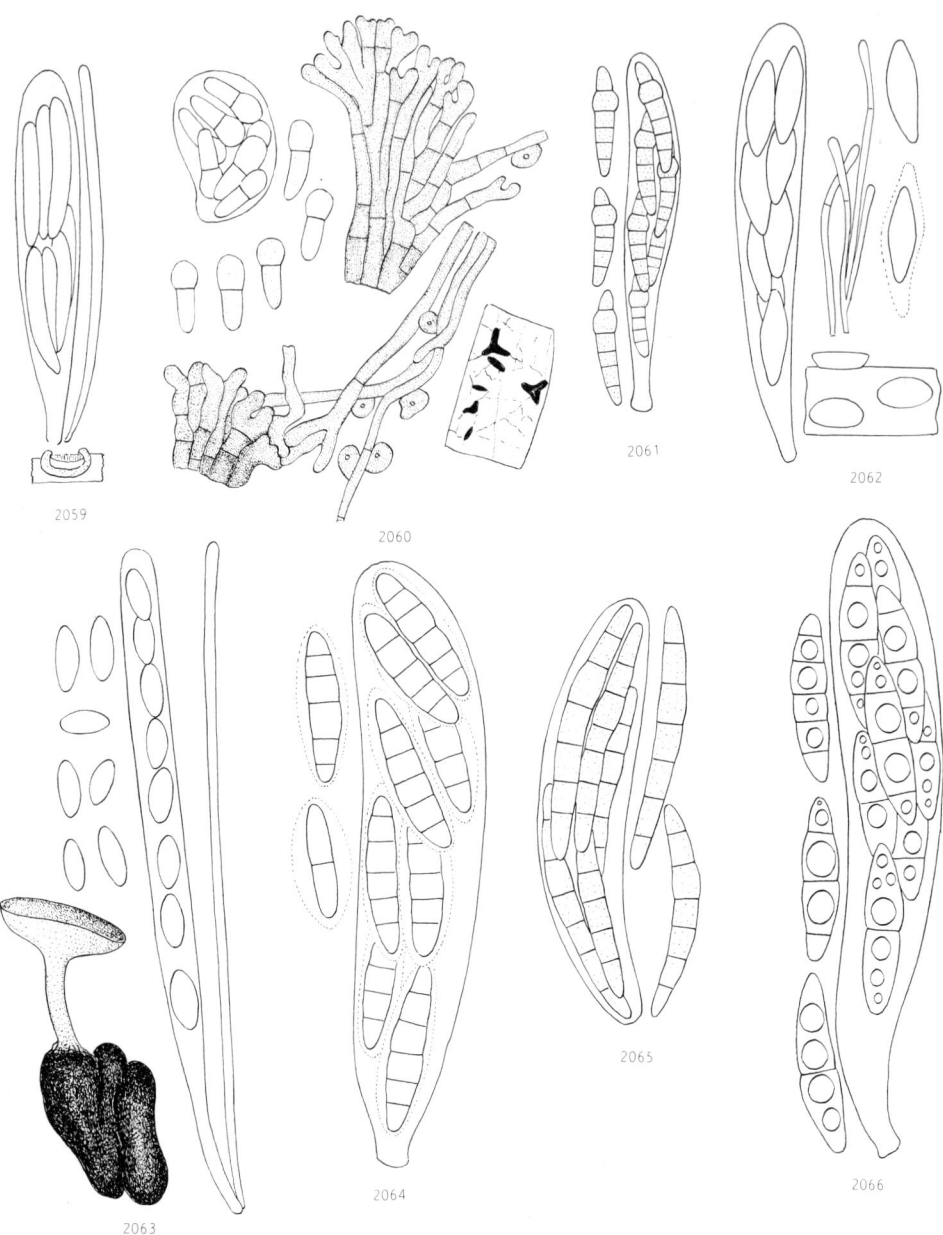

Plate 201. On *Luzula* and *Schoenoplectus*. 2059, *Laetinaevia luzulae*; 2060, *Lembosia luzulae*; 2061, *Leptosphaeria epicalamia*; 2062, *Coleosperma lacustre*; 2063, *Myrioclerotinia scirpicola*; 2064, *Leptosphaeria scirpina*; 2065, *L. sowerbyi*; 2066, *Metasphaeria mosana*.

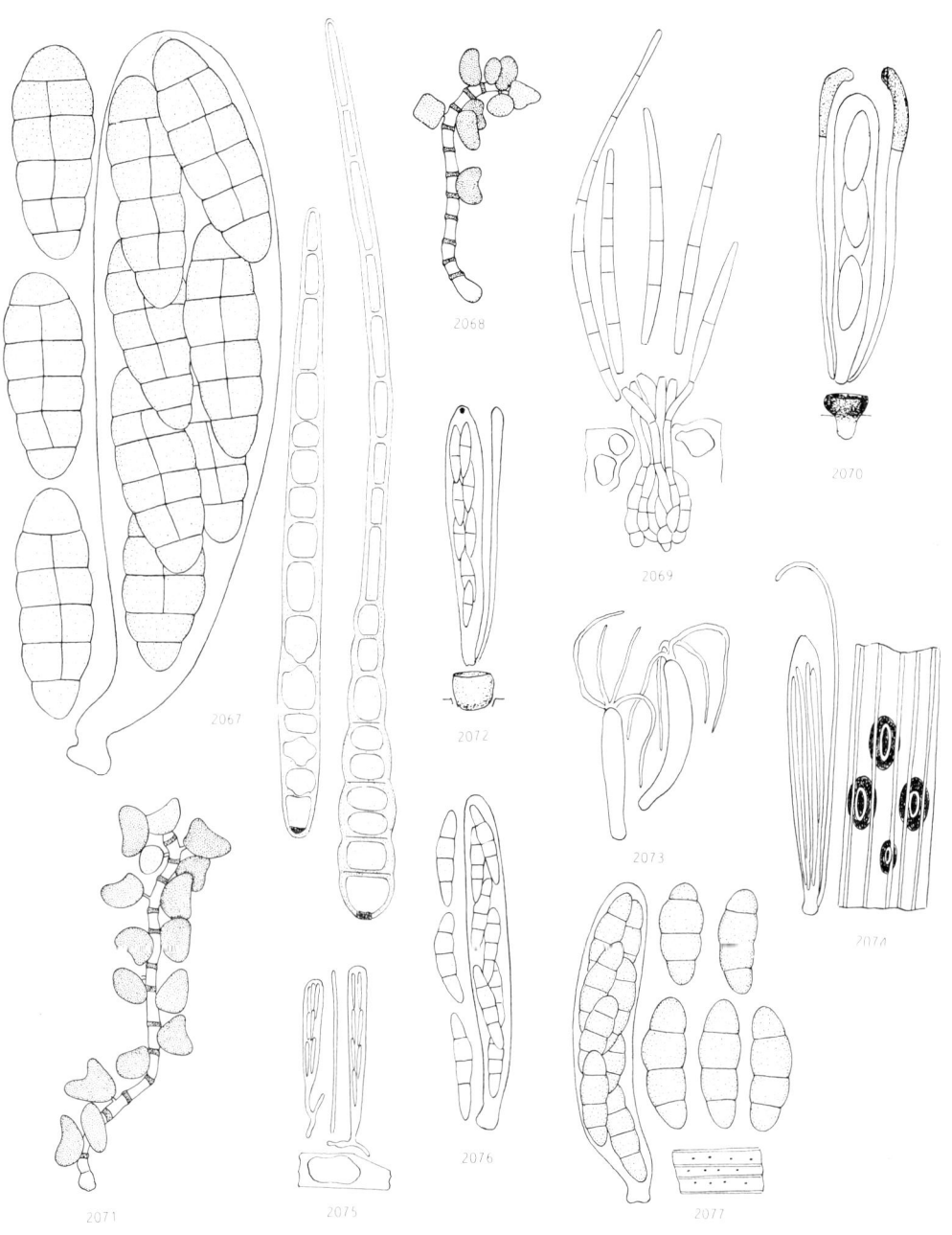

Plate 202. On *Schoenoplectus, Schoenus, Scirpus, Trichophorum* and *Typha*. 2067, *Pyrenophora scirpicola*; 2068, *Arthrinium curvatum* var. *minus*; 2069, *Pseudocercosporella scirpi*; 2070, *Drepanopeziza schoenicola*; 2071, *Arthrinium* state of *Pseudoguignardia scirpi*; 2072, *Dibeloniella trichophoricola*; 2073, *Tiarosporella paludosa*; 2074, *Lophodermium typhinum*; 2075, *Orbilia arundinacea*; 2076, *Leptosphaeria typhae*; 2077, *L. typharum*.

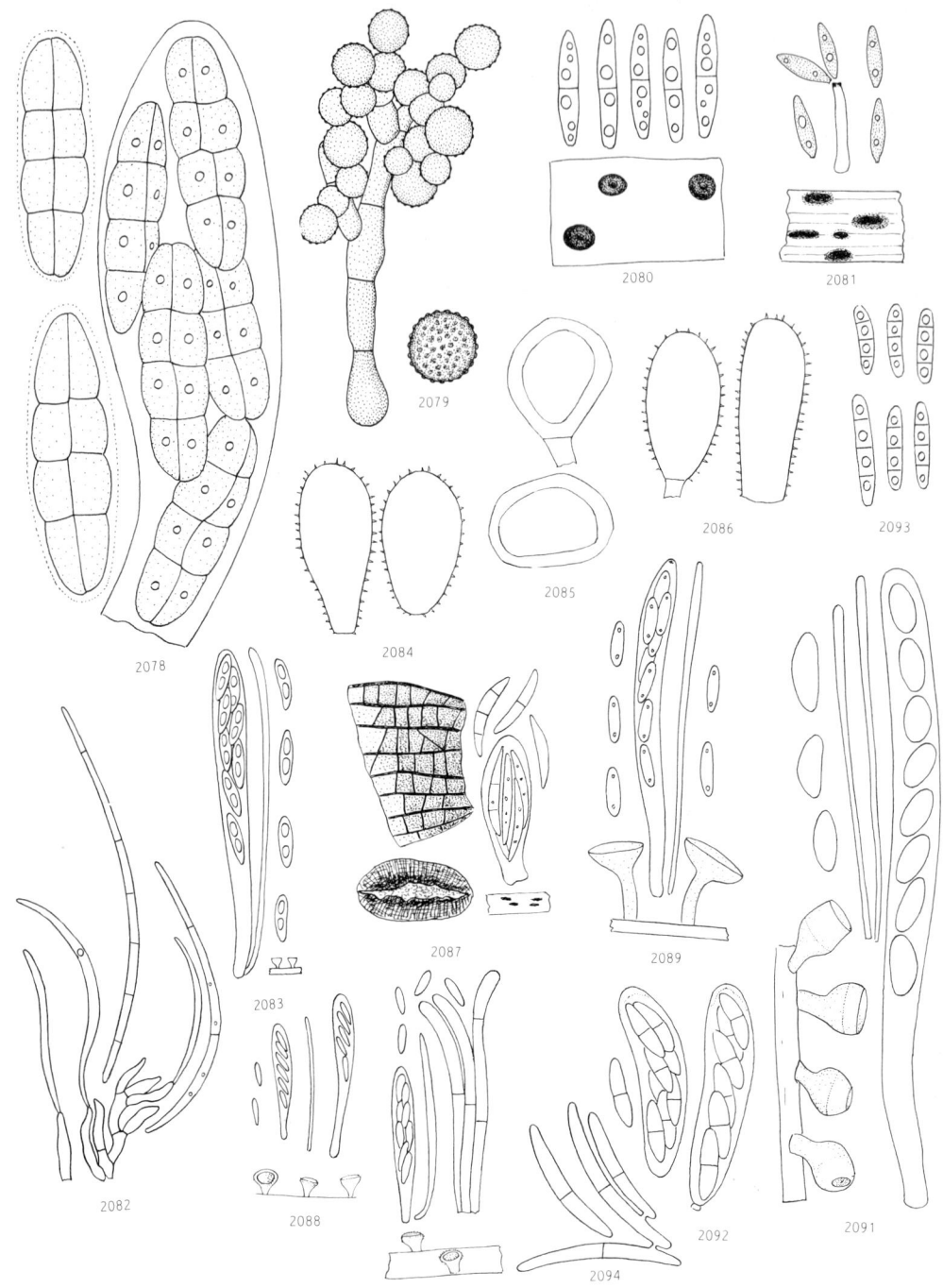

Plate 203. On *Typha*, ferns and horsetails. 2078, *Pyrenophora typhicola*; 2079, *Periconia typhicola*; 2080, *Ascochyta typhoidearum*; 2081, *Hymenopsis typhae*; 2082, *Cercosporella filicis-feminae*; 2083, *Pezizella campanulaeformis*; 2084, *Milesina blechni*; 2085, *Hyalopsora polypodii*; 2086, *Milesina kriegeriana*; 2087, *Leptopeltis filicina*; 2088, *Pezizella chrysostigma*; 2089, *Hymenoscyphus rhodoleucus*; 2090, *Psilachnum inquilinum*; 2091, *Stamnaria persoonii*; 2092, *Mycosphaerella equiseti*; 2093, *Stagonospora equiseti*; 2094, *Titaeospora equiseti*.

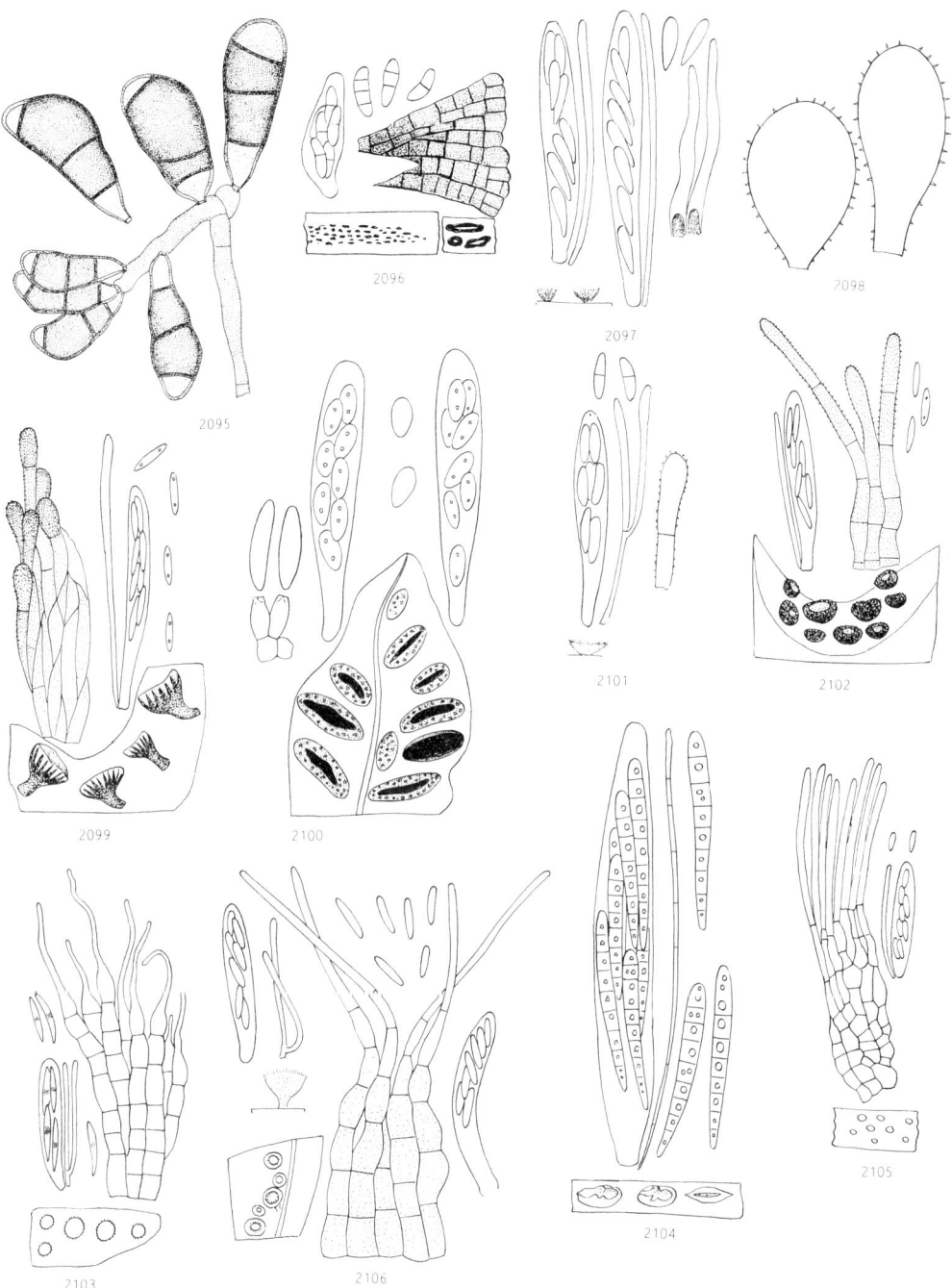

Plate 204. On *Ophioglossum, Osmunda, Phyllitis* and *Pteridium*. 2095, *Curvularia crepinii*; 2096, *Leptopeltis nebulosa*; 2097, *Unguicularia aspera*; 2098, *Milesina scolopendrii*; 2099, *Crocicreas cyathoideum* var. *pteridicola*; 2100, *Cryptomycina pteridis*; 2101, *Dasyscyphus pteridialis*; 2102, *D. pteridis*; 2103, *Hyaloscypha flaveola*; 2104, *Melittosporium pteridinum*; 2105, *Micropodia pteridina*; 2106, *Microscypha grisella*.

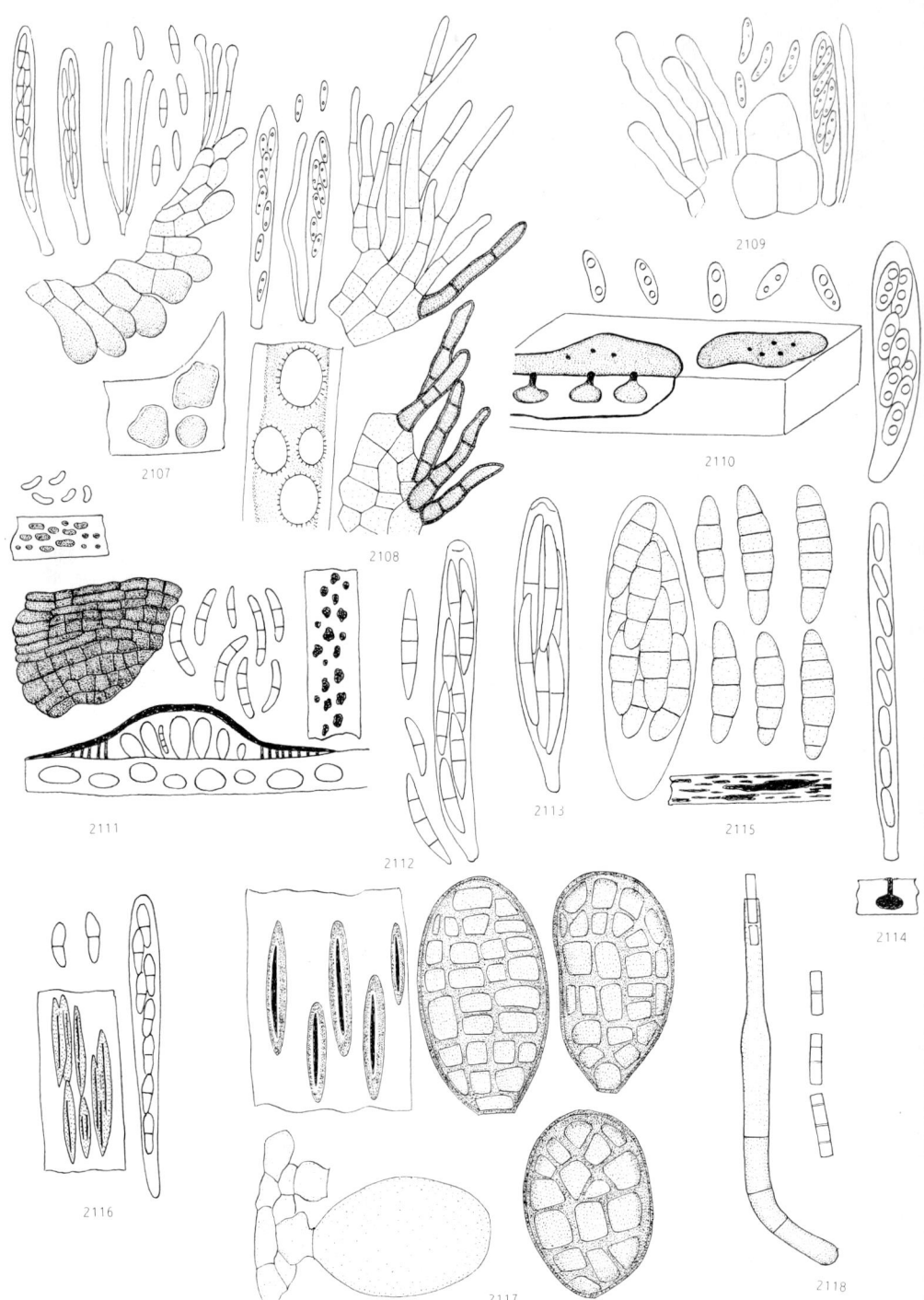

Plate 205. On *Pteridium*. 2107, *Mollisia pteridina*; 2108, *M. pteridis*; 2109, *Psilachnum pteridigenum*; 2110, *Diaporthopsis pantherina*; 2111, *Leptopeltis litigiosa*; 2112, *Monographos fuckelii*; 2113, *Mycosphaerella pteridis*; 2114, *Phomatospora endopteris*; 2115, *Rhopographus filicinus*; 2116, *Scirrhia aspidiorum*; 2117, *Camarographium stephensii*; 2118, *Chalara pteridina*.

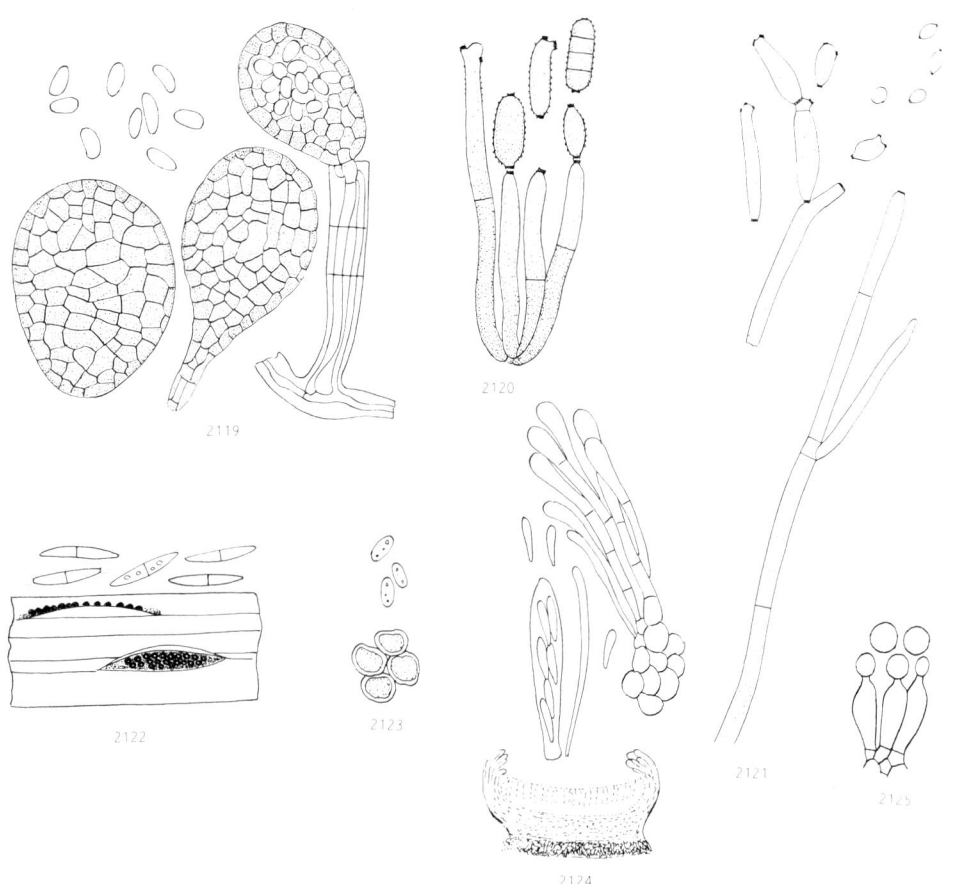

Plate 206. On rusts and powdery mildews. 2119, *Ampelomyces quisqualis*; 2120, *Cladosporium aecidiicola*; 2121, *C. uredinicola*; 2122, *Eudarluca caricis*; 2123, *Hainesia rubi*; 2124, *Micropodia oedema*; 2125, *Tuberculina persicina*.

FUNGUS INDEX

Numbers in italics refer to the illustrations on Plates 1–206

Acanthonitschkea tristis *314*, 21, 82
Acanthophiobolus helicosporus *1971*, 531
Acanthostigmella genuflexa *1900*, 507
Acarosporium 207
Acladium 47
Acremoniella atra *1268*, 288
Acremonium alternatum *1921*, 512
– apii *1358*, 311
Acroconidiella tropaeoli *1696*, 439
Acrogenospora 28
– sphaerocephala *168*, 47
Acrospeira mirabilis *443*, 106
Acrospermum compressum *1704*, 442
– graminum *1757*, 460
– pallidulum *1507*, 361
Actidium hysterioides *819*, 185
– nitidum *661*, 151
Actinocladium rhodosporum *169*, 48
Actinonema 229
Actinopeltis palustris *1901*, 507
Actinoscypha muelleri *1946*, 525
Actinothyrium graminis *1894*, 472, 505
Aegerita viridis 48, 91
Aglaospora profusa *1029*, 227
Agyrium rufum *735*, 168
Albugo candida *1409*, 326
– lepigoni 428
– tragopogonis 420, 435
Allophylaria basalifusca *885*, 203
– claviliformis *1482*, 355
– crystallifera *310*, 81
– macrospora *1451*, 347
– sublicoides *1473*, 351
Alternaria 466, 550
– alternata *1269*, 288, 467
– anagallidis *1345*, 306
– brassicae *1385*, 319
– brassicicola *1386*, 320
– cheiranthi *1424*, 330
– cinerariae *1659*, 421
– cucumerina *1434*, 339
– dauci *1438*, 342
– dennisii *1660*, 421
– dianthi *1442*, 343
– dianthicola *1443*, 343
– linicola *1556*, 381
– petroselini *1601*, 398
– porri *1338*, 304
– radicina *1439*, 342
– ramulosa *1668*, 424
– raphani *1643*, 412
– solani *1669*, 382, 425
– sonchi *1675*, 428
– tenuissima *1270*, 289, 467
– tomato *1561*, 382
– violae *1724*, 449

– zinniae *1730*, 450
Alysidium 48
– resinae *170*, 48
Amblyosporium botrytis *173*, 48
Amerenographium metableticum *1839*, 481
Amerosporium 229
Ampelomyces quisqualis *2119*, 571
Amphicytostroma 261
Amphisphaerella xylostei *702*, 160
Amphisphaeria bufonia *963*, 215
– melanommoides *1929*, 516
– millepunctata *70*, 21
– umbrina *1203*, 267
– vibratilis 196
Ampulliferina fagi *533*, 123
Anavirga laxa *444*, 71, 106
Anisogramma virgultorum *383*, 97
Antennatula 187
– pinophila *289*, 76
Anthostoma amoenum *561*, 129
– decipiens *423*, 103
– dryophilum *964*, 216
– melanotes *1204*, 139, 267
Anthostomella alchemillae 302
– appendiculosa *1065*, 236
– arenaria *1876*, 495
– caricis *1972*, 532
– chionostoma *1816*, 475
– clypeoides *1461*, 348
– conorum *787*, 179
– formosa *757*, 173
– formosa var. taxi *1161*, 258
– fuegiana *2006*, 540
– leptospora *2007*, 540
– limitata *1973*, 532
– lugubris *1829*, 479
– pedemontana 174
– phaeosticta *1830*, 479
– punctulata *1974*, 532
– rubicola *1066*, 236
– sabiniana 174
– scotina *2008*, 540
– tomicoides *1975*, 532
– tomicum *1870*, 494
– tumulosa *1976*, 532
Anthracoidea arenariae 523
– caricis *1942*, 523
– scirpi 558
– subinclusa *1943*, 523
Anungitea fragilis *276*, 71
Aphanomyces cochlioides 318
Apiognomonia errabunda *529*, 70, 122
Apioporthe vepris *1067*, 236
Apiorhynchostoma curreyi *71*, 21
Apiospora montagnei *1793*, 467
Apomelasmia 444
Aporhytisma urticae *1714*, 444
Aposphaeria 33, 67

787

788 Fungus Index

– ulmicola 269
Apostemidium fiscellum *1119*, 247
– fiscellum var. submersum 247
– guernisaci 247
– leptospora *860*, 194
Appendiculella calostroma *1068*, 236
Arachnocrea stipata *275*, 70
Arachnopeziza aurata *1*, 3
– aurelia *938*, 211
– candido-fulva 3
– eriobasis *886*, 203
– obtusipila *671*, 154
Arachnophora fagicola *541*, 125
Arachnoscypha aranea *436*, 105
Arnium apiculatum *1361*, 312
– olerum *1387*, 320
Arthrinium 467, 557
– caricicola *1984*, 534
– curvatum var. minus *2068*, 556
– cuspidatum *2048*, 551
– fuckelii *1985*, 534
– morthieri *1986*, 535
– phaeospermum *1792*, 467
– puccinioides *1987*, 535
– sporophleum *1988*, 535
– urticae *1709*, 443
Arthrobotrys conoides *1271*, 289
– oligospora *1272*, 289
Arthrobotryum stilboideum *980*, 48, 218
Arwidssonia empetri *512*, 119
Ascobolus foliicola *1227*, 276
– lignatilis 3
Ascochyta 332, 400, 502
– acori 300
– agrostis 477
– anisomera 389
– anthoxanthi 504
– armoraciae 313
– avenae 472, 486
– avenae var. confusa 486
– boltshauseri 398
– chenopodii 330
– cynosuricola 490
– dahliicola 341
– desmazieresii 504
– equiseti 564
– fabae 447
– festucae 497
– glaucii 365
– gracilispora 492
– hordei 502
– hordeicola 502
– humuli 370
– kabatiana 153
– lathyri 378
– leptospora 472
– leptospora var. acuta 504
– leptospora var. major 504
– mercurialis 387
– metulispora 137
– missouriensis 504
– molleriana 344
– orobi 392
– phaseolorum 399
– pisi 400
– plantaginis *1608*, 401
– primulae 408

– psammae 481
– pseudacori *1533*, 373
– pteridis 569
– rhodesii *1809*, 472
– senecionicola 421
– sodalis 537
– sonchi 428
– stellariae 430
– straminea 520
– subalpina 490
– trifolii 436
– typhoidearum *2080*, 557, 561
– viburni 273
– vincae 448
– violae 449
– vulgaris 160
Ascochytella grossulariae *1027*, 227
Ascochytula deformis 255
– obiones *1518*, 366
– symphoricarpi 257
– ulicis 266
Ascocorticium anomalum *800*, 182
Ascocoryne cylichnium *554*, 3, 127
– sarcoides *555*, 3, 127
– solitaria 3
Ascodichaena rugosa *556*, 127
Ascotremella faginea *2*, 3
Aspergillus niger *1339*, 304
Asterina veronicae *1718*, 446
Asteroma 87
– coryli *468*, 110
– impressum 440
– microspermum 95
Asteromassaria macrospora *562*, 129
Asteromella 321
Asterosporium asterospermum *591*, 135
Atopospora betulina *375*, 94
Aulographina eucalypti *515*, 119
Aulographum hederae *621*, 70, 143
Aureobasidium 122
– pullulans *306*, 80

Bactridium flavum *1137*, 134, 252
Bactrodesmiella masonii *542*, 125
Bactrodesmium abruptum *174*, 49
– atrum *175*, 49
– betulicola *399*, 99
– longisporum *362*, 91
– obovatum *176*, 49
– pallidum *177*, 49
– spilomeum *178*, 49
– submoniliforme *981*, 218
Balanium stygium *1144*, 254
Belemnospora epiphylla *277*, 71
– pinicola *767*, 175
Belonium hystrix *1892*, 505
– incurvatum 454
– nigromaculatum *1534*, 373
– psammicola *1825*, 478
Belonopsis filispora *1858*, 487
– graminea *1814*, 474
– iridis *1535*, 373
Beltrania querna *925*, 209
Beltraniella pirozynskii *686*, 157
Berlesiella nigerrima *72*, 21
Bertia moriformis *73*, 21
– moriformis var. multiseptata *74*, 21

Betulina fuscostipitata *371*, 68, 94
Bispora antennata *179*, 49
– betulina *180*, 49
Bisporella citrina *3*, 4
– fuscocincta *941*, 213
– pallescens 4
– subpallida *4*, 4
– sulfurina *5*, 4
Blennoria buxi *417*, 102
Bloxamia bohemica *768*, 175
– leucophthalma *1212*, 269
Blumeriella jaapii *861*, 194
Bombardia bombarda *75*, 21
Bostrichonema alpestre *1618*, 404
Botryobasidium aureum *171*, 48
– candicans *172*, 48
– conspersum *167*, 47
Botryosphaeria abietina 75
– dothidea *1037*, 229
– festucae *1758*, 460
– foliorum *1162*, 258
– hoffmannii *563*, 129
– hyperborea 119
– melanops *965*, 216
– obtusa *76*, 21
– quercuum *564*, 130
– rhodorae *997*, 221
– ribis *1021*, 226
– stevensii 139
Botryosporium longibrachiatum *1273*, 289
– pulchrum *1710*, 443
Botryotinia calthae *1403*, 324
– convoluta 373
– draytonii 364
– fuckeliana *271*, 68
– globosa *1340*, 304
– narcissicola 390
– polyblastis 390
– porri 305
– sphaerosperma 305
– squamosa *1341*, 305
Botryotrichum piluliferum *1274*, 289
Botrytis 71, 364, 373
– allii *1342*, 305
– anthophila *1687*, 437
– byssoidea 305
– cinerea *1275*, 289
– elliptica *1552*, 68, 379
– fabae *1720*, 447
– galanthina *1504*, 359
– paeoniae *1593*, 395
– tulipae *1697*, 439
Brachydesmiella biseptata *614*, 49, 134, 141
Brachysporiella laxa *534*, 123
Brachysporium bloxami *181*, 50
– britannicum *182*, 50
– dingleyae *1005*, 50, 222
– masonii *183*, 50
– nigrum *184*, 50
– obovatum *185*, 50
Bremia lactucae *1544*, 376
Broomella vitalbae *459*, 108
Brunchorstia 183
Buergenerula biseptata *1977*, 532
– typhae *1914*, 510
Bulgaria inquinans *942*, 4, 127, 213
Bulgariella pulla *380*, 95

Burenia inundata *1359*, 311
Byssolophis sphaerioides *77*, 22

Cacumisporium capitulatum *186*, 50
Cainiella johansonii *505*, 118
Calcarisporium arbuscula *187*, 51
Calloria neglecta *1699*, 441
Calonectria erubescens *639*, 146
– hederae *622*, 143
– pseudopeziza *78*, 22
Calosphaeria cyclospora *966*, 216
– dryina *967*, 216
– parasitica 130
– pulchella *869*, 196
– wahlenbergii *384*, 97
Calospora arausiaca *968*, 216
– platanoides *315*, 82
Calycella scolochloae *1737*, 454
Calycellina caricina 525
– chlorinella *1228*, 276, 433
– indumenticola *1105*, 244
– leucella *372*, 94
– ochracea *6*, 4, 127
– phalaridis *1899*, 507
– populina *841*, 68, 190
– punctiformis *887*, 203
– rivelinensis *888*, 203
– spiraeae *1483*, 355
Calyptella capula 425, 440
Camarographium stephensii *2117*, 569
Camarops lutea *408*, 22, 101
– microspora *353*, 90
– polysperma *354*, 90
– tubulina 76
Camarosporium 153, 227
– aequivocum *1369*, 314
– ambiens *330*, 85
– feurichii *1923*, 512
– lauri *698*, 158
– obiones *1519*, 366
– oreades *935*, 210
– orni *615*, 142
– pini *795*, 180
– propinquum *485*, 112
– rosae *1043*, 230
– rubicolum *1098*, 241
– salicinum *1141*, 253
– spartii 117
– tiliae *1180*, 262
Camposporium cambrense *543*, 51, 71, 125
– pellucidum *544*, 51, 71, 125
Candelabrum spinulosum *445*, 71, 106
Capnobotrys dingleyae *1164*, 259
Capnodium salicinum *1110*, 245
Capnophialophora 152
Caryospora callicarpa *969*, 216
Catenularia 24, 51, 229
– piceae *739*, 169
Catinella olivacea *7*, 4
Caudospora taleola *970*, 216
Cenangium acuum *742*, 170
– ferruginosum *801*, 182
– graddonii *346*, 88
Cephalotheca clarkii *1871*, 494
– sulfurea *79*, 22
Ceratocladium microspermum *582*, 134
Ceratocystiopsis falcata 22

790 Fungus Index

Ceratocystis coerulescens *820*, 185
– moniliformis *565*, 130
– piceae *821*, 185
– ulmi *1205*, 267
Ceratosphaeria lampadophora *80*, 22
Ceratosporella stipitata *424*, 103
Ceratosporium fuscescens *652*, 51, 149
Ceratostomella ampullasca *81*, 22
– cirrhosa *82*, 22
Cercophora caudata *83*, 23
Cercoseptoria handelii *1006*, 222
– oxyriae *1592*, 394
Cercospora apii *1360*, 311
– armoraciae *1367*, 313
– beticola *1380*, 318
– chenopodii *1425*, 330
– comari *1626*, 406
– depazeoides *1145*, 254
– ferruginea *1370*, 314
– juncicola *2049*, 551
– loti *1558*, 381
– mercurialis *1577*, 387
– microsora *1177*, 262
– murina *1725*, 449
– nigellae *1583*, 391
– physospermi *1604*, 399
– plantaginis *1609*, 401
– radiata *1356*, 310
– resedae *1645*, 413
– rhamni *991*, 220
– scandens *1683*, 432
– violae *1726*, 449
– zebrina *1572*, 385, 437
– zonata *1721*, 447
Cercosporella brassicae 321
– filicis-feminae *2082*, 562
– junci *2050*, 551
– primulae *1630*, 408
Cercosporidium depressum *1349*, 308
– graminis *1794*, 468
Ceriophora palustris *1978*, 532
Ceriospora polygonacearum *1619*, 404
Ceriporiopsis aneirina *253*, 64
Ceuthospora 146
– euonymi *521*, 120
– feurichii *1723*, 448
– hederae *624*, 143
– lauri *876*, 198
– mahoniae *712*, 162
– rhododendri 223
Chaenothecopsis caespitosa *1160*, 258
Chaetapiospora islandica *506*, 118
– rhododendri *998*, 221
Chaetendophragmia britannica *1848*, 484
Chaetochalara bulbosa *642*, 147
– cladii *2013*, 541
Chaetoconis 404
Chaetodiplodia caulina *1426*, 330
Chaetomella acutiseta 80
Chaetomium cochlioides 460
– elatum *1759*, 460
– globosum 460
Chaetopsis grisea *188*, 51
Chaetospermum chaetosporum *1319*, 298
Chaetosphaerella fusca *84*, 23
– fusispora *316*, 82
– phaeostroma *85*, 23

Chaetosphaeria bramleyi *1038*, 229
– callimorpha *1070*, 24, 237
– cupulifera *86*, 24
– inaequalis *87*, 24
– innumera *88*, 24
– myriocarpa *89*, 24
– preussii *90*, 24
– pulviscula *91*, 25
– vermicularioides *92*, 25
Chalara affinis *545*, 72, 125
– aurea *278*, 71
– cladii *2014*, 541
– cylindrica *731*, 72, 167
– cylindrosperma *546*, 72, 125
– fusidioides *769*, 175
– hughesii *926*, 209
– inflatipes *486*, 113
– kendrickii *927*, 209
– ovoidea *583*, 134
– pteridina *2118*, 569
– spiralis *547*, 125
– urceolata *1276*, 289
Chalaropsis thielavioides *1440*, 342
Cheiromycella microscopica *831*, 187
Cheirospora botryospora *634*, 109, 145
Chitonospora ammophilae *1831*, 479
Chlorencoelia versiformis *8*, 5
Chloridium 25, 51
Chlorociboria aeruginascens *943*, 213
– aeruginosa *944*, 213
Chloroscypha sabinae *655*, 151
– seaveri 115, 260
Chromelosporium carneum *279*, 72
– ochraceum *280*, 72, 289
Chromocrea aureoviridis *93*, 25, 65
Chrysomyxa abietis 166
– empetri *513*, 119
– pirolata 409
– rhododendri 166, 220
Ciboria acerina *722*, 165
– amentacea *342*, 87
– aschersoniana *1947*, 525
– batschiana *939*, 211
– betulae *379*, 95
– caucus 192
– juncorum 545
– viridifusca *343*, 87
Ciborinia candolleana *889*, 204
– hirtella *890*, 204
Ciboriopsis tenuistipes *1452*, 347
Cicinnobolus cesatii 571
Ciliolarina laricina *802*, 154, 182
Circinotrichum britannicum *687*, 157
Cirrenalia lignicola *189*, 51
Cistella dentata 5
– geelmuydenii 5
– perparvula 5
Cladosporium 198, 343, 374
– aecidiicola *2120*, 571
– allii-cepae *1343*, 305
– britannicum *190*, 51
– chlorocephalum *1594*, 395
– cladosporioides *1277*, 290, 468
– cucumerinum *1435*, 339
– herbarum *1278*, 290, 468
– macrocarpum *1279*, 290, 468
– magnusianum *1582*, 390

Fungus Index   791

– orchidis *1437*, 341
– phlei *1904*, 508
– sphaerospermum *1280*, 290
– staurophorum *770*, 175
– uredinicola *2121*, 571
– variabile *1676*, 429
Clasterosporium caricinum *1989*, 535
Clathrosphaerina 72, 106
Claussenomyces atrovirens *9*, 5
– olivaceus 168
– prasinulus *10*, 5
Claviceps microcephala 511
– nigricans 543
– purpurea *1760*, 460
Clavidisculum caricis *1948*, 525
– microsporum *381*, 95
Clonostachys compactiuscula *584*, 51, 134
Clypeosphaeria mamillana 25
– notarisii *1071*, 237
Coccomyces arctostaphyli 92
– coronatus *891*, 68, 204
– dentatus *892*, 121, 204
– juniperi *656*, 151
– leptideus *1219*, 270
Cochliobolus sativus *1939*, 519
Codinaea 51, 237
– britannica *281*, 72, 125, 357
– fertilis *548*, 72, 126
– hughesii *549*, 126
– setosa *643*, 147
– simplex *282*, 72
Coleophoma cylindrospora *645*, 73, 148
– empetri *1009*, 73, 223
Coleosperma lacustre *2062*, 555
Coleosporium tussilaginis 169, 420, 427, 440
Coleroa alchemillae *1331*, 302
– chaetomium *1072*, 237
– circinans *1512*, 363
– potentillae *1627*, 406
Colletotrichum 252, 282
– circinans 305
– coccodes *1562*, 383
– dematium *1320*, 298
– fuscum 344
– graminicola *1810*, 472
– helicis 144
– holci *1886*, 501
– liliacearum 346
– lindemuthianum *1602*, 399
– lini *1557*, 381
– malvarum 384
– orbiculare 340
– trichellum *625*, 144
– trifolii 385
– typhae 561
Colpoma crispum *736*, 168
– quercinum *945*, 213
Coniella castaneicola *433*, 73, 104
– fragariae *1500*, 358
Coniochaeta ligniaria *1187*, 25, 264
– malacotricha *822*, 186
– pulveracea *94*, 25
– velutina 25
Coniosporium 31, 51
Coniothyrium 115, 238
– hellebori 367
– ilicis 148

– obiones 366
– psammae *1840*, 481
– pteridis 569
– slaptoniense 149
– sphaerospermum 266
– tamaricis 258
– wernsdorffiae 230
Conoplea fusca *2051*, 551
Conostroma 213
Cordana crassa *363*, 91
– pauciseptata *191*, 51
Cordella clarkii *1990*, 535
Coremiella cubispora *1281*, 290
Coronophora angustata *95*, 26
– gregaria *1153*, 26, 256
Coryne 3, 51, 127
Coryneopsis tamaricis 258
Corynespora biseptata *585*, 51, 134
– cambrensis *1156*, 257
– cassiicola *1436*, 340
– cespitosa *400*, 99
– foveolata *1849*, 484
– olivacea *1178*, 262
– proliferata *1213*, 269
– pruni *1165*, 259
– smithii *653*, 52, 149
Corynesporopsis quercicola *982*, 218
– uniseptata *688*, 157
Coryneum 98, 100, 217
– compactum *1215*, 269
– elevatum *989*, 219
– neesii *990*, 219
Costantinella micheneri *192*, 52
– terrestris *193*, 52
Creopus gelatinosus *96*, 26
– spinulosus *1761*, 461
Cristulariella depraedens *307*, 80
Crocicreas amenti *1118*, 247
– complicatum *1182*, 116, 263
– coronatum *1229*, 137, 276
– culmicolum *1738*, 454
– cyathoideum *1230*, 276
– cyathoideum var. cacaliae *1588*, 393
– cyathoideum var. pteridicola *2099*, 566
– dolosellum *599*, 137
– furvum *1815*, 474
– gramineum var. incertellum 503
– megalosporum *1949*, 525
– megalosporum var. gramineum *1866*, 493
– starbaeckii *1638*, 411
– stramineum 454
– subhyalinum *293*, 68, 78
– tomentosum *1739*, 454
Cronartium flaccidum 180, 395, 439
– ribicola *1016*, 180, 225
Crumenulopsis pinicola *803*, 182
– sororia *804*, 182
Cryptocline cinerescens *936*, 210
– paradoxa *626*, 143
– phacidiella 198
Cryptocoryneum condensatum *194*, 52
– rilstonii *195*, 52
Cryptodiaporthe aesculi *333*, 85
– aubertii *723*, 165
– castaneae *453*, 107
– galericulata *566*, 130
– hranicensis *1172*, 261

- hystrix *317*, 82
- lebiseyi *318*, 83
- populea *852*, 192
- pyrrhocystis *478*, 111
- salicella *1127*, 250
- salicina *1128*, 250
Cryptodiscus pallidus *11*, 5
- rhopaloides *12*, 5, 336
Cryptomycella 566
Cryptomyces maximus *1120*, 248
Cryptomycina pteridis *2100*, 566
Cryptosphaerella annexa 139, 250
Cryptosphaeria eunomia *605*, 139
- populina *853*, 192
Cryptosphaerina fraxini 140
Cryptosporella compta 130
- hypodermia *1206*, 268
- populina 193
Cryptosporiopsis 104, 113, 163, 215
Cryptosporium tami *1684*, 432
Cryptostroma corticale *329*, 84
Cryptovalsa protracta *606*, 140
- suaedicola 431
Cucurbidothis pithyophila *290*, 77
Cucurbitaria berberidis *368*, 93
- dulcamarae 425
- elongata *1030*, 227
- laburni 153
- rhamni *992*, 220
- rubefaciens *1129*, 250
- spartii *501*, 117
Cudoniella acicularis *946*, 6, 213
- clavus *347*, 6, 89
- clavus var. grandis *347A*, 6, 89
- junciseda 545
- rubicunda 178
- tenuispora 6
Cumminsiella mirabilissima *711*, 162
Curvularia crepinii *2095*, 565
- inaequalis *1938*, 519
- protuberata *1874*, 494
- trifolii f. sp. gladioli *1516*, 364
Cyathicula littoralis *1381*, 318
Cyclothyrium 84
Cylindrium elongatum *928*, 72, 209
Cylindrocarpon 36, 52, 132
Cylindrocladium 143
Cylindrocolla 441
Cylindrosporium 321
Cylindrotrichum clavatum *689*, 157
- ellisii *1885*, 499
- oligospermum *1282*, 290
Cymadothea trifolii *1688*, 437
Cyphellopsis anomala *270*, 67
Cytoplacosphaeria rimosa *1924*, 513
Cytospora 41, 67
- abietis 156
- aquifolii 149
- carphosperma 262
- clypeata 241
- euonymi 120
- intermedia 219
- lantanae 273
- lauri 158
- laurocerasi 199
- occulta 92
- oxyacanthae 114
- pinastri 177
- prunorum 198
- rosarum 230
- salicis 253
- subclypeata 223
- tamaricis 258
- taxi 260

Dactylaria chrysosperma *983*, 218
- junci *2052*, 551
- lepida *732*, 167
- obtriangularia *690*, 157
- purpurella *984*, 218
Dactylella arnaudii *1521*, 368
Dactylospora stygia *602*, 138
Dactylosporium macropus *196*, 52
Daldinia concentrica *607*, 26, 140
- vernicosa 264
Darluca filum 571
Dasyscyphus acerinus *294*, 78
- acutipilus *1740*, 455
- acutus *1741*, 455
- acuum *743*, 166, 170
- aeruginellus *1484*, 355
- albotestaceus *1742*, 455
- apalus *2025*, 546
- barbatus 159
- bicolor *947*, 213
- bicolor var. rubi 232
- brevipilus *557*, 6, 95, 127
- callimorphus 525
- calyculiformis *474*, 111
- capitatus *893*, 204
- caricis *1950*, 525
- carneolus *1743*, 455
- carneolus var. longisporus *1744*, 456
- castaneicola *437*, 105
- castaneus *1453*, 347
- cerinus *13*, 6
- ciliaris *894*, 68, 205
- clandestinus *1049*, 232, 277
- clavigerus 347
- clavisporus 546
- controversus *1745*, 456
- corticalis *14*, 6
- coruscatus *895*, 68, 205
- crystallinus *948*, 213
- deflexus *1198*, 68, 266
- diminutus *2026*, 546
- dumorum *1050*, 232
- fascicularis *15*, 7
- fugiens *2027*, 546
- fuscescens *896*, 205
- fuscescens var. fagicola *536*, 123
- grevillei *1231*, 277
- humuli 370
- imbecillis *2022*, 543
- imbecillis var. cladii *2005*, 539
- luzulinus 553
- misellus *1051*, 233
- mollissimus *1232*, 277
- mughonicola 171
- nidulus *1233*, 277
- niveus *949*, 213
- nudipes *1485*, 355
- nudipes var. minor *1454*, 347
- palearum *1746*, 456

– palearum var. niger 496
– papyraceus *16*, 7, 183
– patulus 205
– pteridialis *2101*, 566
– pteridis *2102*, 567
– pudibundus *1121*, 248
– pulveraceus *17*, 7
– pulverulentus *744*, 171
– pygmaeus *1183*, 7, 263
– radotinensis *295*, 78
– rehmii *2028*, 546
– rhodoleucus 456
– rhytismatis *296*, 78
– salicariae *1568*, 384
– soppittii *897*, 205
– sulphurellus *724*, 165
– sulphureus *1234*, 277
– sydowii 544
– tenuissimus *1747*, 456
– virgineus *18*, 7, 68, 123, 277
Deightoniella arundinacea *1922*, 512
Delphinella abietis *287*, 76
Dematioscypha dematiicola *950*, 7, 214
Dematophora 286
Dencoeliopsis johnstonii 96
Dendrodochium 35, 52
– citrinum *832*, 187
– pinastri 188
Dendrostilbella 5, 52
Dendryphiella infuscans *1283*, 291
– vinosa *1284*, 291
Dendryphion 395
– comosum *1285*, 291
– nanum *1286*, 291, 320
Dendryphiopsis 33, 52
Dennisiella babingtonii *999*, 221
Dennisiodiscus prasinus *1881*, 498
Dermea ariae *1151*, 256
– balsamea *878*, 199
– cerasi *862*, 195
– padi *863*, 195
– prunastri 195
– tulasnei 138
Desmazierella acicola *745*, 171
Diapleella clivensis *1249*, 281
Diaporthe arctii *1362*, 312
– arctii var. artemisiae *1371*, 314
– beckhausii 273
– carpini *425*, 103
– chailletii 317
– circumscripta 253
– crataegi *493*, 114
– decedens *479*, 112
– decorticans 197
– detrusa *369*, 93
– eres *97*, 26, 268
– eumorpha 448
– fibrosa *993*, 220
– hederae *627*, 144
– impulsa *1154*, 256
– inaequalis *1188*, 117, 264
– incarcerata *1039*, 229
– leiphemia *971*, 216
– lirella *1495*, 356
– nobilis 156
– nucleata 264
– oncostoma *1031*, 227

– pardalota *703*, 26, 109, 160
– perniciosa *870*, 163, 197
– phaseolorum 399
– phaseolorum var. sojae 447
– pulla *628*, 144
– pustulata 83
– rudis *670*, 26, 153, 161
– sarmenticia 370
– sarothamni 117
– sarothamni var. dulcamarae *1670*, 425
– strumella *1022*, 226
– syngenesia *596*, 136
– tessella 250
– varians *319*, 83
– velata *1173*, 261
Diaporthopsis angelicae 308
– pantherina *2110*, 568
Diatrype bullata *1130*, 250
– disciformis *567*, 26, 130
– stigma *98*, 27, 130
Diatrypella favacea *385*, 27, 90, 97, 112, 130
– quercina *972*, 217
Dibeloniella trichophoricola *2072*, 558
Dichomera mutabilis 189
– saubinetii *331*, 67, 85
Dicytochaeta querna *929*, 209
Dictyopolyschema pirozynskii *740*, 169
Dictyosporium elegans *1795*, 468
– toruloides *197*, 53
Dictyotrichiella pulcherrima *99*, 27
Didymaria kriegeriana *1664*, 423
– linariae *1554*, 380
Didymascella tetraspora *657*, 151
– thujina *1171*, 260
Didymella applanata *1073*, 237
– bryoniae *1394*, 322
– cerastii *1420*, 329
– commanipula *1657*, 419
– corni *461*, 109
– exigua *1407*, 325
– exitialis *1887*, 502
– hebridensis 476
– holosteae *1078*, 430
– lingulicola 332
– lophospora 568
– lycopersici *1563*, 383
– pinodes 400
– prominula 568
– proximella *1979*, 532
– refracta 558
– scotica 479
Didymopleela cladii *2009*, 540
Didymosphaeria conoidea *1253*, 282
– oblitescens *1074*, 237
– rubicola *1075*, 237
– superapplanata *454*, 107
Digitodesmium elegans *198*, 53
Digitosporium 182
Dilophospora 472
Dimerium meliolicola *1069*, 236
Dinemasporium 465
Diplocarpa bloxamii *19*, 7
Diplocarpon earlianum *1501*, 358, 406
– mespili *488*, 113, 116
– rosae *1034*, 229
Diplocladiella scalaroides *1194*, 265
Diplococcium spicatum *199*, 53

Diplodia 38, 67
– ilicicola *646*, 148
– lantanae 273
– laurina 158
– ligustri 158
– lonicerae 161
– mamillana *462*, 109
– melaena *1216*, 270
– rubi *1099*, 241
– rudis 153
– taxi 260
– tiliae 262
– ulicis 266
Diplodina 83, 85, 130, 250
– galii *1508*, 361
Diplorhinotrichum candidulum *200*, 53
– juncicola *2053*, 551
Diplosporonema delastrei *1665*, 423
Discogloeum veronicae *1719*, 446
Discohainesia oenotherae *434*, 104
Discosia artocreas *378*, 73, 95
Discosphaerina fagi *530*, 70, 122
Discosporina deplanata *426*, 103
Discosporium 192
Discula 92, 122
Disculina 91
Ditopella ditopa *355*, 90
Doassansia alismatis *1333*, 303
– limosellae 380
– sagittariae *1653*, 416
Doratomyces microsporus *1287*, 292, 484
– nanus *1288*, 292
– purpureofuscus *1388*, 330
– stemonitis *1289*, 292
Dothichiza 256
Dothidea puccinioides *1189*, 118, 264
– sambuci *1143*, 254
Dothiora pyrenophora *1155*, 256
– ribesia *1023*, 226
– sphaerioides *854*, 193
– taxicola *1162*, 259
– thujae 260
Dothiorella 147, 243
– candollei *418*, 102
– fraxinea 142
Dothistroma 174
Drechslera 475, 486, 497
– avenacea *1856*, 486
– biseptata *1796*, 468
– cactivora *1395*, 322
– dematioidea *1797*, 468
– fugax *1819*, 477
– iridis *1536*, 373
– phlei *1905*, 509
– poae *1927*, 514
– siccans *1891*, 504
Drepanopeziza populi-alba *842*, 190
– populorum *843*, 190
– punctiformis *844*, 190
– ribis *1018*, 225
– salicis *1106*, 244
– schoenicola *2070*, 557
– sphaerioides *1107*, 245
– triandrae *1116*, 246
Duebenia compta 277
Dumontinia tuberosa *1348*, 307
Durella atrocyanea *20*, 7, 116

– commutata *951*, 7, 214
– connivens *558*, 7, 128
– macrospora *952*, 7, 214
– suecica 178

Echinobotryum 292
Echinula asteriadiformis *1052*, 233
Ectostroma iridis *1537*, 373
Elsinoe veneta 237
Embellisia allii *1344*, 305
– phragmospora *1934*, 517
Enchnoa infernalis *973*, 217
– lanata *386*, 97
Encoelia fascicularis *851*, 138, 192
– fuckelii *864*, 195
– furfuracea *475*, 89, 111
– glaberrima *427*, 103
– glauca *476*, 111
– siparia *1200*, 267
– tiliacea 261
Endoconospora cerastii *1421*, 329
Endophragmia alternata *283*, 54, 73
– atra *1711*, 443
– biseptata *201*, 54
– boewei *790*, 179
– boothii *1090*, 54, 240
– catenulata *550*, 126
– cesatii *1620*, 404
– coronata *1166*, 259
– elliptica *1290*, 54, 292
– glanduliformis *401*, 54, 99, 134
– hyalosperma *1291*, 54, 292
– nannfeldtii *202*, 54
– parva *1091*, 240
– pinicola *771*, 175
– prolifera *1497*, 54, 357
– stemphylioides *586*, 134
– taxi *1167*, 259
– uniseptata *203*, 55
– verruculosa *587*, 55, 135
Endophragmiella cambrensis *204*, 55
– corticola *205*, 55
– fagicola *551*, 126
– fallacia *206*, 55
– lauri *691*, 157
– ovoidea *207*, 55
– pallescens *208*, 55
– resinae *741*, 169
– subolivacea *209*, 55
– suttonii *402*, 99
Endophyllum euphorbiae-sylvaticae 353
– sempervivi 420
Endoxyla operculata *823*, 186
Endoxylina pini *824*, 186
Entomosporium 113
Entorrhiza aschersoniana 545
– casparyana 545
– scirpicola 542
Entyloma achilleae 300
– calendulae 323
– calendulae f. bellidis 318
– calendulae f. dahliae 341
– calendulae f. hieracii 370
– chrysosplenii 333
– crastophilum 500
– eryngii 351
– fergussoni 389

– ficariae 411
– fuscum 395
– helosciadii 311
– henningsianum 417
– linariae 380
– matricariae 439
– microsporum *1636*, 411
– ossifragi 391
– serotinum 431
Ephelina lugubris 413
Epibelonium gaeumannii *916*, 208
Epichloe typhina *1762*, 461
Epicoccum purpurascens *1292*, 292, 469
Epithyrium 184
Erinella discolor *1700*, 441
Eriopeziza caesia *953*, 214
Eriospora leucostoma *2000*, 537
Erysiphe aquilegiae 311
– artemisiae 314
– asperifoliorum 341
– betae 318
– biocellata 365, 387
– cichoracearum *1374*, 315
– circaeae 334
– cruchetiana 393
– cruciferarum 320
– depressa 312
– fischeri 421
– galeopsidis 359, 365
– galii 361
– graminis *1763*, 461
– heraclei *1522*, 368
– hyperici 371
– knautiae 375
– limonii *1553*, 380
– pisi 400
– polygoni 405
– ranunculi *1639*, 411
– sordida *1610*, 401
– tortilis *463*, 109
– trifolii 161, 437
– ulmariae 356
– urticae 442
– verbasci 445
Eudarluca caricis *2122*, 571
Eupropolella arundinariae *1845*, 483
– britannica *865*, 195
– celata *1951*, 525
Eutypa acharii *320*, 27, 83
– flavovirens *100*, 27, 130
– lata *101*, 27
– leioplaca *568*, 131
– spinosa *569*, 131
Eutypella acericola *321*, 27, 83
– prunastri *871*, 197
– sorbi 256
– stellulata *1207*, 268
Excipularia fusispora *460*, 108
Exobasidium japonicum 220
– vaccinii 220, 270
Exosporium tiliae *1179*, 262

Farlowiella carmichaeliana *102*, 28
Farysia thuemenii *1944*, 523
Fenestella fenestrata *103*, 28
– salicis *1131*, 250
– vestita *322*, 28, 83

Foveostroma 195, 256
Fragosphaeria purpurea *570*, 131
– reniformis *104*, 28
Frommea obtusa *1624*, 406
Fuckelia 225
Fujimyces oödes *777*, 177
Fulvia fulva *1564*, 383
Fumago vagans 80
Fumagospora 245
Fusariella hughesii *1293*, 293
– sarniensis *1474*, 352
Fusarium *1294*, 55, 293, 426, 469
– oxysporum f. tulipae 440
Fusicladium 95, 201
Fusicoccum 226
– noxium 219
– ulmi 270
Fusidium aeruginosum *930*, 73, 210
– griseum *284*, 73

Gabarnaudia betae *1382*, 318
Gaeumannomyces graminis 461, 532
– graminis var. avenae *1855*, 461
– graminis var. graminis *1764*, 462, 485
– graminis var. tritici *1765*, 462
Gelasinospora cerealis *1936*, 518
Gemmamyces piceae *730*, 167
Geniculosporium 30, 39, 55, 102, 113
Gibbera elegantula 272
– myrtilli *1222*, 272
– salisburgensis *1470*, 350
– vaccinni 272
Gibberella buxi *409*, 101
– cyanogena *1389*, 320
– pulicaris *708*, 28, 117, 161
– zeae *1766*, 462, 469
Gibellina cerealis *1937*, 518
Gliocladium 426
– luteolum *210*, 55
– roseum *211*, 55
Gliomastix 484
– luzulae *1295*, 293
Gloeosporidiella 225
– nobilis 158
– turgida *616*, 142
Gloeosporium rhododendri *1010*, 223
– tiliae 262
Gloeotinia granigena *1890*, 503
Glomerella cingulata *1250*, 282
– montana *1928*, 515
Gloniella moliniae *1893*, 505
Gloniopsis praelonga *1076*, 28, 238
Glonium lineare *105*, 28
Glyphium elatum *704*, 160
Gnomonia cerastris *303*, 79
– comari 406
– fragariae *1502*, 358
– fructicola *1503*, 358
– gnomon *469*, 110
– leptostyla *654*, 149
– rubi *1077*, 238
– salina 516
– setacea *376*, 70, 79, 85, 87, 94
Gnomoniella rubicola *1078*, 238
– tubiformis *339*, 87
Godronia callunigera 323
– cassandrae f. callunae *1397*, 323

Fungus Index

- fuliginosa *1122*, 248
- ribis *1019*, 225
- uberiformis *1020*, 225
- urceolus 96
- Gonatobotrys simplex *1296*, 293
- Gonatobotryum fuscum *985*, 135, 218
- Gonytrichum 24, 56
  - macrocladum *1851*, 484
- Gorgoniceps aridula *805*, 183
  - boltonii 563
  - charnwoodensis *438*, 105
  - micrometra 546
- Graddonia coracina *21*, 8
- Grahamiella dryadis *507*, 118
- Graphium calicioides *212*, 56
  - putredinis *1390*, 320
  - smaragdinum *1007*, 222
- Gremmeniella abietina *806*, 183
- Griphosphaeria corticola *1079*, 109, 229, 238
- Guignardia acerifera 79
  - aesculi 85
  - cytisi *504*, 117
  - fulvida 381
  - istriaca 243
  - miribelii *420*, 102
  - philoprina *640*, 143, 147
  - punctoidea *917*, 208
- Gymnosporangium clavariiforme *487*, 113, 150
  - confusum 113, 116, 150, 165
  - cornutum 150, 255
  - sabinae 150, 200
- Gyrothrix podosperma *1991*, 535
  - verticillata *1712*, 443

- Habrostictis rubra *1201*, 263, 267
- Hadrotrichum virescens *1821*, 477
- Haematomma elatina *807*, 183
- Haglundia elegantior *954*, 214
  - perelegans *955*, 214
- Hainesia 104
  - rubi *2123*, 571
- Haligena spartinae *1930*, 516
- Hansfordia pulvinata *1195*, 265
- Hapalocystis berkeleyi *840*, 189
  - bicaudata *1208*, 268
- Hapalosphaeria deformans *1100*, 241
- Haplariopsis fagicola *552*, 73, 126
- Haplobasidion thalictri *1684A*, 434
- Haplographium 7, 56
- Harpographium 38, 56, 214
- Helicoma 56
- Helicomyces roseus *214*, 56
- Helicoon ellipticum *215*, 56
  - fuscosporum *1092*, 240
- Helicosporium 40, 56, 132, 467
- Helminthosporium microsorum *986*, 218
  - solani *1671*, 425
  - velutinum *216*, 56
- Helotium consobrinum *1647*, 415
  - humile *439*, 105, 123
- Hendersonia 466
  - acicola 177
  - culmicola 498
  - culmiseda *1925*, 513
  - fiedleri var. symphoricarpi *1159*, 258
- Hendersoniopsis 91, 92
- Henicospora minor *1093*, 73, 240

- Hercospora tiliae *1174*, 261
- Herpotrichia herpotrichoides *1080*, 28, 238
  - juniperi *662*, 151
  - macrotricha *1251*, 28, 282
  - parasitica *291*, 77
- Herpotrichiella moravica *106*, 28
  - pilosella *107*, 29
- Heteroconium chaetospira *218*, 57
  - tetracoilum *219*, 57
- Heteropatella 308
  - antirrhini *1357*, 310
  - valtellinensis *1444*, 343
- Heterosphaeria patella *1350*, 308
- Heyderia abietis *727*, 166
  - pusilla *746*, 171
- Hobsonia mirabilis *220*, 57
- Hormiactella asetosa *833*, 188
- Hormotheca robertiani *1513*, 363
- Hyalinia rubella *1123*, 248
- Hyalopeziza ciliata *297*, 68, 78
  - spinicola *440*, 105
- Hyalopsora aspidiotis 569
  - polypodii *2085*, 562
- Hyaloscypha cladii 540
  - flaveola *2103*, 567
  - herbarum 355
  - hyalina *956*, 8, 214
  - hyalina f. 1 *1053*, 233
  - lachnobrachya *298*, 69, 78
  - lachnobrachya var. araneocincta *373*, 69, 94
  - laricionis *747*, 171
  - lectissima *1054*, 233
  - leuconica *672*, 8, 154
  - paludosa *2023*, 544
  - pseudopuberula *898*, 205
  - pygmaea *899*, 205
  - quercina 214
  - quercina var. resinacea 256
  - richonis 233
  - salicina 233
  - scintillans *1952*, 526
  - stevensonii *808*, 183
  - velenovskii *673*, 154
  - zalewskii *446*, 72, 106
- Hyalotricha corticola *725*, 165
  - niveocincta *1478*, 353
  - salicicola *1108*, 245
- Hymenopsis typhae *2081*, 561
- Hymenoscyphus albidus *600*, 137
  - albopunctus *334*, 69, 86, 121
  - calyculus *22*, 8
  - caudatus *272*, 69, 78, 86
  - citrinulus 526
  - conscriptus 248
  - cortisedus 96
  - epiphyllus *900*, 69, 205
  - equisetinus 564
  - fagineus *537*, 124
  - fructigenus *473*, 69, 110
  - herbarum *1235*, 278
  - imberbis *23*, 8
  - immutabilis *845*, 69, 191
  - infarciens 153
  - laetus *24*, 8
  - limonium 278
  - lutescens *781*, 178
  - magnificus 526

Fungus Index

- nitidulus 493
- phyllogenus *846*, 69, 191
- phyllophilus *525*, 69, 121
- pileatus *1236*, 278, 456
- repandus *1455*, 347
- rhodoleucus *2089*, 564
- robustior *1748*, 457
- rokebyensis *538*, 124
- rumicis *1648*, 415
- salicellus *1124*, 248
- scutula *1237*, 278
- scutula f. alba 415
- scutula var. fucatus 405
- scutula var. solani *1672*, 425
- scutula var. suspecta *1749*, 457
- separabilis 233
- serotinus 128
- subferrugineus 8, 248
- sublateritius 278
- subpallescens *311*, 81
- vernus *25*, 8
- vitellinus *1486*, 355
- vitigenus 9
Hypocrea argillacea *108*, 29
- citrina *109*, 29
- pilulifera 549
- rufa *110*, 29, 65
- schweinitzii 29
- splendens *111*, 30
Hypocreopsis lichenoides *1132*, 250
Hypoderma alpinum *1953*, 526
- commune *1238*, 278
- hederae *618*, 142
- ilicinum *901*, 206
- rubi *1055*, 234
- scirpinum 555
Hyponectria buxi *410*, 101
- cookeana *918*, 208
Hypospilina bifrons 208
Hypoxylon chestersii 140
- cohaerens *571*, 131
- confluens *974*, 30, 217
- fragiforme *572*, 131
- fraxinophilum *608*, 140
- fuscum *480*, 30, 97, 112
- howeianum *481*, 30, 83, 97, 112
- mammatum *1133*, 250
- multiforme *387*, 30, 97
- nummularium *573*, 131
- rubiginosum *609*, 30, 140
- rutilum 30, 131
- serpens *112*, 30
- serpens var. effusum 251
- udum *975*, 217
Hysterium acuminatum 31
- angustatum *113*, 31
- insidens *114*, 31
- pulicare *115*, 31
Hysterographium elongatum 251
- fraxini *610*, 141
- mori *116*, 31
Hysteropezizella diminuens 526
- dowardensis *1954*, 526
- exigua *2029*, 547
- foecunda 558
- hebridensis 558
- melatephroides 505

- olivacea 527
- pusilla *2030*, 547
- rehmii *2031*, 547
- valvata *1826*, 479
Hysterostegiella fenestrata 555

Incrupila melatheja 234
- viridipilosa *312*, 81
Isariopsis carnea *1550*, 378
Isothea rhytismoides *508*, 118
- saligna *1111*, 245
Isthmolongispora minima *692*, 157
Itersonilia pastinacae *1597*, 396

Junctospora pulchra *772*, 176

Kabatia periclymeni *707*, 161
Kabatiella caulivora *1689*, 437
- microsticta *1617*, 402
Kabatina thujae 260
Kalmusia sarothamni *502*, 117
Karstenula rhodostoma *597*, 137
Keissleriella culmifida *1767*, 462
- gallica *1649*, 415
- linearis 511
- ocellata 372
- pinicola *825*, 186
- subalpina *1398*, 323
Khuskia oryzae 484
Klasterskya acuum *758*, 174
Kriegeriella minuta *663*, 152
- mirabilis *759*, 174
Kuehneola uredinis *1045*, 231

Lachnellula calyciformis *879*, 199
- occidentalis *674*, 155
- pseudofarinacea *809*, 183
- resinaria 168
- subtilissima *810*, 183
- willkommii *675*, 155
Laetinaevia carneoflavida *1701*, 441
- luzulae *2059*, 553
- pustulata *902*, 206
Lamproconium desmazieresii *1181*, 262
Lanzia coracina *903*, 206
- stellariae *1679*, 430
Laquearia sphaeralis *603*, 138
Lasiobotrys lonicerae *705*, 160
Lasiosphaeria canescens *117*, 32
- caudata *118*, 32
- hirsuta *119*, 32
- ovina *120*, 32
- phyllophila *121*, 32
- spermoides *122*, 32
Lasiostictis fimbriata *782*, 178
Leiosphaerella vexata *464*, 109
Lembosia luzulae *2060*, 554
Lembosina aulographoides *1000*, 221
Lepteutypa hippophaes *635*, 146
Leptographium lundbergii *834*, 188
Leptopeltis filicina *2087*, 563
- litigiosa *2111*, 568
- nebulosa *2096*, 565
- pteridis 568
Leptosphaeria acuta *1705*, 442
- agnita *1475*, 352
- albopunctata 511, 516

– ammophilae *1832*, 480
– arundinacea *1915*, 511
– avenaria 485
– baldingerae 511
– caricicola 533
– caricis *1980*, 533
– centaureae 328
– cesatiana 345
– cladii *2010*, 540
– coniothyrium *1081*, 238
– culmifraga *1768*, 462
– culmifraga var. propinqua *1884*, 499
– derasa 421
– dolioloides *1428*, 332
– doliolum *1252*, 282
– epicalamia *2061*, 554
– epicarecta 533
– eustoma *1769*, 462
– fuckelii *1770*, 463
– galiorum 361
– gloeospora 282
– graminis *1771*, 463, 511
– grandispora 511
– haematites 108
– herpotrichoides *1772*, 463
– jaceae *1414*, 328
– juncicola 549
– juncina *2038*, 549
– libanotis *1351*, 309
– linearis *1903*, 508
– luctuosa *1773*, 463
– lycopodina 565
– macrospora *1254*, 283
– maculans *1335*, 303, 320
– marina *1931*, 516
– maritima 517
– marram *1833*, 480
– modesta *1656*, 419
– mosana *1902*, 507
– nardi *1895*, 506
– niesliana *1551*, 378
– nigrans *1774*, 463
– nodorum *1775*, 463
– ogilviensis 283
– oraemaris 517
– pelagica 517
– planiuscula 427
– pontiformis *1817*, 475
– praetermissa 238
– purpurea *1255*, 283
– raphani *1644*, 413
– scirpina *2064*, 556
– scitula *1509*, 361
– silenes-acaulis 423
– sowerbyi *2065*, 556
– tritici 463
– typhae *2076*, 560
– typharum *2077*, 560
– vectis *1538*, 373
Leptosphaerulina myrtillina 272
– trifolii *1690*, 437
Leptospora rubella *1256*, 283
Leptostroma 177, 234
– eupatorii 352
– juncacearum *2055*, 552
– spiraeinum *1498*, 357
Leptothyrina 143

– rubi *1101*, 242
Leptothyrium 204
– ilicinum *937*, 210
Leptotrochila brunellae *1632*, 409
– cerastiorum *1422*, 329
– medicaginis 386
– radians 325
– ranunculi *1640*, 411
– repanda 406
– verrucosa *1510*, 361
Letendraea helminthicola *217*, 56
Leucoscypha erminea *374*, 94
Leucostoma curreyi *677*, 155
– niveum *855*, 193
– persoonii 799, 256
Leveillula taurica *1520*, 341, 367
Libertella betulina 100
– faginea *592*, 136
Lidophia graminis *1811*, 472
Linodochium 173
Linospora ceuthocarpa 191
Linostomella sphaerosperma *123*, 32
Lirula macrospora 167
Lobatopedis foliicola *931*, 123, 210
Lopadostoma gastrinum *1209*, 268
– turgidum *574*, 131
Lophiostoma alpigenum var. juncinum *2039*, 549
– angustilabrum *1190*, 264, 284
– arundinis *1916*, 511
– caudatum *1917*, 511
– caulium *1257*, 284
– fuckelii *1082*, 239
– fuckelii var. pulveraceum *1083*, 239
– hysterioides *1084*, 239
– macrostomoides *1134*, 251
– nucula *124*, 33
– origani var. rubidum *1258*, 284
– semiliberum *1776*, 463
– vagabundum *1259*, 284
– viridarium *1085*, 239
Lophium mytilinum *826*, 186
Lophodermella conjuncta *748*, 171
– sulcigena 171
Lophodermium arundinaceum *1750*, 457
– caricinum *1955*, 527
– conigenum *783*, 178
– culmigenum *1751*, 457
– gramineum *1752*, 457
– hedericola *619*, 143
– hysterioides *489*, 113
– juniperinum *658*, 151
– maculare 271
– melaleucum *1220*, 271
– neesii 146
– petiolicolum *904*, 206
– piceae *728*, 167
– pinastri *749*, 172
– pini-excelsae 172
– seditiosum *750*, 172
– typhinum *2074*, 559
– vagulum *994*, 221
Lophomerum ponticum *995*, 221
Loramyces juncicola *2019*, 543

Macrophoma falconeri 223
Macrophomina phaseolina *1603*, 399
Mamiania coryli *470*, 110

Fungus Index 799

– fimbriata *421*, 102
Mariannaea elegans *221*, 57
Marssonina 149, 229, 245, 246, 358
– daphnes 118
– panattoniana 376
Massaria inquinans *323*, 83
Massarina eburnea *576*, 132
– emergens *856*, 193
Mastigosporium album *1823*, 478
– deschampsiae *1875*, 494
– muticum *1864*, 491
– rubricosum *1822*, 477
Megaloseptoria mirabilis *733*, 167
Melampsora allii-fragilis 244
– allii-populina 189
– amygdalinae 244
– capraearum 154, 244
– epitea var. epitea 120, 154, 225, 244, 341, 418
– epitea var. reticulatae 244, 418
– euphorbiae 353
– hypericorum 371
– larici-pentandrae 244
– larici-populina 154, 190
– lini 380
– lini var. liniperda 381
– populnea 154, 181, 190, 387
– ribesii-viminalis 244
– salicis-albae 244
– vernalis 418
Melampsorella caryophyllacearum 75, 329
– symphyti 431
Melampsoridium betulinum 86, 93, 154
Melanconis alni *356*, 90
– chrysostroma *428*, 103
– flavovirens *482*, 112
– modonia *455*, 107
– spodiae *429*, 103
– stilbostoma *388*, 98
– thelebola *357*, 90
Melanconium 90, 92
– hederae 145
  atromaticum *130*, 103
Melanomma fuscidulum *125*, 33
– pulvis-pyrius *126*, 33
– subdispersum *389*, 98
Melanopsamma pomiformis *127*, 33
Melanospora chionea *760*, 174
Melanotaenium cingens 380
– endogenum *1506*, 361
– hypogeum 375
– lamii *1545*, 376
Melasmia 79, 245
Melastiza asperula 172
Meliola niessliana *1223*, 272
Melittosporiella pulchella *26*, 9
Melittosporium propolidoides *27*, 9
– pteridinum *2104*, 567
Meloderma desmazieresii *752*, 172
Melogramma campylosporum 33, 104
– spiniferum *576*, 132
Melomastia mastoidea *128*, 33, 160
Menispora 25
– ciliata *222*, 57
– glauca *223*, 58
– tortuosa 58
Meria laricis *679*, 155
Merostictis seriata *1956*, 527

Metacapnodium juniperi *664*, 152
Metacoleroa dickiei *1555*, 380
Metasphaeria cumana f. macrospora *1981*, 533
– mosana *2066*, 556
Micraspis strobilina *784*, 178
Microdiscula phragmitis 513
Microgloeum 195
Micronectriella agropyri 475
Micropeziza cornea *1957*, 527
– karstenii *1753*, 458
– poae *1754*, 458
Micropodia oedema *2124*, 571
– pteridina *2105*, 567
Microscypha ellisii *1958*, 527
– ellisii var. eriophori 544
– enrhiza 636, 146
– grisella *2106*, 567
Microsphaera alphitoides *919*, 208
– astragali 316
– baeumleri 447
– berberidis *370*, 93
– divaricata 137
– euonymi *522*, 121
– euonymi-japonici 121
– grossulariae *1024*, 226
– lonicerae 160
– mougeotii *710*, 162
– penicillata 87
– viburni 273
Microsphaeropsis olivacea 268, 67
Microstroma juglandis 150
Microthelia incrustans *129*, 33
Microthyrium ciliatum *641*, 147
– ciliatum var. hederae *623*, 70, 143
– cytisi var. cytisi *617*, 142
– cytisi var. ulicis *1191*, 265
– cytisi var. ulicis-gallii *1192*, 265
– fagi *531*, 122
– gramineum *1834*, 480
– ilicinum *920*, 104, 208, 480
– inconspicuum *532*, 122
– lauri *683*, 156
– macrosporum *411*, 101
– microscopicum *921*, 70, 122, 208
– pinophyllum *761*, 115, 174
– versicolor *1086*, 239
Miladina lecithina *348*, 89
Milesina blechni *2084*, 75, 562
– carpatorum 563
– dieteliana 566
– kriegeriana *2086*, 75, 563
– muraraie 562
– scolopendrii *2098*, 566
– vogesiaca 566
– whitei 566
Miyagia pseudosphaeria *1674*, 428
Mollisia amenticola *344*, 88
– aquosa *28*, 9
– caespiticia *29*, 10
– caricina *1959*, 458, 527, 547
– chionea *1960*, 527
– cinerea *30*, 10
– cinerella *31*, 10
– clavata *1056*, 234
– coerulans *1476*, 352
– dactyligluma 528
– dilutella *1456*, 347

– discolor *465*, 109
– discolor var. longispora *957*, 10, 215
– fallax *785*, 178
– fusco-paraphysata 558
– fuscostriata *1487*, 355
– humidicola *1961*, 528
– hydrophila *1908*, 509
– juncina *2032*, 547
– junciseda 547
– ligni *32*, 10
– lycopi *1565*, 383
– melaleuca *33*, 10
– mutabilis *1867*, 493
– palustris *2033*, 458, 547
– poaeoides *1755*, 458
– polygoni *1621*, 405
– polygoni var. rumicis 415
– pteridina *2107*, 567
– pteridis *2108*, 567
– rabenhorstii 206
– ramealis *382*, 10, 89, 96
– spectabilis *905*, 206
– stromatica 234
– typhae 559
– ventosa *34*, 10
Mollisina acerina *299*, 69, 79
– rubi *1057*, 69, 234
Mollisiopsis dennisii *1184*, 263
– lanceolata *1488*, 355
Monascostroma innumerosa *2040*, 549
Monilia 114, 162, 271
Monilinia baccarum *1221*, 271
– fructigena *713*, 162
– johnsonii *492*, 114
– laxa 195
– linhartiana 116
– megalospora 271
– mespili 165
– oxycocci 271
– urnula 271
Monochaetia karstenii *1011*, 223
Monodictys antiqua *224*, 58
– castaneae *225*, 58
– lepraria *226*, 58
– levis *1297*, 293
– melanopa *720*, 164
– paradoxa *403*, 99
– putredinis *227*, 58
Monographella nivalis *1777*, 463
Monographos fuckelii *2112*, 568
Monostichella 245
Morenoina arundinariae *1846*, 483
– chamaecyparidis *457*, 108
– clarkii *1087*, 239
– epilobii *1462*, 348
– fimbriata *2041*, 550
– minuta *2042*, 550
– paludosa *2043*, 550
– phragmitis *1918*, 511
– rhododendri *1001*, 222
– websteri *1847*, 483
Mycocentrospora acerina *1298*, 293, 342
– cantuariensis *1529*, 371
Mycosphaerella allicina 464
– brassicicola *1391*, 321
– brunneola 338
– buxicola 101

– carinthiaca 437
– clymenia *706*, 160
– crataegi 114
– depazeiformis 394
– dianthi *1445*, 343
– elodis 372
– equiseti *2092*, 564
– euphorbiae *1479*, 353
– fagi 123
– fragariae 358
– hedericola 143
– iridis *1539*, 374
– isariophora *1680*, 430
– ligustri 158
– lineolata *1778*, 464
– macrospora *1540*, 374
– mori *721*, 165
– octopetalae *509*, 118
– oedema 266
– osborniae 314
– peregrina 414
– perexigua 533
– podagrariae 301
– primulae 408
– pteridis *2113*, 568
– punctiformis *922*, 71, 123, 209
– pyri 201
– recutita 464
– rhododendri *1002*, 222
– ribis 226
– sentina 201
– slaptoniensis 114
– superflua *1708*, 442
– ulmi 266
Myriangium duriaei *611*, 141
Myrioconium 555
Myriosclerotinia curreyana *2034*, 548
– dennisii 544
– duriaeana 528
– scirpicola *2063*, 555
– sulcata *1962*, 528
Myrothecium 36, 58
– atroviride *1798*, 469
– carmichaelii *1685*, 434
– cinctum *1799*, 469
– masonii *1800*, 469
– roridum *1727*, 449
Mytilinidion acicola *665*, 152
– decipiens 152
– gemmigenum *678*, 155
– mytilinellum *762*, 174, 186
– rhenanum *827*, 186
– scolecosporium 186
Myxocyclus polycistis *407*, 100
Myxosporium rosae 230

Naemacyclus caulium *1702*, 441
– minor *753*, 172
Naemospora 136
Naevala perexigua *906*, 206
Naeviopsis tithymalina *1480*, 353
Nectria aquifolii *649*, 148
– arenula *1260*, 284
– aurantiaca 268
– boothii *1584*, 392
– brassicae 321
– cinnabarina *324*, 35, 84

– citrino-aurantia 251
– coccinea *577*, 35, 132
– coryli 252
– desmazieresii *412*, 101
– ditissima 132
– ellisii *1261*, 285
– episphaeria *130*, 35
– flavo-viridis *131*, 35
– fuckeliana *738*, 169
– galligena *717*, 35, 141, 164
– hederae *629*, 144
– inventa *1262*, 285
– keithii 321
– leptosphaeriae *1706*, 442
– magnusiana *132*, 35
– mammoidea *133*, 35
– mammoidea var. rubi 239
– pallidula *134*, 36
– peristomialis 148
– peziza *135*, 36
– pinea *828*, 187
– punicea 220
– punicea var. ilicis *650*, 148
– purtonii *136*, 36
– ralfsii *137*, 36
– sinopica *630*, 145
– solani 426
– veuillotiana *138*, 36
– viridescens *139*, 36
– wegeliniana *140*, 36
Nectriella consolationis *684*, 156
– dacrymycella *1541*, 374
Nectriopsis solani 426
Nematogonium ferrugineum *1214*, 269
Neobulgaria lilacina *35*, 10
– pura *559*, 128
Neohendersonia kickxii *593*, 136
Neopeckia fulcita *141*, 37
Neottiospora caricina *2001*, 538
Niesslia exilis *763*, 174
– exosporioides *1982*, 533
– ilicifolia 147
Nigrospora 484
– sphaerica *1852*, 484
Niptera *1641*, 412
– excelsior *1909*, 510
– exsiliens 215
– lacustris 548
– melatephra *2035*, 548
– muelleri-argoviensis *907*, 206
– phaea 528
– pilosa *1963*, 529
– pulla 529
– ramincola *36*, 11
Nitschkia brevispina *142*, 37
– collapsa *143*, 37
– confertula *612*, 141
– cupularis *144*, 37
– grevillei *145*, 37
– parasitans *146*, 38
Nodulisporium 30, 58, 97, 100, 113, 131
Nyssopsora echinata 388

Ocellaria ocellata *1125*, 249
Ochropsora ariae 255, 307
Oedemium 23
Ohleria rugulosa *147*, 38

Oidiodendron flavum *644*, 147
– griseum *228*, 59
– rhodogenum *229*, 59
– tenuissimum *230*, 59
Oidiopsis 367
Oidium 80, 320
– chrysanthemi 332
Olpidium brassicae 321
Ombrophila ambigua *1882*, 498
Oncopodiella trigonella *632*, 59, 145
Oncospora pinastri *839*, 188
Oomyces carneo-albus *1872*, 494
Ophiobolus acuminatus *1263*, 285
– cirsii 326
– erythrosporus 285
– fruticum *1589*, 393
– typhae 560
– unlasporus *1379*, 317
Ophiosphaerella herpotricha 464
Ophiovalsa betulae *390*, 98
– corylina *483*, 112
– suffusa *358*, 91
Oramasia hirsuta *1138*, 252
Orbilia alnea *37*, 11
– arundinacea *2075*, 559
– auricolor *38*, 11
– coccinella 11
– comma *1202*, 267
– curvatispora *39*, 11
– cyathea *40*, 12
– euonymi 12
– leucostigma *811*, 12, 183
– luteo-rubella *41*, 12
– luzularum 553
– occulta 12
– sarraziniana *42*, 12
– vinosa 12
– xanthostigma *43*, 12
Otthia spiraeae *148*, 38
Ovulariopsis 110
Ovulinia azaleae *996*, 221

Pachnocybe albida *231*, 59
– ferruginea *232*, 59
Pachydisca ascophanoides 249
Pachyella babingtonii *44*, 12
Paradendryphiopsis cambrensis *987*, 219
Paradidymella clarkii *1088*, 240
– holci *1779*, 464
– tosta *1463*, 348
Paraphaeosphaeria michotii *2011*, 464, 540
– rusci *1104*, 243
Parapleurotheciopsis ilicina *932*, 210
– inaequiseptata *516*, 119
Parasympodiella clarkii *835*, 188
Passalora bacilligera *340*, 87
Passeriniella discors *1932*, 517
Patellaria atrata *45*, 12
Patellariopsis atrovinosa *494*, 115
Patellina caesia *796*, 180
Patinellaria sanguinea *46*, 12
Peltasterinostroma rubi 242
Penicillium corymbiferum 364
– expansum 164
– gladioli 364
Periconia atra *1992*, 536
– britannica *1801*, 470

– byssoides *1299*, 294
– cambrensis *233*, 59
– circinata *1940*, 519
– cookei *1300*, 294
– curta *1993*, 536
– digitata *1994*, 537
– funerea *1995*, 537
– glyceriicola *1802*, 470
– hispidula *1803*, 470
– igniaria *1804*, 470
– laminella *1996*, 537
– macrospinosa *1941*, 519
– minutissima *1301*, 294, 471
– typhicola *2079*, 560
Perisporium kunzei *1392*, 321
Peroneutypa heteracantha *149*, 38
Peronospora affinis 359
– agrestis 446
– alsinearum 430
– alta *1611*, 401
– antirrhini 310
– aparines 362
– arborescens 385, 395
– arenariae 388
– calotheca 362
– campestris 313
– chenopodii 330
– chlorae 319
– conferta 329
– conglomerata 363
– cristata 385
– cyparissiae 354
– debaryi 441
– dentariae 326
– destructor 305
– digitalis 344
– dipsaci 345
– erodii 351
– euphorbiae 354
– farinosa *1427*, 331
– ficariae 412
– fragariae 358
– galligena 306
– gei 363
– grisea 446
– honckenyae 370
– lamii 376
– lepidii 339
– lepigoni 429
– leptoclada 367
– leptosperma *1571*, 385, 439
– linariae 340
– lotorum 381
– minor 317
– myosotidis 389
– niessliana 303
– oerteliana 408
– parasitica *1410*, 326
– paula 329
– polygoni 405
– potentillae 407
– radii 332
– ranunculi 412
– rubi 232
– rumicis 415
– schachtii 319
– scleranthi 419

– sordida 391, 419
– sparsa 228
– symphyti 432
– thlaspeos-arvensis 435
– trifoliorum *1691*, 382, 438
– valerianae 444
– valerianellae 445
– valesiaca 354
– verbasci 445
– viciae *1722*, 448
– violacea 375
– violae 450
Perrotia flammea *47*, 13
– phragmiticola *1910*, 510
Persiciospora masonii *976*, 217
Pesotum 267
Pestalotiopsis disseminata *2018*, 542
– funerea *497*, 115
– guepinii *1012*, 223
– monochaetioides *458*, 108, 115
– neglecta *523*, 121
– oxyanthi 166
– sydowiana *1013*, 223
– versicolor 230
Pezicula acericola *313*, 81
– alba 163
– carnea 81
– carpinea *431*, 104
– cinnamomea *958*, 13, 215
– corticola *714*, 163
– coryli *477*, 111
– corylina 111
– frangulae *598*, 137
– houghtonii *866*, 196
– livida *812*, 184
– malicorticis 163
– myrtillina 271
– paradoxa 111
– rubi *1058*, 234
– scoparia 116
– sepium *495*, 115
Peziza ampliata 13
– apiculata *560*, 13, 128
– micropus *48*, 13
– repanda *49*, 13
– varia *50*, 13
Pezizella albohyalina *349*, 89
– alniella *345*, 88
– campanulaeformis *2083*, 562
– chionea *786*, 179
– chrysostigma *2088*, 563
– discreta *1239*, 279
– eburnea *1756*, 458
– effugiens *1622*, 405
– eriophori 544
– fagi *526*, 122
– gemmarum *847*, 191
– glareosa 279
– incerta 529
– leucostigma 14
– nigro-corticata *1964*, 529
– parilis *51*, 14, 89
– punctoidea *1457*, 347
– roburnea *908*, 206
– rubescens *909*, 207
– subtilis *729*, 167
– turgidella 458, 529

– vulgaris *52*, 14
Pezizellaster serrata *1152*, 256
– tami 371
Phacidiella 249
Phacidiopycnis 200
Phacidiostroma multivalve *637*, 146
Phacidium abietinum 76
– infestans 172
– lacerum *754*, 172
Phaeangella ulicis *1185*, 264
Phaeangellina empetri *514*, 119
Phaeocryptopus gaeumannii *880*, 199
Phaeohelotium extumescens *53*, 15
– flexuosum *54*, 15
– italicum *55*, 15
– lilacinum *56*, 15
– monticola *57*, 15
– nobilis 15
– purpureum *813*, 184
– subcarneum 15
– trabinellum *58*, 15
– umbilicatum 15
Phaeoisaria clavulata *234*, 60
– clematidis *235*, 60
Phaeostalagmus cyclosporus *236*, 60
– peregrinus *791*, 60, 179
– tenuissimus *237*, 60
Phaeostromella 192
Phialea sp. 1 *273*, 70
– fumosella 173
– strobilina 168
Phialina parenchymatosa *910*, 207
– ulmariae *1489*, 355
Phialocephala fumosa *447*, 106
– fusca *1196*, 265
– truncata *448*, 106
Phialophora asteris 316
– cinerescens *1446*, 343
Phleogena faginea *595*, 136
Phloeospora 165, 266, 301
– aceris *309*, 80
– heraclei 368
Phloeosporella 195
Phlyctema 163
Phoma 237, 442
– allostoma 260
– ammophilae *1841*, 481
– apiicola 311
– arundinacea 513
– complanata 299, 368
– destructiva 383
– epitricha 565
– exigua var. foveata 426
– exigua var. linicola 381
– hedericola 144
– herbarum 299
– idaei 242
– jacquiniana 434
– limitata 201
– longissima 368
– macrocapsa *1578*, 388
– medicaginis 386
– medicaginis var. pinodella 401
– minutula 161
– nebulosa 299
– polygramma 401
– pseudacori 374

– rubella 336
– samararum 138
– sambuciphila 255
– suaedae 431
– thujana 260
Phomatospora arenariae *1835*, 480
– berkeleyi *1780*, 464
– dinemasporium *1781*, 465
– endopteris *2114*, 568
– gelatinospora 222
– therophila *2044*, 550
Phomopsis 83, 226, 227
– albicans 372
– alnea 92
– ambigua 201
– brachyceras 159
– cacti 322
– caryophylli 344
– cirsii 336
– conorum 177
– controversa 142
– corni 109
– crustosa *647*, 148
– cryptica 161
– cucurbitae 340
– demissa 108
– diachenii 396
– digitalis 344
– durandiana 416
– elliptica 362
– epilobii 350
– eupatoriicola 352
– euphorbiae 354
– glandicola 211
– hieracii 370
– hysteriola 299
– importata 162
– iridis 374
– juniperovora 152
– lavandulae 379
– morphaea 395
– mulleri 242
– oblita 314
– oblonga 270
– occulta 189
– piceae f. obiones 366
– platanoides 85
– polygonorum 405
– pseudotsugae 199
– pterophila 138
– putator 193
– quercella 219
– ramealis 121
– revellans 113
– rusci 243
– salicina 253
– sclerotioides 340
– scobina 142
– silenes 423
– stictica *419*, 102
– subordinaria 401
– tamicola 432
– tinea 274
– verbasci 445
– viticola 274
Phragmidium acuminatum 231
– bulbosum *1046*, 231

– fragariae *1625*, 406
– fusiforme 228
– mucronatum *1032*, 228
– potentillae 406
– rosae-pimpinellifoliae 228
– rubi-idaei *1047*, 231
– sanguisorbae 407
– tuberculatum *1033*, 228
– violaceum *1048*, 232
Phragmonaevia hysterioides *1965*, 529
Phragmoporthe conformis *359*, 91
Phragmotrichum rivoclarinum *366*, 92
Phyllachora dactylidis *1863*, 491
– graminis *1782*, 465
– junci *2045*, 550
– sylvatica *1878*, 497
Phyllactinia guttata *471*, 110, 123
Phyllosticta 147, 258
– cucurbitacearum 340
– hypoglossi 243
Physalospora alpestris 533
– empetri 119
– miyabeana 252
– vitis-idaeae 272
Physoderma alfalfae 386
Phytophthora citricola 371
– erythroseptica 426
– infestans 383, 426
– porri 306
Piggotia 267
– coryli *472*, 110
Pilatia foliorum 70
Pilidium acerinum *286*, 74
– concavum *1469*, 350
Pirottaea brevipila *1415*, 328
– exilispora *1859*, 487
– nigrostriata *1523*, 368
– veneta *1546*, 377
Pithomyces chartarum *1805*, 471
Pithya cupressina *659*, 151
Placuntium andromedae 92
Plagiosphaera immersa 442
Plagiostoma devexa *1623*, 405
– inclinata *304*, 79
– pustula *923*, 209
– tormentillae 407
Plasmodiophora brassicae 321
Plasmopara densa 391
– epilobii 346
– nivea *1328*, 301
– pusilla 363
– pygmaea 307
– ribicola *1017*, 225
– viticola *1226*, 274
Platychora ulmi *1199*, 266
Platystomum compressum *150*, 38
Plectania melastoma 234
Pleiochaeta setosa *709*, 161
Plejobolus arenarius 480
Pleomassaria siparia *391*, 98
Pleospora ambigua 413
– androsaces *1666*, 424
– aquatica *2046*, 550
– bjoerlingii *1383*, 319
– calvescens *1377*, 317
– cytisi *503*, 117
– epilobii *1464*, 348

– herbarum *1264*, 285, 465
– infectoria *1783*, 466
– media 316
– palustris *2020*, 543
– papaveracea *1595*, 395
– phaeocomoides *1265*, 286
– rubelloides *1784*, 466
– rubicunda *1785*, 466
– scrophulariae *1266*, 286
– shepherdiae *496*, 115
– spartinae 517
– straminis *1786*, 466
– triglochinicola *1695*, 438
– vagans *1787*, 466
– valesiaca *2047*, 550
– vitalbae 108
Pleurophoma pleurospora *1142*, 253
Pleurophragmium acutum *1707*, 442
– parvisporum *1302*, 294
Pleurotheciopsis bramleyi *364*, 60, 91
– pusilla *449*, 106
Pleurothecium recurvatum *238*, 61
Ploettnera exigua *1059*, 235
– solidaginis *1375*, 316
Pocillum cesatii *911*, 207
Podosphaera aucupariae 257
– clandestina *490*, 114
– leucotricha *718*, 164
– myrtillina 272
– tridactyla *872*, 197
Podostroma alutaceum *151*, 38
Poetschia cratincola *59*, 15
Pollaccia 192, 246
Polydesmia pruinosa *60*, 115
Polymorphum 127
Polyscytalum berkeleyi *1713*, 443
– fagicola *535*, 73, 123
– fecundissimum *285*, 73
– gracilisporum *693*, 157
– hareae *517*, 120
– pustulans *1673*, 426
– truncatum *518*, 120
– verrucosum *836*, 188
Polyspora lini 381
Polystigma fulvum 197
– rubrum *873*, 197
Polystigmina 197
Polythrincium 437
Potebniamyces pyri 163, 200
Propolis phacidioides 92
Propolomyces versicolor *61*, 16
Prosthecium auctum *360*, 91
Prosthemium stellare *367*, 92
Protocrea delicatula 38
– farinosa *152*, 38
Protomyces macrosporus *1329*, 301
– pachydermus 433
Protomycopsis leucanthemi 332
Protoventuria arxii *1003*, 222
– straussii *1471*, 350
Psammina bommeriae *1842*, 481
Pseudaegerita viridis *361*, 91
Pseudocenangium 173
Pseudocercospora deightonii *773*, 176
– myrticola *726*, 166
– rubi *1094*, 241
Pseudocercosporella capsellae *1393*, 321

Fungus Index 805

– herpotrichoides *1806*, 471
– scirpi *2069*, 556
Pseudoguignardia scirpi *2071*, 557
Pseudohelotium alaunae *1868*, 493
– pineti *755*, 173
Pseudolachnea hispidula *1321*, 299
Pseudomassaria chondrospora *1175*, 261
– corni *466*, 110
– sepincolaeformis *1040*, 230
– thistletonia *1004*, 222
– vaccinii 273
Pseudombrophila deerata 279
Pseudomicrodochium aciculare *450*, 106
– candidum *774*, 176
– cylindricum *451*, 106
Pseudonectria rousseliana *413*, 101
Pseudopatellina conigena *797*, 180
Pseudoperonospora humuli *1530*, 371
– urticae *1698*, 441
Pseudopeziza calthae 324
– medicaginis *1573*, 386
– trifolii *1692*, 438
Pseudophacidium callunae *1399*, 323
– piceae *737*, 168
Pseudorobillarda phragmitis *1926*, 513
Pseudoseptoria donacis *1812*, 473
– stomaticola 473
Pseudospiropes 98, 100
– hughesii *588*, 61, 135, 141
– nodosus *239*, 61
– obclavatus *240*, 61
– rousselianus *1303*, 61, 295
– simplex *241*, 61
– subuliferus *242*, 62
Pseudostypella translucens *751*, 172
Pseudovalsa lanciformis *392*, 98
– longipes *977*, 217
– umbonata *978*, 217
Psilachnum asemum *1966*, 530
– auranticolor *456*, 107
– granulosellum *1967*, 530
– inquilinum *2090*, 564
– lateritio-album *1968*, 330
– pteridigenum *2109*, 567
– rubrotinctum *1490*, 356
Pteridiospora scoriadea *393*, 98
Pteroconium intermedium *1853*, 484
Puccinia acetosae 414
– adoxae *1326*, 301
– aegopodii 301
– albescens 301
– albulensis 445
– allii 303
– angelicae 308
– annularis 433
– antirrhini 310
– apii 311
– arenariae *1366*, 313
– argentata 301, 373
– asparagi 315
– behenis 422
– betonicae 319
– bistortae 308, 338, 403
– brachypodii var. arrhenatheri *1844*, 482
– brachypodii var. brachypodii 486
– brachypodii var. poae-nemoralis 451
– brunellarum-moliniae 408, 505

– bulbocastani 322
– bupleuri 322
– buxi 100
– calcitrapae 327
– calthae 324
– calthicola *1402*, 324
– campanulae 325
– cancellata 545
– caricina var. caricina 225, 440, 521
– caricina var. magnusii 521
– caricina var. paludosa 397, 521
– caricina var. pringsheimiana 521
– caricina var. ribesii-pendulae 521
– caricina var. ribis-nigri-lasiocarpae 521
– caricina var. ribis-nigri-paniculatae 521
– caricina var. uliginosa 396, 521
– caricina var. urticae-acutae 521
– caricina var. urticae-acutiformis 521
– caricina var. urticae-flaccae 521
– caricina var. urticae-hirtae 521
– caricina var. urticae-inflatae 521
– caricina var. urticae-ripariae 521
– caricina var. urticae-vesicariae 521
– chaerophylli *1354*, 309
– chrysanthemi 331
– chrysosplenii *1430*, 333
– circutae 334
– circaeae 334
– cladii 539
– clintonii var. sylvaticae 397
– cnici 335
– cnici-oleracei *1322*, 300
– commutata 444
– conii 338
– convolvuli 324
– coronata *1731*, 136, 220, 451
– crepidicola 339
– cyani 327
– deschampsiae 492
– difformis 360
– dioicae var. arenariicola 327, 521
– dioicae var. dioicae 335, 521
– dioicae var. extensicola 315, 521
– dioicae var. schoeleriana 420, 521
– dioicae var. silvatica 521
– elymi 478, 495
– epilobii 346
– eriophori 427, 558
– eutremae 337
– fergussonii 448
– festucae 159, 496
– galii-cruciatae 360
– galii-verni 360
– gentianae 362
– gladioli 364
– glechomatis 365
– glomerata 420
– graminis *1732*, 452
– heraclei 367
– hieracii 369, 432
– hieracii var. hypochaeridis 372
– hieracii var. piloselloidarum 370
– holcina 500
– hordei 394, 501
– horiana 331
– hydrocotyles 371
– hysterium 435

- iridis 373
- kusanoi 483
- lagenophorae 420
- lapsanae *1548*, 377
- libanotis 422
- liliacearum *1591*, 394
- longicornis 483
- longissima 503
- luzulae 553
- maculosa 389
- magnusiana *1906*, 410, 509
- major 339
- malvacearum *1570*, 384
- menthae 387
- microsora 521
- nemoralis 386, 505
- nitida 301
- obscura 317, 553
- opizii 375, 521
- oxalidis 394
- oxyriae 394
- pazschkei var. jueliana 418
- pazschkei var. pazschkei 418
- pelargonii-zonalis 397
- phragmitis *1907*, 413, 414, 509
- physospermi 399
- pimpinellae 400
- poarum 440, 513
- polemonii 402
- polygoni-amphibii 403
- polygoni-amphibii var. convolvuli 362, 403
- porri *1336*, 303
- pratensis 500
- primulae 408
- prostii 439
- pulverulenta *1450*, 346
- punctata *1505*, 361
- punctiformis 335
- pygmaea var. ammophilina *1824*, 478
- pygmaea var. pygmaea 489
- recondita f. sp. agropyrina 474
- recondita f. sp. agrostidis 476
- recondita f. sp. borealis 433
- recondita f. sp. bromina 488
- recondita f. sp. perplexans 410, 478
- recondita f. sp. persistens 433, 474
- recondita f. sp. recondita 306, 452, 515
- recondita f. sp. triseti 517
- recondita f. sp. tritici 518
- ribis 225
- rugulosa 398
- saniculae 417
- saxifragae 418
- schroeteri 390
- scirpi 555
- septentrionalis 403, 434
- sessilis *1897*, 315, 506
- smyrnii 424
- striiformis *1733*, 452
- striiformis var. dactylidis 491
- tanaceti 314
- thesii 434
- thymi 435
- tumida 338
- umbilici 419, 440
- variabilis 432
- veronicae *1716*, 445
- veronicae-longifoliae 446
- vincae 448
- violae 449
- virgae-aurea 427

Pucciniastrum agrimoniae 302
- areolatum 168, 194
- circaeae 334
- epilobii 76, 346
- guttatum 361
- pyrolae 409
- vaccinii 270

Pycnofusarium rusci 243
Pycnopeziza pachyderma *912*, 207
Pycnostysanus azaleae *1008*, 222
Pycnothyrium 568
Pyrenochaeta fallax *1715*, 444
- ilicis *648*, 148
- lycopersici 383
Pyrenopeziza adenostylidis *1416*, 328
- arctii *1363*, 312
- arenivaga *1827*, 479
- artemisiae *1372*, 315
- benesuada *350*, 89
- brassicae 321
- carduorum *1433*, 336
- chailletii *1524*, 368
- chamaenerii *1458*, 347
- compressula 393
- depressuloides 312
- digitalina *1448*, 344
- escharodes *1060*, 235
- foliicola *335*, 86
- fuckelii *1109*, 245
- fuscescens 530
- inornata *1364*, 312
- lychnidis *1667*, 424
- lycopincola *1566*, 383
- maculans 506
- mercurialis *1579*, 388
- millegrana *1491*, 356
- nervicola *435*, 104
- pastinacae *1598*, 397
- petiolaris *300*, 70, 79, 122
- plantaginis *1612*, 401
- plicata *1352*, 309
- pulveracea *1492*, 356
- revincta *1240*, 279
- rubi *1061*, 235
- salicis 249
- subplicata *1585*, 392
- thalictri *1686*, 434
- urticicola *1703*, 442
Pyrenophora avenae *1857*, 486
- bromi *1862*, 489
- dictyoides *1879*, 497
- erythrospila *1820*, 477
- graminea *1888*, 502
- japonica *1889*, 502
- scirpicola *2067*, 556
- trichostoma *1788*, 466
- tritici-repentis *1818*, 475
- typhicola *2078*, 560
Pyxidiophora caulicola 359
- petchii 321

Quaternaria dissepta *1210*, 269
- quaternata *578*, 132

Rabenhorstia 261
Ramularia 358, 442
– adoxae *1327*, 301
– aequivoca 412
– agrestis *1728*, 450
– ajugae *1330*, 302
– alba 378
– alborosella *1423*, 329
– anthrisci *1355*, 309
– archangelicae 309
– arenariae *1581*, 388
– ari 315
– armoraciae *1368*, 314
– aromatica *1325*, 300
– arvensis *1628*, 407
– asteris *1376*, 316
– bellunensis *1429*, 332
– beticola *1384*, 319
– bistortae 405
– calcea *1517*, 365
– calthae *1404*, 324
– cardamines *1411*, 326
– cardui *1412*, 327
– centaureae *1417*, 328
– centranthi *1419*, 328
– cicutae *1431*, 334
– circaeae *1432*, 334
– cirsii 337
– cochleariae 337
– cupulariae *1634*, 409
– cynarae 341
– destructiva 166
– deusta 378
– didyma *1642*, 412
– doronici 345
– epilobiana 349
– filaris *1605*, 399
– filaris var. lappae *1365*, 312
– gei *1515*, 364
– geranii 363
– hellebori 367
– heraclei *1525*, 368
– holci-lanati 300
– keithii 385
– knautiae *1543*, 375
– lactea *1729*, 450
– lamiicola *1547*, 377
– lapsanae *1549*, 377
– lychnicola 424
– lysimachiarum *1567*, 384
– macrospora *1408*, 325
– meliloti 387
– menthicola *1576*, 387
– montana *1468*, 349
– obducens 397
– onobrychidis *1587*, 392
– parietariae *1596*, 396
– pastinacae *1599*, 397
– picridis *1606*, 400
– pratensis *1650*, 416
– primulae *1631*, 408
– pruinosa *1661*, 421
– purpurascens *1600*, 397
– rhabdospora *1613*, 402
– rhei *1646*, 413
– rosea *1117*, 247
– rubella *1651*, 416

– sambucina *1146*, 254
– scelerata 412
– schultzeri *1559*, 382
– scrophulariae *1658*, 419
– senecionis *1662*, 421
– sphaeroidea *1560*, 382
– stachydis *1677*, 429
– succisae 431
– sylvestris 345
– taraxaci 433
– ulmariae 357
– valerianae 445
– vallisumbrosae 390
– variabilis 344
– winteri *1590*, 393
Readeriella mirabilis *519*, 120
Rebentischia unicaudata 109
Rhabdocline pseudotsugae *881*, 199
Rhabdospora cirsii 337
– inaequalis *1157*, 257
– lupini 161
– pleosporoides 299
– tanaceticola 332
Rhamphoria bevanii *325*, 84
– pyriformis *979*, 218
Rhizina undulata *814*, 184
Rhizodiscina lignyota *62*, 16
Rhizosphaera kalkoffii *734*, 168
Rhodesia subtecta 481
Rhopographus filicinus *2115*, 568
Rhynchosporium alismatis *1334*, 303
– orthosporum *1865*, 471, 491
– secalis *1807*, 471
Rhytisma acerinum *301*, 79
– salicinum 245
Roesleria pallida *63*, 16
Rosellinia aquila *153*, 39
– buxi *414*, 101
– desmazieresii *1135*, 252
– mammiformis *631*, 39, 145
– necatrix *1267*, 286
– obliquata *788*, 179
– thelena *154*, 39
Rosenschoeldia abundans *1633*, 409
Rotula graminis *1808*, 471
Rutstroemia americana *441*, 105
– bolaris 104
– calopus *1883*, 498
– conformata *336*, 86
– echinophila *442*, 105
– firma *959*, 215
– fruticeti *1062*, 235
– henningsiana *2024*, 544
– hercynica *1459*, 348
– juniperi *660*, 151
– lindaviana *1911*, 510
– luteovirescens *302*, 79
– maritima *1828*, 479
– petiolorum *527*, 122
– plana 542
– rhenana *715*, 163
– rubi 235
– sydowiana *913*, 207

Saccothecium sepincola *1041*, 230, 240
Sarcophoma 102
Sarcopodium circinatum *1304*, 295

– tortuosum *243*, 62
Sarcoscypha coccinea *64*, 16
Sarcotrochila alpina *676*, 155
Sarea difformis *815*, 184
– resinae *292*, 77
Schizonella melanogramma *1945*, 523
Schizothyrioma aterrima *1324*, 300
– ptarmicae *1323*, 300
Schizothyrium pomi *2012*, 541
– speireum *1089*, 240
Schizoxylon berkeleyanum *1241*, 279
– friabilis *960*, 215
Schroeteria delastrina *1717*, 446
Scirrhia aspidiorum *2116*, 569
– pini *764*, 174
– rimosa *1919*, 512
Sclerophoma 177
Sclerotinia bulborum 389
– eleocharidis 542
– gregoriana 558
– minor 376
– sclerotiorum *1242*, 279
– spermophila 438
– trifoliorum *1693*, 438
– trifoliorum var fabae 448
Sclerotium cepivorum 306
– tuliparum 440
Scolecobasidium echinophilum 107
Scoleconectria cucurbiticola *829*, 187
Scolecosporiella 560
Scolicosporium macrosporium *594*, 136
Scopinella caulincola *940*, 211
– solani *789*, 179
Scopulariopsis brevicaulis *1305*, 295
Scotiosphaeria endoxylinae 186
Scutellinia scutellata *1126*, 16, 249
Scutomollisia fimbriomarginata *1860*, 487
– integromarginata *1861*, 487
– operculata 542
– puncta *1896*, 506
– stenospora *2036*, 548
Scutoscypha fagi *528*, 122
Seimatosporium 238
– arbuti *1014*, 224
– foliicola *669*, 153
– hypericinum *1532*, 372
– macrospermum *1217*, 270
– mariae *1015*, 224
– rosarum *1044*, 231
– vaccinii 273
Seiridium cardinale *498*, 115
– intermedium *1218*, 270
Selenophoma 117
Selenosporella curvispora *2054*, 552
Septocylindrium bellocense 445
– leucum *792*, 179
– magnusianum 436
Septocyta ruborum *1102*, 242
Septofusidium herbarum *1306*, 295
Septogloeum carthusianum *524*, 121
Septonema fasciculare *837*, 188
– secedens *404*, 100
Septoria 485
– alpicola 350
– anemones 307
– apiicola 311
– armeriae 313

– azaleae 224
– betae 319
– caricicola 538
– caricis 538
– castaneicola 104
– chamaecysti 367
– chelidonii 330
– chrysanthemella 333
– chrysanthemi 333
– clematidis 109
– convolvuli 325
– cornicola *467*, 110
– dianthi 344
– digitalis 345
– epilobii 350
– euonymi 121
– ficariae 412
– galeopsidis 360
– gladioli 364
– glaucis 365
– hederae 144
– hippocastani 86
– hydrocotyles 371
– hyperici 372
– lactucae 376
– lamii 377
– lavandulae 379
– leucanthemi 333
– lunariae 382
– lychnidis 424
– lysimachiae 384
– matricariae 439
– minuta 554
– mougeotii 370
– obesa 333
– oenotherae 392
– orobina 393
– paeoniae 395
– passerinii 503
– petroselini 398
– polygonorum 405
– quercicola 211
– rosae 231
– rubi *1103*, 242
– saponariae 417
– scabiosicola 375
– senecionis-silvaticae 421
– slaptoniensis 266
– socia 333
– sorbi 257
– stachydis 429
– stellariae *1618*, 430
– tanaceti 333
– tormentillae 407
– tritici 519
– urticae 444
– virgaureae 427
Septoriella junci *2056*, 552
Septosporium bulbotrichum *1158*, 257
Septotrullula bacilligera *405*, 62, 100, 135
Sesquicillium buxi *415*, 102
– candelabrum 176
Seynesiella juniperi *666*, 152
Sillia ferruginea *484*, 112
Sirococcus strobilinus *798*, 180
Sirothyriella 175
Spadicoides atra *680*, 62, 156

– bina *244*, 62
– grovei *245*, 62
Spermospora lolii 504
– poagena 514
Sphacelia 460
Sphaceloma 237
Sphacelotheca hydropiperis 404
– inflorescentiae 404
Sphaerellipsis 571
Sphaeridium candidum 176
Sphaerognomonia carpinea *422*, 71, 103
Sphaeropsis sapinea *779*, 177
Sphaerotheca alchemillae *1332*, 302
– dipsacearum 345
– epilobii *1465*, 349
– erigerontis-canadensis *1472*, 351
– euphorbiae 354
– ferruginea *1655*, 417
– fugax 363
– fuliginea 446
– fusca *1449*, 345
– macularis *1531*, 371
– melampyri 391
– mors-uvae *1025*, 226
– pannosa *1042*, 230
– pannosa var. persicae 197
– plantaginis *1614*, 402
– volkartii 118
– xanthii *1396*, 323
Sphaerulina myriadea *924*, 209
– pedicellata *1933*, 517
Spilocaea 164
– pyracanthae *882*, 200
Spilopodia melanogramma *1580*, 388
– nervisequia *1615*, 402
Splanchnonema ampullaceum *1176*, 261
– argus *394*, 98
– foedans *1211*, 269
– pupula *326*, 84
Spondylocladiopsis cupulicola *553*, 126
Spongospora subterranea 427
– subterranea f. sp. nasturtii 391
Sporidesmium adscendens *416*, 102
– altum *1147*, 63, 254
– anglicum *246*, 63
– aturbinatum *1148*, 63, 254
– cambrense *1197*, 265
– cladii *2015*, 541
– clarkii *1095*, 241
– cookei *1149*, 63, 255
– coronatum *247*, 63
– doliiforme *793*, 179
– eupatoriicola *1477*, 352
– folliculatum *248*, 63
– hormiscioides *589*, 63, 135
– larvatum *1168*, 259
– leptosporum *1150*, 64, 255
– paludosum *2016*, 541
– pedunculatum *249*, 64
– pseudobambusae 485
– rubi *1096*, 241
– socium *633*, 64, 145
– vagum *250*, 64
– wroblewskii *341*, 87
Sporomega degenerans 271
Sporonema 325, 362, 386
– diamandidis *799*, 180

Sporoschisma juvenile *251*, 64
– mirabile *252*, 64
Sporotrichum 64
Stachybotrys 33, 64
– atra *1307*, 295
– cylindrospora *1526*, 369
– dichroa *1308*, 296
– oenanthes *1586*, 392
Stachylidium bicolor *1309*, 296
Stagonospora 517
– anglica 538
– aquatica 556
– aquatica var. luzulicola 554
– arenaria var. minor *1877*, 496
– atriplicis *1378*, 317
– calystegiae 325
– caricinella 538
– caricis *2002*, 538
– curtisii 390
– cylindrica 513
– elegans 513
– equiseti *2093*, 565
– fragariae 359
– gigaspora 539
– heleocharidis *2021*, 543
– innumerosa *2057*, 553
– junciseda *2058*, 553
– luzulae 554
– macropycnidia 539
– meliloti *1574*, 386
– paludosa *2003*, 539
– socia 553
– suaedae 431
– subseriata *1813*, 473
– vitensis *2004*, 539
Stagonosporopsis salicorniae 417
Stamnaria persoonii *2091*, 564
Stegonosporium pyriforme *332*, 85
Stegopeziza dumeti *1063*, 235
– lauri *685*, 157
– quercea *914*, 207
Stemphylium 286, 438
– sarciniforme *1694*, 438
– vesicarium *1310*, 296
Stenella lythri *1569*, 384
Sterigmatobotrys macrocarpa *1169*, 259
Stictis elevata 548
– elongatispora *1969*, 530
– friabilis 65, 16
– pusilla 530
– radiata 16
– stellata *1243*, 279
Stigmatea conferta *1224*, 273
Stigmina carpophila *877*, 199
– glomerulosa *668*, 152
– tinea *1225*, 274
Stilbospora 268
– macrosperma *432*, 104
Stioclettia luzulina 554
Stomiopeltis betulae *395*, 98
– cupressicola *499*, 115
– dryadis *510*, 119
– juniperina *667*, 152
– pinastri *765*, 175
Strasseria geniculata *780*, 177, 189
Strigopodia resinae *830*, 187
Stromatinia gladioli 365

810   Fungus Index

– narcissi 390
– serica 366
Strossmayeria basitricha *961*, 215
Subramaniomyces fusisaprophyticus *933*, 210
Subulispora britannica *934*, 210
– minima *694*, 158
Sydowia polyspora *778*, 152, 177
Sydowiella ambigua *396*, 99
– depressula 240
– fenestrans *1466*, 349
Symphyosira rosea *1607*, 400
Symphyosirinia angelicae *1353*, 309
– galii *1511*, 362
– heraclei *1527*, 369
Sympodiella acicola *775*, 176
Synchytrium anemones 308
– endobioticum 427
– mercurialis 388
– succisae 431
– taraxaci 433

Taeniolella alta *365*, 91
– breviuscula *254*, 64
– faginea *590*, 135
– plantaginis *1616*, 402
– pulvillus *988*, 219
– rudis *255*, 64
– stilbospora *1139*, 65, 252
Taeniolina centaurii *1418*, 328
– scripta *256*, 65
Tapesia cinerella 323
– evilescens *1912*, 510
– fusca *351*, 16, 89
– knieffii *1913*, 510
– lividofusca 17
– melaleucoides *1400*, 323
– rosae *1035*, 229
Taphridium umbelliferarum *1528*, 369
Taphrina amentorum 88
– athyrii 562
– betulae 94
– betulina 94
– bullata 200
– caerulescens 207
– carpini 103
– crataegi 114
– deformans *867*, 196
– filicina 563
– johanssonii 192
– padi 196
– populina *848*, 191
– potentillae 407
– pruni 196
– sadebackii 86
– tosquinetii *337*, 86
– ulmi 267
– vestergrenii 563
– wiesneri 196
Tarbertia juncina 551
Teichospora obducens *613*, 141
Terriera cladophila 271
Tetraploa aristata *2017*, 471, 541
Thaxteriella pezicula *213*, 56
Thecaphora deformans 263, 378
– seminis-convolvuli *1406*, 325
– trailii 335
Thecotheus rivicola *352*, 89

Thedgonia ligustrina *699*, 159
Therrya pini *816*, 184
Thielaviopsis 130
– basicola *1441*, 342
Thuemenella britannica 218
Thyridaria rubronotata *327*, 39, 84
Thyriostroma spiraeae *1499*, 357
Thyronectria berolinensis *1026*, 226
Thyrostromella myriana *1838*, 481
Thysanophora penicillioides *288*, 76
– taxi *1170*, 259
Tiarospora perforans *1843*, 481
Tiarosporella paludosa *2073*, 559
Tilletia holci 500
– lolii 503
– menieri *1898*, 507
– olida 487
– sphaerococca 476
– tritici *1935*, 518
Titaeospora equiseti *2094*, 565
Topospora 226, 248, 323
Torrendiella ciliata *1064*, 235
Torula herbarum *1311*, 296, 471
Trachyspora intrusa 302
Tranzschelia anemones *1346*, 307, 434
– discolor *858*, 194, 307
– pruni-spinosae *859*, 194
Trematosphaeria anglica *155*, 39
– callicarpa 40
– circinans *1575*, 366, 386
– clarkii *1873*, 494
– heterospora *1542*, 374
– pertusa *156*, 40
Triblidium caliciiforme *962*, 215
– octosporum *604*, 138
Trichobelonium asteroma *1970*, 530
– obscurum *1401*, 323
Trichocladium asperum *257*, 65
– opacum *258*, 65
Trichoderma 25, 29
– hamatum *259*, 65
– koningii *1312*, 297
– polysporum *260*, 65
Trichodiscus diversipilus *915*, 94, 208
– heterotrichus *1869*, 493
– pinicola *817*, 184
– ulicicola *1186*, 264
– virescentulus *274*, 70
Trichonectria hirta *157*, 40
Trichophaeopsis bicuspis *66*, 17
Trichosphaerella decipiens *1193*, 265
Trichosphaeria notabilis *158*, 40
– pilosa *159*, 40
Trichothecium roseum *261*, 66
Trichothyrina alpestris *1789*, 467
– ammophilae *1836*, 480
– cupularum *539*, 124
– fimbriata *500*, 116
– nigroannulata *1790*, 71, 467
– norfolciana *1983*, 534
– pinophylla *766*, 116, 175
– salicis *1112*, 246
Tricladium castaneicola *452*, 73, 107
Trimmatostroma betulinum *406*, 66, 100
– salicis *1140*, 253
– scutellare *681*, 156, 179
Triphragmium filipendulae 354

– ulmariae *1481*, 354
Triposporium elegans *1097*, 66, 241
Trisulcosporium acerinum *308*, 80
Trochila craterium *620*, 143
– ilicina *638*, 146
– laurocerasi *868*, 196
Troposporella fumosa *857*, 194
– monospora *776*, 152, 176
Trullula melanochlora *1028*, 227
Truncatella angustata *269*, 67
– hartigii 261
– laurocerasi 242
Tubercularia 66, 84
Tuberculina maxima 571
– persicina *2125*, 571
– sbrozzii 572
Tubeufia cerea *160*, 40
– helicoma *579*, 132
– helicomyces *1791*, 467
– parvula *1837*, 480
Tympanis alnea 89
– conspersa *716*, 163
– hypopodia *818*, 184
– laricina 155
– ligustri *700*, 159

Ulocladium atrum *1313*, 297, 472
– botrytis *1314*, 297
– chartarum *1315*, 297, 472
– consortiale *1316*, 298
Uncinia foliicola *338*, 70, 87
Uncinula aduncta 246
– bicornis *305*, 80
– necator 274
– prunastri 197
Unguicularia aspera *2097*, 565
– cirrhata *1244*, 280
– costata *2037*, 549
– dilatopilosa *1460*, 348
– incarnatina 280
– millepunctata *1245*, 280
– scrupulosa *67*, 17
– ulmariae *1493*, 356
Unguiculella eurotioides *1246*, 280
– hamulata *1247*, 280
– robergei *701*, 159
Urceolella crispula *1248*, 280
– trichodea *756*, 173
Uredinopsis filicina 570
Uredo goodyerae 366
– oncidii 429
– quercus *884*, 202
Urocystis agropyri *1734*, 453
– anemones *1347*, 307
– cepulae 304
– colchici 337
– eranthidis 350
– ficariae 411
– filipendulae 354
– fischeri 523
– floccosa 367
– gladiolicola 364
– junci 545
– occulta 515
– primulicola 408
– ranunculi *1637*, 411
– sorosporoides 434

– trientalis 435
– violae 449
Uromyces acetosae 414
– aecidiiformis 379
– airae-flexuosae 492
– ambiguus *1337*, 303
– anthyllidis 310, 370, 436
– appendiculatus 398
– armeriae 313
– behenis 423
– betae 318
– chenopodii 430
– colchici 337
– dactylidis 410, 452
– dianthi 343
– ervi 447
– erythronii 351
– fallens 436
– ficariae *1635*, 410
– gageae 359
– gentianae 362
– geranii 363
– inaequialtus 423
– junci 409, 545
– limonii 379
– lineolatus 318, 365, 391, 557
– minor 436
– muscari 346, 388, 419
– pisi-sativi 116, 142, 153, 353, 400
– polygoni-aviculare 403
– rumicis 410, 415
– salicorniae 417
– scrophulariae 419
– scutellatus 353
– sparsus 428
– transversalis 364
– trifolii 436
– trifolii-repentis 436
– tuberculatus 353
– valerianae 444
– viciae-fabae 447
– viciae-fabae var. orobi 377
Ustilago anomala 404
– avenae *1854*, 482, 485
– bistortarum 404
– bullata 482, 488
– cynodontis 490
– flosculorum 375
– grandis 509
– heufleri 351
– hordei 485, 501
– hypodytes *1735*, 453, 495
– intermedia 418
– kuehneana 415
– longissima *1880*, 498
– marina 542
– nuda 501, 518
– ornithogali 359
– scabiosae 375
– serpens 453
– striiformis *1736*, 453
– succisae *1682*, 431
– tragopogonis-pratensis 435
– utriculosa 404
– vaillantii 331, 389, 419
– vinosa 394
– violacea *1663*, 423

– zeae 520
Ustulina deusta *580*, 40, 132

Valsa ambiens *161*, 41
– auerswaldii 137
– ceratophora 41
– ceuthospora 198
– curreyi 155
– cypri 159
– kunzei 77, 155
– opulina 274
– pini 187
– pruinosa 41
– pustulata 41
– salicina 252
– sordida 193
Valsaria foedans *162*, 41
– insitiva *163*, 41
Valsella adhaerens *397*, 99
– amphoria *581*, 133
– salicis *1136*, 252
Velutarina juniperi 151
– rufo-olivacea *68*, 17
Venturia carpophila *874*, 198
– cerasi *875*, 198
– chlorospora *1113*, 246
– crataegi *491*, 114
– ditricha *377*, 94
– fraxini *601*, 137
– geranii *1514*, 363
– inaequalis *719*, 164
– macularis *849*, 191
– maculiformis *1467*, 349
– minuta *1114*, 246
– palustris *1629*, 407
– pirina *883*, 201
– populina *850*, 192
– rumicis *1652*, 416
– saliciperda *1115*, 246
Veronaea caricis *1997*, 537
– carlinae *1413*, 327
– parvispora *262*, 66
Verpatinia calthicola *1405*, 324
– spiraeicola *1494*, 356
Verticicladium 171

Verticillium *1317*, 285, 298, 383
– albo-atrum 298
– dahliae 298
Vialaea insculpta *651*, 149
Vibrissea flavovirens 215
– truncorum *69*, 17
Virgaria nigra *263*, 66
Virgariella atra *264*, 66
– ovoidea *265*, 66
Volutella 101
– arundinis *1998*, 537
– ciliata *1318*, 298
– melaloma *1999*, 537

Wallrothiella subiculosa *1850*, 484
Wentiomyces *1496*, 356
Wettsteinina dryadis *511*, 119
– niesslii *1920*, 512
Wiesneriomyces javanicus *695*, 158

Xenodochus carbonarius *1654*, 417
Xenotypa aterrima *398*, 99
Xylaria carpophila *540*, 124
– filiformis 71
– hypoxylon *164*, 42
– longipes *328*, 84
– oxyacanthae 114
– polymorpha *165*, 42, 84
Xylohypha ferruginosa *266*, 67
– nigrescens *267*, 67
– ortmansiae *794*, 180
– pinicola *838*, 188

Zalerion arboricola *682*, 156
Zignoella mortieri 187
– ovoidea *166*, 42
– rhytidodes 141
Zoellneria eucalypti *520*, 120
– rosarum *1036*, 229
Zopfia rhizophila *1373*, 315
Zygophiala jamaicensis *1477*, 344
Zygosporium echinosporum *696*, 158
– gibbum *697*, 158
Zythia 358
Zythiostroma 145, 187

HOST INDEX

Abies 75
Acer 77
Achillea 300
Acorus 300
Adder's Tongue 565
Adoxa 300
Aegopodium 301
Aesculus 85
Aethusa 301
Agrimonia 302
Agrimony 302
Agropyron 474
Agrostis 476
Ajuga 302
Alchemilla 302
Alder 86
Alder Buckthorn 136
Alexanders 424
Alfalfa 385
Alisma 303
Alliaria 303
Allium 303
All-seed 330
Almond 194
Alnus 86
Alopecurus 478
Althaea 306
Alyssum 306
Ammophila 478
Anagallis 306
Anchusa 306
Andromeda 92
Anemone 307
Angelica 308
Anisantha 482
Annual Mercury 387
Anthoxanthum 482
Anthriscus 309
Anthyllis 310
Antirrhinum 310
Apium 311
Apple 162
Apricot 194
Aquilegia 311
Arabis 312
Arctium 312
Arctostaphylos 92
Arenaria 313
Armeria 313
Armoracia 313
Arrhenatherum 482
Arrow-head 416
Artemisia 314
Artichoke 340
Arum 315
Arundinaria 483
Ash 137
Asparagus 315
Asplenium 562

Aster 315
Astragalus 316
Athyrium 562
Atriplex 316
Atropa 317
Auricula 408
Autumn Crocus 337
Avena 485
Azalea 220

Ballota 317
Balsam 373
Bamboo 483
Bambusa 486
Barberry 93
Barley 501
Barren Brome 482
Barren Strawberry 406
Bastard Toadflax 434
Bearberry 92
Bear's Foot 367
Bedstraw 360
Beech 121
Beech Fern 569
Beet 318
Bellbine 324
Bell-heather 350
Bellis 317
Bent 476
Berberis 93
Bermuda-grass 490
Berry Catchfly 339
Berula 318
Beta 318
Betonica 319
Betony 319
Betula 93
Bilberry 270
Bindweed 338
Birch 93
Bird-cherry 194
Bird's Foot Trefoil 381
Bistort 402
Bittersweet 425
Bitter Vetch 377
Blackberry 231
Black Bindweed 402
Black Bryony 432
Black Currant 224
Black Horehound 317
Black Mustard 319
Black Nightshade 425
Blackstonia 319
Blackthorn 194
Bladder-seed 399
Blechnum 562
Bluebell 346
Bluebottle 327
Blue Sesleria 515

Bog Asphodel 390
Bog Myrtle 165
Bog-rush 557
Bog Whortleberry 270
Box 100
Brachypodium 486
Bracken 566
Brassica 319
Bristly Ox-tongue 399
Brittle Bladder-fern 562
Briza 488
Broad Bean 446
Broccoli 319
Brome 488
Bromus 488
Brooklime 445
Brookweed 417
Broom 116
Brussels Sprouts 319
Bryonia 322
Buckler-ferns 563
Buckthorn 220
Bugle 302
Bullace 194
Bulrush 555
Bunium 322
Bupleurum 322
Bur Chervil 309
Burdock 312
Burnet Saxifrage 400
Bur-reed 557
Bush Grass 489
Butcher's Broom 243
Butterbur 397
Buttercup 409
Buxus 100

Cabbage 319
Cactaceae 322
Calamagrostis 489
Calamint 322
Calamintha 322
Calendula 322
Calluna 323
Caltha 324
Calystegia 324
Campanula 325
Campion 422
Canadian Fleabane 351
Candytuft 372
Capsella 326
Cardamine 326
Carduus 326
Carex 521
Carlina 327
Carline Thistle 327
Carnation 343
Carpinus 102
Carrot 342

813

Castanea 104
Catchfly 422
Cat's Ear 372
Cat's-tail 508
Cauliflower 319
Celery 311
Centaurea 327
Centaurium 328
Centaury 328
Centranthus 328
Cerastium 329
Chaerophyllum 329
Chamaecyparis 108
Charlock 424
Cheiranthus 330
Chelidonium 330
Chenopodium 330
Cherry Laurel 194
Chickweed 429
Chickweed Wintergreen 435
Chicory 334
Chionodoxa 331
Chives 303
Chrysanthemum 331
Chrysosplenium 333
Cichorium 334
Cicuta 334
Cineraria 420
Cinquefoil 406
Circaea 334
Cirsium 335
Cladium 539
Cleavers 360
Clematis 108
Clinopodium 337
Clove Pink 343
Clover 436
Club-mosses 562
Club-rush 557
Clustered Bellflower 325
Cochlearia 337
Cocksfoot 490
Colchicum 337
Coltsfoot 440
Columbine 311
Colza 319
Comfrey 431
Conium 338
Conopodium 338
Convallaria 338
Convolvulus 338
Cord-grass 516
Cornbine 338
Cornflower 327
Corn Marigold 331
Corn Salad 445
Corn Spurrey 428
Cornus 109
Coronopus 339
Corylus 110
Cotton-grass 543
Couch 474
Cowbane 334
Cowberry 270
Cow Parsley 309
Cowslip 408
Cow-wheat 386

Cranberry 270
Cranesbill 362
Crataegus 113
Creeping Jenny 384
Creeping Ladies' Tresses 366
Creeping Soft-grass 500
Creeping Thistle 335
Crepis 339
Crested Dog's-tail 490
Crested Hair-grass 503
Cross-leaved Heath 350
Crosswort 360
Crowberry 119
Crowfoot 409
Crow Garlic 303
Cuckoo-flower 326
Cuckoo Pint 315
Cucubalus 339
Cucumber 339
Cucumis 339
Cupressus 115
Cydonia 116
Cymbalaria 340
Cynara 340
Cynodon 490
Cynoglossum 341
Cynosurus 490
Cystopteris 562
Cytisus 116

Dactylis 490
Dactylorhiza 341
Daffodil 390
Dahlia 341
Daisy 317
Dame's Violet 369
Dandelion 432
Daphne 118
Darnel 503
Daucus 342
Deadly Nightshade 317
Deer-grass 558
Delphinium 342
Deschampsia 492
Devil's-bit Scabious 431
Dewberry 231
Dianthus 343
Digitalis 344
Diplotaxis 345
Dipsacus 345
Dock 414
Dog's Mercury 387
Dog's Tooth Violet 351
Dogwood 109
Doronicum 345
Douglas Fir 199
Dovedale Moss 418
Downy Birch 93
Dropwort 354
Dryas 118
Dryopteris 563
Duke of Argyll's Tea Plant 162
Dutch Rush 563
Dyer's Greenwood 142
Dyer's Rocket 413

Echium 345

Elder 253
Eleocharis 542
Elm 266
Elymus 495
Empetrum 119
Enchanter's Nightshade 334
Endive 334
Endymion 346
Epilobium 346
Epipactis 350
Equisetum 563
Eranthis 350
Erica 350
Erigeron 351
Eriophorum 543
Erodium 351
Eryngium 351
Erythronium 351
Eucalyptus 119
Euonymus 120
Eupatorium 351
Euphorbia 352
Euphrasia 354
Evening Primrose 392
Everlasting Pea 377
Eyebright 354

Fagus 121
False Acacia 227
False-brome 486
False Oat 482
Fat Hen 330
Felwort 362
Fennel 357
Fescue 496
Festuca 496
Field Bugloss 306
Field Madder 422
Field Maple 77
Field Penny-cress 435
Field Scabious 375
Figwort 419
Filipendula 354
Fire-thorn 200
Flax 380
Fleabane 409
Flowering Currant 224
Fluellen 375
Foeniculum 357
Fool's Parsley 301
Fool's Watercress 311
Forget-me-not 389
Foxglove 344
Fox-tail 478
Fragaria 358
Fragrant Orchid 366
Frangula 136
Fraxinus 137
French Bean 398
Fumaria 359
Fumitory 359
Furze 263

Gagea 359
Galanthus 359
Galeopsis 359
Galium 360

Garlic 303
Garlic Mustard 303
Genista 142
Gentiana 362
Gentianella 362
Geranium 362, 397
Geum 363
Giant Fir 75
Gilliflower 385
Gipsywort 383
Gladdon 373
Gladiolus 364
Glasswort 417
Glaucium 365
Glaux 365
Glechoma 365
Globe Flower 439
Glyceria 498
Goat's-beard 435
Golden Alyssum 306
Golden Oat-grass 517
Golden-rod 427
Golden Saxifrage 333
Goldilocks 409
Good King Henry 330
Goodyera 366
Gooseberry 224
Goosegrass 360
Gorse 263
Goutweed 301
Grape Hyacinth 388
Grape Vine 274
Grass of Parnassus 396
Great Burnet 417
Greater Burnet Saxifrage 400
Greater Celandine 330
Greater Knapweed 327
Green Hellebore 367
Ground Elder 301
Ground Ivy 365
Groundsel 420
Guelder Rose 273
Gymnadenia 366
Gypsophila 366

Hairy Bittercress 326
Hairy Rock Cress 312
Halimione 366
Hard-fern 562
Hardheads 327
Harebell 325
Hare's Ear 322
Hare's Foot 436
Hart's-tongue Fern 566
Hawkbit 379
Hawksbeard 339
Hawkweed 369
Hawkweed Ox-tongue 399
Hawthorn 113
Hazel 110
Heather 323
Hedera 142
Hedge Mustard 424
Helianthemum 366
Helicotrichon 499
Helleborus 367
Hemlock 338

Hemp Agrimony 351
Hemp-nettle 359
Henbit 376
Heracleum 367
Herbaceous Seablite 430
Herb Bennet 363
Herb Paris 396
Herb Robert 362
Hesperis 369
Hieracium 369
Hippocrepis 370
Hippophäe 146
Hogweed 367
Holcus 500
Holly 146
Hollyhock 306
Honesty 382
Honeysuckle 159
Honkenya 370
Hops 370
Hordeum 501
Hornbeam 102
Horse Chestnut 85
Horse-radish 313
Horse-shoe Vetch 370
Horsetails 563
Hound's-tongue 341
Houseleek 420
Humulus 370
Hydrocotyle 371
Hypericum 371
Hypochaeris 372

Iberis 372
Ilex 146
Impatiens 373
Iris 373
Ivy 142
Ivy-leaved Toadflax 340

Jack-by-the-Hedge 303
Jack-go-to-bed-at-noon 435
Jacob's Ladder 402
Japanese Spindle-tree 120
Jasione 374
Jonquil 390
Juglans 149
Juncus 545
Juniper 150
Juniperus 150

Kale 319
Keck 309
Kickxia 375
Kidney-vetch 310
King Cup 324
Knapweed 327
Knautia 375
Knawel 419
Knotgrass 402
Koeleria 503
Kohl-rabi 319

Laburnum 153
Lactuca 375
Ladies' Fingers 310
Lady-fern 562

Lady's Mantle 302
Lady's Smock 326
Lady's Tresses 429
Lamb's Lettuce 445
Lamiastrum 376
Lamium 376
Lapsana 377
Larch 154
Large Bindweed 324
Large Bittercress 326
Larix 154
Larkspur 342
Lathyrus 377
Laurus 156
Laurustinus 273
Lavandula 379
Lavatera 379
Lavender 379
Lawson's Cypress 108
Leek 303
Leontodon 379
Leopard's-bane 345
Lesser Celandine 409
Lettuce 375
Ligustrum 158
Lilium 379
Lily 379
Lily-of-the-Valley 338
Lime 261
Limonium 379
Limosella 380
Linaria 380
Ling 323
Linnaea 380
Linum 380
Listera 381
Locust Tree 227
Lolium 503
London Rocket 424
Lonicera 159
Lords-and-Ladies 315
Lotus 381
Lousewort 397
Love-in-the-Mist 391
Lucerne 385
Lunaria 382
Lupin 161
Lupinus 161
Luzula 553
Lychnis 382
Lycium 162
Lycopersicon 382
Lycopodium 565
Lycopus 383
Lyme-grass 495
Lysimachia 384
Lythrum 384

Mahonia 162
Maiden Pink 343
Maiden's Breath 366
Maize 520
Male Fern 563
Mallow 384
Malus 162
Malva 384
Mangold 318

816 Host Index

Marguerite 331
Marjoram 393
Marram 478
Marsh Andromeda 92
Marsh Helleborine 350
Marsh Marigold 324
Marsh Orchids 341
Marsh Pea 377
Marsh Thistle 335
Marsh Valerian 444
Mat-grass 506
Matricaria 385
Matthiola 385
Meadow-grass 513
Meadow Oat-grass 499
Meadow Rue 433
Meadow Saffron 337
Meadow-sweet 354
Meadow Thistle 335
Meadow Vetchling 377
Meconopsis 385
Medicago 385
Medick 385
Medlar 165
Melampyrum 386
Melancholy Thistle 335
Melilot 387
Melilotus 387
Mentha 387
Mercurialis 387
Mespilus 165
Meu 388
Meum 388
Mezereon 118
Michaelmas Daisy 315
Milfoil 300
Milk Parsley 398
Milk-vetch 316
Mint 387
Moehringia 388
Molinia 504
Monterey Cypress 115
Morus 165
Moschatel 300
Mountain Ash 255
Mountain Avens 118
Mountain Sorrel 394
Mouse-ear Chickweed 329
Mudwort 380
Mugwort 314
Mulberry 165
Mullein 445
Muscari 388
Musk Thistle 326
Mycelis 389
Myosotis 389
Myosoton 389
Myrica 165
Myrrhis 389
Myrtle 166
Myrtus 166

Naked Ladies 337
Narcissus 390
Nardus 506
Narrow-leaved Water Parsnip 318
Narthecium 390

Nasturtium 391, 439
Nigella 391
Nipplewort 377
Noble Fir 75
Norway Maple 77

Oak 201
Oak Fern 569
Oats 485
Odontites 391
Oenanthe 391
Oenothera 392
Old Man's Beard 108
Onion 303
Onobrychis 392
Ononis 393
Ophioglossum 565
Orache 316
Oregon Grape 162
Origanum 393
Ornithogalum 394
Orthilia 394
Osmunda 565
Oxalis 394
Ox-eye Daisy 331
Oxlip 408
Oxyria 394

Paeonia 395
Pansy 448
Papaver 395
Parentucellia 396
Parietaria 396
Paris 396
Parnassia 396
Parsley 398
Parsnip 396
Pasque Flower 307
Pastinaca 396
Pea 400
Peach 194
Pear 200
Pearlwort 416
Pedicularis 397
Pelargonium 397
Pellitory-of-the-Wall 396
Pennywort 371
Peony 394
Pepper Saxifrage 422
Perennial Wall Rocket 345
Periwinkle 448
Persicaria 402
Petasites 397
Petroselinum 398
Petty Whin 142
Peucedanum 398
Phalaris 506
Phaseolus 398
Pheasant's Eye 390
Phleum 508
Phragmites 509
Phyllitis 566
Physospermum 399
Picea 166
Picris 399
Pignut 338
Pimpinella 400

Pine 169
Pineapple-weed 385
Pinus 169
Pisum 400
Plane 189
Plantago 401
Plantain 401
Platanus 189
Plum 194
Poa 513
Poets' Laurel 156
Polemonium 402
Polygonatum 402
Polygonum 402
Polypodium 566
Polypody 566
Polystichum 566
Poplar 189
Poppy 395
Populus 189
Potato 425
Potentilla 406
Poterium 407
Pot Marigold 322
Powdery Mildews 571
Primrose 408
Primula 408
Privet 158
Prunella 408
Prunus 194
Pseudotsuga 199
Pteridium 566
Puccinellia 515
Pulicaria 409
Purple Loosestrife 384
Purple Milk-vetch 316
Purple Moor-grass 504
Purple Small-reed 489
Pyracantha 200
Pyrola 409
Pyrus 200

Quaking-grass 488
Quercus 201
Quince 116

Radish 412
Ragged Robin 382
Ragwort 420
Rampion 325
Ramsons 303
Ranunculus 409
Rape 319
Raphanus 412
Raspberry 231
Red Currant 224
Red Dead-nettle 376
Red Goosefoot 330
Red Rattle 391
Red Valerian 328
Reed 509
Reed Canary-grass 506
Reedmace 559
Reseda 413
Restharrow 393
Rhamnus 220
Rheum 413

Host Index 817

Rhinanthus 413
Rhododendron 220
Rhubarb 413
Rhynchospora 554
Ribes 224
Robinia 227
Rockrose 366
Rosa 228
Rose 228
Rose-root 419
Rough Chervil 329
Rowan 255
Royal Fern 565
Rubia 414
Rubus 231
Rumex 414
Runner Bean 398
Ruscus 243
Rushes 545
Rusts 571
Rye 515
Rye-grass 503

Sagina 416
Sagittaria 416
Sainfoin 392
Salad Burnet 407
Salicornia 417
Salix 243
Salsify 435
Salt-marsh-grass 515
Sambucus 253
Samolus 417
Sand Spurrey 428
Sanguisorba 417
Sanicle 417
Sanicula 417
Saponaria 417
Saw Sedge 539
Saw-wort 422
Saxifraga 418
Saxifrage 418
Scabiosa 418
Scabious 418
Scarlet Pimpernel 306
Scentless Mayweed 439
Schoenoplectus 555
Schoenus 557
Scilla 419
Scirpus 557
Scleranthus 419
Scrophularia 419
Scurvy-grass 337
Sea Arrow-grass 438
Sea Aster 315
Sea Beet 318
Sea Bindweed 324
Sea Buckthorn 146
Sea Holly 351
Sea Lavender 379
Sea Milkwort 365
Sea Pea 377
Sea Pink 313
Sea Purslane 366
Sea Sandwort 370
Sea Spurrey 428
Sea Wormwood 314

Secale 515
Sedges 521
Sedum 419
Self-heal 408
Selinum 420
Sempervivum 420
Senecio 420
Serratula 422
Seseli 422
Sesleria 515
Sheep's Bit 374
Sheep's Sorrel 414
Shepherd's Purse 326
Sherardia 422
Shield-ferns 566
Shrubby Seablite 430
Silaum 422
Silene 422
Silver Birch 93
Silver Fir 75
Silverweed 406
Sinapis 424
Sisymbrium 424
Slender Hare's Ear 322
Slender Thistle 326
Smooth Cat's Ear 372
Smyrnium 424
Snake-root 402
Snapdragon 310
Sneezewort 300
Snowberry 257
Snowdrop 359
Soapwort 417
Solanum 425
Solidago 427
Solomon's Seal 402
Sonchus 427
Sorbus 255
Sorrel 414
Sow-thistle 427
Sparganium 557
Spartina 516
Spear Thistle 335
Spearwort 409
Speedwell 445
Spergula 428
Spergularia 428
Spignel 388
Spike-rush 542
Spinach 429
Spinacia 429
Spindle-tree 120
Spiranthes 429
Spleenwort 562
Spotted Dead-nettle 376
Spotted Orchid 341
Spruce 166
Spurge 352
Spurge Laurel 118
Squill 419
Stachys 420
Stag's-horn Moss 565
Star-of-Bethlehem 394
Stellaria 429
Stemless Thistle 335
Stinging Nettle 440
Stinking Hellebore 367

Stinkweed 345
Stitchwort 429
St John's Wort 371
Stock 385
Stone Bramble 231
Stonecrop 419
Storksbill 351
Strawberry 358
Suaeda 430
Succisa 431
Sugar Beet 318
Sulphur Weed 398
Swede 319
Sweet Bay 156
Sweet Chestnut 104
Sweet Cicely 389
Sweet Flag 300
Sweet Gale 165
Sweet-grass 498
Sweet Vernal-grass 482
Sweet William 343
Sweet Woodruff 360
Swine-cress 339
Sycamore 77
Symphoricarpos 257
Symphytum 431

Tamarisk 258
Tamarix 258
Tamus 432
Tansy 331
Taraxacum 432
Tare 446
Taxus 258
Teasel 345
Teucrium 433
Thalictrum 433
Thelypteris 569
Thesium 434
Thlaspi 435
Three-nerved Sandwort 388
Thrift 313
Thuja 260
Thyme 435
Thyme-leaved Sandwort 313
Thymus 435
Tilia 261
Timothy-grass 508
Tomato 382
Tormentil 406
Totter-grass 488
Tower Mustard 312
Townhall Clock 300
Tragopogon 435
Traveller's Joy 108
Tree Lupin 161
Tree Mallow 379
Trefoil 436
Trichophorum 558
Trientalis 435
Trifolium 436
Triglochin 438
Tripleurospermum 439
Trisetum 517
Triticum 518
Trollius 439
Tropaeolum 439

818 Host Index

Trout Lily 351
Tufted Hair-grass 492
Tulip 439
Tulipa 439
Turnip 319
Tussilago 440
Tutsan 371
Twayblade 381
Typha 559

Ulex 263
Ulmus 266
Umbilicus 440
Urtica 440

Vaccinium 270
Valerian 444
Valeriana 444
Valerianella 445
Verbascum 445
Veronica 445
Vetch 446
Viburnum 273
Vicia 446
Vinca 448
Viola 448
Violet 448
Viper's Bugloss 345
Vitis 274

Wallflower 330
Wall Lettuce 389

Wall Pennywort 440
Wall Rocket 345
Wall-Rue 562
Walnut 149
Wart-cress 339
Water Avens 363
Water Betony 419
Water Chickweed 389
Watercress 391
Water Dropwort 391
Water-pepper 402
Water Plantain 303
Wavy Hair-grass 492
Wayfaring Tree 273
Weasel's Snout 310
Weld 413
Welsh Poppy 385
Welted Thistle 326
Western Red Cedar 260
Wheat 518
Whin 263
White Beak-sedge 554
White Beam 255
White Bryony 322
White Cedar 260
White Dead-nettle 376
Wild Angelica 308
Wild Basil 337
Wild Madder 414
Wild Service Tree 255
Willow 243
Willow-herb 346

Winter Aconite 350
Wintergreen 409
Winter Heliotrope 397
Wood Anemone 307
Wood Avens 363
Wood Bittercress 326
Woodrush 553
Wood Sage 433
Wood Small-reed 489
Wood-sorrel 394
Woody Nightshade 425
Woolly Thistle 335
Wormwood 314
Woundwort 429

Yarrow 300
Yellow Bartsia 396
Yellow Dead-nettle 376
Yellow Flag 373
Yellow Horned Poppy 365
Yellow Loosestrife 384
Yellow Pimpernel 384
Yellow Rattle 413
Yellow Star-of-Bethlehem 359
Yellow Toadflax 380
Yellow-wort 319
Yew 258
Yorkshire Fog 500

Zea 520
Zinnia 450